D1748166

Macromolecular Engineering

Edited by
Krzysztof Matyjaszewski,
Yves Gnanou,
and Ludwik Leibler

1807–2007 Knowledge for Generations

Each generation has its unique needs and aspirations. When Charles Wiley first opened his small printing shop in lower Manhattan in 1807, it was a generation of boundless potential searching for an identity. And we were there, helping to define a new American literary tradition. Over half a century later, in the midst of the Second Industrial Revolution, it was a generation focused on building the future. Once again, we were there, supplying the critical scientific, technical, and engineering knowledge that helped frame the world. Throughout the 20th Century, and into the new millennium, nations began to reach out beyond their own borders and a new international community was born. Wiley was there, expanding its operations around the world to enable a global exchange of ideas, opinions, and know-how.

For 200 years, Wiley has been an integral part of each generation's journey, enabling the flow of information and understanding necessary to meet their needs and fulfill their aspirations. Today, bold new technologies are changing the way we live and learn. Wiley will be there, providing you the must-have knowledge you need to imagine new worlds, new possibilities, and new opportunities.

Generations come and go, but you can always count on Wiley to provide you the knowledge you need, when and where you need it!

William J. Pesce
President and Chief Executive Officer

Peter Booth Wiley
Chairman of the Board

Macromolecular Engineering

Precise Synthesis, Materials Properties, Applications

Edited by
Krzysztof Matyjaszewski, Yves Gnanou, and Ludwik Leibler

Volume 4
Applications

WILEY-VCH Verlag GmbH & Co. KGaA

The Editors

Prof. Dr. Krzysztof Matyjaszewski
Carnegie Mellon University
Department of Chemistry
4400 Fifth Ave
Pittsburgh, PA 15213
USA

Prof. Dr. Yves Gnanou
Laboratoire de Chimie des Polymères Organiques
16, ave Pey-Berland
33607 Pessac
France

Prof. Dr. Ludwik Leibler
UMR 167 CNRS-ESPCI
École Supérieure de Physique
et Chimie Industrielles
10 rue Vauquelin
75231 Paris Cedex 05
France

■ All books published by Wiley-VCH are carefully produced. Nevertheless, authors, editors, and publisher do not warrant the information contained in these books, including this book, to be free of errors. Readers are advised to keep in mind that statements, data, illustrations, procedural details or other items may inadvertently be inaccurate.

Library of Congress Card No.: applied for

British Library Cataloguing-in-Publication Data
A catalogue record for this book is available from the British Library.

Bibliographic information published by the Deutsche Nationalbibliothek
The Deutsche Nationalbibliothek lists this publication in the Deutsche Nationalbibliografie; detailed bibliographic data are available in the Internet at http://dnb.d-nb.de.

© 2007 WILEY-VCH Verlag GmbH & Co. KGaA, Weinheim, Germany

All rights reserved (including those of translation into other languages). No part of this book may be reproduced in any form – by photoprinting, microfilm, or any other means – nor transmitted or translated into a machine language without written permission from the publishers. Registered names, trademarks, etc. used in this book, even when not specifically marked as such, are not to be considered unprotected by law.

Composition K+V Fotosatz GmbH, Beerfelden
Printing betz-druck GmbH, Darmstadt
Bookbinding Litges & Dopf GmbH, Heppenheim
Cover Grafik-Design Schulz, Fußgönheim
Wiley Bicentennial Logo Richard J. Pacifico

Printed in the Federal Republic of Germany
Printed on acid-free paper

ISBN 978-3-527-31446-1

Contents

Preface *XXIII*

List of Contributors *XXV*

Volume 1
Synthetic Techniques

1 **Macromolecular Engineering** *1*
   Krzysztof Matyjaszewski, Yves Gnanou, and Ludwik Leibler

2 **Anionic Polymerization of Vinyl and Related Monomers** *7*
   Michel Fontanille and Yves Gnanou

3 **Carbocationic Polymerization** *57*
   Priyadarsi De and Rudolf Faust

4 **Ionic and Coordination Ring-opening Polymerization** *103*
   Stanislaw Penczek, Andrzej Duda, Przemyslaw Kubisa, and Stanislaw Slomkowski

5 **Radical Polymerization** *161*
   Krzysztof Matyjaszewski and Wade A. Braunecker

6 **Coordination Polymerization: Synthesis of New Homo- and Copolymer Architectures from Ethylene and Propylene using Homogeneous Ziegler–Natta Polymerization Catalysts** *217*
   Andrew F. Mason and Geoffrey W. Coates

7 **Recent Trends in Macromolecular Engineering** *249*
   Damien Quémener, Valérie Héroguez, and Yves Gnanou

8 **Polycondensation** *295*
   Tsutomu Yokozawa

Macromolecular Engineering. Precise Synthesis, Materials Properties, Applications.
Edited by K. Matyjaszewski, Y. Gnanou, and L. Leibler
Copyright © 2007 WILEY-VCH Verlag GmbH & Co. KGaA, Weinheim
ISBN: 978-3-527-31446-1

VI | Contents

9   **Supramolecular Polymer Engineering**   *351*
         G. B. W. L. Ligthart, Oren A. Scherman, Rint P. Sijbesma, and E. W. Meijer

10   **Polymer Synthesis and Modification by Enzymatic Catalysis**   *401*
         Shiro Kobayashi and Masashi Ohmae

11   **Biosynthesis of Protein-based Polymeric Materials**   *479*
         Robin S. Farmer, Manoj B. Charati, and Kristi L. Kiick

12   **Macromolecular Engineering of Polypeptides Using the Ring-opening Polymerization-Amino Acid N-Carboxyanhydrides**   *519*
         Harm-Anton Klok and Timothy J. Deming

13   **Segmented Copolymers by Mechanistic Transformations**   *541*
         M. Atilla Tasdelen and Yusuf Yagci

14   **Polymerizations in Aqueous Dispersed Media**   *605*
         Bernadette Charleux and François Ganachaud

15   **Polymerization Under Light and Other External Stimuli**   *643*
         Jean Pierre Fouassier, Xavier Allonas, and Jacques Lalevée

16   **Inorganic Polymers with Precise Structures**   *673*
         David A. Rider and Ian Manners

Volume 2
Elements of Macromolecular Structural Control

1   **Tacticity**   *731*
         Tatsuki Kitayama

2   **Synthesis of Macromonomers and Telechelic Oligomers by Living Polymerizations**   *775*
         Bernard Boutevin, Cyrille Boyer, Ghislain David, and Pierre Lutz

3   **Statistical, Alternating and Gradient Copolymers**   *813*
         Bert Klumperman

4   **Multisegmental Block/Graft Copolymers**   *839*
         Constantinos Tsitsilianis

5   **Controlled Synthesis and Properties of Cyclic Polymers**   *875*
         Alain Deffieux and Redouane Borsali

6	**Polymers with Star-related Structures** *909* *Nikos Hadjichristidis, Marinos Pitsikalis, and Hermis Iatrou*
7	**Linear Versus (Hyper)branched Polymers** *973* *Hideharu Mori, Axel H. E. Müller, and Peter F. W. Simon*
8	**From Stars to Microgels** *1007* *Daniel Taton*
9	**Molecular Design and Self-assembly of Functional Dendrimers** *1057* *Wei-Shi Li, Woo-Dong Jang, and Takuzo Aida*
10	**Molecular Brushes – Densely Grafted Copolymers** *1103* *Brent S. Sumerlin and Krzysztof Matyjaszewski*
11	**Grafting and Polymer Brushes on Solid Surfaces** *1137* *Takeshi Fukuda, Yoshinobu Tsujii, and Kohji Ohno*
12	**Hybrid Organic Inorganic Objects** *1179* *Stefanie M. Gravano and Timothy E. Patten*
13	**Core–Shell Particles** *1209* *Anna Musyanovych and Katharina Landfester*
14	**Polyelectrolyte Multilayer Films – A General Approach to (Bio)functional Coatings** *1249* *Nadia Benkirane-Jessel, Philippe Lavalle, Vincent Ball, Joëlle Ogier, Bernard Senger, Catherine Picart, Pierre Schaaf, Jean-Claude Voegel, and Gero Decher*
15	**Bio-inspired Complex Block Copolymers/Polymer Conjugates and Their Assembly** *1307* *Markus Antonietti, Hans G. Börner, and Helmut Schlaad*
16	**Complex Functional Macromolecules** *1341* *Zhiyun Chen, Chong Cheng, David S. Germack, Padma Gopalan, Brooke A. van Horn, Shrinivas Venkataraman, and Karen L. Wooley*

Volume 3
Structure-Property Correlation and Characterization Techniques

1	**Self-assembly and Morphology Diagrams for Solution and Bulk Materials: Experimental Aspects** *1387* *Vahik Krikorian, Youngjong Kang, and Edwin L. Thomas*

2 Simulations 1431
Denis Andrienko and Kurt Kremer

3 Transport and Electro-optical Properties in Polymeric Self-assembled Systems 1471
Olli Ikkala and Gerrit ten Brinke

4 Atomic Force Microscopy of Polymers: Imaging, Probing and Lithography 1515
Sergei S. Sheiko and Martin Moller

5 Scattering from Polymer Systems 1575
Megan L. Ruegg and Nitash P. Balsara

6 From Linear to (Hyper) Branched Polymers: Dynamics and Rheology 1605
Thomas C. B. McLeish

7 Determination of Bulk and Solution Morphologies by Transmission Electron Microscopy 1649
Volker Abetz, Richard J. Spontak, and Yeshayahu Talmon

8 Polymer Networks 1687
Karel Dušek and Miroslava Dušková-Smrčková

9 Block Copolymers for Adhesive Applications 1731
Costantino Creton

10 Reactive Blending 1753
Robert Jerome

11 Predicting Mechanical Performance of Polymers 1783
Han E. H. Meijer, Leon E. Govaert, and Tom A. P. Engels

12 Scanning Calorimetry
René Androsch and Bernhard Wunderlich

13 Chromatography of Polymers 1881
Wolfgang Radke

14 NMR Spectroscopy 1937
Hans Wolfgang Spiess

15 High-throughput Screening in Combinatorial Polymer Research 1967
Michael A. R. Meier, Richard Hoogenboom, and Ulrich S. Schubert

Volume 4
Applications

1		**Applications of Thermoplastic Elastomers Based on Styrenic Block Copolymers** *2001*
		Dale L. Handlin, Jr., Scott Trenor, and Kathryn Wright
1.1		Introduction *2001*
1.1.1		Origins and Development *2001*
1.2		Footwear *2008*
1.3		Bitumen Modification *2009*
1.3.1		Paving *2009*
1.3.2		Road Marking *2010*
1.3.3		Roofing *2010*
1.4		Adhesives and Sealants *2011*
1.5		Compounding Applications *2015*
1.5.1		Raw Material Selection *2015*
1.5.2		Processing and Forming *2018*
1.5.3		Automotive *2020*
1.5.4		Wire and Cable *2021*
1.5.5		Medical *2021*
1.5.6		Soft-touch Overmolding *2022*
1.5.7		Ultra-soft Compounds *2022*
1.6		Polymer Modification *2023*
1.7		Viscosity Modification and Gels *2024*
1.7.1		Viscosity Index Improvers *2024*
1.7.2		Oil Gels *2024*
1.8		Emerging Technology in Block Copolymers *2025*
1.8.1		Recycling Compatibilization *2025*
1.8.2		PVC and Silicone Replacement *2025*
1.8.3		Commercial Uses of Other Controlled Polymerized Polymers *2026*
1.8.4		Styrene-Isobutene-Styrene (S–IB–S) *2028*
		List of Abbreviations *2028*
		References *2029*
2		**Nanocomposites** *2033*
		Michaël Alexandre and Philippe Dubois
2.1		Introduction *2033*
2.2		Generalities *2034*
2.2.1		Nanocomposites and Nanohybrids – Definitions *2034*
2.2.2		Key Details *2035*
2.2.3		Main Production Pathways *2036*
2.2.4		Enhanced Properties *2037*
2.3		Layered Silicates *2038*

2.3.1	Nanocomposites Prepared by *In Situ* Polymerization of the Matrix from Anchored Monomers or Initiators 2040
2.3.1.1	Non-controlled Polymerization Processes 2040
2.3.1.2	Controlled Polymerization Process 2043
2.4	Extrapolation to Other Nanofillers 2057
2.4.1	Other Layered Nanoparticles 2057
2.4.2	Carbon Nanotubes 2059
2.4.3	Isometric Nanoparticles 2061
2.5	Current and Potential Applications 2065
2.5.1	Automotive Industry 2065
2.5.2	Packaging Industry 2066
2.5.3	Cable and Wire Industry 2067
2.6	Conclusion 2067
	References 2068

3	**Polymer/Layered Filler Nanocomposites: An Overview from Science to Technology** 2071
	Masami Okamoto
3.1	Introduction 2071
3.2	Historical Point of View 2072
3.3	Structure of Layered Filler and Its Modification 2073
3.4	Preparative Methods and Structure of PLFNCs 2075
3.4.1	Intercalation of Polymer or Pre-polymer from Solution 2075
3.4.2	*In situ* Intercalative Polymerization Method 2076
3.4.3	Melt Intercalation Method 2076
3.5	Interlayer Structure of OMLFs and Intercalation 2077
3.6	Structure and Characterization of PLFNCs 2079
3.7	Nobel Compounding Process 2083
3.8	Control of Nanostructure Properties 2086
3.8.1	Intercalation During Crystallization and Confinement 2086
3.8.2	Multiscale Micromechanical Modeling 2088
3.8.3	Flexibility of a Single Clay Layer 2090
3.8.4	Higher-order Structure Development and Crystallization Controlled by Nano-filler Surfaces 2091
3.8.5	Flocculation Control and Modulus Enhancement 2095
3.9	Melt Rheology of PLFNCs 2097
3.9.1	Linear Viscoelastic Properties 2097
3.9.2	Steady Shear Flow 2102
3.9.3	Elongational Flow and Strain-induced Hardening 2105
3.9.4	Alignment of Silicate Layers 2107
3.9.5	Electrorheology 2109
3.10	Materials Properties of PLFNCs 2110
3.10.1	Thermal Stability 2110
3.10.2	Fire Retardant Properties 2111
3.10.3	Gas Barrier Properties 2113

3.10.4	Ionic Conductivity	*2115*
3.10.5	Optical Transparency	*2115*
3.10.6	Physicochemical Phenomena	*2116*
3.10.6.1	Biodegradability	*2116*
3.10.6.2	Photodegradation	*2119*
3.10.6.3	Pressure–Volume–Temperature (PVT) Behavior	*2121*
3.11	Computer Simulation	*2121*
3.12	Processing Operations	*2123*
3.12.1	Foam Processing Using sc-CO_2	*2124*
3.12.2	Electrospinning Processing	*2127*
3.12.3	Porous Ceramic Materials via PLSNCs	*2128*
3.13	Future Prospects	*2130*
	References	*2130*
4	**Polymeric Dispersants**	*2135*
	Frank Pirrung and Clemens Auschra	
4.1	Dispersant Application Fields	*2135*
4.2	Pigment Dispersants Function and Requirements	*2136*
4.2.1	Dispersant Application Criteria	*2136*
4.2.2	Design of Dispersants by Stabilization Mechanism	*2138*
4.2.3	Characteristic Data of Pigment Dispersants	*2139*
4.2.4	Chemical Architecture of High Molecular Weight Dispersants	*2140*
4.2.5	Anchoring Groups	*2141*
4.3	Classical Polymeric Pigment Dispersants	*2142*
4.3.1	Polyurethane-based Dispersants	*2142*
4.3.1.1	Polyisocyanate Resins	*2142*
4.3.1.2	Cross-linkers	*2144*
4.3.1.3	Monofunctional Side-chains	*2145*
4.3.1.4	Introduction of Anchoring Groups	*2147*
4.3.2	Polyacrylate-based Dispersants	*2148*
4.3.2.1	Architecture of Polyacrylate Dispersants	*2150*
4.3.2.2	Design of the Polyacrylate Backbone	*2151*
4.3.2.3	Introduction of Anchoring Groups	*2153*
4.3.2.4	Dispersants Based on SMA Resins	*2154*
4.3.3	Polyester-based Dispersants	*2156*
4.3.3.1	Polymeric Amine Anchoring Blocks	*2156*
4.3.3.2	Polyester Side Chains and Dispersant Synthesis	*2157*
4.3.4	Polyether-based Dispersants	*2161*
4.4	New Technologies and Trends in Polymeric Dispersants	*2165*
4.4.1	Dispersants Made by Group Transfer Polymerization (GTP)	*2167*
4.4.2	Dispersants Made by Controlled Free Radical Polymerization (CFRP)	*2170*
4.4.2.1	Atom Transfer Radical Polymerization (ATRP)	*2170*
4.4.2.2	Nitroxyl-mediated Radical Polymerization (NMP)	*2172*
4.4.2.3	Radical Addition and Fragmentation Transfer (RAFT)	*2176*

4.4.3	Dispersants Based on Macromonomers *2177*
4.4.4	Dispersants Based on Dendrimers and Hyperbranched Polymers *2177*
	References *2178*

5 Polymeric Surfactants *2181*
Henri Cramail, Eric Cloutet, and Karunakaran Radhakrishnan

5.1	Introduction *2181*
5.2	Main Surfactants Used for Heterogeneous Polymerizations in Aqueous Media *2183*
5.2.1	Synthesis and Main Uses of PEO-based Copolymers *2183*
5.2.1.1	PEO-*b*-PPO and PEO-*b*-PBO Block Copolymers *2183*
5.2.1.2	PS-*b*-PEO and Derivatives *2186*
5.2.1.3	PMMA-*b*-PEO *2191*
5.2.1.4	Miscellaneous Nonionic Amphiphilic PEO-based Block Copolymers *2192*
5.2.2	Synthesis and Main Uses of PAA- and PMAA-based Copolymers *2193*
5.2.2.1	PMMA-*b*-PAA and PMMA-*b*-PMAA *2194*
5.2.2.2	P*n*BA-*b*-PAA *2194*
5.2.2.3	PS-*b*-PAA and PS-*b*-PMAA *2196*
5.2.3	Synthesis and Main Uses of Sulfonate Block Copolymers *2196*
5.2.4	Cationic and Cationizable Block Copolymers *2198*
5.2.5	Miscellaneous *2200*
5.3	Main Surfactants Used for Heterogeneous Polymerizations in Nonaqueous Dispersions *2204*
5.3.1	PEO- and PPO-based Copolymers *2205*
5.3.2	Polyolefin-based Copolymers *2206*
5.3.3	Silicon-based Copolymers *2207*
5.3.4	(Meth)acrylic-based Copolymers *2208*
5.3.5	*In Situ* Formation of Block Copolymer Stabilizers *2209*
5.4	Main Surfactants Used for Heterogeneous Polymerizations in scCO$_2$ *2211*
5.4.1	Silicon-based Stabilizer *2211*
5.4.2	Fluorinated Copolymers *2213*
5.4.3	Polycarbonates *2215*
5.5	Conclusion *2216*
	List of Abbreviations *2216*
	References *2218*

6 Molecular and Supramolecular Conjugated Polymers for Electronic Applications *2225*
Andrew C. Grimsdale and Klaus Müllen

| 6.1 | Introduction *2225* |

6.2	Control of Properties by Substitution in Conjugated Polymers	*2229*
6.3	Controlling Optical Properties via Increased Planarity	*2232*
6.4	Minimization of Defects in Polymers by Synthetic Design	*2234*
6.5	Controlling Chain Packing of Conjugated Polymers	*2236*
6.6	Self-assembly of Supramolecular Polymers by Hydrogen Bonding *2240*	
6.7	Supramolecular Assembly of Discotic PAHsG*2242*	
6.8	Optimizing Intra- and Intercolumnar Order in PAHs	*2245*
6.9	Carbon Nanotubes and Similar Materials from Ordered Graphite Mesophases *2249*	
6.10	Prospects for Molecular Electronics	*2254*
6.11	Conclusion *2256*	
	Acknowledgments *2257*	
	References *2257*	

7   **Polymers for Microelectronics** *2263*
Christopher W. Bielawski and C. Grant Willson

7.1	Introduction *2263*	
7.2	Conjugated Polymers *2264*	
7.2.1	Overview *2264*	
7.2.2	General Considerations *2265*	
7.2.3	Polyacetylene *2267*	
7.2.4	Polyarenes: Polyphenylenevinylene, Polythiophene and Polypyrrole *2270*	
7.2.5	Polyaniline *2272*	
7.2.6	Outlook *2273*	
7.3	Polymeric Dielectrics *2275*	
7.3.1	Overview *2275*	
7.3.2	Polyimides *2276*	
7.3.3	Polybenzoxazoles *2277*	
7.3.4	Polybenzocyclobutanes *2278*	
7.3.5	Polysilsesquioxanes *2280*	
7.3.6	Extended Aromatic Networks *2280*	
7.3.7	Fluorinated Polymers *2281*	
7.3.8	Outlook *2281*	
7.4	Resists in Microlithographic Applications *2282*	
7.4.1	Overview *2282*	
7.4.2	Photolithography *2283*	
7.4.2.1	Background *2283*	
7.4.2.2	Negative Photoresists *2285*	
7.4.2.3	Positive Photoresists *2286*	
7.4.3	Electron Beam Lithography *2287*	
7.4.4	Immersion Lithography *2289*	
7.4.5	Functional Resists *2289*	
7.4.6	Outlook *2290*	

7.5	Conclusions	*2290*
	References	*2291*
8	**Applications of Controlled Macromolecular Architectures to Lithography** *2295*	
	Daniel Bratton, Ramakrishnan Ayothi, Nelson Felix, and Christopher K. Ober	
8.1	Introduction	*2295*
8.2	Specialized Nanofabrication	*2295*
8.2.1	Definition and Scope	*2295*
8.2.2	Microcontact Printing (μCP)	*2296*
8.2.3	Scanning Probe Lithography (SPL)	*2300*
8.2.4	Particle Replication in Non-wetting Templates (PRINT)	*2302*
8.3	Macromolecules as Resists for Photolithography	*2303*
8.3.1	Introduction	*2303*
8.3.2	Deep-UV (248-nm) Resists	*2305*
8.3.3	193-nm Resists	*2307*
8.3.4	157-nm Resists	*2308*
8.3.5	Extreme UV (13.4-nm) Resists	*2309*
8.3.6	Dendrimers and Hyperbranched Polymers as Photoresists	*2310*
8.3.7	Molecular Glass (MG) Photoresists	*2310*
8.3.8	Supercritical CO_2 Processing	*2312*
8.4	Block Copolymers	*2313*
8.4.1	Introduction	*2313*
8.4.2	Engineering Considerations for Using Block Copolymers as Lithographic Masks	*2315*
8.4.3	Self-assembly in Bulk Versus Thin Film	*2315*
8.4.4	Block Copolymers as Lithographic Resists	*2316*
8.4.5	Block Copolymer Thin-film Characterization	*2316*
8.4.6	Synthesis of Block Copolymers by Direct Polymerization	*2317*
8.4.7	Block Copolymers Functionalized by Post-polymerization Modification	*2317*
8.4.8	Block Copolymers as Templates for Advanced Lithography	*2318*
8.4.9	Block Copolymers as Resists for 193-, 157- and 13.4-nm Lithography	*2320*
8.4.10	Block Copolymers with a Fluoroalkyl Block for Supercritical CO_2 Development	*2322*
8.4.11	Block Copolymers as Sequestering Agents and Additives	*2322*
8.5	Two-photon Lithography	*2323*
8.6	Conclusion	*2325*
	Acknowledgments	*2325*
	References	*2325*

9	**Microelectronic Materials with Hierarchical Organization** *2331*	
	G. Dubois, R. D. Miller and James L. Hedrick	
9.1	Introduction *2331*	
9.2	Porous Organosilicates from a Nucleation and Growth Phase Separation Process *2335*	
9.3	Role of Functionality: Nitrogenous Porogens *2337*	
9.4	Porous Structures via Nucleation and Growth: Role of Macromolecular Architecture *2342*	
9.5	Organic Nanoparticle Formation *2352*	
9.6	Block Copolymer Templates: Role of Macromolecular Architecture *2353*	
9.7	Outlook *2360*	
	References *2362*	

10	**Semiconducting Polymers and their Optoelectronic Applications** *2369*	
	Nicolas Leclerc, Thomas Heiser, Cyril Brochon, and Georges Hadziioannou	
10.1	Introduction *2369*	
10.1.1	From Synthetic Metals to Semiconducting Polymers *2369*	
10.2	Synthesis of Conjugated Polymers *2371*	
10.2.1	Homopolymers *2371*	
10.2.2	Random and Alternating Copolymers *2379*	
10.2.3	Conjugated Block Copolymers *2381*	
10.3	Applications in Optoelectronics *2384*	
10.3.1	The Requirements and Design of Polymer Light-emitting Diodes *2384*	
10.3.1.1	State-of-the-Art *2384*	
10.3.1.2	Single-layer PLEDs *2385*	
10.3.1.3	Double-layer PLEDs *2386*	
10.3.1.4	Polymer Blend PLEDs *2387*	
10.3.2	The Requirements and Design of Polymer Laser Materials *2388*	
10.3.2.1	Amplified Spontaneous Emission in Polymer Films *2389*	
10.3.2.2	Polymer Laser *2389*	
10.3.2.3	Polymer Blends *2391*	
10.3.2.4	Electrically Pumped Polymer Laser? *2391*	
10.3.3	The Requirements and Design of Polymer Photovoltaic Materials *2392*	
10.3.4	Photovoltaic Performance of Donor–Acceptor Block Copolymers *2394*	
10.3.5	The Requirements and Design of Polymer Field-effect Transistors *2395*	
10.3.6	The Requirements and Design of Sensors Based on Polymer Semiconductors *2398*	
10.3.7	"Polymer Conductors"/"Synthetic Metals" *2400*	

10.3.8	The Requirements and Design of Thin-film Conducting and Transparent Polymer Materials	2401
10.4	Conclusion	2402
	Acknowledgments	2404
	References	2404

11 Polymer Encapsulation of Metallic and Semiconductor Nanoparticles: Multifunctional Materials with Novel Optical, Electronic and Magnetic Properties 2409

Jeffrey Pyun and Todd Emrick

11.1	Introduction	2409
11.1.1	Polymeric Surfactants and Encapsulating Coatings	2410
11.1.1.1	Types of Polymeric Surfactants	2410
11.1.1.2	Approaches to Nanoparticle Functionalization with Polymers	2412
11.2	Noble Metal Nanocomposites: the Case of Au	2414
11.2.1	Structure and Properties of Metallic Nanoparticles	2414
11.2.2	Synthesis of Au Nanoparticles	2415
11.2.2.1	Gold Nanoparticles with Small Molecule Surfactants	2415
11.2.2.2	Synthesis of Gold Nanoparticles in Linear Polymer Surfactants	2416
11.2.2.3	Branched and Dendritic Polymers	2422
11.2.2.4	Conjugated Polymers	2423
11.3	Semiconductor Nanoparticles: Cadmium Selenide Quantum Dots	2424
11.3.1	Structure and Properties	2424
11.3.2	Synthesis of CdSe Semiconductor Nanoparticles	2425
11.3.2.1	Small Molecule Precursors and Surfactants	2425
11.3.2.2	Polymer Ligands for Quantum Dots	2427
11.3.2.3	Dendritic Polymers	2433
11.4	Metallic Magnetic Nanoparticles: the Case of Co	2434
11.4.1	Fundamental Terms and Classifications of Magnetic Materials	2434
11.4.2	Linear Polymeric Surfactants	2437
11.4.3	Small Molecule Surfactants	2441
11.4.4	End-functional Polymeric Surfactants	2443
11.5	Perspectives	2444
	References	2444

12 Polymeric Membranes for Gas Separation, Water Purification and Fuel Cell Technology 2451

Kazukiyo Nagai, Young Moo Lee, and Toshio Masuda

12.1	Introduction	2451
12.2	Gas Separation Membranes	2452
12.2.1	Gas Separation Mechanisms	2452
12.2.2	Gas Transport in Non-porous Polymer Membranes	2454
12.2.3	Gas Permeability and Selectivity	2459

12.2.4	Reverse-selective Polymer Membranes	*2463*
12.3	Water Purification Membranes	*2466*
12.3.1	Water Purification Mechanisms	*2466*
12.3.2	Non-porous Nanofiltration Polymer Membranes	*2467*
12.3.3	Pervaporation Polymer Membranes	*2469*
12.4	Membranes for Fuel Cells	*2472*
12.4.1	Conventional Technologies	*2472*
12.4.2	Membrane Performance Relevant for Fuel Cell Applications	*2473*
12.4.3	Aromatic Polymers with Sulfonic Acid Groups in the Their Main Chains	*2473*
12.4.3.1	Processability	*2474*
12.4.3.2	Water Sorption Behavior	*2475*
12.4.3.3	Proton Conductivity	*2477*
12.4.3.4	Fuel Permeability	*2478*
12.4.3.5	Hydrolytic Stability and Oxidative Radical Stability	*2480*
12.4.4	Modified Aromatic Polymer Containing Sulfonic Acid Groups	*2481*
12.4.5	Modified Aromatic PBIs	*2485*
12.5	Conclusion	*2487*
	Acknowledgment	*2487*
	References	*2487*
13	**Utilization of Polymers in Sensor Devices**	*2493*
	Basudam Adhikari and Alok Kumar Sen	
13.1	Introduction	*2493*
13.2	Different Analytes and Their Sensing Principles	*2494*
13.2.1	Chemical Analytes	*2494*
13.2.1.1	Gases and Volatile Chemicals	*2494*
13.2.1.2	Hydrogen Ion	*2504*
13.2.1.3	Selective Ions	*2505*
13.2.1.4	Alcohols	*2511*
13.2.1.5	Drugs	*2512*
13.2.1.6	Amines	*2513*
13.2.1.7	Surfactant	*2513*
13.2.1.8	Insecticides and Fungicides	*2513*
13.2.1.9	Aromatic Compounds	*2515*
13.2.1.10	Hydrazine	*2515*
13.2.1.11	Humidity	*2516*
13.2.2	Biosensor Analytes	*2519*
13.2.2.1	Glucose	*2521*
13.2.2.2	Urea	*2524*
13.2.2.3	Creatinine	*2524*
13.2.2.4	Amino acids	*2525*
13.2.2.5	Pyruvate	*2525*
13.2.2.6	Cholesterol	*2526*
13.2.2.7	Peroxide	*2526*

13.2.2.8	Lactate 2527
13.2.2.9	Fructose 2527
13.2.2.10	Foodstuffs and Drinks 2528
13.2.2.11	Environmental Pollutants 2529
13.2.2.12	DNA-sensitive Infectious Agents and Pollutants 2529
13.3	Measurement of Sense of Taste 2530
13.4	Trends in Sensor Research 2532
13.5	Conclusion 2533
	References 2534

14	**Polymeric Drugs** 2541
	Tamara Minko, Jayant J. Khandare, and Sreeja Jayant
14.1	Introduction 2541
14.2	Polymeric Drug Delivery 2542
14.2.1	Water-soluble Polymers 2542
14.2.2	Classification of Water-soluble Polymeric Drugs 2543
14.2.2.1	Linear Polymer (LP) Systems 2544
14.2.2.2	Linear Polymer–Linear Spacer (LPLS) Systems 2545
14.2.2.3	Linear Polymer–Branched Spacer (LPBS) Systems 2546
14.2.2.4	Branched Polymer (BP) Systems 2547
14.2.3	Advantages of Polymeric Drugs 2547
14.2.4	Future Directions 2548
14.3	Prodrug Approach 2548
14.3.1	Prodrug Approach First Type Prodrugs 2548
14.3.2	Second Type Prodrugs 2549
14.3.3	Tumor-activated Prodrugs (TAP) 2550
14.3.4	Drug Delivery Systems 2550
14.3.5	Antedrugs 2551
14.4	Controlling Properties and Function of Polymeric Drugs by Molecular Architecture 2551
14.4.1	Design and Synthesis of Polymeric Prodrugs 2552
14.4.2	Polymeric Bioconjugates in Drug Delivery 2555
14.4.2.1	Dextran and Its Prodrug Conjugates 2555
14.4.2.2	PEG and Its Prodrug Conjugates 2558
14.4.2.3	Prodrug Conjugates of *N*-(2-Hydroxypropyl)methacrylamide (HPMA) Copolymer 2561
14.4.2.4	Dendrimers and Their Conjugates 2564
14.4.2.5	Applications of Molecular Modeling in Prodrug Conjugates 2567
14.4.3	Crucial Aspects in Bioconjugation 2569
14.4.3.1	Reactivity of the Polymer and Biocomponent 2569
14.4.3.2	Use of an Appropriate Method for Preparation of Bioconjugates 2569
14.4.3.3	Steric Hindrance 2570
14.4.4	Future Directions 2571
14.5	Controlled Drug Release 2571

14.5.1	Advantages of Controlled Drug Release	2571
14.5.2	Types of Controlled Drug Release	2572
14.5.3	Future Directions	2575
14.6	Anticancer Polymeric Drugs	2575
14.6.1	General Considerations	2576
14.6.2	Tumor-targeted Anticancer Polymeric Drugs	2576
14.6.2.1	Passive Targeting	2576
14.6.2.2	Passive Targeting Active Targeting	2578
14.6.3	Proapoptotic Anticancer Polymeric Drugs	2581
14.6.4	Future Directions in Targeted Proapoptotic Anticancer Polymeric Drugs	2583
	References	2587

15 **From Biomineralization Polymers to Double Hydrophilic Block and Graft Copolymers** *2597*
Helmut Cölfen

15.1	Biomineralization	2597
15.2	Active Biomineralization Polymers	2600
15.2.1	Proteins	2601
15.2.2	Polysaccharides	2603
15.2.3	Glycoproteins	2603
15.3	The Role of Biomineralization Polymers and Reduction to Their Functional Units	2604
15.4	Double-hydrophilic Block Copolymers (DHBCs) and Double-hydrophilic Graft Copolymers (DHGCs)	2605
15.4.1	Concept of DHBCs and DHGCs	2605
15.4.2	Synthesis of Double Hydrophilic Block Copolymers	2608
15.4.2.1	Living Anionic Polymerization	2608
15.4.2.2	Living Cationic Polymerization	2612
15.4.2.3	Group Transfer Polymerization (GTP)	2612
15.4.2.4	Radical Polymerization	2612
15.4.2.5	Polymerization with Macro Initiators	2612
15.4.2.6	Coupling of Two Polymeric Blocks	2614
15.4.2.7	Polymer Analogous Reactions	2615
15.4.3	Synthesis of Double Hydrophilic Graft Copolymers (DHGCs)	2615
15.4.4	Applications of DHBCs and DHGCs	2616
15.4.4.1	DHBC Superstructures	2616
15.4.4.2	DHBCs as Novel Surfactants	2620
15.5	Conclusion	2631
	List of Abbreviations	2632
	References	2635

16	**Applications of Polymer Bioconjugates** *2645*	
	Joost A. Opsteen and Jan C. M. van Hest	
16.1	Introduction; Control at the Molecular Level Over Synthetic and Biological Macromolecules *2645*	
16.2	Synthetic Methodologies *2646*	
16.2.1	Controlled Polymerization Methods *2646*	
16.2.2	Polypeptide Synthesis *2647*	
16.2.3	Conjugation Methods *2649*	
16.3	Pharmaceutical Applications *2650*	
16.3.1	Anti-cancer Therapeutics *2651*	
16.3.2	Application of Polymer–Peptide/Protein/Antibody Conjugates as Therapeutics *2655*	
16.3.3	Well-defined Macromolecular Therapeutic Architectures *2656*	
16.4	Bioactive Surfaces *2662*	
16.4.1	Bioconjugate Arrays *2662*	
16.4.2	Cell Adhesive Coatings *2666*	
16.4.3	Non-fouling Coatings *2669*	
16.4.4	Antimicrobial Coatings *2670*	
16.5	Smart Materials *2672*	
16.5.1	Smart Polymer Bioconjugates *2672*	
16.5.2	Self-assembled Structures/Supramolecules; Towards Nanotechnology *2675*	
16.6	Conclusions and Outlook *2678*	
	References *2679*	
17	**Gel: a Potential Material as Artificial Soft Tissue** *2689*	
	Yong Mei Chen, Jian Ping Gong, and Yoshihito Osada	
17.1	Introduction *2689*	
17.2	Molecular Design of Robust Gels *2690*	
17.2.1	Topological Gel *2691*	
17.2.2	Nanocomposite Gel *2691*	
17.2.3	Double Network Gel *2692*	
17.2.4	Robust Gels from Bacterial Cellulose *2696*	
17.3	Surface Friction and Lubrication of Gels *2698*	
17.3.1	Complexity of Gel Friction *2699*	
17.3.2	Template Effect on Gel Friction *2702*	
17.3.3	Low Friction by Dangling Chains *2704*	
17.4	Polymer Gels as Scaffolds for Cell Cultivation *2706*	
17.4.1	Modulation of Cell Growth on Various Gels *2706*	
17.4.2	Effect of Charge on Cell Growth *2710*	
17.5	Application of Gels as Substitutes for Biological Tissues *2712*	
17.5.1	Robust Gels with Low Friction – an Excellent Candidate as Artificial Cartilage *2713*	
17.5.2	Wearing Properties of Robust DN Gels *2715*	
	References *2716*	

18	**Polymers in Tissue Engineering** *2719*	
	Jeffrey A. Hubbell	
18.1	Introduction and Historical Perspective *2719*	
18.1.1	Tissue Engineering and Its Challenges for Macromolecular Engineering *2719*	
18.1.1.1	Cells *2720*	
18.1.1.2	Biomolecules *2721*	
18.1.1.3	Materials *2724*	
18.1.1.4	Examples of Tissue Engineering Constructs *2725*	
18.1.2	History of Polymers in Tissue Engineering *2726*	
18.2	Degradable Polymer Scaffolds *2727*	
18.3	Degradable Polymer Matrices *2732*	
18.4	Biofunctionalization and Drug Release *2735*	
18.5	Outlook *2739*	
	References *2740*	

IUPAC Polymer Terminology and Macromolecular Nomenclature *2743*
R. F. T. Stepto

Subject Index *2747*

Preface

Macromolecular Engineering: From Precise Macromolecular Synthesis to Macroscopic Materials Properties and Applications aims to provide a broad overview of recent developments in precision macromolecular synthesis and in the design and applications of complex polymeric assemblies of controlled sizes, morphologies and properties. The contents of this interdisciplinary book are organized in four volumes so as to capture and chronicle best, on the one hand, the rapid advances made in the control of polymerization processes and the design of macromolecular architectures (Volumes I and II) and, on the other, the noteworthy progress witnessed in the processing methods – including self-assembly and formulation – to generate new practical applications (Volumes III and IV).

Each chapter in this book is a well-documented and yet concise contribution written by noted experts and authorities in their field. We are extremely grateful to all of them for taking time to share their knowledge and popularize it in a way understandable to a broad readership. We are also indebted to all the reviewers whose comments and remarks helped us very much in our editing work. Finally, Wiley-VCH deserves our sincere acknowledgements for striving to keep the entire project on time.

We expect that specialist readers will find *Macromolecular Engineering: From Precise Macromolecular Synthesis to Macroscopic Materials Properties and Applications* an indispensable book to update their knowledge, and non-specialists will use it as a valuable companion to stay informed about the newest trends in polymer and materials science.

November 2006
Pittsburgh, USA
Bordeaux, France
Paris, France

Krzysztof Matyjaszewski
Yves Gnanou
Ludwik Leibler

Macromolecular Engineering. Precise Synthesis, Materials Properties, Applications.
Edited by K. Matyjaszewski, Y. Gnanou, and L. Leibler
Copyright © 2007 WILEY-VCH Verlag GmbH & Co. KGaA, Weinheim
ISBN: 978-3-527-31446-1

List of Contributors

Volker Abetz
GKSS Research Centre Geesthacht
GmbH
Institute of Polymer Research
Max-Planck-Straße 1
21502 Geesthacht
Germany

Basudam Adhikari
Indian Institute of Technology
Materials Science Centre
Polymer Division
Kharagpur 721302
India

Takuzo Aida
University of Tokyo
School of Engineering
Department of Chemistry
and Biotechnology
7-3-1 Hongo, Bunkyo-ku
Tokyo 113-8656
Japan

Michaël Alexandre
Materia Nova Research Centre asbl
Parc Initialis
1 avenue Nicolas Copernic
7000 Mons
Belgium

Xavier Allonas
University of Haute Alsace, ENSCMu
Department of General
Photochemistry, UMR 7525 CNRS
3 Alfred Werner street
68093 Mulhouse Cedex
France

Denis Andrienko
Max Planck Institute for Polymer
Research
Ackermannweg 10
55128 Mainz
Germany

René Androsch
Martin Luther University
Halle-Wittenberg
Institute of Materials Science
06099 Halle
Germany

Markus Antonietti
Max Planck Institute of Colloids
and Interfaces
Colloid Department
Research Campus Golm
14424 Potsdam
Germany

Macromolecular Engineering. Precise Synthesis, Materials Properties, Applications.
Edited by K. Matyjaszewski, Y. Gnanou, and L. Leibler
Copyright © 2007 WILEY-VCH Verlag GmbH & Co. KGaA, Weinheim
ISBN: 978-3-527-31446-1

List of Contributors

Clemens Auschra
CIBA Specialty Chemicals, Inc.
Research and Development
Coating Effects
Schwarzwaldallee 215
4002 Basel
Switzerland

Ramakrishnan Ayothi
Cornell University
Department of Materials Science
and Engineering
Bard Hall
Ithaca, NY 14853-1501
USA

Vincent Ball
Institut National de la Santé
et de la Recherche Médicale
INSERM Unité 595
11 rue Humann
67085 Strasbourg Cedex
France
and
Université Louis Pasteur
Faculté de Chirurgie Dentaire
1 place de l'Hôpital
67000 Strasbourg
France

Nitash P. Balsara
University of California
Department of Chemical Engineering
and Materials Sciences
and Environmental Energy
Technologies Divisions
Lawrence Berkeley National
Laboratory
Berkeley, CA 94720
USA

Nadia Benkirane-Jessel
Institut National de la Santé
et de la Recherche Médicale
INSERM Unité 595
11 rue Humann
67085 Strasbourg Cedex
France
and
Université Louis Pasteur
Faculté de Chirurgie Dentaire
1 place de l'Hôpital
67000 Strasbourg
France

Christopher W. Bielawski
The University of Texas at Austin
Department of Chemistry
and Biochemistry
Austin, TX 78712
USA

Hans G. Börner
Max Planck Institute of Colloids
and Interfaces
Colloid Department
Research Campus Golm
14424 Potsdam
Germany

Redouane Borsali
Université Bordeaux 1
CNRS, ENSCPB
Laboratoire de Chimie des Polymères
Organiques
16 avenue Pey-Berland
33607 Pessac
France

Bernard Boutevin
Ingénierie et Architectures
Macromoléculaires
Institut Gerhardt, UMR 5253
Ecole Nationale Supérieure Chimie
de Montpellier
8 rue de l'Ecole Normale
34296 Montpellier
France

Cyrille Boyer
Ingénierie et Architectures
Macromoléculaires
Institut Gerhardt, UMR 5253
Ecole Nationale Supérieure Chimie
de Montpellier
8 rue de l'Ecole Normale
34296 Montpellier
France

Daniel Bratton
Cornell University
Department of Materials Science
and Engineering
Bard Hall
Ithaca, NY 14853-1501
USA

Wade A. Braunecker
Carnegie Mellon University
Department of Chemistry
4400 Fifth Avenue
Pittsburgh, PA 15213
USA

Cyril Brochon
Laboratoire d'Ingénierie des
Polymères pour les Hautes
Technologies
UMR 7165 CNRS
Ecole Européenne Chimie Polymères
Matériaux
Université Louis Pasteur
25 rue Becquerel
67087 Strasbourg
France

Manoj B. Charati
University of Delaware
Department of Materials Science
and Engineering
201 DuPont Hall
and
Delaware Biotechnology Institute
15 Innovation Way
Newark, DE 19716
USA

Bernadette Charleux
Université Pierre et Marie Curie
Laboratoire de Chimie des Polymères
4, Place Jussieu, Tour 44, 1er étage
75252 Paris Cedex 5
France

Yong Mei Chen
Hokkaido University
Section of Biological Sciences
Faculty of Science
Laboratory of soft & wet matter
North 10, West 8
060-0810 Sapporo
Japan

Zhiyun Chen
Washington University in Saint Louis
Center for Materials Innovation and
Department of Chemistry
One Brookings Drive
St. Louis, MO 63130-4899
USA

Chong Cheng
Washington University in Saint Louis
Center for Materials Innovation and
Department of Chemistry
One Brookings Drive
St. Louis, MO 63130-4899
USA

Eric Cloutet
Université Bordeaux 1
Laboratoire de Chimie des Polymères
Organiques
Unité Mixte de Recherche
(UMR 5629) CNRS
ENSCPB
16 avenue Pey-Berland
33607 Pessac Cedex
France

Geoffrey W. Coates
Cornell University
Department of Chemistry
and Chemical Biology
Baker Laboratory
Ithaca, NY 14853
USA

Helmut Cölfen
Max Planck Institute of Colloids
and Interfaces
Colloid Chemistry
Research Campus Golm
Am Mühlenberg 1
14476 Potsdam-Golm
Germany

Henri Cramail
Université Bordeaux 1
Laboratoire de Chimie des Polymères
Organiques
Unité Mixte de Recherche
(UMR 5629) CNRS
ENSCPB
16 avenue Pey-Berland
33607 Pessac Cedex
France

Costantino Creton
Laboratoire PPMD
ESPCI
10 rue Vauquelin
75231 Paris
France

Ghislain David
Ingénierie et Architectures
Macromoléculaires
Institut Gerhardt, UMR 5253
Ecole Nationale Supérieure Chimie
de Montpellier
8 rue de l'Ecole Normale
34296 Montpellier
France

Priyadarsi De
University of Massachusetts Lowell
Polymer Science Program
Department of Chemistry
One University Avenue
Lowell, MA 01854
USA

Gero Decher
Institut Charles Sadron
(C.N.R.S. UPR 022)
6 rue Boussingault
67083 Strasbourg Cedex
France
and
Université Louis Pasteur
Faculté de Chimie
1 rue Blaise Pascal
67008 Strasbourg Cedex
France

Alain Deffieux
Université Bordeaux 1
CNRS, ENSCPB
Laboratoire de Chimie des Polymères Organiques
16 avenue Pey-Berland
33607 Pessac
France

Timothy J. Deming
University of California, Los Angeles
Department of Bioengineering
420 Westwood Plaza
7523 Boelter Hall
Los Angeles, CA 90095
USA

G. Dubois
IBM Almaden Research Center
650 Harry Road
San Jose, CA 95120
USA

Philippe Dubois
Université de Mons-Hainaut
Matériaux Polymères et Composites
Place du Parc 20
7000 Mons
Belgium

Andrzej Duda
Center of Molecular and Macromolecular Studies
Polish Academy of Sciences
Department of Polymer Chemistry
Sienkiewicza 112
90-363 Łodz
Poland

Karel Dušek
Academy of Sciences
of the Czech Republic
Institute of Macromolecular Chemistry
Heyrovského nám. 2
162 06 Praha
Czech Republic

Miroslava Dušková-Smrčková
Academy of Sciences
of the Czech Republic
Institute of Macromolecular Chemistry
Heyrovského nám. 2
162 06 Praha
Czech Republic

Todd Emrick
University of Massachusetts Amherst
Department of Polymer Science
and Engineering
120 Governors Drive
Amherst, MA 01003
USA

Tom A. P. Engels
Eindhoven University of Technology
Department of Mechanical Engineering
P.O. Box 513
5600 MB Eindhoven
The Netherlands

Robin S. Farmer
University of Delaware
Department of Materials Science
and Engineering
201 DuPont Hall
and
Delaware Biotechnology Institute
15 Innovation Way
Newark, DE 19716
USA

Rudolf Faust
University of Massachusetts Lowell
Polymer Science Program
Department of Chemistry
One University Avenue
Lowell, MA 01854
USA

Nelson Felix
Cornell University
Department of Materials Science
and Engineering
Bard Hall
Ithaca, NY 14853-1501
USA

Michel Fontanille
Université Bordeaux 1
Laboratoire de Chimie des Polymères
Organiques
ENSCPB
16 avenue Pey-Berland
33607 Pessac
France

Jean Pierre Fouassier
University of Haute Alsace,
ENSCMu
Department of General
Photochemistry, UMR 7525 CNRS
3 Alfred Werner street
68093 Mulhouse Cedex
France

Takeshi Fukuda
Kyoto University
Institute for Chemical Research
Uji, Kyoto 611-0011
Japan

François Ganachaud
Ecole Nationale Supérieure de Chimie
de Montpellier
Laboratoire de Chimie
Macromoléculaire
8 rue de l'Ecole Normale
34296 Montpellier Cedex 5
France

David S. Germack
Washington University in Saint Louis
Center for Materials Innovation and
Department of Chemistry
One Brookings Drive
St. Louis, MO 63130-4899
USA

Yves Gnanou
Université Bordeaux 1
Laboratoire de Chimie des Polymères
Organiques
ENSCPB
16 avenue Pey-Berland
33607 Pessac
France

Jian Ping Gong
Hokkaido University
Section of Biological Sciences
Faculty of Science
Laboratory of soft & wet matter
North 10, West 8
060-0810 Sapporo
Japan

Padma Gopalan
University of Wisconsin – Madison
Department of Materials Science
and Engineering
1117 Engineering Research Building
1500 Engineering Drive
Madison, WI 53706
USA

Leon E. Govaert
Eindhoven University of Technology
Department of Mechanical
Engineering
P.O. Box 513
5600 MB Eindhoven
The Netherlands

Stefanie M. Gravano
University of California, Davis
Department of Chemistry
One Shields Avenue
Davis, CA 95616-5295
USA

Andrew C. Grimsdale
Nanyang Technological University
School of Materials Science and
Engineering
50 Nanyang Avenue
Singapore 639798

Nikos Hadjichristidis
University of Athens
Department of Chemistry
Panepistimiopolis Zografou
15771 Athens
Greece

Georges Hadziioannou
Laboratoire d'Ingénierie des
Polymères pour les Hautes
Technologies
UMR 7165 CNRS
Ecole Européenne Chimie Polymères
Matériaux
Université Louis Pasteur
25 rue Becquerel
67087 Strasbourg
France

Dale L. Handlin, Jr.
Kraton Polymers
700 Milam
North Tower
Houston, TX 77002
USA

James L. Hedrick
IBM Almaden Research Center
650 Harry Road
San Jose, CA 95120
USA

Thomas Heiser
Université Louis Pasteur
Institut d'Electronique du Solide
et des Systèmes
UMR 7163, CNRS
23 rue du Loess
67087 Strasbourg
France

Valérie Héroguez
Université Bordeaux 1
Laboratoire de Chimie des Polymères
Organiques
ENSCPB
16 avenue Pey-Berland
33607 Pessac
France

Richard Hoogenboom
Eindhoven University of Technology
and Dutch Polymer Institute (DPI)
Laboratory of Macromolecular
Chemistry and Nanoscience
P.O. Box 513
5600 MB Eindhoven
The Netherlands

Jeffrey A. Hubbell
Ecole Polytechnique Fédérale
de Lausanne
Institute of Bioengineering
1015 Lausanne
Switzerland

Hermis Iatrou
University of Athens
Department of Chemistry
Panepistimiopolis Zografou
15771 Athens
Greece

Olli Ikkala
Helsinki University of Technology
Department of Engineering Physics
and Mathematics
and Center for New Materials
P.O. Box 2200
02015 Hut
Espoo
Finland

Woo-Dong Jang
The University of Tokyo
School of Engineering
Department of Chemistry
and Biotechnology
7-3-1 Hongo, Bunkyo-ku
Tokyo 113-8656
Japan

Sreeja Jayant
Rutgers, The State University
of New Jersey
Department of Pharmaceutics
160 Frelinghuysen Road
Piscataway, NJ 08854-8020
USA

Robert Jerome
University of Liège
Center for Education and Research
on Macromolecules (CERM)
Sart-Tilman, B6a
4000 Liège
Belgium

Youngjong Kang
Massachusetts Institute of Technology
Department of Materials Science
and Engineering and
Institute for Soldier Nanotechnologies
Cambridge, MA 02139
USA

Jayant J. Khandare
Rutgers, The State University
of New Jersey
Department of Pharmaceutics
160 Frelinghuysen Road
Piscataway, NJ 08854-8020
USA

Kristi L. Kiick
University of Delaware
Department of Materials Science
and Engineering
201 DuPont Hall
and
Delaware Biotechnology Institute
15 Innovation Way
Newark, DE 19716
USA

Tatsuki Kitayama
Osaka University
Department of Chemistry
Graduate School of Engineering
Toyonaka, Osaka 560-8531
Japan

Harm-Anton Klok
Ecole Polytechnique Fédérale
de Lausanne (EPFL)
Institut des Matériaux
Laboratoire des Polymères
STI – IMX – LP, MXD 112
(Bâtiment MXD), Station 12
1015 Lausanne
Switzerland

Bert Klumperman
Eindhoven University of Technology
Laboratory of Polymer Chemistry
P.O. Box 513
5600 MB Eindhoven
The Netherlands

Shiro Kobayashi
Kyoto Institute of Technology
R & D Center for Bio-based Materials
Matsugasaki, Sakyo-ku
Kyoto 606-8585
Japan

Kurt Kremer
Max Planck Institute for Polymer
Research
Ackermannweg 10
55128 Mainz
Germany

Vahik Krikorian
Massachusetts Institute of Technology
Department of Materials Science
and Engineering and
Institute for Soldier Nanotechnologies
Cambridge, MA 02139
USA

Przemyslaw Kubisa
Center of Molecular and
Macromolecular Studies
Polish Academy of Sciences
Department of Polymer Chemistry
Sienkiewicza 112
90-363 Łodz
Poland

Alok Kumar Sen
Indian Institute of Technology
Materials Science Centre
Polymer Division
Kharagpur 721302
India

Jacques Lalevée
University of Haute Alsace, ENSCMu
Department of General
Photochemistry, UMR 7525 CNRS
3 Alfred Werner street
68093 Mulhouse Cedex
France

Katharina Landfester
University of Ulm
Department of Organic Chemistry III
– Macromolecular Chemistry
and Organic Materials
Albert-Einstein-Allee 11
89081 Ulm
Germany

Philippe Lavalle
Institut National de la Santé
et de la Recherche Médicale
INSERM Unité 595
11 rue Humann
67085 Strasbourg Cedex
France
and
Université Louis Pasteur
Faculté de Chirurgie Dentaire
1 place de l'Hôpital
67000 Strasbourg
France

Nicolas Leclerc
Université Louis Pasteur
Laboratoire d'Ingénierie des
Polymères pour les Hautes
Technologies
UMR 7165 CNRS
Ecole Européenne Chimie Polymères
Matériaux
25 rue Becquerel
67087 Strasbourg
France

Young Moo Lee
Hanyang University
School of Chemical Engineering
College of Engineering
Seoul 133-791
Korea

Ludwik Leibler
Matière Molle et Chimie
UMR 167 CNRS
ESPCI
10 rue Vauquelin
75005 Paris
France

G. B. W. L. Ligthart
DSM Campus Geleen
Performance Materials
PO Box 18
6160 MD Geleen
The Netherlands

Wei-Shi Li
ERATO-SORST Nanospace Project
Japan Science and
Technology Agency (JST)
National Museum of Emerging
Science and Innovation
2-41 Aomi, Koto-ku
Tokyo 135-0064
Japan

Pierre Lutz
Institut Charles Sadron
6 rue Boussingault
67083 Strasbourg Cedex
France

Ian Manners
University of Bristol
Department of Chemistry
Cantock's Close
Bristol BS8 1TS
UK

Andrew F. Mason
IBM Almaden Research Center
650 Harry Road
San Jose, CA 95120
USA

Toshio Masuda
Kyoto University
Department of Polymer Chemistry
Graduate School of Engineering
Katsura Campus
Kyoto 615-8510
Japan

Krzysztof Matyjaszewski
Carnegie Mellon University
Department of Chemistry
4400 Fifth Avenue
Pittsburgh, PA 15213
USA

Thomas C. B. McLeish
University of Leeds
IRC in Polymer Science
and Technology
Polymers and Complex Fluids
Department of Physics
and Astronomy
Leeds LS2 9JT
UK

Michael A. R. Meier
Eindhoven University of Technology
and Dutch Polymer Institute (DPI)
Laboratory of Macromolecular
Chemistry and Nanoscience
P.O. Box 513
5600 MB Eindhoven
The Netherlands

E. W. Meijer
Eindhoven University of Technology
Laboratory of Macromolecular
and Organic Chemistry
P.O. Box 513
5600 MB Eindhoven
The Netherlands

Han E. H. Meijer
Eindhoven University of Technology
Department of Mechanical
Engineering
P.O. Box 513
5600 MB Eindhoven
The Netherlands

R. D. Miller
IBM Almaden Research Center
650 Harry Road
San Jose, CA 95120
USA

Tamara Minko
Rutgers, The State University
of New Jersey
Department of Pharmaceutics
160 Frelinghuysen Road
Piscataway, NJ 08854-8020
USA

Martin Moller
RWTH Aachen
Institut für Technische
und Makromolekulare Chemie
Pauwelsstraße 8
52056 Aachen
Germany

Hideharu Mori
Yamagata University
Faculty of Engineering
Department of Polymer Science
and Engineering
4-3-16, Jonan
Yonezawa 992-8510
Japan

Klaus Müllen
Max Planck Institute
for Polymer Research
Ackermannweg 10
55128 Mainz
Germany

Axel H. E. Müller
University of Bayreuth
Macromolecular Chemistry II
95440 Bayreuth
Germany

Anna Musyanovych
University of Ulm
Department of Organic Chemistry III
– Macromolecular Chemistry
and Organic Materials
Albert-Einstein-Allee 11
89081 Ulm
Germany

Kazukiyo Nagai
Meiji University
Department of Applied Chemistry
1-1-1 Higashi-mita, Tama-ku
Kawasaki 214-8571
Japan

Christopher K. Ober
Cornell University
Department of Materials Science
and Engineering
310 Bard Hall
Ithaca, NY 14853-1501
USA

Joëlle Ogier
Institut National de la Santé
et de la Recherche Médicale
INSERM Unité 595
11 rue Humann
67085 Strasbourg Cedex
France
and
Université Louis Pasteur
Faculté de Chirurgie Dentaire
1 place de l'Hôpital
67000 Strasbourg
France

Masashi Ohmae
Kyoto University
Department of Materials Chemistry
Graduate School of Engineering
Katsura, Nishikyo-ku
Kyoto 615-8510
Japan

Kohji Ohno
Kyoto University
Institute for Chemical Research
Uji, Kyoto 611-0011
Japan

Masami Okamoto
Advanced Polymeric Materials
Engineering
Graduate School of Engineering
Toyota Technological Institute
2-12-1 Hisakata, Tempaku
Nagoya 468-8511
Japan

Joost A. Opsteen
Radboud University Nijmegen
Institute for Molecules and Materials
Toernooiveld 1
6525 ED Nijmegen
The Netherlands

Yoshihito Osada
Hokkaido University
Section of Biological Sciences
Faculty of Science
Laboratory of soft & wet matter
North 10, West 8
Sapporo 060-0810
Japan

Timothy E. Patten
University of California, Davis
Department of Chemistry
One Shields Avenue
Davis, CA 95616-5295
USA

Stanislaw Penczek
Centre of Molecular
and Macromolecular Studies
Polish Academy of Sciences
Department of Polymer Chemistry
Sienkiewicza 112
90-363 Łódz
Poland

Catherine Picart
Université de Montpellier II
Laboratoire de Dynamique
des Interactions Membranaires
Normales et Pathologiques
(C.N.R.S. UMR 5235)
Place Eugène Bataillon
34095 Montpellier Cedex
France

Frank Pirrung
CIBA Specialty Chemicals, Inc.
Research and Development
Coating Effects
Schwarzwaldallee 215
4002 Basel
Switzerland

Marinos Pitsikalis
University of Athens
Department of Chemistry
Panepistimiopolis Zografou
15771 Athens
Greece

Jeffrey Pyun
University of Arizona
Department of Chemistry
1306 E. University Boulevard
Tucson, AZ 85721
USA

Damien Quémener
Université Bordeaux 1
Laboratoire de Chimie des Polymères
Organiques
ENSCPB
16 avenue Pey-Berland
33607 Pessac
France

Karunakaran Radhakrishnan
University of Akron
Institute of Polymer Science
170, University Avenue
Akron, OH 44325
USA

Wolfgang Radke
Deutsches Kunststoff-Institut
Darmstadt
Schlossgartenstraße 6
65289 Darmstadt
Germany

David A. Rider
University of Toronto
Department of Chemistry
80 St. George Street
M5S 3H6 Toronto, Ontario
Canada

Megan L. Ruegg
University of California
Department of Chemical Engineering
Berkeley, CA 94720
USA

Pierre Schaaf
Institut Charles Sadron
(C.N.R.S. UPR 022)
6 rue Boussingault
67083 Strasbourg Cedex
France
and
Ecole Européenne de Chimie
Polymères et Materiaux
25 rue Bequerel
67087 Strasbourg Cedex 2
France

Oren A. Scherman
University of Cambridge
Department of Chemistry
Lensfield Road
Cambridge CB2 1EW
UK

Helmut Schlaad
Max Planck Institute of Colloids
and Interfaces
Colloid Department
Research Campus Golm
14424 Potsdam
Germany

Ulrich S. Schubert
Eindhoven University of Technology
and Dutch Polymer Institute (DPI)
Laboratory of Macromolecular
Chemistry and Nanoscience
P.O. Box 513
5600 MB Eindhoven
The Netherlands

Bernard Senger
Institut National de la Santé
et de la Recherche Médicale
INSERM Unité 595
11 rue Humann
67085 Strasbourg Cedex
France
and
Université Louis Pasteur
Faculté de Chirurgie Dentaire
1 place de l'Hôpital
67000 Strasbourg
France

Sergei S. Sheiko
University of North Carolina
at Chapel Hill
Department of Chemistry
Chapel Hill, NC 27599-3290
USA

Rint P. Sijbesma
Eindhoven University of Technology
Laboratory of Macromolecular
and Organic Chemistry
P.O. Box 513
5600 MB Eindhoven
The Netherlands

Peter F. W. Simon
GKSS Research Centre
Geesthacht GmbH
Institute of Polymer Research
Max-Planck-Straße
21502 Geesthacht
Germany

Stanislaw Slomkowski
Center of Molecular and
Macromolecular Studies
Polish Academy of Sciences
Department of Polymer Chemistry
Sienkiewicza 112
90-363 Łodz
Poland

Hans Wolfgang Spiess
Max-Planck-Institute for Polymer
Research
Spectroscopy
P.O. Box 3148
55021 Mainz
Germany

Richard J. Spontak
North Carolina State University
Departments of Chemical and
Biomolecular Engineering and
Materials Science and Engineering
Raleigh, NC 27695
USA

R. F. T. Stepto
The University of Manchester
School of Materials
Materials Science Centre
Polymer Science
and Technology Group
Grosvenor Street
Manchester M1 7HS
UK

Brent S. Sumerlin
Southern Methodist University
Department of Chemistry
3215 Daniel Avenue
Dallas, TX 75240-0314
USA

Yeshayahu Talmon
Technion – Israel Institute
of Technology
Department of Chemical Engineering
32000 Haifa
Israel

M. Atilla Tasdelen
Istanbul University
Department of Chemistry
Maslak
Istanbul 34469
Turkey

Daniel Taton
Université Bordeaux 1
Laboratoire de Chimie des Polymères
Organiques
LCPO CNRS
ENSCPB
16 avenue Pey-Berland
33607 Pessac
France

Gerrit ten Brinke
University of Groningen
Laboratory of Polymer Chemistry
Materials Science Centre
Nijenborgh 4
9747 AG Groningen
The Netherlands

Edwin L. Thomas
Massachusetts Institute of Technology
Department of Materials Science
and Engineering
and
Institute for Soldier Nanotechnologies
Cambridge, MA 02139
USA

Scott Trenor
Kraton Polymers
Houston, TX 77002
USA

Constantinos Tsitsilianis
University of Patras
and FORTH/ICEHT
Department of Chemical Engineering
Karatheodori 1
26504 Patras
Greece

Yoshinobu Tsujii
Kyoto University
Institute for Chemical Research
Uji, Kyoto 611-0011
Japan

Jan C. M. van Hest
Radboud University Nijmegen
Institute for Molecules and Materials
Toernooiveld 1
6525 ED Nijmegen
The Netherlands

Brooke A. van Horn
Washington University in Saint Louis
Center for Materials Innovation and
Department of Chemistry
One Brookings Drive
St. Louis, MO 63130-4899
USA

Shrinivas Venkataraman
Washington University in Saint Louis
Center for Materials Innovation and
Department of Chemistry
One Brookings Drive
St. Louis, MO 63130-4899
USA

Jean-Claude Voegel
Institut National de la Santé
et de la Recherche Médicale
INSERM Unité 595
11 rue Humann
67085 Strasbourg Cedex
France
and
Université Louis Pasteur
Faculté de Chirurgie Dentaire
1 place de l'Hôpital
67000 Strasbourg
France

C. Grant Willson
The University of Texas at Austin
Departments of Chemistry
and Biochemistry
and Chemical Engineering
Austin, TX 78712
USA

Karen L. Wooley
Washington University in Saint Louis
Center for Materials Innovation and
Department of Chemistry
One Brookings Drive
St. Louis, MO 63130-4899
USA

Kathryn Wright
Kraton Polymers
700 Milam
North Tower
Houston, TX 77002
USA

Bernhard Wunderlich
University of Tennessee
Department of Chemistry
200 Baltusrol Road
Knoxville, TN 37992-3707
USA

Yusuf Yagci
Istanbul University
Department of Chemistry
Maslak
Istanbul 34469
Turkey

Tsutomu Yokozawa
Kanagawa University
Department of Material
and Life Chemistry
Rokkakubashi, Kanagawa-ku
Yokohama 221-8686
Japan

1
Applications of Thermoplastic Elastomers Based on Styrenic Block Copolymers*

Dale L. Handlin, Jr., Scott Trenor, and Kathryn Wright

1.1
Introduction

The emergence of anionic, living polymerization in the 1950s allowed the synthesis of polymers with controlled molecular weight, polydispersity and monomer distribution. A key consequence of controlled molecular architecture was the distinct phase separation and self-assembling morphologies of polymers containing discrete blocks of styrene and isoprene or butadiene. The self-reinforcing nature of such styrenic block copolymers was recognized for its broad commercial potential as a new class of polymers; styrenic thermoplastic elastomers (TPEs).

More than 40 years after the commercial introduction of styrenic block copolymer TPEs, the annual global consumption of these polymers exceeds 10^9 kg. This consumption includes scores of different polymer structures tailored to meet the specific needs of numerous markets and applications. Applications as diverse as the roads we drive on, the cars in which we drive, the shoes we wear and the personal hygiene utensils we use are all enhanced by the properties of styrenic block copolymer TPEs.

This chapter reviews the discovery of the first styrenic block copolymer TPEs and their existing and developing applications. The reader is encouraged to pursue one of the many more comprehensive reviews both from the early days of development [1–3] and more recent reviews [4–8].

1.1.1
Origins and Development

The origins of styrenic block copolymers (SBCs) can be found in the US Government's Synthetic Rubber (GR-S) program, which began during World War II. At that time, Germany was well advanced in the synthesis of rubber and the

---

* A List of Abbreviations can be found at the end of this chapter.

Macromolecular Engineering. Precise Synthesis, Materials Properties, Applications.
Edited by K. Matyjaszewski, Y. Gnanou, and L. Leibler
Copyright © 2007 WILEY-VCH Verlag GmbH & Co. KGaA, Weinheim
ISBN: 978-3-527-31446-1

USA was concerned that the inability to secure a rubber supply would cripple the military effort. Between 1942 and 1945, this program spent $700 million to build 51 synthetic rubber plants in the USA, all run by different companies. This program was successful in developing a synthetic rubber called GR-S, Government Rubber–Styrene, the first large-volume commercial polymer produced in the USA. This rubber was generally a random or tapered copolymer of styrene and butadiene which we now refer to as SBR. In 1955, with the goals of the GR-S program complete, the program was terminated. The government-owned plants were sold to the various operating companies through a bidding process. Shell Chemical Company and Philips Chemical, both butadiene suppliers, bought the plants that they were operating. The Shell-owned plant in Torrance, CA, was of particular importance because it combined styrene, butadiene, isoprene and rubber compounding, which allowed the eventual synthesis and processing of block copolymers. At the final program review in 1955, Stavely at Firestone presented his work on the synthesis of high-cis-1,4 polyisoprene using alkyllithium initiators. Based on Stavely's results, Porter at Shell began lithium alkyl polymerization of polyisoprene. He developed a process to make 93% cis polyisoprene; the closest thing to natural rubber yet developed. Shell started a rapid development program to commercialize anionic homopolyisoprene that "imitated Nature" with the intention of producing polyisoprene on a large scale for tires. At about the same time, Szwarc, Levy and Milkovich announced the synthesis of block copolymers using anionic initiators [9]. The single-page letter in which they reported the synthesis of styrene–butadiene diblocks and the ability to synthesize ABA and other types of block copolymers was published in the *Journal of the American Chemical Society* in 1956. After the completion of his PhD under Maurice Morton at the University of Akron, Milkovich was hired by Shell to help commercialize the first anionic polyisoprene in 1959. Commercial production of anionic polyisoprene rubber began in Torrance, CA, then was further expanded to a dedicated plant in 1961.

Shell was the first, but not the only, company to commercialize synthetic isoprene. Firestone, who did much of the original work, also commercialized anionic polyisoprene, as did Asahi. At the same time, Zelinski and others at Philips developed and commercialized high-cis polybutadiene by anionic polymerization. Zelinski described the period between 1955 and 1960 as being like "Thanksgiving and Christmas all rolled into one" [10]. The ability to generate a wide variety of AB, $(AB)_x$, ABA, and ABC polymers made this an exciting time for polymer chemistry. Most major chemical companies were exploring block copolymers in the late 1950s and early 1960s. Given the excitement about anionic polymerization in industry and academia, why was Shell Chemical Company the ultimate developer of styrenic block copolymers? The answer is a simple one: the large market for synthetic rubber was automotive tires. Goodyear and B.F. Goodrich had found that they could use Ziegler–Natta catalysts to make very high (95–98%) cis polyisoprene, which was closer to natural rubber than anionic polyisoprene and worked well in tires. Firestone and Philips commercialized anionic cis polybutadiene and Philips also commercialized tapered polysty-

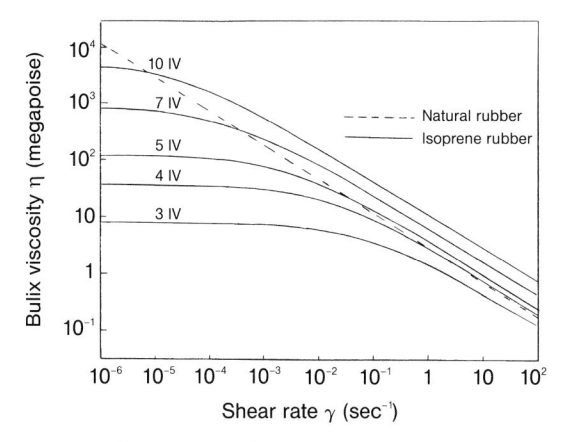

Fig. 1.1 Bulk viscosities of isoprene and natural rubber at 100 °C.

rene–butadiene, both of which were useful in tires. Anionic polyisoprene, which is only 91–93% cis, did not have the green strength (needed to make good sheet), build tack (surface stickiness needed to keep the plies together before molding so that no bubbles form between plies) or hot tear strength (needed for performance under driving conditions) to succeed in tires. Facing the potential failure of a large chemical plant in 1961, Legge at Shell put together a group including Milkovich and Holden to attack the problems plaguing commercial polyisoprene [11]. Although styrenic block copolymers had been known since 1956 and even patented by Porter in 1958 [12], the focus had been on new compositions while their potential as elastomeric materials remained undiscovered. Milkovich, part of the original block polymer chemistry team and an expert in anionic synthesis, shared an office with Holden, a rheologist. Their initial objective was to develop a polymer that could be added to anionic polyisoprene to provide an infinite zero shear rate viscosity comparable to that of natural rubber. As shown in Fig. 1.1, natural rubber is both very high in molecular weight and lightly cross-linked so that even at very low shear rates it does not flow, which imparts green strength. Anionic polyisoprene, however, behaves as a typical thermoplastic with a Newtonian plateau at low shear rates.

Therefore, even if the molecular weight was increased to match the viscosity of natural rubber at typical application shear rates, flow would still occur at low shear rates. Milkovich and Segovia made a range of tapered, random, diblock and triblock polymers to test as rheology modifiers for synthetic polyisoprene. According to Segovia's Shell laboratory notebook, the first SIS with interesting mechanical properties was made on 6 September 1961 [13] (Fig. 1.2).

Initially, only the rheological properties of these so-called "double-ended block copolymers" were measured. Holden confirmed that the addition of 25% styrene–isoprene–styrene (SIS) triblock copolymer to an anionic polyisoprene provided the infinite zero shear viscosity which was needed to mimic the flow curve of natural rubber over a wide range of shear rates (Fig. 1.3). Further, the

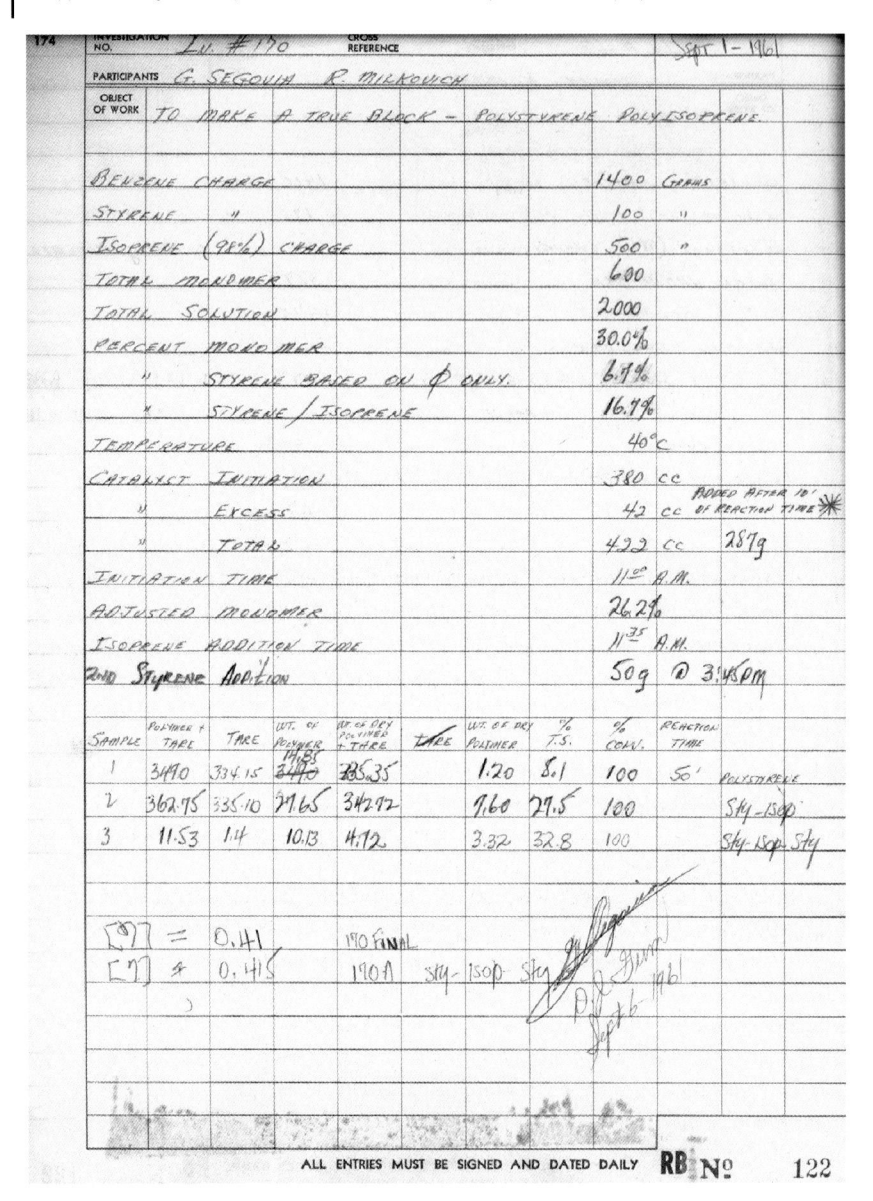

Fig. 1.2 Page 174 of G. Segovia's laboratory notebook at Shell Chemical Company showing the synthesis of the first SIS triblock with commercial potential.

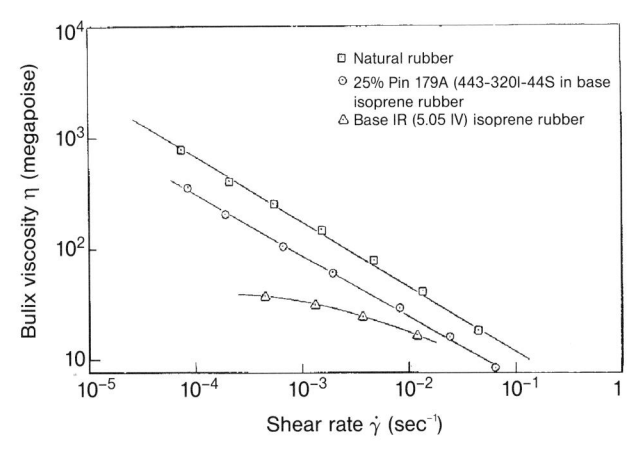

Fig. 1.3 Bulk viscosity of isoprene rubber at 100 °C blended with block copolymer.

activation energy of flow measurements indicated that the flow mechanism was based on the phase separation of the polystyrene blocks from the polyisoprene blocks.

On 29 November 1961, an SIS was made in the pilot plant with molecular weights 30 000–90 000–20 000 whose pure mechanical properties were measured by Hendricks and Holden on 6 December 1961 [14, 15]. This was the first styrenic block copolymer that was recognized as an elastomer without vulcanization. SBS and SI coupled star polymers were made and evaluated over the next few months. For the first time, the potential of these materials outside of the area of polyisoprene flow modification was recognized.

The unique rheology and mechanical properties of SIS and SBS block copolymers led Meier, a polymer physicist in Shell's Emeryville laboratory, to develop the confined chain model of block copolymer phase separation, which was later published in 1969 [16]. Meier reviewed the development of block copolymer theory in Chapter 8 of Ref. [4]. Understanding how the SIS and SBS block copolymers behaved, and to some extent what they were useful for, Milkovich and Holden filed for a US Patent in 1962 [17]. This was the first process patent for commercially interesting ABA block copolymers.

While the addition of SIS block copolymers to anionic polyisoprene did improve the green strength and build tack, it did not improve the hot tear properties. Hence the block copolymer effort to produce a large-volume synthetic tire rubber, as conceived, was a failure. The group at Shell had created polymers with outstanding properties, but not ones suited for the target application. This was a pivotal point in polymer development. Two events helped launch block copolymers in a commercial direction: first, Shell was in the footwear business with SBR compounds and, therefore, knew the application requirements, and second, a German company, Desma, introduced a new direct injection machine for footwear called the Desma 10. It had a single extruder for the injection of natural rubber shoe soles into 10 molding stations arranged on a table that rotated slowly enough for the rubber sole to cure and be removed before the mold reached the extruder again. Hendricks and Danforth realized that if the injected compounds were thermoplastic elastomers, the table could rotate rapidly because no cure time would be required. The reduced cycle time provided the driving force for the commercialization of styrenic thermoplastic elastomers in the footwear market. The first trials were performed in 1964, and in 1965 Scotty Shoes in Massachusetts bought the first truckload of a styrenic block copolymer compound. The 10 000 lb (ca. 4500 kg) of compound was made 30 lb (ca. 13.5 kg) at a time on a Banbury mixer at the original government-sponsored Torrance plant [18]. This footwear application lead to the announcement of the first commercial block copolymers at an ACS Symposium on 3 May 1965, in Miami Beach [19].

With the advent of "earth shoes" with their large TPE soles, volumes of SBS block copolymers soared, driving the construction of the first dedicated block copolymer plant by Shell Chemical in Belpre, OH, in 1971. Although Philips had been working on block copolymer since the early 1960s, it was the burgeoning footwear market that encouraged their first commercial polymers in 1973 [5]. Further discussion of the footwear application can be found in Section 1.2.

Whereas footwear applications accelerated, adhesives, asphalt modification and compound applications were slow to take hold. In the 1960s, adhesives were primarily solvent based. Although SIS polymers offered good tack and excellent strength without curing, the commercial application method in the 1960s was solvent based. The application of SIS from solvent allowed lower solvent levels, but this was not enough of an advantage to displace natural rubber. As described in Section 1.4, it was not until the early 1980s, with the advent of hot melt coating, a process that offered not only the elimination of solvent but also significant application rate increases, that the adhesives market began to emerge as a large commercial application.

The history of block copolymer application in the road and roofing markets was a similar story. Although the first trials were in 1967, application was limited. In the 1980s, however, European commercial roofing began to shift from concrete to metal, which required a more flexible roofing felt, thus encouraging the use of SBS polymers for better elasticity. In the USA, the SHRP Roads Program in the early 1990s helped to standardize requirements for paving asphalt, which led to dramatic increases in SBS-modified asphalts, especially in climates where temperatures varied greatly through the year. More details about bitumen modification can be found in Section 1.3.

Footwear also played a prominent role in the development of hydrogenated block copolymers. One of the prominent features of SBS footwear was good wet traction. Therefore, SBS-soled shoes were frequently used in boating and other outdoor applications. In the late 1960s, it was found that the foxings, that is, the upper edges of the soles, often cracked after prolonged outdoor use. At the same time, fundamental research into hydrogenation chemistry at Shell's Emeryville laboratory was producing fully hydrogenated SIS polymers, referred to as poly(cyclohexane–ethylene propylene–cyclohexane) or C-EP-C polymers. These polymers were being trialed by the National Institutes of Health as materials for artificial hearts [20]. They had the high strength, good elasticity and low thrombogenicity required to make a good heart wall. These polymers were trialed in footwear, but could not be compounded with oils normally used to adjust the hardness of the sole. When SIS and SBS polymers were selectively hydrogenated, the polystyrene blocks were oil resistant [21]. Hence the development of selective hydrogenation, in which only the rubber block of the polymer was hydrogenated, gave rise to a new family of polymers which were not only more environmentally stable but also had greater strength.

Although these selectively hydrogenated polymers had excellent properties in footwear compounds, they did not succeed in footwear applications because it proved too difficult to glue the hydrogenated polymer sole to the uppers. However, these polymers found an immediate place in automotive compounds, as described in Section 1.5.3. SBS compounds had already been trialed in a number of automotive applications with at least one spectacular failure. In 1965, black horn pads were made for Fords cars. After prolonged exposure to the sun, the SBS degraded, causing the pads to crack [18]. With the development of selectively hydrogenated block copolymers, however, UV- and thermally stable compounds were now readily formulated. From 1967 to 1970, a series of compounds were developed using oil and polypropylene with SEBS polymers. The oil and polypropylene allowed tuning of rheology and hardness over a wide range and at reduced cost. The polypropylene also provided extended temperature performance. Marketplace demand in the USA was provided by the regulation passed in 1972 which required all cars to be equipped with a bumper capable of absorbing a 5 mile per hour (mph) impact without damage. The first successful commercial automotive application of hydrogenated block copolymers as a fascia material for the 1974 AMC Matador bumper. Applications such as Jeep CJ fenders, sound-damping panels, air ducts and others soon followed, as described in Section 5.3.

Despite these high-profile applications, a lesser known application, viscosity index improvers (VIIs) for motor oil modification, discussed in Section 1.7.1, actually drove the commercialization of hydrogenated block copolymers. This was a purely internal application to Shell. In 1969, a group began to screen a wide variety of block copolymers for the modification of motor oils. St. Clair found that hydrogenated styrene–isoprene diblocks (S-EP) were not only effective as VIIs, but allowed the generation of the first 10W/50 motor oil [22]. Although it was not clear that a 10W/50 motor oil was needed, it gained rapid market acceptance. This application drove the conversion of the first dedicated synthetic polyisoprene plant built in 1961 to a hydrogenated block copolymer plant in 1974. With this conversion, Shell ceased altogether the polyisoprene production that had led to the invention of commercial block copolymers in the USA, but continued limited production in Europe.

It is important to note that from their introduction in 1965, it took 20 years for styrenic block copolymers to reach 10^8 kg per year. This is a typical time frame for the development of a complex technology that creates significant change. As you consider the emerging applications reviewed in Section 1.8, it is helpful to remember that while many of these technologies have existed for some time, their applications are still developing much as SIS and SBS were developing in the 1960s.

1.2
Footwear

Footwear became the first commercial application for styrenic block copolymers with the advent of the direct injection molder for shoe soles in the 1960s as described in Section 1.1.1 [23–25]. In fact, the first application patent for styrenic block copolymers was for a "footwear assembly" [26]. The driving force for using SBCs was the greatly reduced cycle time, which increased production rates. In the USA, polyurethanes and PVC have largely replaced SBCs in the footwear industry. However, footwear remains a significant market in Europe and South America and has become the largest SBC market in Asia. Because shoe soles require relatively stiff and cheap elastomers, they are generally filled compounds containing SBS linear or radial block copolymers with styrene contents between 30 and 50%. A typical footwear compound consists of mixtures of SBS polymers, oil, polystyrene and up to 50% filler (talc and/or calcium carbonate). Occasionally, poly(ethylene–vinyl acetate) is added to increase the UV, ozone and solvent resistance of the shoe sole. SBS-based footwear has excellent skid resistance, low temperature flexibility and fatigue life. However, it lacks the abrasion resistance under rapid loading that is required for high-performance footwear because of the increase in energy absorption at elevated temperature associated with the polystyrene domain glass transition temperature (T_g).

The high level of filler in SBC footwear, excellent traction and fast cycle times have made SBCs the first choice in low-cost footwear since the "earth shoes" of the 1970s. Today, with the inclusion of either chemical blowing agents (azodicarbonamides) or physical blowing agents, the density of the sole can be reduced by as much as 30–40%. By varying the concentration of either blowing agent through the thickness of the sole, a manufacturer can vary the density and the energy absorption to the sole. When filler is not used, however, the inherent clarity of block copolymers makes it possible to make clear footwear and clear gel inserts for shoes.

1.3
Bitumen Modification

1.3.1
Paving

The modification of bitumen is the largest market for SBCs in the USA and Europe. The bitumen modification market is typically split into two segments: roofing and paving. Bitumen, a product of crude oil refining, is complex and varies with the crude oil source, age of the oil well and processing conditions. Hence, unlike modifiers in adhesive formulations, manufacturers have little control over its composition. Its primary components are asphaltenes and maltenes, which in practice are differentiated by solubility in n-heptane [27, 28]. Asphaltenes, insoluble in n-heptane, are poly-condensed aromatic and polar molecules with molecular weights up to a few thousand grams per mole and hydrogen to carbon ratios of 1.1–1.2. Their physical structure consists of mostly aromatic sheets with paraffinic side-chains. They possess some polarity in the form of ketones, alcohols, ethers and carboxylic acids. The asphaltenes are typically immiscible in either the polystyrene or polybutadiene blocks. Maltenes, soluble in n-heptane, consist of three basic types of molecules, saturates, aromatics and resins [27]. Saturates are typically soluble in the polybutadiene block whereas the aromatics are slightly miscible with the polystyrene blocks at ambient temperature and become increasingly miscible at processing temperatures.

The saturate/aromatic balance is particularly important in formulating high-quality paving materials. If too little aromatic fraction is present, the strong phase separation of the SBC will prevent it from dissolving in the asphalt at processing temperatures of 160–190 °C. Too much aromatic fraction will reduce the strength of the styrene phase association at ambient temperatures, thus reducing the upper service temperature of the modified bitumen. During blending, a range of grafting, chain scission and coupling reactions can occur, which aid in the compatibilization of the SBS and bitumen mixture.

Other than in high process temperature or direct UV exposure applications, bitumen modification uses SBS block copolymers because of the lower cost of the base polymers and their propensity to cross-link during degradation rather than losing continuity by chain scission as polyisoprene midblocks do. These SBS polymers usually contain 30% styrene and are generally linear and radial polymers high molecular weight, typically 100 000–500 000 g mol^{-1} [23]. The high styrene content and radial architecture help to retain hard segment continuity at low polymer levels. When properly blended, the polybutadiene phase is able to absorb up to 10 times its weight in maltenes. At concentrations of 3% polymer, a continuous SBS network swollen by the soluble bitumen components begins to form. On increasing the polymer concentration to 5% or greater, the SBS polymer forms a continuous polymer network. Typical concentrations of SBS in roads are 3–6% based on bitumen (the bitumen concentration

in roads is approximately 5 wt%), where a number of beneficial effects are encountered, including:
- reduced permanent deformation at elevated temperatures and high traffic loading;
- reduced fatigue and low-temperature cracking because of increased ductility of the asphalt binder at low temperatures;
- reduced aggregate loss because of increased binder toughness over wide temperature ranges [28].

1.3.2
Road Marking

A small, but growing, market segment is the modification of road marking paints with SBCs. Thermoplastic road markings are a mixture of glass beads, pigment and filler held together by a binder. Historically, binders of the thermoplastic road markings consist of low molecular weight petroleum-based resins or rosin derivatives which can be mixed with plasticizers to reduce the T_g of the binder and thus toughen the road marking. The addition of SEBS polymers to the road markings improves the mechanical properties of the binder, which leads to an improvement in the erosion resistance and thus the lifetime of the road marking. Including as little as 2 wt% of an SEBS polymer to the road marking paint increases the lifetime of the paint from 1 500 000 to over 4 000 000 wheel passages. The retention of the glass beads is also improved with the inclusion of the SEBS polymer.

1.3.3
Roofing

Roofing membranes and waterproofing coatings require higher elasticity than paving bitumen, therefore they typically contain from 8 to 20% SBS [23, 25]. At this SBS level, the membranes perform more like swollen elastomers than bitumen. Some of the benefits of replacing un-modified roofing materials with SBS modified roofing materials include:
- increased resistance to flow at elevated surface temperatures, which helps prevent the membrane from flowing on highly sloped roofs;
- increased puncture resistance because of increased toughness;
- increased low-temperature flexibility;
- increased resistance to damage by expansion and contraction of the roof during daily cyclic heating cycles.

1.4
Adhesives and Sealants

The use of styrenic block copolymers in adhesives and sealants was first developed by researchers at Shell in the 1960s and continues to be among the most important applications today. SBCs are most often used in pressure-sensitive adhesives (PSAs), which represent a significant fraction of the adhesives market. A typical PSA must possess the following five material properties [29]:
- aggressive and permanent tack;
- adhesion without the need of more than finger or hand pressure;
- no activation by water, heat or solvent;
- a strong holding force;
- sufficient cohesiveness and elasticity to allow removal from smooth surfaces without leaving a residue.

These requirements seem simple enough to achieve, although in practice they require a material to possess contradictory properties. For example, for a PSA to adhere by applying light pressure, it must possess a low enough stiffness to conform to the surface in the time-scale of the application [29–34]. However, in order to exert a strong holding force, there must be sufficient resistance to relaxation to provide the desired mechanical strength.

There are three predominant classes of PSAs available today: solvent-based acrylics, aqueous emulsions (typically acrylic based) and hot melt adhesives (typically SIS and SBS based). Of these three classes, hot melt adhesives are experiencing the largest growth [29, 30, 35–43]. The growth of this class of adhesives compared with emulsion systems is partly because of the tendency for emulsion polymers to absorb water due to the surfactants incorporated during polymerization and to the lower shear holding power of emulsion polymers due to a heterogeneous network morphology [44, 45]. Acrylic copolymers are currently the primary alternative to SBCs in the pressure-sensitive adhesive market. Although acrylics have better specific adhesion to polar materials, adhesive manufacturers cannot easily formulate them for varied adhesion applications. Along with the reduction in the use of solvents, the ease of formulation as described in more detail below has led to the increased popularity of hot melt adhesives based on SBCs. In addition, the inherent purity required for anionic synthesis has allowed FDA clearance of most SBCs, which permits their use in sensitive applications such as diaper and personal care adhesives.

Hot melt adhesives are also attractive due to their ease of use and high production speeds promoted at least in part by the absence of solvents (if desired, SBC adhesives can also be applied from solvent using similar formulations). Since tackifiers and plasticizers are necessary to make the SBC hot melts usable as PSAs, a single hot melt base polymer can be formulated to provide an almost limitless range of properties. For these reasons, hot melt PSAs constitute the largest volume of the PSA market [29]. That said, traditional hot melts do have a few drawbacks. For example, unless the isoprene or butadiene blocks are hydrogenated, residual unsaturations in the polymer backbone are susceptible to oxidation, which can lead to discoloration and embrittlement. Because of relaxations of the physical cross-links provided by the polystyrene domains, cohesive strength is reduced at high temperatures [35, 37, 46]. Butadiene-based SBCs can be cross-linked to increase the high-temperature shear, but the cross-linking typically decreases the tack by increasing stiffness [36, 47–49].

SBC adhesives typically do not have strong specific interactions with substrates such as polar or acid–base interactions. The van der Waals forces of these non-polar materials provide very weak molecular level adhesion. The principal characteristic of SBCs that makes them useful as adhesives is their ability to conform to the surface and to dissipate energy uniformly throughout the bulk of the adhesive layer during peel deformation. The ability of a material to conform (adhesive creep compliance) quickly to a surface is generally referred to as "tack" and is inversely proportional to its modulus. Dahlquist first studied the adhesive creep compliance necessary for adhesive bond formation in the late 1960s [31, 44, 50]. He determined that a 1-s creep compliance greater than 10^{-5} Pa^{-1} (at the use temperature) was necessary for an adhesive to conform to the substrate. As the compliance increases above that value, the adhesive becomes less elastic and cannot deform to wet the substrate properly. On the other hand, if the compressive modulus is too low, the adhesive can flow out of the bond region and no bond is formed.

The most common SBCs used in adhesives are SIS polymers with molecular weights between 100 000 to 200 000 g mol^{-1} and 10–20 wt% polystyrene. Isoprene is the preferred midblock since it has a higher entanglement molecular weight than polybutadiene or the hydrogenated midblocks. The higher entanglement molecular weight translates to the lowest rubber modulus of the commercial SBCs. The low styrene content is desired to assure that the styrene domains will be predominantly spherical, which minimizes the hard phase contribution to the adhesive's modulus. The low styrene block molecular weight also provides a lower order–disorder transition temperature, which can be further reduced with diluents during the adhesive formulation to allow for practical melt temperatures of 150–170 °C.

As mentioned previously, SBCs are typically diluted with oil, diblock copolymers and tackifying resins to improve the strength of adhesion at the desired temperature use range. Although this may seem counterintuitive, recall that the strength of the observed adhesion is strongly related to its ability to dissipate the energy exerted during debonding because the specific interfacial adhesion of these hydrocarbon systems is low [51]. Most pressure-sensitive adhesives are thus formulated to increase the T_g to near −10 °C from −60 °C by blending with tackifying resins (polyisoprene-soluble oligomers) with T_gs above room temperature. This allows the maximum in the loss modulus to approach room temperature and thus provide maximum energy absorption in the adhesive at room temperature. Figure 1.4 shows the elastic modulus and $\tan\delta$ of an SBC polymer with and without tackifying resin (solid and dashed lines, respectively) as a function of temperature. Note that the $\tan\delta$ peak is shifted to higher tempera-

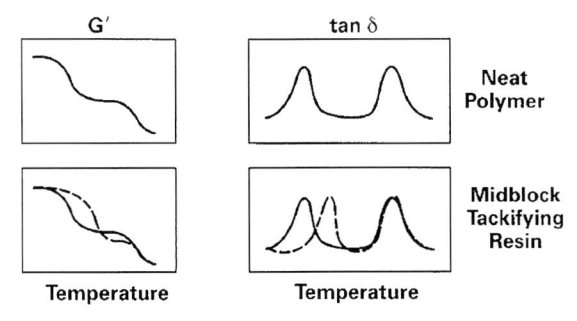

Fig. 1.4 Effect of tackifying resin on shear modulus and tan δ.

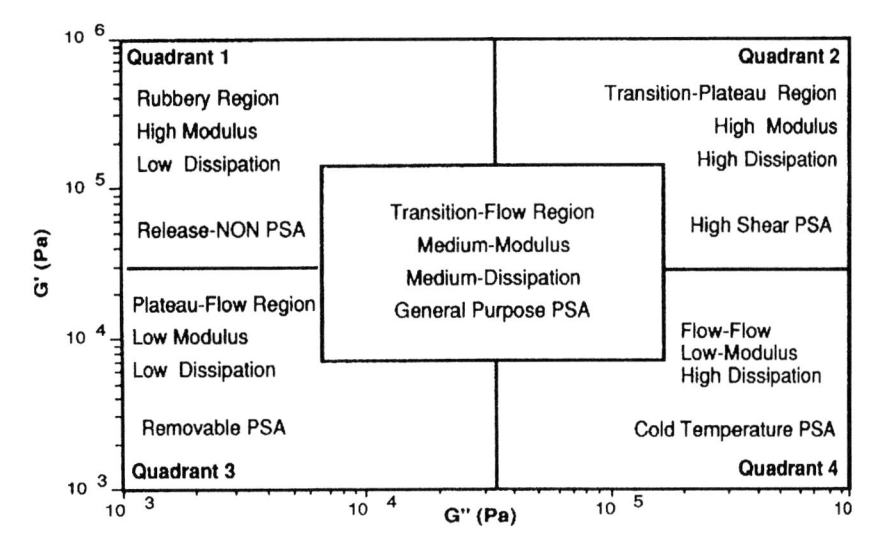

Fig. 1.5 Chang's concept of viscoelastic windows of PSAs related to the rheological characteristics of the adhesive. Reprinted from [32].

tures with the addition of tackifying resins. Tackifying resins are rigid aliphatic oligomers such as oligomerized cyclopentadiene that serve several functions: to increase the T_g of the midblock, perhaps add polarity to increase adhesion either to the backing or the substrate and reduce the melt viscosity, strength and cost of the adhesive formulation. By changing the tackifiers and oils, an SBC adhesive formulation can easily be formulated for a targeted use temperature.

The general type of adhesive formulated can be predicted using the Viscoelastic Windows concept first described by Chang and coworkers [32–34]. The combinations of the values of the dynamic loss (G') and shear (G'') moduli, as measured from the rheological master curves of the PSA, are divided into four regions plus a general-purpose region each describing the PSA properties (Fig. 1.5) [32]. For example, an adhesive with 8×10^4 Pa $< G' < 1 \times 10^6$ Pa and 7×10^3 Pa $< G'' < 1 \times 10^6$ Pa (at the use temperature) is classified as a general-purpose PSA because the viscoelastic properties of the adhesive place it in the transition flow region (possessing medium modulus and medium energy dissipation during peel). If G' and G'' are shifted to lower temperatures, the adhesive then can be used as a cold temperature PSA. A properly balanced adhesive will have enough strength to distribute the stress from the surface uniformly throughout the adhesive so that it will flow with maximum viscous dissipation at a stress just less than the stress necessary to debond the adhesive from the surface.

A wide variety of other additives have been developed to precisely tailor the modulus, strength and T_g of the soft and hard phases. These fall into three categories: plasticizers, midblock resins (previously described) and endblock resins. Plasticizers are generally diblock copolymers and oils chosen to match the polarity of the rubber block. Diblocks are typically used in combination with or in the place of oils because they do not migrate to the surface to form a weak boundary layer and free chain ends increase surface tack. The plasticizers reduce the modulus of the rubber phase, the viscosity of the formulation during melt processing, the strength of the bulk adhesive and the cost of the adhesive.

The endblocks of block copolymer-based PSAs can also be modified with aromatic molecules. For example, adding poly(phenylene oxide) (PPO) polymers and oligomers modifies the properties of the endblock. PPO modification increases in the endblock T_g, which increases the ultimate service temperature of the adhesive formulation. The addition of low molecular weight PPO can also reduce the melt viscosity of the formulation by plasticizing the endblocks during processing. While those advantages seem highly desirable, endblock modification is much less prevalent than midblock modification, due in part to the cost of the PPO and the associated odor. Further details about specific formulations for adhesive tapes were reviewed by Ewins et al. [52].

SBS polymers are often used as cross-linkable adhesives when higher service temperatures are desired. The polybutadiene midblock is much more readily cross-linked by peroxides, UV light (with the addition of photoinitiators) and electron beam radiation than the polyisoprene midblock. The 1,2-butadiene content of the butadiene block can be adjusted to increase its reactivity. Radial and branched architectures are also used to reduce the number of reactions necessary to form a continuous network because cross-linking tends to reduce tack by increasing modulus.

Block copolymers with mixed butadiene and isoprene midblocks(S–I/B–S) are also useful for adhesive compounding [53]. These polymers combine the advantages of both butadiene and isoprene monomers in one material. Although adhesive formulations based on the mixed midblock polymer require a slightly more polar midblock resin than the traditional SIS polymers, they can be more easily formulated for use at lower temperatures due to a lower midblock T_g. The mixed midblock polymers are easier to cross-link than polyisoprene due to the butadiene in the midblock.

For adhesives that require long-term stability in the melt or that will be exposed to UV light, ozone or other oxidative environments, the hydrogenated versions of SBS and SIS (SEBS or SEPS, respectively) are used. Although the saturated rubber blocks are inherently more stable than their unsaturated precursors, they are also higher modulus and more expensive, reducing their use in commercial tape formulations.

1.5
Compounding Applications

Very few applications utilize "neat" block copolymers without any other formulating ingredients. Blending other components, or compounding as it is typically called, is usually necessary to achieve optimal processability within a desired performance window. SBC compounds are a loose class of materials designed to meet a specific set of end-use requirements by blending block copolymers with a wide range of ingredients such as oil, polyolefins, engineering thermoplastics, fillers, stabilizers and others. Hydrogenated SBCs are particularly attractive from a material selection viewpoint due to their versatility and the wide range of properties that can be achieved using various formulating techniques. SEBS-based compounds can be formulated to achieved hardnesses ranging from 30 Shore 00 to 60 Shore D. Hardness scales increase from the softest scale, Shore OO, to Shore O, Shore A and Shore D. Tensile strengths up to 17 MPa (2500 psi), elongations up to 750% and tear strengths up to 97 kN m^{-1} (500 pli) can also be achieved, depending on hardness. SEBS-based formulations can easily have use temperature ranges of –50 to 100 °C and SBS-based formulations can extend down to –70 °C. SBCs are good electrical insulators with a volume resistivity of 10^{14}–10^{16} Ω cm and hydrogenated SBCs have good and thermal and weather stability. In addition, SBC formulations are printable and easily colored. As a result of this wide formulating latitude, SBCs are the material of choice for a wide range of compounded applications in the footwear, medical, soft-touch, automotive and toy markets. Regardless of the application, raw material selection and processing conditions are critical to achieving a consistent compound that meets the performance requirements.

1.5.1
Raw Material Selection

SBC compounds are based on either hydrogenated block copolymers (SEBS or SEPS) or unsaturated block copolymers (SBS or SIS). For the purposes of this discussion, hydrogenated SBCs will only be referred to as SEBS unless noted otherwise, although in most cases the midblock could be hydrogenated isoprene or mixtures of monomers. Hydrogenated SBCs are preferred for most high-performance technical compounds due to superior weatherability and mechanical performance Additional compounding ingredients are chosen to provide desired performance attributes including hardness, tensile strength and elongation, tear strength, compression set, weatherability, cost, surface appearance and color.

Typical compounding ingredients include oil, one or more polyolefins, fillers, endblock resins and additives. Proper selection of raw materials is important to ensure performance and proper processability.

Processing oils and polyolefins are the two most common formulating ingredients used in SBC compounds. Oils improve processability while decreasing hardness and cost of the formulation. Some high molecular weight SEBS copolymers are not melt processable without the addition of oil to enhance flow. Paraffinic white oils are most compatible with SEBS-based compounds whereas naphthenic oils are more compatible with SBS/SIS-based formulations. It is also important to consider color, stability, availability, volatility and FDA and USP compliance when selecting an oil. Polyolefins such as polypropylene (PP) and polyethylene (PE) can be used to promote flow and generate high-strength compounds. Polyolefins generally increase hardness, modulus, tensile strength and tear strength but decrease elongation, resiliency and other rubbery characteristics. Due to its higher melting-point, PP is typically the polyolefin of choice for SEBS formulations, although HDPE, LDPE and LLDPE can also be used. Ethylene vinyl acetate (EVA) can be used but special attention to odor is needed.

Contour diagrams can be generated using statistical design of experiments and are extremely useful to the formulator for determining the optimum formulating range to achieve the desired property set. Contour diagrams are often based on 100 parts of SBC with lines representing formulations of equal performance. As an example, the contour diagram in Fig. 1.6 represents Shore A hardness for the formulating range of 25–125 phr homopolypropylene, 0–160 phr paraffinic oil and 100 parts high molecular weight SEBS with 33% polystyrene content (PSC) (phr indicates parts per hundred parts of rubber). Lines represent formulations with equal hardness. The utility of contour diagrams is realized when trying to determine a starting point formulation to ful-

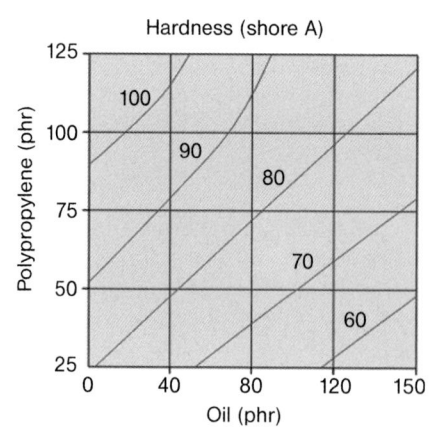

Fig. 1.6 Hardness contour diagram for SEBS–oil–PP formulation based on 100 phr high molecular weight SEBS.

fill a complicated property specification. A single formulation must often meet multiple requirements (e.g. hardness, tensile strength, tear strength, elongation, abrasion resistance). Contour diagrams can be generated for each property and overlayed to determine if a formulation with those ingredients exists that meets all requirements. If so, the formulator can glean some insight into the flexibility of the applicable formulating space and the cost–performance trade-offs.

Mixing is best accomplished when the viscosities of the SEBS–oil blend and polyolefin are matched appropriately at the mixing temperature to achieve isoviscous mixing resulting in a macroscopic interpenetrating network (IPN). The unique rheology of SBCs also allows the formation of IPNs away from isoviscous points by freezing the morphology when shear is stopped. The formation of an IPN for SEBS compounds has been described elsewhere in detail [23, 54]. In a mechanically stabilized IPN, a co-continuous morphology exists where each phase has three-dimensional connectivity; therefore, both phases contribute their respective physical characteristics to the overall blend performance.

Endblock resins and endblock plasticizers are compatible with the PS endblocks and are occasionally used in SEBS compound formulations in order to achieve desired properties. An endblock resin serves to reinforce the endblocks typically by increasing the endblock T_g. Endblock plasticizers serve to improve flow of a formulation and typically decrease the endblock T_g. The effect of endblock resins and plasticizers on shear storage modulus (G') and $\tan\delta$ are illustrated in Fig. 1.7. The solid lines represent the response of a neat block copolymer and the dashed lines represent the response of a modified block copolymer. Poly(phenylene oxide) (PPO) is miscible with the PS domains and serves as an endblock resin when compounded efficiently. Low molecular weight unsaturated styrenic resins can be used in SEBS compound formulations as midblock plasticizers. Raw materials such as these promote flow and allow the formulator to remove some oil from a formulation in oil sensitive applications. The use of endblock resins will also increase compound stiffness.

Many different types of fillers and additives are currently available. There is really no limitation on the type of fillers used in SBS or SEBS compound formulations. Fillers serve to reduce cost, adjust surface appearance and feel and

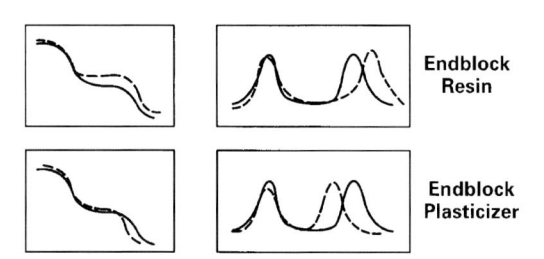

Fig. 1.7 Illustration of dynamic mechanical data for an unmodified block copolymer (solid lines) relative to that modified with an endblock resin and endblock plasticizer (dashed lines).

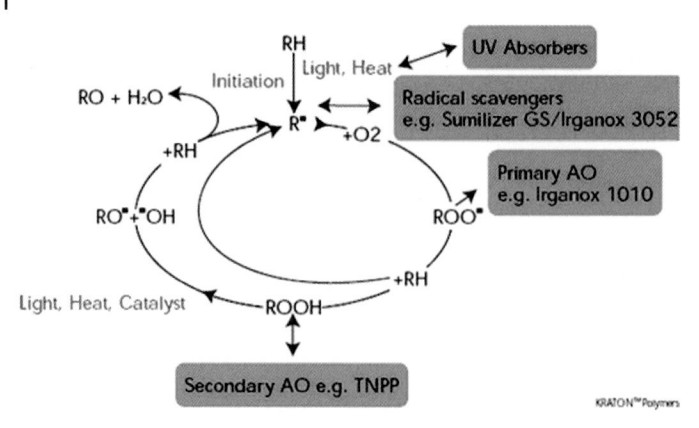

Fig. 1.8 Degradation pathway of elastomers.

enhance oil absorption; however, the addition of too much filler leads to property deterioration. In general, mineral fillers offer little reinforcing effects in SEBS- or SBS-based compounds. Filler addition reduces tensile strength and elongation while hardness and stiffness increase only slightly upon addition of 100–200 phr filler. A wide range of other additives are commonly used in SBC compounds. These can include antioxidants, UV absorbers, antiozonants, fire retardants, pigments, dispersants and color concentrates. There are a variety of antioxidant types: primary, secondary or radical scavenging agents. Figure 1.8 shows the degradation pathway of elastomers and the role of various antioxidant types. A combination of these is often used due to synergistic effects. The choice of the antioxidant package will depend on the nature of the compound (natural or black), the intended application (outdoor or indoor) and the ageing requirements (UV, Florida test, etc.).

Gas fading or pinking can be a major obstacle when selecting the optimal stabilizer package for color-matched parts. Although phenolic antioxidants are very effective stabilizers for long-term heat aging, they are prone to gas fading (primarily due to fork lift nitroxide emissions). Gas fading can be minimized by eliminating the use of a basic HALS in conjunction with a phenolic antioxidant. Many suppliers currently have non-basic HALS systems to choose from.

1.5.2
Processing and Forming

Because of the complex rheology of block copolymers, the use of proper blending and processing techniques is critical to generate the desired compound properties and maintain good reproducibility and quality control. Critical processing aspects are oil incorporation via dry blending or direct injection, metering and blending of dry components, screw and equipment design, temperature, volumetric throughput, pelletization/finishing, dust application and packaging.

1.5 Compounding Applications

Each of these areas will be discussed separately below with emphasis on compounding with hydrogenated SBCs; however, differences for unhydrogenated SBC compounding will be noted when necessary.

To achieve a homogeneous compound, the formulating ingredients should be dry-blended prior to thermal mixing. Various dry blenders can be utilized effectively, such as ribbon blenders, high-speed mixers or tumbler blenders. Most hydrogenated SBCs are supplied in crumb or powder form, which has inherent internal porosity through which oil is absorbed and diffuses into the crumb interior. Once the oil has been incorporated into the crumb, other dry ingredients can be added directly to the dry blender for incorporation.

SBCs form phase-separated structures below their characteristic order–disorder transition temperature (ODT) [23, 55, 56]. At temperatures above the ODT, the block copolymer becomes disordered just as a homopolymer melt. However, due to the strong thermodynamic driving force for phase separation, the ODT for high molecular weight SEBS polymers is often above the decomposition temperature. Although the addition of oil to an SEBS polymer serves to depress the ODT, most high molecular weight SEBS compound formulations are still processed below the ODT of the oiled SEBS polymer and will not flow without the presence of shear.

In fact, infinite viscosity at zero shear rate (illustrated in Fig. 1.9) drives IPN formation. The IPN morphology is formed in high-shear zones during mixing and phase growth is prevented after removal of the shear stress by the infinite viscosity of the SEBS phase [25].

Because of this characteristic rheology, it is necessary to use the appropriate processing equipment and screw designs in order to achieve the optimal compound morphology. While Banbury mixers and Farrel continuous mixers can be utilized to adequately compound many SBC formulations, the twin screw ex-

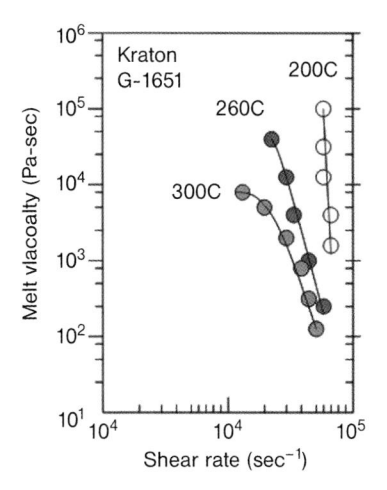

Fig. 1.9 Shear viscosity data for Kraton G-1651.

truder is the most preferred processing technique due to its versatility. Co-rotating intermeshing screws with an L/D of at least 40 and a compression ratio >2 are ideal for SEBS compounding. Some SBS or SIS formulations can even be processed on a single screw extruder equipped with a mixing head. While the specific formulation will determine the optimal processing temperatures, SEBS-based formulations are typically processed >230 °C, whereas 200 °C is often the maximum temperature for SBS and SIS formulations due to degradation tendencies. As more oil is added to the formulation, temperature profiles must be lowered due to lower compound viscosity and to ensure proper feeding by preventing excess lubrication at the feed throat. Once mixing is accomplished, finishing can be achieved either by strand cutting or dicing of harder formulations or under-water pelletization for softer formulations. Soft formulations generally require dusting to prevent pellet blocking during shipping and storage. Common dusting agents include silica, talc, magnesium carbonate, stearates, polyethylene wax and antioxidants.

These fundamental formulating and processing techniques can be utilized to make a wide range of formulations applicable to many market segments. Some of these will be briefly highlighted below.

1.5.3
Automotive

As mentioned earlier, the 1972 5 mph bumper law generated the first applications for SBC formulations in the automotive market. Bumpers that could withstand the new requirements of the law were made from SEBS–PP–oil formulations. Since then, SEBS-based formulations have found utility as fenders, sound-damping panels, air ducts, dampers, shift boot covers, cup holders and airbag door covers. SEBS-based formulations are typically not applicable in "under-the-hood" applications because they cannot meet the high temperature and oil resistance requirements. Product specifications that are of prime importance to most automotive applications include hardness, tensile strength, compression set, tear strength, scratch/abrasion resistance, UV stability, surface appearance and creep performance. Most automotive formulations are based on high molecular weight SEBS polymers in order to meet these often stringent requirements. SEBS compounds containing the correct stabilizer package can exhibit UV performance equal to or better than that of EPDM-based TPVs. An SEBS compound of similar performance exhibited no visible surface cracks and a ΔE less than 1 after 2 years of simulated Kalahari outdoor exposure [57].

Replacement of PVC window encapsulation seals with olefinic-based materials is a current trend in the automotive industry. SEBS compounds are suitable for this type of application due to ease of injection molding, which is critical to minimize glass breakage. Compounds made with high-flow SEBS polymer have lower viscosity without loss of physical properties such as compression set, tensile strength and elongation at break [58]. Although in a very early development stage, it is believed that weather seals based on SBC formulations can meet the end-of-life recycle requirements. SBC formulations may ultimately replace part of the EPDM foamed weather seal market in the long term [59]. One of the benefits of SEBS compounds for automotive is easy colorability, which is critical in an industry where color matching requirements are stringent.

1.5.4
Wire and Cable

Flame-retardant SEBS compounds were developed in the 1970s for electrical wire and cable sheathing. Cold temperature flexibility was the main advantage that SEBS formulations brought to the market place over incumbent materials. These formulations are often based on high molecular weight SEBS polymers in order to achieve the abrasion, flex crack and tensile requirements. In addition, high molecular weight SEBS polymers have enough extendibility to incorporate the large amounts of flame retardant necessary to achieve the V-0 flammability requirements. Traditional flame retardants include halogenated resins in conjunction with antimony oxide. Legislative and environmental pressures are pushing the industry towards non-halogenated flame retardants, which include magnesium hydroxide, phosphite esters, graphite and nitrogen–phosphite types.

1.5.5
Medical

Although fully hydrogenated SEPS polymers were trialed in medical applications in the mid-1960s, significant applications for SBS- and SEBS-based compounds did not develop until the 1980s. Most medical compounds are based on either SEBS–PP–oil, SBS–oil or SBS–oil–PS depending on the application requirements. The most important requirement for material selection is approval for medical use in the application country. In the USA this means US Pharmacopeia and FDA approval. Most medical applications require long qualification lead times, which make raw material changes extremely difficult. Physical properties of most concern to the formulator include sterilization requirements, resilience, tensile strength, tear strength, transparency and colorability. Most SEBS–PP–oil compounds are sterilizable by gamma radiation, beta radiation, ethylene oxide or steam. Radiation is best for SBS-based formulations as degradation and warpage can occur when using ethylene oxide or steam sterilization techniques. Medical applications for molded and extruded goods based on SBC technologies include clear flexible tubing, surgical draping, port caps, syringe bulbs, syringe tips, nipples and physical therapy equipment.

In addition to general molded and extruded goods, SBCs are also used as tie layers in co-extruded films for blood and IV bags, removable protective coatings, heat-sealable films and reclosable packaging constructions. When formulating a tie layer, it is important to match the rheology of the outer materials while balancing functionality. SEBS is most often recommended as the material of choice as a tie layer between PP or PE whereas SBS is often recommended for PS or

HIPS in ABA or ABC constructions. Maleated SEBS can be used as a tie layer between PC and EVOH.

1.5.6
Soft-touch Overmolding

Soft-touch overmolding became a value-added market trend only in the early 1990s and involves molding a soft skin layer on top of a rigid engineering thermoplastic. Soft-touch overmolds serve to provide highly frictional and waterproof surfaces, vibration damping, impact resistance, insulation from heat and electricity and cushioning for improved ergonomics [60]. Consumers often perceive a rubbery look and feel to be of higher value than a "cold", slick plastic feel. Soft-touch overmolding can be achieved either by insert molding or two-component (two-shot) molding. Two-component molding utilizes equipment containing two independent injection units, each shooting a different material. This type of set-up involves two independent runner systems with a rotating mold and is preferred because of faster cycle times compared with insert molding. In addition, the substrate is not exposed to humidity and airborne contamination in a two-component process.

There are many types of thermoplastic elastomers for a fabricator to choose from when selecting a soft-touch overmold material. Styrenic block copolymer formulations are among the most widely used in the marketplace. This is largely due to the wide hardness range, ease of colorability and good UV stability in addition to low polarity, which minimizes the need for drying prior to use. Low polarity also makes SBC solutions less prone to staining [60], but also reduces oil resistance. Lower polarity often makes it difficult for SBC formulations to adhere to polar substrates such as polyamide. When this is required, it is often necessary to incorporate a maleic anhydride functionalized SEBS in the formulation [61].

1.5.7
Ultra-soft Compounds

At the turn of the 21st century, ultra-soft compounding began to gain popularity. These types of materials are often in the Shore 00 hardness range and are on the verge of being considered stiff "oil gels" (see Section 1.7.2). These types of formulations often contain 300–500 phr oil relative to 100 phr SEBS polymer. High oil content and low viscosity facilitate low processing temperatures. Formulating parameters to be considered include low tack, low to no oil bleed out, clarity, colorability and secondary processability. Molding is often challenging as cycle times are long and demolding can be difficult. Applications include cushions, squeezables, mattresses and toys [62].

1.6
Polymer Modification

SBCs are frequently used as impact modifiers of thermoplastic polymers. Many plastics are susceptible to brittle failure due to an inherent lack of energy-absorbing processes throughout the bulk of the material. To increase toughness, energy-dissipating mechanisms must be introduced to limit crack propagation. This can often be achieved by dispersing high-strength rubber polymer particles with a low glass transition temperature such as SBS or SEBS in the thermoplastic matrix. These particles act as stress concentrators that initiate localized energy-absorbing mechanisms. For each plastic, optimum energy dissipation requires a different rubber loading, particle size and particle distribution. This optimum can be achieved with the right degree of compatibility and rheology between the rubber and the plastic. With their wide range of rheology and polymer compositions, SEBS and SBS polymers can provide the right balance of compatibility with a variety of thermoplastic substrates to achieve optimal particle sizes for enhanced toughening [63].

Traditional SBS polymers serve as good tougheners for crystal, general-purpose and high-impact polystyrene resins, but with a significant loss of clarity. SBS polymers with 75% polystyrene content serve to maintain clarity of polystyrene resins while enhancing ductility. These types of formulations can be used to make clear, water-white, disposable packaging, film and molded parts. The addition of small amounts of SBS (<10%) to low-density polyethylene can improve both impact resistance and stress crack resistance. Addition of SBS to high-density polyethylene provides improvements in impact strength, flexibility and film tear and impact strengths [63].

Blending SEBS or SBS polymers with polypropylene can significantly increase the impact strength, particularly at low temperatures. Low molecular weight SBS or SEBS polymers tend to give the best results due to their ability to form a fine dispersion [63]. SEBS polymers have been developed with rubber blocks specifically designed to improve toughness of polypropylene while maintaining good clarity [64]. SEBS polymers can also be used to improve the ductility of polycarbonate in thick sections [≥¼; in (ca. 6 mm)]. High molecular weight SEBS polymers are unique modifiers for PC as they form blends highly resistant to delamination, particularly at SEBS concentrations <10% [63]. Addition of a small amount of high molecular weight SEBS also serves to improve environmental stress crack resistance and to retard the embrittlement of PC due to physical aging.

Maleic anhydride functionalized SEBS polymers can be effective modifiers for polar engineering thermoplastics such as polyamides and polyesters. Nylon 6,6 can be efficiently toughened with ∼20% of an SEBS polymer with ∼1.5% maleation. Nylon 6 requires a lower level of functionality to create the particle size needed for efficient toughening. The saturated midblock of maleated SEBS polymers allow them to stand up to the severe processing conditions, heat and UV aging tests necessary for polyesters such as PBT. Notched Izod impact strength is significantly increased by the addition of 20 and 30% of a maleated SEBS polymer to PET and PBT, respectively [65].

1.7
Viscosity Modification and Gels

1.7.1
Viscosity Index Improvers

Low concentrations of hydrogenated SBCs are very effective rheology modifiers for aliphatic oils in a variety of systems. One of the applications allows, in part, for the existence of multi-grade automobile oils [22]. Engine oil manufacturers use SEP diblocks and star hydrogenated polyisoprene (ethylene propylene, PEP) polymers to modify the viscosity of motor oils. The high solubility of the elastomer block and relative insolubility of the styrene block coupled with precise molecular weight control allow SEP diblocks to form dynamic micelles of very high apparent molecular weight. This highly effective thickening mechanism allows the use of lower viscosity base oils, which results in a system with a flatter viscosity/temperature curve than can be achieved with single-grade oils. Hydrogenated polyisoprene-containing polymers are chosen over the hydrogenated polybutadiene analogues due the their preferred degradation mechanism in the harsh engine environments [65, 66]. SEPs and PEPs degrade via chain scission at high shear rates and in the presence of oxygen, whereas the SEBs degrade with somewhat more cross-linking, leading to increased deposits in the engine. Star polymers are valued for their high shear stability because the scission of one arm results in little overall change in viscosity relative to linear polymers.

The concentration of the hydrogenated SBCs used as modifiers typically varies between 0.4 and 4 wt%. This weight range typically puts the polymer concentration very near the overlap concentration between the semi-dilute and dilute region of the polymer in the oil (C^*). Polymer loading below C^* leads to lower thickening efficiency due to the lack of entanglements. Conversely, concentrations higher than C^* lead to excessive entanglements, which result in unacceptable shear degradation in the engine [67]. SBCs, including hydrogenated polybutadiene-containing polymers, are fairly efficient thickeners for aliphatic oils in other applications such as personal care.

1.7.2
Oil Gels

Further increasing the concentration of SBCs above 6 to as high as 15 wt% in various mineral oils can lead to the formation of thixotropic greases or gels for a number of uses, including cable filling compounds, clear candles and shock-absorbing gels in sporting equipment. Formulations based on triblocks can form strong gels, whereas the use of diblocks leads to greases or weak gels. The use of SBCs in cable filling, the largest of the above markets, was first developed by Bell Laboratories in the late 1970s [68, 69]. Typical formulation include as little as 6% hydrogenated SBC and may include other olefins such as polyethylene. The resulting gel is pumped into optical fiber cables above its gel point during fabrication. The compound lubricates the fibers during installation to reduce breakage and prevent water and air ingress during service.

1.8 Emerging Technology in Block Copolymers

1.8.1 Recycling Compatibilization

Recycling of waste and regrind streams and also after-market parts is becoming the norm in many industries, including packaging and automotive. Phase segregation leading to poor mechanical properties of waste and recycle streams is common because of incompatibility and/or lack of interfacial tension. Low-level addition of SBCs to such recycle streams can serve to compatibilize the blend, resulting in useful mechanical and physical characteristics. Only 5% of an SBS can sufficiently upgrade PS–polyolefin recycle streams while the addition of 5–10% of a maleated SEBS can upgrade polyolefin–polyester, polyolefin–polyamide and ABS–polyolefin recycle streams [70].

1.8.2 PVC and Silicone Replacement

Replacement of silicone and flexible PVC (f-PVC) has become an increasing trend in the TPE market segment. Silicone and f-PVC are both used in the personal care and medical industries where clarity is required. In addition, silicone offers high temperature performance and chemical resistance. F-PVC has a unique physical property set in terms of flexibility, strength, kink resistance and low tack. However, silicone is very expensive and f-PVC has emerging health and environmental concerns. SBC-based compounds are being developed to replace both silicone and f-PVC in the medical and personal care market segments [71]. Compound formulations are often based on polypropylene using high-flow SEBS copolymers. SEBS polymers with modified midblock structures can generate compounds with exceptional transparency in blends with polypropylene [64]. Compound rheology during final film or part formation is key to achieving a smooth surface which reduces surface scattering. Processing aids or special die coatings can also be utilized to enhance surface smoothness. Specific applications for these types of SBC compounds include baby bottle nipples and medical tubing [71].

1.8.3
Commercial Uses of Other Controlled Polymerized Polymers

With the growing complexity of advanced technology, the demand for specifically tailored polymers for individual applications has increased dramatically. Traditionally, anionic polymerization methodologies offer the ability to control accurately the monomer composition and the resulting polymer architecture [72–74]. However, the reactive nature of the anionic propagating species renders it incompatible with many monomer families, including most (meth)acrylates and functional monomers. This incompatibility necessitates specialty initiators or additives for polymerization. Controlled free radical polymerization methods including stable free radical polymerization (SFRP), atom transfer radical polymerization (ATRP), reversible addition fragmentation transfer polymerization (RAFT) and nitroxide-mediated polymerization (NMP), all described in Chapter 5 in Part I, have all recently emerged in academic research as potential alternatives to anionic polymerization [75–81]. These techniques have been slow to find wide-ranging commercial applications, partially due to restrictive manufacturing hurdles such as the need for high monomer concentration, relatively slow reaction rates and the removal of residual monomer.

An ATRP process to produce block and comb copolymers, which are used as dispersants and dispersible inorganic or organic pigments, has been developed [82–84]. Researchers found that the addition of salt-forming components, such as mono-, bi- or tricyclic sulfonic, carboxylic and phosphonic acids, to polymers containing amino groups as repeat units in a hydrophilic polymer block produces pigment dispersions having improved properties for use as dispersants.

Anionic polymers based on acrylates are finding new applications. Leibler's group has researched the use of styrene-*b*-butadiene-*b*-methyl methacrylate and styrene-*b*-butadiene-*b*-(methyl methacrylate-*stat-tert*-butyl methacrylate) triblock copolymers as tougheners for both thermoplastic and thermosetting systems [85–87]. Examples of thermoplastic toughening include the toughening of poly(phenylene oxide), an application that currently uses traditional SBC, poly(vinylidene fluoride) (PVDF) and poly(vinyl chloride), which are not efficiently toughened with traditional SBCs. The polar block of the S-*b*-MMA provides increased compatibility with PVC and PVDF.

The S-*b*-MMA and S-*b*-MMA–*t*BMA and the hydrolyzed version of S-*b*-MMA–*t*BMA (S-*b*-MMA–methacrylic acid) triblocks have also been studied as tougheners for epoxy systems. The triblocks have been shown to toughen effectively epoxy systems based on the diglycidyl ether of bisphenol-A and various amine hardeners. While typical SBCs phase separate from the epoxies during curing (a phenomena that can be reduced with the use of anhydride-functionalized SBCs), methacrylic acid dispersed in the MMA block of these polymers reacts quickly with the oxirane rings in the epoxies [85–87].

Triblocks based on MMA–*n*BA–MMA triblock copolymers have found application in the adhesives market [88]. Since the entanglement molecular weight (M_e) of the *n*-butyl acrylate block is much higher than that of either polyiso-

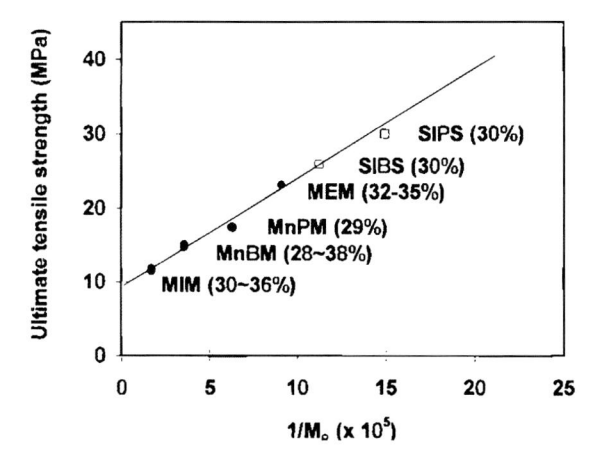

Fig. 1.10 Effect of ultimate tensile strength as a function of M_e of the elastomeric phase [90].

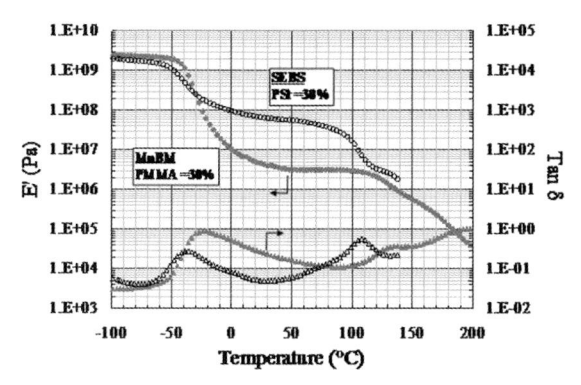

Fig. 1.11 DMA of SEBS and MMA–nBA–MMA block copolymers [88].

prene or polybutadiene, the MMA–nBA–MMA block copolymers are softer and have lower tensile strengths than their analogous SIS and SBS (Figs. 1.10 and 1.11) [89–91]. Although the (meth)acrylic block copolymers may be weaker than their SBC counterparts, there is considerable interest in these polymers as pressure-sensitive adhesives, where the higher tensile strengths of SIS and SBS are not necessarily needed [88, 92]. The (meth)acrylate block copolymers also have a slightly higher use temperature range due to the higher T_g of MMA compared with styrene (Fig. 1.11). Also, unlike SBS and SIS, the MMA–nBA–MMA block copolymers are completely saturated and are therefore more resistant to UV and oxidative degradation without a separate hydrogenation step.

1.8.4
Styrene-Isobutene-Styrene (S–IB–S)

Although traditional SBS or SEBS SBCs are suitable for some packaging and sealant applications, they are often overlooked or found unsuitable for applications requiring low gas permeability coefficients. This is largely due to the inherent high diffusivity of gaseous molecules through the midblock, which is only exacerbated upon addition of oil. Styrenic block copolymers based on a polyisobutene midblock have substantially reduced permeability and do not require hydrogenation for UV and thermal stability. Cationic polymerization of S–IB–S block copolymers has been actively investigated for more than 20 years to overcome problems of chain transfer and low process temperatures [93–96]. The most common synthetic strategy is based on living cationic polymerization initiated with a bifunctional initiator making the polyisobutene center block, followed by a polystyrene end-capping reaction [97]. Compared with the more traditional SEBS and SEPS copolymers. S–IB–S compounds typically exhibit lower hardness, less elasticity, slightly better heat aging resistance and lower gas permeability. In addition, they offer the formulator an opportunity to use significantly less or no oil than traditional SEBS polymers to reach the same compound hardness due to the softness of the isobutene midblock [97]. S–IB–S copolymers have also been evaluated in blends with TPUs to provide enhanced barrier properties and vibration-damping characteristics for overmolding applications [98].

List of Abbreviations

f-PVC	flexible poly(vinyl chloride)
IPN	interpenetrating network
OBO	oil bleed-out
ODT	order–disorder transition
PEP	poly(ethylene propylene)
phr	parts per hundred rubber
PSA	pressure-sensitive adhesive
PSC	polystyrene content (%)
SBC	styrenic block copolymer
SBR	styrene–butadiene rubber
SBS	styrene–butadiene–styrene block copolymer
SEB	styrene–ethylene/butylene diblock copolymer
SEBS	styrene–ethylene/butylene–styrene block copolymer
SEP	styrene–ethylene/propylene diblock copolymer
SEPS	styrene–ethylene/propylene–styrene block copolymer
S–IB–S	styrene–isobutene–styrene block copolymer
SIBS	styrene–isoprene/butadiene–styrene block copolymer
SIS	styrene–isoprene–styrene block copolymer

TPE   thermoplastic elastomer
TPV   thermoplastic vulcanizate
UV   ultraviolet
VII   viscosity index improver

References

1 Moacanin, J., Holden, G., Tschoegl, N. W. (Eds.). *Block Copolymers*, Interscience, New York, **1969**.
2 Burke, J. J., Weiss, V. (Eds.). *Block and Graft Copolymers*, Syracuse University Press, Syracuse, NY, **1973**.
3 Aggarwal, S. L. (Ed.). *Block Copolymers*, Plenum Press, New York, **1970**.
4 Legge, N. R., Holden, G., Schroeder, H. E. (Eds.). *Thermoplastic Elastomers, a Comprehensive Review*, Hanser, Munich, **1987**.
5 Hsieh, H. L., Quirk, R. P. *Anionic Polymerization*, Marcel Decker, New York, **1996**.
6 McGrath, J. E. (Ed.). *Anionic Polymerization, Kinetics, Mechanisms and Synthesis*, American Chemical Society, Washington, DC, **1981**.
7 Noshay, A., McGrath, J. E. *Block Copolymers, Overview and Critical Survey*, Academic Press, New York, **1977**.
8 Holden, G. *Thermoplastic Elastomers*, Hanser, Munich, **2000**.
9 Szwarc, M., Levy, M., Milkovich, R. *Journal of the American Chemical Society* **1956**, *78*, 2656–2657.
10 Zelinski, R. Personal communication.
11 Legge, N. R. In *Proceedings of the 131st Meeting of the Rubber Division, American Chemical Society*, Montreal, **1987**, pp. G79–G115.
12 Porter, L. *US Patent 3 149 182*, **1964**.
13 Segovia, G. *Shell Laboratory Notebook*, 122, **1966**.
14 Holden, G. *Shell Laboratory Notebook*, **1961**.
15 Hendricks, W. *Shell Laboratory Notebook*, **1961**.
16 Meier, D. J. *Journal of Polymer Science, Part C: Polymer Symposia* **1969**, *26*, 81–98.
17 Milkovich, R., Holden, G., to Shell Chemical. *US Patents 3 265 765, 2 231 635*, **1966**.
18 Danforth, R. L. Personal communications.
19 Holden, G. In *ACS Symposia* Miami, FL, **1965**.
20 Bishop, E. T., Porter, L. M., Eisenhut, W. O. *Development of Segmented Elastomers for use in Blood Pumps and Oxygenators*, **1971**.
21 Winkler, L. E., to Shell Oil. *US Patent 3 700 748*, **1972**.
22 St. Clair, D. J., Evans, D. D., to Shell Oil. *US Patent 3 772 196*, **1973**.
23 Bening, R., Korcz, W., Handlin, D. In *Modern Styrenic Polymers: Polystyrene and Related Polymers*, Priddy, S. A. (Ed.). Wiley, New York, **2003**.
24 *KRATON^{TM} Polymers Fact Sheet*, Kraton Polymers, Houston, TX, **2001**.
25 Handlin, D. In *Encyclopedia of Materials Science and Technology*, Elsevier, Amsterdam, **2001**.
26 Hendricks, W. R., Danforth, R. L., to Shell Oil. *US Patent 3 589 036*, **1971**.
27 Reed, J., Whiteoak, D. In *The Shell Bitumen Handbook*, 5th edn., Thomas Tedford Publishing, **2003**.
28 Reed, J., Whiteoak, D. In *The Shell Bitumen Handbook*, 5th edn., Thomas Tedford Publishing, **2003**.
29 Everaerts, A. I., Clemens, L. M. In *Surfaces, Chemistry and Applications*, Chaudhury, M., Pocius, A. V. (Eds.), Elsevier, New York, **2002**, Vol. 2.
30 Satas, D. In *Handbook of Pressure Sensitive Adhesive Technology*, Satas, D. (Ed.), Van Nostrand Reinhold, New York, **1989**.
31 Dahlquist, C. A. In *Handbook of Pressure Sensitive Adhesive Technology*, 2nd edn., Satas, D. (Ed.), Van Nostrand Reinhold, New York, **1989**.

32 Chang, E. P. *Journal of Adhesion* **1991**, *34*, 189–200.
33 Chang, E. P. *Journal of Adhesion* **1997**, *60*, 233–248.
34 Yang, H. W. H., Chang, E. P. *Trends in Polymer Science* **1997**, *5*, 380–384.
35 Schumacher, K.-H., Sanborn, T. *Adhesives and Sealants Industry* **2001**, *8*, 42–44.
36 Paul, C. W. In *Surfaces, Chemistry and Applications*, Chaudhury, M., Pocius, A. V. (Eds.), Elsevier, New York, **2002**, Vol. 2.
37 Schumacher, K.-H., Sanborn, T. In *Annual Meeting of the Adhesion Society*, Williamsburg, VA, **2001**, pp. 165–167.
38 Malik, R. *RadTech Report* **2001**, *15*, 25–31.
39 Malik, J., Clarson, S. J. *International Journal of Adhesion and Adhesives* **2002**, *22*, 283–289.
40 Malik, R. *Adhesives Age* **2002**, *45*, 35.
41 Malik, J., Goldslager, B. A., Clarson, S. J. *Surface Engineering* **2003**, *19*, 121–126.
42 Dobmann, A. *Adhesive Technology* **2000**, *17*, 14–16.
43 Dobmann, A. *Adhesives Age* **2002**, *45*, 26.
44 Tobing, S. D., Klein, A. *Journal of Applied Polymer Science* **2001**, *79*, 2230–2244.
45 Tobing, S. D., Klein, A. *Journal of Applied Polymer Science* **2001**, *79*, 2558–2564.
46 Schumacher, K.-H., Dussterwald, U., Fink, R. *Adhesive Technology* **2000**, *17*, 18–20.
47 Kim, J. K., Kim, W. H., Lee, D. H. *Polymer* **2002**, *43*, 5005–5010.
48 Staeger, M., Finot, E., Brachais, C. H., Auguste, S., Durand, H. *Applied Surface Science* **2002**, *185*, 231–242.
49 Han, C. D., Kim, J., Baek, D. M. *Journal of Adhesion* **1989**, *28*, 201–230.
50 Dahlquist, C. A. *Adhesion: Fundamentals and Practice*, McLaren, London, **1966**.
51 Gay, C., Leibler, L. *Physical Review Letters* **1999**, *82*, 936–939.
52 Ewins, E. E., Clair, D. J. S., Erickson, J. R., Korcz, W. H. In *Handbook of Pressure Sensitive Adhesive Technology*, Satas, D. (Ed.), Van Nostrand Reinhold, New York, **1989**.
53 Keyzer, N. D., Vermunicht, G. Presented at the Afera Technical Seminar, Brussels, **2004**.
54 Gergen, W., Lutz, R., Davison, S. In *Thermoplastic Elastomers*, 2nd edn., Holden, G., Legge, N., Quirk, R. (Eds.), Hanser, Munich.
55 Bates, F., Fredrickson, G. *Physics Today* **1999**, *52*, 32–38.
56 Matsen, M. W., Schick, M. *Macromolecules* **1994**, *27*, 6761–6767.
57 de Groot, H. In *TPE 2003*, **2003**.
58 Muyldermans, X. Presented at the International Body Engineering Conference and Exhibition and Automotive and Transportation Technology Congress, Session Glass and Glazing Systems (Part I), Paris, **2002**.
59 Eller, R. In *TPE 2004*, Brussels, **2004**, pp. 97–107.
60 Varma, R., Liu, D., Venkataswamy, K. In *ANTEC 2003*, Nashville, TN, **2003**, pp. 3051–3058.
61 Mace, J.-M., Moerenhout, J., to Shell Oil. US Patent 5750268, **1998**.
62 Pearce, R., Vancso, G. J. *Macromolecules* **1997**, *30*, 5843–5848.
63 KRATON^{TM} *Polymers for Modification of Thermoplastics*, Kraton Polymers, Houston, TX, **2000**.
64 Yang, B. Presented at SPE TPE TOPCON 2005, Expanding Materials, Applications and Markets, Akron, OH, **2005**.
65 Oshinski, A. J., Keskkula, H., Paul, D. R. *Journal of Applied Polymer Science* **1996**, *61*, 623–640.
66 Crossland, R. K., Clair, D. J. S., to Shell Oil. US Patent 4032459, **1977**.
67 Ver Strate, G., Struglinski, M. J. *ACS Symposium Series* **1991**, *462*.
68 Sabia, R. A., to Bell Telephone Laboratories. US Patent 4176240, **1979**.
69 Mitchell, D. M., Sabia, R. In *29th International Wire and Cable Symposium*, Fort Monmouth, NJ, **1980**, pp. 15–25.
70 Maes, C. Kraton Polymers Internal Communication.
71 Gu, J., Castile, T., Venkataswamy, K. In *ANTEC 2003 – Proceedings of the 61st Annual Technical Conference and Exhibition*, Nashville, TN, **2003**, pp. 3034–3040.
72 Frontini, G. L., Elicabe, G. E., Meira, G. R. *Journal of Applied Polymer Science* **1987**, *33*, 2165–2177.
73 Webster, O. W. *Science* **1991**, *251*, 887–893.
74 Morton, M. *Anionic Polymerization: Principles and Practice*, Academic Press, New York, **1983**.

75 Georges, M. K., Veregin, R. P. N., Kazmaier, P. M., Hamer, G. K. *Macromolecules* **1993**, *26*, 2987–2988.
76 Matyjaszewski, K. (Ed.). *ACS Symposium Series*, **2000**, *768*.
77 Matyjaszewski, K., Xia, J. H. *Chemical Reviews* **2001**, *101*, 2921–2990.
78 Matyjaszewski, K. In *Advances in Controlled/Living Radical Polymerization*, American Chemical Society, Washington, DC, **2003**, Vol. 854.
79 Greszta, D., Mardare, D., Matyjaszewski, K. *Macromolecules* **1994**, *27*, 638–644.
80 Wang, J. S., Matyjaszewski, K. *Macromolecules* **1995**, *28*, 7901–7910.
81 Wang, J. S., Matyjaszewski, K. *Macromolecules* **1995**, *28*, 7572–7573.
82 Auschra, C., Eckstein, E., Zink, M.-O., Muhlebach, A., to Ciba Specialty Chemicals. *US Patent 6849679*, **2005**.
83 Muhlebach, A., Rime, F., Auschra, C., Eckstein, E., to Ciba Specialty Chemicals. *US Patent Application 20030166755*, **2003**.
84 Mühlebach, A., Rime, F., Auschra, C., Eckstein, E., to Ciba Specialty Chemicals, *US Patent 6936656*, **2005**.
85 Rebizant, V., Venet, A.-S., Tournilhac, F., Girard-Reydet, E., Navarro, C., Pascault, J.-P., Leibler, L. *Macromolecules* **2004**, *37*, 8017–8027.
86 Ritzenthaler, S., Court, F., Girard-Reydet, E., Leibler, L., Pascault, J. P. *Macromolecules* **2003**, *36*, 118–126.
87 Rebizant, V., Abetz, V., Tournilhac, F., Court, F., Leibler, L. *Macromolecules* **2003**, *36*, 9889–9896.
88 Hamada, K., Morishita, Y., Kurihara, T., Ishiura, K. In *Proceedings of the 28th Annual Meeting of the Adhesion Society*, Mobile, AL, **2005**, pp. 53–57.
89 Moineau, C., Minet, M., Teyssie, P., Jerome, R. *Macromolecules* **1999**, *32*, 8277–8282.
90 Tong, J.-D., Jerome, R. *Macromolecules* **2000**, *33*, 1479–1481.
91 Tong, J. D., Moineau, G., Leclere, P., Bredas, J. L., Lazzaroni, R., Jerome, R. *Macromolecules* **2000**, *33*, 470–470.
92 Paul, C. W. Pressure Sensitive Tape Council Tech XXVII Global Conference, pp. 247–257.
93 Kwon, Y., Faust, R. *Advances in Polymer Science* **2004**, *167*, 107–135.
94 Fang, T. R., Kennedy, J. P. *Polymer Bulletin* **1983**, *10*, 82–89.
95 Kennedy, J. P., Melby, E. G. *Journal of Polymer Science, Polymer Chemistry Edition* **1975**, *13*, 29–37.
96 Mishra, M. K., Kennedy, J. P. In *Desk Reference of Functional Polymers*, Arshady, R. (Ed.), American Chemical Society, Washington, DC, **1997**.
97 Marshall, D. In *ANTEC 2004*, Chicago, IL, **2004**, pp. 4177–4181.
98 Marshall, D. Presented at SPE TPE TOPCON 2005, Expanding Materials, Applications and Markets, Akron, OH, **2005**.

2
Nanocomposites

Michaël Alexandre and Philippe Dubois

2.1
Introduction

The introduction of filler particles in polymeric materials, resulting in the production of so-called filled polymers or polymeric composite, is a technique that has been used for a very long time in order to modify and/or improve their properties or more practically to reduce the materials' cost. In addition to cost lowering, micron-size inorganic particles are usually introduced in polymers to increase their stiffness, to act as pigments or to modulate the materials' barrier properties or flame-retardant behavior. Nevertheless, addition of particulate fillers may also impart drawbacks to the composites obtained, such as opacity or brittleness.

Since the late 1980s, a new class of materials, resulting from the association of polymers with particles having at least one of their dimensions in the range of a few nanometers (and commonly named "nanoparticles"), has given rise to many advances in the design of new materials that are generally known under the generic name of "polymer nanocomposites". These nanocomposites exhibit markedly improved properties, highly dependent on the chemical nature but also the shape and size of the dispersed nanoparticles. A key factor to obtaining substantially improved properties lies in the ability to disperse properly these nanoparticles, which usually tend to aggregate strongly, as a result of very large specific surface areas and important particle–particle interactions. The use of dedicated macromolecular structures, acting as compatibilizers and either covalently grafted onto the particle surface or strongly interacting with the nanofiller through ionic or other non-covalent bondings, has therefore been developed to optimize the nanoparticles dispersion within a given polymer matrix. Accordingly, macromolecular engineering processes have been approached as unique tools for synthesizing controlled (functionalized) polymeric topologies, allowing the production of highly potent nanocomposite materials.

2.2
Generalities

2.2.1
Nanocomposites and Nanohybrids – Definitions

A *nanocomposite* can be defined as a material resulting from the association of particles having at least one of their dimensions in the range of a few nanometers, dispersed or organized in a continuous phase [1]. In the frame of this chapter, the continuous phase is a polymer matrix.

One can distinguish three types of nanocomposites, depending on the number of spatial dimensions of the particles which are in the nanometer range. When the three dimensions are in the order of one to a few nanometers, the particles can be defined as isometric nanoparticles. They can be metallic (gold, silver, platinum, copper, etc.), oxides (silica, alumina, zinc oxide, titanium dioxide, magnetic iron and cobalt oxides, etc.), but also metal compounds ($CaCO_3$, CdS, CdSe, etc.) or even carbonaceous (carbon blacks, fullerenes). When two dimensions are in the nanometer range and the third one is longer, forming an elongated structure (1D nanoparticles), the resulting nanoparticles can be defined as nanotubes or whiskers, as for example carbon nanotubes, cellulose or chitin whiskers and some specific clays such as sepiolite or palygorskite, which are currently being studied as reinforcing materials with exceptional properties. Finally, when only one dimension is in the nanometer range, the resulting nanoparticles occur in the form of sheets (2D nanoparticles), with dimensions around 1 nm thick and hundreds to thousands of nanometers long and wide. This family of nanoparticles covers also a large variety of chemical structures, from naturally occurring layered (alumino)silicates such as clays, through starch-based platelets or graphite nanosheets, to numerous synthetic layered particles (layered double hydroxides, zirconium phosphates, layered titanates and titanoniobates, molybdenum disulfides, etc.).

A *nanohybrid* can be defined as a nanocomposite for which there exists a relatively strong chemical bond between the nanoparticles and the macromolecular chains that compose (at least partly) the continuous phase. This chemical bond is usually a covalent or an ionic bond. It can be created either by the chemical reaction (or ion exchange) of functionalities borne by polymer chains with the nanoparticle surface ("grafting onto") or by the direct polymerization of a polymer precursor (monomer) from chemical functionalities located on the surface of the nanoparticle ("grafting from"). Another possibility for creating a nanohybrid consists in a copolymerization reaction involving polymerizable moieties anchored onto the nanoparticle surface with "free" monomer in solution or in bulk. In this case, this can be defined as a "grafting through" technique.

2.2.2
Key Details

Probably one of the first industrially prepared nanocomposites consisted in the dispersion of carbon blacks within rubber for tire manufacturing in the mid-1910s, but the use of the term "nanocomposite" only appeared in the scientific literature in the 1980s, for ceramic-based matrices. The first occurrence dealing with the term nanocomposite in relation to a polymer matrix appeared in 1989 in conference proceedings [2] and in 1991 as a paper [3], for the preparation of ultrafine copper particles, prepared in a poly(2-vinylpyridine) matrix by thermal decomposition at 125 °C in the solid state of a Cu^{2+}–formate–poly(2-vinylpyridine) complex. Copper nanoparticles with an average diameter of 3.5 nm were produced and homogeneously dispersed in the polymer matrix. Since 1991, the number of publications dealing with polymer-based nanocomposites has increasing to reach more than 1700 papers in 2005 (based on a survey in *Chemical Abstracts*), as depicted in Fig. 2.1.

Interestingly, it appears that about 43% of the papers published in 2005 deal with clay-based nanocomposites, 19% concern silica and about 9% concern carbonaceous nanoparticles (carbon black, graphite, carbon nanotubes and nanofibers).

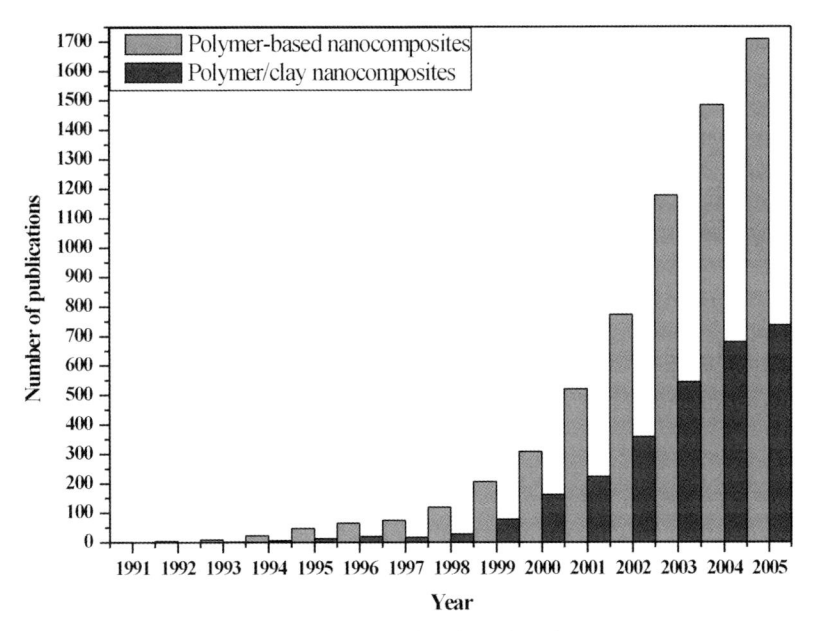

Fig. 2.1 Statistics on publications dealing with polymer-based nanocomposites and clay/polymer-based nanocomposites using SciFinder on the *Chemical Abstracts* database.

Therefore, amongst all the available nanoparticles, clays (and more generally layered silicates) have been and are still the most widely investigated, probably because the starting clay materials are easily available (natural occurrence) and because their intercalation chemistry has already been studied for a long time.

2.2.3
Main Production Pathways

Numerous production pathways have been investigated in order to promote the dispersion of nanoparticles in a polymer matrix. These pathways depend on the nanoparticle properties (chemical nature and surface reactivity, thermal stability, state of aggregation or entanglement for flexible nanotubes, etc.), but also on the polymer matrix characteristics (chemical reactivity towards the nanoparticle surface, ability to melt and physical properties in the molten state, ability to be solubilized, etc.). Despite the wide variety of nanoparticles and polymers, these dispersion techniques can be classified in three main categories:

- Mixing the polymer in the molten state or a solution of the polymer with the nanoparticles (or a suspension of the nanoparticles). In particular, this method achieves nanocomposite preparation through melt blending using various processing devices that allow one to apply some shear to a polymer melt (mixers, extruders, roll-mill, etc.) and to disperse the nanoparticles. With this process, good nanoparticle dispersion can be achieved, especially if the polymer–nanoparticle interactions are larger than the particle–particle interactions. In order to reach this balance of interactions, a chemical modification of the nanoparticles surface, aimed at lowering their surface energy, is often needed to promote good particle–polymer interactions and to weaken particle–particle interactions [4]. This is a very common method used to disperse silica, alumina and calcium carbonate nanoparticles but also investigated to favor the dispersion of layered and acicular clays, carbon black, carbon nanotubes and carbon nanofibers. If, through melt blending, a chemical bond can be created between a reactive polymer matrix and some functional groups located at the nanoparticle surface, this process can also lead to the formation of a nanohybrid material by the so-called "grafting onto" process.
- Creating the nanoparticles within a polymer matrix or diblock-based nanoreactors, starting from soluble reactants. This method has been used to produce metallic nanoparticles (Cu, Ag, Au, etc.) by thermally or chemically induced reduction of metal salts or destruction of soluble metal(0) complexes or to promote the formation of metal oxides, sulfides or other salts by the controlled formation of nanoclusters [5–7]. Under the same process type can be gathered the hydrothermal synthesis of synthetic clays in water-based polymer gels. In such a process, particularly adapted to water-soluble polymers, the polymer gels often act as a template for layer formation [8]. This method, highly dependent on the type of nanoparticle to be synthesized, will not be considered in this chapter.

- Creating the polymer matrix around the nanoparticles, either in solution or in the monomer(s) bulk. Depending on whether a chemical bond is created between the nanoparticles and the growing polymer chains ("grafting from" or "grafting through" processes) or not, a nanohybrid or a nanocomposite is formed, respectively. In the case where no chemical bonds are formed between the nanoparticles and the polymer matrix, good dispersion is insured only if the nanoparticles can disperse homogeneously in the initial monomer or the monomer solution. Better dispersion ability can be promoted by modification of the nanoparticle surface by surface agents. In the case of nanohybrid preparation and especially by the "grafting from" process, that is, the production of polymer chains from polymerization initiators anchored on the nanoparticle surface, the polymer chains growing away from the nanoparticle surface directly promote good dispersion of the nanoparticles within the polymer matrix [5, 9].

It is also possible to combine two of the aforementioned general processes to improve the dispersion of nanoparticles within a polymer matrix. For example, a masterbatch can be prepared using the process involving the matrix polymerization around nanoparticles by the "grafting from" technique in order first to produce nanoparticles individually coated by a thin polymer layer and then to disperse this masterbatch within a polymer matrix by further melt-blending.

2.2.4
Enhanced Properties

The improvement or modification of properties that may occur when nanoparticles are dispersed in a polymeric matrix can cover a very broad range of materials properties and is related to various factors such as the chemical nature of the dispersed nanoparticles, their relative content, their size, their shape and their organization within the polymer matrix.

Table 2.1 gives an overview of some properties acquired by the formation of polymer-based nanocomposites as a function of the nanoparticles' characteristics.

It is interesting that some kinds of nanoparticles allow several properties to be improved at the same time, offering the opportunity to tailor the final behavior of the nanocomposite and its domain of applications. For example, layered clays can impart some mechanical stiffening (without significant loss of ductility) to a polymer matrix while insuring also an improvement of the flame-retardant properties [10] or the fluid barrier performances. Moreover, the intrinsic properties of the polymers used can also participate in the modulation of the nanocomposite properties. For example, the ability of diblock copolymers to "nanostructure" used in conjugation with nanoparticles possessing nonlinear optical or conductive properties can promote the nano-organization of the nanoparticles within the materials, giving rise to interesting behaviors (dichroism, diffraction gratings, preparation of microelectronic devices) using relatively low-cost materials [4, 5, 11].

Table 2.1 Enhanced properties or properties acquired by a polymer matrix through nanocomposite formation.

Property	Nanoparticle characteristics	Nanoparticles
Stiffness	Intrinsic stiffness, high aspect ratio	Phyllosilicates, acicular clays, carbon nanotubes and nanofibers, expanded graphite, cellulose whiskers
Barrier to fluids	Layered structure	Phyllosilicates, expanded graphite
Flame retardancy	Layered structure	Phyllosilicates, expanded graphite, layered double hydroxides
	Elongated structures	Acicular clays, carbon nanotubes
	Ceramization	Polyhedral oligosilsesquioxanes
(Ultra) high refractive index	Refractive index, size	Lead sulfide, iron sulfides, zinc sulfide, titanium dioxide, silicon
(Ultra) low refractive index	Refractive index, size	Gold
Non-linear optical properties	Chemical nature, size	Metal, metal sulfide, metal selenide, metal oxides, etc.
Electric conductivity	Chemical nature	Carbon nanotubes, carbon black, expanded graphite, gold, etc.
Magnetism	Chemical nature	Iron oxides, cobalt, nickel, ferrites, perovskites

2.3
Layered Silicates

As discussed in Section 2.2.2, layered silicates or clays represent the class of nanoparticles most often studied and reported in the literature. The layered silicates commonly used in the preparation of polymeric nanocomposites belong to the structural family known as the 2:1 phyllosilicates. Their crystal lattice consists of two-dimensional layers where a central octahedral sheet of alumina or magnesia is fused to two external silica tetrahedra by the tip so that the oxygen atoms of the octahedral sheet also belong to the tetrahedral sheets. These layers are approximately 1 nm thick and their lateral dimensions may vary from 30 nm to several microns depending on the particular silicate. These layers organize to form stacks with a regular van der Waals gap between them called the interlayer spacing or gallery. Isomorphic substitution within the layers (for example, Al^{3+} replaced by Fe^{2+} or by Mg^{2+} or Mg^{2+} replaced by Li^+) generates negative charges at the particle surface that are counterbalanced by the sorption of highly hydrated alkali or alkaline earth metal cations located in the interlayer spacings. Undoubtedly, the most often used clay is montmorillonite. In order to render these hydrophilic clays more organophilic, the hydrated cations of the interlayer spacings can be exchanged with organic molecules such as alkylammonium or alkylphosphonium cations. When those hydrated cations are exchanged with organic cations such as more bulky alkylammoniums, it results in a larger

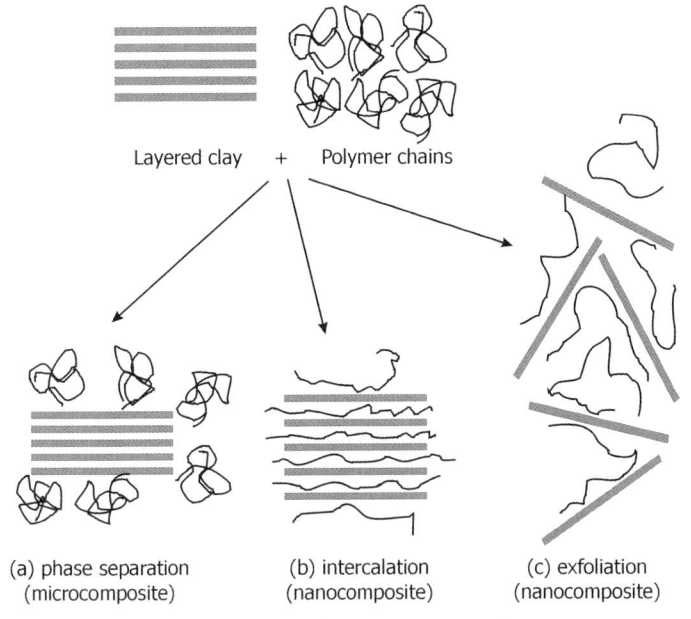

Fig. 2.2 Scheme of different types of composite arising from the interaction of layered silicates and polymers: (a) phase-separated microcomposite; (b) intercalated nanocomposite; (c) exfoliated nanocomposite.

interlayer spacing. The as-modified clay (also named organoclay) being organophilic, its surface energy is lowered and then more compatible with organic polymers. These polymers are then able to penetrate the galleries, at least to some extent.

Depending on the nature of the components used (layered silicate, organic cation and polymer matrix) and the method of preparation, three types of composites may be obtained when a clay is associated with a polymer (Fig. 2.2). When the polymer is unable to intercalate between the silicate sheets, a phase-separated composite (Fig. 2.2a) is obtained, whose properties stay in the same range as traditional microcomposites. Beyond this classical composite, two types of nanocomposite can be obtained. One is an intercalated structure (Fig. 2.2b) in which a single, extended layer of polymer chains is intercalated between the silicate nanoplatelets resulting in a well-ordered multilayer morphology, alternating polymer and inorganic layers. When the silicate layers are completely and uniformly dispersed in a continuous polymer matrix, an exfoliated or delaminated structure is obtained (Fig. 2.2c).

Several strategies have been used in order to prepare polymer-layered silicate nanocomposites. They can be grouped under three main processes [12]:
- *Exfoliation–adsorption:* the layered silicate is exfoliated into single layers using a solvent in which the polymer is soluble. It is well known that such layered

silicates, owing to the weak forces that stack the layers together, can be easily dispersed in a suitable solvent. The polymer then adsorbs on the delaminated sheets and, when the solvent is evaporated (or the dispersion is precipitated), the sheets collapse, trapping the polymer usually to form an ordered multilayer structure. Therefore, this technique mainly leads to intercalated nanocomposites.

- *Melt intercalation:* the layered silicate is mixed with the polymer matrix in the molten state. Under these conditions and if the layer surfaces are sufficiently compatible with the chosen polymer, the polymer can spread into the interlayer spacings and form either an intercalated or an exfoliated nanocomposite. In this technique, no solvent is required, which makes this process more environmentally friendly.
- *In situ intercalative polymerization:* in this technique, the layered silicate is swollen in the monomer (or a monomer solution) and the polymer formation occurs between the sheets. Polymerization can be initiated by either heat or radiation, by the diffusion of a suitable initiator or by an organic catalyst or initiator fixed through cation exchange inside the interlayer before the swelling by the monomer. It is precisely this process that will be further explored in this chapter. Indeed, *in situ* intercalative polymerization showed high efficiency in terms of organoclay delamination in the so-grown polymer matrix, allowing remarkable performances to be attained by the recovered nanocomposites. Furthermore, and interestingly enough, numerous nanohybrid materials have been synthesized via this technique with the possibility of controlling/tailoring the molecular parameters of the *in situ* polymerized chains surface-grafted onto the clay nanolayers.

2.3.1
Nanocomposites Prepared by *In Situ* Polymerization of the Matrix from Anchored Monomers or Initiators

2.3.1.1 Non-controlled Polymerization Processes

The huge interest in layered silicate-based nanocomposites started with the work reported by the Toyota research team [13, 14] on the *in situ* intercalative polymerization of ε-caprolactam from montmorillonite-type clays organomodified with protonated 12-aminolauric acid [$+H_3N(CH_2)_{11}COOH$]. In this process, the organomodified clays are first swollen by the ε-caprolactam monomer at 100 °C and then, by increasing the temperature to 250 °C, ring-opening polymerization (ROP) of the lactam monomer proceeds. After 6 h of polymerization, exfoliated nanocomposites were obtained if the amount of clay was less than 15 wt%, as evidenced by XRD and transmission electron microscopy (TEM) analysis. Comparison of the titrated amounts of COOH and NH_2 end-groups led to the conclusion that the COOH end-groups from the montmorillonite organomodifier are responsible for the initiation of polymerization, therefore indicating grafting of the synthesized polyamide-6 from the clay surface through the protonated amine of 12-aminolauric acid. Further studies [15] have demonstrated that intercalative polymerization of ε-caprolactam could be realized without the necessity to render the clay lipophilic since this monomer, in the presence of hydrochloric acid solutions, was able to intercalate the Na-montmorillonite. At high temperature (200 °C) and in the presence of an excess of ε-caprolactam, the modified clay is able to swell, allowing polymerization to proceed at 260 °C when a polymerization accelerator, such as 6-aminocaproic acid, is added. The resulting composite does not present any diffraction peak characteristic of an interlayer spacing in XRD and TEM observations agree with an individual dispersion of the silicate sheets. All the aforementioned *in situ* ROP reactions of lactam monomers trigger the delamination of the (organo)clays in the polyamide matrix and allow substantially improved thermo-mechanical properties to be attained for the so-recovered nanocomposite materials. However, it is worth pointing out that no control over the ROP and therefore the molecular parameters of the polyamide chains (surface-grafted or not) could be achieved.

Several other processes have been used to produce nanocomposites using the *in situ* intercalative polymerization technique promoted by free radical initiators. Using the "grafting through" technique, Akelah and Moet [16] modified the interlayer of Na-montmorillonite by exchanging the sodium cations with (vinylbenzyl)trimethylammonium chloride. These modified fillers were then dispersed in various solvent mixtures able to swell the modified clays and the polymerizations were carried out at 80 °C for 5 h after the addition of styrene and N,N'-azobis(isobutyronitrile) (AIBN). The composites were isolated by precipitation of the colloidal suspension in methanol, filtered off and dried. By this technique, intercalated composites were produced with interlayer spacings varying between 1.72 and 2.45 nm depending on the nature of the solvent used.

Wang et al. [17] studied the free radical polymerization of styrene and MMA in the presence of Cloisite 10 A, a commercial montmorillonite organomodified by benzyl dimethyl(hydrogenated tallow-alkyl)ammonium cations or in the presence of a montmorillonite modified by styryl hexadecyldimethylammonium, an organic modifier bearing a polymerizable moiety. The efficiency for layered silicate exfoliation was tested for polymerization carried out in bulk, in solution, in suspension or in emulsion. Whatever the polymerization process, the non-reactive clay leads to the formation of intercalated nanocomposites while the clay bearing polymerizable moieties allows the preparation of exfoliated nanocomposites, especially when styrene is polymerized in bulk or in the case of an emulsion process for both MMA and styrene polymerization. Note that the solution process favors, for both monomers, the formation of an intercalated morphology for the resulting nanocomposites.

Another technique which promotes layered silicate exfoliation consists in the use of montmorillonite organomodified with onium cations bearing polymerization initiators.

Fan et al. [18] exchanged Na^+-montmorillonite with monocationic and bicationic free radical initiators based on an azo function (Fig. 2.3).

Intercalation of the initiators has been probed by XRD analysis. The clay intercalated with the bicationic initiator is characterized by an interlayer distance

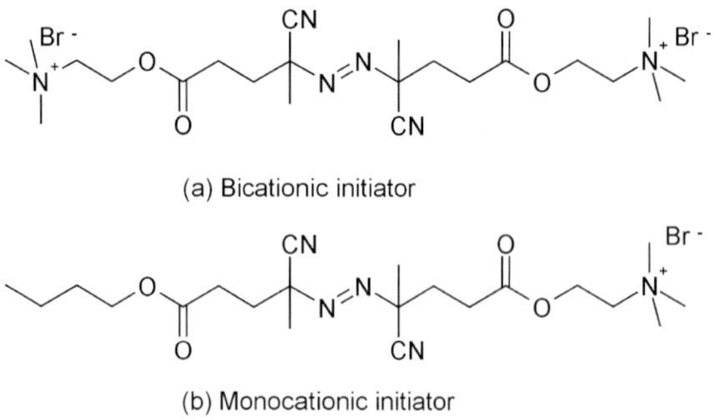

(a) Bicationic initiator

(b) Monocationic initiator

Fig. 2.3 (a) Bicationic and (b) monocationic initiators used by Fan et al. [18, 19] for modifying the montmorillonite gallery and initiating the radical polymerization of acrylic and styrenic monomers.

of 1.52 nm whereas the clay exchanged with monocationic initiator shows an interlayer distance of 2.18 nm. To explain this difference, the authors proposed that the bicationic initiator, which can interact with two anionic sites, may interact either with the same clay nanolayer or interact with two different but neighboring nanoplatelets. The combination of these two possibilities makes the intercalated structure less ordered (as attested by broadening of the X-ray diffraction peak), but also tends to tighten more strongly the clay platelets with one another. These two clays were dispersed for 6–10 h in THF before starting polymerization of styrene at 60 °C for 72 h. Attention was paid to keep the initiator and monomer concentrations constant, allowing the molecular weights of the polystyrene (PS) produced to be compared.

In terms of surface-bound PS, polymer obtained from anchored monocationic initiators has a much higher M_n (51 000 g mol^{-1}) than the one measured for the bicationic initiator (16 000 g mol^{-1}), indicating that the polymerization is more successful with the monocationic initiator. One of the reasons may be that the monocationic initiator-based clays platelets can be much more easily dispersed/ swollen in monomer whereas the bicationic initiator-based clay platelets remain much more associated together through the possible bridges that may occur between clay platelets. This is further attested by XRD analysis of the crude recovered nanocomposites, where total exfoliation is observed for the monocationic initiator system whereas a diffraction peak at a similar angle value to that determined for the starting organomodified clay is recorded for the bicationic system (Fig. 2.3 a). This result indicates that this last nanocomposite still contains an intercalated clay structure. In further work [19], the authors tested their monocationic free radical initiator (see Fig. 2.3 b) for the polymerization of MMA either in bulk, in solution or in suspension. Their study showed that exfoliation can be achieved in bulk and in solution while some intercalation is maintained in suspension.

2.3.1.2 Controlled Polymerization Process

Several "controlled" polymerization processes have been investigated for the preparation of layered silicate-based nanocomposites, mainly using the "grafting from" technique. Hence various controlled radical processes (see below) such as atom-transfer radical polymerization, nitroxide-mediated polymerization, reversible addition fragmentation chain transfer, but also "living" anionic polymerization, metallocene-catalyzed and controlled polymerization and ROP of lactones have be used to produce the nanocomposites. Special focus is given to this last process, which does not require the synthesis of complicated or sensitive initiators, since it can be activated from chemical functions as simple as alcohols.

Ring-opening polymerization of cyclic esters Pantoustier and coworkers [20, 21] synthesized poly(ε-caprolactone) (PCL)-based nanocomposites using *in situ* intercalative polymerization of ε-caprolactone (ε-CL) in the presence of various organomodified clays. This synthetic approach involves the dispersion of the organomodified clay platelets in the liquid monomer followed by bulk polymerization as catalyzed/initiated by selected organometallic compounds such as aluminum or tin alkoxides.

Depending on the nature of the filler, different composite morphologies can be obtained. When clays, either non-organomodified or organomodified with ammonium cations bearing only alkyl chains, were used, essentially intercalated nanocomposites were recovered, as highlighted by the presence of intercalation peaks upon XRD analysis. Only a small amount of exfoliation could be observed by TEM. For the nanocomposites prepared with a commercially available organo clay modified by a functional ammonium cation bearing two primary hydroxyl groups (Cloisite 30B), XRD analysis gives evidence for a fully delaminated structure, as attested by the total absence of a diffraction peak at low diffraction angle. These different observations were also supported by TEM analyses. This behavior can be explained by the presence of hydroxyl functions cationically anchored at the clay surface, which can participate in the polymerization reaction. Initiation of the ε-CL polymerization from the clay surface accounts for the tethering of the *in situ*-grown chains on the clay surface and for clay delamination (Fig. 2.4).

Similar behavior has been observed for other lactone monomers such as L,L-lactide and 1,4-dioxan-2-one. Indeed, Paul et al. [22] successfully exfoliated Cloisite 30B in poly(L,L-lactide) (PLA) by polymerizing L,L-lactide in bulk at 120 °C for 48 h, using triethylaluminum in a molar equivalent with respect to the amount of hydroxyl groups borne by Cloisite 30B. Lee et al. [23] polymerized L-lactide in solution in xylene at 135 °C for 3–7 days, starting from Cloisite 30B activated by tin(II) bis(2-ethylhexanoate) [Sn(Oct)$_2$]. For both processes, exfoliation was observed by XRD and by TEM analysis. Huang et al. [24] studied the *in situ* intercalative polymerization of 1,4-dioxan-2-one catalyzed by triethylaluminum in bulk at 50 °C for 20 h, in the presence of sodium montmorillonite and different montmorillonites modified with octadecyltrimethylammonium and 2-hydroxyethylhexadecyldimethylammonium cations. For this monomer, in-

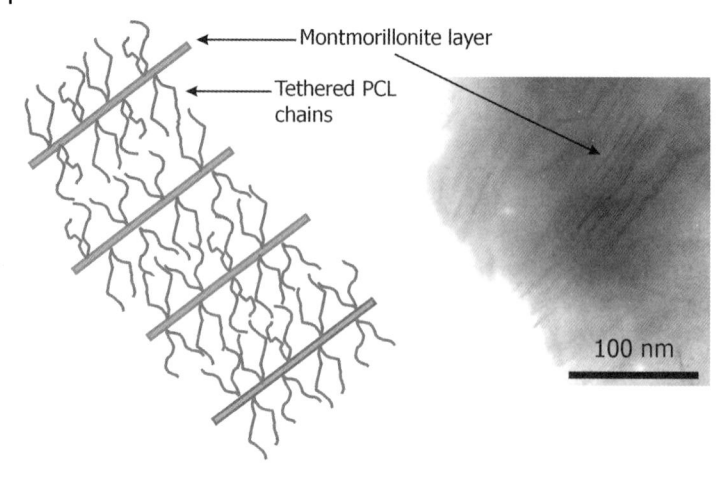

Fig. 2.4 Scheme of exfoliation promoted by polymer growth from tethered initiator and TEM representing the phenomenon for PCL polymerized from hydroxyl groups borne by ammonium cations organomodifying montmorillonite (3 wt% inorganics).

tercalated nanocomposites were obtained as determined by XRD. Unfortunately, the degree of possible exfoliation was not determined by TEM.

Evidence for control over ring-opening polymerization The controlled synthesis of PCL–montmorillonite nanohybrids has been performed by ROP of ε-CL using either $Sn(Oct)_2$ in catalytic amount at 100 °C [21] or triethylaluminum ([Al]/[OH] = 1) at room temperature [25]. Accordingly, the ε-CL polymerization is promoted by the hydroxyl groups (OH) available at the surface of Cloisite 30B, which allows the prediction and control of the molecular weight of the PCL chains produced as shown in Table 2.2.

Whatever the catalyst/initiator, a continuous decrease in the molar masses is observed on increasing the clay content. As far as the tin-based catalyst is concerned, the number-average molar masses measured by size-exclusion chromatography (SEC) are in fair agreement with the theoretical values expected from the initial monomer-to-alkoxide molar ratio. Such behavior arises from the fact that all the polyester chains are initiated by the hydroxyl groups available at the clay surface while this tin derivative behaves as the catalyst added in a tiny amount. Therefore, the molar masses of the tethered chains are a function of the quantity of hydroxyl groups and hence depend on the monomer-to-clay ratio. In order to modulate and control the molar masses of the PCL chains for a given amount of organoclay, the authors modified the relative amount of hydroxyl functions that are tethered to the clay.

Accordingly, montmorillonites were surface modified with a predetermined mixture of non-functional and monohydroxyl-functionalized alkylammonium cations [25–27]. The hydroxyl functions at the clay surface were activated into

Table 2.2 In situ polymerization of ε-CL in the presence of various amounts of Cloisite 30B by catalytic activation with Sn(Oct)$_2$ ([ε-CL]:[Sn]=300) at 100°C for 24 h or promoted by in situ-formed aluminum alkoxides ([AlEt$_3$]=[OH]) at room temperature for 24 h.

Co-initiator	Cloisite 30B content (wt.%)	Conversion (%)	M_n th.[a] (g mol^{-1})	M_n exp.[b] (g mol^{-1})	PDI[b]
Sn(Oct)$_2$	3	97	36 500	37 200	1.6
	5	89	19 700	13 900	1.5
	8	92	12 300	8 000	1.4
	15	51	3 400	4 800	1.7
AlEt$_3$	1	92	116 000	85 300	2.0
	3	91	39 500	41 700	1.8
	5	91	22 700	27 400	1.9
	10	72	8 200	6 000	–[c]

a) M_n th. = [ε-CL]$_0$/[OH]$_0$ × conversion × MW ε-CL.
b) Determined by size-exclusion chromatography; M_n expressed in PCL equivalent.
c) Bimodal.

Sn(II or IV) or Al(III) alkoxide initiators for lactone polymerization, so yielding surface-grafted PCL chains. The surface-grafted PCL chains were analyzed by SEC after extraction using LiCl in solution in THF for cation-exchange reaction with the ammonium cations. The PCL molar masses, measured as a function of the hydroxyl content of the clay, are given in Table 2.3.

In the case of Sn(Oct)$_2$ used in a catalytic amount ([ε-CL]$_0$:[Sn]=300), the molar masses of the tethered PCL were found to decrease with increasing content of hydroxyl groups available at the clay surface. When dibutyltin(IV) dimethoxide in a catalytic amount ([ε-CL]:[Sn]=300) was used, the molar masses of PCL at different clay surface hydroxyl contents show a relatively constant value. This can be explained by the fact that the Sn(IV) compound acts not only as a catalyst but also as a polymerization co-initiator through its methoxide functions. Hence the relatively high content of Sn(IV) combined with the high activity of Sn(IV) alkoxides for promoting intermolecular transesterification reactions can explain the relatively limited effect of the clay surface hydroxyl content on the molecular mass of the PCL chains for this series of nanocomposites. When converting the clay surface hydroxyl groups (for clay organomodified by 100% monohydroxyl-based ammonium cations) into aluminum trialkoxide using a stoichiometric amount of triethylaluminum ([OH]:[Al]=3), remarkable polymerization control can be achieved with experimental molar masses (M_n^{exp}=21 000 g mol^{-1}) very close to the theoretical values (M_n^{th}=19 800 g mol^{-1}) and a very low polydispersity index of 1.2.

As far as PLA polymerization is concerned [22], under the conditions reported in the previous section, PLA chains with a relatively narrow molecular weight

Table 2.3 In situ polymerization of ε-CL initiated by $Sn(Oct)_2$ at 100 °C for 24 h, by $Bu_2Sn(OMe)_2$ at room temperature for 24 h or promoted by in situ-formed aluminum alkoxides ([OH]:[AlEt$_3$] = 3) at room temperature for 24 h in the presence of MMT-$(CH_2CH_2OH)_x$ (3 wt% of inorganics): effect of the co-initiator and the hydroxyl content within the organoclay ([ε-CL]:[Sn] = 300).

Co-initiator	MMT-$(CH_2CH_2OH)_x$ (OH content, x%)	Conversion (%)	M_n exp.[a] (g mol^{-1})	PDI[a]
$Sn(Oct)_2$	100	93	28 000	1.8
	75	96	36 500	1.8
	50	97	47 000	1.9
	25	98	53 500	1.9
	0	98	56 000	2.0
$Bu_2Sn(OMe)_2$	100	98	15 500	1.8
	75	98	16 500	2.0
	50	98	18 000	1.9
	25	99	16 500	2.2
	0	98	18 500	2.0
AlEt$_3$	100	52	20 000	1.2

a) Determined by size-exclusion chromatography; M_n expressed in PCL equivalent.

distribution (PDI $\approx$ 1.5) were obtained, attesting to some level of polymerization control.

The synthesized PCL nanohybrids were also characterized by small-angle X-ray diffraction and transmission electron microscopy. For the nanocomposites obtained from clays with no or small amounts (<50%) of hydroxyl functions available at their surface, intercalation with some extent of exfoliation could be observed. For the nanocomposites obtained from clays exchanged with hydroxyl-functionalized ammoniums in quantities $\geq$50%, the extent of exfoliation was greatly increased and was complete for the clays bearing 100% hydroxyl-based ammoniums cations. Indeed, fully exfoliated PCL organoclay nanocomposites were recovered.

Preparation of surface-grafted diblocks The control over the ROP of ε-CL as initiated from the hydroxyl functions available on the aforementioned organoclays, after adequate activation by AlEt$_3$, was extrapolated to promote the block copolymerization of ε-CL and l,l-lactide (LA) monomers in solution (Fig. 2.5) [28].

It has been shown that adjusting the experimental parameters allows the process to be under control. Indeed, the growing copolyester chains are effectively surface-anchored on the clay and their average length and surface-grafting density can readily be tuned up by the content of hydroxyl functions available on the inorganic platelets. These new organic–inorganic *nanohybrid* materials pres-

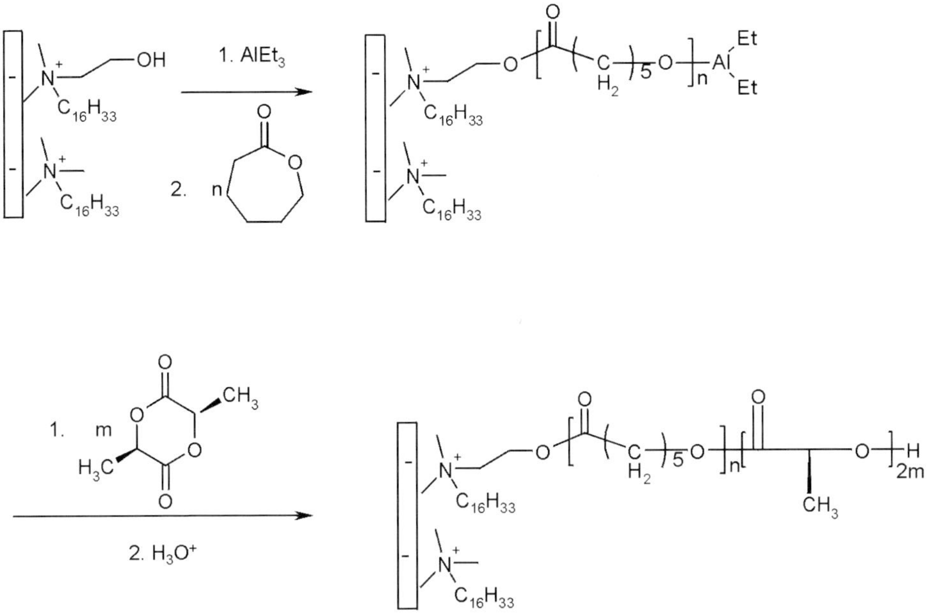

Fig. 2.5 Scheme illustrating the *in situ* block copolymerization of ε-CL and L,L-LA from controlled amounts of hydroxyl groups borne by ammonium cations organomodifying montmorillonite.

ent fully exfoliated morphologies and the grafted copolymers display double semicrystalline and blocky structures (without any trace of transesterification), as confirmed by DSC and NMR analyses, respectively.

Influence of controlled grafting on nanocomposite/nanohybrid properties Gorrasi and coworkers evaluated the water vapor barrier properties of the PCL nanohybrid (3 wt% inorganics) using a microgravimetric method and compared them with unfilled PCL and other nanocomposites obtained by melt blending PCL with Cloisite 30B [29, 30]. The evolution of the water zero concentration diffusion coefficients D_0 for each type of PCL nanocomposite is shown in Fig. 2.6.

Clearly, the exfoliation promoted by the PCL grafting reaction on to the organoclay nanoplatelets strongly reduces the diffusion of water through the samples, with respect to the unfilled PCL sample and the intercalated melt blended nanocomposite. The effect was optimum for the sample prepared from a clay organomodified by 100% monohydroxylated alkylammonium cations, where no water absorption could be measured by this technique. Increasing the hydroxyl content further (Cloisite 30B; 100% dihydroxylated alkylammonium cations) allows the recovery of some water absorption and is characterized by the lowest measurable D_0 value.

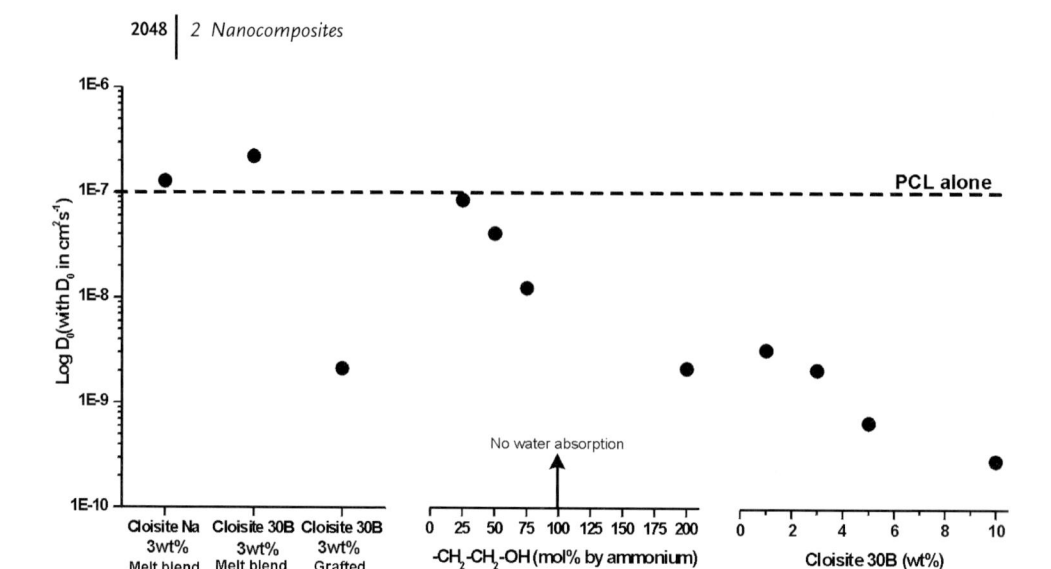

Fig. 2.6 Evolution of the logarithm of the zero diffusion coefficient for water (D_0) in PCL (nano)composites as measured by microgravimetry for (left) a microcomposite (Cloisite Na), an intercalated nanocomposite (Cloisite 30B, melt blend) and an exfoliated nanocomposite (Cloisite 30B, grafted); (center) exfoliated nanocomposites prepared from clays organomodified by various amounts (in mol%) of ammonium cations bearing hydroxyl functions used for initiating ε-CL polymerization (3 wt.% of inorganics as a constant); (right) exfoliated nanocomposites with various clay contents, prepared from Cloisite 30B.

Interestingly, direct permeability measurements of He, H_2 and CO_2 were measured on the same series of samples by Gain et al. [31]. The relative permeability (P_n/P_m, where P_n is the permeability of the nanocomposite and P_m the permeability of the PCL matrix, expressed in barrer) as a function of the grafting density is shown in Fig. 2.7.

The same general behavior is observed for all three gases, with a rapid decrease in the relative permeability with increasing grafting density until a coverage of 100% of monohydroxylated ammonium cations. Then, a leveling off of this effect is observed on further increasing the polyester grafting density by substituting 100% of dihydroxylated ammonium cations for 100% of monohydroxylated ammonium cations. Globally, a reduction by one-third of the gas permeation can be achieved for only 3 wt% of inorganics in the nanohybrids. This effect is mainly due to the increase in the tortuosity pathway of the gas molecules arising from the improved exfoliation observed at higher grafting density.

Karaman et al. [32] prepared PCL-grafted nanocomposites starting from clays organomodified with 100 and 60% monohydroxylated ammoniums, using $Sn(Oct)_2$ as polymerization catalyst, leading to an exfoliated (3 wt% clay) and an intercalated (8.2 wt% clay) nanocomposite, respectively. Stretching calorimetry studies were used to evaluate the mechanical work and heat effects during stretching. It was found that for the exfoliated nanocomposite based on ca.

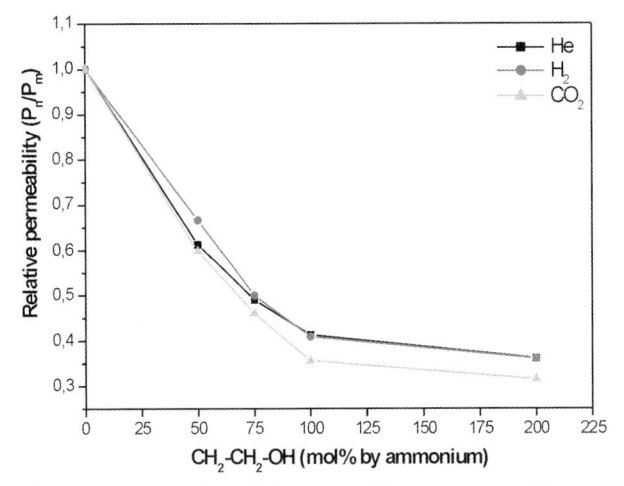

Fig. 2.7 Evolution of the relative permeability to helium, hydrogen and carbon dioxide for pure PCL and nanocomposites prepared from clays organomodified with various amounts (in mol%) of ammonium cations bearing hydroxyl functions initiating ε-CL polymerization (3 wt.% of inorganics as a constant).

100% grafted PCL chains, interfacial debonding was significantly less important than for the nanocomposite where the PCL chains were grafted on only 60% of the available clay surface, indicating that the ionic links between the clay surface and the grown polymer chains are sufficiently strong to influence the adhesive properties at the interface.

Use of aliphatic polyester-grafted nanoclay as masterbatches for nanocomposite preparation Owing to the well-known miscibility of PCL with numerous polymers, highly filled nanohybrids (synthesized by the previously discussed grafting ROP) have been added as masterbatches in commercial thermoplastic matrices known for their miscibility with PCL. A PCL–clay "masterbatch" (25 wt% clay) was prepared in bulk by *in situ* intercalative polymerization of CL initiated by the hydroxyl groups borne by the Cloisite 30B clay, using $Sn(Oct)_2$ as ROP catalyst [21]. This "masterbatch" was melt blended with a chlorinated polyethylene (CPE) (Tyrin 3652) at 175 °C for 10 min in order to reach 3 wt% clay content. For the sake of comparison, a direct melt blend of Cloisite 30B and CPE was prepared under the same conditions. TEM micrographs of these two composite materials (Fig. 2.8) show a much better dispersion of the clay nanoplatelets with extensive exfoliation when the PCL "masterbatch" is used in comparison with the simple melt blend with the commercial organo-clay.

When comparing the mechanical properties (Table 2.4), a large increase in the Young's modulus with only a slight decrease in the tensile strain at break is observed for the exfoliated nanocomposite compared with both the unfilled matrix and the intercalated sample, indicating a large effect of exfoliation using the PCL "masterbatch".

CPE + 3wt% Cloisite 30B

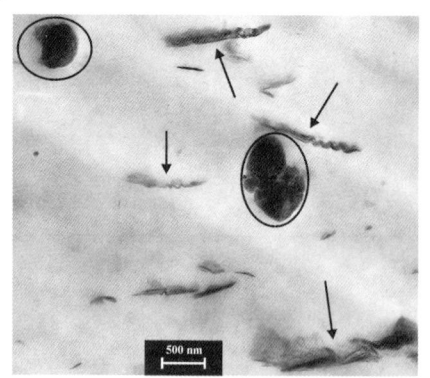

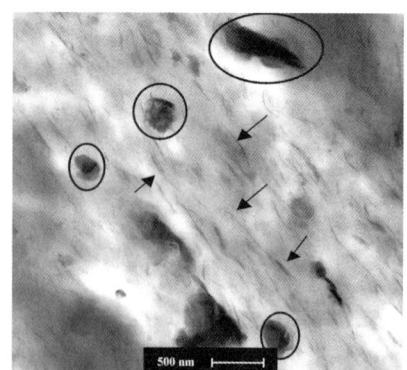

CPE + 3wt% Cloisite 30B grafted with PCL

Fig. 2.8 TEM microphotographs of nanocomposites based on 3 wt.% clay dispersed in chlorinated polyethylene: (a) using Cloisite 30B (arrows show large stacks of intercalated clay); (b) using a 75 wt.% rich PCL-grafted masterbatch (arrows show some exfoliated clay layers). The large circled particles are additives of the native CPE matrix (talc).

Table 2.4 Tensile properties of chlorinated polyethylene (CPE) and nanocomposites arising from the dispersion of 3 wt.% of Cloisite 30B (CPE + CL30B) or the dispersion of a masterbatch prepared by the *in situ* grafting of PCL (75 wt.%) at the surface clay to give a 3 wt.% exfoliated nanocomposite (CPE + masterbatch).

Blend	Tensile stress at break (MPa)	Tensile strain at break (%)	Young's modulus (MPa)
CPE	16.0 ± 0.7	1302 ± 29	4.3 ± 0.3
CPE + CL30B	14.7 ± 0.9	1219 ± 38	8.1 ± 0.8
CPE + masterbatch	8.8 ± 0.6	1111 ± 63	14.3 ± 3.0

The same effect can be observed when PCL or PVC commercial matrices are used [33]. For instance, the thermal stability of PCL nanocomposites as a function of clay content was investigated by thermogravimetry (TGA). A significant enhancement of thermal stability was recorded for the thermoplastic matrices filled with the "PCL-grafted organo-clay" nanohybrids, added as a "masterbatch" to the commercial PCL matrix by melt blending.

Other controlled polymerization processes

Radical processes Table 2.5 gathers information about the controlled radical polymerization processes used to prepare nanocomposites based on layered silicates.

Table 2.5 Clay-based nanocomposites prepared using controlled radical polymerization processes.

Entry	Polymerization process	Clay organomodifier	Catalyst or radical promoter and controller	Monomer(s)	Solvent	T (°C)	Ref.
1	ATRP		CuBr with tetradentate amine ligand	MMA	Acetone	60	32
2	ATRP		CuBr with bis(pyridylmethyl)amine C17 ligand	Styrene, MMA, BuA	–	110, 50–90, 70	33
3	ATRP		CuBr with bis(pyridylmethyl)amine C17 ligand	1. Styrene 2. BuA	–	110, 70	34
4	NMP	TEMPO-based alkoxyamine with benzoate	–	Styrene	–	125	35
5	NMP	4-vinylbenzyl ammonium C15	TEMPO / Benzoyl peroxide	Styrene	–	120	36

Table 2.5 (continued)

Entry	Polymerization process	Clay organomodifier	Catalyst or radical promoter and controller	Monomer(s)	Solvent	T (°C)	Ref.
6	RAFT	(structure)	2,2′-Azobis(isobutyronitrile) (AIBN)	Styrene	DMF	110	37
7	RAFT	(structure)	AIBN/ (structure)	BuA, MMA	–	60	38
		(structure)	AIBN/ (structure)	Styrene	–	110	
		(structure)	AIBN/ (structure)	MMA	–	60	

Atom-transfer radical polymerization (ATRP) processes have been studied by Böttcher et al. [34] and Zhao and coworkers [35, 36]. Böttcher et al. carried out MMA polymerization in acetone dispersion as promoted from an ammonium bromoisobutyrate initiator anchored at the surface of the clay platelets (entry 1 in Table 2.5). The control over the polymerization was attested by the linear first-order kinetics recorded for the monomer consumption ($\ln[M]/[M]_0$) versus time) and the production of PMMA with relatively low polydispersity indices (around 1.1 up to 52% conversion). The intercalation peak of the resulting nanocomposites was shifted towards lower 2θ values with conversion, with a broadening of the peak until complete disappearance, attesting to a continuous PMMA growth within the interlayer spacings that promotes platelet separation until extensive destructuring of the clay stacking. The same organomodifier was investigated by Zhao and coworkers in ATRP of styrene, butyl acrylate (BA) and MMA in bulk [35]. Using the catalytic systems in Table 2.5 (entry 2), either in bulk or in a solvent, good polymerization control was achieved, with relatively good prediction of the molar masses and low polydispersity indices for the *in situ*-grown polymer chains (< 1.4). Better exfoliation was achieved with the ester-based polymers (MMA, BA) than with PS, for which a consequent amount of intercalated stacks was observed by TEM. Interestingly, poly(styrene-*b*-butyl acrylate) block copolymers (PSBA) were also anchored at the surface of clay platelets using the same technique [36], again with a good level of polymerization control (Table 2.5, entry 3). TEM analysis performed after staining the PSBA copolymer-based samples with hydrazine–osmium tetraoxide shows 2–5-nm dark dots of poly(butyl acrylate) domains organized around the clay platelets. Such small-sized PBA phases are explained by the fact that the block copolymer chains remain immobilized on the surface of the clay platelets and cannot develop larger domains.

Weimer et al. used nitroxide-mediated polymerization (NMP) to prepare clay nanoplatelets grafted with PS chains with defined molecular parameters [37]. The controlled polymerization was carried out in pure styrene by dispersing a clay previously modified by an ammonium cation bearing a nitroxide function (Table 2.5, entry 4). Exfoliation was confirmed by XRD and TEM analyses. Polymerization control was assessed through the good agreement between the experimental and theoretical number-average molecular masses of the PS chains (24 400 g mol^{-1} for the theoretical value compared with 21 500 g mol^{-1} for the experimental value) together with a low polydispersity index (PDI = 1.3). Polymerization "livingness" was assessed by both an efficient chain-extension experiment and chain length controlled by combining various quantities of the nitroxide-containing ammonium cations with a non-functional ammonium cation for clay surface cationic organomodification.

NMP was also used by Xu et al. using a "grafting-through" technique [38] (Table 2.5, entry 5). A montmorillonite (3.5 and 3.9 wt%), organomodified with *N,N*-dimethylhexadecyl-(4-vinylbenzyl)ammonium cations was dispersed in styrene in the presence of AIBN as initiator and 2,2,6,6-tetramethylpiperidinoxy (TEMPO) as the nitroxide for polymerization control. The polymerization reaction was carried out at 120 °C for 48 h. Under these conditions, 8.8 and 6.6 wt% of PS were found to be tethered to the clay surface through copolymerization with the vinyl-functionalized ammonium cations. Even though relatively low, this amount of tethering allows an essentially exfoliated structure to be obtained, as attested by XRD and TEM analyses. Interestingly, the extracted PS chains were characterized by relatively low PDI (1.2 and 1.3) and number-average molar masses increasing with monomer conversion.

A reversible addition–fragmentation transfer (RAFT) polymerization mechanism has also been studied. Zhang et al. [39] investigated styrene polymerization in DMF at 110 °C using AIBN as the radical promoter and a sodium montmorillonite exchanged with a cationic RAFT agent as shown in Table 2.5, entry 6. Under these conditions, styrene polymerization occurred in a controlled way, as attested by the linear relationship of the number–average molar masses of extracted PS chains with conversion, together with relatively low PDI ranging from 1.09 to 1.33. Upon styrene polymerization, clay stack destructuring and exfoliation were revealed by XRD analysis, that is, by a gradual shift towards low diffraction angles until disappearance of the intercalation peak with styrene conversion, and also by TEM. Part of the PS chains remained grafted on the clay nanolayers through degenerative transfer of growing chains with the RAFT agent. Indeed, in this system, the polymerization of styrene was initiated by AIBN and mostly took place in solution. Fractional precipitation allowed selective recovery of the clay platelets grafted with PS and proved to be characterized by a clay loading of 20 wt%.

Salem and Shipp [40] used RAFT polymerization to promote the formation of intercalated and exfoliated nanocomposites by the "grafting through" technique (Table 2.5, entry 7). Polystyrene exfoliated nanocomposites were obtained from the polymerization of styrene in bulk at 110 °C in the presence of montmorillonite organomodified with N,N-dimethylhexadecyl-(4-vinylbenzyl)ammonium (VB16) cations, using AIBN as initiator and 2-(2-cyanopropyl) dithiobenzoate as the RAFT agent. A linear evolution of the number-average molar masses with monomer conversion was observed throughout the polymerization and the experimental M_ns were in good agreement with the theoretical values. About 20 wt% of the PS chains were attached on the montmorillonite nanolayers. Montmorillonite organomodified with VB16 (VB16-clay) showed a propensity to delaminate readily in styrene, allowing the preparation of truly exfoliated nanocomposites as determined by XRD and TEM. The same organomodified clay was also used to prepare PBA- and PMMA-based nanocomposites, starting from the monomer in bulk and with the polymerization initiated at 60 °C using AIBN as the initiator and 4-cyano-4-methyl-4-thiobenzoylsulfanylbutyric acid as the RAFT agent. With these monomers, control of the polymerization was effective even if the VB16-clay delamination was not initially obtained in the monomer. While the evolution of the diffraction peak in XRD with monomer conversion leads to total clay destructuring in BA polymerization after 10 h, intercalation is still observed for PMMA, even after 48 h of polymerization and 98% conversion. In this last case, only 3 wt% of polymer chains were effectively grafted to the clay nanosheets, indicating that the poor dispersion of the clay within the monomer reduces drastically the ability of the growing chains to react with the unsaturated functions anchored on the clay surface. However, changing VB16 cations for 2-methacryloyloxyethylhexadecyldimethylammonium cations, more compatible with MMA monomer, allows an exfoliated nanocomposite with controlled M_n and low PDI (1.24) to be obtained.

In conclusion, whatever the process used, it always starts with the synthesis of an ammonium cation bearing either a suitable initiator group or a polymerizable moiety, followed by organomodification of the clay and then polymerization of the chosen monomer(s). Polymerization can be carried out in bulk or in a solvent. As a rule, initiator anchoring does not impede the control of the polymerization and, as for ROP, the controlled radical polymerization processes from clay surface grafted initiators leads to polymer grafting and greatly favors nanoplatelet exfoliation in the *in situ*-grown polymer matrix as attested by both XRD and TEM.

Ionic-like processes Advincula and coworkers [41, 42] modified a montmorillonite with trimethyl{12-[4-(1-phenylvinyl)phenoxy]dodecyl}ammonium cation (Fig. 2.9) in order to generate an anionic initiator by reaction with *n*-butyllithium in benzene. After careful washing of the clay with dried benzene to remove any excess or unreacted *n*-BuLi, the clay was characterized by a red coloration, indicating the presence of anionic species anchored to the clay surface.

On addition of styrene in benzene, polymerization could occur and PS–clay nanocomposites were produced. Unbound PS chains, produced by residual initiator in solution, were recovered by washing with toluene, while surface-bound PS was recovered after exchange reaction with Li cations in THF. Table 2.6 shows the results of SEC analysis of unbound and bound PS chains, for three polymerizations carried out with various monomer-to-initiator ratios.

A linear correlation between monomer-to-initiator ratio (grams per millimole) and M_n for the bound PS is an indication of a "living" anionic polymerization mechanism. Unbound PS are characterized by a narrower PDI, indicating, according to the authors, that the degree of polymerization control in solution was higher than for bound PS. Bound PS are also characterized by much lower M_n than unbound PS. This observation is explained by the authors by a lower efficiency for initiation/chain growth inside interlayer spacings than at the na-

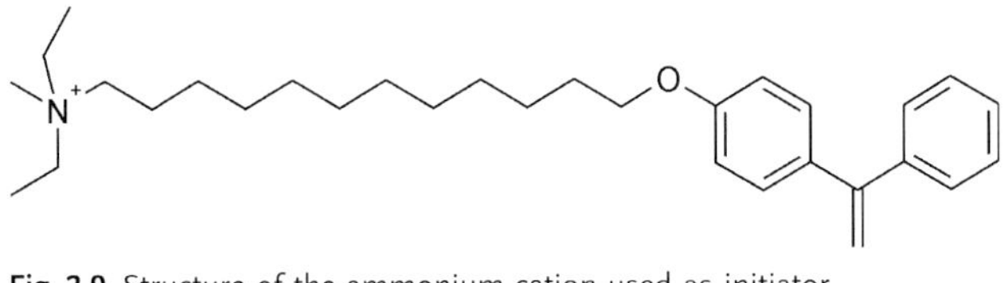

Fig. 2.9 Structure of the ammonium cation used as initiator (after reaction with *n*-BuLi) to promote the controlled anionic polymerization of styrene from an organoclay surface.

Table 2.6 Molecular parameters of bound and unbound PS resulting from the anionic polymerization of styrene from clay nanoplatelets organomodified by the ammonium cation shown in Fig. 2.9.

Styrene/initiator (g mmol^{-1})	M_n (bound PS) (10^3 g mol^{-1})	PDI (bound PS)	M_n (unbound PS) (10^3 g mol^{-1})	PDI (unbound PS)
9.6	7.4	1.3	35.5	1.2
14.6	11.9	1.4	78.6	1.1
19.9	21.8	1.3	58.5	1.3

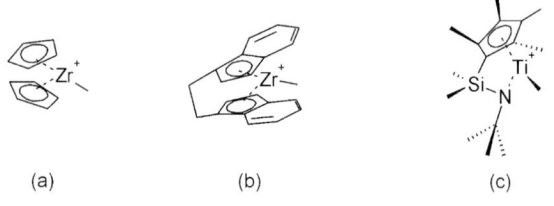

Fig. 2.10 Structure of various metallocenium cations used for the intercalative polymerization of (a) atactic, (b) syndiotactic and (c) isotactic PMMA.

noplatelet perimeter surfaces. Clay exfoliation was not proved, since XRD shows an intercalation peak at even lower interlayer spacing than for the organomodified clay.

Mariott and Chen [43] used montmorillonite intercalated with various metallocenium cations (Fig. 2.10) to control the tacticity of PMMA chains during *in situ* intercalative polymerization of MMA.

Intercalation of the metallocenium was obtained by reaction of the dimethylated derivative of the metallocene catalyst with montmorillonite previously exchanged with a protonated amine [methyl bis(hydrogenated tallow-alkyl)amine]. Whereas atactic PMMA (39.8% *mr* triads) was produced when bis(η^5-cyclopentadienyl)methylzirconium(IV) cation was intercalated (Fig. 2.10a), isotactic PMMA (93% *mm* triads) was obtained using *rac*-ethylenebis(indenyl)methylzirconium(IV) cation (Fig. 2.10b) and syndiotactic PMMA (72% *rr* triads) was recovered when (*tert*-butylamido)dimethyl(tetramethyl-η^5-cyclopentadienyl)silane titanium methyl cation was intercalated (Fig. 2.10c). Both XRD analysis and TEM analyses of the materials produced indicated the formation of essentially exfoliated nanocomposites.

In terms of tacticity control, syndiotactic polystyrene (sPS)–montmorillonite nanocomposites have also been prepared through metallocene-based polymerization of styrene. Bruzaud et al. [44] prepared sPS–montmorillonite (organomodified with alkylammonium or alkylphosphonium cations) nanocomposites via *in situ* "coordination–insertion" polymerization. Polymerizations were carried out in toluene solution with η^5-cyclopentadienyltitanium trichloride–methylaluminoxane (MAO) as the catalytic system. Various protocols (use of commercial or TMA-free MAO, use of sonication) were examined in order to test their influence in terms of nanoclay dispersion. Whatever the protocol used, highly disordered nanocomposites were obtained, as attested by featureless XRD patterns in the area dealing with clay stacking while sPS chains were effectively prepared as confirmed by NMR and DSC.

2.4
Extrapolation to other Nanofillers

"Controlled" polymerization processes have also been studied to prepare nanocomposites based on other types of nanoparticles. Some typical examples will be discussed in the following sections.

2.4.1
Other Layered Nanoparticles

H-magadiite, a synthetic layered silicate possessing reactive hydroxysilyl groups on its basal plane, has been successively reacted with 3-aminopropyldimethylethoxysilane and 2-bromopropionyl bromide to anchor an ATRP initiator on the Si–OH surface [45]. The polymerization of styrene was then carried out in acetone in the presence of CuBr ligated by bipyridine, yielding intercalated and even partially exfoliated PS–magadiite nanocomposites, as attested by XRD and TEM. The organomodification and polymerization process is depicted in Fig. 2.11.

Some control over the polymerization was achieved, as attested by the relatively low PDI (1.3) and steady increase in M_n with polymerization time. Furthermore, nanoindentation experiments revealed a large increase in both Young's modulus (from 6.1 GPa from unfilled PS to 21.5 GPa for PS filled with 5 wt% H-magadiite) and surface hardness (from 0.13 GPa without H-magadiite to 0.30 GPa for 5 wt% H-magadiite).

On another hand, anionic polymerization of styrene initiated by graphite intercalated with potassium proved to give rise to the formation of essentially intercalated nanostructures [46]. Even though the SEC trace of the recovered PS was bimodal, indicating a loss of polymerization control, some "livingness" of the anionic polymerization was demonstrated by chain extension experiments. At 6.5 wt% of graphite, the dielectric constant of PS was increased by a factor of 48, whereas at 8.2 wt% of graphite, the percolation threshold for volume conductivity was reached.

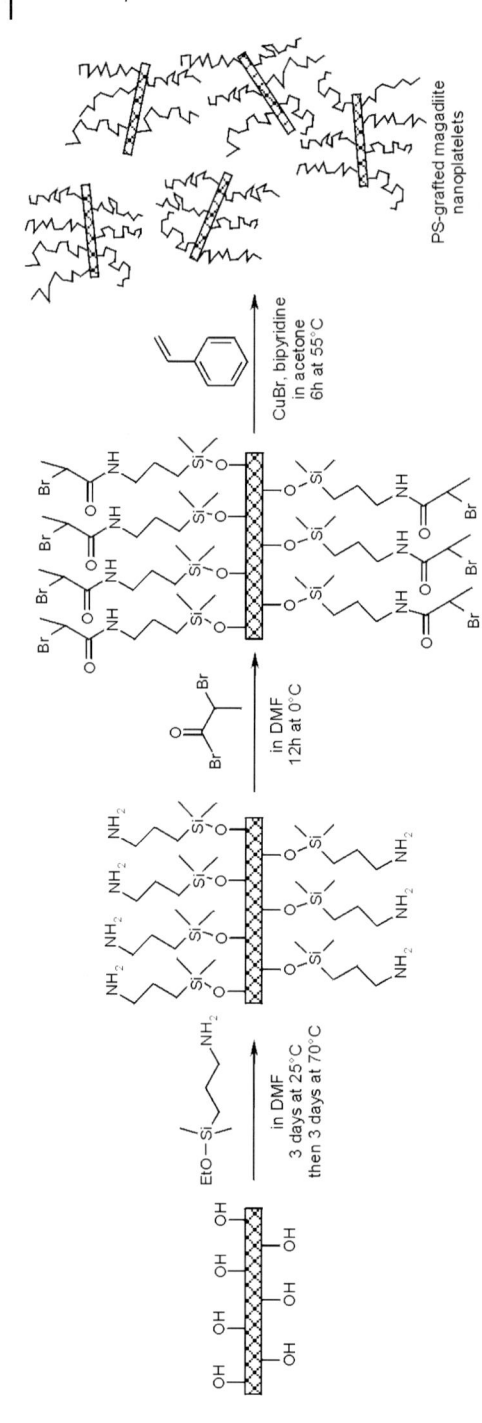

Fig. 2.11 Preparation of H-magadiite–PS nanohybrid through initiator anchoring followed by ATRP of styrene.

2.4.2
Carbon Nanotubes

Several groups have studied "controlled" polymerization processes in order to graft polymers at the surface of carbon nanotubes (CNT). Either "grafting onto" or "grafting from" techniques were used. Jérôme and coworkers [47] used NMP to synthesize alkoxyamine-terminated PS. Alkoxyamine-terminated PCL was also prepared by ROP of ε-CL from a hydroxyl-bearing NMP initiator. This macroinitiator also served for producing nitroxide terminated P(CL-b-S) block copolymer. These procedures allow fine control of the structure of the (co)polymers, which were efficiently grafted on to multiwall carbon nanotubes (MWNTs) by radical addition reaction in toluene at 130 °C for 24 h. Grafting was proved both by TGA and by dispersion tests in organic solvents. The same method was successfully applied by Adronov and co-workers [48] in order to graft PS and poly(tert-butyl acrylate-b-styrene) block copolymers on to single-walled carbon nanotubes (SWNTs). Interestingly, SWNTs grafted with the block copolymer was further treated with trifluoroacetic acid in $CHCl_2$ to hydrolyze the tert-butyl ester into carboxylic acid pendant groups, in order to obtain SWNTs grafted with amphiphilic block copolymers. ATRP was also used by Jérôme and co-workers [49] to prepare poly[methyl methacrylate-co-(1-pyrene)-methyl 2-methyl-2-propenoate] copolymers. The polymerization was initiated by ethyl 2-bromopropionate in the presence of CuBr and 1,1,4,7,10,10-hexamethyl-triethylenetetramine in toluene at 85 °C. This copolymer was able to interact strongly by non-covalent interactions between the pyrene moieties and the carbon nanotube surface. This strong interaction allowed much better CNT dispersion in organic solvents. Finally, Liu et al. [50] studied MWNTs surface-modified through ligand (cyclopentadienyl) exchange with ferrocene, subsequently modified by a methylenestyrene moiety (see Fig. 2.12) to promote further the anionic coupling of living polystyryllithium chains on to the so-modified MWNTs.

For the preparation of polymer–carbon nanotube nanocomposites by the "grafting from" technique, the ATRP process has been essentially investigated. In order to anchor the active bromide (or chloride)-based initiator to the nanotubes, it is necessary first to derivatize the surface of CNTs. Therefore, several techniques have been used to promote the formation of primary alcohols or carboxylic acid chlorides along the CNT surface. Kong et al. [51–53] modified the MWNT surface by a multi-step process in order to anchor covalently bromoisobutyryl ester functions as ATRP initiator. This functionalization method consists in first oxidizing the MWNT surface with concentrated nitric acid to promote the formation of carboxylic acids that are transformed into acid chlorides by reaction with $SOCl_2$, which are further reacted with an excess of ethylene glycol to form anchored hydroxyl groups. These – OH functions are then reacted with 2-bromo-2-isobutyryl bromide to form the ATRP initiator. Using the CuBr–N,N,N',N'',N''-pentamethyldiethylenetriamine catalytic system, they were able to graft PMMA in bulk at 60 °C [53]. The thickness of the polymer layer, as observed by TEM, was well controlled by the monomer to MWNT feed ratio. To

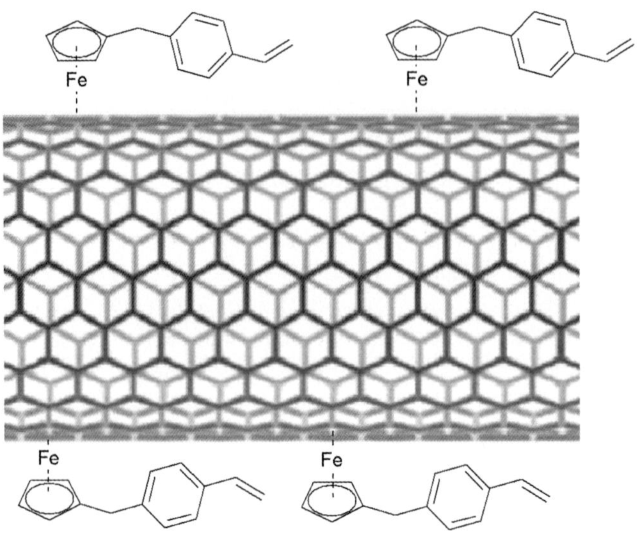

Fig. 2.12 Sketch of styryl moieties immobilized at the surface of carbon nanotubes through ferrocene-like surface functionalization.

highlight the control over the polymerization reaction, the nanotubes so grafted with PMMA chains were further considered as the "macroinitiator" in the polymerization of 2-hydroxyethyl methacrylate (HEMA). The resulting MWNT–PMMA-b-P(HEMA) nanocomposite was characterized by FTIR and ^{1}H NMR spectroscopy. The same procedure was used to graft PS and PS-b-P(tert-butyl acrylate) [51] and MWNT grafted with PS-b-P(acrylic acid) after treatment of the former with CF_3COOH [52]. Yao et al. [54] used a 1,3-dipolar cycloaddition to functionalize SWNTs with phenolic derivatives, grafted along their sidewall. These phenols were further derivatized with 2-bromoisobutyryl bromide to form ATRP initiators that were used to promote MMA or tBA polymerization in DMF suspension by using CuBr–2,2′-dipyridyl as the catalytic system. PMMA chains were recovered by the soft saponification of the phenolic ester bond. PMMA was characterized by very high molar masses that do not vary linearly with polymerization time, revealing a non-controlled process. This absence of control was attributed to some radical scavenging action of the shortened SWNTs used in the study. Baskaran et al. [55] used commercially available MWNTs functionalized with 1% of carboxylic acid functions, which were derivatized in acid chlorides by reaction with $SOCl_2$ and further reacted with hydroxyethyl 2-bromoisobutyrate to anchor the ATRP initiator. Polymerization of MMA and styrene was carried out in bulk at 90 and 100 °C, respectively in the presence of CuBr–N,N,N',N'',N''-pentamethyldiethylenetriamine catalytic system. MWNT–PS, MWNT–PMMA, MWNT–PS-b-PMMA and MWNT–PMMA-co-PS were successfully obtained through this technique.

2.4.3
Isometric Nanoparticles

Nanosized silica-, gold- and magnetic iron-based nanoparticles probably represent the most widely studied isometric nanoparticles on to which polymers have been grafted. Table 2.7 gathers some representative systems involving the "grafting from" methodology [56–71].

Various techniques (ATRP, reverse ATRP, NMP, anionic, ROP, etc.) have been explored with nanosized silicas (entries 1–7, Table 2.7). Whereas most workers focused on the molecular and morphological characterization of their materials, proving the covalent anchoring of polymer chains with controlled molecular parameters and homogeneous coating of the nanoparticles, Bartholome and coworkers [60, 61] (entry 5 in Table 2.7) investigated more deeply the rheological properties of their nanosized silica-based PS nanocomposites. PS chains with M_n between 15 000 and 60 000 g mol^{-1} and PDI around 1.2 were accordingly synthesized. Rheological measurements suggest that the grafted polymer chains on the silica particles create a steric repulsion between the filler particles that prevents the formation of a silica network within the polymer matrix, paving the way for the control of nanoparticle ordering and interdistance and spatial organization.

In addition to nanometric silica particles, other nanoparticles, interesting for their intrinsic properties (optical, magnetic, etc.) have also been grafted with polymers by using "controlled" polymerization processes. For instance, gold nanoparticles have been functionalized by various thiol derivatives, bearing reactive functions able to initiate either ATRP (entries 8 and 9, Table 2.7) or cationic ROP (entry 10, Table 2.7). However, since gold–sulfide bonds are relatively unstable, especially at high temperature, Kotal et al. [67] coated their gold nanoparticles with a thin layer of silica in order to be able to carry out polymerization reactions at higher temperature. These silica-coated gold nanoparticles were surface-treated with [(chloromethyl)phenylethyl]trimethoxysilane in order to anchor an activated chloride which could initiate the MMA polymerization in xylene at 90 °C using CuCl–bipyridine as ATRP catalyst (entry 11 in Table 2.7). Interestingly, the linear increase in monomer conversion {expressed as $\ln([M_0]/[M])$ versus time} with polymerization time and the linear increase in M_n with conversion again attest to the control over the polymerization.

Finally, specific initiators for ATRP (entries 12 and 14, Table 2.7) and coordinative ROP (entry 13, Table 2.7) have been anchored at the surface of various iron oxide-based magnetic nanoparticles to create core–shell nanoparticles while controlling the molecular properties of the polymer chains constituting the shell.

In addition to "grafting from" techniques, Corbierre and coworkers [72, 73] studied the *in situ* preparation of gold nanoparticles by reduction of $HAuCl_4$ in the presence of thiol-terminated PS. The resulting PS-coated gold nanoparticles were further dispersed in PS matrices by solvent casting. For that purpose, PS chains terminated with a thiol function were synthesized by anionic polymeriza-

Table 2.7 Nanocomposites based on isometric nanoparticles prepared using controlled radical polymerization processes.

Entry	Nanofiller	Type of controlled polymerization	Particle size (nm)	Anchored initiator	Catalyst and/or sacrificial initiator	Mono-mer(s)	Solvent	T (°C)	Ref.
1	Silica	ATRP	70	SiO$_2$–O–Si–(CH$_2$)$_2$–C$_6$H$_4$–CH$_2$Cl	bipy / CuCl	Styrene	–	130	56
2			75	SiO$_2$–O–Si–(CH$_2$)$_3$–O–C(O)–CH(Br)CH$_3$	bipy(C$_9$H$_{19}$)$_2$ / CuBr	Styrene	–	110	57
					bipy(C$_9$H$_{19}$)$_2$ / CuBr/CuBr$_2$	MMA	Xylene	90	
3			20	SiO$_2$–O–Si–(CH$_2$)$_3$–O–C(O)–C(CH$_3$)$_2$Br	bipy(C$_9$H$_{19}$)$_2$ / CuBr/CuBr$_2$	Styrene BuA MMA	–	90	58
4		Reverse ATRP	16	SiO$_2$–O–O–C(CH$_3$)$_3$	bipy / CuCl$_2$	MMA	Cyclo-hexane	70	59

#	Method	Size	Initiator (structure)	Reagents	Monomer	Solvent	T (°C)	Ref
5	NMP	13	(SiO₂)–O–Si–[CH₂]₁₀–O–C(=O)–CH(Ph)–O–N(tBu)–P(=O)(OEt)₂	(structure shown)	Styrene	Toluene	110	60, 61
6	Anionic	12–20	(SiO₂)–O–Si–[CH₂]₁₁–C₆H₄–C(=CH₂)Ph + s-BuLi	—	Styrene	Benzene	N/A	62
7	Coord. ROP	7	(SiO₂)–O–Si–[CH₂]₁₀–OH	Triethylaluminum then 2-propanol	ε-CL	Toluene	50	63
			(SiO₂)–O–Si–[CH₂]₃–O–CH₂–CH(OH)–CH₂OH	Triethylaluminum; Aluminum or yttrium tri(isopropylate) and tin(II) bis(2-ethylhexanoate)	ε-CL, lactide			
8	ATRP	50–70	(Au)–S–[CH₂]₁₁–O–C(=O)–C(CH₃)₂–Br	cyclam, CuBr	MMA	Xylene	20	64
9	ATRP	0.8–7	(Au)–S–[CH₂]₁₁–O–C(=O)–C(CH₃)₂–Br	(−)-Sparteine, CuBr	MMA	DMF	40	65
10	Cationic ROP	4	(Au)–S–[CH₂]₁₁–O–SO₂CF₃	—	Oxazolines	CHCl₃	N/A	66

Table 2.7 (continued)

Entry	Nanofiller	Type of controlled polymerization	Particle size (nm)	Anchored initiator	Catalyst and/or sacrificial initiator	Monomer(s)	Solvent	T (°C)	Ref.
11	Silica-coated gold	ATRP	21–29	Au–O–Si–benzyl chloride	2,2′-bipyridine; CuCl	MMA	Xylene	90	67
12	γ-Fe$_2$O$_3$	ATRP	10	Fe$_2$O$_3$–O–C(=O)–C(CH$_3$)$_2$–Br	2,2′-bipyridine; CuBr	Styrene	–	120	68
13		Coord. ROP	5–15	Fe$_2$O$_3$–O–Si–(CH$_2$)$_3$–NH–(CH$_2$)$_2$–NH$_2$	Aluminum tri(isopropylate) Tin(II) bis(2-ethylhexanoate)	ε-CL	Toluene	50	69
14	Fe$_3$O$_4$	ATRP	60	Fe$_3$O$_4$–O–Si–arylsulfonyl chloride	(−)-Sparteine, CuCl	MMA	Diphenyl ether	70	70
15	MnFe$_2$O$_4$	ATRP	9	MnFeO$_4$–O–C(=O)–CH$_2$–Cl	4,4′-dinonyl-2,2′-bipyridine; CuCl	Styrene	Xylene	130	71

tion of styrene and the polystyryl anion was finally titrated with one unit of poly(propylene sulfide) to generate the thiol end-group. When gold nanoparticles were coated with relatively long PS chains ($M_n = 13\,300$ g mol^{-1}), they could be homogeneously dispersed in various PS matrices, independently of their molar mass. However, when the nanoparticles were coated with short-chain PS ($M_n = 2000$ g mol^{-1}), they could only be dispersed in PS matrices with low molar mass. This phenomenon was explained by the authors by a lower grafting density for nanoparticles coated with high-M_n PS chains, forming a loose brush around the nanoparticle that allows the PS from the matrix to mix with the surface-grafted PS chains.

2.5
Current and Potential Applications

In addition to carbon black and nanosized silica particles, whose dispersion in rubbers has been used for many years in the tire industry, commercial applications of polymer nanocomposites are mainly concentrated on clay-based and carbon nanotube-based nanofillers. Currently, these nanocomposites find essential applications in the automotive, packaging and cable and wire industries. The three industrial sectors will be discussed separately in the following sections.

2.5.1
Automotive Industry

Keen interest in polymer nanocomposites started with the discovery, at Toyota Research Laboratories at the end of the 1980s, that the polymerization of ε-caprolactam in the presence of small amounts of montmorillonite (2–5 wt%) led to the production of new materials with interesting mechanical and thermal properties. The resulting 5 wt% clay–polyamide 6 nanocomposite, showing a very large increase in tensile modulus (by 190%) and heat deflection temperature (going from 65 to 152 °C in the final nanocomposite) was used for the fabrication of timing belt covers in automotive engines in the 1990s. This belt cover showed good rigidity, excellent thermal stability and no warp [74]. More recently, the same type of nanocomposite (polyamide 6 filled with 2 wt% montmorillonite) has been used by Maserati in a large engine cover part on its Quattroporte model. This nanocomposite was selected for its good mechanical performance, especially at high temperature, its low weight and its improved surface appearance compared with glass-reinforced polyamide. General Motors has integrated thermoplastic polyolefin (TPO)-clay nanocomposites in several automotive parts, to replace the heavier TPO filled with much larger amounts of talc (20–35 wt%). Between 2002 and 2004, they produced TPO–clay nanocomposites for step assists on the GMC Safari and Chevrolet Astro vans. In 2004, they used a polypropylene–nanoclay nanocomposite on the body side molding of the Chevrolet Impala, and in 2005 the cargo bed (trim, center bridge, sail panel and box rail protector) of the GM Hummer H2 SUT involved more than 7 lb of TPO–nanoclay nanocomposites [75]. A clay-based nanocomposite produced by Putsch under the trade-name Elan XP, where the nanofiller acts as a compatibilizer in a PP (60–80 wt%)–PS(40–20 wt%) blend, has been used for the fabrication of heater vents in Audi and Volkswagen cars as a highly scratch-resistant material for replacement of ABS [76, 77]. PP–clay nanocomposites have also been used in Honda cars to produce seat backs, in order to replace the 20 wt% glass-filled PP that caused processing problems, visual defects and warping [75]. It is also reported that, in (very) small amounts, carbon nanotubes are involved in cars, typically in polyamide matrices encountered in quick connector and filters of the fuel-line system against static electricity [75]. This continually increasing involvement of nanocomposites in cars (and also in other transport engines such as in airplanes or trains) finds its origin in the weight-reducing ability arising from the introduction of very small amounts of nanofillers that tremendously modify the mechanical and thermal resistance of the resulting materials, making them highly competitive with respect to more conventional talc or short glass-fiber filled materials.

2.5.2
Packaging Industry

When nanoclays are properly exfoliated in a polymer matrix, they give rise to the formation of homogeneously dispersed clay nanosheets of very large aspect ratio (from 100 to 1000, depending on the nature of the chosen clay). These nanosheets are impermeable to other molecules and therefore act as "nanobarriers" to their diffusion throughout the nanocomposite material [1]. These nanobarriers make the diffusion pathway of the permeant molecules much more tortuous, decreasing (sometimes by more than an order of magnitude) the permeability of the nanocomposite materials, as compared with the unfilled polymer matrix. This property has been tested commercially for the preparation of packaging with increased passive barrier properties.

Polyamide–clay nanocomposites (Aegis NC from Honeywell, Imperm from Nanocor) are therefore used as internal barrier layers in PET multilayer bottles used to package beer, sodas and flavored alcoholic drinks [78]. The increasing use of clay-based nanocomposites in PET plastic bottles is related to the necessity to increase the shelf-life of the goods and the need to reduce the numbers of layers in multilayer bottles for cost savings and better recycling. Moreover, polyamide–clay nanocomposites are a cost-effective alternative to the more widely used ethylene–vinyl alcohol copolymers. Other developments are also foreseen in film and pouch packaging for food (cheese, meat, fish, etc.).

2.5.3
Cable and Wire Industry

Clay-based polymer nanocomposites are used by several cable and wire manufacturers such as Nexans (France), Cablewerk Eupen (Belgium) and Draka Cable (The Netherlands) for flame-retardant (halogen-free) cable jacketing with good processing and mechanical properties [75]. The introduction of nanoclays in polyolefin-based (PE, EVA) cable sheath improves the flame-retardant rating such as the UL94 test by eliminating dripping and promoting the formation of stable char. For example, by adding 4wt% of nanoclay in PE–EVA compounds, it has been possible to reduce the amount of aluminum or magnesium hydroxides from 65 to 52 wt% in a standard halogen-free formulation while maintaining the desired flame retardancy. This new formulation is characterized by an improvement of the mechanical and surface properties of the resulting sheath and higher extrusion rates. Flame retardancy in polymer–clay nanocomposites comes from a combination of the barrier properties of the clay nanosheets towards the ingress of oxygen and the egress of fuel molecules issued from the polymer degradation, and also from the ability of the clay nanosheets to migrate and accumulate at the surface of the burning materials to form a protective layer (a char) that limits further gas and thermal energy diffusion.

However, despite such successful applications of nanocomposites in industry, several issues still need to be resolved in order to integrate nanoclays (and nanocomposites) more widely in a larger range of applications. One issue relies on the relatively low thermal stability of the ammonium cations commonly used to organomodify natural clays. Their thermal degradation, especially during nanocomposite processing, limits the use of the nanoclay technology to polymers with processing temperatures below 250 °C. Another issue is that, for a given property, such as flame resistance, the use of nanoclay alone is generally not sufficient to compete with current solutions. Major improvements should arise, however, from their combination with other flame-retardant solutions to achieve synergistic behaviors.

2.6
Conclusion

Nanocomposites and nanohybrids based on polymer matrices and nanoparticles are attracting more and more interest in several aspects of everyday life. The modification of properties arising from the dispersion of either nanofillers with high aspect ratio (such as clay nanoplatelets) or nanoparticles displaying intrinsic original properties (for example, electrical, thermal and mechanical properties of carbon nanotubes, optical properties of gold or CdS nanoparticles, magnetic properties of iron or cobalt-based nanoparticles) allows a wide array of new applications to be envisaged. Macromolecular engineering is and will continue to be closely associated with the development of these new materials, since the new sought-after properties can be magnified by the precise nanostructuring of these nanoparticles within the nanomaterials. The preparation of nanohybrids, using controlled "grafting from", "grafting through" and "grafting onto" polymerization techniques, has already proved to improve greatly the homogeneous dispersion of nanoparticles within the polymer matrices. Clearly, manipulating the nature, composition, molecular weight and topology (for example, block copolymers, hyperbranched copolymers) of the so-grafted polymer chains represents another challenge for accessing unique structural organization at the nanolevel of surface-modified nanoparticles in the organic host. Undoubtedly, the combination of controlled and "living" polymerization processes and tailored interfacial chemical reactions with (nano)tools encountered in the production of polymeric (nano)composites will open up new opportunities for key applications in material sciences. In this chapter, several such pioneering approaches have been highlighted together with the first important industrial applications. The involvement of macromolecular engineering in the nanosciences and particularly in the domain of nanocomposites is only in its infancy and can expect a very promising future.

References

1 M. Alexandre, P. Dubois, *Mater. Sci. Eng. R Rep.* **2000**, *28*, 1–63.
2 A. M. Lyons, S. Nakahara, E. M. Pearce, *Mater. Res. Soc. Symp. Proc.* **1989**, *132*, 111–117.
3 A. M. Lyons, S. Nakahara, E. M. Pearce, J. V. Waszczak, *J. Phys. Chem.* **1991**, *95*, 1098–1105.
4 W. Caseri, *Macromol. Rapid Commun.* **2000**, *21*, 705–722.
5 R. Shenhar, T. B. Norsten, V. M. Rotello, *Adv. Mater.* **2005**, *17*, 657–669.
6 J. F. Ciebien, R. T. Clay, B. H. Sohn, R. E. Cohen, *New J. Chem.* **1998**, *22*, 685–691.
7 G. Carotenuto, B. Martorana, P. B. Perlo, L. Nicolais, *J. Mater. Sci.* **2003**, *13*, 2927–2930.
8 K. A. Carrado, *Appl. Clay Sci.* **2000**, *17*, 1–23.
9 J. Pyun, K. Matyjaszewski, *Chem. Mater.* **2001**, *13*, 3436–3448.
10 M. Alexandre, G. Beyer, C. Henrist, R. Cloots, A. Rulmont, R. Jérôme, P. Dubois, *Macromol. Rapid Commun.* **2001**, *22*, 643–646.
11 M. R. Bockstaller, R. A. Mickiewicz, E. L. Thomas, *Adv. Mater.* **2005**, *17*, 1331–1349.
12 C. Oriakhi, *Chem. Br.* **1998**, *34*, 59–62.
13 A. Usuki, Y. Kojima, M. Kawasumi, A. Okada, Y. Fukushima, T. Kurauchi, O. Kamigaito, *J. Mater. Res.* **1993**, *8*, 1179–1184.
14 Y. Fukushima, A. Okada, M. Kawasumi, T. Kurauchi, O. Kamigaito, *Clay Miner.* **1988**, *23*, 27–34.
15 Y. Kojima, A. Usuki, M. Kawasumi, A. Okada, T. Kurauchi, O. Kamigaito, *J. Polym. Sci., Part A: Polym. Chem.* **1993**, *31*, 983–986.
16 A. Akelah, A. Moet, *J. Mater. Sci.* **1996**, *31*, 3589–3596.
17 D. Y. Wang, J. Zhu, Q. Yao, C. A. Wilkie, *Chem. Mater.* **2002**, *14*, 3837–3843.
18 X. Fan, C. Xia, R. C. Advincula, *Langmuir* **2003**, *19*, 4381–4389.
19 X. W. Fan, C. J. Xia, R. C. Advincula, *Langmuir* **2005**, *21*, 2537–2544.
20 N. Pantoustier, B. Lepoittevin, M. Alexandre, D. Kubies, C. Calberg, R. Jérôme, P. Dubois, *Polym. Eng. Sci.* **2002**, *42*, 1928–1937.
21 N. Pantoustier, M. Alexandre, P. Degée, D. Kubies, R. Jérôme, C. Henrist, A. Rulmont, P. Dubois, *Compos. Interfaces* **2003**, *10*, 423–433.

22 M. A. Paul, M. Alexandre, P. Degée, C. Calberg, R. Jérôme, P. Dubois, *Macromol. Rapid Commun.* **2003**, *24*, 561–566.
23 S. Lee, C. H. Kim, J. K. Park, *J. Appl. Polym. Sci.* **2006**, *101*, 1664–1669.
24 F.-Y. Huang, Y.-Z. Wang, X.-L. Wang, K.-K. Yang, Q. Zhou, S.-D. Ding, *J. Polym. Sci., Part A: Polym. Chem.* **2005**, *43*, 2298–2303.
25 B. Lepoittevin, N. Pantoustier, M. Alexandre, C. Calberg, R. Jérôme, P. Dubois, *Macromol. Symp.* **2002**, *183*, 95–102.
26 B. Lepoittevin, N. Pantoustier, M. Alexandre, C. Calberg, R. Jérôme, P. Dubois, *J. Mater. Chem.* **2002**, *12*, 3528–3532.
27 B. Lepoittevin, N. Pantoustier, M. Devalckenaere, M. Alexandre, D. Kubies, C. Calberg, R. Jérôme, P. Dubois, *Macromolecules* **2002**, *35*, 8385–8390.
28 E. Pollet, C. Delcourt, M. Alexandre, P. Dubois, *Macromol. Chem. Phys.* **2004**, *205*, 2235–2244.
29 G. Gorrasi, M. Tortora, V. Vittoria, E. Pollet, M. Alexandre, P. Dubois, *J. Polym. Sci., Part B: Polym. Phys.* **2004**, *42*, 1466–1475.
30 G. Gorrasi, M. Tortora, V. Vittoria, E. Pollet, B. Lepoittevin, M. Alexandre, P. Dubois, *Polymer* **2003**, *44*, 2271–2279.
31 O. Gain, E. Espuche, E. Pollet, M. Alexandre, P. Dubois, *J. Polym. Sci., Part B: Polym. Phys.* **2005**, *43*, 205–214.
32 V. M. Karaman, E. G. Privalko, V. P. Privalko, D. Kubies, R. Puffr, R. Jérôme, *Polymer* **2005**, *46*, 1943–1948.
33 B. Lepoittevin, N. Pantoustier, M. Devalckenaere, M. Alexandre, C. Calberg, R. Jérôme, C. Henrist, A. Rulmont, P. Dubois, *Polymer* **2003**, *44*, 2033–2040.
34 H. Böttcher, M. L. Hallensleben, S. Nuss, H. Wurm, J. Bauer, P. Behrens, *J. Mater. Chem.* **2002**, *12*, 1351–1354.
35 H. Zhao, S. D. Argoti, B. P. Farrell, D. A. Shipp, *J. Polym. Sci., Part A: Polym. Chem.* **2004**, *42*, 916–924.
36 H. Zhao, D. A. Shipp, *Chem. Mater.* **2003**, *15*, 2693–2695.
37 M. W. Weimer, H. Chen, E. P. Giannelis, D. Y. Sogah, *J. Am. Chem. Soc.* **1999**, *121*, 1615–1616.
38 L. Xu, S. Reeder, M. Thopasridharan, J. Ren, D. A. Shipp, R. Krishnamoorti, *Nanotechnology* **2005**, *16*, S514–S521.
39 B. Q. Zhang, C. Y. Pan, C. Y. Hong, B. Luan, P. J. Shi, *Macromol. Rapid Commun.* **2006**, *27*, 97–102.
40 N. Salem, D. A. Shipp, *Polymer* **2005**, *46*, 8573–8581.
41 X. Fan, Q. Zhou, C. Xia, W. Cristofoli, J. Mays, R. Advincula, *Langmuir* **2002**, *18*, 4511–4518.
42 Q. Y. Zhou, X. W. Fan, C. J. Xia, J. Mays, R. Advincula, *Chem. Mater.* **2001**, *13*, 2465–2467.
43 W. R. Mariott, E. Y. X. Chen, *J. Am. Chem. Soc.* **2003**, *125*, 15726–15727.
44 S. Bruzaud, Y. Grohens, S. Ilinca, J.-F. Carpentier, *Macromol. Mater. Eng.* **2005**, *290*, 1106–1114.
45 C.-P. Li, C.-M. Huang, M.-T. Hsieh, K.-H. Wei, *J. Polym. Sci., Part A: Polym. Chem.* **2005**, *43*, 534–542.
46 M. Xiao, L. Sun, J. Liu, Y. Li, K. Gong, *Polymer* **2002**, *43*, 2245–2248.
47 P. Petrov, F. Stassin, C. Pagnoulle, R. Jérôme, *Chem. Commun.* **2003**, 2904–2905.
48 Y. Liu, Z. Yao, A. Adronov, *Macromolecules* **2005**, *38*, 1172–1179.
49 X. Lou, C. Detrembleur, V. Sciannamea, C. Pagnoulle, R. Jérôme, *Polymer* **2004**, *45*, 6097–6102.
50 I. C. Liu, H.-M. Huang, C.-Y. Chang, H.-C. Tsai, C.-H. Hsu, R. C.-C. Tsiang, *Macromolecules* **2004**, *37*, 283–287.
51 H. Kong, C. Gao, D. Yan, *Macromolecules* **2004**, *37*, 4022–4030.
52 H. Kong, C. Gao, D. Yan, *J. Mater. Chem.* **2004**, *14*, 1401–1405.
53 H. Kong, C. Gao, D. Yan, *J. Am. Chem. Soc.* **2004**, *126*, 412–413.
54 Z. Yao, N. Braidy, G. Botton, A. Adronov, *J. Am. Chem. Soc.* **2003**, *125*, 16015–16024.
55 D. Baskaran, J. W. Mays, M. S. Bratcher, *Angew. Chem.* **2004**, *43*, 2138–2142.
56 T. von Werne, T. E. Patten, *J. Am. Chem. Soc.* **1999**, *121*, 7409–7410.
57 T. von Werne, T. E. Patten, *J. Am. Chem. Soc.* **2001**, *123*, 7497.
58 J. Pyun, S. Jia, T. Kowalewski, G. D. Patterson, K. Matyjaszewski, *Macromolecules* **2003**, *36*, 5094–5104.
59 Y.-P. Wang, X.-W. Pei, X.-Y. He, K. Yuan, *Eur. Polym. J.* **2005**, *41*, 1326–1332.

60 C. Bartholome, E. Beyou, E. Bourgeat-Lami, P. Chaumont, N. Zydowicz, *Macromolecules* **2003**, *36*, 7946–7952.
61 C. Bartholome, E. Beyou, E. Bourgeat-Lami, P. Cassagnau, P. Chaumont, L. David, N. Zydowicz, *Polymer* **2005**, *46*, 9965–9973.
62 Q. Zhou, S. Wang, X. Fan, R. Advincula, J. Mays, *Langmuir* **2002**, *18*, 3324–3331.
63 M. Joubert, C. Delaite, E. Bourgeat-Lami, P. Dumas, *J. Polym. Sci., Part A: Polym. Chem.* **2004**, *42*, **1976–1984**.
64 T. K. Mandal, M. S. Fleming, D. R. Walt, *Nanoletters* **2002**, *2*, 3–7.
65 K. Ohno, K.-m. Kho, Y. Tsujii, T. Fukuda, *Macromolecules* **2002**, *35*, 8989–8993.
66 R. Jordan, N. West, A. Ulman, Y.-M. Chou, O. Nuyken, *Macromolecules* **2001**, *34*, 1606–1611.
67 A. Kotal, T. K. Mandal, D. R. Walt, *J. Polym. Sci., Part A: Polym. Chem.* **2005**, *43*, 3631–3642.
68 Y. Wang, X. Teng, J.-S. Wang, H. Yang, *Nanoletters* **2003**, *3*, 789–793.
69 C. Flesch, E. Bourgeat-Lami, S. Mornet, E. Duguet, C. Delaite, P. Dumas,
J. Polym. Sci., Part A: Polym. Chem. **2005**, *43*, 3221–3231.
70 E. Marutani, S. Yamamoto, T. Ninjbadgar, Y. Tsujii, T. Fukuda, T. M., *Polymer* **2004**, *45*, 2231–2235.
71 C. R. Vestal, Z. J. Zhang, *J. Am. Chem. Soc.* **2002**, *124*, 14312–14313.
72 M. K. Corbierre, N. S. Cameron, M. Sutton, S. G. J. Mochrie, L. B. Lurio, A. Ruhm, R. B. Lennox, *J. Am. Chem. Soc.* **2001**, *123*, 10411–10412.
73 M. K. Corbierre, N. S. Cameron, M. Sutton, K. Laaziri, R. B. Lennox, *Langmuir* **2005**, *21*, 6063–6072.
74 A. Okada, A. Usuki, *Mater. Sci. Eng. C* **1995**, *3*, 109–115.
75 L. M. Sherman, *Plast. Technol.* **2004**, *feature article*, November.
76 J. H. Schut, *Plast. Technol.* **2006**, *feature article*, February.
77 B. H. Naitove, *Plast. Technol.* **2006**, *feature article*, January.
78 J. A. Grande, *Plast. Technol.* **2005**, *feature article*, August.

3
Polymer/Layered Filler Nanocomposites: An Overview from Science to Technology

Masami Okamoto

3.1
Introduction

A decade of research has shown that nanostructured materials have the potential to significantly impact growth at every level of the world economy in the 21st century. This new class of materials is now being introduced in structural applications, such as gas barrier film, flame retardant product, and other load-bearing applications.

Of particular interest are the recently developed nanocomposites consisting of a polymer and layered silicate because they often exhibit remarkably improved mechanical and various other properties [1] when compared with pure polymer or conventional composites (both micro- and macro-composites). A primary development in polymer/layered silicate nanocomposites, a Nylon 6/layered silicate hybrid [2] reported by Toyota Central Research & Development Co. Inc. (TCRD), was successfully prepared by *in situ* polymerization of ε-caprolactam in a dispersion of montmorillonite (MMT). The silicate can be dispersed in liquid monomer or a solution of monomer. It has also been possible to melt-mix polymers with layered silicates, avoiding the use of organic solvents. The latter method permits the use of conventional processing techniques such as injection molding and extrusion. The extensive literature in nanocomposite research has been covered in recent reviews [3–5].

Continued progress in nanoscale controlling, as well as an improved understanding of the physicochemical phenomena at the nanometer scale, have contributed to the rapid development of novel nanocomposites. This chapter presents current research on polymer/layered filler nanocomposites (PLFNCs) with the primary focus on recent advances from basic science to technology.

3.2
Historical Point of View

Earlier attempts at preparing polymer/layered filler composites are found in almost half-a-century old patent literatures [6, 7]. In such cases, incorporation of 40 to 50 wt% clay mineral (bentonite, hectorite, etc.) into a polymer was attempted but the results were unsatisfactory. The maximal modulus enhancement was only around 200%, although the clay loading was as much as 50 wt%. The failure was obvious because they failed to achieve good dispersion of clay particles in the matrix, in which silicate minerals existed as agglomerated tactoids. Such a poor dispersion of the silicate particles could improve the material rigidity, but certainly sacrificed the strength, the elongation at break and the toughness of the materials [6, 7].

A prime reason for the impossibility of improving the tactoids dispersion into well-dispersed exfoliated monolayers of the silicate is obviously due to the intrinsic incompatibility of hydrophilic layered silicates with hydrophobic engineering plastics. One attempt at circumventing this difficulty was made by Unitika Ltd. [8] about 30 years ago in preparing Nylon 6/layered silicate composites (not nanocomposites) via *in situ* polymerization of ε-caprolactam with montmorillonite (MMT), but the results turned out to be not very good.

The first major breakthrough occurred in 1987, when Fukushima and Inagaki of TCRD, via their detailed study on polymer/LS composites, persuasively demonstrated that *lipophilization* by replacing inorganic cations in galleries of the native clay with alkylammonium surfactant successfully made them compatible with hydrophobic polymer matrices [9]. The modified clay was thus called lipophilized clay, organophilic clay or simply organo-clay (organoclay). Furthermore, they found that the lipophilization enabled one to expand silicate galleries and exfoliate the silicate layers into single layers of 1 nm thickness.

Six years later, in 1993, Usuki, Fukushima and their colleagues at TCRD successfully prepared, for the first time, *exfoliated* Nylon 6/MMT hybrid via *in situ* polymerization of ε-caprolactam, in which alkylammonium-modified MMT was thoroughly dispersed in advance [2, 10].

Apart from this, the intercalation of small molecules into silicate galleries has been found by researchers when studying Mayan archeological sites [11]. Maya blue was used in Mesoamerica and colonial Mexico as late as the 20th century. The Maya blue color is resistant to diluted mineral acids, alkalis, moderate heat and even biocorrosion. This blue color contains clay (mainly MMT clay and palygorskite $(Mg_5(Si,Al)_8O_{20}(OH)_{28}H_2O)$, see Table 3.1) and indigo molecules $(C_{16}H_{10}N_2O_2)$. Intercalation of indigo molecules into MMT galleries and/or encapsulation in the pores of palygorskite might explain the corrosion resistance in the extreme condition of the rain forest. Up to now Maya blue paint has been understood as an origin of intercalation, and recognized as an ancient nanostructured material.

Table 3.1 Clay mineral (phyllosilicates) classification.

Type	Group	Groupoid	Species	Tetra-hedron	Octa-hedron	Interlayer cation
2:1 $Si_4O_{10}(OH)_2$	pyrophyllite Talc ($x \sim 0$)	di. tri.	Pyrophyllite Talc	Si_4 Si_4	Al_2 Mg_3	– –
	smectite ($0.25 < x < 0.6$)	di.	Montmorillonite	Si_4	$(Al_2, Mg)_2$	Na, Ca, H_2O
		di.	Hectorite	Si_4	$(Mg_2, Li)_2$	Na, Ca, H_2O
		di.	Beidellite	$(Si, Al)_4$	Al_2	Na, Ca, H_2O
		tri.	Saponite	$(Si, Al)_4$	Mg_3	Na, Ca, H_2O
	vermiculite ($0.25 < x < 0.9$)	di.	Vermiculite	$(Si, Al)_4$	$(Al, Mg)_2$	K, Al, H_2O
		tri.	Vermiculite	$(Si, Al)_4$	$(Mg, Al)_3$	K, Mg, H_2O
	mica ($x \sim 1$)	di.	Muscovite	$Si_3 \cdot Al$	Al_2	K
			Paragonite	$Si_3 \cdot Al$	Al_2	Na
	brittle mica ($x \sim 2$)	tri.	Phlogopite	$Si_3 \cdot Al$	$(Mg, Fe^{2+})_3$	K
			Biotite	$Si_3 \cdot Al$	$(Fe^{2+}, Mg)_3$	K
2:1:1 $Si_4O_{10}(OH)_8$	chlorite (large variation of x)	di.	Donbassite	$(Si, Al)_4$	Al_2	$Al_2(OH)_6$
		di.-tri.	Sudoite	$(Si, Al)_4$	$(Al, Mg)_2$	$(Mg, Al)_3(OH)_6$
		tri.	Clinochlore	$(Si, Al)_4$	$(Mg, Al)_3$	$(Mg, Al)_3(OH)_6$
			Chamosite	$(Si, Al)_4$	$(Fe, Al)_3$	$(Fe, Al)_3(OH)_6$
1:1 $Si_2O_5(OH)_4$	kaolin-mineral	di.	Kaolinite	Si_2	Al_2	–
			Halloysite	Si_2	Al_2	H_2O
	Serpentinite ($x \sim 0$)	tri.	Chrysotile	Si_2	Mg_3	–
Needle	Sepiolite Palygorskite ($x \sim 0$)	tri.	Sepiolite	Si_{12}	Mg_8	$(OH_2)_4 \cdot H_2O$
			Palygorskite	Si_8	Mg_8	$(OH_2)_4 \cdot H_2O$
Amorphous (low crystalline)			Imogolite	SiO_3OH	$Al(OH)_3$	–
			Allophane	(1-2) $SiO_2 \cdot$ (5-6) H_2O		
			Hisingerite	$SiO_2–Fe_2O_3H_2O$		

x indicates degree of isomorphous substitution;
di. indicates dioctahedral, tri. indicates trioctahedral.

3.3
Structure of Layered Filler and Its Modification

The commonly used clays for the preparation of PLFNCs belong to the same general family of phyllosilicates. Their crystal structure consists of layers made up of two silica tetrahedra fused to an edge-shared octahedral sheet of either aluminum or magnesium hydroxide. The layer thickness is around 1 nm and the lateral dimensions of these layers may vary from 30 nm to several μm and even larger, depending on the particular layered silicate. Stacking of the layers

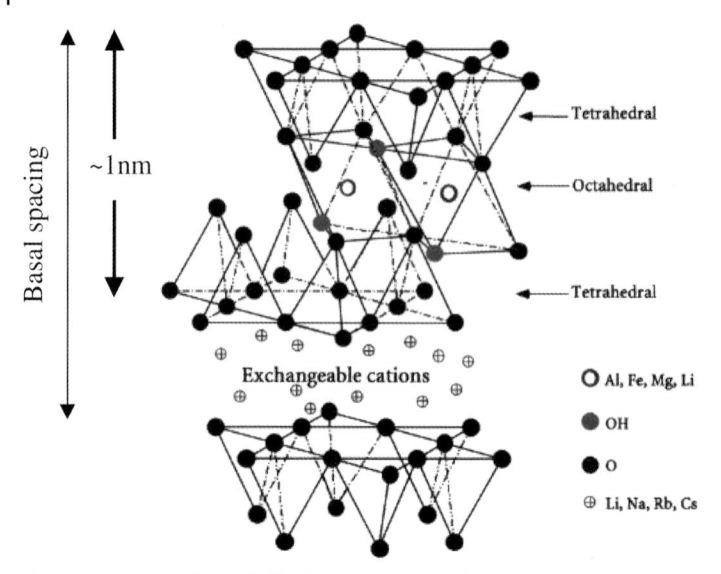

Fig. 3.1 Structure of 2:1 phyllosilicates (montmorillonite).

leads to a regular van der Waals gap between the layers called the interlayer or gallery. Isomorphic substitution within the layers (for example, Al^{3+} replaced by Mg^{2+} or by Fe^{2+}, or Mg^{2+} replaced by Li^+) generates negative charges that are counterbalanced by alkali and alkaline earth cations situated inside the galleries, as shown in Fig. 3.1 and Table 3.1.

The most commonly used layered silicates are montmorillonite (MMT), hectorite and saponite having different chemical formulae, respectively, $M_x(Al_{2-x}Mg_x)Si_4O_{10}(OH)_2$, $M_x(Mg_{3-x}Li_x)Si_4O_{10}(OH)_2$ and $M_x(Si_{4-x}Al_x)Si_4O_{10}(OH)_2$ ($x=0.25–1.0$). The type of clay is characterized by a moderate surface charge (cation exchange capacity) (CEC of 80–120 mequiv/100 g) and layer morphology. These clays are only compatible with hydrophilic polymers, such as poly(ethylene oxide) (PEO), poly(vinyl alcohol) (PVA). To improve compatibility with other polymer matrices, one must convert the normally hydrophilic silicate surface to organophilic, which makes possible intercalation of many engineering polymers. Generally, this can be done by ion-exchange reactions with cationic surfactants including primary, secondary, tertiary, and quaternary alkyl ammonium or alkylphosphonium cations. The role of alkylammonium or alkylphosphonium cations in the organosilicates is to lower the surface energy of the inorganic host and improve the wetting characteristics with the polymer matrix, and results in a larger interlayer spacing. One can evaluate a Na^+ density of 0.7 Na^+/nm^2, i.e., 7000 alkylammonium salt molecules are localized near the individual silicate layers ($\sim 100\times 100$ nm^2) and active surface area ($\sim 700–900$ m^2 g^{-1}, as determined by BET). This result indicates that the organoclay platelets are *hairy* plates. Furthermore, the surface hydroxyl concentration of clays was determined by titration with triethyl aluminum. Assuming that the hydroxyl groups are randomly distributed on the edge surface, one can calculate

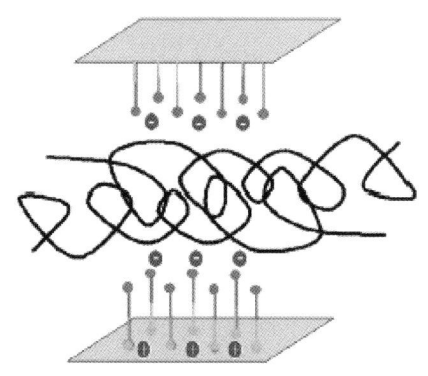

Fig. 3.2 Principle of a repulsion of sheet-like nanoparticles via surface modification with bifunctional molecules. Reprinted with permission from [19], © 2003, Elsevier Science.

a Si–OH density of 5Si–OH/nm^2 [12], i.e., 500 –OH groups are localized near the edge surface of the individual silicate layers ($\sim 1\times 100$ nm^2). This lipophilic–hydrophobic balance is the key issue in the fine dispersion of the organoclay platelets into the polymeric matrices. Additionally, the alkylammonium or alkylphosphonium cations could provide functional groups that can react with the polymer matrix or in some cases initiate the polymerization of monomers to improve the strength of the interface between the inorganic and the polymer matrix [13, 14].

Because the thermal degradation of many organophilic clays begins at temperatures higher than 200 °C, clays with enhanced thermal stability are desired. A recent approach is polymerically modified clays [15–18]. Poly(diallylammonium) salt and oligomeric styrene-based ammonium salt have been prepared and used to produce PLFNCs. The more interesting idea of the introduction of a repulsion between the single clay layers is reported by Fischer (see Fig. 3.2) [19]. The cations located in between the clay platelets are ion-exchanged by one of the functional groups of these organic molecules, e.g., an ammonium group, leaving another functional group, which can be positively or negatively charged, present on the clay layers. 4-Amino-1-naphthalenesulfonic acid is one of the candidates.

3.4
Preparative Methods and Structure of PLFNCs

3.4.1
Intercalation of Polymer or Pre-polymer from Solution

This is based on a solvent system in which polymer or pre-polymer is soluble and the silicate layers are swellable. The layered silicate is first swollen in a solvent, such as water, chloroform or toluene etc. When the polymer and layered silicate solutions are mixed, the polymer chains intercalate and displace the solvent within the interlayer of the silicate. Upon solvent removal, the intercalated structure remains, resulting in PLFNCs.

3.4.2
In situ Intercalative Polymerization Method

In this method, the organoclay is swollen within the liquid monomer or monomer solution so that the polymer formation can occur in between the intercalated sheets. Polymerization can be initiated either by heat or radiation, by the diffusion of a suitable initiator or by an organic initiator or catalyst fixed through cation exchange inside the interlayer before the swelling step by the monomer.

3.4.3
Melt Intercalation Method

Since the possibility of direct melt intercalation was first demonstrated [20], melt intercalation has become a method of preparation of the intercalated polymer/layered silicate nanocomposites. This process involves annealing, statically or under shear, a mixture of the polymer and OMLFs above the softening point of the polymer. During annealing, the polymer chains diffuse from the bulk polymer melt into the nano-galleries between the layered fillers.

In order to understand the thermodynamic issue associated with the formation of nanocomposites, Vaia et al. have applied a mean-field statistical lattice model and found that conclusions based on the mean field theory agreed well with the experimental results [21, 22]. The entropy loss associated with confinement of a polymer melt does not prohibit nanocomposite formation because an entropy gain associated with the layer separation balances the entropy loss of polymer intercalation, resulting in a net entropy change near to zero. Thus, from the theoretical model, the outcome of nanocomposite formation via polymer melt intercalation depends on energetic factors, which may be determined from the surface energies of the polymer and OMLF.

Nevertheless, we have often faced the problem that the nanocomposite shows fine and homogeneous distribution of the nanoparticles in the polymer matrix (e.g., poly(L-lactide) (PLA)) without a clear peak shift of the mean interlayer spacing of the (001) plane, as revealed by wide-angle X-ray diffraction (WAXD) analysis [23]. Furthermore we sometimes encounter a decrease in the interlayer spacing compared with that of pristine OMLF, despite very fine dispersion of the silicate particles. To the best of our knowledge, the mechanism of direct melt intercalation in nanocomposite formation is not very well explored in the literature. For this reason, information on the structure of the surfactant (intercalant)–polymer interface is necessary to understand the intercalation kinetics that can predict final nanocomposite morphology and overall material properties.

3.5
Interlayer Structure of OMLFs and Intercalation

Lagaly has suggested a paraffin-type layer structure for the intercalants in the case of highly surface charged clay minerals [24]. This model was derived from WAXD analysis assuming all-trans conformation of the intercalants. Vaia et al. [25] have shown, by using Fourier transform infrared spectroscopy (FTIR), which is sensitive to the gauche/trans conformer ratio in alkyl chains and the lateral chain–chain interactions, that alkyl chains can vary from liquid-like to solid-like, with the liquid-like structure dominating as the interlayer density or chain length decreases, or as the temperature increases. In addition, for the longer chain length intercalants, the intercalants in the layered silicate can show a thermal transition akin to melting or liquid-crystalline to liquid-like transitions upon heating. The gauche content was found to decrease with increasing intercalant concentration. The chains adopt an essentially all-trans conformation when intercalant concentration is high (more than the cation-exchange capacity (CEC)) [26]. Osman [27] reported that the all-trans conformation of the alkyl chains was preferentially adopted when using high surface charge density clay (vermiculite).

On the other hand, Paul et al. [28] reported, using molecular simulations, disordered conformation of the alkyl chains (containing hydroxyl-ethyl units) with a tendency to lie on the silicate surfaces. A systematic investigation is required using OMLFs with different types of intercalant and nano-fillers with different surface charge density. Furthermore, the correlation between the intercalant–polymer interface and the melt intercalation is still sparse, as mentioned before. A better understanding of the correlation and subsequent preparation of the nanocomposites is of fundamental importance in controlling the nanoscale structure.

Okamoto et al. [29] reported on the interlayer structure of OMLFs with respect to the number per area (surface charge density) and size of intercalant chains for nanocomposite formation. In the study they presented an interdigitated layer structure model of the OMLF, where the intercalants are oriented with some inclination to the host layer in the interlayer space. Details regarding this model and explanation are presented in [29]. The illustration of a model of interlayer structure of some intercalant (N-(cocoalkyl)-N,N-[bis(2-hydroxyethyl)]-N-methyl ammonium cation: qC$_{14}$(OH) in gallery space of layered titanate (HTO: H$_{1.07}$Ti$_{1.73}$O$_{3.95}$ · 0.5H$_2$O) is shown in Fig. 3.3. For nano-fillers with high surface charge density (1.26 e$^-$ nm^{-2}, see Table 3.2), the intercalants can adopt a configuration where the alkyl chains (with all-trans conformation) are tilted under the effect of the van der Waals forces, which decreases the chain–chain distance. For this reason the angle a should be directly related to the packing density of the alkyl chains. The value of a decreases until close contact between the chains is attained, giving an increasing in the degree of the crystallinity of the intercalants within the nano-galleries.

This interdigitated structure may imply that the intercalants can adopt a different orientation angle when polymer molecules penetrate into the galleries.

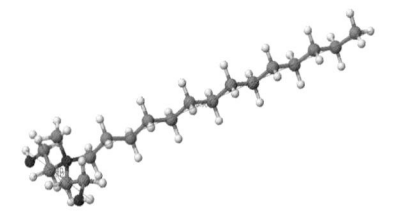

N-(coco alkyl)-N,N-[bis(2-hydroxyethyl)]-N-methyl ammonium [qC$_{14}$(OH)]

HTO-qC$_{14}$(OH)

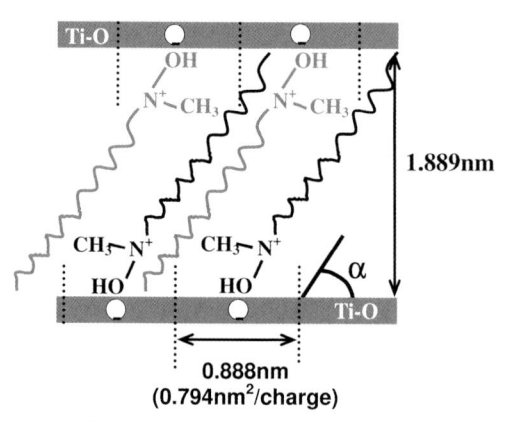

Fig. 3.3 Illustration of a model of interlayer structure of intercalant N-(cocoalkyl)-N,N-[bis(2-hydroxyethyl)]-N-methyl ammonium cation (qC$_{14}$(OH)) in the gallery space of layer titanate (HTO). The average distance between exchange sites is 0.888 nm calculated by surface charge density of 1.26 e$^-$ nm^{-2}. For qC$_{14}$(OH), the obtained molecular length, thickness and width are 2.09 nm, 0.881 nm and 0.374 nm, respectively (see upper panel). The tilt angle α of the intercalants can be estimated by the combination of the interlayer spacing, molecular dimensions and the loading amount of intercalants when the alkyl chains adopt an all-trans conformation. Reprinted from [29], © 2006, Wiley-VCH.

Apparently, the interdigitated layer structure provides a balance between polymer chains penetration and the different orientation angle of the intercalants. It is necessary to understand the meaning of the interlayer expansion in the intercalation, estimated by WAXD analysis.

From this result, the entropic contribution of the intercalants, which leads to the entropy gain associated with the layer expansion after intercalation of the small molecules and/or polymer chains, may not be significant due to the interdigitated layer structure. Presumably the penetration takes place by pressure drop within the nano-galleries, nano-capillary action, generated by the two platelets.

Table 3.2 Characteristic parameters of nano-fillers. Reprinted with permission from [29], © 2006, Wiley-VCH.

Parameters	HTO	syn-FH	MMT
Chemical formula	$H_{1.07}Ti_{1.73}O_{3.95} \cdot 0.5H_2O$	$Na_{0.66}Mg_{2.6}Si_4O_{10}(F)_2$	$Na_{0.33}(Al_{1.67}Mg_{0.33})Si_4O_{10}(OH)_2$
Particle size/nm	~100–200	~100–200	~100–200
BET area/m^2 g^{-1}	~2400	~800	~700
CEC[a]/meq/100 g	~200 (660)	~120 (170)	~90(90)
e$^-$/charge nm^{-2}	1.26	0.971	0.708
Density/g cm^{-3}	2.40	2.50	2.50
Refractive index (n_D^{20})	2.3	1.55	1.55
pH	4–6	9–11	7.5–10

a) Methylene blue adsorption method. The values parentheses are calculated from the chemical formulae of the nano-fillers.

3.6
Structure and Characterization of PLFNCs

Depending on the degree of penetration of the matrix into the OMLF galleries, nanocomposites are obtained with structures ranging from intercalated to exfoliate. Polymer penetration, resulting in finite expansion of the silicate layers, produces intercalated nanocomposites consisting of well-ordered multilayers with alternating polymer/inorganic layers and a repeat distance of a few nanometers (intercalated, see Fig. 3.4) [30]. On the other hand, extensive polymer penetration resulting in disorder and eventual delamination of the silicate layers produces *near to* exfoliated nanocomposites consisting of individual silicate layers dispersed in the polymer matrix (exfoliated) [31]. Under some conditions, the intercalated nanocomposites exhibit flocculation because of the hydroxylated edge–edge interaction of the silicate layers (intercalated-and-flocculated). The length of the oriented collections, in the range 300–800 nm, is far larger than original clay (mean diameter ≅ 150 nm) [32, 33]. Such flocculation presumably is governed by an interfacial energy between the polymer matrix and the organoclays and is controlled by ammonium cation–matrix polymer interaction. The polarity of the matrix polymer is of fundamental importance in controlling the nanoscale structure.

The structure of nanocomposites has typically been established using wide-angle X-ray diffraction (WAXD), small-angle X-ray scattering (SAXS) analysis and transmission electron microscope (TEM) observation. Due to its ease and availability WAXD is most commonly used to probe the PLFNC structure and sometimes to study the kinetics of the polymer melt intercalation. By monitoring the position, shape, and intensity of the basal reflections from the distributed silicate layers, the nanocomposite structure, either intercalated or exfoliated, may

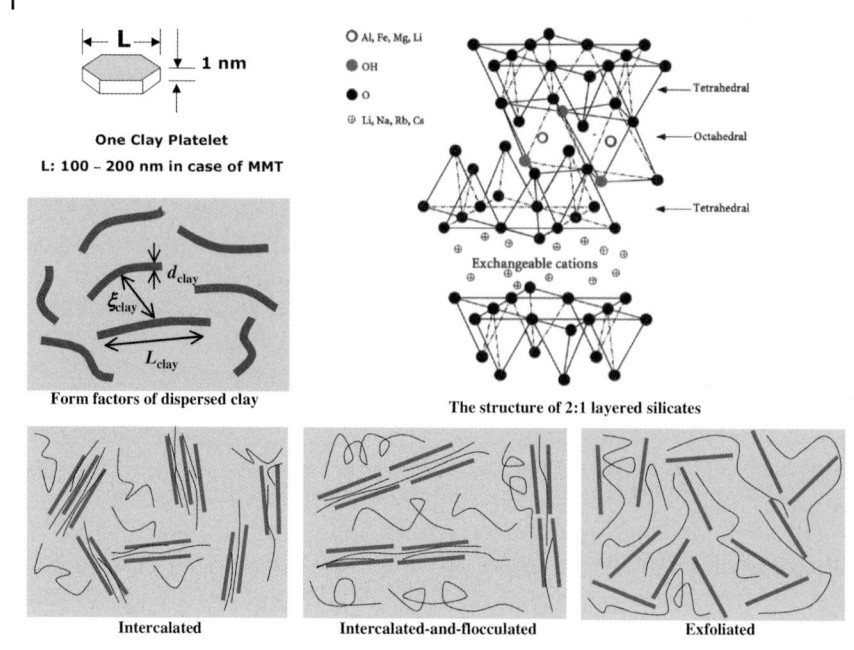

Fig. 3.4 Schematic illustration of three different types of thermodynamically achievable PLFNCs. Reprinted from [30], © 2003, American Chemical Society.

be identified. For example, in the case of exfoliated nanocomposites, the extensive layer separation associated with the delamination of the original silicate layers in the polymer matrix results in the eventual disappearance of any coherent X-ray diffraction from the distributed silicate layers. On the other hand, for intercalated nanocomposites, the finite layer expansion associated with the polymer intercalation results in the appearance of a new basal reflection corresponding to the larger gallery height.

WAXD offers a convenient method to determine the interlayer spacing of the silicate layers in the original layered silicates and in the intercalated nanocomposites (within 1–5 nm), however, little can be said about the spatial distribution of the silicate layers or any structural inhomogeneities in the nanocomposites. Additionally, some layered silicates initially do not exhibit well-defined basal reflection. Thus, peak broadening and intensity decreases are very difficult to study systematically. Therefore, conclusions concerning the mechanism of nanocomposites formation and their structure based solely on WAXD patterns are only tentative.

On the other hand, TEM allows a qualitative understanding of the internal structure, spatial distribution of the various phases, and defect structure through direct visualization. However, special care must be exercised to guarantee a representative cross-section of the sample. The WAXD patterns and corre-

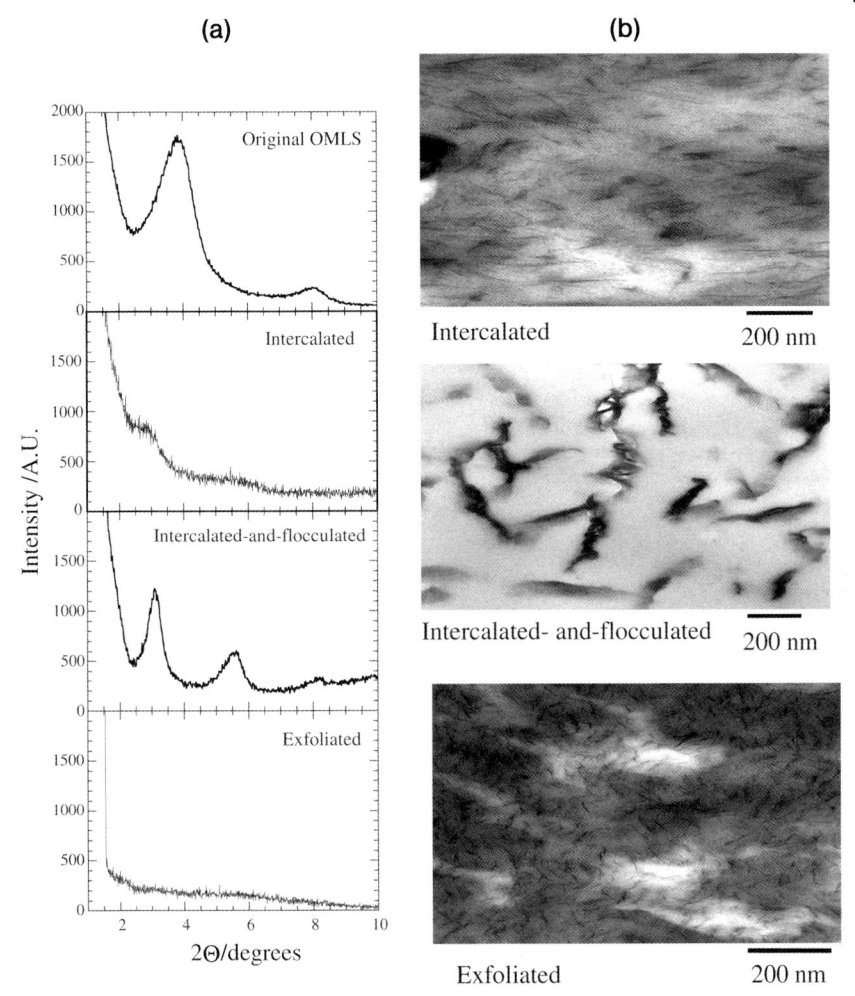

Fig. 3.5 (a) WAXD patterns and (b) TEM images of three different types of nanocomposites. Reprinted from [3], © 2003, Elsevier Science.

sponding TEM images of three different types of nanocomposites are presented in Fig. 3.5.

Solid-state nuclear magnetic resonance (NMR) method to quantify the level of clay exfoliation is also a very important facet of nanocomposite characterization. The main objective in solid-state NMR measurement is to connect the measured longitudinal relaxation, T_1^H, of proton (and ^{13}C nuclei) with the quality of clay dispersion. The extent of and the homogeneity of the dispersion of the silicate layers within the polymer matrix are very important for determining physical properties. The surfaces of naturally occurring layered silicates such as MMT are mainly made of silica tetrahedra while the central plane of the layers

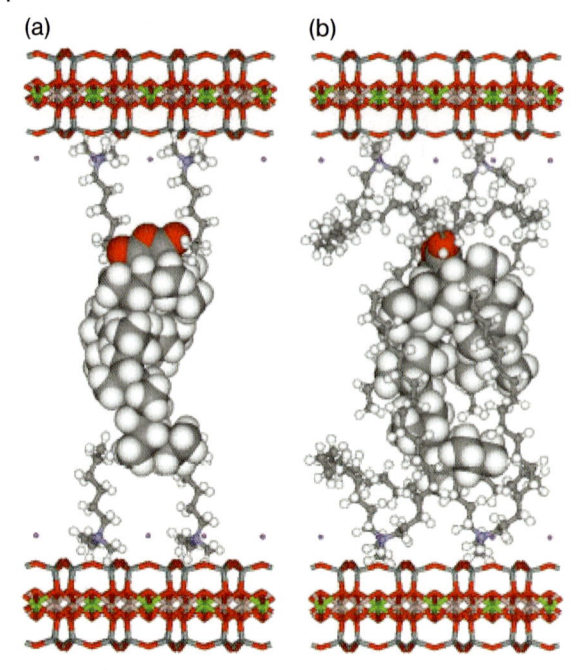

Fig. 3.6 Three-component model used for basal spacing simulations, consisting of two layers of MMT with K$^+$ cations (stick model), four molecules of trimethylammonium cation (a) or dimethylstearylammonium cation (b) (stick and ball model), and one molecule of maleated PP (PP-MA) (ball model). Reprinted from [36], © 2004, Elsevier Science.

contains octahedrally coordinated Al^{3+} (see Fig. 3.1 and Table 3.1) with frequent non-stoichiometric substitutions, where an Al^{3+} is replaced by Mg^{2+} and, somewhat less frequently, by Fe^{3+}. Typical concentrations of Fe^{3+} (spin=5/2) in naturally occurring clays produce nearest-neighbor Fe–Fe distances of about 1.0–1.4 nm [34], and at such distances, the spin exchange interaction between the unpaired electrons on different Fe atoms is expected to produce magnetic fluctuations in the vicinity of the Larmor frequencies for protons or ^{13}C nuclei [34]. The spectral density of these fluctuations is important because the T_1^H of protons (and ^{13}C nuclei) within about 1 nm of the clay surface can be directly shortened. For protons, if that mechanism is efficient, relaxation will also propagate into the bulk of the polymer by spin diffusion. Thus, this paramagnetically induced relaxation will influence the overall measured T_1^H to an extent that will depend both on the Fe concentration in the clay layer and, more importantly, on the average distances between clay layers. The latter dependence suggests a potential relationship between measured T_1^H values and the quality of the clay dispersion. If the clay particles are stacked and poorly dispersed in the polymer matrix, the average distances between polymer/layered filler interfaces are greater, and the average paramagnetic contribution to T_1^H is weaker. VanderHart et al. [35] also employed the same arguments in order to understand the stability of a particular OMLF under different processing conditions.

Compared to OMLFs, nanocomposite structure is difficult to model using atomic scale molecular dynamics (MD) because the intercalated polymer chain conformation is complex and can hardly be the equilibrium state. However, Pricl et al. [36] explored and characterized the atomic scale structure to predict binding energies and basal spacing of PLFNCs based on polypropylene (PP) and maleated (MA) PP (PP-MA), MMT, and different alkylammonium ions as intercalants (see Fig. 3.6). From a global interpretation of all these MD simulation results, they concluded that intercalants with smaller volume are more effective for clay modification as they improve the thermodynamics of the system by increasing the binding energy. On the other hand, intercalants with longer tails are more effective for intercalation and exfoliation processes, as they lead to higher basal spacing. Additional information is necessary to predict a more reasonable nanostructure of PLFNCs; some literature [37–40] related to the confined polymer chains within the silicate galleries by using coarse-grained MD simulation has appeared.

3.7
Nobel Compounding Process

Hasegawa and Usuki [41] reported a novel compounding process using the Na^+-MMT slurry and demonstrated the preparation of a Nylon 6 nanocomposite, where the silicate layers exfoliate and homogeneously disperse at nanometer level. The greatest merit of this compounding process is that the PLFNC consisted of Nylon 6 and Na^+-MMT is prepared without any surfactants of the clay minerals and additives. However, it is difficult to prepare a completely exfoliated nanocomposite by this method (see Fig. 3.7). Originally the perfect exfoliation of the silicate platelets may be impossible due to the strong interaction between hydroxylated edge–edge groups, the clay platelets are sometimes flocculated in polymer matrices as reported by van Olphen [42] (see Section 3.8.5).

Complete exfoliation of the OMLFs is also not feasible after melt intercalation, although appropriate shear is applied during melt compounding. Recently, an *in situ* polymerization method used supercritical CO_2 (sc-CO_2) as a processing aid to achieve a uniform distribution in a PLFNC at high MMT clay loading (~40 wt%) [43]. Zhao et al. [44] presented unambiguous evidence for sc-CO_2-mediated intercalation of poly(ethylene oxide) (PEO) into Na^+-MMT compared with polymer intercalation in solution (water). They successfully intercalated PEO into clay via a sc-CO_2-mediated process. This mechanism is probably similar to that in polymer melts. Therefore, they speculated that the sc-CO_2-mediated intercalation is an enthalpically driven process, deriving from a favorable intercalation between PEO and MMT. The scCO_2-mediated process is also conducted during the melt compounding process. Ozisik et al. [45] examined

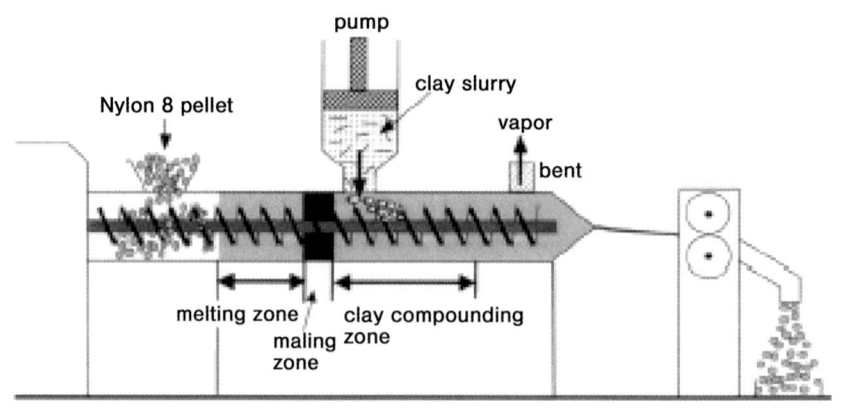

Fig. 3.7 Schematic representation depicting a compounding process for preparing the Nylon 6 nanocomposites using the clay slurry. Nylon 6 was put into the extruder at 2 kg h^{-1} and melted in the melting zone. The clay slurry was pumped into the cylinder of the extruder at 2 kg h^{-1} and compounded with the melting Nylon 6 in the compounding zone at 240–250 °C. The screw rotation speed was 200 rpm. The residence time in the cylinder was approximately 10 min. Sealing zone was set by using a sealing ring to prevent water back flow for the hopper. The water of the slurry was removed from the vent by vacuum to obtain the nanocomposites. Reprinted from [41], © 2003, Elsevier Science.

the effect of the sc-CO_2 fed to the tandem extruder on the dispersion of OMLFs with different intercalants into Nylon 6 matrix. In the absence of sc-CO_2, pressure improved the MMT clay delamination by reducing the free volume of the polymer and increasing the interaction between chains and ultimately increasing the viscosity. Using sc-CO_2 did not improve the clay dispersion due to decreasing the melt viscosity.

Another interesting approach for the delamination of OMLFs is the use of ultrasound in the preparation of PLFNCs. Lee et al. [46] reported the effect of the *in situ* ultrasound on the polymer/MMT melt phase. They found an effective method to enhance the dispersion, intercalation and exfoliation of OMLFs in thermoplastic-based nanocomposites. The same experiment was done by Guo et al. [47] for PP-based nanocomposite preparation. The schematic representation of this technology is shown in Fig. 3.8. The maximum power output and frequency of the ultrasonic generator are 300 W and 20 kHz, respectively. They described the fine dispersion of silicate layers in the PP matrix after ultrasonic treatment (100 W). However, the ultrasonic oscillations exhibited little effect on the delamination of OMLFs, as revealed by transmission electron microscopy (TEM) observation.

Thus, compounding with the assistance of sc-CO_2 fluids and ultrasonication did not improve the state of the nano-filler dispersion once a critical morphology was established. We have to develop more innovative compounding processes, especially in the preparation of the nanocomposites possessing exfoliated layered fillers. Despite its importance, the pressure drop (Δp) within the nano-

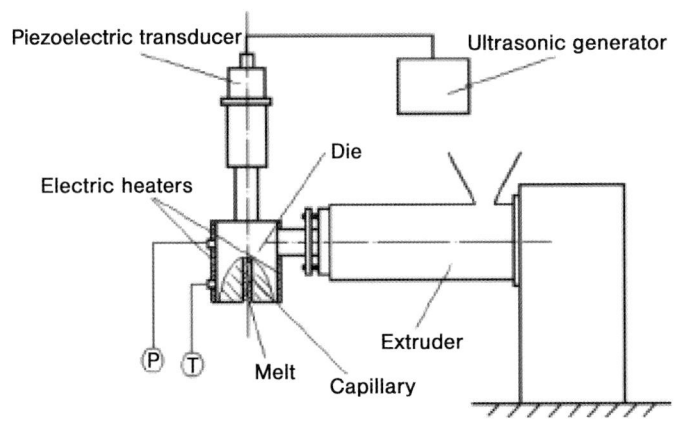

Fig. 3.8 Schematic diagram of ultrasonic oscillations extrusion system. Reprinted from [47], © 2006, Elsevier Science.

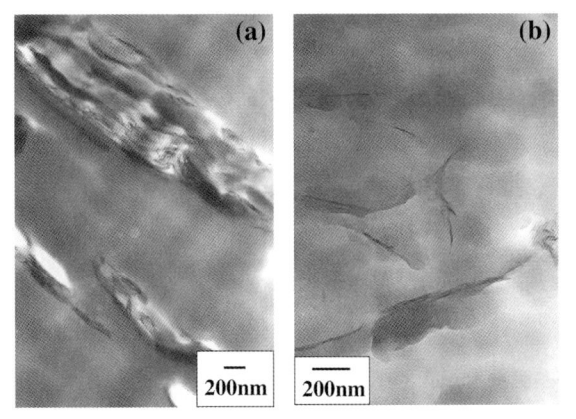

Fig. 3.9 Bright field TEM images of (a) mixture before solid-state processing and (b) processed mixture. Both samples are prepared by annealing at 300 °C for 30 s (without shear processing). The dark entities are the cross-section and/or face of intercalated-and-stacked silicate layers, and the bright areas are the matrix. Reprinted from [50], © 2006, Wiley-VCH.

galleries [29], which makes the polymer penetration more difficult, has not yet been discussed fully. Okamoto [50] reported the estimated value of Δp (~ 24 MPa), which is much larger than the shear stress (0.1 MPa) during melt compounding [29].

Another challenge is the exfoliation of talc fillers by solid-state shear processing using the pan-type mill to prepare PP/talc nanocomposites [48]. Although the delamination of talc fillers was not achieved in the nanocomposite, as revealed by TEM images, no indication of the layer correlation was observed in

WAXD profiles. Solid-state shear processing may be an innovative technique to delaminate the layered fillers because the pressure drop (Δp) within the nano-galleries makes the polymer penetration more difficult [49]. In this regard, very recently, Saito and Okamoto [50] reported the use of solid-state processing for the preparation of poly(p-phenylenesulfide) (PPS)-based nanocomposites having finely dispersed layered fillers.

The mixture of PPS and OMLF (95/5 wt.) was subjected to the processing using a thermostatted hot-press at 150 °C, below T_m of PPS (i.e., PPS is still in the solid-state), and applying pressures of 33 MPa for 30 s. The mixture exhibited disorder and delaminated layer structure with a thickness of 10–20 nm into the PPS matrix. On the contrary, nanocomposite prepared by melt compounding at 300 °C for 3 min showed large stacked silicate layers in the PPS matrix. The processing led to delamination of the silicate layers and attained the discrete dispersion. This approach can be extended to prepare polymeric nanocomposites with fine dispersion of the nano-fillers in overcoming the pressure drop within the nano-galleries (see Fig. 3.9).

3.8
Control of Nanostructure Properties

3.8.1
Intercalation During Crystallization and Confinement

Maiti et al. [51] reported that in the PP/MMT nanocomposites (PPCN in Fig. 3.10), at high crystallization temperature ($T_c \geq 110$ °C), where the crystallization rate is low enough to solidify the system, the intercalation should be anticipated in the melt state during crystallization. The driving force of the intercalation originates from strong hydrophilic interaction between the maleic anhydride (MA) group and the polar clay surfaces [52]. With increasing T_c, the small peak and shoulder shift toward the smaller angle region in the PPCNs, suggesting that the extent of intercalation takes place with crystallization [51].

Figure 3.10 shows quantitatively the diffraction peak from the (001) planes ($d_{(001)}$) of the clay gallery, as a function of T_c, obtained from their respective Bragg reflections. Here, in the case of PPCN2 (including 2 wt% of MMT), the peak is not prominent. The dotted line shows the effect of annealing on the $d_{(001)}$ value of organoclay. The $d_{(001)}$ increases with T_c for both PPCN4 and PPCN7.5 systems and PPCN4 always exhibits a significantly higher value than that of PPCN7.5. These results imply that intercalation proceeds at T_c and increases with decreasing clay content. Further decrease in clay content from 4 to 2 wt% in PPCN2 leads to a partially exfoliated system, as discussed in the literature [52]. That is, the PPCN with low clay content crystallized at high T_c (≥ 110 °C) exhibits a higher amount of intercalation than that with high clay content crystallized at any T_c.

At high T_c (≥ 110 °C) (low crystallization rate), the melt state exists for quite a long time and PP-g-MA chains have enough time to intercalate before crystalli-

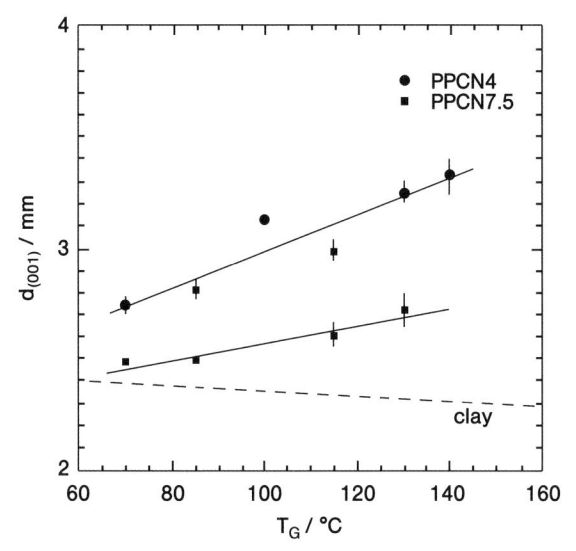

Fig. 3.10 T_c dependence of the interlayer spacing of PPCN4 and PPCN7.5. The broken line shows the annealing effect on organoclay. Reprinted from [51], © 2002, American Chemical Society.

zation can occur in the bulk. Then, enhanced intercalation is produced. The extent of intercalation is strongly dependent on the time of the molten state. In other words, the intercalated PPCNs are not equilibrated. By decreasing the clay content in the nanocomposites, the virtual gallery space in the silicate layers decreases and consequently, the PP-g-MA molecules try to fit, through interaction, in the minimum space, causing higher intercalated species. For sufficiently low clay content, a system, like PPCN2 having less gallery space, is partially exfoliated due to the high number density of the tethering junction.

There are two possible ways of ordering polymer chains inside the silicate gallery: (i) polymer molecules escape from the gallery and crystallize outside (*diffuse out*) or (ii) molecules may penetrate into the silicate gallery when they are in the molten state (*diffuse in*). When PPCN4 is directly crystallized from the melt at 70 °C for two different times of 30 min and 17 h, the interlayer spacing is the same (2.75 nm). If PPCN4 melt is annealed at 150 °C, just above T_m (=145 °C) for sufficiently long time and then subsequently crystallized at 70 °C for 30 min, the interlayer spacing increases to 2.96 nm. Furthermore, when PPCN4 is crystallized from the melt at 30 °C, where the crystallization rate is slow enough, the interlayer spacing becomes 3.08 nm. All these experiments indicate that the extent of intercalation is strongly dependent on the time of the molten state and ordering of polymer chains occurs through a *diffuse-in* mechanism. In other words, a slower crystallization rate makes a more intercalated species as molten polymer molecules have sufficient time to diffuse into the silicate gallery. Based on the WAXD and TEM micrographs, the nature of interca-

Table 3.3 Dynamic storage modulus of PP-g-MA and PPCNs at $T = 50\,°C$ crystallized at different temperatures. Reprinted from [51], © 2002, American Chemical Society.

System	T_c /°C	$G' \times 10^{-8}$/Pa	Increase (%)
PP-g-MA	70	2.92	9.9
	130	3.21	
PPCN2	70	4.79	
	130	4.50	
PPCN4	70	5.16	30.6
	130	6.74	
PPCN7.5	70	7.49	13.3
	130	8.49	

lation has been represented by Maiti and Okamoto [51, 53]. Thus, by suitably crystallizing the PPCNs we can control the fine structure (*confined orientation*) of the PLFCNs.

According to Khare's prediction [54], the confinement of polymer chains increases the viscosity and mechanical properties of the system significantly. One can expect some difference in mechanical properties with the change in the degree of intercalation in the PP-based nanocomposites vis-à-vis the clay content and T_c (see Table 3.3). It is clear from Table 3.3 that, for a particular T_c, the dynamic stage modulus (G') increases with increasing clay content. The PP-g-MA crystallized at 130 °C exhibit a 9.9% increase in G' compared to the sample crystallized at 70 °C. The PPCN7.5 and PPCN4 show 13.3 and 30.6% increases, respectively, under the same conditions. The effect of T_c on G' is in the order of PP-MA < PPCN7.5 < PPCN4. It may be recalled that the T_c dependence of $d_{(001)}$ showed the order of intercalation PPCN7.5 < PPCN4 in Fig. 3.10. This implies that much higher efficiency of the intercalation for the reinforcement is attained in the PPCN4. For PPCN2, owing to the partial exfoliation, the degree of intercalation decreases and hence the modulus decreases compared to the low T_c condition (70 °C). Here, it should be mentioned that the crystallinity increases a little with increasing T_c for both PP-g-MA and PPCNs and the extent is almost the same for all systems. So, it is believed that it is not the crystallinity but rather the degree of intercalation that affects the storage modulus.

3.8.2
Multiscale Micromechanical Modeling

Sheng and Boyce [55] reported a multiscale modeling strategy to account for the hierarchical morphology of the nanocomposite: at a length scale of thousands of microns, the structure is one of high aspect ratio particles within a matrix; at the length scale of microns, the clay particle structure is either exfoliated clay

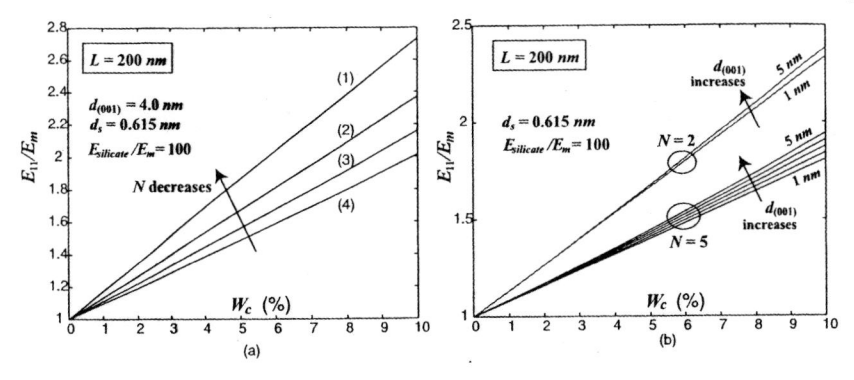

Fig. 3.11 Effect of clay structural parameters (N, $d_{(001)}$) on the macroscopic modulus predicted by Mori–Tanaka model. (a) Effect of N at fixed $d_{(001)} = 4.0$ nm. (b) Effect of $d_{(001)}$ at two fixed values $N = 2$ and $N = 5$. Reprinted from [55], © 2004, Elsevier Science.

sheets of nanometer level thickness or stacks of parallel clay sheets separated from one another by interlayer galleries of nanometer level height, and the matrix, if semi-crystalline, consists of fine lamella, oriented with respect to the polymer/clay interfaces. Models of various representative volume elements of the underlying structure of the clay filled polymer are constructed. Figure 3.11 shows the influence of internal clay structural parameters (the average number of silicate layers per clay stack: N, $d_{(001)}$) on the macroscopic modulus of the PLFNC. The enhancement of modulus (E_{11}/E_m) is plotted as a function of clay content (W_c) and N at fixed $d_{(001)}$. The strong dependence of the modulus on N is clearly demonstrated; at a fixed W_c, the modulus increases with decreasing N; the amount of increase gradually expands as $N \to 1$. On the other hand, the effect of $d_{(001)}$ on the modulus for two different values of N ($N=2$ and 5) is rather small, and depends on the specific value of N. This increment is rather negligible when N is small, however, when the nanocomposite is highly intercalated (e.g., $N=5$), an increase of a few nanometers in $d_{(001)}$ can cause a considerable increase in modulus.

In the case of Nylon 6, the transcrystallization behavior induced by the nanofiller (clay) is taken into account by modeling a layer of matrix surrounding the particle to be highly textured and therefore mechanically anisotropic. Micromechanical models (numerical and analytical) based on the "effective clay particle" were employed to calculate the overall elastic modulus of the amorphous and semi-crystalline PLFNCs and to compute their dependence on the matrix and clay properties as well as internal clay structural parameters. The proposed modeling technique captures the strong modulus enhancements observed in elastomer/MMT nanocomposites as compared with the moderate enhancements observed in glassy and semi-crystalline PLFNCs. For the case where the matrix is semi-crystalline (like Nylon 6), the proposed approach captures the effect of transcrystallized matrix layers in terms of composite modulus enhancement, however, this effect is found to be surprisingly minor in comparison with the composite-level effects of stiff particles in a matrix. The reason for this is discussed in Section 8.4.

Note that in order to determine the nanocomposite modulus the modulus of nano-clay of 400 GPa is employed. Most nanocomposite researchers believe that the nano-clay has a high modulus of 170 GPa [1]. But this value is not absolutely acceptable even in the case of a mono-layered clay sheet.

3.8.3
Flexibility of a Single Clay Layer

A large degree of flexibility of the mono-layered clay sheet is reported [56]. Two transmission electron microscope (TEM) images are shown (see Fig. 3.12). One arises from the clay layers that appear as about 150–200 nm curved sheets. When viewed edge-on as in Fig. 3.12, several 3–5 nm stacked sheets are apparent. The curved nature of the sheet is observed, for it is well known that smectite clay sheets have a large degree of flexibility [56]. Sato and Kawamura [57] reported the study of the flexibility of smectite clay minerals by using molecular dynamics (MD) simulations. They took into account the quantitative understanding of the mechanical behavior of a single clay layer in a completely exfoliated state. The repeating unit of a layer is taken to be $a_0 = 0.52$ nm and $b_0 = 0.902$ nm with formula of $2Na_{1/3}Al_2[Si_{11/3}Al_{1/3}]O_{10}(OH)_2$ which corresponds

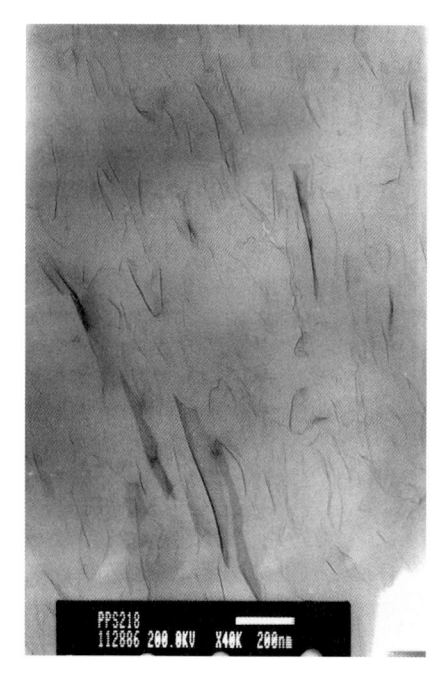

Fig. 3.12 Bright field TEM images of PPS-based nanocomposite: loading filler = 5.3 wt%. Unpublished data from M. Okamoto.

to that of beidellite (see Table 3.1). When the size of the basic cell ($A=9.3$ nm, $B=2.6$ nm, and $C=5$ nm) (A-type cell) is reduced by 3–40% in the A-direction, the stationary structure of a clay layer is obtained as a curved sheet with a 2:1 smectite-type layer. In such a curved state, the layer experiences an external stress of 0.5–0.7 GPa. The layer structure of a clay fractures when the size of the same basic cell is reduced by more than 40%. This value is much lower than that of moscovite (~ 2 GPa) which is also reported by the same authors [58]. The simulation was also done by reducing the size of the basic cell ($A=3.1$ nm, $B=10.7$ nm, and $C=5$ nm) (B-type) in the B-direction. The clay layer is found to be more flexible along the A-direction than along the B-direction. When the microscopic structure of a curved clay layer is examined, it is concluded that the main origin of the flexibility lies in the change in the Si–O–Si angle in the silicate tetrahedral sheets rather than in the change in bond lengths. These simulation results agree with the atomic force microscopy (AFM) observations [59].

3.8.4
Higher-order Structure Development and Crystallization Controlled by Nano-filler Surfaces

The formation of the γ-form in the presence of clay in Nylon 6/MMT nanocomposite is well known [60]. The essential difference between the γ-form and the a-form is the molecular packing; in the a-form hydrogen bonds are formed between anti-parallel chains while the molecular chains have to twist away from the zigzag planes to form the hydrogen bonds among the parallel chains in the γ-form giving rise to less inter-chain interaction as compared to the a-form.

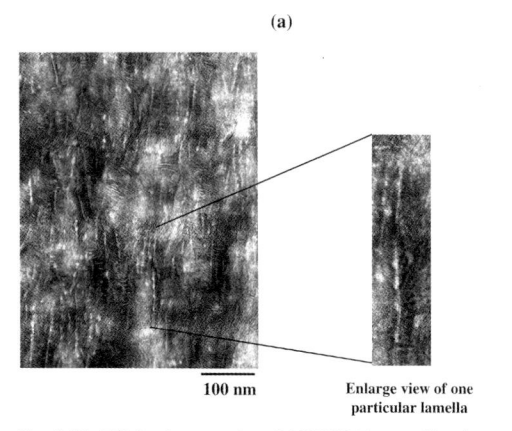

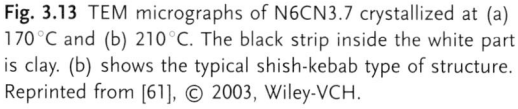

Fig. 3.13 TEM micrographs of N6CN3.7 crystallized at (a) 170°C and (b) 210°C. The black strip inside the white part is clay. (b) shows the typical shish-kebab type of structure. Reprinted from [61], © 2003, Wiley-VCH.

The lamellar morphology and distribution of MMT particles in the Nylon 6-based nanocomposite (N6CN3.7) (MMT = 3.7 wt%), crystallized at 170 and 210 °C, have been reported by Okamoto and Maiti [61] (see Fig. 3.13). The white strips (Fig. 3.13a) represent the discrete lamellar pattern and, after a close look, a black clay particle is clearly observed inside the lamella. In other words, lamellar growth occurs on both sides of the clay particles i.e. the clay particle is sandwiched by the formed lamella. This is a unique observation of lamellar orientation on the clay layers. In the semi-crystalline polymer the stacked lamellar orientation generally occurs.

The lamellar pattern at high T_c (Fig. 3.13b) is somehow similar but along with the sandwiched structure, branched lamellae are formed which originate from the parent sandwiched lamella. There are no clay particles found inside the branched lamellae and the γ-phase having irregular chain packing with distortion (γ^*-phase) is formed, as revealed by wide-angle X-ray diffraction (WAXD) which one can observe only in the case of high T_c crystallized N6CN3.7. This epitaxial growth (γ^*-phase) on the parent lamella forms the *shish-kebab* type of structure, which virtually enhances the mechanical properties of the nanocomposites like a bone material which consists of collagen fibrils reinforced with tiny mineral particles, a few nanometers in thickness [62].

From this sandwiched structure the accurate determination of the long spacing and lamellar thickness of N6NC3.7 from small angle X-ray scattering (SAXS) is questionable [63, 64]. It has to be remembered that Nylon 6 has the highest capability of forming hydrogen bonding to form a hydrogen-bonded sheet. Pseudohexagonal packing is favored with hydrogen-bonding between the silicate layers and Nylon 6, as a result the induction time of N6CN3.7 becomes very short, as compared to neat Nylon 6. Once one molecular layer is nucleated on the clay surface, other molecules may form hydrogen-bonding to the molecule already hydrogen-bonded to the silicate surface, giving rise to the discrete lamellar structure on both sides of the clay. The nucleation and growth process is demonstrated in Fig. 3.14, following direct observation by TEM [61]. This unique mechanism can well explain the higher crystallization rate of PLFNCs along with the morphology and developed internal structure. This sandwiched structure (each silicate layer is strongly covered by polymer crystals) makes the

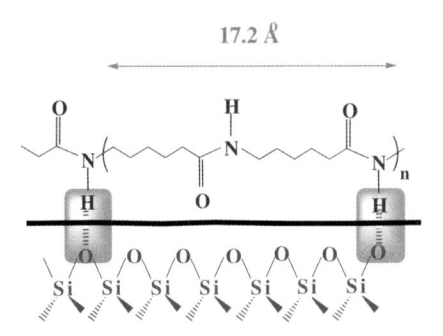

Fig. 3.14 Schematic view of the nucleation and growth mechanism in N6CN3.7. Reprinted from [61], © 2003, Wiley-VCH.

system very rigid and as a result the heat distortion temperature (HDT) increases to 80 °C but the surrounding excess amorphous part (lower crystallinity of N6CN3.7 as compared to neat Nylon 6) can easily retain the polymeric properties like impact strength, ultimately making an improved/perfect system in PLFNCs.

Kim and Kressler [65] also reported that the fine lamellae of Nylon 12 crystals are oriented perpendicular to the Nylon 12/MMT interface, i.e., on planes lying normal to the injection molding direction. This interfacial ordering may be a result of the crystallization process and is similar to the well-known transcrystallization [66]. The nanocomposites consist of a nanostructured network, finely dispersed and uniformly oriented silicate layers are aligned perpendicular to lamellae, and the two materials are strongly bonded to each other (see Fig. 3.15). In these Nylon-based nanocomposites, the nano-clay particles with the sandwiched-and-network (shish-kebab) structure have a high function as micro-void initiation sites, which are necessary for high toughness during deformation. Fratzl et al. [67] reported that the mechanical behavior of the collagen mineral nanocomposite in bone depends crucially on both components, mineral and wet collagen, and on their interaction (see Fig. 3.16).

In the case of polyvinylidene (PVDF)-based nanocomposites, the formation of the β-form has been observed [68, 69]. Shah et al. [69] reported a remarkable order of magnitude enhancement in toughness of the nanocomposites (see Fig. 3.17). They postulated that nucleation of the fiber-like β-phase (more ductile than the α-phase) on the surface of individual silicate layers leads to a structure conducive to plastic flow under applied stress [70–72]. Energy dissipation could

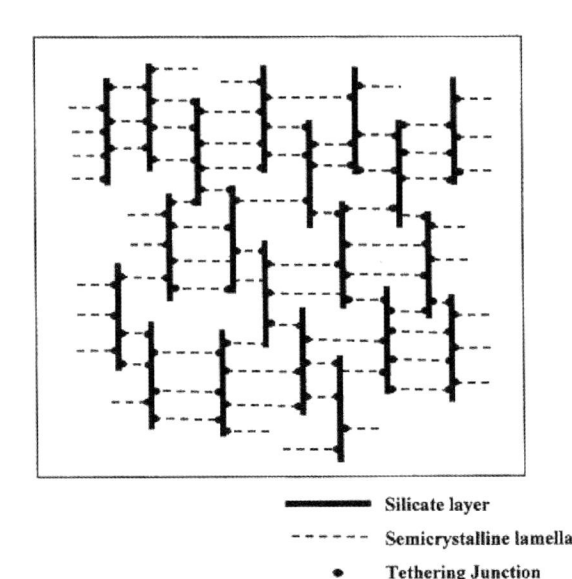

Fig. 3.15 Nanostructured network in Nylon 12/MMT nanocomposite. Reprinted from [65], © 2001, Elsevier Science.

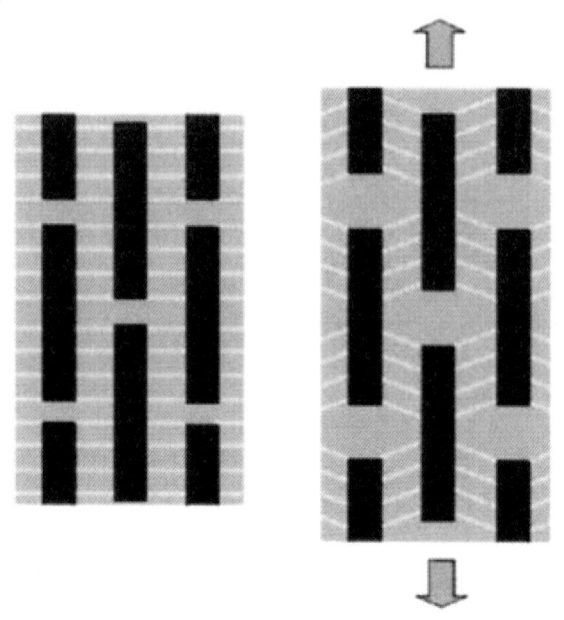

Fig. 3.16 Schematic arrangement of mineral and organic phase in a mineralized collagen fibril subjected to tensile load. Mineral particles are shown in dark gray and should be imagined as platelets viewed edge on. The horizontal white lines in the light gray matrix (left part of figure) are not indicating any physical reality, they are just drawn to visualize the shear deformation in the matrix between the particles as a consequence of tensile deformation of the tissue (right part of figure). Reprinted from [67], © 2004, The Royal Society of Chemistry.

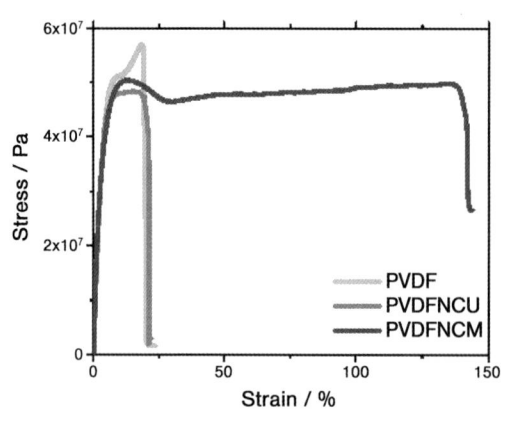

Fig. 3.17 Stress–strain curves for neat PVDF (α-phase crystallite), PVDFNCU (microcomposite by using MMT) and PVDFNCM (nanocomposite by using organoclay) showing the dramatic increase in elongation at break for the nanocomposite. Reprinted from [69], © 2004, Wiley-VCH.

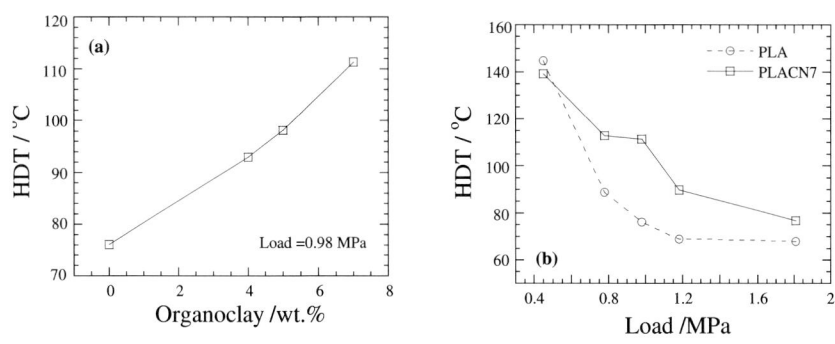

Fig. 3.18 (a) Organoclay (wt%) dependence of HDT of neat PLA and various PLACNs. (b) Load dependence of HDT of neat PLA and PLACN7. Reprinted from [73], © 2003, Elsevier Science.

be further enhanced due to the presence of more mobile β crystallites which have the potential to act like rigid fillers surrounded by the crystalline phase of PVDF. Thus the crystallization control by silicate surfaces may provide not only a new approach for toughening of polymers but also the way towards a novel approach for the design of new materials.

In the case of polyester systems, Yamada et al. [73] examined the HDT of various polylactide (PLA)-based nanocomposites (PLACNs) with different load conditions. With PLACN (MMT = 5 wt%), there is marked increase in HDT with an intermediate load of 0.98 MPa, from 76 °C for the neat PLA up to 111 °C for PLACN (see Fig. 3.18). In the case of high load (1.81 MPa), however, it is very difficult to achieve high HDT enhancement without strong interaction between the polymer matrix and organoclay as for Nylon systems [61]. So the improvement of HDT with intermediate load (0.98 MPa) originates from the better mechanical stability of the PLACNs due to mechanical reinforcement by the dispersed clay particles, a higher value of the degree of crystallinity χ_c and intercalation. This is qualitatively different from the behavior of Nylon systems, where the MMT layers stabilize in a different crystalline phase (γ-phase) [61] with strong hydrogen bonding between the silicate layers and Nylon 6 as a result of the discrete lamellar structure on both sides of the clay (see Figs. 3.13 and 3.14). Nylon-based nanocomposites have been successfully prepared without a strategy for designing materials with the desired properties of the PLFNCs.

3.8.5
Flocculation Control and Modulus Enhancement

Most nanocomposite researchers obdurately believe that the preparation of a completely exfoliated structure is the ultimate target for better overall properties. However, these significant improvements are not observed in all nanocomposite systems, including systems where the silicate layers are near to exfoliated [74].

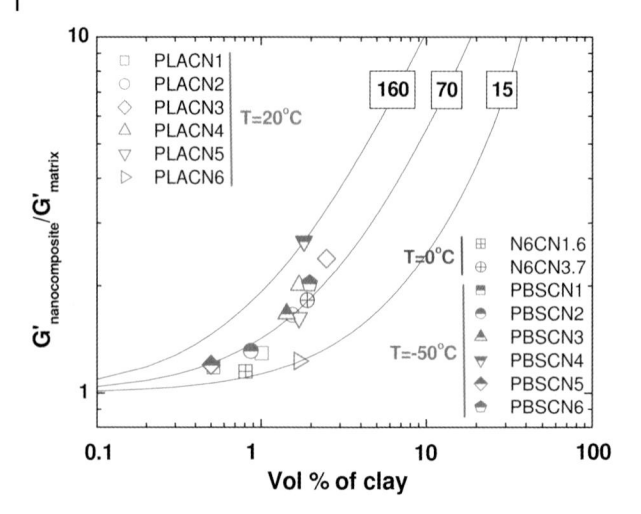

Fig. 3.19 Plots of $G'_{nanocomposite}/G'_{matrix}$ vs. vol% of clay for various nanocomposites. The Einstein coefficient k_E is shown with the number in the box. The lines show the calculated results from Halpin and Tai's theory with various k_E.

While, from the barrier property standpoint, the development of exfoliated nanocomposites is always preferred, Nylon-based nanocomposite systems are completely different from other nanocomposite systems, as discussed before.

In Fig. 3.19, Okamoto summarized the clay content dependence of the dynamic storage modulus (G') of various types of nanocomposites obtained at well below the T_g of the matrices. The Einstein coefficient (k_E), derived by using Halpin and Tai's theoretical expression modified by Nielsen, is shown in the figure, and represents the aspect ratio (L_{clay}/d_{clay}) of dispersed MMT particles without intercalation. From this figure, it is clearly observed that poly(butylene succinate) (PBS)-based nanocomposites (PBSCNs) show very high increment in G' compared to other nanocomposites having the same content of clay in the matrix. PPCNs are well known for intercalated systems, N6CNs are well-established exfoliated nanocomposites, PLACNs are becoming established intercalated-and-flocculated nanocomposites, while PBSCNs are intercalated-and-extended flocculated nanocomposite systems [75, 76]. Due to the strong interaction between hydroxylated edge–edge groups, as mentioned above, the MMT particles are sometimes flocculated in the polymer matrix. As a result of this flocculation the length of the clay particles increases enormously and hence the overall aspect ratio increases. For the preparation of high molecular weight PBS, di-isocyanate [OCN-(C_6H_{12})-NCO] type end-groups are generally used as a chain extender. These isocyanate end group chain extenders form urethane bonds with hydroxyl-terminated low molecular weight PBS, and each high molecular weight PBS chain contains two such kind of bonds. These urethane type bonds lead to strong interaction with silicate surface, by forming hydrogen

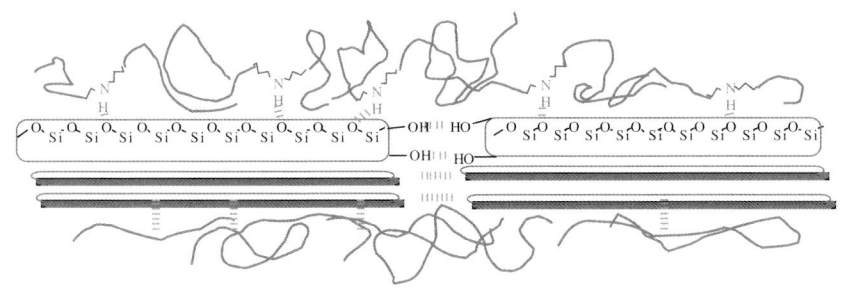

Fig. 3.20 Formation of hydrogen bonds between PBS and clay, which leads to the flocculation of the dispersed silicate layers. Reprinted from [75], © 2003, American Chemical Society.

bonds, and hence strong flocculation (see Fig. 3.20). For this reason, the aspect ratio of dispersed clay particles is much higher in the case of PBSCNs than with all nanocomposites, and hence the high enhancement of the modulus.

This behavior can be explained with the help of the classical rheological theory of suspension of conventional filler reinforced systems. According to this theory [77], the rotation of filler is possible when the volume fraction of clay $\phi_{filler} < \phi_{critical} \cong$ (aspect ratio)$^{-1}$. All PBSCNs studied here follow this relation except PBSCN4 (MMT = 3.6 wt%), in which $\phi_{filler} \gg$ (aspect ratio)$^{-1}$. For this reason, in PBSCN4, rotation of dispersed intercalated with flocculated stacked silicate layers is completely hindered and only translational motion is available, hence they exhibit a very high modulus. This behavior is clearly observed in dynamic storage modulus measurements in the molten state [75]. In the case of N6CN3.7 (MMT = 3.7 wt%) we can see the same high increment in G' as found for PBSCNs. The development of the flocculated structure occurs even though N6CNs are well-established exfoliated nanocomposite systems.

3.9
Melt Rheology of PLFNCs

3.9.1
Linear Viscoelastic Properties

The measurement of the rheological properties of the PLFNCs in the molten state is crucial to gain a fundamental understanding of the nature of the processability and the structure–property relationships for these materials. Dynamic oscillatory shear measurements of polymeric materials are generally performed by applying a time-dependent strain of $\gamma(t) = \gamma_0 \sin(\omega t)$ and the resultant shear stress is $\sigma(t) = \gamma_0 [G' \sin(\omega t) + G'' \cos(\omega t)]$, with G' and G'' being the storage and loss modulus, respectively.

Generally, the rheology of polymer melts depends strongly on the temperature at which the measurement is carried out. It is well known that for thermorheological simplicity, isotherms of storage modulus ($G'(\omega)$), loss modulus ($G''(\omega)$) and complex viscosity ($|\eta^*|(\omega)$) can be superimposed by horizontal shifts along the frequency axis:

$$b_T G'(a_T\omega, T_{ref}) = b_T G'(\omega, T);$$

$$b_T G''(a_T\omega, T_{ref}) = b_T G''(\omega, T)$$

$$|\eta^*|(a_T\omega, T_{ref}) = |\eta^*|(\omega, T)$$

where a_T and b_T are the frequency and vertical shift factors, and T_{ref} is the reference temperature. All isotherms measured for pure polymer and for various nanocomposites can be superimposed along the frequency axis.

In the case of polymer samples, it is expected, at the temperatures and frequencies at which the rheological measurements were carried out, that the polymer chains should be fully relaxed and exhibit characteristic homopolymer-like terminal flow behavior (i.e. the curves can be expressed by a power law of $G' \propto \omega^2$ and $G'' \propto \omega$).

The rheological properties of *in situ* polymerized nanocomposites with end-tethered polymer chains were first described by Krisnamoorti and Giannelis [78]. The flow behavior of PCL- and Nylon 6-based nanocomposites differed extremely from that of the corresponding neat matrices, whereas the thermorheological properties of the nanocomposites were entirely determined by the behavior of the matrices [78]. The slope of $G'(\omega)$ and $G''(\omega)$ versus $a_T\omega$ is much smaller than 2 and 1, respectively. Values of 2 and 1 are expected for linear mono-dispersed polymer melts, and the large deviation, especially in the presence of a very small amount of layered silicate loading, may be due to the formation of a network structure in the molten state. However, such nanocomposites based on the *in situ* polymerization technique exhibit fairly broad molar mass distribution of the polymer matrix, which hides the structurally relevant information and impedes the interpretations of the results.

To date, the melt state linear dynamic oscillatory shear properties of various kinds of nanocomposites have been examined for a wide range of polymer matrices including Nylon 6 with various matrix molecular weights [79], PS [80], polystyrene (PS)-polyisoprene (PI) block copolymers [81, 82], poly(ε-caprolactone) (PCL) [83], PP [84–86], PLA [87, 88], PBS [75, 89], and so on [90].

The linear dynamic viscoelastic master curves for the neat PLA and various PLACNs are shown in Fig. 3.21 [88]. The linear dynamic viscoelastic master curves were generated by applying the time–temperature superposition principle and shifted to a common temperature T_{ref} using both the frequency shift factor a_T and the modulus shift factor b_T. The moduli of the PLFNCs increase with increasing clay loading at all frequencies ω. At high ωs, the qualitative behavior of $G'(\omega)$ and $G''(\omega)$ is essentially the same and unaffected by frequency. However, at low frequencies $G'(\omega)$ and $G''(\omega)$ increase monotonically with increasing clay content. In the low frequency region, the curves can be expressed

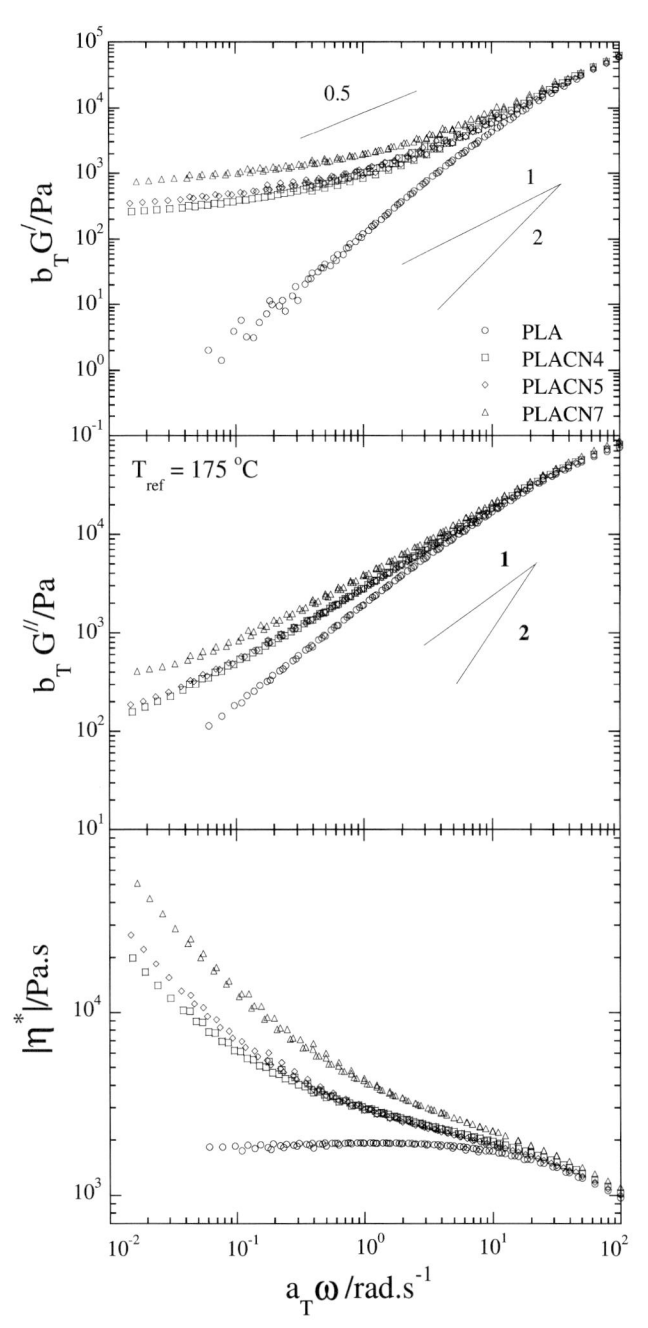

Fig. 3.21 Reduced frequency dependence of storage modulus, loss modulus and complex viscosity of neat PLA and various PLACNs. Reprinted from [88], © 2003, Elsevier Science.

Table 3.4 Terminal slopes of G' and G" vs. $a_T\omega$ for PLA and various PLACNs. Reprinted from [88], © 2003, Elsevier Science.

System	G'	G"
PLA	1.3	0.9
PLACN4	0.2	0.5
PLACN5	0.18	0.4
PLACN7	0.17	0.32

by the power-law of $G'(\omega) \propto \omega^2$ and $G''(\omega) \propto \omega$ for neat PLA, suggesting that this is similar to those of the narrow M_w distribution homopolymer melts. On the other hand, for $a_T < 5$ rad s^{-1}, the viscoelastic response [particularly $G'(\omega)$] for all the nanocomposites displays significantly diminished frequency dependence as compared to the matrices. In fact, for all PLACNs, $G'(\omega)$ becomes nearly independent at low $a_T\omega$ and exceeds $G''(\omega)$, characteristic of materials exhibiting a pseudo-solid-like behavior [78]. The values of the terminal zone slopes of both neat PLA and PLACNs are estimated in the lower $a_T\omega$ region (<10 rad s^{-1}), and are presented in Table 3.4. The lower slope values and the higher absolute values of the dynamic moduli indicate the formation of a "spatially-linked" structure in the PLACNs in the molten state [81]. Because of this structure or high geometric constraints, the individual stacked silicate layers are incapable of freely rotating and hence, by imposing small $a_T\omega$, the relaxations of the structure are prevented almost completely. This type of prevented relaxation due to the highly geometric constraints of the stacked and intercalated silicate layers leads to the presence of the pseudo-solid-like behavior observed in PLACNs. This behavior probably corresponds to the shear-thinning tendency, which appears strongly in the viscosity curves ($a_T\omega < 5$ rad s^{-1}) ($|\eta^*|$) vs. $a_T\omega$). Such features depend strongly on the shear rate in the dynamic measurement because of the formation of the shear-induced alignment of the dispersed clay particles [91].

The temperature-dependent frequency shift factor (a_T, Williams–Landel–Ferry type [92]) used to generate the master curves shown in Fig. 3.21 are shown in Fig. 3.22. The dependence of the frequency shift factors on the silicate loading suggests that the temperature-dependent relaxation process observed in the viscoelastic measurements is somehow affected by the presence of the silicate layers [78]. In the case of N6CN3.7, with the hydrogen-bonding to the already formed molecule hydrogen-bonded to the silicate surface, the system exhibits a large value of flow activation energy (estimated from the slope in Fig. 3.22a), nearly one order higher in magnitude than that of neat Nylon 6 [93].

The shift factor b_T shows large deviation from a simple density effect although it would be expected that the values would not vary far from unity [92]. One possible explanation is an internal structure development occurring in PLACNs during measurement (shear process). The alignment of the silicate

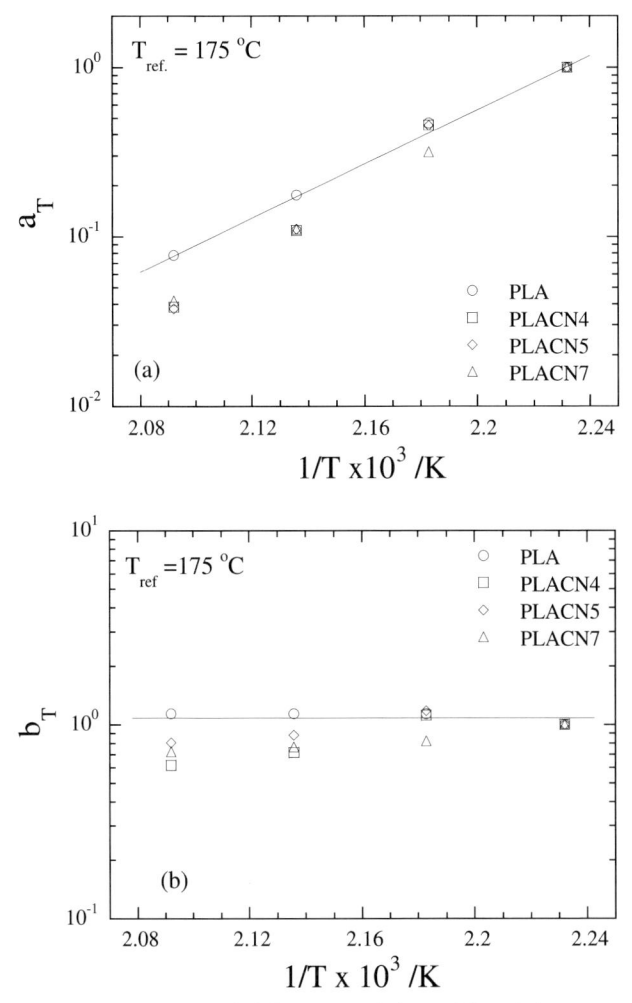

Fig. 3.22 (a) Frequency shift factor a_T and (b) modulus shift factor b_T as a function of temperature. Reprinted from [88], © 2003, Elsevier Science.

layers probably helps PLFNC melts to withstand the shear force, thus leading to the increase in the absolute values of $G'(\omega)$ and $G''(\omega)$.

Galgali and his colleagues [85] have also shown that the typical rheological response in nanocomposites arises from frictional interactions between the silicate layers and is not due to the immobilization of confined polymer chains between the silicate layers. They have also shown a dramatic decrease in the creep compliance for the PP-based nanocomposite with 9 wt% of MMT. They showed a dramatic three-order of magnitude drop in the zero shear viscosity beyond the apparent yield stress, suggesting that the solid-like behavior in the quiescent state is a result of the percolated structure of the layered silicate.

Ren et al. [81] measured the viscoelastic behavior of a series of nanocomposites of disordered PS-PI block copolymer and MMT. Dynamic moduli and stress relaxation measurements indicate solid-like behavior for nanocomposites with more than 6.7 wt% MMT, at least on a time scale of the order of 100 s. According to them, this solid-like behavior is due to the physical jamming or percolation of the randomly distributed silicate layers at a surprisingly low volume fraction due to their anisotropic nature. The fact that alignment of the silicate layers by large shear stresses results in a more liquid-like relaxation behavior supports the percolation argument.

3.9.2
Steady Shear Flow

The steady shear rheological behavior of neat PBS and various PBSCNs is shown in Fig. 3.23. The steady viscosity of PBSCNs is enhanced considerably at all shear rates with time, and at a fixed shear rate increases monotonically with increasing silicate loading [76]. On the other hand, all intercalated PBSCNs exhibit strong rheopexy behavior, and this becomes prominent at low shear rates, while neat PBS exhibits a time-independent viscosity at all shear rates. With increasing shear rates, the shear viscosity attains a plateau after a certain time, and the time required to attain this plateau decreases with increasing shear rates. The possible reasons for this type of behavior may be due to the planar alignment of the clay particles towards the flow direction under shear. When the shear rate is very slow (0.001 s^{-1}), clay particles take longer to attain complete planar alignment along the flow direction, and this measurement time (1000 s) is too short to attain such alignment and hence strong rheopexy behavior is shown. On the other hand, under high shear rates (0.005 s^{-1} or 0.01 s^{-1}) this measurement time is sufficient to attain such alignment, and hence, nanocomposites show time-independent shear viscosity after a certain time.

Figure 3.24 shows the shear rate dependence of viscosity for neat PBS and the corresponding nanocomposites measured at 120 °C. While the neat PBS exhibits almost Newtonian behavior at all shear rates, nanocomposites exhibited non-Newtonian behavior. At very low shear rates, the shear viscosity of nanocomposites initially exhibits some shear-thickening behavior, corresponding to the rheopexy behavior observed at very low shear rates (see Fig. 3.23). After that all nanocomposites show very strong shear thinning behavior at all shear rates and this behavior is analogous to the results obtained in the case of dynamic oscillatory shear measurements [73]. Additionally, at very high shear rates, the viscosities of nanocomposites are comparable to that of neat PBS. These observations suggest that the silicate layers are strongly oriented towards the flow direction at high shear rates, and shear thinning behavior at high shear rates is dominated by that of neat polymer.

The PLFNC melts always exhibit significant deviation from the Cox–Merz relation [94], while all neat polymers nicely obey the empirical Cox–Merz relation, which requires that for $\dot{\gamma}=\omega$, the viscoelastic data should obey the relationship

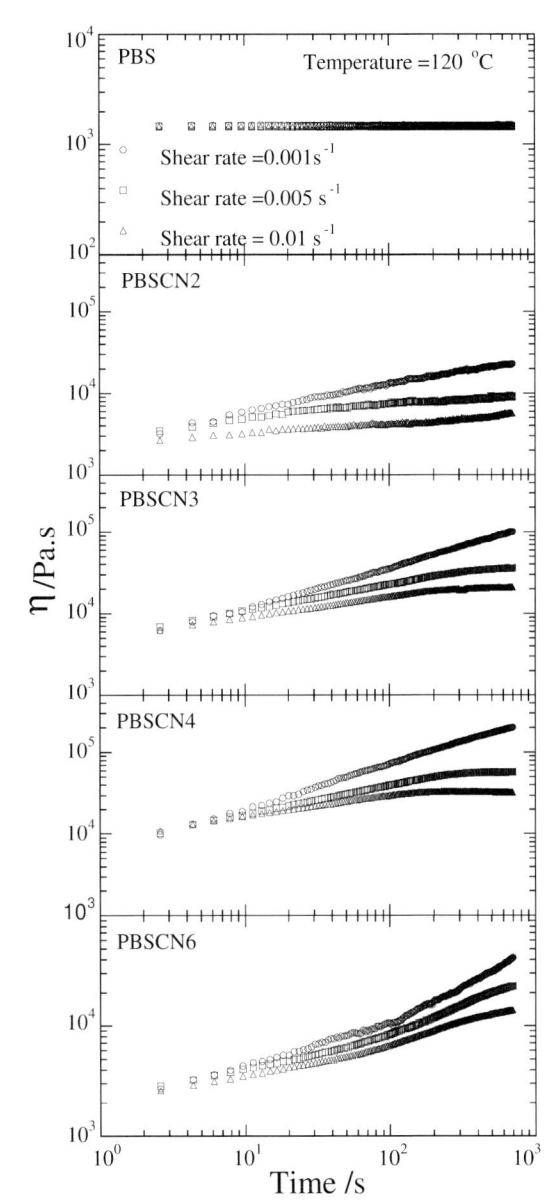

Fig. 3.23 Time variation of shear viscosity for PBSCN. Reprinted from [75], © 2003, American Chemical Society.

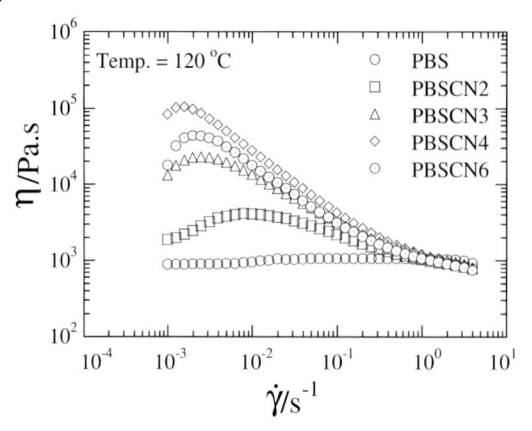

Fig. 3.24 Shear viscosity as a function of shear rates for the shear rate sweep test. Reprinted from [75], © 2003, American Chemical Society.

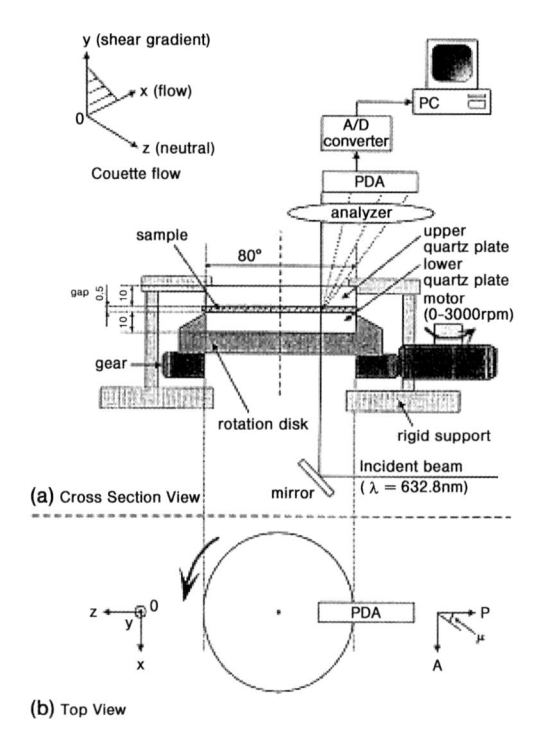

Fig. 3.25 A schematic illustration of Rheo-SALS apparatus: (a) cross-sectional and (b) top views. Reprinted from [95], © 2000, The Society of Rheology, Japan.

$\eta(\dot{\gamma}) = |\eta^*|(\omega)$. We believe there are two possible reasons for the deviation from the Cox–Merz relation in the case of nanocomposites: first, this rule is only applicable for homogeneous systems like homopolymer melts but nanocomposites are heterogeneous systems. For this reason this relation is obeyed in the case of neat polymer [75]. Second, the structure formation is different when nanocomposites are subjected to dynamic oscillatory shear and steady shear measurements.

Okamoto et al. [95] constructed a unique rheo-optical device, an angle light scattering apparatus (Rheo-SALS), which enables us to perform time-resolved measurements of light intensity scattered from the internal structure developed under shear flow. Figure 3.25 shows a schematic illustration of the apparatus: Plane polarized light normal to the O–x-axis (flow direction) was applied vertically to the parallel-plate type shear cell along the velocity gradient (O–y-axis). Scattering profiles were observed either under Vv mode (depolarized geometry in which the optical axis of the analyzer was set parallel to that of the polarizer) or Hv (the cross-polarized geometry with the two axes being set perpendicular to the two axes) optical alignment at an azimuthal angle μ of $0°$. They reported the time variation of the mean-square density fluctuation $\langle \eta^2 \rangle$, the mean-square anisotropy $\langle \delta^2 \rangle$ and the relevant value of the correlation distance (ξ_η and ξ_δ) upon imposition/cessation of steady shear flow at both low shear rate ($\dot{\gamma}$) ($\cong 0.5$ s^{-1}) and high $\dot{\gamma}$ ($\cong 60$ s^{-1}).

3.9.3
Elongational Flow and Strain-induced Hardening

Okamoto et al. [96] first conducted an elongation test on PP-based nanocomposites (PPCN4) in the molten state at constant Hencky strain rate $\dot{\varepsilon}_0$ using an elongation flow optorheometer [97] and they also attempted to control the alignment of the dispersed silicate layers with nanometer dimensions of an intercalated PPCNs under uniaxial elongational flow.

Figure 3.26 shows double logarithmic plots of transient elongational viscosity $\eta_E (\dot{\varepsilon}_0; t)$ against time t observed for Nylon 6/LS system (N6CN3.7) and PPCN4 (MMT=4 wt%) with different Hencky strain rates $\dot{\varepsilon}_0$ ranging from 0.001 s^{-1} to 1.0 s^{-1}. The solid curve represents the time development of three-fold shear viscosity, $3\eta_0 (\dot{\gamma}; t)$, at 225 °C with a constant shear rate $\dot{\gamma} = 0.001$ s^{-1}. In $\eta_E (\dot{\varepsilon}_0; t)$ at any $\dot{\varepsilon}_0$, N6CN3.7 melt shows a weak tendency towards *strain-induced hardening* as compared to that of PPCN4 melt. A strong behavior of strain-induced hardening for the PPCN4 melt originated from the perpendicular alignment of the silicate layers to the stretching direction, as reported by Okamoto et al. [96].

From TEM observation (see Fig. 3.13), the N6CN3.7 forms a fine dispersion of the silicate layers of about 100 nm in L_{clay}, 3 nm thickness in d_{clay} and ξ_{clay} of about 20–30 nm between them. The ξ_{clay} value is an order of magnitude lower than the value of L_{clay}, suggesting the formation of a spatially-linked structure of the dispersed clay particles in the Nylon 6 matrix. For the N6CN3.7 melt, the silicate layers are so densely dispersed into the matrix that they are difficult to

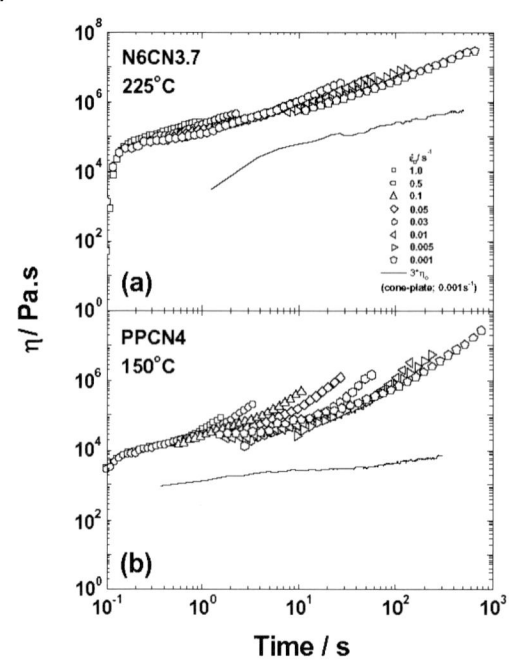

Fig. 3.26 Time variation of elongational viscosity $\eta_E(\dot{\varepsilon}_0; t)$ for (a) N6CN3.7 melt at 225 °C and for (b) PPCN4 at 150 °C. The solid line shows three times the shear viscosity, $3\eta_E(\dot{\gamma}; t)$, taken at a low shear rate $\dot{\gamma} = 0.001$ s^{-1} on a cone-plate rheometer. Reprinted from [1], © 2003, Rapra Technology Ltd.

align under elongational flow. Under flow fields, the silicate layers might move translationally, but not rotationally, in such a way that the energy loss becomes a minimum. This tendency was also observed in PPCN7.5 melt having a higher content of MMT (=7.5 wt%) [93].

On the other hand, one can observe two features for the shear viscosity curve. First, the extended Trouton rule, $3\eta_0(\dot{\gamma}; t) \cong \eta_E(\dot{\varepsilon}_0; t)$, does not hold for both N6CN3.7 and PPCN4 melts, in contrast to the melt of ordinary homopolymers. The latter, $\eta_E(\dot{\varepsilon}_0; t)$, is more than 10 times larger than the former, $3\eta_0(\dot{\gamma}; t)$. Second, again unlike ordinary polymer melts, $3\eta_0(\dot{\gamma}; t)$ of the N6CN3.7 melt increases continuously with t, never showing a tendency to reach a steady state within the time span (600 s or longer) examined here. This *time-dependent thickening* behavior may be called *anti-thixotropy* or *rheopexy*. Via slow shear flow ($\dot{\gamma} = 0.001$ s^{-1}), $3\eta_0(\dot{\gamma}; t)$ of N6CN3.7 exhibits a much stronger rheopexy behavior, almost two orders of magnitude higher than that of PPCN4. This reflects the fact that the shear-induced structural change involves a process with an extremely long relaxation time, as for other PLFNCs having rheopexy behavior [75], especially under weak shear field.

3.9.4
Alignment of Silicate Layers

The orientation of silicate layers and Nylon 6 crystallites in injection-molded N6CN using WAXD and TEM has been examined [98, 99]. Kojima and his colleagues found in the sample three regions of different orientations as a function of depth. Near the middle of the sample, where the shear forces are minimal, the silicate layers are oriented randomly and the Nylon 6 crystallites are perpendicular to the silicate layers. In the surface region, shear stresses are very high, so both the clay layers and the Nylon 6 crystallites are parallel to the surface. In the intermediate region, the clay layers, presumably due to their higher aspect ratio, still orient parallel to the surface and the Nylon 6 crystallites assume an orientation perpendicular to the silicate. Medellin-Rodriguez et al. [100] reported that the molten N6CN samples showed planar orientation of silicate layers along the flow direction, which is strongly dependent on shear time as well as clay loading, reaching a maximally orienting level after being sheared for 15 min with $\dot{\gamma}=60\ \text{s}^{-1}$.

Okamoto and his colleagues conducted TEM observation for the sheared N6CN3.7 with $\dot{\gamma}=0.0006\ \text{s}^{-1}$ for 1000 s [1]. The edges of the silicate layers lying along the z-axis (marked with the arrows (A)) or parallel alignment of the silicate edges to the shear direction (x-axis) (marked with the arrows (B)) rather than random orientation in the Nylon 6 matrix is observed, but in fact, one cannot see these faces in this plane (Fig. 3.27). Here, it should be emphasized that the planar orientation of the silicate faces along the x–z plane does not take place prominently. For the case of rapid shear flow, the commonly applicable conjecture of the planar orientation of the silicate faces along the shear direction was first demonstrated to be true by Kojima and his colleagues [98].

In uniaxial elongational flow (converging low) for a PPCN4, the formation of *a house-of-cards* structure is found by TEM analysis [96]. The perpendicular (but *not* parallel) alignment of disk-like clay particles with large anisotropy toward the flow direction might sound unlikely but this could be the case, especially under an elongational flow field, in which the extentional flow rate is the square of the converging flow rate along the thickness direction, if the assumption of *affine* deformation without volume change is valid. Obviously under such conditions, the energy dissipation rate due to viscous resistance between the disk surface and the matrix polymer is minimal when the disks are aligned perpendicular to the flow direction.

Moreover, Lele and his colleagues [86] recently reported an *in situ* Rheo-X-ray investigation of flow-induced orientation in syndiotactic PP/layered silicate nanocomposite melt.

Bafna et al. [101] developed a technique to determine the three-dimensional (3D) orientation of various hierarchical organic and inorganic structures in PLFNCs. They studied the effect of compatibilizer concentration on the orientation of various structures in PLFNCs using 2D small angle X-ray scattering (SAXS) and 2D WAXD in three sample/camera orientations. Reflections and

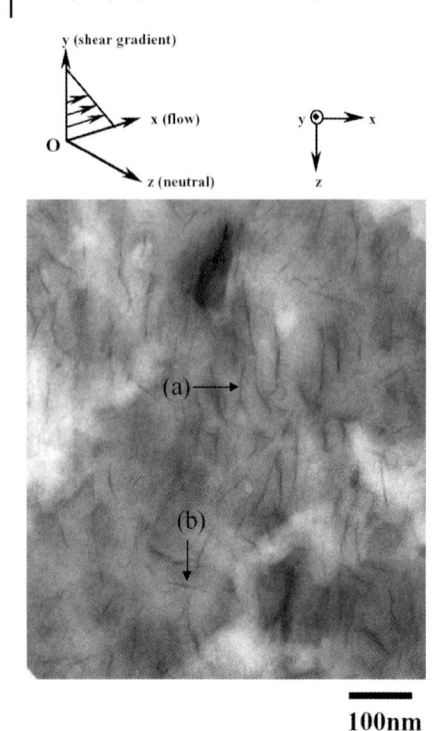

Fig. 3.27 TEM micrograph in the x–z plane showing N6CN3.7 sheared at 225 °C with $\dot{\gamma}=0.0006\ \text{s}^{-1}$ for 1000 s. The x-, y- and z-axes correspond respectively to flow, shear gradient and neutral direction. Reprinted from [1], © 2003, Rapra Technology Ltd.

orientation of six different structural features were easily identified: (a) clay clusters/tactoids (0.12 µm), (b) modified/intercalated clay stacking period (002) (2.4–3.1 nm), (c) stacking period of unmodified clay platelets (002) (1.3 nm), (d) MMT-clay (110) and (020) planes, normal (b) and (c), (e) polymer crystalline lamellae (001)(19–26 nm), long period ((001) is an average crystallographic direction), and (f) polymer unit cell (110) and (200) planes. The corresponding identified reflections are presented in Fig. 3.28. A 3D study of the relative orientation of the above mentioned structures was carried out by measuring three projections of each sample. Quantitative data on the orientation of these structural units in the nanocomposite film were determined through calculations of the major axis direction cosines and through a ternary, direction-cosine plot called a "Wilchinsky triangle" [102], previously proposed in lamellar orientation studies [103]. It allows a direct comparison of average preferred orientations for different structural features. In this way it is conceptually more useful than stereographic projections involving orientation density maps for a single WAXD reflection.

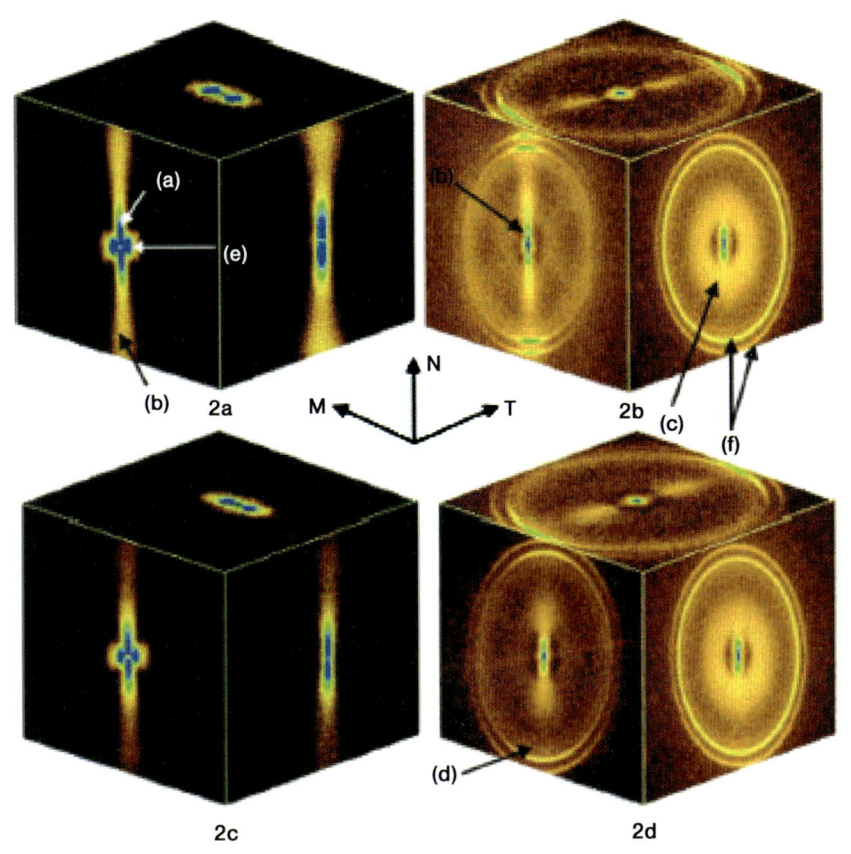

Fig. 3.28 2D SAXS (a and c) and WAXS (b and d) patterns for orientation MN (left face), NT (right face) and MT (top face) of films HD603 (a and b) and HD612 (c and d). The numbers in parentheses represent the reflections from the following: (a) clay tactoids, (b) modified/intercalated clay (002) plane, (c) unmodified clay (002) plane, (d) clay (110) and (020) plane, (e) polymer crystalline lamellar, (f) polymer unit cell (110) plane (inner ring) and (200) plane (outer ring). Reprinted from [101], © 2003, Elsevier Science.

3.9.5
Electrorheology

Electrorheological fluids (EFR)s, sometimes referred to as "smart fluids", are suspensions of polarizable particles dispersed in insulating media. A mismatch in conductivity or dielectric constant between the dispersed particle and the continuous medium phase induces polarization upon application of an electric field. The induced particle dipoles under the action of an electric field tend to attract neighboring particles and cause the particles to form fibril-like structures, which are aligned to the electric field direction.

Among various materials [104–107], semiconducting polymers are one of the novel intrinsic ER systems since they have the advantage of a wide range of working temperature, reduced abrasion of device, low cost, and relatively low current density. As a result, development of a high-performance ER fluid, followed by conducting polymer optimization and tuning, has been the subject of considerable interest for practical applications as a new electromechanic interface. Nevertheless, the yield stress and modulus of ER fluids are lower than those of magnetorheological fluids. Thus the performance of conducting polymer-based ER fluids is still insufficient for the successful development of specific application devices.

On this basis of this information, Kim et al. [104] first introduced a nanocomposite as an EFR using polyaniline (PANI)/MMT nanocomposites with intercalated structure. Though PANI/clay-intercalated nanocomposites are a new material for application as ER materials, yield stresses of the system showed less than 100 Pa at 1.2 kV mm^{-1} (20 wt% suspensions). This value is a little lower than the yield stress of a pure PANI particle system [105]. In other words, no synergistic effect of clay on yield stress was shown.

Park and his colleagues [106] have observed remarkable enhancement of yield stress for electrorheological fluids in PANI-based nanocomposites of clay. In their further study [107], they fabricated three kinds of EFRs with different contents of PANI-based nanocomposite and pure PANI particles in order to investigate the effect of nanocomposite particles on the enhancement of yield stress more systematically. They observed that there is an optimum content ratio between nanocomposite and pure PANI particles to produce minimum yield stress.

3.10
Materials Properties of PLFNCs

3.10.1
Thermal Stability

The thermal stability of polymeric materials is usually studied by thermogravimetric analysis (TGA). The weight loss due to the formation of volatile degradation products is monitored as a function of temperature ramp. When the heating is operated under an inert gas flow, a non-oxidative degradation occurs while the use of air or oxygen allows one to follow the oxidative degradation of the samples. Generally the incorporation of MMT-clay in the polymer matrix enhances the thermal stability by acting as a superior insulator and mass transport barrier to the volatile products generated during decomposition.

Blumstein [108] first reported the improved thermal stability of a PLFNC that combined poly(methyl methacrylate) (PMMA) and MMT. He showed that PMMA intercalated (d-spacing increase of 0.76 nm) between the galleries of MMT clay-resisted thermal degradation under conditions that would otherwise completely degrade pure PMMA. These PMMA-based nanocomposites were prepared by free radical polymerization of MMA intercalated in the MMT. TGA data reveal that nanocomposites with both linear PMMA and cross-linked PMMA intercalated into MMT layers have a 40–50 °C higher decomposition temperature. Blumstein argues that the stability of the PMMA-based nanocomposite is due not only to its different structure but also to restricted thermal motion of the PMMA in the gallery.

Recently, there have been many reports concerned with the improved thermal stability of nanocomposites prepared with various types of organoclay and polymer matrices [109–111]. Very recently, Zanetti and his colleagues [109] conducted detailed TG analyses of nanocomposites based on poly(ethylen-co-vinyl acetate) (EVA). The inorganic phase was fluorohectorite or MMT, both exchanged with octadecylammonium cation. According to them the deacylation of EVA in nanocomposites is accelerated and may occur at temperatures lower than those for the pure polymer or corresponding microcomposite due to catalysis by the strongly acid sites created by thermal decomposition of the silicate modifier. These sites are active when there is an intimate contact between the polymer and the silicate. Slowing down of the volatilization of the deacylated polymer in nitrogen may be due to the labyrinth effect of the silicate layers in the polymer matrix [110].

In air, the nanocomposite presents a significant delay of weight loss that may derive from the barrier effect due to diffusion of both the volatile thermo-oxidation products to the gas phase and oxygen from the gas phase to the polymer. According to Gilman et al. this barrier effect increases during volatilization owing to ablative reassembly of the reticulum of the silicate on the surface [111].

Many researchers believe the role of nano-filler in the nanocomposite structure might be the main reason for the difference in the TGA results of these systems compared to those for previously reported systems. The MMT-clay acts as a heat barrier, which could enhance the overall thermal stability of the system, as well as assisting in the formation of char after thermal decomposition. Thereby, in the beginning stage of thermal decomposition, the clay could shift the decomposition temperature higher. However, after that, this heat barrier effect would result in a reverse thermal stability. In other words, the stacked silicate layers could hold accumulated heat that could be used as a heat source to accelerate the decomposition process, in conjunction with the heat flow supplied by the external heat source.

3.10.2
Fire Retardant Properties

The Cone calorimeter is the most effective bench-scale method for studying the fire retardant properties of polymeric materials. Fire-relevant properties, measured by the Cone calorimeter, such as heat release rate (HRR), peak HRR, and smoke and CO yield, are vital to the evaluation of the fire safety of materials.

Gilman reviewed the flame-retardant properties of nanocomposites in detail [112, 113].

Table 3.5 Cone calorimeter data of various polymers and their nanocomposites with organoclay. Reprinted from [113], © 2000, American Chemical Society.

Sample (structure)	Residue yield (%) (±0.5)	Peak HRR (kW m^{-2}) (Δ%)	Mean HRR (kW m^{-2}) (Δ%)	Mean H_c MJ kg^{-1}	Mean SEA m^2 kg^{-1}	Mean CO yield kg/kg
Nylon 6	1	1010	603	27	197	0.01
N6 nanocomposite 2% (delaminated)	3	686 (32)	390 (35)	27	271	0.01
N6 nanocomposite 5% (delaminated)	6	378 (63)	304 (50)	27	296	0.02
PS	0	1120	703	29	1460	0.09
PS-silicate mix 3% (immiscible)	3	1080	715	29	1840	0.09
PS-nanocomposite 3% (intercalated/ delaminated)	4	567 (48)	444 (38)	27	1730	0.08
PSw/DBDPO/ Sb$_2$O$_3$) 30%	3	491 (56)	318 (54)	11	2580	0.14
PP-MA	5	1525	536	39	704	0.02
PP-MA nano- composite 2% (intercalated/ delaminated)	6	450 (70)	322 (40)	44	1028	0.02
PP-MA nano- composite 4% (intercalated/ delaminated)	12	381 (75)	275 (49)	44	968	0.02

Heat flux, 35 kW m^{-2}; H_c, specific heat of combustion; SEA, specific extinction area. Peak heat release rate, mass loss rate, and SEA data, measured at 35 kW m^{-2}, are reproducible to within ±10%. The carbon monoxide and heat of combustion data are reproducible to within ±15%.

Table 3.5 lists the cone calorimeter data of three different kinds of polymer and their nanocomposites with MMT. We can see that all Nylon 6-based nanocomposites reported here show reduced flammability. Peak HRR is reduced by 50–75% for Nylon 6, PS, and PP-g-MA nanocomposites [113]. According to the authors, the MMT must be nanodispersed for it to affect the flammability of the nanocomposites. However, the MMT-clay need not be completely delaminated for it to affect the flammability of the nanocomposite. In general, the nanocomposite flame-retardant mechanism involves the buildup of high-performance carbonaceous-silicate chars on the surface during burning; this insulates the underlying material and slows the mass loss rate of decomposition products.

For a PPCN with 4 wt% MMT-clay [113], there is a 75% reduction in flammability compared to the neat matrix (see Fig. 3.29).

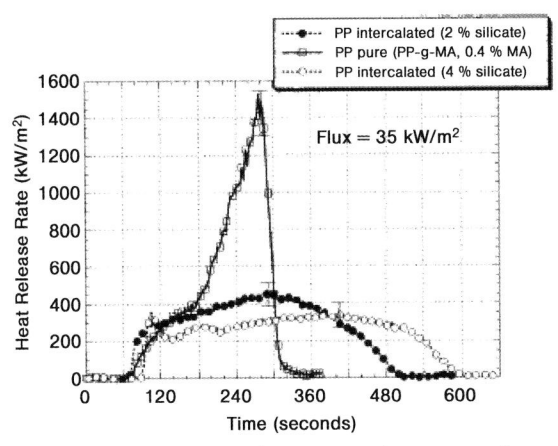

Fig. 3.29 Heat release rate during cone-calorimetry combustion of heat PP-g-MA and PPCNs. Reprinted from [113], © 2000, American Chemical Society.

3.10.3
Gas Barrier Properties

Nano-fillers are believed to increase the barrier properties by creating a maze or "tortuous path" (see Fig. 3.30) that retards the progress of the gas molecules through the matrix resin. The direct benefit of the formation of this type of path is clearly observed in polyimide/MMT nanocomposite which shows dramatically improved barrier properties with simultaneous decrease in thermal expansion coefficient [114, 115]. The polyimide-based nanocomposites reveal a several-fold reduction in the permeability of small gases, e.g. O_2, H_2O, He, CO_2, and the organic vapor ethylacetate with the presence of a small fraction of organoclay. For example, at 2 wt% MMT loading, the permeability coefficient of water vapor was decreased tenfold for the synthetic mica relative to pristine polyimide. By comparing nanocomposites made with layered silicates of various aspect ratios the permeability was noted to decrease with increasing aspect ratio.

The O_2 gas permeability was measured for the exfoliated PLA/fluorohectorite nanocomposites prepared by Sinha Ray and Yamada [116]. The relative perme-

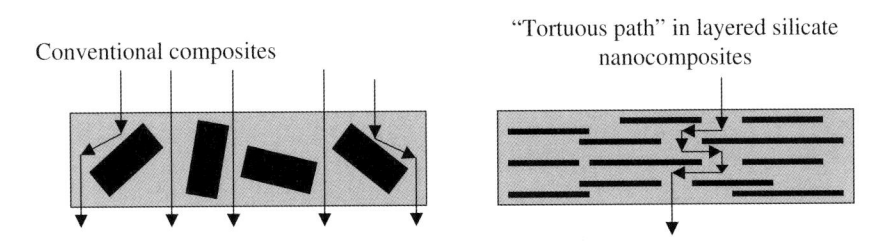

Fig. 3.30 Formation of tortuous path in polymer/nano-filler system.

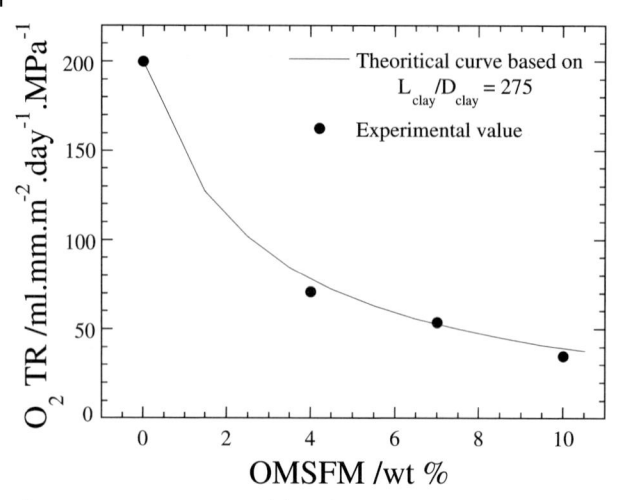

Fig. 3.31 Oxygen gas permeability of neat PLA and various PLACNs as a function of organoclay content measured at 20 °C and 90% relative humidity. The filled circles represent the experimental data. Theoretical fits based on Nielsen tortuosity model. Reprinted from [116], © 2003, American Chemical Society.

ability coefficient value, i.e. P_{PLACN}/P_{PLA} where P_{PLACN} and P_{PLA} stand for the nanocomposite and pure PLA permeability coefficients, respectively, has been plotted as a function of the wt% of mica. The curve fitting was achieved by using Nielsen's theoretical expression allowing the prediction of gas permeability as a function of the length and width of the filler particles as well as their volume fraction within the PLA-matrix (see Fig. 3.31).

In the Nielsen model [117], assuming platelets of length ($\cong L_{clay}$) and width ($\cong D_{clay}$) of the clay, which are dispersed parallel in the polymer matrix, then the tortuosity factor (τ) can be expressed as

$$\tau = 1 + (L_{clay}/2D_{clay})\phi_{clay} \qquad (1)$$

where ϕ_{clay} is the volume fraction of dispersed clay particles. Therefore, the relative permeability coefficient (P_{PLFNC}/P_{Neat}) is given by

$$P_{PLFNC}/P_{Neat} = \tau^{-1} = 1/[1 + (L_{clay}/2D_{clay})\phi_{clay}] \qquad (2)$$

where P_{PLFNC} and P_{Neat} are the permeability coefficients of PLFNC and neat polymer, respectively.

The H_2O-vapor permeability for the polyurethane urea (PUU)/MMT nanocomposites is presented by Manias and his colleagues in terms of P_c/P_o, i.e. the permeability coefficient of the nanocomposite (P_c) relative to that of the neat PUU (P_o) [118]. The nanocomposite formation results in a dramatic decrease in

H_2O-vapor transmission through the PUU sheet. The solid lines are based on the argument of the tortuosity model for the aspect ratio of 300 and 1000. A comparison between the experimental values and the theoretical model prediction suggests a gradual change in the effective aspect ratio of the filler (L_{clay}/D_{clay}).

3.10.4
Ionic Conductivity

Solvent-free electrolytes are of much interest because of their charge-transport mechanism and their possible applications in electrochemical devices. With this in mind, Vaia and his colleagues [119] have considered the preparation of polyethylene oxide (PEO)/MMT nanocomposites to fine tune the ionic conductivity of PEO. An intercalated nanocomposite prepared by melt intercalation of PEO (40 wt%) into Li^+-MMT (60 wt%) has been shown to enhance the stability of the ionic conductivity at lower temperature when compared to the more conventional PEO/$LiBF_4$ mixture. This improvement in conductivity is explained by the fact that PEO is not able to crystallize when intercalated, hence eliminating the presence of crystallites, non-conductive in nature. The higher conductivity at room temperature compared to conventional PEO/$LiBF_4$ electrolytes with a single ionic conductor character makes these nanocomposites promising new electrolyte materials.

Okamoto and his colleagues [120] have reported the correlation between the internal structure and ionic conductivity behavior of PMMA/MMT and PS/MMT nanocomposites having various dispersed morphology of the clay layers by using an impedance analyzer in the temperature range 90–150 °C. The nanocomposites with finer dispersion of the clay layers exhibit higher ionic conductivity than the other systems such as PMMA/MMT nanocomposite with stacking layer structure. The activation energy of the conductivity in finer dispersed morphology systems becomes larger than the other systems and the corresponding organoclay solids.

3.10.5
Optical Transparency

Although layered silicates are 1 µm in lateral size, they are just 1 nm thick. Thus, when single layers are dispersed in a polymer matrix, the resulting nanocomposite is optically clear in the visible region. Manias and his colleagues have reported the UV/visible transmission spectra of pure poly(vinyl alcohol) (PVA) and PVA/Na^+-MMT nanocomposites with 4 and 10 wt% MMT [121]. The spectra show that the visible region is not affected at all by the presence of the silicate layers and retains the high transparency of the PVA. For the UV wavelengths, there is strong scattering and/or absorption, resulting in very low transmission of the UV light. This behavior is not surprising as the typical MMT lateral sizes are 50–1000 nm.

Like PVA, various other polymers also show optical transparency after nanocomposites preparation with organoclay [122].

3.10.6
Physicochemical Phenomena

3.10.6.1 Biodegradability

Another most interesting and exciting aspect of nanocomposite technology is the significant improvements of biodegradability of biodegradable polymers after nanocomposite preparation with organoclay. Aliphatic polyesters are among the most promising materials for the production of environmentally friendly biodegradable plastics. Biodegradation of aliphatic polyester is well known, in that some bacteria degrade it by producing enzymes, which attack the polymer. Tetto and his colleagues [123] first reported some results about the biodegradability of nanocomposites based on PCL, where they found that the PCL/MMT nanocomposites showed improved biodegradability compared to pure PCL. According to them, the improved biodegradability of PCL after nanocomposite formation may be due to the catalytic role of the organoclay in the biodegradation mechanism. But it is still unclear how the clay increases the biodegradation rate of PCL.

Recently, Yamada and Okamoto et al. [124–126] first reported the biodegradability of neat PLA and PLA-based nanocomposites prepared with trimethyl octadecylammonium-modified MMT ($3C_1C_{18}$-MMT) including a detailed mechanism. The used compost was prepared from food waste and tests were carried out at a temperature of $58\pm2\,°C$. Figure 3.32a shows the real picture of the recovered samples of neat PLA and PLACN4 ($3C_1C_{18}$-MMT=4 wt%) from compost with time. The decreased molecular weight M_w and residual weight percentage R_w of the initial test samples with time are also reported in Fig. 3.32b. The biodegradability of neat PLA is significantly enhanced after PLSNC preparation. Within one month, both the extent of M_w and the extent of weight loss are at almost the same level for both PLA and PLACN4. However, after one month a sharp change occurs in the weight loss of PLACN4, and within two months, it is completely degraded in compost. The degradation of PLA in compost is a complex process involving four main phenomena, namely: water absorption, ester cleavage and formation of oligomer fragments, solubilization of oligomer fragments, and finally diffusion of soluble oligomers by bacteria [127]. Therefore, the factor, which increases the hydrolysis tendency of PLA, ultimately controls the degradation of PLA. They expect that the presence of terminal hydroxylated edge groups of the silicate layers may be one of the factors responsible for this behavior. In the case of PLACN4, the stacked (~ 4 layers) and intercalated silicate layers are homogeneously dispersed in the PLA matrix (from TEM image [126]) and these hydroxy groups start heterogeneous hydrolysis of the PLA matrix after absorbing water from compost. This process takes some time to start. For this reason, the weight loss and degree of hydrolysis of PLA and PLACN4 are almost same up to one month (see Fig. 3.32b). However, after one

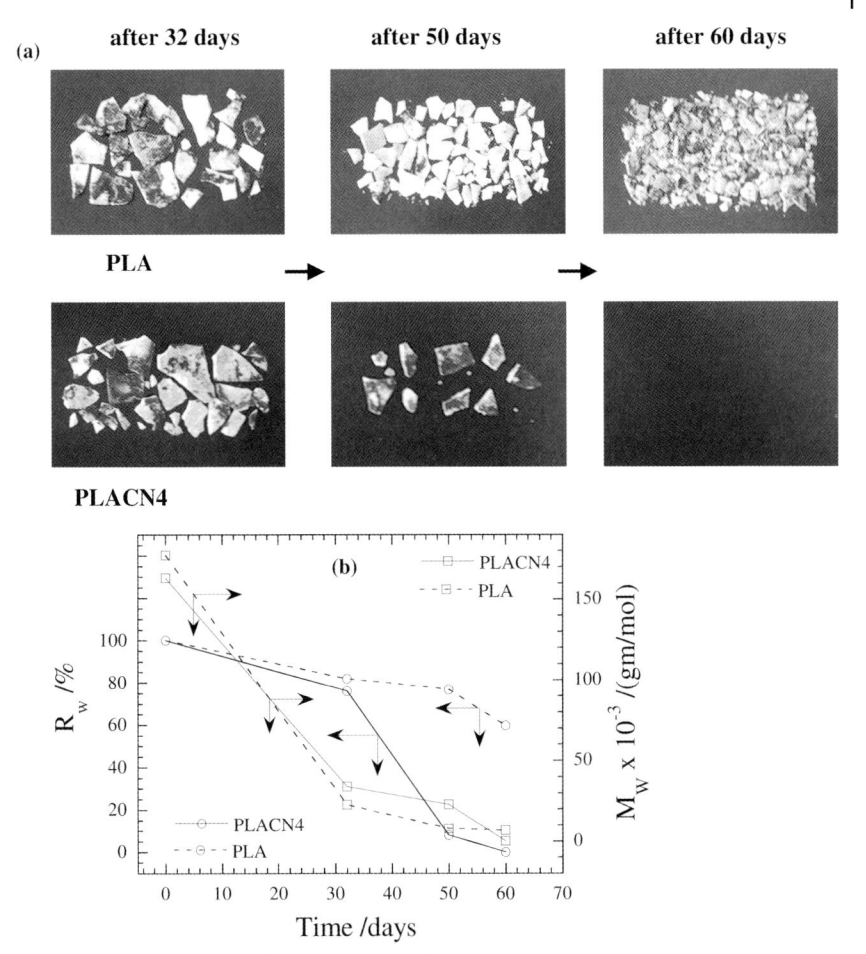

Fig. 3.32 (a) Real picture of biodegradability of neat PLA and PLACN4 recovered from compost with time. Initial shape of the crystallized samples was $3\times10\times0.1$ cm^3. (b) Time dependence of residual weight, R_w, and of matrix, M_w, of PLA and PLACN4 under compost at 58 ± 2 °C. Reprinted from [124], © 2002 American Chemical Society.

month there is a sharp weight loss in case of PLACN4 compared to that of PLA. That means one month is a critical value to start heterogeneous hydrolysis, and due to this type of hydrolysis the matrix becomes very small fragments which disappear with compost. This assumption was confirmed by conducting the same type of experiment with PLACN prepared by using the dimethyl dioctadecyl ammonium salt-modified synthetic mica which has no terminal hydroxylated edge group, and the degradation tendency was almost the same as with neat PLA [125].

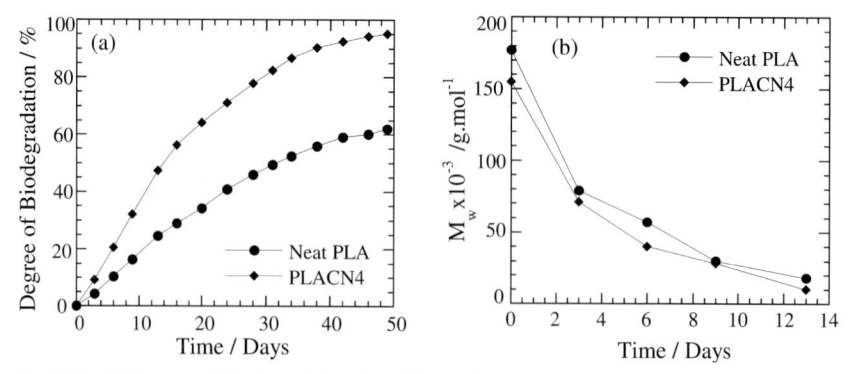

Fig. 3.33 (a) Degree of biodegradation (i.e. CO_2 evolution), and (b) time-dependent change of matrix M_w of neat PLA and PLACN4 (MEE clay = 4 wt%) under compost at $58 \pm 2\,°C$. Reprinted from [125], © 2004, Wiley-VCH.

They also conducted a respirometric test to study the degradation of the PLA matrix in a compost environment at $58 \pm 2\,°C$. For this test the compost used was made from bean-curd refuse, food waste, and cattle feces. Unlike weight loss, which reflects the structural changes in the test sample, CO_2 evolution provides an indicator of the ultimate biodegradability of PLA in PLACN4 (prepared with (N(cocoalkyl)N,N-[bis(2-hydroxyethyl))]-N-methylammonium-modified synthetic mica), i.e. mineralization, of the samples. Figure 3.33 shows the time dependence of the degree of biodegradation of neat PLA and PLACN4, indicating that the biodegradability of PLA in PLACN4 is enhanced significantly. The presence of organoclay may thus cause a different mode of attack on the PLA component, which might be due to the presence of hydroxy groups. Details of the mechanism of biodegradability are presented in the relevant literature [125, 126].

K. Okamoto and M. Okamoto also investigated the biodegradability of neat PBS before and after nanocomposite preparation with three different types of organoclay. They used alkylammonium or alkylphosphonium salts for the modification of pristine-layered silicates, and these surfactants are toxic for microorganisms [128].

Figure 3.34a shows the real pictures of recovered samples of neat PBS and various nanocomposites from the compost after 35 days. It may be clearly observed that many cracks have appeared in the nanocomposite samples compared to neat PBS. This observation indicates the improved degradability of nanocomposites in compost. This kind of fracture has an advantage for biodegradation because it makes it easy to mix with compost and creates much more surface area for further attack by microorganisms. It should be noted here that the extent of fragmentation is directly related to the nature of the organoclay used for the nanocomposite preparation. They also conducted the GPC (gel permeation chromatography) measurement of recovered samples from compost, and found that the extent of molecular weight loss was almost the same for all

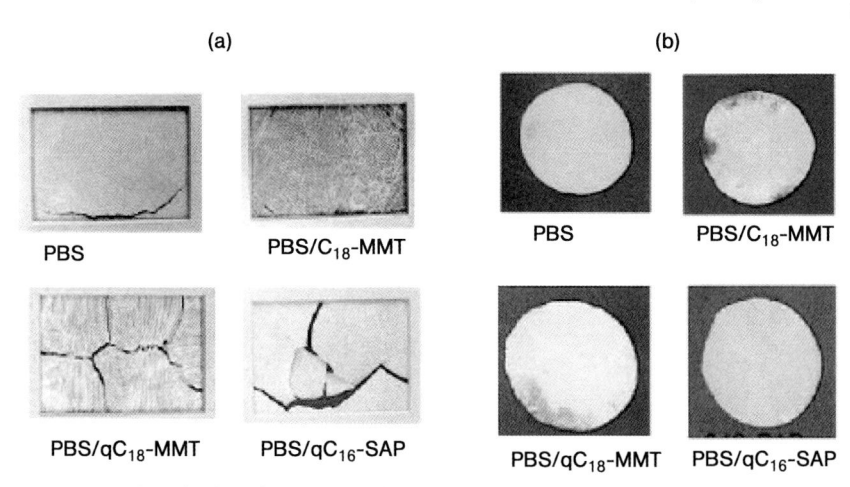

Fig. 3.34 Biodegradability of neat PBS and various nanocomposites sheets (a) under compost, and (b) under soil field. Reprinted from [128], © 2003, John Wiley & Sons, Inc.

samples. This result indicates that the extent of hydrolysis of PBS in the pure state or OMLS-filled systems is the same in compost.

Except for the PBS/qC$_{16}$-SAP (n-hexadecyl tri-n-butyl phosphonium cation-modified saponite) system, the degree of degradation is not different for other samples. This observation indicates that MMT or alkylammonium cations, and at the same time other properties, have no effect on the biodegradability of PBS. The accelerated degradation of a PBS matrix in the presence of qC$_{16}$-SAP may be due to the presence of alkylphosphonium surfactant. This kind of behavior is also observed in the case of PLA/MMT nanocomposite systems.

They also observed the nature of the degradation of PBS and various nanocomposites under a soil field. This experiment was conducted for 1, 2, and 6 months. After 1 and 2 months, there is no change in the nature of the surface of the samples, but after 6 months black or red spots appeared on the surface of the nanocomposite samples. Figure 3.34b represents results of the degradation of neat PBS and various nanocomposite sheets recovered from a soil field after 6 months. They reported that these spots on the sample surface are due to fungus attack, because when these parts were put into the slurry we observed clear growth of fungus. These results also indicate that nanocomposites exhibit the same or a higher level of biodegradability compared with the PBS matrix.

3.10.6.2 Photodegradation

Hiroi, Okamoto et al. [129] first reported the photodegradability of neat PLA and the corresponding PLA-based nanocomposite prepared by using organically modified layered titanate as a new nano-filler. One of the features of this material is its photocatalytic reactivity similar to that of titania (TiO$_2$). The photo-

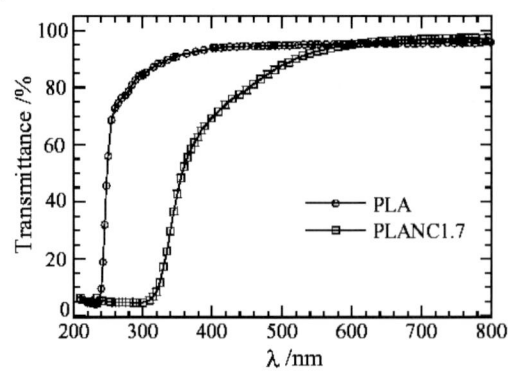

Fig. 3.35 UV/VIS transmission spectra of neat PLA and nanocomposite (PLANC1.7). Reprinted from [129], © 2004, Wiley-VCH.

Table 3.6 GPC results of sample recovered from weathermeter after 300 h. Reprinted from [129], © 2004, Wiley-VCH.

Samples	$M_w \times 10^{-3}$ (g mol^{-1})	M_w/M_n	$M_w/M_w^{0\ a)}$
PLA	198	1.53	0.94
PLANC1.7	93.7	1.89	0.68
PLANC3.9	86.3	1.86	0.76

a) M_w^0 is molecular weight before test.

catalytic reactions of anatase-TiO_2, such as evolution of hydrogen gas from water or oxidative degradation of organic compounds, have attracted intense research interest because of their possible application to the conversion of solar energy into chemical energy [130].

Figure 3.35 shows the UV/vis transmission spectra of pure PLA and nanocomposite (PLANC1.7). The spectra show that the VIS region (> ~400 nm) is changed with increasing absorbence by the presence of titanate layers compared with neat PLA. For UV wavelengths, there is strong absorption up to 320 nm, resulting in 0% transmittance. This significant change in the spectra may indicate the occurrence of the photodegradation of the PLA matrix. To confirm this, some preliminary experiments were conducted on the photodegradation of PLANCs under a sunshine weathermeter at 60 °C. After 300 h, there was no change in the nature of the sample surfaces of neat PLA, however, the surface color of nanocomposite samples became yellow and/or light brown. Table 3.6 shows the GPC measurement of recovered samples from the test. The drop in M_w accompanied by broadening of M_w/M_n indicates that enhancement of degradation of PLA in the titanate-filled system has occurred.

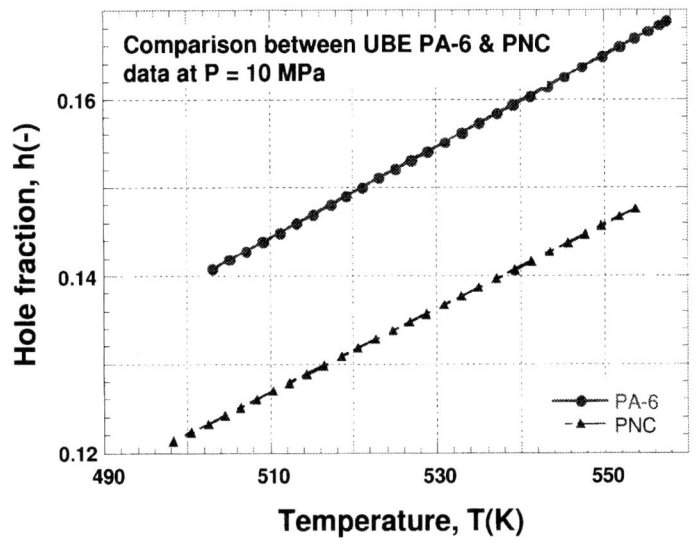

Fig. 3.36 Hole fraction versus temperature for N6-based nanocomposite. Reprinted from [131], © 2001, VSP.

3.10.6.3 **Pressure–Volume–Temperature (PVT) Behavior**

In Nylon 6-based nanocomposites (N6CNs) the reduction of free volume by 12 to 17% was reported through a pioneering effort by Utracki et al. [131]. Figure 3.36 shows the temperature dependence of the free volume reduction by the addition of nanoclay particles. They speculated that the N6 chains are absorbed on the solid clay surface and the first few absorbed layers are immobilized. This loss of mobility translates, in their model, into a free volume loss. The effect was simulated by a large adhesion ratio $\varepsilon_{12}^*/\varepsilon_{11}^*$ ($=313.54$ kJ mol^{-1}/32.09 kJ mol^{-1}) which is an energetic interaction parameter between polymer–clay and polymer–polymer. They also reported the value of the free volume reduction for the polystyrene-based nanocomposite system [132]. In case of 4 wt% organoclay loading, the reduction is much lower (4 to 6%) than that of N6-based nano-composites due to the weak interaction between polymer and silicate surfaces.

Some literatures are also available [133] related to the PVT dependencies of commercial PP melt and its nano-composites including small amount of a compatibilizer. Another recent approach is positron annihilation lifetime spectroscopy for evaluation of the free volume fraction in PLFNCs [134].

3.11
Computer Simulation

Ginzburg and Balazs developed simple models that describe the liquid crystalline ordering in the polymer–platelet systems [135–138]. They combined density

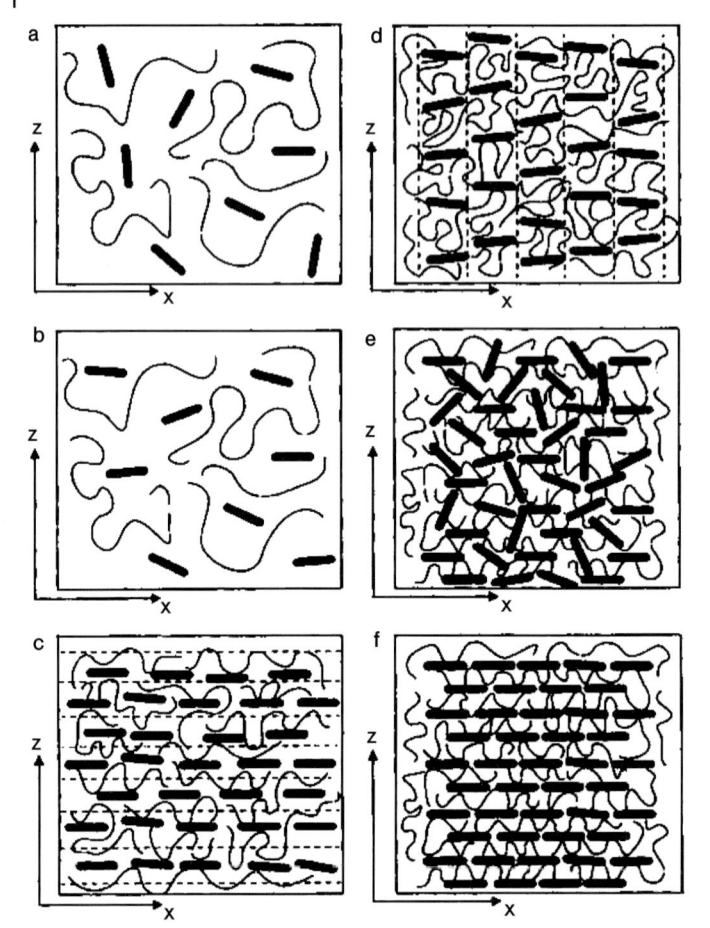

Fig. 3.37 Mesophases of oblate uniaxial particles dispersed in a polymer: (a) isotropic (I), (b) nematic (N), (c) smectic (S), (d) columnar (Col), (e) plastic solid (house-of-cards) (PS), and (f) crystal (Cr). The nematic director **n** in ordered phases is aligned along the Z axis, the disks lie in the X–Y plane. Dashed lines show smectic layers (c) and columns (d). Reprinted from [136], © 2000, American Chemical Society.

functional theory (DFT) with a self-consistent field model (SCF) to calculate the phase behavior of thin, oblate colloidal particles that are coated with surfactants and dispersed in a polymer melt. These coated particles represent organoclay sheets. By intergrafting the two methods, they investigated the effect of the surfactant characteristics (grafting density, ρ_{gr} and length, N_{gr}) and the polymer–intercalant interaction energy on the polymer/clay phase diagram (see Fig. 3.37) [136]. Depending on the values of these critical parameters and the clay volume fraction, ϕ, the system can be in an isotropic or nematic phase (exfoliated sys-

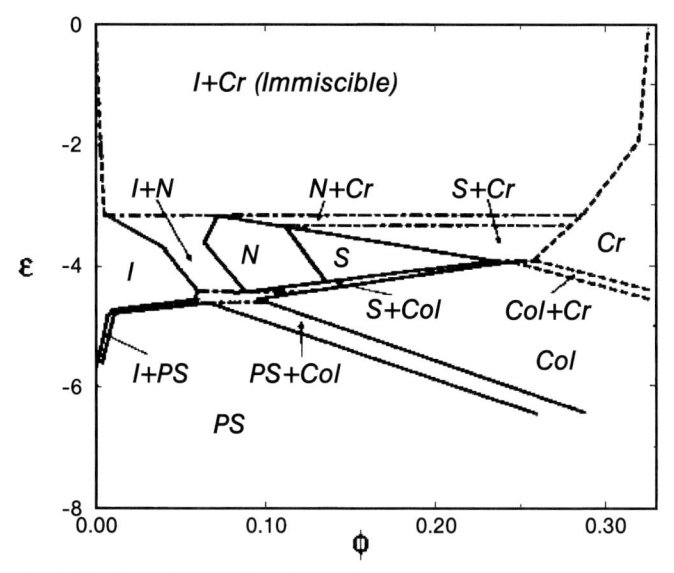

Fig. 3.38 Phase diagram for polymer/clay mixture. Here ε is the sticker-clay adhesion energy. Reprinted from [137], © 2000, Wiley-VCH.

tem). The system can also form a smectic, crystal, columnar, or house-of-cards plastic solid as well as a two-phase immiscible mixture. Using this model they isolated conditions that lead to the stabilization of the exfoliated nanocomposite system and to the narrowing of the immiscible two-phase regions.

Furthermore, they extended their study to take end-functionalized polymer chains into consideration [137]. The parameter ε characterizes the interaction between the terminal functional group and the surface. For small negative ε, the system is completely immiscible. As the sticker/surface attraction is increased, the system exhibits isotropic and nematic morphologies at relatively low clay volume fractions ($\phi < 0.12$). The latter structures correspond to thermodynamically stable, exfoliated nanocomposite. Finally, when the surface/sticker attraction is increased even further, the system exhibits a plastic solid at low ϕ, and a columnar phase at very high ϕ (see Fig. 3.38). The resulting phase diagrams can be used as guidelines for nanocomposite formation with thermodynamically stable morphologies.

3.12
Processing Operations

Flow-induced internal structural change occurs in both shear and elongational flow, but the changes almost differ from each other, as judged from the above results on $\eta_E(\dot{\varepsilon}_0; t)$ and $3\eta_0(\dot{\gamma}; t)$ (see Fig. 3.26). Thus, with these rheological features of the PLFNCs and the characteristics of each processing operation, the decision as to which process type should be selected for a particular nanocomposite for the enhancement of its mechanical properties can be made.

For example, the strong strain-induced hardening in $\eta_E(\dot{\varepsilon}_0;t)$ is requisite for withstanding the stretching force during the processing, while the rheopexy in $3\eta_E(\dot{\gamma}_0;t)$ suggests that for such PLFNC a promising technology is the processing in a confined space such as injection molding where shear force is crucial.

3.12.1
Foam Processing Using sc-CO_2

Very recently, the first successful nanocomposite foam, processed by using supercritical CO_2 as a physical foaming agent, appeared through a pioneering effort by Okamoto and his colleagues [139, 140].

Figure 3.39 shows the typical results of SEM images of the fracture surfaces of the intercalated polycarbonate (PC)/layered silicate nanocomposites (PCCNs) and PC/SMA blend (matrix) without clay foamed at 160 °C under different isobaric saturation conditions of supercritical CO_2 (10, 14 and 18 MPa) [141, 142]. PC/SMA foams exhibit the polygon closed-cell structures having pentagonal and hexagonal faces, which express the most energetically stable state of polygon cells. Such a foam structure was obtained probably because these foams belong to the polymeric foams having high gas phase volume (>0.6). Obviously, under

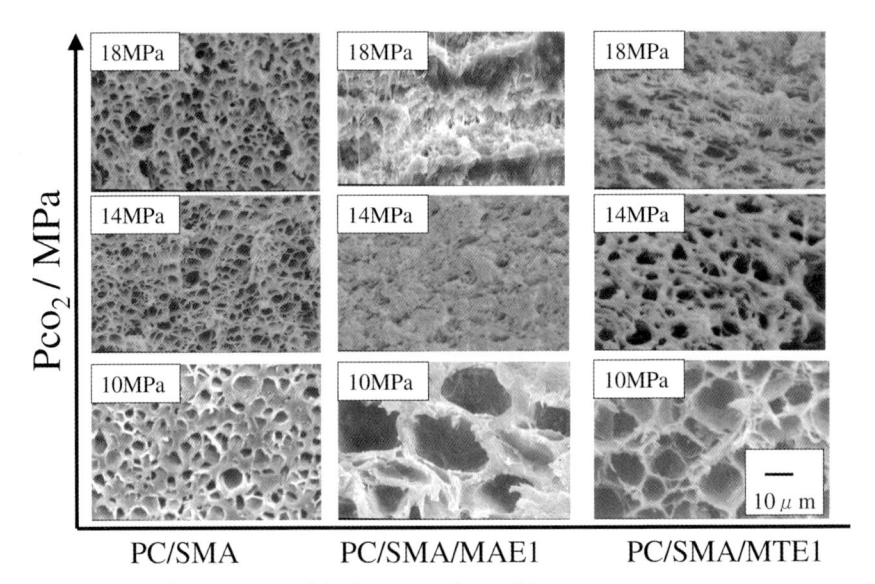

Fig. 3.39 Typical SEM images of the fracture surfaces of the PCCNs and PC/SMA blend without clay foamed at 160 °C under different isobaric conditions (10, 14 and 18 MPa). Reprinted from [142], © 2006, Wiley-VCH.

low saturation CO_2 pressure (~10 MPa) both PCCN foams exhibit large cell size, indicating that the dispersed clay particles hinder CO_2 diffusion by creating a maze or a more tortuous path, as discussed in the literature [1]. However, high CO_2 pressure (~18 MPa) provides a large supply of CO_2 molecules, which can subsequently form a large population of cell nuclei upon depressurization. The PC/SMA/MAE1 ($2C_{12}C_{18}$-fluorohectrite) (including 1 wt% of organoclay) foam shows smaller cell size, i.e., larger cell density compared to PC/SMA foam, suggesting that the dispersed clay particles act as nucleating sites for cell formation and lowering of d with clay. The incorporation of nano-filler hinders CO_2 diffusion and simultaneously induces heterogeneous nucleation because of a lower activation energy barrier compared to homogeneous nucleation. The characterization of the interfacial tension between bubble and matrix was carried out by using modified classical nucleation theory [143].

In Table 3.7, the values of the interfacial tension of the systems are summarized [142]. We can see that the PC/SMA system has a slightly large value (17.3 mJ m^{-2})

Table 3.7 Interfacial tension of the systems calculated using Suh and Colton's theory.

System	P_{CO_2}/MPa	γ/mJ m^{-2}
PC/SMA-CO_2	10	17.3
PC/SMA/MAE-CO_2 [a]	10	14.6
PC/SMA/MTE-CO_2 [a]	10	16.0

[a] MAE: dimethyl dioctadecyl ammonium cation-modified synthetic fluorohectorite.
MTE: methyl trioctyl ammonium cation-modified synthetic fluorohectorite.

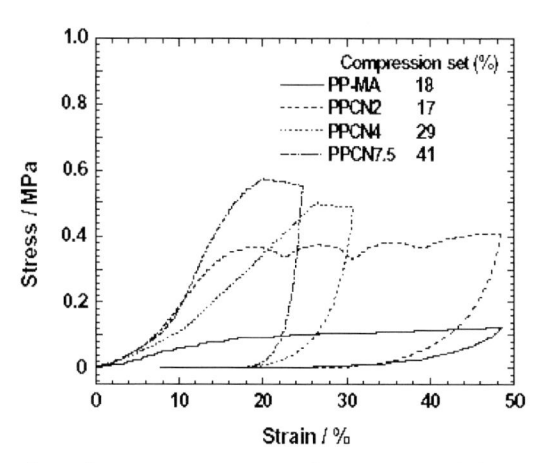

Fig. 3.40 Stress–strain curves and strain recovery behavior of the PP-based nanocomposites (PPCNs). Reprinted from [1], © 2003, Rapra Technology Ltd.

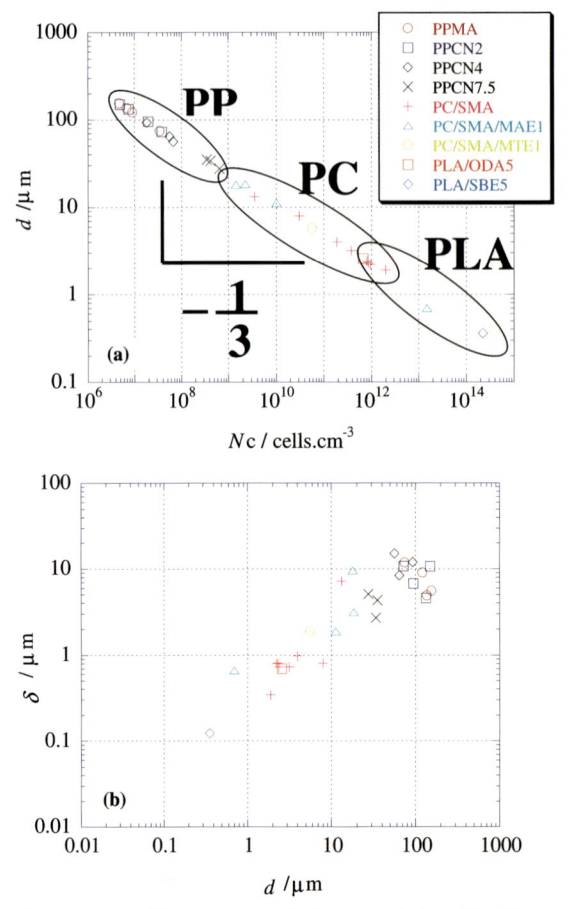

Fig. 3.41 (a) Cell size versus cell density and (b) cell wall thickness (δ) versus cell size for PCCN systems. For the comparison, here we show other nanocomposite foams obtained by a series of our recent studies [142, 146]; PP: polypropylene-based, PLA: polylactide-based nanocomposite foams.

compared to that of PC/SMA/MAE1 (14.6 mJ m^{-2}) and PC/SMA/MTE1 (C$_{13}$C$_8$-fluorohectrite) (16.0 mJ m^{-2}) with low CO$_2$ pressure (10 MPa). These estimated values of γ are in good agreement with that of another poly-(methyl methacrylate) (PMMA)–CO$_2$ system (10–20 mJ m^{-2}) [144]. The value for the PC/SMA system decreases with increasing CO$_2$ pressure, as expected, while PC/SMA/MAE1 show a constant value with increasing pressure. This trend reflects the relative importance of heterogeneous nucleation, which dominates over homogeneous nucleation in the event that the amount of CO$_2$ available for bubble nucleation is limited.

Figure 3.40 shows the stress–strain curves and the strain recovery behavior of the PP-based nanocomposite (PPCN) foams [139] in the compression mode at a constant strain rate of 5% min^{-1}. The nanocomposite foams exhibit a high modulus compared to neat PP-g-MA foam. The residual strain is 17% for PPCN2 (including 2 wt% of organoclay) as well as neat PP foam, providing the excellent strain recovery and the energy dissipation mechanism, probably with the house-of-cards structure formation in the cell wall, which enhance the mechanical properties of the nanocomposites like a spruce wood which is close to a right-handed helix [145].

Figure 3.41 shows the cell size (d), the cell density (N_c) and the mean cell wall thickness (δ) relations of the nanocomposite foams obtained by a series of recent studies [142, 146]. In the case of nanocomposite foams, the cell wall thickness becomes 2–6 times thicker than that of neat polymer foams due to the spherical cell shape [140] caused by the high modulus of the materials during processing. The controlled structure of the nanocomposite foams from microcellular ($d \cong 20$ μm and $N_c \cong 1.0 \times 10^9$ cell cm^{-3}) to nanocellular ($d \cong 200$ nm and $N_c \cong 1.0 \times 10^{14}$ cell cm^{-3}).

Recently, some literature has also become available [147–150] related to the reactive extrusion foaming of various nanocomposites.

3.12.2
Electrospinning Processing

Fibers and nanofibers of N6CN (diameter 100–500 nm) were electrospun from 1,1,1,3,3,3-hexafluoro-2-propanol (HFIP) solution and collected as non-woven fabrics or as aligned yarns [151]. The electrospinning process resulted in highly aligned MMT particles and Nylon 6 crystallites. Cylindrical-shaped fibers and nanofibers and ribbon-shaped fibers were also found in the products (see Fig. 3.42). Using polystyrene (PS)/MMT and poly(methyl methacrylate) (PMMA)/MMT nanocomposite solutions, the same approach was conducted [152, 153]. The fiber diameters of PS-based nanocomposite were adjusted from 4 μm to

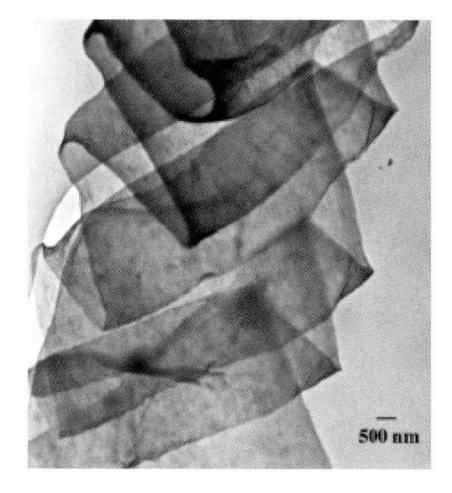

Fig. 3.42 TEM micrograph of a ribbon-shaped nanofiber. Reprinted from [151], © 2002, Elsevier Science.

Fig. 3.43 SEM images of electrospun PS fibers with 4 wt% MMT at different PS concentration: (a) 20 wt%, (b) 15 wt%, (c) 10 wt%, (d) 7.5 wt%, and (e) 5 wt%. Reprinted from [152], © 2006, American Chemical Society.

150 nm by changing the tetrahydrofuran (THF) and dimethylformamide (DMF) solution concentration (see Fig. 3.43). The addition of MMT to the spinning solution produced fibers with a highly aligned MMT layer structure at low MMT content (4 wt%). The agglomerated tactoids of MMT were observed at higher concentration. The shear modulus of the fiber was enhanced and the glass transition temperature was increased by 20 °C due to the MMT dispersion. Electrospinning can be expected to align other nano-fillers such as carbon nanotubes.

3.12.3
Porous Ceramic Materials via PLSNCs

A new route for the preparation of porous ceramic material from thermosetting epoxy/clay nanocomposite was first demonstrated by Brown et al. [154]. This route offers attractive potential for diversification and application of the PLFNCs. Okamoto and coworkers have reported the results on the novel porous ceramic material via burning of the PLA/MMT system (PLACN) [155]. In the PLACN containing 3 wt% inorganic clay. The SEM image of the fracture surface of porous ceramic material prepared from simple burning of the PLACN in a furnace up to 950 °C is shown in Fig. 3.44. After complete burning, as seen in the figure, the PLACN becomes a white mass with porous structure. The bright lines in the SEM image correspond to the edge of the stacked silicate layers. In the porous ceramic material, the silicate layers form a house-of-cards structure,

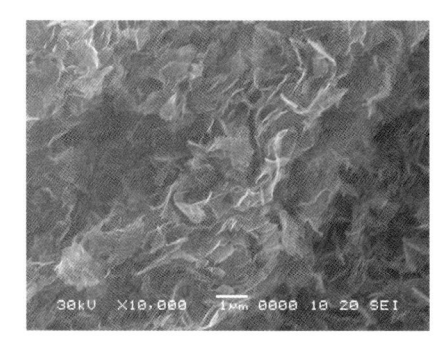

Fig. 3.44 SEM image of porous ceramic material after coating with platinum layer (∼10 nm thickness). Reprinted from [155], © 2002, American Chemical Society.

which consists of large plates having length ∼1000 nm and thickness ∼30–60 nm. This implies that the further stacked platelet structure is formed during burning. The material exhibits an open-cell type structure having a 100–1000 nm diameter void, BET surface area 31 $m^2 g^{-1}$ and low density of porous material of 0.187 g ml^{-1} estimated by the buoyancy method. The BET surface area value of MMT is 780 $m^2 g^{-1}$ and that of the porous ceramic material is 31 $m^2 g^{-1}$, suggesting that about 25 MMT plates are stacked together. When MMT is heated above 700 °C (but below 960 °C) first all the OH groups are eliminated from the structure and thus MMT is decomposed into a non-hydrated aluminosilicate. This transformation radically disturbs the crystalline network of the MMT, and the resulting diffraction pattern is indeed often typical of an amorphous (or non-crystalline) phase. The estimated rough value of the compression modulus (K) is in the order of ∼1.2 MPa, which is five orders of magnitude lower than the bulk modulus of MMT (∼10^2 GPa) [1]. In the stress–strain curve, the linear deformation behavior is nicely described in the early stage of the deformation, i.e., the deformation of the material closely resembles that of ordinary polymeric foams [156]. This open-cell type porous ceramic material consisting of the house-of-cards structure is expected to provide the strain recovery and excellent energy dissipation mechanism after unloading in the elastic region up to 8% strain, probably each plate bends like a leaf spring. This porous ceramic material is a new material possessing elasticity and is very lightweight. This new route for the preparation of a porous ceramic material via burning of nanocomposites can be expected to pave the way for a much broader range of applications of the PLSNCs. This porous ceramic material closely relates to an excellent insulator property for flame retardancy of PLFNCs [1]. The flame behavior must derive from the morphological control of the shielding properties of the graphitic clay created during polymer ablation.

3.13
Future Prospects

Development of the PLFNCs is one of the latest evolutionary steps of polymer technology. The PLFNCs offer attractive potential for diversification and application of conventional polymeric materials. Some PLFNCs are already commercially available and applied in industrial products.

Biodegradable polymers based on nanocomposites have a great deal of future promise for potential applications as high-performance biodegradable materials. These are entirely new types of materials based on plant and natural materials (organoclay). When disposed of in compost, they are safely decomposed into CO_2, water, and humus through the activity of microorganisms. The CO_2 and water will become corn or sugarcane again through plant photosynthesis. Undoubtedly, their unique properties, originating from the controlled nanostructure, pave the way to a much broader range of applications (already commercially available through Unitika Ltd., Japan), and open a new dimension for plastics and composites. The major impact will be at least a decade away.

References

1 Okamoto M., *Polymer/Layered Silicate Nanocomposites*, Rapra Review Report No 163, p. 166, Rapra Technology Ltd., London (2003).
2 Usuki A., Kojima Y., Okada A., Fukushima Y., Kurauchi T., Kamigaito O., *J. Mater. Res.* 8, 1174 (1993).
3 Ray S.S., Okamoto M., *Prog. Polym. Sci.* 28, 1539 (2003).
4 Gao F., *Mater. Today*, 7, 50 (2004).
5 Usuki A., Hasegawa N., Kato M., *Adv. Polym. Sci.*, 179, 135 (2005).
6 National Lead Co., *US Patent* 2 531 396 (1950).
7 Union Oil Co., *US Patent* 3 084 117 (1963).
8 Unitika Ltd., *Japanese Kokai Patent* 109 998 (1976).
9 Fukushima Y., Inagaki S., *J. Inclusion Phenom.* 5, 473 (1987).
10 Kojima Y., Usuki A., Kawasumi M., Okada A., Fukushima Y., Karauchi T., Kamigaito O., *J. Mater. Res.* 8, 1185 (1993).
11 Jose-Yacaman M., Rendon L., Arenas J., Puche M.C.S., *Science*, 273, 223 (1996).
12 Hermosin M.C., Cornejo J., *Clays Clay Miner.*, 34, 591 (1986).
13 Krishnamoorti R., Vaia R.A., Giannelis E.P., *Chem. Mater.* 8, 1728 (1996).
14 Sinha Ray S., Yamada K., Okamoto M., Ogami A., Ueda K., *Chem. Mater.*, 15, 1456 (2003).
15 Su S.P., Jiang D.D., Wilkie C.A., *Polym. Degrad. Stab.*, 83, 321 (2004).
16 Su S.P., Jiang D.D., Wilkie C.A., *Polym. Degrad. Stab.*, 83, 333 (2004).
17 Vuillaume P.Y., Glinel K., Jonas A.M., Laschewsky A., *Chem. Mater.*, 15, 3625 (2003).
18 Acosta E.J., Deng Y., White G.N., Dixon J.B., McInnes K.J., Senseman S.A., Frantzen A.S., Simanek E.E., *Chem. Mater.*, 15, 2903 (2003).
19 Fischer H., *Mater. Sci. Eng.*, C23, 763 (2003).
20 Vaia R.A., Ishii H., Giannelis E.P., *Chem. Mater.* 5, 1694 (1993).
21 Vaia R.A., Giannelis E.P., *Macromolecules* 30, 7990 (1997).
22 Vaia R.A., Giannelis E.P., *Macromolecules* 30, 8000 (1997).
23 Sinha Ray S., Yamada K., Okamoto M., Ueda K., *J. Nanosci. Nanotechnol.*, 3, 503 (2003).

24 Lagaly G., *Clay Miner.*, 16, 1 (1970).
25 Vaia R. A., Teukolsky R. K., Giannelis E. P., *Chem. Mater.* 6, 1017 (1994).
26 Li Y. Q., Ishida H., *Langmuir*, 19, 2479 (2003).
27 Osman M. A., *J. Mater. Chem.*, 13, 2359 (2003).
28 Paul D. R., Zeng Q. H., Yu A. B., Lu G. O., *J. Colloid Interface Sci.*, 292, 462 (2005).
29 Yoshida O., Okamoto M., *Macromol. Rapid Commun.*, 27, 751 (2006).
30 Sinha Ray S., Okamoto K., Okamoto M., *Macromolecules*, 36, 2355 (2003).
31 Giannelis E. P., Krishnamoorti R., Manias E., *Adv. Polym. Sci.* 138, 107 (1999).
32 Okamoto M., Morita S., Kim Y. H., Kotaka T., Tateyama H., *Polymer* 42, 1201 (2001).
33 Sinha Ray S., Maiti P., Okamoto M., Yamada K., Ueda K., *Macromolecules* 35, 3104 (2002).
34 Yang D. K., Zax D. B., *J. Chem. Phys.* 110, 5325 (1991).
35 VanderHart D. L., Asano A., Gilman J. W., *Chem. Mater.*, 13, 3796 (2001).
36 Totha R., Coslanicha A., Ferronea M., Fermeglia M., Pricl S., Miertus S., Chiellini E., *Polymer*, 45, 8075 (2004).
37 Sinsawat A., Anderson K. L., Vaia R. A., Farmer B. L., *J. Polym. Sci. Part B: Polym. Phys.*, 41, 3272 (2003).
38 Kuppa V., Menakanit S., Krishnamoorti R., Manias E., *J. Polym. Sci. Part B: Polym. Phys.*, 41, 3285 (2003).
39 Zeng Q. H., Yu A. B., Lu G. Q., Standish R. K., *Chem. Mater.*, 15, 4732 (2003).
40 Sheng N., Boyce M. C., Parks D. M., Rutledge G. C., Abes J. I., Cohen R. E., *Polymer*, 45, 487 (2004).
41 Hasegawa N., Okamoto H., Kato M., Usuki A., Sato N., *Polymer*, 44, 2933 (2003).
42 van Olphen, H. *An Introduction to Clay Colloid Chemistry*, Wiley, New York (1977).
43 Zerda A. S., Caskey C., Lesser A., *Macromolecules*, 36, 1603 (2003).
44 Zhao Q., Samulski E. T., *Macromolecules*, 36, 6967 (2003).
45 Yang K., Ozisik R., *Polymer* 47, 2849 (2006).
46 Lee E. C., Mielewski D. F., Baird R. J., *Polym. Eng. Sci.*, 44, 1773 (2004).
47 Zhao L., Li J., Guo S., Du Q., *Polymer* 47, 2460 (2006).
48 Shao W., Wang Q., Li K., *Polym. Eng. Sci.*, 45, 451 (2005).
49 Yoshida O., Okamoto M., *J. Polym. Eng.*, in press (2006).
50 Saito T., Okamoto M., Hiroi R., Yamamoto M., Shiroi T., *Macromol. Rapid. Commun.*, 27, 1472 (2006).
51 Maiti P., Nam P. H., Okamoto M., Kotaka T., Hasegawa N., Usuki A., *Macromolecules*, 35, 2042 (2002).
52 Nam P. H., Maiti P., Okamoto M., Kotaka T., Hasegawa N., Usuki A., *Polymer*, 42, 9633 (2001).
53 Maiti P., Nam P. H., Okamoto M., Kotaka T., Hasegawa N., Usuki A., *Polym. Eng. Sci.*, 42, 1864 (2002).
54 Khare R., Pablo J. J. de, Yethiraj A., *Macromolecules* 29, 7910 (1996).
55 Sheng N., Boyce M. C., Parks D. M., Rutledge G. C., Abes J. I., Cohen R. E., *Polymer*, 45, 487 (2004).
56 Carrado K. A., in *Handbook of Layered Materials*, Auerbach S. M., Carrado K. A., Dutta P. K. (Eds.), Marcel Dekker, New York (2004), pp. 1–38.
57 Sato H., Yamagishi A., Kawamura K., *J. Phys. Chem. B* 105, 7990 (2001).
58 Seo Y. S., Ichikawa Y., Kawamura K., *Mater. Sci. Res. Int.* 5, 13 (1999).
59 Tamura K., Setsuda H., Taniguchi M., Yamagishi A., *Langmuir*, 15, 6915 (1999).
60 Medellin-Rodriguez F. J., Burger C., Hsiao B. S., Chu B., Vaia R. A., Phillips S., *Polymer*, 42, 9015 (2001).
61 Maiti P., Okamoto M., *Macromol. Mater. Eng.*, 288, 440 (2003).
62 Landis W. J., *Connect. Tissue Res.*, 35, 1 (1996).
63 Lincoln D. M., Vaia R. A., Wang Z. G., Hsiao B. S., *Polymer*, 42, 1621 (2001).
64 Lincoln D. M., Vaia R. A., *Macromolecules*, 37, 4554 (2004).
65 Kim G. M., Lee D. H., Hoffmann B., Kressler J., Stoppelmann G., *Polymer*, 42, 1095 (2001).
66 Stern T., Wachtel E., Marom G., *J. Polym. Sci., Part B Polym. Phys.*, 35, 2429 (1997).

67 Fratzl P., Gupta H.S., Paschalis E.P., Roschger P., *J. Mater. Chem.*, 14, 2115 (2004).
68 Priya L., Jog J.P., *J. Polym. Sci. Part B: Polym. Phys.*, 40, 1682 (2002).
69 Shah D., Maiti P., Gunn E., Schmidt D.F., Jiang D.D., Batt C.A., Giannelis E.P., *Adv. Mater.*, 16, 1173 (2004).
70 Kader M.A., Nah C., *Polymer*, 45, 2237 (2004).
71 Buckley J., Cebe P., Cherdack D., Crawford J., Ince B.S., Jenkins M., Pan J., Reveley M., Washington N., Wolchover, *Polymer*, 47, 2411 (2006).
72 Dillon D.R., Tenneti K.K., Li C.Y., Ko F.K., Sics I., Hsiao B.S., *Polymer*, 47, 1678 (2006).
73 Sinha Ray S., Yamada K., Okamoto M, Ueda K., *Polymer*, 44, 857 (2003).
74 Chen J.S., Poliks M.D., Ober C.K., Zhang Y., Wiesner U., Giannelis E.P., *Polymer*, 43, 4895 (2002).
75 Sinha Ray S., Okamoto K., Okamoto M., *Macromolecules*, 36, 2355 (2003).
76 Okamoto K., Sinha Ray S., Okamoto M., *J. Polym. Sci. Part B: Polym. Phys.*, 41, 3160 (2003).
77 Utracki L.A., *Polymer Alloys and Blends: Thermodynamics and Rheology*, Hasser Publishers, New York (1990).
78 Krishnamoorti R., Giannelis E.P., *Macromolecules* 30, 4097 (1997).
79 Fornes T.D., Yoon P.J., Keskkula H., Paul D.R., *Polymer* 42, 9929 (2001).
80 Hoffman B., Dietrich C., Thomann R., Friedrich C., Mulhaupt R., *Macromol. Rapid Commun.* 21, 57 (2000).
81 Ren J., Silva A.S., Krishnamoorti R., *Macromolecules* 33, 3739 (2000).
82 Mitchell C.A., Krishnamoorti R., *J. Polym. Sci. Part B Polym. Phys.* 40, 1434 (2002).
83 Lepoittevin B., Devalckenaere M., Pantoustier N., Alexandre M., Kubies D., Calberg C., Jerome R., Dubois P., *Polymer* 43, 1111 (2002).
84 Solomon M.J., Almusallam A.S., Seefeld K.F., Somwangthanaroj S., Varadan P., *Macromolecules* 34, 1864 (2001).
85 Galgali G., Ramesh C., Lele A., *Macromolecules* 34, 852 (2001).
86 Lele A., Mackley M., Galgali G., Ramesh C., *J. Rheol.* 46, 1091 (2002).
87 Sinha Ray S., Maiti P., Okamoto M., Yamada K., Ueda K., *Macromolecules* 35, 3104 (2002).
88 Sinha Ray S., Yamada K., Okamoto M., Ueda K., *Polymer* 44, 6631 (2003).
89 Okamoto K., Sinha Ray S., Okamoto M., *J. Polym. Sci. Part B: Polym. Phys.* 41, 3160 (2003).
90 Krishnamoorti R., Yurekli K., *Curr. Opin. Colloid Interface Sci.* 6, 464 (2001).
91 Okamoto M., Taguchi H., Sato H., Kotaka T., Tatayama H., *Langmuir* 16, 4055 (2000).
92 Williams M.L., Landel R.F., Ferry J.D., *J. Am. Chem. Soc.* 77, 3701 (1955).
93 Nam P.H., Master Thesis, Toyota Technological Institute (2001).
94 Cox W.P., Merz E.H., *J. Polym. Sci.*, 28, 619 (1958).
95 Okamoto M., Sato H., Taguchi H., Kotaka T., *Nippon Rheology Gakkaishi*, 28, 201 (2000).
96 Okamoto M., Nam P.H., Maiti P., Kotaka T., Hasegawa N., Usuki A., *Nano Lett.*, 1, 295 (2001).
97 Kotaka T., Kojima A., Okamoto M., *Rheol. Acta*, 36, 646 (1997).
98 Kojima Y., Usuki A., Kawasumi M., Okada A., Kurauchi T., Kamigaito O., Kaji K., *J. Polym. Sci. Part B: Polym. Phys.*, 33, 1039 (1995).
99 Yalcin B., Cakmak M., *Polymer*, 45, 2691 (2004).
100 Medellin-Rodriguez F.J., Burger C., Hsiao B.S., Chu B., Vaia R.A., Phillips S., *Polymer*, 42, 9015 (2001).
101 Bafna A., Beaucage G., Mirabella F., Mehta S., *Polymer*, 44, 1103 (2003).
102 Roe R., *Methods of X-ray and Neutron Scattering in Polymer Science*, Oxford University Press, New York (2000), p. 199.
103 Bafna A., Beaucage G., Mirabella F., Skillas G., Sukumaran S., *J. Polym. Sci., Part B: Polym. Phys.* 39, 2923 (2001).
104 Kim J.W., Kim S.G., Choi H.J., Jhon M.S., *Macromol. Rapid Commun.*, 20, 450 (1999).
105 Kim J.W., Noh M.H., Choi H.J., Lee D.C., Jhon M.S., *Polymer*, 41, 1229 (2000).

106 Park J.H., Lim Y.T., Park O.O., *Macromol. Rapid Commun.*, 22, 616 (2001).
107 Lim Y.T., Park J.H., Park O.O., *J. Colloid Interface Sci.*, 245, 198 (2002).
108 Huang X., Brittain W.J., *Macromolecules* 34, 3255 (2001).
109 Zanetti M., Camino G., Thomann R., Mulhaupt R., *Polymer* 42, 4501 (2001).
110 Camino G., Sgobbi R., Colombier S., Scelza C., *Fire Mater.* 24, 85 (2000).
111 Gilman J.W., Ksahiwagi T., Giannelis E.P., Manias E., Lomakin S., Lichtenhan J.D., Jones P., in *Chemistry and Technology of Polymer Additives*, Ch. 14, Al-Malaika S., Golovoy A., Wilkie C.A. (Eds.), Blackwell Science, Oxford (1999).
112 Gilman J.W., *Appl. Clay Sci.* 15, 31 (1999).
113 Gilman J.W., Jackson C.L., Morgan A.B., Harris Jr R., Manias E., Giannelis E.P., Wuthenow M., Hilton D., Phillips S.H., *Chem. Mater.* 14, 3776 (2002).
114 Yano K., Usuki A., Okada A., Kurauchi T., Kamigaito O., *J. Polym. Sci. Part A Polym. Chem.* 31, 2493 (1993).
115 Yano K., Usuki A., Okada A., *J. Polym. Sci. Part A Polym. Chem.* 35, 2289 (1997).
116 Sinha Ray S., Yamada K., Okamoto M., Ogami A., Ueda K., *Chem. Mater.* 15, 1456 (2003).
117 Nielsen L., *J. Macromol. Sci. Chem.* A1(5), 929 (1967).
118 Xu R., Manias E., Snyder A.J., Runt J., *Macromolecules* 34, 337 (2001).
119 Vaia R.A., Vasudevan S., Krawiec W., Scanlon L.G., Giannelis E.P., *Adv. Mater.* 7, 154 (1995).
120 Okamoto M., Morita S., Kotaka T., *Polymer* 42, 2685 (2001).
121 Strawhecker K.E., Manias E., *Chem. Mater.* 12, 2943 (2000).
122 Manias E., Touny A., Strawhecker K.E., Lu B., Chung T.C., *Chem. Mater.* 13, 3516 (2001).
123 Tetto J.A., Steeves D.M., Welsh E.A., Powell B.E., *ANTEC'99*, 1628 (1999).
124 Sinha Ray S., Okamoto M., Yamada K., Ueda K., *Nano. Lett.*, 2, 1093 (2002).
125 Sinha Ray S., Yamada K., Ogami A., Okamoto M., Ueda K., *Macromol. Rapid Commun.*, 23, 943 (2002).
126 Sinha Ray S., Yamada K., Okamoto M., Ueda K., *Macromol. Mater. Eng.*, 288, 936 (2003).
127 Liu J.W., Zhao Q., Wan C.X., *Space Med. Med. Eng.* 14, 308 (2001).
128 Okamoto K., Sinha Ray S., Okamoto M., *J. Polym. Sci. Part B: Polym. Phys.*, 41, 3160 (2003).
129 Hiroi R., Sinha Ray S., Okamoto M., Shiroi T., *Macromol. Rapid. Commun.*, 25, 1359 (2004).
130 Fujishima A., Honda K., *Nature*, 37, 238 (1972).
131 Simha R., Utracki L.A., Garcia-Rejon A., *Composite Interfaces*, 8, 345 (2001).
132 Tanoue S., Utracki L.A., Garcia-Rejon A., Tatibouet J., Cole K.C., Kamal M.R., *Polym. Eng. Sci.*, 44, 1046 (2004).
133 Utracki L.A., Simha R., *Macromolecules*, 37, 10123 (2004).
134 Wang Y., Wu Y., Zhang H., Zhang L., Wang B., Wang Z., *Macromol. Rapid Commun.*, 25, 1973 (2004).
135 Ginzburg V.V., Balazs A.C., *Macromolecules*, 32, 5681 (1999).
136 Ginzburg V.V., Singh C., Balazs A.C., *Macromolecules*, 33, 1089 (2000).
137 Ginzburg V.V., Balazs A.C., *Adv. Mater.*, 12, 1805 (2000).
138 Ginzburg V.V., Gendelman O.V., Manevitch L.I., *Phys. Rev. Lett.*, 86, 5073 (2001).
139 Okamoto M., Nam P.H., Maiti M., Kotaka T., Nakayama T., Takada M., Ohshima M., Usuki A., Hasegawa N., Okamoto H., *Nano Lett.*, 1, 503 (2001).
140 Nam P.H., Okamoto M., Maiti P., Kotaka T., Nakayama T., Takada M., Ohshima M., Hasegawa N., Usuki A., *Polym. Eng. Sci.*, 42(9), 1907 (2002).
141 Mitsunaga M., Ito Y., Okamoto M., Sinha Ray S., Hironaka K., *Macromol. Mater. Eng.*, 288, 543 (2003).
142 Ito Y., Yamashita M., Okamoto M., *Macromol. Mater. Eng.*, 291, 773 (2006).
143 Colton J.S., Suh N.P., *Polym. Eng. Sci.*, 27, 485 (1987).
144 Goel S.K., Beckman E.J., *Polym. Eng. Sci.*, 34, 1137 (1994).
145 Fratzl P., *Curr. Opin. Colloid Interface Sci.*, 8, 32 (2003).
146 Ema Y., Ikeya M., Okamoto M., *Polymer*, 47, 5350 (2006).

147 Cao X., Lee L. J., Widya T., Macosko C., *Polymer*, 46, 775 (2005).
148 Chandra A., Gong S., Turng L. S., Gramann P., Cordes H., *Polym. Eng. Sci.*, 45, 52 (2005).
149 Taki K., Yanagimoto T., Funami E., Okamoto M., Ohshima M., *Polym. Eng. Sci.*, 44, 1004 (2004).
150 Strauss W., D'Souza N. A., *J. Cellular Plastics*, 40, 229 (2004).
151 Fong H., Liu W., Wang C. S., Vaia R. A., *Polymer*, 43, 775 (2002).
152 Ji Y., Li B., Ge S., Sokolov J. C., Rafailovich M. H., *Langmuir*, 22, 1321 (2006).
153 Wang M., Hsieh A. J., Rutledge G. C., *Polymer*, 46, 3407 (2005).
154 Brown J. M., Curliss D. B., Vaia R. A., *Proc. of PMSE*, Spring Meeting, San Francisco, California (2000), p. 278.
155 Sinha Ray S., Okamoto K., Yamada K., Okamoto M., *Nano. Lett.*, 2, 423 (2002).
156 Gibson L. J., Ashby M. F. (Eds.), *Cellular Solids*, Pergamon Press, New York (1988), p. 8.

4
Polymeric Dispersants

Frank Pirrung and Clemens Auschra

4.1
Dispersant Application Fields

The term "polymeric dispersant" is not connected to a scientifically precise definition. In a broad sense, it can comprise all kinds of functional polymers, which, based on a designed interfacial activity, have found practical use as additives to create effects on the stabilization and rheology of disperse systems, such as in paints, inks, plastics, textiles, adhesives, construction materials, agrochemicals, cosmetics, lubricants and fuels. In a more specific definition, the term "polymeric dispersant" relates to applications in which solid particles are to be dispersed and stabilized in a liquid system. Here by far the main application is to disperse color pigments and particulate fillers for use in coatings. This chapter will focus on this main application area of polymeric pigment dispersants used in coatings and closely related applications such as inks.

Within the coatings industry, major sub-segments include automotive coatings, industrial coatings, architectural coatings and decorative coatings. Especially automotive coatings have been a major driver for the development and growth of modern pigment dispersants, because this industry has continuously defined increasing technical requirements combined with economic demands. A prominent example is the use of high-performance organic pigments, which has steadily increased since the mid-1980s and which was technically enabled only by the introduction of specially designed pigment dispersants [1].

4.2
Pigment Dispersants Function and Requirements

4.2.1
Dispersant Application Criteria

In coatings, pigment dispersants are use for two main tasks. The first is to work as a "processing aid" to facilitate and rationalize the manufacture of pigment dispersions, which are used as tinting pastes for paints. A well-designed pigment dispersant will help to speed up the pigment grind process by improving pigment wetting and by stabilizing the resulting dispersed particle suspension. Another important aspect is good control on millbase rheology, which allows the manufacture of easily processable tinting pastes with higher pigment loads.

The second main task is to work as a "performance additive" to achieve and guarantee the best possible color properties of a pigment in the final coating. The optical quality of a pigmented paint is critically related to the capability to realize a stable pigment dispersion in the liquid paint and to preserve the optimum dispersion into the cured paint film. Any deviation from an ideal dispersion of primary pigment particles, e.g. by flocculation during processing, application or curing of the paint can results in a loss of color strength and usually leads to optical paint defects such as reduced gloss, seeding, settling, flooding or floating of pigments.

The specific advantages of polymeric dispersants for organic pigments can be understood by considering the effects which are caused by the small particle size and the relative nonpolar nature of particles. Depending on the intended end use as a transparent or more opaque pigment, the dimensions of the primary crystals of organic pigments are typically in the range 50–500 nm. This results in a much larger total particle surface to be wetted, as compared with traditional inorganic pigments with particle sizes typically well above 1 µm.

Second, organic pigment crystals usually have a reduced adsorption capability for organic molecules, due to non-uniform surfaces. On a single crystal, faces of high and low polarity can be distinguished, whereas only the high-polarity surfaces will allow good adsorption of stabilizing compounds.

Third, Brownian motion has more influence on flocculation with smaller particles: a classical surfactant-like dispersant dynamically wets and de-wets the pigment surface, hence the Brownian contribution to flocculation is more prominent and additional stabilization mechanisms such as steric hindrance become necessary [2].

Finally, the organic pigment surfaces have fewer (or no) ionic groups compared with inorganic particles, so that stabilizing interactions with the resin polymers used in the paints and inks are much weaker. Organic pigments contain mainly chromophore groups consisting of C-, H-, O- and N-atoms that are usually conjugated and bear optional substituents [3]. Industrially important classes of organic pigments include copper phthalocyanines, azo condensation pigments, quinacridones and diketopyrrolopyrroles. The pigment producer usually tries to enhance the wettability and resin adsorption capability of the pigments by post-treatment of the surface with polar pigment derivatives or other compounds which modify the particle surface for better dispersibility in the intended end-use formulations. Such compounds for surface treatments are typically polycyclic molecules carrying one or several acid or amino groups and corresponding salts thereof. Due to the growing demand for organic pigments with small particle sizes and difficult-to-wet surfaces, the stabilization of the finely dispersed particles is a challenge.

For practical use, a dry pigment powder first has to be dispersed to form a liquid dispersed pigment concentrate, which is then used as a tinting paste in the formulation of the final paint. The manufacture of the pigment dispersion is done in appropriate high-shear milling equipment in the presence of solvent and dispersant. This millbase recipe can in addition contain low molecular weight resins to support the formation of a stable pigment dispersion. Depending on the quality of pigment dispersion, the pigment particles can exist in different states of agglomeration (Fig. 4.1). Ideally, the pigment dispersion process tries to achieve a stable dispersion of primary particles and to prevent the flocculation of these particles at any later stage of the paint processing (Fig. 4.2).

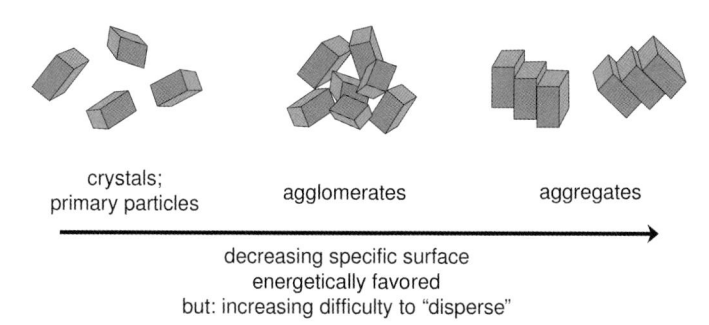

Fig. 4.1 Schematic presentation of different states of interaction of pigment particles.

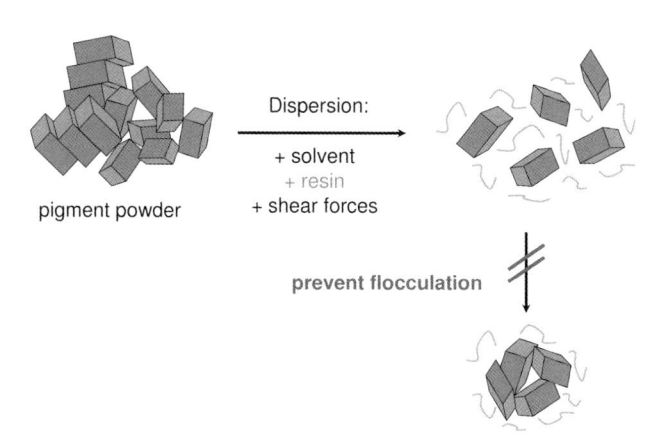

Fig. 4.2 Schematic presentation of pigment dispersion process.

4.2.2
Design of Dispersants by Stabilization Mechanism

The stabilization of a particle dispersed in a liquid medium can take place by either steric or electrostatic mechanisms (Fig. 4.3). Steric stabilization implies that dispersant polymers are adsorbed on the particle surface and at the same time polymers extend from the surface to form a shell of well-solvated polymer chains. Such an array provides efficient particle stabilization by osmotic forces which effectively prevent further approach of individual particles. The minimum molecular weight necessary for effective steric stabilization [4] to occur lies around 5000 g mol^{-1}. In case of electrostatic stabilization mechanisms [5], also lower molecular weight polyelectrolytes have been found to be effective, because in this case the stabilizing effect is first built up by a surface charge. The adsorption of dissociating salt structures on the particle's surface causes the build-up of a diffuse electric double layer, which provides an electrostatic repulsion if two particles approach each other. This mechanism only works in liquids with high dielectric constant and therefore in practice is only relevant for aqueous formulations. For aqueous coatings only anionic polyelectrolyte dispersants are relevant, because the usual aqueous coatings resins are anionically stabilized emulsion polymers.

Steric stabilizing polymeric dispersants are the most important product class for solvent-based applications, especially when organic pigments are used in high-quality paints.

Both stabilization mechanisms are based on the principle that adsorbed dispersants increase the energy of repulsion and thereby the potential energy barrier (E_a), which can prevent the formation of agglomerates, if E_a it is larger than the kinetic energy of the particles (Fig. 4.4). Adsorbed and solvated polymers provide better stabilization by a stronger and long-range repulsion than low molecular weight surfactants.

Therefore, an important aspect of the polymeric dispersants is their high molecular weight. Such structures are important not only to accommodate all the necessary anchoring groups, but also to ensure a sufficient distance between the particles in the dispersed stage. However, there are practical limits on the molecular weight of the dispersant since paint and ink formulations are liquid media and all components of this solution have to show sufficient solubility.

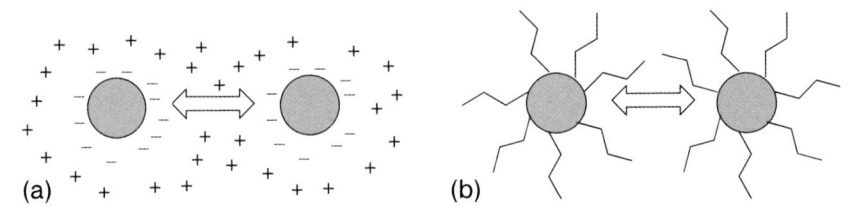

Fig. 4.3 Stabilization of particles. (a) Electrostatic and (b) steric principles.

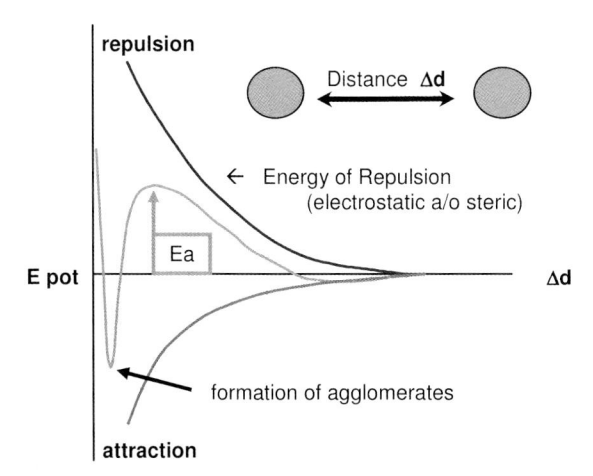

Fig. 4.4 Schematic potential energy diagram. The curve with the label "Ea" represents the total potential energy of a two-particle system.

Another factor which limits the molecular weight of dispersants is the formation of bridges and interactions between particles or with the resin or other dispersant molecules, leading to an undesirable increase in viscosity and formation of a pseudoplastic structure. On the other hand, too low molecular weight polymers are unable to build up sufficient steric stabilization. From experience, the best results are obtained with molecular weights (M_n) between 5000 and 25 000 g mol^{-1}, measured by standard gel permeation chromatography (GPC) against defined standards, such as linear polystyrene.

4.2.3
Characteristic Data of Pigment Dispersants

GPC gives useful information on molecular weight and its distribution. Nevertheless, as a relative method which provides only apparent molecular weights according to the hydrodynamic volume in a given solvent, it can be only a rough criterion for comparison of different dispersants, especially when chemically and structurally different polymers are compared.

For the practical physical characterization and quality control of polymeric dispersants, several other parameters are determined. These include solid content [6], acid value [7], amine value [8], color [9] and specific gravity [10]. The acid value and amine value usually give good indications about the type (aminic or acidic) and amount of pigment-affinic anchoring groups present in the dispersant polymer.

4.2.4
Chemical Architecture of High Molecular Weight Dispersants

The state-of-the-art described in the following sections shows that different polymer architectures have proven to be suitable for effective pigment stabilization (Fig. 4.5):
- random copolymers: polyacrylates;
- graft copolymers: e.g. modified styrene–maleic anhydride copolymers;
- comb copolymers: e.g. polyacrylate copolymers with macromonomers;
- block copolymers: polyurethanes, polyesters and polyacrylates;
- hybrid structures, block, comb, branched: polyurethanes and polyesters.

Some of the possible monomer arrangements are shown schematically in Fig. 4.6.

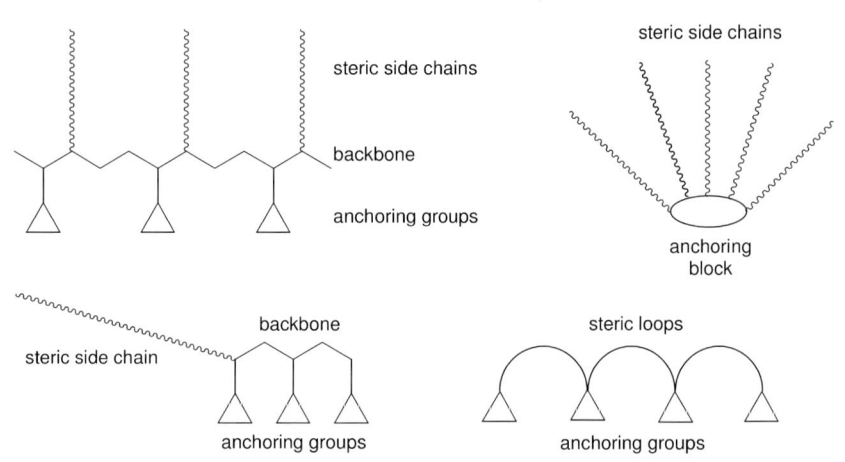

Fig. 4.5 Polymer architecture.

Random (linear)	ABACDBBCAC...
Block (linear)	AAAAAA-BBBBBB...
Gradient (linear)	AAAAAA-B-AAA-BB-AA-BBBB-A-BBBBBB...
Hybrid	AAAABBBCACA-XXXYYZYYZ...
Random (branched)	AABCBAADAB.....
	B
	C
	Y A
	X A
	X C
	Y
Grafted	ABACDBBAACA....

(A, B, C, D and X, Y, Z = monomer units)

Fig. 4.6 Monomer arrangements.

All these polymer architectures have the common feature that they are composed of defined regions of functionality. These can be identified as the backbone of the polymer carrying (a) pigment-affinic (or anchoring) groups, (b) steric side-chains and (c) groups to improve the compatibility with the resin system. Often, monomer units can be identified that simply build up the backbone of the polymer and their main function is to give space to the pendant functional groups. For practical reasons, the steric side-chains and the compatibility-enhancing groups cannot be absolutely distinguished.

Polar polymers play the dominant role for aqueous-based systems. The amount of polar side-chains (i.e. polyether chains) on the polymer backbone determines whether the dispersant is emulsifiable or totally soluble in water. Additionally, water compatibility can be achieved by deprotonation of free carboxylic acid groups on the polymer backbone by amines or caustic solutions.

4.2.5
Anchoring Groups

Anchoring groups provide adsorption to the pigment surface and determine on the type of pigment on which the dispersant can be used. Anchoring principles can be based on acid–base reactions (ionic forces), hydrogen bonding, complex formation, polar interaction (dipole–dipole) and even London dispersion forces. A wide range of chemistries can be used for anchoring groups [11] and can be classified into the following categories:
- acidic anchoring groups: carboxyl, sulfate and phosphate, including their neutralized forms;
- electroneutral groups: polyether, aromatic rings (substituted phenyl and naphthyl moieties) and heterocyclic derivatives thereof;
- aminic anchoring groups (and their quaternized forms): typical examples aniline, indole, imidazole, imide, morpholine, oxazoline, piperazine, polyethylenimine, pyridine, trialkylamines and triazole.

These functional moieties are usually chemically connected through spacers to the backbone of the polymeric dispersant.

Aminic groups, often combined with aromatic rings, have proven to be effective anchoring groups in polymeric dispersants, as these moieties can readily form hydrogen bridges, cause dipole–dipole interactions and induce London dispersive forces (listed in reducing strength) between the dispersant and the organic pigment surface. As these forces lead to much weaker adsorption than the ionic forces that predominate on inorganic surfaces, dispersants for organic particles should contain a larger number of anchoring groups on each polymeric molecule. They can be distributed block-wise or statistically over the polymeric chain to ensure complete adsorption of the dispersant. These anchoring groups also function on inorganic surfaces of pigments and extenders, since they can be polarized so that they also interact with the charged surface.

In order to obtain dispersants with a wide application range with regard to different pigment types, combinations of the mentioned anchoring categories can be used. For the best results, practice has shown that several different anchoring groups per polymer molecule should be incorporated to achieve stable adsorption on the pigment surface.

4.3
Classical Polymeric Pigment Dispersants

4.3.1
Polyurethane-based Dispersants

The term polyurethane-based (PUR) dispersants describes a family of products that appeared on the market on an industrial scale in the 1980s [12]. They were mainly designed for solvent-based coating systems. Due to their high molecular weights, in the range 5000–25 000 g mol^{-1} (M_n against polystyrene standards) and the presence of polar functional groups, they are supplied as 25–60% solutions in organic solvents (esters, aromatics, alcohols), in order to obtain free-flowing delivery forms at room temperature. Attempts to develop universal dispersants for use in both solvent- and water-based systems were not very successful in the market, due to the presence of organic co-solvents. The polyurethanes chemically embrace polymers which are not only composed of urethane linking groups, but also a large number of ester and ether functionalities, which build up the polymeric backbone. A general structure is shown in Fig. 4.7.

4.3.1.1 Polyisocyanate Resins
The main element is the polyisocyanate prepolymer, indicated with a circular symbol, which can be a technical NCO group-containing resin, as they are used as hardeners in two-component polyurethane lacquers. They are characterized

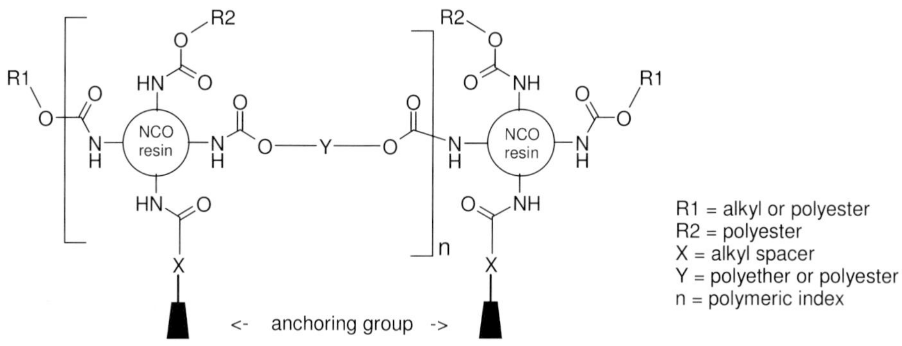

R1 = alkyl or polyester
R2 = polyester
X = alkyl spacer
Y = polyether or polyester
n = polymeric index

Fig. 4.7 Schematic structure of a polyurethane dispersant.

by an oligomeric structure, bearing between two and six reactive NCO groups on each molecule, and are produced by addition reactions of the monomeric isocyanates [13]. Typical examples of suitable prepolymers are polyisocyanates, which are available through addition of asymmetric diisocyanates to polyols (Fig. 4.8) or by a biuret reaction from diisocyanates (Fig. 4.9). Another important class are oligomers, which are the result of cyclization of diisocyanates with iso-

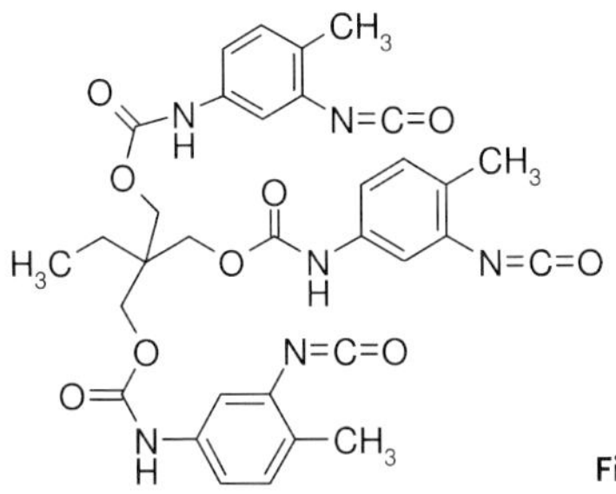

Fig. 4.8 Addition product of diisocyanate to a polyol.

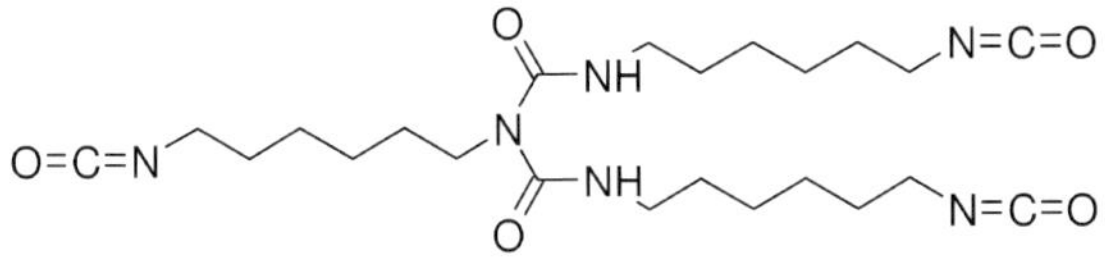

Fig. 4.9 Biuret reaction product from diisocyanate.

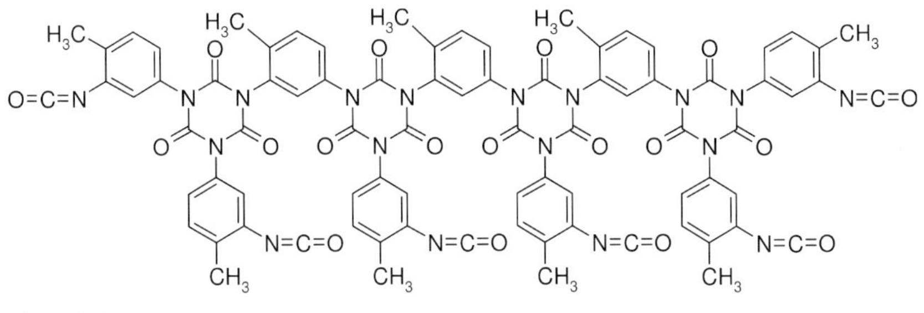

Fig. 4.10 Aromatic isocyanurate.

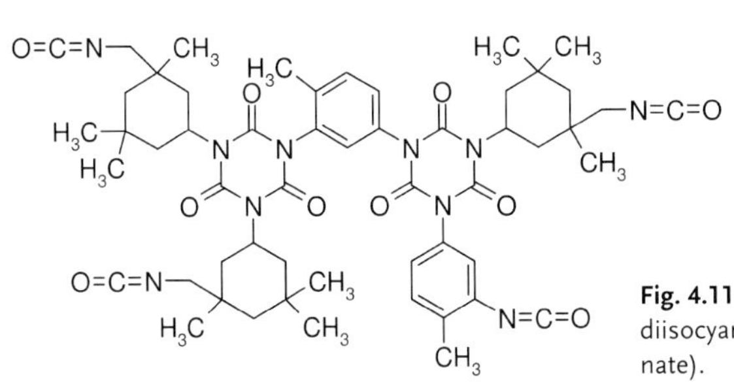

Fig. 4.11 Mixed isocyanurate (tolylene diisocyanate and isophorone diisocyanate).

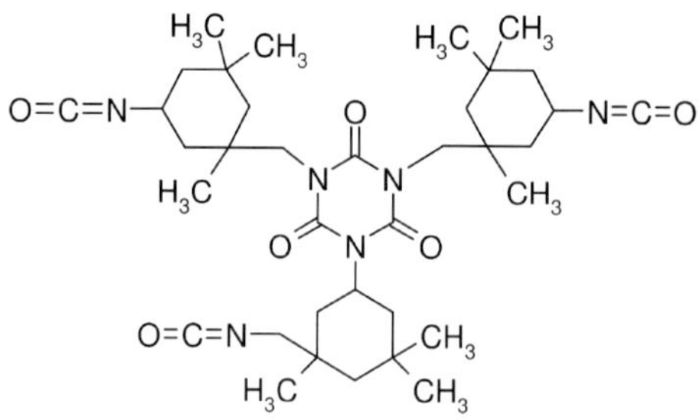

Fig. 4.12 Aliphatic isocyanurate.

cyanurate base structures (Figs. 4.10–4.12). The diisocyanate units may be aliphatic or aromatic in nature, but can also be combined to obtain mixed oligomers. These versatile building blocks are available on a large scale and in their delivery forms typically are not pure compounds, but a mixture of formulas with similar structures indicated above with multiple NCO functionality. Most of these products are supplied as solutions in inert solvents, such as alkyl acetates, due to the intrinsic viscosities to give better handling.

4.3.1.2 Cross-linkers

The repetitive element in the schematic polymeric structure is generated by interconnecting the oligomeric NCO-resin units by difunctional cross-linkers (Y in Fig. 4.7) with, for example, molecular weights between 500 and 4000 g mol^{-1} (M_n). These cross-linkers can be based on dihydroxy-functional polyethers (ethylene oxide, propylene oxide, block copolymers thereof and polytetrahydrofuran), and also polymeric hydroxycarboxylic acids and dicarboxylic acids [12, 14]. Figure 4.13 shows schematically a few examples of cross-linkers.

The use of diamines is usually limited, due to their fast addition reaction to the NCO groups and uncontrolled gel formation. The polarity of the backbone and hence compatibility with coating formulations can be adjusted from polar to nonpolar by using either polyethers, mixed polyester–polyethers or just polyester moieties (e.g. Fig. 4.14) [15].

The amount of cross-linker to build up the polymeric backbone can vary within certain limits and is dependent on the application. Since the addition to the NCO-oligomer is statistically determined, a certain degree of branching on the NCO-bearing moieties cannot be avoided and is typical for this kind of disper-

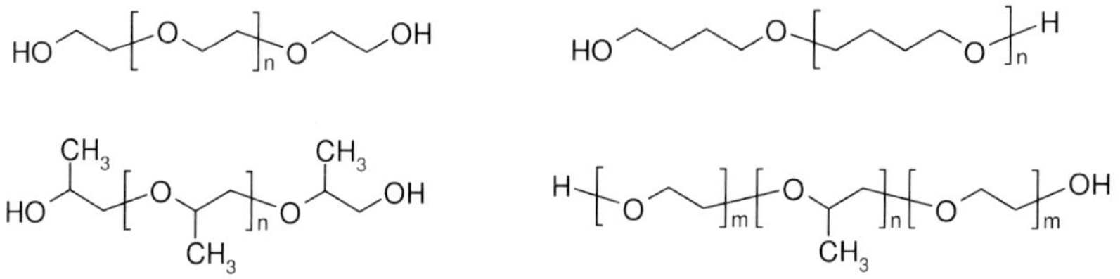

Fig. 4.13 Cross-linkers based on polyether structures.

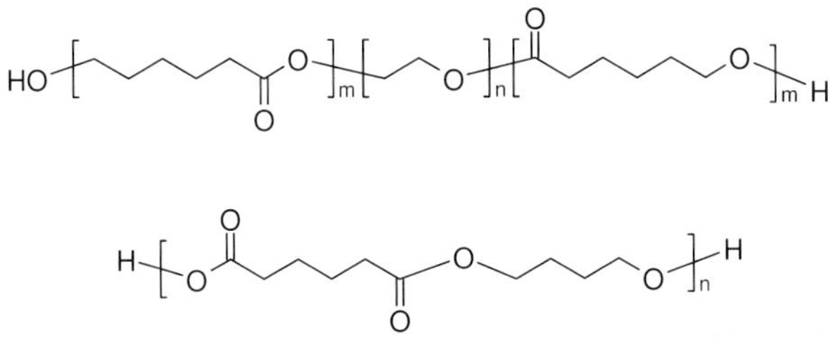

Fig. 4.14 Cross-linkers based on polyesters and polyester-block-polyethers.

sant. The branching feature can contribute to an efficient covering of the pigment surface by a dense brush of adsorbed polymers, which will lead to favorable low viscosities in the formulated pigment paste. Trifunctional cross-linkers, based on triol-initiated polycaprolactone chains, have also been described, leading to an even higher degree of branching in the polymer backbone [15].

4.3.1.3 Monofunctional Side-chains

In addition to the difunctional cross-linkers, monofunctional polymeric chains with molecular weights between 500 and 5000 g mol^{-1} need to be introduced into the backbone to induce sufficient steric hindrance (osmotic pressure) around the polymer for effective particle stabilization. These can be monofunctional polyethers, polyesters and mixed structures.

A valuable tool to obtain well-defined monohydroxy-functional polyesters is ring-opening polymerization (ROP) of lactones. For this purpose, monohydroxy initiators (e.g. longer aliphatic alcohols and monofunctional polyethers) or amines are reacted with lactones [e.g. (substituted) δ-valerolactone and ε-caprolactone] at temperatures between 120 and 180 °C. The reaction is efficiently accelerated in the presence of transition metal catalysts (Sn, Zr, Ti, Zn complexes) or protic acids (p-toluenesulfonic acid) to activate the acyl–oxygen bond [16].

Lactone polymerization proceeds via a ring-opening process and can lead to high molecular weight polyesters with low polydispersity (typically $M_n/M_w < 1.5$) at high reaction rates, and good control of terminating units can be achieved. ROP of lactones is an addition polymerization, where one end of a growing polymer chain acts as a reactive center that can react with additional cyclic monomers to propagate the chain. Selective acyl–oxygen cleavage of the lactone ring is the first step in the sequence. The resulting hydroxy end-group on the growing chain reacts further according to the same ring-opening mechanism and the polymer formation proceeds step by step (Fig. 4.15) [17].

The starting alcohols or amines can be chosen from a wide variety of readily available compounds. Examples of useful alcohols for this reaction are fatty alcohols such as oleyl alcohol, synthetic C_{10}–C_{18} oxo alcohols, cyclohexanol, 2-ethylhexanol, phenoxyethanol and monofunctional polyethers such as monomethoxy polyethylene and monobutoxy polypropylene glycols (M_n 300–1000 g mol^{-1}).

Monohydroxy polyacrylates constitute another class of monofunctional polymers that can be used as steric side-chains. They are accessible, for example, through free radical polymerization of acrylic and methacrylic acid esters in the presence of mercapto-functional alcohol regulators, such as 2-mercaptoethanol and -propanol (Fig. 4.16) [18, 19].

The resulting polymers have molecular weights between 500 and 5000 g mol^{-1} (M_n), which can be set by adjusting the concentrations of regulator versus initiator (typically peroxides and diazo compounds) and the reaction temperature. The selection of the alkyl ester on the monomer is very flexible and can vary from short- to long-chain or branched alkyl groups, and also moieties containing heteroatoms (polyethers). Vinylic monomers (e.g. styrene) can be copolymerized into these chains. Some examples of monofunctional chains from the patent literature are shown in Fig. 4.17.

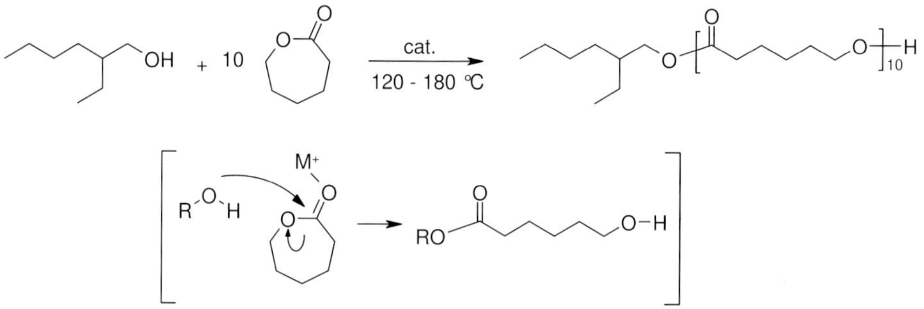

Fig. 4.15 Ring-opening polymerization (ROP) of lactones.

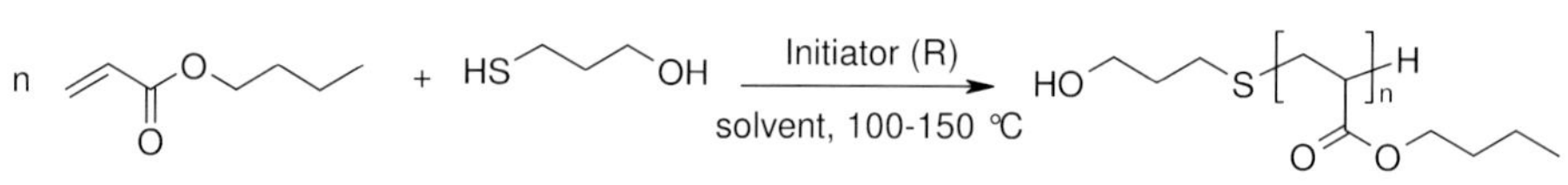

Fig. 4.16 Monohydroxy-functional polyacrylate chains.

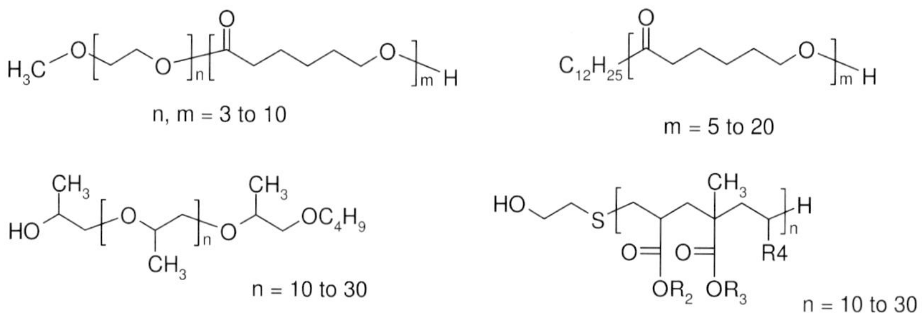

Fig. 4.17 Monofunctional polymeric chains for steric hindrance.

Defined polyoxyalkylene monoalkyl ethers can be used without modification, provided that they have sufficient molecular weights. However, from practical experience it has been found that a nonpolar character of the pending chains has a positive effect on the final viscosity of highly filled pigment preparations. It is assumed that nonpolar chains exhibit fewer interactions (dipole–dipole, hydrogen bridges) with polar resin polymers and other dispersant chains when their concentration is increased. As a consequence, the solubility of the neat polyurethane dispersant is sometimes low compared with other dispersant chemistries such as polyacrylates. Therefore, the choice of solvent is more critical to provide a usable product delivery form (typical useful solvents include xylene, toluene and ethyl acetate). These solvents, however, do not always fulfill the strict environmental requirements of modern paint systems, making these dispersants less widely accepted.

Both the monofunctional side-chains and the difunctional cross-linker polymers are reacted with the NCO-containing building blocks in defined stoichiometric ratios to avoid gel formation or insufficient molecular weight for steric stabilization in the final application. To obtain a polymer structure according to Fig. 4.7 which can provide good steric hindrance, it is optimal that each NCO-resin unit carries at least one of the long-chain monohydroxy-functional intermediates. Depending on the technology described in the patents, the sequence of addition of the various reactive components to the NCO-resin can be either stepwise or in one portion. For the addition reaction to the NCO group, well-known catalysts such as dibutyltin dilaurate can be used.

4.3.1.4 Introduction of Anchoring Groups

A very important component and distinctive element of PUR dispersants is the anchoring groups for the pigment surface, because they determine the range of pigments on which the dispersant can be used. Each producer tends to focus on their own technology and choice of functional group. Often a combination of several anchoring moieties is applied, providing a wider applicability to commercial pigments from various sources and surface treatments.

In order to connect the anchoring molecules covalently to the polymer backbone via the remaining NCO groups, it is necessary that these molecules carry at least one active hydrogen atom (Zerewitinoff hydrogen atom [20]) that reacts preferentially with isocyanates. To avoid unwanted cross-linking, usually not more than one active hydrogen is found in these monomers.

It has been reported that groups containing tertiary or aromatic nitrogen atoms with pK_a values between 5 and 12 (as a 0.01 M solution in water) are suitable for pigment affinity [19]. These aminic structures obviously impart basicity to the polymer and allow salt formation, since many of these moieties can also be used in the form of their organic or inorganic salts or by subsequent quaternization by alkylation reactions, e.g. reaction with carboxylic acids, alkyl halides or sulfates.

As basic structural elements of the anchoring monomers, piperazines, pyrimidines, pyrroles, morpholines, pyridines, pyrrolidines, imidazoles, benzimida-

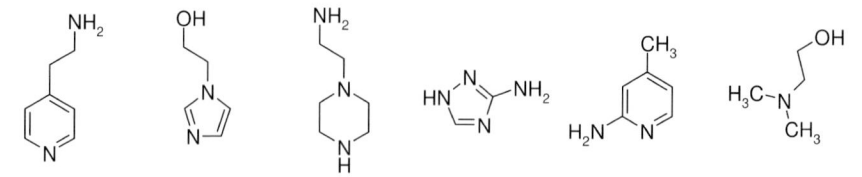

Fig. 4.18 Examples of reactive anchoring moieties.

zoles, benzothiazoles, triazoles and triazine heterocycles and also trialkylamines are mentioned in the patent literature [12, 21]. All of these bear reactive substituents for isocyanates, such as hydroxy- and aminoalkyl chains, as illustrated in Fig. 4.18.

The expensive and sometimes hazardous nature of the anchoring monomers makes it necessary to ensure complete chemical incorporation of the anchoring groups into the backbone. The complete incorporation will also improve the storage stability of such polymers, since many of the anchoring monomers are crystallizing solids and in addition are prone to cause side-reactions with other coating components (yellowing, short pot-life) if present in the free form.

In the case of the introduction of amino functional anchoring groups, urea bonds in the polymer will be generated, which are more polar and easily form inter- and intramolecular hydrogen bridges. Therefore, products containing urea bonds usually are delivered in lower solid contents (25–40 wt.%) for easy handling, as compared with those polymers that are composed of only hydroxy-reactive building blocks (solid contents up to 60 wt.%).

Suitable solvent environments showing good solubility parameters for all educts and products are aromatics (xylene, alkylbenzenes), acetates (ethyl and butyl acetate, methoxypropyl acetate) and other polar aprotic solvents (DMF, NMP) [22].

Although PUR dispersants are mainly designed for solvent-based formulations, there are a few examples in the patent literature of polyisocyanate adducts for aqueous systems.

In one specific example [23], dihydroxy-functional polyethers (M_n 1000–1500 g mol^{-1}) are reacted with 2,4-toluylene diisocyanate until a defined conversion and degree of cross-linking, followed by reaction of the remaining NCO groups with N,N-dimethylethanolamine for its pigment-affinic properties. The polyisocyanate addition products obtained are water-soluble, as they contain 30–95 wt.% of hydrophilic polyether chains based on ethylene oxide.

4.3.2
Polyacrylate-based Dispersants

Polyacrylate dispersants include classical polyelectrolytes for inorganic pigments and fillers (e.g. calcium carbonate) that are used in water-based systems [24, 25].

Their mode of action is based on the zeta potential of inorganic particles [26], which shows the degree to which electrostatic stabilization is possible. The purpose of polyelectrolyte dispersing agents in aqueous media is mainly to shift the (net) zeta potential at the particle/liquid interface to a negative value in order to eliminate the attractive forces. By increasing the charge on the surface, approaching particles experience a stronger repulsion. This form of electrostatic stabilization is very efficient for systems consisting of only one type of pigment with the same charge.

Examples of compounds that act on charge repulsion are polyphosphates and polyacrylates (Fig. 4.19). Both systems are neutralized or partly neutralized to achieve water solubility. They bear a negative charge that increases the total surface charge of the particle, when adsorbed on the surface.

Such polyacrylates generally have a molecular weight between 1000 and 10000 g mol^{-1} with a wide variation of chain lengths ($M_w/M_n > 2$).

The products that are commercially most successful as dispersants are based on poly(acrylic acid) and copolymers of acrylic acid and AMPS (2-acrylamido 2-methylpropanesulfonic acid). The polymers are obtained by free radical solution polymerization in the free acid form; however, they are delivered neutralized with amines or alkali metal hydroxides in aqueous or alcoholic solutions.

An improvement of this concept led to product forms with a narrower molecular weight distribution with polydispersities below 1.5. This can be achieved by fractionation techniques in which the partly neutralized lower molecular weight oligomers and the higher molecular weight fractions show different solubilities in defined organic and aqueous environments, by means of which they can be separated from each other [27]. The precise split into higher and lower molecular weight fractions can be selected by altering the process conditions, degree of neutralization, type of neutralization agent and choice and amount of organic solvent (typically C_1–C_5 alcohols or C_3–C_4 ketones). The fractions (usually two) find separate applications in coating preparations.

As pigments and extenders carry different charges, it is practically impossible to stabilize combinations of different pigments by the above method of charge stabilization. For combinations of pigments, it is necessary to use polymers that can – in addition to providing an electrostatic effect – also build up a steric stabilization around the pigment particles to avoid their flocculation. In addition, the concept of charge stabilization only applies to aqueous systems, whereas in organic solvent-based coatings, good stabilization has to rely only on steric principles.

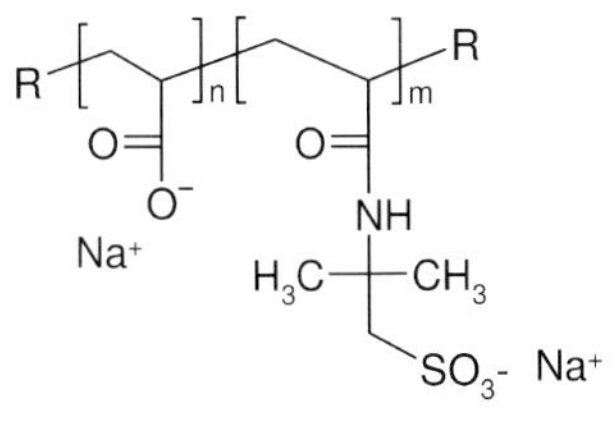

Fig. 4.19 Polyelectrolyte dispersant.

4.3.2.1 Architecture of Polyacrylate Dispersants

Polyacrylates which work according to a steric repulsion mechanism are available for water- and solvent-based paint systems and can also be used on complex pigment mixtures. Such polyacrylates possess distinct functional regions that can be distinguished as anchoring group, polymer backbone and steric hindrance-inducing side-chains. The molecular weights can typically be found in the range 8000–20 000 g mol^{-1} with polydispersities between 2 and 4.

These polymers are produced according to classical free radical copolymerization techniques in solution: reaction of acrylic, methacrylic and vinylic monomers in the presence of a thermally labile initiator, such as an organic peroxide or diazo compound. The concept is generalized in Fig. 4.20. In some cases, non-functional alkanethiols (e.g. dodecanethiol, mercaptopropionate esters) are added to regulate the molecular weights and molecular weight distributions.

Due to the chemistry, polyacrylic dispersants mostly have a linear structure with a robust C–C backbone and pendant functionalities depending on the monomers employed. The part of the acrylic copolymer chain not carrying anchoring groups can act as a steric barrier. With random copolymers the adsorption situation can be visualized in the form of extending loops into the medium, as indicated in Fig. 4.21.

The following product classes have been developed and are commercially available:

1. Polyacrylates as 40–70% solutions in organic solvents (e.g. butyl acetate, 2-butanol, glycol esters and ethers). This is the first generation of products designed for solvent-based coating systems.
2. Polyacrylates containing grafted polar side-chains (polyethers) in polar organic solvents. These polymer solutions can be (self-)emulsified in aqueous formulations, but can also be used in organic media.
3. Polyacrylates in polar solvents (alcohols, e.g. methoxypropanol), containing carboxylic acid groups in their backbone. By addition of base (amines, caustic

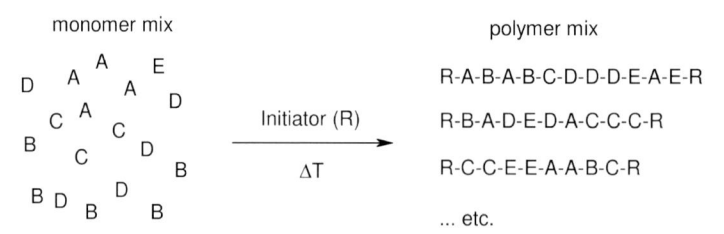

Fig. 4.20 Random polymerization of acrylic monomers.

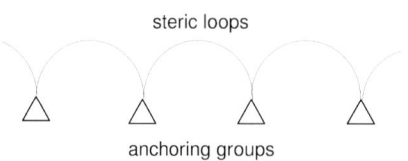

Fig. 4.21 Steric loops on polymeric dispersant.

solutions) they become fully water soluble, while the original non-neutralized form can be used for solvent-based systems. A recognized disadvantage of this kind of product for water-based systems is the presence of volatile organic solvents. In addition, in the case of non-volatile amines or alkali metal hydroxides, an enhanced water sensitivity of the cured film can be detected.

4. Polyacrylates with pre-neutralized carboxylic acid groups, delivered in water. These can be used as delivered directly in aqueous systems; however, at lower pH (<5), they have the risk of becoming water insoluble, with loss of performance.
5. Polyacrylates containing large amounts (>40 wt.% on dry polymer) of polar side-chains, which make the entire polymeric chain water soluble. These polymers do not need additional carboxylate formation for full water solubility and allow wider use in aqueous systems with pH values ranging from ca. 3 to 10. The polar side-chains typically are long-chain polyethylene glycol ethers with molecular weights up to 1500 g mol^{-1}.
6. Polyacrylates that are prepared directly in aqueous environment by classical emulsion polymerization [28]. The final polymers are by nature water-insoluble acrylic dispersions. They contain hydrophilic side-chains, which are designed to unfold once the dispersant is adsorbed on the pigment surface and compatibilize the pigment with the polar medium.
7. Polyacrylates on the edge of water solubility in VOC-free [29] polar glycol ethers, which can also be used in polar solvent-based systems. These products find special application in universal pigment concentrates for tinting both aqueous and solvent-borne base coats.

The overview of products shows that polar polymers are relevant for aqueous-based systems.

The amounts, lengths and end-groups of polar side-chains (i.e. polyethylene oxide or polyethylene–polypropylene oxide) on the polymeric backbone determine whether the dispersant is emulsifiable or totally soluble in water. Alternatively, water compatibility can be achieved by deprotonation of free carboxylic acid groups – introduced into the polymeric backbone via (meth)acrylic acid as comonomer – by amines or caustic solutions.

However, the use in aqueous systems has some risk of hydrolysis of the ester functions from the C–C backbone and subsequent loss of steric hindrance performance, hence these dispersants have limited suitability for aqueous environments with very low (<3) or very high (>10) pH values.

4.3.2.2 Design of the Polyacrylate Backbone

From the patent literature, a wide range of monomers can be found as basic building blocks for polyacrylic dispersants [30]. In order to obtain resin-compatible structures, a choice of acrylate esters and the corresponding methacrylate monomers is readily available from industry and virtually all compatibilities can be achieved from nonpolar to polar. Figure 4.22 shows a classification of mono-

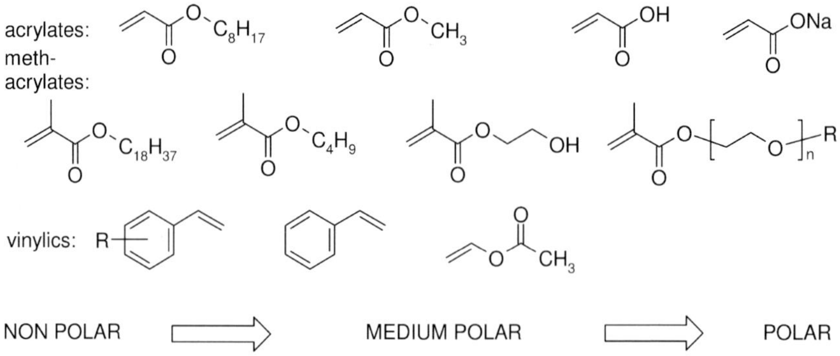

Fig. 4.22 Classification of monomers for the polymeric backbone.

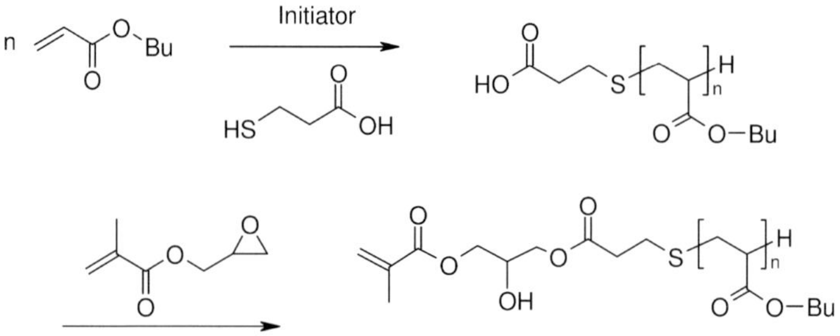

Fig. 4.23 Synthesis of macromonomers via chain-transfer agents.

mers over the polarity regions. The nature of the alkyl ester of the polymerizable (meth)acrylate group determines the polarity and the T_g of the final dispersant backbone. For their intended function, the amounts of these monomers can reach up to 80 mol% of the total amount of monomers in the composition. Vinylic monomers such as styrene, vinyltoluene and -methylstyrene are often used in combination with the above. The introduction of the aromatic ring allows a broader solubility in resin systems, as compared with polymeric chains that only contain acrylic ester moieties [30].

Grafted structures are accessible through the use of macromonomers, where parts (up to 25 wt.%) of the conventional monomers mentioned above are substituted [31]. Suitable macromonomers are copolymers of ethylenically unsaturated monomers and have typical molecular weights between 5000 and 10 000 g mol^{-1}. The defined monofunctionality can be obtained by using appropriate chain-transfer agents such as mercaptopropionic acid, leading to carboxy-terminated polyacrylic chains, which in a subsequent step are reacted with glycidyl methacrylate, as shown in Fig. 4.23. These long-chain comonomers provide the polymer with a branched structure, which is claimed to reduce viscosity more efficiently in highly filled pigment preparations.

4.3.2.3 Introduction of Anchoring Groups

The introduction of anchoring groups into the polymeric chain is essential for its function as a pigment dispersant. Similarly to PUR dispersants, compounds with aliphatic and aromatic amines have proven to be very effective. The following options are practiced:

1. Direct copolymerization of reactive monomers bearing pigment-affinic groups (examples are shown in Fig. 4.24). These monomers can be vinylic or acrylic functional and their relative amounts can be up to 20 mol% based on the total amount of monomers. Useful examples are 2- and 4-vinylpyridine, 1-vinylimidazole, 2-methyl-N-vinylimidazole, vinylpyrrolidone, vinylcarbazole, vinylquinoline, N,N-dimethylaminoethyl (meth)acrylate and *tert*-butylaminoethyl acrylate [30, 32–34].

2. Incorporation of defined anchoring moieties by condensation reactions. For this approach, transaminolysis [35] and transesterification [36] are valuable tools. Initially, the polymeric backbone without anchoring groups is prepared by conventional random solution copolymerization and, subsequently, amino- or primary alcohol-functional compounds are introduced at random on to the backbone by partial cleavage of the accessible esters on the polymeric backbone (Fig. 4.25). For practical reasons, low-boiling alcohols (methanol, ethanol and butanol) are liberated and removed from the reaction medium. Catalysts (e.g. titanium complexes) are employed to ensure high conversions.

3. Post-modification, by addition reaction of anchoring moieties containing reactive –OH, –NH$_2$, –COOH or –SH groups on to acid, epoxy or isocyanate groups on the polymer backbone [30], as shown in Fig. 4.26. These groups become accessible in the polymeric backbone by copolymerizing glycidyl methacrylate, acrylic acid or 2-hydroxyethyl acrylate, the last subsequently being modified with

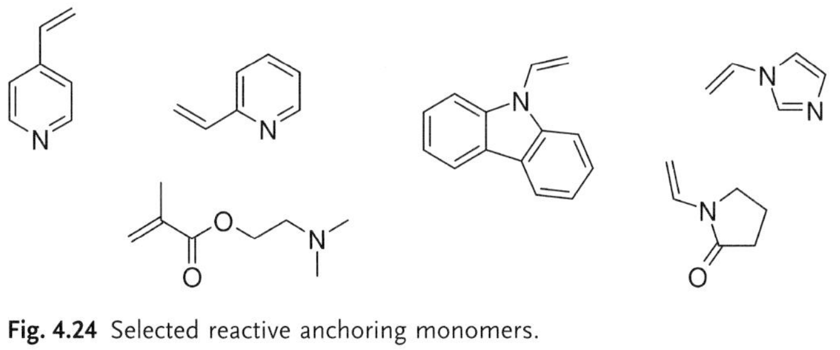

Fig. 4.24 Selected reactive anchoring monomers.

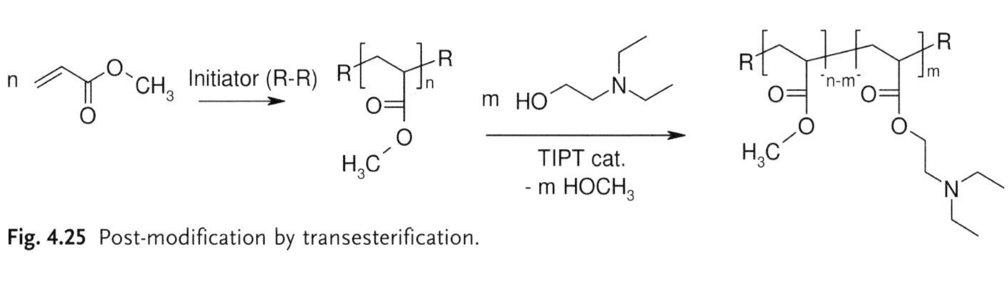

Fig. 4.25 Post-modification by transesterification.

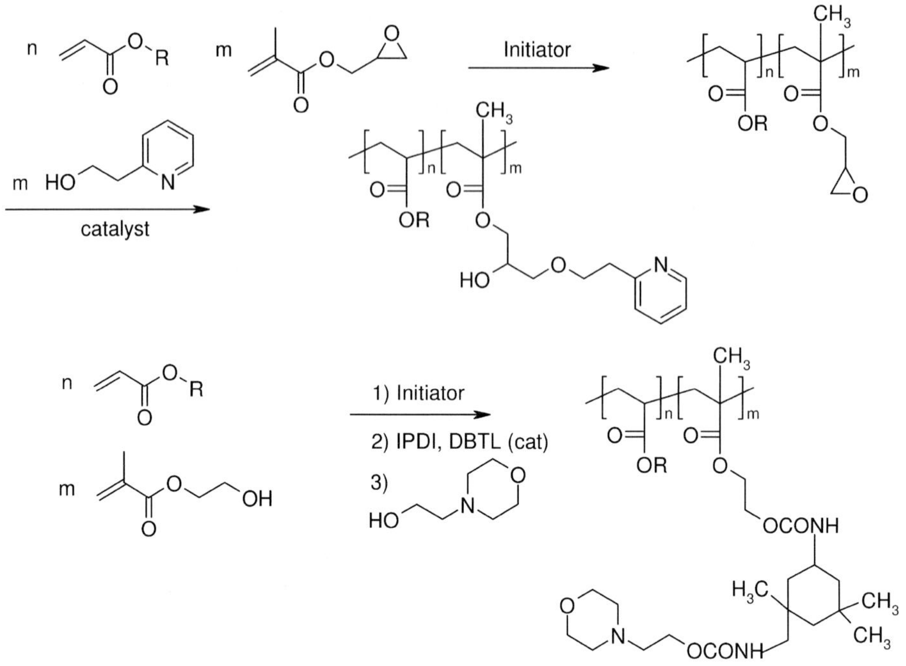

Fig. 4.26 Post-modification by addition reactions.

one equivalent of a diisocyanate (e.g. isophorone diisocyanate). A highly selective addition of primary alcohols to the secondary NCO group of isophorone diisocyanate can be achieved by dibutyltin dilaurate catalysis [37].

Typical reagents for post-incorporation of the anchoring group include those used for polyurethane dispersants (see Fig. 4.18) and others described in the literature [30].

4.3.2.4 Dispersants Based on SMA Resins

A special class of vinylic-type copolymers used as polymeric dispersants is based on addition products between styrene–maleic anhydride copolymers (SMA resins) and alcohols or amines [38]. Styrene and maleic anhydride can be copolymerized in a bulk process using free radical initiators to yield an oligomeric alternating block copolymer with a hydrolytically stable C–C backbone, as illustrated schematically in Fig. 4.27.

The copolymerization of styrene with maleic anhydride creates a material with a higher glass transition temperature than polystyrene (T_g 125–155 °C) and is chemically reactive with nucleophilic groups.

In practice, available SMA resins contain up to 50 mol% of maleic anhydride and some grades also contain small amounts of butadiene or ethylene as a comonomer. Products ranging from 1 mol of maleic anhydride to 1 mol of styrene up to 3 mol of styrene, with molecular weights (M_n) from 1500 to 4000 g mol^{-1} are commonly used in the coating and printing ink industries [39]. These oligomers can be used as a starting point for the addition of steric side-chains and anchoring groups. Monohydroxy- or amino-functional polyethers and other aliphatic alcohols can be introduced into this backbone by a ring-opening addition (Fig. 4.28). Eventually formed acidic groups can be neutralized with ammonia or organic amines to afford good water solubility of the polymer.

The use of polyether-functional alcohols as steric side-chains opens up the possibility of designing such polymers for universal use properties, being compatible in both solvent- and water-based coatings. Short-chain aliphatic end-groups, in combination with long-chain polyethers ($n > 10$), lead to products with full water solubility, without the need to neutralize the carboxylic function with amines. Stabilization of both inorganic an organic pigments is usually effective.

The use of polyalkylene glycol monoalkyl monoamines (Fig. 4.29) [40] leads to products with a hydrolytically stable amide or imide function, in addition to water solubility. Generally, the polyether amines are made by aminating a polyether polyol with ammonia in the presence of an Ni, Cu or Cr catalyst [41].

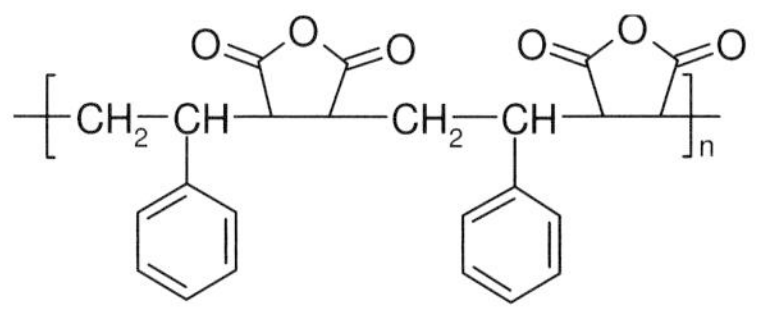

Fig. 4.27 SMA resins.

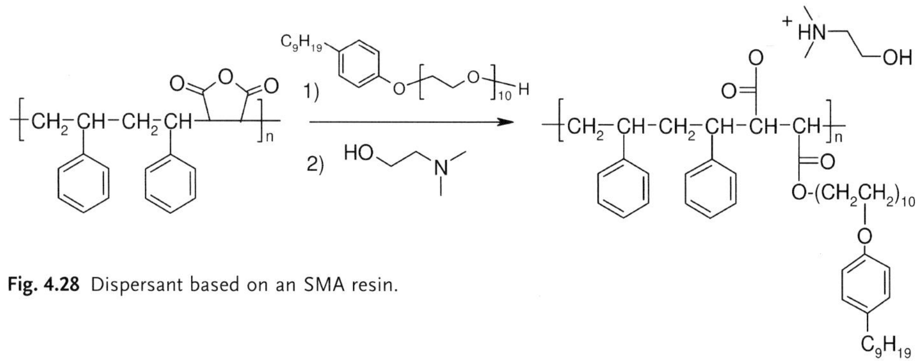

Fig. 4.28 Dispersant based on an SMA resin.

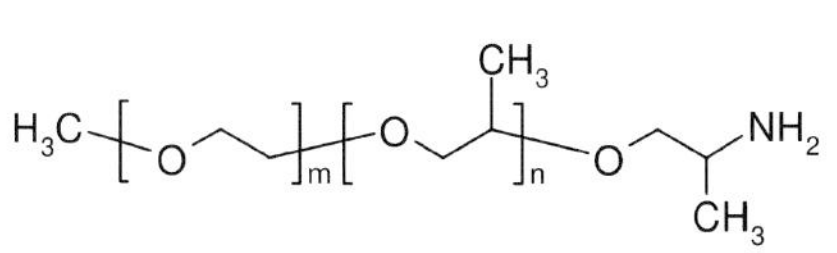

Fig. 4.29 Amino-functional polyether ($m = 1–49$; $n = 3–39$).

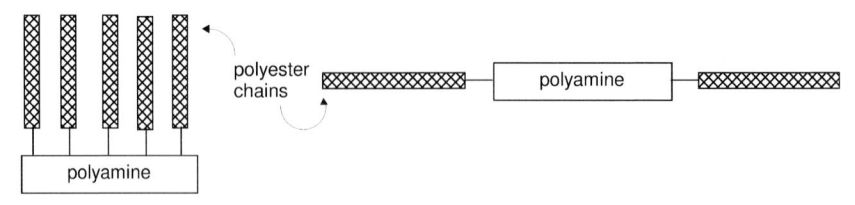

Fig. 4.30 Architecture of polyester dispersants.

4.3.3
Polyester-based Dispersants

This class of polymeric dispersants mainly revolves around amino- and amido-functional block copolymers of oligo- and polyamines with polyesters and named after the polyester blocks being the main component of the formulation. These materials started to appear commercially in the 1980s [42, 43].

These polymers can best be described as block, graft or comb copolymers with an amino-functional core (main chain) and at least two pendant monofunctional polyester chains (Fig. 4.30).

4.3.3.1 Polymeric Amine Anchoring Blocks

The polyamine acts as the pigment-affinic block. Practical importance is attributed to polyethylenimine (PEI), which is available on an industrial scale in a variety of molecular weights and degrees of branching. Work described by Meyers and Royer in 1977 led to the commercial availability of a variety of products based on PEI technology [44].

PEI is formed by protic- or Lewis acid-catalyzed polymerization of ethylenimine (aziridine) to give a highly branched hydrophilic three-dimensional matrix (Fig. 4.31) [45, 78]. About 25% of the resultant amines are primary, 50% are secondary and 25% are tertiary. Each of these types of products is delivered in the form of viscous liquids or aqueous solutions. Solid contents range from 10 to 100%, depending on the molecular weights, which are available from a few hundred up to several hundred thousand g mol^{-1}.

Fig. 4.31 Synthesis of PEI.

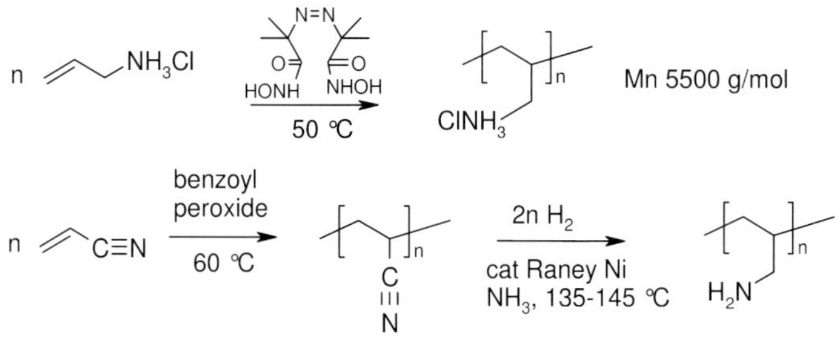

Fig. 4.32 Industrial synthesis of polyallylamine.

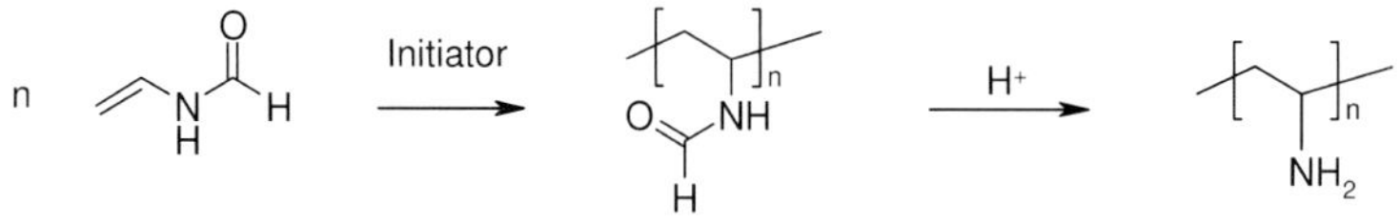

Fig. 4.33 Polyvinylamine.

Low-temperature polymerization of aziridine affords linear PEI, which can be separated from branched isomers by precipitation [46]. Technical grades of oligomeric PEI (diethylenetriamine, tetraethylenepentamine, pentaethylenehexamine, etc.) are obtained by condensation of dichloroethane with ammonia or ethylenediamine [47].

In another example, the aminic anchoring block is polyallylamine [48]. Industrial-scale production can be achieved by either azo-type initiated free radical polymerization of inorganic salts of (mono)allylamine (e.g. its hydrochloride or sulfate) [49] yielding high degrees of polymerization, or by hydrogenation of polyacrylonitrile (Fig. 4.32) [50]. Direct polymerization of (mono)allylamine can only be done with difficulty, leading to low molecular weight products, due to the self-termination of allyl compounds. This allylic degradative chain transfer is common to allylic monomers [51].

Polyvinylamine [55], which is commercially available in various molecular weights, is less commonly used as a building block. It is obtained by free radical polymerization of N-vinylformamide [52] to yield water-soluble polyvinylformamide (PNVF) [53], followed by direct acidic or basic hydrolysis to polyvinylamine. The process is illustrated in Fig. 4.33. PNVF is probably the most practical precursor for the preparation of polyvinylamine and is a less toxic alternative than polyacrylamide and other cationic, water-soluble polymers.

4.3.3.2 Polyester Side Chains and Dispersant Synthesis

Side-chains are connected to the aminic core polymer in order to enhance compatibility of the polar polyamine with the resin medium and to induce steric hindrance around the pigment particle after its adsorption. Defined monofunctional polyesters have evolved to be of practical importance for this type of dispersant. Linkage of the polyester chains takes place by reaction:

- between the polyester and defined carboxylic acid end-groups and the polyamine by amide formation;
- by Michael addition of the primary amines of the polyamine to acrylic functions of polyester-based macromonomers [54];
- of salt linkages between the polyamine and the acidic side-chain; in this case, carboxy-functional, phosphoric and sulfuric acid esters have been described [55, 56].

Defined linear polyesters with one carboxylic acid group at one end are accessible mainly by the following routes.

1. Polymerization of hydroxycarboxylic acids such as 12-hydroxystearic acid (12-HSA) [57]. The polyester is obtained by thermal condensation of 12-HSA or related compounds (ricinoleic acid, castor oil fatty acid, etc.) including their mixtures in the region of 160–200 °C, to the desired molecular weight between M_n 500 and 5000 g mol^{-1} (Fig. 4.34). The resulting polyester is carboxy functional on one end and hydroxy functional on the other, the latter group optionally being end-capped through esterification with monocarboxylic acids such as lauric, stearic and oleic acid [58].

    The consecutive reaction of PEI (M_n 50 000 g mol^{-1}) with these polyesters leads to mixtures of amide and salt linkages. Due to the highly nonpolar character of poly(12-HSA), dispersants obtained in this way are considered suitable for applications based on nonpolar organic media, such as impact printing inks.

2. Ring-opening polymerization (ROP) of cyclic lactones (typically ε-caprolactone and δ-valerolactone), initiated with a monofunctional carboxylic acid (Fig. 4.35). According to the literature, this results in linear carboxy-functional poly(carbonylalkyleneoxy) chains [42]. The most suitable catalysts for this polymerization are tetrabutyl titanate [42] and zirconium butoxide [54].

    Using carboxylic acids instead of nucleophilic alcohols as initiators for the ROP, as in the case of the previously described polycaprolactones as steric chains for polyurethane dispersants, active hydrogens are now absent and the mechanism is suggested to involve propagation through a carbonium ion,

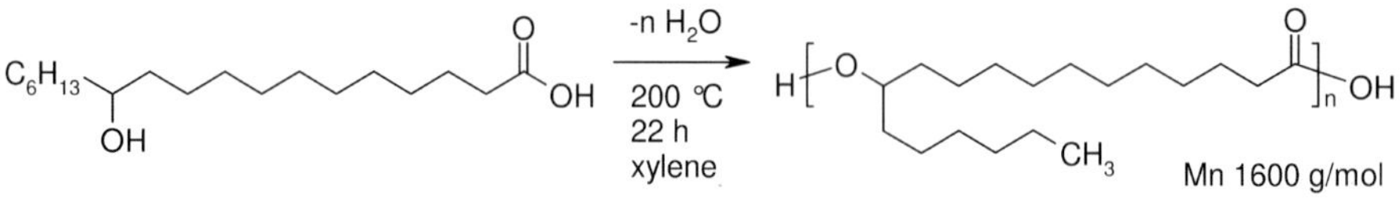

Fig. 4.34 Poly(12-hydroxystearic acid).

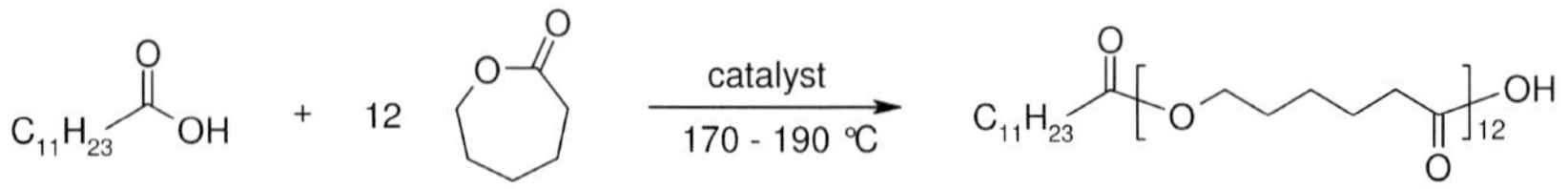

Fig. 4.35 Polyester acid by ROP of lactones.

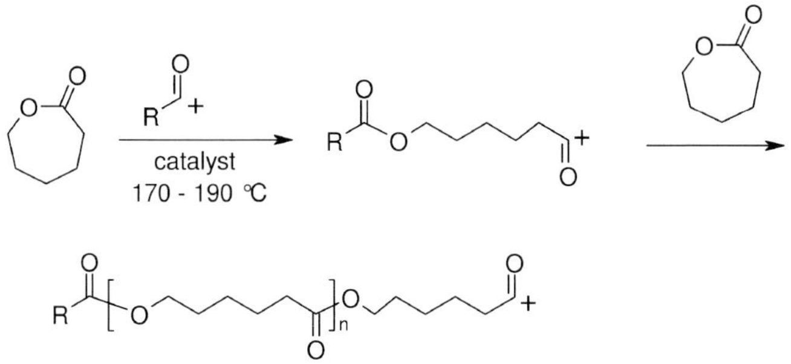

Fig. 4.36 ROP in the absence of active hydrogen initiators.

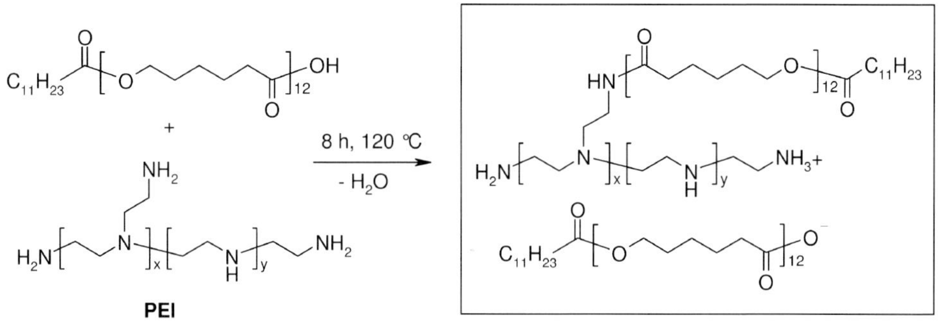

Fig. 4.37 PEI-based polyester dispersant.

with acyl–oxygen cleavage being the prime mode of ring opening, shown in Fig. 4.36 [59]. Alkyl–oxygen cleavage takes place in unique cases only, as with three-membered rings [60].

For dispersant synthesis, these carboxyl functional polyesters are then reacted with PEI (e.g. M_n 10000 g mol^{-1}) in such a way that practically a mixture of amide and ionic salt bonds is obtained. One possible synthesis is shown in Fig. 4.37. The complete amidization reaction of such polyesters with the sterically hindered amines in the branched PEI core would require long reaction times and high temperatures (above 200 °C), which is known to cause strong discoloration of the product even in an inert (N2) atmosphere. If mild conditions are chosen (low temperatures and short reaction times), almost exclusively salt bonds are formed; more severe conditions lead to larger amounts of the preferred amide bonds [42]. In order to find a compromise and for practical reasons, a certain number of salt bonds remain. However, the presence of unreacted primary and secondary amines for two-component polyurethanes and epoxy resin systems is unwanted.

As the relative amount of polyester chain in the final dispersant outweighs the amount of polyamine in the formulations cited in the literature usually by factor of 10, the resulting adducts have melting-points between 50 and 60 °C (e.g. pure poly–caprolactone, m.p. 63 °C) [45], rendering them practically undesirable for many paint production processes, where liquid components are preferred. In addition, the storage stability of the formulated paint may become a problem if the dispersant shows a tendency to separate out of the liquid formulation by crystallization.

The incorporation of comonomers into the homogeneous polycaprolactone chains is one of the ways to overcome this drawback of the first generation of polyester dispersants. More recent publications focus mainly on the improvement of the crystallization behavior [61]. For discrimination of suitable monomer compositions, the resulting dispersant is exposed to low-temperature conditions to assess its sensitivity towards solidification. A general guideline describes that the material should survive 10 °C for 5 days in a solution of organic solvent. δ-Valerolactone has been identified as a useful comonomer [61]. Another approach has been published in which trimethylene carbonates (e.g. 5,5-dimethyltrimethylene carbonate) [62] and lactic and glycolic acid [63] were selectively copolymerized into a polycaprolactone chain.

The regularity in the polymer can also be disturbed by incorporating a mixture of structurally different polyesters into the same dispersant, such as the combination of a poly(ricinoleic acid) and a polylactone chain, which tend to show less tendency to solidify in this combination [64].

3. Polyesters that are obtained from combined condensation and addition reactions of ε-caprolactone and 12-HSA, which subsequently form a condensate of or a salt with a polyallylamine [48]. The polyesters can be a defined block copolymer, by first building the hydroxy-end-capped poly(hydroxycarboxylic acid), followed by ROP of a lactone [65]. A source to hydroxy-functional polyesters with defined carboxylic function has been described starting from lactic acid as initiator for ROP of caprolactone, which subsequently is reacted with PEI [66].

Polyallylamines with grafted polyester chains play an important role as dispersants for high-performance dispersants for electronic applications. These polyesters (M_n below 10 000 g mol^{-1}) originate from ROP of lactones by hydroxy acid-functional initiators and contain a free carboxylic acid [67]. The polyesters can also contain amide bonds by incorporating lactams or amine building blocks. The molecular weights of the polyallylamines used are up to 20 000 g mol^{-1}. The reaction conditions are chosen so that the majority (more than 65%) of the primary amines form amides and the remainder salt bonds

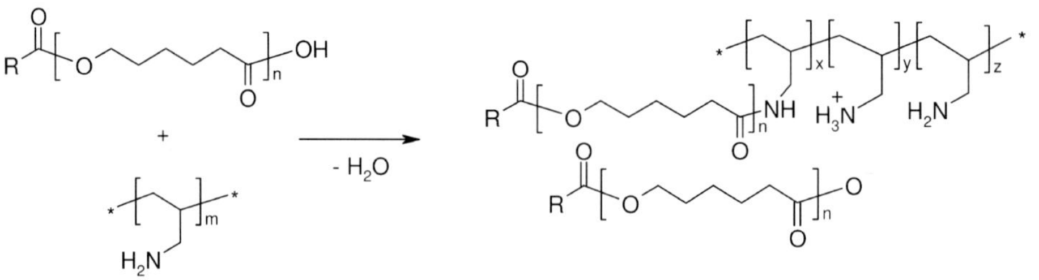

Fig. 4.38 Polyester-grafted polyallylamines.

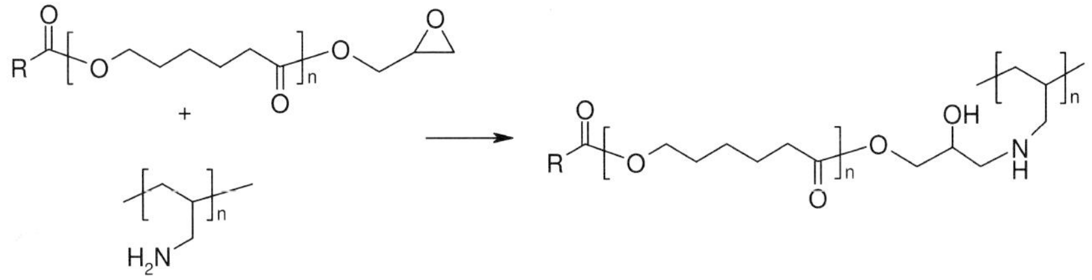

Fig. 4.39 Functionalization of polyallylamine by glycidyl macromonomers.

(Fig. 4.38). Depending on the ratios of the reactants, the final dispersant can also contain free amino groups.

In the literature [67], the linear structure of the polyallylamine core is claimed to be advantageous over the branched PEI and is explained by better accessibility of the amino groups as anchoring groups at the site of adsorption on the pigment surface. Amines in a highly branched polyamine are expected to suffer from steric hindrance during the adsorption process.

In a special case, these polyesters were attached to the allylamine through a glycidyl ester-modified macromonomer (Fig. 4.39) [68].

4. Polyesters based on polycaprolactone or poly(hydroxycarboxylic acid)s condensed with PEI and subsequently reacted with cyclic five- or six-membered anhydrides (such as maleic, succinic and phthalic anhydride) on residual amine and hydroxy functionality to yield carboxylic acid-modified dispersants [69]. The advantages of the these structures over similar unmodified polyester dispersants [42, 57] are described as not showing a tendency to discolor and inhibiting acid-catalyzed reactions such as melamine cross-linking or reacting with isocyanates. Therefore, they can be used in acid-catalyzed and isocyanate processes without the adverse effects of amino-functional polyesters.

The final polyester-based dispersants synthesized in either of the ways described above can be used as obtained after the functionalization of the polyamine with the polyester, being macromolecular structures containing mixtures of amide and salt bonds and in some cases are subjected to further reactions, such as quaternization by N-alkylation with alkyl sulfates (e.g. dimethyl sulfate), alkyl halides (e.g. methyl bromide) or further neutralization with selected acids [54].

In addition to polyamines, non-polymeric amines carrying polyester chains have been reported in the patent literature as versatile pigment dispersants for special applications, such as for grinding crude copper phthalocyanine pigments [70]. In such a case, polyesters based on 12-hydroxystearic acid and M_n of ca. 1600 g mol^{-1} are condensed with N,N-dimethylaminopropylamine, followed by quaternization of the amino group by dimethyl sulfate (Fig. 4.40).

4.3.4
Polyether-based Dispersants

There are numerous examples of surfactant-type polyether dispersants, which are characterized by a low molecular weight (typically M_n below 1000 g mol^{-1}),

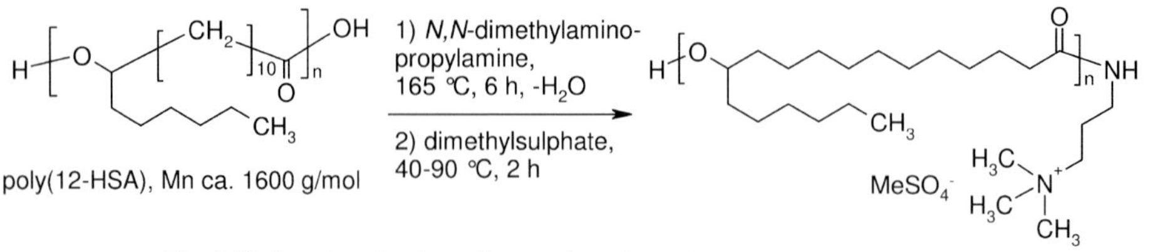

Fig. 4.40 Functionalization of non-polymeric amines.

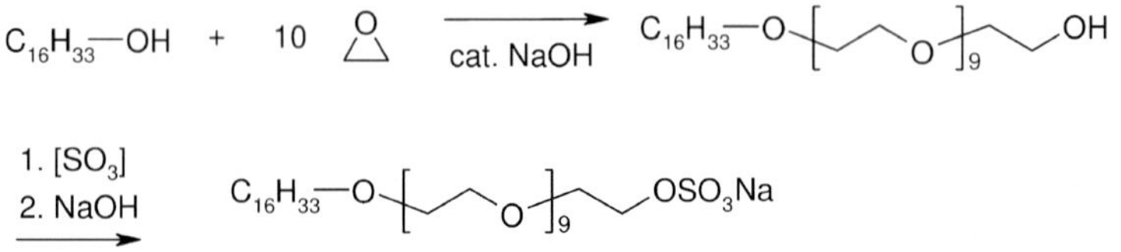

Fig. 4.41 Synthesis of surfactant by ethoxylation and subsequent modification.

water compatibility, water sensitivity and migrational behavior in the cured paint film, good wetting of substrate and inorganic particles, and foam stabilization in the liquid aqueous paint. Classical examples are nonylphenol ethoxylates and C_8–C_{18} alkyl ethoxylates, which can be optionally end-functionalized with acid groups (sulfate or phosphate) or quaternary amines. The products can be classified according to their functional end-groups as electroneutral, anionic or cationic products. Application-wise, they are popular for low-cost coating systems, mainly to wet and disperse inorganic materials ($CaCO_3$ fillers and TiO_2 and Fe_2O_3 pigments). These dispersants have been extensively described in various literature reviews and will not be discussed in this chapter [71]. Synthesis of e.g. an alkyl ethoxylate is generally illustrated in Fig. 4.41. Alkali metal hydroxide-catalyzed anionic ROP of ethylene (or propylene) oxide, starting with an alkyl alcohol, leads to the monohydroxy-functional polymer. Further reaction yields a sulfate surfactant.

This product group not only finds application as wetting and dispersants for coatings, but in general also can be used as detergents or emulsifiers for the formulation of various industrial and cosmetic products.

However, there are only few examples of higher molecular weight ($M_n > 2000$ g mol^{-1}) polyether-based dispersants [72]. The good water compatibility of polyether chains makes these structures especially interesting for very polar organic systems and water-based applications, where the previously described product classes based on e.g. polyesters and polyurethanes may show limited solubility.

For the targeted higher molecular weight of polyether dispersants, the use of very long-chain aliphatic alcohols as a starting point for the ethoxylation becomes indispensable, in order to maintain the optimum ratio between the hydrophobic and hydrophilic parts in the chain (HLB value) [73]. Natural fatty al-

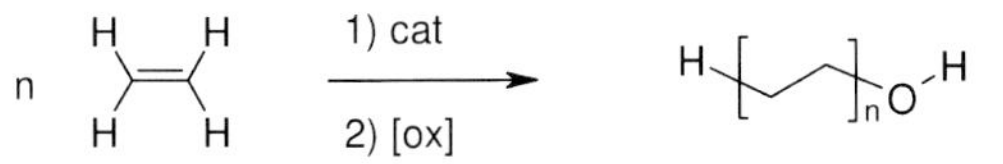

Fig. 4.42 Synthesis of long-chain synthetic alcohols from ethylene.

cohols are not readily available above C_{20}, therefore linear synthetic C_{20}–C_{50} alcohols are used, which are accessible through oligomerization of ethylene and end-group functionalization (Fig. 4.42) [74]. Since they are oligomers, they have a molecular weight distribution themselves and are used as the starting point of ethoxylation chemistry.

The polarity and molecular weight of the final dispersant can be adjusted by the length of the aliphatic alcohol and the degree of ethoxylation. Useful molecular weights (M_n) range from 2000 to 5000 g mol^{-1} and the final adducts are solid materials as pure substances at room temperature.

Difunctional dispersants with phosphate or sulfate anchoring groups and polyether main building blocks have been described as suitable dispersants for a variety of inorganic materials in water-based systems [75]. The basic element consists of a polymeric diol, derived from either pure polyethylene glycols or ABA block copolymers of ethylene oxide and propylene oxide, where the propylene glycol moiety constitutes the middle block (B). The polyethers employed are described as having molecular weights up to 6000 g mol^{-1} and which are further reacted with hydroxycarboxylic acid derivatives to generate a polyester elongation on both sides (monomers, e.g. ε-caprolactone, 12-hydroxystearic acid and ricinoleic acid). The conditions and mechanisms of formation of these polyester moieties are similar to those described elsewhere in this chapter. The resulting symmetric dihydroxy-functional hybrid block copolymers (polyester–polyether–polyester) are finally converted into diphosphate esters by reacting the diol with polyphosphoric acid, P_2O_5 or $POCl_3$, a shown in Fig. 4.43.

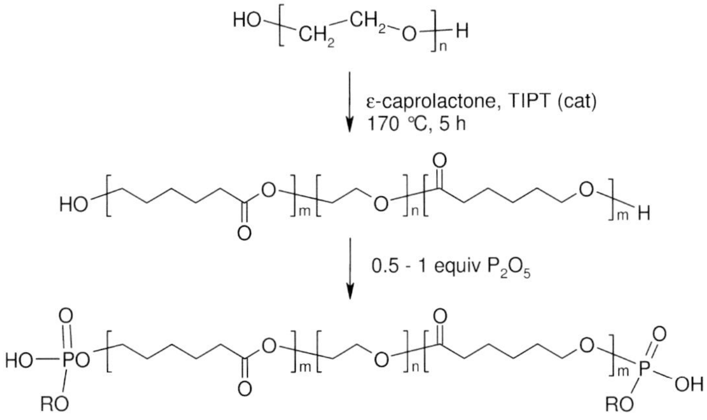

Fig. 4.43 Polyester–polyether-based phosphate dispersants.

4 Polymeric Dispersants

The corresponding linear monofunctional polyether dispersants with phosphoric ester groups are obtained in an analogous way starting from polyethylene glycol monomethyl ethers, which are elongated by ROP of lactones [76]. Molecular weights for the most effective product forms, however, are below 2000 g mol^{-1}.

Monoalkyl polyether dispersants without additional polyester elongation have been mentioned, containing both phosphoric and carboxylic acid groups as pigment anchors at one end of the molecule [77].

In a recent publication, the combination of polyether chains and polyethylene imines was described [78]. It was aimed at providing block copolymers of the above structural elements, which can be used as dispersants and detergents in aqueous media. The polymeric structure is obtained by grafting up to 50 mol of alkylene oxides (e.g. ethylene and propylene oxide) to the reactive nitrogens of a PEI (M_n between 500 and 2000 g mol^{-1}), leading to OH-terminated polyether side-chains. In a subsequent step, these OH groups are end-functionalized by capping with e.g. an anhydride (Fig. 4.44), an epoxy compound such as allyl glycidyl ether or an alkyl isocyanate. It is shown that the hydrophobic modification of the polyglycol chain enhances the surface activity, as compared with the unmodified polar adducts known from industry [79].

Although in Fig. 4.44 the reaction of ethylene oxide with a PEI is schematically simplified, it should be emphasized that primary N-atoms can react two times with alkylene oxides leading to two pendant polyglycols, while available secondary N-atoms are also functionalized once in this way, yielding finally a highly branched structure.

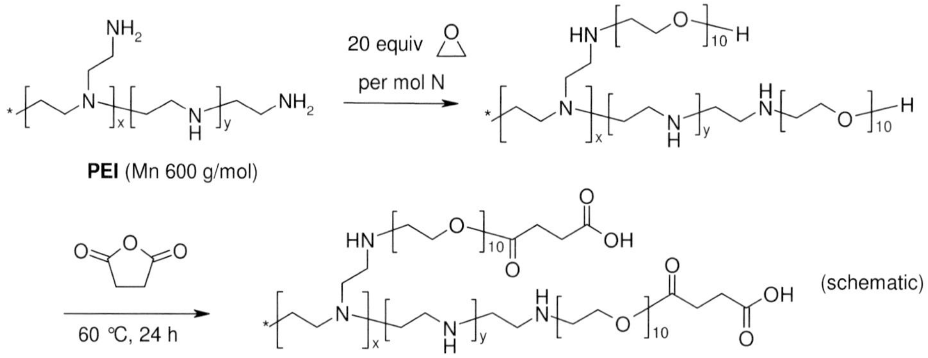

Fig. 4.44 Polyalkylenimine alkylene oxide copolymers.

4.4
New Technologies and Trends in Polymeric Dispersants

In the coatings industry, environmental issues such as the replacement of heavy metal-based pigments and the trend to high solids coatings, and also economic aspects, continue to increase the demands for modern pigment dispersants. Such trends in coatings and the demands from other important application fields such as ink jet technology and electronic applications, e.g. the high quality requirements for color filter production, have stimulated intensive research for new polymeric pigment dispersants in recent years.

Research efforts have been directed towards improved control of polymer architecture by new techniques of "living" polymerization, which are characterized by the fulfillment of certain kinetic criteria of the polymerization reaction and the absence of polymer chain-terminating side-reactions [80].

"Living" polymerization provides possibilities for preparing a multitude of polymers which are well defined in terms of molecular dimensions, polydispersity, topology, composition and functionalization. During recent decades, many living systems based on anionic and cationic mechanisms have been developed; however, they did not lead to commercial dispersants [81]. In more recent years, controlled free radical polymerization has attracted much interest for the synthesis of novel dispersants with improved performance based on optimized polymer architecture.

The very low concentration of active growing polymer chains is the key to obtaining a controlled free radical polymerization in order to suppress unwanted side-reactions. This creates a "living" polymer chain, which can further grow by monomer addition. The "living" character allows precise control of molecular weight and distribution and permits the synthesis of block copolymers by sequential polymerization of different monomers. Block copolymers, especially those with an AB structure (Fig. 4.45), belong to the most efficient architectures for pigment dispersants [82].

Stabilization by relatively short-chain AB block copolymers (M_n typically below 15 000 g mol^{-1}) can be superior to stabilization generated by loops and grafted tails of long-chain random copolymers, partly because of the possibility of achieving stronger and well-ordered adsorption of AB block copolymers with well-solvated stabilizer blocks extending from the pigment surface. In analogy with the classical dispersant design, the part of the polymer responsible for (electro)steric hindrance must be long enough to induce steric stabilization and be compatible and soluble in the resin/solvent medium, while the block de-

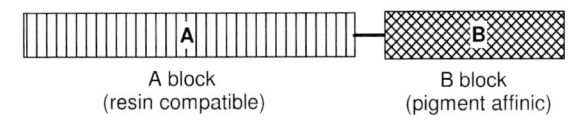

Fig. 4.45 Design of AB block copolymer dispersants.

signed for adsorption on the pigment surface must be available and show strong affinity for this surface through the use or introduction of relevant pigment-affinic anchoring groups similar to those discussed in previous sections. This anchoring block must be large enough to cause irreversible interaction with the pigment surface, but on the other hand not too large to adsorb on more than one particle. By design, the polarity of the soluble steric block determines the suitability of the final dispersant for either nonpolar solvent-based systems or for polar solvent- or even aqueous-based systems.

In addition to good structural control via "living" polymerization, these methods also provide products with a low polydispersity, typically below 1.5. This desirable side-effect minimizes fractions of very high and very low molecular weights, as can be found typically in random copolymers at levels of the order of several weight percent or more (Fig. 4.46). For dispersant performance, higher molecular weight fractions of the polymers tend to cause a strong increase in viscosity in the formulated pigmented systems. This is ascribed to the numerous interactions of long chains containing polar groups with other pigment particles, leading to flocculation in an extreme case and entanglements of the long chain with other coating components such as resins, whereas the lower molecular weight fractions fail to build up the required bulk around the pigment particle for steric stabilization and will therefore not contribute to the overall performance of the additive. For these reasons, dispersants based on "living" polymers have good potential for better performance at lower use levels.

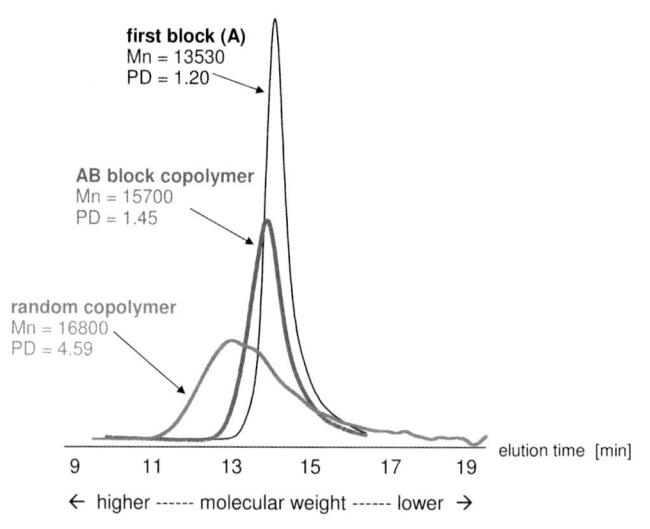

Fig. 4.46 GPC traces of an AB block and a random copolymer at similar M_n, based on the same monomer composition.

4.4.1
Dispersants Made by Group Transfer Polymerization (GTP)

The pioneering work of Webster's group on GTP opened the path for the living polymerization of a range of vinylic monomers [83]. This technique can be used to produce methacrylic block copolymers with M_n generally between 1500 and 20000 g mol^{-1} and is applied to the synthesis of dispersants for applications in paints and ink jet applications [84]. GTP involves initiators able to split into two fragments, of which one starts the polymerization by reacting with an acrylic monomer and the other fragment continuously transfers to the other end of the growing polymer chain. GTP does not allow for polymerization of acrylic monomers without a substituent on the ethylenically unsaturated β-atom, which makes the process specially suited for a range of commercially available methacrylates.

Although the mechanism of monomer addition [85] at first glance appears to be anionic in nature, the intermediate species are end-capped and therefore are stable enough to avoid the cryogenic techniques typical of a classical anionic polymerization (0 to $-40\,°C$). Nevertheless reaction temperatures for GTP are significantly lower than those for classical acrylic polymerizations and limit around room temperature or slightly above.

Figure 4.47 shows a simplified initiation process for GTP. The most commonly used initiators are based on silylketene acetal structures, which are made by trimethylsilane addition to methacrylic esters. For polymerization to take place they are combined with catalytic amounts of nucleophilic anions (e.g. a salt of an oxyanion such as tetrabutylammonium m-chlorobenzoate or CsF$_2$), which is dosed into the reaction medium over a prolonged period of time during the monomer addition to ensure sufficient polymerization events, as the catalyst is slowly consumed. The catalyst initiates silylketene decomposition starting the addition to a monomeric species. For thermal control, the monomers are added to the reaction mixture over a longer period (several hours). At the end of the monomer addition, the reactive Me$_3$Si group on the growing end of the polymer chain is removed by quenching the process with an alcohol. Due to the inherent reactivity of the silylketene acetals, this reaction needs to be carried out under strict exclusion of humidity and nucleophilic impurities and requires polar aprotic solvents such as ethylene glycol dimethyl ether (glyme), 1-methoxy-2-propylacetate or THF. Monomer conversions by GTP are generally very

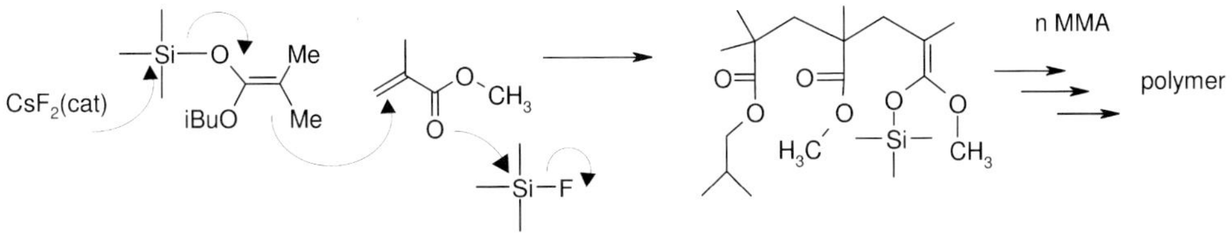

Fig. 4.47 Simplified mechanism of group transfer polymerization.

high and conversions of over 99% have been reported [86], hence the polymerization proceeds until almost total monomer consumption in each step. When building up block structures, monomers can be sequentially polymerized without purification of the sensitive intermediates.

A useful polymer composition for pigment dispersants is based on an MMA–GMA block copolymer as shown in Fig. 4.48 [87]. The epoxy functions, being arranged at one end of the polymer, can subsequently be opened with carboxylic acids such as 4-nitrobenzoic acid and 3,5-dinitrobenzoic acid to introduce the pigment-affinic groups. Amino functions have been successfully incorporated in the same way via p-aminobenzoic acid and typically 90% conversion of the acid can be obtained after 4 h of reaction at 130 °C in polar aprotic solvents. In the application, the basic aromatic amines tend to act as more effective anchoring groups than neutral nitro-aromatics. However, aromatic amines possess a strong risk of yellowing under outdoor exposure. Attempts to overcome this by introducing aliphatic amines to the epoxy function also have been described, but this did not afford very effective dispersants [87].

Better performance characteristics for the anchoring B block were obtained through the introduction of tertiary aliphatic amino groups, for instance by the use of dimethylaminoethyl methacrylate as monomer [88]. For the attached steric A block, methyl methacrylate, butyl methacrylate, 2-ethylhexyl methacrylate and combinations of these monomers were employed. Optionally, the alkylation of the tertiary amines on the basic polymer structure leads to high-performance dispersants for pigments with difficult to wet surfaces. Useful alkylation agents mentioned for this purpose are dimethyl sulfate, benzyl halides and methyl toluenesulfonate. Figure 4.49 shows a typical representative of this dispersant class.

Recently, the introduction of heterocyclic aromatic rings has become available for GTP, through the epoxy-functional intermediates shown in Fig. 4.48 [89].

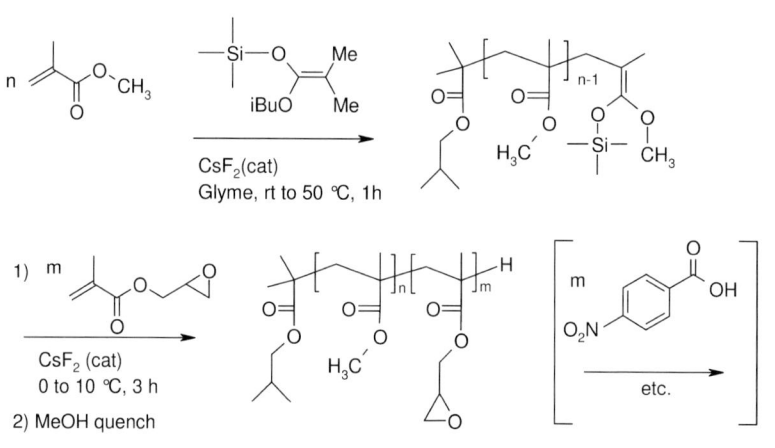

Fig. 4.48 AB block copolymer (MMA–GMA) for further modification via GTP.

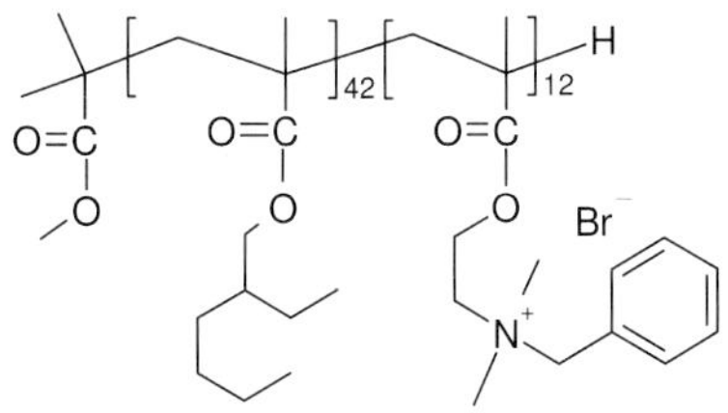

Fig. 4.49 GTP pigment dispersant based on a quaternized 2-EHA–DMAEMA block copolymer.

The literature mentions 2-methylimidazole, 4-methylimidazole and morpholine as moieties for anchoring, which can be added to the epoxide-groups.

The synthesis of ABC block copolymers has been described, in which the B block comprises the anchoring units and both the A and C blocks function primarily as steric stabilizer blocks [90]. Monomers used for the A and C blocks are alkyl and aryl methacrylates, with the addition that the C block also contains hydroxy-functional monomers. This modification was done with the intention of introducing reactive groups into the dispersant, which are able to cross-link with the coating composition into which it will be incorporated, such as a two-component polyurethane system. In this way, the dispersant molecule becomes an integral part of the cured film matrix to avoid intercoat adhesion problems, when a clear coat is applied over a pigmented base coat containing the dispersant. For the anchoring block B, dialkylaminoethyl methacrylates are used as in previous cases, with the option for quaternization. The molecular weight of the total ABC block copolymer can be found in the same region as for AB blocks, below 20 000 g mol^{-1}.

Worth mentioning for this example is that the incorporation of hydroxy-functional monomers via GTP is far from trivial and needs protection of the OH function prior to the polymerization by silane blockage, an example being trimethylsiloxyethyl methacrylate. The silyl group is removed after polymerization in the quenching step. Nevertheless, distillation of the volatile solvents, deblocked protection groups and residual monomers is required at the end of the process. A typical monomer sequence for a final ABC dispersant is shown in Fig. 4.50, within each block, A and C, the sequence of monomers being in random order.

Despite powerful control of the polymer architecture, GTP has found only limited industrial use. Disadvantages of this method are the stringent requirements on the purity of the reactants and the limited availability and difficult handling of the reactive silylketene acetal initiators, leading to relatively high

[MMA$_{15}$–BMA$_{13}$]-b-[DMAEMA$_{10}$]-b-[MMA$_{18}$–BMA$_{15}$–HEMA$_3$]

Fig. 4.50 ABC block copolymer via GTP (40 mol% alkylation of amino groups in B block by benzyl chloride not shown; MMA = methyl methacrylate, BMA = butyl methacrylate, DMAEMA = N,N-dimethylamino methacrylate and HEMA = 2-hydroxyethyl methacrylate).

processing costs. Also, the range of functional monomers is relatively limited to structures without acidic hydrogen atoms or special monomers with protected groups.

4.4.2
Dispersants Made by Controlled Free Radical Polymerization (CFRP)

In recent years, promising new techniques for the controlled free radical polymerization (CFRP) of vinylic monomers have been developed. The most prominent methods are atom-transfer radical polymerization (ATRP) [91] and nitroxyl-mediated controlled free radical polymerization (NMP) [92]. Both techniques allow for the "living" polymerization of a broad range of acrylic monomers under conditions which are attractive for industrial use. The basic working principle of these techniques of CFRP is similar. A growing polymer chain with its active radical chain end is reversibly capped by a terminating species. In the case of ATRP, a halogen-terminated polymer chain is reversibly activated by a Cu(I) complex. In the case of NMP-mediated CFRP, a stable nitroxyl radical reversibly caps the growing polymer chain. This is illustrated in Fig. 4.51.

4.4.2.1 Atom Transfer Radical Polymerization (ATRP)

ATRP has been used for the synthesis of new block copolymer-type and comb copolymer-type pigment dispersants [93]. Figure 4.52 shows an example of the synthesis of an acrylic block copolymer pigment dispersant with aminic anchoring groups by ATRP.

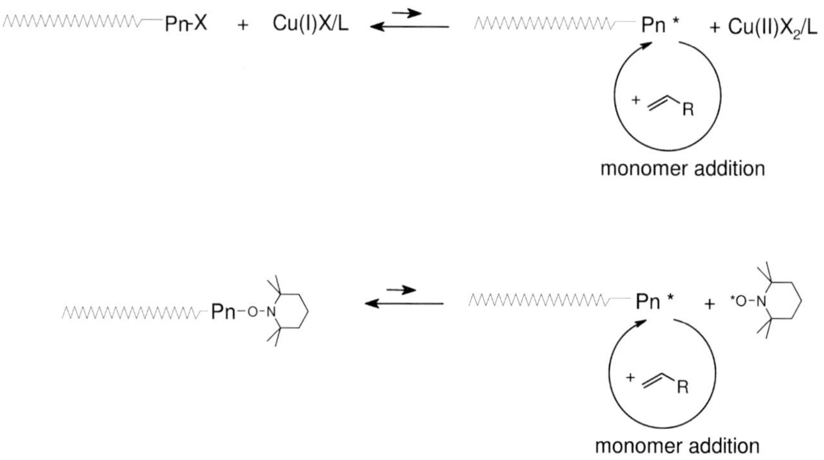

Fig. 4.51 Mechanism of controlled free radical polymerization (X=halogen. L=ligand, Pn=polymer chain with *n* repeat units).

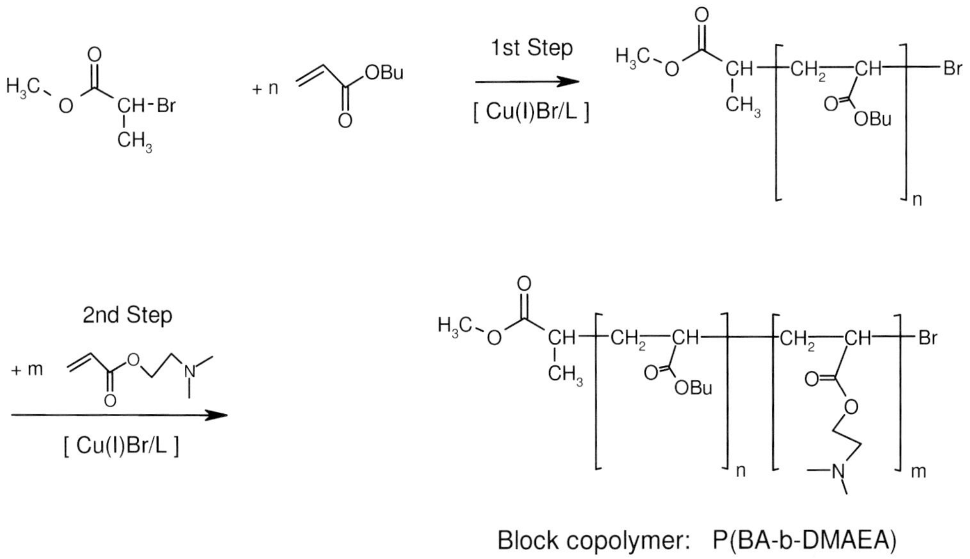

Fig. 4.52 Synthesis of block copolymer dispersant P [BA-b-DMAEA] by ATRP.

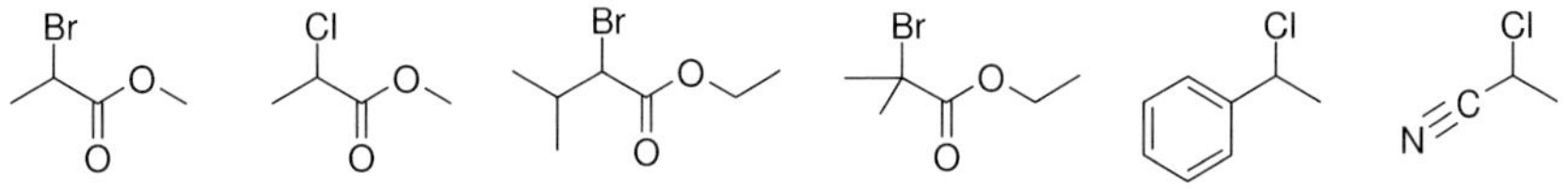

Fig. 4.53 Examples of ATRP initiators.

Examples of suitable polymerization initiators, which contain a radically transferable atom or group X, and detailed kinetic and mechanistic studies are available in the literature [94] and some of the materials are shown in Fig. 4.53. These α-halo ketones, esters and other halogenated compounds are readily available and are activated by a Cu salt (CuBr or CuCl) stabilized by a bidentate aminic ligand, such as bipyridine or ethylenediamine derivatives.

Monomer selection for ATRP is less restrictive than in the case of GTP [95]. The direct polymerization of hydroxy-functional monomers is possible. In addition, both acrylic and methacrylic, and also aromatic vinyl, monomers can be used to construct the polymer, which enlarges the pool of available monomers. Unfortunately, acidic monomers cannot be reacted and therefore any acidic functions in the polymeric backbone have to be generated indirectly, e.g. by hydrolysis of incorporated *tert*-butyl acrylate. For pigment affinity in dispersants, ATRP allows the use of basic (tertiary amine) monomers.

The transition between the two amphiphilic blocks (each based on either hydrophilic or hydrophobic monomers) does not necessarily have to be abrupt. Gradient block copolymers have been synthesized by ATRP [96], which show advantages in reduced foam stabilization in the formulated paint, as the surface activity for the air/liquid interface is reduced with the smooth transition of the polarity along the chain. Further, gradient polymers tend to show less haze in a final coating application, as the tendency to form micelle structures in the liquid paint is minimized as a result of the reduced surfactant character. For the preparation of the gradient along the AB block, the monomers of the second block are dosed into the polymerizing reaction mixture containing monomers for the first block at a defined rate chosen such that the conversion of the first monomer has been completed before the end of the dosing process. However, the deliberate dilution and concentration gradient of the anchoring moieties in the pigment-affinic block are expected to lead to less efficient products and a higher risk of pseudoplastic behavior in the pigment paste.

Although the Cu(I) catalyst and its ligands appear to be relatively cheap raw materials, catalyst removal (copper salts) remains a challenge for industrial scale-up. The same applies to halide impurities. Although on a small laboratory scale purification techniques such as precipitation, filtration (e.g. by column chromatography), centrifugation and extraction have all been shown to provide final products with low Cu levels, in the ppm range, such techniques are relative expensive for industrial applications.

4.4.2.2 Nitroxyl-mediated Radical Polymerization (NMP)

Nitroxyl-mediated controlled free radical polymerization broadens the basis for future developments of pigment dispersants by CFRP, as functional monomers and conditions suitable for industrial production become available.

The suitability of TEMPO for the polymerization of styrenic monomers has been known for a long time; however, this polymerization regulator is unsuitable for acrylic monomers (Fig. 4.54) [92].

Only recently, new nitroxide-based (NOR) polymerization regulators have been developed which are able to effect controlled NMP of a wide range of acrylic monomers [97]. Acrylates require regulators with greater steric hindrance than styrenic compounds. A general structure for such regulators and some representative compounds are shown in Fig. 4.55. At least two of the R_1–R_4 groups need to be larger than methyl. Regulators with two ethyl groups adjacent to the N-atom proved to be especially suitable for acrylates and acrylamides, and also for styrenic monomers. Their use for both acrylates and methacrylates has been reported [92]. The nitroxides are readily prepared by oxidation of the corresponding secondary amine or hydroxylamine. The R group is a moiety that is bound to the oxygen via a C-atom and can be introduced by adding a C-radical to the N–O residue.

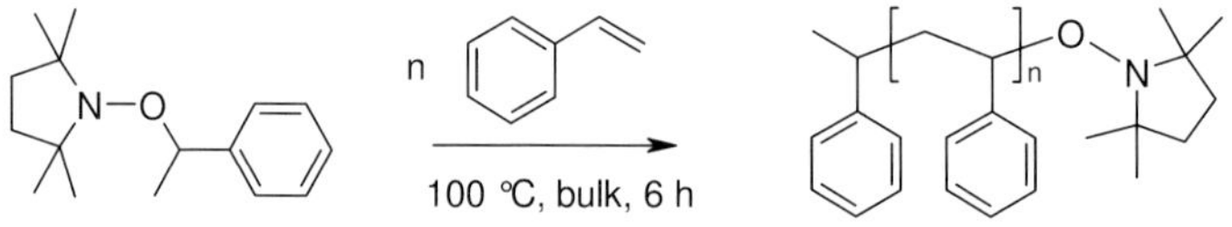

Fig. 4.54 NMP polymerization of styrene by TEMPO.

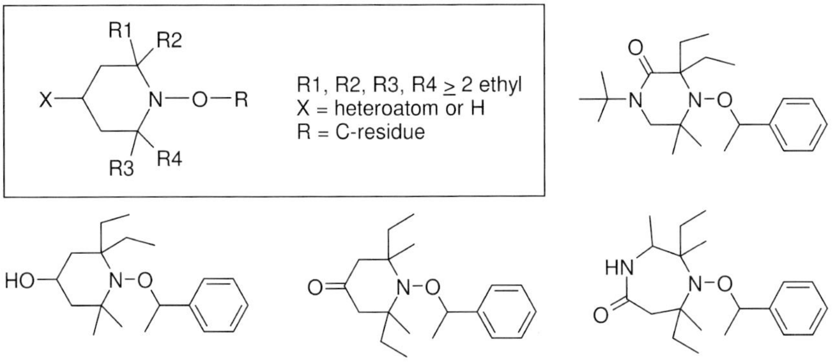

Fig. 4.55 NOR polymerization regulators.

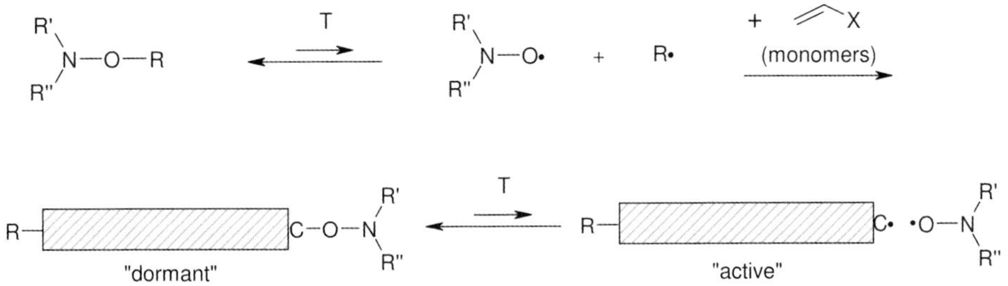

Fig. 4.56 Schematic NMP polymerization scheme using NOR regulators.

From a mechanical point of view, the polymerization proceeds through an equilibrium between "dormant" and "active" species, as illustrated in Fig. 4.56. The direction of this equilibrium is controlled by the reaction temperature. At a minimum temperature of e.g. 120–130 °C, some of the six-membered NOR regulators show a significant concentration of active species, so that effective chain growth can take place. Polymerization proceeds via addition of monomers to the active polymer radicals which are formed on the growing chain by a reversible termination process.

Compounds such as those shown in Fig. 4.55 are useful for industrial applications because the reaction temperature for activation of the C–O bonds is around 100–150 °C. In this temperature range, most common monomers are still below their boiling-points. The reaction is not sensitive towards impurities present in technical-grade monomers, such as humidity and inhibitors, which makes this process very robust. For coating applications, there is no specific need to remove or convert regulator fragments after the process, as in the case of ATRP or GTP, because the fragments of the NOR polymerization regulator are readily compatible with coating resins.

In the course of the work on the development of NOR regulators, structure–activity relationships in the polymerization of n-butyl acrylate, as shown in Fig. 4.57, were established, which are in accordance with semiempirical molecu-

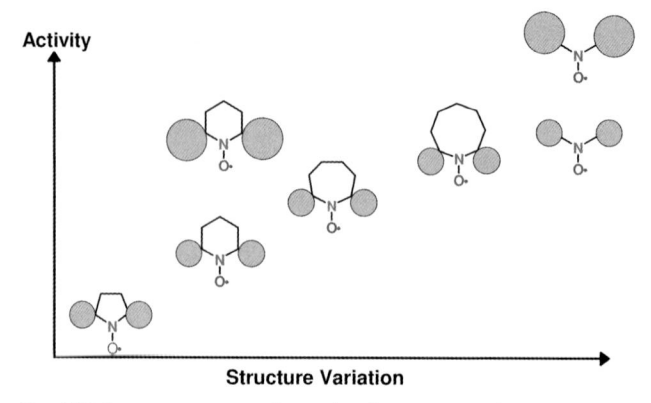

Fig. 4.57 Structure–activity relationship for NOR as polymerization regulators (model system n-butyl acrylate, bulk).

Fig. 4.58 Dispersant synthesis through NOR-mediated CFRP.

lar orbital calculations [98]. The activity increases and therefore the activation temperatures required decrease, with larger nitroxide rings and open structures, while the same effect is found by increasing the steric bulk around the N-atom at constant ring size.

For dispersant design, AB block copolymers can be produced by sequential addition of acrylic monomers as shown in Fig. 4.58 with an n-butyl acrylate–dimethylaminoethyl acrylate block structure. The process conditions are tuned such that conversion of the monomers in each step is optimized with respect to time, temperature and avoidance of side-reactions. One of the unwanted events is termination of growing chains by elimination of the regulator fragment or recombination of free radicals to "dead" chains. The amount of "dead" versus "living" chains can be expressed by the term *block efficiency*, which is the ratio between "living" chains and total amount of chains in a reaction mass. In industrial reality, it is feasible to reach up to 80% conversion of monomers with a block efficiency of more than 95% by choosing the right process conditions. Conversions lower than 100% are not a limiting factor for NOR technology, since excess monomers can be easily removed by distillation at the point of desired conversion and reutilized in subsequent polymerizations. However, monomer removal in industrial practice cannot always be achieved to the full 100% and a certain low level of residuals of the A block monomers is accepted, before addition of the B block monomers. As a consequence, the second block will contain certain low amounts of the A block monomers.

Monomers used for dispersant synthesis include those for random acrylic copolymers. For AB block structured dispersant synthesis, monomers for the steric block (A) are polymerized separately, followed by addition of anchoring moieties for block B. For the latter, dialkylaminoacrylates and vinylic amino aromatics, such as 2- and 4-vinylpyridine, are well suited. Tertiary aliphatic amines can be easily transformed into quaternary salt structures showing enhanced properties for selected pigment classes.

For polar solvent systems and water-based applications, polyether-based monomers are incorporated into the A block whereas for more nonpolar systems, acrylic alkyl esters are highly useful.

Figure 4.59 shows an example of the superior application performance of acrylic AB block copolymer dispersants made by controlled free radical polymer-

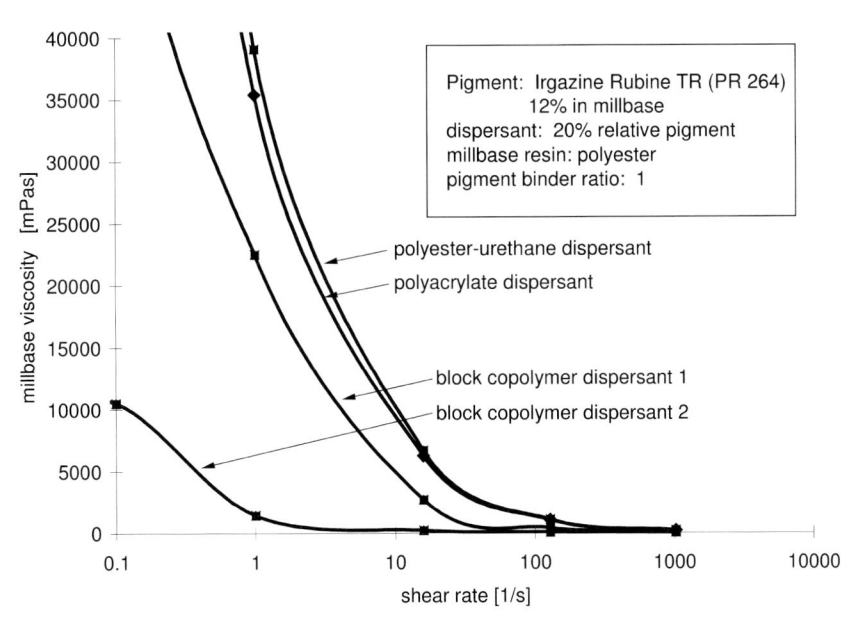

Fig. 4.59 Rheological performance of block copolymer dispersants in a millbase of a transparent organic pigment. Block copolymers 1 and 2 are acrylic block copolymers with different pigment-affinic groups.

ization. This example demonstrates that the viscosity of pigment concentrates can be well controlled by properly selected block copolymer dispersants. The performance with respect to specific pigments can be optimized by changing the pigment-affinic groups. Such tailor-made dispersants allow the use of higher pigment loads in the grinding step and offer improved flocculation stability in paint formulations.

4.4.2.3 Radical Addition and Fragmentation Transfer (RAFT)

RAFT was developed as a method for controlled polymerization of acrylic monomers, involving the addition of a growing chain with a radical end to other chain ends capped with a dithio ester [99]. After addition, the resulting stabilized radical will fragment back into the same radical that had been captured before or it liberates a new radical fragment. Both released fragments can then add a monomeric species (Fig. 4.60), usually a methacrylate.

In the literature, there appear to be no examples of commercial pigment dispersants based on RAFT technology, although excellent molecular architecture control and narrow molecular weight distributions are obtained. This is mainly because part of the chain ends are capped with dithio esters, to afford smelly and deep-yellow products, which need to be further processed, e.g. by peroxide treatment. Unfortunately, the availability of the RAFT regulators on a technical scale is limited and involves a multi-step synthesis.

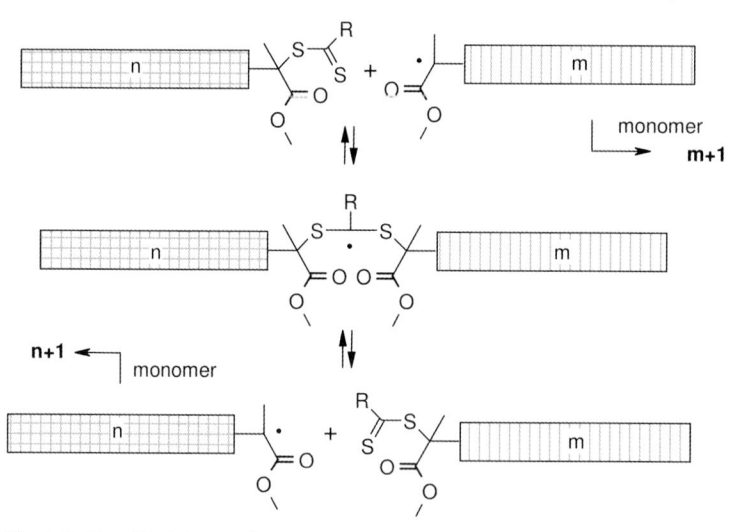

Fig. 4.60 Simplified RAFT polymerization scheme.

4.4.3
Dispersants Based on Macromonomers

Macromonomers can be synthesized by different well-known routes. ATRP offers a new, convenient way to synthesize macromonomers avoiding the use of mercaptans; see, for example, Fig. 4.61 [100]. Copolymerization of such macromonomers with ethylenically unsaturated monomers leads to comb polymers with interesting dispersing properties. It is assumed that the arrangement of polymeric chains with polar properties in the grafts and nonpolar properties in the main backbone (or the inverse situation) leads to increased polarity differences within the molecular structure as opposed to linear AB blocks and therefore increased amphiphilic character of the dispersant.

The reactive poly(butyl acrylate)s obtained according to Fig. 4.61 can be copolymerized with other monomers in a classical fashion (e.g. initiated by AIBN) to afford the desired comb structures. Preferred comonomers with pigment-affinic properties include acrylic acid, 1-vinyl-2-pyrrolidone, N,N-dimethylacrylamide and N,N-dimethylaminoethyl acrylate.

4.4.4
Dispersants Based on Dendrimers and Hyperbranched Polymers

The patent literature describes one case in which dendrimers were selected as defined building blocks to attach anchoring groups and steric side-chains for applications in solvent-based coatings [101]. Dendrimers with polyester backbones are commercially available materials based on 2,2-dimethylolpropionic acid [102], which has the unique functionality of one –COOH and two –OH groups. The shell of this dendrimer consists of a large number of OH groups, where the multi-functionality depends on the type of core used and the number of generations. For dispersant design, part of the outer –OH groups were functionalized with pigment anchoring groups, while to the other part compatibility- and steric hindrance-enhancing moieties were attached (Fig. 4.62). Options for the anchoring groups are the classical nitrogen-containing structures, and also carboxylic, phosphoric and sulfonic esters. Specific examples for the steric side-chains are EO/PO-type polyethers, polyesters based on polycaprolactone and fatty acids (stearic, 12-hydroxystearic and ricinoleic acid).

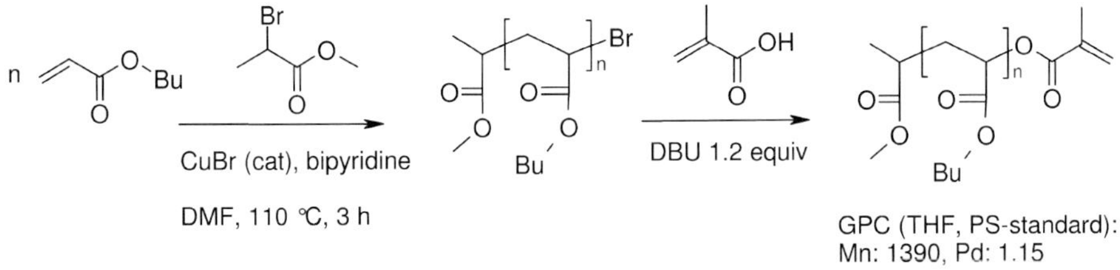

Fig. 4.61 Synthesis of macromonomers by ATRP.

X = polyol *(pentaerythritol)*

B = building block *(2,2-DMPA)*

n = 2 - 5

A = anchoring group *(phthalic anhydride)*

R = steric side chain *(stearic acid)*

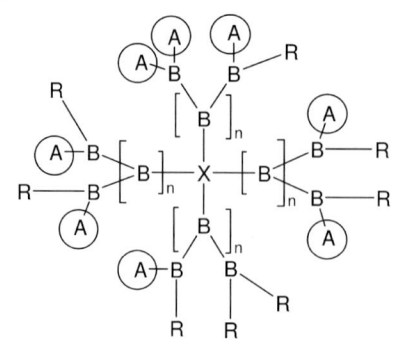

Fig. 4.62 Dendritic dispersant (schematic).

The technical advantage of the use of a dendrimer as base structure is given by its three-dimensional spherical architecture. This provides that the molecule on one hand is highly flexible to orient its anchoring groups and steric chains into the desired directions and on the other hand occupies a smaller hydrodynamic volume as compared with linear structures with similar molecular weights. This feature allows low, Newtonian-like viscosities in the millbase of pigment preparations.

References

1 (a) Schofield, J.D. In *Handbook of Coating Additives*, Calbo L.J. (Ed.), Marcel Decker, New York, **1992**, pp. 71–104; (b) Pirrung, F.O.H., Quednau, P.H., Auschra, C. *Chimia* **2002**, 56, 170–176.
2 (a) Bieleman J.H. In *Lackharze*, Stoye, D., Freitag, W. (Eds.), Carl Hanser, Munich, **1996**, p. 647; (b) Bieleman, J.H. In *Lackadditive*, Bieleman J.H. (Ed.), Wiley-VCH, Weinheim, **1998**, p. 67.
3 Herbst, W. Hunger, K. *Industrial Organic Pigments*, VCH, Weinheim, **1997**.
4 Verwey, E.J., Overbeek, J.T.G. *Theory of Stability of Lyophobic Colloids*, Elsevier, Amsterdam, **1948**.
5 Napper, D.H. *Polymeric Stabilization of Colloidal Dispersions*, Academic Press, London, **1983**.
6 E.g. according to ISO 3251 (in %).
7 E.g. according to DIN 53402 (in mg KOH g^{-1}).
8 E.g. according to DIN 16945 (in mg KOH g^{-1}).
9 E.g. according to ISO 4630 (in Gardner units).
10 E.g. according to DIN 51757 (in g cm^{-3}).
11 Van den Haak, H.J.W. *J. Coat. Technol.* **1997**, 69, 137–142.
12 (a) EP 270126, **1986** (Byk-Chemie); (b) EP 318999, **1987** (Byk-Chemie); (c) EP 438836, **1989** (EFKA Chemicals).
13 For methods, see for example Bartl, H., Falbe, J. (Eds.), *Houben-Weyl, Methoden der Organischen Chemie*, Vol. E 20, Makromolekulare Stoffe, Georg Thieme, Stuttgart, **1987**, pp. 1735–1751.
14 EP 520586, **1991** (EFKA Chemicals).
15 EP 154678, **1984** (Byk-Chemie).
16 Brode, G.L., Koleske, J.V. *J. Macromol. Sci. Chem.* **1972**, A6, 1109–1144.
17 Cerrai, P.,Tricoli, M., Andruzzi, F., Paci, M. *Polymer* **1989**, 30, 338.
18 US 4032698, **1975** (Du Pont).
19 US 4070388, **1976** (DuPont).
20 (a) Karrer, P. *Organic Chemistry*, English Translation, Elsevier, Amsterdam, **1938**, p. 135. (Zerewitinoff Hydrogen: active hydrogen that can be determined by treatment with the Grignard reagent methylmagnesium iodide, which reacts

with any acidic hydrogen atom to form methane. This gas can be determined quantitatively); (b) Saunders and Frisch. *Polyurethanes: Chemistry and Technology*, Interscience, New York, **1963** (Part I) and **1964** (Part II).
21 EP 956153, **1996** (EFKA Additives).
22 EP 318999, **1987** (Byk-Chemie).
23 US 5969002, **1996** (Bayer).
24 Hond, R. *Verfkroniek* **1994**, *6*, 44.
25 EP 129329, **1983** (Allied Colloids).
26 Connah, M. *Verfkroniek* **1995**, *2*, 17–19.
27 (a) EP 127388, **1983** (Allied Colloids); (b) EP 185458, **1984** (Allied Colloids).
28 EP 1562696, **2002** (EFKA Additives).
29 EU Directive 2004/42/EG, **2004**:: VOC=volatile organic compounds, i.e. b.p. <250°C at 1013 hPa.
30 (a) EP 311157, **1987** (EFKA Chemicals); (b) US 5688858, **1987** (EFKA Chemicals).
31 EP 1081169, **1999** (Byk-Chemie).
32 US 6680355, **1998** (BASF).
33 (a) DE 2934642, **1978** (PPG Industries); (b) US 4302561, **1980** (PPG Industries).
34 US 5134187, **1988** (Kansai Paint).
35 EP 879860, **1997** (Byk-Chemie).
36 EP 595129, **1992** (Th. Goldschmidt).
37 Spyrou, E. *Farbe Lack* **2000**, *10*, 126–130.
38 US 6423785, **1999** (Goldschmidt).
39 E.g. Brydson, J.A. *Plastics Materials*, Elsevier, Amsterdam, **1999**, p. 450.
40 Examples of polyalkylene glycol monoalkyl monoamines are the so-called Jeffamine M series of polyethers available from Huntsman Corporation.
41 (a) US 4766245, **1988** (Texaco); (b) US 6465606, **1993** (Huntsman).
42 EP 208041, **1985** (ICI).
43 EP 158406, **1984** (ICI).
44 Meyers, W. E., Royer, G. P. *J. Am. Chem. Soc.* **1977**, *99*, 6141.
45 Zhuk, D. S., Gembitskii, P. A., Kargin V. A. *Russ. Chem. Rev.* **1965**, *34*, 515–527.
46 Gembitskii, P. A., Chmarin, A. I., Kleshcheva, N. A., Zhuk, D. S. *Polym. Sci. USSR* **1978**, *20*, 1695–1702.
47 Kern, W., Brenneisen, E. *J. Prakt. Chem.* **1941**, *159*, 193.
48 JP 2003-186584 (USPA 20040266911), **2003** (Toyo Ink).
49 US 4528347, **1983** (Nitto Boseki).
50 US 2456428, **1944** (Shell).
51 Schildkecht, C. E. In *Allyl Compounds and their Polymers*, Wiley-Interscience, New York, **1973**, pp. 29–30.
52 (a) US 6965052, **2003** (University of Pittsburgh); (b) US 4942259, **1988** (Air Products and Chemicals)
53 Badesso, R. J., Nordquist, A. F., Pinschmidt, R. K., Jr., Sagl, D. J. In *Hydrophilic Polymers: Performance with Environmental Acceptance*, Glass, E. (Ed.), American Chemical Society, Washington, DC, **1995**, p. 489.
54 (a) US 6787600, **1999** (Lubrizol Corporation); (b) EP 713894, **1994** (Daicel Chemical Industries); (c) US 6583213, **1998** (Avecia); (d) JP 09157374, **1995** (Daicel Chemical Industries).
55 WO 2002/85507, **2001** (Avecia).
56 EP 893155, **1998** (Byk-Chemie).
57 (a) WO 2000/24503, **1998** (Avecia); (b) WO 2000/46313, **1999** (Avecia).
58 US 4224212, **1978** (ICI).
59 Cherdron, H., Ohse, H. Korte, F. *Makromol. Chem.* **1962**, *56*, 179.
60 Yamashita, Y., Tsuda, J., Okada, M., Ivatsuki, S. *J. Polym. Sci., Part A-1* **1966**, *4*, 2121.
61 US 6197877, **1996** (Zeneca).
62 (a) WO 2001/80987, **2000** (Avecia); (b) WO 2002/85507, **2001** (Avecia).
63 (a) WO 98/19784, **1996** (Zeneca); (b) WO 99/49963, **1998** (Avecia).
64 WO 2001/21298, **1999** (Avecia).
65 WO 2004/78333, **2003** (Lubrizol).
66 JP 8010601, **1994** (Daicel Chemical Industries).
67 EP 768321, **1995** (Ajinomoto).
68 (a) JP 12204281, **2000** (Ajinomoto); (b) JP 987537, **1997** (Ajinomoto).
69 US 6878799, **2001** (King Industries).
70 WO 95/17473, **1993** (Zeneca).
71 E.g. Stache, H., Kosswig, K. (Eds.). *Tensid-Taschenbuch*, Carl Hanser, Munich, **1990**.
72 (a) JP 2000-265098, **2000** (Fuji); (b) JP 2001-384798, **2001** (Maekawa S., Tanaka H.).
73 Davies, J. T., Rideal, E. K. *Interfacial Phenomena*, 2nd edn., Academic Press, New York, **1963**.
74 US 5376170, **1993** (Petrolite).
75 (a) WO 95/34593, **1994** (Zeneca); (b) US 5785894, **1994** (Zeneca); (c) US 5981624, **1996** (Zeneca).

76 (a) EP 417490, **1990** (Byk-Chemie); (b) US 6562897, **1998** (Avecia).
77 WO 2005/085261, **2004** (EFKA Additives).
78 EP 1518882, **2003** (Nippon Shokubai).
79 US 4597898, **1982** (Procter and Gamble).
80 (a) Johnson, A. F., Mohsin, M. A., Meszena, Z. G., Graves-Morris, P. *JMS Rev. Macromol. Chem. Phys.* **1999**, *C39*, 527; (b) Fischer H., *J. Polym. Sci., Part A: Polym. Chem.* **1999**, *37*, 1885.
81 (a) Webster, O.W. *Science* **1991**, *251*, 887; (b) Duivenvoorde, F. L., van Nostrum, C. F., Laven, J., van der Linde, R. *J. Coat. Technol.* **2000**, *72*, 145.
82 Jakubauskas, H. L. *J. Coat. Technol.* **1986**, *58*, 71.
83 (a) Webster, O.W., Hertler, W. R., Sogah, D.Y., Farnham, W. B., RajanBabu, T.V. *J. Am. Chem. Soc.* **1983**, *105*, 5706; (b) Sogah, D.Y., Hertler, W. R., Webster, O.W., Cohen, G.W. *Macromolecules* **1987**, *20*, 1473; (c) US 4417034, **1982** (Du Pont).
84 (a) Spinelli, H. J. *Prog. Org. Coat.* **1996**, *27*, 255; (b) US 4812517, **1987** (Du Pont); (c) WO 99/03938, **1998** (Du Pont).
85 Webster, O. W. *Adv. Polym. Sci.* **2004**, *167*, 1–34.
86 US 6306521, **2000** (Du Pont).
87 US 4656226, **1985** (Du Pont).
88 US 4755563, **1986** (Du Pont).
89 US 6316564, **1999** (Du Pont).
90 US 6413306, **1999** (Du Pont).
91 (a) Matyjaszewski, K., Xia, J. *Chem Rev.* **2001**, *101*, 2921; (b) WO 96/30421, **1995** (Carnegie-Mellon University).
92 US 4581429, **1984** (CSIRO).
93 (a) WO 00/40630, **2000** (Ciba SC); (b) WO 01/51533, **2001** (Ciba SC); (c) WO 01/44376, **2001** (PPG); (d) WO 01/44388, **2001** (PPG); (e) WO 01/44389 **2001** (PPG).
94 (a) WO 96/30421, **1995** (Carnegie-Mellon University); (b) WO 98/01480, **1996** (Carnegie-Mellon University).
95 WO 00/40630, **2000** (Ciba SC).
96 EP 1416019, **2003** (Byk-Chemie).
97 (a) Zink, M. O., Kramer, A., Nesvadba, P. *Macromolecules* **2000**, *33*, 8106; (b) GB 2335190, **2000** (Ciba SC); (c) GB 2342649, **2000** (Ciba SC).
98 Moad, G., Rizzardo, E. *Macromolecules* **1995**, *28*, 8722.
99 Mayadunne, R. T. A., Rizzardo, E., Chiefari, J., Chong, Y. K., Moad, G., Thang, S. H. *Macromolecules* **1999**, *32*, 6977.
100 US 6936656, **2000** (Ciba SC).
101 WO 02/057004, **2001** (EFKA Additives).
102 (a) Johansson, M., Malmström, E., Hult, A. *Eur. Coat. J.* **2002**, *7–8*, 26–33; (b) SE 503 342, **1994** (Perstorp).

5
Polymeric Surfactants *

Henri Cramail, Eric Cloutet, and Karunakaran Radhakrishnan

5.1
Introduction

Surface-active polymers or polymeric surfactants today represent a very large and important class of materials that find many applications as emulsifiers, dispersion stabilizers, compatibilizers and wetting agents [1]. A variety of polymeric surfactants – reactive or not –, ionic or neutral, composed of linear or nonlinear arrangements of chemically different polymeric chains such as block, graft and star copolymers, and also end-functionalized oligomers (see Scheme 5.1), have been reported. Among them, block copolymers and in particular amphiphilic block copolymers have a predominant place.

Research on block copolymers as polymeric surfactants has been ongoing for a long time and the colloidal behaviors of these materials have been analyzed from both experimental and theoretical points of view by many research groups [2–7]. In most cases, the different blocks are incompatible, leading to the formation of well-defined self-assembled structures both in bulk and in selective solvents. Their specific colloidal properties make them very attractive for a plethora of applications.

The surface activity of polymeric surfactants differs from the classical surface-active behavior of low molar mass surfactants. It is now well established that polymeric surfactants have lower diffusion coefficients [8, 9] and lower critical micelle concentrations (CMC) than common surfactants such as sodium dodecyl sulfate (SDS). Typically, the CMC values of polymeric surfactants [10–12] are in the concentration range between 10^{-9} and 10^{-4} mol L^{-1} whereas those of low molar mass and ionic surfactants range from 10^{-3} to 1 mol L^{-1}. The low CMCs values are advantageous for many applications, since only traces of polymer are required to form micelles and, for instance, dilution, which is problematic in the case of classical surfactants, does not alter polymeric micelles.

---

* A List of Abbreviations can be found at the end of this chapter.

Macromolecular Engineering. Precise Synthesis, Materials Properties, Applications.
Edited by K. Matyjaszewski, Y. Gnanou, and L. Leibler
Copyright © 2007 WILEY-VCH Verlag GmbH & Co. KGaA, Weinheim
ISBN: 978-3-527-31446-1

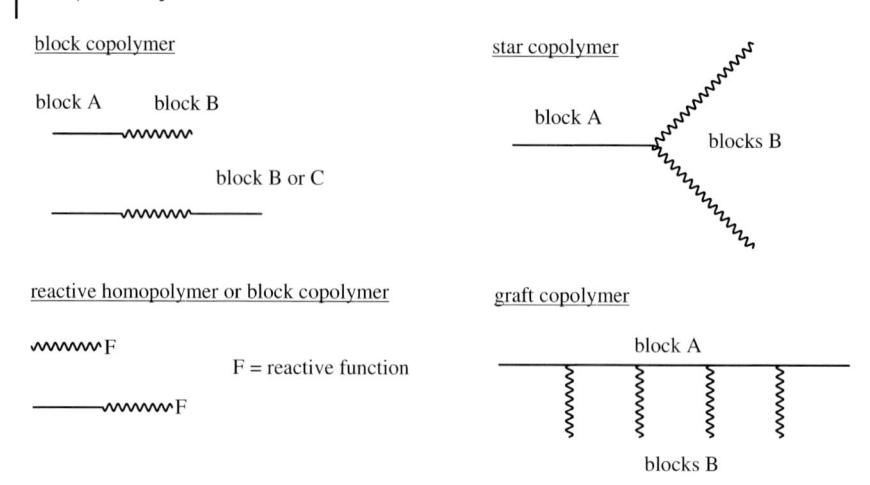

Scheme 5.1 Typical structures of polymeric surfactants.

It must also be pointed out that polymeric surfactants are generally strongly attached to the colloidal particles either by physical adsorption or covalent chemical bonding and stabilize the colloidal particles by steric stabilization, whereas low molar mass surfactants stabilize them by electrostatic interaction. Consequently, polymeric surfactants do not desorb so easily from the colloidal particles, avoiding the drawbacks of low molar mass surfactants, which gather, during the application of the polymer dispersions as film or bulk material, in hydrophilic spots. In addition, the shorter range steric stabilization enables more latex particles to be packed in a given volume and hence high solid content latexes with relatively low viscosity can be prepared. Further, amphiphilic block copolymers can offer novel property profiles at interfaces in comparison with their low molar mass analogues.

The objective of this chapter is to give a general overview of the main polymers used as surfactants in heterogeneous polymerizations, with special emphasis on the synthesis of these well-defined macromolecular structures. The continuous progresses in controlled/living ionic and free-radical polymerizations have allowed the synthesis of macromolecular surfactants, with controlled composition, dimension, functionality and architecture. Specific attention will be also paid towards reactive polymeric surfactants that nowadays represent an important class of materials in the field [13–16].

Many possibilities were available to us to present such a wide domain of research. It appeared to us that a discrimination of polymeric surfactants according to the main families used was the most convenient method to help the reader in finding a suitable polymeric surfactant capable of being applied in an appropriate way for a given polymerization system. Therefore, this chapter has been divided into two main sections. In the first, the synthesis of surfactants – reactive or not – used for aqueous emulsion and dispersion polymerizations is presented. Some colloidal features of the described surfactants and some selected latex characteristics prepared in their presence are also covered. In the second and third parts, emphasis is put on the main polymeric families used as steric stabilizers for nonaqueous dispersion polymerization, which also represents an important domain for latex synthesis. A special focus on stabilizers used in supercritical carbon dioxide (scCO$_2$), as dispersant medium, completes this section at the end of the chapter.

The number of publications in the field of polymeric surfactants is so large that an exhaustive analysis of all of them in a single chapter is not feasible, which led us to make choices. We generally selected the more recent investigations that focus on polymeric-type surfactants. We deliberately did not cover surfactants of low molar mass such as common ionic surfactants and inisurf (both initiator and surfactant) or transsurf (both transfer agent and surfactant) agents.

Although data on emulsion, mini-emulsion, suspension and dispersion polymerizations in the presence of surfactants will be given when discussing a specific surfactant, the reader will find more precise features on these different polymerization techniques in Chapter 13 in Volume 1.

5.2
Main Surfactants Used for Heterogeneous Polymerizations in Aqueous Media

Heterogeneous radical polymerization in aqueous media is one of the most important industrial processes for the production of polymers. In this chapter, we will mainly focus on the most effective polymeric surfactants used in aqueous medium as the dispersant phase. Three main types of surfactants can be discriminated: (i) the nonionic copolymers based on poly(ethylene oxide) (PEO), (ii) the anionic copolymers based on poly(acrylic acid) (PAA) and poly(methacrylic acid) (PMAA) and (iii) the copolymers containing cationic or cationizable monomer units, e.g. poly[amino(methyl)acrylate]s.

5.2.1
Synthesis and Main Uses of PEO-based Copolymers

PEO-based (co)polymers are an important class of surfactants thanks to the solubility of PEO in water and in a wide range of organic solvents, and also to its non-toxic nature. The effects of the architecture and composition of PEO-based block copolymers on their associative properties in aqueous solution have been widely studied [17, 18].

5.2.1.1 PEO-*b*-PPO and PEO-*b*-PBO Block Copolymers
PEO-*b*-PPO and PEO-*b*-PPO-*b*-PEO block copolymers are commonly used in emulsion and mini- and micro-emulsion polymerizations. PEO-*b*-PPO block copolymers are obtained by anionic polymerization and are available commercially (Pluronic, from BASF, Synperonics from ICI) with various compositions and ar-

chitectures (blocks, grafts, star blocks, etc.). Despite the great interest in and wide applications of such amphiphilic copolymers, their polydispersity in terms of both molar mass and composition impede precise and useful correlations between structure and colloidal properties of the surfactant. Therefore, the search for well-defined and high molar mass PEO-b-PPO block copolymers is still challenging. The relatively ill-defined structure of such copolymers comes from the lack of control of PO anionic polymerization, explained by an important transfer reaction to the monomer [19]. Many attempts to control the anionic polymerization of PO have been made by several groups [20, 21]. The recent results from Deffieux and co-workers are probably one of the most effective routes to achieve control of PO anionic polymerization [22]. The authors reported that addition of trialkylaluminum, playing the role of a monomer activator, to an alkali metal alkoxide initiator permits the synthesis, in hydrocarbon media, of high molar mass PPO (up to 20×10^3 g mol^{-1}). These results led the same group to perform the anionic sequential copolymerization of ethylene oxide and propylene oxide via the monomer activation mechanism. Well-defined tapered-like and di- and triblock ethylene oxide–propylene oxide copolymers of high molar mass were readily prepared in the presence of triisobutylaluminum as additive to conventional anionic initiators [23, 24]. as shown in Scheme 5.2.

This method for preparing well-defined and high molar mass di- and triblocks copolymers based on EO and PO will certainly open up novel vistas towards the investigation of the colloidal properties of such block copolymers and uses in many domains.

PPO-b-PEO block copolymers with various PEO:PPO ratios have been used as surfactants for the elaboration of PS and PMMA latexes by free-radical polymerization [25–27]. The surfactant is generally located at the interface between the aqueous and the organic phases but, in some cases, clouding of the aqueous

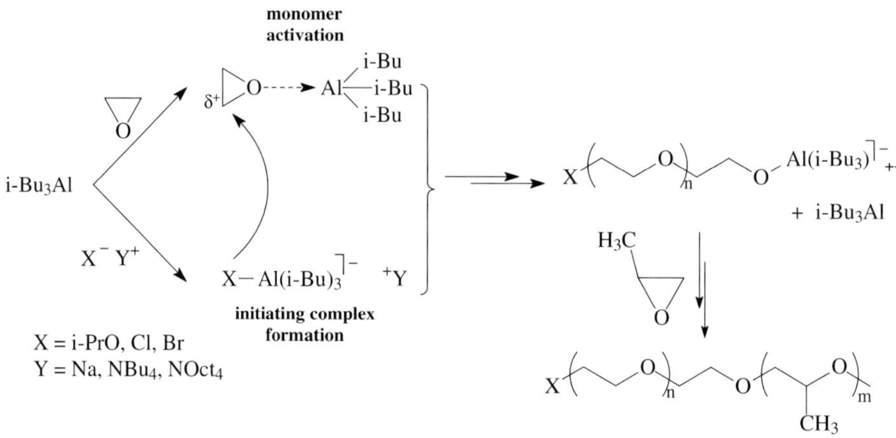

Scheme 5.2 Synthesis of well-defined and high molar mass PEO-b-PPO block copolymers by anionic polymerization (monomer activation mechanism).

5.2 Main Surfactants Used for Heterogeneous Polymerizations in Aqueous Media

phase can be observed, leading to the phase inversion phenomenon and to the formation of bimodal latex particles.

Guyot and coworkers synthesized a series of reactive PPO-*b*-PEO block copolymers end-capped with methacrylic, vinylic, allylic or maleic moieties and used them as both surfactants and monomers (also called surfmers) for the preparation of calibrated PS and acrylic latex particles [28–31]. Scheme 5.3 shows the method for the preparation of maleic-ended macromonomer based on PEO and PPO block copolymers.

The authors observed that the colloidal stability is generally improved thanks to the reactive surfactant that remains covalently linked to the particles. They also found that some of the surfactant might be buried within the particles as it is reactive throughout the polymerization process. This phenomenon may be seen as a problem in terms of the final latex properties.

The same group has developed a strategy for the synthesis of non-reactive and reactive PEO-*b*-PBO block copolymers. The reactive copolymers functiona-

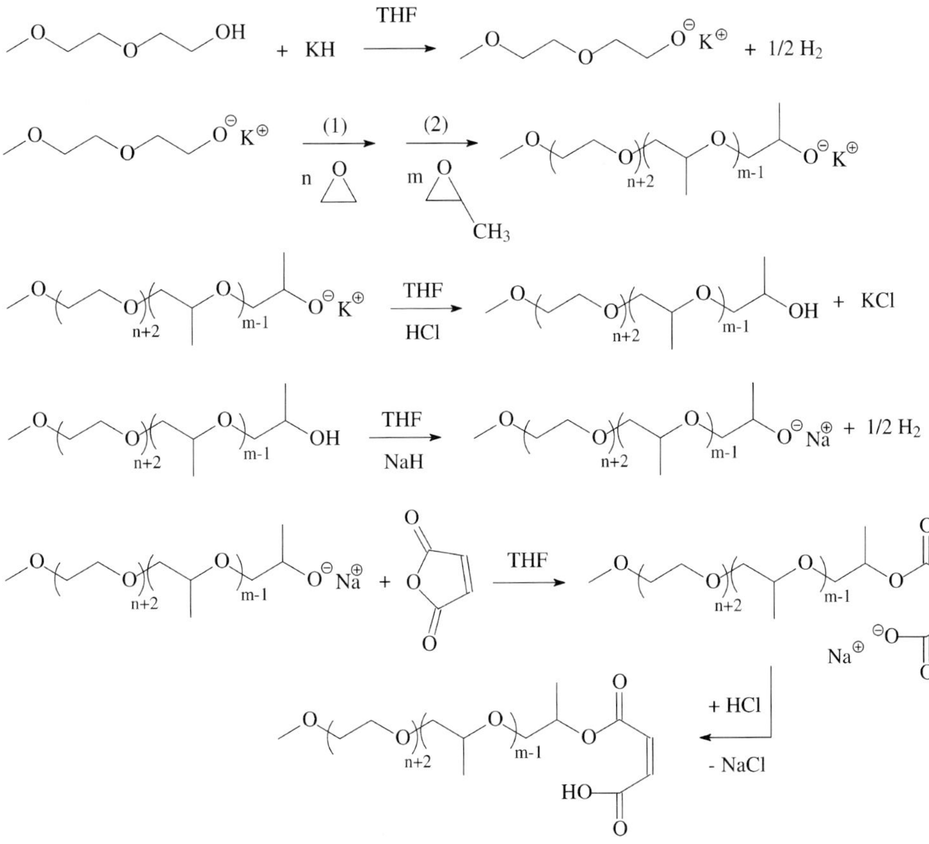

Scheme 5.3 Synthesis of maleic-type ended PEO-*b*-PPO block copolymer.

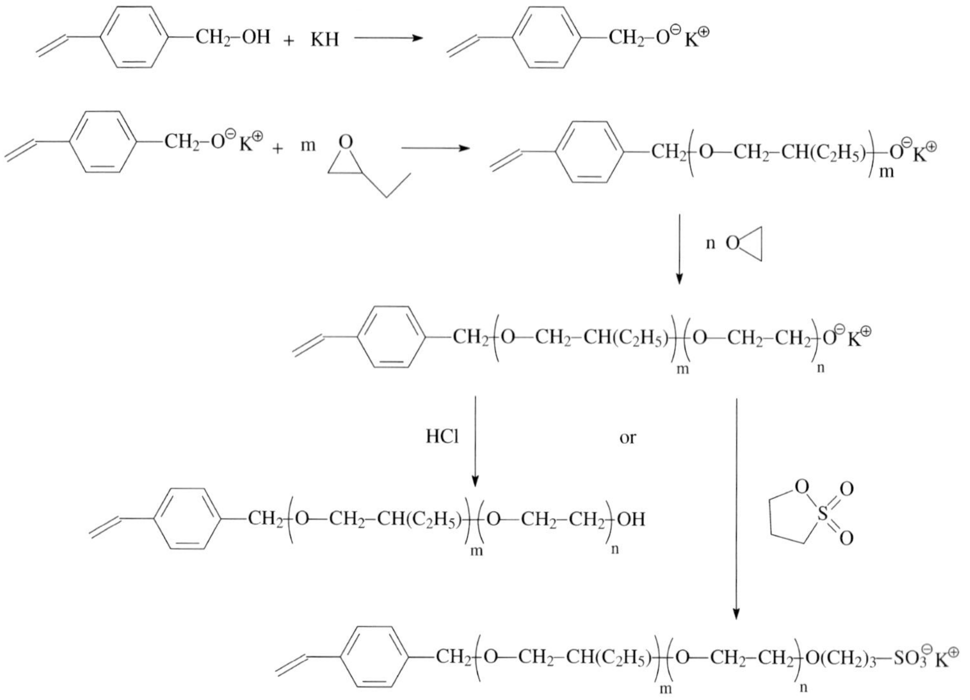

Scheme 5.4 Synthesis of nonionic and anionic styrenyl-ended PBO-b-PEO block copolymers.

lized with a styrenyl group were obtained using 4-vinylbenzyl alcohol as the initiator for the ring-opening polymerization (ROP) of the cyclic monomers. Nonionic and ionic surfactants were prepared according to the terminating agent, as depicted in Scheme 5.4 [32]. The key characteristics of seeded emulsion polymerization of MMA with these surfactants (both nonionic and ionic) are lesser coagulum with the anionic-type surfactants and more stable latexes with the reactive type [33].

5.2.1.2 PS-b-PEO and Derivatives

Polystyrene-b-poly(ethylene oxide) copolymers are widely used in heterogeneous polymerization in aqueous media. Such copolymers can be synthesized by living anionic polymerization or a combination of living anionic and controlled radical polymerizations.

Synthesis of PS-b-PEO by anionic route Lithium-based anionic polymerization initiators are generally inefficient for the ring-opening polymerization of EO due to the formation of a strong lithium–oxygen bond formed during the initiation step. However, this was overcome (a) by use of addition of certain promoters such as strong polar solvents such as dimethyl sulfoxide with either potas-

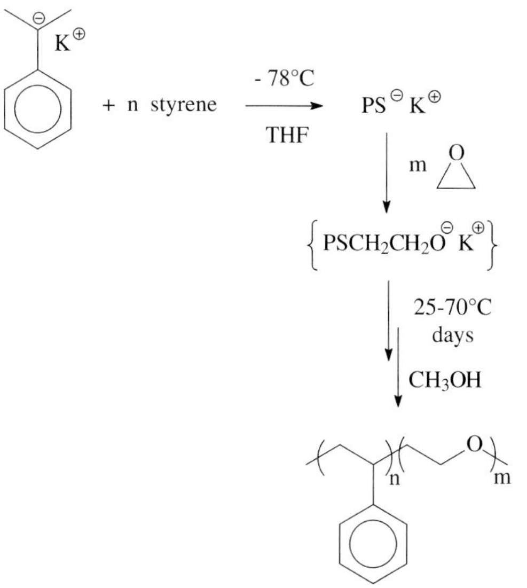

Scheme 5.5 One-pot synthesis of PS-b-PEO block copolymers using potassium-based anionic initiator.

sium *tert*-butoxide, potassium *tert*-amyloxide or potassium 2,6-di-*tert*-butylphenoxide [34], (b) replacement of lithium by potassium [35] or (c) using a potassium-based initiator from the beginning [36]. Scheme 5.5 illustrates a one-pot synthesis to prepare PS-*b*-PEO block copolymers using cumylpotassium as an initiator.

A more recent investigation in the field was reported by Deffieux and co-workers [24]. They were able to copolymerize styrene and ethylene oxide using lithium-based initiators. They found that EO polymerization occurs readily from polystyryllithium, in the presence of a trialkylaluminum that activates EO monomer, in hydrocarbon solvent.

Synthesis of PS-*b*-PEO by CRP route The recent breakthrough developments in the field of controlled radical polymerization (CRP) have allowed the synthesis of such block copolymers by alternative approaches. Various CRP methods such as atom-transfer radical polymerization (ATRP), nitroxide-mediated radical polymerization (NMP) and reversible addition–fragmentation chain transfer (RAFT) process are commonly used [37]. All these methods take advantage of a rapid dynamic equilibration established between a minute amount of growing free radical species and a large majority of dormant species to control the radical polymerization. Due to the very small number of active chain ends at a given point in time, the intermolecular coupling termination reactions – the most common termination reaction in conventional free radical polymerization – is drastically reduced, resulting in polymers with a relatively narrow molecular weight distribution.

In general, the synthetic protocol for the synthesis of PS-*b*-PEO block copolymer involves the PEO macro-initiator by the reaction of PEO-OH with a suitably

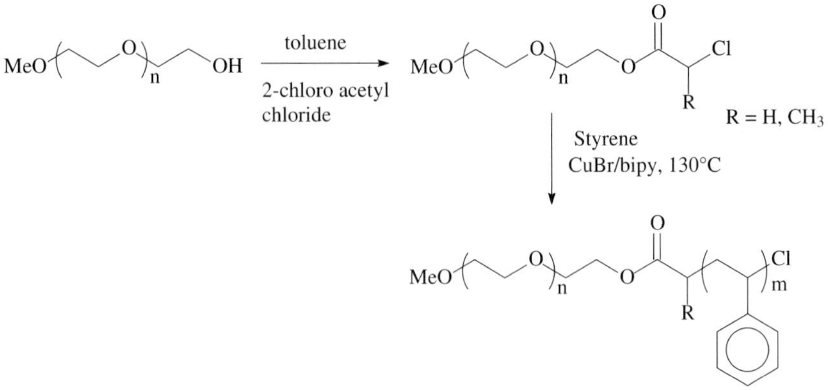

Scheme 5.6 Synthesis of PEO-b-PS copolymer by ATRP from macro-initiator PEO.

functionalized initiator followed by CRP of styrene. Reining et al. [38, 39], Jankova et al. [40] and Beaudoin et al. [41] have reported the synthesis of such block copolymers by CRP. An illustration of this route to PS-b-PEO block copolymer is shown in Scheme 5.6. The two-stage synthetic procedure involves refluxing MeO-PEO-OH with 2-chloroacetyl chloride in toluene, isolation of the macro-initiator formed and synthesis of the block copolymer in bulk using CuBr–bipyridine catalyst at 130 °C.

A special note should be made about the difference in end-groups in PS-b-PEO copolymers synthesized by anionic and CRP techniques as chemical anchoring of such surfactants to latexes is of significant importance. Block copolymers synthesized by the anionic method may have hydroxyl end-groups (i.e. PS-b PEO-OH), whereas those synthesized by CRP have specific groups representative of the type of CRP technique used (e.g. halogen groups in the case of ATRP, Scheme 5.6). Synthesis of block copolymers having hydroxyl end-groups by CRP will require the use of protection/deprotection chemistries or other organic chain-end modification reactions.

Some applications of PS-b-PEO and PEO-b-PS-b-PEO as surfactants Styrene emulsion polymerizations with well-defined PEO-b-PS block copolymers have been studied in detail by Riess and Piirma over several decades [1, 7]. In the emulsion polymerization of styrene with PS-b-PEO surfactants, the absence of an induction period for particle formation and a high initial rate of polymerization are worthy of special mention. It has also been pointed out that to obtain latex particles of similar size, a larger amount of polymeric surfactant compared with low molecular weight surfactants needs to be added to the system.

Riess found that the use of thiols, although primarily used as chain transfer agents, is necessary to have a stable latex in the free radical emulsion polymerization of styrene with PS-b-PEO surfactants and that among block copolymers based on PS and PEO, PEO-b-PS-b-PEO triblock copolymers were the most efficient stabilizers [42–43].

It is worth mentioning that in the emulsion polymerization of MMA or nBA using a series of PS-b-PEO block copolymers as surfactants, it has also been observed that most of the surfactants desorb from the latex due to large incompatibility between the PMMA or PnBA and PS [42].

The latter surfactants were also used to stabilize micro-emulsions [44] in the presence of a co-surfactant or to prepare colloidal dispersions of conductive polymers [45–46]. Landfester et al. also reported a series of polyaddition reactions involving tetraepoxides with various diamines, dithiols or bisphenols, in mini-emulsions using numerous stabilizers including PS-b-PEO block copolymers [47]. Latex particles with a broad size distribution between 30 and 600 nm were obtained.

Synthesis and use of reactive PS-b-PEO as surfactant The *in situ* formation of PS-b-PEO block copolymers is also another interesting possibility that was investigated. For that purpose, end-functionalized PEO must first be prepared. The surface-active properties of PEO macromonomers and their role in radical polymerization (of styrene and other monomers) in disperse media have been widely investigated [48]. Shay et al. [49] detailed the use of a PEO end-capped with a methylstyrene unit by urethane chemistry for the dispersion polymerization of styrene in polar media (Scheme 5.7). Shear rheological measurements showed that the stability of the latex can be correlated with the stabilizer concentration; the higher the surface coverage of the latex by the surfactant, the higher is the stability.

A similar approach was followed by Tauer and Yildiz [50]. They used symmetrical PEO-azo initiators with either terminal hydroxyl or styrenyl end-groups in the emulsion polymerization of styrene.

Gibanel and coworkers [51, 52] studied the dispersion polymerization of styrene in water–alcohol using PS-b-PEO as surfactant having various chain ends (Scheme 5.8). These surfactants were synthesized by a combination of anionic polymerization and termination of the anionic chain end using appropriate electrophilic unsaturated groups as well as organic transformations [53].

Analysis of the purified latex by ^{1}H NMR spectroscopy showed that the incorporation of these surfactants in the latex depends on the chain length of the individual blocks. Apparently, a good surfactant is an amphiphilic polymer capable of adsorbing on the particle surface. Interestingly, no specific effect was observed in the presence of polymerizable end-groups on the surfactants. On the other hand, styrenyl end-capped PS-b-PEO resulted in particles of smaller size

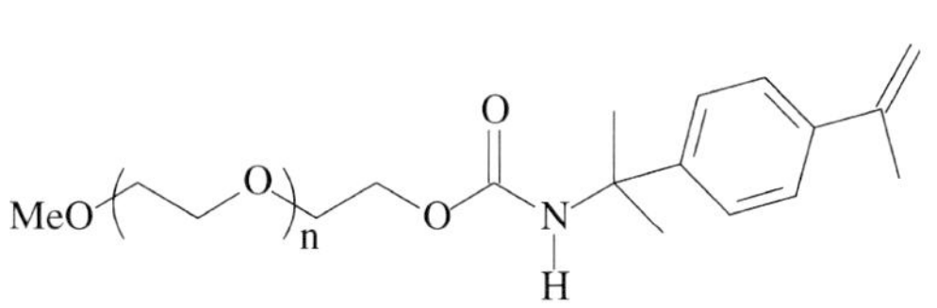

Scheme 5.7 Macromonomer based on PEO developed by Shay et al. [49].

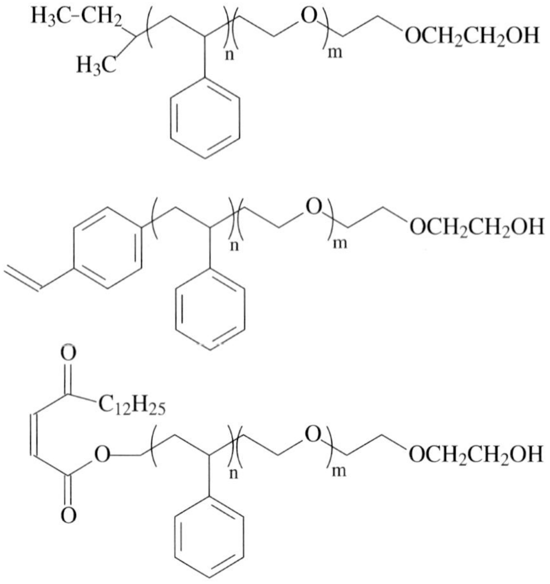

Scheme 5.8 Structure of various PS-b-PEO reactive block copolymers.

(~200 nm compared with ~350 nm). This study also showed that amphiphilic macromonomers produced latexes of better stability over hydrophilic macromonomers.

It is worth mentioning that PS-b-PEO block copolymers having norbornenyl reactive end-groups have also been reported for the aqueous suspension [54] and mini-emulsion [55] ROP of cyclooctadiene in water to give 1,4-PB particles in the size ranges 20 µm and 220 nm, respectively.

Synthesis of styrenyl end-capped PS–(PEO)$_2$ stars (Scheme 5.9) and use in emulsion and dispersion polymerizations of styrene have been reported by Gnanou and coworkers [56–57]. These gemini-type surfactants have higher CMC values than their linear analogues. Also, no induction period for particle/polymer formation was observed in contrast to linear analogues. This absence of induction period may be due to the faster exchange between the gemini-type dispersant molecules present in the medium and those engaged in the micelles.

Amphiphilic star-shaped and dendrimer-like copolymers based on a PS core and PEO corona were also prepared using a core first method combining ATRP and anionic polymerization [58]. The surface properties of the stars were compared with those of linear diblock analogues, demonstrating similar behavior.

Peleshanko et al. also described the synthesis of asymmetric hetero-arm PEO-b-PS star block polymers having terminal hydroxyl groups by a combination of ATRP and anionic polymerization techniques and protection/deprotection chemistries [59]. These polymers, developed for different applications, could be also excellent candidates as surfactants.

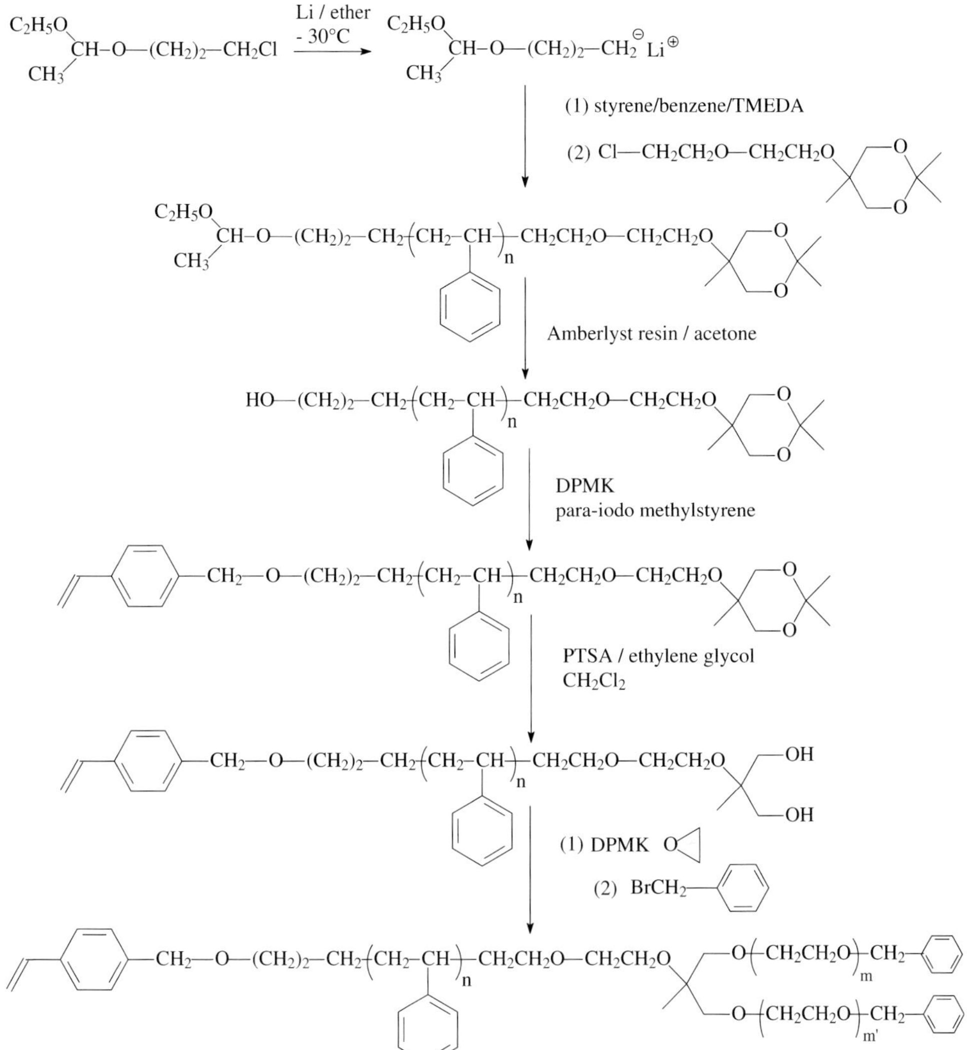

Scheme 5.9 Synthesis of PS–(PEO)$_2$ star block copolymer bearing a styrenyl end-group.

5.2.1.3 PMMA-*b*-PEO

PMMA-*b*-PEO amphiphilic block copolymers are also commercially available (Goldschmidt). Such copolymers were recently synthesized by ATRP from MeO-PEO-OH [38, 60], as shown in Scheme 5.10.

Wu and Piirma have also reported the synthesis of such block copolymers by chain end coupling [61].

PMMA-*b*-PEO di- and triblock copolymers have been thoroughly investigated for the emulsion polymerization of styrene and MMA [1]. Recent studies have

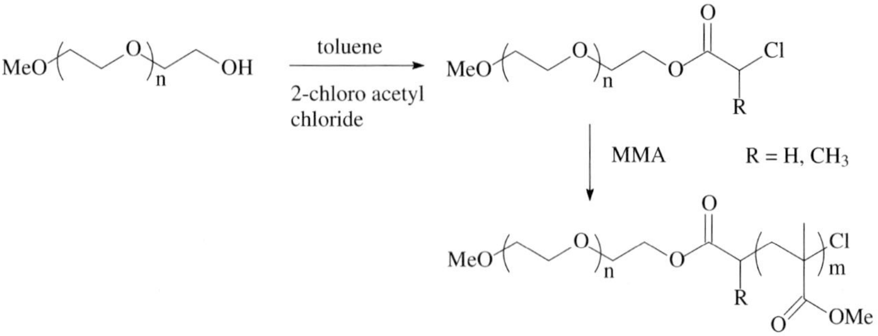

Scheme 5.10 Synthesis of MeO-PEO-b-PMMA by ATRP.

revealed that block copolymers having molar masses of 1000 and 3000 g mol^{-1} for PMMA and PEO, respectively, are the most efficient in the emulsion polymerization of the above monomers [7]. PMMA-g-PEO graft copolymers were also reported to be very effective surfactants for inverse emulsion polymerization of acrylamide [62]. PMMA-b-PEO block copolymers were also tested as co-surfactants for the mini-emulsion terpolymerization of MMA, nBA and AA and for the preparation of microgels and microlatexes with potential applications in various formulations [7].

5.2.1.4 Miscellaneous Nonionic Amphiphilic PEO-based Block Copolymers

Block copolymers of polybutadiene (PB) or polyisoprene (PI) with PEO are also an important class of materials that may find applications as surfactants. Föster and Krämer found that the phosphazene base tBuP$_4$ (Scheme 5.11) allowed the synthesis of PB-b-PEO and PI-b-PEO block copolymers in THF by simply adding ethylene oxide to the living PB or PI chain ends [63]. They prepared narrow disperse polymers over a broad range of molecular weights with good control of block lengths for an exact 1:1 stoichiometry of t-BuP$_4$:s-BuLi.

Interestingly, Deffieux and coworkers were also able to prepare PI-b-PEO and PB-b-PEO block copolymers by adding EO monomer together with trialkylaluminum to polyisoprenyllithium or polybutadienyllithium growing chains, respectively, in a hydrocarbon medium as already described for the synthesis of PS-b-PEO.

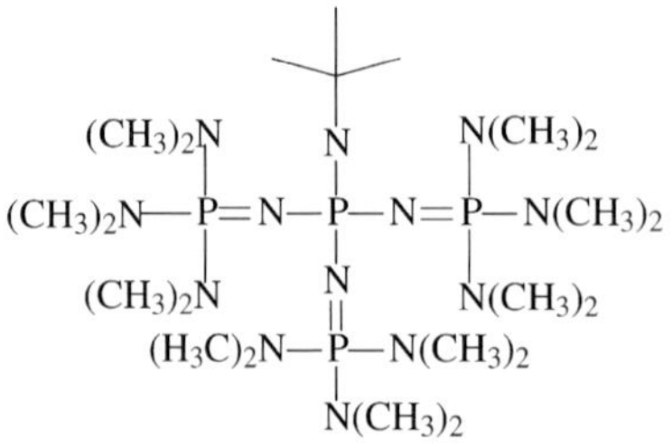

Scheme 5.11 Structure of the phosphazene base tBuP$_4$.

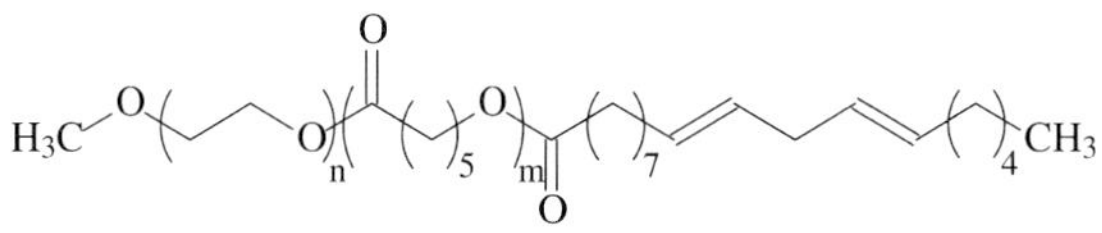

Scheme 5.12 PEO-*b*-PCL block copolymer end-capped with linoleic group for use as cross-linkable surfactant.

Ito and coworkers also described the synthesis of PEO-*b*-poly(styrene oxide) block copolymers having vinylbenzyl ether end-groups by sequential anionic polymerization of ethylene oxide and styrene oxide with potassium 2-methoxyethoxide, then end-capped by *p*-vinylbenzyl bromide. Such a macromonomer has been successfully used for the emulsion and dispersion polymerization of styrene [64].

Amphiphilic di- and triblock copolymers consisting of PEO as central hydrophilic segment and poly(ε-caprolactone) (PCL) as hydrophobic segment(s) were prepared by ROP then end-functionalized by esterification with linoleic acid, which contains reactive double bonds (Scheme 5.12) [65]. These copolymers were used as surfactants for PMMA latex preparations. The best results were obtained with diblock copolymers for which less coagulum, smaller particle size and more stable latexes were obtained. Interestingly, upon drying and film formation, the surfactants can cross-link oxidatively, preventing surfactant migration.

5.2.2
Synthesis and Main Uses of PAA- and PMAA-based Copolymers

Well-defined amphiphilic block copolymers containing PAA or PMAA (specifically PS-*b*-PAA and PMMA-*b*-PAA) as the hydrophilic block have been widely used by several groups as surfactants in heterogeneous polymerizations. Such block copolymers are typically synthesized in two steps. The first step involves the synthesis of the corresponding well-defined P*t*BA or P*t*BMA block copolymer precursor by anionic or CRP methods. This precursor is subjected to selective hydrolysis to remove the *tert*-butyl groups to obtain the PAA- or PMAA-based amphiphilic block copolymer, as illustrated in Scheme 5.13.

The hydrolysis of *tert*-butyl groups is usually carried out by reaction with trimethylsilyl iodide at room temperature and then with aqueous HCl [66], with *p*-toluenesulfonic acid in toluene under reflux conditions [67], with HCl in diox-

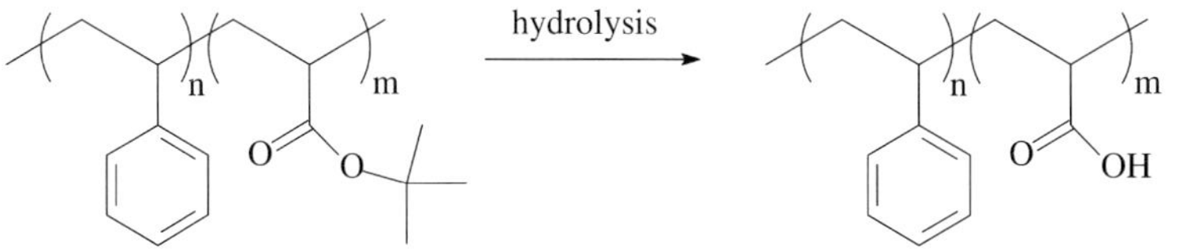

Scheme 5.13 Synthesis of PS-*b*-PAA block copolymers.

ane under reflux conditions [68] or with trifluoroacetic acid at room temperature [69]. Zhang and Ruckenstein [70] proposed an alternative method involving elimination of a 1-(alkoxy)ethyl group and synthesis of PMMA-b-PMAA copolymers by anionic [71] and ATRP [72] methods. In addition, protected monomers such as BzMA [73] and NPMA [74] have also been used in place of tBMA to obtain the target block copolymers.

5.2.2.1 PMMA-b-PAA and PMMA-b-PMAA

PMMA-b-PMAA, PMMA-b-PAA and PMA-b-PAA copolymer synthesis by various methods, namely anionic [6, 66], ATRP [68, 75] as well as GTP [73, 76] are well documented.

Liu et al. used PMMA-b-PMAA as a dispersant in the micro-emulsion polymerization of MMA [77]. It was observed that the block copolymer chain length plays a key role in determining the number of polymer chains per micelle before and after polymerization. Under same external conditions, the number of particles (N_w) was relatively unaffected when block copolymers with longer PMMA chain lengths were used.

PMMA-b-PAA has been used in the emulsion polymerization of MMA and MMA–nBuA copolymers and it was found that block copolymers with PAA chain lengths of equal or greater length than the hydrophobic PMMA chains are extremely efficient as stabilizers down to block copolymer to monomer ratios of less than 1 wt.% [78].

Spinelli reported that poly[EHMA-b-(EHMA-r-MAA)] block copolymers (EHMA=ethylhexyl methacrylate) synthesized by GTP are excellent surfactants for the emulsion polymerization of acrylates [79]. High solid contents (40–50 wt.%), small particle size and minimal coagulum were the key characteristics of the latexes prepared with these block copolymers. The amount of methacrylic acid in the block copolymer plays a key role in determining the size of the particles and amount of coagulum. An increase in MAA content resulted in smaller lattices with lower coagulum, probably due to higher MAA content block copolymers forming smaller micelles.

In addition, it is also important to recall that statistical copolymerization of MAA and PEGMA by conventional free radical chemistry also yields high molar mass and polydisperse surfactants whose surface activity is pH dependent, exploited advantageously to break emulsions [80].

5.2.2.2 PnBA-b-PAA

Recent advances in RAFT have allowed the synthesis of acrylic acid block copolymers directly [81]. CRP using reversible addition fragmentation chain transfer involves the reaction of a di-/trithioester chain transfer agent (CTA), $S=C(Z)–SR$, with a primary radical derived from an initiator or a propagating polymer chain, forming a new CTA and eliminating a radical capable of reinitiating the polymerization.

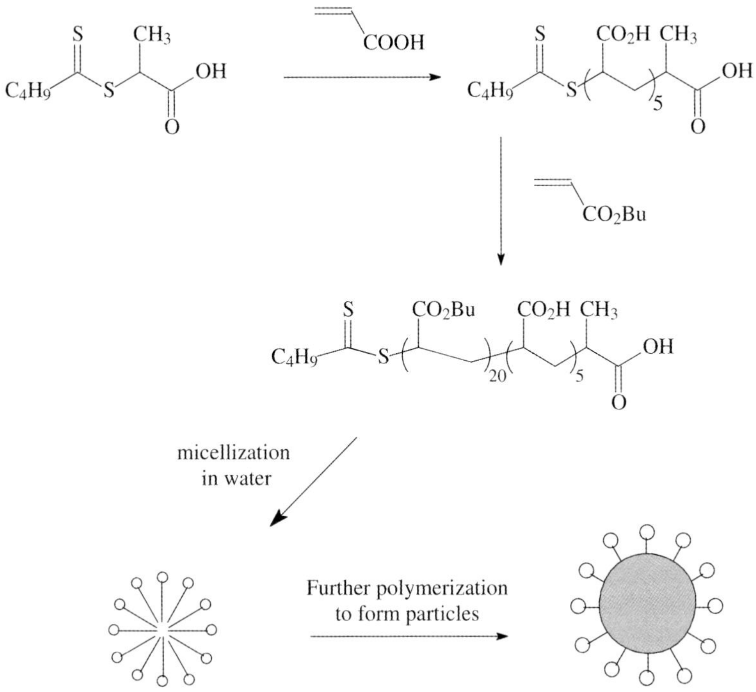

Scheme 5.14 Synthesis and self-assembly of P*n*BA-*b*-PAA RAFT agent towards acrylic particle formation.

RAFT has been used to synthesize P*n*BA-*b*-PAA copolymer and employed in the emulsion polymerization of *n*BA–MMA copolymer [82–83]. The studies revealed that these block copolymers could be used as stabilizers for obtaining latexes up to 50 wt.% solid content without any coagulum. Recent investigations [84] have shown that polymerization under RAFT control can be implemented in an *ab initio* emulsion polymerization. The method consists in the formation of an amphipathic trithiocarbonate RAFT agent to produce an initial diblock copolymer with hydrophilic and hydrophobic components of degrees of polymerization chosen so that they can self-assemble into micelles. When polymerization is continued by feeding of the hydrophobic monomer, the chains cannot desorb from these micelles and continued polymerization results in the formation of stable latex particles. The synthesis of core–shell latex particles composed of poly(acrylic acid)-*b*-poly(butyl acrylate)-*b*-polystyrene triblocks copolymers could be prepared according to this technique. The synthesis and self-assembly of these diblock copolymers is shown in Scheme 5.14.

5.2.2.3 PS-b-PAA and PS-b-PMAA

As already introduced, PS-b-PAA and PS-b-PMAA block copolymers can be prepared by anionic [67, 85, 86] or ATRP [87–89] methods followed by hydrolysis of the *tert*-butyl groups.

The elegant work done by Eisenberg and co-workers on morphological studies of these block copolymers in water is worth special mention [90–93]. They observed a range of morphologies, namely spherical, rod-like, vesicular and large complex micelles filled with reverse micelles, depending on the chain length of PAA.

Charleux and coworkers described the direct synthesis of controlled poly(styrene-*co*-acrylic acid)s of various compositions by nitroxide-mediated random copolymerization [94]. Recently, the same group reported the easy preparation of well-defined amphiphilic gradient poly(styrene-*co*-acrylic acid) copolymer and their use as efficient stabilizers (at a concentration of 3–4 wt.%) for the emulsion polymerization of styrene and mixtures of methyl methacrylate and butyl acrylate [95]. More recently, Billon and coworkers reported an alternative route to PS-b-PAA block copolymers by direct nitroxide-mediated polymerization using the PS block as a macro-initiator [96]. This approach allowed the preparation of copolymers with broader range of molar masses and hydrophilic–lipophilic balance (HLB).

Emulsion polymerization studies of styrene in the presence of PS-b-PAA have shown that block copolymers act as efficient stabilizers up to 0.5 wt.% with respect to monomer. Block copolymers having 10 styrene units and a maximum of 50 acrylic acid units gave the best results. The tri- and star-block copolymers with external hydrophilic blocks behaved like block copolymers whereas triblocks with internal hydrophilic blocks behaved differently [97]. It has also been shown that block copolymers with high acid content behaved like low molecular weight surfactants [98].

Y-shaped polystyrene-b-[poly(*tert*-butyl acrylate)]$_2$ block copolymers were also prepared by ATRP using a multi-step route involving protecting-group chemistry. Hydrolysis of the *tert*-butyl groups led to the formation of amphiphilic PS–(PAA)$_2$ Y-shaped block copolymers [99]. Such copolymers with specific shapes should be good candidates as surfactants.

5.2.3
Synthesis and Main Uses of Sulfonate Block Copolymers

The benefits of long polyelectrolyte chains as stabilizers for colloidal particles were predicted theoretically by Pincus [100]. In that sense, sulfonate block copolymers represent an important class of surfactants. One main feature of these copolymers containing a strong polyelectrolyte moiety is their capacity to stabilize emulsions over a wide range of pH and temperature, in contrast to carboxylated analogues or PEO-based copolymers. First, the pioneering work of Teyssié and coworkers should be mentioned [101–102]. They synthesized by "living" anionic polymerization poly(alkyl acrylate-b-sulfonated glycidyl methacrylate) block copolymers which proved to be efficient surfactants for the preparation and stabilization of acrylic latexes in aqueous medium. Interestingly, replacement of the alkyl group (methyl) by an allyl group leads to a covalent linkage of the emulsifiers with the particles.

The synthesis, in a two-step procedure using a mixed water–ethylene glycol solvent, of amphiphilic block copolymers composed of an ionic PSS and hydrophobic polystyrene, using TEMPO-mediated living radical polymerization, was described by Vairon and coworkers [103]. Low molar mass TEMPO-capped PSS is synthesized first and then used as a macro-initiator for styrene polymerization in heterogeneous conditions, as illustrated in Scheme 5.15. The control of the polymerization is ensured by TEMPO located between the continuous and dispersed phases.

Partly sulfonated (10–30%) PB and PI and their block copolymers with polystyrene also represent a family of very effective reactive stabilizers in emulsion polymerization, comparable to low molar mass surfactants such as sodium dodecyl sulfate (SDS), as reported by Tauer and coworkers [104, 105]. They demonstrated that such surfactants are also reactive stabilizers and covalently linked to the particles when nonionic water-soluble free-radical initiators are used. Indeed, coagulum is mainly observed with ionic initiators, explained by a change of the solution state of the polyelectrolyte stabilizer under these conditions. An increase in the latex stability against salt was obtained by using partly sulfonated polybutadiene-b-poly(ethylene glycol) block copolymers, which combine electrostatic and steric stabilizations. The same group demonstrated that long polyelectrolyte chains contribute via electrosteric stabilization to an extraordinary stability of colloidal particles against electrolytes, as verified experimentally by means of poly(ethylethylene)-b-poly(styrene sulfonate) block copolymers [106].

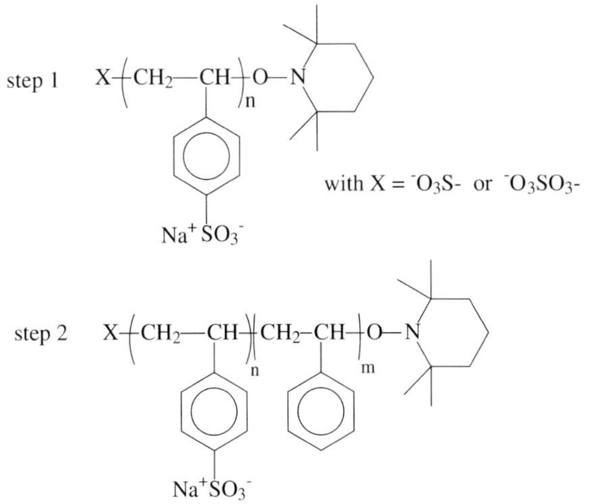

Scheme 5.15 Structure of amphiphilic block copolymer PSS-b-PS prepared by CRP.

5.2.4
Cationic and Cationizable Block Copolymers

Cationic copolymers are also an important class of surfactants. Mention should be made of the investigations of Teyssié and co-workers, who prepared PMMA latexes by radical polymerization in the presence of poly(methyl methacrylate)-b-poly[2-(dimethylamino)ethyl methacrylate] (PMMA-b-PDMAEMA), used as a cationic surfactant after quaternization [107].

Lieske and Jaeger also reported the original polymerization of diallyldimethylammonium chloride (DADMAC) in the presence of macro azo initiators obtained from azobisisobutyronitrile (AIBN) and poly(ethylene glycol) yielding poly(ethylene glycol)-b-poly(diallyldimethylammonium chloride) block copolymers. The latter copolymers were found to be efficient stabilizers for the emulsion polymerization of styrene [108].

Amphiphilic block copolymers composed of a polystyrene hydrophobic block and a cationic poly(vinylbenzyltriethylammonium chloride) hydrophilic block were synthesized by controlled radical polymerization using ATRP and nitroxide-mediated methodologies [109, 110]. Such copolymers were found to be effective electrosteric stabilizers in the free-radical emulsion polymerization of styrene yielding small PS particles (90 nm) and stable latexes [87].

Recently, Charleux and co-workers reported the synthesis of well-defined cationic amphiphilic comb-like copolymers in the mini-emulsion polymerization of styrene [111, 112]. The copolymers were prepared by controlled free-radical copolymerization of styrene and vinylbenzyl chloride using either the reversible addition–fragmentation chain-transfer method or TEMPO-mediated polymerization. The benzyl chloride moieties were modified by two different long alkyl chain tertiary amines to yield the amphiphilic copolymers, as illustrated in Scheme 5.16.

It is remarkable to note that a small amount of these copolymers was sufficient to stabilize the latex (0.5–2 wt.% vs monomer) thanks to their high structural quality.

Armes and coworkers [113] developed a series of well-defined pH-responsive cationic surfactants based on tertiary amine methacrylate monomers, synthesized in the presence of oxyanionic initiators (Scheme 5.17).

Selected macromonomers based on 2-(dimethylamino)ethyl methacrylate (DMAEMA) were used to prepare sub-micrometer-sized and micrometer-sized polystyrene latexes via aqueous emulsion and nonaqueous dispersion polymerization. In emulsion conditions, the acidity of the medium was fixed around pH 2 to ensure the full protonation of the DMAEMA units, preventing the precipitation of the macromonomer under the polymerization conditions (60 °C). Armes and co-workers also synthesized, by GTP, cationic block copolymers composed of MMA and DMAEMA. Such copolymers were proved to be efficient steric stabilizers for the synthesis of polystyrene latexes. The pH-dependent surface activity of such latexes enabled them to be used as stimulus-responsive particulate emulsifiers for the preparation of oil-in water emulsions [114]. Similar results were ob-

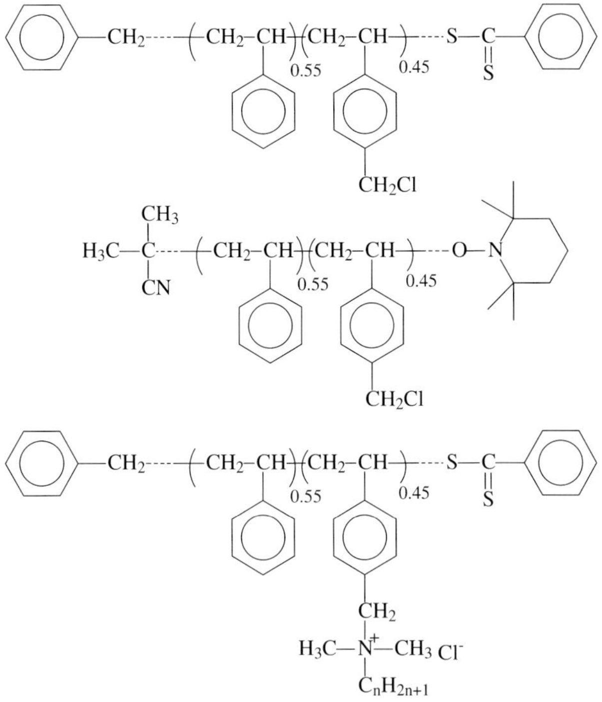

Scheme 5.16 Amphiphilic cationic block copolymers.

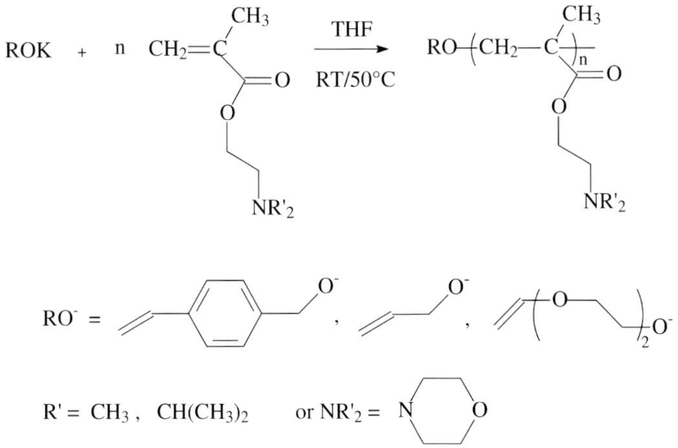

Scheme 5.17 Synthesis of pH-responsive cationic polymers based on tertiary amine methacrylate.

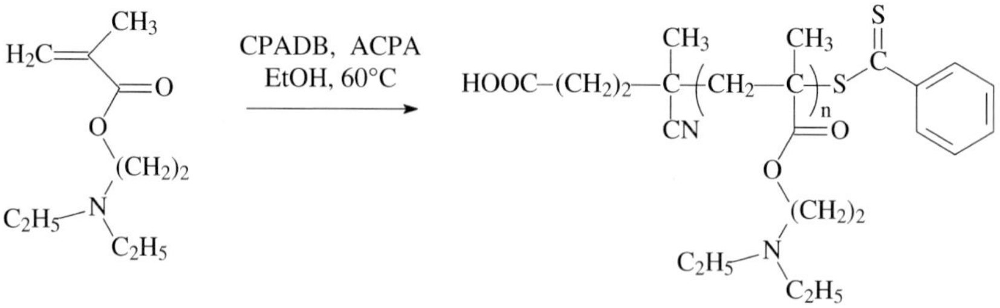

Scheme 5.18 Synthesis of macro-RAFT agent based on 2-(diethylamino)ethyl methacrylate.

tained with poly(propylene oxide)-b-poly[2-(diethylamino)ethyl methacrylate], readily prepared by ATRP [115]. Haddleton and co-workers also described the synthesis of Y-shaped block copolymers composed of MMA and DMAEMA [116].

Armes and co-workers, in collaboration with Charleux and coworkers, used similar DMAEMA-based cationic diblock copolymers and macromonomers thereof as stabilizers for the mini-emulsion polymerization of styrene [117]. The best results in terms of latex stability were obtained with the 95% quaternized diblock macromonomer, which was demonstrated to be strongly adsorbed on the surface of the latex particles.

Recently, Cai and Armes reported the synthesis via ATRP of well-defined Y-shaped zwitterionic block copolymers, namely poly[2-(diethylamino)ethyl methacrylate]-b-[poly(succinyloxyethyl methacrylate)]$_2$ [118]. As already described by the same group with similar linear block copolymers, such Y-shaped copolymers exhibit the so-called 'schizophrenic' micellization in aqueous medium, with respect to pH [119].

Also worth noting is the batch emulsion polymerization of styrene stabilized by a hydrophilic macro-RAFT agent based on protonated 2-(diethylamino)ethyl methacrylate [120]. The reaction leads to a stable latex composed of well-defined polyelectrolyte. The synthesis of the RAFT agent is shown in the Scheme 5.18.

5.2.5
Miscellaneous

Simionescu et al. [121] detailed the synthesis of PDMS-b-poly(N-acetyliminoethylene) triblock copolymers (Scheme 5.19) by cationic ROP of 2-methyl-2-oxazoline with chloromethylphenyl end-capped PDMS and employed them in the emulsion polymerization of styrene and acrylates. The results of these studies indicated the involvement of the surfactant in the initiation process, thus consider-

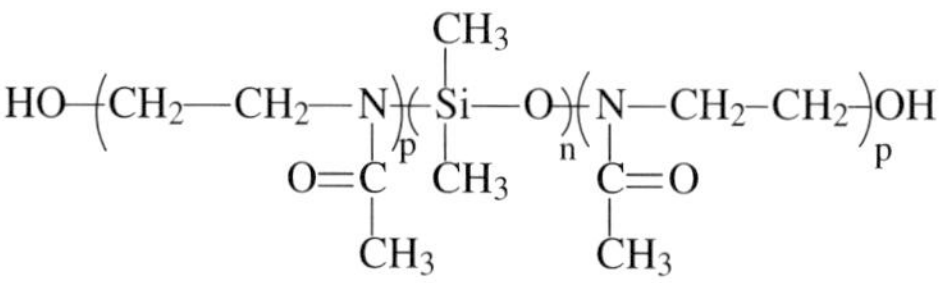

Scheme 5.19 Structure of PDMS-b-Poly(N-acetyliminoethylene).

ably influencing the conversion and the final molecular weight. The same group recently described the use of maleic acid-terminated poly(N-acetylethylenimine) macromonomer as co-surfactant for the production of nanoparticles of MMA and nBA copolymers by micro-emulsion [122].

The controlled synthesis of amphiphilic poly(methacrylate)-g-(polyester/polyether) graft terpolymer is also worth mentioning [123]. The authors prepared the graft copolymer in two steps by combining ATRP and coordination–insertion ROP. Typically, they first carried out the ATRP copolymerization of MMA, 2-hydroxyethyl methacrylate (HEMA) and poly(ethylene glycol) methyl ether methacrylate (PEGMA). In the second step, the ROP of ε-caprolactone or L,L-dilactide (LA) was performed in the presence of poly(MMA-co-HEMA-co-PEGMA) together with a tin- or an aluminum-based catalyst. The resulting graft copolymers and the precursor terpolymer proved to be efficient surfactants.

Block copolymers composed of methyl vinyl ether (MVE) and isobutyl vinyl ether (IBVE) are also interesting materials due to their thermo-adjustable amphiphilic properties. Du Prez and coworkers synthesized well-defined PMVE-b-PIBVE block copolymers by "living" cationic polymerization (Scheme 5.20) and investigated their colloidal properties as a function of the copolymer composition and temperature [124]. They found that below the LCST, the block copolymers have emulsifying properties similar to or even better than those of the commercial PEO-b-PPO block copolymers.

Poly(vinyl alcohol) (PVOH) and PVOH-containing block copolymers also represent an important class of surfactants. Generally, vinyl acetate is first polymerized by conventional free radical polymerization followed by ATRP-mediated polymerization of a second monomer [125, 126]. The polymerization of VAc by controlled radical polymerization has been investigated by many research groups [127–131]. Recently, Jérme and coworkers were able to prepare well-defined PVAc and related block copolymers with PS, PEA and PMMA by combination of cobalt-mediated radical polymerization and ATRP. The derivatization of PVAc into PVOH was achieved by methanolysis without breaking the link between the two blocks. The self-association of amphiphilic PS-b-PVOH in a water–THF mixture (4:1) resulted in the formation of vesicles [133].

Recently, Laschewsky and coworkers prepared a series of amphiphilic block copolymers constituted of a constant PnBA hydrophobic block and of various hydrophilic blocks (nonionic and ionic) exhibiting different hydrophilicities (Scheme 5.21) [134–136]. They chose the RAFT route since it can be applied to a wide range of monomers and is also tolerant towards a wide range of experimental conditions.

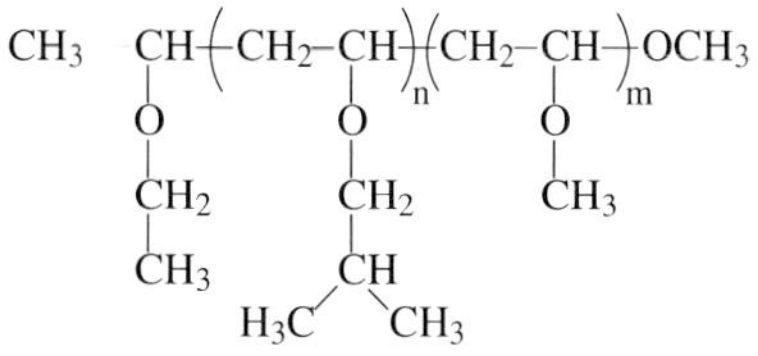

Scheme 5.20 Structure of PIBVE-b-PMVE block copolymer prepared by "living" cationic polymerization.

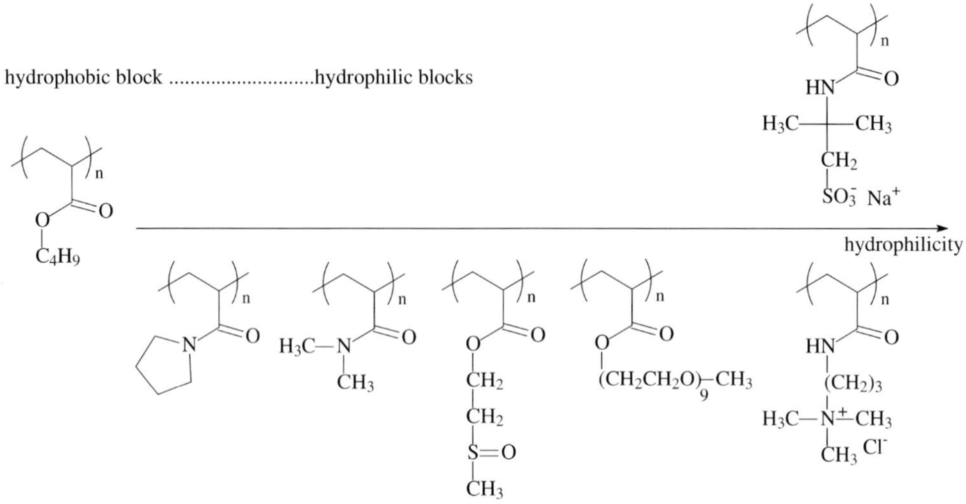

Scheme 5.21 Series of amphiphilic block copolymers prepared by RAFT and exhibiting different HLB.

In this methodology, P*n*BA macro-RAFT agent was first synthesized by polymerization of butyl acrylate with benzyldithiophenyl acetate and AIBN and subsequently used for the polymerization of various hydrophilic monomers.

Using a similar procedure, Sanderson and coworkers investigated the preparation of dithiobenzoate end-capped polystyrene oligomers in bulk for water-borne organic dispersions [137].

Tauer and coworkers [138] used polystyrene-*b*-polyglutamate block and star copolymers as surfactants in the emulsion polymerization of styrene. The two-step block copolymer synthesis involves the ROP of γ-benzyl-L-glutamate-*N*-carboxyanhydride by ω-amino-functional polystyrene synthesized by anionic polymerization as illustrated in Scheme 5.22.

The authors demonstrated that star-shaped stabilizers yield smaller particles (70–130 nm) with a broad distribution compared with the linear analogues (100–220 nm).

Polymeric surfactants based on amphiphilic polysaccharides are also of growing interest since polysaccharides provide biocompatibility, biodegradability and other specific properties [139–141]. The preparation of neutral polymeric surfactants derived from dextran by chemical modification and a grafting method is worth mentioning [142–144]. Racles and Hamaide [145] also described the synthesis of nonionic surfactants constituted of monosaccharide-functionalized polysiloxanes bearing either terminal or pendant mannose moieties (Scheme 5.23). Such polymers were used as surfactants for the preparation of PCL particles (200 nm diameter) by nano-precipitation in water.

The synthesis of well-defined block copolymers and other architectures based on glycopolymers should be of great interest in the field. Recent studies by

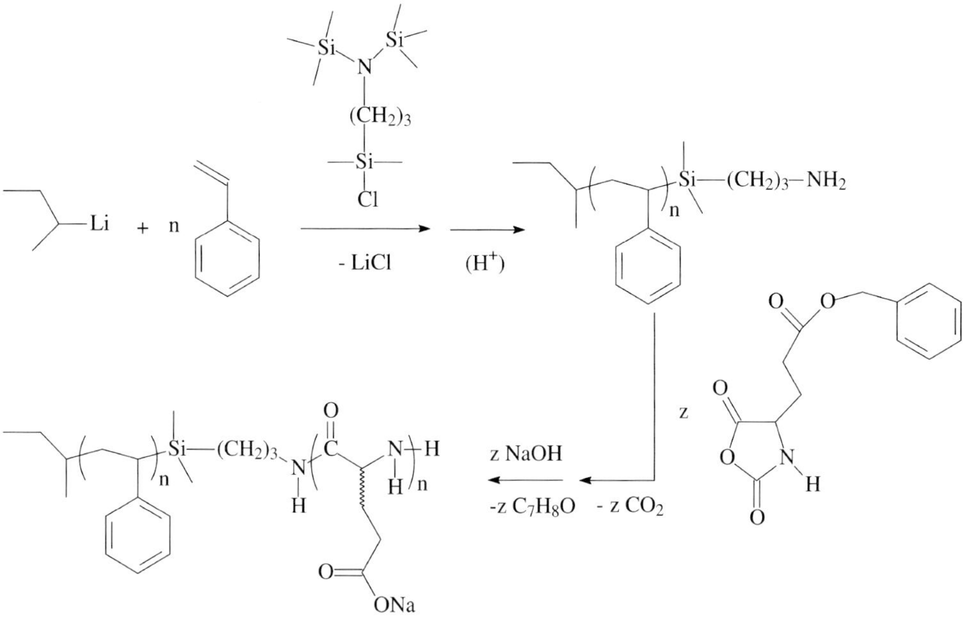

Scheme 5.22 Synthesis of polystyrene-b-poly(glutamate) block copolymer.

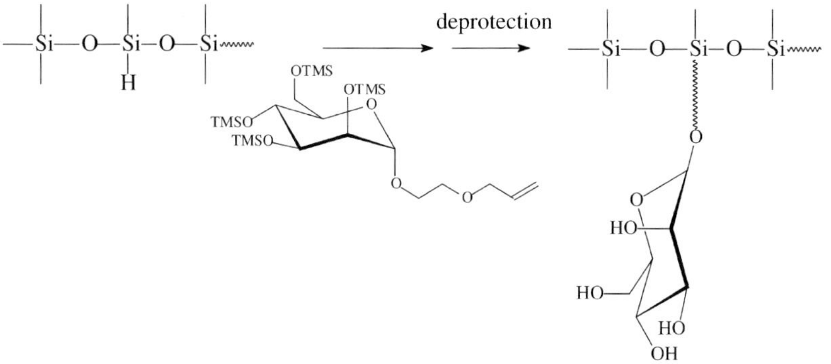

Scheme 5.23 Synthesis of surfactant based on polysiloxane bearing mannose moieties.

RAFT provided access to such copolymers, for example combining poly(acrylol glucosamine) (PAGA) and polyNIPAM, without protecting group chemistry, as shown in Scheme 5.24 [146, 147].

Another class of polymers that has attracted increasing interest in the last decade is the double hydrophilic block copolymers (DHBCs). Such block copolymers consists of two water-soluble blocks of different chemical nature. The amphiphilicity of DHBCs can be revealed as soon as the chains interact with a substrate (typically a mineral substrate) or even totally modified by varying the pH, the tem-

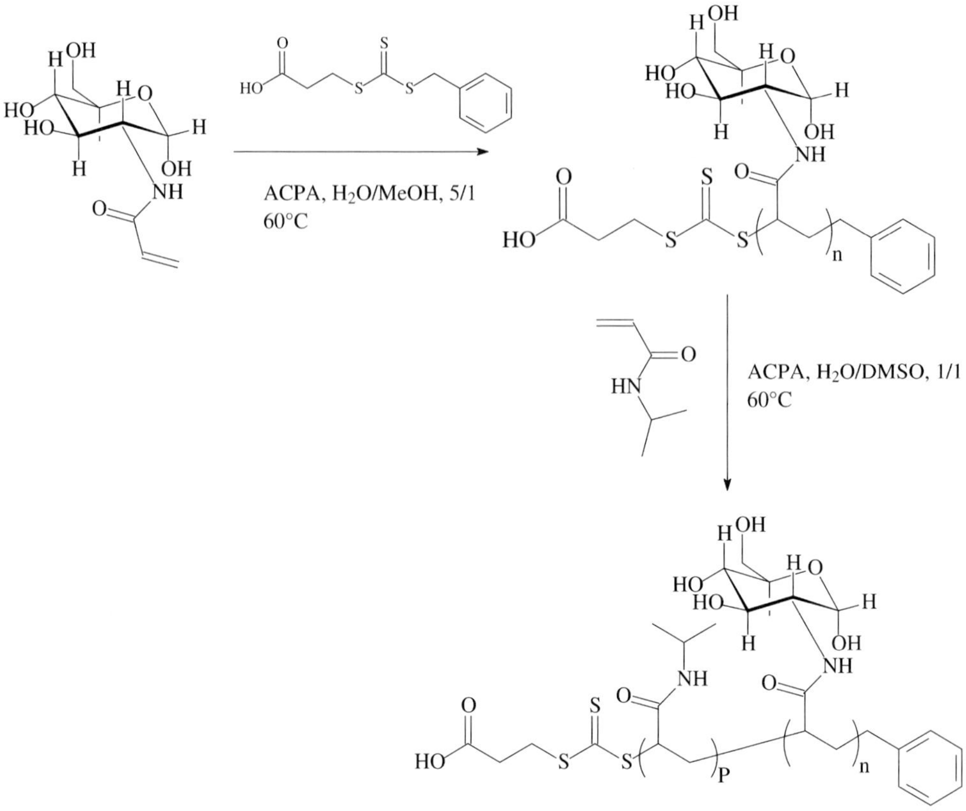

Scheme 5.24 Synthesis of PAGA-b-PNIPAM using the RAFT method.

perature or the ionic strength of the solution. In this respect, these copolymers show versatile behaviors that make them very attractive for many applications and in particular as crystal growth modifiers or novel drug carrier systems, domains of investigation that are outside the scope of this chapter. Readers are directed to an interesting review by Cölfen [148]. Additionally, it is interesting to mention studies by Tauer et al., who prepared, via radical heterophase polymerization, some original DHBCs, i.e. PSS-b-PMAA, PEG-b-PNIPAM-b-PMMA, PDEAEMA-b-PNIPAM-b-PMAA and related particles [149].

5.3
Main Surfactants Used for Heterogeneous Polymerizations in Nonaqueous Dispersions

Nonaqueous heterogeneous polymerizations are generally designed as dispersion polymerizations. In most cases, the reaction mixture is initially homogeneous – this is not always the case; sometimes one monomer may be partially or totally insoluble – then becomes heterogeneous along with the formation of the insoluble growing polymer chain in the continuous phase. The phase behaviors of the initial and final systems are governed by the solubility of each component (monomers, catalyst, polymer), quantified to a first approximation by the solubility parameter value (δ) as defined by Hansen [150]. The closer are the δ values of the monomer and the solvent, the better is the miscibility. The flocculation of the chains is prevented by the presence of a steric stabilizer and, generally, polymeric particles of micrometer size are obtained. Therefore, in nonaqueous dispersions (NAD), the choice of a well-designed copolymer having one block compatible with the continuous phase and another block having affinity with the growing polymeric chains is crucial. The steric stabilizer may be also formed *in situ* with the help of a reactive homopolymer (or macromonomer), as already discussed. Illustrations of the main steric (reactive or not) stabilizers specifically used in NAD are given hereafter, including polar and nonpolar media and scCO$_2$.

5.3.1
PEO- and PPO-based Copolymers

The synthesis of related block copolymers was described in the first part of this chapter. The versatility of PEO makes PEO-based block copolymers also attractive materials in nonaqueous dispersions. Indeed, the dispersion polymerization of styrene in methanol using PS-*b*-PEO as a stabilizer was reported by Riess and coworkers [151]. AA and MAA were also polymerized in glycol media in the presence of the same steric stabilizer. You et al. reported the polymerization of ethyl methacrylate (EMA) and lauryl methacrylate (LMA) monomers in propylene glycol using PEO-*b*-PS-*b*-PEO block copolymers [152, 153]. They observed that the stabilization ability of these triblock copolymers was correlated with their molar mass and conformation in propylene glycol. The polymerization of styrene and acrylonitrile was also investigated in polyether glycol in the presence of PS-*b*-PPO block copolymers [154].

In contrast to all the examples given through this review, which mostly deal with free-radical polymerization, very few studies have been reported concerning nonaqueous dispersions prepared by polyaddition (polycondensation) in the presence of block copolymers as stabilizers, although this area has been investigated for a fairly long time, as described by Barrett [155]. The development of nonaqueous dispersions, i.e. (o/o) suspension polycondensation has been reported in a patent [156]. Typically, the authors use PEO-*b*-PDMS block copolymers as stabilizers to synthesize polyurethane and polyurea microcapsules. Sivaram and coworkers also investigated this area and reported that PI-*b*-PEO block copolymers were efficient stabilizers for the synthesis of PUR particles of micrometer size in paraffin oil [157, 158].

5.3.2
Polyolefin-based Copolymers

The discovery in the two last decades of novel catalysts allowing the controlled/ living polymerization of olefins is probably one of the more recent important breakthroughs in the field of polymer chemistry [159–162]. The direct synthesis of copolymers [163] and end-functionalized polyolefins [164–166] is now feasible. Nevertheless, the polyolefin-based (co)polymers used as surfactants so far have principally been obtained from hydrogenation of the corresponding unsaturated (co)polymer prepared by living anionic polymerization [167]. Such (co)polymers, known under the trade mark Kraton, were introduced by Shell in the 1960s. A typical structure of a saturated copolymer obtained from PS-b-PI block copolymer is shown in Scheme 5.25.

PS-b-poly(ethylene-co-propylene) copolymers, PS-b-PEP, are essentially used as surfactants in nonpolar media such as aliphatic hydrocarbon solvents, for which the polyolefin block has affinity. The free-radical dispersion polymerization of MMA in the presence of PS-b-PEP (Kraton G1701), as a stabilizer, reported by Stejskal and coworkers should be mentioned [168, 169]. The authors characterized precisely the formation of micelles and also the dynamics of PMMA particle formation by dynamic light scattering. The same group also investigated the GTP of MMA in n-heptane in the presence of such a stabilizer [170]. Mülhaupt and coworkers also reported the use of PS-b-PEP as a surfactant in the controlled radical TEMPO-mediated polymerization of styrene in decane [171]. They found a broad particle size distribution, explained by the high reaction temperature.

PS-hydrogenated PI was used for the dispersion polymerization of 1-vinylpyrrolidone [172]. Finally, Landfester et al. reported the use of poly(ethylene-co-butene)-b-poly(ethylene oxide) (KLE3729 from Goldschmidt, Essen, Germany;

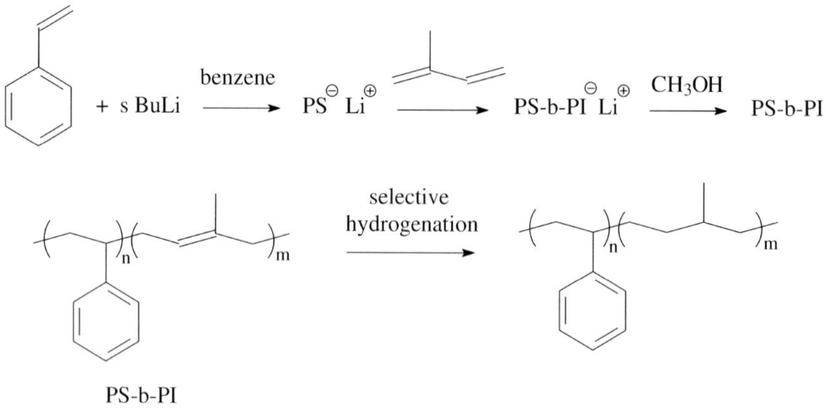

Scheme 5.25 Synthesis of PS-b-PI block copolymer by anionic polymerization followed by selective hydrogenation of the PI block.

$M_w = 6600$ g mol^{-1}, 44 wt.% EO) for nonaqueous inverse mini-emulsions of hydroxyethyl methacrylate, acrylamide or acrylic acid in cyclohexane or hexadecane as the nonpolar continuous phase [173]. The stability of the latex was ensured by the addition of water or a salt, used as a "lipophobe". Small and narrow distributed latexes in the range 50–200 nm were obtained.

5.3.3
Silicon-based Copolymers

Block copolymers having one silicon block are also an important family of steric stabilizers. These copolymers are generally prepared by living anionic polymerization, allowing the control of the copolymer composition [174, 175]. The siloxane monomer has to be carefully purified and its polymerization performed at low temperature must be discontinued before complete conversion to avoid secondary reactions such as back-biting reactions (Scheme 5.26).

Free-radical dispersion polymerization of styrene was performed in n-alkanes using PS-b-PDMS block copolymers as steric stabilizers. Interesting investigations monitored by SANS experiments with deuterated copolymers enabled Higgins et al. to measure the packing and conformation of the stabilizing chains at the particle–liquid interface [176]. They observed that the deuterated PS block was dispersed into the core of the PS particles, while a small phase separation was only revealed at the surface of the PMMA particles, in comparative investigations. The anionic dispersion polymerization of styrene and DVB was also investigated in hydrocarbon media [177–179].

The synthesis of triblock copolymers via GTP by sequential addition of first MMA and then trimethylsilyl methacrylate to a PDMS-based macro-initiator was also reported [180]. After removal of the silyl protecting group, such a terpolymer was found to be an efficient stabilizer for MMA polymerization in scCO$_2$. Remarkably, the latex formed can be redispersed in water via a charge stabilization mechanism.

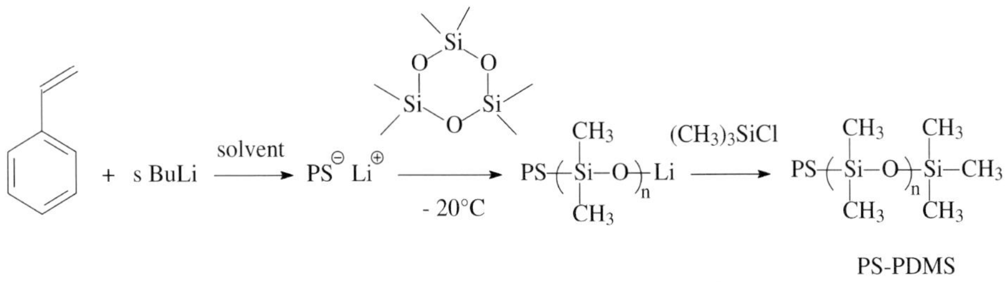

Scheme 5.26 Synthesis of PS-b-PDMS block copolymer by living anionic polymerization.

5.3.4
(Meth)acrylic-based Copolymers

A series of PMMA-based copolymers were also investigated as steric stabilizers for nonaqueous dispersion polymerization. The main route to such copolymers is anionic polymerization, as depicted in Scheme 5.27 for PMMA-b-PtBA block copolymers.

Anionic polymerization of methacrylate monomer requires a low temperature, a polar solvent and a relatively less active and sterically hindered initiator to avoid (or limit) the reaction of active anions with the carbonyl group of the (meth)acrylic monomer [181].

PMMA-b-PtBA block copolymers were used as effective stabilizers for the dispersion polymerization of MMA and nBuA in ethanol, a medium in which PtBA is preferentially soluble. Colloidal particles in the size range 60–100 nm could be prepared, as reported by Kuentz and Riess [7].

Poly[2-(dimethylamino)ethyl methacrylate]-b-poly(alkyl methacrylate) block copolymers were also tested as stabilizers for styrene polymerization in alcoholic media (1-butanol to 1-octanol). Near monodisperse, micrometer sized polystyrene particles were produced [182].

The anionic polymerization of MMA was also performed in heptane–THF medium in the presence of PNMA-b-PMMA and PDMS-b-PMMA block copolymers [7].

It is also worth mentioning the preparation of poly(ethylene terephthalate) (PET) particles by a dispersion technique from a solution of bis(hydroxyethyl) terephthalate in high-boiling petroleum. A graft copolymer consisting of a PMMA backbone with pendant poly(12-hydroxystearic acid) soluble side-chains was used as the stabilizer [155].

Slomkowski et al. also reported interesting work dealing with the preparation of biodegradable nano- and microparticles composed of polylactides, Poly(L-Lc) or PCL with controlled surface properties [183]. A series of di- and triblock copoly-

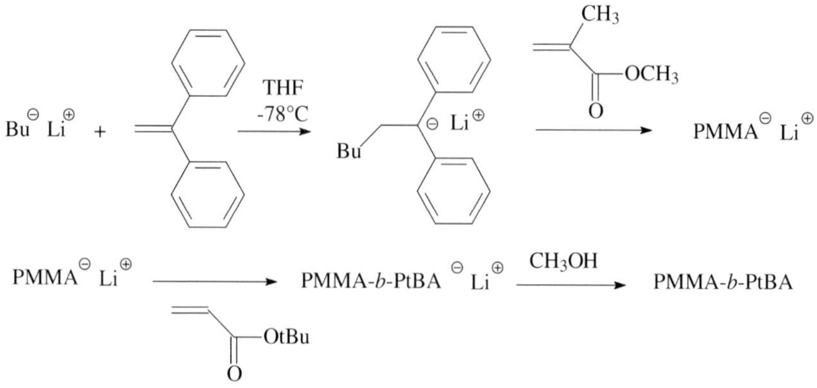

Scheme 5.27 Synthesis of PMMA-b-PtBA block copolymers by anionic polymerization.

mers were synthesized and used as surfactants. Typically, poly(L-Lc) microspheres (∼2 µm) were obtained in heptane–1,4-dioxane (4:1, v/v) in the presence of PCL-b-PODMA-b-PDMAEMA. Thanks to the presence of the PDMAEMA block, in which amino groups are protonated at low pH, the particles could be transferred into water in a second stage.

5.3.5
In Situ Formation of Block Copolymer Stabilizers

As already reported in the case of heterogeneous polymerization in aqueous media, the formation and stabilization of latex particles may be accomplished by *in situ* formation of the block copolymer that serves as surfactant. A similar approach has been investigated for dispersion polymerization in organic dispersant media where the block copolymer plays the role of a steric stabilizer.

The anionic dispersion polymerization of 1,4-divinylbenzene (DVB) was reported by Okay and Funke [184]. Highly cross-linked 1,4-divinylbenzene particles with a shell of poly(4-*tert*-butylstyrene) (P*t*BS) chains were prepared in *n*-heptane at 50 °C. "Living" P*t*BS chains with various chain lengths were used as initiators of the DVB polymerization and also as steric stabilizers. The mechanism of particle formation were studied and the conditions governing the transition from the microgel to the macrogel regime analyzed. It is worth mentioning that a similar approach was already described by Barrett for the anionic dispersion polymerization of styrene [155].

The synthesis of polyurethane microspheres of core–shell type, by *in situ* formation of the steric stabilizer, has been also reported. Previous investigations by Graham and coworkers should be mentioned [185, 186]. More recently, Sivaram and coworkers prepared a reactive polymer with a long hydrophobic acrylate ester moiety and bearing two primary hydroxyl groups (Scheme 5.28). Dispersion polymerizations were carried out in paraffin oil and PUR particles with an average diameter in the range 10–50 µm were obtained [157, 158].

The preparation of core–shell nanoparticles by self-assembly and step-growth polymerization was also described by Knauss and Clark [187]. In that case, functional PEO was used as a steric stabilizer for the preparation of PUR nanoparticles in toluene as the dispersant phase.

Cramail and coworkers also investigated the synthesis of PUR core–shell particles [188–194] following a similar approach. They first prepared a series of macro-

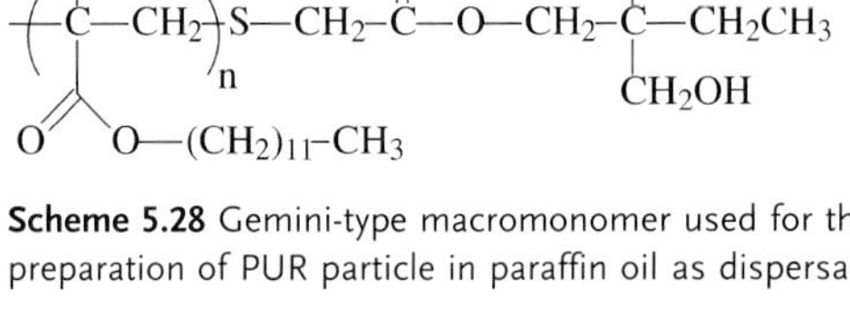

Scheme 5.28 Gemini-type macromonomer used for the preparation of PUR particle in paraffin oil as dispersant medium.

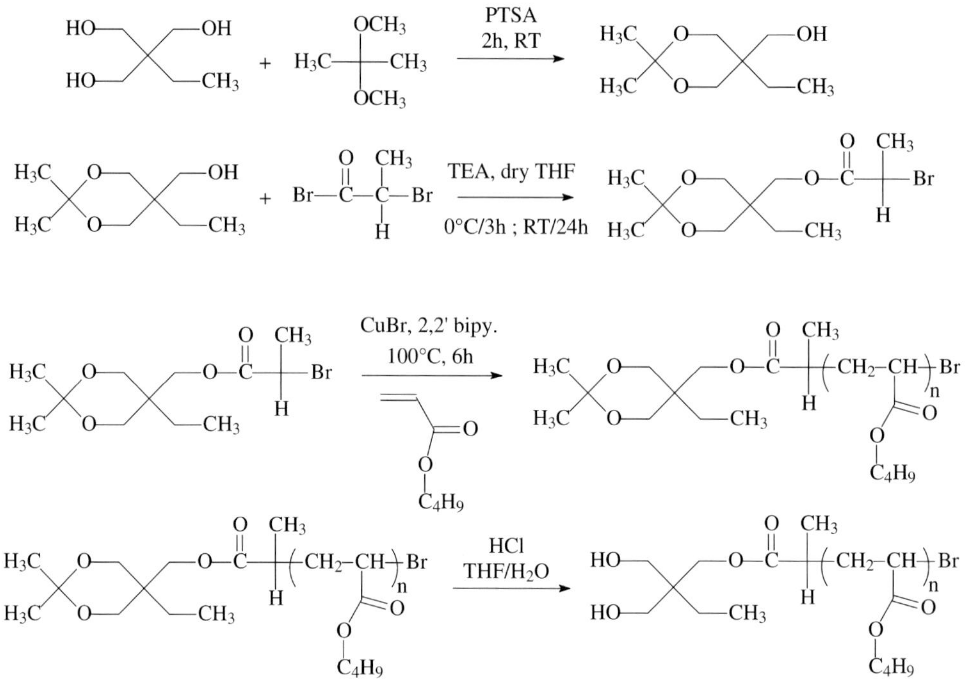

Scheme 5.29 Synthesis by ATRP of ω,ω'-hydroxy-PnBA used as steric stabilizer for PUR latex particle synthesis.

monomers, such as α- or ω,ω'-hydroxy-terminated PS, PB, PnBA, PtBA, PIB and PDMS. Dispersion polymerizations, in cyclohexane at 60 °C, of diols and diisocyanates, in the presence of the macromonomer and DBTDL as a catalyst, yielded well-defined particles of micrometer size. The best results in terms of particle size calibration were obtained with low-T_g macromonomers. It is worth noting that the use of an ω,ω'-hydroxy macromonomer leads to the *in situ* formation of grafted copolymers. The latter were more efficient stabilizers than block copolymers arising from monohydroxyl macromonomers. Indeed, the concentration of ω,ω'-hydroxy macromonomer needed to obtain stable and calibrated PUR latex particles was found to be much lower than that of a monohydroxyl macromonomer. ω,ω'-Hydroxy-terminated macromonomers based on PnBA were synthesized by ATRP. The two hydroxyl functions could be attached to the polyacrylate chains from either initiation or termination reactions. An illustration of a synthetic route to obtain ω,ω'-hydroxy-PnBA is given in Scheme 5.29.

A typical SEM analysis of micrometer-size and well-calibrated PUR particle synthesized under the conditions described above is given in Fig. 5.1.

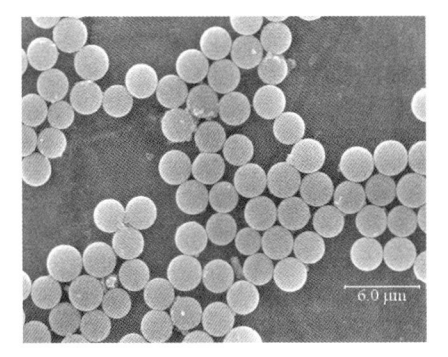

Fig. 5.1 SEM image of PUR particles synthesized in cyclohexane in the presence of PDMS-OH (4700 g mol^{-1}) macromonomer.

5.4
Main Surfactants Used for Heterogeneous Polymerizations in scCO$_2$

In the last decade, the use of supercritical carbon dioxide (scCO$_2$) as a polymerization medium has been the focus of considerable academic and industrial research. scCO$_2$ is very attractive for environmentally friendly processing techniques since it is inexpensive and non-flammable and its critical point is easily accessible ($T_c = 31.1$ °C, $P_c = 73.8$ bar). Moreover, scCO$_2$ represents a real alternative to volatile organic compounds since its physical properties (viscosity and density) can be readily modulated with both the pressure and the temperature. scCO$_2$ is a nonpolar medium that behaves like aliphatic solvents; it is a poor solvent for most polymers of high molar mass, except silicon-based polymers, fluoropolymers and polycarbonates. The use of steric polymeric stabilizers composed of "CO$_2$-philic and CO$_2$-phobic" sequences is therefore required to perform polymerization in this dispersant medium. Some selected recent studies are discussed in the following.

5.4.1
Silicon-based Stabilizer

Silicon polymers are attractive surfactants since they are soluble in CO$_2$ but also less expensive than their fluorinated counterparts. Moreover, silicones are soluble in a range of conventional organic solvents, allowing easy characterization of the reaction products. De Simone and coworkers first reported the dispersion polymerization of MMA in CO$_2$ using a commercially available methacrylate-terminated polydimethylsiloxane (PDMS) macromonomer [195]. High PMMA molecular weights with regular spherical particle morphologies were obtained with high yields provided that a high stabilizer concentration was used (3.5–16% w/w vs monomer). It was shown that only a fraction of the macromonomer was actually copolymerized with the PMMA and the role of the non-grafted PDMS was not clear. O'Neill et al. also investigated the dispersion polymerization of MMA in scCO$_2$ using PDMS macromonomer stabilizers [196]. DeSimone and

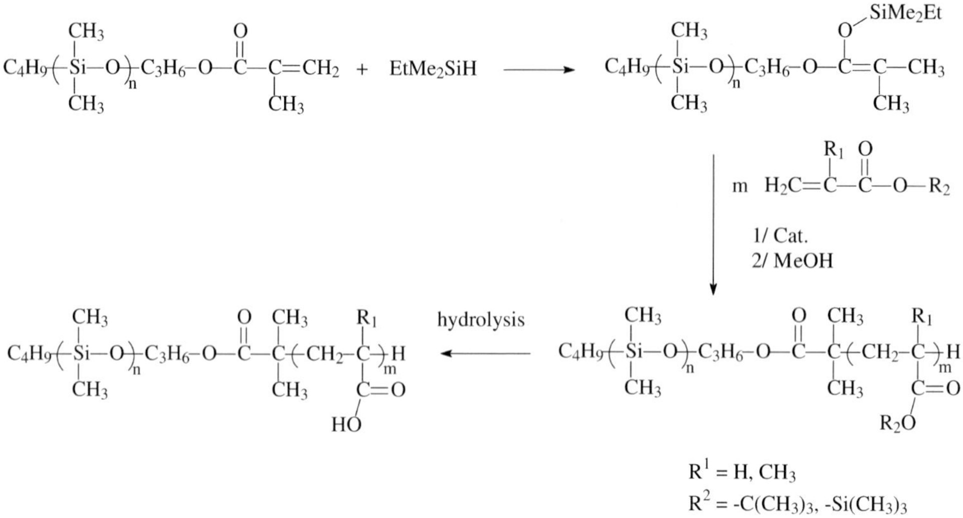

Scheme 5.30 Synthesis of PDMS-*b*-PMA or PDMS-*b*-PAA.

coworkers developed PDMS-based block copolymers for dispersion polymerizations of styrene and MMA. For instance, PS-*b*-PDMS block copolymers were synthesized (according to the procedure described by Saam et al. [197]) and used as stabilizers to prepare PS particles by dispersion polymerization [198]. The particle sizes, molecular weights and molecular weight distributions were found to be influenced significantly by the density of the CO_2 continuous phase (i.e. the pressure) and also by the presence of helium.

Johnston and co-workers described the synthesis of polydimethylsiloxane-*b*-poly(methacrylic acid) (PDMS-*b*-PMA) block copolymers (Scheme 5.30), which were efficiently used as stabilizers for the dispersion polymerization of MMA [199, 200]. The PDMS-*b*-PMA stabilizer is particularly interesting because it allows the formation of water-dispersible PMMA powders. This is due to the "ambidextrous" nature of the surfactant, which can provide stabilization in water and in CO_2. The authors showed that more regular PMMA particles were formed when mixtures of PDMS-based surfactants were used, i.e. mixtures of PDMS-*b*-PMA and commercially available graft copolymer PDMS-g-poly(pyrrolidonecarboxylic acid).

Chambon et al. performed PUR particle synthesis in $scCO_2$ (30–40 MPa) as dispersant medium in the presence of monohydroxyl end-capped PDMS [201]. They observed that higher molar mass PUR and macromonomer incorporation were achieved in $scCO_2$ compared with data obtained in cyclohexane. Nevertheless, a broader particle size distribution was obtained in $scCO_2$. More interestingly, use of DBTDL was not necessary in $scCO_2$ to obtain PUR particles in good yield and with a short reaction time.

5.4.2
Fluorinated Copolymers

During the last decade, fluorinated polymers have been extensively used with the development of polymerizations conducted in $scCO_2$. The first report on the synthesis by radical polymerization of a "CO_2-philic" fluorinated polyacrylate was described by DeSimone et al. (Scheme 5.31) [202]. They took advantage of the amphiphilic nature of the poly(1,1-dihydroperfluorooctyl acrylate) (PFOA) homopolymer formed ($M_n = 1.1 \times 10^4$ or 2.0×10^5 g mol^{-1}) to effect the dispersion polymerization of MMA to high conversions (>90%) and high degrees of polymerization (>3000) [203]. Similarly, they described that the methacrylate analogue [i.e. poly(1,1-dihydroperfluorooctyl methacrylate) (PFOMA)] could be effectively synthesized and used for the preparation of PS particles in the micrometer size range (2.9–9.6 µm) [204].

The success achieved with such fluorinated (meth)acrylate homopolymers brought about a number of studies to design their block copolymer counterparts, which were shown to be of great interest as surfactants for dispersion polymerizations. Most of the reports have dealt with the synthesis of copolymers (statistical, block or graft) containing a CO_2-philic component and a CO_2-phobic (CO_2-insoluble) anchoring component. The most common lipophilic anchoring components have been PS and PMMA. Concerning PS-based block copolymers, Lacroix-Desmazes et al. recently described the synthesis of fluorinated block copolymers by nitroxide-mediated radical polymerization (Scheme 5.32) [205]. The block copolymers are composed of a PS block and either a fluorinated polystyrene or a fluorinated polyacrylate and are aimed to act as surfactants for $scCO_2$ applications.

Further, Ma and Lacroix-Desmazes described the synthesis of hydrophilic/CO_2-philic poly(ethylene oxide)-b-poly(1,1,2,2-tetrahydroperfluorodecyl acrylate) (PEO-b-PFDA) block copolymers via RAFT or ATRP in the presence of degenerative transfer agents or a macroinitiator based on PEO [206]. Such block copolymers showed potential as surfactants for water-in-CO_2 emulsions.

Regarding PMMA-based copolymers, Lepilleur and Beckman [207] described the synthesis of poly(methyl methacrylate-co-hydroxyethyl methacrylate)-g-poly(perfluoropropylene oxide) poly(MMA-co-HEMA)-g-poly(PFPO) graft copolymers

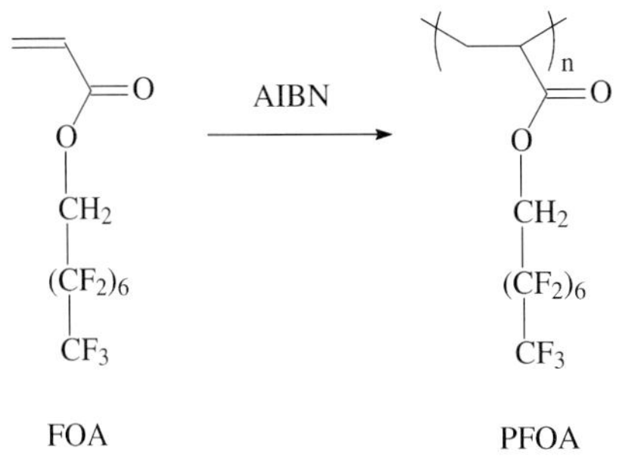

Scheme 5.31 Synthesis of PFOA by conventional radical polymerization in $scCO_2$.

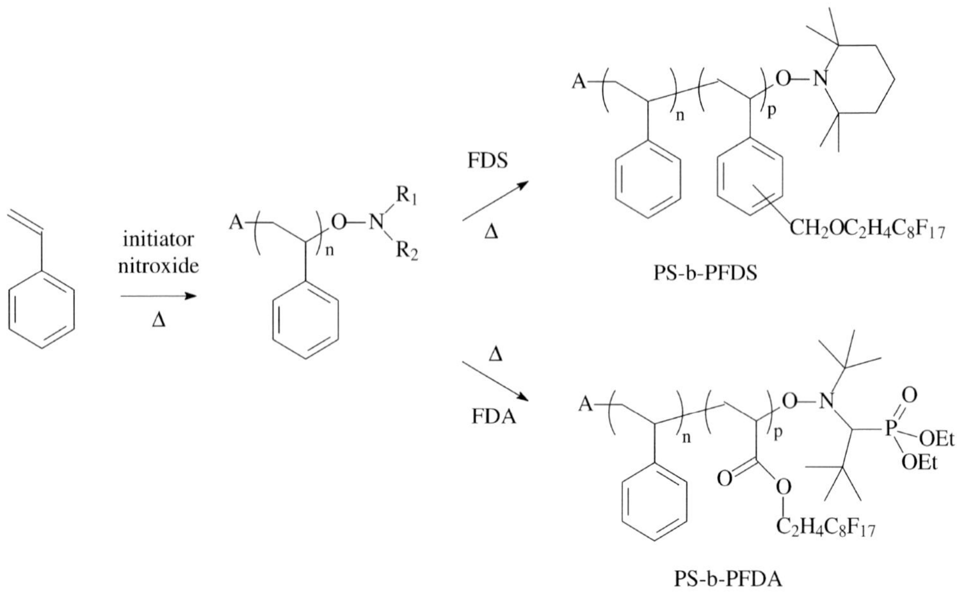

Scheme 5.32 Synthesis of PS-b-PFDS and PS-b-PFDA by the CRP route.

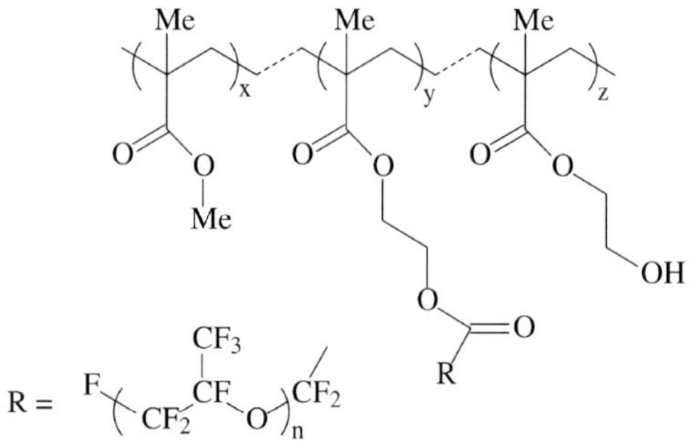

Scheme 5.33 Structure of poly(MMA-co-HEMA)-g-poly(PFPO) used as surfactant for MMA polymerization in scCO$_2$.

for application as stabilizers for the dispersion polymerization of methyl methacrylate in scCO$_2$ (Scheme 5.33).

The authors first synthesized poly(MMA-co-HEMA) block copolymers by free radical polymerization using AIBN and then performed grafting by coupling between PFPO-acid chloride and HEMA.

5.4.3
Polycarbonates

Currently, there are tremendous efforts to find non-fluorinated CO_2-philic (co)-polymers that could be use as surfactants [208]. Beckman and coworkers have reported that a series of poly(ether–carbonate) copolymers (Scheme 5.34) readily dissolve in CO_2 at low pressure [209].

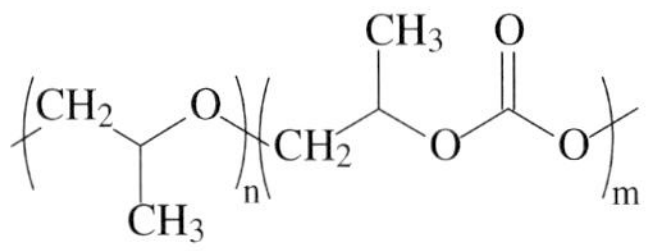

Scheme 5.34 Structure of poly(ether carbonate) obtained by copolymerization of PO and CO_2.

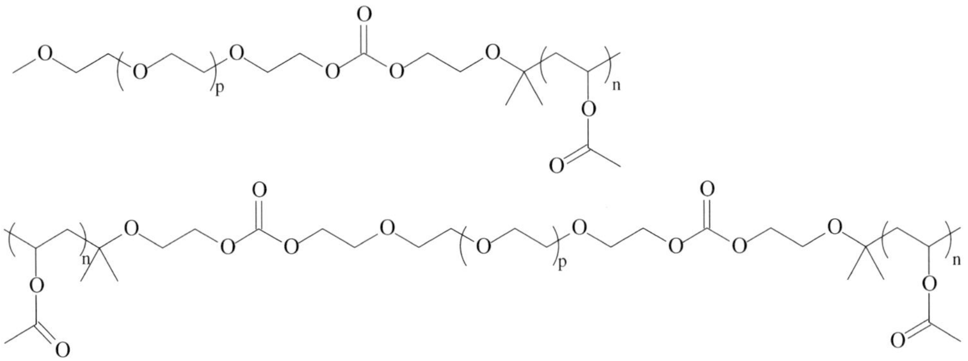

Scheme 5.35 Structure of block copolymers composed of PVAc and PEO segments.

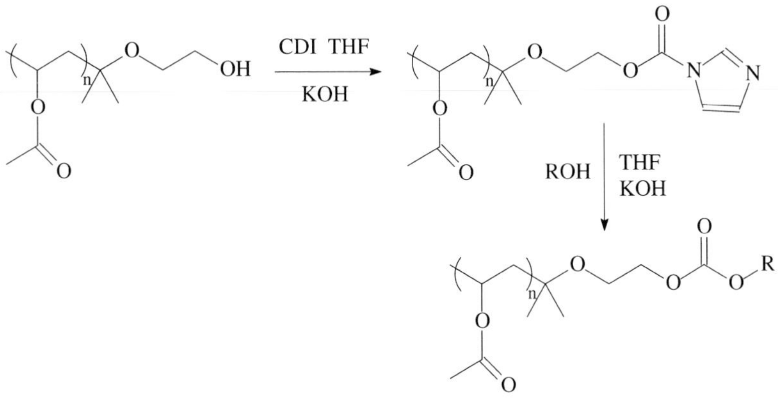

Scheme 5.36 Route to the synthesis of PVAc-b-PEG.

Tan and Cooper recently described the synthesis of block copolymers constituted of oligo(vinyl acetate) and poly(ethylene oxide) segments (Scheme 5.35) [210]. First, monohydroxy-functionalized oligo(vinyl acetate)s (PVAc) were obtained by free-radical polymerization in the presence of a chain transfer agent (2-isopropoxyethanol) and fractionated by supercritical fluid extraction. Then, these oligomers were allowed to react with poly(ethylene glycol) diols (PEG) through a carbonyldiimidazole (CDI) coupling (Scheme 5.36).

For example, a PVAc-b-PEG-b-PVAc triblock surfactant (with $n \approx 60$ and $p \approx 30$) was found to emulsify up to 97% v/v CO_2 in water and to form a uniform, opaque emulsion which was stable for at least 48 h. The remarkable stability of these emulsions was further demonstrated by polymerization of the continuous aqueous phase to give porous, cross-linked polyacrylamide materials.

5.5
Conclusion

As has been illustrated through out this review, the search for novel polymeric surfactants with well-defined structure, composition, architecture and functionality is still the focus of interest of a large community involved in polymer chemistry and physics. It is worth mentioning that real breakthroughs, in terms of polymeric surfactant synthesis, could be achieved thanks to controlled polymerization methods, essentially anionic and free-radical polymerizations. It can be visualized that the controlled radical polymerization method, which not only allows one to polymerize a wide range of monomers but also permits the synthesis of amphiphilic block copolymers directly in water, will further be advanced towards the synthesis of polymeric surfactants with specific characteristics.

The search for environmentally friendly processing techniques is absolutely crucial in the field nowadays. In that sense, it can be speculated that surfactants based on natural renewable resources, which correspond today to "niche" markets, will gain more importance (and volume) in the future. Indeed, biocompatible and biodegradable surfactants, derived from natural oils and fats and incorporating sugar- and polyol-type moieties, could progressively be developed to answer consumer demands in terms of greener and more powerful materials.

List of Abbreviations

ACPA	4,4'-azobis(4-cyanopentanoic acid)
AIBN	2,2'-azobisisobutyronitrile
ATRP	atom-transfer radical polymerization
nBA	n-butyl acrylate
tBA	*tert*-butyl acrylate
BzMA	benzyl methacrylate
CDI	carbonyldiimidazole

CMC	critical micelle concentration
CPADB	(4-cyanopentanoic acid)-4-dithiobenzoate
CRP	controlled radical polymerization
CTA	chain transfer agent
DADMAC	diallyldimethylammonium chloride
DBTDL	dibutyltin dilaurate
DHBC	double hydrophilic block copolymer
DVB	divinylbenzene
EA	ethyl methacrylate
EO	ethylene oxide
EG	ethylene glycol
EHMA	ethylhexyl methacrylate
EMA	ethyl methacrylate
FOA	1,1′-dihydroperfluorooctyl acrylate
GTP	group transfer polymerization
HEMA	hydroxyethyl methacrylate
HLB	hydrophilic–lipophilic balance
IBVE	isobutyl vinyl ether
LCST	lower critical solubility temperature
LMA	lauryl methacrylate
MAA	methacrylic acid
MMA	methyl methacrylate
NAD	nonaqueous dispersion
NMP	nitroxide-mediated polymerization
PAA	poly(acrylic acid)
PAGA	poly(acrylol glucosamine)
PB	polybutadiene
PnBA	poly(n-butyl acrylate)
PBO	poly(butene oxide)
PtBS	poly(tert-butylstyrene)
PCL	poly(ε-caprolactone)
PDMAEMA	poly(dimethylaminoethyl methacrylate)
PDMS	poly(dimethylsiloxane)
PEB	poly(ethylenebutene)
PEG	poly(ethylene glycol)
PEGMA	poly(ethylene glycol methacrylate)
PEO	poly(ethylene oxide)
PEP	poly(ethylene-co-propylene)
PET	poly(ethylene terephthalate)
PFDA	poly(1,1,2,2-tetrahydroperfluorodecyl acrylate)
PFDS	poly(1,1,2,2-tetrahydroperfluorodecylstyrene)
PFOA	poly(1,1′-dihydroperfluorooctyl acrylate)
PFOMA	poly(1,1′-dihydroperfluorooctyl methacrylate)
PFOP	poly(perfluoropropylene oxide)
PI	polyisoprene

PLA	poly(lactic acid)
PMAA	poly(methacrylic acid)
PMMA	poly(methyl methacrylate)
PNIPAM	poly(N-isopropylacrylamide)
PNMA	poly(nonyl methacrylate)
PO	propylene oxide
PPO	poly(propylene oxide)
PS	polystyrene
PSS	poly(styrene sulfonate)
PUR	polyurethane
PVAc	poly(vinyl acetate)
PVME	poly(vinyl methyl ether)
PVOH	poly(vinyl alcohol)
RAFT	reversible addition–fragmentation transfer
ROP	ring-opening polymerization
SANS	small-angle neutron scattering
$scCO_2$	supercritical carbon dioxide
SDS	sodium dodecyl sulfate
SEM	scanning electron microscopy
TEMPO	2,2,6,6-tetramethylpiperidine-N-oxy
VAc	vinyl acetate

References

1 I. Piirma, *Polymeric Surfactants*, Surfactant Science Series 42, Marcel Dekker, New York, **1992**.
2 Z. Tuzar, P. Kratochvil, in *Surface and Colloid Science*, E. Matijevic (Ed.), Plenum Press, New York, **1993**.
3 N. Hadjichristidis, S. Pispas, G. A. Floudas (Eds.), *Block Copolymers: Synthetic Strategies, Physical Properties and Applications*, Wiley-VCH, Weinheim, **2003**.
4 P. Alexandridis, B. Lindman, *Amphiphilic Block Copolymers: Self Assembly and Applications*, Elsevier, Amsterdam, **2000**.
5 G. Riess, Ph. Dumas, G. Hurtrez, Block copolymer micelles and assemblies, in *MML Series*, Vol. 5, R. Arshady, A. Guyot (Eds.), Citus Books, London, **2002**, p. 69.
6 G. Riess, *Prog. Polym. Sci.* **2003**, *28*, 1107–1170.
7 G. Riess, C. Labbe, *Macromol. Rapid Commun.* **2004**, *25*, 401–435.
8 S. Creutz, J. Van Stam, F. C. De Schryver, R. Jérme, *Macromolecules* **1998**, *31*, 681–689.
9 Y. Y. Wom, H. T. Davis, F. S. Bates, *Macromolecules* **2003**, *36*, 681–689.
10 A. Jada, B. Siffert, G. Riess, *Colloids Surf. A* **1993**, *75*, 203–209.
11 S. Antoun, J.-F. Gohy, R. Jérme, *Polymer* **2001**, *42*, 3641–3648.
12 Y. Chang, J. D. Bender, M. V. B. Phelps, H. R. Allcock, *Biomacromolecules* **2002**, *3*, 1364–1369.
13 A. Guyot, K. Tauer, J. M. Asua, S. Van Es, C. Gauthier, A. C. Hellgren, D. C. Sherrington, A. Montoya-Goni, M. Sjoberg, O. Sindt, F. Vidal, M. Unzue, H. Schoombrood, E. Shipper, P. Lacroix-Desmazes, *Acta Polym.* **1999**, *50*, 57–66.
14 A. Guyot, *Macromol. Symp.* **2002**, *179*, 105–132.
15 A. Guyot, *Adv. Colloid Interface Sci.* **2004**, *108–109*, 3–22.

16 S. Liu, S. P. Armes, *Curr. Opin. Colloid Interface Sci.* **2001**, *6*, 249–256.
17 B. Chu, *Langmuir* **1995**, *11*, 414–421.
18 C. Booth, D. Attwood, *Macromol. Rapid Commun.* **2000**, *21*, 501–537.
19 R. P. Quirk, G. M. Lizarraga, *Macromol. Chem. Phys.* **2000**, *201*, 1395–1404.
20 M. Akatsuka, T. Aida, S. Inoue, *Macromolecules* **1994**, *27*, 2820–2825.
21 W. Braune, J. Okuda, *Angew. Chem. Int. Ed.* **2003**, *42*, 64–68.
22 C. Billouard, S. Carlotti, Ph. Desbois, A. Deffieux, *Macromolecules* **2004**, *37*, 4038–4043.
23 V. Rejsek, D. Sauvanier, C. Billouard, Ph. Desbois, A. Deffieux, S. Carlotti, *Macromolecules*, submitted.
24 P. Desbois, A. Deffieux, S. Carlotti, C. Billouard, *Patent WO 2004 104068*, **2004**.
25 Y. C. Chen, V. Dimonie, M. S. El-Aasser, *J. Appl. Polym. Sci.* **1992**, *45*, 487–499.
26 Y. C. Chen, V. Dimonie, M. S. El-Aasser, *J. Appl. Polym. Sci.* **1992**, *46*, 691–706.
27 I. T. Kim, P. F. Luckham, *Colloids Surf.* **1992**, *68*, 243–259.
28 E. T. Schipper, O. Sindt, T. Hamaide, P. Lacroix Desmazes, B. Mueller, A. Guyot, M. J. W. A. Van den Enden, F. Vidal, J. J. G. S. Van Es, A. L. German, A. Montoya Goni, D. C. Sherrington, H. A. S. Schoonbrood, J. M. Asua, M. Sjoeberg, *Colloid Polym. Sci.* **1998**, *276*, 402–411.
29 O. Soula, R. Pétiaud, M. F. Llauro, A. Guyot, *Macromolecules* **1999**, *32*, 6938–6943.
30 A. Guyot, I. Capek, *Macromol. Symp.* **2002**, *179*, 105–132.
31 M. Dufour, A. Guyot, *Colloid Polym. Sci.* **2003**, *281*, 97–104.
32 O. Soula, A. Guyot, *Langmuir* **1999**, *15*, 7956–7962.
33 O. Soula, A. Guyot, N. Williams, J. Grade, T. Blease, *J. Polym. Sci., Part A: Polym. Chem.* **1999**, *37*, 4205–4217.
34 R. P. Quirk, J. Kim, C. Kaush, M. Chun, *Polym. Int.* **1996**, *39*, 3–10.
35 M. A. Hillmyer, F. S. Bates, *Macromolecules* **1996**, *29*, 6994–7002.
36 N. Ekizoglu, N. Hadjichristidis, *J. Polym. Sci., Part A: Polym. Chem.* **2001**, *39*, 1198–1202.
37 K. Matyjaszewski, *Controlled/Living Radical Polymerization: Progress in ATRP, NMP and RAFT*, American Chemical Society, Washington, DC, Vol. 768, **2000**.
38 B. Reining, H. Keul, H. Höcker, *Polymer* **1999**, *40*, 3555–3563.
39 B. Reining, H. Keul, H. Höcker, *Polymer* **2002**, *43*, 7145–7154.
40 K. Jankova, X. Chen, J. Kops, W. Batsberg, *Macromolecules* **1998**, *31*, 538–541.
41 E. Beaudoin, P. E. Dufils, D. Gigmes, S. Marque, C. Petit, P. Tordo, D. Bertin, *Polymer* **2006**, *47*, 98–106.
42 G. Riess, *Colloids Surf. A* **1999**, *153*, 99–110.
43 G. Riess, *Polym. Adv. Technol.* **1995**, *6*, 497–508.
44 M. Antonietti, R. Basten, S. Lohmann, *Macromol. Chem. Phys.* **1995**, *196*, 441–466.
45 S. Sapurina, J. Stejskal, Z. Tuzar, *Colloids Surf. A* **2001**, *180*, 193–198.
46 J. Stejskal, Conducting polymer nanospheres and nanocomposites, in *MML Series*, Vol 5, R. Arshady, A. Guyot (Eds.), Citus Books, **2002**, Ch 6–7, p. 195.
47 K. Landfester, F. Tiarks, H-P. Hentze, M. Antonietti, *Macromol. Chem. Phys.* **2000**, *201*, 1–5.
48 I. Capek, *Adv. Colloid Interface Sci.* **2000**, *88*, 295–357.
49 J. S. Shay, R. J. English, R. J. Spontak, C. M. Balik, S. A. Khan, *Macromolecules* **2000**, *33*, 6664–6671.
50 K. Tauer, U. Yildiz, *Macromolecules*, **2003**, *36*, 8638–8647.
51 S. Gibanel, V. Heroguez, J. Forcada, Y. Gnanou, *Macromolecules* **2002**, *35*, 2467–2473.
52 S. Gibanel, V. Heroguez, J. Forcada, *J. Polym. Sci., Part A: Polym. Chem.* **2002**, *40*, 2819–2827.
53 A. Bcsi, J. Forcada, S. Gibanel, V. Héroguez, M. Fontanille, Y. Gnanou, *Macromolecules*, **1998**, *31*, 2087–2097.
54 A. Chemtob, V. Héroguez, Y. Gnanou, *Macromolecules* **2004**, *37*, 7619–7627.
55 D. Quemener, V. Héroguez, Y. Gnanou, *Macromolecules* **2005**, *38*, 7977–7982.
56 S. Gibanel, J. Forcada, V. Heroguez, M. Schappacher, Y. Gnanou, *Macromolecules* **2001**, *34*, 4451–4458.
57 S. Gibanel, V. Heroguez, J. Forcada, *J. Polym. Sci., Part A: Polym. Chem.* **2001**, *39*, 2767–2776.

58 F. Raju, D. Taton, J. L. Logan, P. Massé, Y. Gnanou, R. S. Duran, *Macromolecules* **2003**, *36*, 8253–8259.
59 S. Peleshanko, J. Jeong, V. V. Shevchenko, K. L. Genson, Y. Pikus, M. Ornatska, S. Petrash, V. V. Tsukruk, *Macromolecules* **2004**, *37*, 7497–7506.
60 H. Keul, A. Neumann, B. Reining, H. Höcker, *Macromol. Symp.* **2000**, *161*, 63–72.
61 W. Wu, I. Piirma, *Polym. Bull.* **1993**, *31*, 531–538.
62 Z. Xu, C. Yi, S. Cheng, L. Feng, *J. Appl. Polym. Sci.* **2001**, *79*, 528–534.
63 S. Föster, E. Krämer, *Macromolecules* **1999**, *32*, 2783–2785.
64 H. Imai, S. Kawaguchi, K. Ito, *Polym. J. (Tokyo)*, **2003**, *35*, 528–534.
65 B. Tan, D. W. Grijpma, T. Nabuurs, J. Feijen, *Polymer* **2005**, *46*, 1347–1357.
66 C. K. Smith, G. Liu, *Macromolecules*, **1996**, *29*, 2060–2067.
67 J. P. Hautekeer, S. K. Varshney, R. Fayt, C. Jacobs, R. Jérme, Ph. Teyssié, *Macromolecules*, **1990**, *23*, 3893–3898.
68 Q. Ma, K. L. Wooley, *J. Polym. Sci. Polym. Chem.* **2000**, *38*, 4805–4820.
69 C. Burguière, C. Chassenieux, B. Charleux, *Polymer* **2003**, *44*, 509–518.
70 H. Zhang, E. Ruckenstein, *Macromolecules* **1998**, *31*, 7575–7580.
71 E. Ruckenstein, H. Zhang, *Macromolecules* **1998**, *31*, 9127–9131.
72 J. Yuan, Y. Shi, Z. Fu, W. Yang, *Polym. Int.* **2006**, *55*, 360–364.
73 S. P. Rannard, N. C. Billingham, S. P. Armes, J. Mykytiuk, *Eur. Polym. J.* **1993**, *29*, 407–414.
74 Y. Liu, L. Wang, C. Pan, *Macromolecules* **1999**, *32*, 8301–8305.
75 J. Qiu, B. Charleux, K. Matyjaszewski, *Prog. Polym. Sci.*, **2001**, *26*, 2083–2134.
76 J. Kriz, B. Masar, H. Pospisil, J. Plestil, Z. Tuzar, M. A. Kiselev, *Macromolecules* **1996**, *29*, 7853–7858.
77 T. Liu, H. Schuch, M. Gerst, B. Chu, *Macromolecules* **1999**, *32*, 6031–6042.
78 T. Rager, W. H. Meyer, G. Wegner, K. Mathauer, W. Mächtle, W. Schrof, D. Urban, *Macromol. Chem. Phys.* **1999**, *200*, 1681–1691.
79 H. J. Spinelli, *Prog. Org. Coat.* **1996**, *27*, 255–260.
80 B. Drescher, A. B. Scranton, J. Klier, *Polymer* **2001**, *42*, 49–58.
81 Y. K. Chong, T. P. T. Le, G. Moad, E. Rizzardo, S. H. Thang, *Macromolecules* **1999**, *32*, 2071–2074.
82 N. Gaillard, A. Guyot, J. Claverie, *J. Polym. Sci., Part A: Polym. Chem.* **2003**, *41*, 684–698.
83 C. J. Ferguson, R. J. Hughes, B. T. T. Pham, B. S. Hawkett, R. G. Gilbert, A. K. Serelis, C. H. Such, *Macromolecules* **2002**, *35*, 9243–9245.
84 C. J. Ferguson, R. J. Hughes, D. Nguyen, B. T. T. Pham, R. G. Gilbert, A. K. Serelis, C. H. Such, B. S. Hawkett, *Macromolecules* **2005**, *38*, 2191–2204.
85 C. Ramireddy, Z. Tuzar, K. Prochazka, S. E. Webber, P. Munk *Macromolecules* **1992**, *25*, 2541–2545.
86 A. Desjardins, A. Eisenberg, *Macromolecules* **1991**, *24*, 5779–5790.
87 C. Burguiere, S. Pascual, B. Coutin, A. Polton, M. Tardi, B. Charleux, K. Matyjaszewski, J.-P. Vairon, *Macromol. Symp.* **2000**, *150*, 39–44.
88 K. A. Davis, B. Charleux, K. Matyjaszewski, *J. Polym. Sci., Part A: Polym. Chem.*, **2000**, *38*, 2274–2283.
89 W. Guojian, Y. Deyue, *J. Appl. Polym. Sci.* **2001**, *82*, 2381–2386.
90 X. F. Zhong, S. K. Varshney, A. Eisenberg, *Macromolecules*, **1992**, *25*, 7160–7167.
91 L. Zhang, A. Eisenberg, *Science* **1995**, *268*, 1728–1731.
92 O. Terreau, C. Bartels, A. Eisenberg, *Langmuir* **2004**, *20*, 637–645.
93 A. Choucair, C. Lavigueur, A. Eisenberg, *Langmuir* **2004**, *20*, 3894–3900.
94 L. Couvreur, B. Charleux, O. Guerret, S. Magnet, *Macromol. Chem. Phys.* **2003**, *204*, 2055–2063.
95 C. Lefay, B. Charleux, M. Save, C. Chassenieux, O. Guerret, S. Magnet, *Polymer* **2006**, *47*, 1935–1945.
96 G. Laruelle, J. François, L. Billon, *Macromol. Rapid Commun.* **2004**, *25*, 1839–1844.
97 C. Burguiere, S. Pascual, C. Bui, J. P. Vairon, B. Charleux, K. A. Davis, K. Matyjaszewski, I. Betremieux, *Macromolecules* **2001**, *34*, 4439–4450.
98 C. Burguière, C. Chassenieux, B. Charleux, *Polymer* **2003**, *44*, 509–518.

99 R. Francis, B. Lepoittevin, D. Taton, Y. Gnanou, *Macromolecules* **2001**, *34*, 4451–4458.
100 P. Pincus, *Macromolecules* **1991**, *24*, 2912–2919.
101 L. Leemans, R. Fayt, Ph. Teyssié, *J. Polym. Sci.* **1990**, *28*, 1255–1262.
102 L. Leemans, R. Fayt, Ph. Teyssié, N.C. de Jaeger, *Macromolecules* **1991**, *24*, 5922–5925.
103 M. Bouix, J. Gouzi, B. Charleux, J.P. Vairon, P. Guinot, *Macromol. Rapid Commun.* **1998**, *19*, 209–213.
104 K. Tauer, A. Zimmermann, *Macromol. Rapid Commun.* **2000**, *21*, 825–831.
105 K. Tauer, A. Zimmermann, H. Schlaad, *Macromol. Chem. Phys.* **2002**, *203*, 319–327.
106 K. Tauer, H. Müller, L. Rosengarten, K. Riedelsberger, *Colloids Surf. A* **1999**, *153*, 75–88.
107 L. Leemans, R. Jérme, Ph. Teyssié, *Macromolecules* **1998**, *31*, 5565–5571.
108 A. Lieske, W. Jaeger, *Macromol. Chem. Phys.* **1998**, *199*, 255–260.
109 W. Jaeger, U. Wendler, A. Lieske, J. Bohrish, C. Wandrey, *Macromol. Symp.* **2000**, *161*, 87–96.
110 W. Jaeger, U. Wendler, A. Lieske, J. Bohrish, *Langmuir* **1999**, *15*, 4026–4032.
111 M. Save, M. Manguian, C. Chassenieux, B. Charleux, *Macromolecules*, **2005**, *38*, 280–289.
112 M. Manguian, M. Save, C. Chassenieux, B. Charleux, *Colloid. Polym. Sci.* **2005**, *284*, 142–450.
113 S.F. Lascelles, F. Malet, R. Mayada, N.C. Billingham, S.P. Armes, *Macromolecules* **1999**, *32*, 2462–2471.
114 J.I. Amalvy, G.F. Unali, Y. Li, S. Granger-Bevan, S.P. Armes, B.P. Binks, J.A. Rodrigues, C.P. Whitby, *Langmuir* **2004**, *20*, 4345–4354.
115 S. Liu, S.P. Armes, *Angew. Chem. Int Ed.* **2002**, *41*, 1413–1416.
116 S.A.F. Bon, K. Ohno, D.M. Haddleton, *ACS Symp. Series* **2000**, *780*, 148–161.
117 L. Houillot, J. Nicolas, M. Save, B. Charleux, Y. Li, S.P. Armes, *Langmuir* **2005**, *21*, 6726–6733.
118 Y. Cai, S.P. Armes, *Macromolecules* **2005**, *38*, 271–279.
119 V. Bueten, S. Liu, J.V.M. Weaver, X. Bories-Azeau, Y. Cai, S.P. Armes, *React. Funct. Polym.* **2006**, *66*, 157–165.
120 M. Manguian, M. Save, B. Charleux, *Macromol. Rapid Commun.* **2006**, *27*, 399–404.
121 C.I. Simionescu, G. David, V. Alupei, M. Rusa, A. Ioanid, B.C. Simionescu, *Angew. Macromol. Chem.* **1998**, *255*, 17–21.
122 C. Babac, G. Guven, G. David, B.C. Simionescu, E. Piskin, *Eur. Polym. J.* **2004**, *40*, 1947–1952.
123 I. Ydens, P. Degee, J. Libiszowski, A. Duda, S. Penczek, P. Dubois, *ACS Symp. Ser.* **2003**, *854*, 283–298.
124 B. Verdonck, E.J. Goethals, F.E. Du Prez, *Macromol. Chem. Phys.* **2003**, *204*, 2090–2098.
125 H.J. Paik, M. Teodorescu, J. Xia, K. Matyjaszewski, *Macromolecules* **1999**, *201*, 1189–1199.
126 D. Batt-Coutot, D.M. Haddleton, A.P. Jarvis, R.L. Kelly, *Eur. Polym. J.* **2003**, *39*, 2243–2252.
127 T. Otsu, T. Matsunaga, T. Doi, A. Matsumoto, *Eur. Polym. J.* **1995**, *31*, 67–78.
128 M.C. Iovu, K. Matyjaszewski, *Macromolecules* **2003**, *36*, 9346–9354.
129 M. Destarac, D. Charmot, X. Franck, S.Z. Zard, *Macromol. Rapid Commun.* **2000**, *21*, 1035–1039.
130 M. Wakioka, K.Y. Baek, T. Ando, M. Kamagaito, M. Sawamoto, *Macromolecules* **2002**, *35*, 330–333.
131 A. Favier, C. Barner-Kowollik, T.P. Davis, M.H. Stenzel, *Macromol. Chem. Phys.* **2004**, *205*, 925–936.
132 J. Bernard, A. Favier, T.P. Davis, C. Barner-Kowollik, M.H. Stenzel, *Polymer* **2006**, *47*, 1073–1080.
133 A. Debuigne, J-R. Caille, N. Willet, R. Jérme, *Macromolecules* **2005**, *38*, 9488–9496.
134 M. Mertoglu, S. Garnier, A. Laschewsky, K. Skrabania, J. Storsberg, *Polymer* **2005**, *46*, 7726–7740.
135 S. Garnier, A. Laschewsky, *Macromolecules* **2005**, *38*, 7580–7592.
136 S. Garnier, A. Laschewsky, *Langmuir* **2006**, *22*, 4044–4053.

137 J. J. Voslo, D. de Wet-Roos, M. P. Tonge, R. D. Sanderson, *Macromolecules* **2002**, *35*, 4894–4902.
138 H. Kukula, H. Schalaad, K. Tauer, *Macromolecules* **2002**, *35*, 2538–2544.
139 C. Wollenweber, A. V. Makievski, R. Miller, R. J. Daniels, *Colloid Interface Sci.* **2000**, *117*, 419–426.
140 K. Holmberg, B. Jönsson, B. Kronberg, B. Lindman, *Surfactants and Polymers in Aqueous Solution*, Wiley, New York, **2002**, Ch. 12, pp. 261–276.
141 Th. F. Tadros, A. Vandamme, B. Levecke, K. Booten, C. V. Stevens, *Adv. Colloid Interface Sci.* **2004**, *108–109*, 207–226.
142 C. Nouvel, C. Frochot, V. Sadtler, P. Dubois, E. Dellacherie, J. L. Six, *Macromolecules* **2004**, *37*, 4981–4988.
143 E. Rotureau, C. Chassenieux, E. Dellacherie, A. Durand, *Macromol. Chem. Phys.* **2005**, *206*, 2038–2046.
144 E. Rotureau, M. Leonard, E. Marie, E. Dellacherie, T. A. Camesano, A. Durand, *Colliids Surf. A* **2006**, in press.
145 C. Racles, T. Hamaide, *Macromol. Chem. Phys.* **2005**, *206*, 1757–1768.
146 L. Albertin, M. H. Stenzel, C. Barner-Kowollik, L. John R. Foster, T. P. Davis, *Macromolecules* **2005**, *38*, 9075–9084.
147 J. Bernard, X. Hao, T. P. Davis, C. Barner-Kowollik, M. H. Stenzel, *Biomacromolecules* **2006**, *7*, 232–238.
148 H. Cölfen, *Macromol. Rapid Commun.* **2001**, *22*, 219–252.
149 K. Tauer, V. Khrenov, N. Shirshova, N. Nassif, *Macromol. Symp.* **2005**, *226*, 187–201.
150 C. M. Hansen, *Hansen Solubility Parameters: a Users's Handbook*, CRC Press, Boca Raton, FL, **2000**.
151 C. L. Winzor, Z. Mrazek, M. A. Winnik, M. D. Croucher, G. Riess, *Eur. Polym. J.* **1994**, *30*, 121–128.
152 X. You, V. L. Dimonie, A. Klein, *J. Appl. Polym. Sci.* **2001**, *80*, 1951–1962.
153 X. You, V. L. Dimonie, A. Klein, *J. Appl. Polym. Sci.* **2001**, *80*, 1963–1975.
154 R. P. Quirk, G. M. Lizzaraga, J. E. Davis, G. Aviles, *Polym. Prepr.* **1999**, *40*, 1003–1004.
155 K. E. J. Barrett, *Dispersion Polymerization in Organic Media*, Wiley, New York, **1975**.
156 N. Yabuuchi, T. Imamura, T. Mukae, K. Ishii, *US Patent 5 603 986*, **1997**.
157 L. S. Ramanathan, P. G. Shukla, S. Sivaram, *Pure Appl. Chem.* **1998**, *70*, 1295–1299.
158 L. S. Ramanathan, D. Baskaran, P. G. Shukla, S. Sivaram, *Macromol. Chem. Phys.* **2002**, *203*, 998–1002.
159 S. D. Ittel, L. K. Johnson, M. Brookhart, *Chem. Rev.* **2000**, *100*, 1169–1203.
160 G. W. Coates, P. D. Hustad, S. Rartz, *Angew. Chem. Int. Ed.* **2002**, *41*, 2236–2257.
161 H. Makio, N. Kashiwa, T. Fujita, *Adv. Synth. Catal.* **2002**, *344*, 477–493.
162 M. Mitani, J. Saito, S-I. Ishii, Y. Nakayama, H. Makio, N. Matsukawa, S. Matsui, J-I. Mohri, R. Furuyama, H. Terao, H. Bando, H. Tanaka, T. Fujita, *Chem. Rec.* **2004**, *4*, 137–158.
163 H. Zhang, K. Nomura, *J. Am. Chem. Soc.* **2005**, *127*, 9364–9365.
164 J. Imuta, N. Kashiwa, Y. Toda, *J. Am. Chem. Soc.* **2002**, *124*, 1176–1177.
165 R. Godoy Lopez, C. Boisson, F. D'Agosto, R. Spitz, F. Boisson, D. Bertin, P. Tordo, *Macromolecules* **2004**, *37*, 3540–3542.
166 R. Godoy Lopez, C. Boisson, F. D'Agosto, R. Spitz, F. Boisson, D. Gigmes, D. Bertin, *Macromol. Rapid Commun.* **2006**, *27*, 173–181.
167 M. P. Mc Grath, E. D. Sall, S. J. Tremont, *Chem. Rev.* **1995**, *95*, 381–398.
168 J. Stejskal, D. Hlavata, C. Sikora, C. Konak, J. Pletsil, P. Kratochvil, *Polymer* **1992**, *33*, 3675–3685.
169 B. Hirzinger, M. Helmstedt, J. Stejskal, *Polymer* **2000**, *41*, 2883–2891.
170 A. D. Jenkins, D. Maxfield, C. G. Dos Santos, D. R. M. Walton, J. Stejskal, P. Kratochvil, *Makromol. Chem. Rapid Commun.* **1992**, *13*, 61–63.
171 M. Hölderle, M. Baumert, R. Mülhaupt, *Macromolecules* **1997**, *30*, 3420–3422.
172 D. Horak, M. Krystüfek, J. Spevacek, *J. Polym. Sci., Part A: Polym. Chem.* **2000**, *38*, 653–663.
173 K. Landfester, M. Willert, M. Antonietti, *Macromolecules* **2000**, *33*, 2370–2376.
174 K. Almdal, K. Mortensen, A. J Ryan, F. S. Bates, *Macromolecules* **1996**, *29*, 5940–5947.
175 V. Bellas, H. Iatrou, N. Hadjichristidis, *Macromolecules* **2000**, *33*, 6993–6997.

176 J. S. Higgins, P. E. Tomlins, J. V. Dawkins, G. G. Maghami, S. A. Shakir, *Polym. Commun.* **1988**, *29*, 122–125.
177 J. V. Dawkins, G. Taylor, in *Polymer Colloids*, R. M. Fitch (Ed.), Plenum Press, New York, **1980**, p. 447.
178 M. A. Awan, V. L. Dimonie, M. S. El-Aasser, *J. Polym. Sci. Polym. Chem.* **1996**, *34*, 2633–2649.
179 M. A. Awan, V. L. Dimonie, M. S. El-Aasser, *J. Polym. Sci. Polym. Chem.* **1996**, *34*, 2651–2664.
180 G. Li, M. Z. Yates, K. P. Johnson, K. T. Lim, S. E. Webber, *Macromolecules*, **2000**, *33*, 1606–1612.
181 R. D. Allen, T. E. Long, J. E. Mac Grath, *Polym. Bull.* **1986**, *15*, 127–134.
182 F. L. Baines, S. Dionisio, N. C. Billingham, S. P. Armes, *Macromolecules* **1996**, *29*, 3096–3102.
183 S. Slomkowski, M. Gadzinowski, S. Sosnowski, C. De Vita, A. Pucci, F. Ciardelli, W. Jakubowski, K. Matyjaszewski, *Macromol. Symp.* **2005**, *226*, 239–252.
184 O. Okay, W. Funke, *Macromolecules* **1990**, *23*, 2623–2628.
185 N. B. Graham, J. Mao, *Colloids Surf.s A* **1996**, *118*, 211–220.
186 N. B. Graham, A. Cameron, *Pure Appl. Chem.* **1998**, *70*, 1271–1275.
187 D. M. Knauss, S. L. Clark, *Polym. Prepr.* **2002**, *43*, 324–325.
188 E. Cloutet, B. Radhakrishnan, H. Cramail, *Polym. Int.* **2002**, *51*, 978–985.
189 B. Radhakrishnan, E. Cloutet, H. Cramail, *Colloid Polym. Sci.* **2002**, *280*, 1122–1130.
190 P. Chambon, B. Radhakrishnan, E. Cloutet, E. Papon, H. Cramail, *Macromol. Symp.* **2003**, *199*, 47–57.
191 B. Radhakrishnan, P. Chambon, E. Cloutet, H. Cramail, *Colloid Polym. Sci.* **2003**, *281*, 516–530.
192 P. Chambon, E. Cloutet, H. Cramail, *Macromolecules* **2004**, *37*, 5856–5859.
193 P. Chambon, E. Cloutet, H. Cramail, *Macromol. Symp.* **2005**, *226*, 227–238.
194 B. Radhakrishnan, G. Balerdi, E. Cloutet, H. Cramail, *Macromol. Symp.* **2005**, *229*, 56–65.
195 K. A. Shaffer, T. A. Jones, D. A. Canelas, J. M. DeSimone, *Macromolecules* **1996**, *29*, 2704–2706.
196 M. L. O'Neill, M. Z. Yates, K. P. Johnston, C. D. Smith, S. P. Wilkinson, *Macromolecules* **1998**, *31*, 2838–2847.
197 J. C. Saam, D. J. Gordon, S. Lindsey, *Macromolecules* **1970**, *3*, 1–4.
198 D. A. Canelas, J. M. DeSimone, *Macromolecules* **1997**, *30*, 5673–5682.
199 M. Z. Yates, G. Li, J. J. Shim, S. Maniar, K. P. Johnston, K. T. Lim, S. Webber, *Macromolecules* **1999**, *32*, 1018–1026.
200 K. T. Lim, S. Webber, K. P. Johnston, *Macromolecules* **1999**, *32*, 2811–2815.
201 P. Chambon, E. Cloutet, H. Cramail, T. Tassaing, M. Besnard, *Polymer* **2005**, *46*, 1057–1066.
202 J. M. De Simone, Z. Guan, C. S. Elsbernd, *Science* **1992**, *257*, 945–947.
203 J. M. De Simone, E. E. Maury, Y. Z. Menceloglu, J. B. McClain, T. J. Romack, J. R. Combes, *Science* **1994**, *265*, 356–359.
204 H. Shiho, J. M. De Simone, *J. Polym. Sci., Part A: Polym. Chem.* **1999**, *37*, 2429–2437.
205 P. Lacroix-Desmazes, P. Andre, J. M. DeSimone, A. V. Ruzette, B. Boutevin, *J. Polym. Sci., Part A: Polym. Chem.* **2004**, *42*, 3537–3552.
206 Z. Ma, P. Lacroix-Desmazes, *J. Polym. Sci., Part A: Polym. Chem.* **2004**, *42*, 2405–2415.
207 C. Lepilleur, E. J. Beckman, *Macromolecules* **1997**, *30*, 745–756.
208 E. J. Beckman, *Chem Commun.* **2004**, 1885–1888.
209 T. Sarbu, T. Styranec, E. J. Beckman, *Nature* **2000**, *405*, 165–168.
210 B. Tan, A. I. Cooper, *J. Am. Chem. Soc.* **2005**, *127*, 8938–8939.

6
Molecular and Supramolecular Conjugated Polymers for Electronic Applications

Andrew C. Grimsdale and Klaus Müllen

6.1
Introduction

Organic electronics is one of the most exciting and important emerging technologies. The performance of organic semiconductors – both small molecules and polymers, as the active materials in electronic devices such as field-effect transistors (FETs) [1], solar cells [2] and light-emitting diodes (LEDs) [3] – has reached the point where some products have entered the commercial marketplace and others are being developed. One of the most scientifically exciting aspects of this field is the ability to control the electronic properties of an organic compound by synthetic design, where not only the properties of individual molecules, but also those of ensembles of molecules, i.e. their supramolecular properties, have to be controlled. Understanding and manipulating the self-assembly of organic molecules and polymer chains is thus a vital part of this field. The formation and properties of supramolecular assemblies of conjugated organic systems have recently been reviewed by Meijer and coworkers [4]. The aspect of organic electronics that has most gripped the public imagination, however, is the potential for achieving the ultimate in miniaturization by constructing devices in which a single molecule or polymer chain acts as the active electronic component – truly molecular electronics – although here the use of a well-defined aggregate is perhaps a more realistic target. In this chapter, we intend to show how the molecular and supramolecular optoelectronic properties of organic polymers can be tuned by synthetic design and to illustrate some of the progress that has been made towards the "Holy Grail" of viable molecular electronics. This is not a comprehensive review of conjugated polymers in general, but an overview of some aspects of the electronic and device behavior of conjugated polymers. For space reasons, the synthesis, processing and device aspects of various classes of conjugated polymers will all have to be considered together, with no separate synthetic section, so that for a more detailed discussion of the synthetic aspects the reader is referred to the cited literature.

Before introducing the classes of polymers that we will discuss, let us first consider what the requirements are that such materials must satisfy, as it is these parameters which will guide the design of the polymers. These criteria are themselves dependent upon the nature of the device in which the materials are to be used. Thus in FETs, the primary criterion for a material is that it must possess high charge carrier mobility as the switching speed of the transistor is dependent upon the mobility of the charges flowing through it. It is also desirable that the barriers to charges flowing between the electrodes and the material be low. In LEDs, by contrast, high charge carrier mobility is not essential, but instead, since these devices require opposite charge to combine to produce light, efficient and balanced charge injection is needed to obtain high efficiency (as measured in terms of light produced compared with charges injected). This is sometimes achieved by using multiple layers, which can raise questions as to processing of materials (see below). The other primary device parameters in LEDs are color purity and color stability, which depend on the optical properties of the active materials. The main performance criterion in solar cells is the efficiency with which they convert light into current. Here both the optical properties (how well the active materials absorb light) and the electrical properties (how well the charges move through the material to be collected at the electrodes) of the active materials are important.

A complicating factor in solar cells is that whereas FETs and LEDs require only a single active material (although blends of materials can be used), solar cells generally require both an electron-donating and an electron-accepting component for efficient operation. In all these devices, the stability and lifetime of the device (and hence of the materials within it) are also crucial parameters, especially for commercial applications. Although FETs, LEDs and solar cells are currently the most important organic electronic devices, other devices such as memories and sensors are also being investigated and have their own specific performance criteria.

The requirements for materials intended for device applications can be divided into three areas. First, there are the electronic properties, which are determined by the energy levels of the highest occupied molecular orbital (HOMO) and the lowest unoccupied molecular orbital (LUMO). The difference in the energy levels of the HOMO and LUMO in an organic material – often referred to as the bandgap by analogy with inorganic semiconductors – determines its inherent electrical properties, i.e. whether the undoped material is insulating, semiconducting or conducting. Optical properties such as the color of the material (a consequence of its absorption spectrum) and of its luminescence are also dependent on this bandgap. It should be noted that whereas the bandgap is the primary determinant of the absorption and emission maxima in isolated molecules or polymer chains as seen in dilute solution, the solid-state optical properties are also profoundly affected by intermolecular processes such as aggregation, which may produce red or blue shifts in the absorption or emission spectra and/or reduce the luminescence efficiency. The energy levels of the HOMO and LUMO determine whether the material will be one that is readily reduced, i.e. accepts electrons (n-type material) or one that is readily oxidized, i.e. accepts holes (p-type material). In devices, the energy differences between these orbitals and the relevant electrodes determine the efficiency of injection of charge into the material and so are crucial to the overall device performance. Next one must consider how the material is to be processed to form thin films. Small molecules can be deposited by vacuum deposition techniques, but this is expensive and is not possible for polymers. Solution processing is much cheaper but obviously requires that the material be soluble or that there be a soluble precursor material which can be readily converted to the final material by a method that does not damage the film or underlying substrates. If one is to make multilayer devices then the layers need to possess orthogonal solubilities so that an underlying layer will not be disturbed by the deposition of a later layer or the earlier layers must be rendered insoluble, e.g. by thermally or photochemically induced cross-linking. Furthermore, solution processing works better for amorphous materials such as polymers, since crystalline materials tend to form multi-crystalline films in which the grain boundaries cause problems, e.g. in charge transport, which seriously degrade device performance. Finally, the stability of the material towards oxidation, particularly in the presence of charge and/or light, determines the lifetime of devices using it.

Now, let us introduce the classes of molecules that we will discuss. First come the standard covalently linked conjugated oligomers and polymers. Here one has a backbone of alternating single and double (or triple) bonds as exemplified by polyacetylene (**1**, Fig. 6.1), by polyarylenes such as poly(*p*-phenylene) (PPP, **2**) or polyheteroarylenes such as polythiophene (**3**) and by mixed systems such as poly(*p*-phenylenevinylene)s (PPVs, **4**) or poly(*p*-phenyleneethynylene)s (PPEs, **5**) in which aromatic units are linked by alkene or alkyne moieties. A key feature of these materials is that their π-electrons are readily delocalized, so that it was long believed that materials with extended conjugated π-systems might be intrinsically conductive. However, all the above systems are at best semiconducting in their undoped (electrically neutral) state. Upon introducing charges on the chain (doping), usually by oxidation, their conductivity is dramatically enhanced, so that they turn into conductors, as was first demonstrated for polyacetylene (**1**) [5], which discovery earned Heeger, MacDiarmid and Shirakawa the Nobel Prize in Chemistry in 2000. Since then, high conductivity has been demonstrated for other conjugated systems such as polythiophene (**3**), polypyrrole (**6**) and polyaniline (PANI) (**7**), but in all cases only after extensive doping, and

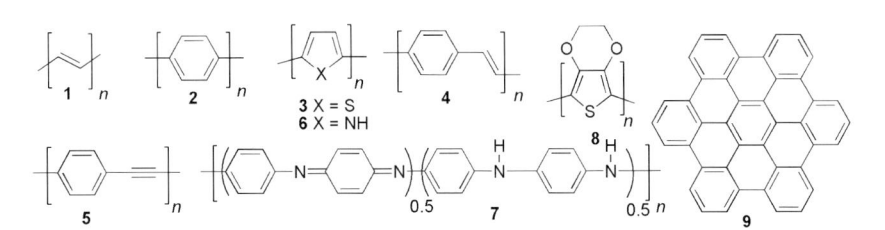

Fig. 6.1 Some representative conjugated systems.

the question as to whether an undoped conjugated polymer can be intrinsically conducting remains open. Polyaniline is a special case among conducting polymers in that it can exist in a variety of oxidation states with varying amounts of benzenoid and quinoidal units, which become conductive upon doping with acid. The highest conductivity is obtained upon protonation of the emeraldine form, shown in structure **7**.

To date, the applicability of most conducting polymers has been hindered by their instability in the (oxidized) conductive form – only polyaniline (**7**), which is doped by protonation rather than oxidation, is stable in the conducting form. A further aspect limiting applications is the difficulty in processing most conducting polymers. Polyacetylene (**1**), polythiophene (**3**) and polypyrrole (**6**) are all insoluble and so films of them must be prepared *in situ*. In the case of **1** this can be done by using thermal conversion of a soluble precursor [6], while films of **3** and **5** are readily obtainable by electropolymerization [7]. The processability can be improved by substitution, e.g. PEDOT (**8**) is soluble in organic solvents and can be processed as an aqueous suspension when doped with poly(sodium 4-styrylsulfonate) (PSS). This suspension is now available commercially and is widely used by academic and industrial researchers as a hole injecting layer in LEDs and solar cells. Polyaniline is readily prepared by oxidation of aniline and can be processed from acidic solvents such as cresol or as a complex with acids such as camphorsulfonic acid (CSA).

Such complexes are commercially available and polyaniline is acquiring applications, e.g. in antistatic coatings. However, there still remain a number of problems with materials such as PEDOT:PSS and PANI:CSA, which will need to be overcome before they can be more widely applied. Accordingly, we will not discuss further conducting polymers, which have been reviewed elsewhere [7–9], but will instead address the use of the semiconducting undoped materials. Even without giving a full account of the chemistry of these various classes of conjugated materials, it can readily be seen that, for example, PPEs are rigidly linear systems whereas the flexible vinylene units in PPVs allow them to form twisted, coiled chain conformations, which affects their solid-state packing and consequent optical and electronic properties. Also, the vinylene units in PPVs display much different chemistry from the benzene rings, e.g. they are much more reactive towards electrophiles such as oxygen, which means that the chemistry of PPVs is very different to that of PPPs. Similarly, in addition to having a different topology, the thiophene rings in **3** are much more electron rich than the benzene rings in **2**, so that polythiophenes display very different physical and chemical behavior to PPPs. The relative electron densities also produce differences in stability, especially towards attack by electrophiles such as oxygen.

We will first examine how the properties of conjugated polymers can be controlled by means of attaching groups as side-chains to them. Here both the nature and the position of the substituents can be used to tune the desired properties, as exemplified by derivatives of **3** and **4**. Next we show how the planarity of polyphenylenes determines their optical properties. Then we will consider the problem of chemical defects in the polymer chains (e.g. missing bonds) affecting their optical and electrical properties and how these can be minimized by synthetic design. Finally, we will discuss the control of the supramolecular order (specifically the chain packing) of conjugated polymers. This leads naturally on to the second class of materials, which are supramolecular polymers in which the monomer units are linked together by non-covalent bonds. These materials have the great advantage over covalently linked polymers that since they are held together by dynamic non-covalent forces and so molecules are continually moving into and out of the assemblies, any structural imperfections (defects) are self-healing at or near ambient temperatures, as the assembly strives to attain thermodynamic equilibrium, whereas defects in covalently linked polymers are reparable (if at all) only by the much more strenuous processes of making or breaking bonds.

Here we will first discuss materials in which conjugated units are linked together by hydrogen bonds. Next we will describe the formation of supramolecular structures by π-stacking. This is exemplified by two-dimensional polycyclic aromatic hydrocarbons (PAHs) such as hexa-*peri*-hexabenzocoronene (HBC, **9**), which form extended ordered columns with very high charge carrier mobilities along the axis of the columns. A distinguishing feature of these materials is that, in contrast to the situation in covalently linked polymers, the 'monomer' units are aligned perpendicular to the direction of conjugation. Here we will compare methods for controlling both the intra- and the intercolumnar order in HBCs. Then we will discuss how such order can be taken advantage of in new methods for preparing carbon nanotubes and similar materials. We conclude with a brief overview of prospects for making functional devices utilizing single chemical units (molecules or polymer chains).

There are a number of basic ways in which one can aim to manipulate the properties of conjugated molecules and polymers. One of the most obvious is by size, as a larger conjugation length means a shift in absorption and emission spectra towards longer wavelengths as the bandgap decreases. This, however, reaches a limit – the so-called effective conjugation length (ECL) – after which further extension of the conjugated system produces no further change in the spectrum. Other obvious methods are by changing the topology of the π-system, for example anthracene versus phenanthrene, or by incorporating heteroatoms, e.g. pyridine as against benzene. However, the most versatile way to modify easily the properties of an organic material is to attach substituents, and it is to this we now turn.

6.2
Control of Properties by Substitution in Conjugated Polymers

The most basic effect of substituents is to improve processability, e.g. by enhancing solubility or fusibility. This is of particular importance for polymers as these are most readily processed in solution or in the melt. For example, PPV (**4**) is an insoluble, infusible material. Thin films for devices are therefore pro-

Fig. 6.2 PPV precursor polymer and some substituted PPV derivatives.

duced by processing the soluble precursor polymer **10** (Fig. 6.2), which is then converted to **4** by heating under vacuum [10]. (The synthesis of PPVs will be discussed in more detail later.) By contrast, the long alkyl chains in MEH-PPV (**11**) make it soluble in common organic solvents, enabling thin films to be produced by spin-casting or other solution-based techniques. The second most obvious effect of substituents is to alter the orbital energies of the compound. For example, unsubstituted PPV has a bandgap of 2.25 eV and hence produces yellow–green luminescence. Therefore, the electron-donating alkoxy groups in MEH-PPV raise the energy of the HOMO compared with PPV and so reduce the bandgap so that whereas PPV displays yellow–green luminescence, MEH-PPV is a red–orange emitter [11]. Other solubilizing substituents such as alkyl as in BuEH-PPV (**12**) [12] and silyl as in DMOS-PPV (**13**) [13] are not strongly electron donating and therefore do not reduce the bandgap significantly, so these polymers are green emitters.

Substituents can also affect the optical properties of conjugated polymers in more subtle ways. Thus, whereas 2,5-dialkoxy substitution as in **11** red shifts the emission of a PPV derivative, the 2,3-dialkoxy-substituted polymer **14** has its emission blue shifted compared with PPV, because the steric repulsion between the substituents on adjacent units twists the polymer backbone and so reduces the conjugation [14]. Such effects are also seen in other classes of conjugated polymers such as polyphenylenes (see below) or poly(2,5-thiophene)s with solubilizing substituents at the 3-positions. Here two adjacent 3-substituted thiophenes can be coupled either 2:2 (head-to-head, HH), 2:5 (head-to-tail, HT) or 5:5 (tail-to-tail, TT) (Fig. 6.3). Head-to-head coupling induces blue shifts in ab-

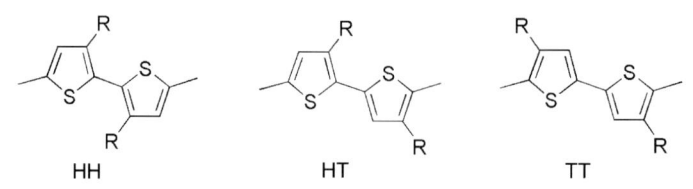

Fig. 6.3 Head-to-head, head-to-tail and tail-to-tail dyads in polythiophenes.

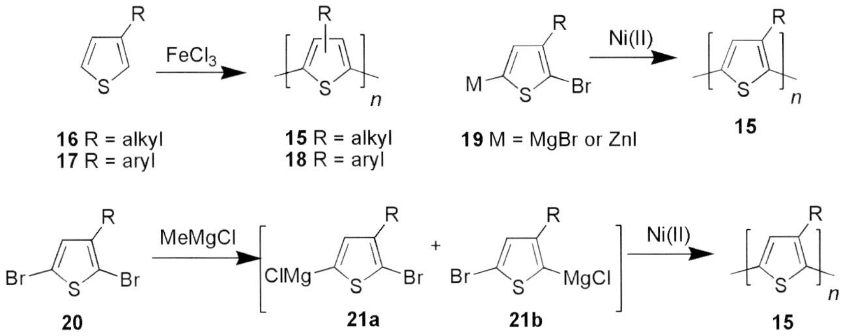

Scheme 6.1 Routes to poly(3-alkylthiophene)s.

sorption and emission spectra as the steric repulsion of the substituents causes an out-of-plane torsion of the polymer chain [15].

The only totally regioregular polythiophene possible is one with all HT links and so the amount of HT links in a polythiophene is called the degree of regioregularity. The control of the regioregularity of polythiophenes offers an excellent example of how properties can be tuned by synthetic design. A more detailed overview of the synthesis of polythiophenes can be found in the review by McCullough [16]. The simplest route to poly(3-alkylthiophene)s (P3ATs, **15**, Scheme 6.1) is to polymerize oxidatively a 3-alkylthiophene (**16**), most commonly using iron(III) chloride. Here the oxidation generates a cation, which then undergoes Friedel–Crafts addition to another thiophene. Since the alkyl group neither stabilizes a cation more at the 2- than at the 5-position, nor provides a major steric barrier to attack at the 2-position, it is not surprising that the resulting polymers are regiorandom with between 52 and 80% HT links depending on the substrate and reaction conditions [15, 16]. If one oxidizes a 3-arylthiophene (**17**) then the aryl stabilizes a cation at the 2-position via resonance and also offers a larger barrier to substitution at the 2-position, so that the resulting polymers (**18**) are 94% HT [17]. To obtain regioregular P3ATs (**15**), polymerizations utilizing transition metal-mediated cross-couplings of 5-metallo-2-halothiophenes (**19**) were developed [16]. Here the choice of catalyst was found to be decisive in obtaining high (>98%) regioregularity, which was attributed to the degree of steric hindrance about the metal center in the key carbon–carbon bond-forming step. More recently, it has been shown that similarly high regioregularity (97%) can be obtained by a Grignard metathesis (GRIM) method [18]. Here treatment of a 2,5-dihalo-3-alkylthiophene (**20**) with 1 equiv. of methylmagnesium halide produces a mixture of the thiophene Grignard reagents **21a** and **21b** (83:17), but the much higher steric demands for the aryl–aryl coupling than in oxidative coupling mean that HH coupling is kinetically and thermodynamically so disfavored that the product (**15**) is still regioregular [19].

The effect of achieving complete regioregularity is to increase considerably the conjugation length – a change from 7 to 40 thiophene rings has been determined for R=$C_{12}H_{25}$ – which is indicated by a marked red shift in the absorption and emission maxima and a reduction of 0.3–0.4 eV in the bandgap [16]. This reflects a much greater planarity of the polymer chain, which, as will be discussed later, also improves the degree of order in the solid state.

6.3
Controlling Optical Properties via Increased Planarity

Red shifts in the optical spectra with increasing planarity are also observed in PPPs, but here the planarization is achieved by linking adjacent phenylene rings with methine bridges. In a single-stranded PPP (**22**, Fig. 6.4) [20], steric repulsion between the solubilizing alkyl or alkoxy groups induces severe torsion between adjacent rings [21], which results in limited conjugation along the chain so that such materials typically emit in the violet ($\lambda_{max} \approx 405$ nm). At the other extreme, double-stranded ladder-type PPPs (LPPPs, **23**) in which all the rings are bridged, are blue–green emitters ($\lambda_{max} = 460$ nm) with very small Stokes shifts, indicative of a highly rigid structure with little difference in geometry between the ground and excited states [22]. By linking only some of the rings – so-called "step-ladder" polymers – the absorption and emission maxima can be tuned so that polyfluorenes (PFs, **24**) [23] and polyindenofluorenes (PIFs, **25**) [24] are blue–violet emitters ($\lambda_{max} = 420$–430 nm), whereas poly(ladder-type

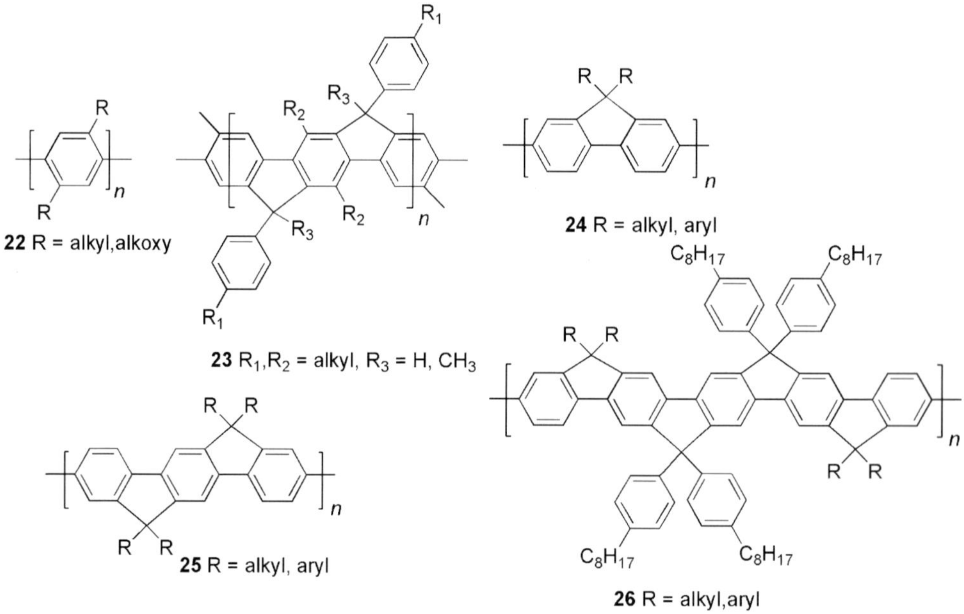

Fig. 6.4 Polyphenylenes with different degrees of bridging.

pentaphenylenes) (26) produce blue emission (λ_{max}=445 nm) [25]. As would be expected, the polymers' Stokes shifts decrease as the amount of bridging, and hence their rigidity, increase.

Interestingly, the effective conjugation length for emission in all the bridged materials seems to be around 12–15 benzene rings, but that for absorption decreases from 24 benzene rings for PFs to 12 for LPPPs. This suggests that the excited states are relatively planar in all these materials, so that the effect of rigidification is most notable in the ground states, which is consistent with the much larger red shift seen in the absorption than in the emission spectra. Anomalously, the LPPP 23b has a smaller persistence length (8 nm) than the ladder-type pentaphenylene polymer 26 (25 nm) [26]. The most obvious explanation for this is that the LPPP contains chemical defects, i.e. that there are some missing bridges. To see how this might occur, let us briefly consider the synthesis of 23 (Scheme 6.2) [22, 27]. First a precursor polymer 27 is made by a Suzuki polycondensation [28] of a dihalobenzene 28 and a benzenebisboronate 29. This is then converted to a polycarbinol 30 either by reduction (30a) or by addition of an organolithium (30b). The methine bridges are then formed by a polymer-analogous Friedel–Crafts alkylation using boron trifluoride. Although no defects are detectable by NMR, this has a sensitivity of at best 1–2% and there is no way to remove the defective polymer chains. By contrast, in the synthesis of the step-ladder polymers 24–26 the bridges are prepared during synthesis of the monomers and so rigorous purification of the monomers and intermediate compounds should prevent any such defects appearing in the final polymers.

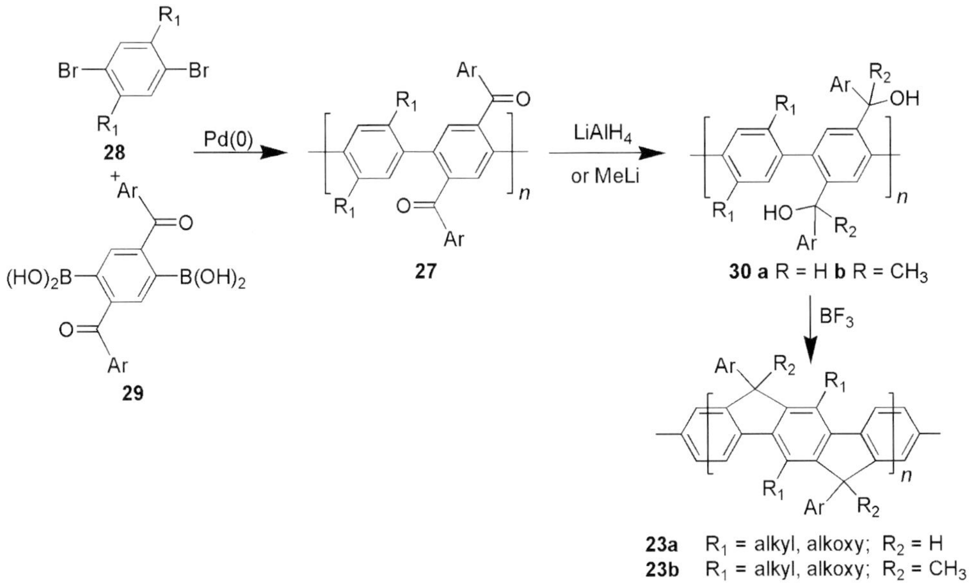

Scheme 6.2 Synthesis of LPPPs.

The above example, in addition to showing the effects of planarity on the optical properties of polyphenylenes, also illustrates the importance of chemical defects in determining polymer properties and of the need to design a synthesis so as to minimize them, to which topic we now turn.

6.4 Minimization of Defects in Polymers by Synthetic Design

The synthesis of PPVs (more detailed discussions can be found in the reviews by Denton and Lahti [29] and by Scherf [27]) provides an excellent example of how minimization of defects is essential for achieving optimal performance from conjugated polymers in devices. The most widely used methods for making PPVs are base-promoted reactions of α,ω-*p*-xylylene compounds **31**, most commonly the dihalides (Gilch synthesis [30]) or the disulfonium salts (Wessling synthesis [10]). These form quinodimethane intermediates **32**, which predominantly couple via a radical mechanism to produce precursor polymers **33** (Scheme 6.3). These are either converted *in situ* to the conjugated polymer, e.g. MEH-PPV (**11**), by further base-promoted eliminations or are isolated and subsequently thermally converted to the final product, e.g. PPV (**4**). It has been shown that tolane (**34**) and bisbenzyl (**35**) moieties are produced during Gilch (and presumably also Wessling) polymerizations through non-regioregular coupling of the quinodimethane intermediates and that high levels of these defects significantly degrade the performance and lifetime of LEDs using these polymers [31–33].

It has also been demonstrated that the amount of defects is determined by the nature of the substituents R_1 and R_2 [32]. When both are alkoxy as in **31 a**, then there is no discrimination as to which halide is removed and so both pos-

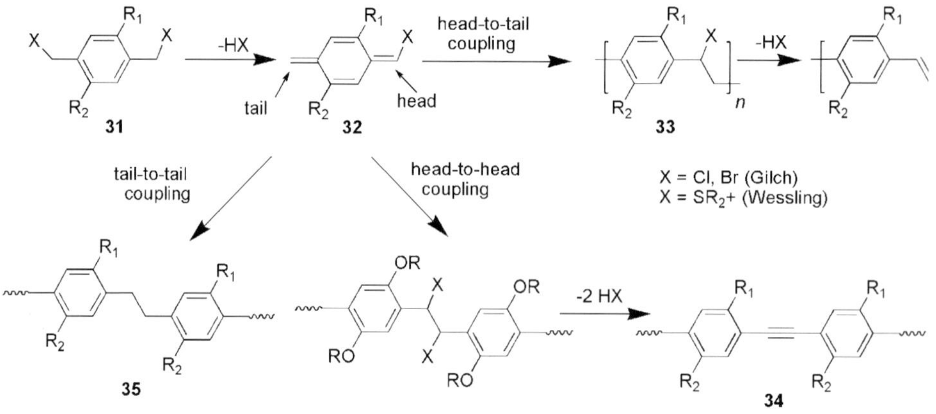

Scheme 6.3 Formation of defects during synthesis of PPVs via quinodimethanes.

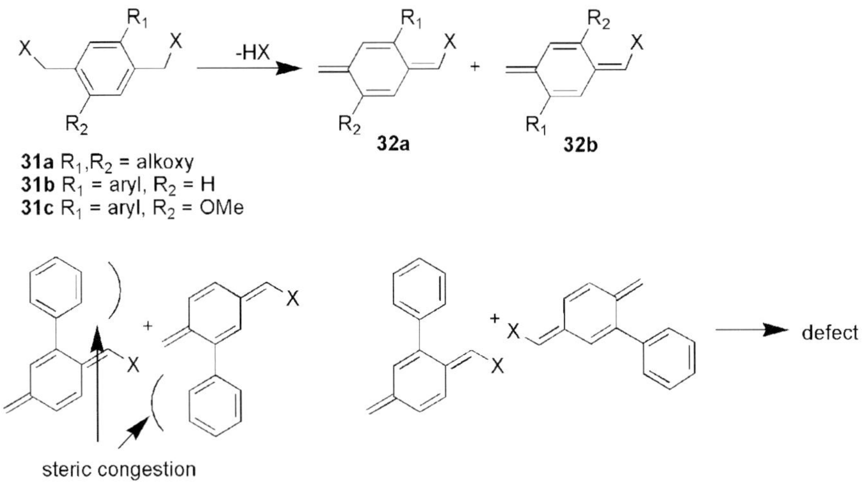

Scheme 6.4 Explanation for high defect levels in aryl-PPVs.

sible quinodimethanes **32a** and **32b** are formed equally, but there is no steric barrier to head-to-tail coupling so defect formation is low (Scheme 6.4). Both quinodimethanes are also equally likely if R_1 is an aryl ring and R_2 is hydrogen (**31b**), but here there is a large barrier to head-to-tail coupling between the different isomers, which means that a lot of defects are formed. The lowest defect levels were found when the two substituents were an aryl and a methoxy (**31c**) as now **32a** was formed preferentially due to the higher acidity of the benzyl protons α- to the methoxy group and there was no barrier to head-to-tail coupling. As a result, polymers made from such monomers exhibited the best performance in LEDs.

One could also try to avoid these defects by using a different synthetic method. For example, MEH-PPV (**11**), produced by the Horner polycondensation of a bisaldehyde and a bisphosphonate, appears to be defect free, but this method takes longer and gives much lower molecular masses than the Gilch route [34].

Such a trade-off between synthetic simplicity and polymer quality is also seen for the synthesis of PPEs (**35**) [35]. Here the experimentally simpler method involves Hagihara–Sonogashira coupling of a dialkyne **36** with a diiodide **37** (Scheme 6.5). This route, however, produces a large amount of diyne defects due to alkyne–alkyne coupling and also leaves fluorescence-quenching halides as end-groups. By contrast, metathesis polymerization of a dialkyne **38** produces defect-free polymers with well-defined end-groups, but this method is currently more limited in the range of materials that can be successfully polymerized and also, unlike the other method, cannot be used to make alternating copolymers.

Defects can have a number of deleterious effects on the optical and electrical properties of organic materials. Often they suppress emission by providing sites for non-radiative decay. Sometimes the defects themselves are emissive, as for example the ketone defects seen in PFs (see below), and so their emission can

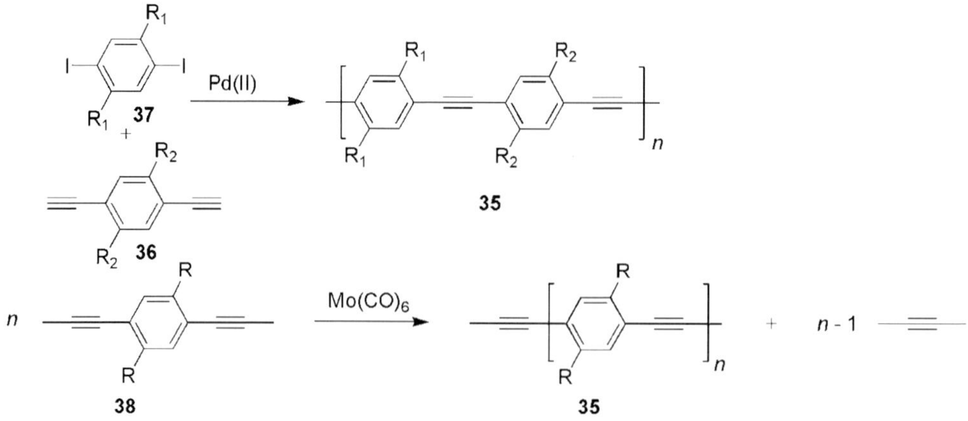

Scheme 6.5 Alternative routes to PPEs.

contaminate or even suppress the inherent emission from the material. The effects of defects are most pronounced in the solid state, as here inter-chain interactions can promote exciton migration to defect sites. One can therefore minimize the effects of defects by suppressing aggregation and other interactions of polymer chains. On the other hand, strong inter-chain interactions are useful for obtaining high charge carrier mobility, which is a desirable property for FETs and other applications [1]. As a result, controlling the inter-chain interactions of conjugated polymers so as to suppress undesirable and promote beneficial interactions is an essential element in optimizing their behavior.

6.5
Controlling Chain Packing of Conjugated Polymers

The packing of polymer chains depends crucially on the nature and pattern of the substituents. Thus regioregular poly(3-alkylthiophene)s (**15**), in addition to showing greater planarity, also display much closer and more regular chain packing than their regiorandom counterparts. One effect is that their conductivity is much higher, typically 100–200 S cm^{-1} as against 0.1–1 S cm^{-1} for irregular polymers, with values as high as 1000–2000 S cm^{-1} being recorded for some samples [16]. The charge carrier mobility of regioregular P3ATs in FETs is also several orders of magnitude higher than for the corresponding irregular polymers, with a highest value of 0.1 cm^2 V^{-1} s^{-1} being reported for regioregular P3HT (**15** R=C$_6$H$_{13}$) [36, 37]. This high mobility was found to correlate with a lamellar layer structure of the polymer film with the π-systems normal to the substrate [37]. Polythiophenes have rather high-lying HOMOs so that they are readily oxidized, which causes the FET performance to drop noticeably over time when devices using P3HT are operated under ambient atmosphere. Phenylene-based materials such as PFs are much more stable towards oxidation but

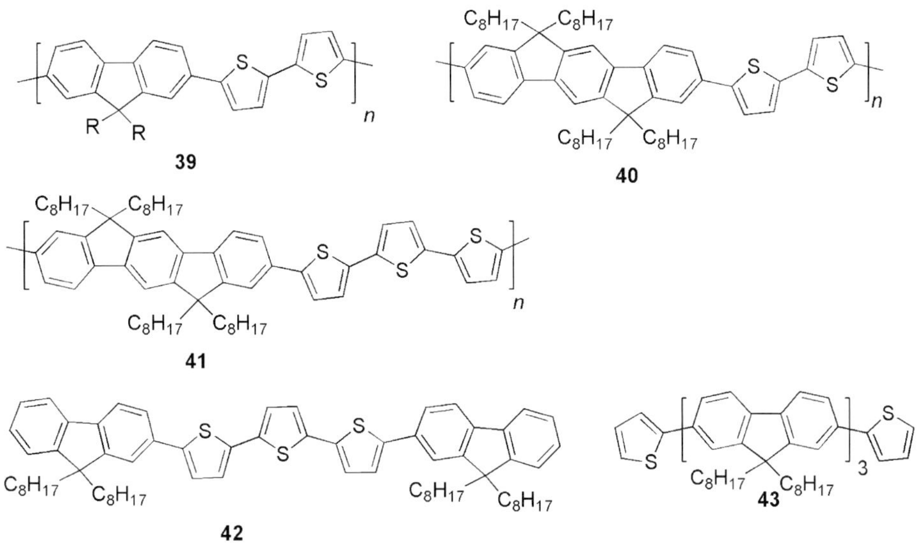

Fig. 6.5 Phenylene–oligothiophene copolymers and oligomers.

their mobility values are also lower ($\sim 10^{-4}$ cm^2 V^{-1} s^{-1}) [38]. Phenylene–thiophene copolymers might be expected to display better stability than polythiophenes but higher mobility than polyphenylenes. Indeed, the fluorene–bithiophene copolymer F8T2 (**39**, Fig. 6.5) has been used to make FETs with mobilities of up to 10^{-2} cm^2 V^{-1} s^{-1}. The key to obtaining this high value is to take advantage of a thermotropic liquid crystalline state to orient the polymer chains, by heating the material to 280 °C to bring it into the ordered mesophase and then rapidly cooling it to freeze in the order [39]. The sizes of the blocks in such copolymers greatly influence their charge carrier mobilities, as illustrated by the indenofluorene–oligothiophene copolymers **40** and **41**. The bithiophene copolymer (PIF-T2, **40**) has a mobility of less than 10^{-5} cm^2 V^{-1} s^{-1}, whereas the value for terthiophene copolymer PIF-T3 (**41**) is around 10^{-4} cm^2 V^{-1} s^{-1} [40]. To understand these results, the polymer packing has been studied by a joint atomic force microscopy (AFM)/molecular modeling approach.

Hence molecular modeling indicates that both F8T2 (**39**) and PIF-T3 (**41**) possess low-energy linear chain conformations, which can close pack with the oligothiophenes stacking on top of the oligophenylenes to form the fibrils visible in AFM images of their thin films (Fig. 6.6) [41].

By contrast, in PIF-T2 (**40**) interactions between the β-hydrogens on the thiophenes and the neighboring hydrogens on the indenofluorene units produce bent chains which cannot close pack, thus leading to amorphous films. Films of the model compound **42** display ordered structures, whereas the inverse molecule **43** produces completely amorphous films [42]. This suggests that the π-stacking of the terthiophenes is stronger than of the terfluorenes, possibly due to the greater planarity of the former. Ordered nanostripes of the oligomer **42**

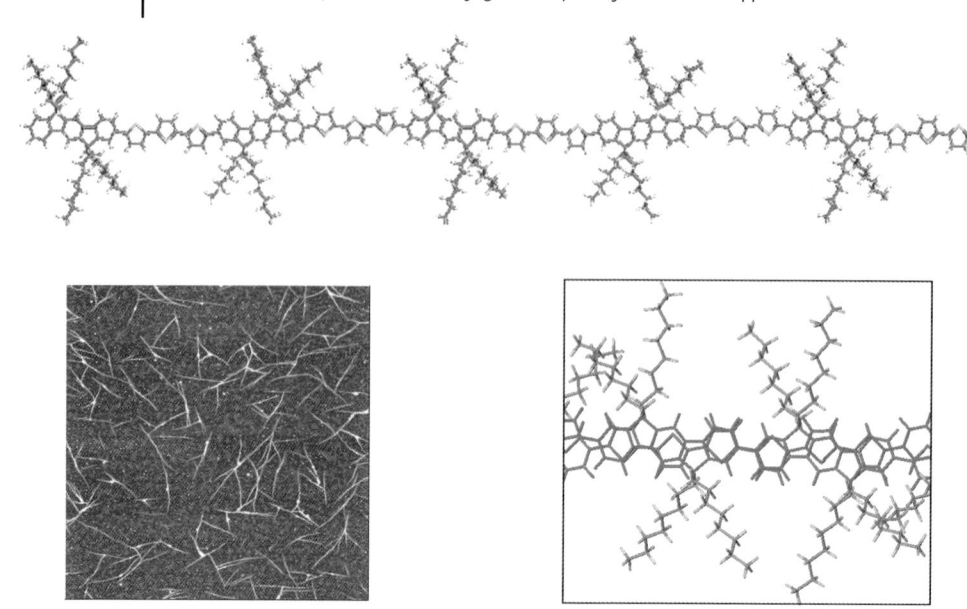

Fig. 6.6 AFM image of a deposit of and simulations of the chain conformation and packing of the copolymer **41**. Reproduced with permission of Wiley-VCH from [150].

have been deposited by a lithographically controlled wetting process between two electrodes to construct nanostructured FETs. These devices exhibit charge carrier mobilities orders of magnitude higher than corresponding devices utilizing spin-cast or vacuum-deposited films of **42** [43].

This combined AFM/modeling approach has also proved useful in understanding the optoelectronic behavior of fluorene **24** and indenofluorene **25** homopolymers. The degree of order in the morphology correlates with the amount of green emission seen in solid-state PL or upon operation of an LED [44]. Polymers bearing *n*-octyl substituents such as **24a** [45] and **25a** [44] (Fig. 6.7) can pack closely (calculated inter-chain distance 0.5 nm), show fibrillar morphologies and exhibit the rapid appearance of a green emission band. The PIF **25b** [44] with branched 2-ethylhexyl chains and polymers such as **24c** [46] and **25c** [47], with even bulkier aryl units, cannot close pack, as shown by both modeling and X-ray scattering experiments, and display featureless morphologies [48]. All display stable blue PL, but whereas the EL emission from **25b** rapidly turns green, only blue EL was seen from **24c** [46] and **25c** [47]. These differences in packing also explain the different phase-forming behaviors of the *n*-alkyl, branched-alkyl and aryl-substituted PFs and PIFs. Thus *n*-alkyl PFs and PIFs typically form two nematic liquid crystalline (LC) phases, whereas ethylhexyl-substituted polymers exhibit only a single nematic phase and no LC phases have been detected for the aryl-substituted polymers [24, 49, 50].

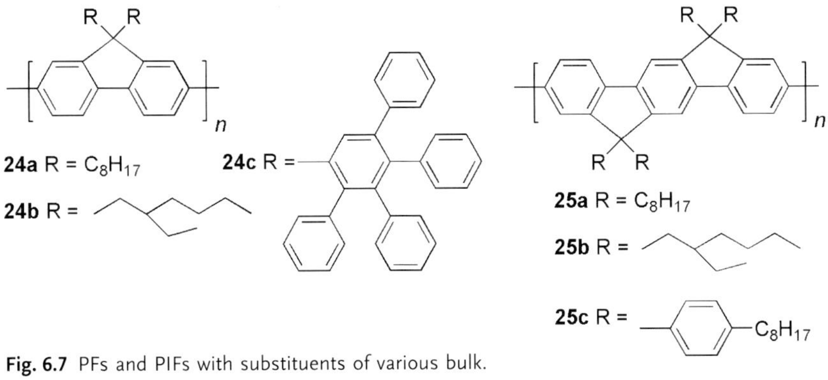

Fig. 6.7 PFs and PIFs with substituents of various bulk.

Since emissive ketone defects have recently been identified as the major source of the long-wavelength emission seen from PFs [51], PIFs [48] and LPPPs [52], these differences in emission behavior are explained by the change in chain packing reducing exciton migration to defect sites. Furthermore, the ketone defects in alkyl-substituted PFs (and presumably in PIFs) have been shown to arise from the presence of bridgehead hydrogens due to incomplete alkylation [51]. It is noteworthy that the blue emission from the LPPP **23a** with bridgehead hydrogens is extremely unstable with long-wavelength emission appearing very rapidly in the solid state, whereas the methyl-substituted MeLPPP **23b** shows stable blue–green emission [22]. Better purification of dialkylfluorene monomers to remove monoalkylated material has been found to retard significantly the appearance of green emission from LEDs using PFs [53]. Similarly, improving the alkylation of indenofluorene so as to avoid underalkylation was found to convert the solid-state PL of **25a** from green to blue [48]. The extreme sensitivity of emission color towards lower energy defect sites is illustrated by the complete suppression of the solid-state blue PL from **25c** by only 0.3 wt% of a green-emitting dopant [48].

The extra stability of the blue EL from aryl-substituted polymers such as **25c**, however, does not just reflect reduced exciton migration, but also differences in the synthetic route. Here the bridges are made by a Friedel–Crafts addition similar to that used as the last step in making LPPPs (Scheme 6.2), which, as with MeLPPP, precludes the presence of bridgehead hydrogens. It is also probable, although as yet not definitively proven, that diaryl bridges are more resistant to oxidation than dialkyl bridges.

Another way to control the packing of conjugated materials is to incorporate them within rod–coil block copolymers. Such systems are well known to be able to form a wide range of morphologies ranging from sheets of mushroom-like aggregates [54], through spherical, vesicular, cylindrical and lamellar aggregates [55–57], to porous membranes [58, 59]. The relative sizes of the two blocks profoundly affect the packing. Thus, in fluorene–ethylene oxide block copolymers (PF–PEO), when the volume ratio of the PEO is below 0.4, self-assembly of the rods produces fibrillar morphologies, whereas higher ratios of PEO lead to unstructured aggregates being formed as the long-range packing of the rods is suppressed [60]. The morphology of films of rod–coil copolymers can also be controlled by means of the solvent. Thus, when a poly(phenylene-*block*-ethylene oxide) (PPP–PEO) copolymer was deposited from solvents that are good for the PP block, a fibrillar morphology was observed due to stacking of the conjugated rods, whereas when it was deposited from methanol, which is a good solvent for the coils and bad for the rods, round, apparently hollow, deposits were formed [45]. The film morphology from such copolymers is also dependent on the substrate. Thus, copolymers (OPV–PPO) with phenylenevinylene oligomers attached to poly(propylene oxide) chains form nanofibers [61]. Copolymers with large PO coils form large, unaligned fibers when deposited on polyimide substrates, whereas deposition on mica produces alignment of the fibers over ranges of several tens of micrometers. By contrast, copolymers with small PO coils do not form aligned fibers on mica.

So far, we have dealt with covalently linked polymers which can form supramolecular assemblies by means of interactions between their π-systems. However, supramolecular polymers can be formed in which non-covalent forces induce self-assembly of conjugated systems, and we now turn to the first example of these.

6.6
Self-assembly of Supramolecular Polymers by Hydrogen Bonding

Hydrogen bonding is one of the most important non-covalent methods by which molecules can assemble and is widely used in nature for this purpose, e.g. in forming the famous DNA double-helix. It can also be used to self-assemble optically and electronically active units to form complicated superstructures. Thus, the ureidopyrimidinones in molecule **44** (Fig. 6.8) bind the molecules together in solution to form a supramolecular polymer in which the monomer units are linked by hydrogen bonds [62]. Molecule **45** forms dimers through hydrogen bonding, which then produce helices by π–π-stacking [63]. In the final structure, there is thus a hierarchy of order with π-stacking giving higher level order to units already ordered by hydrogen bonding. Similarly, hydrogen-bonded donor–acceptor–donor triads of oligomers **46** with the perylenediimide **47** form helical stacks in solution which can be deposited as chiral fibers [64, 65]. A photovoltaic cell constructed using this triad showed poor results because the aggregates tend to organize themselves laterally on the surface, which is not favorable for charge transfer to the electrodes [65]. The self-assembly of DNA base-pairs can be used to produce supramolecular systems, e.g. hybridization of perylene dyes with oligonucleotides attached has been used to prepare linear assemblies containing equally spaced dye units [66].

These hydrogen-bonded systems represent one approach towards the construction of supramolecular polymers. In these examples, however, the overlap

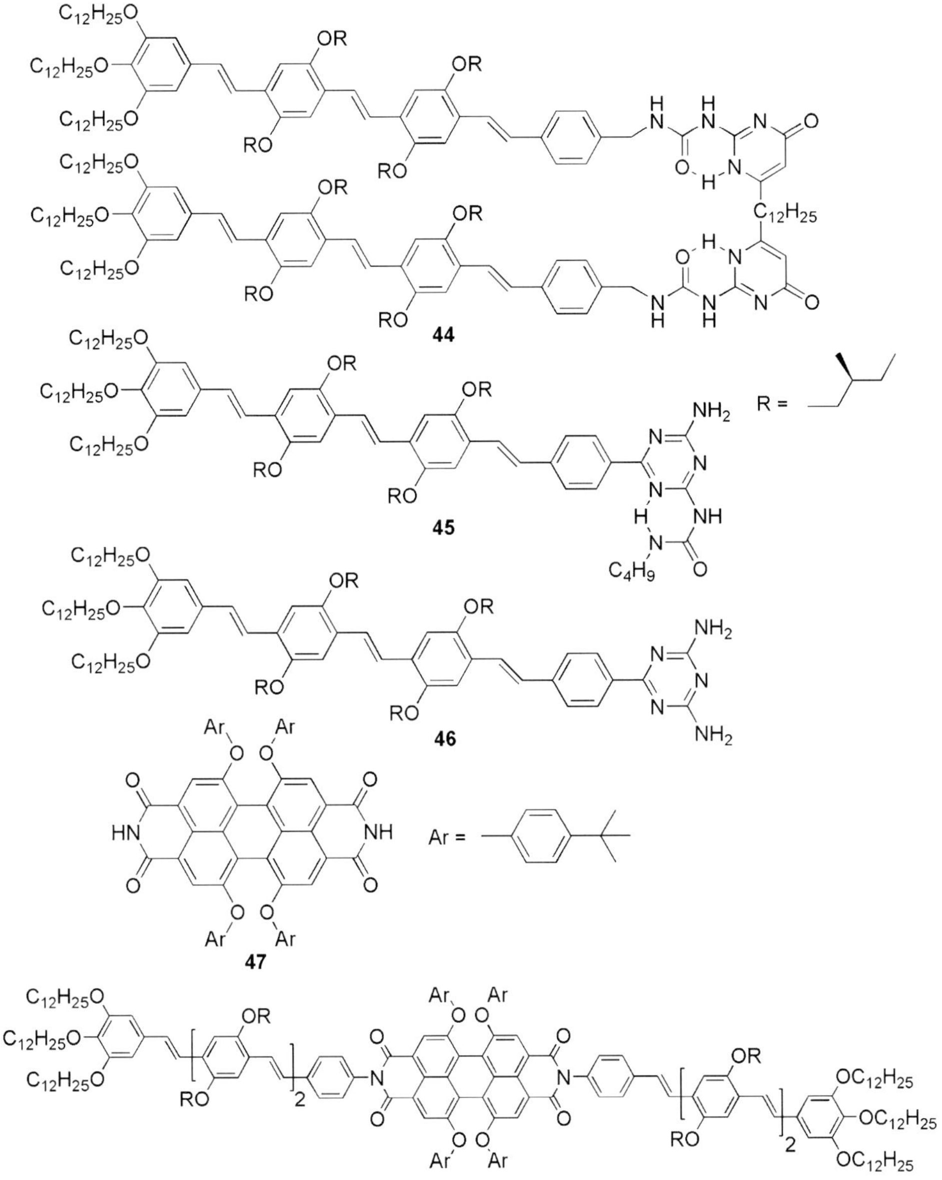

Fig. 6.8 OPVs with quadruple H-bonding end-groups.

between the π-systems is not optimal. If one wants to use such supramolecular systems as the active components in solar cells or FETs, then one needs high charge carrier mobilities, which in turn require a much greater delocalization of electrons between the stacked conjugated systems. The π-stacking of discotic PAHs to form columns offers a more efficient overlap between the adjacent π-systems than the above systems, with the additional advantage that transverse shifts of individual components are less detrimental to the overall electron delocalization. The dynamic nature of such systems, with individual discs being able to leave and re-enter the columns, also provides a self-healing process to minimize the effects of any defects in the columns. Accordingly, we now turn our attention to such materials.

6.7
Supramolecular Assembly of Discotic PAHs

Planar PAHs [67] can be looked upon as molecular subunits of graphite consisting of fused benzene rings. They can be divided into two types according to the model of Clar, in which the π-electrons are distributed among the benzene rings so as to form the greatest possible number of "full" rings, i.e. π-electron sextets [68]. In all-benzenoid PAHs (also known as two-dimensional polyphenylenes), all the π-electrons in the PAH can be distributed so that each benzene ring contains either six or no π-electrons (which means that such molecules must have a number of ring carbon atoms divisible by six). All-benzenoid PAHs are more stable, less reactive and have larger HOMO–LUMO gaps than isomeric compounds that do not satisfy this criterion [67].

Efficient routes have been developed to a large variety of processable graphene molecules of various sizes and molecular topologies as illustrated in Fig. 6.9 (substituents not shown). These molecules range in diameter from 1.4 nm (**9**) to 3.5 nm (**50**), representing a considerable increase in the size of the conjugated π-systems. As the size of the discs increases, their absorption spectra red shift markedly [69]. The extra π-center in **51** strongly perturbs the π-electron cloud, so that this material shows bands not seen in the UV–visible absorption spectra of the more symmetrical **9**, **49** or **50** [70]. The loss of symmetry also enhances the intensity of the lowest band of the a-transition. Of the skeletons shown, the C_{42} (HBC, **9**) [71], C_{96} superphenalene (**49**) [72] and C_{222} (**50**) [73] are all-benzenoid, whereas the C_{44} (**51**) [70] is not. In all cases the PAHs are made in good yield by oxidative cyclohydrogenation of a soluble three-dimensional polyphenylene dendrimer precursor. This simple and efficient route is suitable for the large-scale synthesis of all-benzenoid PAHs, which are now obtained as structurally perfect nano-sized 2D-π-systems. PAHs bearing solubilizing groups such as alkyl groups are solution and melt processable. Large, unsubstituted PAHs such as **50** are insoluble, but thin films can be obtained by ultra-high-vacuum deposition at high temperatures (550–650 °C) due to their excellent thermal stability.

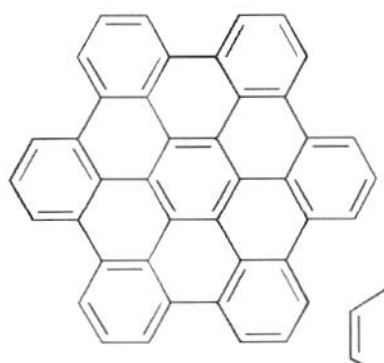

Fig. 6.9 Some representative PAH skeletons.

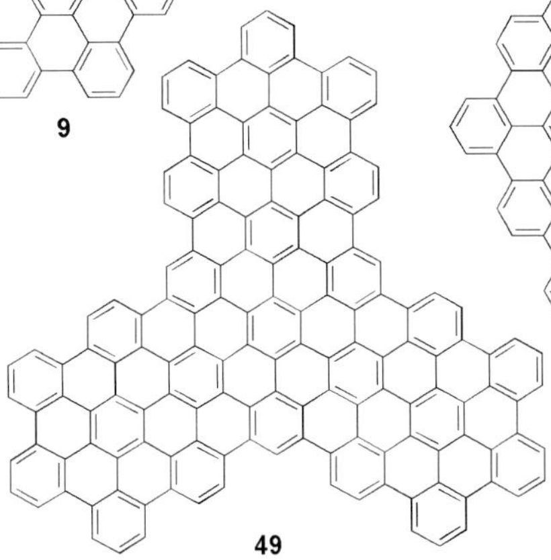

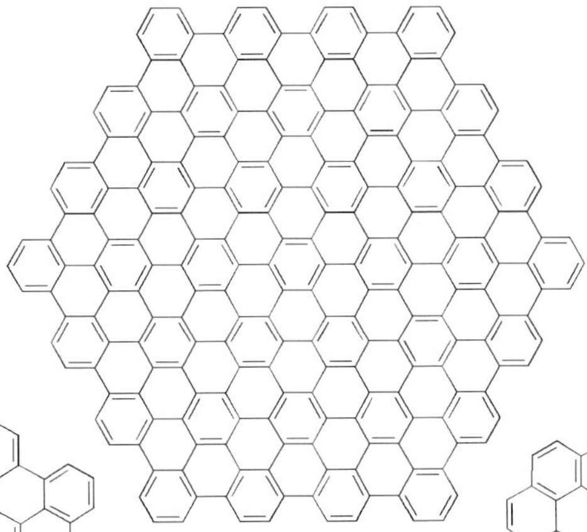

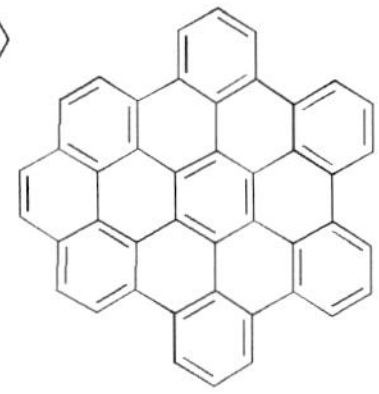

Soluble discotic graphite molecules such as alkyl-substituted HBCs can form thermotropic liquid crystalline phases, in which the molecules align into ordered nano-sized columns [74]. While the π-interaction of the aromatic cores leads to self-assembly into the columnar structures, the side-chains are partially organized in the core periphery, producing nano-phase separated arrays. The supramolecular order is strongly dependent on the thermal behavior of the compound. Thus, in the crystalline state, the discs arrange in a herringbone structure which is characterized by molecular tilting with respect to the columnar axis. During heating to the liquid crystalline phase, the molecular arrangement changes to an orthogonal intracolumnar packing of the discs about 0.35 nm apart, which allows for efficient overlap of the π-orbitals on adjacent discs, so producing high charge-carrier mobility along the axis of the column [69, 74–76]. Each column thus acts as a supramolecular polymer in which the movement of charge along the column between the discotic "monomer" units corresponds to the charge transport along a conjugated polymer chain.

As a result of the direction of greatest charge mobility being perpendicular to the plane of the discs, the highest efficiencies in electronic devices will be attained when the columns are aligned perpendicular to the electrodes, which in FETs means the discs must be edge-on to the substrate (heterotropic alignment) and for a solar cell that they be face-on (homeotropic alignment). Whereas the former is the favored orientation for discs deposited from solution, it has been shown that the latter can be achieved by cooling a layer of an HBC from the isotropic phase between two substrates [74, 77]. Here again the size of the molecules matters. While discotic mesophases have been obtained for a large number of disc-type molecules [74], it is the size of the PAHs which promotes aniso-

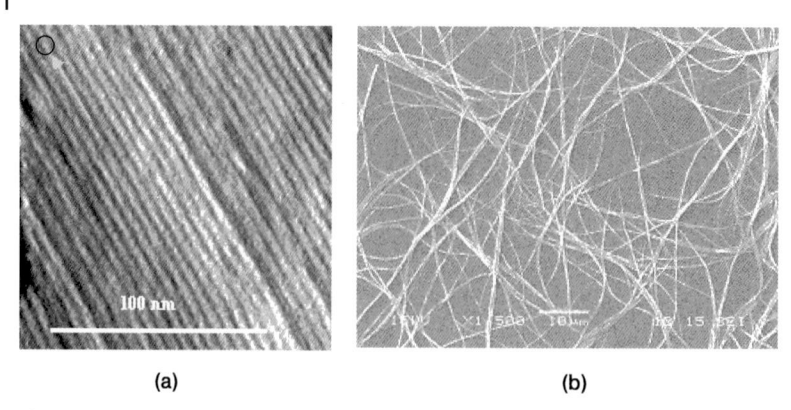

Fig. 6.10 Zone-cast films (a) and microfibers (b) of HBCs cast from THF solution on to glass substrates. Reproduced with permission of Wiley-VCH from [150].

54 R = $C_{12}H_{25}$

55 R = —⟨⟩—$OC_{12}H_{25}$

56 R = —⟨⟩—N(—⟨⟩—$C_{12}H_{25}$)(—⟨⟩—$C_{12}H_{25}$)

52 R = alkyl

53 R = —⟨⟩—$C_{12}H_{25}$

Fig. 6.11 Discotic liquid crystalline HBCs.

tropic LC phases stable over a broad temperature range and to increased charge-carrier mobilities [69].

Two methods have been used for producing aligned films of HBCs. In the first, the material is spin-cast upon a pre-aligned polytetrafluoroethylene (PTFE) layer [78], whereas the second utilizes zone-casting from solution [79]. In this simple procedure, a solution is deposited through a nozzle on to a moving substrate, with both nozzle and substrate being thermally stabilized. The concentration and temperature gradients suffice to align the columnar nanostructures uniaxially along the deposition direction. By varying the casting conditions, either continuous films with a "quasi-single-crystalline", long-range ordered surface layer (Fig. 6.10a) or highly oriented microfibers (Fig. 6.10b) can be obtained. Zone crystallization from the melt has also been used to obtain highly aligned films [80].

Efficient FETs have been constructed using films of HBCs **52** (Fig. 6.11), which were aligned between the electrodes both by means of a PTFE pre-alignment layer [78] ($\mu = 10^{-3}$ cm^2 V^{-1} s^{-1}) and by zone-casting ($\mu = 10^{-2}$ cm^2 V^{-1} s^{-1}) [81]. A blend of the aryl-substituted HBC **53** with an electron-accepting perylene diimide dye showed high efficiency in a photovoltaic device with an external quantum efficiency (EQE) of 34% being seen at 460 nm [82]. Here the partial phase separation of the two components to provide separate percolation pathways for holes and electrons seems to be the main reason for the good efficiency. Thus, optimizing the performance of FETs and photovoltaic cells with PAHs as active components requires increasingly complex supramolecular hierarchies.

6.8
Optimizing Intra- and Intercolumnar Order in PAHs

The charge transport in columns of HBCs can be optimized by attaining the maximum degree of intracolumnar order and thus best overlap of orbitals. One way of increasing the order of the columns is to replace the alkyl substituents in **52** with (di)arylethyne units as in **54–55** (Fig. 6.11) [83]. Compound **55** displays helical stacking with each successive disc rotated by 15° about the columnar axis. Very high charge-carrier mobility is also achieved by attaching hole-accepting triarylamine units as in **56** [84]. Here the HBC cores and triarylamine units provide two coaxial pathways for charges to move.

Individual columns of discs up to several hundred nanometers in length have been observed in spin-cast films of **53** by AFM [83]. By contrast, drop-cast films of **55** contain nanoribbons several micrometers in length, which are interpreted as being arrays of 5–6 columns. Alignment of such HBC fibers or columns between two electrodes would permit the fabrication of a nano-FET.

The above materials illustrate how one can optimize intracolumnar order in HBCs, but the control of intercolumnar order is equally important if one is to use more than a single column in a device. One method to achieve this is to combine one- and two-dimensional structural motifs by preparing HBCs with rigid conjugated linkers, e.g. the HBC dimers and trimers **57–59** (Fig. 6.12). Such compounds might be blended with other HBCs and thus link together the columns in which they are incorporated. These conjugated links might also allow charge to be transferred between columns, thus bypassing any defects in the columns. Compounds **57–58** self-assemble into ordered columnar mesophases, as shown by X-ray diffraction experiments. By contrast, the non-linear

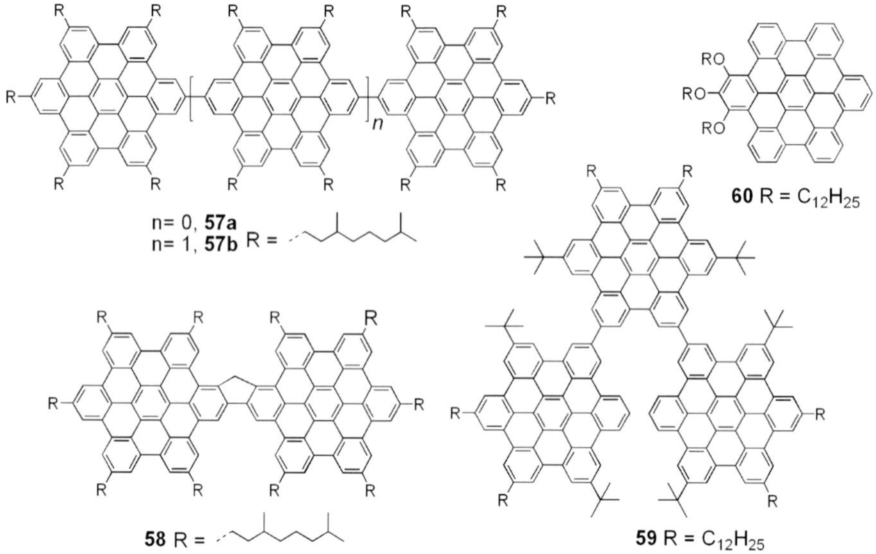

Fig. 6.12 Oligomers of HBC and "unwrapped" HBC **60**.

trimer **59** does not form an ordered columnar array, presumably because the HBCs in this material are twisted even more out-of-plane by steric interactions than the rings in the others [85]. The "unwrapped" HBC molecule **60** displays a stable columnar liquid crystalline phase with a low isotropization temperature (190 °C) [86] and can be used as the active component in field-effect transistors. Thereby the interactions between unsubstituted aromatic cores should allow for better intercolumnar charge transport.

Another way to combine one- and two-dimensional structural motifs is to prepare triply stranded graphite ribbons **61** by dehydrogenation of highly phenylated polyphenylenes **62** (Scheme 6.6) [87]. These materials are insoluble and, due to the presence of isomeric substructures in the precursor polymers, which lead to kinks in the ribbons, they are not totally regular in structure. As a result, TEM images of the ribbons show a mixture of domains containing ordered graphite layers and of disordered areas. If the solubility of the materials can be improved, then such a route may offer a way to producing long graphitic molecules, which might act as organic wires, without the need for very high temperatures, as are generally required in making extended graphitic materials such as nanotubes (see below).

An alternative method to produce order between columns is to attach groups to the outside of the columns, enabling them to bind to each other, to surfaces or to other compounds which might be used to help order the columns. Here both covalent and non-covalent bonds can be used to bind the columns. For example, the order of the columns of the HBCs **63** (Fig. 6.13) can be locked in by thermally polymerizing the acrylate or methacrylate end-groups within the tem-

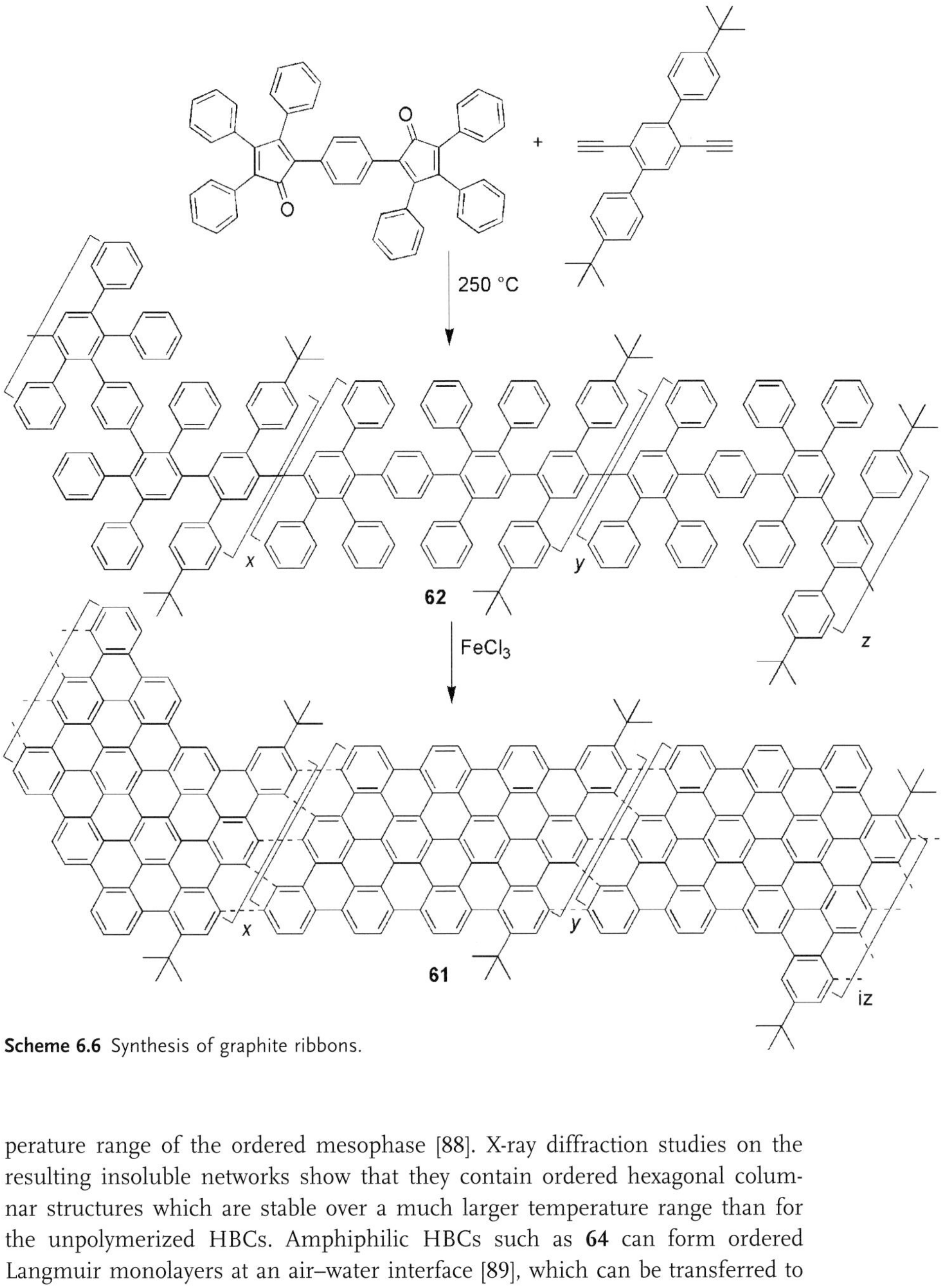

Scheme 6.6 Synthesis of graphite ribbons.

perature range of the ordered mesophase [88]. X-ray diffraction studies on the resulting insoluble networks show that they contain ordered hexagonal columnar structures which are stable over a much larger temperature range than for the unpolymerized HBCs. Amphiphilic HBCs such as **64** can form ordered Langmuir monolayers at an air–water interface [89], which can be transferred to quartz substrates to produce ordered Langmuir–Blodgett films. Complexation of **64** with an L-lysine–ethylene oxide block copolymer **65** produces a hexagonal lattice of the helical block copolymer strands, each of which is surrounded symmetrically by six HBC columns which form a hexagonal sublattice [90].

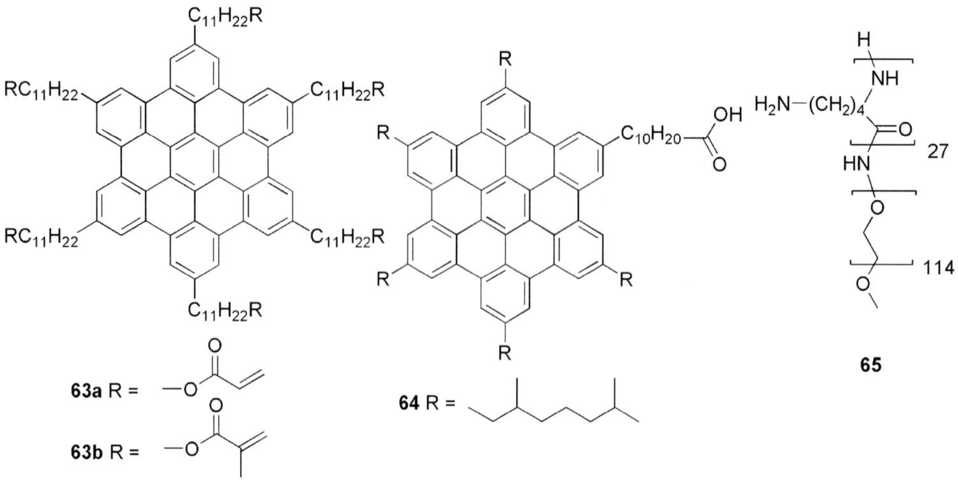

Fig. 6.13 HBCs whose intra- or intercolumnar order can be preserved by cross-linking or complexation.

As was discussed above, optimizing the mobilities by the intra- and intercolumnar order of HBCs is a major research goal. If one attaches bulky groups to an HBC, on the other hand, then one can restrict the stacking of the discs (Fig. 6.14). Such compounds cannot be used in existing electronic devices but exemplify how the self-assembly of molecules is controllable by means of varying the substituents. The dendronized HBC **66** cannot stack and so exists as isolated discs. Attaching the dendrons via an intermediate phenylene ring reduces the steric hindrance to stacking so that **67a** forms stable dimers, whereas **67b** displays an equilibrium between monomer and dimer which can be studied by NMR [91]. This demonstrates some more ways by which the assembly of nano-objects can be controlled, besides those previously mentioned such as hydrogen bonding.

In the hexa-*peri*-hexabenzocoronene (HBC) derivative **68**, chromophores are attached to the six dendron substituents. As HBCs are good hole-accepting units and perylene imides are good electron acceptors, one sees efficient electron transfer from the HBC core to the dyes [92]. An important motivation for HBC–peryleneimide hybrids is the combination of light-induced charge separation with charge-carrier transport in supramolecular architectures.

A similar molecule **69** has an HBC donor core surrounded by six alkyl chains bearing electron-accepting anthraquinone (AQ). Current–potential curves were separately recorded by a combination of scanning tunneling microscopy (STM) and scanning tunneling spectroscopy for donor and acceptor components of the molecule. These showed characteristic differences in their slope, indicating that the tunneling currents were strongly enhanced by the resonant involvement of the HBC HOMO and the quinone LUMO. Remarkably, given that this material was designed primarily with photovoltaic activity in mind, transistor activity has

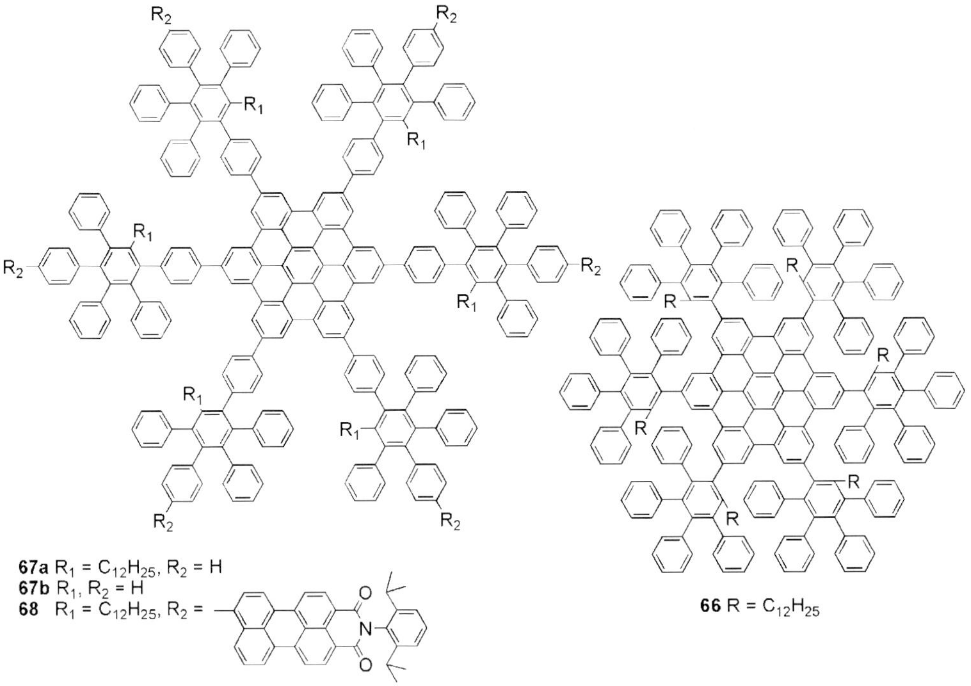

Fig. 6.14 HBCs with dendron substituents.

67a $R_1 = C_{12}H_{25}$, $R_2 = H$
67b $R_1, R_2 = H$
68 $R_1 = C_{12}H_{25}$, $R_2 = $ [perylene diimide group]

66 $R = C_{12}H_{25}$

been demonstrated for a single molecule of **69** on a surface, in which both the HBC and anthraquinone lie flat on the surface (Fig. 6.15). When 9,10-dimethoxyanthracene (DMA, **70**) was added, it formed a charge-transfer complex with the anthraquinone, whose dipole changed the substrate work function by ~120 meV and so induced a detectable change in the *I–V* characteristics of the HBC core as measured by an STM tip, thus producing a chemical gate effect [93], with a nanometer-sized gate [94]. Although it is certainly a fascinating experiment to demonstrate a FET function at a nanometer-length scale, the currents involved are much too small in order – at present – to be of practical value in a real FET.

In addition to their electronic properties, HBCs have another role to play in the development of molecular electronics, as compounds from which carbonaceous materials such as carbon nanotubes can be made.

6.9
Carbon Nanotubes and Similar Materials from Ordered Graphite Mesophases

Since their discovery by Ijima [95], carbon nanotubes (CNTs) have become perhaps the best-known of a broad family of carbonaceous materials which have a long history (soot was used as a colorant by the earliest humans) and have prov-

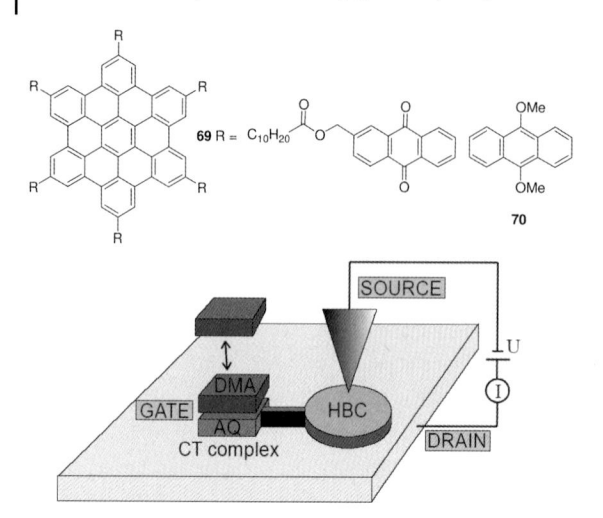

Fig. 6.15 A single-molecule chemical FET. Reproduced with permission of Wiley-VCH from [150].

en of great industrial importance. As a result, they have been the subject of several reviews [96–99]. Nanotubes can be thought of as being individual graphene sheets which have been rolled up and the edges joined together and, depending on how the sheet has been rolled up, can be produced as so-called armchair, zigzag or (chiral) helical structures [97]. These isomers have different electronic properties. CNTs may comprise either a single rolled-up graphite sheet (single-walled nanotubes, SWNTs) or an array of such nanotubes concentrically nested like the rings in a tree trunk (multiwalled nanotubes, MWNTs). The processing and characterization of CNTs present much the same problems as are encountered with large PAHs. CNTs are generally produced as insoluble bundles which can be broken up with ultrasound. Some contaminants such as amorphous carbon can be burnt off at 300 °C in air whereas others can be removed by microfiltration or by size-exclusion chromatography with surfactants being used to stabilize the raw nanotube bundles. Functionalization of nanotubes has been used to assist in processing and to manipulate their properties. Techniques such as X-ray diffraction, AFM, electron microscopy – both scanning (SEM) and tunneling (TEM) – and Raman scattering being extensively used in studying and characterizing nanotubes.

Nanotubes (and the related fullerenes) have become of intense academic and industrial interest recently because their electronic properties mean that they have considerable potential for use in electronic devices or as chemical sensors [97, 98]. Their most significant electronic characteristic is that they can be either metallic (conducting) or semiconducting, depending on their structure. As a result, nanotubes have been investigated as active materials in devices such as electromechanical actuators, supercapacitors and field-emitted electron sources.

A single semiconducting SWNT has even been used to make a nano-FET by bridging the gap between two electrodes [100].

A problem with CNTs is that they are insoluble and so hard to process, which can be overcome by blending them with polymers (conjugated or otherwise), which also acts as a way to purify the nanotubes. In particular, conjugated polymers help to solubilize CNTs by wrapping around them through interactions between their π-systems [101]. Blends of CNTs with polymers are being used as conductive plastic components in the automobile industry [98], while blending them with conjugated polymers has been shown to improve the electrical performance of PLEDs [102, 103]. Because the electrical transport and thermopower properties of non-metallic CNTs are very sensitive to the presence of substances that affect the amount of injected charges, nanotubes are of interest for use in chemical sensing applications and they are already being used as scanning probe tips for AFMs [98].

The outstanding physical characteristics of nanotubes, in particular their high tensile strength, have created interest in using fibers of CNTs or composites of nanotubes with polymers as structural materials. The ability of CNTs to incorporate metal ions makes them attractive candidates as electrodes for lithium ion batteries, although their current performance is not yet competitive with that of existing materials [98].

The main obstacle to the widespread use of CNTs remains the difficulties in obtaining these materials, particularly the SWNTs, in both high yield and high purity (the latter being in part due to the problems in processing them) [98, 99]. The standard methods for preparing carbon nanotubes such as electric arc discharge of graphite electrodes [95], laser ablation of graphite [104] or disproportionation of carbon monoxide over a metal catalyst [105], typically involve high temperatures or pressures. One method that is particularly attractive for the synthesis of aligned CNTs is chemical vapor deposition from the gas phase, in which a hydrocarbon gas is passed over a catalyst in a furnace [106]. The development of new, mild methods for preparing CNTs, especially aligned materials, has thus been of major interest within the materials science community.

As a first step towards this goal, a new milder method for preparing carbon nanoparticles has been developed using a two-stage pyrolysis of an alkyl HBC (52 R=$C_{12}H_{25}$) [107]. The concept is that the first stage of the pyrolysis occurs at 400 °C, at which temperature the HBCs are still in an ordered liquid crystalline mesophase, a sort of carbomesophase. This results in the formation of large PAH intermediates (detectable by MALDI-TOF-MS) which preserve some mesomorphic order, as shown by polarizing optical microscopy. In the second step, pyrolysis at 800 °C results in mild graphitization to produce carbon nano- and microparticles with graphitic nanostructures, including fibers and rods (Fig. 6.16).

Pyrolysis on mica afforded a mixture of films coated on the surface and discrete nanoparticles, whose preferred shape appeared to be a hexagonally symmetric zigzag, e.g. large micro-objects with cross-sections of 1–10 mm and lengths in the micrometer range (Fig. 6.16c) and root-like micro-objects with cross sections of 400 nm^{-1} μm and lengths of > 30 μm (Fig. 6.16d).

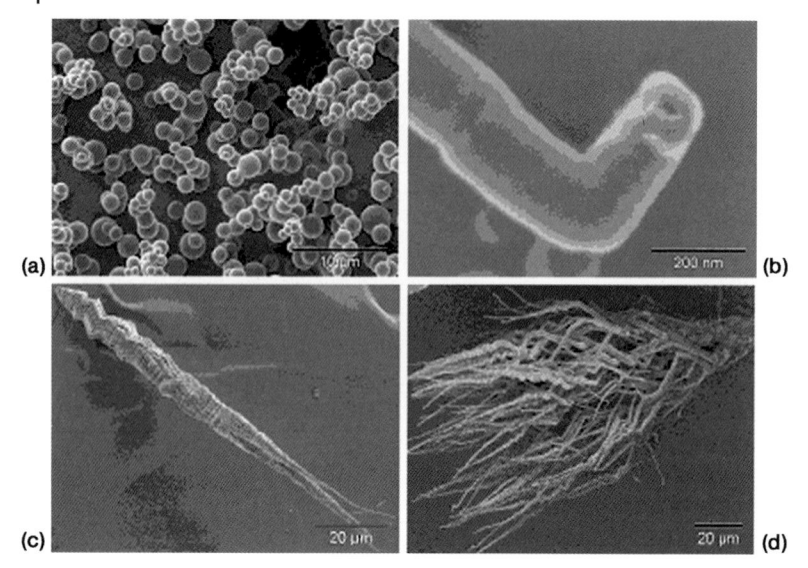

Fig. 6.16 (a) Spherical and (b) rod-like particles obtained by pyrolysis of HBC on quartz. (c, d) Micro-objects formed by pyrolysis of HBC on mica. Reproduced with permission of Wiley-VCH from [150].

Since metal particles are known to catalyze the formation of the aryl–aryl bonds in CNTs, an organometallic complex can serve in a pyrolysis as both a carbon source and a catalyst precursor and so the solid-state pyrolysis of organometallic compounds such as metallo-dehydro [n]annulenes has been developed by a number of groups as a new mild method for making CNTs, primarily MWNTs [108–111]. The pyrolysis of an HBC–cobalt complex **71** in the solid state produces carbon nanotubes (Fig. 6.17). Nearly quantitative yields of uniform either "bamboo-shaped" or straight nanotubes are obtained. The shape and yield of the carbon nanostructures can be finely controlled by regulation of the heating process. This provides an alternative high-yielding synthesis of carbon nanotubes. The precursor **71** (Fig. 6.17) contains both an as-ready graphite carbon source (HBC) and a catalyst precursor and at high temperatures the graphite islands form tube-like graphene structures on the cobalt catalyst units.

The next progression is to use the ordering of graphene molecules such as HBC to induce order in the nanotube pyrolysis products. Thus a well-controlled pyrolysis of HBC molecules has been performed in a porous alumina membrane which acts as a template [112]. The HBC molecules were introduced to the nanochannels within the alumina template by a simple wetting process, where it is important that the HBC discs adopt an edge-on arrangement of the template surface. Pyrolysis then furnishes carbon nanotubes. After dissolution of the template, high-resolution transmission electron microscopy (Fig. 6.17) showed that the carbon nanotubes were formed with graphite layers perpendic-

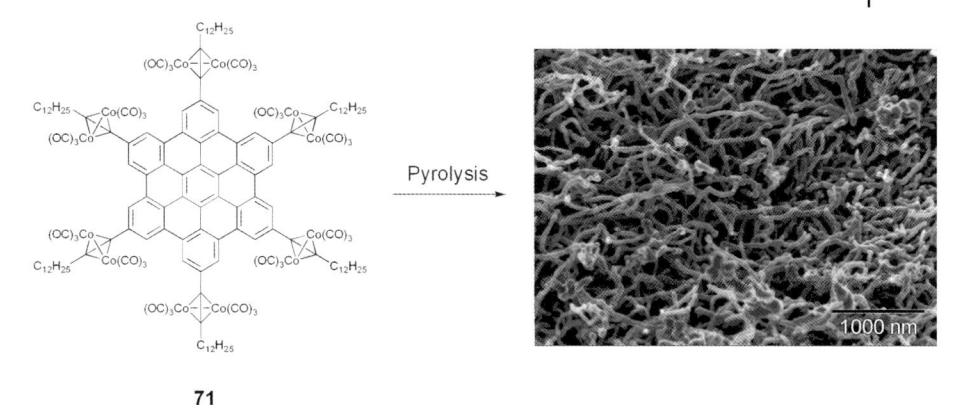

6.9 Carbon Nanotubes and Similar Materials from Ordered Graphite Mesophases

Fig. 6.17 High-yielding synthesis of carbon nanotubes by pyrolysis of an HBC–cobalt complex in the solid-state. Reproduced with permission of Wiley-VCH from [150].

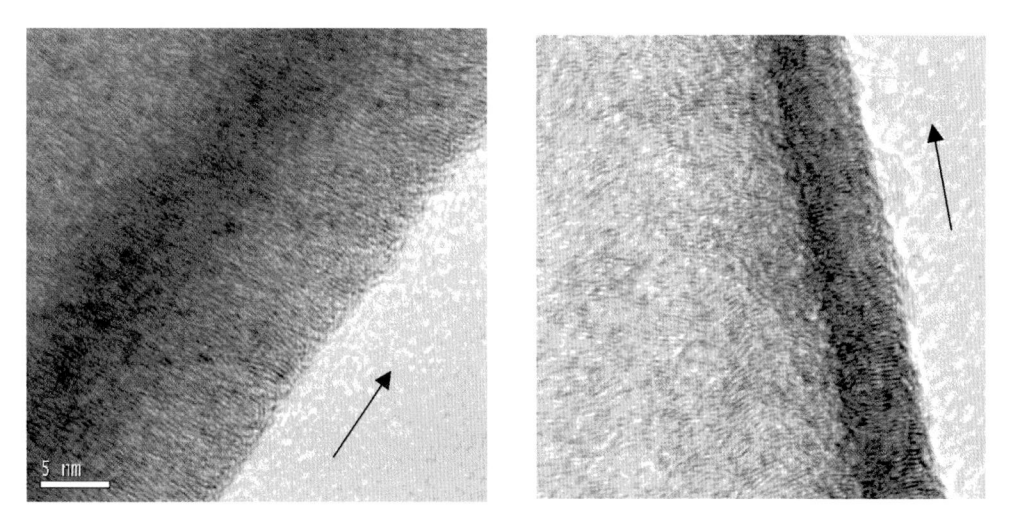

Fig. 6.18 TEM pictures of the nanotubes showing the aligned graphene layers. Reproduced with permission of Wiley-VCH from [150].

ular to the nanotube axis, i.e. matching the alignment of the original HBCs. It was found that the ordered pre-organization of the liquid crystalline HBC molecules on the surface of the membrane played a key role in obtaining carbon nanotubes with controlled graphene layer orientation (Fig. 6.18). This special alignment of the graphene sheets is expected to induce significantly different properties in these nanotubes compared with those made by the previously known methods.

By further development of these concepts, it is aimed to produce nanotubes with controllable size, morphology and electronic properties. For example, pyrolysis of materials containing heteroatoms could be used to produce selectively n- or p-type materials.

The further development of these carbon nanotubes will depend in part upon finding ways to process them. This is particularly important if one wishes to utilize individual tubes rather than bundles or aggregates of tubes for electronic devices. (Individual unfunctionalized tubes are not useful as reinforcing agents for polymers as they tend to aggregate in such media.) The question of whether one can utilize individual nanotubes or fibers or even molecules for electronic devices is one of the hottest topics in current materials science and it is to that issue that we now turn to conclude this chapter.

6.10
Prospects for Molecular Electronics

There are a number of issues that need to be considered in designing a truly molecular electronic device, i.e. one in which each active component contains only a single chemical unit (a molecule or polymer chain). Of these, the most basic is how one gets energy into the system and how one gets it out. This is less of a problem for optical than for electrical devices, as light can be detected and manipulated at very low levels, so that detection and analysis of single photons are now routinely performed [113, 114]. Also, light input and measurement of light output does not require direct contact with a material, whereas inputting charge necessitates close and regular contact between the molecule and an electrode, as of course does collecting any current produced. Care must of course be taken that the length scale of the device compared with the wavelength of light being used does not cause problems associated with photonic effects such as diffraction. Further, the electrical or optical properties of individual molecules may well differ from those of a bulk material as not only are there intermolecular interactions in the latter which affect these properties, but also in the bulk one sees an average response from many molecules, which should remain relatively consistent between experiments and over time (except for the effects of fatigue due to, e.g., oxidation), while individual molecules may produce a variety of responses reflecting individual energy states, etc., at any given time.

There has been a considerable amount of work into exploring the optical and electronic properties of individual molecules by means of single-molecule spectroscopy (SMS) [115, 116] and STM [117, 118]. STM can be used for the visualization of individual molecules after their immobilization on a surface, often as part of an ordered monomolecular layer. By combining STM with scanning tunneling spectroscopy, it is possible to record current–potential curves for individual molecules or even parts of molecules [119]. As we have seen above (Fig. 6.15), it is even possible to use an STM tip as one electrode in a single-molecule FET. The STM tip has also been used to manipulate nano-objects, e.g. metal atoms or organic molecules, and even to perform nanochemistry by making and breaking individual chemical bonds [119] or initiating a polymerization [120].

There has been much interest in using individual conjugated molecules as electronic components, e.g. rectifiers [121], or as so-called molecular wires [122–126]. The main obstacle is the irreproducibility of many of the results from studies on the electrical properties of individual molecules, the methods for achieving which have recently been reviewed [127]. Most commonly the molecules are tested as a self-assembled monolayer (SAM), e.g. by scanning a SAM on a gold electrode or other conductive surfaces with an STM or conducting AFM tip. To attach the molecules to the conductive surface, groups such as isonitriles or thiols are used as so-called "alligator clips" and the position of these anchoring groups on the molecule has been shown to affect the electric current detected [128]. There has also been interest in developing single-molecule transistors using organic materials, in which the organic component would act like a quantum dot [129]. Such devices have been demonstrated using individual carbon nanotubes [130], C_{60} molecules [131], rotaxanes [132] and fullerene dimers [133]. Recently, a thiol-endcapped PPP **72** (Fig. 6.19) of ~ 8 nm average length has been deposited between two electrodes in a field-effect transistor (FET) structure. Upon altering the gate voltage, weak transistor behavior was observed showing that, in principle, isolated conjugated polymer chains could be used in single-molecule FETs, although the transistor properties will need considerable improvement [134].

A much more immediately promising approach to molecular electronics is to use individual polymer chains in bio- or chemical sensors. Already sensors based on self-assembled monolayers are one of the most important emerging nanotechnologies [135]. As shown above, conjugated polymers can form ordered monolayers and chemical sensors utilizing conjugated polymers as polymers may show greater sensitivity than small molecules [136–138]. The principle behind most of this work is that interactions between the analyte and the polymer induce a change in the polymer chain conformation affecting its electrical and optical properties, which are detected as changes in the conductivity, redox behavior (measured by cyclic voltammetry), absorption or fluorescence of the poly-

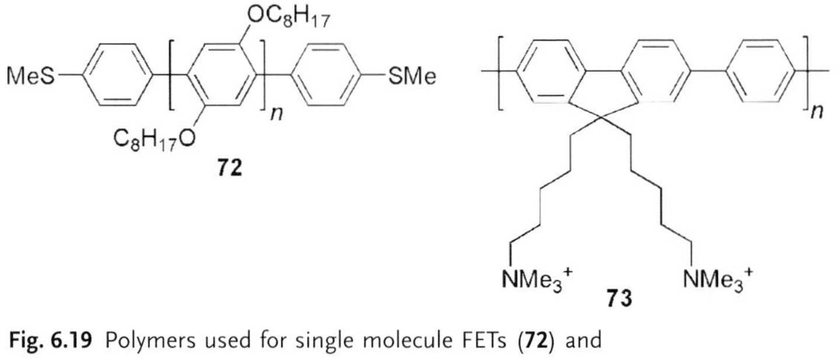

Fig. 6.19 Polymers used for single molecule FETs (**72**) and DNA detection (**73**).

mer. The materials that can be detected by such methods include halide ions [139, 140], chemical warfare agents [141], proteins [142, 143], nucleic acids [144–147] and even bacteria [148]. The sensitivity of these methods is such that as few as 220 DNA molecules in 180 µL (2.4×10^{-18} M) can be detected [145]. Another DNA-sensing approach uses a polycationic fluorene–phenylene copolymer 73 (Fig. 6.19) and a peptide nucleic acid (PNA) strand bearing a chromophore dye [149]. The PNA can complex with a strand of DNA complementary to it to form a polyanion, which is then electrostatically attracted to the copolymer. In this circumstance, energy transfer can occur from the polymer to the dye, leading to intense fluorescence from the dye, whereas in the presence of non-complementary DNA such complexation and energy transfer cannot occur. Since fluorescence can be detected at very low, even single-photon, levels it might in principle be possible to use single polymer chains as sensors. For the moment this appears to be the most likely method for achieving viable molecular electronic devices. However, the rate at which new advances in the probing of molecular properties and of reliably and reproducibly manipulating materials at the nano-level are being made may soon invalidate that statement. One thing is certain: whatever the future holds in the world of nanotechnology, chemistry will play a leading role in its development.

6.11
Conclusion

In this chapter, we have provided an overview of how the molecular and supramolecular properties of organic polymers, both covalently and non-covalently bound (supramolecular polymers), can be tuned by synthetic design. We have endeavored to illustrate ways in which such materials can be made, processed and studied and also to present examples of their applications, both current and potential. In particular, we have sought to indicate how the design of molecules and, thanks to advances in the understanding of molecular assembly processes, of ordered ensembles of molecules, has brought the realization of the dream of truly molecular electronics closer to fruition.

In the first sections we showed how the electronic and optical properties of covalently linked conjugated polymers can be tuned by means of controlling the nature and position of substituents, as exemplified by the dramatic differences between regioregular and regiorandom polythiophenes and/or the planarity of the polymer backbone, as demonstrated for phenylene-based polymers. A key factor in optimizing the properties of conjugated polymers is the minimization of potential chemical defects in the structure by means of careful choices of substrate and synthetic method. Greater understanding of the structures of potential defects and the mechanisms for their formation has permitted synthetic chemists to develop materials with greatly improved performance in electronic devices. Also assisting in the latter process has been the probing of the relationships between the structure, the packing behavior and the optoelectronic properties of polymers. Here we have shown how a combined AFM/molecular modeling approach can be used to elucidate information about polymer packing and to understand differences in device performance between structurally closely related polymers.

Next, we demonstrated how non-covalent forces can be used to form supramolecular polymers. First, hydrogen-bonded assemblies of oligo-phenylenevinylenes were presented. Then the assembly of discotic PAHs into ordered columnar mesophases was introduced and it was shown how the order both within and between columns could be optimized by varying the pattern and nature of the substituents and how this relates to device performance. Next, the use of such ordered systems as the starting materials for the preparation of novel carbonaceous materials was discussed. Finally, we presented a brief overview of the prospects for the development of electronic devices using individual molecules or ensembles of molecules as their active components. Already as we have shown there have been a number of examples of the electronic behavior of single molecules being measured in device-like setups, but the results indicate that although molecular electronics is no longer the stuff of science fiction, there is still some distance to go before it can be considered anything more than an intellectual curiosity. What is clear is that ongoing studies into the synthesis and properties of conjugated molecules offer the prospects for significant advances both in fundamental science and in developing new technology.

Acknowledgments

We thank all our coworkers and colleagues who contributed to the research we have presented here and in the preparation of this chapter. We also gratefully acknowledge funding from the Deutsche Forschungsgemeinschaft through Sonderforschungsbereiche 625. We thank Wiley-VCH for permission to reproduce images from our recent review on nanomaterials [150].

References

1 Dimitrakopoulos, C.D., Malenfant, P.R.L. *Adv. Mater.* **2002**, *14*, 99–117.
2 Brabec, C.J., Sariciftci, N.S., Hummelen, J.C. *Adv. Funct. Mater.* **2001**, *11*, 15–26.
3 Kraft, A., Grimsdale, A.C., Holmes, A.B. *Angew. Chem. Int. Ed.* **1998**, *37*, 402–428.
4 Hoeben, F.J.M., Jonkheijm, P., Meijer, E.W., Schenning, A.P.H.J. *Chem. Rev.* **2005**, *105*, 1491–1546.
5 Chiang, C.K., Fincher, C.R., Jr., Park, Y.W., Heeger, A.J., Shirakawa, H., Louis, E.J., Gau, S.C., MacDiarmid, A.G. *Phys. Rev. Lett.* **1977**, *39*, 1098–1101.
6 Martens, J.H.F., Pichler, K., Marseglia, E.A., Friend, R.H., Cramail, H., Khosravi, E., Parker, D., Feast, W.J. *Polymer* **1994**, *35*, 403–414.
7 Naarman, H. *Plast. Eng.* **2005**, *70*, 737–778.
8 MacDiarmid, A.G., Epstein, A.J. *Mater. Res. Soc. Symp. Proc.* **1994**, *328*, 133–144.
9 MacDiarmid, A.G. *Angew. Chem. Int. Ed.* **2001**, *40*, 2581–2590.
10 Wessling, R.A. *J. Polym. Sci., Polym. Symp.* **1985**, *72*, 55–66.

11 Braun, D., Heeger, A. *J. Appl. Phys. Lett.* **1991**, *58*, 1982–1984.
12 Andersson, M. R., Yu, G., Heeger, A. J. *Synth. Met.* **1997**, *85*, 1275–1276.
13 Hwang, D.-H., Kim, S. T., Shim, H.-K., Holmes, A. B., Moratti, S. C., Friend, R. H. *J. Chem. Soc., Chem. Commun.* **1996**, 2241–2242.
14 Martin, R. E., Geneste, F., Riehn, R., Chuah, B. S., Cacialli, F., Friend, R. H., Holmes, A. B. *Chem. Commun.* **2000**, 291–292.
15 Rasmussen, S. R., Straw, B. D., Hutchinson, J. E. *ACS Symp. Ser.* **1999**, *735*, 347–367.
16 McCullough, R. D. *Adv. Mater.* **1998**, *10*, 93–116.
17 Andersson, M. R., Selse, D., Berggren, M., Järvinen, H., Hjertberg, T., Inganäs, O., Wennerström, O., Österholm, J. E. *Macromolecules* **1994**, *27*, 6503–6506.
18 Loewe, R. S., Khersonsky, S. M., McCullough, R. D. *Adv. Mater.* **1999**, *11*, 127–130.
19 Loewe, R. S., Ewbank, P. C., Liu, J., Zhai, L., McCullough, R. D. *Macromolecules* **2001**, *34*, 4324–4333.
20 Schlüter, A. D., Wegner, G. *Acta Polym.* **1993**, *44*, 59–69.
21 Park, K. C., Dodd, L. R., Levon, K., Kwei, T. K. *Macromolecules* **1996**, *29*, 7149–7154.
22 Scherf, U. *J. Mater. Chem.* **1999**, *9*, 1853–1864.
23 Neher, D. *Macromol. Rapid Commun.* **2001**, *22*, 1365–1385.
24 Setayesh, S., Marsitzky, D., Müllen, K. *Macromolecules* **2000**, *33*, 2016–2020.
25 Jacob, J., Sax, S., Piok, T., List, E. J. W., Grimsdale, A. C., Müllen, K. *J. Am. Chem. Soc.* **2004**, *126*, 6987–6995.
26 Somma, E., Loppinet, B., Fytas, G., Setayesh, S., Jacob, J., Grimsdale, A., Müllen, K. *Colloid Polym. Sci.* **2004**, *282*, 867–873.
27 Scherf, U. *Top. Curr. Chem.* **1999**, *201*, 163–222.
28 Schlüter, A. D. *J. Polym. Sci. Part A: Polym. Chem.* **2001**, *39*, 1533–1556.
29 Denton, F. R., Lahti, P. M. *Plast. Eng.* **1998**, *48*, 61–102.
30 Gilch, H. G., Wheelwright, W. L. *J. Polym. Sci. Part A-1* **1966**, *4*, 1337–1349.
31 Becker, H., Spreitzer, H., Ibrom, K., Kreuder, W. *Macromolecules* **1999**, *32*, 4925–4932.
32 Becker, H., Spreitzer, H., Kreuder, W., Kluge, E., Schenk, H., Parker, I., Cao, Y. *Adv. Mater.* **2000**, *12*, 42–48.
33 Becker, H., Gelsen, O., Kluge, E., Kreuder, W., Schenk, H., Spreitzer, H. *Synth. Met.* **2000**, *111/112*, 145–149.
34 Pfeiffer, S., Hörhold, H.-H. *Macromol. Chem. Phys.* **1999**, *200*, 1870–1878.
35 Bunz, U. H. F. *Chem. Rev.* **2000**, *100*, 1605–1644.
36 Sirringhaus, H., Tessler, N., Friend, R. H. *Science* **1998**, *280*, 1741–1744.
37 Sirringhaus, H., Brown, P. J., Friend, R. H., Nielsen, M. M., Bechgaard, K., Langeveld-Voss, B. M. W., Spiering, A. J. H., Janssen, R. A. J., Meijer, E. W., Herwig, P., de Leeuw, D. M. *Nature* **1999**, *401*, 685–688.
38 Redecker, M., Bradley, D. D. C., Inbasekaran, M., Woo, E. P. *Appl. Phys. Lett.* **1998**, *73*, 1565–1567.
39 Sirringhaus, H., Wilson, R. J., Friend, R. H., Inbasekaran, M., Wu, W., Woo, E. P., Grell, M., Bradley, D. D. C. *Appl. Phys. Lett.* **2000**, *77*, 406–408.
40 Sonar, P., Grimsdale, A. C., Müllen, K., Surin, M., Lazzaroni, R., Leclère, P., Pinto, J., Chua, L.-L., Sirringhaus, H., Friend, R. H., *J. Mater. Chem.*, submitted.
41 Surin, M., Sonar, P., Grimsdale, A. C., Müllen, K., Lazzaroni, R., Leclère, P. *Adv. Funct. Mater.* **2005**, *15*, 1426–1434.
42 Surin, M., Sonar, P., Grimsdale, A. C., Müllen, K., De Feyter, S., Habuchi, S., Sarzi, S., Van der Auweraer, M., De Schryver, F. C., Cavallini, M., Moulin, J.-F., Biscarini, F., Fermoní, C., Lazzaroni, R., Leclère, P. *J. Chem. Mater.*, ASAP.DOI 10.103a/b6101132a.
43 Cavallini, M., Stoliar, P., Mouloin, J. F., Biscarini, F., Surin, M., Leclère, P., Lazzaroni, R., Nielsen, M., Breiby, D. W., Sonar, P., Grimsdale, A. C., Müllen, K. *Nano Lett.* **2005**, *5*, 2422–2425.
44 Grimsdale, A. C., Leclère, P., Lazzaroni, R., MacKenzie, J. D., Murphy, C., Setayesh, S., Silva, C., Friend, R. H., Müllen, K. *Adv. Funct. Mater.* **2002**, *12*, 729–733.

45 Leclère, P., Hennebicq, E., Calderone, A., Brocorens, P., Grimsdale, A. C., Müllen, K., Brédas, J. L., Lazzaroni, R. *Prog. Polym. Sci.* 2003, *28*, 55–81.

46 Setayesh, S., Grimsdale, A. C., Weil, T., Enkelmann, V., Müllen, K., Meghdadi, F., List, E. J. W., Leising, G. *J. Am. Chem. Soc.* 2001, *123*, 946–953.

47 Jacob, J., Zhang, J., Grimsdale, A. C., Müllen, K., Gaal, M., List, E. J. W. *Macromolecules* 2003, *36*, 8240–8245.

48 Keivanidis, P. E., Jacob, J., Oldridge, L., Sonar, P., Carbonnier, B., Baluschev, S., Grimsdale, A. C., Müllen, K., Wegner, G. *ChemPhysChem* 2005, *6*, 1650–1660.

49 Teetsov, J., Fox, M. A. *J. Mater. Chem.* 1999, *9*, 2117–2122.

50 Grell, M., Knoll, W., Lupo, D., Meisel, A., Miteva, T., Neher, D., Nothofer, H.-G., Scherf, U., Yasuda, A. *Adv. Mater.* 1999, *11*, 671–675.

51 Scherf, U., List, E. W. J. *Adv. Mater.* 2002, *14*, 477–487.

52 Lupton, J. M. *Chem. Phys. Lett.* 2002, *365*, 366–368.

53 Craig, M. R., de Kok, M., Hofstraat, J. W., Schenning, A. P. H. J., Meijer, E. W. *J. Mater. Chem.* 2003, *13*, 2861–2862.

54 Zubarev, E. R., Pralle, M. U., Sone, E. D., Stupp, S. I. *J. Am. Chem. Soc.* 2001, *123*, 4105–4106.

55 Jenekhe, S. A., Chen, X. L. *J. Phys. Chem.* 2000, *104*, 6332–6335.

56 Wang, H., Wang, H. H., Urban, V. S., Littrell, K. C., Thiyagarajan, P., Yu, L. *J. Am. Chem. Soc.* 2000, *122*, 6855–6861.

57 Vriezma, D. M., Kros, A., de Gelder, R., Cornelissen, J. J. L. M., Rowan, A. E., Nolte, R. J. M. *Macromolecules* 2004, *37*, 4736–4739.

58 Widawski, G., Rawiso, M., François, B. *Nature* 1994, *369*, 387–389.

59 De Boer, B., Stalmach, U., Nijland, H., Hadziioannou, G. *Adv. Mater.* 2000, *12*, 1581–1583.

60 Surin, M., Marsitzky, D., Grimsdale, A. C., Müllen, K., Lazzaroni, R., Leclère, P. *Adv. Funct. Mater.* 2004, *14*, 708–715.

61 Wang, H., You, W., Jiang, P., Yu, L., Wang, H. H. *Chem. Eur. J.* 2004, *10*, 986–993.

62 El-ghayoury, A., Schenning, A. P. H. J., van Hal, P. A., van Duren, J. K. J., Janssen, R. A. J., Meijer, E. W. *Angew. Chem. Int. Ed.* 2001, *40*, 3660–3663.

63 Schenning, A. P. H. J., Jonkheijm, P., Peeters, E., Meijer, E. W. *J. Am. Chem. Soc.* 2001, *123*, 409–416.

64 Schenning, A. P. H. J., Jonkheijm, P., Hoeben, F. J. M., van Herrikhuyzen, J., Meskers, S. C. J., Meijer, E. W., Herz, L. M., Daniel, C., Silva, C., Philipps, R. T., Friend, R. H., Beljonne, D., Miura, A., De Feyter, S., Zdanowska, M., Uji-i, H., De Schryver, F. C., Chen, Z., Würthner, F., Mas-Torrent, M., den Boer, D., Durkut, M., Hadley, P. *Synth. Met.* 2004, *147*, 43–48.

65 Würthner, F., Chen, Z., Hoeben, F. J. M., Osswald, P., You, C.-C., Jonkheijm, P., van Herrikhuyzen, J., Schenning, A. P. H. J., van der Schoot, P. P. A. M., Meijer, E. W., Beckers, E. H. A., Meskers, S. C. J., Janssen, R. A. J. *J. Am. Chem. Soc.* 2004, *126*, 10611–10618.

66 Abdalla, M. A., Bayer, J., Rädler, J. O., Müllen, K. *Angew. Chem. Int. Ed.* 2004, *43*, 3967–3970.

67 Watson, M. D., Fechtenkötter, A., Müllen, K. *Chem. Rev.* 2001, *101*, 1267–1300.

68 Clar, E. *The Aromatic Sextet*, Wiley, London, 1972.

69 Debije, M. G., Piris, J., de Haas, M. P., Warman, J. M., Tomovic, Z., Simpson, C. D., Watson, M. D., Müllen, K. *J. Am. Chem. Soc.* 2004, *126*, 4641–4645.

70 Wang, Z., Tomovic, Z., Kastler, M., Pretsch, R., Negri, F., Enkelmann, V., Müllen, K. *J. Am. Chem. Soc.* 2004, *126*, 7794–7795.

71 Herwig, P. T., Kayser, C. W., Müllen, K., Spiess, H. W. *Adv. Mater.* 1996, *8*, 510.

72 Tomovic, Z., Watson, M. D., Müllen, K. *Angew. Chem. Int. Ed.* 2004, *43*, 755–758.

73 Simpson, C. D., Brand, J. D., Berresheim, A. J., Przybilla, L., Räder, H.-J., Müllen, K. *Chem. Eur. J.* 2002, *8*, 1424–1429.

74 Simpson, C. D., Wu, J., Watson, M. D., Müllen, K. *J. Mater. Chem.* 2004, *14*, 494–504.

75 van de Craats, A. M., Warman, J. M., Fechtenkötter, A., Brand, J. D., Harbison, M. A., Müllen, K. *Adv. Mater.* 1999, *11*, 1469–1472.

76 van de Craats, A. M., Warman, J. M., Müllen, K., Geerts, Y., Brand, J. D. *Adv. Mater.* **1998**, *10*, 36–38.
77 Liu, C.-y., Fechtenkötter, A., Watson, M. D., Müllen, K., Bard, A. J. *Chem. Mater.* **2003**, *15*, 124–130.
78 van de Craats, A. M., Stutzmann, N., Bunk, O., Nielsen, M. M., Watson, M., Müllen, K., Chanzy, H. D., Sirringhaus, H., Friend, R. H. *Adv. Mater.* **2003**, *15*, 495.
79 Tracz, A., Jeszka, J. K., Watson, M. D., Pisula, W., Müllen, K., Pakula, T. *J. Am. Chem. Soc.* **2003**, *125*, 1682–1683.
80 Pisula, W., Kastler, M., Wasserfallen, D., Pakula, T., Müllen, K. *J. Am. Chem. Soc.* **2004**, *126*, 8074–8075.
81 Pisula, W., Menon, A., Stepputat, M., Lieberwirth, I., Kolb, U., Tracz, A., Sirringhaus, H., Pakula, T., Müllen, K. *Adv. Mater.* **2005**, *17*, 684–689.
82 Schmidt-Mende, L., Fechtenkötter, A., Müllen, K., Moons, E., Friend, R. H., MacKenzie, J. D. *Science* **2001**, *293*, 1119–1122.
83 Wu, J., Watson, M. D., Zhang, L., Wang, Z., Müllen, K. *J. Am. Chem. Soc.* **2004**, *126*, 177–186.
84 Wu, J., Baumgarten, M., Debije, M. G., Warman, J. M., Müllen, K. *Angew. Chem. Int. Ed.* **2004**, *43*, 5331.
85 Wu, J., Watson, M. D., Tchebotareva, N., Wang, Z., Müllen, K. *J. Org. Chem.* **2004**, *69*, 8194–8204.
86 Wang, Z., Watson, M. D., Wu, J., Müllen, K. *Chem. Commun.* **2004**, 336–337.
87 Wu, J., Gherghel, L., Watson, M. D., Li, J., Wang, Z., Simpson, C. D., Kolb, U., Müllen, K. *Macromolecules* **2003**, *36*, 7082–7089.
88 Brand, J. D., Kübel, C., Ito, S., Müllen, K. *Chem. Mater.* **2000**, *12*, 1638–1647.
89 Laursen, B. W., Nørgaard, K., Reitzel, N., Simonsen, J. B., Nielsen, C. B., Als-Nielsen, J., Bjørnholm, T., Sølling, T. I., Nielsen, M. M., Bunk, O., Kjaer, K., Tchebotareva, N., Watson, M. D., Müllen, K., Piris, J. *Langmuir* **2004**, *20*, 4139–4146.
90 Thünemann, A. F., Kubowicz, S., Burger, C., Watson, M. D., Tchebotareva, N., Müllen, K. *J. Am. Chem. Soc.* **2003**, *125*, 352–356.

91 Wu, J., Fechtenkötter, A., Gauss, J., Watson, M. D., Kastler, M., Fechtenkötter, C., Wagner, M., Müllen, K. *J. Am. Chem. Soc.* **2004**, *126*, 11311–11321.
92 Wu, J., Qu, J., Tchebotareva, N., Müllen, K. *Tetrahedron Lett.* **2005**, *46*, i1565–1568.
93 Jäckel, F., Watson, M. D., Müllen, K., Rabe, J. P. *Phys. Rev. Lett.* **2004**, *92*, 1883031–1883034.
94 Jäckel, F., Wang, Z., Watson, M. D., Müllen, K., Rabe, J. P. *Synth. Met.* **2004**, *146*, 269–272.
95 Ijima, S. *Nature* **1991**, *354*, 56.
96 Ajayan, P. M. *Chem. Rev.* **1999**, *99*, 1787–1799.
97 Rao, C. N. R., Satishkumar, B. C., Givindaraj, A., Nath, M. *ChemPhysChem* **2001**, *2*, 78–105.
98 Baughman, R. H., Zakhidov, A. A., de Heer, W. A. *Science* **2002**, *297*, 787–792.
99 Dai, H. *Surf. Sci.* **2002**, *500*, 218–241.
100 Tans, S. J., Verschueren, A. R. M., Dekker, C. *Nature* **1998**, *393*, 49–52.
101 Dalton, A. B., Coleman, J. N., in het Panhuis, M., McCarthy, B., Drury, A., Blau, W. J., Nunzi, J.-M., Byrne, H. J. *J. Photochem. Photobiol. A* **2001**, *144*, 31–41.
102 Curran, S. A., Ajayan, P. M., Blau, W. J., Carroll, D. L., Coleman, J. N., Dalton, A. B., Davey, A. P., Drury, A., McCarthy, B., Maier, S., Strevens, A. *Adv. Mater.* **1998**, *10*, 1091–1093.
103 Fournet, P., O'Brien, D. F., Coleman, J. N., Hörhold, H.-H., Blau, W. J. *Synth. Met.* **2001**, *121*, 1683–1684.
104 Thess, A., Lee, R., Nikolaev, P., Dai, H. J., Petit, P., Robert, J., Xu, C. H., Lee, Y. H., Kim, S. G., Rinzler, A. G., Colbert, D. T., Scuseria, G. E., Tomanek, D., Fischer, J. E., Smalley, R. E. *Science* **1996**, *273*, 483.
105 Dai, H., Rinzler, A. Z., Nikolaev, P., Thess, A., Colbert, D. T., Smalley, R. E. *Chem. Phys. Lett.* **1996**, *260*, 471–475.
106 Dai, H. *Acc. Chem. Res.* **2002**, *35*, 1035–1044.
107 Gherghel, L., Kübel, C., Lieser, G., Räder, H.-J., Müllen, K. *J. Am. Chem. Soc.* **2002**, *124*, 13130–13138.

108 Dosa, P.I., Erben, C., Iyer, V.S., Vollhardt, K.P.C., Wasser, I.M. *J. Am. Chem. Soc.* **1999**, *121*, 10430–10431.

109 Laskoski, M., Steffen, W., Morton, J.G.M., Smith, M.D., Bunz, U.H.F. *J. Am. Chem. Soc.* **2002**, *124*, 13814–13818.

110 Rao, C.N.R., Govindaraj, A. *Acc. Chem. Res.* **2002**, *35*, 998–1007.

111 Iyer, V.S., Vollhardt, K.P.C., Wilhelm, R. *Angew. Chem. Int. Ed.* **2003**, *42*, 4379–4383.

112 Zhi, L., Wu, J., Li, J., Kolb, U., Müllen, K. *Angew. Chem. Int. Ed.* **2005**, *44*, 2120–2123.

113 Novikov, E., Hofkens, J., Cotlet, M., De Schryver, F.C., Boens, N. *Rev. Fluoresc.* **2004**, *1*, 299–340.

114 Bertone, N., Biasi, R., Dion, B. *Proc. SPIE* **2005**, *5726*, 153–163.

115 Baschè, T., Moerner, W.E., Orrit, M., Wild, U.P. *Single Molecule Optical Detection, Imaging and Spectroscopy*, Verlag-Chemie, Weinheim, **1997**.

116 Moerner, W.E., Orrit, M. *Science* **1999**, *283*, 1670–1676.

117 Binning, G., Rohrer, H., Gerber, C., Weibel, E. *Helv. Phys. Acta* **1982**, *55*, 726.

118 Binning, G., Rohrer, H., Gerber, C., Weibel, E. *Appl. Phys. Lett.* **1982**, *40*, 178.

119 Samori, P. *J. Mater. Chem.* **2004**, *14*, 1353–1366.

120 Miura, A., De Feyter, S., Abdel-Mottaleb, M.M.S., Gesquière, A., Grim, P.C.M., Moessner, G., Sieffert, M., Klapper, M., Müllen, K., De Schryver, F.C. *Langmuir* **2003**, *19*, 6474–6482.

121 Aviram, A., Ratner, M.A. *Chem. Phys. Lett.* **1974**, *29*, 277–283.

122 Roncali, J. *Acc. Chem. Res.* **2000**, *33*, 147–156.

123 Tour, J.M., Rawlett, A.M., Kozaki, M., Yao, Y., Jagessar, R.C., Dirk, S.M., Price, D.W., Reed, M.A., Zhou, C., Chen, J., Wang, W., Campbell, I. *Chem. Eur. J.* **2001**, *7*, 5118–5134.

124 Tour, J.M. *Acc. Chem. Res.* **2000**, *33*, 791–804.

125 Robertson, N., McGowan, C.A. *Chem. Soc. Rev.* **2003**, *32*, 96–103.

126 Tour, J.M. *Molecular Electronics: Commercial Insights, Chemistry, Devices, Architecture and Programming*, World Scientific, River Edge, NJ, **2003**.

127 James, D.K., Tour, J.M. *Chem. Mater.* **2004**, *16*, 4423–4435.

128 Mayor, M., Weber, H.B. *Angew. Chem. Int. Ed.* **2004**, *43*, 2882–2884.

129 Kouwenhoven, L. *Science* **1997**, *275*, 1896–1897.

130 Tans, S.J., Verschueren, A.R.M., Dekker, C. *Nature* **1998**, *393*, 49–52.

131 Park, H., Park, J., Lim, A.K.L., Anderson, E.H., Alivisatos, A.P., McEuen, P.L. *Nature* **2000**, *407*, 57–60.

132 Yu, H., Luo, Y., Beverly, K., Stoddart, J.F., Tseng, H.-R., Heath, J.R. *Angew. Chem. Int. Ed.* **2003**, *42*, 5706–5711.

133 Pasupathy, A.N., Park, J., Chang, C., Soldatov, A.V., Lebedkin, S., Bialczak, R.C., Grose, J.E., Donev, L.A.K., Sethna, J.P., Ralph, D.C., McEuen, P.L. *Nano Lett.* **2005**, *5*, 203–207.

134 Lee, J.-O., Lietschnig, G., Wiertz, F.G.H., Struijk, M., Janssen, R.A.J., Egberink, R., Reinhoudt, D.N., Grimsdale, A.C., Müllen, K., Hadley, P., Dekker, C. *Ann. N.Y. Acad. Sci.* **2003**, *1006*, 122–132.

135 Flink, S., van Veggel, F.C.J.M., Reinhoudt, D.N. *Adv. Mater.* **2000**, *12*, 1315–1328.

136 McQuade, D.T., Pullen, A.E., Swager, T.M. *Chem. Rev.* **2000**, *100*, 2537–2574.

137 Kuroda, K., Swager, T.M. *Macromol. Symp.* **2003**, *201*, 127–134.

138 Leclerc, M., Ho, H.A. *Synlett* **2004**, 380–387.

139 Ho, H.A., Leclerc, M. *J. Am. Chem. Soc.* **2003**, *125*, 4412–4413.

140 Kim, T.-H., Swager, T.M. *Angew. Chem. Int. Ed.* **2003**, *42*, 4803–4806.

141 Zhang, S.-W., Swager, T.M. *J. Am. Chem. Soc.* **2003**, *125*, 3420–3421.

142 Kumpumbu-Kalemba, L., Leclerc, M. *Chem. Commun.* **2000**, 1847–1848.

143 McQuade, D.T., Hegedus, A.H., Swager, T.M. *J. Am. Chem. Soc.* **2000**, *122*, 12389–12390.

144 Ho, H.-A., Boissinot, M., Bergeron, M.G., Corbeil, G., Dore, K., Boudreau, D., Leclerc, M. *Angew. Chem. Int. Ed.* **2002**, *41*, 1548–1551.

145 Dore, K., Dubus, S., Ho, H.-A., Levesque, I., Brunette, M., Corbeil, G.,

Boissinot, M., Boivin, G., Bergeron, M.G., Boudreau, D., Leclerc, M. *J. Am. Chem. Soc.* **2004**, *126*, 4240–4244.

146 Ho, H.-A., Leclerc, M. *J. Am. Chem. Soc.* **2004**, *126*, 1384–1387.

147 Ho, H.-A., Boissinot, M., Bergeron, M.G., Corbeil, G., Dore, K., Boudreau, D., Leclerc, M. *ACS Symp. Ser.* **2005**, *888*, 359–367.

148 Disney, M.D., Zheng, J., Swager, T.M., Seeberger, P.H. *J. Am. Chem. Soc.* **2004**, *126*, 13343–13346.

149 Gaylord, B.S., Heeger, A.J., Bazan, G.C. *J. Am. Chem. Soc.* **2003**, *125*, 896–900.

150 Grimsdale, A.C., Müllen, K. *Angew. Chem. Int. Ed.* **2005**, *44*, 5592–5629.

7
Polymers for Microelectronics

Christopher W. Bielawski and C. Grant Willson

7.1
Introduction

Over the past four decades, synthetic polymeric materials have played a critical role in the manufacture of electronic devices and components that have profoundly impacted modern society. The value of these materials stems from the ability to modify and control systematically their physical and chemical properties, including not only their electronic characteristics, but also their solubilities, thermal stabilities, dielectric constants, absorptivities, etc., in a logical manner.

These advantages have allowed the microelectronics industry to grow at an unprecedented pace. Synthetic polymers are currently used as photoresists in lithographic applications for fabricating microprocessors and memory chips, dielectrics and other components in circuit boards, in addition to photovoltaics and organic light-emitting diodes (OLEDs). More recently, they have been used in smart cards, large area displays and high-density storage media. Undoubtedly, the high demands placed on the microelectronics industry will continue to inspire the development of new materials with improved physical properties.

This chapter will survey the current status of advanced polymeric materials and their use in microelectronic devices and related applications. Attention will be directed towards conjugated polymers, particularly those with conductive properties, polymer dielectrics and resist materials for photolithography. The chapter will conclude with some final comments about the field in general and some expectations for the future.

7.2
Conjugated Polymers

7.2.1
Overview

Conjugated polymers have been the focus of intensive research endeavors by a large number of groups over the past three decades [1]. They have found utility in an impressive number of applications in the microelectronics industry, including use in charge storage devices such as batteries and supercapacitors, as precursors for printed circuit board technology and integrated circuits, as a means to protect and house electronic devices, as efficient discharge layers and as photoresists in electron beam lithography [2]. For example, they can be found in coin-type batteries found in common wrist-watches, as key components in electrolytic capacitors, in magnetic storage media, electrostatic loudspeakers and as OLEDs. Annual worldwide sales of conjugated polymers have already surpassed US$ 1 billion and undoubtedly will continue to grow [3].

In general, conjugated polymers are comprised of organic macromolecules with a framework of alternating single and double bonds. More formally, they possess a backbone of σ-bonds, formed from the overlap of sp^2 hybridized orbitals and π-bonds, formed from the overlap between the remaining out-of-plane p_z orbitals. The simplest such chemical structure is polyacetylene; however, a wide variety of other conjugated polymers are known, including those containing heteroatoms (Fig. 7.1). In addition, with the advent and progression of various methods in organic synthesis, a broad range of derivatives are now known.

It is important to note that although conjugated polymers formally possess extensively delocalized electronic structures, they are not inherently conductive. However, charged derivatives (which are accessible via a variety of chemical and/or electrochemical oxidation or reduction methods) are highly conductive and can readily transport charge. This feature, combined with the myriad of chemical methods available to modify systematically molecular structure, pro-

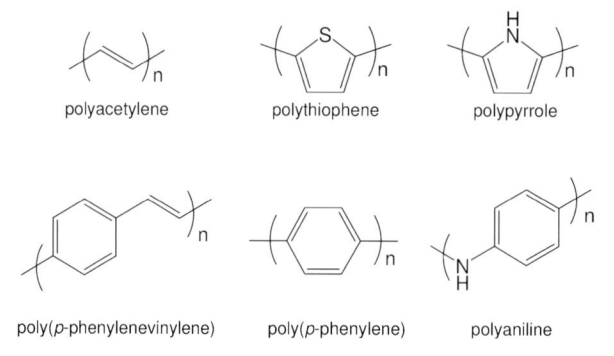

Fig. 7.1 Examples of various types of conjugated polymers.

vides several handles for logically tuning the conductivity of the resultant polymer to meet a specific application and/or need. In addition, since most organic materials are flexible, processable (or can be made processable) and lightweight, conducting polymers are potentially suitable for applications that have not yet been realized. This section will start with a short overview of basic terminology used in the field and some fundamentals of electrical conductivity. Common classes of conjugated polymers will then be surveyed with particular emphasis on utility and potential in various aspects of the microelectronics industry. The final section will conclude the chapter with an outlook and some predictions for the future.

7.2.2
General Considerations

The relationship between conductivity, potential and resistance is known as Ohm's law: $V = IR$, where V = potential in volts, I = current in amps and R = resistance in ohms. Simply put, this equation states that under constant voltage, higher current necessitates lower resistance. The resistance or conductivity through any system or material is typically treated as a wire and therefore will depend on its length (l), cross-sectional area (A) and intrinsic ability to move charge [which is generally defined in terms of resistivity (ρ) with units of ohm cm]. Using these terms, a convenient (and widely used) equation can be constructed that defines conductivity (σ) as the inverse of resistivity ($\sigma = \rho^{-1}$ with units of $\text{ohm}^{-1}\text{ cm}^{-1}$ or S cm^{-1}). The potential of a material for use in an electronic application is based on its conductivity, which covers a considerably large range (over 20 orders of magnitude). Typical metallic conductors such as copper, iron or silver exhibit conductivities on the order of 10^6 S cm^{-1}, whereas typical semiconductors, Si and Ge, display much lower values of 10^{-2} and 10^{-5} S cm^{-1}, respectively. Insulators, plastics and most organic materials exhibit considerably smaller conductivities, with values as low as 10^{-14} S cm^{-1} (diamond).

Equal in importance to conductivity is the band structure of the material [4]. Conducting polymers are believed to possess an extensive, spatially delocalized, band-like electronic structure which stems from the splitting of overlapping molecular orbitals in a manner consistent with the band structure of solid-state semiconductors (Fig. 7.2) [5].

It is generally believed that the mechanistic origins of conductivity in these polymers stems from the motion of charged defects/carriers along the conju-

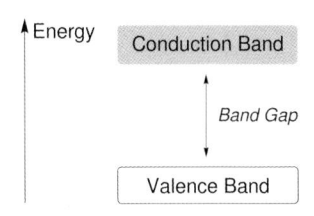

Fig. 7.2 Illustration of the energy difference (i.e. the bandgap) between the valence band and the conduction band in an electronically conductive polymer.

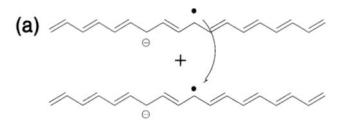

Fig. 7.3 Negatively charged defects in polyacetylene. Combination of two polarons (a) affords neutral polyacetylene and a bipolaron (b).

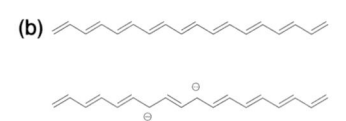

gated framework [6]. These species are either positively charged (p-type) or negatively charged (n-type) and are formed via oxidation or reduction, respectively, of the polymeric material. Figure 7.3 shows a basic description of these processes in the context of n-type carriers (although the concepts also apply equally to p-type carriers). In particular, reduction of the polymer shown initially generates a radical anion, which is referred to as a polaron and comprises both the charge and the accompanying structural distortion (e.g. in this example there is a hybridization change). Once anions and radicals are formed, two polarons can combine to form a lower energy bipolaron. Despite the coulombic repulsion of the two negatively charged ions, theory suggests that one bipolaron is more stable than two polarons. In essence, the structural defects associated with the aforementioned processes can be considered simply as a boundary between two chains of infinite conjugation on either side. Once this boundary is hurdled, each polaron can (in theory) migrate in either direction without affecting the energy of the backbone. This "charge carrier mobility" ultimately defines the conductivity of conjugated polymers.

Conductivities of various materials are defined in terms of their intrinsic mobility (μ) and the number (n) of charge carriers (electrons and/or holes) according to the equation $\sigma = n\mu e$. In general, the bandgaps of conjugated polymers are usually fairly large and the number of charge carriers is very small. Consequently, conjugated organic polymers are in fact insulators in their neutral form (i.e. the search for an intrinsically conducting organic polymer continues). The conductivity of the polymer can be enhanced, however, by introducing more charge carriers into the material. A convenient way to do this is through chemical or electrochemical oxidation (p-doping) or reduction (n-doping). Examination of the previous equation suggests that another method to increase conductivity is to modify the molecular structure of the polymer to optimize electron delocalization properties. This is often the more challenging option because the structure of the bulk material, specifically the role and presence of any structural defects, must be simultaneously considered.

Although at first glance it may seem that the polymer material will conduct electrons only along its conjugated backbone, this is in fact not true. In the solid state (i.e. once they are implemented into a device), conductive organic polymers can conduct in all directions. In fact, it is extremely unlikely (on a kinetic and thermodynamic basis) for a polymer to exist in a fully extended conformation (this is especially true for high molecular weight polymers) and span from one electrode to another. As such, conductivity can occur via intermolecular and intramolecular pathways. This generally implies an electron transfer reaction via the "hopping" of polarons (charged species) from one chain to another (Fig. 7.3). Therefore, polymer crystallinity (packing density, chain alignment, defects, etc.) plays a critical role defining the material's conductivity and the development of new processing methods to maximize, or at least influence, crystallinity is extremely important. As seen below, most conjugated polymers are poorly soluble in common (and even uncommon) organic solvents. Thus, a variety of methods have been employed to increase their solubility. A typical way in which this has been done is through the incorporation of long alkyl chains into the conjugated polymer, which often greatly enhances their solubilities. However, there is usually a trade-off and, while many of these approaches provide soluble polymers, the presence of the solubilizing groups often frustrates chain crystallinity, preventing close packing of chains that would enable interchain electron transfer to occur. In addition to the development of new conjugated polymers, future efforts will continue to be devoted toward addressing these important issues.

7.2.3
Polyacetylene

From either a historical, theoretical or experimental point of view, any discussion of conducting polymers should include polyacetylene. Efforts to prepare this polymer date back to Natta in the 1950s, whose group first attempted to polymerize acetylene [7]. Unfortunately, the products obtained were always black, air-sensitive, insoluble – and uncharacterizable – powders. A key breakthrough came about in the early 1970s when Shirakawa and Ikeda discovered that passing gaseous acetylene over a Ziegler–Natta-type catalyst caused a shiny, copper-colored film of polymeric material to be formed [8]. Subsequent analysis of this material revealed that it was unsaturated with almost exclusively the cis geometric configuration about the double bonds (Fig. 7.4). Heating the material to 150 °C, however, produced its stereoisomeric all-trans form, which had a silver-colored appearance and for all purposes bore a striking resemblance to a mirror. These features were remarkable for a material comprised entirely of carbon and hydrogen. More importantly, they suggested that such polymers might show some metalloid character and, in particular, interesting electronic properties. Unfortunately, it was soon determined that neither the all-cis, the all-trans

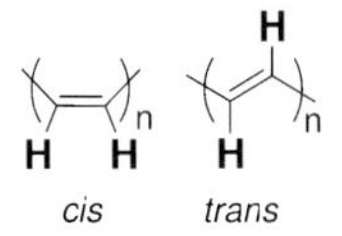

Fig. 7.4 Structures of *cis*- and *trans*-polyacetylene.

or any intermediate material was conductive. For example, cis-polyacetylene exhibits a conductivity of 10^{-10} S cm^{-1} (which is similar to that of glass) and trans-polyacetylene exhibits a conductivity of 10^{-5} S cm^{-1}, which is considerably higher than that of its cis-isomer but still too low for many applications.

A short time later, MacDiarmid, Heeger and colleagues began to investigate the potential of doping polyacetylene as a means to improve its conductivity. In conjunction with Shirakawa, they reported that exposure of polyacetylene to a molecular halogen (Cl_2, Br_2 or I_2) produced a material with conductivities as high as 10^3 S cm^{-1} [9]. This value is well within the conductor range and thus the field of conducting organic polymers was born. For this important work, Shirakawa, MacDiarmid and Heeger shared the 2000 Nobel Prize in Chemistry [10].

Unfortunately, polyacetylene has a few drawbacks, which have hampered its commercial realization. The material is insoluble, brittle and extremely difficult to process. Significant effort continues towards improving the solubility and processability of polyacetylene. The key appears to be finding the right balance between crystallinity and solubility. Planar conjugated polymers tend to be highly crystalline and therefore notoriously insoluble, making them difficult to use. One of the more effective methods for increasing solubility is to decrease crystallinity through the incorporation of flexible side-chains. For example, polymerization of terminal (R–C≡C–H) or internal alkynes (R–C≡C–R) would lead to polyacetylate derivatives with pendant carbon chains at alternating or every carbon atom along the backbone, respectively. While there are methods to accomplish this task effectively, the conductivities of the respective doped materials are often still low. It was determined that the pendant R groups (alkyl, aryl, etc.) cause significant distortion along the polymer backbone, ultimately disrupting conjugation (Fig. 7.5).

A major step forward in polyacetylene chemistry came as the result of an advance in methods used to polymerize acetylene with specially designed variants of known catalyst systems. Under certain conditions, Naarman and coworkers demonstrated that acetylene could be simultaneously polymerized and processed to produce ultimately a highly crystalline polyacetylene with conductivities that rivaled those of copper metal [11]. Alternatively, a variety of generalized strategies to circumvent the processing problems associated with polyacetylene have also been reported. Feast and coworkers demonstrated that post-polymer-

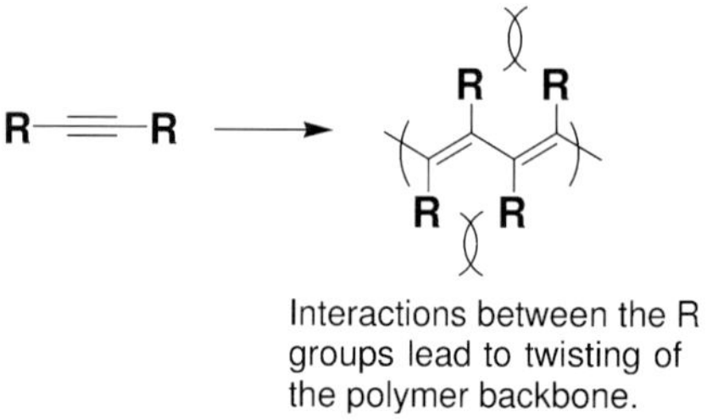

Interactions between the R groups lead to twisting of the polymer backbone.

Fig. 7.5 Steric interactions along the backbone of 1,2-disubstituted polyacetylenes lead to twisting of the polymer chains and decreased conjugation.

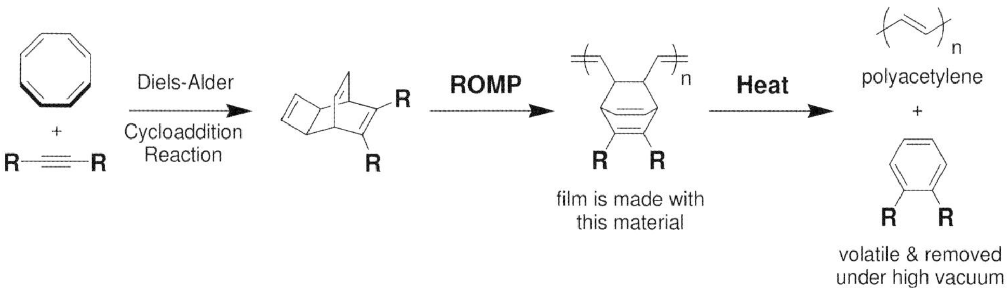

Fig. 7.6 Formation of films of polyacetylene through the polymerization of cycloaddition adduct between cyclooctatetraene and an alkyne followed by treatment with heat under high vacuum.

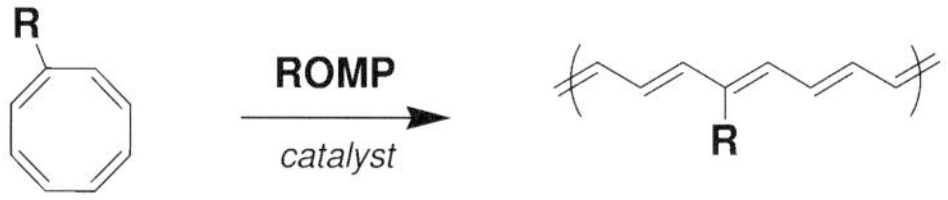

Fig. 7.7 Direct polymerization of derivatized cyclooctatetraenes using a highly active ruthenium catalyst affords soluble forms of polyacetylene.

ization reactions can be used to create highly conjugated polymers after the material has been processed [12]. As shown in Fig. 7.6, an alkyne and cyclooctatetraene were reacted to form a Diels–Alder adduct that was later polymerized using ring-opening metathesis polymerization (ROMP). This provided a highly soluble polymer that was amenable for processing. Subsequent heating resulted in a retro-Diels–Alder reaction, which, when performed under high vacuum, led to loss of a volatile arene and provided a quality polyacetylene film already in place. Although the stereospecificity of the retro-Diels–Alder reaction results in formation of an olefin with a cis (and, as noted above, less conductive) geometry, the high temperature used in this reaction facilitates isomerization to the more thermodynamically stable (and more conductive) trans form.

The development of highly active, well-defined catalysts for mediating ring-opening metathesis polymerizations led by Schrock and Grubbs during the 1980s and 1990s paved new avenues for producing polyacetylene [13]. For example, the ROMP of cyclooctatetraene afforded a one-step method for forming polyacetylene. This material ultimately exhibited high conductivities after doping due to the relatively low number of defects along the polymer backbone. Although the polymeric material was still insoluble, it was later demonstrated that solubility can be drastically improved by adding a single alkyl chain to the cyclooctatetraene monomer [14]. This affords a polyacetylene derivative with an alkyl group on every eighth carbon atom (Fig. 7.7). This material is soluble, processable and can doped to achieve acceptable levels of conductivity, thus creating a good balance between solubility and distortion along the backbone.

Despite these efforts, neither polyacetylene nor any of its derivatives have found a substantial commercial market. Its importance, however, was launching the field of organic conducting polymers. Since the discovery of polyacetylene, a wide variety of other organic polymers have been developed with more desirable combinations of conductivity, processability and stability.

7.2.4
Polyarenes: Polyphenylenevinylene, Polythiophene and Polypyrrole

More processable conducting polymeric materials have been achieved through replacement of the olefins in the polyacetylene (PA) structure described above with aryl groups. This has a general effect of increasing the stability of the system while providing several new handles and synthetic strategies for incorporating solubilizing groups. Poly(p-phenylene) (PPP), poly(p-phenylenevinylene) (PPV), polythiophene (PT) and polypyrrole (PP) are a few examples (structures are shown in Fig. 7.1). Unfortunately, the bandgaps of these materials are generally very large (E_g=4.5, 2.5, 1.7 and 3.0 eV, respectively; for comparison, the E_g of PA is 1.5 eV). This is important because there is a general conception that smaller bandgaps are needed to produce good conductivities since a smaller bandgap generally implies that the merging of the bands can happen more readily. For example, metallic-like conductivities are commonly associated with PT and other small bandgap systems. However, considering the number of factors involved in processing, this concept may not be universally true, especially in doped systems.

In addition to the pursuit of enhancing conductivities, researchers have focused on exploring a number of other unique properties associated with conjugated poly(arene)s. For example, one of the most promising applications is as thin-film materials for use in light-emitting diodes (LEDs) and flat panel displays. When an electric current passes through conducting polymers, light is emitted, a phenomenon known as electroluminescence. In its simplest form, an organic LED (OLED) consists of a layer of conjugated polymer sandwiched between two electrodes. Upon application of an electric current to the device, light is emitted. More formally, holes (positive charges) are created at the anode whereas electrons (negative charges) are created at the cathode. The conductive property of the conjugated polymer permits the holes and electrons to percolate through the material. When they meet, an excited state of the polymer is formed. This state is similar to that formed when a photon is absorbed and, accordingly, relaxation to the ground state releases energy in the form of fluorescence (i.e. light is produced). The color that is emitted in the above process depends strongly on the material used. Since the optical properties of PPV and PT can be tuned to a considerable extent, a broad range of colors can be produced using these polymeric materials. In fact, the displays used in many automobile stereo systems currently utilize these organic polymeric films. A related area in which these polymers have found application has been in photovoltaics, devices that convert light into electricity. As mentioned above, the bandgaps of many semiconducting polymers fall between $E_g=0$ and 2.2 eV, which are comparable to the energies of the photons found in the visible spectrum. One class of conjugated polymers that has found particular success in this area has been PPVs. Early devices constructed of these polymers were capable of generating open-circuit voltages of up to 1.7 V, depending on the electrode employed, with quantum efficiencies of up to 1% upon application of low-intensity light (0.1 mW cm^{-2}) [15]. Since the efficiencies of photovoltaic devices are determined primarily by the ability of the conjugated polymer not only to generate excited states, but also to conduct resulting charges, it becomes critically important to generate these excited states at appropriate distances. If they are too close, they simply recombine; too far, and too much energy is required. Hence it was theorized that an interpenetrating network consisting of a good acceptor and a good donor could improve performance. In particular, blending a functionalized PPV (as the donor) with a soluble PT (as the acceptor), devices with improved quantum efficiencies may be achieved [16].

Another advantage of conjugated polymers containing aromatic groups is derived from their structure, which allows the introduction of solubilizing groups without causing severe changes to the polymers' conjugation (and thus changes in conductivity). Of course, any change will depend on the polymer structure and the group introduced. For example, introduction of a long alkyl chain on to the aromatic rings in PPP will inherently place that group ortho to the main chain, leading to some steric interactions. This effect is less pronounced in PT because the smaller angles associated with five-membered rings effectively position the substituents away from the backbone. In addition, as shown in Fig. 7.8,

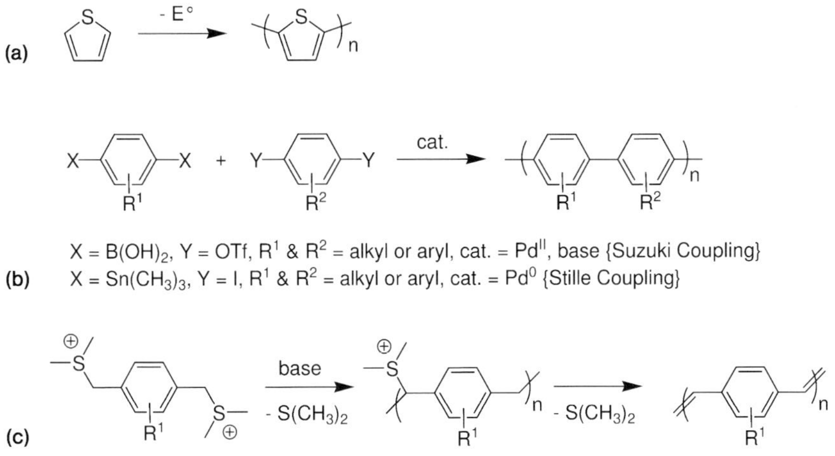

Fig. 7.8 Various synthetic strategies used to prepare conjugated, aromatic polymers. (a) Electrochemical polymerization of thiophene; (b) Pd-catalyzed cross-coupling of disubstituted benzenes; (c) condensation polymerization of a disulfonium-derivatized p-xylene under basic conditions.

there are a large number of synthetic strategies that are currently available to prepare these conjugated aromatic polymers. The most convenient method is electrochemical, which is particularly effective for polymerizing thiophene and pyrrole. This allows a very direct and simple method for polymerizing thin films of thiophene on various electrodes. For PPP and derivatives, condensation-type polymerizations of dihalogenated arenes using various metal-mediated coupling reactions (e.g. Suzuki and Stille type) are particularly effective. Alternatively, sulfonium chemistry can be used to prepare a highly soluble and processable PPP precursor. Similarly to the polyacetylene precursor described above, thermally induced elimination of a volatile sulfide and hydrocarbon provides the desired conjugated polymer. Collectively, these advances allow the design of sophisticated architectures, such as hyperbranched polymers and dendrimers.

Polypyrrole is another conjugated polymer that surged in popularity due to its ability to function as an effective p-doped material after partial oxidation. Like PT, it can be conveniently prepared through electrochemical polymerization of its monomer. An interesting application that may hold promise for this class of conjugated polymers is in advanced microelectronic devices used to deliver electrical stimuli, particularly as it applies toward nerve regeneration. The idea is to introduce a doped PP to an injured site non-invasively and then use an external electromagnetic field to stimulate charge transfer and encourage new nerves to grow.

7.2.5
Polyaniline

Polyaniline (PA) is another conjugated polymer with a variety of interesting characteristics [17]. The polymer contains nitrogen as a key component of the backbone, which is susceptible to oxidation and reduction. PA is produced commercially through the electrochemical polymerization of aniline, although this method introduces some regiochemical defects. Although the material is usually drawn in the "all-para" form, it contains significant amounts of its meta isomer. More recently, Pd-catalyzed aryl amination, developed by Buchwald and Hartwig [18], has been used to produce regiopure polyaniline consisting of either the all-para or the all-meta isomers [19]. While the commercial material typically shows conductivities on the order of 10 S cm^{-1}, polyanilines with conductivities as high as 10^3 S cm^{-1} have been prepared.

Importantly, polyaniline is sufficiently soluble that transparent films can be made from solutions in organic solvents or water. This feature has the interesting consequence that polyaniline does not require harsh chemical or electrochemical doping processes to achieve high conductivities. Instead, its electrochemical properties can be modified by simply adding acid or base. Figure 7.9 shows polyaniline in its various oxidation states. Although the emeraldine base (which is blue or yellow) is not conductive, it can be protonated with acid to its emeraldine salt (which is green), which possesses significant electronic delocalization and, in many respects, behaves as an "organic metal." Notably, simple

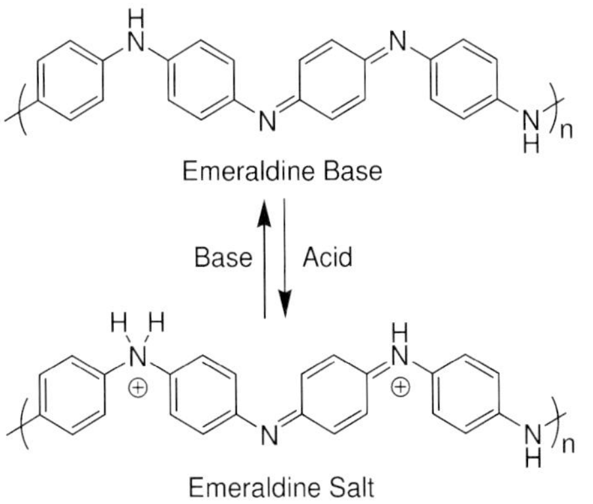

Fig. 7.9 Various oxidized and reduced form of polyaniline. Notably, the emeraldine salt form is conducting whereas the emeraldine base form is not.

treatment of polyaniline with acid or base provides a convenient and relatively simple method to tune both the conductivity and optical properties of this material. This feature, combined with its ability to be processed, makes it one of the most promising of the conducting polymers to date. As such, numerous applications are ultimately envisioned for this material. Particularly, transparent antistatic and thin-film coatings for printed circuit boards or for corrosion protection are relevant and of interest to applications in microelectronics.

More recently, effort has been directed towards developing derivatives of polyaniline that are soluble in their conducting form [20]. Particular attention has been devoted towards their dissolution in aqueous media. For example, water-solubilizing sulfonate groups have been directly incorporated into the polyaniline's backbone through simple electrophilic aromatic substitution reactions (with sulfuric acid), and the product forms a negatively charged material after introduction of base. Polymerization of sulfonated aniline monomers provides an alternative approach to analogous polymers [21]. Recently reported was a new family of water-soluble polyanilines, so-called "PanAquas" [22]. They can be prepared in a single step using a process in which aniline is first coordinated to a polymeric acid template. Upon polymerization, the aniline remains bound to the acid, which prevents an insoluble, cross-linked network from forming. Upon neutralization, the polyaniline can be cleanly isolated from the template.

7.2.6
Outlook

One area of conducting polymers that will continue to attract attention is the development of extremely low bandgap organic polymers. This could lead to a material that is intrinsically conducting, ultimately curtailing the need for doping. This would be a substantial discovery because doping generally accelerates degradation of the material's physical and mechanical properties. In addition, after a material is doped, it generally cannot be processed. Notably, some progress toward this goal has been made. In particular, a new breed of single-walled carbon nanotubes (SWNTs) have been discovered which are inherently conductive. Combined with the significant advances in the processing of these materials that have been developed over the past few years, this discovery should lead to many new advances in this important area.

An area in which conducting polymers may find use in the microelectronics industry is protection against electrostatic charge (ESC) and electrostatic discharge (ESD), which is a particularly challenging and costly problem [23]. Airborne particles can result in electrostatic charges that can reach thousands of volts. This leads to serious problems if these particles become lodged on important electroactive surfaces as they will essentially "short-circuit" or cause physical damage to the device. In addition, as integrated circuit density continues to increase, effects related to ESC and ESD become increasingly important. Currently, conductive materials are extensively used to help combat these effects. For example, they are strategically placed in clean rooms (as mats) to help dissipate charge and are incorporated into packaging materials during transport. These conductors are typically ionic composites, metallic coatings and/or carbon resins. However, these materials are not ideal for protection against ESD. Ionic conductors generally exhibit poor conductivities and, as expected for an ionic material, are strongly dependent on ambient humidity. Metallic coatings are stable towards moisture and air and show high conductivities; however, they are expensive to fabricate and coating of some surfaces can be problematic. Furthermore, contamination is a recurring problem with any carbon-based material and relatively high loadings of carbon are needed to obtain reasonable conductivities. This poses problems with preparing the composite and, in particular, maintaining structural/mechanical properties. In contrast, conducting polymers may offer new alternatives – and advantages over the materials described above – for protection against ESD. As discussed above, such materials can, in theory, be tuned to a wide range of conductivity regimes and can be tailored to address temperature, humidity and stability issues. In addition, the numerous processing and preparation methods enable efficient methods to eliminate contaminants.

Other areas that will continue to attract attention will be in the development of new displays based on organic polymer films. In particular, the development of displays that are highly flexible will continue to attract attention [24]. Other promising applications of conducting polymers in the near future include organic transistors, field-effect transistors and photovoltaics. The last application is particularly promising because although photovoltaics based on conjugated polymers are not as efficient as their silicon analogues, they produce higher voltages. Hence any means to collect radiation more efficiently and transfer it to the polymer could lead to a major breakthrough in this area.

7.3
Polymeric Dielectrics

7.3.1
Overview

The continued improvement in the performance of logic and memory circuits has been so successful that now signal propagation through the interconnecting wiring is becoming speed limiting. Since the speed of signal propagation through this structure depends on the dielectric constants of the insulating medium, there has been increased demand to develop new materials with dielectric constants below 2.5, with 2.0 on the horizon. Since organic materials inherently possess rather low dielectric constants, a wide range of new polymeric materials have been proposed to meet the aforementioned needs. However, in additional to low dielectric constants, these materials must meet a number of requirements for broad utility. Most importantly, they must show outstanding thermal stabilities as the formation of metal interconnects found in a wide range of advanced electronic components often requires temperatures that exceed 400 °C. It is also important for the material to exhibit low water absorptivity and low coefficient of thermal expansion (CTE) to prevent the formation of microcracks. This section will survey a variety of polymeric materials that hold promise in this area, including polyimides, polybenzoxazoles, polybenzocyclobutanes,

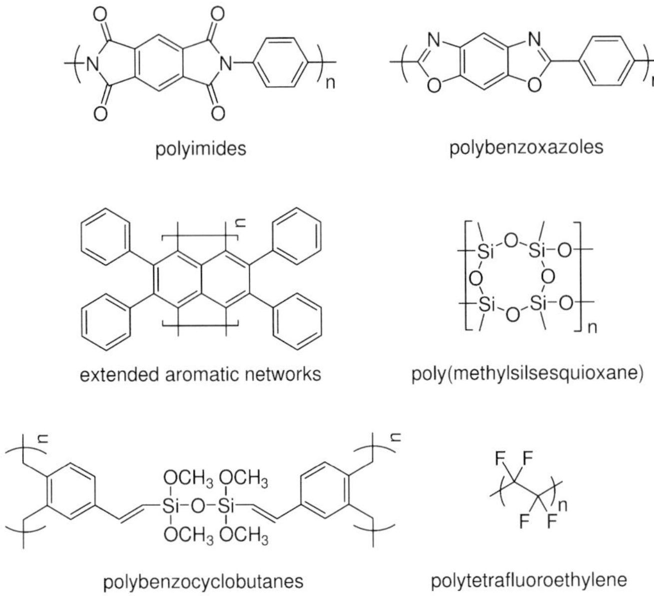

Fig. 7.10 Representative examples of various polymers that show relatively low dielectric constants.

polysilsesquioxanes, extended aromatic networks and fluorinated polymers (Fig. 7.10) [25]. Emphasis will be placed on chemical synthesis, physical properties and techniques available for film deposition of these materials. The section will conclude with an outlook for the field.

7.3.2
Polyimides

The search for new polymeric materials with low dielectric constants that were suitable for use as interlayer dielectrics (ILDs) started with polyimides ($\varepsilon = 3.0$–3.8). These polymers exhibit excellent thermal stabilities (up to 600 °C in some cases) and mechanical properties and low CTEs and can be prepared with sufficiently high purities for use in integrated circuits. However, most polyimides are relatively insoluble and must be prepared from an appropriate, soluble precursor. This is often accomplished by first reacting a dianhydride with a diamine in a polar aprotic solvent such as *N,N*-dimethylacetamide, which affords a soluble and processable polyamic acid. A second step involves thermal treatment which facilitates intramolecular dehydration and ultimately affords the corresponding polyimide. An example of this sequence is shown in Fig. 7.11 [26]. Despite their advantages as dielectric materials, polyimides have a few shortcomings. They are highly polar materials with a strong tendency to absorb water. The adsorption of just 1% (w/w) of water increases the dielectric constant by 0.8. When spin coated, polyimides form anisotropic films with the chains oriented in the substrate plane. This arrangement provides very low coefficients of thermal expansion in the plane of the coating. Their anisotropy also carries over into the polarizability and therefore the dielectric constant. Creative synthetic efforts have successfully minimized the anisotropy of thin films of these materials and they continue to serve in a wide variety of roles, including flexible conductor substrates, on-chip dielectrics and passivation layers.

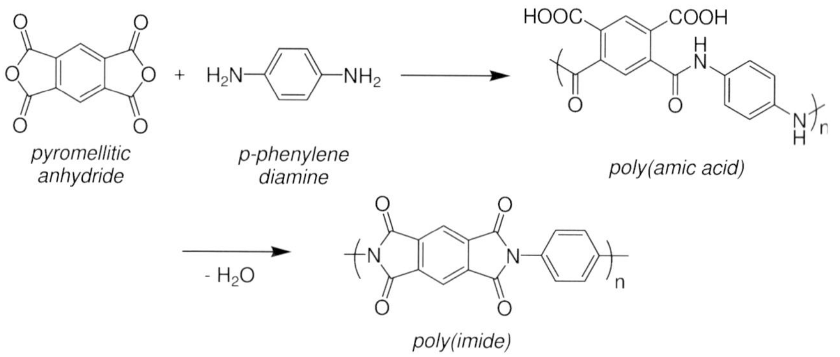

Fig. 7.11 Representative example of the general synthetic approach used to prepare polyimides.

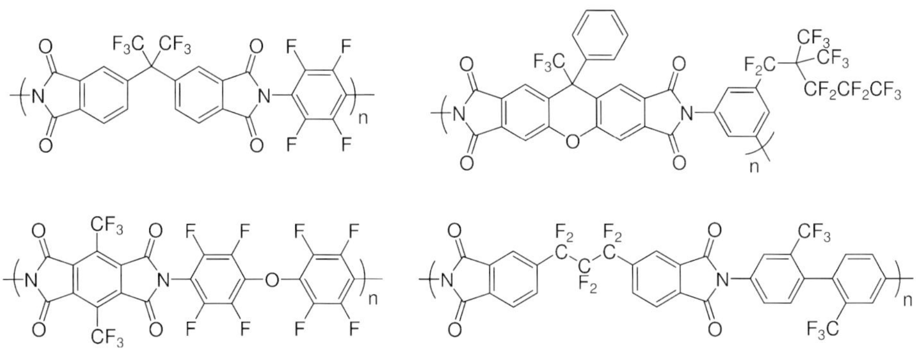

Fig. 7.12 Representative examples of fluorinated polyimides with low dielectric constants.

Two main approaches have been used to decrease the dielectrics of these materials (and reduce their sensitivity towards water). The first method involved introducing a number of bulky hydrocarbon chains into the backbone, which effectively reduced the overall polarity of the polymer (a factor that strongly influences the material's dielectric constant) and increased its hydrophobic content. The second method involved the introduction of fluorinated side-chains (for representative examples, see Fig. 7.12) [27]. Interestingly, although the C–F bond is highly polar, it is also very short, which results in a relatively small dipole. Fluorination appears to have a stronger effect than incorporation of hydrocarbon chains as smaller dielectric constants are generally observed in these materials ($\varepsilon \leq 2.4$) with relatively low water uptakes and minimal CTEs. Through further optimization and derivatization, fluorinated polyimides and other derivatives with slightly lower dielectric constants may be possible [28].

7.3.3
Polybenzoxazoles

Polybenzoxazoles are another class of aromatic polymers that hold potential for low-ε applications. As shown in Fig. 7.13, they are structurally similar to polyimides and share many common physical characteristics. They are relative stable at elevated high temperatures and are often difficult to process and dissolve in common organic solvents due their rigid architectures. For example, polybenzoxazoles prepared from 4,6-diaminoresorcinol and terephthalic acid form high-strength fibers and are processed by spinning lyotropic solutions of these polymers from polyphosphoric acid. Some of these solubility limitations have been overcome through the incorporation of flexible groups into the aromatic groups on the main chains of the polymers. These groups provide an effective means to disrupt packing forces and tune solubility and glass transition temperatures of the polymeric materials. However, processing is generally performed in a two-step process similarly to that for the polyimides noted above.

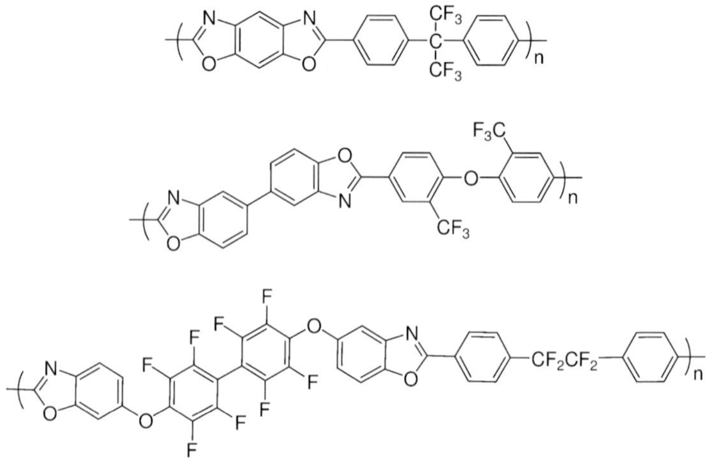

Fig. 7.13 Representative examples of fluorinated polybenzoxazoles.

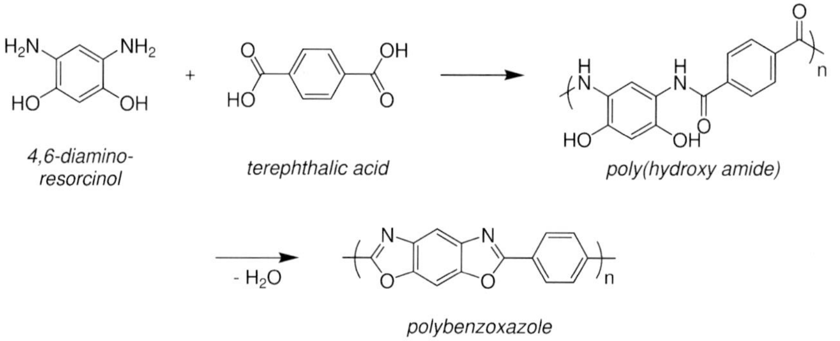

Fig. 7.14 Representative example of the general synthetic approach used to prepare polybenzoxazoles.

As shown in Fig. 7.14, following the preparation of a polyhydroxyamide, subsequent thermal treatment liberates water through an intramolecular cyclization mechanism and affords the respective polybenzoxazole. As in the case of polyimides, low dielectric constants may be achieved by the introduction of fluorine and is often accomplished by reacting fluorinated bis(o-aminophenol)s with bis(carboxylic acid)s.

7.3.4
Polybenzocyclobutanes

During the 1980s and 1990s, Dow Chemical developed and commercialized a new class of polymers based on bicyclo[4.2.0]octa-1,3,5-triene (or benzocyclobutene, BCB) [29]. Cyclotene is a silicon-containing derivative that is currently

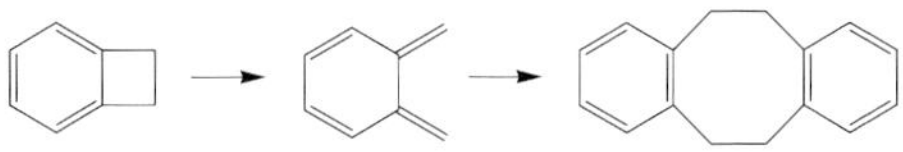

Fig. 7.15 Proposed mechanism of benzocyclobutene dimerization. The first step involves an electrocyclic ring-opening reaction to afford a diene, which subsequently dimerizes via a Diels–Alder cycloaddition reaction.

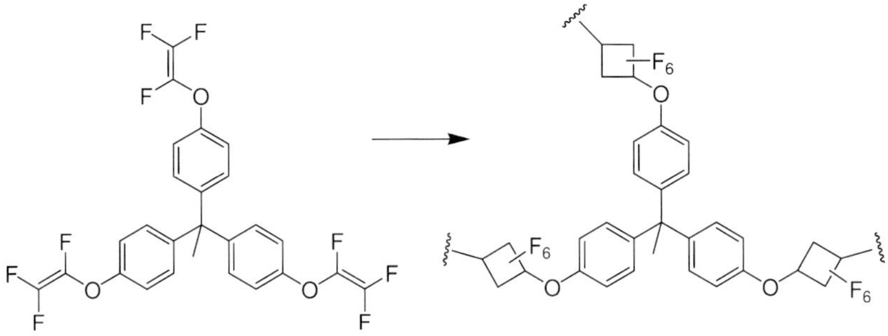

Fig. 7.16 Network formation upon polymerization of a trifunctional fluorinated vinyl ether.

available for microelectronic applications. The polymerization mechanism involves the Diels–Alder cycloaddition reaction (Fig. 7.15). Importantly, the reaction does not emit any volatiles (detrimental in ultra-clean circuit manufacturing environments) and shows minimal shrinkage upon polymerization. The dielectric constants of cured films of these polymers are $\varepsilon = 2.6$–2.7 and show minimal water adsorption (<0.2%) [30]. Unfortunately, thermal stability was found to be a critical issue with Cyclotene and has restricted its use in microchip-based applications.

The lessons learned from fluorinating polyimides were ultimately applied to structurally similar polybenzocyclobutanes, resulting in the development of polyfluorocyclobutane (PFCB)-type materials [31]. As shown in Fig. 7.16, this material is based on a solution of monomer(s) (and/or small oligomers) which are thermally cross-linked during curing. Notably, the monomers contain multiple units of a perfluorovinyl ether which dimerize to ultimately form the [2+2]-cycloaddition product. (Woodward–Hoffman rules state that formal [2+2]-cycloaddition reactions are thermally forbidden, so this monomer may be dimerizing through a different mechanism.) Despite being highly strained, the resultant cyclobutane rings are stable, which has been suggested to result from the ability of the highly polar C–F bonds to remove tension from the four-membered rings. Importantly, the fully cured material has a dielectric constant of $\varepsilon = 2.35$ and exhibits minimal water adsorption. A disadvantage of this material is that, due to the large number of fluorine groups, it does not show good adhesion properties and a promoter layer is required. Furthermore, when compared with some of the polyimides discussed above and its hydrocarbon analogue, PFCB shows limited thermal stability and a relatively high CTE.

7.3.5
Polysilsesquioxanes

The polysilsesquioxanes hold significant promise for use as interlayer dielectric materials. Although they are typically prepared from trialkoxysilanes, the resulting polymeric materials are not cross-linked, but instead are mixtures of ladder-like structures. The curing of these materials results in the loss of various silanol functionalities, but must be performed carefully and under an inert atmosphere as the formation of SiO_2 becomes favorable at elevated temperatures (>500 °C) and occurs with substantial shrinkage. Leading examples are currently polyhydridosilsesquioxane and polymethylsilsesquioxane, which exhibit low dielectric constants ($\varepsilon = 2.6$–2.7) and low water adsorption [32]. An additional advantage of these materials is that their SiO_2 resemblance makes them suitable for use with existing lithography procedures (see below). A current issue with these materials is the high propensity for microcracks to form in their thin films, which can greatly affect their dielectric properties.

7.3.6
Extended Aromatic Networks

Another promising class of low dielectric materials from the Dow Chemical are macromolecular aromatic networks, marketed under the name SiLK [33]. As shown in Fig. 7.17, two synthetic approaches were employed to prepare these materials. In the first approach, the o-alkynyl groups on a variety of o-bis(alkynyl) precursors undergo a Bergman-type cycloaromatization reaction at elevated temperatures to afford radical-based intermediates [34]. These intermediates are

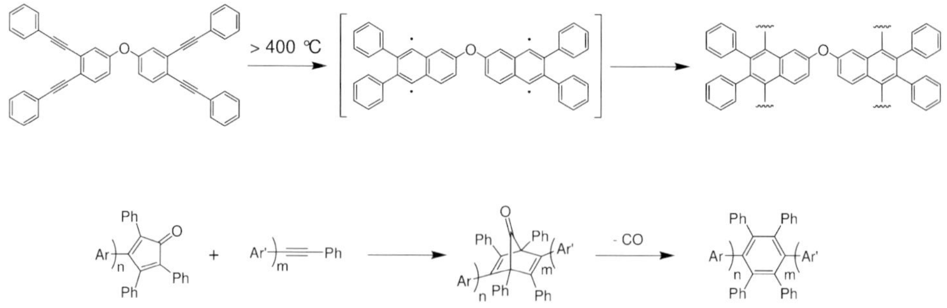

Fig. 7.17 (a) Polymerization of o-dialkylarenes via Bergman cycloaromatization followed by radical coupling; (b) synthesis of macromolecular networks via cycloaddition of multifunctional cyclopentadienones with alkynylarenes followed by extrusion of carbon monoxide.

highly reactive and quickly undergo radical coupling to form an extended aromatic network. The second approach involves a series of Diels–Alder [4+2]-cycloadditions between multifunctional cyclopentadienones and alkynylarenes followed by the expulsion of CO to form an aromatic network. In general, these materials generally exhibit values of ε on the order of 2.6–2.7, are stable up to 500 °C and show low CTEs and low water absorptivity. However, the polymerization mechanism and temperatures involved in their synthesis place limitations on the types of functional groups that can be incorporated into these materials.

7.3.7
Fluorinated Polymers

With $\varepsilon = 1.9$–2.1, polytetrafluorethylene (Teflon) and related copolymers exhibit the lowest dielectric constants of all non-porous organic materials. Unfortunately, Teflon is a highly crystalline material, which makes conventional processing a challenge. Furthermore, it is very insoluble and its decomposition temperature is close to its melting temperature, making melt processing almost impossible. However, a Teflon-based emulsion suitable for preparing films using spin coating and other methods was recently developed. These films show excellent homogeneities and thermal stabilities after drying and removal of the emulsifier. Of course, one of Teflon's main properties is that it shows poor adhesion to most surfaces. Hence a silane adhesive must be used to enhance superficial interaction. Films prepared using this method show dielectric constants as low as $\varepsilon = 1.7$, although the CTE is relatively high compared with some of the previously mentioned polymers and adhesion is often a challenge [35].

7.3.8
Outlook

A large number of significant developments have been made in the last decade towards the development of materials with extremely low dielectric constants. Although some issues regarding the use of these materials in integrated circuits still remain (thermal stability, CTEs, etching characteristics, etc.), these issues will undoubtedly be resolved to meet the growing demand for circuits and other devices with increasing performance. Considering that the dielectric constants of most gases are close to vacuum ($\varepsilon \leq 1$), the development of new porous materials may offer an attractive alternative to low-dielectric films [36]. For example, the introduction of extremely bulky groups along the backbones of polymeric materials may decrease the packing density and increase the free volume. Since it is important to keep the pores disconnected to prevent crack formation, control over this process will be critical. The use of block copolymers or other materials with advanced topologies that show phase separation characteristics may provide a solution to these issues.

7.4
Resists in Microlithographic Applications

7.4.1
Overview

Lithography is one of the oldest techniques known for transferring an image from one substrate to another. It is founded on the basic principle that "oil and water do not mix" and involves relatively simple procedures. In early methods, an image was painted on to a smooth piece of stone (generally limestone) using a thick, greasy substance. The surface of the stone was then treated with water and the parts not protected by the paint became moistened. An oil-based ink was then rolled on to the stone and ultimately retained by areas painted with grease and rejected by moistened areas. A piece of paper was pressed on to the stone, which effectively transferred the image, a process that could be repeated, often several hundred times. The second step in the process was later improved by using acidified aqueous media which reacted with the stone to form a salt layer over the entire surface. This etching process sensitized the painted areas to accept ink and repel water better and simultaneously desensitized the moistened areas so that they accepted water and repelled ink. Collectively, this improved the quality of the transferred image. Although photochemical processes and photoresists have replaced the use of point pressure to define the image and the modern-day stone is a silicon wafer, this "stone writing" process is used to create the fine features and complex patterns found on modern computer chips and in advanced microelectronic devices. The features produced using these techniques are incredibly small and lines with widths smaller than 100 nm can be constructed.

In 1965, Intel founder Gordon Moore predicted that the average transistor density of semiconductor devices would double every 18 months (Fig. 7.18). To meet "Moore's law," a continual reduction in the minimum feature sizes on a microprocessor or a memory chip was needed. As will be discussed below, resolution is inversely related to the wavelength of exposure. Therefore, the use of shorter wavelengths of light is an obvious solution. Indeed, over the years the exposure sources used for patterning photoresists has shifted from the 436- and 365-nm emissions of high-pressure mercury arc lamps to 248- and 193-nm excimer laser-based systems.

Currently, most modern photolithographic methods vary according to the type of radiation used (wavelength, intensity, etc.) as the photoresists employed in these applications respond to the light in a variety of different ways. For example, resists can be patterned with photons of different wavelengths, electron and ion beams, in addition to X-rays. Although photolithography using the longer wavelengths mentioned have dominated manufacturing, the manufacturing process would not exist without masters made using the latter [37]. Each of these photolithographic methods utilize polymers that undergo a change in solubility upon irradiation as a key step. These materials and their associated processes

Moore's Law

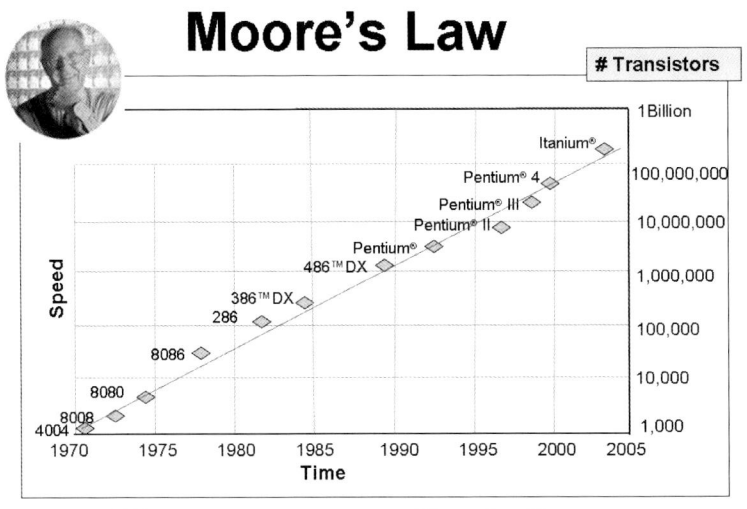

Microprocessor Evolution

Fig. 7.18 Graphical depiction of Moore's law.

have provided service to nearly all aspects of the electronics industry and their continued improvement has been a key reason for their superlative expansion. In the following section, a basic overview of the photolithographic process with a focus on the two types of common resists (negative and positive) will be presented. The discussion will then progress to polymeric materials used in electron beam and immersion lithographies. We will conclude with some comments and trends in recent developments in this field.

7.4.2
Photolithography

7.4.2.1 Background
Photolithography is an optical-based process involved in patterning the integrated circuits that comprise modern computer processors, memory chips and other high-technology devices [38]. More formally, it is a printing process in which the image to be printed is rendered on a flat surface, such as a sheet of zinc, aluminum or silicon. The key component in this process is the photoresist, which generally is a synthetic organic polymer that can be processed and coated on to a relevant substrate or surface. Importantly, the polymeric material must be photoreactive, such that exposure to light changes physical or chemical properties associated with the material, rendering it more soluble or less soluble in some solvent upon exposure to light. In general, since this process involves the making and/or breaking of carbon–carbon bonds, the distinction between whether the material increases or decreases its solubility (in a given solvent) upon exposure to light defines the lithographic process [39].

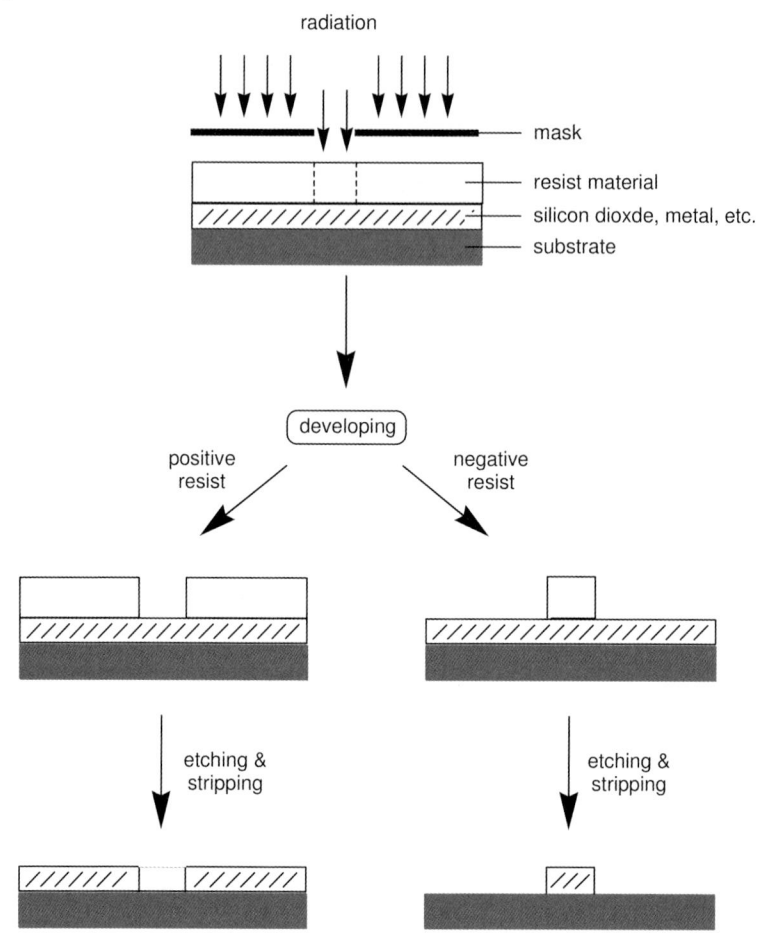

Fig. 7.19 Schematic representation of a photolithographic process using positive (left) and negative (right) photoresists.

A scheme of the basic photolithography process is shown in Fig. 7.19. After coating the photoresist on to a substrate (commonly a silicon wafer), radiation is collimated through a quartz/chrome mask to project the corresponding image on to the surface. This limits exposure of the photoresist material to selected regions, which is effectively the negative of the mask. The photoresist material is designed to respond in one of two ways, depending on its chemical nature. In a negative photoresist, exposure to light renders the material insoluble. Thus, subsequent treatment with solvent removes the unexposed areas of the photoresist and reveals the underlying substrate, which is typically silicon, silicon dioxide, silicon nitride or a metal such as gold. Since the substrate actually contains electronic properties, the overall process is called "negative" because the exposed regions of the substrate were not exposed to light. This is different than a positive photoresist, which increases in solubility upon exposure to light. Thus, solvent treatment reveals areas of the substrate that were exposed to light, providing patterned access to the substrate, which can then be etched or modified selectively. After the image has been transferred, the residual photoresist material is then removed [40]. Historically, the dominant approach in the microelectronics industry employed negative photoresists. However, more recently, positive photoresists have become more common because they can often afford finer features. The chemistry involved in each type of photoresists is discussed below [41].

7.4.2.2 Negative Photoresists

Negative photoresist materials are generally synthetic organic polymers that become less soluble in standard solvents upon exposure to light. Most commonly, this is accomplished through a reaction that links polymer chains to each other (i.e. cross-linking), which results in an insoluble network. One of the first important negative photoresist materials was based on synthetic rubbers such as polyisoprene, which contains olefins throughout its backbone. A solution of the polymer and a photoactivatable cross-linking agent such as a bis(azide) were then cast into a thin film. Photolysis of azides generally results in loss of nitrogen and affords a nitrene which is very reactive and can rapidly insert into C–H bonds or form aziridines with olefins. Thus, the difunctionality inherent to a bis(nitrene) permits two reactions per molecule and, in the presence of polyisoprene (and light), ultimately leads to a cross-linked material. After the material has been irradiated, the surface is then treated with a developer such as methyl ethyl ketone or a chlorinated hydrocarbon, dissolving away polymer chains that have not been cross-linked. A schematic representation of this approach is illustrated in Fig. 7.20. Similar strategies have employed light to generate free radicals, which then initiated the polymerization and cross-linking of a suitable bis(acrylate) monomers.

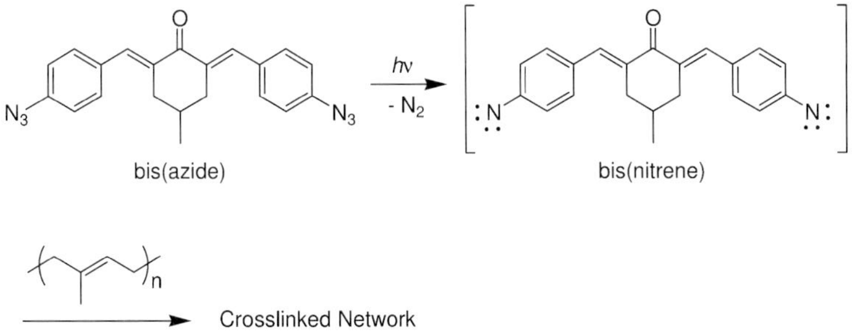

Fig. 7.20 Classical negative photoresist materials and chemistry. Upon irradiation of a bis(azide), it loses nitrogen to form a bis(nitrene), which subsequently reacts with polyisoprene to form an insoluble, cross-linked polymer.

7.4.2.3 Positive Photoresists

A disadvantage of negative photoresists noted above is that developing often swells cross-linked regions, which leads to a reduction in the sharpness of the features. This drawback motivated the development of positive photoresists, which are currently the workhorses of the semiconductor industry. In a positive photoresist, exposure to radiation causes the material to become more soluble, enabling exposed regions to be directly washed away. Fundamentally, this is a more challenging task when compared with negative photoresists because it will inevitably involve a well-defined chain scission reaction or a large molecular change to enhance solubility.

The classical single-component positive resist is poly(methyl methacrylate) (PMMA), which undergoes random chain scissioning when exposed to radiation in the spectral region below about 290 nm. The solubility rate of PMMA is exponentially and inversely proportional to the logarithm of the number-average molecular weight. Hence a change in molecular weight causes a large change in solubility rate. PMMA is an excellent resist that has only two flaws: it is rather insensitive and does not provide an efficient etch in modern reactive ion techniques. A classic example of a two-component resin uses Novolac resins,

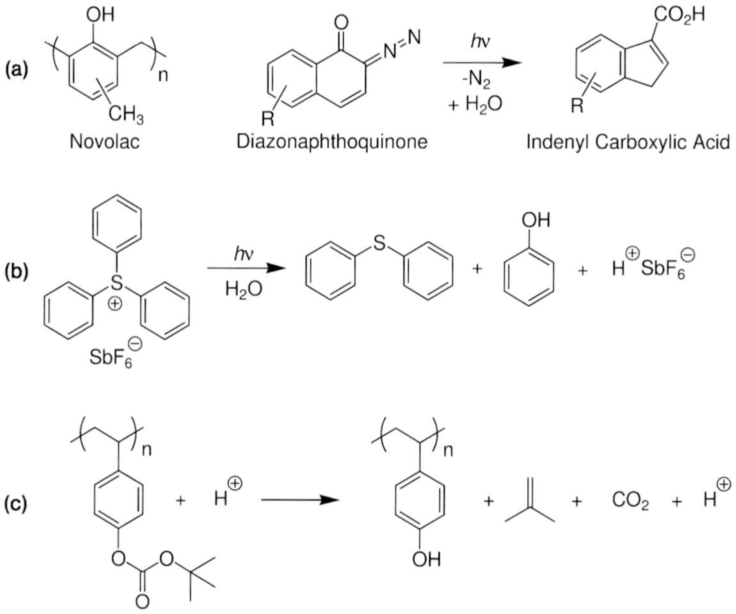

Fig. 7.21 Classical positive photoresist materials and chemistry. (a) The solubility of Novolac resins impregnated with diazonaphthoquinones are greatly increased upon photo-induced rearrangement; (b) triphenylsulfonium hexafluorantimonate fragments into diphenyl sulfide and a protonic acid upon exposure to light; (c) a *tert*-butoxycarbonyl derivative of polystyrene undergoes fragmentation into poly(4-hydroxypolystryene) in the presence of catalytic amounts of acid.

which are soluble in organic solvent and basic, aqueous media. If a film of this material is blended with a photoresponsive dissolution inhibitor, such as hydrophobic diazonaphthoquinone, the dissolution behavior becomes significantly reduced. Photolysis of diazoquinone leads to a Wolff-type rearrangement to afford an indenylcarboxylic acid, which is quickly deprotonated in basic media (Fig. 7.21 a). The increase in the overall polarity of the material leads to rapid dissolution of the exposed region. Although these materials allowed the production of features with 500-nm resolution, even finer detail soon became necessary. The Novolac resin and the diazonaphthoquinone both absorb too strongly below about 290 nm to allow exposure of even very thin films. Hence improvements in resolution by reduction of exposure wavelengths required the development of new imaging chemistry that emerged as chemically amplified systems. In these systems, sulfonium salts (e.g. triphenylsulfonium hexafluoroantimonate), which were known to afford Brønsted acids upon exposure to deep UV radiation (Fig. 7.21 b), were combined with polystyrenes bearing pendant *tert*-butoxycarbonyl (t-BOC) groups, which are relatively insoluble in aqueous media. Exposure of these materials to radiation caused the formation of a latent image of acid which catalyzed the thermolysis of the t-BOC, exposing a soluble phenolate residue. Importantly, the deprotection process is catalytic in acid (Fig. 7.21 c). Although this feature generally translates to extremely high efficiencies, it opens the potential for acid diffusion into non-irradiated areas, which can compromise image resolution. This type of resist is used to make all currently manufactured chips.

7.4.3
Electron Beam Lithography

Electron beam technology is commonly used to fabricate masks for the photolithographic applications noted above and also for the construction of high-resolution, low-volume integrated circuits in laboratories. Relative to photolithography, electron beam lithography is slow because the pattern information is transferred in a serial fashion, pixel by pixel. However, this technique can be used to generate complex patterns with ultra-high resolution. For example, this lithography technique has been shown to pattern features on the order of ~20 nm with the smallest lateral dimensions approaching ~5 nm [42].

In electron beam lithography, a focused beam of electrons is directly scanned over a resist material. A mask is not required as patterns are generated with the aid of a computer. Since most resists are typically insulators, ions (i.e. positive and/or negative charges) can be become trapped and ultimately find their way into the underlying silicon substrate. These trapped ions and associated surface charges can lead to beam deflection, which can cause image distortion. To prevent this problem, the resist material (or the substrate) is often coated with a conducting polymer to dissipate charge. Alternatively, films of carbon or various metal oxides can be used; however, high-temperature evaporative processes are required for film deposition, which erode the performance of the resist material [43].

Sulfonated polystyrene, an ionic polymer with high solubilities in aqueous media, has been used as an efficient charge dissipator in electron beam lithography. This material is easily processable and films of good quality can be made. However, it is not conjugated and therefore its conductivity is relatively low, leading to relatively low charge dissipation and the problems noted above. Development of soluble conjugated polymers would provide an attractive alternative for electron beam lithography. One particularly valuable example is polyaniline [44]. As noted earlier, polyaniline can provide materials with conductivities greater than 10 S cm^{-1}. From an industrial standpoint, these polymeric materials have a number of inherent advantages and therefore are preferred for many applications. For example, they are environmentally stable, relatively soluble and can be easily prepared (in one step) through the electrochemical polymerization of aniline [45]. The aromatic ring provides a powerful handle for further tuning through chemical derivatization to meet any specific need of a certain application [46]. In addition, their properties can be further tuned through dopant variation or through oxidation (or reduction) to one of polyaniline's various oxidation states [47]. The inherent advantages of polyaniline permit its use as a discharge layer in electron beam applications. Generally, polyaniline is incorporated as a conducting layer (usually as a middle layer) in a multi-layered resist material. To convert the polyaniline into its conductive form, it is treated with dilute hydrochloric acid prior to use. Because the resulting polymer is insoluble, subsequent resist materials may be coated on top of the polyaniline layer. Since the entire circuit is grounded and the polymer is conductive, charge is effectively removed, hence charge build-up is minimized. It is important to note that the level of conductivity, which can be conveniently tuned by changing the amount of added oxidant, depends on the structure of the resist and the desired speed of removal. Since polyaniline is more soluble as its free base than its charged derivatives, a one-step process was developed to minimize processing steps. In this method, doping is performed *in situ*, which eliminates the need to use exogenous acid. These acidic and basic films are not compatible with the use of acid-catalyzed, chemically amplified resins.

Another drawback to the approaches discussed above is that, after exposure and development of the resist material, polyaniline remains in unexposed areas. Therefore, to transfer the pattern to underlying substrates, a subsequent step involving ion etching with oxygen is needed. To solve this problem, the conducting polyaniline layer can be placed on top of the resist to permit the simultaneous removal of the polyaniline and the resist. In this respect, common polar aprotic solvents are generally not usable because they can dissolve the resist material. In addition, since industry is moving towards environmentally friendly solvents, compatibility or solubility in aqueous media is strongly preferred. Notably, the water soluble PanAqua materials described above have been found to be very effective at eliminating charge build-up on the resist material.

7.4.4
Immersion Lithography

Current semiconductor fabrication photolithography technology can pattern 90-nm features in full production. Advanced products with minimum features of 65 nm will soon be in production and 37-nm products are in development [48]. Immersion lithography shares many similarities with immersion microscopy, which is often used to characterize cells and other small biological materials [49]. In this technique, it has been shown that increasing the refractive index of the medium between the imaging lens and the imaging substrate can greatly improve resolution. It does so by increasing the depth of focus while also permitting lenses with larger numerical apertures (capture angles) to be used within the imaging system. In fact, switching the medium from air ($n=1.00$) to water ($n=1.47$) results in lenses with numerical apertures approaching 1.3. In addition to finding liquids and polymers that are transparent with high refractive indices, remaining technical challenges associated with this technique involve developing new methods for handling the immersion liquid and designing new catadioptric aspheric lenses. Regardless, this method offers the potential of creating a new high-volume, low-cost technology for fabricating devices with sub-50-nm resolutions.

7.4.5
Functional Resists

A highly sensitive, high-resolution conducting resist material could significantly simplify the process of constructing advanced circuitry and related devices. In particular, with conducting resists, additional charge dissipation would be unnecessary. Considerable attention has been devoted towards the development of conducting resists. Initial investigations focused on the use of polyanilines containing various amounts of onium salts (e.g. triphenylsulfonium hexafluoroantimonate). Upon exposure to radiation, the salts decomposed to produce Brønsted acids, which acidify the polyaniline into its conductive state. However, since the doped polymer is insoluble, the solubility differential is formed between regions which were and were not exposed to light. Subsequently, poly-o-toluidine, a material that is considerably more soluble than its structurally related polyaniline, has been developed and permits a broader range of solvents to be used. Conducting resists based on poly(3-alkylthiophenes) have also been developed based on the bis(azide) cross-linking system discussed above [50]. More recently, methacrylate functional groups have been incorporated into the thiophene framework [51], which provides a handle for polymerization using standard free radical methods. Thus, the materials can be conveniently and efficiently cross-linked through exposure to radiation. As in the aforementioned example, the material must be subsequently doped to facilitate conduction.

7.4.6
Outlook

The semiconductor and microelectronics industries will continue to demand faster and lighter devices. This means that the size of the device will have to continue to decrease. If photolithographic methods are going to continue to be used to construct these devices, then radiation with extremely small wavelengths will have to be used. If history provides any predictions for the future, it is that each step in this process will bring its own set of challenges. For example, in the late 1990s, the semiconducting industry was preparing to employ 157-nm radiation to construct integrated circuits. Initially this caused a major setback as nearly all chemical functionalities absorb strongly at this wavelength. This is highly undesirable as it leads to signal attenuation, diffusion issues and ultimately poor image qualities. In pursuit of polymers suitable for 157-nm based photolithography, an empirical observation was made: integrating fluorinated residues into polymeric materials which have high degrees of hydrogen deficiency resulted in greatly reduced absorptivities at 157 nm. Although a solution was imminent, the semiconductor industry abruptly skipped 157-nm photolithography altogether and set the new goal at 193 nm using immersion techniques. It is expected that new, high refractive index polymeric materials will be required to support this emerging imaging technology. Furthermore, the use of X-rays is still on the horizon [52]. As these short wavelengths of light carry extraordinary amounts of energy, new photoresist materials will be required to manage sensitivities [53]. Likewise, new light sources and optical equipment based on reflection rather than transmission may be needed [54]. Alternatively, imprint lithography, which is based on compression molding and pattern transfer processes, may provide a practical method for fabricating integrated circuits and other electronic devices with extremely small (<10-nm) features [55]. Ultimately, the high demand for processors and other microelectronic devices with increased speeds will require the continuing development and optimization of new materials and tools for their fabrication.

7.5
Conclusions

The discovery and development of synthetic polymers with useful electronic properties have profoundly impacted the microelectronics and other high-technology industries. A wide assortment of these materials are currently used as photoresists in lithographic applications for fabricating microprocessors and memory chips, in photovoltaic and other applications that require conjugated polymers and in polymeric dielectrics used for charge storage. Despite these extraordinary advances, the high demands placed on the microelectronics industry will continue to require new materials with improved physical properties such as ultra-low dielectric constants, low dissipation factors, increased moisture and oxygen compatibilities, high thermal stabilities and enhanced conductivities. Ultimately, new polymeric materials, perhaps with advanced architectures and unprecedented properties, will be needed to sustain growth in the foreseeable future.

References

1 (a) Frommer, J.E., Chance, R.R. in *Encyclopedia of Polymer Science and Engineering*, Vol. 5, Wiley, New York, **1986 AQ4**; (b) Scott, J.C. *Science* **1997**, *278*, 2071.

2 (a) Soane, D.S., Martynenko, Z. (Eds.). *Polymers in Microelectronics*, Elsevier, Amsterdam, **1989**; (b) Harper, C.A. (Ed.) *Electronic Packaging and Interconnection Handbook*, McGraw-Hill, New York, **1991**; (c) Cadenhead, R.L. in *Materials and Electronic Phenomena*, Electronics Materials Handbook, Dostal, C.A. (Ed.), ASM International, Materials Park, OH, **1989**.

3 (a) Alivisatos, A.P., Barbara, P.F., Castleman, A.W., Chang, J., Dixon, D.A., Klein, M.L., McLendon, G.L., Miller, J.S., Ratner, M.A., Rossky, P.J., Stupp, S.I., Thompson, M.E. *Adv. Mater.* **1998**, *10*, 1297.

4 (a) Albright, T.A., Burdett, J.K., Whangbo, M.-H. (Eds), *Orbital Interactions in Chemistry*, Wiley, New York, **1985**; (b) Hoffmann, R. (Ed) *Solids and Surfaces: a Chemist's View of Bonding in Extended Structures*, VCH, New York, **1988**.

5 Heinze, J. *Top. Curr. Chem.* **1990**, *152*.

6 (a) Stewart, M.D., Patterson, K., Somervell, M.H., Willson, C.G. *J. Phys. Org. Chem.* **2000**, *13*, 767; (b) Ito, H. *J. Polym. Sci., Part A: Polym. Chem.* **2003**, *41*, 3863; (c) Willson, C.G., Trinque, B.C. *J. Photopolym. Sci. Technol.* **2003**, *16*, 621.

7 Natta, G., Mazzanti, G., Corradini, P. *Atti Accad. Naz. Lincei Cl. Sci. Fis. Mat. Nat. Rend.* **1958**, *25*, 3.

8 Shirakawa, H., Ikeda, S. *Polym. J.* **1971**, *2*, 231.

9 (a) Chiang, C.K., Fincher, C.R., Park, Y.W., Heeger, A.H., Shirakawa, H., Louis, W.J., MacDiarmid, A.G. *Phys. Rev. Lett.* **1977**, *39*, 1098; (b) Chiang, C.K., Druy, M.A., Gau, S.C., Heeger, A.J., Louis, E.J., MacDiarmid, A.G. *J. Am. Chem. Soc.* **1978**, *100*, 1013.

10 (a) Shirakawa, H. *Angew. Chem. Int. Ed.* **2001**, *40*, 2574; (b) MacDiarmid, A.G. *Angew. Chem. Int. Ed.* **2001**, *40*, 2581; (c) Heeger, A.J. *Angew Chem. Int. Ed.* **2001**, *40*, 2591.

11 Basescu, N., Liu, Z.-X., Moses, D., Heeger, A.J., Naarmann, H., Theophilou, N. *Nature* **1987**, *327*, 403.

12 (a) Edwards, J.H., Feast, W.J. *Polymer* **1980**, *21*, 595; (b) Swager, T.M., Dougherty, D.A., Grubbs, R.H. *J. Am. Chem. Soc.* **1988**, *110*, 2973.

13 (a) Schrock, R.R. *Angew. Chem. Int. Ed.* **2006**, *45*, 3748; (b) Grubbs, R.H. *Angew. Chem. Int. Ed.* **2006**, *45*, 3760.

14 Gorman, C.B., Ginsburg, E.J., Sailor, M.J., Moore, J.S., Jozefiak, T.H., Lewis, N.S., Grubbs, R.H., Marder, S.R., Perry, J.W. *Synth. Met.* **1991**, *41*, 1033.

15 Marks, R.N., Halls, J.J.M., Bradley, D.D.C., Friend, R.H., Holmes, A.B. *J. Phys. Condens. Matter* **1994**, *6*, 1379.

16 (a) Halls, J.J.M., Walsh, C.A., Greenham, N.C., Masseglia, E.A., Friend, R.H., Marrati, S.C., Holmes, A.B. *Nature* **1995**, *376*, 498; (b) Grandstrom, M., Petritsch, K., Arias, A.C., Lux, A., Anderson, M.R., Friend, R.H. *Nature* **1998**, *395*, 257.

17 Chen, S.-A., Fang, Y., Chen, S.-A., Fang, Y. *Synth. Met.* **1993**, *60*, 215.

18 (a) Wolfe, J.P., Wagaw, S., Marcox, J.-F., Buchwald, S.L. *Acc. Chem. Res.* **1998**, *31*, 805; (b) Hartwig, J.F. *Angew. Chem., Int. Ed.* **1998**, *37*, 2046.

19 (a) Kanbara, T., Nakadani, Y., Hasegawa, K. *Polym. J.* **1999**, *31*, 206; (b) Spetseris, N., Ward, R.E., Meyer, T.Y. *Macromolecules* **1998**, *31*, 3158; (c) Goodson, F.E., Hauck, S.I., Hartwig, J.F. *J. Am. Chem. Soc.* **1999**, *121*, 7527.

20 (a) Liao, Y.H., Angelopoulos, M., Levon, K. *J. Polym. Sci., Polym. Chem. Ed.* **1995**, *33*, 2725; (b) Cao, Y., Smith, P., Heeger, A.J. *Synth. Met.* **1992**, *48*, 91; (c) Tzou, K., Gregory, R.V. *Synth. Met.* **1993**, *53*, 365; (d) Yue, J., Wang, Z.H., Cromack, K.R., Epstein, A.J., MacDiarmid, A.G. *J. Am. Chem. Soc.* **1991**, *113*, 2665.

21 (a) DeArmitt, C., Armes, S.P., Winter, J., Uribe, F.A., Gottesdeld, S., Mombourquette, C. *Polymer* **1993**, *34*, 158; (b) Nguyen, M.T., Kasai, P., Miller, J.L., Diaz, A.F. *Macromolecules* **1994**, *27*, 3625.

22 Angelopoulos, M., Patel, N., Shaw, J.M. *Mater. Res. Soc. Symp. Proc.* **1994**, *328*, 173.

23 (a) Law, S.L., Mucha, S., Banks, S. *Elect. Pack. Prod.* **1991**, *31*, 82; (b) Brown, L., Burns, D. *Elect. Pack. Prod.* **1990**, *30*, 50; (c) Jillie, D. *Semicond. Int.* **1993**, *16*, 120.

24 (a) Garnier, F., Horowitz, G., Peng, X., Fichou, D. *Adv. Mater.* **1990**, *2*, 392; (b) Burroughes, J.H., Bradley, D.D.C., Brown, A.R., Marks, R.N., Mackey, K., Friend, R.H., Burns, P.L., Holmes, A.B. *Nature* **1990**, *347*, 539.

25 Maier, G. *Prog. Polym. Sci.* **2001**, *26*, 3.

26 Sroog, C.E. *Prog. Polym. Sci.* **1991**, *16*, 561.

27 (a) Sasaki, S., Nishi, S. in *Polyimides: Fundamentals and Applications*, Ghosh, M.K., Mittal, K.L. (Eds), Marcel Dekker, New York, **1996**; (b) Auman, B.C. in *Polyimide Science and Technology*, Feger, C., Khojasteh, M.M., Htoo, S. (Eds), Technomic, Lancaster, PA, **1993**.

28 (a) Hu, C.-K., Small, M.B., Kaufman, F., Pearson, D.J. *Mater. Res. Soc. Symp. Proc.* **1990**, *225*, 369; (b) Chen, S.T. *Mater. Res. Soc. Symp. Proc.* **1995**, *381*, 141.

29 Kirchhoff, R.A., Bruza, K.J. *Adv. Polym. Sci.* **1994**, *117*, 1.

30 Yang, G.-R., Zhao, Y.-P., Neirynck, J.M., Murarka, S.P., Gutman, R.J. *Mater. Res. Soc. Symp. Proc.* **1997**, *476*, 161.

31 (a) Mills, M.E., Townsend, P., Castillo, D., Martin, S., Achen, A. *Microelect. Eng.* **1997**, *33*, 327; (b) Hendricks, N.H. *Mater. Res. Soc. Symp. Proc.* **1997**, *443*, 3.

32 (a) Hacker, N.P., Davis, G., Figge, L., Krajewski, T., Lefferts, S., Nedbal, J., Spear, R. *Mater. Res. Soc. Symp. Proc.* **1997**, *476*, 25; (b) Bremmer, J.N., Liu, Y., Gruszynski, K.G., Dall, F.C. *Mater. Res. Soc. Symp. Proc.* **1997**, *476*, 37; (b) Kim, S.M., Yoon, D.Y., Nguyen, C.V., Han, J., Jaffe, R.L. *Mater. Res. Soc. Symp. Proc.* **1998**, *511*, 39; (c) Többen, D., Weigand, P., Shapiro, M.J., Cohen, S.A. *Mater. Res. Soc. Symp. Proc.* **1997**, *443*, 195.

33 Martin, S.J., Godschalx, J.P., Mills, M.E., Shaffer, E.O., Townsend, P.H. *Adv. Mater.* **2000**, *12*, 1769.

34 Jones, R.R., Bergman, R.G. *J. Am. Chem. Soc.* **1972**, *94*, 660.

35 Brandrup, J., Immergut, E.H. (Eds) *Polymer Handbook*, 3rd edn, Wiley, New York, **1989**.

36 (a) Hedrick, J.L., Labadie, J.W., Russell, T.P., Hofer, D., Wakharkar, V. *Polymer* **1993**, *34*, 4717; (b) Hedrick, J.L., Carter, K.R., Labadie, J.W., Miller, R.D., Volksen, W., Hawker, C.J., Yoon, D.Y., Russell, T.P., McGrath, J.E., Briber, R.M. *Adv. Polym. Sci.* **1998**, *141*, 1; (c) Hedrick, J.L., Russel, T.P., Sanchez, M., DiPietro, R., Swanson, S., Meccerreyes, M., Jerome, R. *Macromolecules* **1996**, *29*, 3642.

37 Thompson, L.F., Willson, C.G., Bowden, M.J. (Eds) *Introduction to Microlithography. ACS Professional Reference Book*, American Chemical Society, Washington, DC, **1994**.

38 (a) MacDonald, S.A., Willson, C.G., Fréchet, J.M. *Acc. Chem. Res.* **1994**, *27*, 151; (b) Reiser, A. *Photoreactive Polymers: the Science and Technology of Resists*, Wiley, New York, **1989**; (c) MacDonald, E., Iwayanagi, T. (Eds.). *Polymers in Microlithography: Materials and Processes. ACS Symposium Series*, American Chemical Society, Washington, DC, **1989**.

39 (a) Stewart, M.D., Patterson, K., Somervell, M.H., Willson, C.G. *J. Phys. Org. Chem.* **2000**, *13*, 767; (b) Ito, H. *J. Polym. Sci., Part A: Polym. Chem.* **2003**, *41*, 3863; (c) Willson, C.G., Trinque, B.C. *J. Photopolym. Sci. Technol.* **2003**, *16*, 621.

40 Moreau, W.M. (Ed) *Semiconductor Lithography*, Plenum Press, New York, **1988**.

41 Shaw, J.M. in *Imaging for Microfabrication, Imaging Processes and Materials*, Sturge, J., Walworth, V., Shepp, A. (Eds) Van Nostrand Reinhold, New York, **1989**.

42 Kubena, R. L. *Mater. Res. Soc. Symp. Proc.* **1993**, *279*, 567.
43 (a) Todokoro, Y., Takasu, Y., Ohkuma, T. *Proc. SPIE* **1985**, *587*, 179; (b) Kakuchi, M., Hikito, M., Sugita, A., Onose, K., Tamamura, T. *J. Electrochem. Soc.* **1986**, *133*, 1755.
44 (a) Huang, W. S. *Polymer* **1994**, *35*, 4057; (b) Angelopoulos, M., Shaw, J. M., Kaplan, R. D., Perreault, S. *J. Vac. Sci. Technol. B* **1989**, *7*, 1519; (c) MacDiarmid, A. G., Chaing, J. C., Richter, A. F., Epstein, A. J. *Synth. Met.* **1987**, *18*, 285.
45 MacDiarmid, A. G., Chaing, J. C., Richter, A. F., Somasiri, N. L. D., Epstein, A. J. in *Conducting Polymers*, Alcacer, L. (Ed) Reidel, Dordrecht, **1985**.
46 (a) Yue, J., Epstein, A. J. *J. Am. Chem. Soc.* **1990**, *112*, 2800; (b) Liao, Y. H., Angelopoulos, M., Levon, K. *J. Polym. Sci., Polym. Chem. Ed.* **1995**, *33*, 2725.
47 (a) Cao, Y., Smith, P., Heeger, A. J. *Synth. Met.* **1992**, *48*, 91; (b) Angelopoulos, M., Patel, N., Saraf, R. *Synth. Met.* **1993**, *55*, 1552.
48 (a) Hoffnagle, J. A., Hinsberg, W. D., Sanchez, M., Houle, F. A. *J. Vac. Sci. Technol. B* **1999**, *17*, 3306; (b) Switkes, M., Kunz, R. R., Rothschild, M., Sinta, R. F., Yeung, M., Baek, S. Y. *J. Vac. Sci. Technol. B* **2003**, *21*, 2794; (c) Mulkens, J., Flagello, D. G., Streefkerk, B., Graeupner, P. *J. Microlith. Microfab. Microsyst.* **2004**, *3*, 104; (d) Owa, S., Nagasaka, H., Ishii, Y., Hirakawa, O., Yamamoto, T. *Solid State Technol.* **2004**, *47*, 43.
49 Savile Bradbury, B. B. *Introduction to Light Microscopy*, BIOS Scientific, Oxford, **1998**.
50 (a) Cai, S. X., Kanskar, M., Nabity, J. C., Keana, J. F. W., Wybourne, M. N. *J. Vac. Sci. Technol. B* **1992**, *10*, 2589; (b) Abdou, M. S. A., Diaz-Guijada, G. A., Arroyo, M. I., Holdcroft, S. *Chem. Mater.* **1991**, *3*, 1003.
51 Lowe, J., Holdcroft, S. *Macromolecules* **1995**, *28*, 4608.
52 Goethals, A. M., Bisschop, P. D., Hermans, J., Jonckheere, R., Van Roey, F., Van den Heuvel, D., Eliat, A., Ronse, K. *J. Photopolym. Sci. Technol.* **2003**, *16*, 549.
53 Brainard, R. L., Cobb, J., Cutler, C. A. *J. Photopolym. Sci. Technol.* **2003**, *16*, 401.
54 (a) Li, Y., Ota, K., Murakami, K. *J. Vac. Sci. Technol. B* **2003**, *21*, 127; (b) Stulen, R. *Proc. Electrochem. Soc.* **1997**, *97-3*, 515; (c) Larruquert, J. I., Keski-Kuha, R. A. M. *Appl. Opt.* **2002**, *41*, 5398.
55 (a) Chou, S. Y., Krauss, P. R., Renstrom, R. J. *Science* **1996**, *272*, 85; (b) Bratton, D., Yang, D., Dai, J. Y., Ober, C. K. *Polym. Adv. Technol.* **2006**, *17*, 94.

8
Applications of Controlled Macromolecular Architectures to Lithography

Daniel Bratton, Ramakrishnan Ayothi, Nelson Felix, and Christopher K. Ober

8.1
Introduction

The term "lithography" simply means "marking on stone". From its origins in the early 19th century, lithography has evolved into a broad array of techniques that allow the transfer of nano-sized features of arbitrary complexity from a single master to a number of identical reproductions. This powerful technique is already well established as the backbone of the microelectronics industry and is now a pivotal technique in the nanotechnology revolution.

In this chapter, we explore several forms of lithography and its application to polymer science. In one sense, we explore a "top-down" approach, examining the applicability of soft lithography (or specialized nanofabrication) to the production of patterned macromolecular architectures. In a "bottom-up" approach, we also show how the ability to control the structure and functionality of polymers has helped advance the field of photolithography since its inception and describe novel alternatives such as molecular glass resists. Lastly, we examine the properties of block copolymers that permit novel nano-applications enabled through lithography, such as porous films for bioseparation.

The technique of lithography owes much to the production and exploration of different macromolecular architectures and we hope that this chapter will both enlighten and entertain the interested reader.

8.2
Specialized Nanofabrication

8.2.1
Definition and Scope

Specialized nanofabrication refers to a series of techniques used to produce patterns or structures with wide-ranging applications in fields as diverse as from

Macromolecular Engineering. Precise Synthesis, Materials Properties, Applications.
Edited by K. Matyjaszewski, Y. Gnanou, and L. Leibler
Copyright © 2007 WILEY-VCH Verlag GmbH & Co. KGaA, Weinheim
ISBN: 978-3-527-31446-1

biotechnology to microelectronics. In traditional lithographic techniques, radiation is used to transfer a pattern on to a substrate coated with a suitable resist. In specialized nanofabrication, physical contact between a stamp, mold or a probe and a substrate is used to transfer a pattern.

There are two distinct forms of specialized nanofabrication. The first is soft lithography, which is an umbrella term describing the patterning of suitable substrates using stamps or molds in a manner analogous to a printing press [1]. The second is scanning probe lithography (SPL), which uses a probe to manipulate matter on a substrate either by transferring a chemical "ink" or by selectively removing features from an existing layer [2]. The importance of such techniques is growing, as illustrated by the rapid expansion of this field in the past decade. Compared with traditional lithography, specialized nanofabrication requires comparatively simple, inexpensive equipment and often employs milder conditions which allow the direct patterning of delicate species, e.g. biomolecules [3–5]. However, it is less commercially advanced and some techniques, especially certain forms of SPL, may not represent practical commercially exploitable technologies.

Our goal in this section is to summarize recent key work in specialized nanofabrication relating to the patterning and manipulation of macromolecules. For this reason, we restrict ourselves to microcontact printing (µCP), scanning probe lithography (SPL) and particle replication in non-wetting templates (PRINT). As embossing or molding techniques usually involve the patterning of pre-synthesized homopolymers rather than controlled macromolecular structures, they are not covered here and the interested reader is referred to a number of recent excellent reviews for more comprehensive treatment [1, 6].

8.2.2
Microcontact Printing (µCP)

First popularized by Whitesides and coworkers [7], µCP uses an elastomeric stamp to transfer molecules ("ink") on to an appropriate substrate. The ink binds to the substrate either by chemisorption or physisorption in the regions of contact between the stamp and substrate. The pattern is therefore determined by the topography of the stamp. Polydimethylsiloxane (PDMS) is usually used as a stamp material, owing to its high chemical stability and low surface energy [8]. Stamps can be constructed with features of arbitrary complexity through the use of "masters" created by traditional lithography. Typically, a substrate is coated with a photoresist and then patterned. Etching is used to transfer the pattern into the hard substrate. A polymerizable mixture, typically employing disubstituted acrylic or methacrylic PDMS oligomers, is poured on to the mold and cured to form the soft stamp, which is then peeled off the substrate (Fig. 8.1). The stamp is "inked" by soaking it in a solution of the desired ink, resulting in physisorption onto the stamp surface. The stamp is then brought into contact with another substrate, resulting in transfer of the physisorbed ink to the new substrate.

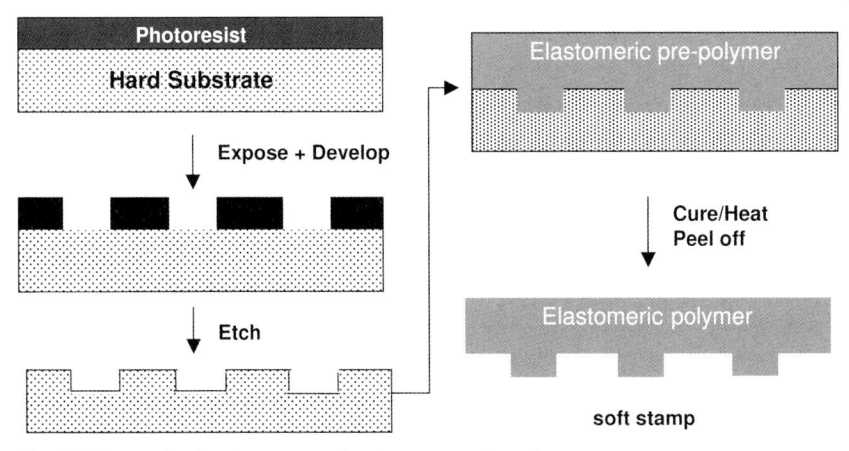

Fig. 8.1 Schematic showing process for the construction of stamps for microcontact printing (µCP). A pattern is first etched on a suitable substrate, then an appropriate polymerizable mixture is poured over the mold and cured. The resulting stamp is peeled off the substrate prior to use.

The application of µCP to controlled macromolecular architectures is perhaps best illustrated through the "bottom-up" patterning of polymer brushes. These materials are macromolecules bound at one end to a substrate such as silica or gold, although in principle any suitably functionalized substrate can be used as an anchor to form a brush [9]. Surface-initiated polymerization (SIP) from bound initiating molecules remains the most versatile method of controlling brush thickness and density. Polymer brushes permit the formation of functionalized surfaces and hence have been used for a range of applications from biotechnology to colloid stabilization. In addition to the "bottom-up" approach offered by SIPs, µCP has also been used to pattern macromolecules in a "top-down" manner, by inking the stamp with macromolecules then applying them to a suitable substrate. These approaches are summarized in Fig. 8.2.

Husemann et al. were the first to employ µCP for SIP [10]. In this approach, an inert thiol molecule, $CH_3(CH_2)_{15}SH$, is used as an ink to form a pattern on gold-coated silica. A functionalized thiol, $HO(C_2H_4O)_2(CH_2)_{11}SH$, is then used to backfill the unpatterned areas on the gold layer. By employing the free hydroxyl groups as initiators for the ring-opening polymerization of ε-caprolactone, polymer brushes were grown in periodic arrays from the gold substrates.

In principle, any controlled polymerization can be used for SIP, although in practice the control and versatility offered by atom transfer free radical polymerization (ATRP) has resulted in this becoming one of the most popular methods for creating smart brushes. Braun and coworkers [11] used the "backfilling" technique to functionalize silicon wafers with an ATRP initiator, after which brushes of poly(N-isopropylacrylamide) were grown (Fig. 8.3). Hamelinck et al. [12] described the synthesis of liquid crystalline polymer brushes. The bromine

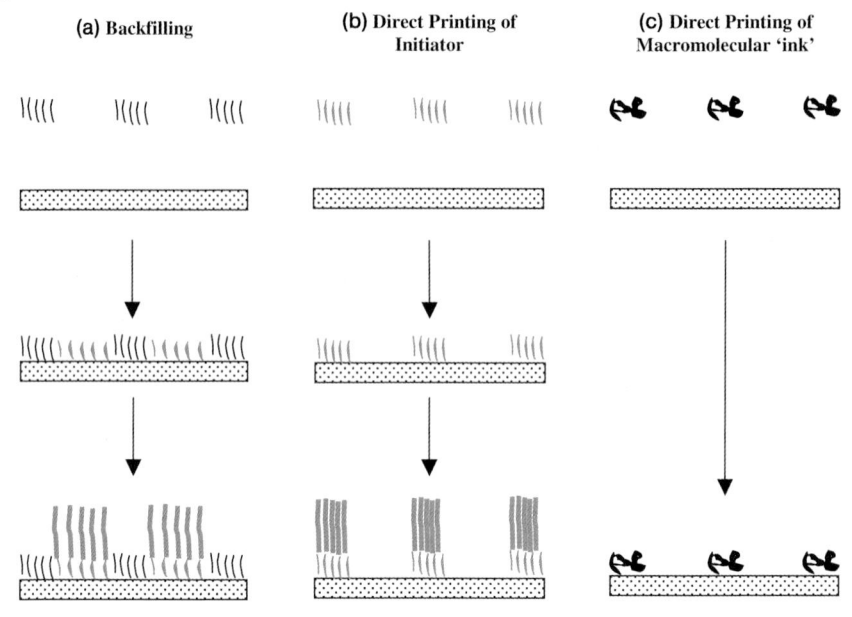

Fig. 8.2 Three approaches to patterning macromolecules using μCP. In (a) an inert "ink" is transferred to a substrate, then the initiator molecules are backfilled into the remaining free reactive sites [10]. Surface-initiated polymerization (SIP) is used to produce a patterned brush. Part (b) shows a more direct technique, whereby initiator molecules are used as an ink and directly transferred to the surface [13], and (c) shows direct printing with a macromolecular ink that results in a patterned substrate [14].

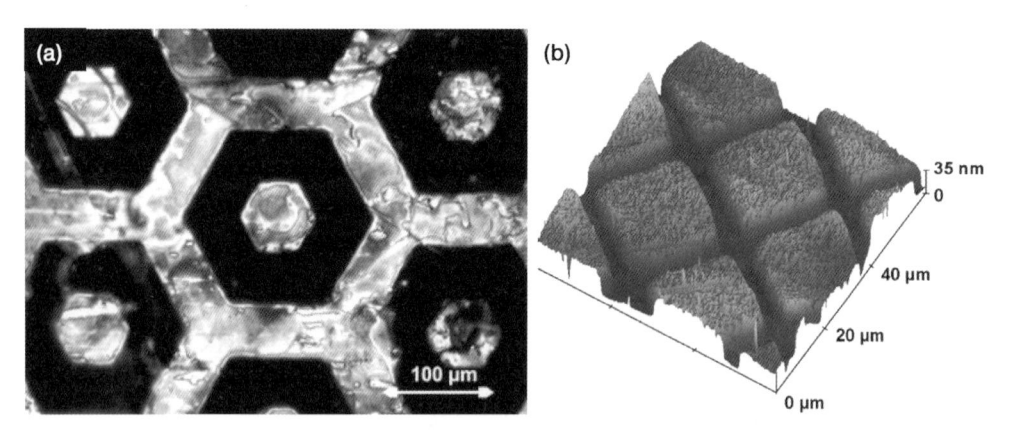

Fig. 8.3 Polymer brushes patterned by microcontact printing. (a) 200-μm wide hexagons of homeotropically aligned liquid crystalline brushes synthesized by the direct printing of an ATRP initiator on to a substrate [12]. Reprinted from P. J. Hamelinck, W. T. S. Huck, Journal of Materials Chemistry **2005**, *15*, 381, with permission from the Royal Society of Chemistry. (b) AFM image of polymer brush grown by ATRP after backfilling unpatterned regions of the substrate [11]. Reprinted with permission from H. Tu, C. E. Heitzman, P. V. Braun, Langmuir **2004**, *20*, 8313. Copyright 2004, American Chemical Society.

or chlorine functionalized initiator was directly patterned using μCP on to glass or silicon (rather than backfilled), then used to initiate the ATRP of acrylate-functionalized mesogens to form regular patterns of homeotropically aligned liquid crystalline brushes.

For efficient pattern transfer using μCP, it is imperative that the "ink" adheres or bonds strongly to the substrate to prevent diffusion and loss of image integrity. For this reason, most studies employ silica as a substrate, due to the variety of species with which it can form covalent bonds. Comparatively few studies deal with the direct initiation of polymer brushes from a polymer substrate. Farhan and Huck [13] modified poly(ethylene terephthalate) (PET) and poly(ethylene naphthalate) (PEN) using plasma oxidation to obtain free hydroxyl groups on the surface. An ATRP initiator could then covalently bind to these otherwise inert materials to allow the growth of patterned brushes. It was noted, however, that the brush thickness was not as well controlled as when silica was used as a substrate, due to higher surface roughness and lower density of initiating molecules on the surface.

In addition to the "bottom-up" approach offered by SIPs, μCP has also been used to produce patterned macromolecules in a "top-down" manner, by inking the stamp with macromolecules then applying them to a suitable substrate. For example, Arrington et al. [14] reported the μCP of a primary amine-terminated polyamidoamine dendrimer (PAMAM) on a silicon oxide substrate. The electrostatic interactions between the ink and substrate proved strong enough to resist sonification. The authors noted an empirical correlation between the thickness of the transferred dendrimer film and the concentration of the ink solution, i.e. more concentrated solutions led to thicker transferred patterns, as measured using atomic force microscopy (AFM). The printing of cross-linkable poly(amidoamine organosilicon) on to glass, silicon and polyelectrolyte layers for organoelectronic applications has also been described [15]. Thomas and coworkers reported the formation of covalently attached macromolecular polymer brushes by employing a reactive diblock copolymer as an ink [16]. In this case, a rod–coil diblock copolymer consisting of styrene and 3-(triethoxysilyl)propyl isocyanate was found to attach covalently to a silica surface by the covalent binding of the silyl groups. Han and coworkers [17] grafted carboxylic acid-terminated poly(2-vinylpyridine) on to a polystyrene brush grown from a silicon substrate, resulting in a patterned mixed brush system that exhibited some interesting "switching" properties when immersed in different solvents, allowing the wettability of the surface to be tuned.

Kern and coworkers reported μCP on silicon oxide of a monolayer of CdS–dendrimer nanocomposites [18]. Once again, the concentration of the ink solution was found to be a critical factor in the formation of a stable pattern. For example, a highly concentrated ink leads to the formation of multilayered structures rather than monolayers. Reinhoudt and coworkers [19] described the applications of μCP to the manufacture of molecular printboards. These were formed by the printing of functionalized calixarene molecules, dendritic wedges and dendrimers on to a self-assembled layer of β-cyclodextrin on gold or silver oxide surfaces. The printed molecules were then used to direct the absorption of other materials, such as fluorescent dyes. These examples clearly illustrate the ability of μCP to permit the formation of smart surfaces for future technology applications.

For the most part, μCP is used to produce micron-sized features, rather than the sub–50-nm structures often encountered in advanced lithographic techniques, e.g. E-beam lithography. However, Huck and coworkers recently introduced a new technique, known as nanocontact printing (nCP) [20]. The low elastic modulus of PDMS can result in pattern deformation upon contact with a substrate, making the production of sub-100-nm structures difficult. In this case, a novel double layer-stamp was constructed from a soft layer of PDMS, which mimics the topography of the master, attached to a layer of harder, higher molecular weight PDMS. The improved mechanical properties of the stamp, combined with the use of a high molecular weight dendrimer ink to prevent diffusion, allowed the large-scale parallel patterning of a silicon substrate with 40-nm features.

8.2.3
Scanning Probe Lithography (SPL)

SPL embraces a suite of techniques which use the interaction between a probe tip and a substrate to form a pattern. Typical examples of the instrumental techniques used include atomic force microscopy (AFM) and scanning tunneling microscopy (STM). SPL can form patterns either by selectively depositing or manipulating material on a surface (e.g. dip pen lithography, DPN) [21] or removing material from the surface (nanoshaving) (Fig. 8.4).

In addition to physical manipulation, SPL can be used to modify a surface chemically, e.g. by selectively oxidizing the surface using an electric field. The

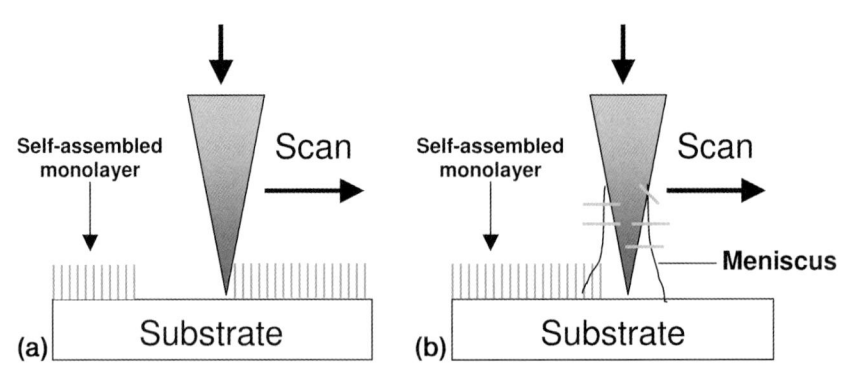

Fig. 8.4 Two applications of scanning probe lithography (SPL) to the patterning of substrates. (a) Nanoshaving, where an AFM tip is used to remove selected areas of self-assembled monomers. (b) Dip pen lithography (DPN), where an AFM tip is used to deposit an "ink" on to a substrate.

resolution offered by such techniques is far in advance of other forms of nanofabrication, e.g. photolithography or soft lithography, offering the potential to manipulate individual atoms or molecules into patterns of arbitrary complexity. SPL has produced some of the most instantly recognizable images in nanofabrication, including the famous IBM logo written in xenon atoms on the surface of a single crystal of nickel [22]. However, the inability of SPL to pattern wide areas in a parallel manner has so far limited its acceptance by industry, although recent efforts to design a parallel version capable of simultaneous patterning may change this in the near future [2].

Fréchet and coworkers have employed dendrimer-based monolayers as resists for SPL [23]. Poly(benzyl ether) dendrimers functionalized with monochlorosilanes were allowed to form a covalently attached monolayer on the surface of a silicon wafer. An AFM tip was then used to apply a voltage to selective regions of the monolayer, resulting in oxidation of the organic layer and the silicon wafer to produce raised features of silicon oxide. This surface could then be selectively etched using hydrofluoric acid (HF) to produce positive tone features. A similar dendrimer functionalized with carboxylic acid side-groups was bound to an amine-functionalized silicon wafer and was found to act as a resist when patterned in an identical manner [24]. Dendrimers can also be employed as positive (by nanoshaving) and negative tone resists (by oxidative scanning) to pattern titanium films [25]. Direct patterning of dendrimers has also been achieved by DPN [26]. Amine-functionalized dendrimers were employed as the "ink" to produce 300-nm lines on silicon and glass substrates. A similar technique was used to pattern amine-functionalized PAMAM dendrimers by covalently attaching them to a reactive thiol monolayer adsorbed to a gold substrate [27].

SPL has also been employed in the nanoscale patterning of polymer brushes (Fig. 8.5). Liu et al. reported the site-specific ring-opening metathesis polymerization (ROMP) by first directly applying thiol-functionalized norbornene molecules on a gold substrate using DPN, then growing brushes by surface-initiated ROMP in conjunction with a palladium catalyst [28]. Nanoshaving has been used to expose selectively a gold substrate through an inert thiol monolayer. Backfilling the exposed substrate with a thiol-functionalized initiator allowed the growth of poly(N-isopropylacrylamide) brushes through ATRP [29, 30].

The patterning of conducting polymers by SPL has been widely reported. Lim and Mirkin used DPN to pattern polypyrrole and poly(anilinesulfonic acid) onto modified positively or negatively charged silicon substrates [31]. Electrostatic interactions between the surface and these water-soluble polymers helped promote adhesion. The smallest feature size obtained was 130 nm, larger than that usually obtained with small-molecule inks and the authors speculated that this was due to rapid diffusion of the macromolecules in the aqueous ink. An alternative technique to direct transfer has been reported by Jang et al. [32]. In this process, a conducting AFM tip is used to apply a voltage to a spin-coated terthiophene polymer. The film is selectively oxidized and cross-linked, converting it from an insulator into a conducting material. This solid oxidative cross-linking resulted in conducting polymer line widths of 120 nm.

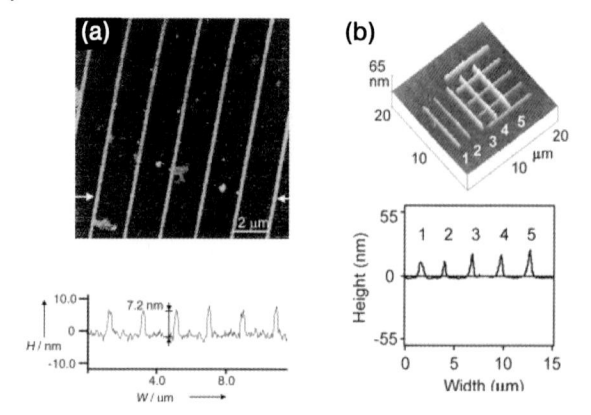

Fig. 8.5 Applications of dip pen lithography to polymer brushes. (a) AFM image of polymer brush grown by ROMP after direct deposition of initiator by DPN [28]. Reprinted from X. Liu, S. Guo, C. A. Mirkin, *Angewandte Chemie, International Edition* **2003**, *42*, 4785, with permission of John Wiley & Sons, Inc. (b) AFM image of polymer brush grown after 'nanoshaving' of inert thiol monolayer followed by backfilling with ATRP initiator [30]. Reprinted with permission from M. Kaholek, W.-K. Lee, B. LaMattina, K. C. Caster, S. Zauscher, *Nano Letters* **2004**, *4*, 373. Copyright 2004, American Chemical Society.

Scanning probe contact lithography (SPCL) is an interesting technique representing a hybrid between μCP and DPN [33]. The authors constructed an AFM tip from PDMS and used it in a similar fashion to a μCP stamp to produce a pattern of an inert thiol monolayer on a gold substrate. The resolution was comparatively low for an SPL technique (images were in the micron rather than nanometer region) but a recent publication reported far higher resolution using a PDMS-coated commercially available Si_3N_4 AFM tip [34]. Using this stamp and an ink composed of PAMAM dendrimers, the authors were able to produce regular 70-nm lines of patterned macromolecules on a silicon substrate.

8.2.4
Particle Replication in Non-wetting Templates (PRINT)

In this section, we do not describe either imprint or step-and-flash imprint lithography (SFIL), as these methods have been reviewed elsewhere [35, 36]. However, PRINT is an interesting extension to imprint lithography recently demonstrated by DeSimone and coworkers, with possible applications in the formation of nanoparticles consisting of controlled macromolecular architectures. The technique depends on the use of cross-linked perfluorinated polyethers (PFPEs) rather than PDMS as a master template. Although PDMS has a variety of desirable characteristics as a material for templates, including high chemical tolerance and low surface energy [8], it suffers from swelling in certain organic solvents, resulting in a reduction in feature integrity. This limits the usefulness of PDMS for the construction of microfluidic devices, for example.

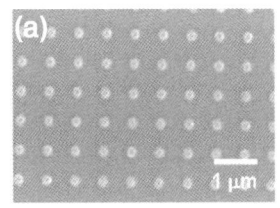

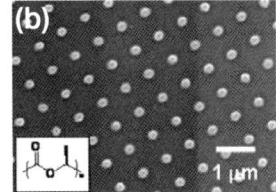

Fig. 8.6 Particle replication in non wetting templates. (a) SEM image of the PFPE master used to generate discrete 200 nm particles of poly(lactic acid), as shown in (b) [39]. Reprinted with permission from J. P. Rolland, B. W. Maynor, L. E. Euliss, A. E. Exner, G. M. Denison, J. M. DeSimone, *Journal of the American Chemical Society* **2005**, *127*, 10096. Copyright 2005, American Chemical Society.

PFPEs, however, have many of the same desirable characteristics, including high chemical tolerance, without the tendency to swell in organic solvents [37, 38]. In addition, PFPEs are non-wetting and it is this feature that PRINT [39] relies upon to produce discrete, isolated structures rather than embossed films. No scum layer is formed in the process, which makes the collection of the structures for further applications extremely facile.

DeSimone and coworkers [39] have used the technique to produce a wide variety of well-defined polymer nanoparticles synthesized using a variety of different chemistries, including thermally initiated ring-opening polymerization, UV-initiated free radical polymerization and acid-catalyzed oxidative coupling (Fig. 8.6). In addition, it was discovered that the inherently mild nature of the process allowed the resulting nanoparticles to be loaded with various biologically active molecules (although presumably this depends on the chemistry used to form the nanoparticles).

PRINT is clearly a versatile soft lithographic tool with a great deal of potential for the production of nanoparticles for medical use and further research will surely produce alternative applications as the technology matures.

8.3
Macromolecules as Resists for Photolithography

8.3.1
Introduction

Despite the recent advances in soft lithography, most processes (μCP, step and flash, etc.) still depend on conventional lithography to generate the initial master or mask. In the semiconductor industry, the ability to manufacture millions of small transistors and devices less than one hundredth the width of a human hair on a single silicon substrate is vital. With each year that passes, the industry finds ways to shrink these devices even further and thus achieve higher device densities on a single chip. This shrinking in size, in accordance with

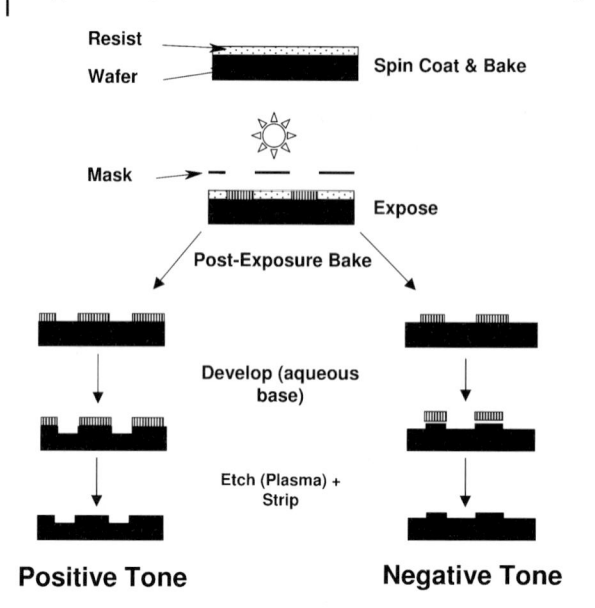

Fig. 8.7 The process of photolithography. The resist is first spin-coated on to a silicon wafer, then baked to remove residual solvent. Exposure results in a solubility switch in the exposed region, rendering it soluble (positive tone) or less soluble (negative tone) in the developer. The pattern is transferred into the substrate by etching and the residual resist is cleaned from the wafer.

Moore's law [40], has contributed to increases in efficiency and demand and thus has fueled the growth of the semiconductor industry.

Photolithography is the method of choice in the manufacture of these small devices. By using UV light and a mask, arbitrary patterns can be made in a photosensitive polymer film. This pattern can then be transferred to the underlying silicon substrate, usually by reactive ion etching, as illustrated in Fig. 8.7.

Primarily, these photosensitive films are comprised of polymers that have specific structures and functionalities to make them an effective photoresist. These photoresist films must have a high enough glass transition temperature (T_g) to withstand any annealing process (typically ~100 °C) and they must also undergo some type of solubility switch upon exposure to UV energy. Typically, the resolution of the smallest features readily obtained is governed by the wavelength of light used during patterning. Therefore, different functionalities can be built into the macromolecule to ensure transparency at the wavelength of radiation in question.

Importantly, a photoresist must be resistant to the anisotropic plasma etch processes that transfer the pattern down to the silicon substrate (Fig. 8.7). This etch resistance is determined by the structure of the photoresist. Two parameters have been found empirically to describe the etch resistance of a specific polymer, namely the Ohnishi parameter [41] and the Ring parameter [42]. The former states that the number of carbon atoms must be maximized while the number of oxygen atoms is minimized in a resist composition. The latter predicts that more cyclic structures will impart a higher etch resistance to the photoresist. These restrictions tend to limit the types of polymers and functionalities that can be used in resist materials, but impressive results with alternative designs have been seen nonetheless.

8.3.2
Deep-UV (248-nm) Resists

The first important development that allowed the patterning of sub-micron features was the development of diazonaphthoquinones (DNQs) as dissolution inhibitors in Novolac-based photoresists [43]. In the unexposed state, DNQ acts as a dissolution inhibitor [44, 45], preventing the free hydroxyl groups in the Novolac resin from being dissolved in the alkaline aqueous development process. However, upon UV exposure, DNQ undergoes a Wolff rearrangement [46], converting the DNQ into a polar molecule and allowing the exposed areas to be dissolved more quickly than the unexposed areas of the photoresist film (Fig. 8.8). This photochemistry provided an excellent positive-tone resist platform for patterning with the 436- and 365-nm UV exposure technologies of the time.

However, as the demand for smaller device features continued, the industry required a shift to 248-nm UV patterning (KrF laser). The previous DNQ–Novolac system was not sensitive enough at this wavelength and resulted in sloping features because the resist matrix absorbed the UV energy too strongly. This necessitated the need for a different type of photoresist that was more sensitive and transparent. The solution relied on the use of photoacid generators (PAGs) in the resist matrix.

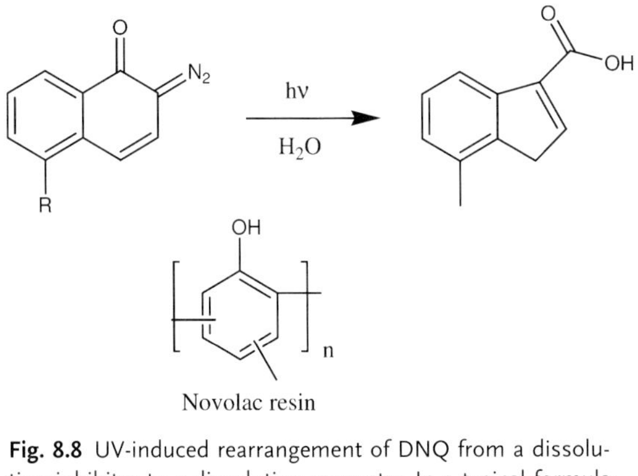

Fig. 8.8 UV-induced rearrangement of DNQ from a dissolution inhibitor to a dissolution promoter. In a typical formulation, DNQ is blended within a matrix of base soluble Novolac resin.

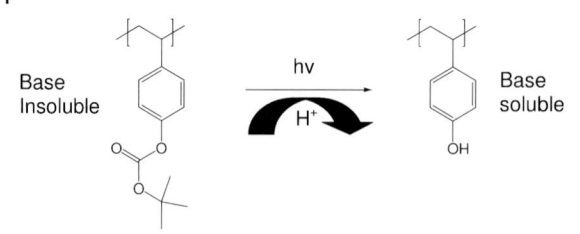

Fig. 8.9 The process of chemical amplification, shown here with tert-butoxycarbonyl-protected polyhydroxystyrene. Exposure to radiation of an appropriate wavelength activates a photoacid generator, releasing catalytic amounts of proton. These diffuse through the resist and induce the solubility switch in the exposed regions of the resist, forming the desired pattern.

Such PAGs generate a proton upon absorption of a photon over a range of UV wavelengths, which goes on to catalyze acid–based reactions in the photoresist film [47]. The PAGs used in photoresist films are primarily based on onium salts [48] and many different structures exist based on the wavelength and performance needed [49–51]. For the most part, acid production mechanisms are complex [49], although the fact that one proton can go on to catalyze hundreds of subsequent reactions has resulted in these resists being termed chemically amplified (CA) resists.

The prefered polymer system for CA resists has been based on low molecular weight (~10 000) polyhydroxystyrene (PHOST) copolymers. They have a structure similar to Novolac resins and the aromatic rings impart plasma etch resistance. The breakthrough was in the use of a reactive acid-labile protecting group, tert-butoxycarbonate group (tBoc) [47]. In this way, the exposed regions of a photoresist film are deprotected and become more soluble in aqueous base. This results in a positive tone resist system where the solubility contrast between the exposed and unexposed regions is very high. This process is illustrated in Fig. 8.9.

However, solubility contrast is not the only concern when designing a photoresist. The photoresist film must also adhere well to the underlying substrate, avoid swelling in insoluble regions and create robust standing features. Because of this, many polymeric CA photoresists are actually copolymers comprised of different monomers with different purposes. These are typified by the IBM APEX, ESCAP and KRS resists [52, 53]. In the APEX-type resist, the PHOST is only partially protected, resulting in a polymer that is more hydrophilic with better adhesion than its fully protected version. The ESCAP-based resists rely on copolymerizing PHOST with tert-butyl acrylate (Fig. 8.10). The deprotected acrylate is more acidic than the deprotected PHOST and therefore imparts better solubility. The KRS-type resists are very similar to the APEX resists, but rely on the less reactive tert-butoxy protecting group [54].

Although the details can vary widely between resist platforms, the critical performance themes remain the same. The right balance must be struck between

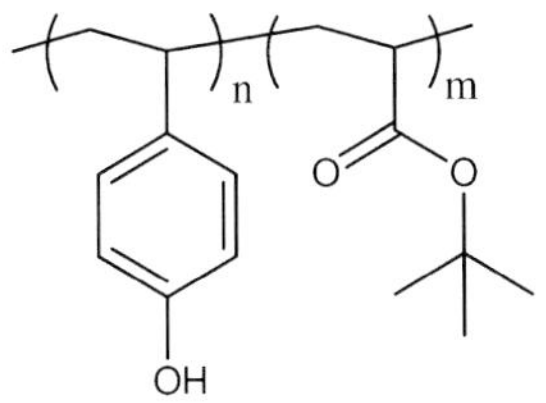

Fig. 8.10 ESCAP-based resist platform featuring an acid-labile group on the acrylate.

solubility contrast, adhesion, swelling and etch resistance. Strategies have included more reactive protecting groups such as tetrahydropyranyl groups [55] and those incorporating acetal bonds [56, 57]. Others have also looked at acid-catalyzed depolymerizations of acetal-based systems [57] and polycarbonates [50]. All are instructive as to the wide variety of chemically amplified chemistries available for resist materials, although the PHOST-based materials have been the most pervasive and features smaller than 200 nm in width have been achieved.

8.3.3
193-nm Resists

In keeping with the progression of Moore's law, features smaller than 150 nm eventually were demanded, which is smaller than can be achieved with 248-nm irradiation. This necessitated the transition to ArF (193-nm) excimer sources. However, since aromatic polymers such as PHOST are highly absorbing at 193 nm, a different polymer platform is needed in order to create suitable photoresists. Acrylate-based polymers systems are comparatively transparent at this wavelength and some of the first patterns were demonstrated with a terpolymer of methyl methacrylate, *tert*-butyl methacrylate and methacrylic acid [58]. The *tert*-butyl methacrylate unit provided an acid-labile group for chemical amplification, while the methacrylic acid group improved solubility and the methyl methacrylate provided mechanical robustness.

Good patterns can be achieved with simple acrylic photoresists, but plasma etch resistance is poor. It was found that if alicyclic groups were attached, etch resistance improved dramatically [59, 60]. These groups include adamantane, lactone, cyclohexane and norbornene groups attached to the polymer. Since alicyclic groups lack conjugation, they are more transparent than aromatic groups at 193 nm while offering similar etch resistance. Etch resistance can also be improved by adding silicon to the polymer. This is achieved by using silsesquioxane-based photoresist platforms, for example, with traditional acid-labile and solubility-enhancing groups attached [61, 62].

Photolithography at 193 nm also has its eventual resolution limits, but these limits are currently being extended by the use of immersion lithography. This involves placing a liquid medium between the final lens of the exposure system and the photoresist layer [63]. Since this liquid has a higher refractive index than air, it allows for a larger depth of focus and process window for higher resolution. Water is the current immersion fluid of choice, which means that photoresists targeted for this technology must be water-resistant and suffer minimal leaching of components into the water. Acrylic systems have shown excellent stability under water and component leaching is alleviated through the use of top barrier coatings [43, 50, 61].

8.3.4
157-nm Resists

Reducing the exposure wavelength from 193 to 157 nm (F_2 laser) can further reduce the achievable feature size resolution. At this shorter wavelength, absorbance issues pose a greater challenge. Fortunately, fluorinated materials have been found to be highly transparent at this wavelength [42]. This allows structures similar to those of 193-nm resists to be used as 157-nm resists, even though conjugated bonds such as the carbonyl bonds in acrylates absorb highly at this wavelength [61]. This is exemplified by the use of the fluorinated acrylate, 2-[4-(2-hydroxyhexafluoroisopropyl)cyclohexane]hexafluoroisopropyl acrylate [64] (Fig. 8.11).

The large volume of fluorine in a photoresist containing this monomer reduces absorption while increasing the acidity of the hydroxyl group and the cyclohexane moiety improves etch resistance. Resolution down to 80 nm is achievable with this system copolymerized with tetrahydropyranyl methacrylate.

The etch resistance of these systems can be improved by introducing more alicyclic groups and can be achieved by using photoresist polymers with a polynorbornene backbone and pendent acid-labile moieties. Also, cyclized poly-

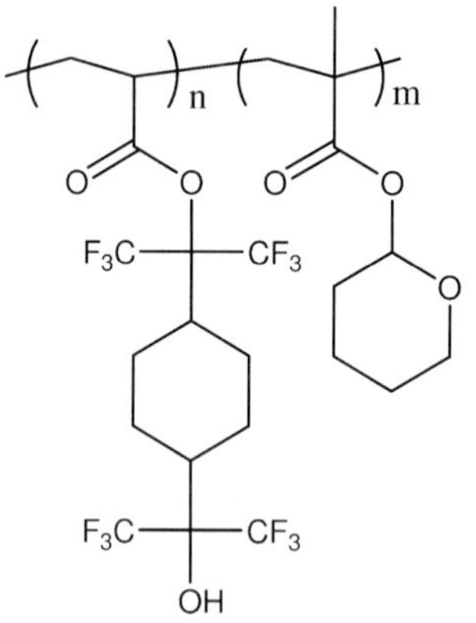

Fig. 8.11 Photoresist targeted for 157-nm transparency. Poly{2-[4-(2-hydroxyhexafluoroisopropyl)cyclohexane]hexafluoroisopropyl acrylate-*co*-tetrahydropyranyl methacrylate}.

dienes increase the glass transition temperature while improving etch resistance. Adamantyl [65] and tetrahydropyranyl protecting groups can also be used to increase etch resistance. Aside from these basic systems, siloxane and silsesquioxane polymers have been found to be highly transparent at 157 nm [42] while offering superior oxygen plasma etch resistance. Although changes in industry strategy now make it unlikely that 157-nm photoresists will be used, many interesting polymers have been developed in this effort and several structures may be used in 193-nm resists.

8.3.5
Extreme UV (13.4-nm) Resists

The ultimate advanced technique optical lithography is extreme ultraviolet (EUV) lithography, which actually operates at 13.4 nm, in the range of soft X-rays [66–69]. This technique is expected to produce feature sizes in the sub-50-nm regime due to the extremely short wavelength of the EUV radiation. However, because of the very short exposure wavelength, almost any material absorbs energy to some degree and energy absorption is no longer dictated by atomic bonding; rather, it is dictated by atomic composition [70]. To date, EUV sources are somewhat weak and every spare photon must be harnessed to maximize the imaging capability of this undoubtedly powerful technique [71–73]. Oxygen and fluorine absorb strongly at this wavelength, which makes designing resists for EUV lithography something of a challenge, given that today's chemically amplified resist systems usually consist of a high percentage of these elements. For example, both Novolac resin and poly(methyl methacrylate) absorb strongly [74–76], which can result in poor side-wall profiles. It follows that resists for EUV lithography should be transparent at this wavelength and have high sensitivity in order to compensate for the weak radiation sources. In addition, line edge roughness (LER), that is, defects in image uniformity, becomes critical at sub-50-nm feature sizes; hence resists must provide low LER values [69].

Carbon, hydrogen, boron and silicon are all elements with high EUV transparency; hence the interest in the design of resists rich in these elements. For example, polysilane-based resists have been employed at this exposure wavelength [77, 78], as have boron-containing polymers [79, 80]. PHOST- [81] and norbornene-based resists [82] have been used with some success, with the high carbon/oxygen ratio rendering them comparatively transparent at this wavelength. Possible solutions to strongly absorbing resist systems include the use of thin resist films, employing strategies such as top-surface imaging [83] or bilayer resist systems [84]. Completely oxygen-free resists have been reported, including poly(trimethylsilylstyrene-*co*-chloromethylstyrene) [79], which acts as a negative tone resist. However, despite promising results, there are no commercially available resists which yet meet all of these criteria for a dedicated resist for EUV lithography, although recent research into novel resists has shown excellent progress (see Section 8.3.7).

8.3.6
Dendrimers and Hyperbranched Polymers as Photoresists

Extending the range of macromolecular architectures investigated for photoresists, suitably functionalized dendrimers and hyperbranched polymers have been employed previously as photoresists. Tully and coworkers reported the synthesis of poly(benzyl ether) and poly(benzyl ester) protected at the periphery with *t*Boc chemically amplifiable groups [85, 86]. Patterns were generated using both 248-nm UV and electron beam radiation, resulting in the generation of sub-100-nm images. It was suggested that the high functional group density at the periphery of the dendrimer (or hyperbranched) molecules together with their unique geometries minimizes chain entanglement and therefore LER. LER is defined as a deviation from a "perfect" fitted edge of a feature produced by photolithography and this parameter becomes increasingly important at sub-50-nm feature sizes [69]. Comparison of a hyperbranched resist with commercially available, linear polymeric resists concluded that these non-linear resists showed comparable sensitivity and resolution to conventional resists and resulted in images with lower LER [87]. Other chemically amplifiable examples have subsequently been reported, using dendrimers based on poly(o-hydroxyamides) [88] and hyperbranched polyesters [56].

Ueda and coworkers have reported dendrimers based on polypropylenimine as alkali-soluble resists, which proved patternable using 365-nm UV radiation in conjunction with diazonaphthoquinone, a photoactive dissolution inhibitor [89]. Photosensitive hyperbranched polyimides have been reported to act as negative tone resists through cross-linking [90].

8.3.7
Molecular Glass (MG) Photoresists

MG photoresists (MGPs) represent a relatively new area of research for next-generation lithography and may one day replace the more traditional polymeric resists. MGPs are small, discrete organic molecules which exist in the form of stable amorphous glasses at room temperature [91]. Typically, small molecules tend to crystallize, but this can be avoided via the incorporation of bulky groups, non-planar structures or a variety of conformers into the molecular design of new materials. MGs are in a state of thermodynamic non-equilibrium (much like amorphous polymers) and show T_gs rather than crystalline melting points.

MGs have been previously investigated for applications such as in molecular electronics [91], but chemically amplifiable groups and etch-resistant moieties can be readily incorporated into MGs to enhance their utility as photoresists. In particular, the small size of these materials compared with amorphous polymeric resists is thought to lead to high-resolution images with lower LER [92].

The synthesis and lithographic characterization of a variety of MGPs for E-beam lithography have been reported by Shirota and coworkers. A suitably functionalized trisubstituted aromatic core has proved especially successful and

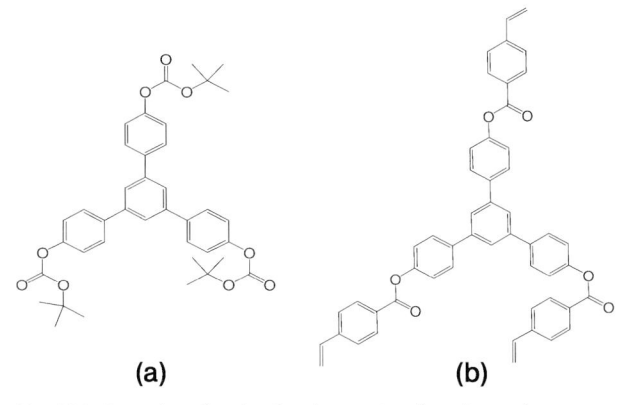

Fig. 8.12 Examples of molecular glass resists for advanced lithography. (a) BCMTPB, positive tone chemically amplified resist [94] and (b) p-VCTPB (m- and o-substituted analogues also reported) negative tone resists [92].

some examples of this species are shown in Fig. 8.12. TsOTPB, a positive tone resist, and AISTPA, a negative tone resist, allowed the formation of 150- and 70-nm lines, respectively [93]. The reported sensitivity was approximately 3 mC cm^{-2} at 50 keV. The VCTPB class of compounds which functioned as negative tone resists proved more sensitive (50 µC cm^{-2} at 50 keV), giving comparable feature sizes [92]. As may be expected, chemically amplifiable MGPs such as BCMTPB and analogues proved considerably more sensitive when exposed in the presence of a photoacid generator [93, 94], with sensitivities <5 µC cm^{-2}. Feature sizes of around 40 nm were obtained with these materials.

Calixarenes are a class of molecules with analogous structures to PHOST or Novolac resins and their naturally irregular structures has made them especially versatile cores for MG photoresists. Matsui and coworkers reported that p-alkylphenol calix[6]arenes function as negative tone resists for E-beam lithography, albeit with low sensitivity, requiring a dose of 800 µC cm^{-2} to form an image [41]. Calix[4]resorcinarenes were employed as a negative tone resist for 365-nm lithography using a polyfunctional benzylic alcohol as a cross-linker in an acid-catalyzed process [95]. P-Chloromethyl-substituted calix[4]arenes also functioned as negative tone E-beam resists in a similar mechanism to that observed with poly(chloromethylstyrene) [96]. Calix[8]arenes were monosubstituted with bulky side-groups such as either toluene-p-sulfonyl chloride to render them amorphous, then blended with DNQ dissolution inhibitors to produce a positive tone image using 248-nm radiation [97]. Ober and coworkers have focused on the design of chemically amplifiable MGPs for EUV and E-beam lithography [15, 98]. Recently, the synthesis and lithographic characterization of chemically amplifiable calix[4]resorcinarene derivatives as positive tone resist for EUV lithography was reported [99]. Images with feature sizes as low as 30 nm with low LER were obtained after exposure and development with aqueous base (Fig. 8.13). In addi-

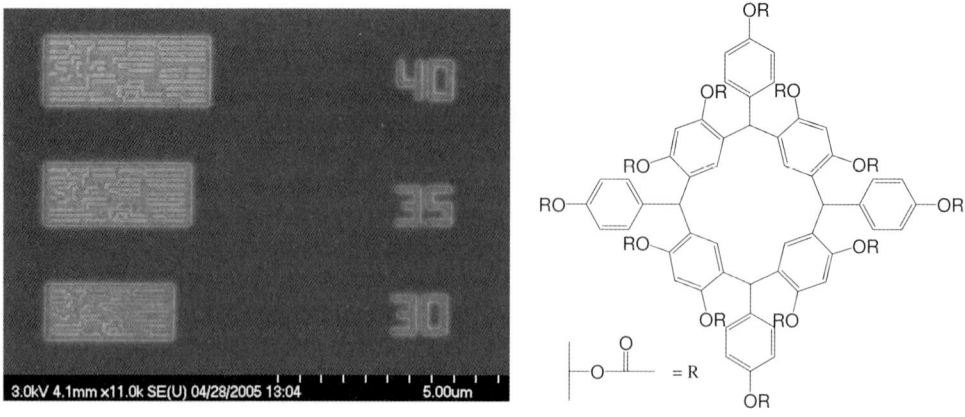

Fig. 8.13 Example of a positive tone molecular glass resist for EUV lithography. 4-Hydroxycalix[4]resorcinarene protected with *t*Boc was successfully employed as a positive tone resist, resulting in 30-nm feature sizes [99].

tion, a suitable cross-linker such as Powderlink™ could be blended with other small crystalline molecules together with a PAG to form an amorphous photoresist formulation. An acid-catalyzed cross-linking reaction gave negative tone images with feature sizes of approximately 50 nm [99].

Several MGPs designed with saturated aliphatic functionalities (rather than aromatic groups) designed for 193-nm lithography have been reported. Sooriyakumaran et al. described the synthesis and lithographic characterization of a resist based on polyhedral oligomeric silsesquioxane (POSS) [100]. POSS is best described as a cage of silicon and oxygen and the reported molecule was functionalized with chemically amplifiable bulky side-groups to raise the naturally low melting point of POSS. Using this resist, the authors reported the production of 120-nm feature sizes using 193-nm lithography with LER comparable to commercially available resists. This resist is also expected to show excellent dry etch resistance thanks to the silicon in the backbone. In addition, an adamantyl core functionalized with cholic acid was reported to have a T_g of 90 °C and allowed the formation of 70-nm patterns using 193-nm radiation [101].

8.3.8
Supercritical CO₂ Processing

It has been reported that to process one silicon wafer in a typical semiconductor fabrication line, more than 1 L of solvent and 1 L of rinse water are sent to waste [102]. The use of any replacement solvent would be highly desirable and supercritical carbon dioxide (scCO$_2$) as a replacement solvent would make the lithography process more environmentally benign. In addition, a supercritical fluid is a single-phase fluid with no phase boundary that eliminates surface tension when drying, thus permitting the retention of fine, closely packed, high aspect-ratio features. This fact has been demonstrated by supercritical drying after aqueous base development [103, 104].

Hence development and drying steps can be effectively combined if the photoresist is developable in $scCO_2$. This can be achieved if typical photoresist monomers are copolymerized with fluorinated methacrylates to create CO_2-soluble photoresists. Polymers containing these fluorinated compounds have been shown to be appreciably soluble in $scCO_2$ and submicron patterns have been developed using this technology [61, 105, 106].

The incorporation of fluorinated moieties can be avoided if cosolvents are added to the supercritical fluid. This is particularly necessary if resists are being processed for extremely short wavelength lithography where fluorine absorbs strongly. It has been shown that adding small amounts of a cosolvent to $scCO_2$ can drastically change its solvating power and thereby allow a wider range of photoresists to be developed in CO_2. Films of poly(glycidyl methacrylate) have been patterned and developed in $scCO_2$ with 2% acetone added and features on the order of 300 nm have been demonstrated [107].

Another way to eliminate the need for fluorination is to re-envision the type of resist used. Due to their small size compared with polymeric resists, molecular glasses have a greater tendency to be soluble in $scCO_2$. Solubility is enhanced by the existence of conjugated groups on the molecule and solubility can be achieved without the need for fluorinated moieties. Using this philosophy, it has been shown that a broad range of calixarene-type resists are soluble in $scCO_2$ [108]. Ober et al. have also demonstrated that a *t*Boc-protected hexa(hydroxyphenyl)benzene resist is soluble in $scCO_2$, with 50-nm resolution shown [109]. This result demonstrates the adaptability of MGs and shows the potential of supercritical processing.

8.4
Block Copolymers

8.4.1
Introduction

Block copolymers are a distinct form of macromolecular architecture which consists of two or more chemically distinct, covalently attached polymer blocks. Microphase separation-induced nanometer-scale structure formation can occur when incompatible blocks are combined together. The use of block copolymer microdomains as lithographic templates is at the early stages of competing with commercial top-down conventional lithographic techniques and is emerging as an alternative route for generating nanoscale patterns. Self-assembly of block copolymers in thin films, along with some kind of physical or chemical assistance, allows one to generate patterns from 100 to 10 nm in a comparatively economical manner compared with the advanced lithographic techniques discussed in the previous sections. In addition to using block copolymers as litho-

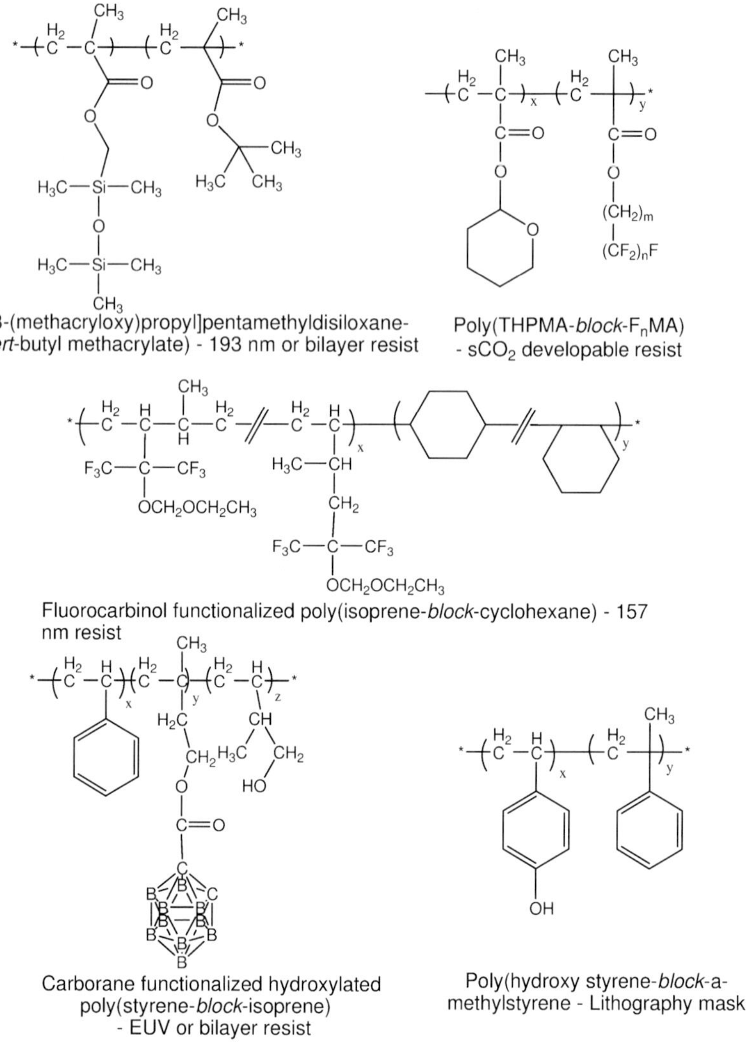

Fig. 8.14 Polymers with controlled architecture studied for use as nano-templates and resists

Poly([3-(methacryloxy)propyl]pentamethyldisiloxane-block-tert-butyl methacrylate) - 193 nm or bilayer resist

Poly(THPMA-block-F_nMA) - sCO_2 developable resist

Fluorocarbinol functionalized poly(isoprene-block-cyclohexane) - 157 nm resist

Carborane functionalized hydroxylated poly(styrene-block-isoprene) - EUV or bilayer resist

Poly(hydroxy styrene-block-a-methylstyrene - Lithography mask

graphic masks, they offer other advantages in terms of resist chemistry. The tendency of block copolymers to migrate to interfaces and their micellization behavior can improve adhesion, the miscibility of various additives and overall resist performance, including resolution and contrast. A detailed description of general aspects of block copolymer science and engineering, block copolymer self-assembly in thin films and block copolymer lithography is detailed elsewhere [110]. This section will mainly focus on recent advances involving the use of block copolymers as a means to generate nanoscale structures and as novel photoresist or resist additives in their own right [111, 112]. The structures of block copolymers designed, synthesized and evaluated as lithographic masks and resists are illustrated in Fig. 8.14.

8.4.2
Engineering Considerations for Using Block Copolymers as Lithographic Masks

In order for the convergence of block copolymer lithography with classical top-down lithographic processes truly to take place, block copolymer resists must be compatible with steps such as spin coating, etch processing and thin-film characterization [113]. A major challenge is to find common industrial lithographic solvents for all block copolymers, since the solvent used in film formation also influences microdomain formation. In general, high-boiling solvents such as toluene and propylene glycol methyl ether acetate (PGMEA) have been used for thin-film formation [114–117]. Block copolymers, like all resists, are expected to form defect-free films over large areas (e.g. an 8-inch wafer) on different substrates including silicon, silicon nitride and other semiconductor industry materials [118]. The thickness of the film (d) can be adjusted by varying the spin speed (R) and solution concentration (C) as both parameters influence the film thickness according to the equation $d \propto C/R^{1/2}$. The composition and architecture of block copolymers play an important role in transferring nanoscale structure to an underlying substrate, hence the importance of controlled polymerization during synthesis. To enhance wet and dry etch resistance, selective metal decoration of one block, block copolymers with a metal containing block or photochemical cross-linking have been used [116, 117, 119].

8.4.3
Self-assembly in Bulk Versus Thin Film

Block copolymer self-assembly is an alternative method of achieving a desired pattern by striking a balance between chemistry and thermodynamics. Block copolymers microphase separate to tens of nanometers depending on the Flory–Huggins interaction parameters between the two blocks and their degree of polymerization. The resulting morphology can be tuned by changing the volume fraction of the components, keeping the interaction parameters constant. Lamellae, cylinders and spheres are the commonly encountered morphologies even though the microstructure can be complicated by the number of blocks and the connectivity between the blocks. The thermodynamics of these structures have been extensively studied in bulk [120, 121].

In thin films, microdomain formation is influenced by the interfacial energies and the geometric constraints introduced by the interfaces. The influence of the interface dominates when the film thickness approaches the microdomain spacing. Commonly encountered problems include submerging of microdomains inside the films, no microdomain formation during thin-film formation and the formation of a different morphology observed in the thin film compared with that in bulk [122, 123]. In general, self-assembly is limited to a few periodic forms of high symmetry such as spheres or cylinders, but some degree of intelligent guidance can be used to achieve preferred structures and long-range ordering depending on the application. Three different types of mechanisms are generally used for orienting block copolymer thin films [119]: (1) controlling the topography of the substrate and thus the film thickness, (2) chemical modification of the surface to change the substrate–polymer interactions and (3) applying external stimuli (electric, thermal, eutectic solidification, crystallization, solvent evaporation, etc.) to induce long-range ordering.

8.4.4
Block Copolymers as Lithographic Resists

Block copolymers are useful in many applications where two or possibly three different polymers are connected together to yield a material with hybrid properties. Materials designed for use as a resist for advanced lithography are expected to meet diverse requirements posed by the particular approach. In addition to resolving features smaller than 100 nm, the resist must have tolerance to both dry and wet etches, develop in environmentally friendly solvents and have uniform adhesion to the substrate, high contrast and high sensitivity to the radiation. The use of block copolymers as a resist material offers the possibility of combining two distinct resist functions such as imaging and etch resistance using separate phases in a single material. Such separation of function may offer benefits in terms of dimensional stability, etch behavior and development characteristics [111].

8.4.5
Block Copolymer Thin-film Characterization

As we move to smaller dimensions, our ability to make small structures is in some cases outstripping our ability to characterize them. Transmission electron microscopy (TEM) has been widely used to study the structure of block copolymer in bulk, but its uses for thin-film characterization has been limited by the demanding sample preparation. In contrast, scanning electron microscopy (SEM) and atomic force microscopy (AFM) are two imaging techniques [119] widely used to analyze block copolymer thin films because of their flexibility with sample preparation and substrate compatibility. Low-voltage SEM is widely used for imaging block copolymer microstructures in thin films. Microdomains that present surface topography, such as those formed by poly(styrene-*block*-methyl methacrylate) [poly(S-*b*-PMMA)], were readily imaged by SEM with standard topography-enhancing operating procedures. However, microdomains of block copolymer systems that are often submerged beneath a surface wetting layer require additional steps such as metal decoration, tuned operating voltage and etch methods to enhance contrast. AFM generates contrast by sensing topographic features or variations in mechanical properties. The contrast obtained by AFM is influenced by both topographic and tip–polymer interactions. The issues include the difficulty involved in decoupling the contrast-influencing parameter and the damage to the AFM tip caused by inexperienced operators. The flexibility in sample preparation and mild operating conditions have made tapping mode AFM a powerful tool for examining the surface of block copolymer films [117].

Small-angle X-ray scattering under grazing incidence (GISAXS) has received much attention in recent times for characterizing the surface of block copolymer thin films [124]. Unlike transmission SAXS measurements, the sampling volume in GISAXS increases because the penetration of the X-ray photon into the substrate is limited when the angle of incidence is close to the critical angle. GISAXS helps to map lateral structure of polymer thin films on a mesoscopic length scale of up to ~100 nm. The intensity obtained in a GISAXS image is a result of the electron density difference between the two phases. Therefore, this technique will prove useful for characterizing nanostructures created in a thin film by monitoring the scattering intensity difference before and after removal of the processable block.

8.4.6
Synthesis of Block Copolymers by Direct Polymerization

Block copolymers with predetermined molecular weight and narrow molecular weight distribution are important for achieving self-assembly with a high degree of long-range order. Living anionic polymerization [125] has been widely used to synthesize polymers for self-assembly because of its excellent control over architecture and molecular weight. Several block copolymers synthesized by anionic polymerization for use as potential lithography masks include poly(styrene-*block*-isoprene), poly(S-*b*-MMA), poly(styrene-*block*-2-vinylpyridine) and poly(styrene-*block*-ferrocinyldimethylsilane). Poly(α-methylstyrene-*block*-4-*tert*-butoxystyrene) [poly(α-MS-*b*-tBuOS)] used in studies of block copolymer lithography was synthesized by living anionic polymerization [117]. Controlled radical polymerization (CRP) [126–128] has emerged as a viable alternative method to produce block copolymers that phase separate to form periodic structures with good order. Block copolymers that do not exhibit microphase separation can still be used as photoresists and can be synthesized using various controlled polymerization techniques including group transfer polymerization (GTP) [105].

8.4.7
Block Copolymers Functionalized by Post-polymerization Modification

Post-polymerization modification is one of the more versatile methods used in polymer synthesis to obtain polymers with varying functionality, which in turn improve the physical or chemical properties of the resulting block copolymer. This method can suffer from incomplete reactions and low yield, although certain reactions such as hydrolysis and hydrogenation proceed by quantitative conversion and high yield. These reactions have been successfully used in block copolymer synthesis and the resulting block copolymers were utilized successfully in microdomain formation. For example, hydrogenation of the double bonds in isoprene in a block copolymer results in a new block copolymer with improved chemical and thermal stability [129]. Similarly, the poly(α-methylstyrene-*block*-4-hydroxystyrene) [poly(α-MS-*b*-HOST)] copolymer used to study photoresist chemistry in block copolymers was synthesized by hydrolysis of poly(α-MS-*b*-*t*BuOS) [117].

Post-polymerization modification allows one to incorporate selectively certain elements into the block copolymer, which in turn improves transparency at a particular wavelength of radiation, etch resistance or solubility. For example, the introduction of boron to a block copolymer improves both etch resistance and transparency at EUV wavelengths [130]. Similarly, modification of block copolymers with hexafluoro-2-propanol groups increases its transparency at 157 nm [131] or selective introduction of fluorinated groups in the block copolymer allows a resist to be developed in scCO$_2$ [105].

8.4.8
Block Copolymers as Templates for Advanced Lithography

The optimal utilization of nanoscopic patterns for fabrication requires spatial and orientational control of block copolymer microdomains. The facile fabrication of large-area periodic functional structures with block copolymers is an important field in its own right. In many applications, such as multifunctional on-chip bioseparation, simple periodic patterning is insufficient and spatial control of the microdomains is necessary to form integrated multilevel structures. A photochemistry-driven process for the fabrication of spatially controlled nanoporous polymer films was developed with a new diblock copolymer–small molecule additive system [117, 132]. The process makes use of a combination of conventional "top-down" photolithography with a block copolymer-assisted "bottom-up" strategy. Poly(α-MS-*b*-HOST) was designed and synthesized with PHOST block functioning as a high-resolution negative-tone photoresist (by cross-linking) and the α-MS block was selectively removed by photodegradation (Fig. 8.15). With a straightforward spin-coating process, a perpendicular cylindrical orientation of the removable block α-MS microdomain was induced over a wide range of film thicknesses on different substrates. A final irradiation step under vacuum produced uniformly sized pores, having a diameter of 18 nm in the patterned area without any collapse, as shown in an AFM height image in Fig. 8.15d. Thus, uniform nanometer-sized pores in submicron-sized photopatterns were generated by merging "top-down" photolithographic imaging with subsequent selective removal of a "bottom-up" self-assembled nanostructure. This represents a novel nanofabrication process for obtaining spatially controlled nanopores. To examine the internal thin film structure on macroscopic length scales and to demonstrate that the AFM images are representative throughout the whole thickness of the film, GISAXS was employed to probe

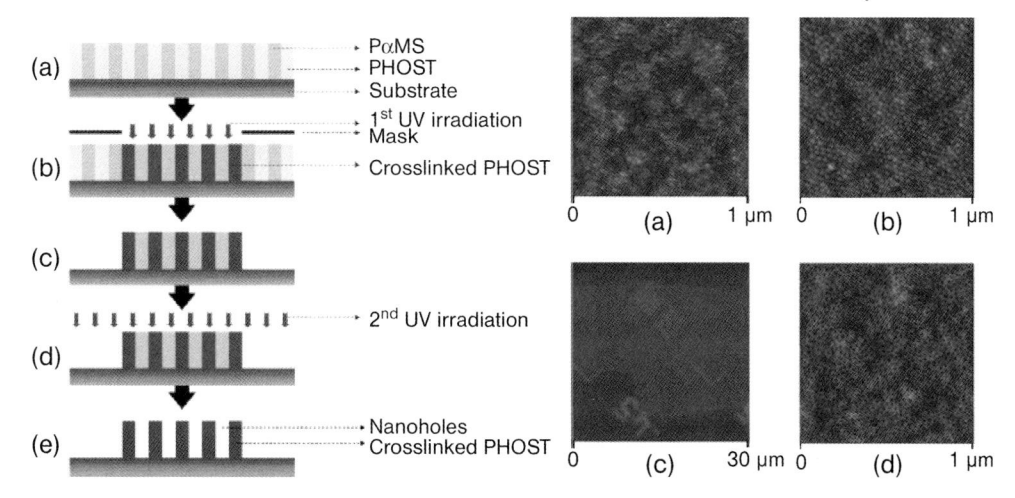

Fig. 8.15 Novel nanofabrication process for obtaining spatially controlled nanopores. (Left) (a) spin-coat poly(α-MS-b-PHOST)–cross-linker–PAG mixture on to a silicon wafer to form vertical cylinders of Pα-MS in the PHOST matrix, (b) irradiate using a 248-nm stepper with a photomask and bake, (c) develop with a mixed solvent to form micron-sized patterns on top of substrate, (d) irradiate using a 365-nm lamp under vacuum and (e) form patterns with nanoporous channels. AFM height images obtained from thin films of Pα-MS-b-PHOST on a silicon wafer. (Right) (a) after spin coating from PGMEA without annealing; (b) after development, the microstructure and its orientation were retained; (c) patterns with feature size as small as 450 nm were obtained after development; (d) further UV irradiation generated nanometer-sized pores on the patterned region [117]. Reprinted with permission from M. Li, K. Douki, K. Goto, X. Li, C. Coenjarts, D. M. Smilgies, C. K. Ober, *Chemistry of Materials* **2004**, *16*, 3800. Copyright 2004, American Chemical Society.

the orientation of the nanostructures in the thin films. This is a technique very well suited to the analysis of self-assembled thin films.

Figure 8.16a shows a two-dimensional GISAXS image taken from a thin film (152 nm thick) before UV removal of the Pα-MS block. The vertical streaks in the GISAXS pattern indicate that the Pα-MS cylinders orientate perpendicular to the substrate with the inter-cylinder distance determined to be 33.5 nm. With complete removal under vacuum of Pα-MS during a second UV irradiation step, vertical streaks with a much stronger scattering intensity appear at the same q_x value in the GISAXS image (Fig. 8.16b). This indicates that there is a much larger density contrast after treatment that was assigned to hollow nanochannels being generated in the polymer film. Moreover, the persistence of the vertical streaks shows that the nanochannels are formed without any significant collapse of the matrix polymer block after the removal of Pα-MS.

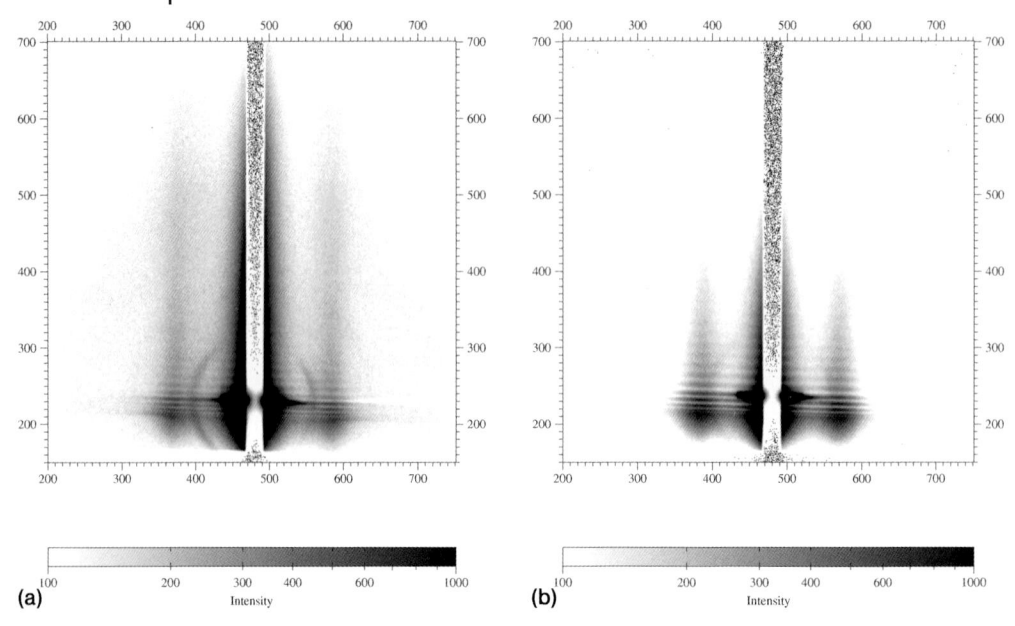

Fig. 8.16 GISAXS images obtained from a thin film of Pa-MS-b-PHOST (152 nm in thickness) on a silicon wafer. X-rays of wavelength 0.155 nm incident at 0.2075° to the film surface were used. (b) Before removal of Pa-MS block. The vertical streaks indicate that the cylinders are standing up on the silicon substrate throughout the film. (b) After removal of Pa-MS block. The dramatic increase in the scattering contrast of the streaks indicates the nanoporous nature of the thin film [117]. Reprinted with permission from M. Li, K. Douki, K. Goto, X. Li, C. Coenjarts, D. M. Smilgies, C. K. Ober, *Chemistry of Materials* **2004**, *16*, 3800. Copyright 2004, American Chemistry Society.

8.4.9
Block Copolymers as Resists for 193-, 157- and 13.4-nm Lithography

Ober et al. have prepared a variety of block and random copolymers containing monomers of *tert*-butyl methacrylate and [3-(methacryloxy)propyl]pentamethyldisiloxane, which they employed as photoresists (Fig. 8.14) [133]. The block copolymers were synthesized by the controlled technique GTP and random copolymers were prepared by conventional radical polymerization. The polymers are highly transparent at 193 nm, making them attractive candidates for ArF lithography. The presence of silicon increases the oxygen reactive ion etch resistance of the resist, making them suitable for the imageable layer of a bi-level resist system. After 193-nm exposure, the block copolymers exhibit superior development behavior in aqueous base to that of the corresponding random copolymers, despite the presence of the long hydrophobic siloxane block. A non-ideal development behavior was observed which may be due to the lower effective concentration of photogenerated acid at the surface of the film compared with the bulk of the film. Line-space patterns 1 μm wide were imaged with block co-

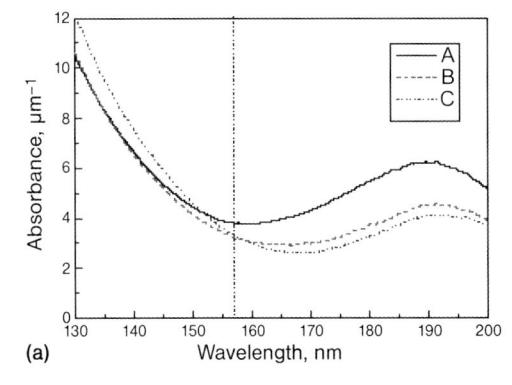

Fig. 8.17 (a) VUV spectra of precursor polymers and resist: (A) poly(IPF$_6$OH-b-CHD); (B) poly(IPF$_6$OH-b-CH); (C) EOM-protected poly(IPF$_6$OH-b-CH) (resist). (b) SEM image of 0.5–1-μm lines (dosage 30.30 mJ cm^2) EOM-protected poly(IPF$_6$OH-b-CH) [131]. Reprinted from Y.C. Bae, C.K. Ober, *Polymer Bulletin* **2004**, *52*, 321. Copyright 2004, with kind permission from Springer Science and Business Media.

polymer resist, whereas a compositionally identical random copolymer suffered from poor solubility and "pinhole" defect formation [133].

A fluorocarbinol functionalized poly(isoprene-*block*-cyclohexane) [poly(IP-b-CH)] copolymer (Fig. 8.14) was synthesized by coupling living anionic polymerization and post-polymerization modification. The final block copolymer exhibited an absorption coefficient of 3.30 μm^{-1} at 157 nm [131]. The VUV spectra of precursor polymers poly(IPF$_6$OH-b-CHD) (A), poly(IPF$_6$OH-b-CH) (B) and the final resist with a chemically amplified ethoxymethyl group (EOM) poly(IP-F$_6$OH-b-CH) (C) are shown in Fig. 8.17a. Exposure was carried out on a 248-nm stepper and feature sizes down to 0.5 μm were resolved prior to optimization of exposure and formulation conditions (Fig. 8.17b).

A boron-containing block copolymer negative-tone resist (Fig. 8.14) for EUV lithography was synthesized by coupling living anionic polymerization and polymer modification chemistry [130]. The amount of boron in the block copolymer was controlled from 0 to 100% by changing the feed ratio. The boron-containing block copolymers showed excellent transparency at EUV wavelengths, with absorption as low as 1.24 μm^{-1} for 30% functionalized block copolymer. The partially modified boron-containing block copolymer (5 mol% unreacted hydroxyl groups) was formulated with both a suitable cross-linker (TMMGU) and a photoacid generator and then imaged using a 248-nm stepper, producing patterns as small as 350-nm lines and spaces. A reactive ion etch study on a block copolymer containing 7.4 wt% boron showed higher oxygen etch resistance than a random copolymer containing 9.9 wt% boron, indicating that the blocky architecture of the resist imparts some genuine advantages in resist performance. This behavior was examined by near-edge-ray absorption fine structure (NEXAFS) and contact angle measurements. These surface-sensitive techniques indicated that the low surface energy drives the boron-containing block to the air/polymer interface and dominates the sample surface, hence the surface of the resist is richer in boron than the bulk film. The high etch resistance of this boron-containing block copolymer has high potential in bilayer resists application.

8.4.10
Block Copolymers with a Fluoroalkyl Block for Supercritical CO_2 Development

Supercritical CO_2 ($scCO_2$) technology has been attracting considerable attention as the feature size required in the microelectronics industry decreases, as discussed in Section 8.3.8. This novel developer exhibits tunable solubility parameters, zero surface tension and high diffusivities and can help mitigate pattern collapse in high aspect ratio features and improve resolution. The penetration of the developer at the substrate/film interface and the strength of anchoring of the resist to the substrate have been identified as important factors in the effectiveness of $scCO_2$ for development. Block copolymers possess the advantage of having two chemically different segments linked covalently, where each component can be tailored to segregate to specific interfaces. Therefore, a combination of block copolymers with $scCO_2$ development can improve adhesion, contrast and resolution, given the correct choice of blocks. We have synthesized series of poly(tetrahydropyranyl methacrylate-*block*-perfluoroalkyl methacrylate) [poly-(THP-*b*-FnMA)] (where $n=1, 2, 3, \ldots$) copolymers by group transfer polymerization and examined their development behavior in $scCO_2$ (Fig. 8.14) [105]. Block copolymers containing 50 wt% of a fluorinated methacrylate component (F3MA and F7MA) were found to be soluble in $scCO_2$ under mild conditions, at temperatures <80 °C and pressures <7000 psi. The polymers were lithographically evaluated using a 193-nm exposure source and were found to have excellent sensitivities (<10 mJ cm^{-2}) with the appropriate photoacid generator combinations. The fluoropolymers were found to resolve features as small as 0.2 µm without any optimization, indicating the remarkable potential of this class of polymer as photoresists combined with $scCO_2$ development.

8.4.11
Block Copolymers as Sequestering Agents and Additives

A typical commercial chemically amplified photoresist is a mixture of several components including the resist, photoacid generator, dissolution inhibitors and bases. The major problem in the design of chemically amplified photoresists for lithography has been the intricate balance required between various parameters such as sensitivity, etch resistance, hydrophobicity and adhesion and the interaction between various resist components. Any additives used in the formulation are expected to increase specific resist performance without interfering with any of the basic functions of the resist. The homogeneity of the formulation is a key issue as phase separation of the components in the thin film can impair resist performance. Block copolymers have been shown to serve as compatibilizers at interfaces, for example, by acting as stabilizers in various polymerization reactions or surfactants in detergency applications [110]. For these reasons, block copolymers have been investigated as additives in photoresists to enhance resist homogeneity and resist performance [134].

Block copolymers of poly[*tert*-butyl methacrylate-*block*-(3-methoxy)propylpentamethyldisiloxane] were studied as photoresists for sub-100-nm patterning and compared with random copolymers having the same silicon content and molecular weight. It was found that these block copolymer photoresists were more sensitive and exhibited superior development characteristics and enhanced resolution compared with equivalent random copolymers. A possible explanation for the better performance of the block copolymer is thought to be the difference in the microstructure of the block and random copolymers, which resulted in a more homogeneous formulation with less PAG aggregation. A detailed investigation of a block copolymer of poly[*tert*-butyl methacrylate-*block*-(3-methoxy)propylpentamethyldisiloxane] was compared with model copolymers of MMA and *t*BMA. Rutherford backscattering (RBS) experiments performed on these block copolymers examined PAG distribution and revealed the importance of the polarities of PAG and the polymer matrix and their mutual interactions on component mixing and phase separation [134].

The ability of block copolymers to act as surfactant-like additives was examined in a "proof-of-concept" experiment performed with a photoresist matrix of THPMA-*block*-IBMA, previously shown to possess an uneven distribution of PAG aggregation [134]. The results demonstrated that the right block copolymer additive promoted the homogeneous distribution of the PAG throughout the resist. Block copolymers were also investigated as additives by adding 2 wt% IBMA-*block*-MMA or *t*BMA-*block*-MMA copolymers to a "zero-thinning" resist formulation (to act as compatibilizers) followed by lithographic evaluation of the resulting blends. The addition of the block copolymers was found to improve greatly the resolution of the photoresist. Ion beam experiments were performed to investigate the mechanism of enhancement in resolution and it was found that the block copolymer additives prevented segregation to the wafer/polymer interface and enhanced resist performance by dispersing the photoacid generator uniformly through the depth of the resist film.

8.5
Two-photon Lithography

As a final example of the possibilities provided by lithographic methods, we briefly describe two-photon lithography. Traditional photolithography-based patterning processes are inherently two-dimensional in nature, due to their reliance on top-down UV patterning. Therefore, fabricating three-dimensional structures becomes a challenging exercise in multi-step processing. However, the use of two-photon excitation lends itself to inherent three-dimensional patterning. This process is achieved by the simultaneous absorption of two photons at the focal point of a laser, where the probability of simultaneous arrival of two photons is much higher [135]. This allows for three-dimensional, spatially resolved exposure points in a thin film. If the focal point is rastered throughout the films, three-dimensional patterns can be fabricated.

Typically infrared laser energies are used so that there is minimal incidental absorption of the laser energy elsewhere in the film. The patterning process[136, 137]. The resulting three-dimensional patterns are typically on the order of microns in size, although features as small as 350 nm have been shown [138]. Although the doses required to fabricate structures can be high, the use of photosensitive conjugated dyes improves the two-photon absorption cross-section of the film and therefore can substantially improve efficiency [136]. The inclusion of photoacid generators substantially reduces writing times [139].

Two-photon lithography lends itself to the fabrication of many varied structures. The usefulness of these structures depends on the material that is being patterned. Radical initiated cross-linking of poly(dimethylsiloxane) resins has been demonstrated [138] (Fig. 8.18), making this process very useful in the fabrication of microfluidic devices. Also, three-dimensional hydrogels have been patterned by the cross-linking of acrylamide resins, leading to the possibility of structured drug delivery devices [140, 141].

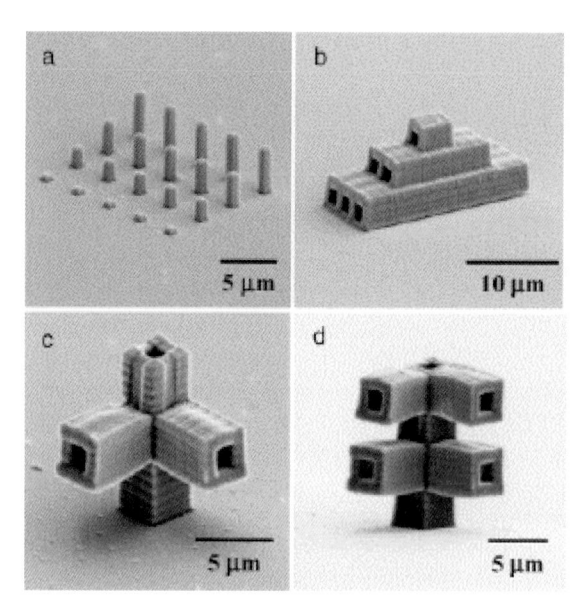

Fig. 8.18 Arbitrarily complex three-dimensional structures fabricated in poly(dimethylsiloxane) using 3D two-photon lithography [138]. Reprinted with permission from C.A. Coenjarts, C.K. Ober, Chemistry of Materials **2004**, 16, 5556. Copyright 2004, American Chemical Society.

8.6
Conclusion

The goals set by the semiconductor industry for continuing the reduction of pattern resolution are breathtaking. Plans are to achieve dimensions of less than 30 nm in just a few years. At the same time, the number of tools and therefore choices for making structures at these length scales are increasing. The result is that lithography and patterning are spreading from the semiconductor industry to many other areas of research, science and technology. For example, the ability to make three-dimensional shapes has possibilities for producing microfluidic devices of use to the biomedical community. It will be interesting to see what the future holds for patterning and the level of control that will be achieved over the next few years.

Acknowledgments

This review was made possible by research support from several organizations. The authors would like to thank the National Science Foundation for financial support, including that provided through the Cornell Center for Materials Research and the Nanobiotechnology Center. The Semiconductor Research Corporation has enabled initial and ongoing efforts in the area of lithography and patterning such as those described here. International Sematech has also supported our efforts in this area. Finally, several companies have provided crucial research support, including Intel, IBM, Air Products, Rohm and Haas Microelectronics and Praxair.

References

1. B. D. Gates, Q. Xu, J. C. Love, D. B. Wolfe, G. M. Whitesides, *Annual Review of Materials Research* **2004**, *34*, 339.
2. D. S. Ginger, H. Zhang, C. A. Mirkin, *Angewandte Chemie, International Edition* **2004**, *43*, 30.
3. R. S. Kane, S. Takayama, E. Ostuni, D. E. Ingber, G. M. Whitesides, *Biomaterials* **1999**, *20*, 2363.
4. G. M. Whitesides, E. Ostuni, S. Takayama, X. Jiang, D. E. Ingber, *Annual Review of Biomedical Engineering* **2001**, *3*, 335.
5. W. Senaratne, L. Andruzzi, C. K. Ober, *Biomacromolecules* **2005**, *6*, 2427.
6. B. D. Gates, Q. Xu, M. Stewart, D. Ryan, C. G. Willson, G. M. Whitesides, *Chemical Reviews* **2005**, *105*, 1171.
7. R. J. Jackman, J. L. Wilbur, G. M. Whitesides, *Science* **1995**, *269*, 664.
8. J. C. McDonald, G. M. Whitesides, *Accounts of Chemical Research* **2002**, *35*, 491.
9. S. G. Boyes, A. M. Granville, M. Baum, B. Akgun, B. K. Mirous, W. J. Brittain, *Polymer Brushes* **2004**, 151.
10. M. Husemann, D. Mecerreyes, C. J. Hawker, J. L. Hedrick, R. Shah, N. L. Abbott, *Angewandte Chemie, International Edition* **1999**, *38*, 647.
11. H. Tu, C. E. Heitzman, P. V. Braun, *Langmuir* **2004**, *20*, 8313.
12. P. J. Hamelinck, W. T. S. Huck, *Journal of Materials Chemistry* **2005**, *15*, 381.

13 T. Farhan, W. T. S. Huck, *European Polymer Journal* **2004**, *40*, 1599.
14 D. Arrington, M. Curry, S. C. Street, *Langmuir* **2002**, *18*, 7788.
15 N. Kohli, P. R. Dvornic, S. N. Kaganove, R. M. Worden, I. Lee, *Macromolecular Rapid Communications* **2004**, *25*, 935.
16 J.-W. Park, E. L. Thomas, *Journal of the American Chemical Society* **2002**, *124*, 514.
17 K. Yu, C. Yang, J. Fu, R. Xing, N. Zhao, Y. Han, *Surface Science* **2004**, *572*, 490.
18 X. C. Wu, A. M. Bittner, K. Kern, *Advanced Materials* **2004**, *16*, 413.
19 T. Auletta, B. Dordi, A. Mulder, A. Sartori, S. Onclin, C. M. Bruinink, M. Peter, C. A. Nijhuis, H. Beijleveld, H. Schoenherr, G. J. Vansco, A. Casnati, R. Ungaro, B. J. Ravoo, J. Huskens, D. N. Reinhoudt, *Angewandte Chemie, International Edition* **2004**, *43*, 369.
20 H.-W. Li, B. V. O. Muir, G. Fichet, W. T. S. Huck, *Langmuir* **2003**, *19*, 1963.
21 R. D. Piner, J. Zhu, F. Xu, S. Hong, C. A. Mirkin, *Science* **1999**, *283*, 661.
22 D. M. Eigler, E. K. Schweizer, *Nature* **1990**, *344*, 524.
23 D. C. Tully, K. Wilder, J. M. J. Fréchet, A. R. Trimble, C. F. Quate, *Advanced Materials* **1999**, *11*, 314.
24 D. C. Tully, A. R. Trimble, J. M. J. Fréchet, K. Wilder, C. F. Quate, *Chemistry of Materials* **1999**, *11*, 2892.
25 M. Rolandi, I. Suez, H. Dai, J. M. J. Fréchet, *Nano Letters* **2004**, *4*, 889.
26 R. McKendry, W. T. S. Huck, B. Weeks, M. Fiorini, C. Abell, T. Rayment, *Nano Letters* **2002**, *2*, 713.
27 G. H. Degenhart, B. Dordi, H. Schoenherr, G. J. Vancso, *Langmuir* **2004**, *20*, 6216.
28 X. Liu, S. Guo, C. A. Mirkin, *Angewandte Chemie, International Edition* **2003**, *42*, 4785.
29 M. Kaholek, W.-K. Lee, S.-J. Ahn, H. Ma, K. C. Caster, B. LaMattina, S. Zauscher, *Chemistry of Materials* **2004**, *16*, 3688.
30 M. Kaholek, W.-K. Lee, B. LaMattina, K. C. Caster, S. Zauscher, *Nano Letters* **2004**, *4*, 373.
31 J.-H. Lim, C. A. Mirkin, *Advanced Materials* **2002**, *14*, 1474.
32 S.-Y. Jang, M. Marquez, G. A. Sotzing, *Journal of the American Chemical Society* **2004**, *126*, 9476.
33 X. Wang, K. S. Ryu, D. A. Bullen, J. Zou, H. Zhang, C. A. Mirkin, C. Liu, *Langmuir* **2003**, *19*, 8951.
34 H. Zhang, R. Elghanian, N. A. Amro, S. Disawal, R. Eby, *Nano Letters* **2004**, *4*, 1649.
35 M. Colburn, T. Bailey, B. I. Choi, J. G. Ekerdt, S. V. Sreenivasan, C. C. Willson, *Solid State Technology* **2001**, *44*, 67.
36 M. D. Stewart, S. C. Johnson, S. V. Sreenivasan, D. J. Resnick, C. G. Willson, *Journal of Microlithography, Microfabrication and Microsystems* **2005**, *4*, 011002/1.
37 J. P. Rolland, R. M. Van Dam, D. A. Schorzman, S. R. Quake, J. M. DeSimone, *Journal of the American Chemical Society* **2004**, *126*, 2322.
38 J. Rolland, E. C. Hagberg, G. M. Denison, K. R. Carter, J. M. De Simone, *Angewandte Chemie, International Edition* **2004**, *43*, 5796.
39 J. P. Rolland, B. W. Maynor, L. E. Euliss, A. E. Exner, G. M. Denison, J. M. DeSimone, *Journal of the American Chemical Society* **2005**, *127*, 10096.
40 P. K. Bondyopadhyay, *Proceedings of the IEEE* **1998**, *86*, 78.
41 J. Fujita, Y. Ohnishi, Y. Ochiai, S. Matsui, *Applied Physics Letters* **1996**, *68*, 1297.
42 R. R. Kunz, S. C. Palmateer, A. R. Forte, R. D. Allen, G. M. Wallraff, R. A. Dipietro, D. C. Hofer, *Proceedings of SPIE – The International Society for Optical Engineering* **1996**, *2724*, 365.
43 R. R. Dammel, *Diazonaphthoquinone-based Resists*, SPIE Press, Bellingham, WA, **1993**, 1.
44 T. Kajita, T. Ota, H. Nemoto, Y. Yumoto, T. Miura, *Proceedings of SPIE – The International Society for Optical Engineering* **1991**, *1466*, 161.
45 K. Honda, J. B. T. Beauchemin, R. J. Hurditch, A. J. Blankeney, K. Kawabe, T. Kokubo, *Proceedings of SPIE – The International Society for Optical Engineering* **1990**, *1262*, 493.
46 O. Sus, *Justus Liebigs Annalen der Chemie* **1994**, *556*, 55.
47 H. Ito, H. D. Truong, S. D. Burns, D. Pfeiffer, W. Huang, M. M. Khojasteh,

P. R. Varanasi, M. Lercel, *Proceedings of SPIE – The International Society for Optical Engineering* **2005**, *5753*, 109.
48 J. V. Crivello, *Polymer Engineering and Science* **1983**, *23*, 953.
49 G. S. Calabrese, A. A. Lamola, R. Sinta, J. W. Thackeray, A. K. Berry, *Polym. Microelektron. Proc. Int. Symp.* **1990**, 435.
50 J. M. J. Fréchet, F. Bouchard, F. Houlihan, B. Kryczka, C. G. Willson, *ACS Polymeric Materials Science and Engineering* **1985**, *53*, 263.
51 T. Sakamizu, H. Shiraishi, H. Yamaguchi, T. Ueno, N. Hayashi, *Japanese Journal of Applied Physics, Part I: Regular Papers and Short Notes* **1992**, *31*, 4288.
52 R. Wood, C. Lyons, J. Conway, R. Mueller, in *Proceedings of KTI Interface '88*, **1988**, p. 341.
53 W. Huang, R. Kwong, A. Katnani, M. Khojasteh, *Proceedings of SPIE – The International Society for Optical Engineering* **1994**, *2195*, 37.
54 D. R. Medeiros, A. Aviram, T. C. Guarnieri, W. Huang, R. Kwong, C. K. Magg, A. P. Mahorowala, W. M. Moreau, K. E. Petrillo, M. Angelopoulos, *IBM Journal of Research and Development* **2001**, *45*, 639.
55 T. Sakamizu, T. Arai, K. Katoh, S.-i. Uchino, F. Murai, Y. Suzuki, H. Shiraishi, *Journal of Photopolymer Science and Technology* **1998**, *11*, 547.
56 Z. Feng, M. Liu, Y. Wang, L. Zhao, *Journal of Applied Polymer Science* **2004**, *92*, 1259.
57 S. Moon, K. Naitoh, T. Yamaoka, *Chemistry of Materials* **1993**, *5*, 1315.
58 R. D. Allen, G. M. Wallraff, W. D. Hinsberg, L. L. Simpson, *Journal of Vacuum Science and Technology, B: Microelectronics Processing and Phenomena* **1991**, *9*, 3357.
59 Y. Kaimoto, K. Nozaki, S. Takechi, N. Abe, *Proceedings of SPIE – The International Society for Optical Engineering* **1992**, *66*, 1672.
60 K. Yamashita, M. Endo, M. Sasago, N. Nomura, H. Nagano, S. Ono, T. Sato, *Journal of Vacuum Science and Technology, B: Microelectronics Processing and Phenomena* **1993**, *11*, 2692.
61 Y. C. Bae, K. Douki, T. Yu, J. Dai, D. Schmaljohann, H. Koerner, C. K. Ober, W. Conley, *Chemistry of Materials* **2002**, *14*, 1306.
62 R. Sooriyakumaran, H. Truong, L. Sundberg, M. Morris, B. Hinsberg, H. Ito, R. Allen, W. Huang, D. Goldfarb, S. Burns, D. Pfeiffer, *Proceedings of SPIE – The International Society for Optical Engineering* **2005**, *5753*, 329.
63 M. Switkes, M. Rothschild, R. R. Kunz, S.-Y. Baek, D. Cole, M. Yeung, *Microlithography World* **2003**, *12*, 4.
64 J.-B. Kim, T.-H. Oh, K. Kim, *Proceedings of SPIE – The International Society for Optical Engineering* **2005**, *5753*, 603.
65 V. R. Vorha, K. Douki, Y. Kwark, X. Liu, C. K. Ober, Y. C. Bae, W. Conley, D. Miller, D. Zimmerman, *Proceedings of SPIE – The International Society for Optical Engineering* **2002**, *4690*, 84.
66 R. L. Brainard, G. G. Barclay, E. H. Anderson, L. E. Ocola, *Microelectronic Engineering* **2002**, *61/62*, 707.
67 R. L. Brainard, J. Cobb, C. A. Cutler, *Journal of Photopolymer Science and Technology* **2003**, *16*, 401.
68 J. L. Cobb, R. L. Brainard, D. J. O'Connell, P. M. Dentinger, *Materials Research Society Symposium Proceedings* **2002**, *705*, 91.
69 H. B. Cao, J. M. Roberts, J. Dalin, M. Chandhok, R. P. Meagley, E. M. Panning, M. K. Shell, B. J. Rice, *Proceedings of SPIE – The International Society for Optical Engineering* **2003**, *5039*, 484.
70 D. Seligson, L. Pan, P. King, P. Pianetta, *Nuclear Instruments and Methods in Physics Research, Section A: Accelerators, Spectrometers, Detectors and Associated Equipment* **1988**, *A266*, 612.
71 N. N. Matsuzawa, H. Oizumi, S. Mori, S. Irie, S. Shirayone, E. Yano, S. Okazaki, A. Ishitani, D. A. Dixon, *Japanese Journal of Applied Physics, Part 1: Regular Papers, Short Notes and Review Papers* **1999**, *38*, 7109.
72 N. N. Matsuzawa, H. Oizumi, S. Mori, S. Irie, E. Yano, S. Okazaki, A. Ishitani, *Microelectronic Engineering* **2000**, *53*, 671.
73 N. N. Matsuzawa, S. Irie, E. Yano, S. Okazaki, A. Ishitani, *Proceedings of SPIE – The International Society for Optical Engineering* **2001**, *4343*, 278.

74 O. R. Wood II, J. E. Bjorkholm, L. Fetter, M. D. Himel, D. M. Tennant, A. A. MacDowell, B. La Fontaine, J. E. Griffith, G. N. Taylor, W. K. Waskiewicz, D. L. Windt, J. B. Kortright, E. K. Gullikson, K. Nguyen, *Journal of Vacuum Science and Technology, B: Microelectronics and Nanometer Structures* **1994**, *12*, 3841.

75 O. R. Wood II, J. E. Bjorkholm, K. F. Dreyer, L. Fetter, M. D. Himel, R. R. Freeman, D. M. Tennant, J. E. Griffith, G. N. Taylor, *OSA Proceedings on Extreme Ultraviolet Lithography*, edited by F. Zernike, D. T. Attwood, Optical Society of America, Washington, DC **1995**, 83–88.

76 G. N. Taylor, R. S. Hutton, S. M. Stein, C. H. Boyce, O. R. Wood, II, B. LaFontaine, A. A. MacDowell, D. R. Wheeler, G. D. Kubiak, A. K. Ray-Chaudhuri, K. W. Berger, D. A. Tichenor, *Proceedings of SPIE – The International Society for Optical Engineering* **1995**, *2437*, 308.

77 J. P. Bravo-Vasquez, Y.-J. Kwark, C. K. Ober, *Proceedings of SPIE – The International Society for Optical Engineering* **2005**, *5753*, 732.

78 C. R. Kessel, L. D. Boardman, S. J. Rhyner, J. L. Cobb, C. C. Henderson, V. Rao, U. Okoroanyanwu, *Proceedings of SPIE – The International Society for Optical Engineering* **1999**, *3678*, 214.

79 J. Dai, C. K. Ober, L. Wang, F. Cerrina, P. F. Nealey, *Proceedings of SPIE – The International Society for Optical Engineering* **2002**, *4690*, 1193.

80 J. Dai, C. K. Ober, S.-O. Kim, P. F. Nealey, V. Golovkina, J. Shin, L. Wang, F. Cerrina, *Proceedings of SPIE – The International Society for Optical Engineering* **2003**, *5039*, 1164.

81 T. Hirayama, D. Shiono, S. Matsumaru, T. Ogata, H. Hada, J. Onodera, T. Arai, T. Sakamizu, A. Yamaguchi, H. Shiraishi, H. Fukuda, M. Ueda, *Proceedings of SPIE – The International Society for Optical Engineering* **2005**, *5753*, 738.

82 Y.-J. Kwark, J.-P. Bravo-Vasquez, C. K. Ober, H. B. Cao, H. Deng, R. P. Meagley, *Proceedings of SPIE – The International Society for Optical Engineering* **2003**, *5039*, 1204.

83 A. Ray-Chaudhuri, G. Kubiak, C. Henderson, D. Wheeler, T. Pollagi, Advanced Electronics Manufacturing Department, Sandia National Laboratories [Technical Report] **1997**, *8285*, 1–13.

84 R. Sezi, M. Sebald, R. Leuschner, *Polymer Engineering and Science* **1989**, *29*, 891.

85 D. C. Tully, A. R. Trimble, J. M. J. Fréchet, *Proceedings of SPIE – The International Society for Optical Engineering* **2000**, *3999*, 1202.

86 D. C. Tully, A. R. Trimble, J. M. J. Fréchet, *Advanced Materials* **2000**, *12*, 1118.

87 M. Williamson, A. R. Neureuther, *Proceedings of SPIE – The International Society for Optical Engineering* **2000**, *3999*, 1189.

88 C. S. Hong, M. Jikei, R. Kikuchi, M. Kakimoto, *Macromolecules* **2003**, *36*, 3174.

89 T. Fujigaya, M. Ueda, *Journal of Photopolymer Science and Technology* **2000**, *13*, 339.

90 S. Makita, H. Kudo, T. Nishikubo, *Journal of Polymer Science, Part A: Polymer Chemistry* **2004**, *42*, 3697.

91 Y. Shirota, *Journal of Materials Chemistry* **2005**, *15*, 75.

92 T. Kadota, H. Kageyama, F. Wakaya, K. Gamo, Y. Shirota, *Materials Science and Engineering, C: Biomimetic and Supramolecular Systems* **2001**, *C16*, 91.

93 M. Yoshiiwa, H. Kageyama, Y. Shirota, F. Wakaya, K. Gamo, M. Takai, *Applied Physics Letters* **1996**, *69*, 2605.

94 T. Kadota, H. Kageyama, F. Wakaya, K. Gamo, Y. Shirota, *Chemistry Letters* **2004**, *33*, 706.

95 T. Nakayama, K. Haga, O. Haba, M. Ueda, *Chemistry Letters* **1997**, *3*, 265.

96 M. Ishida, J.-i. Fujita, T. Ogura, Y. Ochiai, E. Ohshima, J. Momoda, *Japanese Journal of Applied Physics, Part 1: Regular Papers, Short Notes and Review Papers* **2003**, *42*, 3913.

97 T. Nakayama, M. Ueda, *Journal of Materials Chemistry* **1999**, *9*, 697.

98 S. W. Chang, D. Yang, J. Dai, N. Felix, D. Bratton, K. Tsuchiya, Y.-J. Kwark, J.-P. Bravo-Vasquez, C. K. Ober, H. B. Cao, H. Deng, *Proceedings of SPIE – The International Society for Optical Engineering* **2005**, *5753*, 1.

99 S. W. Chang, N. Felix, D. Yang, A. Ramakrishnan, C. K. Ober, *PMSE Preprints* **2005**, *92*, 131.

100 R. Sooriyakumaran, H. Truong, L. Sundberg, M. Morris, B. Hinsberg, H. Ito, R. Allen, W.-S. Huang, D. Goldfarb, S. Burns, D. Pfeiffer, *Journal of Photopolymer Science and Technology* **2005**, *18*, 425.

101 E. K. Kim, J. G. Ekerdt, C. G. Willson, *Journal of Vacuum Science and Technology, B: Microelectronics and Nanometer Structures – Processing, Measurement and Phenomena* **2005**, *23*, 1515.

102 S. L. Wells, J. DeSimone, *Angewandte Chemie, International Edition* **2001**, *40*, 518.

103 H. Namatsu, K. Yamazaki, K. Kurihara, *Journal of Vacuum Science and Technology B* **2000**, *18*, 780.

104 D. L. Goldfarb, J. J. D. Pablo, P. F. Nealey, J. P. Simons, W. M. Moreau, M. Angelopoulos, *Journal of Vacuum Science and Technology, B: Microelectronics Processing and Phenomena* **2000**, *18*, 3313.

105 N. Sundararajan, S. Yang, K. Ogino, S. Valiyaveettil, *Chemistry of Materials* **2000**, *12*, 41.

106 D. Flowers, E. N. Hogan, R. Carbonell, J. M. DeSimone, *Proceedings of SPIE – The International Society for Optical Engineering* **2002**, *4690*, 419.

107 Y. Mao, N. M. Felix, P. T. Nguyen, C. K. Ober, K. K. Gleason, *Journal of Vacuum Science and Technology, B: Microelectronics Processing and Phenomena* **2004**, *22*, 2473.

108 B. F. Graham, A. F. Lagalante, T. J. Bruno, J. M. Harrowfield, R. D. Trengove, *Fluid Phase Equilibria* **1998**, *150–151*, 829.

109 N. M. Felix, Tsuchiya, C. K. Ober, *Advanced Materials* **2006**, *18*, 442.

110 I. W. Hamley, *Developments in Block Copolymer Science and Technology*, Wiley & Sons, Chichester UK **2004**, 1–29.

111 A. H. Gabor, C. K. Ober, *ACS Symposium Series* **1995**, *614*, 281.

112 C. K. Ober, M. Li, K. Douki, K. Goto, X. Li, *Journal of Photopolymer Science and Technology* **2003**, *16*, 347.

113 C. G. Willson, M. J. Bowden, L. F. Thompson, *Introduction to Microlithography*, American Chemical Society, Washington, DC, **1994**.

114 T. Thurn-Albrecht, J. Schotter, G. A. Kastle, N. Emley, T. Shibauchi, L. Krusin-Elbaum, K. Guarini, C. T. Black, M. T. Tuominen, T. P. Russell, *Science* **2000**, *290*, 2126.

115 J. Y. Cheng, C. A. Ross, V. Z. H. Chan, E. L. Thomas, R. G. H. Lammertink, G. J. Vancso, *Advanced Materials* **2001**, *13*, 1174.

116 M. Park, C. Harrison, P. M. Chaikin, R. A. Register, D. H. Adamson, *Science* **1997**, *276*, 1401.

117 M. Li, K. Douki, K. Goto, X. Li, C. Coenjarts, D. M. Smilgies, C. K. Ober, *Chemistry of Materials* **2004**, *16*, 3800.

118 C. T. Black, K. W. Guarini, K. R. Milkove, S. M. Baker, T. P. Russell, M. T. Tuominen, *Applied Physics Letters* **2001**, *79*, 409.

119 C. Harrison, J. A. Dagata, D. H. Adamson, *Developments in Block Copolymer Science and Technology*, Wiley & Sons, Chichester UK **2004**, 295.

120 F. S. Bates, *Science* **1991**, *251*, 898.

121 F. S. Bates, G.-H. Fredrickson, *Physics Today* **1999**, *52*, 32.

122 M. J. Fasolka, A. M. Mayes, *Annual Review of Materials Research* **2001**, *31*, 323.

123 C. S. Henkee, E. L. Thomas, L. J. Fetters, *Journal of Materials Science* **1988**, *23*, 1685.

124 C. M. Papadakis, P. Busch, D. Posselt, D.-M. Smilgies, *Advances in Solid State Physics* **2004**, *44*, 327.

125 R. P. Q. Henry L. Hsieh, *Anionic Polymerization: Principles and Applications*, Marcel Dekker, New York, **1996**.

126 K. Matyjaszewski, J. Xia, *Chemical Reviews* **2001**, *101*, 2921.

127 C. J. Hawker, A. W. Bosman, E. Harth, *Chemical Reviews* **2001**, *101*, 3661.

128 C. Barner-Kowollik, T. P. Davis, J. P. A. Heuts, M. H. Stenzel, P. Vana, M. Whittaker, *Journal of Polymer Science, Part A: Polymer Chemistry* **2003**, *41*, 365.

129 C. Harrison, Z. Cheng, S. Sethuraman, D. A. Huse, P. M. Chaikin, D. A. Vega, J. M. Sebastian, R. A. Register, D. H. Adamson, *Physical Review E: Statistical, Nonlinear and Soft Matter Physics* **2002**, *66*, 011706/1.

130 J. Dai, C. K. Ober, *Proceedings of SPIE – The International Society for Optical Engineering* **2004**, *5376*, 508.

131 Y. C. Bae, C. K. Ober, *Polymer Bulletin* **2004**, *52*, 321.
132 P. Du, M. Li, K. Douki, X. Li, C. B. W. Garcia, A. Jain, D.-M. Smilgies, L. J. Fetters, S. M. Gruner, U. Wiesner, C. K. Ober, *Advanced Materials* **2004**, *16*, 953.
133 A. H. Gabor, L. C. Pruette, C. K. Ober, *Chemistry of Materials* **1996**, *8*, 2282.
134 N. Sundararajan, C. F. Keimel, N. Bhargava, C. K. Ober, J. Opitz, R. D. Allen, G. Barclay, G. Xu, *Journal of Photopolymer Science and Technology* **1999**, *12*, 457.
135 W. Denk, J. H. Strickler, W. W. Webb, *Science* **1990**, *248*, 73.
136 B. H. Cumpston, S. P. Sundaravel, S. Barlow, D. L. Dyer, J. E. Ehrlich, L. L. Erskine, A. A. Ahmed, S. M. Kuebler, I.-Y. S. Lee, D. McCord-Maughon, J. Qin, H. Rockel, M. Rumi, X.-L. Wu, S. R. Marder, J. W. Perry, *Nature* **1999**, *398*, 51.
137 S. Kawata, H.-B. Sun, T. Tanaka, K. Takada, *Nature* **2001**, *412*, 697.
138 C. A. Coenjarts, C. K. Ober, *Chemistry of Materials* **2004**, *16*, 5556.
139 S. J. Ahn, M. Kaholek, W.-K. Lee, B. LaMattina, T. H. LaBean, S. Zauscher, *Advanced Materials* **2004**, *16*, 2141.
140 T. Watanabe, M. Akiyama, K. Totani, S. M. Kuebler, F. Stellacci, W. Wenseleers, K. Braun, S. R. Marder, J. W. Perry, *Advanced Functional Materials* **2002**, *12*, 611.
141 J. D. Pitts, P. J. Campagnola, G. A. Epling, S. L. Goodman, *Macromolecules* **2000**, *33*, 1514.

9
Microelectronic Materials with Hierarchical Organization

G. Dubois, R. D. Miller, and James L. Hedrick

9.1
Introduction

Historically, semiconductor manufacturers have generated new products according to Moore's law, which states that device densities on chips should nearly double every 18–24 months [1]. This has generally been accomplished by a continuous reduction in device dimensions, driven by advances in microlithography. The transistors and other active electronic devices are integrated on a silicon substrate and connected to each other through complex on-chip wiring. This interconnect wiring, on-chip insulation and local packaging are denoted the back-end-of-the line (BEOL), whereas the active devices on the substrate constitute the front-end-of-the line (FEOL) part of the chip. Nowadays, chips contain hundreds of millions of active devices with interconnect wiring distributed over 8–10 vertical wiring levels that approach 10 000 m in cumulative length. With the continual reduction in device dimensions and as device densities increase, chip performance degrades due to signal delays and cross-talk between the conductor lines. Signal propagation delays are characterized by the product of metal resistance times the capacitance of the lines (RC) [2] and this delay depends on the resistivity of the wiring metallurgy, the dielectric constant of the insulating media and the dimensions of the metal lines. For example, the switch to copper metallurgy from aluminum reduced the resistivity of the metal by 30% [3], providing a significant improvement in performance but forcing the introduction of damascene processing [4]. Moreover, the dielectric constant of the on-chip insulating medium has a strong influence on both signal delays and power consumption. Historically, this insulator has been SiO_2 since this has the requisite thermal and mechanical properties to survive integration/fabrication. Unfortunately, it has a dielectric constant of ∼4.0, causing signal delays and cross-talk between lines for 130-nm technology and beyond. The semiconductor industry has found that replacement of this insulator is more difficult than expected and the International Technology Roadmap for Semiconductors has significantly relaxed its targeted dielectric constant values for the foreseeable

Macromolecular Engineering. Precise Synthesis, Materials Properties, Applications.
Edited by K. Matyjaszewski, Y. Gnanou, and L. Leibler
Copyright © 2007 WILEY-VCH Verlag GmbH & Co. KGaA, Weinheim
ISBN: 978-3-527-31446-1

future. The difficulty stems, in part, from the current manufacturing processes that require the insulating materials to be thermally stable, have a high glass transition temperature (T_g), low moisture absorption, good adhesion to metals and passivation coatings, mechanical toughness to prevent crack propagation and mechanical hardness to withstand the rigors of chemical mechanical polishing, dicing and wire bonding. Nonetheless, future chip generations will require a dielectric constant between 2.4 and 2.0 to realize the full benefits of the reduced lithographic features.

Alternatives to SiO_2 with a lower dielectric constant have been developed and can be generally divided into two general classes including organic high-temperature polymers (generally thermosetting materials) and materials that are more inorganic in nature such as organosilicates. The inorganic materials generally possess the hardness and thermal stability for microelectronic fabrication, but are rather fragile and have poor crack resistance. Conversely, the organic materials generally have tough, ductile mechanical properties but are soft. The applicability of organic thermosetting polymers has quickly reached a limit because of their incompatibility with the traditional integration processes developed and optimized for SiO_2 [5]. As a result, the competition for the replacing insulating materials existing has turned to more silica-like materials. For instance, the first improvement in the k value was achieved by replacing some Si–O bonds in silica with the less polarizable Si–F bond, forming F-doped silica glasses (FSGs) (Fig. 9.1). An even lower dielectric constant was obtained by introducing methyl groups within the structure of silica, resulting in carbon-doped oxides (CDOs) (Fig. 9.1). Moreover, the replacement of an oxygen atom with either F or C leads to an increase in the interatomic distances or "free volume" and results in an additional decrease of the dielectric constant.

CDO materials have been successfully introduced for 90-nm technology and above (Fig. 9.2). Future technologies, e.g. at 65 and 45 nm, will require insulating materials which possess a dielectric constant of 2.7 or lower. A class of promising low-dielectric constant materials are methyl silsesquioxanes, having the general formula $(RSiO_{1.5})_n$, where R is an organic substituent. These organosilicates have intrinsic dielectric constants of >2.7, low moisture absorption, exceptional thermal stability ($\sim 500\,°C$) and reasonable moduli. Following the general strategy for lowering the dielectric constant by the incorporation of air

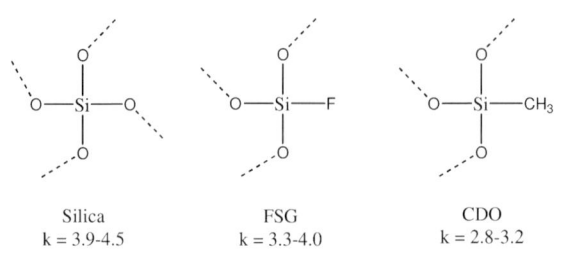

Silica
$k = 3.9\text{-}4.5$

FSG
$k = 3.3\text{-}4.0$

CDO
$k = 2.8\text{-}3.2$

Fig. 9.1 Possible structural variations in silica or organosilicates to facilitate enhanced dielectric performance.

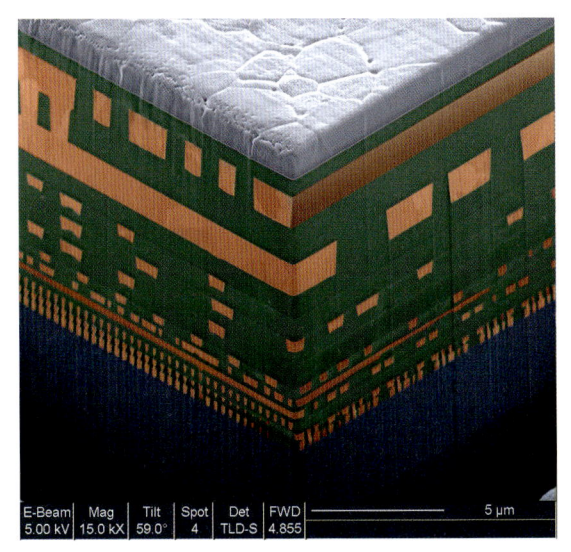

Fig. 9.2 Cross-section of 64-bit high-performance microprocessor chip built in IBM's 90-nm Server-Class CMOS technology with Cu/low-k wiring. Above the transistors, the wiring levels include one W local interconnect, five "1x-scaled" Cu levels in full SiCOH low-k dielectric, three "2x-scaled" Cu/SiCOH levels, two "6x-scaled" Cu levels in FTEOS/SiO$_2$ dielectric and one Al(Cu) terminal pad and wiring level. The minimum M1 Cu linewidths and spaces are 0.12 μm.

via a nanocomposite, it is expected that increasing the volume fraction of air will continuously lower the dielectric constant, while maintaining the other desired properties. This allows the dielectric constant to be continuously tuned, extending the lifetime of the materials and deposition tools, a critical consideration in implementing a new material.

As mentioned one route to materials with dielectric extendibility is the controlled introduction of porosity [i.e. a nanocomposite composed of an insulating material and air ($k=1.01$)] [6]. In this way, the dielectric constant of an existing material can be systematically reduced simply by the addition of more air. It is critical that the pores be small relative to the minimal device feature sizes, pore distribution be narrow, contaminant uptake be minimal and the materials must show high strength and toughness even at low density. As a consequence, the development of new porous insulators and strategies to create porosity has become an active area of research [7]. Interestingly, the method used to create porosity in dielectric materials is strongly related to their mode of application to the wafer. Dielectric films can be deposited from the gas phase by chemical vapor deposition (CVD) or from solution by spin-on methods. Porous carbon-doped ox (CDO) samples deposited by CVD and plasma-enhanced CVD (PE-CVD) have been reported with dielectric constants as low as 2.1 [8], but the mechanism by which the pores are formed is still unclear and not well studied. In contrast, the creation of porosity with spin-on deposited materials is now well understood and can be divided into two main categories. The first includes all materials where the porosity is introduced exclusively through sol–gel processes [9], whereas the second includes materials where the porosity is generated through the use of sacrificial "porogens" that are decomposed during the film thermal processing [10]. The first strategy implies the formation of a rigid silicate network while the solvent is still present. The amount of solvent relative to solid content in the sol after aging and the method of solvent removal will determine the level of porosity. Aerogels are obtained when supercritical drying of the wet gel (in most cases CO_2 is used) is performed after aging [11], whereas controlled thermal curing of the sol leads to the formation of xerogels [12]. Independent of the drying method, it has been shown that not only are the pore sizes in these sol–gel materials difficult to control but the difficulty increases as a function of the porosity level, rending these materials unattractive for microelectronic applications.

In order to allow better control of the pores size and morphology, sacrificial macromolecular porogens have been used. These polymers are thermally stable up to at least 250 °C and remain unaffected during the organosilicate vitrification step. Subsequent decomposition of these porogens at higher temperature leads to the formation of a porous structure (Fig. 9.3).

Two mechanisms have been described to explain the generation of porosity through polymer decompositionporosity: nucleation and growth and templating. In the first case, the porogen is soluble in the hydrophilic organosilicate before curing. When the vitrification of the resin starts, the silicate network becomes more hydrophobic, resulting in a nano-phase separation of the porogen. The stage at which this phase separation occurs determines the average pore size distribution in the material. In contrast, when molecular amphiphiles are used, typically polymeric particles containing a hydrophobic core and a hydrophilic corona, the porogens are compatible in the organosilicate at a pre-curing stage, but not truly soluble (nano-dispersion).

Since these particles are already nano-phase separated before the organosilicate vitrification, the size of the pores obtained after thermal decomposition depends only upon the size of the starting particles, hence the term "templating mechanism".

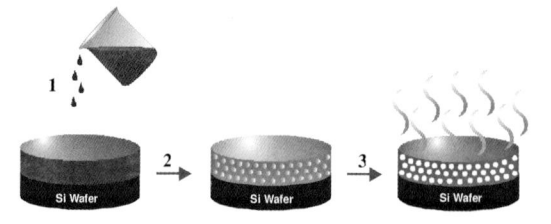

- Blending of thermosetting resin and porogen
- Spin coating of thin films (1)
- Formation of inorganic/organic nanohybrids (2)
- Generation of nanoporosity (3)

Fig. 9.3 General strategy to obtain porous organosilicates via nucleation and growth: spin coating deposition, heat to advance MSSQ molecular weight and effect phase separation of sacrificial polymer and decomposition of organic component to generate porous organosilicate.

9.2
Porous Organosilicates from a Nucleation and Growth Phase Separation Process

One approach to nanoporous organosilicates is based on a miscible polymer pair, where one component is the thermosetting methylsilsequioxane (MSSQ) resin and the other is a sacrificial polymer or porogen designed to generate the porosity upon its removal [10]. Although the polymer pair is initially miscible, the increase in the molecular weight of the MSSQ with curing changes both the molecular weight and solubility parameter of the MSSQ, causing the porogen to phase separate, generating an organic–inorganic hybrid. This phase separation process is generally referred to as a nucleation and growth process. Subsequent heating facilitates network formation and causes the organic component to undergo thermolysis, leaving behind pores with a size and shape templated by the initial hybrid morphology. This approach is similar to that employed by Fréchet and coworkers [13] in the preparation of porous polystyrene beads and monoliths for chromatography, by rubber and thermoplastic toughened thermosetting epoxy resins and the generation of porous epoxies via chemically induced phase separation. The vitrification, phase separation and thermolysis processes can be mapped with a combination of dynamic mechanical and thermogravimetric techniques. Qualitative features of the variation of modulus of the MSSQ resin versus temperature during the initial heating to 450 °C show that the resin softens and becomes fluid at ~70 °C, but begins to vitrify at ~200 °C owing to cross-linking and chain extension reactions (see Fig. 9.5). The porogens are designed to undergo quantitative thermolysis above this vitrification temperature. Samples cured to intermediate temperatures (200 °C) show evidence of a second transition that can be identified as the sacrificial polymer domains, consistent with its phase separation. Control of the pore size and shape is critical, but by this method it is difficult to control the growth of the second phase. Shown in Fig. 9.4 is a typical phase diagram that shows the

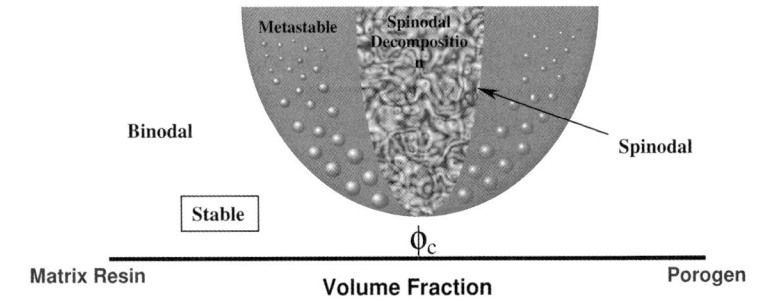

Fig. 9.4 Generic phase diagram of organic–inorganic hybrid mixture that phase separates via the nucleation and growth mechanism. Increasing molecular weight leads to either bimodal or spinodal phase separation and vitrification kinetically limits the growth of the second phase.

variation in structure with increasing porogen concentration. At low porogen contents the hybrid remains homogeneous; however, above a certain concentration, a "porogen-rich" phase nucleates and the growth of this phase depends on the viscosity or vitrification of the polymerizing MSSQ medium. Higher porogen contents result in phase separation via spinodal decomposition that produces a co-continuous structure and higher volume fractions, and inverts the morphology where the porogen is the continuous phase. Clearly, the regime of interest for interlayer dielectric applications is nucleation and growth, where the challenge is to control the growth to limit the size of the pores.

The morphology of the nanohybrid is strongly dependent on the interaction between the porogen and the MSSQ matrix and the molecular weight, macromolecular architecture, functionality and loading level of the porogen are important design features. Therefore, to minimize pore size, it is desirable to postpone the nucleation of the second phase (Fig. 9.4). It is important to maximize the interaction between the organic and inorganic components to delay nucleation of the organic phase, allowing the vitrification of the MSSQ matrix to trap the organic phase kinetically. This has been accomplished through the introduction of functionality in either the backbone or end-groups through the use of complex macromolecular architectures. Poly(ethylene oxide) (PEO) and poly(propylene oxide) (PPO) are examples of functional linear polymers that have been shown to be highly miscible with MSSQ prepolymers and phase separate via

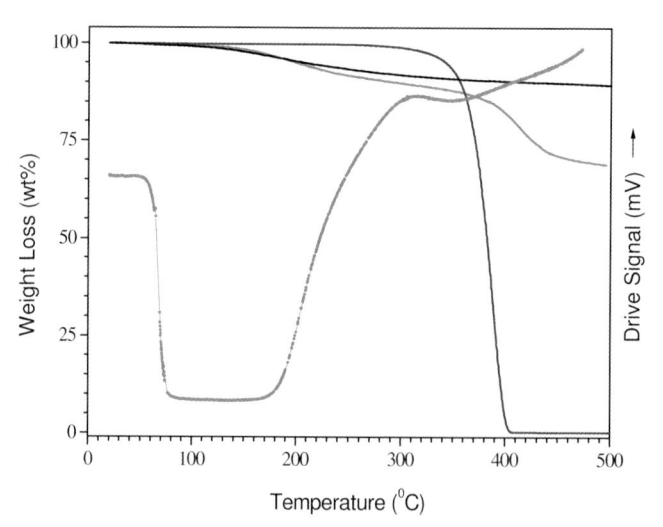

Fig. 9.5 Desired thermal properties for the generation of an inorganic–organic hybrid system that phase separates via nucleation and growth processes. Thermogravimetric analysis (TGA) shows clean and quantitative decomposition of the organic component but only after vitrification of the MSSQ. TGA analysis of MSSQ alone that shows weight loss associated with the condensation byproducts (vitrification). Drive signal or relative modulus as a function of temperature shows the increase in modulus above 200 °C associated with the vitrification process and consistent with the TGA results.

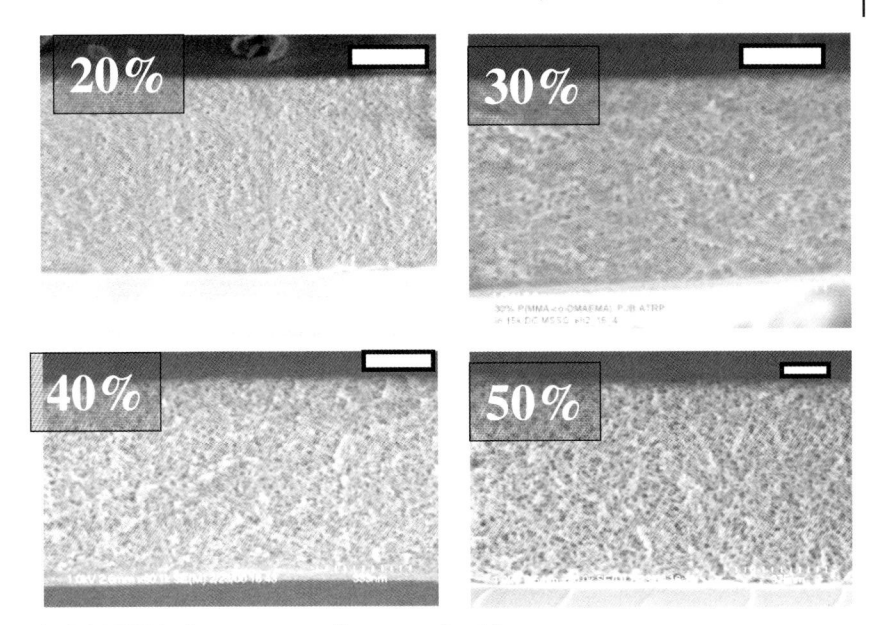

Fig. 9.6 FESEM of porous organosilicates templated from PPO concentrations ranging from 20 to 50%. Pore size increases with increasing porogen composition, a characteristic of nucleation and growth phase separation.

nucleation and growth, resulting in the organosilicate nanostructures with small features [14]. Shown in Fig. 9.6 are FESEM micrographs of porous organosilicates templated from PPO concentrations ranging from 20 to 50%. Characteristics of this phase separation process are pore distributions randomly initiated often with irregular shapes and a dependence of the pore size and shape on loading level. Higher levels of porogen cause phase separation to occur at lower MSSQ conversions, allowing growth of the organic phase that produces progressively larger pore sizes. The inversion in the morphology causes a significant reduction in film thickness and the remaining MSSQ looks like sintered particles.

9.3
Role of Functionality: Nitrogenous Porogens

Nitrogenous macromolecular porogens for generating porosity in silica and organosilicates have an extended history. Such materials have been particularly useful for nucleation and growth processes since nitrogen substitution in organic polymers often delivers characteristics commensurate with the formation of miscible blends with organosilicates, i.e. polarity, hydrogen bonding capability and basicity, which promote strong interactions between the polymeric porogen and the thermosetting matrix material.

Early work with nitrogenous porogens was described by Chujo and coworkers, who found that nanoporous silica could be prepared by adding polar polymers such as polyoxazolines, poly(N-methylpyrrolidinones) and poly(N,N-dimethylacrylamides) to the sol–gel polymerization of tetraethyl orthosilicate (TEOS) and firing the spun films to 600 °C to eliminate the porogen [15]. They also demonstrated that semi-interpenetrating networks derived from the simultaneous hydrolysis/condensation of TEOS and polymerization of N,N-dimethylacrylamide produced nanoporous silica. Similarly, dendritic amidoamines (PAMAM) have been tested as porogens for silica and, although some correlation between pore and porogen size was observed, the relationship was not exact, suggesting some folding of the porogen in the matrix [16].

More recently, in conjunction with collaborators at DSM, we studied some commercial hyperbranched polyester–amides (HYBRANES, DSM) with modified end-groups produced by the condensation of various anhydrides with diisopropanolamine [17] as porogens when mixed with low molecular weight MSSQ. Several of the samples (e.g. HA-1690 and P-1000) formed clear nanocomposites at loading levels up to 70% and nanoporous films after calcination. However, comparison of the refractive indices and dielectric constants of the porous MSSQ films generated from the HYBRANES with those derived from an efficient porogen [i.e. branched poly(propylene oxide), MW $\approx$ 6000] suggested that pore formation from the HYBRANE samples was less efficient. This was also consistent with the observation that TGA analysis of the thermal decomposition of the HYBRANES alone often showed a significant char yield. It is also possible that, because of the high branching levels, phase separation is inefficient, leading to collapse of the smallest pores.

With macromolecular porogens where the chain functionality is only weakly interacting with the organosilicate prepolymer [e.g. polycaprolactone, poly(methyl methacrylate)], miscibility depends strongly on the nature and the number of the chain ends. In these cases, low molecular weight, highly branched porogens (dendrimers, hyperbranched, star derivatives, etc.) are more soluble in the uncured resin and tend to produce nanoscopic porosity upon calcination. In contrast, for polar, reactive or hydrogen bonding polymers for which the functional groups on the polymer interact strongly with the thermosetting resin, miscible blends and nanoporous porosity are generated even using linear polymers. Although we have tested various nitrogenous polymeric porogens, including polyvinylpyridines, polyacrylamides and polydimethylacrylamides, we have extensively studied copolymers of dimethylaminoethyl methacrylate and methyl methacrylate (DMAEMA-co-MMA) (Fig. 9.7). Random copolymers from these monomers can be prepared by free radical polymerization in either the absence or presence of a chain transfer species [2-propanol or tris(trimethylsilyl)silane] to control molecular weights or by atom transfer radical polymerization (ATRP) [18]. The polymer molecular weights ranged from 4000 to 30000 and the polydispersities were 1.1–1.4 after purification by solvent–nonsolvent precipitation. The amount of DMAEMA in the copolymer was varied from 0 to 30 mol%. Samples which contained between 20 and 30% of DMAEMA formed excellent

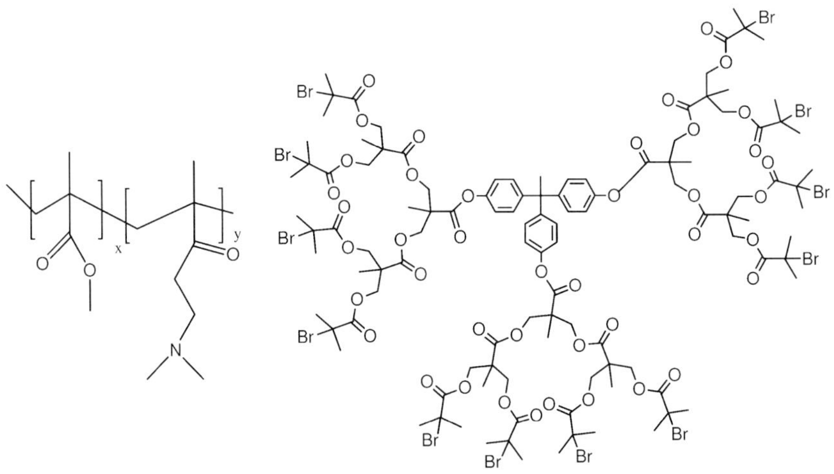

Fig. 9.7 Dendritic initiator containing activated alkyl bromides for the preparation of a star-shaped nitrogenous polymer (methyl methacrylate and dimethylaminoacrylamide) via ATRP.

coatings when formulated in MSSQ even when the MSSQ films alone showed defects [19]. In this case, the presence of porogen itself improved the film quality. Interestingly, although the addition of monomeric base causes rapid gelation of the MSSQ solution, the addition of the basic polymer seems to have no noticeable effect on the solution stability. For samples containing 25–30 mol% of DMAEMA, we obtained clear films and nanoscopic porosity for loading levels up to 70 wt%.

The apparent miscibility of the porogen with the MSSQ resin is dictated by specific polymer–polymer interactions involving the dimethylamino substituents on the porogen and the pendant SiOH functionality on the thermosetting resin [20]. This can be detected by IR studies on the porogen–resin mixture. In this case, we employed two different MSSQ resins, each with the same elemental composition $(MeSiO_{1.5})_n$ and similar molecular weights (1500–2000 g mol^{-1}) and polydispersities. The resins differ structurally in that one has relatively more SiOH functionality than the other as determined by the strength (normalized for thickness) of the SiOH absorptions at 3400 and 930 cm^{-1} in the infrared. We have accordingly denoted these resins MSSQ-HI and MSSQ-LO, respectively. The functional interaction is demonstrated by a 120-cm^{-1} shift in the SiOH stretching frequencies around 3400 cm^{-1} towards higher energies upon addition of the porogen, a value which scales with the amount of the porogen present. Interestingly, the position of the porogen carbonyl absorption is not affected by the addition of the organosilicate resin, suggesting that the dimethylamino substituent is responsible for the shift. The curing (vitrification) of the organosilicate resin is also accelerated by presence of the porogen such that considerable curing is observed at 150 °C. Very little resin curing is observed at this temperature without the porogen. Porogen decomposition begins around 200 °C (~20% weight loss) and is complete by 350 °C. TGA analysis suggests a char residue of 2–4%, a feature that can greatly affect the mechanical properties (see below). The final microstructure of the porous MSSQ is also influenced by the presence and amount of porogen as determined by the relative ratio of the characteristic Si–O–Si stretching bands at 1030 and 1130 cm^{-1} in the porous films. Recently, we have also prepared a variety of highly branched copolymers derived from dendronized initiators such as shown in Fig. 9.7. The copolymer (75:25 MMA–DMAEMA) arms were synthesized in a controlled fashion by ATRP procedures using CuBr and N-(n-propyl)-2-pyridylmethanimine [21]. In this study, we prepared copolymers with 4, 6 and 12 arms for comparison. Preliminary results with low molecular weight MSSQ resins showed exceptional solubility for the 6- and 12-arm derivatives. As expected, all of the branched porogens produced nanoscopic porosity upon calcination, but seemed to provide no measurable advantage over the linear copolymer. In fact, the copolymer derived from the 12-arm initiator was less efficient at generating porosity than the linear material (refractive index comparisons), suggesting that phase separation is less efficient, possibly resulting in small pore collapse. These results suggest that for strongly interacting polymer functionality, the efficacy of porogen branching is mitigated. Possible applications using higher molecular weight, less functionalized resins where the linear polymers are intrinsically less soluble are envisioned.

Small-angle X-ray scattering (SAXS) studies on MSSQ-HI and MSSQ-LO at comparable porogen loading levels (35 wt%) suggest that the average pore size in the latter is larger (7.6 vs. 2.0 nm). This is consistent with earlier nucleation from the latter resin, which has fewer SiOH functionalities to promote miscibility. A similar conclusion was reached in studies of the porogen which contained ~2 mol% of a pyrene-containing monomer (4-pyrenylbutyl methacrylate) incorporated so that the ratio of excimer to monomer fluorescence could be monitored during the formation of the nanocomposite. For the MSSQ-LO resin, the neat porogen limiting ratio was achieved at ~20% loading in the nanohybrid (cured at 180 °C), whereas this was not observed with the MSSQ-HI resin until almost 50% loading. Comparison with films of the neat porogen showed that the limiting ratio observed in MSSQ was very similar to the ratio in the neat porogen film. These data are also consistent with the proposal that the porogen phase separates, curing at lower temperatures for the MSSQ-LO resin, and hence leads to larger pores.

The SAXS studies using this porogen in MSSQ-LO revealed another interesting characteristic of the nucleation and growth process of phase separation [22]. Here the initial studies utilized a PMMA–DMAEMA (75:25) with a polydispersity index (PDI) of ~1.1. As expected, the calculated average pore size increased from 4.0 to 10 nm as the porogen loading increased from 5–50 wt% and the size dispersity also increased substantially (Fig. 9.8). Other effects are also important: at comparable loading levels, higher molecular weight porogens produce larger pores and the nature of the resin can also influence the pore size.

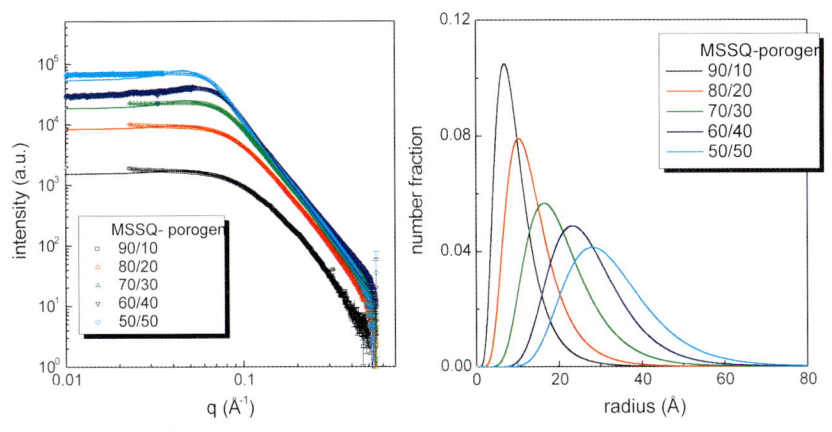

Fig. 9.8 SAXS studies on PMMA–PDMEMMA copolymers in hybrid mixtures with MSSQ. Pore size varies as a function of copolymer composition in this nucleation and growth process, moving from bimodal to spinodal phase separation.

As a result of these studies, we concluded that although polymer blending coupled with a nucleation and growth mechanism is operationally simple and can produce nanoscopic porosity, the pore size and distribution depend on many parameters: (i) porogen structure, molecular weight and polydispersity, (ii) resin structure and (iii) processing conditions.

Finally, a note of caution is required for systems using nitrogenous porogens, particularly those which are also basic. In the deposition of MSSQ films containing the MMA–DMAEMA porogen on oxide surfaces, we have found by both transmission electron microscopy (TEM) and small-angle neutron scattering (SANS) studies using partially deuterated porogen (derived from MMA-d_8) that there is some segregation of the porogen at the oxide surface, leading ultimately to a higher pore density at the interface [23]. This effect is shown in Fig. 9.9, which shows both the TEM and SANS results. The TEM picture suggests increased pore density at the substrate interface. This is confirmed by SANS using deuterated toluene for contrast enhancement. In the latter case, the scattering length density (SLD) is clearly higher both at the film surface/toluene interface as expected but also at the film/substrate interface. It is believed that this results from the strong attraction of the basic porogen to the acidic oxide surface. Such inhomogeneous distributions of porogen and porosity do not occur with non-basic porogens such as poly(propylene oxide). A surprising result of the build-up of the porogen at the interface and the formation of 2–4% char yield upon decomposition of the porogen is that the interfacial fracture energy increases remarkably from ~ 2.6 to >30 J m^{-2} for glasses containing 50% porosity [24]. This most unexpected result has been rationalized by a hypothesis that a powerful molecular bridging mechanism enabled by organic residues at the interface is responsible for the enormous and unexpected increase in surface

Depth profile of pore distribution with d-solvent

TEM showing increased porosity at the interface caused by porogen interaction with the surface

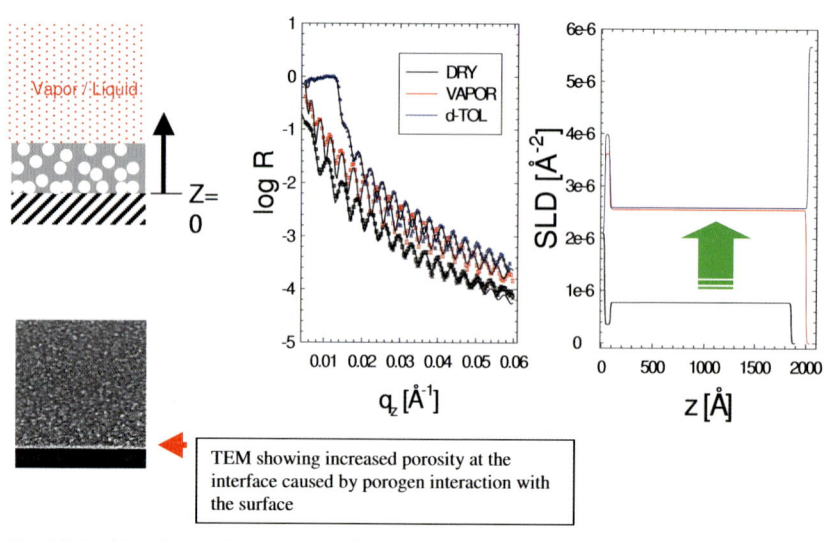

Fig. 9.9 Probing the interface: SANS and TEM results demonstrate the phase separation of the surface-enriched organic component at the oxide/film interface. A manifestation of the basic porogen segregating to the acidic oxide interface.

fracture energy. This suggestion was supported by modeling using bridging mechanics and suggests that controlled decomposition of porogens could lead to improved mechanical properties in fragile glasses.

9.4
Porous Structures via Nucleation and Growth: Role of Macromolecular Architecture

It is well established that not only the elemental composition but also the molecular arrangement of a molecule has a significant impact on the properties of the material. For example, polyethylene is one of the most commonly used bulk polymers and exists both as linear low-density polyethylene (LDPE) and as high-density polyethylene (HDPE); on more recently, synthetic advances have allowed highly branched or amorphous polyethylene, all material which differ only in the nature of the skeletal structure [25]. It is well known that LDPE and HDPE have significantly different mechanical properties due to their different crystalline contents. However, owing to the broad polydispersity of these constitutional isomers, detailed and precise structure–property relationships relating to the effects of branching are difficult. In many respects, dendrimers represent the transition from small organic molecules to polymers and provide a model for the investigation of the solution and physical properties of materials with precisely defined branched structures [26]. Nanostructured organic–inorganic thin

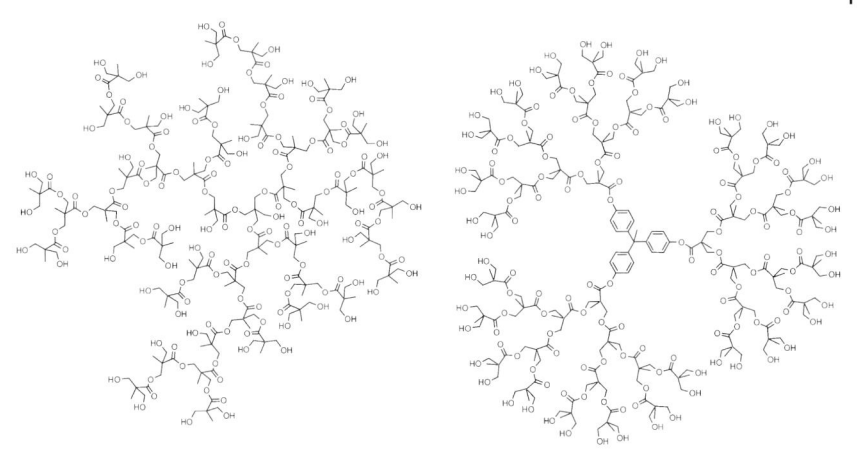

Fig. 9.10 Hydroxyl functional dendritic initiators (4th generation) derived from 2,2′-bis(hydroxymethyl)propionic acid.

films have been demonstrated from branched and dendritic organic templates [27]. The control of the phase morphology in these hybrid systems was attributed to the stronger interactions of the organic polymer with the MSSQ matrix prior to and during the templated vitrification due to the increased number of end-groups and abundant functionality.

Star-shaped polyesters were prepared by the controlled polymerization of ε-caprolactone or lactide initiated from the numerous chain-end hydroxymethyl groups of the analogous dendrimeric and hyperbranched polyesters derived from 2,2′-bis(hydroxymethyl)propionic acid (Fig. 9.10) [28]. The initiators chosen for study were generations 2–5 of the hydroxy-terminated dendritic polyesters, reported by Hult and Fréchet's groups and the corresponding dendrimer generations 1–4 [29]. The number-average molecular weight per arm correlated closely with the monomer to initiator ratio. By varying this ratio or the size of the initiator, molecular weights ranging from 20 000 to 210 000 g mol^{-1} could be obtained for the hyperbranched initiators. In a similar fashion, the initiation of ε-caprolactone from the dendrimeric analogues proved to be extremely facile and gave star polymers with accurate control of molecular weight and narrow polydispersities irrespective of the generation employed. For example, polymerization of 960 molar equivalents of ε-caprolactone with the fourth generation (G-4) dendrimer (48 surface hydroxyl groups) gave a star-shaped polymer in 96% yield at 110 °C (20 h) (M_n = 115 000 g mol^{-1}, PDI = 1.18) (Scheme 9.1).

Similarly, novel poly(methyl methacrylate)s with star-shaped architectures were prepared by controlled radical polymerization starting from dendritic 2-, 4-, 6- and 12-arm initiators (Scheme 9.2) [30]. The initiators were prepared by coupling bromo-functionalized bis-MPA dendrons with the first generation dendrimer. The use of atom transfer radical polymerization (ATRP) allows the synthesis of branched and star polymers with accurate control of molecular weight and poly-

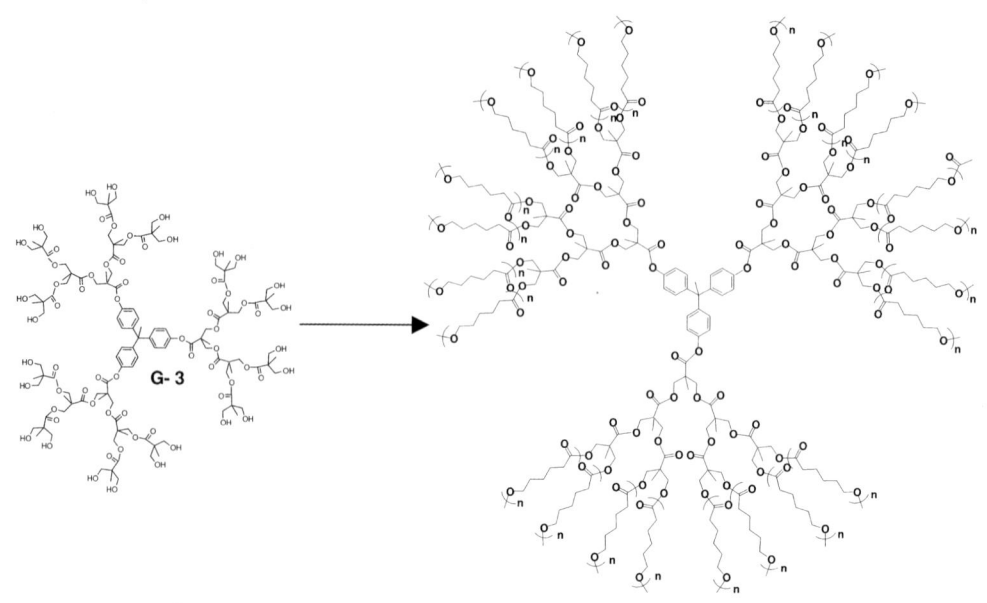

Scheme 9.1 Star-shaped poly(caprolactone) via ROP from 3rd-generation dendrimer.

dispersity. The solubility of the star-shaped copolymers can be tuned by the addition of comonomers such as N,N-dimethylacrylamide or hydroxyethyl methacrylate that add the requisite functionality.

Star-shaped polymers having alternating arms were prepared from miktofunctional initiators using consecutive ATRP and ring-opening polymerization (ROP) techniques in a core-out approach (Scheme 9.3). A building block containing initiating sites for both ATRP and ROP was designed and was coupled to a multifunctional core. In this way, the initiating sites were arranged in an alternating fashion for the synthesis of block copolymers containing alternating arms of PMMA and PCL emanating from a core, providing another level of control to the preparation of miktoarm polymers. Amphiphilic alternating arm copolymers were prepared containing poly(caprolactone) (PCL) and poly(acrylic acid) and related copolymers.

The modification of star-shaped and linear polymers with functional dendrons provided unique materials with abundant functionality. In the case of the star-shaped polymers, two general strategies were investigated where dendrons were selectively placed either in the core of the stars (denoted miktoarm dendritic-linear star polymers) or decorated at the periphery of the star chain-ends. The dendritic-linear miktoarm star polymers were prepared using a tandem "core-in"/"core-out" approach [31] utilizing consecutive convergent dendron attachment and ROP or controlled radical polymerization (CRP) [32]. The rational design of orthogonally protected, multifunctional compounds containing sites for the attachment of dendrons in a convergent approach together with an initiator for

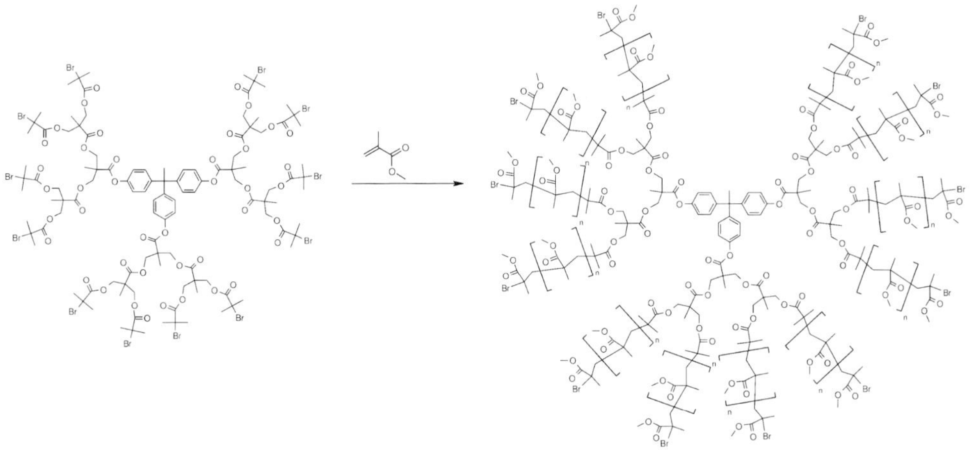

Scheme 9.2 Star-shaped poly(methyl methacrylate) via ATRP 2nd-generation dendrimer.

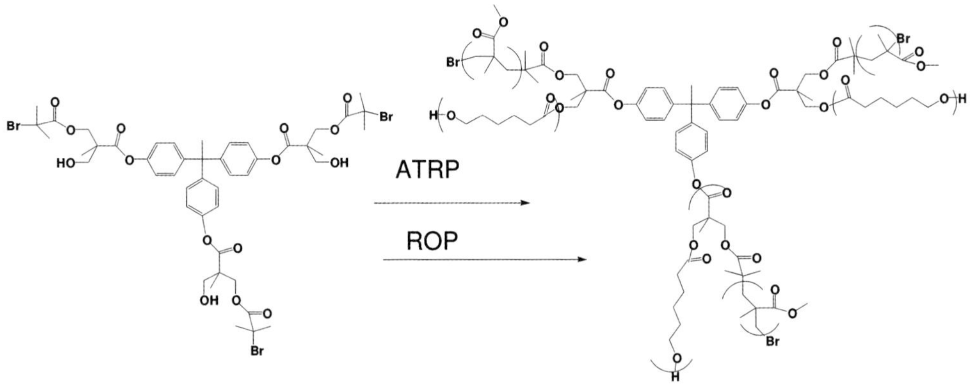

Scheme 9.3 Miktoarm star-shaped copolymer using disparate living polymerization techniques.

ROP enabled the synthesis. Dendritic-linear AB and ABA block copolymers have been prepared, where the A-blocks are hydrophilic dendrimers derived from bis-MPA derivatives and the B-blocks are linear chains of poly(ε-caprolactone) (Fig. 9.11) [33, 34]. Libraries of new initiators composed of the first through third generation protected dendrons and hydroxyl groups protected with either benzyl ether or benzylidene acetal groups were prepared. Deprotection of these hydroxyl groups by catalytic hydrogenolysis yielded the requisite nucleophilic initiators for the controlled ROP of ε-caprolactone in the presence of a suitable organometallic promoter. The nature and size of the arms and dendrons had a significant influence on the hydrodynamic radii and the ability to form self-assembled structures in solution.

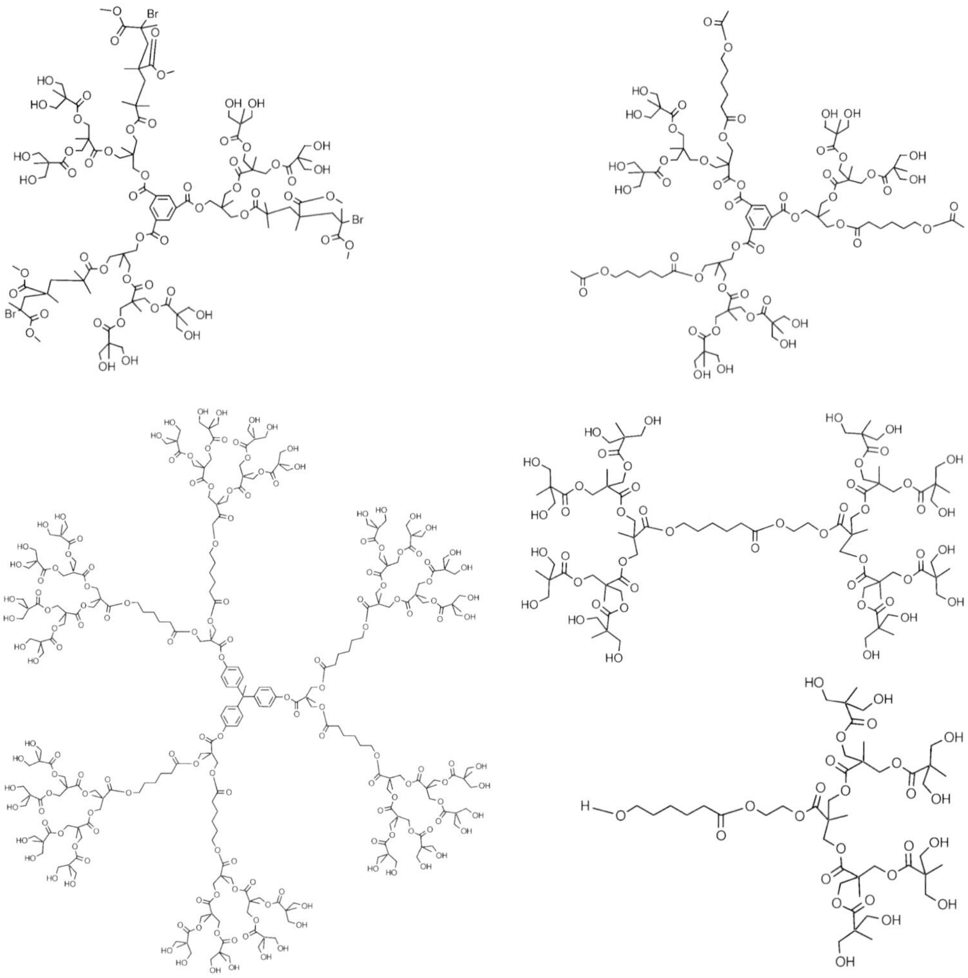

Fig. 9.11 Dendritic-linear hybrids; miktoarm, star-shaped, dumbbell and dendritic-linear hybrids.

In the second approach, Mitsunobu condensation conditions [35] were used to couple hydroxyl-functional star polycaprolactones to the focal point of protected dendrons (G-1–G-2) to produce star polymers with 12, 24 or 48 functional groups upon deprotection. Interestingly, the hydrodynamic volume for the sample functionalized with the third-generation dendron was clearly larger than the others and the extended conformation presumably stems from steric crowding associated with the large dendron on the relatively short polycaprolactone chains [36]. In a definitive study aimed at unraveling the effects of polymer architecture, six topological or constitutional isomers of dendrimer-like star polymers were prepared, each having comparable molecular weights ($\sim 80\,000$ g mol^{-1}),

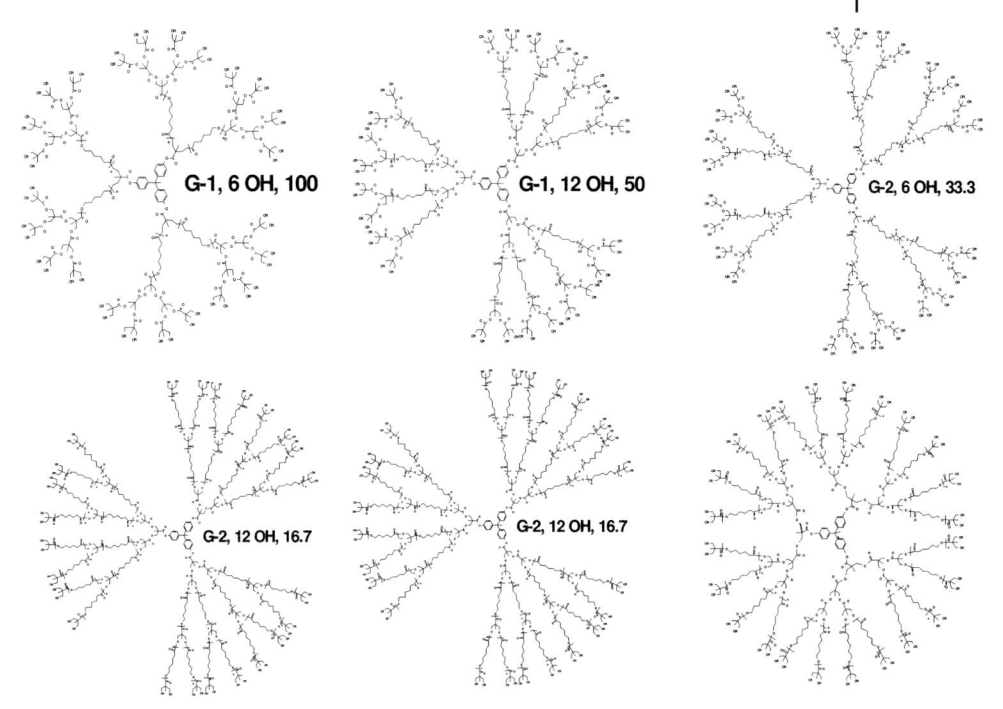

Fig. 9.12 Constitutional isomers of dendritic polymers, having 45 branching junctions, 48 hydroxyl end-groups and a molecular weight of 80 000 g mol^{-1}. Branching is located at chain-ends or core or is distributed throughout the structure. Copolymers permit basic studies of branching in star-shaped polymers.

narrow polydispersities and an identical number of branching points (45) and surface hydroxyl functionalities (48) [37]. The general structures are shown in Fig. 9.12, where the only difference in the samples is in the location of the branching sites. The most pronounced effects on the physical properties, morphology (crystallinity) and hydrodynamic volume were for the polymers in which the branching was distributed evenly throughout the sample in a dendrimer-like fashion. The versatility of this approach has provided the possibility of understanding the relationship between architecture and physical properties. Dynamic light scattering measurements yielded values of the hydrodynamic radius, R_h, which, when combined with values of the radius of gyration, R_g, from SAXS measurements, showed that the radial density distribution of molecular material was consistent with a many-arm star polymer and was not significantly biased toward core-dense or shell-dense distributions. A form of the Cotton–Daoud model for the radial density distribution of star polymers gave good agreement with experimentally determined values. Trends within the radial density distributions were consistent with the designed architectures. In dilute solution, SANS studies of these systems showed that the location of the junctions alters the size of the cores and the conformation of their component polymer chains. Furthermore, in SANS studies of semi-dilute solutions, the intermolecular interactions were affected by the core architecture; the interactions between molecules with more constrained architectures can lead to a liquid-like structuring of the cores. In complementary neutron spin-echo (NSE) spectroscopy studies, the dynamics of the internal chains of the cores were explored and it was shown that the chains are Rouse-like in their behavior. These results reveal that the peculiarities in the statics of these systems are mirrored in the dynamics of the polymeric chains comprising the cores.

Several groups have reported a new approach to hyperbranched polyesters utilizing a single-step self-condensation procedure involving ROP chemistry which is related to the self-condensing vinyl polymerizations previously reported by Fréchet's group and others [38]. The bis(hydroxymethyl)-substituted ε-caprolactone monomer was prepared in a four-step procedure. The ε-caprolactone derivative was generated by the Baeyer–Villiger oxidation of the corresponding cyclohexanone derivative. Monoprotection of 1,4-cyclohexanediol by esterification was accomplished in moderate yields by reaction of 2,2′-bis(phenyldioxymethyl)-propionyl chloride [39]. This compound was oxidized with pyridinium chlorochromate to yield the respective protected hydroxyl and bis(hydroxyl) functional cyclohexanones. The benzyl ester protecting group was readily removed by catalytic hydrogenolysis using Pd/C. This new monomer was either self-polymerized or copolymerized with ε-caprolactone at 110 °C in the presence of a catalytic amount of $Sn(Oct)_2$ (Scheme 9.4) [40].

This general polymerization concept was extended by combining self-condensing vinyl and cyclic ester polymerization through the combination of controlled radical and ring-opening polymerization techniques [41]. The concurrent polymerization of an ABC monomer with a BCD monomer was introduced as a means of expanding the type of branched copolymers and end-group functionality. γ-(ε-Caprolactone) 2-bromo-2-methylpropionate serves as a monomer (AB functionality) for ROP and an activated alkyl bromide moiety serves as an initiator (C functionality) for ATRP of vinyl monomers, once properly activated. The

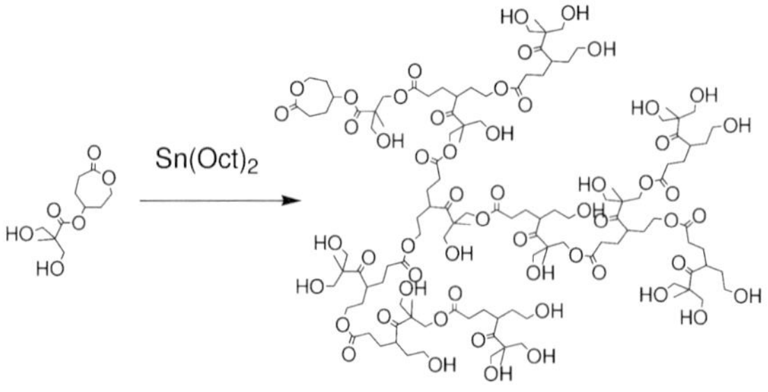

Scheme 9.4 Synthesis of hyperbranched polyesters via self-condensing ROP.

9.4 Porous Structures via Nucleation and Growth: Role of Macromolecular Architecture

other monomer, 2-hydroxyethyl methacrylate, contains an initiating site for ROP (B functionality) and a vinyl group (CD functionality) that will polymerize by ATRP. Different from "self-condensing vinyl polymerizations", where one vinyl monomer containing an initiating site (AB*) is used to prepare highly branched structures, our system employs two monomers (ABC and BCD) that polymerize by different chemistries (AB or CD) to produce branched copolymers (Scheme 9.5). The initiating center for each type of chemistry resides on the opposite monomer, precluding the possibility of self-polymerization of a given monomer, and insures copolymerization and branching.

Dendritic polyesters, prepared from AB_x macromonomers, initiated from either the first (AB_2) or second (AB_4)-generation dendrons of bis-MPA, protected as the benzyl ester, generated narrowly dispersed products with molecular weights (4000–18 000 g mol^{-1}) that closely tracked the monomer to initiator ratio (Scheme 9.6) [42]. The benzyl groups on the initiator were removed by catalytic hydrogenolysis, generating the requisite acid functionality and the a-carboxylic-ω-dihydroxy AB_2 macromonomer. Polymerization of the AB_x macromonomer was accomplished using dicyclohexylcarbodiimide (DCC) and 4-(dimethylamino)-pyridinium 4-toluenesulfonate (DPTS) in methylene chloride. High molecular weight hyperbranched polymers were obtained with the expected broadening in polydispersity associated with the condensation polymerization. ^{1}H NMR analysis of the hyperbranched polymers showed a branching pattern which is similar to those of polymers prepared by a tedious stepwise divergent approach. Hyperbranched copolymers were prepared by the co-condensation of different AB_x macromonomers.

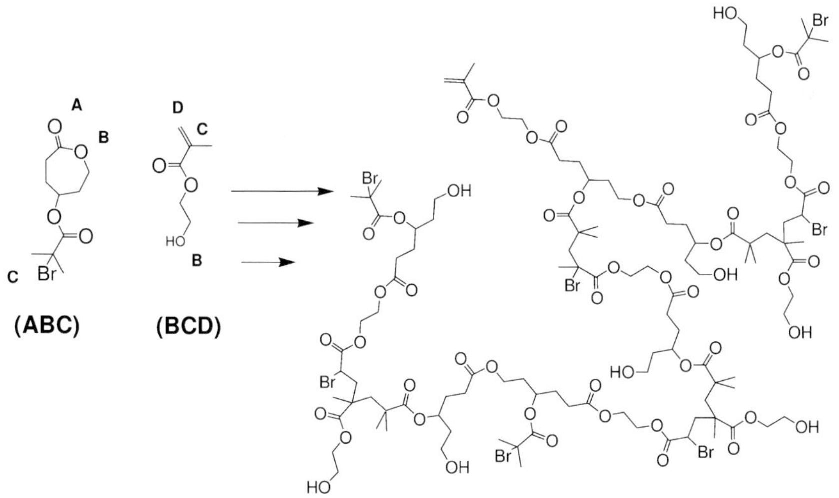

Scheme 9.5 Self-condensing vinyl and cyclic ester polymerization through controlled radical and ring-opening polymerization.

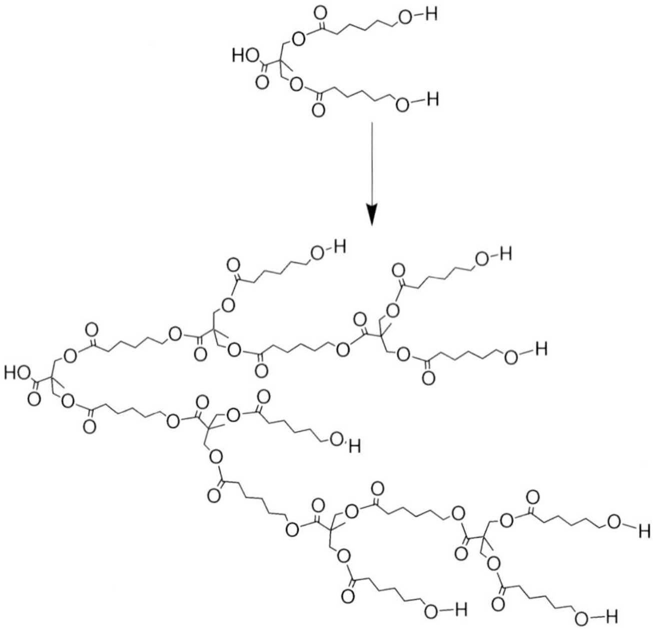

Scheme 9.6 Dendritic polyesters prepared from the polymerization of AB$_x$ macromonomers.

The functionality of the polyesters can be further increased by the incorporation of monomers containing protected functional groups (hydroxyl, bishydroxyl, amino and carboxyl) (Fig. 9.13). Each of the ε-caprolactone derivatives was generated by the Baeyer–Villiger oxidation of the corresponding cyclohexanone derivative. Monoprotection of 1,4-cyclohexanediol by benzylation or esterification was accomplished in moderate yields by reaction of benzyl bromide or 2,2′-bis-(phenyldioxymethyl)propionyl chloride. Each could be oxidized with pyridinium chlorochromate to yield the respective protected hydroxyl and bis(hydroxyl) functional cyclohexanones. The benzyl ether and benzyl ester protecting groups are readily removed by catalytic hydrogenolysis using Pd/C. Alternatively, ethyl 4-ketocyclohexylcarboxylate was hydrolyzed and the free carboxylic acid was re-esterified to the benzyl or *tert*-butyl 4-ketocyclohexylcarboxylate by esterification with either benzyl bromide or *tert*-butanol. These protecting groups were chosen because they are readily cleaved to the respective carboxylic acids under mild conditions. The azacyclohexanone derivative was produced from the commercially available ethylene ketal of 4-piperidone by acetylation with trifluoroacetic anhydride followed by transketalization using *p*-toluenesulfonic acid in excess acetone. The trifluoroacetyl protecting group is easily removed by NaBH$_4$ reduction. Removal of the protecting groups on the aliphatic polyesters yielded the functionalized polymers. Alternatively, Ree and coworkers modified the chain-ends of star-shaped polycaprolactone porogens to facilitate favorable interactions with

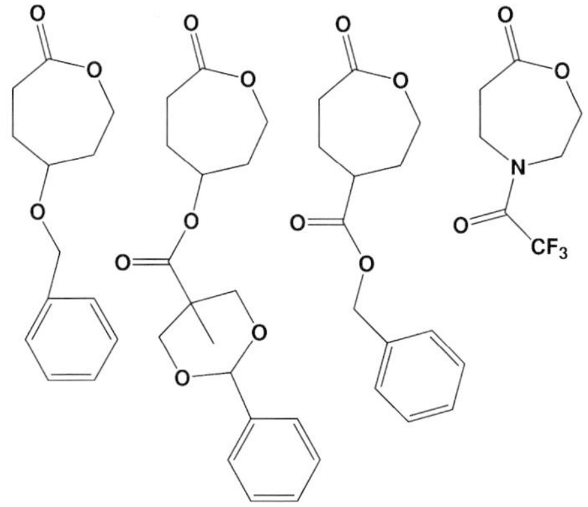

Fig. 9.13 Protected functional monomers designed to tailor the properties of the branched structures.

the organosilicate to minimize aggregation [43]. For example, triethoxysilyl groups were introduced to the chain-end that significantly minimized the increase in pore size for porogen compositions as high as 40%.

The general vision or strategy behind the investigation of novel macromolecular architectures is to generate nanocomposites with MSSQ from the formation of miscible polymer mixtures that are designed to undergo arrested nucleation and phase separation growth upon MSSQ vitrification. The miscibility is believed to result, in part, from the abundant functional groups characteristic of the structures prepared. Mixtures of the MSSQ component and various porogens were co-deposited from solution on silicon wafers by spinning and subsequent baking. The onset of network formation begins at 150 °C and by 250 °C essentially all of the SiOH functional groups are consumed, affording an organosilicate with a T_g and related properties comparable to those of a fully cured material. At these temperatures, the polyester or polymethacrylates are thermally stable. Dynamic mechanical, dielectric and thermal analysis measurements were used to study the morphology of the hybrids and many of the isomeric porogen architectures investigated were miscible with the MSSQ derivatives at low temperatures, provided their molecular weights were low. For each of the hybrids surveyed, a phase-separated morphology was generated during the initial cure (250 °C), but the size scale of phase separation was strongly dependent on the porogen architecture [44]. For example, FESEM micrographs of three representative architectures including linear ($k=2.0$, pore size >100 nm), star-shaped [$k=1.95$, pore size ∼30 nm (SAXS)] and hyperbranched [$k=2.0$, pore size ∼18 nm (SAXS)] poly(caprolactone) are shown in Fig. 9.14. The dramatic differences observed for the isomeric porogens is believed to result from the enhanced solubility of the branched materials, facilitated, in part, from the abundant functional chain-ends and the chain topology. The improved solubility is believed to delay phase separation until the later stages of the cure, minimizing coarsening of the morphology. Thermolysis of the macromolecular template generated nanoporous organosilicate films with dielectric constants that tracked the volume fraction of the template. Similarly, the use of star-shaped polymers

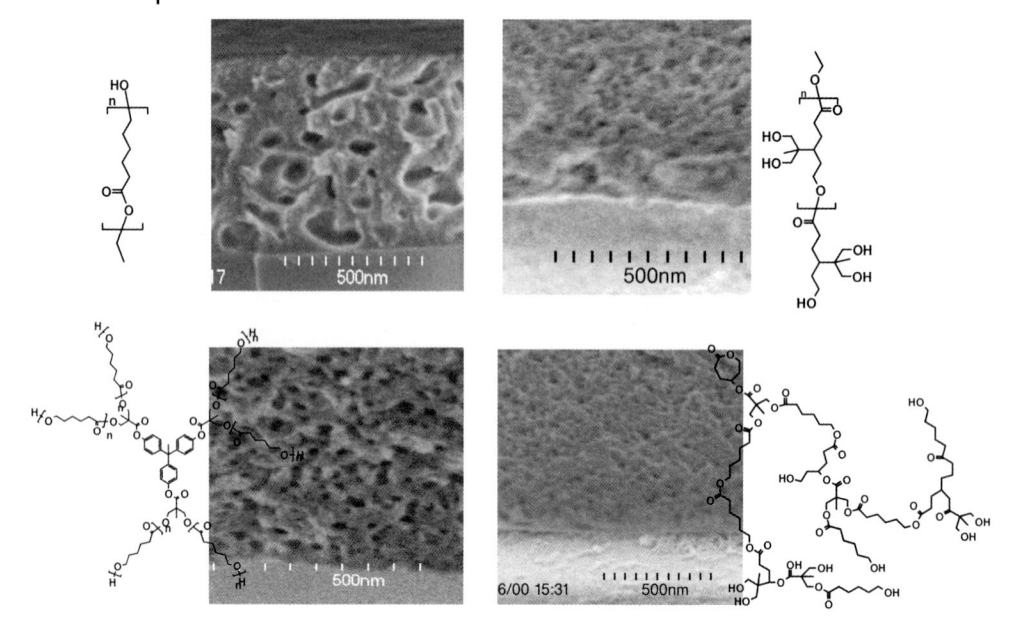

Fig. 9.14 FESEM micrographs of porous organosilicates templated from constitutional isomers of linear, linear with functionality, branched and star-shaped macromolecules, designed to establish the role of functionality and branching.

($k=2.10$, pore size ~ 20 nm) and the results are similar to those observed by other groups [45]. Interestingly, the linear polyester containing a compatibilizing pendent functionality gave comparable results.

9.5
Organic Nanoparticle Formation

A new approach to organic nanoparticle formation has provided a unique and versatile route to shape-persistent macromolecules that have dimensions characteristic of dendritic polymers [46]. This approach relies on the controlled intramolecular cross-linking of a functionalized polymer chain. The versatility of shape control is provided by the possible variables, which include the latent cross-link chemistry/density and polymer type, functionality and architecture. In one example, linear aliphatic polyesters containing pendant acrylate functionalities [47] were predominantly self-cross-linked via a radical mechanism in ultradilute conditions generating primarily single chain nanoparticles with only minor amounts of intermolecular coupling (Fig. 9.15). Hydrodynamic radii varying from 3.8 to 13.1 nm were measured by dynamic light scattering. The size of the particle increased with the molecular weight of the linear precursor polymer for a constant level of cross-linking functionality.

9.6 Block Copolymer Templates: Role of Macromolecular Architecture

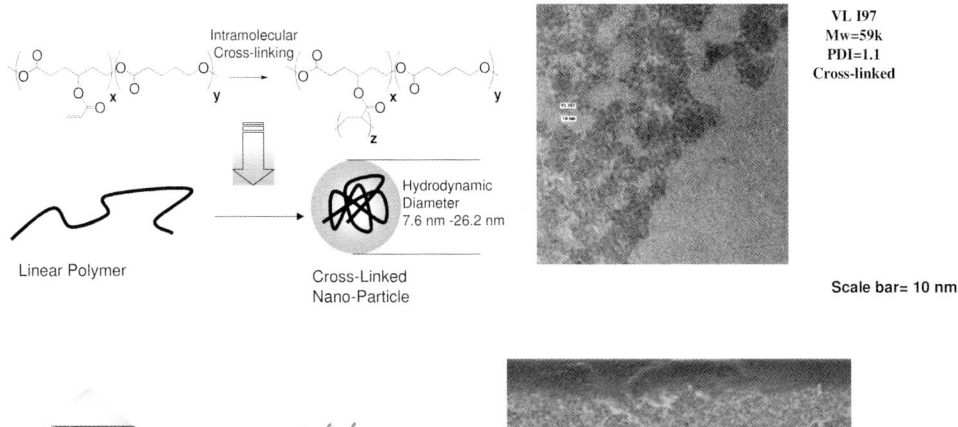

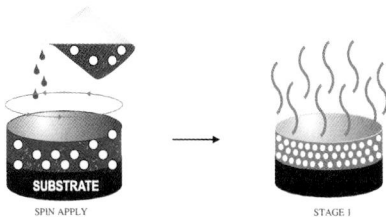

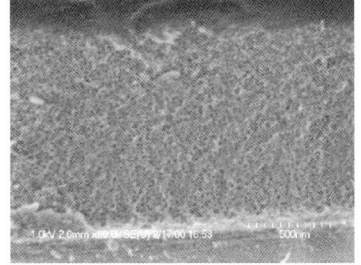

Fig. 9.15 General strategy for organic nanoparticle formation based on the collapse and self-cross-linking of an aliphatic polyester containing pendant acrylate functionalities. General strategy for hybrid formation where nanoparticles are simply dispersed into organosilicate; TEM of nanoparticles and FESEM of porous organosilicate where the porosity was templated by the dispersed particles.

Solutions of MSSQ and the polycaprolactone nanoparticles were deposited and cured to 430 °C to effect network formation and the decomposition of the nanoparticle. Figure 9.15 shows the FESEM of the porous MSSQ with and pores smaller than 10 nm uniformly distributed throughout the film. The average pore size of the porous organosilicate calculated from SAXS measurements was 7.2 nm, which compares favorably with the size of the cross-linked nanoparticles (R_h = 6.5 nm, by dynamic light scattering). The dielectric constant of a porous MSSQ prepared from 20 wt% nanoparticles was 2.1, a considerable reduction from that of the dense film (2.8).

9.6
Block Copolymer Templates: Role of Macromolecular Architecture

Block copolymers are remarkable self-assembling systems that can assume a wide variety of morphologies, including lamellar, hexagonal-packed cylindrical and body-centered cubic micellar structures, depending on the relative volume fractions of the blocks [48]. The use of amphiphilic block copolymers to direct the organization of polymerizing silica has been shown to produce well-ordered mesoporous silica with pore sizes of 75–300 Å [49]. Unlike conventional surfactants, block copolymers offer the possibility of fine-tuning the polymer–solvent phase behavior by adjusting molecular weight, composition and monomer types. The main route to mesoporous silica using macromolecular amphiphiles is through sol–gel chemistry [50, 51] and efforts to prepare continuous films with accessible porosity have been limited, particularly when the assembly has been directed. The molecular architecture of a copolymer chain also has a pronounced effect on the morphology and interfacial activity; however, the use of such complex architectures as structure-directing agents for the synthesis of nanostructured inorganic materials has received little attention.

The porous structures generated by the co-assembly of binary polymer mixtures of a diblock copolymer of polystyrene and poly(ethylene oxide) (PS–PEO)

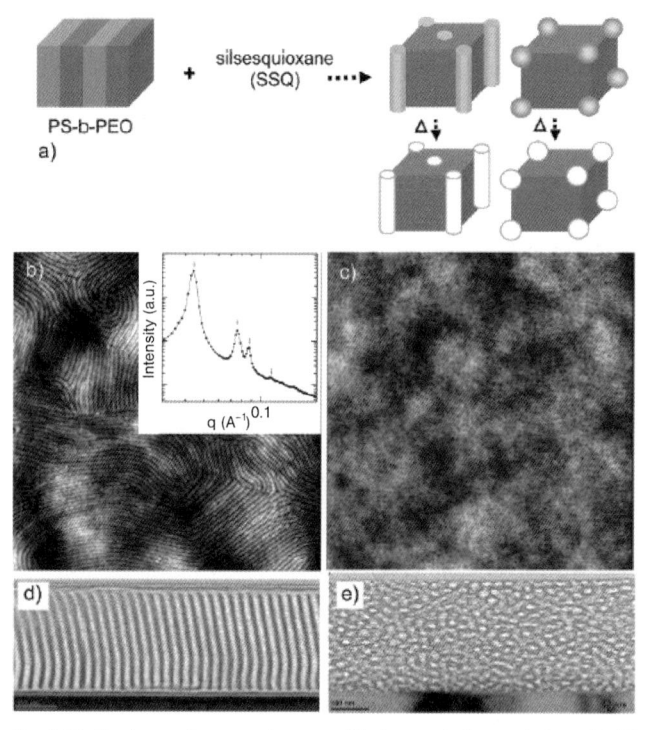

Fig. 9.16 Block copolymer and organosilicate co-assembled structures. (a) Schematic illustration of co-assembled structures. The co-assembly morphology is determined by the relative volume fractions of the PS and PEO + SSQ phases; heating to above 300 °C results in a porous, nanostructured SSQ. (b, c) 1 μm×1 μm height-contrasted AFM images (10-nm height scale) of the porous films containing a cylindrical (b) and a spherical morphology (c). The inset in (b) is the SAXS profile of the film. (d, e) TEM cross-sectional micrographs of films with a cylindrical (d) and a spherical (e) morphology (SSQ occupies darker regions). The film thicknesses are ~285 nm. The molecular weight of PS-b-PEO used for both films is 19 000 g mol^{-1}. Samples for (b)–(e) were annealed at 20 °C in chloroform for 15 h.

and MSSQ, which is selectively miscible with the PEO, is shown in Fig. 9.16. The relative volume fraction of the PS and PEO + MSSQ phases, which determines the morphology of co-assembled nanostructures, is controlled simply by varying the volume fraction of each block of the PS–PEO and the mixing composition of the PS–PEO and MSSQ mixtures (Fig. 9.16a). Cross-linking of the MSSQ occurs above 150 °C and, since this step preserves the structure, the morphology of the final porous film (created by further heating to 450 °C) is nearly identical with that of the nanohybrid. Continuous films containing cylindrical or spherical pores were generated and the porous cylindrical structure could be oriented normal to the surface by tuning the interfacial energy at the air/film and film/substrate interfaces. This process affords highly robust, thermally stable templates with a wide variety of applications.

In a second example, environmentally responsive dendritic-linear block copolymers [54] were used to organize organosilicate vitrificates into nanostructured lamellar morphologies that orient parallel to the surface and, upon thermolysis of the template, produce a porous lamellar structure between organosilicate sheets [55]. The copolymers were prepared by the divergent growth of dendrons, derived from acetal-protected anhydride derivatives of bis-MPA (Fig. 9.17) [56] from a monohydroxy functional poly(ethylene oxide) oligomer (PEG) as described by Fréchet and coworkers. Deprotection of the acetonide groups, followed by functionalization of the surface hydroxyl groups by esterification with heptanoic acid, was accomplished for copolymers with a 4–6 dendron generation. Mixtures of the block copolymers with organosilicates form hybrids with a two-phase structure upon spin coating, generated by self-assembly through the selective solubilization of the PEO with the MSSQ and concurrent collapse of the dendron into a nanostructured morphology. Dynamic mechanical analysis (DMA) and SANS measurements demonstrated that the dendron portion is phase separated in the as-cast film and serves as the macromolecular template to generate the nanostructure morphology X-ray reflectivity (XR, TEM and SAXS). Further heating of hybrid samples to 430 °C removes the block copolymer template, creating a porous thin film. Mixtures containing the higher volume fractions of copolymer manifest a nanoporous lamellar structure that is oriented parallel to the surface and this morphology is maintained over a wide compositional range ($\sim$35–70%). The bulky nature of the dendron affects their ability to pack efficiently into the traditional spherical micelles and, to help relieve some of the packing constraints of the dendrons, elongated micellar structures are generated. This, in turn, may explain the porous lamellar structure (4 nm) obtained between organosilicate sheets (6–9 nm) that persists over a wide compositional range (35–70%). Moreover, the substrate surface SiO_2 causes the lamellar structure to orient parallel over the entire substrate (Fig. 9.17). Interestingly, with this macromolecular template, lamellar morphologies were realized in the co-assembly process at very low compositions of copolymer, a distinctive feature of this architecture.

Related architectures such as dendritic-linear dumbbell-shaped structures are prepared by the divergent growth of dendrons, derived from acetal-protected an-

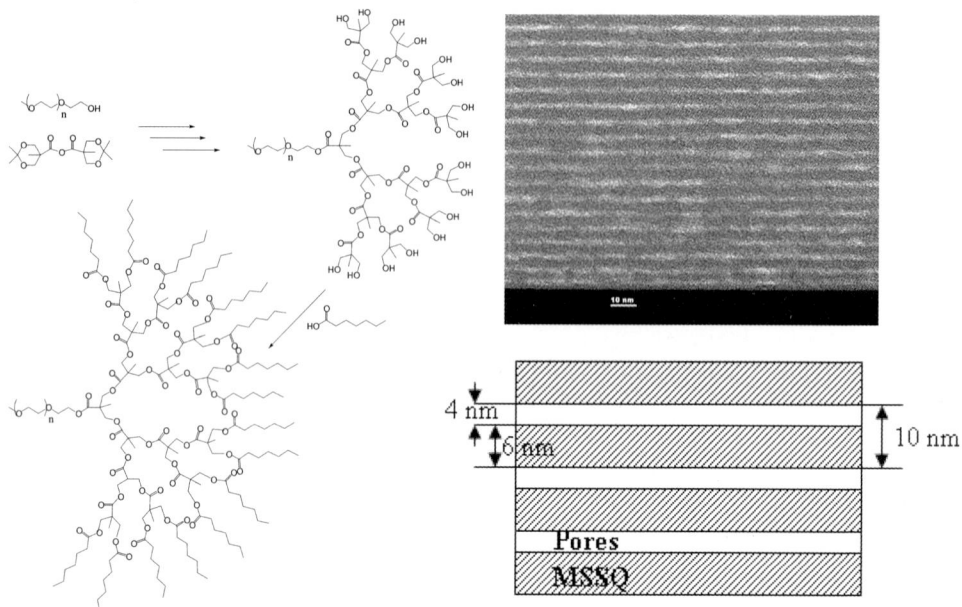

Fig. 9.17 Topologically unsymmetric dendritic-linear diblock copolymers that respond in a unique fashion to their environment were used in organic–inorganic hybrids to template an organosilicate vitrificate into nanostructures. The copolymers were prepared by the divergent growth of dendrons, derived from 2,2′-bis(hydroxymethyl)propionic acid (bis-MPA) and from a monohydroxy functional poly(ethylene oxide) oligomer (PEG). The block copolymers were used to organize organosilicate vitrificates into nanostructured lamellar morphologies that orient parallel to the surface and, upon thermolysis of the template, a perforated porous lamellar structure (4 nm) between organosilicate sheets (6–9 nm) was obtained.

hydride derivatives of bis-MPA from a difunctional PEO oligomer (Fig. 9.18). Deprotection of the acetonide groups, followed by functionalization of the surface hydroxyl groups by esterification with heptanoic acid, was accomplished for copolymers having 4th and 5th dendron generations. Mixtures of the dumbbell-shaped block copolymers with organosilicates form hybrids with a two-phase structure upon spin coating, generated by self-assembly through the selective solubilization of the PEO with the MSSQ and concurrent collapse of the dendron into a nanostructured morphology. Heating the hybrid samples to 430 °C removes the block copolymer template, creating a nanoporous thin film also with a lamellar structure that is oriented parallel to the surface (Fig. 9.18).

Another example of how the architecture of the copolymer has a pronounced effect on the structure-directing ability of the MSSQ vitrificate has been demonstrated using a stimuli-responsive , star-shaped copolymer that creates a nano-sized domain through an MSSQ-mediated collapse of the interior core of the core–corona polymeric structure. The outer corona of the star compatibilizes the insoluble core in the thermosetting resin and suppresses aggregation or pre-

9.6 Block Copolymer Templates: Role of Macromolecular Architecture

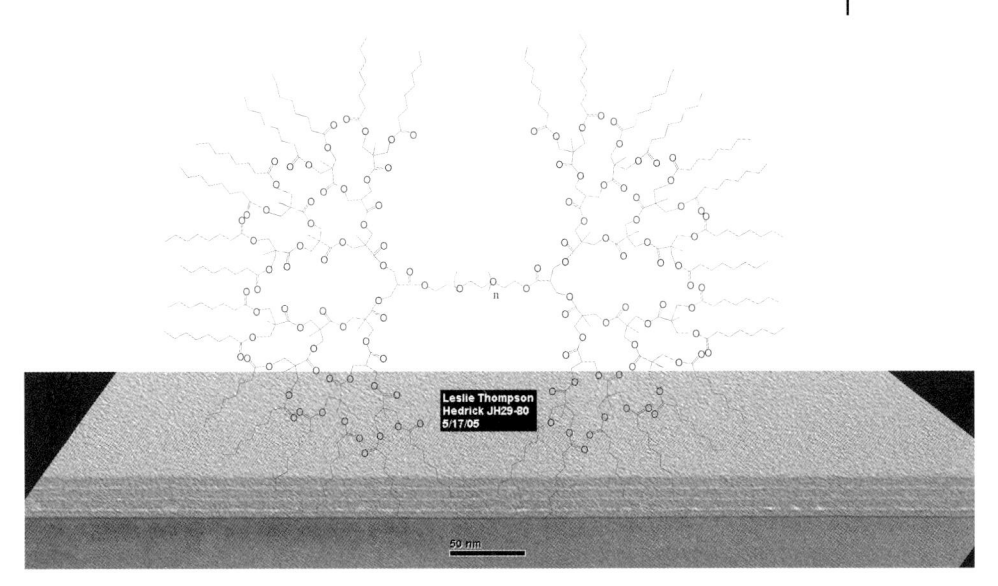

Fig. 9.18 Dendritic-linear dumbbell-shaped copolymers, prepared by divergent growth of dendrons off a PEO scaffold, used to organize vitrifying organosilicate into lamellar nanostructures that orient parallel to the surface.

cipitation of the insoluble interior so that a single polymer molecule templates cross-linking and ultimately generates a single hole (general strategy shown in Fig. 9.19). Importantly, the use of such a macromolecular architecture produces non-interconnected porosity to high volume fractions. The initial unimolecular micellar star polymers were prepared using living ruthenium-catalyzed ring-opening metathesis polymerization (ROMP), anionic or ring-opening polymerizations in tandem with controlled radical processes [57]. Shown in Scheme 9.7 is a general synthetic scheme of an amphiphilic radial block copolymer prepared by tandem controlled ring-opening and atom transfer radical polymerization [58] using dendritic imitators. The amphiphilic copolymers were dissolved in a solution containing MSSQ prepolymer in propylene glycol monomethyl ether and the resulting solution was spun on a silicon wafer to produce thin films that were cured to 430 °C to effect network formation of the MSSQ and decomposition of the sacrificial copolymer template. The refractive indices of the samples decrease predictably from 1.36 to 1.26 and 1.21 with increasing copolymer loading (20 and 40%, respectively) and dielectric constants of 2.32 and 1.95 were measured for the latter samples, consistent with a porous structure. DMA, TEM, SANS and SAXS measurements strongly suggest that the core is phase separated in the as-cast films and serves as a macromolecular template, where one star generates one pore (Fig. 9.20). The porous morphology was strongly dependent on the arm number, molecular weight and type of the interior core and also the volume fraction of copolymer in MSSQ mixtures

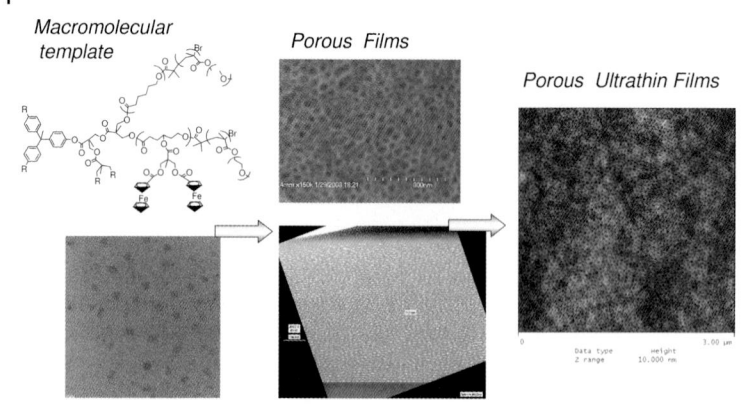

Fig. 9.19 Star-shaped environmentally responsive copolymers create nano-sized domains through a matrix-mediated collapse of the interior core of the core–corona polymeric structure. This approach relies on the outer corona of the star to compatibilize the insoluble core with the thermosetting resin and prevent aggregation such that these individual molecules template the cross-linking of the matrix and ultimately generate a single hole. Shown is TEM of ultra-thin films of the radial copolymer (a 12-arm PCL core with an average DP of 50 per arm) deposited from water was used to investigate the responsiveness in a polar environment and also obtain an estimate of the size in the "dry" state. The contrast in the TEM micrograph of the copolymer, resulting from the ferrocene functionality introduced into the PCL core, confirms the two-phase structure containing a collapsed core with a diameter around 22 nm and $\sim$49 nm spacing between the core centers. FESEM and TEM micrographs of copolymer–MSSQ mixtures heated to 430 °C and porous thin films (AFM micrograph).

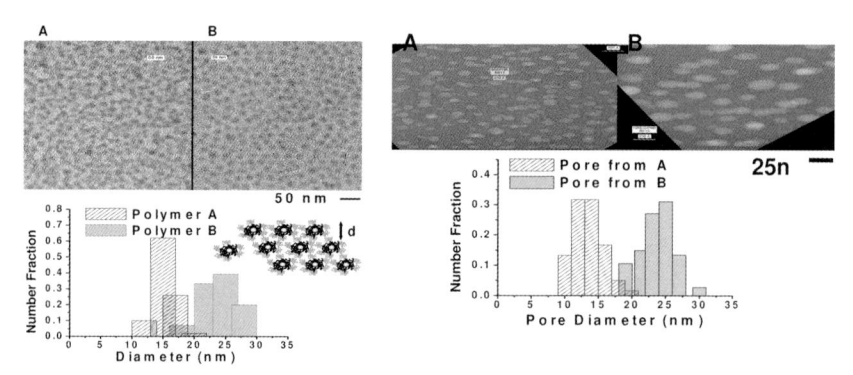

Fig. 9.20 TEM of a low (a) and high (b) molecular weight block copolymer where the core has been selectively stained with core sizes of $\sim$15 and 25 nm. Porous organosilicates generated from the macromolecular template described above. One molecule generates one hole.

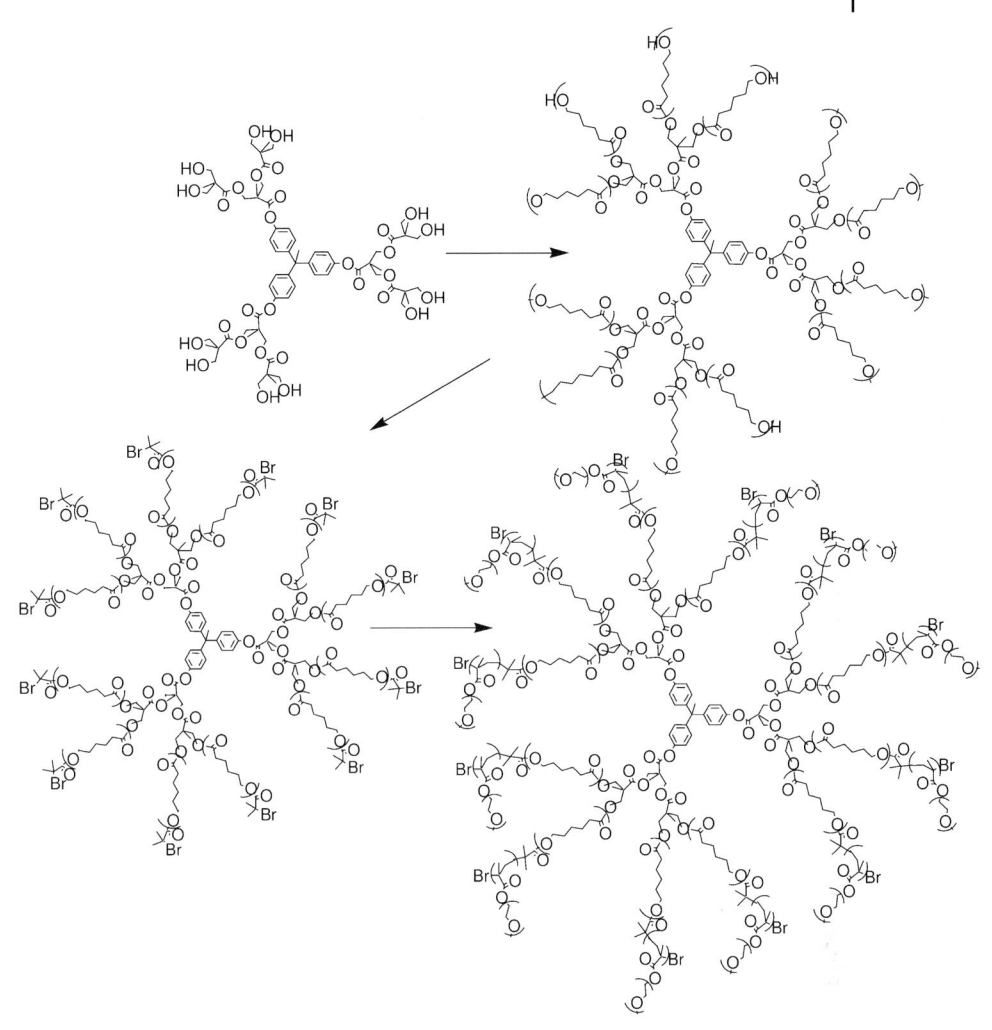

Scheme 9.7 Synthesis of radial block copolymer via ROP and ATRP.

(Fig. 9.21). The bottom-up/core-out approach to star porogens produced porous structures with pores that range in size from 7 to 40 nm, depending on the polymer molecular weight and architecture.

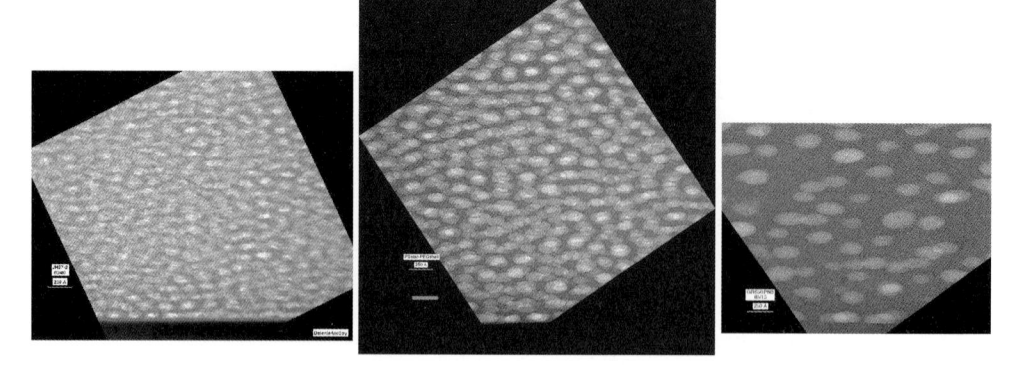

Fig. 9.21 Generality of radial block copolymer-templated organosilicate vitrification. Variation of core from poly(caprolactone), polystyrene and norbornene (from left to right).

9.7
Outlook

It is now well established that the introduction of porosity into organosilicate thin films negatively impacts the mechanical properties of these materials, manifested by a reduction in the Young's modulus and an increase in crack propagation rate in water [59]. Consequently, integration of these ultra-porous materials remains a continuing challenge, because of cracking and damage observed during photo-resist strip, etching, chemical mechanical polishing (CMP), dicing and wire bonding.

These issues have been addressed with limited success by varying the resin and porogen structures, addition of toughening additives [60], control of porous morphologies and post-porosity treatment using UV [61] or E-beam exposure [62]. Although improvements in moduli of 25–50% are often achieved upon post-porosity treatment, this requires a separate processing step and chemical changes are produced in the materials, which can degrade the electrical properties. An equivalent improvement in terms of modulus was obtained by changing the nature of the porogen. As can be seen in Fig. 9.22, while an exponential decay behavior is observed for the Young's modulus as a function of the porogen loading, a $\sim 30\%$ improvement is obtained for the particle porogen at 40 vol% ($k=2.0$).

Independently of the toughening method used, the mechanical properties of most porous organosilicate remain low: 1–1.5 GPa at a dielectric constant of 2.0 (40 vol.% porogen).

In an effort to improve significantly the mechanical properties of low-k materials without using post-porosity treatment methods, we and others have investigated the templated sol–gel polymerization of bridged silicon monomer precursors [63] (Fig. 9.23). These hybrid organic–inorganic films are obtained by the

9.7 Outlook

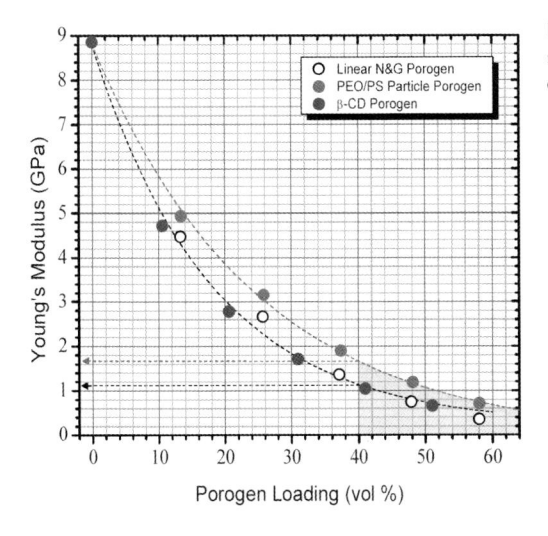

Fig. 9.22 Influence of the porogen structure on the mechanical properties.

bis(triethoxysilyl)methane

1,2-bis(triethoxysilyl)ethane

1,3,5-tris(diethoxysila)cyclohexane

Fig. 9.23 Examples of bridge silicon precursors used for the preparation of low-k films.

sol–gel process through the hydrolysis and polycondensation of organosilicon precursors containing of at least two Si(OR)$_3$ groups with R=Me, Et. Although new to the semiconductor industry, materials obtained from the sol–gel polymerization of bridged silicon precursors have been known for several years [64]. A variety of organic building blocks have been reported for the preparation of bridged polysilsequioxanes, ranging from rigid arylenic [65], acetylenic [66], olefinic [67] and alkylenes incorporating 1–14 methylene groups [68] to functionalized groups, such as amines [69], ethers [70], sulfides [71], phosphines [72], amides [73], ureas [74], carbamates [75] and carbonates [76]. The sol–gel polymerization is generally acid-, base- or fluoride-catalyzed, leading to the rapid formation of polymers that irreversibly form gels and subsequently afford amorphous xerogels and aerogels, upon drying [77]. In this case, the porosity formed is a function of the sol–gel parameters and drying conditions [78].

On the other hand, the synthesis of porous films requires the addition of alkylene oxide polymeric surfactants to sol–gel solutions. Spin coating of these solutions on to a silicon wafer leads after thermal processing to high-quality,

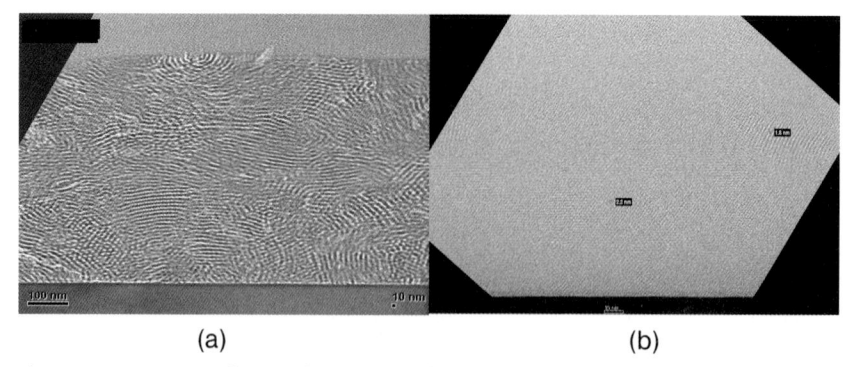

Fig. 9.24 TEM images of porous low-dielectric films obtained after decomposition of P123 (Bayer) (a) or a Brij surfactant

low-dielectric films. TEM images of porous films obtained from a methylene-bridged silicon precursor and alkylene oxide surfactant are presented in Fig. 9.24. As shown, the porogen decomposition leads to the formation of organized cylindrical pores at the wafer surface, the size of which depends on the surfactant structure and molecular weight. The Young's modulus measured for these bridged polysilsequioxanes is exceptionally high, 4–6 times that produced from SSQ derivatives with a similar porogen and 2–3 times that obtained from UV-treated materials deposited by either spin-on or CVD processes. As a result of their excellent thermal, dielectric, electrical and mechanical properties, these materials are promising new candidates for dielectric applications at $k=2.3$ and beyond.

References

1 G. E. Moore, *Electronics*, **1965**, *38*(8), April, 19; (b) G. E. Moore, *IDEM Tech. Dig.*, **1975**, 11.
2 Wilson, S. R. in *Handbook of Multilevel Metallization for Integrated Circuits*, S. R. Wilson, C. J. Tracy, J. L. Freeman (Eds.). Noyes, Park Ridge, NJ, **1993**, Chap. 1.
3 D. Edelstein, J. Heidenreich, R. Goldblatt, W. Cote, C. Uzoh, N. Lustig, P. Roper, T. McDevitt, W. Motsiff, A. Simon, J. Durkovic, R. Wachnik, H. Rathore, R. Schultz, L. Su, S. Luce, J. Slattery *Tech. Dig. IEEE Int. Electron Devices Mtg.* **1997**, 376.
4 V. H. Nguyen, H. van Kranenburg, P. H. Woerlee *Copper for Advanced Interconnect*, Proceedings of the Third International Workshop on Materials Science (IWOMS'99), 2–4 November **1999**, Hanoi.
5 N. Matsunaga, N. Nakamura, K. Higashi, H. Yamaguchi, T. Watanabe, K. Akiyama, S. Nakao, K. Fujita, H. Miyajima, S. Omoto, A. Sakata, T. Katata, Y. Kagawa, H. Kawashima, Y. Enototo, T. Hasegawa, H. Shibata. BEOL International Interconnect Technical Conference (IITC) Proceedings, San Francisco, **2005**.
6 For example: (a) J. L. Hedrick, J. W. Labadie, T. P. Russell, D. Hofer, Wakharker *Polymer* **1993**, *34*, 4717; (b) Y. Charler, J. L. Hedrick, T. P. Russell, W. Wolksen, *Polymer* **1995**, *36*, 48; (e) B. Lee, W. Oh, Y. Hwang, Y.-H. Park,Yoon, K. S. Jin, K. Heo, J. Kin, W. Kim, M. Ree, *Adv. Mater.* **2005**, *17*, 696.
7 M. Ree, J. Yoon, K. Heo, *J. Mater. Chem* **2006**, *16*, 685.
8 (a) A. Grill, V. Patel, K. P. Rodbell, E. Huang, M. R. Baklanov, K. P. Mogilnkov,

M. Toney, H.-C. Kim, *J. Appl. Phys.* **2003**, *94*, 3427; (b) A. Grill, V. Patel, *Appl. Phys. Lett.* **2001**, *79*, 803.

9 For example: (a) D. Zhao, J. Feng, Q. Huo, N. Melosh, G. H. Fredickson, B. F. Chmelka, G. D. Stuckey, *Science* **1998**, *279*, 548; (b) P. J. Bruinsma, N. J. Hess, J. R. Bontha, S. Baskaran, *Mater. Res. Soc. Proc.* **1996**, *443*, 105; (c) Y. Lu, R. Ganguli, C. A. Drewien, M. T. Anderson, C. J. Brinker, W. Gong, Y. Guo, H. Soyez, B. Dunn, M. H. Huang, J. I. Zink, *Nature* **1997**, *389*, 364

10 For example: (a) Q. R. Huang, H.-C. Kim, E. Huang, D. Mecerreyes, J. L. Hedrick, W. Volksen, C. W. Frank, R. D. Miller, *Macromolecules* **2003**, *36*, 7661; (b) Q. R. Huang, W. Volksen, E. Huang, M. Toney, C. W. Frank, R. D. Miller, *Chem. Mater.* **2002**, *14*, 3676; (c) R. D. Miller, R. Beyers, K. R. Carter, R. F. Cook, M. Harbison, C. J. Hawker, J. L. Hedrick, V. Y. Lee, E. Liniger, C. Nguyen, J. Remenar, M. Sherwood, M. Trollsas, W. Volksen, D. Y. Yoon, *Mater. Res. Soc. Proc.* **1999**, *565*, 4.

11 (a) A. N. Husing, U. Schubert, *Angew. Chem. Int. Ed.* **1998**, *37*, 22; (b) L. W. Hrubesh, L. E. Keene, V. R. Latore, *J. Mater. Res.* **1993**, *8*, 1736.

12 (a) T. Ramos, K. Roderick, A. Maskara, D. M. Smith, *Mater. Res. Soc. Proc.* **1997**, *443*, 91; (b) D. M. Smith, J. Anderson, C. C. Cho, G. P. Johnson, S. P. Jeng, *Mater. Res. Soc. Proc.* **1995**, *381*, 261; (c) A. Jain, S. Rogojevic, S. V. Nitta, V. Pisupatti, W. G. Gill, P. C. Wayer Jr., J. L. Plawsky, *Mater. Res. Soc. Proc.* **1999**, *565*, 29.

13 For example: J. A. Tripp, F. Svec, J. M. J. Fréchet, S. Zeng, J. C. Mikkelsen, J. G. Santiago, *Sens. Actuators B* **2004**, *99*, 66; (b) E. C. Peters, F. Svec, J. M. J. Fréchet, *Chem. Mater.* **1997**, *9*, 1898; (c) E. Hilder, F. Svec, J. M. J. Fréchet, *Electrophoresis*, **2002**, *23*, 3934.

14 S. Yang et al. *Chem. Mater.* **2001**, *13*, 2762; (b) S. Yang et al. *Chem. Mater.* **2002**, *14*, 369.

15 Y. Chujo, T. Saegusa, *Adv. Polym. Sci.* **1992**, *100*, 12; (b) T. Saegusa, Y. Chujo, *Makromol. Chem. Macromol. Symp.* **1992**, *64*, 1; (c) R. Tamaki, K. Naka, Y. Chujo *Polym. J.* **1998**, *30*, 60.

16 Y. Chujo, H. Matsuki, S. Kure, T. Saegusa, T. Yazawa, *J. Chem. Soc., Chem. Commun.* **1994**, 635; (b) G. Larsen, E. Lotero, M. Marquez, *J. Phys. Chem. B* **2000**, *104*, 4840; (c) G. Larsen, E. Lotero, M. Marquez, *Chem. Mater.* **2000**, *12*, 1513.

17 P. Froehling, J. Brackman, *Macromol. Symp.* **2000**, *151*, 581

18 K. Matyjaszewski, J. Xia, In *Handbook of Radical Polymerization*, K. Matyjaszewski, T. P. Davis (Eds.). Wiley, Hoboken, NJ, **2002**, Chap. 11.

19 C. J. Hawker, J. L. Hedrick, E. E. Huang, V. Y. Lee, T. Magbitang, D. Mecerreyes, R. D. Miller, W. Volksen, *US Patent 6685983*, **2004**; (b) C. J. Hawker, J. L. Hedrick, E. E. Huang, V. Y. Lee, T. Magbitang, R. D. Miller, Volksen, V. *US Patent 6670285*, **2003**.

20 Q. R. Huang, W. Volksen, E. Huang, M. Toney, C. W. Frank, R. D. Miller, *Chem. Mater.* **2002**, *14*, 3676; (b) Q. R. Huang, H.-C. Kim, E. Huang, D. Mecerreyes, J. L. Hedrick, W. Volksen, C. W. Frank, R. D. Miller, *Macromolecules* **2003**, *36*, 7661; (c) Q. R. Huang, C. W. Frank, D. Mecerreyes, W. Volksen, R. D. Miller, *Chem. Mater.* **2005**, *17*, 1521.

21 N. Gogibus, A. Heise, T. Magbitang, R. D. Miller, unpublished results.

22 E. Huang, M. F. Toney, W. Volksen, D. Mecerreyes, P. Brock, H.-C. Kim, C. J. Hawker, J. L. Hedrick, V. Y. Lee, T. Magbitang, R. D. E. Miller, *Appl. Phys. Lett.* **2002**, *81*, 2232.

23 H.-C. Kim, W. Volksen, R. D. Miller, E. Huang, G. Yang, R. M. Briber, K. Shin, S. K. Satija, *Chem. Mater.* **2003**, *15*, 609.

24 D. A. Maidenberg, W. Volksen, R. D. Miller, R. Dauskardt *Nat. Mater.* **2004**, *3*, 464.

25 L. K. Johnson, C. M. Killian, M. Brookhart, *J. Am. Chem. Soc.* **1995**, *117*, 6414.

26 For example: (a) D. A. Tomalia, J. M. J. Fréchet, *J. Polym. Sci., Part A: Polym. Chem. Ed.* **2002**, *40*, 2719; (b) C. J. Hawker, K. L. Wooley, *Science* **2005**, *309*, 1200; (c) A. W. Bosman, H. M. Jansen, E. W. Meijer, *Chem. Rev.* **1999**, *99*, 1665; (d) S. Hecht, J. M. J. Fréchet, *Angew. Chem. Int. Ed.* **2001**, *40*, 74; (e) H. Frey,

Angew. Chem. Int. Ed. **1998**, *37*, 2193; (f) A. Zhu, P. Bharathi, J. O. White, H. G. Drickamer, J. S. Moore, *Macromolecules* **2001**, *34*, 4606; (g) H. A. Al-Muallem, D. M. Knauss, *J. Polym. Sci., Polym. Chem.* **2001**, *39*, 152; (i) T. Sato, D.-L. Jiang, T. Aida, *J. Am. Chem. Soc.* **1999**, *121*, 10658; (j) V. Percec, W. D. Cho, G. Ungar, D. J. P. Yeardley, *J. Am. Chem. Soc.* **2001**, *123*, 1302; (k) R. Gronheid, J. Hofkens, F. Köhn, T. Weil, E. Reuther, K. Müllen, F. C. De Schryver, *J. Am. Chem. Soc.* **2002**, *124*, 2418; (l) J. M. J. Fréchet, *Science* **1994**, *263*, 1710; (m) J. L. Hedrick, R. D. Miller, C. J. Hawker, K. R. Carter, W. Volksen, D. Yoon, M. Trollsås, *Adv. Mater.* **1998**, *10*, 1049.

27 For example: J. L. Hedrick et al. *Chem. Eur. J.* **2002**, *8*, 3308; (b) B. Lee, H. Y. Park, Y. Hwang, K. S. Jin, W. Oh, M. Ree, *Nat. Mater.* **2005**, *4*, 147; (c) Y. Chujo, H. Matsuki, S. Kure, T. Saegusa, T. Yazawa, *Chem. Commun.* **1994**, 635.

28 M. Trollsås, C. J. Hawker, J. F. Remenar, J. L. Hedrick, M. Johansson, H. Ihre, A. Hult, *J. Polym. Sci., Chem. Ed.* **1998**, *36*, 2793; (b) B. Atthoff, M. Trollss, H. Claesson, J. L. Hedrick, *Macrom. Chem. Phys. Ser.* **1998**, *713*, 127; (c) S. Hecht, N. Vladimirov, J. M. J. Fréchet, *J. Am. Chem. Soc.* **2001**, *123*, 18.

29 (a) H. Ihre, O. L. Padilla De Jesus, J. M. J. Fréchet, *J. Am. Chem. Soc.* **2001**, *123*, 5908; (b) H. Ihre, A. Hult, E. Soderlind, *J. Am. Chem. Soc.* **1996**, *178*, 6388; (c) E. Malmstrom, M. Johansson, A. Hult, *Macromolecules* **1995**, *28*, 1698; (d) M. Malkoch, E. Malmstrom, A. Hult, *Macromolecules* **2002**, *35*, 8307; (d) E. Malmstrom, M. Johansson, A. Hult, *Macromolecules* **1995**, *95*, 137; (e) E. Malmstrom, M. Johansson, A. Hult, *Macromol. Chem. Phys.* **1996**, *197*, 3199; (f) H. R. Ihre, O. L. P. De Jesus, F. C. Szoka, J. M. J. Fréchet, *Bioconj. Chem.* **2002**, *13*, 443.

30 A. Heise, C. Nguyen, R. Malek, J. L. Hedrick, C. W. Frank, R. D. Miller, *Macromolecules* **2000**, *33*, 2346.

31 T. Glauser, C. M. Stancik, M. Moller, S. Voytek, A. P. Gast, J. L. Hedrick, *Macromolecules* **2002**, *35*, 5774.

32 (a) M. Trollss, J. L. Hedrick, *J. Am. Chem. Soc.* **1998**, *120*, 4644; (b) M. Trollsås, H. Claesson, B. Atthoff, J. L. Hedrick, *Angew. Chem. Int. Ed.* **1998**, *37*, 3132; (c) M. Trollsås, J. L. Hedrick, D. Mecerreyes, Ph. Dubois, R. Jérôme, H. Ihre, A. Hult, *Macromolecules* **1998**, *31*, 2756; (d) M. Trollsâs, M. A. Kelly, H. Claesson, R. Siemens, J. L. Hedrick, *Macromolecules* **1999**, *32*, 4917; (e) Y. Gnanou, D. Taton, *Macromol. Symp.* **2001**, *174*, 333; (f) D. Seebach, G. F. Herrmann, U. Lengweiler, B. M. Bachmann, W. Amrein, *Angew. Chem. Int. Ed. Engl.* **1996**, *35*, 2795.

33 A. Wursch, M. Möller, T. Glauser, L. S. Lim, S. B. Voytek, J. L. Hedrick, C. W. Frank, J. G. Hilborn, *Macromolecules* **2001**, *34*, 6601.

34 (a) V. Percec, C. Ahn, G. Ungar, D. Yeardley, M. Möller, S. Sheiko, *Nature* **1998**, *391*, 161; (b) V. Percec, D. Schlueter, G. Ungar, S. Z. D. Cheng, A. Zhang, *Macromolecules* **1998**, *31*, 1745; (c) V. Percec, C.-H. Ahn, W.-D. Cho, A. M. Jamieson, J. Kim, T. Leman, M. Schmidt, M. Gerle, M. Möller, S. Prokhorova, S. Sheiko, S. Z. D. Cheng, A. Zhang, G. Ungar, D. J. P. Yeardley, *J. Am. Chem. Soc.* **1998**, *120*, 8619; (d) J. M. J. Fréchet, I. Gitsov, T. Monteil, S. Rochat, J. Sassi, C. Verlati, D. Yi, *Chem. Mater.* **1999**, *11*, 1267; (e) J. P. Kampf, C. W. Frank, E. E. Malmström, C. J. Hawker, *Langmuir* **1999**, *15*, 227; (f) I. Gitsov, K. L. Wooley, J. M. J. Fréchet, *Angew. Chem. Int. Ed. Engl.* **1992**, *31*, 1200; (g) I. Gitsov, K. L. Wooley, C. J. Hawker, P. Ivanova, J. M. J. Fréchet, *Macromolecules* **1993**, *26*, 5621; (h) I. Gitsov, P. Ivanova, J. M. J. Fréchet, *Macromol. Rapid Commun.* **1994**, *15*, 387; (i) J. C. M. van Hest, D. A. P. Delnoye, M. W. P. L. Baars, C. Elissen-Roman, M. H. P. van Genderen, E. W. Meijer, *Chem. Eur. J.* **1996**, *2*, 1616; (j) J. C. M. van Hest, D. A. P. Delnoye, M. W. P. L. Baars, M. H. P. van Genderen, E. W. Meijer, *Science* **1995**, *268*, 1592; (h) P. T. Hammond et al. *Macromolecules*, **2002**, *35*, 231.

35 Mitsunobo, *Synthesis* **1981**, 1.

36 M. Trollsås, H. Claesson, B. Atthoff, J. L. Hedrick, J. A. Pople, A. P. Gast, *Macromol. Symp.* **2000**, *153*, 87.

37 M. Trollsås, B. Atthoff, J. L. Hedrick, J. A. Pople, A. P. Gast, *Macromolecules* **2000**, *33*, 6423.

38 (a) J. M. J. Fréchet, M. Henmi, I. Gitsov, S. Aoshima, M. R. Leduc, R. B. Grubbs, *Science* **1995**, *269*, 1080; (b) C. J. Hawker, J. M. J. Fréchet, R. G. Grubbs, J. Dao, *J. Am. Chem. Soc.* **1995**, *117*, 10763; (c) S. G. Gaynor, S. Edelman, K. Matyjaszewski, *Macromolecules* **1996**, *29*, 1079; (d) P. Lu, J. K. Paulasaari, W. P. Webber, *Macromolecules* **1997**, *30*, 7024; (e) K. Matyjaszewski, S. G. Gaynor, M. Kulfan, Podwika, *Macromolecules* **1997**, *30*, 5192; (f) K. Matyjaszewski, J. Pyun, S. Gaynor, *Macromol. Rapid Commun.* **1998**, *19*, 665; (g) M. W. Weimer, J. M. J. Fréchet, I. Gitsov, *J. Polym. Sci., Part A: Polym. Chem.* **1998**, *36*, 955.

39 M. Trollsås, V. Lee, D. Mecerreyes, P. Lowenhielm, M. Möller, R. D. Miller, J. L. Hedrick, *Macromolecules* **2000**, *33*, 4619.

40 (a) M. Liu, N. Vladimirov, J. M. J. Fréchet, *Macromolecules* **1999**, *32*, 6881; (b) M. Trollss, P. Lowenhielm, V. Lee, M. Möller, R. D. Miller, J. L. Hedrick, *Macromolecules* **1999**, *32*, 9062.

41 D. Mecerreyes, M. Trollsås, J. L. Hedrick, *Macromolecules* **1999**, *32*, 8753.

42 (a) M. Trollsås, J. L. Hedrick, D. Mecerreyes, P. Dubois, R. Jérôme, H. Ihre, A. Hult, *Macromolecules* **1997**, *30*, 8508; (b) M. Trollss, J. L. Hedrick, *Macromolecules* **1998**, *31*, 4390; (c) M. Trollsås, B. Atthoff, H. Claesson, J. L. Hedrick, *Macromolecules* **1998**, *31*, 3439; (d) M. Trollsås, J. L. Hedrick, D. Mecerreyes, P. Dubois, R. Jérôme, *J. Polym. Sci., Part A: Polym. Chem. Ed.* **1998**, *36*, 3187.

43 B. Lee, W. Oh, J. Yoon, Y. Hwang, J. Kin, B. G. Landes, M. Ree, *Macromolecules* **2006** in press.

44 (a) C. Nguyen, K. R. Carter, C. J. Hawker, J. L. Hedrick, R. Jaffy, R. D. Miller, J. Remenar, H. Rhee, P. Rice, M. Toney, D. Yoon, *Chem. Mater.* **1999**, *11*, 3080; (b) C. Nguyen, C. J. Hawker, R. Miller, J. L. Hedrick, J. G. Hilborn, *Macromolecules* **2000**, *33*, 4281; (c) D. Mecerreyes, E. Huang, T. Magbitang, W. Volksen, C. J. Hawker, V. Lee, R. D. Miller, J. L. Hedrick, *High Perform. Polym.* **2001**, *13*, 11; (d) J. L. Hedrick, C. J. Hawker, M. Trollsås, J. Remenar, D. Y. Yoon, R. D. Miller, *Mater. Res. Symp. Proc.* **1998**, *519*, 65.

45 (a) B. H. Bolze, M. Ree, H. S. Youn, S.-H. Chyu, K. Char, *Langmuir* **2001**, *17*, 6683; (b) B. Lee, W. Oh, J. Yoon, Y. Hwang, M. Ree, S.-H. Chu, K. Char, *Polymer* **2004**, *44*, 2519.

46 D. Mecerreyes, V. Lee, C. J. Hawker, J. L. Hedrick, A. Wurssch, W. Volksen, T. Magbitang, E. Huang, R. D. Miller, *Adv. Mater.* **2001**, *13*, 204.

47 (a) D. Mecerreyes, R. D. Miller, J. L. Hedrick, C. Detrembleur, R. Jérme, *J. Polym. Sci., Part A: Polym. Chem.* **2000**, *38*, 870; (b) D. Mecerreyes, J. Humes, R. D. Miller, J. L. Hedrick, C. Detremblur, P. Lecomte, R. Jérme, *Macromol. Rapid Commun.* **2000**, *21*, 779.

48 For example: (a) E. L. Thomas *Polymer*, **2003**, *44*, 6725; (b) F. Bated et al. *Macromolecules*, **2003**, *36*, 782; (c) T. P. Russell et al. *Adv. Mater.* **2002**, *14*, 1373.

49 For example: (a) M. Templin, A. Franck, A. Du Chesne, H. Leist, T. Ahang, R. Ulrich, V. Schadler, U. Wiesner, *Science* **1997**, *278*, 1795; (b) D. Zhao, J. Feng, Q. Huo, N. Melosh, G. H. Frederickson, B. F. Chmelka, G. D. Stucky *Science* **1998**, *279*, 548; (c) E. F. Connor, L. K. Sundberg, H.-C. Kim, J. J. Cornelissen, T. Magbitang, P. M. Rice, V. Y. Lee, C. J. Hawker, W. Volksen, J. L. Hedrick, R. D. Miller *Angew. Chem. Int. Ed.* **2003**, *42*, 3785; (d) Y. Y. Yuan, P. Hentze, W. M. Arnold, B. K. Marlow, M. Antonietti, *Nano-Lett* **2002**, *2*, 1359; (e) A. Jain, U. Wiesner, *Macromolecules* **2004**, *37*, 5665; (f) P. Simon, R. Ulrich, H. W. Spiess, U. Wiesner, *Chem. Mater.* **2001**, *13*, 3464; (g) T. Magbitang et al. *Adv. Mater.* **2005**, *17*, 1031; (h) C. J. Brinker et al. *Langmuir* **2003**, *19*, 7295; (i) K. Yu, A. Hurd, A. Eisenburg, C. J. Brinker, *Langmuir* **2001**, *17*, 796; (j) Kim et al. *J. Phys. Chem. B* **2002**, *106*, 2552; (h) J. N. Cha, G. D. Stucky, D. E. Morse, T. J. Deming, *Nature* **2000**, *403*, 289; (i) P. Yang, T. Deng, D. Zhao, P. Feng, D. Pine, B. F. Chmelka, G. M. Whitesides, G. D. Stucky, *Science*

1998, *282*, 2244; (j) A. Sellinger, P. M. Weiss, A. Nguyen, Y. Lu, R. A. Assink, C. J. Brinker, *Nature*, **1998**, *394*, 256; (k) Y. Lu, H. Fan, T. Ward, T. Reiker, C. J. Brinker, *Nature* **1999**, *398*, 223.

50 For example: (a) C. T. Kresge, M. E. Leonowicz, W. J. Roth, J. C. Vatuli, J. S. Beck *Nature* **1992**, *359*, 710; (b) J. S. Beck et al. *J. Am. Chem. Soc.* **1992**, *114*, 10834; (c) Q. Huo et al. *Nature* **1994**, *368*, 317; (d) M. E. Davies et al. *Nature* **1988**, *331*, 698; (e) Q. Huo et al. *Chem. Mater.* **1994**, *6*, 1176; (f) P. Yang, T. Deng, D. Zhao, P. Feng, D. Pine, B. F. Chmelka, G. M. Whitesides, G. D. Stucky, *Science* **1998**, *282*, 2244–2246; (g) Y. Chujo, H. Matsuk, S. Kure, T. Saegusa, T. Yazawa, *J. Chem. Soc.,Chem Commun.* **1994**, 635; (h) D. L. Gin, W. Gu, B. A. Pindzola, W.-W. Zhou, *Acc. Chem. Res.* **2001**, *34*, 973; Y. Zhou, J. H. Schattka, M. Antonietti, *NanoLett* **2004**, *4*, 477; (i) B.-H. Han, S. Poarz, M. Antonietti, *Chem. Mater.* **2001**, *13*, 3915; (j) A. Okabe, T. Fukushima, K. Ariga, T. Aida, *Angew. Chem. Int. Ed.* **2002**, *41*, 3414; (k) M. Kimura, K. Wada, K. Ohta, K. Hanabusa, H. Shirai, N. Kobayashi, *J. Am. Chem. Soc.* **2001**, *123*, 2438.

51 (a) M. Templin, A. Franck, A. Du Chesne, H. Leist, T. Ahang, R. Ulrich, V. Schadler, U. Wiesner, *Science* **1997**, *278*, 1795; (b) D. Zhao, J. Feng, Q. Huo, N. Melosh, G. H. Frederickson, B. F. Chmelka, G. D. Stucky, *Science* **1998**, *279*, 548; (c) Y. Y. Yuan, P. Hentze, W. M. Arnold, B. K. Marlow, M. Antonietti, *NanoLett* **2002**, *2*, 1359; (d) A. Jain, U. Wiesner, *Macromolecules* **2004**, *37*, 5665; (e) P. Simon, R. Ulrich, H. W. Spiess, U. Wiesner, *Chem. Mater.* **2001**, *13*, 3464; (f) T. Magbitang et al. *Adv. Mater.* **2005**, *17*, 1031; (g) C. J. Brinker et al. *Langmuir* **2003**, *19*, 7295; (h) K. Yu, A. Hurd, A. Eisenburg, C. J. Brinker, *Langmuir*, **2001**, *17*, 796; (i) H.-C. Kim et al. *J. Phys. Chem. B* **2002**, *106*; (j) J. N. Cha, G. D. Stucky, D. E. Morse, T. J. Deming, *Nature* **2000**, *403*, 289; (k) P. Yang, T. Deng, D. Zhao, P. Feng, D. Pine, B. F. Chmelka, G. M. Whitesides, G. D. Stucky, *Science* **1998** *282*, 2244; (l) A. Sellinger, P. M. Weiss, A. Nguyen, Y. Lu, R. A. Assink, C. J. Brinker, *Nature*, **1998**, *394*, 256; (m) Y. Lu, H. Fan, T. Ward, T. Reiker, C. J. Brinker *Nature* **1999**, *398*, 223.

54 For example: (a) J. P. Kampf, C. W. Frank, E. E. Malmstrom, C. J. Hawker *Science* **1999**, *283*, 1730; (b) J. P. Kampf, C. W. Frank, E. E. Malmstrom, C. J. Hawke *Langmuir* **1999**, *15*, 227; (b) V. Percec, B. Barboiu, C. Grigoras, T. K. Bera, *J. Am. Chem. Soc.* **2003**, *125*, 6503.

55 H.-C. Kim, T. Magbitang, V. Y. Lee, J. N. Cha, H.-L. Wang, W. R. Chung, R. D. Miller, G. Dubois, W. Volksen, J. L. Hedrick, *Angew. Chem.* in press.

56 H. Ihre, O. L. Padilla De Jesus, J. M. J. Fréchet, *J. Am. Chem. Soc.* **2001**, *123*, 5908.

57 (a) E. F. Connor, L. K. Sundberg, H.-C. Kim, J. J. Cornelissen, T. Magbitang, P. M. Rice, V. Y. Lee, C. J. Hawker, W. Volksen, J. L. Hedrick, R. D. Miller, *Angew. Chem. Int. Ed.* **2003**, *42*, 3785; (b) T. Magbitang et al. *Adv. Mater.* **2005**, *17*, 1031.

58 For example: (a) K. Matyjaszewski (Ed.), *ACS Symp. Ser.* **1998**, *685*; (b) T. E. Patten, J. Xia, T. Abernathy, K. Matyjaszewski, *Science* **1996**, *272*, 866; (c) M. Kato, M. Kamigaito, M. Sawamoto, T. Higashimura, *Macromolecules* **1995**, *28*, 1721; (d) C. Granel, P. Dubois, R. Jérôme, P. Teyssié *Macromolecules* **1996**, *29*, 8576; (e) V. Percec, B. Barboiu, *Macromolecules* **1995**, *28*, 7970; (f) C. J. Hawker, A. N. Bosman, E. Harth, *Chem. Rev.* **2001**, *101*, 3661.

59 M. W. Lane, X. H. Liu, T. M. Shaw, *IEEE Trans. Device Mater. Reliab.* **2004**, *4*, 142.

60 G. Dubois et al. *Abstr. Pap. Am. Chem. Soc.* **227**, Pt. 2, U554.

61 A. S. Lucas, M. L. O'Neil, J. L. Vincent, R. N. Vrtis, M. D. Bitner, E. J. Karwacki, *US Patent Application 2004175957A1*.

62 T. C. Chang, T. M. Tsai, P. T. Liu, C. W. Chen, T. Y. Tseng, *Thin Solid Films* **2004**, *469/470*, 383.

63 Y. Lu, H. Fan, N. Doke, D. A. Loy, R. A. Assink, D. A. LaVan, C. J. Brinker, *J. Am. Chem. Soc.* **2000**, *122*, 5258; K. Landskron, B. D. Hatton, D. D. Perovic, G. Ozin, *Science* **2003**, *302*, 266; B. D. Hatton, K. Landskron, W. Whitnall, D. D. Perovic, G. A. Ozin, *Adv. Funct. Mater.* **2005**, *15*, 823; G. Dubois, T. Magbitang, W. Volksen, E. E. Simonyi, R. D. Miller,

IEEE Proc. IITC **2005**, 226; G. Dubois, T. Magbitang, W. Volksen, E. E. Simonyi, R. D. Miller, *Proc. 22nd International VMIC* **2005**, 31.

64 D. A. Loy, K. J. Shea, *Chem. Rev.* **1995**, *95*, 1431; K. J. Shea, D. A. Loy, *Chem. Mater.* **2001**, *13*, 3306.

65 K. J. Shea, D. A. Loy, O. W. Webster, *J. Am. Chem. Soc.* **1992**, *114*, 6700; K. J. Shea, D. A. Loy, O. W. Webster, *Polym. Mater. Sci. Eng.* **1990**, *63*, 281; R. J. P. Corriu, J. J. E. Moreau, P. Thepot, M. W. C. Man, *Chem. Mater.* **1992**, *4*, 1217.

66 K. J. Shea, D. A. Loy, O. W. Webster, *J. Am. Chem. Soc.* **1992**, *114*, 6700; R. J. P. Corriu, J. J. E. Moreau, P. Thepot, M. W. C. Man, *Chem. Mater.* **1992**, *4*, 1217; B. Boury, R. J. P. Corriu, *Adv. Mater.* **2000**, *12*, 989; B. Boury, R. J. P. Corriu, P. Delord, V. Le Strat, *J. Non-Cryst. Solids* **2000**, *265*, 41; B. Boury, R. J. P. Corriu, V. Le Strat, *Chem. Mater.* **1999**, *11*, 2796; B. Boury, P. Chevalier, R. J. P. Corriu, P. Delord, J. J. E. Moreau, M. W. C. Man, *Chem. Mater.* **1999**, *11*, 281; B. Boury, R. J. P. Corriu, V. Le Strat, P. Delord, *New J. Chem.* **1999**, *23*, 531; P. Chevalier, R. J. P. Corriu, P. Delord, J. J. E. Moreau, M. W. C. Man, *New J. Chem.* **1998**, *22*, 423; P. Chevalier, R. J. P. Corriu, J. J. E. Moreau, M. W. C. Man, *J. Sol–Gel Sci. Technol.* **1997**, *8*, 603.

67 D. A. Loy, J. P. Carpenter, S. A. Yamanaka, M. D. McClain, J. Greaves, S. Hobson, K. J. Shea, *Chem. Mater.* **1998**, *10*, 4129; R. J. P. Corriu, J. J. E. Moreau, P. Thepot, M. W. C. Man, *J. Mater. Chem.* **1994**, *4*, 987.

68 J. H. Small, K. J. Shea, D. A. Loy, *J. Non-Cryst. Solids* **1993**, *160*, 234; H. W. Oviatt, K. J. Shea, J. H. Small, *Chem. Mater.* **1993**, *5*, 943; D. A. Loy, G. M. Jamison, B. M. Baugher, E. M. Russick, R. A. Assink, S. Prabaker, K. J. Shea, *J. Non-Cryst. Solids* **1995**, *186*, 44; D. A. Loy, J. P. Carpenter, S. A. Myers, R. A. Assink, J. H. Small, *J. Am. Chem. Soc.* **1996**, *118*, 8501; S. A. Myers, R. A. Assink, D. A. Loy, K. J. Shea, *J. Chem. Soc., Perkin Trans. 2* **2000**, 545; D. A. Loy, J. P. Carpenter, T. M. Alam, R. Shaltout, P. K. Dorhout, J. Greaves, J. H. Small, K. J. Shea, *J. Am. Chem. Soc.* **1999**, *121*, 5413.

69 C. Li, T. Glass, G. L. Wilkes, *J. Inorg. Organomet. Polym.* **1999**, *9*, 79.

70 D. A. Loy, J. V. Beach, B. M. Baugher, R. A. Assink, K. J. Shea, J. Tran, J. H. Small, *Chem. Mater.* **1999**, *11*, 3333.

71 S. Kohjiya, Y. Ikeda, *Rubber Chem. Technol.* **2000**, *73*, 534.

72 J.-P. Bezombes, C. Chuit, R. J. P. Corriu, C. Reye, *Can. J. Chem.* **2000**, *78*, 1519; R. J. P. Corriu, C. Hoarau, A. Mehdi, C. Reye, *Chem. Commun.* **2000**, 71; J.-P. Bezombes, C. Chuit, R. J. P. Corriu, C. Reye, *J. Mater. Chem.* **1999**, *9*, 1727; M. Jurado-Gonzales, D. Li Ou, B. Ormsby, A. C. Sullivan, J. R. H. Wilson, *Chem. Commun.* **2000**, 67; A. Aliev, D. Li Ou, B. Ormsby, A. C. Sullivan, *J. Mater. Chem.* **2000**, *10*, 2758.

73 C. Guizard, P. Lacan, *New J. Chem.* **1994**, *18*, 1097.

74 N. N. Vlasova, A. E. Pestunovich, Y. N. Pozhidaev, A. I. Kirillov, M. G. Voronkov, *Izv. Sib. Otd. Akad. Nauk SSSR, Ser. Khim. Nauk* **1989**, *2*, 106.

75 R. Shaltout, D. A. Loy, M. D. McClain, S. Prabakar, J. Greaves, K. J. Shea, *Polym. Prepr.* **2000**, *41*, 508.

76 D. A. Loy, J. V. Beach, B. M. Baugher, R. A. Assink, K. J. Shea, J. Tran, J. H. Small, *Chem. Mater.* **1999**, *11*, 3333.

77 C. J. Brinker, G. W. Scherer, *Sol–Gel Science: the Physics and Chemistry of Sol-Gel Processing*. Academic Press, San Diego, **1990**.

78 G. Cerveau, R. J. P. Corriu, E. Framery, *Chem. Mater.* **2001**, *13*, 3373; G. Cerveau, R. J. P. Corriu, *Coord. Chem. Rev.* **1998**, *178–180*, 1051.

10
Semiconducting Polymers and their Optoelectronic Applications

Nicolas Leclerc, Thomas Heiser, Cyril Brochon, and Georges Hadziioannou

10.1
Introduction

10.1.1
From Synthetic Metals to Semiconducting Polymers

Research on "synthetic metals" evolved from a rather esoteric occupation to a lively field of activity ever since the discovery, in 1977, of electrical conductivity in the simple hydrocarbon polymer polyacetylene upon oxidation or reduction (doping) [1, 2] (Fig. 10.1). Note that conductivity $\sigma = e\mu N$ [S m^{-1}] = [W^{-1} m^{-1}] expresses the amount of charge that flows under a given electric field. The charge is carried by particles, the electrons. Hence the conductivity is the product of the electron charge (e), the mobility of the charge-carrying electrons (μ in [m^2 V^{-1} s–1]) *and* the number of charge carriers per volume (N in [m^{-3}]). Highly doped conjugated polymers, or "conducting polymers", although not the only

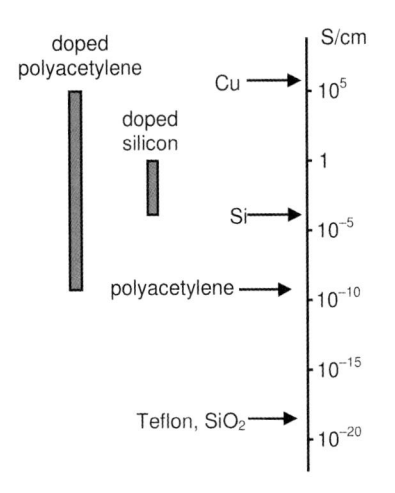

Fig. 10.1 Conductivity of polyacetylene compared with that of several major materials, such as copper (metal), intrinsic and extrinsic (doped) silicon (semiconductor), silicon oxide and polytetrafluoroethylene (Teflon) (as insulators). Polyacetylene in its pure state.

Macromolecular Engineering. Precise Synthesis, Materials Properties, Applications.
Edited by K. Matyjaszewski, Y. Gnanou, and L. Leibler
Copyright © 2007 WILEY-VCH Verlag GmbH & Co. KGaA, Weinheim
ISBN: 978-3-527-31446-1

class of materials known today as "synthetic metals", now form the most widely investigated group. They combine flexibility and ease of processing, common to plastics, with metal-type electrical conductivity and can be used as electrostatic materials for antistatic clothing or electromagnetic shielding.

The more recent dramatic developments in the field of conjugated polymers took off in 1990, after the discovery of the electroluminescence (EL) of poly(phenylenevinylene) (PPV), which opened the way to the development of polymer light-emitting diodes [3]. As a consequence, research during the 1990s in this highly interdisciplinary area has focused on the semiconductor properties of undoped polymers rather than on metal conduction and on the development of organic semiconductor devices. Today, derivatives of polythiophene (PT), poly(p-phenylene) (PPP), poly(p-phenylenevinylene) (PPV) and polyfluorene (PF) are the major candidates for use as active material in field-effect transistors, light-emitting diodes (LEDs), photodetectors, photovoltaic cells, sensors and lasers in solution and in solid state [4, 5]. Some applications have already been commercialized, while others have been shown to be technically feasible (plastic solar cells, electronic circuitry).

The electronic and optical properties of these semiconducting polymers are mostly due to the alternation of simple and double bonds (or π-conjugation) along the polymer backbone. Conjugation leads to highly delocalized π-orbitals, which reduce the HOMO–LUMO gap below a few eV and make the conjugated polymer chain behave, to a first approximation, as a one-dimensional semiconductor [6]. This enables the polymer to absorb and emit light in the visible spectrum, a property which is of fundamental importance for optoelectronic devices. The π-conjugated systems contribute to the structural organization through the π-π interactions on top of the van der Waals interactions between neighboring molecules, thus determining the molecular ordering in the solid state and strongly influencing the intermolecular charge transport and solid-state optical properties. Due to the relatively large HOMO–LUMO gap (~ 2 eV), the thermally induced (or intrinsic) free charge carrier densities in pure semiconducting polymers are usually very low. In most devices, charge carriers are generated either optically, by photon absorption-induced HOMO–LUMO transitions (optical doping), electronically, by electron transfer from the Fermi level of a metal electrode into the LUMO band (n-type doping) or from the HOMO band to the metal Fermi level (p-type doping), or chemically, by introducing reducing agents (n-type semiconductor) or oxidizing agents (p-type semiconductor).

In addition to their semiconductor properties, polymers also provide a way to obtain patterned structures by means of inexpensive techniques such as spin casting, photolithography, inkjet printing, soft lithography, screen printing and micromolding on to almost any type of substrate, including flexible ones [4, 7]. The main advantage offered by polymers over the traditional semiconductor materials is the versatility of processing methods, which allows a polymer to be obtained in virtually any desired shape and in composite form with many other materials. Deposition as a thin film over a macroscopically large area is particularly attractive. Where classical polymer processing could be used, the processing cost should be low. For this to become a reality, the parent conjugated polymers, which are highly intractable because of their conjugated, inflexible backbone, have to be derivatized without degrading their optoelectronic properties. With clever synthetic chemistry, impressive progress has already been made on this aspect [8]. An appropriate and well-defined chemical structure is a prerequisite for the control of ultimate properties, but it does not end there. The properties of polymer materials depend sensitively on the details of their processing history and each step has to be carefully carried out so as to achieve the desired result.

The present chapter focuses on the design and synthesis of polymer semiconductors for application in polymer light-emitting diodes (PLEDs), polymer lasers, polymer photovoltaics (PVs), polymer field-effect transistors (PFETs), sensors and transparent conducting thin polymer films. Section 10.2 deals with the synthesis of conjugated homopolymers and different kinds of conjugated copolymers, and Section 10.3 describes in more detail the applications of these materials in semiconductor devices.

10.2
Synthesis of Conjugated Polymers

10.2.1
Homopolymers

Polyacetylene (PA) (Fig. 10.2) represents the historic parent of this class of materials. It was the first plastic material to exhibit metallic conductivity, a discovery which was rewarded with the Nobel Prize in Chemistry in 2000 [1].

Acetylene was first polymerized by Natta and coworkers in 1958 [9]. What was obtained at that time was an intractable powdered black compound. The synthetic procedure was very similar to the synthesis of polyethylene from titanium catalysts. In fact, Ti(OProp)$_4$ easily reacts with Al(Et)$_3$ to give a Ti–C bond; the polymer chains grow upon insertion of the monomers in the created Ti–C bonds. The work of Shirakawa [1, 10] followed the procedure of Natta, but the catalyst concentration was much higher so that a highly viscous catalytic "soup" remained on the wall of the reaction flask. The diffusion of acetylene in the system in vacuum led to the formation of a shiny, free-standing film which had to be washed several times in order to remove the catalyst residues. There are several variations to this synthetic approach and readers may refer to reviews [11, 12] for details.

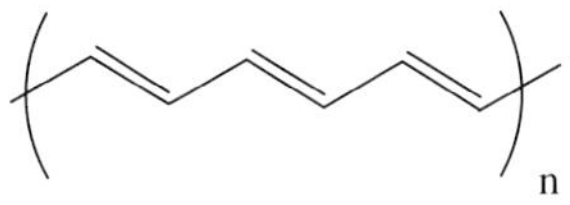

Fig. 10.2 Polyacetylene.

The major drawbacks of polyacetylene are related to its insolubility and to the impossibility of treating the polymer after synthesis: once polymerized the polymer should be taken as it is. Attempts to solubilize PA were made by different groups with two main approaches: the precursor route and the grafting approach, which will be illustrated below for other conjugated polymers. The discovery in the early 1990s of the light emission from conjugated polymers in their semiconducting form and the fact that the electronic structure of PA does not allow light emission prompted the development of many new structures of conjugated polymers which have been conceived and synthesized the last 15 years.

Most investigated conjugated polymers other than PA can be divided into three subgroups: poly(arylenevinylene)s, polyaryls and poly(aryleneethynylene)s. Among these, poly(arylenevinylene)s, and in particular PPV, are the most studied and well understood homopolymers. For each of them, low solubility and processability needed to be addressed and led to the introduction of either precursor or grafting approaches.

The Wessling method, which consists in the preparation of a water-soluble sulfonium precursor of PPV, was one of the first precursor routes to be reported [13]. After deposition of a sulfonium thin film, a thermal conversion to the final PPV was made by heating at 220–250 °C under vacuum (Fig. 10.3). Several modifications to this general procedure have been made, changing the conditions of the transformation process [14–16] and/or the nature of the chemical species to be eliminated [17–19].

In the grafting approach, the introduction of alkyl or alkoxy groups on the phenyl ring aims at the synthesis of a soluble PPV derivative, which can be processed into thin polymeric films by means of wet deposition techniques. The first grafting procedure of PPV is based on the Gilch reaction, using dihalomethyldialkoxy monomers which are converted to a soluble PPV derivative by means of an excess of base. One of the most popular PPVs synthesized through this method was

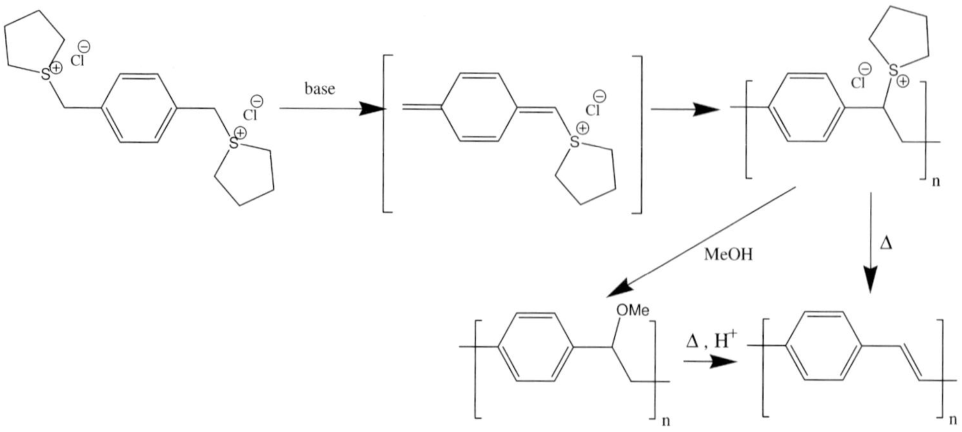

Fig. 10.3 Wessling–Zimmerman route to PPV.

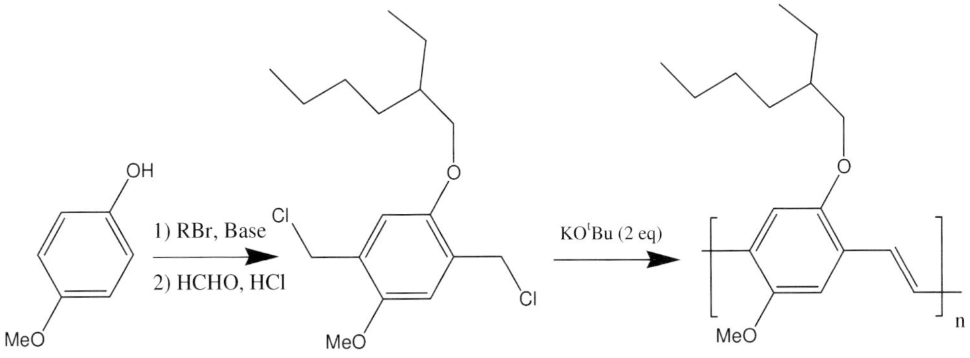

Fig. 10.4 Gilch route to MEH–PPV.

poly[2-methoxy-5-(2′-ethylhexyloxy)-1,4-phenylenevinylene] (MEH-PPV), which is fluorescent in the red–orange part of the visible spectrum [20, 21] (Fig. 10.4). There are, however, two limitations to this method: it does not work for all aromatic systems and it produces some defects in the conjugated backbone that affect the electro-optical properties of the polymer.

To overcome these limitations, polycondensation reactions, such as the Horner–Wittig or the Knoevenagel routes, can be used. The polymers obtained in this way are generally of much lower molecular weight than the materials obtained by the Gilch method, but the lack of defects in the backbone yields better electro-optical properties.

Knoevenagel condensation between a bisnitrile phenyl moiety and a dialdehyde allows the preparation of poly(2,5,2′,5′-tetrahexyloxy-7,8′-dicyanodi-p-phenyle nevinylene) (CN-PPV) (Fig. 10.5). This polymer exhibits a large bathochromic shift (PL maximum at 560 nm in the yellow region) compared with the blue-emitting non-cyano substituted parent [22].

Horner and Wittig polycondensations with aryl biphosphonium salts and aryl biphosphonate salts, respectively, which are parent methods, have been widely used to synthesize PPV derivatives [23], but also other polymers such as poly(carbazolevinylene) (PCV) [24] (Fig. 10.6) and poly(thienylenevinylene) (PTV) [25]. Comparative studies have shown the Horner condensation to be superior in that it produces higher molecular weight materials with all double bonds being trans,

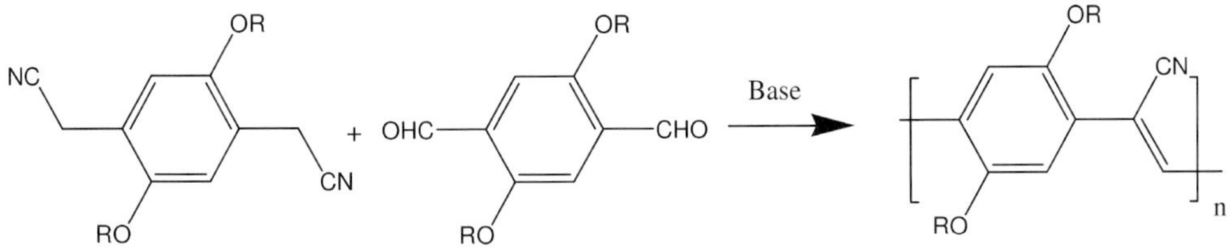

Fig. 10.5 Knoevenagel polycondensation route to CN–PPV.

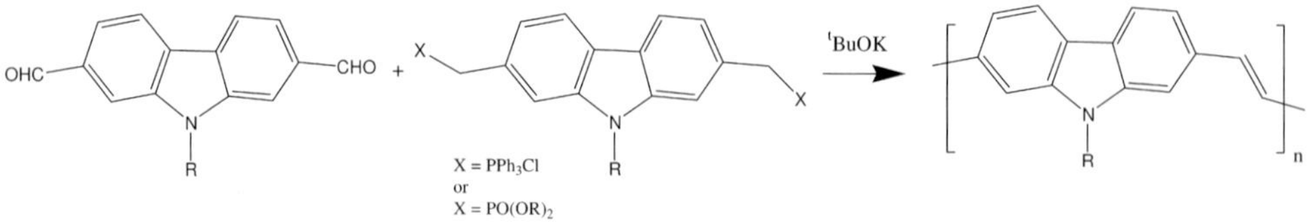

Fig. 10.6 Wittig and Horner routes to PCV.

whereas the Wittig route results in a higher proportion of *cis*-vinylene units, which degrades the electro-optical properties of the material [26].

Another direct route for poly(arylenevinylene) synthesis is the Heck coupling method, which is a transition metal-mediated reaction. The coupling of dihalophenyl with a divinylphenylene offers a direct route to obtain PPV ("AB" approach in Fig. 10.7a), similar to the self-coupling of vinylbromobenzene ("AA/BB" approach in Fig. 10.7b) [27]. As with the condensation method discussed above, the molecular weights are generally lower than those obtained by the Gilch route, but the materials are relatively defect free.

Cross-coupling of dihaloarenes with bisstannylethane (Stille reaction) also allows the synthesis of poly(arylenevinylene)s. A poly(pyridinevinylene) has been made by this route [28] and also a poly(thienylenevinylene) (PTV) [29] (Fig. 10.8).

Polyarylenes are the second subclass of highly investigated conjugated polymers. Poly(*p*-phenylene) (PPP) [30], a blue-emitting material, and poly(3-alkylthiophene)s (P3ATs) [31, 32] for their general electro-optic properties (red–orange emitters, but also efficient active layers in OFET and PV devices) are the most popular examples of this class.

Because of similar drawbacks to PPV and PA (insolubility and unprocessability), PPP was essentially synthesized by precursor routes. Not surprisingly, the

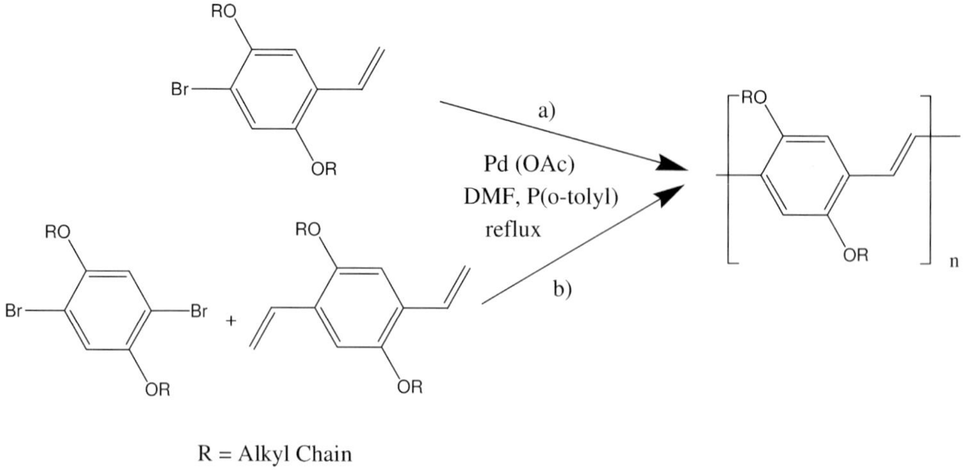

R = Alkyl Chain

Fig. 10.7 PPV synthesis by Heck reaction: (a) AB approach and (b) AA/BB approach.

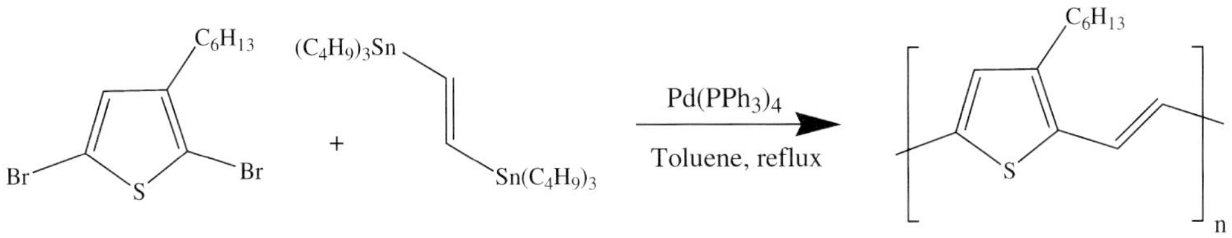

Fig. 10.8 Synthesis of PTV by the Stille route.

polymers so obtained had many defects in their conjugated backbone, including some ortho bonds. To overcome this problem, Yamamoto and coworkers developed reductive coupling, from the dihalomonomer, using bis(cyclooctadiene)-nickel(0) and 2,2′-bipyridyl, and applied it to the synthesis of PPPs, P3ATs and PFs [33] (Fig. 10.9).

P3ATs have been extensively studied for their electro-optical properties and their applications in PFET, PV and PLED devices. As 3-alkylthiophene is an asymmetric monomer, the molecular structures obtained in the polymerization process depend on the ratios between head-to-tail (HT), tail-to-tail (TT) and head-to-head (HH) connections (Fig. 10.9) and lead to different electro-optical properties. Due to steric hindrance between side-groups, the head-to-head con-

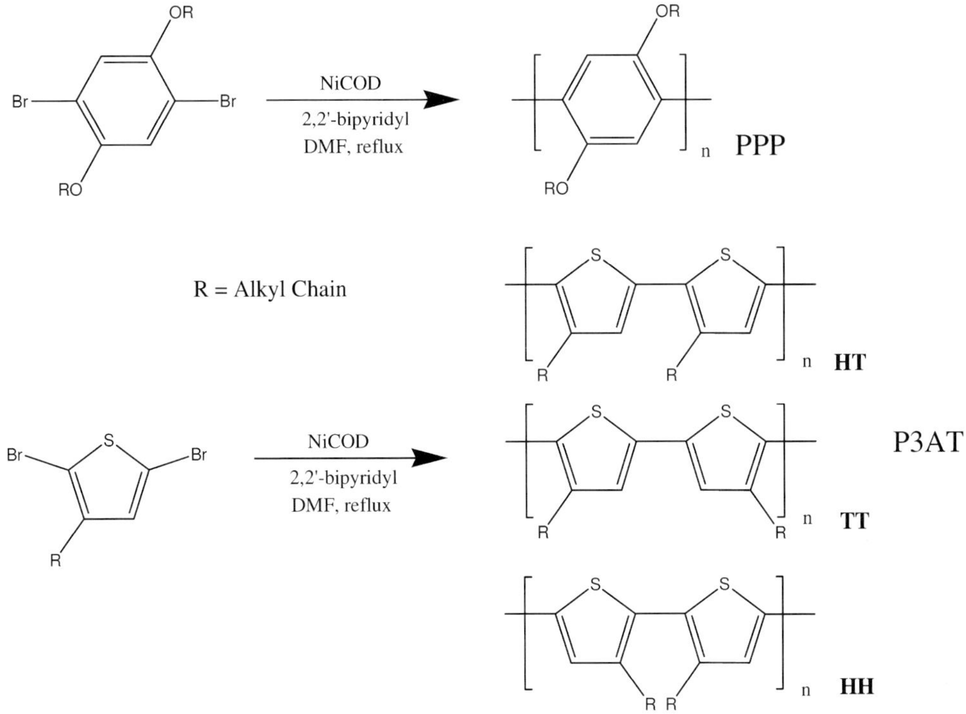

Fig. 10.9 PPP and P3AT synthesis by the Yamamoto route.

figuration reduces the planarity of the molecule, interrupts the conjugation length and affects the molecular ordering in the solid state. This in turn will affect the thin-film electro-optical and conduction properties. With the Yamamoto synthesis, the preferred head-to-tail configuration is obtained with a yield of approximately 90%, as determined from ^{1}H NMR analysis [34]. In order to have even better control of the regioregularity of P3HT, McCullough and Lowe reported in 1992 a synthetic procedure based on a Kumada reaction between the Grignard compound selectively in the 5-position of a 2,5-dihalogenated-3-alkylthiopene [32] (Fig. 10.10).

Other synthetic approaches have been introduced for the preparation of PATs. Oxidative coupling, essentially with iron(III) chloride, is one of them. Wegner's group used this route to synthesize regiorandom P3ATs [35]. This procedure has become very popular since the commercialization of poly(3,4-diethyloxy)-thiophene (PEDOT), a PAT derivative. PEDOT, doped with polystyrenesulfonic acid as counterion, is commercially available as a blue suspension in water and is often used to form a transparent conducting layer on indium–tin oxide (ITO). PEDOT is obtained through an oxidative coupling using $FeCl_3$ as oxidizing agent [36] (Fig. 10.11). This method has the advantage of avoiding the introduction of halogen atoms on the aromatic rings during the monomer preparation procedure. However, purification of the final product is necessary in order to avoid luminescence quenching by metal residues.

In comparison with oxidative coupling, transition metal-mediated cross-coupling reactions are much more difficult to use, but are of great interest for the synthesis of polyarenes. A major advantage of these methods results from their high versatility and their wide tolerance towards many functional groups. The most widely used cross-coupling reaction is the Suzuki coupling, which generally gives high yields and high molecular weights. Other methods such as Stille,

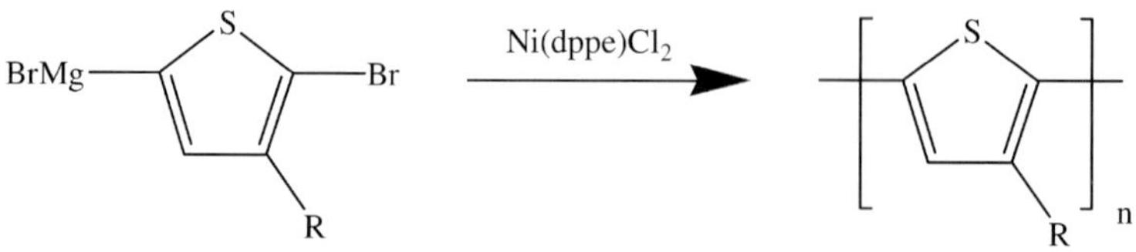

Fig. 10.10 Kumada synthesis of P3AT.

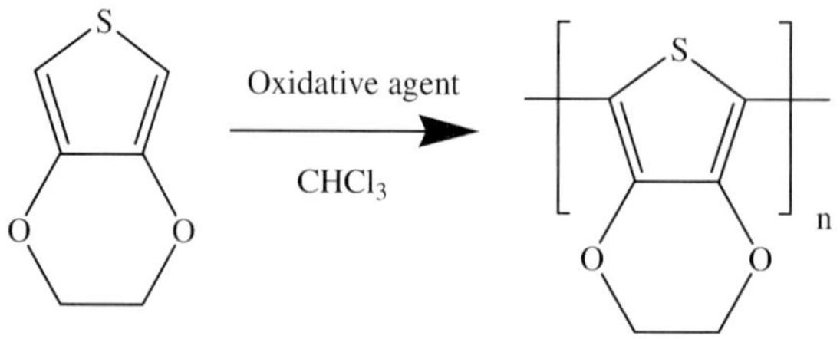

Fig. 10.11 PEDOT synthesis by oxidative coupling.

Negishi, Kumada or Yamamoto couplings (see below) are well known and generally give lower molecular weights. As for the polycondensation reactions, both AB and AA/BB approaches are often used.

The synthesis of PFs is a good example of a cross-coupling reaction. Leclerc and coworkers [37] were the first to introduce the palladium-catalyzed Suzuki coupling between 2,7-dibromofluorene-9-carboxylates and fluorene derivatives bearing diboronic moieties in order to obtain a well-defined poly(2,7-fluorene) derivative (Fig. 10.12 a). PFs may be considered as PPP derivatives with absorption spectra shifted towards higher wavelengths. The CH_2 bridge connecting two adjacent phenylene rings induces indeed a planarization of the molecule which enhances the conjugation length and reduces the HOMO–LUMO gap. One of the advantages of Suzuki coupling over other palladium-catalyzed cross-coupling reactions, such as the Stille reaction, is the easier purification of boronic acid or ester starting compounds as compared with the stannyl monomer used in the Stille reaction.

In order to have a soluble PF, the 9-position is generally substituted with two chains (alkyl functionalized). However, when this substitution is not quantitative, it may lead in the presence of oxygen to the formation of peroxide, which subsequently evolves into a keto derivative, as shown in Fig. 10.12 b [38].

This oxidation process has been found to be responsible for photoluminescence quenching and for the introduction of a new band in the PL spectrum at 520 nm. For this reason, several groups have focused their research on the use of polycarbazole (PC). Carbazole is a fluorene derivative in which the C-9 atom is replaced by a nitrogen atom. The latter is more stable towards oxidation and is easily substituted with a wide variety of functional groups. However, until recently, only poorly conjugated poly(N-alkyl-3,-6-carbazole) derivatives could be synthesized. Morin and Leclerc recently reported the first synthesis of poly(N-alkyl-2,7-carbazole) [39] (Fig. 10.13). This polymer had been synthesized either by Yamamoto [40], Kumada [41] (well-defined oligomers were obtained in that case) or Suzuki [41] reactions. The emission of PCs is in the same range as PFs, 420–460 nm. As expected, the fully aromatic feature of carbazole makes it much more stable than fluorene.

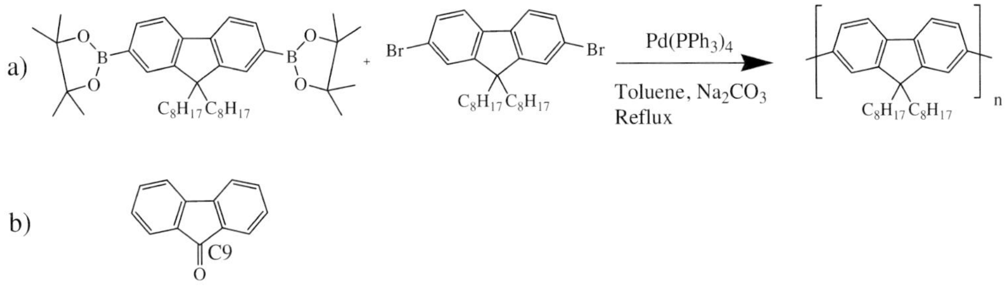

Fig. 10.12 (a) First polyfluorene synthesized by the Suzuki method; (b) fluorenone.

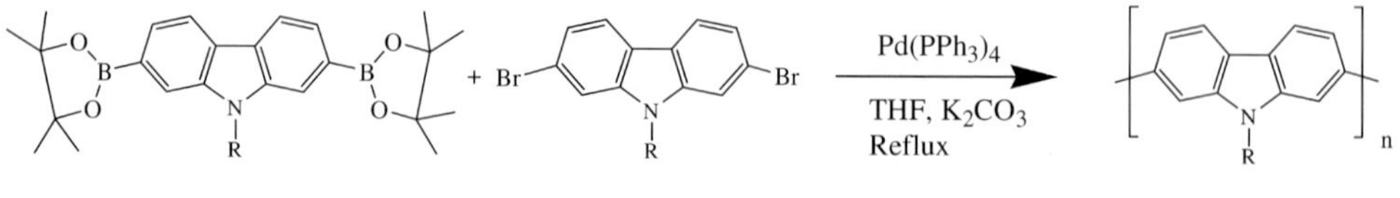

Fig. 10.13 Suzuki synthesis of 2,7-polycarbazole.

A third subfamily of conjugated polymers comprises poly(aryleneethynylene)s (PAEs). They are more rigid than PAVs and have a tendency towards aggregation. For this reason, they have been less investigated than other conjugated polymers. Two routes can be used to prepare PAEs: Hagihara–Sonogashira coupling and alkyne metathesis. The first method involves Pd catalysis, with a copper cocatalyst, between dihaloarenes and an aryldiyne. The poly(phenyleneethynylene)s (PPEs), structurally closest to PPVs, are the most studied polymers of this class. Giesa and Shulz reported the first improved synthesis of soluble PPE with Pd–CuI-catalyzed coupling [42] (Fig. 10.14a). The PL spectrum of this polymer is in the range 445–455 nm (green emitting).

Yamamoto and coworkers reported the synthesis of regiorandom poly(3-hexylthienylvinyleneethynylene) (P3HTE) by a similar route [43]. Pang used a variant of the previous Yamamoto coupling to prepare a regioregular P3HTE [44]. Surprisingly, the photoluminescence wavelength of regioregular P3HTE was identical with that of regiorandom P3HTE. The regioregularity effect in P3HTEs appeared to be much smaller than that in P3HTs. This reduced effect could be explained in terms of the reduced steric interactions between substituents on adjacent thiophene rings and of the reduced interactions between the alkyl side-chain and the sp^2 lone pair on the sulfur atoms of neighboring thiophenes. However, the regioregularity has a significant effect on the fluorescence quantum yield, since the regioregular P3HTE exhibits a ca. 10 times stronger fluorescence than regiorandom P3HTE. The high fluorescence yield in the case of regioregular P3HTE may be related to the closer molecular stacking which is allowed by the HT linkage.

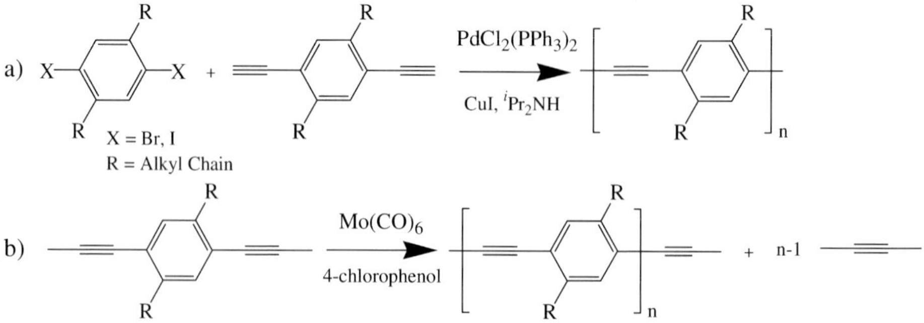

Fig. 10.14 (a) Hagahira–Sonogashira synthesis of PPE and (b) alkyne metathesis of PPE.

The Hagihara–Sonogashira coupling has the disadvantage, however, of allowing the presence of fluorescence-quenching halogens as end-groups and the formation of diyne defects [45].

To circumvent this problem, the metathesis, using molybdenum– or tungsten–carbyne complexes, has been developed by Bunz and coworkers [46]. It has the advantage of producing defect-free materials with well-defined end-groups (Fig. 10.14b). The materials obtained are of higher quality, but this procedure suffers from greater experimental difficulty and is not applicable to all alkynes.

10.2.2
Random and Alternating Copolymers

Many studies have focused on the fine-tuning of properties of conjugated materials such as processability, morphology and the related absorption/emission spectra, charge injection barriers and charge transport (mobility). As discussed above, this can be done to some extent by varying the homopolymer substituents. Another, most effective way to achieve that goal is to introduce different elementary units into the polymer, that is, the synthesis of copolymers.

General rules can be used as guidelines for these investigations. For instance, the more electron-rich the arylene is, the more red-shifted the emission will be. Thus phenylene-based materials are blue- or blue–green-emitting materials, whereas polymers based on highly electron-rich thiophene units produce orange–red emission. Since copolymers show emission spectra which are intermediate between the corresponding homopolymers, it is straightforward to imagine some new copolymers in order to fine-tune the emission color.

From a chemical point of view, three families of copolymers can be highlighted: random, alternating and block copolymers. The last family will be discussed in more details in the next part.

Random copolymers can be easily achieved by classical polycondensation methods. For example, some derivatives of carbazole-based copolymers were synthesized by a nickel-catalyzed Yamamoto coupling reaction. Leclerc and coworkers have synthesized a copolymer by Yamamoto coupling between an equimolar mixture of 1,4-dichloro-2-(2′-ethylhexyloxy)benzene and N-ethylhexyl-2,7-dichlorocarbazole [47]. This copolymer shows an emission in the blue region (416 nm) with a slight blue shift in comparison with the N-ethylhexyl-2,7-carbazole homopolymer. Shim and coworkers have recently synthesized a fluorene–carbazole-based copolymer by reductive Yamamoto coupling between an equimolar mixture of the electron-withdrawing 3,6-bis{1-[4-bromobenz(1-cyanovinylene)]}-9-(2-ethylhexyl)carb azole and the electron-donating 9,9-bis(2-ethylhexyl)-fluorene monomer, in order to use exciplex formation for white emission [48]. Exciplex emission is formed by recombination of electrons and holes at the interface between electron acceptor and donor molecules [49] (Fig. 10.15).

However, these AA-type polycondensation coupling reactions are not completely satisfactory in the case of copolymer synthesis because of their lack of control of the regioregularity and chain composition of copolymer(s). For this

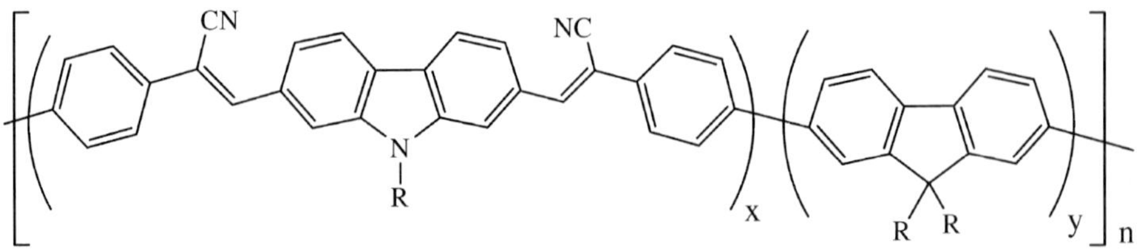

R = Ethylhexyl

Fig. 10.15 White electroluminescent polymer obtained by the Yamamoto route.

reason, most recent studies are based on AA/BB and AB approaches via cross-coupling reactions (CCRs). Among these CCRs, the Suzuki reaction is the most widely used method because boronic esters and acids are easier to handle and purify than the Grignard and stannyl compounds used in Kumada and Stille reactions [50]. Moreover, a wide variety of functional groups are accepted. Nevertheless, in agreement with the Carothers equation, the stoichiometric ratio between two different functional groups in the case of the AA/BB approach should be respected to achieve high molar mass copolymers [51].

The use of an AA/BB approach allows the incorporation of electron-accepting units, e.g. oxadiazoles, triarylamines and quinoxalines, in a well-defined alternating copolymer. This allows better control of the electronic properties of the material. Zhu and coworkers described the synthesis of a novel series of electron-deficient oxadiazole-, quinoline- and quinoxaline-containing conjugated copolymers based on fluorene with the purpose of combining both the high electron affinities of the former units and the blue light-emitting and high quantum efficiencies of the latter unit in the same polymer [52]. They obtained

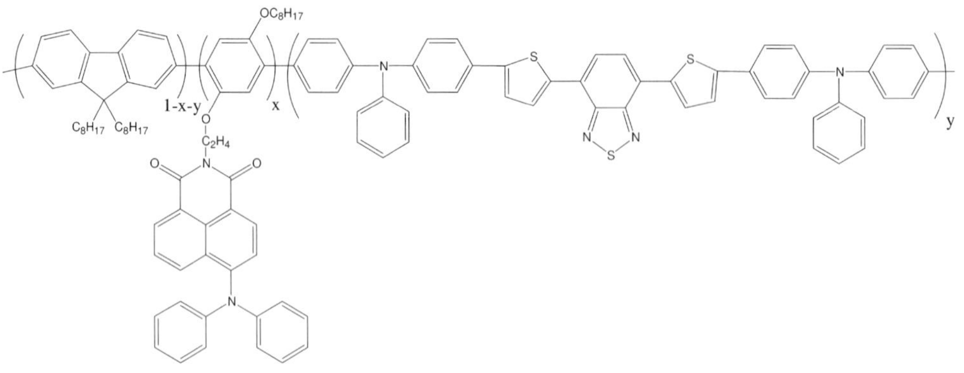

Fig. 10.16 Fluorene-based white electroluminescent copolymer synthesized by the Suzuki method.

strong blue photoluminescent thin films (around 414–476 nm) together with a low-lying LUMO energy level, which allowed a lower electron injection barrier.

Recently, Wang and coworkers reported the synthesis of fluorene-based white electroluminescent copolymers with simultaneous blue, green and red emission. This was achieved by Suzuki copolymerization between a side-chain green-emitting component, a red-emitting component and a blue-emitting component [53]. Partial energy transfer and charge trapping from the blue PF backbone to the green and red chromophores led to individual emission from three emissive species and the generation of white-light emission with three primary colors (Fig. 10.16).

10.2.3
Conjugated Block Copolymers

Another strategy to enhance the properties of conjugated polymers is to synthesize block copolymers. In this way, it is possible to combine specific properties of each block in the same material. A relatively poorly investigated aspect of polymeric materials in electronic applications is the ability to self-assemble. Self-assembly allows morphology control at a size scale previously inaccessible, but potentially important to the optimization of light-emitting and photovoltaic devices. The efficiency of organic electronic devices depends strongly on the morphology of the active layer. In photovoltaic devices, for instance, blends [54–63] are widely used to form an interpenetrated network of donor and acceptor domains. However, the morphology of blended films is not perfectly controlled and the stability of the active layer is problematic. In contrast, block copolymers are known to phase separate in ordered microphases due to the strong incompatibility of the different blocks [64]. Specifically tailored polymers can self-organize into mesomorphic structures with features on the nanometer scale. For these reasons, block copolymers based on conjugated polymers or oligomers have been widely investigated [65–73].

Block copolymers consisting of two conjugated rigid segments have been studied by Sun and coworkers. They synthesized various multiblock copolymers [71, 74] by the condensation reaction of two different α,ω-end-functionalized conjugated polymers. When one of the two conjugated blocks is an electron acceptor and the other an electron donor, the final copolymer can be used as an active layer for photovoltaic devices. There are, however, two major drawbacks for this approach: the final copolymer exhibits poor processability and the control of the morphology is not trivial with these rigid copolymers.

A possible way to circumvent these drawbacks consists in combining a conjugated rigid block with a nonconjugated flexible block. These so-called rod–coil block copolymers can provide a better processability and enhance the ability of the material to self-organize.

Due to the large difference in the chemical nature of the two blocks, the synthesis of a conjugated rod–coil copolymer is not as simple as that of a classical coil–coil block copolymer. There are two possible approaches to synthesizing such copolymers: the convergent way and the divergent way.

In the convergent way, the block copolymer is obtained by reaction on a conjugated polymer with one or two coil segments in order to obtain di- or triblock copolymers. This connection can occur by a simple quenching of a living coil polymer, synthesized by anionic polymerization, with an end-functionalized conjugated oligomer [75–77]. It is also possible to make a simple condensation reaction between an end-functionalized coil polymer and various conjugated polymers [76–81]. For example, poly(p-phenylene-b-ethylene glycol) diblocks (1 in Fig. 10.17) were obtained by Mullen and coworkers [77] The control of the condensation reaction yield is crucial and a purification step is often necessary to remove unreacted homopolymers. In either case, the rod block is synthesized by a polycondensation reaction and must have one or two reactive functions at the chain ends. The chemical nature of these chain ends strongly affects the formation of rod–coil di- or triblock copolymers.

For the convergent way, the different blocks are made by successive polymerization steps. In this approach, a simple and elegant route is first to polymerize successively two coil blocks by living anionic polymerization and then convert one of the two blocks into a conjugated structure [82] (2 in Fig. 10.17). Anionic living polymerization is known to result in well-defined homopolymers and block copolymers [83]. This method, however, is very restrictive with regard to the choice of the rod block and the existence of permanent defects in conjugation.

The use of a fully conjugated macromolecular initiator (macroinitiator) will bypass this difficulty. Indeed, it is possible to convert the rod block into a macroinitiator and then polymerize the second block directly on to it. This route has been used in conjunction with anionic living polymerization by Marsitzky et al. to polymerize ethylene oxide with a polyfluorene macroinitiator as the conjugated rod block [84]. Anionic polymerization, which is a popular route towards block copolymers because it yields nearly monodisperse molecular weight distributions, requires drastic reaction conditions (very high sensitivity to impurities and functional groups) and cannot be used with a wide variety of monomers,

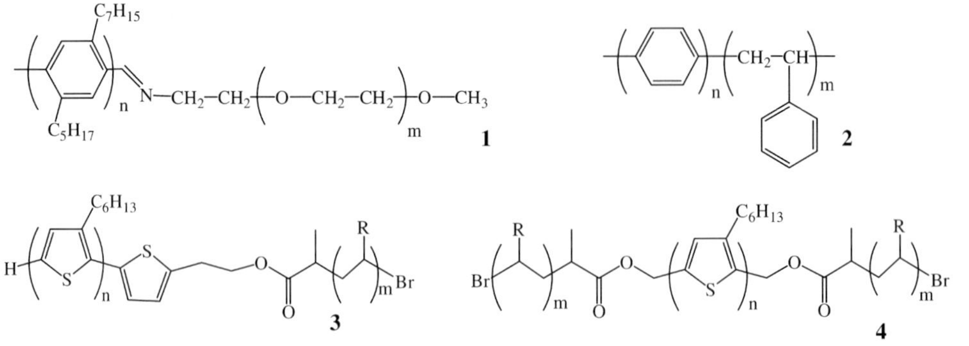

Fig. 10.17 Examples of rod–coil block copolymers.

particularly those with functional groups. In fact, controlled/"living" polymerization techniques are more versatile than anionic polymerization. The two main controlled/"living" radical polymerization techniques are nitroxide-mediated radical polymerization (NMRP) [85] and atom-transfer radical polymerization (ATRP) [86]. These are compatible with a wide range of monomers (acrylates, styrene, etc.), including functionalized derivatives, and result in narrow polydispersity molecular weight control, which is necessary for controlled self-assembly. Furthermore, (macro)initiators for NMRP or ATRP are very stable and it is possible to purify and characterize these compounds before (re)initiating the radical polymerization. Various rod–coil block copolymers have been synthesized by polymerization of styrene or (meth)acrylate derivatives either by ATRP with regioregular PT [87, 88], PF (**3** and **4**, respectively, in Fig. 10.17) [89] or oli-

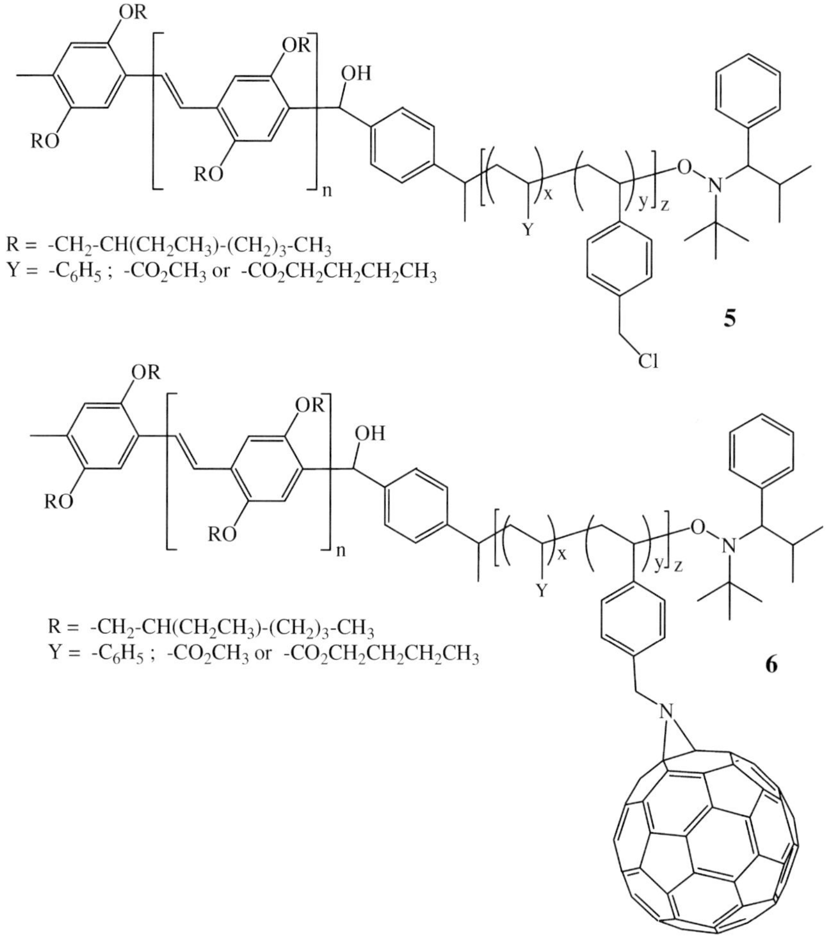

R = -CH$_2$-CH(CH$_2$CH$_3$)-(CH$_2$)$_3$-CH$_3$
Y = -C$_6$H$_5$; -CO$_2$CH$_3$ or -CO$_2$CH$_2$CH$_2$CH$_2$CH$_3$

5

R = -CH$_2$-CH(CH$_2$CH$_3$)-(CH$_2$)$_3$-CH$_3$
Y = -C$_6$H$_5$; -CO$_2$CH$_3$ or -CO$_2$CH$_2$CH$_2$CH$_2$CH$_3$

6

Fig. 10.18 Examples of donor–acceptor rod–coil copolymers for photovoltaic applications.

gomers [90, 91] as macroinitiators to reach diblock or triblock rod–coil copolymers. By NMRP, PPP [70, 92] diblock copolymers (**5** in Fig. 10.18) were first obtained. Then, other macroinitiators based on other polymers, such as PTV [93] or PF [94], were developed for di- and tri-block copolymers. The use of conjugated macroinitiators for controlled/"living" radical polymerization is actually one of the most efficient routes for the synthesis of the targeted rod–coil di- and triblock copolymers.

Di- and triblock rod–coil copolymers have good self-organization capacities. For this reason, they exhibit very interesting photophysical properties both in solution and in thin solid films. Moreover, they show good processability and are interesting candidates for PLED. Furthermore, in the case of a conjugated polymer block which behaves as an electron donor/hole conductor, the coil block can be functionalized with an electron acceptor group (such as fullerenes) in order to obtain self-organized donor–acceptor materials for photovoltaic applications (**6** in Fig. 10.18) [70, 92].

Various well-defined conjugated rod–coil di- and triblock copolymers and rod–rod copolymers have been designed. This new class of copolymers has good tunability regarding the length and nature of each block. These parameters of the design are a good basis for the study of self-assembly and optoelectronic properties of these new materials. However, the solubility of these materials can be poor, especially with rod–rod copolymers. Presently many unexplored design parameters, such as the nature of each block, must be considered for improving the morphology and optoelectronic properties of these copolymers and many research groups are working on these aspects.

10.3
Applications in Optoelectronics

10.3.1
The Requirements and Design of Polymer Light-emitting Diodes

10.3.1.1 State-of-the-Art

The development of LEDs based on polymers started only in 1990 and has continued really fast. We are already well above the level needed for daytime operation. The lifetime, at a reasonable output, is about 10 000 h. Within 10 years it has developed into a viable material and a viable product. During the same period there has also been a development of inorganic materials, the nitrides. Such competition benefits both parties. Nonetheless, at the moment, it is possible to make full-color displays with polymers as the only active material. There is also a parallel development of molecular organic solids. The difference between organic solids and polymers is mainly in the processing technology: in organic solids one uses the same technology for preparing the device as for gallium arsenide, that is, ultra-high vacuum deposition techniques, lithographic techniques, etc. This is costly, but the advantage is that this technology already exists in manufacturing plants, so one can transfer it easily from one material to another. On the other side, polymers provide a lot more flexibility in manufacturing techniques, for instance, they can be applied like a coating on the wall. However, we are not yet able to make materials that perform well enough to have them applied like paint. Imagine how simple manufacturing could be if paint technology was used to make devices.

10.3.1.2 Single-layer PLEDs

The most basic polymer light-emitting diodes consist essentially of one organic layer sandwiched between two electrodes (Fig. 10.19a). During device operation, positive and negative charge carriers are injected into the organic layer. Electroluminescence occurs when electron–hole pairs (or excitons) are formed and exciton recombination takes place via photon emission. The cathode needs to be a low-workfunction metal (calcium or aluminum), the Fermi level of which should be as close as possible to the LUMO level of the organic layer. This is a necessary condition for achieving a low electron injection barrier (and thus a low operating voltage). Similarly, the anode consists of a high-workfunction me-

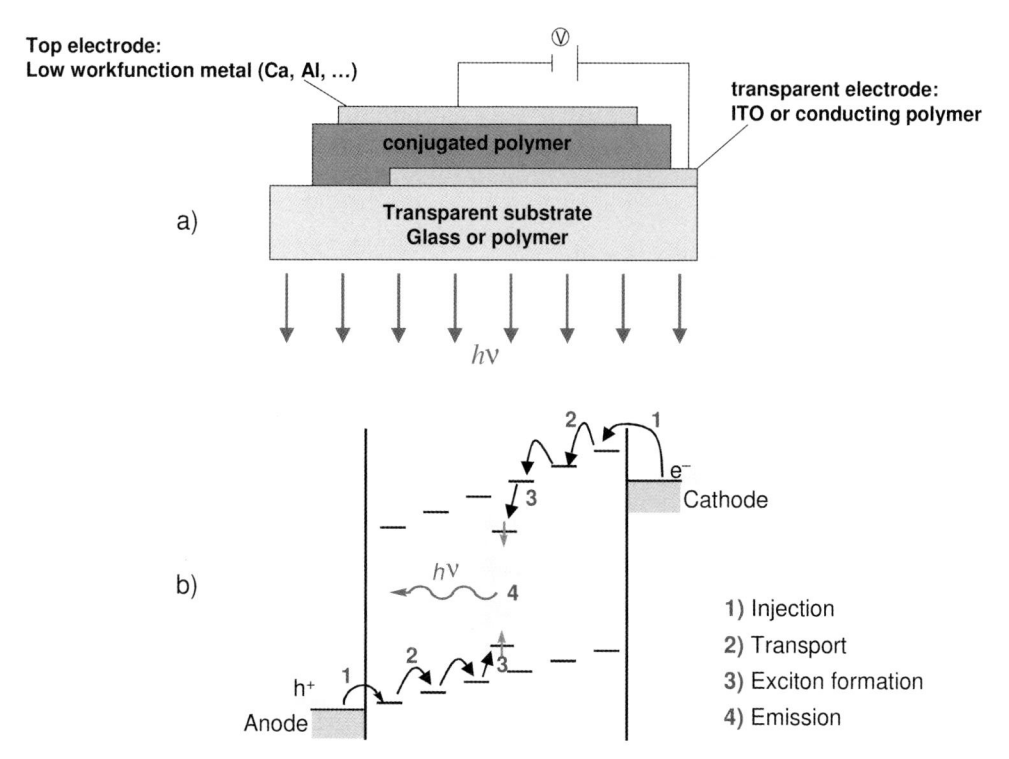

Fig. 10.19 Schematic representations of (a) a PLED device and (b) electroluminescence mechanism.

tal, the Fermi level of which should match the organic HOMO level. Indium–tin oxide (ITO), which is a conducting transparent ceramic, is the most widely used anode material. Its transparency is necessary to allow emitted light to escape the device.

The voltage is put across the device by lowering the potential of the cathode with respect to the anode. In most cases, the organic layers behave as an intrinsic semiconductor and no space charge appears in the film upon application of the voltage. As a consequence, the voltage drops uniformly across the layer, as schematized in Fig. 19b. Under operation, electrons and holes, injected by the cathode and the anode, respectively, will drift across the layer by "hopping" from one site to another. The electrons and holes may then interact on a given molecule in a process called exciton formation. The exciton undergoes relaxation and will eventually recombine by emitting a photon.

To make such a device efficient, the polymer layer must allow balanced charge transport and a high exciton formation rate. If the current is indeed dominated by one charge carrier type, most of the majority charges will just cross the device without participating in the electroluminescence. For achieving balanced charge transport, the injection barriers at the metal/organic interfaces need to be reasonably close, a condition which is difficult to achieve in practice. The threshold voltage at which electroluminescence occurs depends essentially on the highest injection barrier.

It is equally important to have mobilities of the same magnitude for both charge carriers. In the opposite case, exciton formation will indeed occur close to the metal electrode which injects the slow carriers and leads to luminescence quenching (due to metal-induced defects).

The use of ITO as a transparent electrode causes several problems. Contamination of the emissive layer through diffusion of oxygen impurities may occur during device operation and enhance the photochemical degradation of the organic semiconductor. Since ITO is not a stoichiometric compound, its electrochemical properties are uncertain and vary from batch to batch. Also, the interface roughness may be at the origin of localized short-circuits and enhance the leakage current.

10.3.1.3 Double-layer PLEDs

Some of the drawbacks of single-layer PLEDs can be circumvented by using additional organic layers. In case of an n-type emissive layer, one possible option is to include a hole-transporting layer (HTL) between the anode and the emissive electron-transporting layer (ETL) (combination of an emissive HTL and an electron-transporting layer is also possible). The HTL can be either a semiconducting polymer, a hopping transporter such as PVK or even a doped, conducting polymer. Balanced charge transport can then be obtained through an appropriate choice of the HOMO and LUMO levels of both organic layers, which independently will adjust the injection barriers at both metal/organic interfaces. Moreover, at the organic/organic interface a barrier for electron (hole) injection into the HTL (ETL) layer will appear, which may be used to counter a difference in mobilities or injection barriers at both electrodes. A conducting HTL will enhance the electric field across the emitting layer and so help the injection of negative charges and assist their transport across the ETL. Most important in the use of two organic layers is that the exciton formation and recombination will take place at or near the organic/organic interface, well away from the deleterious metal electrodes. The organic/organic interface energetics will establish in which layer excitons are formed and influence the color of the emitted light versus applied voltage. Intermolecular excitons (or exciplex) may also form at the interface and result in color shifts.

An additional bonus of the HTL is that it prevents contamination of the emitting layer by the ITO. One of the materials that is commonly used between the anode and the emissive layer is sulfonated polystyrene-doped PEDOT. It is deposited as a very thin layer. It prevents contamination and defines the Fermi level of the anode more reliably. The higher work function of PEDOT–PSS compared with ITO further allows the hole injection barrier to be reduced.

10.3.1.4 Polymer Blend PLEDs

Another possibility for engineering PLED devices is to use blends. In this case, a small amount of a chromophore (guest), which can be either a small molecular dye or a luminescent polymer, is added to the polymer matrix (host). The host must be a charge-transporting polymer which allows for exciton formation and energy transfer (mostly by the Förster mechanism) from the host to the emitter (dye) [95, 96]. To achieve high electroluminescence efficiency, the emission spectra of the host matrix must overlap with the absorption spectra of the dye.

Among the possible materials to be used as host, poly(vinylcarbazole) (PVK) is of particular interest: it is stable and forms good films. An example of using PVK and Si-PPV (a polymer of limited conjugation) is illustrated in Fig. 10.20.

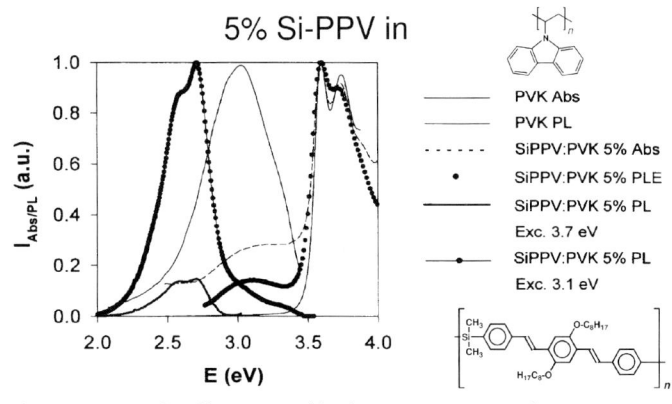

Fig. 10.20 Example of host–guest blend using Förster transfer.

Since the HOMO–LUMO gap of PVK is larger than that of Si-PPV, energy transfer is possible. For this particular system, the Förster transfer has been studied by the absorption/photoluminescence experiments shown in Fig. 10.20. The PVK absorbs at short wavelengths and emits in the violet region, around 3 eV. When blended with 5% of the Si-PPV depicted, the absorption characteristics are almost identical with the superposition of the absorption spectra of both components. The emission of the blend, however, is dominated by the fluorescence of Si-PPV, peaking at 2.7 eV. (similar to the emission that occurs when exciting at the wavelength of Si-PPV maximum absorption, $\sim$ 3.1 eV.). The emission spectrum is also found to be comparable to that of Si-PPV in solution, which indicates that no aggregation of chromophores occurs in the blend. The dispersion of chromophores reduces the fluorescence quenching by non-radiative recombination processes that frequently occur in condensed phases.

10.3.2
The Requirements and Design of Polymer Laser Materials

According to some scientists, the natural follow-up to the polymer LED is the polymer laser. Let us first consider the general principle. What is used here is light amplification by stimulated emission. The medium is a material that efficiently supports stimulated emission and it is continuously pumped by an external source to maintain an inverted population. The medium is positioned within a resonator, a cavity between two mirrors, so the light travels back and forth. The trick is that one mirror is a fully reflective mirror and the other is only partially reflective. Through the latter mirror, 1–2% of the light that hits it is transmitted and escapes.

The light within the resonator is a standing wave and its wavelength is selected by the length of the cavity (a Fabry–Perot interferometer) in combination with the emission spectrum of the medium. The emission characteristics define to a large extent in which narrow wavelength range gain may be experienced. The interferometer serves to sharpen this output line considerably (the width of the wavelength distribution goes down to fractions of a nanometer). The output light is coherent in space and time, highly monochromatic, polarized and highly directional (limited beam divergence).

The intensity gain over one round trip in the cavity is gain minus losses:

$$\frac{I_{i+1}}{I_i} = R_1 \exp[2L(g-a)] R_2 > 1$$

where g=gain coefficient, a=loss coefficient and R_1, R_2 are the factors for the reflectivity of the two mirrors, one being 1 and the other only slightly less. If the whole product on the right-hand side is still larger than 1, the lasing will work.

The >100-fold extension of the pathlength of interaction between the light beam and the excited states dramatically increases the amplification.

Polymer lasers, which are still under development, represent a major challenge to the organic semiconductor device community. There are several requirements for polymers to be used for lasing: (i) a four-level system, which facilitates the achievement of population inversion; (ii) high chromophore density, which favors a high density of excited states; (iii) high cross-section for stimulated emission, a molecular property tuned through chemical engineering; (iv) high luminescence efficiency, with a low probability of nonradiative decay, which makes it easier to achieve population inversion and hence lowers the threshold for the occurrence of amplification. A major issue is the final point, (v) low excited-state absorption: once the polymer is excited, new species are created and these have their own, different absorption spectra. As the excited population grows, the excited-state absorption increases. It usually lies at longer wavelengths than the ground-state absorption and interferes much more with the emitted light (loss processes).

Many of the desired properties can be adjusted with chemical tuning of a given group of molecules. These properties and the related materials performances also depend strongly on structure and disorder at the chain level and at the level of the bulk solid.

10.3.2.1 Amplified Spontaneous Emission in Polymer Films

Lasing in conjugated polymers has been observed to start at a concentration of excited states of $\sim 10^{17}$ cm^{-3}, ca. 1 per 10 000 chromophore units. This is possible due to the principles of the four-level system.

Mirrorless lasing does not require a cavity feedback system to observe laser-like phenomena. As soon as amplification occurs, it shows up as spectral narrowing. The spectrum of emission changes as one wavelength is amplified at the expense of others. The higher the excitation energy is (pulse power, not photon energy), the narrower the spectrum becomes.

In the example shown in Fig. 10.21a, a laser operating at 355 nm was simply pumping a thin Si-PPV polymer film. The path of interaction is created by waveguiding within the thin film. The stimulated wave propagates in the plane of the film. If the length over which amplification takes place is large enough, light emitted from the edges of the film will be considerably narrowed compared with normal photoluminescence. The emission peaks at 450 nm (Fig. 10.21b). The triggering of the stimulated emission is not done by a probe beam but by spontaneous emission (luminescence) of the polymer material [amplified spontaneous emission (ASE)]. Since there are no resonator structures to define the wavelength better, the peaks from ASE are much less narrow than they are in the output of a real laser.

10.3.2.2 Polymer Laser

Hadziioannou and coworkers [97], Heeger and coworkers [98] and Friend and coworkers [99] have demonstrated the use of semiconducting polymers for real lasing, by employing a resonator structure. Two mirrors, one 99% reflective (imper-

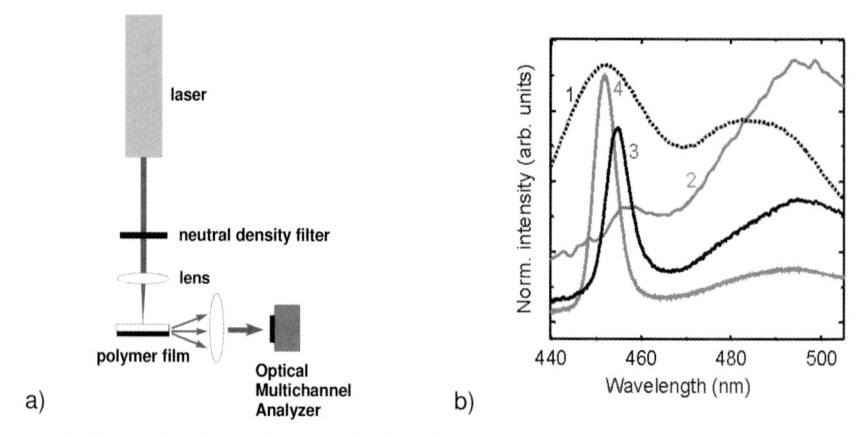

Fig. 10.21 (a) Mirrorless lasing assembly in order to observe the ASE phenomenon and (b) emission spectra of an Si–PPV copolymer for increasing excitation energy at 355 nm.

fections) and the other 95%, were used. To be able to change the distance between the mirrors, a grease which was refractive-index-matched to the polymer was used on one side of the film (Fig. 10.22a). Figure 10.22b (A) shows the normal luminescence of dioctyloxy-PPV in (dotted line) and the spectrum that was observed by pumping a thin film on glass (solid line). The line narrowing due to ASE in the film is very obvious. Using the resonator structure, one could go further.

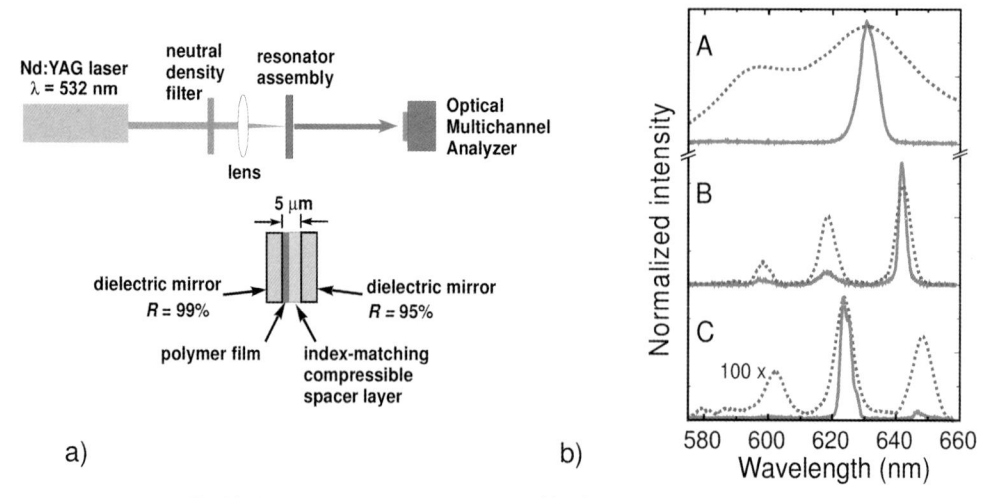

Fig. 10.22 (a) Resonator structure assembly. (b) Emission spectra of a dioctyloxy-PPV on glass (A) and in a planar optical cavity resonator structure (B, C; different lengths). Dotted lines, below excitation threshold energy; solid lines, above threshold.

At low laser power, when the emission spectrum is broad, the mode structure of the resonator was observed. The resonator selects three lines from the envelope of the spectrum. At high laser intensity, the most intense of these peaks is favored and grows at the expense of the others. The line narrows further compared with the linewidth dictated by the resonator. When the resonator is tuned by changing its length (by moving a mirror), one sees that the wavelengths of the resonator shift and that at some point it leads to the selection of another mode as the one that receives most gain at high excitation energy (it is closest to the emission maximum) [100] [Fig. 10.22 b (B and C)].

These results demonstrate that optically induced lasing in polymers is possible.

10.3.2.3 Polymer Blends

As for the PLEDs, one could also use a blend here, chromophores in an absorbing matrix [101]. The characteristics are Förster transfer, resulting in an improved four-level system, and reduced quenching due to molecular dispersion of the chromophores and therefore a higher efficiency. This would imply that a lower excitation density is required to obtain net amplification and hence a lower threshold intensity of the excitation source. In the example shown in Fig. 10.23, the reduced overlap between adsorption and emission in the blend of two PPVs as compared with the separate components is clearly visible.

10.3.2.4 Electrically Pumped Polymer Laser?

The question has been from the start: could one devise an electrically pumped polymer laser? An optically pumped laser, light in–light out, might not bring so many advantages over the existing situation. The electrical polymer laser would have advantages similar to those of the polymer LED: flexible, large area, etc. It requires a large current density to generate enough excitations. This is a major problem. The current itself can be sustained by the material if the mode of operation is pulsed rather than continuous, so that heat can be conducted away.

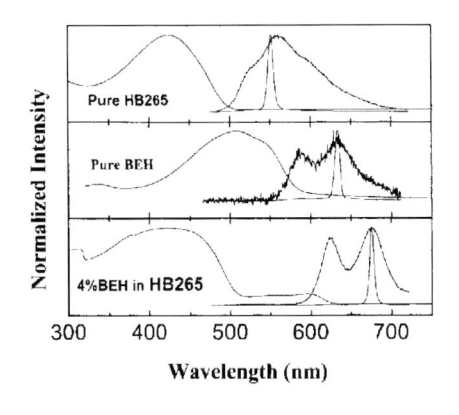

Fig. 10.23 Absorption, spontaneous emission and ASE for pure HB265 (a soluble PPV), BEH–PPV and 4 wt% BEH–PPV in HB265 (PPV derivative obtained from Covion).

The main problem is apparently in the loss factors: the charge carrier-induced absorptions. To create the density of excited states required, a lot of charge carriers are running around. The absorption spectrum of the polarons is much more to the red (even into the IR) region than that of the ground-state species. The total absorption ranges all the way from the ground-state absorption to the IR and a large proportion of the emitted radiation is reabsorbed. Other loss processes have to do with the presence of metal electrodes interfaced with the polymer film. These electrodes hamper the waveguiding in the polymer film, so if waveguiding is essential to the gain process, additional organic layers must be incorporated.

10.3.3
The Requirements and Design of Polymer Photovoltaic Materials

Photovoltaic cells (PVs) are either light sensor (photodetector) or energy conversion (solar cell) devices. Our emphasis will be on the latter application; in that arena, a well-performing organic material would have to compete with amorphous silicon with regard to energy conversion efficiency and fabrication costs. The potential of the polymer clearly lies in the promise of large-area, mechanically flexible, active coatings fabricated by inexpensive processing techniques.

A basic requirement for a photovoltaic material is photoconductivity, i.e. that charges are generated upon illumination. Subsequently, then, these charges must drift (move in an electric field) towards electrodes for collection. In an organic molecular material, photoexcitation does not directly yield free charge carriers. Due to the low dielectric constant of organics, an electron in the excited state is bound to its vacancy (hole) fairly strongly, the binding energy being several tenths of an electronvolt [102]. This bound electron–hole pair is called an exciton. Escape from the Coulomb attraction is promoted by offering an energetically favorable pathway to an electron-accepting molecule. This is the donor–acceptor (D–A) concept, which is commonly applied to organic photovoltaic materials [103, 104]. Dissociation of the exciton, via rapid electron transfer (< 200 fs), leaves a positively charged donor molecule and a negatively charged acceptor molecule. These are cation and anion species, respectively, stabilized by charge delocalization within their conjugated systems and by polarization of their environment. Exciton dissociation occurs at the interface between donor and acceptor species. Although it is not *a priori* evident what the nature of this interface should be in terms of scale and geometry, optimization within the D–A concept is likely to imply that this interface be made large and easily accessible for the excitons generated. Since excitons have a finite lifetime, they also have a finite diffusion range. Hence the requirement of accessibility naturally leads to constraints for the geometry of the interface. A spatially distributed interface with a correlation length of 10 nm would be compatible with the evolution of the exciton: the exciton would have a higher probability of reaching the interface and dissociate than of decaying in another way, e.g. radiatively. After dissociation, the charges must be further separated and transported each through its own phase so as to avoid recombination before the electrode is reached. This D–A concept has recently been implemented by using interpenetrating polymer blends of donor and acceptor homopolymers sandwiched between two asymmetric contacts (two metals with different workfunctions) for photovoltaic devices [104, 105]. This bi-continuous network of donor–acceptor heterojunctions facilitates simultaneously the efficient exciton dissociation and balanced bipolar charge transport throughout the whole volume of the device.

The performance of this type of device is very sensitive to the morphology of the blend, since exciton dissociation as described above occurs only at the donor/acceptor interface and charge transport depends on the geometry and composition of the phases. A system of spatially distributed donor–acceptor heterojunctions may contain imperfections that degrade performance, such as fully dispersed domains and cul-de-sac-type discontinuities of the donor and/or acceptor phases. Furthermore, the extended interface area is accompanied by an enlarged phase boundary volume in which donor and acceptor species are molecularly mixed. This is likely to result in increased energy level disorder, causing an increase of the charge trap density and a reduction of the electron and/or hole mobility. Hence, although exciton dissociation may be enhanced, transport is impeded by the circuitous geometry of the interface. For these and other reasons, the overall device performance can be lower than expected. The transport problem is altogether absent in a double-layer structure with a single, planar donor–acceptor heterojunction. Since, however, the exciton diffusion range is typically ~ 10 nm and hence shorter than the light absorption depth, a properly dimensioned and structured network of donor–acceptor heterojunctions should be more efficient in terms of the exciton dissociation than the simple double-layer structure. It is not obvious how to impose such a nanometer-size and regularly interpenetrating morphology on a mixture of donor and acceptor homopolymers: they would probably either mix molecularly or phase separate into nearly pure components. This is exactly the point at which block copolymers provide the answer, because of their ability to self-organize. Block copolymers are a well-known class of compounds that can show well-ordered, regular morphologies through microphase separation, which is in line with the strategy outlined above [106]. The composition of the components is fixed due to the synthesis, the morphology of phase-separated films can be controlled (by the interaction parameters, block lengths and ratios) and the dimensions of the domains are of the order of several nanometers. Therefore, this class of materials seems destined to yield much higher performances in photovoltaic devices than the homopolymer blend. The simplest molecule would be a photovoltaic diblock copolymer consisting of a block with donor functionality linked to a block with acceptor properties. The electronic functions could be either in the main chain of the blocks or in substituents. Microphase separation would produce a suitable geometry at the proper scale that could be fine-tuned via the lengths of the blocks. In this context, cylindrical and bi-continuous interpenetrating morphologies are the most appropriate.

10.3.4
Photovoltaic Performance of Donor–Acceptor Block Copolymers

Based on the general principles discussed above, a self-structuring block copolymer has been designed and synthesized [68]. It is a diblock copolymer consisting of a semiconducting rod block and a coil block densely functionalized with acceptor moieties (e.g. C_{60}). The synthesis of such a rod–coil polymer poses several challenges. To obtain a well-defined material providing a bi-continuous interpenetrated microphase-separated structure throughout its volume, good control over the length of both blocks and their volume ratio is desirable. Furthermore, the incorporation of a sufficient amount of acceptor moieties into one block of the diblock copolymer is necessary to satisfy two requirements for the efficient operation as a photovoltaic material: (i) creating an accessible donor/acceptor interface at which dissociation of excitons into separate charge carriers is promoted, thus reducing the probability of decay along other routes, of which luminescence is one; (ii) providing separate pathways for transport of holes (via the rod block) and electrons (via the coil block), thus reducing the recombination probability. A reduction in the photoluminescence lifetime relative to the non-acceptor-functionalized diblock copolymer has been observed and constitutes a first indication that an effective donor/acceptor interface has been created by incorporating the electron acceptor [69].

When comparison is made between the D–A block copolymer and the blend of the donor and acceptor homopolymers of the corresponding blocks, an enhancement in photovoltaic performance is observed on going from the blend to the block copolymer [67]. This superior response can be seen in measurements of current versus voltage (I–V) (Fig. 10.24) on devices from the blend and the block copolymer under monochromatic illumination of 1 mW cm^{-2} at 458 nm. The enhancement is mainly due to the larger D–A interface and the higher continuity of transport pathways for charges in the block copolymer.

Although both blend and block copolymers show almost complete quenching of the fluorescence in the solid state, the collection efficiencies (the ratio of collected electrons and absorbed photons) obtained are significantly smaller than unity. This could be attributed to several processes, among which the following two seem to be the most important. First, exciton dissociation upon photoexcitation could be neither the only nor the main energy deactivation pathway. Energy transfer could compete with the dissociation or the charge separation might not be effective [107]. Second, in either system, blend or block copolymer, mixing of donor and acceptor at the molecular scale may introduce increased energy level disorder in both phases, which results in an increase in the charge trap density with a significant increase in the space-charge field and, consequently, in a reduction of the electron and/or hole mobility [108].

The results obtained with a C_{60}-modified PPV–PS block copolymer, which still leave ample room for optimization as far as microstructure is concerned, seem to validate the block copolymer strategy towards self-assembling materials for photovoltaic applications. Other strategies have recently been pursued with success making use of the self-assembling properties of liquid-crystal materials [109].

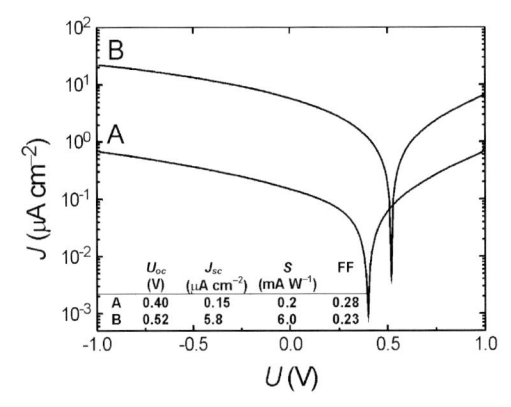

[Graph showing J (μA cm⁻²) vs U (V) with curves A and B, and table:]

	U_{oc} (V)	J_{sc} (μA cm⁻²)	S (mA W⁻¹)	FF
A	0.40	0.15	0.2	0.28
B	0.52	5.8	6.0	0.23

Fig. 10.24 Photovoltaic response of a PPV-b-P(S-stat-C_{60}MS) donor–acceptor block copolymer (B) compared with a blend of donor homopolymer and acceptor polymer (A) under monochromatic illumination of 458 nm. U_{oc}, open-circuit voltage; J_{sc}, short-circuit current density; S, sensitivity; FF, fill factor. The fill factor is defined as the maximum electrical power $(IV)_{max}$ that can be extracted from a photovoltaic diode divided by the product of maximum current I_{sc} and maximum voltage U_{oc} [67].

The record photovoltaic power conversion efficiency in polymer semiconductors is ~5%. This record was achieved by optimizing the network structure of donor and acceptor homopolymer blends via trial and error in the solvent casting and post-deposition annealing of polymer films [55]. In our opinion, the scientific and technological challenges for highly efficient polymer photovoltaic materials and devices are the following: (i) obtaining chemically ultra-pure polymer materials; (ii) optimizing the light absorption of the active polymer for efficient harvesting of the solar spectrum; (iii) making a judicious choice of donor and acceptor polymer pairs for the most efficient exciton dissociation and charge transport; (iv) understanding and optimizing the electrode/polymer and donor polymer/acceptor polymer interfaces for better charge extraction/injection; and (v) optimizing/controlling the nanostructuring of the active material. Most efficiently and elegantly, the nanostructuring can be accomplished through self-organization of matter rather than by manipulation or machining, wherever possible.

10.3.5
The Requirements and Design of Polymer Field-effect Transistors

Conjugated polymers were used as semiconductor layer in field-effect transistors already in the early 1980s [110]. Although the initial devices showed fairly low performances, the continuous progress in understanding and control of polymer semiconductors has allowed impressive improvements in device performance, up to the point that makes polymer field-effect transistors (PFETs) today a possible substitute for amorphous silicon transistors in low-cost and large-scale electronic products. Since PFETs are produced at low temperatures

(<200 °C) and most likely at very low cost, they are of particular interest to flexible electronic devices such as radiofrequency disposable identification tags or electronic driving circuits in flexible displays.

In a PFET, the intrinsic polymer semiconductor forms a conducting channel between two metal electrodes (the "source" and "drain" contacts; Fig. 10.25). The charge carrier density within the polymer film, which is separated by a thin dielectric layer from the conducting substrate (or "gate" contact), can be modulated by applying an appropriate voltage V_{GS} between source and gate. A positive (negative) gate voltage induces an accumulation of negative (positive) charge carriers in the first few monolayers of the polymer film. As a result, the current I_D flowing from source to drain under a fixed source–drain voltage, V_{DS}, is controlled by V_{GS} and is either "n-type" (electron current, $V_{GS}>0$) or "p-type" (hole current, $V_{GS}<0$). The device performances are best described in terms of the channel transconductance defined by $g_m = \frac{\partial I_D}{\partial V_{GS}}(V_{GS})\big|_{V_{DS}}$

In the linear regime (low V_{DS}),

$$G_m \cong \frac{WC_i}{L}\mu V_{DS}$$

where W and L are the channel width and length, respectively, C_i the insulating layer capacitance per unit area and µ the field-effect charge carrier mobility. This relationship highlights the influence of the charge carrier mobility on the device performance, and also the importance of the device geometry (through W and L) and composition. C_i depends indeed on the dielectric constant and thickness of the insulating layer and fixes the amount of charge accumulated in the channel for a given gate voltage.

Another important device characteristic is the current ratio, "I_{on}/I_{off}", in the saturation regime, that is, when $V_{DS} > V_{th} - V_{GS}$, where V_{th} is the device-specific "threshold" gate voltage above which a significant accumulation of carriers in the channel occurs. I_{on} and I_{off} are the drain currents with or without a gate voltage, respectively. High I_{on}/I_{off} values are a prerequisite for reduced energy consumption and high signal-to-noise ratios. Both field-effect mobility and current ratio have been increased by several orders of magnitude since the first demonstration of a PFET and are rapidly approaching values typical of amorphous silicon transistors [111–113].

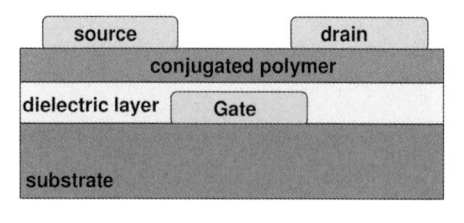

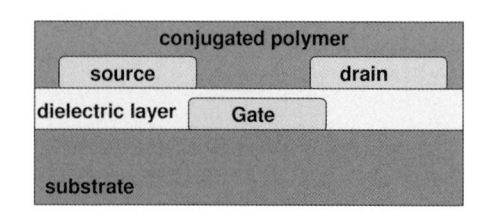

a) top-contact

b) bottom-contact

Fig. 10.25 (a) Top-contact PFET and (b) bottom-contact PFET.

From a material science point of view, the device performance relies on:
- the positive and/or negative charge carrier transport through the polymer film near the interface with the insulating layer;
- the electronic quality of the semiconductor/insulator interface;
- the purity of the semiconductor;
- the charge injection barriers at the source and drain electrodes.

High field-effect mobilities can only be achieved if the molecular ordering in the first few nanometer thickness allows efficient intramolecular and intermolecular charge transport in a direction parallel to the surface. This can be achieved through molecular engineering by synthesizing molecules which, in the solid state, allow strong π–π intermolecular interactions along the appropriate direction. The two orders of magnitude increase in hole mobility observed in P3HT thin films, on going from a regiorandom to a regioregular molecule, illustrates this point well [112, 114, 115]. The regular head-to-tail arrangement of the alkyl groups in the rr-P3HT allows a better in-plane molecular packing (see paragraph 10.2.1), leading to the formation of two-dimensional crystalline lamellas whose π-stacking axes are mostly parallel to the substrate. Also, hole mobilities as large as $0.6 \, \text{cm}^2 \, \text{V}^{-1} \, \text{s}^{-1}$ have recently been reported on a liquid-crystalline semiconducting thieno[3,2-b]thiophene polymer, which is characterized by a highly organized morphology with large crystalline domains [113].

Progress with n-type PFETs has been impeded by a frequently observed electron trapping mechanism which resulted in reduced channel conduction. Although the origin of these effects is not completely understood yet, Friend and coworkers showed recently that electron trapping by hydroxyl groups present at the semiconductor/dielectric interface is a major cause [116]. By using a hydroxyl-free gate dielectric, they obtained good electron conduction in most conjugated polymers, with electron mobilities reaching values up to $10^{-2} \, \text{cm}^2 \, \text{V}^{-1} \, \text{s}^{-1}$.

In general, to achieve good transistor performance, special attention needs to be paid to the semiconductor/dielectric interface. It must simultaneously be exempt of electronically active trapping centers and allow the formation of a well-ordered polymer layer with the appropriate morphology. The currently best operating PFETs use silicon oxide as a dielectric and highly doped silicon as gate electrode [113]. It is obvious that similar performances on more application-oriented dielectric materials (organic solution-processed dielectrics, for instance) can only be reached if the semiconductor/dielectric interface is of sufficient electronic quality and if it allows the polymer film to maintain adequate morphology.

The chemical purity of semiconductors is another point of major importance, since unintentional chemical doping (through oxidation of the polymer, for instance) will reduce the I_{on}/I_{off} ratio. This is a major disadvantage of polymers compared with molecular semiconductors, seeing that standard purification procedures can no longer be used.

Finally, the semiconductor–metal contacts are also essential to the device operation. Source and drain contacts should have low charge carrier injection barriers in order to reduce the operating voltage. This in turn requires the metal Fermi level to be close either to the polymer HOMO level, in the case of p-type conduction, or to the LUMO level, for n-type conduction. The former case is most often achieved by using gold as a high electron workfunction metal, whereas low workfunction metals such as Ca or Al are more appropriate for n-type PFETs. In general, the quality of the semiconductor–metal contacts is found to depend on the procedure used to fabricate the transistor structure. Bottom-contact transistors (Fig. 10.25 b) generally suffer from a larger contact resistance than the equivalent top-contact structure. The presence of an uncontrolled oxide layer and an alternation of the semiconductor growth mechanism close to the electrodes, leading to a larger defect density, are possibly at the origin of the reduced performances.

Further progress in the field of PFETs strongly relies on our ability to continually improve our control over the semiconductor morphology and interface quality, in particular in devices obtained using procedures such as inkjet printing techniques and dielectric materials, compatible with large-scale, low-cost device production.

10.3.6
The Requirements and Design of Sensors Based on Polymer Semiconductors

Sensors with the ability to detect chemicals, e.g. volatile compounds and biological species, including DNA and proteins, have been given increasing attention by the scientific and industrial communities. Miniaturization is demanded for all types of sensors, because of the need for better portability, higher sensitivity, lower power dissipation and better device integration.

Conjugated polymers offer a great number of opportunities to couple analyte receptor interactions, and also non-specific interactions, into observable responses.

Conjugated polymer-based sensors have been formulated into a variety of schemes [117]:
- Conductometric sensors yield changes in electrical conductivity in response to analyte interactions.
- Potentiometric sensors rely on analyte-induced changes in the system's chemical potential.
- Colorimetric sensors refer to changes in the polymer absorption properties.

The last kind of sensor is perfectly illustrated by the work of Leclerc's group [118]. In order to detect specific nucleic acids for the diagnosis of infections, identification of genetic mutations and forensic analyses, they used a fast and simple electrostatic approach based on a cationic-soluble polythiophene which allows the direct detection and specific identification of less than 10 molecules of the negatively charged complementary DNA or RNA target without any chemical reaction on the probes or targets. The cationic water-soluble polythiophene can form stoichiometric polyelectrolyte complexes (coacervates) with anionic oligonucleotide probes (duplex). These stoichiometric complexes tend to form aggregates in aqueous solutions. As a result of aggregation and of the close proximity and stiffening of the polythiophene chains, the intrinsic fluorescence of the polymer is red shifted and quenched. However, the fluorescence intensity increases again through the hybridization process with the perfectly matched complementary nucleic acid strand. The blue shift and increase in fluorescence are due to a helical structure of the polythiophene backbone that wraps with greater affinity around the anionic phosphate backbone of double-stranded nucleic acids (as compared with its affinity for single-stranded nucleic acids) combined with better solubilization of the resulting nonstoichiometric polyelectrolyte complex (triplex), which limits interchain quenching.

Aptamers [119], synthetic oligonucleotides with characteristic 3-D structures, represent a new and interesting class of ligands to bind a protein selectively without using antibodies extracted from animals. In order to detect the blood-clotting enzyme thrombin, Plaxco and coworkers [120] realized a potentiometric sensor, using a methylene blue-labeled aptamer probe as an electro-active molecular beacon. Signal generation occurs upon a large, binding-induced conformational change in a redox-modified, electrode-attached aptamer. The authors presumed that the conformational change (unfolding to folding G-quartet conformation) due to the recognition event between thrombin and aptamer inhibits the electron transfer from the methylene blue to the gold electrode.

Concerning the conductometric type of sensor, an example is given by the work of Lee and Swager on the chemical vapor detection in an elementary device in which a polymer thin film is sandwiched between two electrodes [121]. The analyte-induced reductions in conjugation length of the conjugated polymer backbone or induced energy mismatches between adjacent redox-active sites yield a reduced conductivity of the conducting layer. In order to enhance the sensitivity of the system, they used molecular insulated nanowires as active material instead of standard conducting polymer thin films. In general, analyte–polymer interactions lead to changes in the charge carrier local energy landscape which alters the electrical conductivity of the layer. However, π–π interactions between adjacent conjugated backbones provide multiple pathways to the charge-carrying species. The resulting three-dimensional electronic network allows charge carriers, encountering a local analyte-induced energy barrier, to find alternative routes to the collecting electrodes. This restricts the conductivity response to the presence of the analyte and, therefore, the detection limit of the device. If insulated nanowires are used as sensing and charge-transporting media, intramolecular transport will dominate. The resulting one-dimensional electronic network leaves no escape to the free carrier in case of a broken pathway (Fig. 10.26) and, correspondingly, improves the detection limit. In addition, the phenyl canopy, which in this case was used as a molecular "scaffold", increases the intermolecular separation, thereby facilitating the diffusion of the analyte towards the conjugated backbone and improving the sensor's response.

In conclusion, conjugated polymers have attracted wide theoretical interest and practical applications in sensor technology. The advantages offered by the

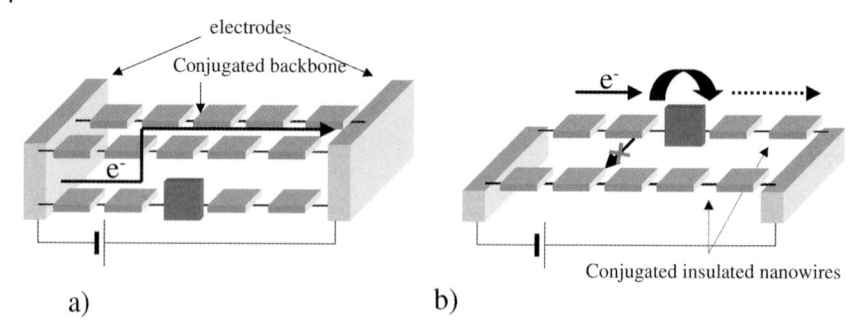

Fig. 10.26 Electronic pathway in thin film (a) and molecular nanowire devices (b). The dark square represents an analyte-induced local perturbation or energy barrier.

polymeric materials to sensor technology are the common ones, i.e. their relatively low cost and simple, low-thermal budget processing techniques and compatibility with different substrates, but they also bring specific benefits for sensor technology, such as the wide choice of their molecular structure and the possibility of incorporating side-chains or grains with specific behavior into the bulk material or on the surface, allowing the production of films with specific physical and chemical properties, including sensing behavior. The last, but no less important, feature is the biocompatibility of many of these materials.

10.3.7
"Polymer Conductors"/"Synthetic Metals"

After the discovery that polyacetylene can reach high electrical conductivities, the field of conducting polymers, which is often referred to as synthetic metals, has attracted the interest of many scientists and engineers in academia and industry [122]. From the start, much of the combined research effort of industrial and academic researchers was focused on developing materials that are stable in the conducting state, easily processable and relatively simple to produce at low cost. Of the many conducting polymers that have been developed over the past 30 years, those based on polyanilines, polypyrroles, polythiophenes, polyphenylenes and poly(p-phenylenevinylene)s have attracted the most attention [122]. Of these, polyaniline is the best because it allows the fabrication of processable conductive materials at relatively low cost and in large amounts. Unfortunately, due to the possible presence of benzidine moieties in the polymer backbone, which might yield toxic (carcinogenic) products upon degradation, many industrial and academic groups have limited their research in polyaniline chemistry. In the mid-1980s the development of a polythiophene derivative, poly(3,4-ethylenedioxythiophene), often abbreviated to PEDT or PEDOT, with its high conductivity (~ 300 S cm^{-1}), its high transparency in the thin-film configuration and its very high stability in the oxidized state, gave a new impetus to the subfield of polymeric conducting materials. Initially, PEDOT was found to be an insoluble polymer. This problem was elegantly circumvented by the use of a water-soluble polyelectrolyte, poly(styrenesulfonic acid) (PSS), as the charge-balancing dopant during polymerization to yield PEDT–PSS. This combination resulted in a water-soluble polyelectrolyte system with good film-forming properties, high conductivity (~ 10 S cm^{-1}), high visible light transmissivity and excellent stability [123]. With this new system, now known under its commercial name Baytron P (P stands for polymer), numerous applications have been envisaged, beyond its simple use as an antistatic thin film for photographic applications, electrode material in capacitors, material for through-hole plating of printed circuit boards, transparent electrodes, and more are expected in the emerging technology of plastic electronics.

10.3.8
The Requirements and Design of Thin-film Conducting and Transparent Polymer Materials

Conducting polymers are highly absorbing due to the presence of a high density of polarons. Films with thickness in the micrometer range are therefore already intensely colored and fairly dark. There are several applications for which a conducting and highly transparent film is desirable or required. Examples are antistatic coatings (especially those for photographic films) and transparent electrodes. To obtain better transparency, the amount of conducting polymer in the films must be reduced to no more than a few volume-%, but at the same time a fair level of conductivity should be maintained. A concept to realize just that is to dilute the conducting polymer molecularly, to make a percolating system in a controlled manner. This is done by coating transparent spheres with a thin layer of conducting polymer and then forcing these coated spheres into intimate contact by film formation. Hence there are ample pathways, of mesoscopic width (~ 1 nm), of pure conductive polymer available to the charge carriers. In such a system, the decrease in conductivity would be given by the overall dilution factor of the conducting polymer. In a regular blend of the polymers, percolation would cease at much higher conducting polymer fractions.

Huijs and coworkers [124] used the approach of core–shell latex spheres with polypyrrole and polythiophene as a shell and an acrylate core. The conducting polymer was formed *in situ* around the latex spheres by oxidative coupling in an aqueous system. The polypyrrole-based materials were the most successful, due to the morphology that developed during synthesis.

In Fig. 10.27, the conductivity of a (dried) PPy-coated poly(butyl methacrylate) latex is compared with that of a mixture of PBMA latex and PPy, for various PPy fractions. For the coated latex, the resulting conductivity is between 0.1 and 1 S cm^{-1} even at PPy fractions of only 1%. For the blend, the percolation was found to be very poor already at medium PPy content. Film transparency is no longer determined by photoabsorption (which is very low because of the composition) but by light scattering from residual voids. The spheres are in the micrometer range so their complete fusion is essential, for both transparency and me-

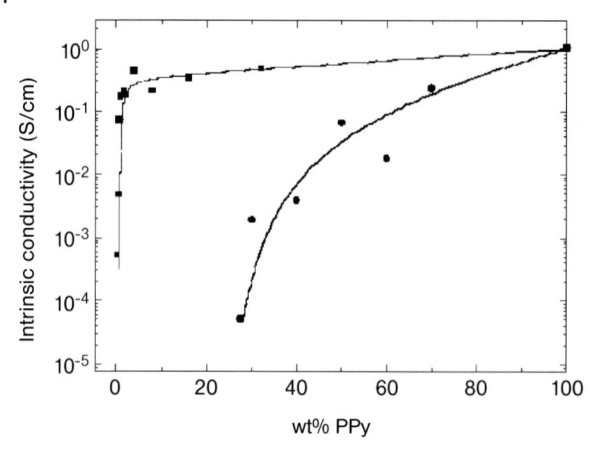

Fig. 10.27 Conductivity of (a) thin film formed from PPy-coated poly(butyl methacrylate) latex and (b) thin film from a mixture of the PBMA latex and PPy, for various PPy fractions.

chanical coherence. A T_g of the latex polymer not far above room temperature is beneficial. Fusion of the spheres was further promoted by coalescence agents, in order to eliminate high-temperature treatments that might degrade the conduction properties of the PPy shell.

The polymer blend method was used earlier to obtain transparent and conductive thin polymer films. For this, poly(aniline–camphorsulfonic acid) (PANI–CSA) blends with poly(methyl methacrylate) (PMMA) and nylon exhibited fairly high conductivities (1–10 S cm^{-1}) at relatively low ($\sim$1%) loading levels of the conducting polymer in the host [125]. This high conductivity is attributed to the formation of a continuous percolated network of the conducting polymer in the matrix of the host.

The most promising and commercially available conductive and transparent thin films were obtained from spin casting of the aqueous poly(3,4-ethylenedioxythiophene)–poly(styrenesulfonic acid) (PEDOT–PSS) (Baytron P) solution which had a high degree of mechanical integrity with conductivities ranging between 1 and 10 S cm^{-1}. These films are highly stable and can be treated for up to 1000 h at 100 °C with no change in conductivity [123].

10.4
Conclusion

Semiconducting polymers are relatively new and promising materials for the development of new technologies such as plastic electronics. Moreover, they involve novel basic phenomena, phenomena that had been predicted to exist or treated theoretically but were not found in actual materials until these polymers came along. They show these phenomena and make them experimentally accessible. One prominent aspect of the field of semiconducting polymer materials is that it is highly interdisciplinary. This field has evolved through the collaboration of solid-state physicists, polymer chemists, theoretical physicists and chemists, polymer engineers, physical engineers and even electrical engineers. The field of research is now approximately 30 years old and very dynamic.

Due to the ability of semiconducting polymers to conduct electronic charge, these materials have been considered to replace conventional materials in solid-state devices such as transistors, light-emitting diodes and solar cells, radiofrequency identification tags (RFIDs), etc. [126]. The driving force for the development of organic-based devices is their easy and cheaper fabrication as compared with their inorganic counterparts. However, organic materials, unlike crystalline semiconductors, are not crystalline solids but amorphous or polycrystalline materials. Phenomena such as charge transport are different than usually observed with inorganic semiconductor materials and this results in a different performance behavior of the devices fabricated with these materials, which has to be accounted for when considering different applications. In particular, plastic chips will probably never be as fast or as miniaturized as their silicon counterparts and, therefore, it seems most unlikely that plastic transistors will replace the silicon type in microprocessors. On the other hand, they will eventually be successfully employed as switching devices for active matrix flat-panel displays, RFIDs [127], wearable electronics [128] and disposable chemical sensors [118]. The goal of plastic electronics is to identify the right applications for the right materials and *vice versa*, to create a material with the best properties tailored for a certain application.

In this chapter, the birth and the development of the field of polymer semiconductor materials have been described, together with their applications in devices and systems in present and future technologies. The various families of conjugated polymers, their optoelectronic properties and the synthetic methodologies to produce them have been discussed. A few optoelectronic devices and optoelectronic polymer material systems have been described with regard to their optical, electrical and/or electronic functions and also the requirements that the polymer semiconducting materials must fulfill. These requirements concern the optoelectronic properties and processability characteristics for the realization of the highest possible performance devices and systems, based on today's fundamental understanding of the optical and electronic phenomena in conjugated polymers. The field of semiconducting polymers is undergoing a rapid evolution. Due to space limitations, it was impossible to include all possible applications and the latest developments in the field; therefore, for a more comprehensive description of the field, readers might consult the most recent literature [129, 130].

Acknowledgments

The present chapter reviews examples, for the most part, of published and unpublished work of Professor Georges Hadziioannou's former research team at Groningen University, The Netherlands, and the present research team at the Université Louis Pasteur, Strasbourg, France. Professor Georges Hadziioannou is grateful to the numerous contributions of his former and present PhD students, postdoctoral fellows, visiting scientists, research associates and colleagues around the world.

References

1 H. Shirakawa, E.J. Lewis, A.G. MacDiarmid, C.K. Chiang, A. Heeger, *J. Chem. Soc., Chem. Commun.*, **1977**, 578.
2 A.J. Heeger, *Angew. Chem. Int. Ed.*, **2001**, 40, 2591.
3 J.H. Burroughes, D.D.C. Bradley, A.R. Brown, R.N. Marks, K. MacKay, R.H. Friend, P.L. Burn, A.B. Holmes, *Nature*, **1990**, 347, 539.
4 A.J. Heeger, *MRS Bull.*, **2001**, 26, 900.
5 R.H. Friend, J.H. Burroughes, T. Shimoda, *Phys. World*, **1999**, 12 (6), 35.
6 A.J. Heeger, *J. Phys. Chem. B*, **2001**, 105, 8476.
7 Z. Bao, *Adv. Mater.*, **2000**, 12, 227.
8 A. Kraft, A.C. Grimsdale, A.B. Holmes, *Angew. Chem. Int. Ed.*, **1998**, 37, 402.
9 G. Natta, G. Mazzanti, P. Corradini, *Atti Accad. Naz. Lincei Rend. Cl. Sci. Fis., Mat. Nat.*, **1958**, 25, 3.
10 H. Shirakawa, S. Ikeda, *Synth. Met.*, **1980**, 1, 175.
11 A. Bolognesi, M. Catellani, S. Destri, in *Comprehensive Polymer Science*, Vol 4, Ed. G. Allen, Pergamon Press, Oxford, **1988**, Chapt. VIII.
12 W.J. Feast, J. Tsibouklis, K.L. Power, L. Groeneudaal, E.W. Meijer, *Polymer*, **1996**, 37, 1506.
13 R.A. Wessling, *J. Polym. Sci., Polym. Symp.*, **1985**, 72, 55.
14 P.L. Burn, D.D.C. Bradley, R.H. Friend, D.A. Halliday, A.B. Holmes, R.W. Jackson, A. Kraft, *J. Chem. Soc., Perkin Trans. 2*, **1992**, 3225.
15 R.O. Garay, U. Baier, C. Bubeck, K. Mullen, *Adv. Mater.* **1993**, 5, 561.
16 J. Gmeiner, S. Karg, M. Meier, W. Riess, P. Strohriegl, M. Schwoer, *Acta Polym.*, **1993**, 44, 201.
17 A. Beerden, D. Vandenzande, J. Gelan, *Synth. Met.*, **1992**, 52, 387.
18 M. Herold, J. Gmeiner, W. Riess, M. Schwoerer, *Synth. Met.*, **1996**, 76, 109.
19 F. Louwet, D. Vardenzande, J. Gelan, J. Mullens *Macromolecules*, **1995**, 28, 1330.
20 A.J. Heeger, D. Braun, *PCT Int. Patent Appl. WO 92/16.023*, **1992**.
21 A.J. Heeger, D. Braun, *Chem. Abstr.*, **1992**, 118, 157401j.
22 M.R. Pinto, B. Hu, F.E. Karasz, L. Akcelrud, *Polymer*, **2000**, 41, 2603.
23 S. Pfeiffer, H.H. Hörhold, *Macromol. Chem. Phys.*, **1999**, 200, 1870.
24 N. Leclerc, A. Michaud, K. Sirois, J.F. Morin, M. Leclerc, *Adv. Funct. Mater.*, **2006**, 13, 1694.
25 C. Martineau, P. Blanchard, D. Rondeau, J. Delaunay, J. Roncali, *Adv. Mater.* **2002**, 14, 283.
26 K.L. Brandon, P.G. Bentley, D.D.C. Bradley, D.A. Drunmer, *Synth. Met.*, **1997**, 91, 305.
27 Y. Liu, P. Lathi, F. La, *Polymer*, **1998**, 39, 5241.
28 M.J. Marsella, D.K. Fu, T.M. Swager, *Adv. Mater.*, **1995**, 7, 145.
29 J. Hou, C. Yang, J. Qiao, Y. Li, *Synth. Met.*, **2005**, 150, 297.
30 G. Grem, G. Leditzky, B. Ullrich, G. Leising, *Adv. Mater.*, **1992**, 4, 36.
31 M.R. Andersson, M. Berggren, G. Gustaffson, T. Hejertberg, O. Inganäs,

O. Wennertström, *Synth. Met.*, **1995**, *71*, 2183.

32 R. D. McCullough, R. D. Lowe, *J. Chem. Soc., Chem. Commun.*, **1992**, 70.

33 T. Yamamoto, A. Morita, Y. Miyazaki, T. Maruyama, H. Wakayama, Z.-H. Zhou, Y. Nakamura, T. Kanbara, *Macromolecules*, **1992**, *25*, 1214.

34 P. C. Stein, A. Bolognesi, M. Catellani, S. Destri, L. Zetta, *Synth. Met.*, **1991**, *41*, 559.

35 M. Leclerc, F. Martinez Diaz, G. Wegner, *Makromol. Chem.*, **1989**, *190*, 3105.

36 R. Corradi, S. P. Armes, *Synth. Met.*, **1997**, *84*, 453.

37 (a) M. Ranger, M. Leclerc, *J. Chem. Soc., Chem. Commun.*, **1997**, 1597; (b) M. Ranger, D. Rondeau, M. Leclerc, *Macromolecules*, **1997**, *30*, 7686.

38 E. J. W. List, R. Guentner, P. S. Scanducci de Freitas, U. Scherf, *Adv. Mater.*, **2002**, *14*, 374.

39 J. F. Morin, M. Leclerc, *Macromolecules*, **2001**, *34*, 4680.

40 J. F. Morin, M. Leclerc, *Macromolecules*, **2002**, *35*, 8413.

41 U. Geissler, M. L. Hallensleben, A. Rienecker, N. Rohde, *Polym. Adv. Technol.*, **1997**, *8*, 87.

42 R. Giesa, R. C. Shulz, *Macromol. Chem. Phys.*, **1993**, *191*, 857.

43 T. Yamamoto, M. Takagi, K. Kizu, T. Maruyama, K. Kubota, H. Kanbara, T. Kurihara, T. Kaino, *J. Chem. Soc., Chem. Commun.*, **1993**, 797.

44 J. Li, Y. Pang, *Macromolecules*, **1997**, *30*, 7487.

45 M. Moroni, J. L. LeMoigne, S. Luzatti, *Macromolecules*, **1994**, *27*, 562.

46 (a) K. Weiss, A. Michel, E. M. Auth, U. H. F. Bunz, T. Mangel, K. Mullen, *Angew. Chem.*, 1997, *36*, 506; (b) U.H.F. Bunz, *Chem. Rev.*, **2000**, *100*, 1605.

47 J. F. Morin, P. L. Boudreault, M. Leclerc, *Macromol. Rapid Commun.* **2002**, *23*, 1032.

48 N. S. Cho, D.-H. Hwang, B.-J. Jung, J. Oh, H. Y. Chu, H.-K. Shim, *Synth. Met.*, **2004**, *143*, 277.

49 M. Mazzeo, D. Pisignano, F. Della Sala, J. Thompson, R. I. R. Blyth, G. Gilgli, R. Cingolani, *Appl. Phys. Lett.*, **2003**, *82*, 334.

50 A. D. Schlüter, *J. Polym. Sci., Part A: Polym. Chem.*, **2001**, *39*, 1533.

51 J. M. G. Cowie, *Polymers: Chemistry and Physics of Modern Materials*, International Textbooks, Aylesbury, **1973**, pp. 4–5.

52 X. Zhan, Y. Liu, X. Wu, S. Wang, D. Zhu, *Macromolecules*, **2002**, *35*, 2529.

53 J. Liu, Q. Zhou, Y. Cheng, Y. Geng, L. Wang, D. Ma, X. Jing, F. L. Wang, *Adv. Mater.*, **2005**, *17*, 2974.

54 C. J. Brabec, N. S. Saraciftci, J. C. Hummelen, *Adv. Funct. Mater.*, **2001**, *11*, 15.

55 J.-L. Bredas, J. Cornil, A. J. Heeger, *Adv. Mater.* **1996**, *8*, 447.

56 G. Yu, J. Gao, J. C. Hummelen, F. Wudl, A. J. Heeger, *Science*, **1995**, *270*, 1789.

57 D. Gebeyehu, C. J. Brabec, F. Padinger, T. Fromhertz, J. C. Hummelen, B. Badt, H. Schindler, N. S. Sariciftci, *Synth. Met.*, **2001**, *118*, 1.

58 F. Padinger, R. S. Rittberger, N. S. Sariciftci, *Adv. Funct. Mater.*, **2003**, *13*, 85.

59 H. Hoppe, N. Arnold, N. S. Sariciftci, D. Meissner, *Sol. Energy Mater. Sol. Cells*, **2003**, *80*, 105.

60 T. Martens, J. D'Haen, T. Munters, Z. Beelen, L. Goris, J. Manca, M. D'Olieslaeger, D. Vanderzande, L. de Schepper, R. Andriessen, *Synth. Met.*, **2003**, *138*, 243.

61 C. W. T. Bulle-Lieuwma, J. K. J. van Duren, X. Yang, J. Loos, A. B. Sieval, J. C. Hummelen, R. A. J. Janssen, *Appl. Surf. Sci.*, **2004**, *231–232*, 274.

62 H. Hoppe, M. Niggemann, C. Winder, J. Kraut, R. Hiesgen, A. Hinsch, D. Meissner, N. S. Sariciftci, *Adv. Funct. Mater.*, **2004**, *14*, 1005.

63 L. Schmidt-Mende, A. Fechtenkötter, K. Müllen, R. H. Friend, J. D. MacKenzie, *Physica E*, **2002**, *14*, 263.

64 F. S. Bates, G. H. Frederickson, *Phys. Today*, 1999, *52*, 32.

65 S. A. Jenekhe, X. L. Chen, *Science*, **1999**, *283*, 372.

66 M. Lee, B. Cho, W. Zin, *Chem. Rev.*, **2001**, *101*, 3869.

67 B. de Boer, U. Stalmach, P. F. van Hutten, C. Meltzer, V. V. Krasnikov, G. Hadziioannou, *Polymer*, **2001**, *42*, 9097.

68 U. Stalmach, B. de Boer, C. Videlot, P. F. van Hutten, G. Hadziioannou, *J. Am. Chem. Soc.*, **2000**, *122*, 5464.

69 U. Stalmach, B. de Boer, A. D. Post, P. F. van Hutten, G. Hadziioannou, *Angew. Chem. Int. Ed.*, **2001**, *40*, 428.
70 R. A. Segalman, C. Brochon, G. Hadziioannou, in *Organic Photovoltaics: Mechanisms, Materials and Devices*, Eds. S.-S. Sun, N. S. Sariciftci, Francis and Taylor, CRC Press, New York, **2005**, pp. 403–420.
71 S.-S. Sun, *Sol. Energy Mater. Sol. Cells*, **2003**, *79*, 257.
72 F. C. Krebs, O. Hagemann, M. Jørgensen, *Sol. Energy Mater. Sol. Cells*, **2004**, *83*, 211.
73 S. M. Lindner, M. Thelakkat, *Macromolecules*, **2004**, *37*, 8832.
74 C. Zhang, S. Choi, J. Haliburton, T. Cleveland, R. Li, S.-S. Sun, A. Ledbetter, C. Bonner, *Macromolecules*, **2006**, *39*, 4317.
75 W. J. Li, H. B. Wang, L. P. Yu, T. L. Morkved, H. M. Jaeger, *Macromolecules*, **1999**, *32*, 3034.
76 D. Marsitzky, T. Brand, Y. Geerts, M. Klapper, K. Mullen, *Macromol. Rapid Commun.*, **1998**, *19*, 385.
77 G. N. Tew, M. U. Pralle, S. I. Stupp, *J. Am. Chem. Soc.*, **1999**, *121*, 9852.
78 M. A. Hempenius, B. M. W. Langeveld-Voss, J. van Haare, R. A. J. Janssen, S. S. Sheiko, J. P. Spatz, M. Moller, E. W. Meijer, *J. Am. Chem. Soc.*, **1998**, *120*, 2798.
79 V. Francke, H. J. Rader, Y. Geerts, K. Mullen, *Macromol. Rapid Commun.*, **1998**, *19*, 275.
80 H. Kukula, U. Ziener, M. Schops, A. Godt, *Macromolecules*, **1998**, *31*, 5160.
81 J. Genzer, E. Sivaniah, E. J. Kramer, J. G. Wang, H. Korner, K. Char, C. K. Ober, B. M. DeKoven, R. A. Bubeck, D. A. Fischer, S. Sambasivan, *Langmuir*, **2000**, *16*, 1993.
82 P. Leclere, V. Parente, J. L. Bredas, B. Francois, R. Lazzaroni, *Chem. Mater.*, **1998**, *10*, 4010.
83 M. Szwarc, *J. Polym. Sci. Part A, Polym. Chem.*, **1998**, *36*, IX–XV.
84 D. Marsitzky, M. Klapper, K. Mullen, *Macromolecules*, **1999**, *32*, 8685.
85 C. J. Hawker, A. W. Bosman, E. Harth, *Chem. Rev.*, **2001**, *101*, 3661.
86 Matyjaszewski, K. J. H. Xia, *Chem. Rev.*, **2001**, *101*, 2921.
87 M. C. Iovu, M. Jeffries-EL, E. E. Sheina, J. R. Cooper, R. D. McCullough, *Polymer*, **2005**, *46*, 8582.
88 J. Liu, E. Sheina, T. Kowalewski, R. D. McCullough, *Angew. Chem. Int. Ed.*, **2002**, *41*, 329.
89 Lu, S., Q. L. Fan, S. J. Chua, W. Huang, *Macromolecules*, **2003**, *36*, 304.
90 P. Tosolakis, J. Kallitsis, A. Godt, *Macromolecules*, **2002**, *35*, 5758.
91 C. Chochos, J. Kallitsis, V. Gregoriou, *J. Phys. Chem. B*, **2005**, *109*, 8755.
92 M. H. van der Veen, B. de Boer, U. Stalmach, K. I. van de Wetering, G. Hadziioannou, *Macromolecules*, **2004**, *37*, 3673.
93 K. van de Wetering, C. Brochon, C. Ngov, G. Hadziioannou, *Macromolecules*, **2006**, *39*, 4289.
94 G. Klaener, M. Trollas, A. Heise, M. Husemann, B. Atthoff, C. J. Hawker, J. L. Hedrick, R. D. Miller, *Macromolecules*, **1999**, *32*, 8227.
95 F. Garten, A comparison of the electrical properties of polymer LEDs based on poly(thiophene)s and PPV derivatives, PhD Thesis, Rijksuniversiteit Groningen, **1998**.
96 T. J. Wang, C. H. Chen, S. M. Chang, Y. J. Tzeng, Y. C. Chao, *Microwave Opt. Technol. Lett.* **2003**, *38*, 406.
97 H. J. Brouwer, V. V. Krasnikov, A. Hilberer, G. Hadziioannou, *Adv. Mater.*, **1996**, *8*, 935.
98 F. Hide, M. A. Díaz-García, B. J. Schwartz, M. R. Andersson, Q. Pei, A. J. Heeger, *Science*, **1996**, *27*, 273.
99 N. Tessler, G. J. Denton, R. H. Friend, *Nature*, **1996**, *382*, 695.
100 H. J. Brouwer, Semiconducting polymers for light-emitting diodes and lasers: a structural, photophysical and electrical study of PPV-type alternating copolymers and oligomers, PhD Thesis, Rijksuniversiteit Groningen, **1998**.
101 M. D. McGehee, A. J. Heeger, *Adv. Mater.*, **2000**, *12*, 1655.
102 B. A. Gregg, *J. Phys. Chem. B*, **2003**, *107*, 4688.
103 N. S. Sariciftci, L. Smilowitz, A. J. Heeger, F. Wudl, *Science*, **1992**, *258*, 1474.

104 G. Yu, J. Gao, J.C. Hummelen, F. Wudl, A.J. Heeger, *Science*, **1995**, *270*, 1789.
105 J.J.M. Halls, C.A. Walsh, N.C. Greenham, E.A. Marseglia, R.H. Friend, S.C. Moratti, A.B. Holmes, *Nature*, **1995**, *376*, 498.
106 F.S. Bates, G.H. Fredrickson, *Annu. Rev. Phys. Chem.*, **1990**, *41*, 525.
107 J.-F. Eckert, J.-F. Nicoud, J.-F. Nierengarten, S.-G. Liu, L. Echegoyen, F. Barigelletti, N. Armaroli, L. Ouali, V. Krasnikov, *J. Am. Chem. Soc.*, **2000**, *122*, 7467.
108 L. Ouali, V.V. Krasnikov, U. Stalmach and G. Hadziioannou, *Adv. Mater.*, **1999**, *11*, 1515.
109 L. Schmidt-Mende, A. Fechtenkötter, K. Müllen, E. Moons, R.H. Friend, J.D. MacKenzie, *Science*, **2001**, *293*, 1119.
110 (a) F. Ebisawa, T. Kurokawa, S. Nara, *J. Appl. Phys.*, **1983**, *54*, 3255; (b) A. Tsumura, H. Koezuka, T. Ando, *Appl. Phys. Lett.*, **1986**, *49*, 1210.
111 C.D. Dimitrakopoulos, P.R.L. Malenfant, *Adv. Mater.*, **2002**, *14*, 99.
112 R.J. Kline, M.D. McGehee, M.F. Toney, *Nat. Mater.*, **2006**, *5*, 222.
113 I. Mucculloch, M. Heeney, C. Bailey, K. Genevicius, I. Macdonald, M. Shkunov, D. Sparrowe, S. Tierney, R. Wagner, W. Zhang, M.L. Chabinyc, R.J. Kline, M.D. Mcgehee, M.T. Toney, *Nat. Mater.*, **2006**, *5*, 328.
114 Z. Bao, A. Dobabalapur, A.J. Lovinger, *Appl. Phys.Lett.*, **1996**, *69*, 4108.
115 H. Sirringhaus, N. Tessler, R.H. Friend, *Science*, **1998**, *280*, 1741.
116 L.L. Chua, J. Zaumseil, J.F. Chang, E.C.-W. Ou, P.K.-H. Lo, H. Sirringhausm, R.H. Friend, *Nature*, **2005**, *434*, 194.
117 (a) D.T. McQuade, A.E. Pullen, T.M. Swager, *Chem. Rev.*, **2000**, *100*, 2537; (b) C. Bartic, G. Borghs, *Anal. Bioanal. Chem.*, **2006**, *384*, 354; (c) L. Wang, D. Fine, D. Sharma, L. Torsi, A. Dodabalapur, *Anal. Bioanal. Chem.*, **2006**, *384*, 310; (d) J.T. Mabeck, G.G. Malliaras, *Anal. Bioanal. Chem.*, **2006**, *384*, 343.
118 (a) K. Doré, S. Dubus, H.-A. Ho, I. Lévesque, M. Brunette, G. Corbeil, M. Boissinot, G. Boivin, M.G. Bergeron, D. Boudreau, M. Leclerc, *J. Am. Chem. Soc.*, **2004**, *126*, 4240; (b) H.-A. Ho, K. Doré, M. Boissinot, M.G. Bergeron, R.M. Tanguay, D. Boudreau, M. Leclerc, *J. Am. Chem. Soc.*, **2005**, *127*, 12673.
119 M.T. Bowser, *Analyst*, **2005**, *130*, 128.
120 Y. Xiao, A.A. Lubin, A.J. Heeger, K.W. Plaxco, *Angew. Chem. Int. Ed.*, **2005**, *44*, 5456.
121 D. Lee, T.M. Swager, *Chem. Mater.*, **2005**, *17*, 4622.
122 (a) T.A. Skotheim, R.L. Elsenbaumer, J.R. Reynolds (Eds.), *Handbook of Conducting Polymers*, 2nd edn., Marcel Dekker, New York, **1998**; (b) H.S. Nalwa (Ed.), *Handbook of Organic Conductive Molecules and Polymers*, Vols. 1–4, Wiley, Chichester, **1997**.
123 (a) Bayer AG, European Patent 339 340, **1988**; (b) F. Jonas, L. Schrader, *Synth. Met.*, **1991**, *41–43*, 831; (c) G. Heywang, F. Jonas, *Adv. Mater.*, **1992**, *4*, 116; (d) I. Winter, C. Reece, J. Hormes, G. Heywang, F. Jonas, *Chem. Phys.*, **1995**, *194*, 207; (e) L. Groenendaal, F. Jonas, D. Freitag, H. Pielartzik, J.R. Reynolds, *Adv. Mater.*, **2000**, *12*, 481; (f) S. Kirchmeyer, K. Reuter, *J. Mater. Chem.*, **2005**, *15*, 2077–2088.
124 (a) F.M. Huijs, Transparent conducting polymer thin films, PhD Thesis, Rijksuniversiteit Groningen, **2000**; (b) F.M. Huijs, F.F. Vercauteren, B. de Ruiter, D. Kalicharan, G. Hadziioannou, *Synth. Met.*, **1999**, *102*, 1151; (c) F.M. Huijs, J. Lang, D. Kalicharan, F.F. Vercauteren, J.J.L. van der Want, G. Hadziioannou, *J. Appl. Polym. Sci.*, **2001**, *79*, 900; (d) F.M. Huijs, F.F. Vercauteren, G. Hadziioannou, *Synth. Met.*, **2002**, *125*, 395.
125 Y. Cao, G.M. Treacy, P. Smith, A. Heeger, *J. Appl. Phys. Lett.*, **1992**, *60*, 2711.
126 (a) N.J. Greenham, R.H. Friend, Semiconductor device physics of conjugated polymers, in *Solid State Physics*, Eds. H. Ehrenreich, F. Spaepen, Academic Press, New York, **1995**; (b) G. Horowitz, *J. Mater. Res.*, **2004**, *19*, 1946.
127 D. Voss, *Nature*, **2000**, *407*, 442.

128 (a) G. P. Collins, *Sci. Am.*, **2004**, 76;
(b) J. B. Lee, V. Subramanian, *IEEE Trans. Electron Devices*, **2005**, *52*, 269.

129 G. Hadziioannou, P. F. van Hutten (Eds.), *Semiconducting Polymers: Chemistry, Physics and Engineering*, Wiley-VCH, Weinheim, **2000**.

130 G. Hadziioannou, G. Malliaras (Eds.), *Semiconducting Polymers: Chemistry, Physics and Engineering*, 2nd edn., Wiley-VCH, Weinheim, **2006**, in press.

11
Polymer Encapsulation of Metallic and Semiconductor Nanoparticles: Multifunctional Materials with Novel Optical, Electronic and Magnetic Properties

Jeffrey Pyun and Todd Emrick

11.1
Introduction

The synthesis of novel organic–inorganic hybrid materials is a compelling and exceptionally important area of research in modern materials chemistry and nanoscience, due to the intriguing and unique properties of materials at the nanoscale [1–8]. Combining the rich diversity of organic materials with the unique optical, magnetic and electronic properties of metals and semiconductors creates a vast realm of possibilities in the design of novel structures possessing precisely programmed properties and performance. While a large number of different hybrid nanocomposite materials have been synthesized in the past decade, a common theme in the preparation of these materials is interfacial control of well-defined building blocks on the molecular, nanoscopic and mesoscopic length scales. There is more than ever before an urgent need for fundamental insight into structure–property correlations in such composite systems. Gaining such insight requires higher levels of synthetic sophistication, processing techniques and characterization associated with well-defined and controlled nanoparticle–polymer composites.

The hybridization of organic polymers and inorganic colloids has been pursued as a wide range of approaches have been developed to prepare precisely structured materials [9]. Extensive efforts have focused on preparing either well-defined organic polymers or inorganic nanoparticles. The advantages of combining organic polymers with inorganic colloids is now widely recognized, as the inherently modular nature of polymer chemistry allows for installation of diverse chemical functionality [10], which has been used both to synthesize and to functionalize inorganic nanoparticles.

This chapter describes the synthesis and functionalization of metallic and semiconductor nanoparticles using synthetic polymers. The various types of polymeric coatings and surfactants that have been investigated will be discussed. In particular, linear polymers (including homopolymers, random copolymers and block copolymers), branched polymers and conjugated polymers that

Macromolecular Engineering. Precise Synthesis, Materials Properties, Applications.
Edited by K. Matyjaszewski, Y. Gnanou, and L. Leibler
Copyright © 2007 WILEY-VCH Verlag GmbH & Co. KGaA, Weinheim
ISBN: 978-3-527-31446-1

Polymeric Surfactants
Linear Polymers. Dendrimers

Inorganic Colloids
Metals, Semiconductors

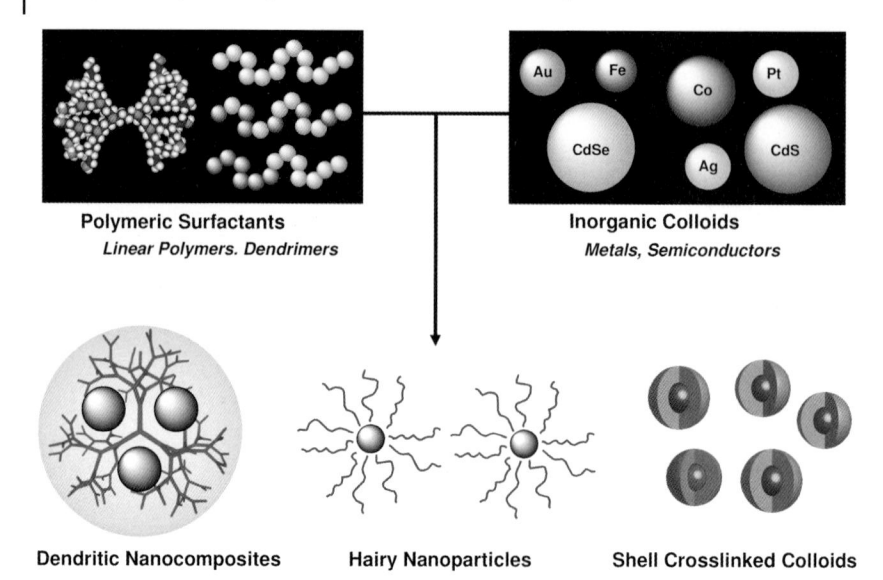

Dendritic Nanocomposites   **Hairy Nanoparticles**   **Shell Crosslinked Colloids**

Scheme 11.1 Examples of hybrid nanocomposite materials prepared by encapsulation or coating of inorganic colloids with organic polymers.

coordinate or adsorb to inorganic colloidal surfaces will be described (Scheme 11.1). The use of these polymers in place of small molecule surfactants for the functionalization and encapsulation of inorganic colloids from noble metals (Au), semiconductors (CdSe) and magnetic nanoparticles (Co) will be covered. Readers are directed to additional sources for reviews on synthetic approaches toward similar nanocomposite materials via electrostatic layer by layer deposition [11–14], dispersions of colloids into polymerizable matrices and other fundamental aspects of nanoparticles [8, 15].

11.1.1
Polymeric Surfactants and Encapsulating Coatings

While a variety of conventional small molecule surfactants have been used to prepare inorganic colloids, polymers offer distinctive advantages when used to replace these surfactants, due to the tunable composition of polymers that permits the introduction of multiple functional groups on the same molecule. Additionally, the extended nature of polymers on the nanoscale is inherently well suited for stabilization of inorganic colloidal dispersions in a variety of media.

11.1.1.1 Types of Polymeric Surfactants
A number of different linear (co)polymers have been used to prepare and functionalize inorganic nanoparticles. Polymeric materials are ideal candidates to synthesize and encapsulate inorganic colloids, as a wide range of functionality

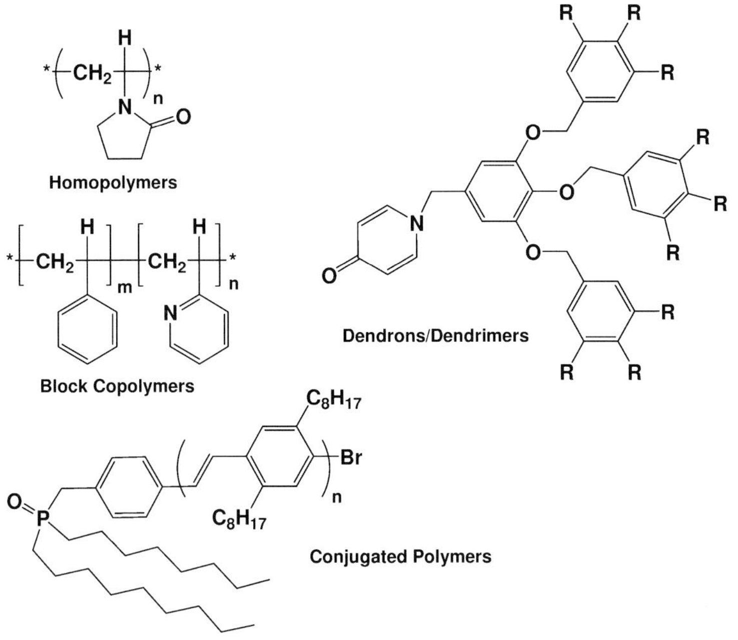

Scheme 11.2 Examples of polymeric surfactants: linear homopolymers, block copolymers, conjugated and branched/dendritic polymers.

can be installed on the polymer backbone and thus is presented on the nanoparticle shell. Furthermore, polymeric surfactants may allow tuning of nanoparticle size and morphology in a manner not accessible when using small molecule surfactants. By variation of polymer molecular weight, composition and architecture, a wide range of hybrid nanocomposites have been prepared, containing various types of insulating or conjugated polymers. While the type of polymeric surfactant employed varies depending on the targeted application, a few general structural features are required. In particular, polymer surfactants for nanoparticles must be designed to include covalently bound and/or coordinating functionality that serves to anchor the polymer chains to the nanoparticle surface. Various types of polymer materials have been developed where these ligating moieties are located along the polymer backbone and at the chain-ends.

Polymer ligands for encapsulation of inorganic nanoparticles have been varied widely and include conjugated polymers, branched polymers and dendritic materials. Typically, divergently prepared dendrimers, such as polyamidoamine (PAMAM) dendrimers and star polymers, have been used as nanoreactors to template the growth of colloids within the branched polymer interior. Convergently prepared dendrons have also been used in this fashion, where functionality placed at the dendritic focal point is used for ligation to the colloid surface (Scheme 11.2). Various examples of functionalization and assembly with biological macromolecules have also been described [16, 17].

11.1.1.2 Approaches to Nanoparticle Functionalization with Polymers

A number of different strategies have been reported for functionalizing inorganic nanoparticles with polymer materials, including the dispersion and polymerization of colloids in cross-linkable polymer matrices, the encapsulation of colloids in latex particles by emulsion polymerization and simple blending of the polymer and nanoparticle components. This chapter will focus on composite materials that combine inorganic nanoparticles with structurally well-defined polymers that serve either as encapsulating shells or as stabilizers in colloidal dispersions. Functionalization approaches most relevant to the preparation of these materials include ligand exchange from preformed particles, surface-initiated polymerization from functional colloidal initiators and *in situ* particle reactions in the presence of polymer ligands (Scheme 11.3).

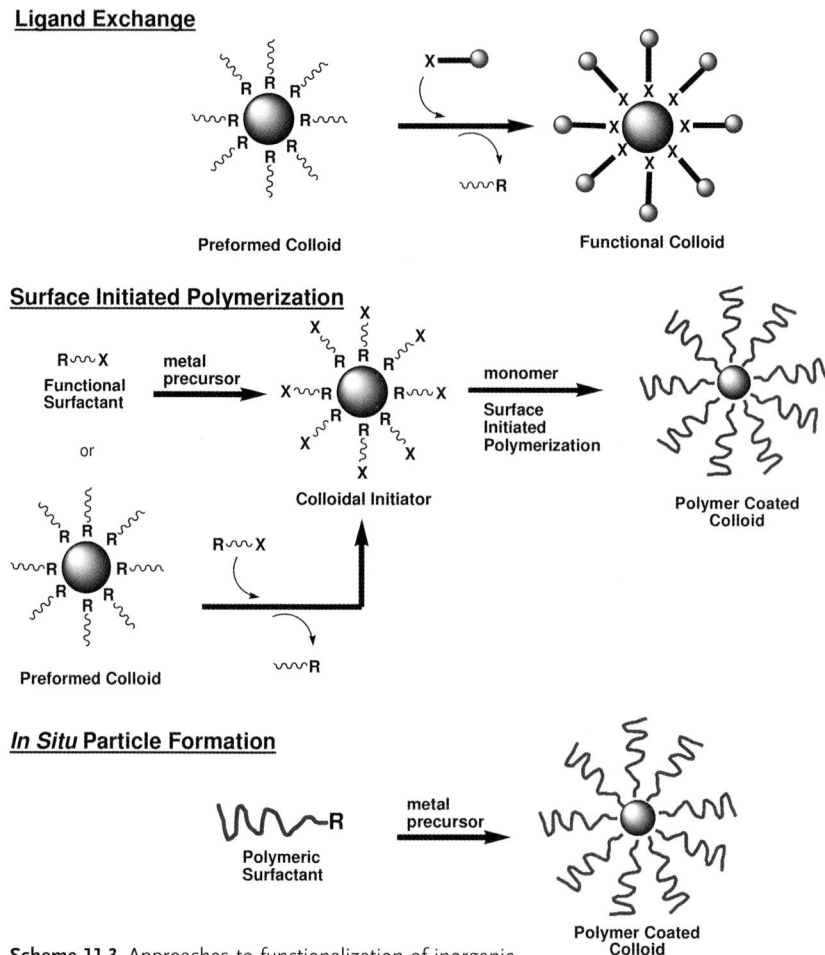

Scheme 11.3 Approaches to functionalization of inorganic colloids with polymers.

Ligand exchange methods provide a general approach to surface modification of pre-synthesized colloids and involve replacing the ligands used in the initial nanoparticle synthesis with new ligands that contain the desired chemical functionality. In many cases, the chemical and thermal conditions required for nanoparticle growth limit the range of functionality that can be used in the growth step. In ligand exchange, the nanoparticle surface and the incoming functional ligand must be appropriately paired to promote ligand exchange of bound capping agents. To this end, a wide range of small molecule surfactants and end-functional polymers have been used in ligand exchange to form organic shells around inorganic nanoparticles, allowing for tunable properties in the resulting hybrid nanocomposites. This approach has proven successful in the surface modification of gold [18, 19] and semiconductor nanoparticles [20–25] and has also been applied to a number of different self-assembling and optically active materials. Functionalization of other metal and metal oxide colloids by ligand exchange is not as well established as that of gold and semiconductor particles, but represents an area of growing interest and activity.

Surface-initiated polymerization methods provide a different approach for attachment of polymers to nanoparticles, in which the nanoparticles are first functionalized with moieties capable of controlling polymerization processes, then used in the polymerization chemistry needed to grow the tethered chains. Due to the extensive development of living and controlled polymerizations [26–29], these types of surface-initiated processes can be used to grow well-defined functional polymers from nanoparticle surfaces. In addition, a few examples of step-growth polymerization from nanoparticle surfaces have also been reported. Living or controlled chain-growth polymerizations, such as ring-opening metathesis polymerization (ROMP) [30, 31] and a range of controlled radical processes, such as atom transfer radical polymerization (ATRP) [27, 32–34], nitroxide-mediated polymerization (NMP) [35–38] and reversible addition fragmentation chain-transfer polymerization (RAFT) [39–42], have been widely utilized to prepare various types of core–shell hybrid materials. Typically, colloidal "macroinitiators" are synthesized by *in situ* particle reactions with functionalized surfactants that are capable of subsequent polymer growth from the nanoparticle surfaces. This method is distinct from the synthesis and use of monomer-functionalized nanoparticles, which participate in polymerization chemistry as cross-linking agents due to their multifunctional nature.

In situ nanoparticle formation: functional polymers, with chain-end or pendent functionality, have also been used as replacements for small molecule surfactants to support the growth of a wide range of nanoparticle compositions. While such functional polymers do not necessarily require well-defined structures in terms of polydispersity, the successful synthesis and stabilization of well-defined nanoparticles does require the presence of ligating moieties on the polymer. Nanoparticle reactions performed in the presence of these polymeric surfactants typically employ metal salt precursors and reducing agents [15] or metal carbonyl complexes [43]. Although this approach is successful in many cases, one general limitation lies in the potential complications when attempting to embed desired, non-surface ligating, functionality within the polymer materials used in the nanoparticle synthesis.

11.2
Noble Metal Nanocomposites: the Case of Au

11.2.1
Structure and Properties of Metallic Nanoparticles

The interesting properties of metallic materials, including electrical and thermal conductivity, optically characteristic luster and malleability, arise from the highly delocalized electronic structure throughout the extended metallic network [44]. The combined overlap of atomic orbitals gives rise to closely spaced molecular orbitals, which form bands that span a range of energy levels. As for molecular systems, the electrons in the highest occupied molecular orbital (HOMO) reside in a collection of overlapping orbitals that form the *valence band*, in which the highest energy electrons occupy the *Fermi level*. The energy levels at and above the lowest unoccupied molecular orbital (LUMO) form the *conduction band* and the distance between the valence and conduction bands is defined as the *bandgap*. The absence of a bandgap in metals allows facile promotion of valence electrons to the conduction band, where the electrons are delocalized and highly mobile and thus able to provide high electrical and thermal conductivity (Scheme 11.4) [45].

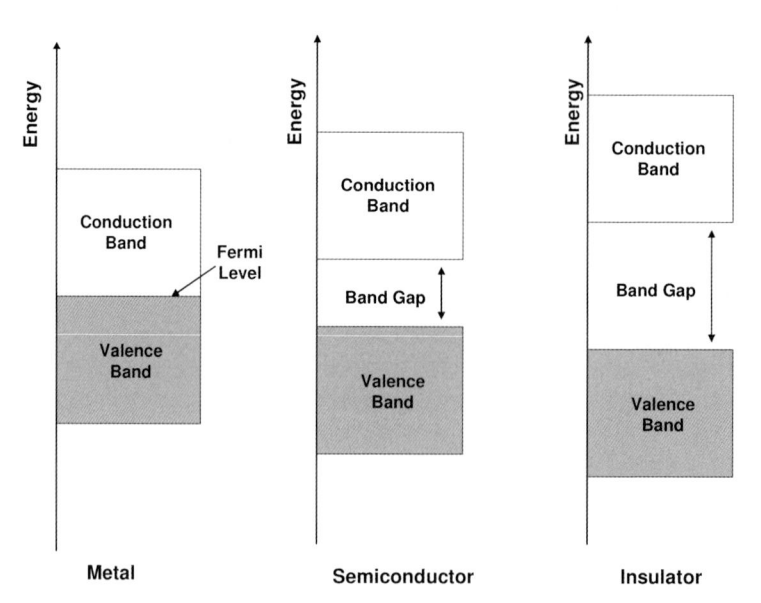

Scheme 11.4 General electronic structure of metals, semiconductors and insulators.

As the dimensions of metals are reduced to particles of nanoscale dimensions, deviation from bulk properties is observed. Such effects are encountered when the dimensions of the metallic nanoparticle are smaller than the mean free path of the valence electrons [46]. Below a critical particle size, the onset of a bandgap has been observed, which significantly alters the electronic and optical properties of the material. The metal-to-insulator transition arising from these effects was confirmed by electrochemical impedance measurements on colloidal assemblies [47, 48]. The optical properties of these materials are most evident in the brilliant color of colloidal dispersions of certain metals (Au, Ag), a consequence of surface plasmon resonance. This is induced by the excitation of surface bound valence electrons on the metal surface upon interaction of the surface with light. For gold nanoparticles (5–20 nm), a surface plasmon band is observed near 530 nm. The surface plasmon band is attributed to the frustrated environment of electrons on the metal surface in size regimes below their mean free paths. Upon irradiation with visible light, the influence of the induced electromagnetic field creates coherent oscillation of excited electrons in the conduction band and the onset of a strong absorption band [44, 46].

11.2.2
Synthesis of Au Nanoparticles

While colloidal gold has been utilized for centuries in applications ranging from dyes to medical curatives, interest in gold nanoparticles continues today in research spanning both state-of-the-art electronic and medical applications. A number of different systems have been developed using small molecule and macromolecular surfactants to synthesize and stabilize gold colloids in both aqueous and organic media. Organic soluble gold nanoparticles, a more recent development, have led in part to a renaissance in novel synthetic hybrid nanocomposites that combine polymers and gold colloids. While the scope of such systems is very broad, the strategies utilized to prepare Au–polymer nanocomposites can be traced back to the approaches described in Scheme 11.3. In the majority of these systems, gold precursor salts such as hydrogen tetrachloroaurate ($HAuCl_4$) are combined with reducing agents (e.g. citrate salts, sodium borohydride, hydrazine) and surfactant stabilizers to give nanoparticles that are then used in some fashion to prepare the desired composites [49].

11.2.2.1 Gold Nanoparticles with Small Molecule Surfactants
The seminal efforts of Faraday in the late 1800s are well worth noting in the context of Au nanoparticle synthesis. Faraday reported the synthesis of Au colloidal materials using a biphasic solvent system in the presence of small molecule surfactants [50]. Since then, a variety of synthetic methods that afford gold nanoparticles have been developed, for example by Turkevich et al. [51], using water-soluble thiols as surfactants and a sodium citrate reducing agent, in an aqueous preparation of electrostatically stabilized gold colloidal dispersions. Na-

noparticle diameters arising from this method afforded particles tunable over the 5–30-nm diameter range. This method is used widely today to prepare Au colloidal precursors that possess weakly bound ligands. An alternative and extremely popular synthesis of gold nanoparticles is found in the biphasic system reported by Brust, Schiffrin et al. [52], where $AuCl_4^-$ anions in the aqueous phase are transferred to the organic phase (e.g. toluene) using the phase transfer catalyst tetra-n-octylammonium bromide, where they are reduced by $NaBH_4$. Gold nanoparticles formed in the organic phase are stabilized by dodecanethiol, which functions effectively as a ligand for the gold surface. The Brust–Schiffrin process readily yields very small gold nanoparticles, typically in the 1–3-nm diameter range. Moreover, single-phase processes have also been developed to prepare thiol-stabilized gold nanoparticles of well-defined size [53].

The methods described above provide hydrophilic or hydrophobic gold nanoparticles that are well suited for certain applications, but not nearly universal in scope. In recent years, synthetic methods designed to tailor gold nanoparticles with particular functional ligands have proven very useful for expanding the capabilities of these nanoparticles. For example, direct reduction of gold salts in the presence of functional surfactant molecules gives gold nanoparticles with functionality in the ligand periphery [18, 54, 55]. This approach is relevant for functional groups that are compatible with the synthetic conditions employed. Thus, the concept of ligand exchange or so-called "place exchange" of functional surfactants on to preformed gold nanoparticles has been extensively investigated by Murray and coworkers [19, 56] as a general approach to nanoparticle functionalization and recently by Rotello and coworkers for the preparation of gold nanoparticles tailored to control protein activity [57].

11.2.2.2 Synthesis of Gold Nanoparticles in Linear Polymer Surfactants

A variety of polymer surfactants have been used to prepare gold nanoparticles, including elegant research activities in the areas of hybrid materials obtained by modern techniques of living polymerization, regioregular step-growth polymerization [58] and dendrimer [59, 60] synthesis. These developments have opened the door to a rich diversity of polymeric materials that have been used to synthesize novel Au-nanocomposites.

Linear homopolymers for *in situ* nanoparticle preparation Water-soluble homopolymers have been used extensively as surfactants in the preparation and stabilization of Au colloids. Typical examples include polyvinylpyrrolidone (PVPy) and poly(ethylene oxide) (PEO) for the stabilization of colloidal Au dispersions [61, 62]. The biological applications of polymer-stabilized Au nanoparticles to label cells for histological studies has also generated considerable interest. Extensive investigations by Mayer and Mark [63] concluded that polymers possessing ligating moieties along the backbone, such as poly(2-vinylpyridine) (P2VP), are effective stabilizers for gold nanoparticles. Gold nanoparticles were obtained from water–ethanol mixtures, using either UV irradiation or potassium borohydride

(KBH$_4$) to reduce the HAuCl$_4$ precursor. These studies are consistent with the findings of Miyak and coworkers [64], where thiols or nitriles as pendent groups were introduced to the P2VP backbone as random copolymer surfactants. In both cases, uniform Au colloids possessing particle sizes smaller than 5 nm were prepared. In a nonaqueous synthesis of Au colloids, P2VP-stabilized nanoparticles prepared in formamide, which functions as both the solvent and reducing agent, gave uniform Au colloids 20–35 nm in diameter [65]. The recent work of Sun and Xia [66] demonstrated that nonspherical nanoparticles, such as metallic Au (and Ag) "nanoboxes", could be prepared in PVPy with tunable nanoparticle morphology under optimized conditions. Thiol-terminated PEO was also used to prepare Au colloids using a modified Brust–Schiffrin procedure [67]. PEO thiol as the stabilizing surfactant proved superior to prior studies that used PEO without a ligand chain-end. While all of these examples demonstrate the versatility of water-soluble homopolymers as surfactants for gold nanoparticles, variations in polymer composition, functionality and molecular weight control all are advantageous with regard to control over nanoparticle size and, as shown subsequently, in self-assembly.

End-functionalized polystyrene, prepared by living anionic polymerization, has also been used in an *in situ* synthesis of gold nanoparticles [68, 69]. This strategy further highlights the advantage of using well-defined polymer ligands that are tethered strictly from the chain-end, such that the thickness of the corona layer is controlled by the degree of polymerization of the polystyrene chains. It was further demonstrated that PS-coated Au colloids could be dispersed in blends with free PS due to the compatibilizing shell on the nanoparticle.

The use of thiol-terminated polymers for encapsulation of gold nanoparticles was extended by McCormick and coworkers [70] using reversible addition fragmentation chain-transfer (RAFT) polymerization, a convenient technique for obtaining thiol-terminated polymers (Fig. 11.1). In these systems, the growth of well-defined polymers is mediated by a thioester chain-transfer agent that shuttles between active and dormant states. Reduction of thioester-terminated RAFT-derived polymers with NaBH$_4$ in the presence of HAuCl$_4$ gave Au nanoparticles with the desired polymer corona. Moreover, the functional group tolerance of controlled radical polymerization allowed the synthesis of multiple functional polymers used in the preparation of the Au nanoparticles. Using this versatile approach, uniform Au nanoparticles of $\sim$20-nm diameter, possessing water-soluble polyelectrolyte, polyacrylamide and block copolymer coatings, were prepared. It is worth noting that functionalization of Au nanoparticles has also been achieved by ligand exchange using poly(methacrylic acid) [71] and polyphosphazene-based [72] polymer ligands.

Surface-initiated polymerization from Au nanoparticles has also been used effectively to prepare well-defined core–shell nanocomposite materials. Nguyen and coworkers [73] demonstrated ROMP from norbornene-functionalized Au nanoparticles to tether polynorbornene derivatives to the nanoparticle surface (Scheme 11.5). Using a modified Brust–Schiffrin method, norbornene-functionalized Au colloids were synthesized, then used in the sequential surface-initiated

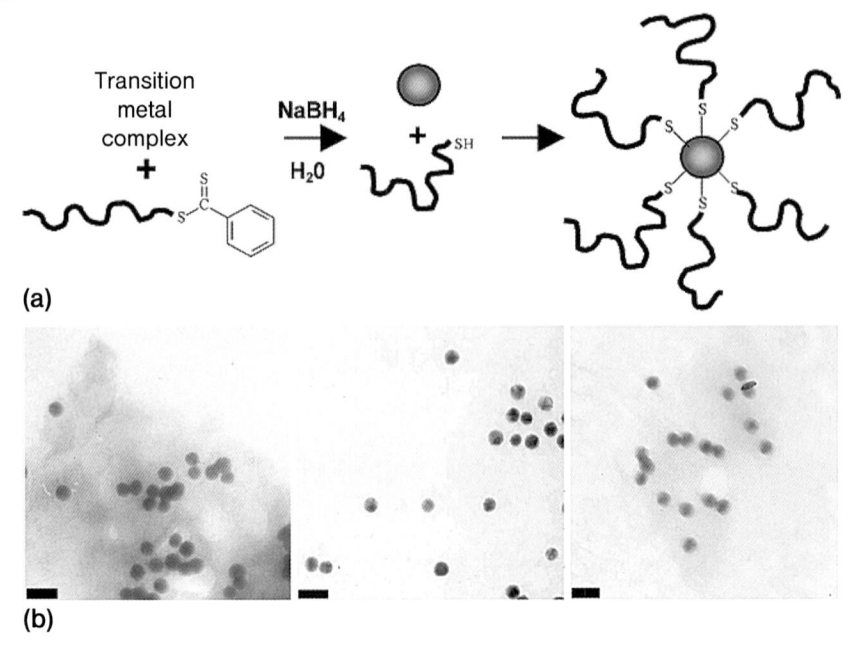

Fig. 11.1 Top: general synthetic scheme to prepare polymer-coated Au nanoparticles from macromolecular surfactants prepared from RAFT. Bottom: TEM images of polymer-stabilized Au colloids prepared from this process. Scale bar=40 nm. Reproduced with permission from [70], Copyright 2002 American Chemical Society.

ROMP of ferrocene-functionalized norbornenes. The attachment of the desired block copolymers to the Au shell was confirmed using NMR spectroscopy and cyclic voltammetry.

Hallensleben and coworkers [74] showed that surface-initiated polymerization from Au nanoparticles can also be accomplished using the controlled radical polymerization technique ATRP. In ATRP, copper(I) catalysts mediate the surface-initiated polymerization and the ability to perform this polymerization at room temperature is key for suppressing detachment of the thiol-terminated polymers from the gold surface. In this study, the polymer-encapsulated Au nanoparticles obtained following the polymerization were characterized by transmission electron microscopy (TEM), atomic force microscopy (AFM) and size-exclusion chromatography (SEC). Cleavage of the tethered chains from the gold surface and subsequent SEC characterization revealed that polymers of predictable molar mass were obtained by this method. Similar studies were performed by Walt and coworkers [75], using copper(I)–cyclam complexes for the ATRP of MMA from polydisperse gold nanoparticles.

Several other studies on linear nanoparticle-polymer composites should be noted. Detailed kinetic studies of the surface-initiated ATRP of MMA from Au nanoparticle initiators were reported by Fukuda and coworkers [76]. Polymeriza-

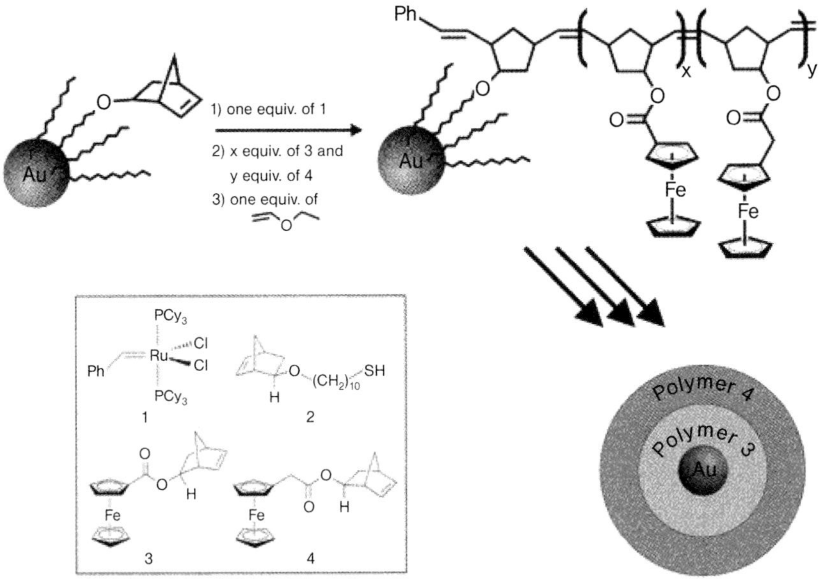

11.2 Noble Metal Nanocomposites: the Case of Au

Scheme 11.5 Synthesis of ferrocenyl functional Au nanoparticles via sequential surface-initiated ROMP. Reproduced with permission from [73], Copyright 1999 American Chemical Society.

tions were performed at 40 °C using a homogeneous DMF solution of a copper(I)–sparteine catalyst complex. First-order monomer consumption and a linear increase of molar mass with conversion were observed up to $M_n = 30\,000$ g mol^{-1}, indicating the controlled nature of this grafting-from polymerization. Nanocomposites prepared from these materials were investigated using Langmuir–Blodgett film techniques [77]. The preparation of Au colloids coated with thermoresponsive polymers was conducted using RAFT polymerization, for example for the growth of poly(N-isopropylacrylamide) from Au nanoparticle surfaces [78, 79]. Finally, polyoxazoline-coated Au nanoparticles were synthesized using living cationic polymerization from the nanoparticle surface. Hydroxyl-terminated alkanethiols were used as monolayers on the gold nanoparticles, then converted into triflates for subsequent polymerization of 2-oxazoline. Chain-end termination with amines was employed to modify the surface properties of the composite materials obtained by this method [80].

Linear block copolymers Block copolymers have proven useful as both "nanoreactors" and templates for the synthesis of nanoparticles and the formation as composite films, respectively. The early work of Cohen and coworkers [81] demonstrated that well-defined block copolymers prepared by ROMP can be loaded with Au precursors and cast into a microphase separated film to template the formation of nanoparticles. This general strategy was applied to a wide range of inorganic nanoparticle materials [82]. A recent report by Thomas and coworkers demonstrates that blending of Au colloids with block copolymers to form phase

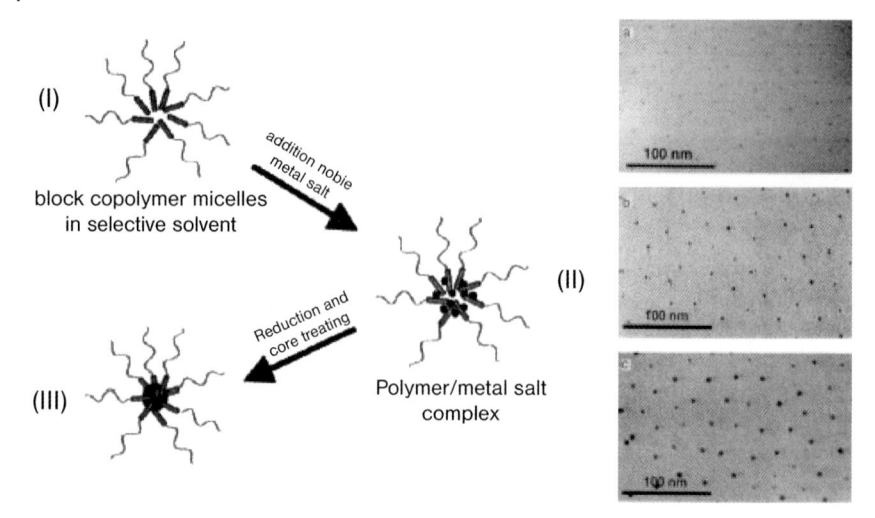

Fig. 11.2 Synthesis of Au colloids using amphiphilic block copolymers as micellar templates. TEM images show that sizes of Au nanoparticles can be tuned by ratio of block copolymer to metal precursor used in the particle synthesis. Reproduced with permission from [85], Copyright 1996 Wiley-VCH.

separated films is a versatile route to nanocomposite materials formed by hierarchical assembly [83].

The concept of using amphiphilic block copolymers as micellar templates for Au nanoparticles was demonstrated elegantly by Möller and coworkers (Fig. 11.2) [84–86]. In these systems, well-defined block copolymers of polystyrene-*block*-poly(ethylene oxide) (PS-*b*-PEO) or polystyrene-*block*-poly(2-vinylpyridine) (PS-*b*-P2VP), prepared by living anionic polymerization, were dissolved in nonpolar solvents such as toluene to induce polymer reverse micelle formation. Au salts were directed into the relatively hydrophilic interior of the reverse micelle and inorganic colloids were formed by the addition of reducing agent. Au nanoparticles prepared by this method were found to be relatively uniform in size, with tunable diameters in the range 2–6 nm. These block copolymer-encapsulated nanoparticles proved stable and could be patterned on to semiconductor surfaces. Control over polymer thickness allows these materials to serve as "self-masks", by their ordered deposition and assembly of nanoparticle arrays, followed by plasma treatment to remove the organic component [87, 88]. The use of block copolymer surfactants was also independently conducted by Antonietti et al. using PS-*b*-PVP block copolymer surfactants to prepare Au nanoparticles [89].

An alternative approach to functionalize Au nanoparticles with amphiphilic block copolymers was recently developed by Kang and Taton [90, 91]. In this system, well-defined block copolymers of poly(styrene-*block*-acrylic acid) (PS-*b*-PAA) and poly(methyl methacrylate-*block*-acrylic acid) (PMMA-*b*-PAA) were prepared by ATRP, then mixed with preformed Au colloids in a mutually good solvent

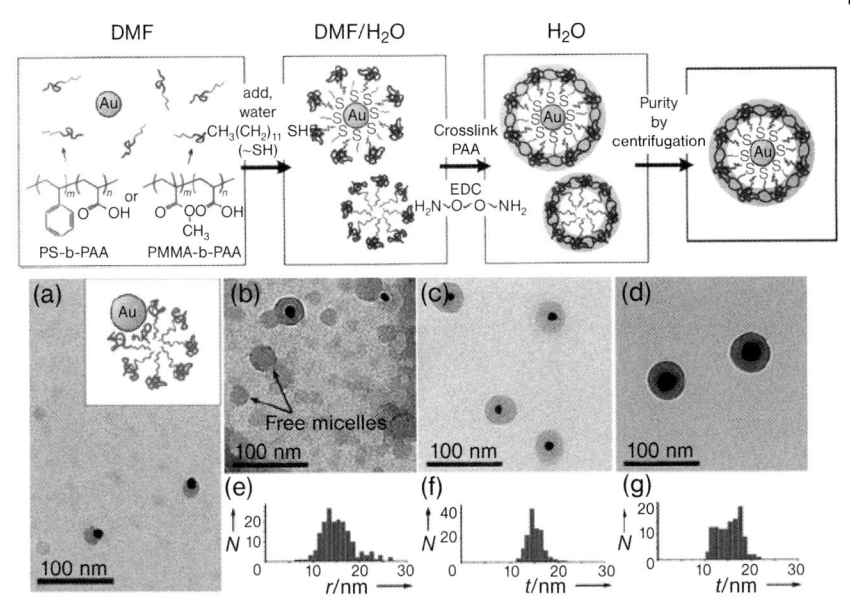

Fig. 11.3 General scheme for the encapsulation of Au colloids with block copolymers and cross-linking of the surfactants into shell-cross-linked nanoparticles. TEM imaging confirms the formation of core–shell structures where the inner dark core is assigned to the Au nanoparticle and the polymer shell the lighter corona. Reproduced with permission from [90], Copyright 2004 Wiley-VCH.

system, such as DMF. Addition of water leads to the encapsulation of the Au nanoparticles within the interior of block copolymer micelles (Fig. 11.3). This technique provided an application of the conditions used to form the "crew-cut" micelles described previously by Zhang and Eisenberg [92] and does not require the block copolymer surfactant to possess ligating groups to bind and stabilize colloidal surfaces, but rather functions by physical adsorption of the polymer chains on the nanoparticle surface.

The authors further demonstrated the ability to cross-link the polymer shell of these encapsulated nanoparticles by the addition of a water-soluble carbodiimide and a diamine cross-linker, in an adaptation of chemistry developed by Wooley and coworkers for shell-cross-linked nanoparticles (SCKs) [93–97]. By control of the block copolymer chain length and composition, shell-cross-linked Au nanoparticles possessing tunable polymeric shells and functionality were achieved. This versatile system also permitted the assembly and permanent linking of polymer-coated Au colloids into one-dimensional plasmonic chains by optimization of the coupling conditions [98].

Synthetic block copolymers and Au nanoparticles have also been utilized to form hierarchical assemblies using various noncovalent interactions between surfactants and nanoparticles. In the "bricks and mortar" approach developed

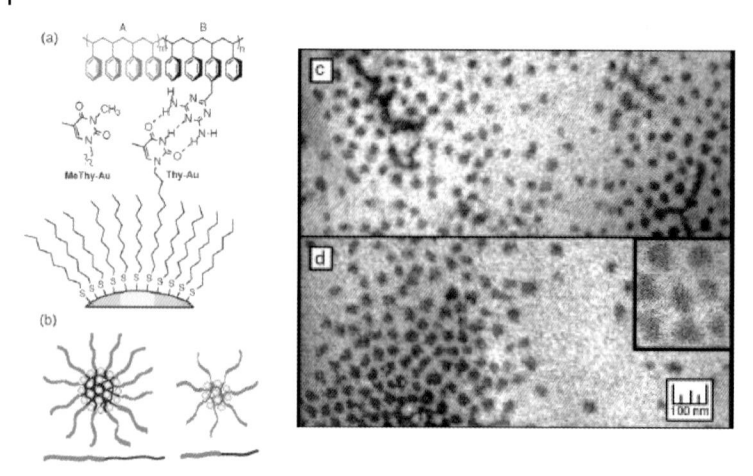

Fig. 11.4 "Bricks and mortar" approach to nanoparticle–polymer assemblies. Block copolymers with triazene side-chain groups are blended with thymine-decorated Au colloids. TEM imaging confirms the formation ordered hierarchical assemblies from the noncovalent interactions between the block copolymer and the functional nanoparticle. Reproduced with permission from [99], Copyright 2002 American Chemical Society.

by Rotello and coworkers [99, 100], thymine functional alkanethiol-capped Au nanoparticles were blended with block copolymers containing PS and PS–triazine segments, which form complementary hydrogen bonding pairs and induce aggregation (Fig. 11.4). An attractive feature of this system is the controllable incorporation of the solubilizing PS segment on the block copolymer, which limits (by sterics) the aggregation number of colloids and polymers. Using this approach, the authors successfully prepared uniform polymer–nanoparticle assemblies, as confirmed by analysis of TEM images and light scattering data.

11.2.2.3 Branched and Dendritic Polymers

Various types of branched polymers have also been investigated as templates/ligands for inorganic nanoparticles. Building on the living polymerization concepts described above, star copolymers composed of block copolymer arms were synthesized using living anionic polymerization, then used as static nanoreactor templates for the preparation of Au nanoparticles. Advincula and coworkers [101] demonstrated the use of PS-b-P2VP star block copolymers to prepare well-defined Au nanoparticles bound to the pyridine interior of the polymeric surfactant. Similar success was achieved for PEO-b-poly(ε-caprolactone) (PCL) star block copolymers, as reported by Schubert and coworkers [102].

Dendritic surfactants have also been used extensively for nanoparticles, as seen for example in the work of Crooks and coworkers [103–105] on dendrimers such as PAMAM and PPI to prepare dendrimer-encapsulated nanoparticles

(DENs) with metal and bimetallic colloids. In these systems, the dendrimer serves as both a template and nanoreactor to load metal precursor salts and reduce to form nanoparticles within the dendritic interior. The preparation of Au DENs using PAMAM dendrimers was reported initially by Esumi et al. [106], where a strong dependence on dendrimer generation and Au nanoparticle size was observed. Using a similar approach, Amis and coworkers [107] prepared Au–PAMAM DENs as model systems to probe fundamental aspects of templated synthesis of Au colloids within dendritic hosts. Solid-state and dilute solution characterization of these materials was conducted by TEM, small-angle X-ray scattering (SAXS) and small-angle neutron scattering (SANS) techniques. Optimization of the reaction conditions to prepare Au–PAMAM DENs was carried out by Crooks and coworkers and led to the finding that "magic number" Au-to-dendrimer feed ratios result in highly uniform Au nanoparticles [108]. Reports by other groups demonstrated that novel nanocomposites could also be prepared using thiophene-modified PAMAM [109] and polypropylenimine (PPI) dendrimers [110].

Poly(benzyl ether) dendrons, prepared by convergent methods [59], have also been used as polymeric ligands in *in situ* approaches to prepare Au nanoparticles. Functional dendrons possessing pyridone [111] and thiol [112] ligating groups at the focal point were synthesized and used effectively to mediate the growth of Au nanoparticles and form stable colloidal dispersions. The perfectly defined size and controlled chain-end functionality of dendrimers imparts a particularly high level of precision to these composite nanostructures.

11.2.2.4 Conjugated Polymers

The preparation of metal nanoparticles with conjugated polymers is an intriguing combination of disparate compositions, with both useful electronic and optical properties. The synthesis of such hybrid inorganic–organic electronic materials has been investigated widely and reviewed elsewhere [113]. One of the fundamental challenges in working with typical conjugated polymers is their intractable nature, which complicates simple blending approaches to incorporate metal nanoparticles. The general approach commonly used to circumvent this problem is the dispersion of metal nanoparticles within solutions of monomer precursors, such as pyrrole, aniline or thiophene, which is then polymerized around the nanoparticle inclusions to prepare the final conductive nanocomposite. An elegant example of the preparation of such composites was reported by Feldheim and coworkers [114], in which channels of porous alumina membranes were filled with metal precursor salts and pyrrole monomers, followed by polymerization within the template to form nanowire and nanoparticle chain-like structures. Alternatively, soluble conjugated polymers can be used to prepare Au nanoparticle composites, as demonstrated by Naka and coworkers for polydithiafulvene–Au nanoparticle composites [115, 116]. A similar approach was reported by Advincula and coworkers [117] to synthesize nanocomposite thin films by the reduction of metal precursor salts using a sexithiophene-functionalized poly(4-vinylpyridine) copolymer. Regioregular poly(3-hexylthiophene) was also demonstrated by Zhai and McCullough [58, 118] to stabilize effectively Au nanoparticle dispersions.

11.3
Semiconductor Nanoparticles: Cadmium Selenide Quantum Dots

11.3.1
Structure and Properties

Semiconductors are intriguing materials that possess many of the characteristic properties of both metals and insulators, as derived from the intermediate electronic band structure (i.e. narrow bandgap between the valence and conduction bands). Excitation of electrons into the conduction band gives rise to electron–hole pairs or excitons, that are loosely bound before charge dissociation. The intriguing properties of this class of materials has been applied to a number of different areas in microelectronic and display technologies [44, 45, 119, 120].

The impact of semiconductor nanoparticle size on electronic properties has been studied extensively, revealing an electronic nature intermediate between

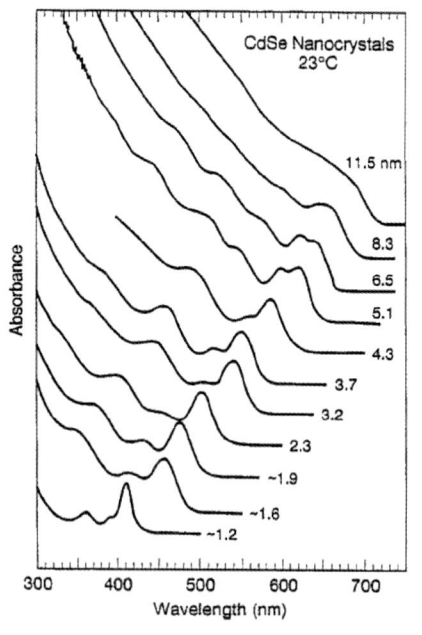

Fig. 11.5 Size dependence of CdSe nanocrystals due to quantum confinement. As the nanocrystal diameter increases, the bandgap decreases. Reproduced with permission from [130], Copyright 1998 Elsevier Science.

that of bulk and molecular materials [121–129]. Such unique properties of semiconductor nanocrystals, or "quantum dots", arise from quantum confinement of excitons within nanoparticles. As the nanoparticle dimensions are reduced below the Bohr exciton diameter, these quantum size effects are manifest in increasing bandgap with decreasing size. These bandgaps are strongly size dependent, as noted in the solution-state photoluminescence emission wavelengths of quantum dots (Fig. 11.5) [130–137]. This is an exciting feature of semiconductor nanoparticles, as size-tunable photoluminescence emission carries important implications across a range of applications, from light-emitting displays through optoelectronic devices to fluorescent tags in biological detection assays [21, 138–141].

Several different types of semiconductor nanoparticles have been synthesized. II–VI quantum dots composed of CdS, CdSe and CdTe are prepared by the reaction of Group II precursors, in the form of metal alkyls, metal oxides or organic salts, with organophosphine or organosilyl chalcogenides in nonvolatile coordinating solvents [e.g. trioctylphosphine oxide (TOPO)]. III–V quantum dots composed of InP and InAs are prepared similarly, from metal and chalcogenide precursors in high-boiling, coordinating solvents. Other semiconductor nanoparticles, such as PbSe and PbS, have also been prepared and studied, in part for their photoluminescence emission wavelengths approaching or in the infrared region [8]. While these and other types of quantum dots are all of current interest, this section will focus on CdSe quantum dots, especially on their blending and surface passivation, with polymers.

11.3.2
Synthesis of CdSe Semiconductor Nanoparticles

11.3.2.1 Small Molecule Precursors and Surfactants

Bottom-up chemical synthesis of semiconductor nanoparticles has evolved to give well-defined, high-quality particles that possess impressive size uniformity, especially when prepared in the 2–8-nm diameter range. Pioneering studies by Brus and coworkers [142] provided methods for the room temperature synthesis of semiconductor nanoparticles using water-in-heptane inverse micelles for the synthesis. This inverse micellar approach gave nanoparticles with good size control from ~2 to 4.5 nm in diameter by variation of the water-to-surfactant ratio. Despite the success of these initial studies, improvements were needed, as the particles proved difficult to isolate due to irreversible aggregation of the dried powders.

Alternative quantum dot syntheses developed around procedures involving rapid injection of cadmium-based organometallic precursors, especially dimethylcadmium, into hot coordinating solvents such as TOPO, containing selenium metal [20, 143]. An attractive feature of this synthetic approach is the temporal separation of nucleation and growth processes, which provides not only a controllable average size, but also a very narrow size distribution of the particles. In these reactions, TOPO, and also other phosphorus-based compounds

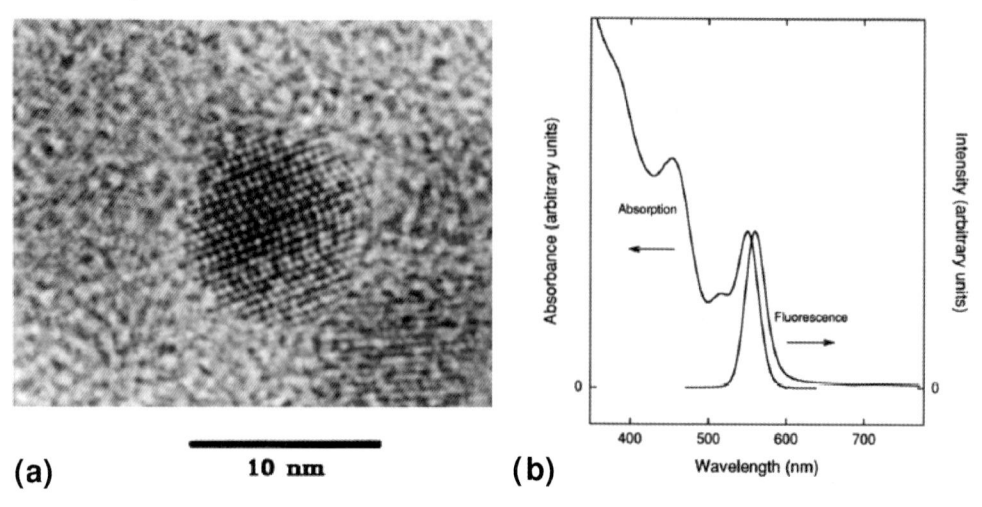

Fig. 11.6 (a) Bright field TEM of 8-nm CdSe nanocrystal prepared by rapid injection of CdSe precursors into hot TOPO/TOP. (b) Absorbance and fluorescence spectra of 3.5-nm CdSe quantum dots. Reproduced with permission from [20], Copyright 1993 American Chemical Society.

such as phosphonic acids, serve as surfactants that control quantum dot growth by surface passivation and, moreover, prevent nanoparticle aggregation in solution and the solid state following the synthesis. Size-selective precipitation, for example from 1-butanol–methanol mixtures, gives highly uniform CdS, CdSe and CdTe nanoparticles in the range ∼1.5–10 nm. High-magnification TEM of the nanocrystals reveals their atomic lattice morphology (Fig. 11.6a). Moreover, the effect of quantum confinement was manifest in the solution optical properties of these nanoscale materials, with red-shifting visible and photoluminescence spectra as a function of size. Solution quantum yields, however, were reduced by surface detects or uncoordinated sites on the nanocrystal periphery that facilitate charge trapping and nonradiative decay pathways. Inorganic passivation of quantum dots has largely overcome these quantum yield inefficiencies and solution quantum yields of semiconductor nanoparticle samples can be very high (>75%). These passivated materials are core–shell structures, for example CdSe nanocrystals with overcoat layers of higher bandgap material, such as CdS or ZnS. Preparative methods for such core–shell materials, as reported for example by Alivisatos and coworkers, and also Bawendi and co-workers, have led to very bright quantum dot fluorophores for many applications and are now available commercially [144, 145].

In recent years, improvements to nanoparticle preparations have been developed utilizing less toxic and more stable precursors [146–148]. Cadmium sources such as CdO, $CdCO_3$ and $Cd(OAc)_2$ were used in conjunction with various carboxylic acids (e.g. myristic acid or oleic acid) to prepare surfactant metal salts which proved to be excellent precursors to high-quality CdSe nanocrystals.

Mixtures of TOPO and long-chain aliphatic amines were also employed as surfactants to stabilize semiconductor nanoparticle surfaces during and following their synthesis. Although these are high-temperature methods, the procedure is generally more convenient than earlier procedures employing pyrophoric compounds such as dimethylcadmium. The use of both aliphatic amines and TOPO as surfactants resulted in significant improvement of room-temperature quantum yields without the need for an inorganic protective coating. The authors attribute the high quantum yields to a combination of factors, including an improved passivation of the nanocrystal surface by aliphatic amine surfactants and control over the cadmium-to-selenium ratios (the photoluminescence emission bright point increased as the Cd:Se ratio deviated from 1:1).

11.3.2.2 Polymer Ligands for Quantum Dots

The ligand environment around quantum dots, like other nanoparticles, is critical to stabilization of the particles against aggregation and compatibilization of the particles in polymer materials and thin films [25]. In addition, functionality on the polymer backbone can be exploited to tailor the nanoparticles further with functional organic shells. Some success has been achieved by blending preformed semiconductor nanoparticles with polymers such as certain polymethacrylates [149] and conjugated polymers [138, 139]. The straightforward nature of this blending approach is compromised by the typical incompatibility of the ligand coverage of quantum dots with the polymer matrix, limiting the scope of dispersed nanocomposites that can be obtained.

Linear homopolymers Polymers can be end-capped with the type of functionality (e.g. amines, thiols, phosphine oxides) found in small molecule ligands, making them suitable for functionalization of quantum dots. The synthetic accessibility of well-defined, end-functional polymers through living controlled polymerizations has opened up many options for introducing nanoparticles into polymer materials. An early example was reported by Hedrick and coworkers, who demonstrated the effectiveness of thiol-terminated poly(ε-caprolactone) (PCL) to prepare polymer-coated CdS nanoparticles [150]. Well-defined PCL was prepared by tin-catalyzed ring-opening polymerization of ε-caprolactone, followed by chain-end functionalization with the desired thiol. CdS nanoparticles were synthesized in the presence of this polymer using a modified inverse emulsion methodology with Cd(II) salts and thiourea precursors.

Attachment of chain-end functionalized polymers to CdSe nanocrystals has also been reported by Skaff and Emrick, for example by ligand-exchange procedures that give PEGylated quantum dots with solubility in both water and organic solvents, owing to the amphiphilicity of PEO (Scheme 11.6) [23]. Pyridine-terminated PEO, prepared by Mitsunobu coupling of 4-hydroxypyridine and hydroxyl-terminated PEO, was used in this study. The pyridine end-group binds to the quantum dots and the exposed PEO chain provides aqueous solu-

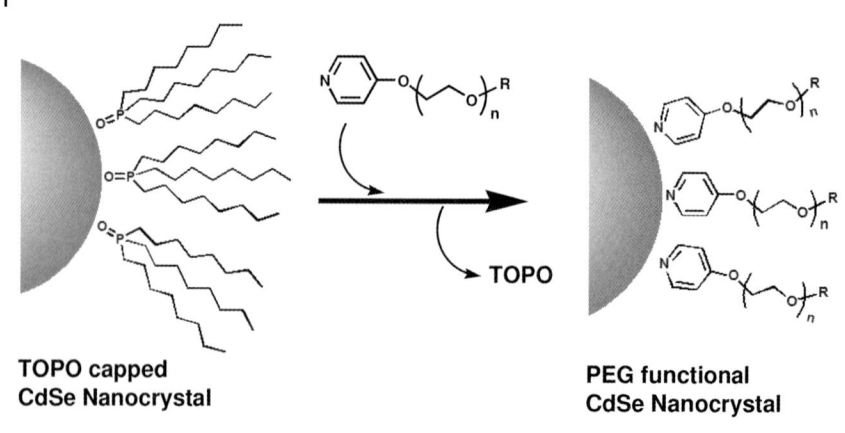

Scheme 11.6 Functionalization of CdSe quantum dots with PEG surfactant chains via a ligand-exchange approach from TOPO-capped precursors.

bility. Such PEO coverage on quantum dots provides potential applicability for applications in biology.

Surface-initiated polymerization using a variety of controlled and living polymerization techniques from semiconductor colloidal initiators by Emrick and coworkers [22, 23, 151] has been developed to allow the preparation of a wide range of polymer-encapsulated materials. In early studies, CdSe quantum dots were functionalized with ruthenium alkylidenes, making them suitable for ROMP by following introduction of cyclic olefin monomers. This synthesis utilized ligand exchange to convert the conventional TOPO coverage on the nanoparticles to a vinylbenzene coverage, which was followed by a metathesis reaction with ruthenium alkylidene complexes to prepare semiconductor nanoparticle initiators for ROMP (Scheme 11.7) [22]. A critical step in the synthesis was the use of di-*n*-octylphosphine oxide (DOPO), which serves as a valuable precursor to "TOPO-like" ligands. Subsequent surface-initiated ROMP using cyclooctene and substituted cyclooctene afforded nanoparticle-polymer core–shell structures that, importantly, maintained the inherent absorption and photoluminescent properties of the quantum dots.

This approach to functionalized quantum dots was extended to surface-initiated controlled radical polymerization, such as nitroxide-mediated polymerization (NMP) and reversible addition–fragmentation chain-transfer polymerization (RAFT). For NMP, a DOPO-functionalized alkoxyamine based on 2,2,6,6-tetramethyl-1-piperidinyloxy (TEMPO) was synthesized and attached to the quantum dots by ligand exchange. Dissolution of the colloidal initiator in styrene, followed by surface-initiated NMP of styrene and styrenic derivatives, gave the desired nanocomposites. Gel permeation chromatographic analysis of the polystyrene, following its removal from the nanoparticle surface, showed that the grafted chains could be prepared from $M_n = 15\,000$ to $120\,000$ g mol^{-1}, with poly-

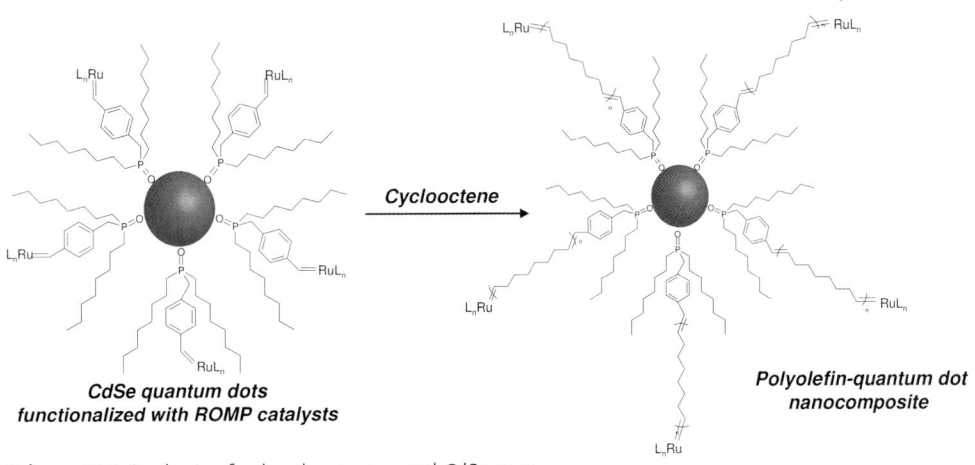

Scheme 11.7 Synthesis of polycyclooctene-coated CdSe nanocrystals prepared by surface-initiated ROMP [22].

dispersities from 1.3 to 1.5 [24]. To access a wider range of monomers and polymers, a surface-initiated RAFT system was developed from CdSe nanocrystals with trithiocarbonate-containing ligands [151]. The use of RAFT is particularly advantageous, as methacrylate and other functional vinylic monomers are amenable to this process. Furthermore, the lower temperatures (ca. 60 °C) used in surface-initiated RAFT polymerization suppress the contribution of side-reactions, such as thermal self-initiation of styrene, which is more readily observed at the higher temperatures (ca. 125 °C) of NMP.

Linear block copolymers Amphiphilic block copolymer micellar templates have been used in the preparation of nanoparticles, including CdS quantum dots. Early work by Moffitt and coworkers used polystyrene-*block*-poly(acrylic acid) diblock copolymers, both in solution as micelles and as microphase-separated films, to load cadmium acetate salts. The cadmium loading is followed by treatment with H_2S to give the desired CdS–diblock copolymer nanocomposite [152, 153]. Douglas and coworkers used a similar approach to synthesize CdS nanoparticles in the interior of polystyrene-*block*-poly(2-vinylpyridine) micelles [154]. The nonpolar polystyrene block allows these reactions to be performed in organic solvents. On the other hand, "doubly hydrophilic" diblock copolymers, such as poly(ethylene oxide)-*block*-polyethylenimine (PEO-*b*-PEI), can also be used to grow CdS quantum dots, as demonstrated by Antonietti and coworkers [155]. In this system, the PEO segment is the hydrophilic stabilizing segment, while the amines on the PEI passivate the semiconductor nanoparticle surface. Although these methods are valuable for dispersing the quantum dots, the nanoparticles obtained through this approach generally do not possess the exceptional crystallinity and optical properties relative to quantum dots obtained by high-temperature methods. Hence a variety of other methods have been explored in which

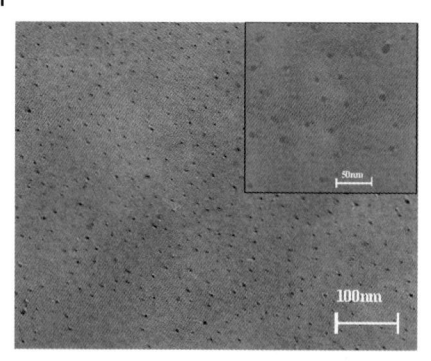

Fig. 11.7 TEM image of CdSe quantum dots deposited into nanoporous diblock copolymer templates by electrophoretic deposition [150].

quantum dots of the highest possible quality are prepared first, followed by blending or surface functionalization with diblock copolymers.

Schrock and coworkers demonstrated that polynorbornene diblock copolymers prepared by ROMP could be blended with preformed semiconductor nanoparticles to form nanocomposite films with ordered placement of quantum dots into specific phase-separated domains [156]. Polynorbornene diblock copolymers containing phosphine or phosphine oxide pendant groups in one of the blocks were prepared and shown to sequester CdSe nanocrystals selectively into the phosphorus-containing domain, either in the bulk or in thin films. Diblock copolymers have been used in other ways as "bottom-up" methods towards materials with hierarchical order. Using a sequential process, CdSe nanocrystals coated with either PEO or mercaptoundecanoic acid were deposited on a pre-assembled diblock copolymer template [157–159]. Porous arrays of periodically spaced hollow cylinders were prepared from polystyrene-*block*-poly(methyl methacrylate) diblock copolymer templates, by UV degradation of PMMA cylinders dispersed in a PS matrix. Such patterned nanoparticle arrays can be achieved by dip coating the nanoporous template into a solution of TOPO-coated CdSe nanocrystals or by electrophoretic deposition of carboxylate-coated CdSe nanocrystals (Fig. 11.7). CdSe nanorods, when coated with PEO, can also be arranged spatially using diblock copolymer templates.

Linear conjugated polymers The synthesis of nanocomposites composed of quantum dots and organic conjugated polymers has been investigated as a route to toward electronically active hybrid materials. These nanocomposites are particularly intriguing for the evaluation of electronic communication at the interface of the inorganic and organic phases. Such composites are of interest for providing new materials for display applications, by taking advantage of the light-emitting properties of each of the components, and also in photovoltaic applications.

Alivisatos and coworkers [138, 139] investigated blends of CdSe nanoparticles with poly(3-hexylthiophene) (P3HT) for photovoltaic applications. In this nano-

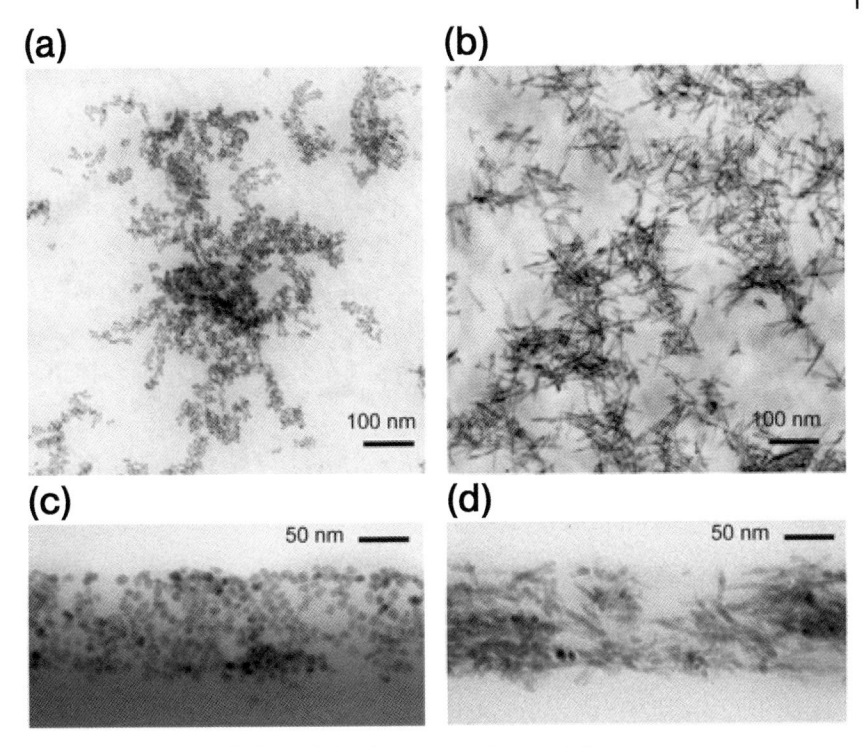

Fig. 11.8 TEM image of CdSe–polythiophene composite material cast form a binary solvent system. Images (a) and (b) are thin films of 7-nm spheres and 7×60-nm rods, respectively, in poly(3-hexylthiophene); images (c) and (d) are cross-sections of 10-nm spheres and 7×60-nm rods, respectively, in poly(3-hexylthiophene). Reproduced with permission from [138], Copyright 2002 American Association for the Advancement of Science.

composite, the nanoparticles serve as n-type semiconductors, whereas P3HT is the p-type organic semiconductor phase. Gross phase separation between the inorganic and organic semiconductors was mitigated by optimization of solvent mixtures (i.e. chloroform and pyridine) used in spin-coating procedures. In these quantum dot–P3HT nanocomposites, interparticle contacts from continuous aggregated domains were surmised to form a semi-percolating network for charge transport (Fig. 11.8). Organic photovoltaic cells generated from these composites materials exhibited an external quantum efficiency of 54% and a monochromatic power conversion efficiency of 6.9%. Although these efficiencies are lower than those of pure inorganic solar cells such as silicon, the "spin-on" application of such films holds much potential for the future of flexible polymer-based nanocomposites.

This approach to semiconductor nanocomposites was advanced further by Fréchet and coworkers, by enhancing the miscibility of CdSe quantum dots in

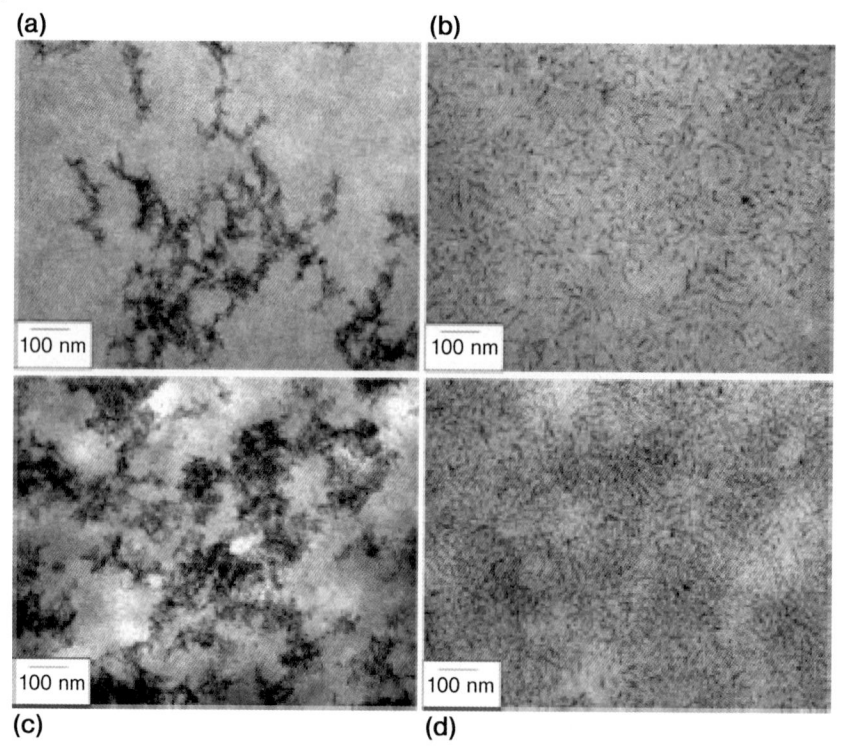

Fig. 11.9 TEM images of four films consisting of CdSe (20 wt%) and P3HT (a); CdSe (20 wt%) and amine-terminated P3HT (b), CdSe (40 wt%) and P3HT (c) and CdSe (40 wt%) and amine-terminated P3HT (d). Reproduced with permission from [141], Copyright 2004 American Chemical Society.

P3HT thin films [140]. Oligothiophenes with phosphonic acid chain-ends were synthesized and used in ligand–exchange chemistry as an electronically active surfactant for the quantum dots. Charge and energy transfer from the semiconductor nanoparticle to the ligand was observed when oligothiophenes of sufficient conjugation lengths were used as encapsulants. This approach was extended to longer polymers using a modification of the McCullough method for polythiophene, where P3HT was end-capped with amines, then blended in solution with pyridine-capped semiconductor nanoparticles and spin-coated into thin films [141]. TEM images of these (Fig. 11.9) confirmed that the presence of the functional polymer significantly improved nanoparticle dispersion. This enhanced dispersion also enhanced the AM 1.5 power efficiencies (PE) of photovoltaic devices made from these nanocomposite films ($PE_{\text{dispersed film}}=1.6\%$ at 40 wt.% CdSe vs. $PE_{\text{aggregated film}}=0.6\%$ at the same CdSe loading).

An alternative approach to CdSe-conjugated polymer nanocomposites was reported by Emrick and coworkers [160], by application of surface-initiated step-growth polymerization to prepare quantum dots encapsulated with poly(p-phe-

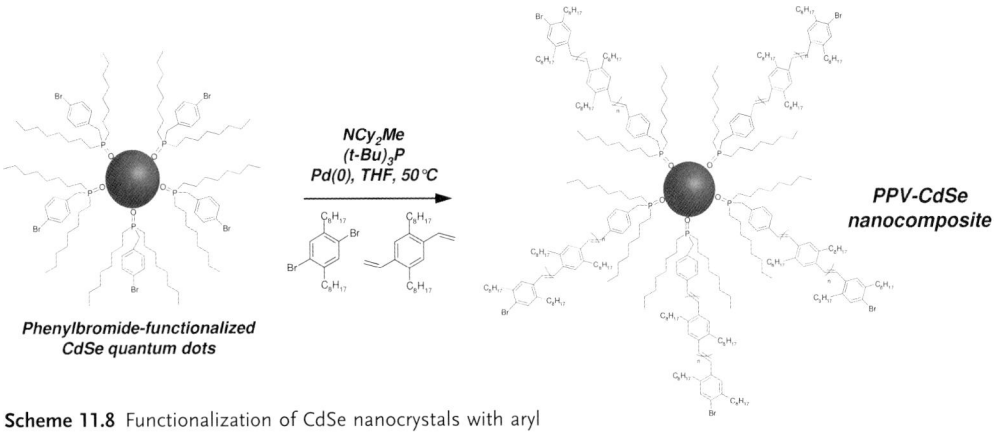

Scheme 11.8 Functionalization of CdSe nanocrystals with aryl bromide–phosphine oxide ligands and copolymerization with small molecule monomers to afford PPV-coated quantum dots [160].

nylenevinylene) (PPV) shells. Functionalization of CdSe nanoparticles with aryl bromides was achieved first by the synthesis of a novel phenyl bromide-functionalized phosphine oxide based on DOPO alkylation, then by copolymerization of divinyl- and dibromobenzene derivatives to give PPV-coated CdSe quantum dots (Scheme 11.8). TEM images of thin films cast from these PPV-quantum dot composites confirmed that the presence of the PPV shell imparted miscibility of the quantum dot in the conjugated polymer matrix. The authors further noted the impact of dispersion vs. aggregation in these nanocomposites and the dominance of photoluminescence emission from the quantum dots in the dispersed case, seen as a result of enhanced energy transfer between the polymer and nanoparticle, enabled by the dispersed morphology.

11.3.2.3 Dendritic Polymers

Convergently-synthesized dendrons have also been investigated as well-defined surfactants for the passivation and functionalization of semiconductor nanoparticles. Dendritic encapsulation also provides amplification of functionality at the dendron periphery for additional chemical modifications of the nanoparticles. Peng and coworkers [161] demonstrated the use of dendritic ligands to passivate CdSe semiconductor nanoparticles and impart improved chemical, thermal and photochemical stability to the quantum dots. Various thiol-functionalized dendron ligands were synthesized with polar and nonpolar groups at the periphery to modify the solubility of the CdSe nanocrystals after dendritic ligand exchange. The same authors synthesized third-generation (G-3) polyamide dendron ligands with vinyl terminal groups at the periphery. Functionalization of CdSe nanocrystals with these dendrons by ligand exchange, followed by cross-metathesis reactions with ruthenium–alkylidene complexes, resulted in shell cross-linked "dendritic boxes" around the quantum dot core [162].

Oligothiophene dendrons have also been reported as surfactants for the encapsulation of CdSe nanocrystals [163]. These electroactive dendrons were synthesized using a convergent approach by successive Stille coupling reactions, where n-hexyl solubilizing groups were installed at the dendron periphery and a phosphonic acid group was placed at the focal point. Absorption and fluorescence spectroscopy confirmed photoexcited charge transfer between the semiconductor interfaces, while single layer devices fabricated from these materials exhibited modest power efficiencies ($PE \approx 0.3\%$).

11.4
Metallic Magnetic Nanoparticles: the Case of Co

11.4.1
Fundamental Terms and Classifications of Magnetic Materials

Magnetic materials derived from Fe, Co or Ni metals fall into a special class of materials that are responsive to external fields arising from the cooperative interactions of unpaired electronic spins of atoms in a crystalline lattice [44]. While all materials are responsive to magnetic fields, most substances exhibit weak susceptibility to magnetic induction as dictated by their chemical composition. Noble metals, such as Au and Cu, have paired electron spins in the valence and conduction bands of the solid and hence do not exhibit significant magnetic properties. The way in which certain materials respond to a magnetic field has been described broadly by the following classifications [45]:

1. *Diamagnetism*: for a diamagnetic substance, the interaction under an applied field is weakly repulsive. The magnetic susceptibility (χ) is a unitless ratio of the induced magnetization (M) and the applied field (H) ($\chi = M/H$) and is indicative of how effectively an applied field induces a magnetic dipole in a material. For a diamagnetic substance, $\chi < 0$ ($\sim 10^{-5}$–10^{-6}), which is typical of most organic and biological materials (Scheme 11.9a).
2. *Paramagnetism*: for a paramagnetic substance, the interaction under an applied field is weakly attractive in the direction of the magnetic poles. For a paramagnetic substance, $\chi > 0$ ($\sim 10^{-3}$–10^{-5}) (Scheme 11.9b).
3. *Ferromagnetism*: for a ferromagnetic substance, the electronic spins on different metal centers are coupled into a parallel alignment that persists over thousands of atoms in *magnetic domains*. Ferromagnetic nanoparticles below a critical size (dependent on composition) can be single-domain nanocrystals. Below the *Curie temperature* (T_C), the net magnetic moment can be very high as the forces of magnetic attractions are larger than thermal fluctuations (kT). The *Curie temperature* for bulk ferromagnetic metals, such as Co and Fe, is hundreds of degrees higher than room temperature (for cobalt $T_C = 1131\,°C$ and iron $770\,°C$).

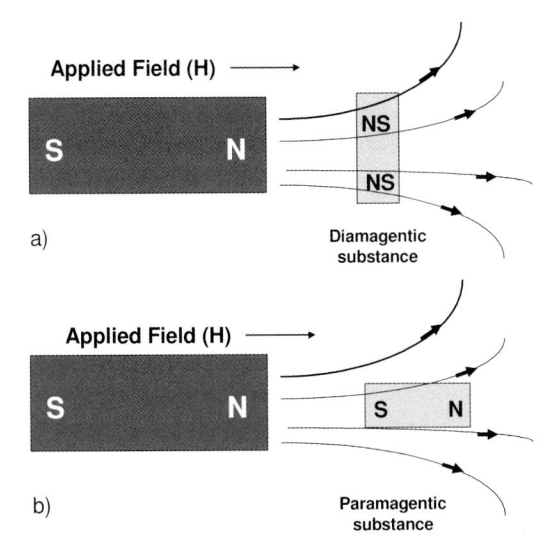

Scheme 11.9 (a) For diamagnetism, the induced dipole is repulsive to the applied field, which causes rotation of the material to minimize unfavored antiparallel alignment.
(b) For paramagnetism, the induced dipole is in the same direction as applied field and the net force is attractive.

4. *Antiferromagnetism*: in these materials, the electronic spins of neighboring atoms are fixed in an antiparallel alignment, resulting in a low magnetic moment. Antiferromagnetism is often observed in paramagnetic substances below a critical temperature known as the *Néel temperature*. The coupling of spins from metal centers in the crystalline lattice usually occurs through spin polarization from a bridging atom or from a ligand through a process known as *superexchange*. Cobalt oxide (CoO) is an example of a well-known antiferromagnetic material.
5. *Ferrimagnetism*: ferrimagnetic substances are similar to ferromagnetic materials as a strong net dipole moment is present in the material despite the antiparallel alignment of nonequivalent spin moments. These interactions also arise from *superexchange*. Iron oxides, such as magnetite (Fe_3O_4), are examples of ferrimagnetic materials.

The relationship of the induced magnetization (M) for various substances as a function of the applied magnetic field strength (H) is often plotted to ascertain the magnetic properties of the material (Fig. 11.10 a). For a ferromagnetic substance, an applied field magnetizes the material by the alignment of the unpaired spins to a saturation magnetization value (M_s). When the field is removed, the magnetization of the material does not decay at the same rate as the reduction of the field, hence a magnetic *hysteresis* is observed. When the applied field is removed (i.e. $H=0$), the remnant magnetization (M_r) value is indi-

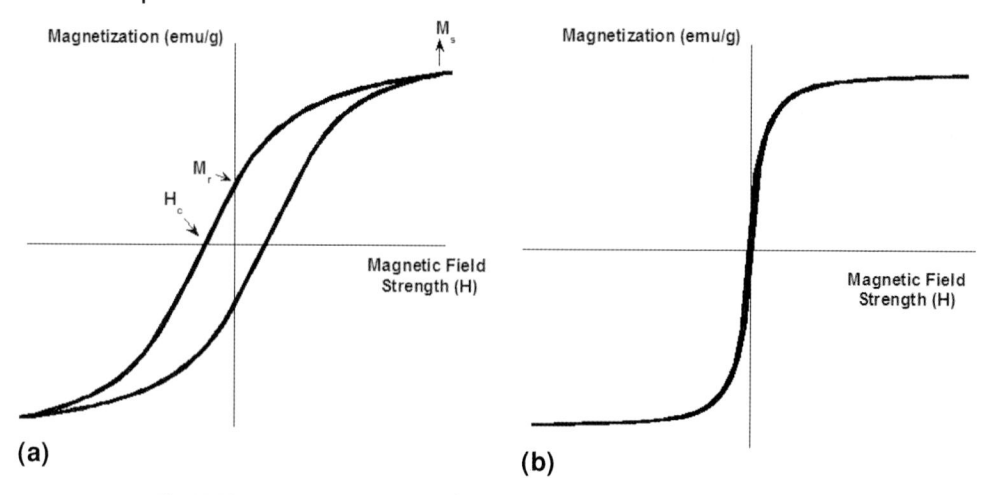

Fig. 11.10 Magnetization curves of M vs. H for (a) ferromagnetic and (b) superparamagnetic materials. Note that in the ferromagnetic case, a hysteresis is present where retention of the induced magnetic moment is observed removal of the applied field.

cative of the magnetic "memory" of the field-induced alignment at M_s. If the applied field is reversed (negative field), the M_r value is driven to zero at an applied field strength referred to as the coercivity (H_c). Increasing the negative field direction forces the materials to be magnetized at a negative M_s value, generating a symmetric plot of M vs. H for both field directions. "Hard" ferromagnetic materials exhibit weak dependences of M vs. H, as characterized by high M_r and H_c values, which yield wide, square-like hysteresis loops. Hard magnets are required for magnetic storage media. "Soft" ferromagnets exhibit a strong dependence of M vs. H and hence typically possess lower M_r and H_c values and are characterized by a narrow hysteresis loop [44, 164].

As for noble metals and semiconductors, quantum confinement effects on bulk properties are readily manifest in *superparamagnetic* materials. When the dimensions of bulk ferromagnets are reduced to critical nanoscale dimensions, the materials are no longer capable of retaining a strong dipole moment at certain temperatures in the absence of an external magnetic field. Due to the small size of nanoparticles, the reduced number of aligned spins is no longer sufficient to retain a net dipole moment that can compete with thermal fluctuations (i.e. magnetic dipolar forces weaker than kT). Hence superparamagnetic materials do not exhibit a hysteresis in M vs. H plots, indicating that while dipoles can easily be aligned and magnetized to a high M_s value, complete and rapid loss of the magnetization is observed when the field is removed (Fig. 11.10b). At a critical temperature, referred to as the *blocking temperature* (T_B), the dipole moments in the nanoparticle align and couple to generate a large net induced magnetization. Above T_B, thermal fluctuations dominate and the material rapidly loses coupling of unpaired spins.

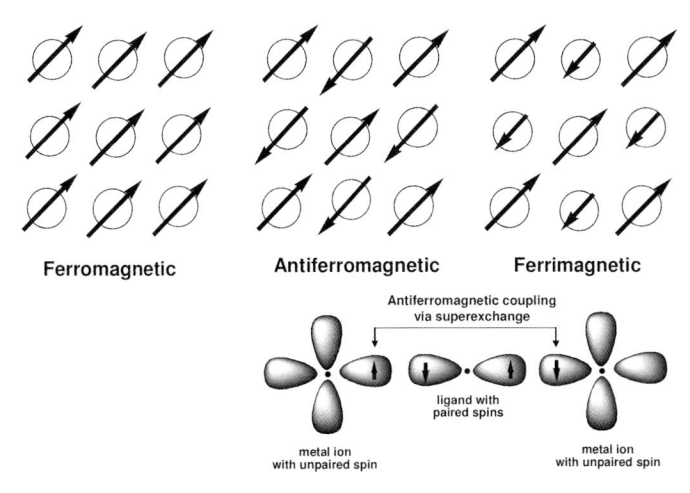

Ferromagnetic **Antiferromagnetic** **Ferrimagnetic**

Scheme 11.10 Parallel spin alignments in ferromagnetic materials and antiparallel spins in both antiferromagnetic and ferrimagnetic materials. Superexchange between metal centers via spin polarization with a paired spin ligand gives rise to antiparallel alignments.

In the synthesis of well-defined ferromagnetic nanoparticles, many metals require high annealing temperatures to form crystalline phases that are ferromagnetic. Solution wet-chemical approaches to synthesized magnetic nanoparticle often yield colloids of uniform particle size and morphology, but form crystalline phases that exhibit inferior magnetic properties. In the case of cobalt, hcp Co and fcc Co are commonly observed, where the hcp-phase possesses enhanced M_s and H_c values [165–167]. The epsilon (ε) phase of Co [168] has also been observed and has inferior magnetic properties relative to the hcp or fcc Co phases. High-temperature annealing of well-defined ε-Co particle arrays afforded either hcp or fcc nanoparticles of the same particle size but enhanced magnetic properties [168, 169]. However, in the preparation of nanocomposites with polymeric components, this high-temperature annealing step cannot be employed, requiring other methodologies to obtain such desired nanocomposites.

11.4.2
Linear Polymeric Surfactants

The synthesis of colloidal cobalt nanoparticles was initially achieved by the thermolysis of dicobaltoctacarbonyl (Co_2CO_8) in the presence of linear polymeric surfactants. The use of small molecule surfactants was developed after the initial reports using polymeric systems and will be discussed in later sections. The seminal report by Thomas at the Chevron Group demonstrated the effective use of poly(methyl methacrylate-*random*-ethyl acrylate-*random*-N-vinylpyrrolidone) [p(MMA-*r*-EA-*r*-VPy] terpolymers as stabilizers in the preparation of ferromag-

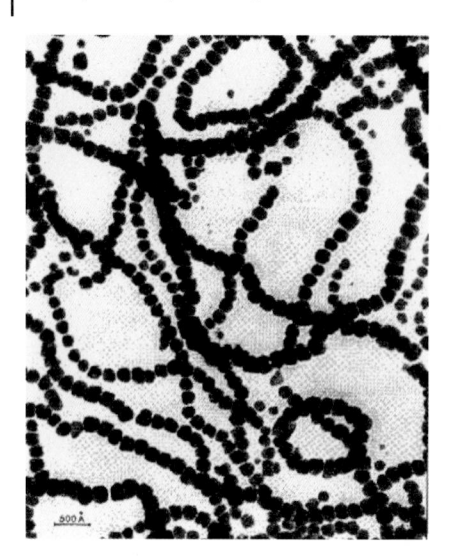

Fig. 11.11 TEM image of ferromagnetic cobalt colloids prepared by the thermolysis of Co_2CO_8 using PVP terpolymer stabilizers. Reproduced with permission from [165], Copyright 1966 American Institute of Physics.

netic cobalt colloids by the thermolysis of Co_2CO_8 in refluxing toluene [165]. Polymer-stabilized nanoparticles with tunable sizes in the range 2–30 nm were synthesized by variation of the surfactant-to-metal carbonyl ratios. Cobalt single-domain nanocrystals were synthesized, with the majority of colloids possessing the fcc phase, as determined from XRD. TEM of nanoparticles deposited from solution confirmed that colloids possessing uniform size and morphology were obtained, where 1-D nanoparticle chains were also observed from the dipolar association between ferromagnetic cores (Fig. 11.11). Fcc cobalt nanoparticles larger than 10 nm in diameter were found to organize into 1-D assemblies, indicative that the superparamagnetic limit of these colloids is below 10 nm. In similar studies, Hess and Parker [170] also reported the synthesis of cobalt colloids using a library of both condensation and chain growth-derived polymers as stabilizers.

Later studies by Smith and coworkers [171] at Xerox investigated the use of polydienes and polystyrene-*block*-poly(4-vinylpyridine) (PS-*b*-P4VP) block copolymers as surfactants in the preparation of metallic iron nanoparticles. Iron pentacarbonyl $[Fe(CO)_5]$ precursors were used to form metallic colloids by thermolysis at high temperatures ($\sim 150\,^\circ$C). Well-defined Fe colloids of tunable sizes in the range 10–20-nm diameter were synthesized by this approach and TEM confirmed that 1-D nanoparticle chains formed from association of ferromagnetic colloids. Oxidation studies using Fe colloids confirmed that the formation of oxide layers on the nanoparticle shell proceeded at ambient conditions that compromised the magnetic properties of the colloidal material. This was the first study to confirm with electron microscopy and diffraction that passivating oxide layers readily formed around ferromagnetic Fe colloidal cores. These authors subsequently probed the mechanism of nanoparticle formation using polymeric surfactants and proposed that polar groups on certain polymers facilitated the thermolysis of metal carbonyl precursors by nucleating the growth of colloids [172].

Mechanistic studies on the thermolysis reactions of Co_2CO_8 were conducted in a viscous polymeric medium (i.e. polystyrene) and followed the consumption of the precursor and formation of various intermediate clusters (e.g. Co_4CO_{12}) on the pathway to forming Co metal colloids [173]. The multiple possible pathways towards various Co intermediates complicated a precise mechanistic study from the dimeric precursor to the final nanoparticle product. Recent mechanistic studies on small molecule surfactants for the thermolysis of Co_2CO_8 using attenuated total reflectance–Fourier transform infrared (ATR–FTIR) yielded similar conclusions [174].

Bronstein and Antonietti investigated the use of PS-b-PVP surfactants in the thermolysis of Co_2CO_8 and found that ferromagnetic cobalt nanoparticles could also be formed using these copolymer surfactants, as reported by Smith and coworkers for Fe nanoparticles. These ferromagnetic cobalt colloids were also observed to form extended 1-D chains when cast on to TEM grids [175]. Ferromagnetic cobalt colloids were also synthesized from polystyrene-block-poly(2-vinylpyridine) copolymer surfactants with similar results [176].

Using a different class of well-defined polymeric surfactants, Riffle and coworkers [177, 178] demonstrated that functional polydimethylsiloxane (PDMS) copolymers were effective surfactants to synthesize fcc ferromagnetic and superparamagnetic cobalt colloids. ABA triblock copolymers with a ligating (nitrile-containing) B-block were prepared using anionic ring-opening polymerization, in which a difunctional macroinitiator of poly[(3-cyanopropyl)methylsiloxane] (PCPMS) was chain extended with hexamethylcyclotrisiloxane to yield PDMS-b-PCPMS-b-PDMS copolymers. Cobalt nanoparticles with polysiloxane shells were then prepared by *in situ* thermolysis with Co_2CO_8 in refluxing toluene, where the polymer shell allowed efficient dispersion of magnetic colloids into the PDMS carrier fluids (Fig. 11.12). These magnetic PDMS fluids have potential applications as nontoxic stimuli-responsive fluids for biomedical materials and surgical tools for retinal repair.

Similar approaches to the preparation of block copolymer–cobalt nanocomposite materials were reported by Tannenbaum and coworkers [179], in which cobalt nanoparticles were formed using only a PS-b-PMMA surfactant in the solution thermolysis of Co_2CO_8. Thin films cast from this mixture revealed that cobalt nanoparticles preferentially sequestered into the PMMA domains. An alternative approach to preparing cobalt–nanocomposite materials was elegantly demonstrated by Grubbs and coworkers [180, 181], in which alkyne-functionalized block copolymers were complexed with metal carbonyls to incorporate cobalt precursors in the polymeric surfactant (Fig. 11.13). Block copolymers carrying pendant alkyne groups were prepared by the sequential nitroxide-mediated polymerization of styrene and 4-phenylethynylstyrene. Treatment of the block

Fig. 11.12 (a) Synthesis of polysiloxane-coated cobalt nanoparticles using PDMS–PCMPS-b-PDMS copolymer surfactants. (b) TEM images of cobalt nanoparticles examining the kinetics of the reaction with respect to particle size and morphology. Reproduced with permission from [178], Copyright 2002 Elsevier.

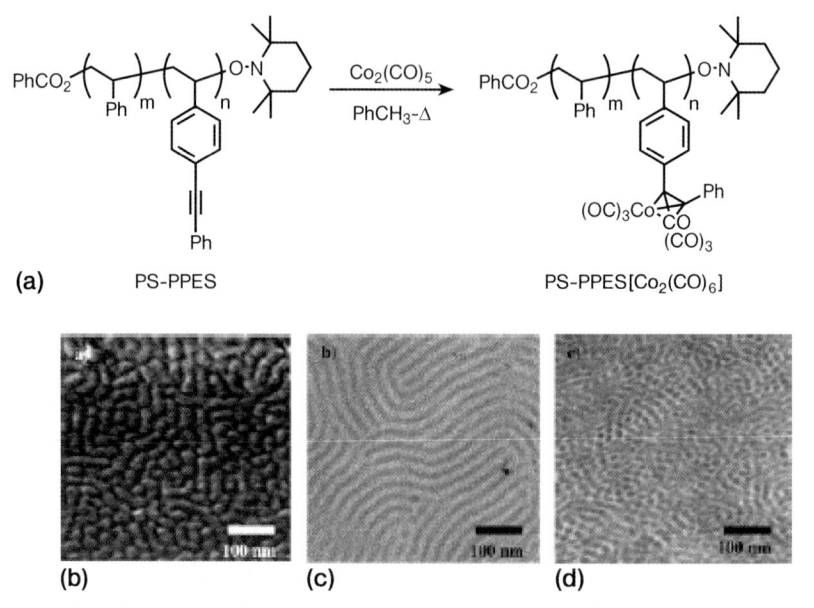

Fig. 11.13 (a) Synthesis of block copolymer–cobalt carbonyl adducts via ligand exchange of Co_2CO_8 on to polystyrene-block-poly(4-phenylethynylstyrene). (b) AFM phase image of PS_{181}–$PPES_{56}$ $[Co_2(CO)_6]_{50}$ dip-coated from toluene solution on freshly cleaved mica; (c) TEM image of PS_{181}–$PPES_{56}$ $[Co_2(CO)_6]_{50}$ drop cast from toluene solution on to a C-coated Cu grid; (d) TEM image of microtomed bulk PS_{173}–$PPES_{67}$ $[Co_2(CO)_6]$. Reproduced with permission from [180], Copyright 2004 American Chemical Society.

copolymer with Co_2CO_8 resulted in the formation of an alkynyl–cobalt adduct that was incorporated into the polymer surfactant. The cobalt carbonyl-bound polymeric precursor was then cast into thin films, where AFM and TEM provided visual evidence for selective sequestration of cobalt complexes into the domains of the alkyne-containing copolymer segments. High-temperature pyrolysis of these films yielded ceramic material composed of magnetic nanoparticles within a carbonaceous matrix.

11.4.3
Small Molecule Surfactants

A significant breakthrough in the preparation of magnetic cobalt nanoparticles was the finding that small molecule surfactants, such as TOPO, oleic acid and aliphatic amines used in the synthesis of semiconductor nanocrystals, are also applicable to the preparation of cobalt nanoparticles. In the seminal report of Dinega and Bawendi [168], TOPO was used in the thermolysis of Co_2CO_8 in refluxing toluene to give particles exhibiting a new metastable crystalline phase, termed ε-Co. As for the case in quantum dots, the lability of TOPO permits an Ostwald ripening mechanism to proceed. TEM images indicated that the cobalt

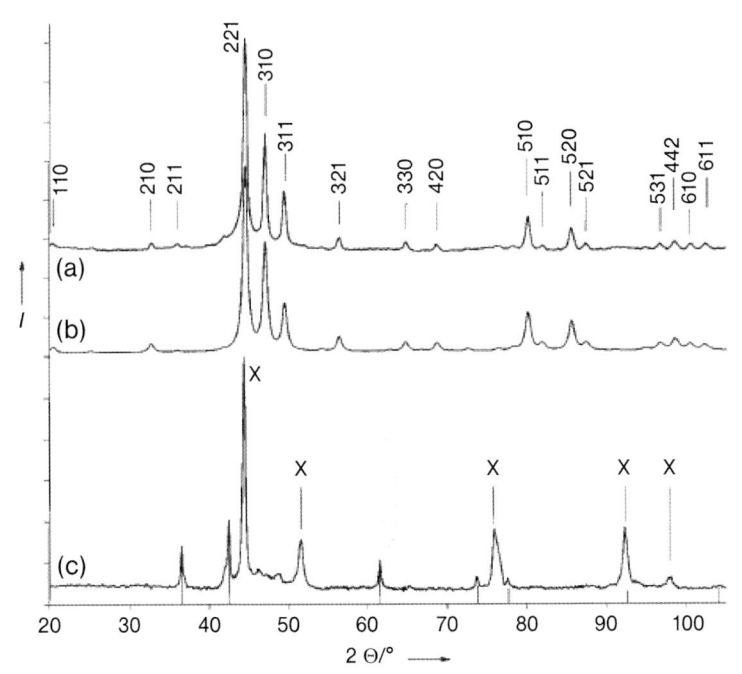

Fig. 11.14 XRD powder diffraction patterns of (a) as synthesized ε-Co nanoparticles; (b) simulation of ε-Co phase; (c) fcc Co nanoparticles obtained by annealing particles from (a) at 500 °C. Reproduced with permission from [168], Copyright 1999 Wiley-VCH.

nanoparticles obtained in this process possessed a distribution of sizes with an average size of 20 nm. XRD of the as-synthesized cobalt nanoparticles revealed distinctive diffraction patterns from the hcp or fcc phases (Fig. 11.14).

Using a different synthetic procedure, Sun and Murray prepared well-defined metallic colloids of uniform size and morphology [169]. A mixed surfactant cocktail of oleic acid and trialkylphosphines was used to stabilize the nanoparticles generated by the reduction of cobalt(II) chloride ($CoCl_2$) using superhydride ($LiBEt_3$) in dioctyl ether at 200 °C. Highly uniform ε-Co nanoparticles, in the range 2–11 nm in diameter, were synthesized using this approach. Subsequently hcp and fcc Co nanoparticles were obtained by annealing the ε-Co particle arrays at 300 and 500 °C, respectively.

An improved and facile synthesis of ε-Co nanoparticles possessing tunable, uniform particle size and morphology was developed by Alivisatos and coworkers [166], using a mixture of oleic acid, TOPO or aliphatic amines in the thermolysis of Co_2CO_8. In contrast to the method of Dinega and Bawendi, thermolysis reactions were performed at high temperatures (185 °C) in 1,2-dichlorobenzene, which provided temporal resolution of nucleation and growth stages of the reaction, allowing the preparation of nearly monodisperse nanoparticles. The mixture of oleic acid and TOPO ligands permitted efficient passivation of magnetic colloids, while still allowing for equilibration of metal species through Ostwald ripening. Nanoparticle powders obtained from these reactions proved stable and redispersible in solvents, due to the presence of the tightly binding oleic acid surfactant layer. By variation of the Co_2CO_8 precursor and the surfactants, both uniform superparamagnetic and ferromagnetic colloids were synthesized (Fig. 11.15). The authors later reported the use of an aliphatic amines and

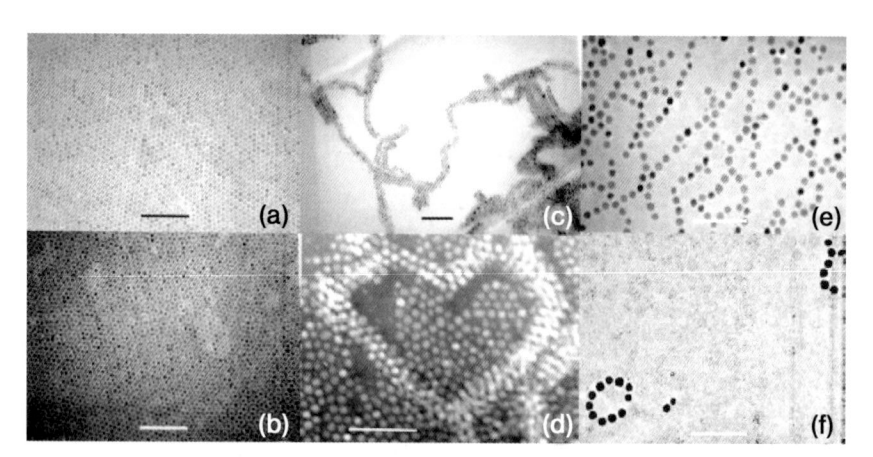

Fig. 11.15 TEM of oleic acid–TOPO-capped cobalt nanoparticle arrays, where images (a)–(d) are superparamagnetic ($D=12$ nm) and images (e) and (f) are ferromagnetic cobalt colloids ($D=16$ nm). Reproduced with permission from [166], Copyright 2001 American Association for the Advancement of Science.

TOPO surfactants to afford exquisite control of particle morphology, allowing the synthesis of hcp Co particles, disks and rods [167].

11.4.4
End-functional Polymeric Surfactants

Whereas a number of different nanocomposite materials have been prepared using magnetic cobalt nanoparticles and polymers, methodologies using well-defined polymeric surfactants that yield high-quality ferromagnetic colloids have not been developed extensively. In previous systems to form ferromagnetic cobalt nanoparticles using polymers, surfactants with a broad range of molar mass and random sequence compositions of ligating groups were employed in thermolysis reactions of Co_2CO_8 at relatively low temperatures ($\sim 110\,°C$). Under these conditions, hybridization with the polymer surfactant was achieved. However, nanoparticles with a wide distribution of size and morphology were often obtained, in contrast to the high temperature approaches ($\sim 185\,°C$) using small molecule surfactants, such as oleic acid and TOPO, that gave narrow size distribution samples.

To overcome these challenges, Pyun and coworkers [182] synthesized well-defined end-functional polymeric surfactants bearing either amine or phosphine oxide ligands to mediate the growth of cobalt nanoparticles (Fig. 11.16). These polymers were meant to mimic the function of aliphatic amines and TOPO in the high-temperature thermolysis of Co_2CO_8 in refluxing 1,2-dichlorobenzene. Polymer surfactants were synthesized using functional alkoxyamine initiators for nitroxide-mediated polymerization, yielding well-defined polystyrenes of precise molar mass and composition ($M_n=5000$, $M_w/M_n=1.09$). Using these polymers, fcc ferromagnetic cobalt nanoparticles of precise size ($D=15\pm1.5$ nm;

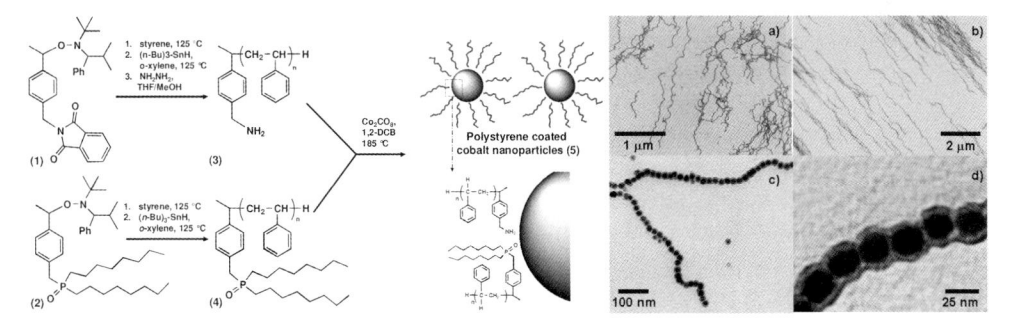

Fig. 11.16 Synthesis of polystyrenic surfactants using nitroxide mediation polymerization and cobalt nanoparticles. TEM images of ferromagnetic pS-coated cobalt nanoparticles: (a) self-assembled by deposition from toluene dispersions on to carbon-coated copper grids; (b) cast from toluene dispersion and aligned under a magnetic field (100 mT); (c) self-assembled single nanoparticle chains; (d) high-magnification image visualizing cobalt colloidal core (dark center) and pS surfactant shell (light halo). Reproduced with permission from [182], Copyright 2006 American Chemical Society.

$M_s = 38$ emu g^{-1}; $H_c = 100$ Oe at 20 °C) were synthesized. TEM images confirmed the expected ferromagnetic behavior of these colloidal cobalt nanoparticles, as evidenced by the formation of 1-D nanoparticle chains spanning several microns in length. The deposition of 1-D assemblies was controlled easily by the application of a weak magnetic field during the casting of colloids on surfaces.

The general area of magnetic nanoparticle functionalization remains largely unexplored. Ligand exchange and surface-initiated polymerizations have been conducted using superparamagnetic iron oxide [183–186], iron–platinum (FePt) [187] and spinel ferrite [188] colloids. However, these functionalization methodologies have not been applied to metallic cobalt nanoparticle systems or other ferromagnetic nanoparticles. Additionally, the marriage of conjugated polymers and dendrimers to cobalt and other ferromagnetic colloids is also an area of opportunity that has not been widely explored.

11.5
Perspectives

Polymer–nanoparticle composites as an area of research and development is exceedingly bright, with tremendous potential for new and innovative developments. The current landscape is ideal for continued exploration, as synthetic techniques to prepare both well-defined polymers and inorganic nanoparticles are now readily available. As discussed in this chapter, the synthetic approach to control the interface between these types of materials is essential for the creation of novel chemical compositions and multifunctional materials. With increasing demands for alternative energy technologies, new materials capable of solar energy conversion and hydrogen generation/storage are needed, as are their characterization and formulation into useful devices. Both semiconductor nanoparticles and electroactive polymers are expected to be important components in these types of materials. In the area of microelectronics, the need for magnetic storage media of higher density and larger capacities is ever present, requiring improved materials with enhanced magnetic properties on smaller dimensions. Flexible magnetic storage is also an important application in need of new materials that inherently require hybridization of polymers and magnetic colloids. Moreover, methodologies to prepare and characterize novel magnetic nanocomposite films will be an area in need of materials research. In all of these applications, the ability to combine useful, often mutually exclusive properties into a composite material is the critical challenge that synthetic materials chemistry, physics and engineering will need to address.

References

1 Novak, B. M., *Adv. Mater.* **1993**, *5*, 422–433.
2 Loy, D. A., Shea, K. J., *Chem. Rev.* **1995**, *95*, 1431–1442.
3 Wen, J., Wilkes, G. L., *Chem. Mater.* **1996**, *8*, 1667–1681.
4 Foerster, S., Antonietti, M., *Adv. Mater.* **1998**, *10*, 195–217.

5 Pyun, J., Matyjaszewski, K., *Chem. Mater.* **2001**, *13*, 3436–3448.
6 Simon, P. F. W., Ulrich, R., Spiess, H. W., Wiesner, U., *Chem. Mater.* **2001**, *13*, 3464–3486.
7 Antonietti, M., Ozin, G. A., *Chem. Eur. J.* **2004**, *10*, 28–41.
8 Burda, C., Chen, X., Narayanan, R., El-Sayed, M. A., *Chem. Rev.* **2005**, *105*, 1025–1102.
9 Shenhar, R., Norsten, T. B., Rotello, V. M., *Adv. Mater.* **2005**, *17*, 657–669.
10 Hawker, C. J., Wooley, K. L., *Science* **2005**, *309*, 1200–1205.
11 Caruso, F., *Adv. Mater.* **2001**, *13*, 11–22.
12 Kotov, N. A., *MRS Bull.* **2001**, *26*, 992–997.
13 Hammond, P. T., *Adv. Mater.* **2004**, *16*, 1271–1293.
14 Tang, Z., Kotov, N. A., *Adv. Mater.* **2005**, *17*, 951–962.
15 Cushing, B. L., Kolesnichenko, V. L., O'Connor, C. J., *Chem. Rev.* **2004**, *104*, 3893–3946.
16 Storhoff, J. J., Mirkin, C. A., *Chem. Rev.* **1999**, *99*, 1849–1862.
17 Rosi, N. L., Mirkin, C. A., *Chem. Rev.* **2005**, *105*, 1547–1562.
18 Hostetler, M. J., Green, S. J., Stokes, J. J., Murray, R. W., *J. Am. Chem. Soc.* **1996**, *118*, 4212–4213.
19 Hostetler, M. J., Templeton, A. C., Murray, R. W., *Langmuir* **1999**, *15*, 3782–3789.
20 Murray, C. B., Norris, D. J., Bawendi, M. G., *J. Am. Chem. Soc.* **1993**, *115*, 8706–8715.
21 Pathak, S., Choi, S., Arnheim, N., Thompson, M. S., *J. Am. Chem. Soc.* **2001**, *123*, 4103–4104.
22 Skaff, H., Ilker, M. F., Coughlin, E. B., Emrick, T. S., *J. Am. Chem. Soc.* **2002**, *124*, 5729–5733.
23 Skaff, H., Emrick, T. S., *Chem. Commun.* **2003**, 52–53.
24 Sill, K., Emrick, T. S., *Chem. Mater.* **2004**, *16*, 1240–1243.
25 Kalyuzhny, G., Murray, R. W., *J. Phys. Chem. B* **2005**, *109*, 7012–7021.
26 Coates, G. W., *Chem. Rev.* **2000**, *100*, 1223–1252.
27 Matyjaszewski, K., Xia, J., *Chem. Rev.* **2001**, *101*, 2921–2990.
28 Hawker, C. J., Bosman, A. W., Harth, E., *Chem. Rev.* **2001**, *101*, 3661–3688.
29 Hadjichristidis, N., Pitsikalis, M., Pispas, S., Iatrou, H., *Chem. Rev.* **2001**, *101*, 3747–3792.
30 Buchmeiser, M. R., *Chem. Rev.* **2000**, *100*, 1565–1604.
31 Trnka, T. M., Grubbs, R. H., *Acc. Chem. Res.* **2001**, *34*, 18–29.
32 Patten, T. E., Xia, J., Abernathy, T., Matyjaszewski, K., *Science* **1996**, *272*, 866–868.
33 Matyjaszewski, K., Patten, T. E., Xia, J., *J. Am. Chem. Soc.* **1997**, *119*, 674–680.
34 Patten, T. E., Matyjaszewski, K., *Adv. Mater.* **1998**, *10*, 901–915.
35 Hawker, C. J., *J. Am. Chem. Soc.* **1994**, *116*, 11185–11186.
36 Hawker, C. J., *Acc. Chem. Res.* **1997**, *30*, 373–382.
37 Malmström, E. E., Hawker, C. J., *Macromol. Chem. Phys.* **1998**, *199*, 923–935.
38 Benoit, D., Chaplinski, V., Braslau, R., Hawker, C. J., *J. Am. Chem. Soc.* **1999**, *121*, 3904–3920.
39 Chiefari, J., Chong, Y. K., Ercole, F., Krstina, J., Jeffery, J., Le, T. P. T., Mayadunne, R. T. A., Meijs, G. F., Moad, C. L., Moad, G., Rizzardo, E., Thang, S. H., *Macromolecules* **1998**, *31*, 5559–5562.
40 Chong, Y. K., Le, T. P. T., Moad, G., Rizzardo, E., Thang, S. H., *Macromolecules* **1999**, *32*, 2071–2074.
41 Moad, G., Chong, Y. K., Postma, A., Rizzardo, E., Thang, S. H., *Polymer* **2005**, *46*, 8458–8468.
42 Moad, G., Rizzardo, E., Thang, S. H., *Aust. J. Chem.* **2005**, *58*, 379–410.
43 Green, M., *Chem. Commun.* **2005**, 3002–3011.
44 Klabunde, K. J., Nanoscale Materials in Chemistry, 2nd ed. John Wiley and Sons, **2001**.
45 Shriver, D. F., Atkins, P. W., Langford, C. H., *Inorganic Chemistry.* Freeman, New York, **1990**.
46 El-Sayed, M. A., *Acc. Chem. Res.* **2001**, *34*, 257–264.
47 Schoen, G., Simon, U., *J. Colloid Polym. Sci.* **1995**, *273*, 202.
48 Schoen, G., Simon, U., *J. Colloid Polym. Sci.* **1995**, *273*, 101.
49 Daniel, M.-C., Astruc, D., *Chem. Rev.* **2004**, *104*, 293–346.

50 Faraday, M., *Philos. Trans.* **1857**, *147*, 145–181.
51 Turkevitch, J., Stevenson, P. C., Hillier, J., *Discuss. Faraday Soc.* **1951**, *11*, 55–75.
52 Brust, M., Walker, M., Bethell, D., Schiffrin, D. J., Whyman, R. J., *J. Chem. Soc., Chem. Commun.* **1994**, 801–801.
53 Brust, M., Fink, J., Bethell, D., Schiffrin, D. J., Kiely, C. J., *J. Chem. Soc., Chem. Commun.* **1995**, 1655–1656.
54 Ingram, R. S., Hostetler, M. J., Murray, R. W., *J. Am. Chem. Soc.* **1997**, *119*, 9175–9178.
55 Templeton, A. C., Wuelfing, W. P., Murray, R. W., *Acc. Chem. Res.* **2000**, *33*, 27–36.
56 Templeton, A. C., Hostetler, M. J., Kraft, C. T., Murray, R. W., *J. Am. Chem. Soc.* **1998**, *120*, 1906–1911.
57 Hong, R., Fisher, N. O., Verma, A., Goodman, C. M., Emrick, T., Rotello, V. M., *J. Am. Chem. Soc.* **2004**, *126*, 739–743.
58 McCullough, R. D., *Adv. Mater.* **1998**, *10*, 93–116.
59 Grayson, S. M., Fréchet, J. M. J., *Chem. Rev.* **2001**, *101*, 3819–3867.
60 Fréchet, J. M. J., *J. Polym. Sci., Part A: Polym. Sci.* **2003**, *41*, 3713–3725.
61 Hirai, H., *J. Macromol. Sci.* **1979**, *A13*, 633–649.
62 Hayat, M. A., *Colloidal Gold: Principles, Methods and Applications.* Academic Press, New York, **1989**.
63 Mayer, A. B. R., Mark, J. E., *Eur. Polym. J.* **1999**, *34*, 103–108.
64 Teranishi, T., Kiyokawa, I., Miyake, M., *Adv. Mater.* **1998**, *10*, 596–599.
65 Han, M. Y., Quek, C. H., Huang, W., Chew, C. H., Gan, L. M., *Chem. Mater.* **1999**, *11*, 1144–1147.
66 Sun, Y., Xia, Y., *Science* **2002**, *298*, 2176–2179.
67 Wuelfing, P. W., Gross, S. M., Miles, D. T., Murray, R. W., *J. Am. Chem. Soc.* **1998**, *120*, 12696–12697.
68 Bockstaller, M. R., Kolb, R., Thomas, E. L., *Adv. Mater.* **2001**, *13*, 1783–1786.
69 Corbierre, M. K., Cameron, N. S., Sutton, M., Mochrie, S. G. J., Lurio, L. B., Rühm, A., Lennox, R. B., *J. Am. Chem. Soc.* **2001**, *123*, 10411–10412.
70 Lowe, A. B., Sumerlin, B. S., Donovan, M. S., McCormick, C. L., *J. Am. Chem. Soc.* **2002**, *124*, 11562–11563.
71 Mangeney, C., Ferrage, F., Aujard, I., Marchi-Artzner, V., Jullien, L., Ouari, O., El Djouhar Re'kaï, E. D., Laschewsky, A., Vikholm, I., Sadowski, J. W., *J. Am. Chem. Soc.* **2002**, *124*, 5811–5821.
72 Walker, C. H., St. John, J. V., Wisian-Neilson, P., *J. Am. Chem. Soc.* **2001**, *123*, 3846–3847.
73 Watson, K. J., Zhu, J., Nguyen, S. B. T., Mirkin, C. A., *J. Am. Chem. Soc.* **1999**, *121*, 562–563.
74 Nuss, S., Böttcher, H., Wurm, H., Hallensleben, M. L., *Angew. Chem. Int. Ed.* **2001**, *40*, 4016–4018.
75 Mandal, T. K., Fleming, M. S., Walt, D. R., *NanoLett* **2002**, *2*, 3–7.
76 Ohno, K., Hoh, K.-M., Tsuji, Y., Fukuda, T., *Macromolecules* **2003**, *35*, 8989–8993.
77 Ohno, K., Koh, K., Tsuji, Y., Fukuda, T., *Angew. Chem. Int. Ed.* **2003**, *42*, 2751–2754.
78 Shan, J., Nuopponen, M., Jiang, H., Kauppinen, E., Tenhu, H., *Macromolecules* **2003**, *36*, 4526–4533.
79 Zhu, M., Wang, L., Exarhos, G. J., Li, A. D. Q., *J. Am. Chem. Soc.* **2004**, *126*, 2656–2657.
80 Jordan, R., West, N., Chou, Y.-M., Nukyen, O., *Macromolecules* **2001**, *34*, 1606.
81 Ng Cheong Chan, Y., Schrock, R. R., Cohen, R. E., *Chem. Mater.* **1992**, *4*, 24–27.
82 Clay, R. T., Cohen, R. E., *Supramol. Sci.* **1995**, *2*, 183–191.
83 Bockstaller, M. R., Lapetnikov, Y., Margel, S., Thomas, E. L., *J. Am. Chem. Soc.* **2003**, *125*, 5276–5277.
84 Spatz, J. P., Roescher, A., Sheiko, S., Krausch, G., Möller, M., *Adv. Mater.* **1995**, *7*, 731–735.
85 Spatz, J. P., Roescher, A., Möller, M., *Adv. Mater.* **1996**, *8*, 337–340.
86 Spatz, J. P., Mössmer, S., Möller, M., *Chem. Eur. J.* **1996**, *12*, 1552–1556.
87 Spatz, J. P., Herzog, T., Mössmer, S., Ziemann, P., Möller, M., *Adv. Mater.* **1999**, *11*, 149–153.
88 Spatz, J. P., Mössmer, S., Hartmann, C., Möller, M., Herzog, T., Krieger, M., Boyen, H., Ziemann, P., Kabius, B., *Langmuir* **2000**, *16*, 407–415.

89 Antonietti, M., Wenz, E., Bronstein, L., Seregina, M., *Adv. Mater.* **1995**, *7*, 1000–1005.

90 Kang, Y., Taton, T. A., *Angew. Chem. Int. Ed.* **2004**, *44*, 409–412.

91 Kang, Y., Taton, T. A., *Macromolecules* **2005**, *38*, 6115–6121.

92 Zhang, L., Eisenberg, A., *Science* **1995**, *268*, 1768.

93 Zhang, Q., Remsen, E. E., Wooley, K. L., *J. Am. Chem. Soc.* **2000**, *122*, 3642–3651.

94 Wooley, K. L., *J. Polym. Sci., Part A: Polym. Sci.* **2000**, *38*, 1397–1407.

95 Ma, Q., Remsen, E. E., Kowalewski, T., Wooley, K. L., *J. Am. Chem. Soc.* **2001**, *123*, 4627–4628.

96 Qi, K., Ma, Q., Remsen, E. E., Clark, J. C. G., Wooley, K. L., *J. Am. Chem. Soc.* **2004**, *126*, 6599–6607.

97 Joralemon, M. J., O'Reilly, R., Hawker, C. J., Wooley, K. L., *J. Am. Chem. Soc.* **2005**, *127*, 16892–16899.

98 Kang, Y., Erickson, K. J., Taton, T. A., *J. Am. Chem. Soc.* **2005**, *127*, 13800–13801.

99 Frankamp, B. L., Uzun, O., Ilhan, F., Boal, A. K., Rotello, V. M., *J. Am. Chem. Soc.* **2002**, *124*, 892–893.

100 Boal, A. K., Ilhan, F., DeRouchy, J. E., Thurn-Albrecht, T., Russell, T. P., Rotello, V. M., *Nature* **2000**, *404*, 746.

101 Youk, J., Park, M., Locklin, J., Advincular, R., Yang, J., Mays, J., *Langmuir* **2002**, *18*, 2455–2458.

102 Filali, M., Meier, M. A. R., Schubert, U. S., Gohy, J., *Langmuir* **2005**, *21*, 7995–8000.

103 Lemon, B. I., Crooks, R. M., *J. Am. Chem. Soc.* **2000**, *122*, 12886.

104 Crooks, R. M., Zhao, M., Sun, L., Chechik, V., Yeung, L. K., *Acc. Chem. Res.* **2001**, *34*, 181–190.

105 Scott, R. W. J., Wilson, O. M., Crooks, R. M., *J. Phys. Chem. B* **2005**, *109*, 692–704.

106 Esumi, K., Suzuki, A., Aihara, N., Usui, K., Torigoe, K., *Langmuir* **1998**, *14*, 3157.

107 Grohn, F., Kim, G., Bauer, B. J., Amis, E. J., *Macromolecules* **2001**, *34*, 2179–2185.

108 Kim, Y.-G., Oh, S.-K., Crooks, R. M., *Chem. Mater.* **2004**, *16*, 167–172.

109 Deng, S., Locklin, J., Patton, D., Baba, A., Advincula, R. C., *J. Am. Chem. Soc.* **2005**, *127*, 1744–1751.

110 Sun, X., Luo, Y., *Mater. Lett.* **2005**, *59*, 4048–4050.

111 Wang, R., Yang, J., Zheng, Z., Carducia, M. D., Jiao, J., Seraphin, S., *Angew. Chem. Int. Ed.* **2001**, *40*, 549–552.

112 Li, D., Li, J., *Colloids Surf. A* **2005**, *257/258*, 255–259.

113 Sih, B. C., Wolf, M. O., *Chem. Commun.* **2005**, 3374–3384.

114 Marinakos, S. M., Brousseau, L. C., III, Jones, A., Feldheim, D. L., *Chem. Mater.* **1998**, *10*, 1214.

115 Zhou, Y., Itoh, H., Uemura, T., Naka, K., Chujo, Y., *Chem. Commun.* **2001**, 613.

116 Zhou, Y., Itoh, H., Uemura, T., Naka, K., Chujo, Y., *Langmuir* **2002**, *18*, 277.

117 Patton, D., Locklin, J., Meredith, M., Xin, Y., Advincula, R., *Chem. Mater.* **2004**, *16*, 5063.

118 Zhai, L., McCullough, R. D., *J. Mater. Chem.* **2004**, *14*, 141–143.

119 El-Sayed, M. A., *Acc. Chem. Res.* **2004**, *37*, 326–333.

120 Skaff, H., Emrick, T. S., *Semiconductor Nanoparticles: Synthesis, Properties and Integration into Polymers for the Generation of Novel Composite Materials*. Kluwer Academic/Plenum Publishers, New York, **2004**.

121 Bawendi, M. G., Wilson, W. L., Rothberg, L., Carroll, P. J., Jedju, T. M., Steigerwald, M. L., Brus, L. E., *Phys. Rev. Lett.* **1990**, *65*, 1623–1626.

122 Brus, L. E., Szajowski, P. F., Wilson, W. L., Harris, T. D., Schuppler, S., Citrin, P. H., *J. Am. Chem. Soc.* **1995**, *117*, 2915–2922.

123 Nirmal, M., Dabbousi, B. O., Bawendi, M. G., Macklin, J. J., Trautman, J. K., Harris, T. D., Brus, L. E., *Nature* **1996**, *383*, 802–804.

124 Chen, C. C., Herhold, A. B., Johnson, C. S., Alivisatos, A. P., *Science* **1997**, *276*, 398–401.

125 Empedocles, S. A., Bawendi, M. G., *Science* **1997**, *278*, 2114–2117.

126 Empedocles, S. A., Bawendi, M. G., *Acc. Chem. Res.* **1999**, *32*, 389–396.

127 Nirmal, M., Brus, L. E., *Acc. Chem. Res.* **1999**, *32*, 407–414.
128 Shim, M., Wang, C. J., Guyot-Sionnest, P., *J. Phys. Chem. B* **2001**, *105*, 2369–2373.
129 Shim, M., Wang, C. J., Norris, D. J., Guyot-Sionnest, P., *MRS Bull.* **2001**, *26*, 1005–1008.
130 Chestnoy, N., Hull, R., Brus, L. E., *J. Phys. Chem.* **1986**, *85*, 2237–2242.
131 Brus, L. E., *J. Phys. Chem. Solids* **1998**, *59*, 459–465.
132 Alivisatos, A. P., Harris, T. D., Brus, L. E., Jayaraman, A., *J. Chem. Phys.* **1988**, *89*, 5979–5982.
133 De Schryver, F. C., *Pure Appl. Chem.* **1998**, *70*, 2147.
134 Li, L. S., Hu, J. T., Yang, W. D., Alivisatos, A. P., *NanoLett* **2001**, *1*, 349–351.
135 Marcus, M. A., Flood, W., Steigerwald, M., Brus, L. E., Bawendi, M. G., *J. Chem. Phys.* **1991**, *95*, 1572–1576.
136 Rabani, E., Hetenyi, B., Berne, B. J., Brus, L. E., *J. Chem. Phys.* **1999**, *110*, 5355–5369.
137 Rossetti, R., Ellison, J. L., Gibson, J. M., Brus, L. E., *J. Chem. Phys.* **1984**, *80*, 4464–4469.
138 Huynh, W. U., Dittmer, J. J., Alivisatos, A. P., *Science* **2002**, *295*, 2425–2427.
139 Huynh, W. U., Peng, X. G., Alivisatos, A. P., *Adv. Mater.* **1999**, *11*, 923.
140 Milliron, D., Alivisatos, A. P., Pitois, C., Edder, C., Fréchet, J. M. J., *Adv. Mater.* **2003**, *15*, 58–61.
141 Liu, J., Tanaka, T., Sivula, K., Alivisatos, A. P., Fréchet, J. M. J., *J. Am. Chem. Soc.* **2004**, *126*, 6550–6551.
142 Steigerwald, M. L., Alivisatos, A. P., Gibson, J. M., Harris, T. D., Kortan, R., Muller, A. J., Thayer, A. M., Duncan, T. M., Douglass, D. C., Brus, L. E., *J. Am. Chem. Soc.* **1988**, *110*, 3046–3050.
143 Rogach, A. L., Talapin, D. V., Shevchenko, E. V., Kornowski, A., Haase, M., Weller, H., *Adv. Funct. Mater.* **2002**, *12*, 653–664.
144 Peng, X. G., Schlamp, M. C., Kadavanich, A. V., Alivisatos, A. P., *J. Am. Chem. Soc.* **1997**, *119*, 7019–7029.
145 Dabbousi, B. O., Rodriguez-Viejo, J., Mikulec, F. V., Heine, J. R., Mattoussi, H., Ober, R., Jensen, K. F., Bawendi, M. G., *J. Phys. Chem. B* **1997**, *101*, 9463–9475.
146 Qu, L. H., Peng, Z. A., Peng, X. G., *NanoLett* **2001**, *1*, 333–337.
147 Peng, Z. A., Peng, X. G., *J. Am. Chem. Soc.* **2002**, *124*, 3343–3353.
148 Qu, L. H., Peng, X. G., *J. Am. Chem. Soc.* **2002**, *124*, 2049–2055.
149 Lee, J., Sundar, V. C., Heine, J. R., Bawendi, M. G., Jensen, K. F., *Adv. Mater.* **2000**, *12*, 1102–1105.
150 Carrot, G., Scholz, S. M., Plummer, C. J. G., Hilborn, J. G., Hedrick, J. L., *Chem. Mater.* **1999**, *11*, 3571–3577.
151 Skaff, H., Emrick, T. S., *Angew. Chem. Int. Ed.* **2004**, *43*, 5383–5386.
152 Moffitt, M., McMahon, L., Pessel, V., Eisenberg, A., *Chem. Mater.* **1995**, *7*, 1185–1192.
153 Wang, C. W., Moffitt, M. G., *Langmuir* **2004**, *20*, 11784–11796.
154 Zhao, H. Y., Douglas, E. P., Harrison, B. S., Schanze, K. S., *Langmuir* **2001**, *17*, 8428–8433.
155 Qi, L., Colfen, H., Antonietti, M., *NanoLett* **2001**, *1*, 61–65.
156 Fogg, D. E., Radzilowski, L. H., Blanski, R., Schrock, R. R., Thomas, E. L., *Macromolecules* **1997**, *30*, 417–426.
157 Misner, M. J., Skaff, H., Emrick, T. S., Russell, T. P., *Adv. Mater.* **2003**, *15*, 221–224.
158 Zhang, Q., Xu, T., Butterfield, D., Misner, M. J., Ryu, D. Y., Emrick, T. S., Russell, T. P., *NanoLett* **2005**, *5*, 357–361.
159 Zhang, Q., Gupta, S., Emrick, T. S., Russell, T. P., *J. Am. Chem. Soc.* **2006**, *128*, 3898–3899.
160 Skaff, H., Sill, K., Emrick, T. S., *J. Am. Chem. Soc.* **2004**, *126*, 11322–11325.
161 Wang, Y. A., Li, J. J., Chen, H., Peng, X. G., *J. Am. Chem. Soc.* **2002**, *124*, 2293–2298.
162 Guo, W., Li, J. J., Wang, A., Peng, X. G., *J. Am. Chem. Soc.* **2003**, *125*, 3901.
163 Locklin, J., Patton, D., Deng, S., Baba, A., Millan, M., Advincula, R., *Chem. Mater.* **2004**, *16*, 5187–5193.
164 Huber, D. L., *Small* **2005**, *1*, 482–501.
165 Thomas, J. R., *J. Appl. Phys.* **1966**, *37*, 2914–2915.
166 Puntes, V. F., Krishnan, K. M., Alivisatos, A. P., *Science* **2001**, *291*, 2115–2117.

167 Puntes, V. F., Zanchet, D., Erdonmez, C. K., Alivisatos, A. P., *J. Am. Chem. Soc.* **2002**, *124*, 12874–12880.
168 Dinega, D. P., Bawendi, M. G., *Angew. Chem. Int. Ed.* **1999**, *38*, 1788–1791.
169 Sun, S., Murray, C. B., *J. Appl. Phys.* **1999**, *85*, 4325–4330.
170 Hess, P. H., Parker, P. H., *J. Appl. Polym. Sci.* **1966**, *10*, 1915.
171 Griffiths, C. H., O'Horo, M. P., Smith, T. W., *J. Appl. Phys.* **1979**, *50*, 7108–7115.
172 Smith, T. W., Wychick, D., *J. Phys. Chem.* **1980**, *84*, 1621–1629.
173 Tannenbaum, R., *Inorg. Chim. Acta* **1994**, *227*, 233–240.
174 Lagunas, A., Jimeno, C., Font, D., Sola, L., Pericas, M. A., *Langmuir* **2006**, *22*, 3823–3829.
175 Platonova, O. A., Bronstein, L. M., Solodovnikov, S. P., Yanovskaya, I. M., Obolonkova, E. S., Valetsky, P. M., Wenz, E., Antonietti, M., *Colloid Polym. Sci.* **1997**, *275*, 426–431.
176 Diana, F. S., Lee, S., Petroff, P. M., Kramer, E. J., *NanoLett* **2003**, *3*, 891–895.
177 Stevenson, J. P., Rutnakornpituk, M., Vadala, M., Esker, A. R., Riffle, J. S., Charles, S. W., Wells, S., Dailey, J. P., *J. Magn. Magn. Mater.* **2001**, *225*, 47–58.
178 Rutnakornpituk, M., Thompson, M. S., Harris, L. A., Farmer, K. E., Esker, A. R., Riffle, J. S., Connolly, J., St. Pierre, T. G., *Polymer* **2002**, *43*, 2337–2348.
179 Tadd, E. H., Bradley, J., Tannenbaum, R., *Langmuir* **2002**, *18*, 2378–2384.
180 Miinea, L. A., Sessions, L. B., Ericson, K. D., Glueck, D. S., Grubbs, R. B., *Macromolecules* **2004**, *37*, 8967–8972.
181 Sessions, L. B., Miinea, L. A., Ericson, K. D., Glueck, D. S., Grubbs, R. B., *Macromolecules* **2005**, *38*, 2116–2121.
182 Korth, B. D., Keng, P., Shim, I., Bowles, S. E., Tang, C., Kowalewski, T., Nebesny, K. W., Pyun, J., *J. Am. Chem. Soc.* **2006**, in press.
183 Boal, A. K., Das, K., Gray, M., Rotello, V. M., *Chem. Mater.* **2002**, *14*, 2628–2636.
184 Wang, Y., Teng, X., Wang, J.-S., Yang, H., *NanoLett* **2003**, *3*, 789–793.
185 Ninjbadgar, T., Yamamoto, S., Fukuda, T., *Solid State Sci.* **2004**, *6*, 879–885.
186 Gravano, S. M., Dumas, R., Liu, K., Patten, T. E., *J. Polym. Sci., Part A: Polym. Sci.* **2005**, *43*, 3675–3688.
187 Hong, R., Fischer, N. O., Emrick, T. S., Rotello, V. M., *Chem. Mater.* **2005**, *17*, 4617–4621.
188 Vestal, C. R., Zhang, Z. J., *J. Am. Chem. Soc.* **2002**, *124*, 14312–14313.

12
Polymeric Membranes for Gas Separation, Water Purification and Fuel Cell Technology

Kazukiyo Nagai, Young Moo Lee, and Toshio Masuda

12.1
Introduction

A membrane is an interphase between two adjacent phases acting as a selective barrier, regulating the transport of substances between the two compartments and is employed for specific function(s) including separation of gases and liquids, ions or biological matter. Membrane separation technology enjoys various advantages over conventional methodologies such as isothermal operation at low temperatures, no requirement for additives, low energy consumption and ease of integration into other separation or reaction processes. Historically, membrane science and technology have found a wide array of industrial applications for gas separation and water purification [1–4] and are currently being applied for fuel cell applications, particularly as polymer electrolyte membranes.

Research activity in the field of membrane separation enjoys a long history of more than 100 years. However, a commercial process was successfully launched in 1960s for the desalination of sea water by reverse osmosis and development of a very thin membrane to provide a high water flux (i.e. highly efficient pure water production) was a technical breakthrough for this commercialization [1]. Furthermore, the development of asymmetric polymer membranes led to a dramatic cost reduction of the membrane separation process. Therefore, membrane separation technology has been applied for reverse osmosis, ultrafiltration, microfiltration, pervaporation and gas separation for the last four decades.

Since the early stages of membrane engineering, water purification membranes have contributed to the production of drinking water and rinse water in the electronics industry (e.g. semiconductor manufacture). This technique also finds applications for in-house water purification and huge-scale systems in cities. The present-day applications of ultrafiltration, microfiltration and nanofiltration include the fermentation, medical, chemical, food and beverage industries. On the other hand, the pervaporation process is still undergoing development in the industrial sector for water purification and liquid separation and is especially attractive for the production of pure ethanol as a biofuel.

Macromolecular Engineering. Precise Synthesis, Materials Properties, Applications.
Edited by K. Matyjaszewski, Y. Gnanou, and L. Leibler
Copyright © 2007 WILEY-VCH Verlag GmbH & Co. KGaA, Weinheim
ISBN: 978-3-527-31446-1

The first commercial equipment making use of gas separation membranes was installed in 1980 for hydrogen gas separation from nitrogen, in ammonia purge gas streams [1, 2, 4–7]. To date, this technique has gained multidimensional significance in areas such as air separation, nitrogen purification, natural gas separation, flue gas purification (i.e. carbon dioxide removal from air), dehydration and recovery volatile organic compounds (VOCs). In particular, many countries have focused on both the production of hydrogen as a clean energy source and the removal of carbon dioxide to prevent global warming. In Japan, in 2002, an oxygen enrichment membrane system was installed in air conditioners to maintain appropriate oxygen concentrations in rooms to create a comfortable living environment.

The quest for clean and efficient sources of electric power has intensified research on proton-exchange membrane fuel cells (PEMFCs) in the past decade and the fuel cell is expected to be an efficient energy system for future applications. One of the developing core technologies in fuel cell systems is an electrolyte membrane. The current approach to developing proton exchange membranes mainly targets non-perfluorinated polymers with a thermostable polymer backbone.

The properties of a membrane depend strongly on the chemical structure and microstructure of the polymer, both of which are substantially affected by the molecular weight of the polymer, presence of impurities, membrane formation process, membrane thickness and membrane pretreatment. Therefore, the synthesis of novel polymers with well-defined structure as 'designed' membrane materials will not only contribute to the development of new membrane materials but will also lead to significant advances in the science and technology of membranes. This chapter reviews the fundamentals of membrane science and the structure–property relationships of polymer membranes employed for gas separation, water purification and fuel cell technology and provides a comprehensive overview of the development of polymeric membranes having advanced or novel functions in the various membrane separation processes for liquid and gaseous mixtures (gas separation, reverse osmosis, pervaporation, nanofiltration, ultrafiltration, microfiltration) and in other important applications of membranes such as fuel cell systems.

12.2
Gas Separation Membranes

12.2.1
Gas Separation Mechanisms

Gas separation using a polymer membrane is attributed to differences in the molecular size of the gas and/or interaction between the gas molecules and the membrane, in a gaseous mixture. For the separation of binary gas mixture through a membrane, there are five basic separation mechanisms, as illustrated in Fig. 12.1 [1, 2, 8–11]. These mechanisms are primarily classified according to

12.2 Gas Separation Membranes

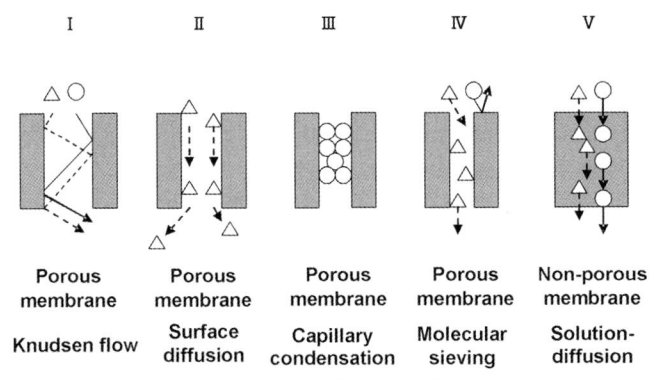

Fig. 12.1 Schematic representation of basic mechanisms for gas separation through membranes.

I	II	III	IV	V
Porous membrane	Porous membrane	Porous membrane	Porous membrane	Non-porous membrane
Knudsen flow	Surface diffusion	Capillary condensation	Molecular sieving	Solution-diffusion

the type of membrane: porous membranes (Types I–IV) and non-porous membranes (Type V). Porous membranes are characterized by the presence of permanent pores across a membrane. As the diameter of gas molecules ranges from 0.2 to 0.5 nm [12], the maximum pore size is about 2 nm for separations based on the differences between the Knudsen flow-rates of two gases (Type I). When there is a specific interaction between gas molecules and the surface inside a pore, surface diffusion takes place (Type II). Condensable gases/vapors tend to condense inside a pore (Type III). These mechanisms result in moderate gas selectivity while molecular sieving behavior is expected to provide high selectivity (Type IV). It is difficult, however, to prepare well-ordered angstrom-size (i.e. gas size) pores in the polymer membranes.

Gas transport through porous membranes with an average pore diameter of 1–5 nm is affected by the both Knudsen diffusion and surface diffusion. Capillary condensation can also occur for the transport of condensable gases, such as carbon dioxide. So far, there have been two types of gas transport model studies: one assumes that the pore structure is an ideal tube-like structure, while the other postulates that the evaluated material is distorted and is completely different from the first supposition. However, with the rapid development of visual analytical techniques, such as scanning electron microscopy (SEM), transmission electron microscopy (TEM) and atomic force microscopy (AFM), photoimages of many of the porous membranes are revealing structures completely different from those proposed in the models.

A simplified transport model is valid for all porous membranes irrespective of their actual microporous structure. The gas permeance, Q, through a porous membrane can be described by

$$Q\sqrt{MT} = ae^{\beta(V_c\sqrt{T_c})} \tag{1}$$

where M is the molecular weight of the gas, T is the absolute temperature, V_c is the critical volume of the gas and T_c is the critical temperature of the gas

[13]; α is the Knudsen diffusion factor and β is the surface diffusion factor, both of which are independent of the *real* geometric pore structures. Equation 1 allows the relative evaluation of gas transport properties of porous membranes regardless of their actual microporous structure.

12.2.2
Gas Transport in Non-porous Polymer Membranes

In general, polymer membranes for gas separation are designed on the basis of a solution–diffusion mechanism (Type V), using differences in the molecular properties of the gases such as diameter, shape or volume for gas diffusion and condensability or polarity for gas solution [1, 2, 7–9, 14, 15]. Gas molecules dissolve in the surface of a *dense* membrane and the dissolved molecules diffuse through the transient gaps between polymer chains.

According to the solution–diffusion mechanism, permeation (P) of gas A through a polymer membrane is a product of gas solubility (S) and gas diffusivity (D):

$$P_A = S_A D_A \tag{2}$$

The selectivity of gas A over gas B is defined as the ratio of their permeability coefficients, P_A/P_B:

$$a_{A/B} = \frac{P_A}{P_B} = \left(\frac{S_A}{S_B}\right)\left(\frac{D_A}{D_B}\right) \tag{3}$$

Selectivity is thus described as the product of the ratios of solubilities (S_A/S_B) and diffusivities (D_A/D_B). The first term is the solubility selectivity and the second is the diffusivity selectivity. To enhance the gas permeation selectivity, an increase in solubility selectivity and/or diffusivity selectivity is a prerequisite.

Gas separation membranes are required to display both high gas permeability and permeation selectivity. The relationship between gas permeability and permeation selectivity is presented in Fig. 12.2 [16]. In general, highly permeable polymers tend to show low selectivity and *vice versa*. Robeson proposed the upper bound line (i.e. the limit of separation properties) from these data in 1991 [16, 17]. Over the past 15 years, substantial research efforts have been directed to overcoming the limit imposed by the upper bound, for the development of gas separation polymers [18].

The relationship between gas diffusivity and gas solubility of common glassy polymers is presented in Fig. 12.3 [19], where dotted lines denote the carbon dioxide permeability [cm^3 (STP) cm cm^{-2} s^{-1} $cmHg^{-1}$]. These data were collected from carbon dioxide permeation experiments at 35 °C, at a feed pressure of 10 atm. Glassy polymers exhibit a narrow range of solubility coefficients compared with diffusion coefficients; for instance, the carbon dioxide solubility coefficients of most glassy polymers fall in the range 1–10 cm^3 (STP) cm^{-1} (poly-

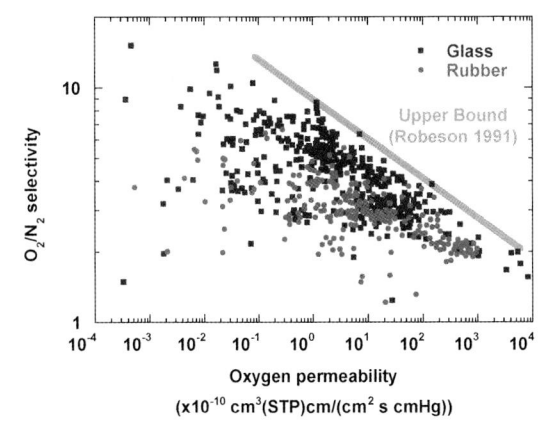

Fig. 12.2 Permselectivity vs permeability for the O_2–N_2 gas pair in rubbery and glassy polymers [16]. Reproduced with permission from Elsevier © 1991.

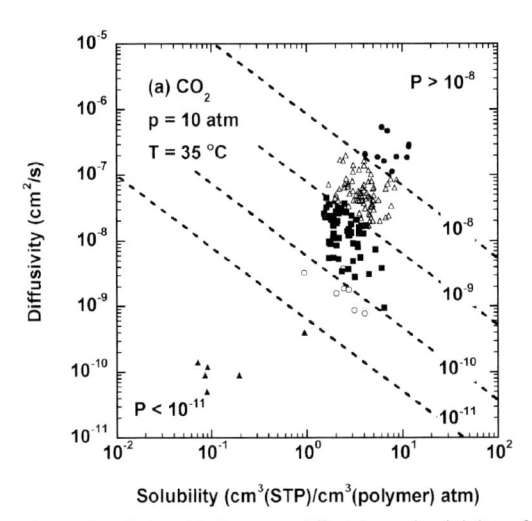

Fig. 12.3 Relationship between diffusivity and solubility of carbon dioxide in glassy polymers at 35 °C [19]. Gas permeability: $P > 10^{-8}$ (●); $10^{-8} > P > 10^{-9}$ (△); $10^{-9} > P > 10^{-10}$ (■); $10^{-10} > P > 10^{-11}$ (○); and $P < 10^{-11}$ (▲). Reproduced with permission from Elsevier © 2005.

mer) atm^{-1}, whereas the diffusivity varies from 10^{-6} to 10^{-11} cm^2 s^{-1}. Hence the carbon dioxide permeability depends more on the diffusivity than solubility and the same behavior is observed for the permeation of other gases. As a consequence, diffusivity selectivity is of prime significance in determining the gas selectivity and the fact that the diffusivity selectivity dominates the upper bound has been theoretically demonstrated by Freeman, as shown in Fig. 12.2.

When there is no or little interaction between the polymer and the gas molecules, gas permeation selectivity is primarily dependent on diffusivity selectivity rather than solubility selectivity [1, 2, 7–9, 14, 15]. This is because the gas solubility is strongly dependent on the gas condensability (e.g. boiling temperature, critical temperature, Lennard–Jones force constant) and is independent of the chemical structure [20, 21].

Notwithstanding the primary structure of the polymer, diffusion of the gas molecules across a polymer membrane is correlated to its fractional free volume (FFV) and this relation holds when there is no or little interaction between the gas molecules and the polymer chain:

$$D = A_D \exp\left(-\frac{B_D}{FFV}\right) \quad (4)$$

where A_D and B_D are adjustable parameters. The free volume theory is also applicable for gas permeation [22]:

$$P = A_P \exp\left(-\frac{B_P}{FFV}\right) \quad (5)$$

where A_P and B_P are adjustable parameters.

Figure 12.4 presents the carbon dioxide diffusivity and permeability of common glassy polymers as a function of FFV [19]. The diffusivity and permeability tend to decrease with increasing reciprocal FFV as expected from Eqs. 4 and 5. However, these data are scattered; as the carbon dioxide diffusivity varies from 10^{-6} to 10^{-8} cm^2 s^{-1} at a 1/FFV value of 5.5 (i.e. FFV=0.18), as shown in . 12.4a and better linear correlation coefficients were reported only for the structurally related polymers, such as polycarbonates, polysulfones and polyarylates. This behavior illustrates the limitations of the use of Eqs. 4 and 5 to estimate gas diffusion and permeability coefficients and suggests that accurate correlations can only be obtained for structurally related polymers.

These less accurate correlations are due to the distribution of free volume space, which can be estimated by positron annihilation lifetime spectroscopy (PALS) [23]. PALS data are analyzed using parameter τ_n (ns) as space size and parameter I_n (%) as the amount of space τ_n. The space size increases with an increasing n value and for most polymers it lies between τ_1 and τ_3 levels. So far, the τ_4 level space has only been reported for some substituted polyacetylenes [e.g. poly(1-trimethylsilyl-1-propyne) [poly(TMSP)] and fluorinated polymers [e.g. poly[2,2-bis(trifluoromethyl)-4,5-difluoro-1,3-dioxole-co-tetrafluoroethylene] [poly(TFE/PDD)] [23–25]. Gas diffusivity is correlated with the reciprocal of $\tau_3^3 \times I_n$ (ns^3 %), which is the total volume of τ_3 level space in a polymer membrane. Gas diffusion in poly(TMSP) and poly(TFE/PDD) is described in terms of $\tau_4^3 \times I_n$ and/or $\tau_3^3 \times I_n + \tau_4^3 \times I_n$ (ns^3 %), while the connectivity of τ_1–τ_4 level space across a polymer membrane is also an important factor in determining the gas diffusion.

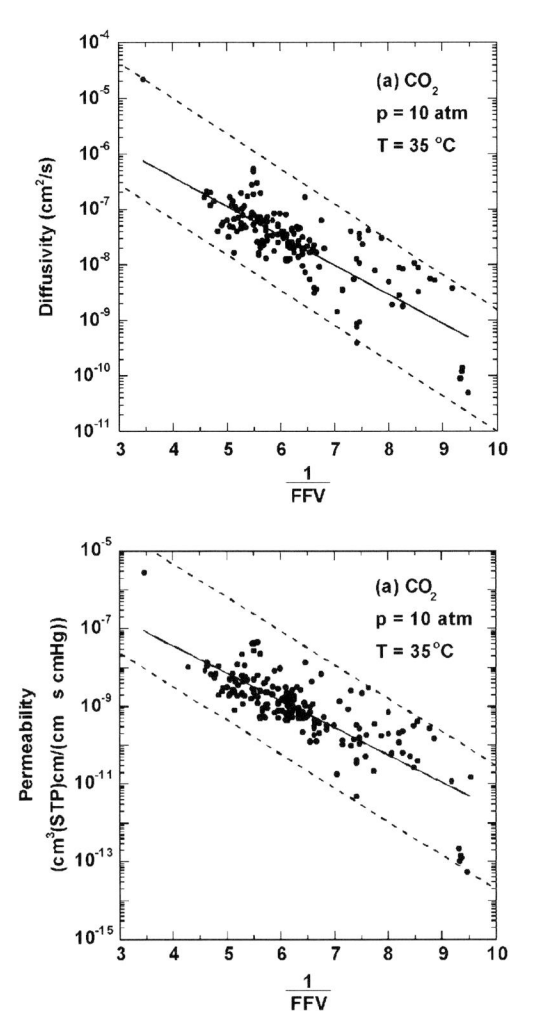

Fig. 12.4 Diffusivity and permeability of carbon dioxide in glassy polymers at 35 °C as a function of reciprocal fractional free volume (1/FFV) [19]. Lines represent least-squares fit to the data using Eqs. 4 and 5. Reproduced with permission from Elsevier © 2005.

As mentioned previously, the classical free volume theory does not take into account the effects of interaction between the gas molecules and the polymer chains; however, the dielectric constant of polymer membranes is correlated with FFV, which is strongly influenced by the polarity of polymers. Based on the Clausius–Mossotti equation, gas diffusivity and gas permeability defined as functions of the dielectric constant (ε) are given as

$$D = A \exp\left(\frac{-B}{1-a}\right) \tag{6}$$

$$P = A' \exp\left(\frac{-B}{1-a}\right) \tag{7}$$

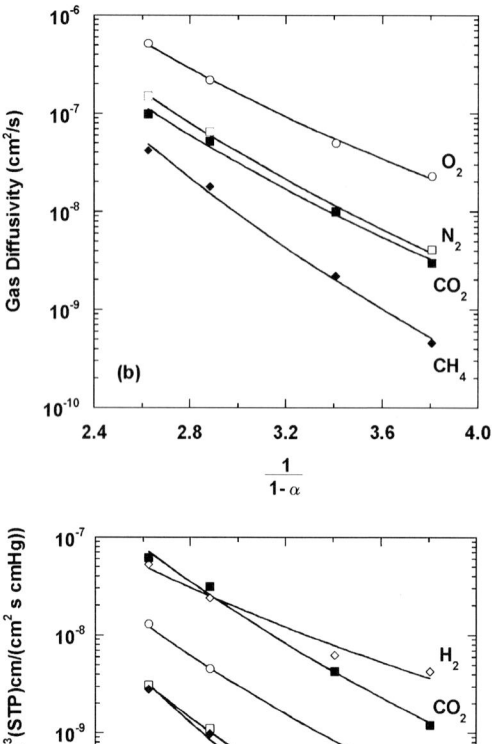

Fig. 12.5 Diffusivity and permeability of various gases in fluorine-containing polyimides at 30 °C as a function of reciprocal $1 - a$. Gases: hydrogen ($\diamond$), nitrogen ($\square$), oxygen ($\circ$), methane ($\blacklozenge$) and carbon dioxide ($\blacksquare$). Lines represent least-squares fit to the data using Eqs. 6 and 7.

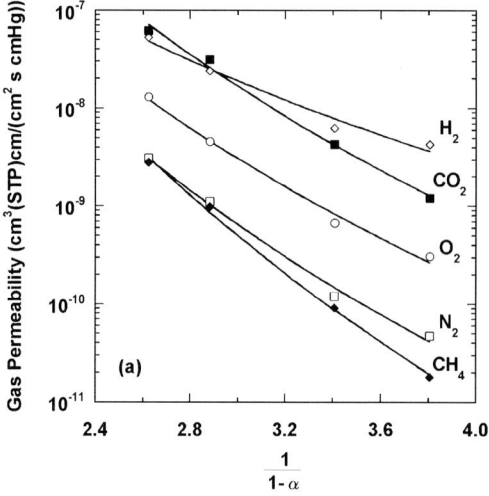

where

$$a = 1.3 \frac{V_W}{P_{LL}} \frac{\varepsilon - 1}{\varepsilon + 2}$$

A, A', B, B' are adjustable constants, V_W is the specific van der Waals volume and P_{LL} is the molar polarization [26].

The gas diffusivity and gas permeability of a family of 5,5'-[2,2,2-trifluoro-1-(trifluoromethyl)ethylidene]bis-1,3-isobenzofurandione (6FDA)-based polyimides as a function of the reciprocal of $1 - a$ are plotted in Fig. 12.5. The $1 - a$ values of 6FDA-based polyimides are not simply equal to their *FFV* values, as a takes into account the interaction between gas molecules and polymer chains, hence the $1-a$ values are 1.6-2.2 times larger than FFV.

12.2.3
Gas Permeability and Selectivity

The oxygen permeability coefficients (P_{O_2}) of some representative polymers are summarized in Table 12.1 [27–34]. Poly(1-trimethylsilyl-1-propyne) [poly(TMSP)] is the most gas-permeable polymeric material and its oxygen permeability is about 1×10^{-6} cm^3 (STP) cm cm^2 s cmHg at room temperature. On the other hand, polyacrylonitrile is the least permeable with an oxygen permeability of about 1×10^{-14} cm^3 (STP) cm cm^2 s cmHg. Different polymer structures give rise to different spaces in polymer segment aggregates (i.e. free volume) and a small change in the primary polymer structure brings forth a significant difference in the gas permeation properties, thus leading to applications of polymeric materials ranging from gas permeation membranes to packaging films. As explained previously, this permeability difference is due to the differences in gas diffusivity and gas solubility.

Substituted polyacetylenes are a class of highly gas-permeable glassy polymers [$P_{O_2} > 1000$ barrer; 1 barrer $= 1\times 10^{-10}$ cm^3 (STP) cm cm^2 s cmHg] and have been

Table 12.1 Oxygen permeability coefficients (P_{O_2}) of various polymer membranes.

Polymer	Temperature (°C)	P_{O_2} [$\times 10^{-10}$ cm^3 (STP cm cm^{-2} s^{-1} cmHg^{-1})]	Ref.
Poly(1-trimethylsilyl-1-propyne)	30	13 000	27
Poly[2,2-bis(trifluoromethyl)-4,5-difluoro-1,3-dioxole-co-tetrafluoroethylene]	35	960	28
Poly(dimethylsiloxane)	35	800	29
Poly(1,1-dihydroperfluorooctyl acrylate)	23	130	34
6FDA-TMPD[a]	35	110	30
Poly(propylene oxide)	30	83	31
Poly(bis-trifluoroethoxyphosphazene)	35	77	32
Natural rubber	30	23	33
Polystyrene	30	2.6	33
Polyethylene (density $=0.964$ g^{-3})	30	0.40	33
Poly(vinyl chloride)	30	0.045	33
Nylon 6	30	0.038	33
Poly(ethylene terephthalate)	30	0.035	33
Poly(vinylidene chloride)	30	0.0053	33
Polyacrylonitrile	30	0.0003	33

a) Polyimide synthesized using 2,2-bis(3,4-dicarboxyphenyl)hexafluoropropane dianhydride (6FDA) and 2,4,6-trimethyl-1,3-phenylenediamine (TMPD).

most extensively investigated as membrane-forming materials aiming at practical applications [4, 35–40]. These studies are motivated by the extremely high gas permeability of poly(TMSP) [36, 41], whose oxygen permeability (P_{O_2} = 4000–9000 barrer) is about 10 times larger than that of poly(dimethylsiloxane). Poly(TMSP) has enjoyed substantial prominence and been the focus of considerable attention from membrane scientists not only because of its high permeability but also owing to its ability to give a free-standing thin film. The gas permeation mechanism in poly(TMSP) is different from that of poly(dimethylsiloxane) (i.e. the dual-mode mechanism vs the solution-diffusion mechanism) [36]. According to the dual-mode sorption model, a polymer consists of a continuous chain matrix, along with microvoids (holes), frozen in the matrix; and these microvoids, present in both discrete and continuous domains, are caused by the non-equilibrium thermodynamic state of glassy polymers. Hence the solubility of a gas molecule in glassy polymers can be expressed in terms of Henry's law of solubility (dissolution in continuous chain matrix) and Langmuir-type sorption (sorption in microvoids) [19].

Examples of highly gas-permeable substituted polyacetylenes are listed in Table 12.2 [36, 42–56]. The P_{O_2} values and oxygen/nitrogen permselectivities ($P_{O_2}/$

Table 12.2 Oxygen permeability coefficients (P_{O_2}) and P_{O_2}/P_{N_2} of substituted polyacetylenes.

$\pm C=C\pm$ $R^1 R^2$		P_{O_2} (barrer[a])	P_{O_2}/P_{N_2}	Ref.
R^1	R^2			
Me	SiMe$_3$	4×10^3–9×10^3	1.8	36
Me	SiEt$_3$	860	2.0	42, 43
Me	SiMe$_2$Et	500	2.2	43–45
Me	SiMe$_2$-i-C$_3$H$_7$	460	2.7	43, 44
Me	GeMe$_3$	7800		46
Me	i-C$_3$H$_7$	2700	2.0	47
Ph	C$_6$H$_4$-p-SiMe$_3$	1100–1550	2.1	48, 49
Ph	Ph	910	2.2	50
Ph	C$_6$H$_4$-m-SiMe$_3$	1200	2.0	49
Ph	C$_6$H$_4$-m-GeMe$_3$	1100	2.0	51
Ph	C$_6$H$_4$-p-t-C$_4$H$_9$	1100	2.2	52
β-Naphthyl	C$_6$H$_4$-p-SiMe$_3$	3500	1.8	53
β-Naphthyl	Ph	4300	1.6	53
C$_6$H$_4$-p-F	C$_6$H$_4$-p-SiMe$_3$	2900	1.5	54
C$_6$H$_4$-p-F	Ph	3000	1.4	54
C$_6$H$_3$-m,p-F$_2$	C$_6$H$_4$-p-SiMe$_3$	3600	1.5	54
C$_6$H$_3$-m,p-F$_2$	Ph	3800	1.3	54
H	C$_6$H$_2$-2,4,5-(CF$_3$)$_3$	780	2.1	55
H	C$_6$H$_3$-2,5-(CF$_3$)$_2$	450	2.3	55
H	C$_6$H$_3$-o,p-(SiMe$_3$)$_2$	470	2.7	56

a) 1 barrer = 1×10^{-10} cm^3 (STP) cm cm^{-2} s^{-1} cmHg^{-1}.

P_{N_2}) (25 °C) of more than 100 substituted polyacetylene derivatives have been determined so far [36]. Among substituted polyacetylenes, those possessing spherical substituents such as *t*-Bu, Me$_3$Si and Me$_3$Ge groups display high gas permeability, whereas bulkier substituents such as Ph$_3$Si, *i*-Pr$_3$Si and adamantyl groups are rather unfavorable for high gas permeability. A majority of the less permeable polyacetylenes possess long *n*-alkyl groups. Moreover, the presence of the phenyl group as a main substituent results in considerably lower gas permeability. For the sake of comparison, the P_{O_2} values of (P_{O_2}/P_{N_2}) commercially available oxygen-permeable polymer membranes at 25 °C are as follows [40, 57]: poly(dimethylsiloxane) 600 barrer (2.0); poly(4-methyl-1-pentene) 32 barrer; natural rubber 23 barrer (2.3); and poly(oxy-2,6-dimethylphenylene) 15 barrer (5). The extraordinarily high gas permeability of polyacetylenes is attributed to the high free volume, unusual free volume distribution and the high local mobility of spherical substituents, which is presumably derived from their low cohesive energy structure, stiff main chain and spherical substituents.

High gas permeability is not a characteristic feature of poly(TMSP) only; poly(diphenylacetylene)s with spherical ring-substituents such as *p*-SiMe$_3$, *m*-SiMe$_3$, *m*-GeMe$_3$ and *p-t*-C$_4$H$_9$ also exhibit fairly high gas permeability coefficients [37, 49, 51, 52]. Whereas poly(diphenylacetylene) is insoluble in any solvent, its derivatives with bulky ring substituents are usually soluble in common organic solvents such as toluene and chloroform and afford free-standing membranes by solution casting. Poly(diphenylacetylene) derivatives show good thermal stability [onset temperature of weight loss (T_0) > 400 °C] and possess film-forming ability. The ease of modifying the ring substituents provides an opportunity to tune both the permeability and the solubility and glass transition temperature of the resulting polymers. The gas permeability of poly(diphenylacetylene)s displays a significant dependence on the shape of ring substituents and, generally, those with spherical ring substituents such as *t*-Bu, Me$_3$Si and Me$_3$Ge groups exhibit very large P_{O_2} values of up to 1000–1500 barrer, which is about one-quarter of that of poly(TMSP) and approximately twice as large as that of poly(dimethylsiloxane). Poly(phenylacetylene)s usually show lower gas permeability than do poly(diphenylacetylene)s, which is probably due to the stiffer structure of the latter.

Poly(diphenylacetylene) membrane has been prepared by the desilylation of poly[1-phenyl-2-*p*-(trimethylsilyl)phenylacetylene] membrane catalyzed by trifluoroacetic acid [58]. The prepared polymer membrane shows high thermal stability, insolubility in any solvent and high gas permeability (e.g. the oxygen permeability at 25 °C is 910 barrer) [50]. The high gas permeability of poly(diphenylacetylene) seems to arise from the presence of many microvoids in the polymer matrix generated due to the stiff polymer chain composed of alternating double bonds and the steric effect of the spherical pendent groups. In a similar way, poly(diphenylacetylene)s having various silyl groups, such as *i*-PrMe$_2$Si, Et$_3$Si, and *n*-C$_8$H$_{17}$Me$_2$Si, are soluble in common organic solvents and their membranes can be converted into poly(diphenylacetylene) membranes by desilylation; however, their oxygen permeability coefficients (120–3300 barrer) are different from one another irrespective of their very similar polymer structures. Moreover, it has been observed that the removal of the bulkier silyl groups leads to a greater increment in the oxygen permeability.

Poly[1-aryl-2-*p*-(trimethylsilyl)phenylacetylene]s [aryl=naphthyl, fluorenyl, phenanthryl] are soluble in common organic solvents and afford free-standing membranes [53, 59, 60]. These Si-containing polymer membranes are desilylated to give the corresponding membranes of poly(1-aryl-2-phenylacetylene)s. Both of the starting and desilylated polymers show very high thermal stability and high gas permeability; for instance, the T_0 and P_{O_2} values of poly(1-β-naphthyl-2-phenylacetylene) are 470 °C and 4300 barrer, respectively [53]. Among disubstituted polyacetylenes, poly(diphenylacetylene)s with silyl groups and fluorine atoms have proved to be the best candidates as highly gas-permeable materials [54]. The *FFV* of poly[1-*p*-fluorophenyl-2-*p*-(trimethylsilyl)phenylacetylene] is 0.28 and appreciably large; cf. 0.26 of poly[1-phenyl-2-*p*-(trimethylsilyl)-phenylacetylene] [54]. The P_{O_2} of poly[1-*p*-fluorophenyl-2-*p*-(trimethylsilyl)phenylacetylene] is as high as 2900 barrer, which is about twice that of poly[1-phenyl-2-*p*-(trimethylsilyl)phenylacetylene]. Therefore, the incorporation of fluorine atoms into poly[1-phenyl-2-*p*-(trimethylsilyl)phenylacetylene] generally results in a significantly enhanced gas permeability.

Disubstituted acetylenes with hydroxy groups cannot be polymerized because of the deactivation of Ta and Nb catalysts caused by the presence of polar moieties such as hydroxy groups [35, 39]. In contrast, 1-phenyl-2-*p*-(*tert*-butyldimethylsiloxy)phenylacetylene (**1a**), a protected monomer, polymerizes to give a high molecular weight polymer [61](Scheme 12.1). This polymer (**2a**) is soluble in common organic solvents and provides a free-standing membrane. Desilylation of the membrane **2a** yields a poly(diphenylacetylene) membrane having free hydroxy groups (**3a**), which furnishes the first example of a highly polar substituent-carrying poly(diphenylacetylene). Polymer **3a** is insoluble in nonpolar solvents such as toluene and chloroform, unlike the starting polymer (**2a**).

The P_{O_2} value of **2a** at 25 °C is 160 barrer (Table 12.3), which is relatively small compared with other poly(diphenylacetylene) derivatives and other gases

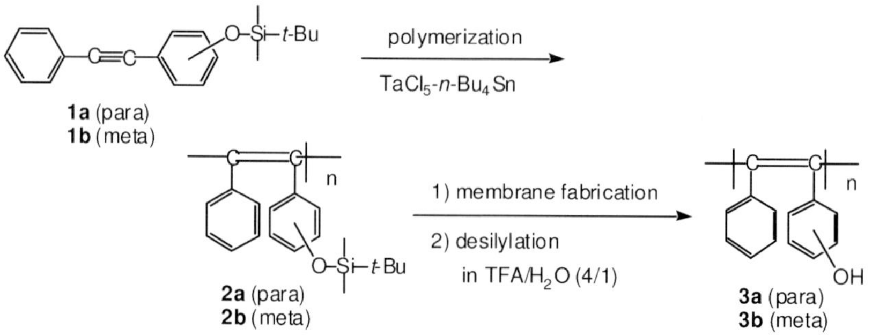

Scheme 12.1 Synthesis of *t*-BuMe$_2$SiO-containing poly(diphenylacetylene)s and OH-containing poly(diphenylacetylene)s.

Table 12.3 Gas permeability coefficients (P) of polymers **2** and **3** [a].

Polymer	P (barrer[b])						P_{O_2}/P_{N_2}	P_{CO_2}/P_{CH_4}	P_{CO_2}/P_{N_2}
	H_2	He	CO_2	O_2	N_2	CH_4			
2a [c]	330	170	810	160	50	160	3.2	5.1	16.2
2b [c]	380	210	880	190	67	170	2.8	5.2	13.1
3a [d]	56	38	110	8.0	2.4	2.3	3.3	47.8	45.8
3b [d]	86	46	130	15	5.1	9.6	2.9	13.5	25.5

a) P values measured at 25 °C.
b) 1 barrer = 1×10^{-10} cm^3 (STP) cm cm^{-2} s^{-1} cmHg^{-1}.
c) Methanol-conditioned.
d) Hexane-conditioned.

show similar behavior, whereas the permeability of **2b**, having a *m*-hydroxy group, is somewhat higher for every gas than that of **2a**. The values of desilylated polymers **3a** and **3b** are 8.0 and 15 barrer, respectively, which demonstrate a significant decrement in gas permeability upon desilylation, probably owing to the decrease in *FFV*. Generally, polymers bearing hydroxy groups such as poly(vinyl alcohol) (P_{O_2}=0.00665 barrer) exhibit very low gas permeability and can be utilized as gas-barrier membranes [43, 57]. When this fact is taken into account, the relatively high gas permeability of **3a** and **3b** suggests fairly sparse structures, in other words, large free volumes, as are common for sterically crowded substituted polyacetylenes. The P_{O_2}/P_{N_2} of **2a** and **3a** are 3.2 and 3.3, respectively, which are very close to each other and a similar tendency is observed in the case of desilylated derivatives **2b** and **3b**. In contrast, the separation factors of carbon dioxide and methane (P_{CO_2}/P_{CH_4}) and of carbon dioxide and nitrogen (P_{CO_2}/P_{N_2}) for polymers **3a** and **3b** are 14–48 which are appreciably large [18, 62], indicating a remarkable improvement in the CO_2 separation performance, as a result of desilylation. The increment in the separation factor for two gas pairs (P_{CO_2}/P_{CH_4}, P_{CO_2}/P_{N_2}) upon desilylation is fairly large for a para-substituted monomer in comparison with its meta-substituted counterpart. It is especially noteworthy that the P_{CO_2}/P_{CH_4} value for **3a** is located above the Robeson's upper bound [16]. The methane permeability (P_{CH_4}) undergoes a remarkable decrease upon desilylation, not only in comparison with CO_2 but also other small-sized gases (H_2, He), which is probably due to the bulky size and nonpolar character of methane gas.

12.2.4
Reverse-selective Polymer Membranes

Polymer membranes exhibit a general tendency to show small gas molecule selectivity, i.e. the small molecules permeate faster through a membrane relative to the larger ones due to the strong size-sieving effect. Most of the substituted polyacetylenes tend to show this behavior; however, some highly gas-permeable substituted polyacetylenes exhibit an interesting feature of large gas molecule selectivity in contrast to the general size-seizing effect [10, 11, 36, 63]. Substituted polyacetylenes furnish a rare example of a class of polymer membranes whose size-sieving ability can be switched from small gas molecules to larger ones by an appropriate selection of substituents.

In the case of poly(TMSP) and poly(4-methyl-1-pentyne) [poly(MP)] membranes, the blocking effect, resulting from gas condensation (Type III in Fig. 12.1), was observed in the separation of n-butane and methane from their mixture [36]. Condensable gas molecules occupy the free volume space in the polymer matrix, thus preventing gas diffusion through the membrane. According to PALS analysis, only the τ_4 level space contributes to the blocking behavior, thus suggesting that the polymers having the τ_4 level space possibly experience this effect. The selectivity of a binary gas mixture in common polymeric membranes [e.g. polysulfone as a glassy polymer and poly(dimethylsiloxane) as a rubbery polymer] is usually either equal to or lower than that observed with pure components; in contrast, substituted polyacetylenes exhibit a much higher n-butane/methane selectivity for the binary gas mixture than that obtained form pure gas measurements. For instance, the permeability coefficients of n-butane and methane in an n-butane–methane mixture through a poly(TMSP) membrane are about 30 and 90% lower than those for pure n-butane and methane, respectively, while selectivity in the binary gas mixture is six times higher than that determined for pure gases [64]. As a result, poly(TMSP) provides a selectivity value of 30 for n-butane–methane binary gas mixture, the highest ever displayed by any known polymer membrane for this binary mixture, with much higher permeability also. In the case of poly(TMSP), SF_6–He, SF_6–N_2, C_{3+} (alkanes higher than propane)–CH_4 and chlorofluorocarbon–N_2 mixtures serve as other examples of the binary gas mixtures for which such a blocking effect has been observed [64–67].

When impermeable filler is added to a polymer material, its gas permeability is reduced. One of the transport models for impermeable filler-containing polymer membranes is derived from Maxwell's equation [68]:

$$P_c = P_p \left(\frac{1 - \phi_f}{1 + \frac{\phi_f}{2}} \right) \tag{8}$$

where P_c is the permeability of the composite membrane, P_p is that of the pure polymer and ϕ_f is the volume fraction of the filler.

Unlike common polymers, an increase in the fumed silica content in poly(MP) leads to an increase in the τ_4 level space, thus resulting in the enhancement of mixed-gas selectivity and n-butane permeability [69].

Fumed silica powder acts as a reinforcing filler in polymer materials such as silicone rubber. The contact of polymer matrix on silica surface strongly affects the interaction between the polymer segments and the silica surface and, in

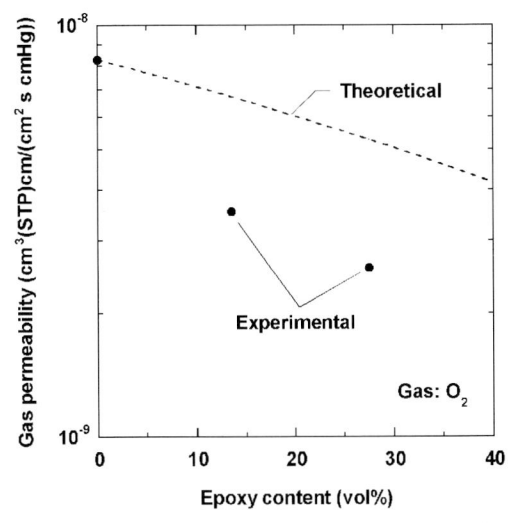

Fig. 12.6 Experimental (●) and theoretical (– – –) values for oxygen permeability coefficients at 30 °C as a function of filler in poly(propylene oxide) membranes containing epoxy resin as a filler [71]. Theoretical permeability coefficients were determined using Eq. 8. Reproduced with permission from the Polymer Society of Japan © 2005.

turn, gas solution behavior. For instance, the solubility in poly(dimethylsiloxane) (PDMS) membranes containing silica powder, bearing hydrophilic silanol groups on the surface, is smaller than the combined solubility in individual components (i.e. solubility in PDMS membranes without silica plus adsorption on silica powder) [70]. On the other hand, the solubility in PDMS membranes containing silica powder, having hydrophobic trimethylsiloxy groups on the surface, is larger than the combined solubility of these components. Interestingly, in the case of poly(propylene oxide) (PPO) membranes containing epoxy resin as an organic filler, the gas permeability decreases with increase in the filler content (Fig. 12.6) [71]. The permeability reduces to 45–50% of the theoretical value, estimated using Eq. 8 and is presumably due to the reduction in the molecular motion of polymer segments around the interface of epoxy resin filler.

It is argued that the filler interacts strongly with polymer segments to decrease the polymer chain mobility and occupies the free volume space leading to a decrease in the free volume of the polymer membrane. Unlike common polymers, the free volume space in poly(MP) increases upon addition of silica powder, which may be by virtue of enhanced free volume space in the polymer matrix regions and/or the large τ_4 level space created around the poly(MP)/silica interface; however, it has not been revealed yet whether the added silica powder substantially affects the chain mobility of poly(MP).

12.3
Water Purification Membranes

12.3.1
Water Purification Mechanisms

Water purification is the most successful industrial application of membrane technology since the 1960s [1]. The membrane-based processes for water purification can be classified into reverse osmosis, electrodialysis, microfiltration, ultrafiltration, nanofiltration and pervaporation. Reverse osmosis, microfiltration, ultrafiltration and electrodialysis are well-developed technologies and polymeric materials used for these membrane applications are conventional; for instance, cellulose acetate and polyamide for reverse osmosis, Teflon, poly(vinylidene fluoride), polypropylene, polyethylene, polycarbonate and cellulose acetate for microfiltration and cellulose acetate, polysulfone, poly(ether sulfone), polyimide, polyacrylonitrile, poly(vinyl alcohol) and poly(phenylene sulfide) for ultrafiltration [1, 72]. On the other hand, pervaporation and nanofiltration are still undergoing rapid development and a variety of polymers are being designed to serve various purposes.

One of the ways to classify the water purification membranes is primarily based on the pore size, which is correlated with transport and separation mechanisms. Figure 12.7 presents the pore sizes of membranes and the penetrant size along with the sizes of some small molecules. The pore size of the water

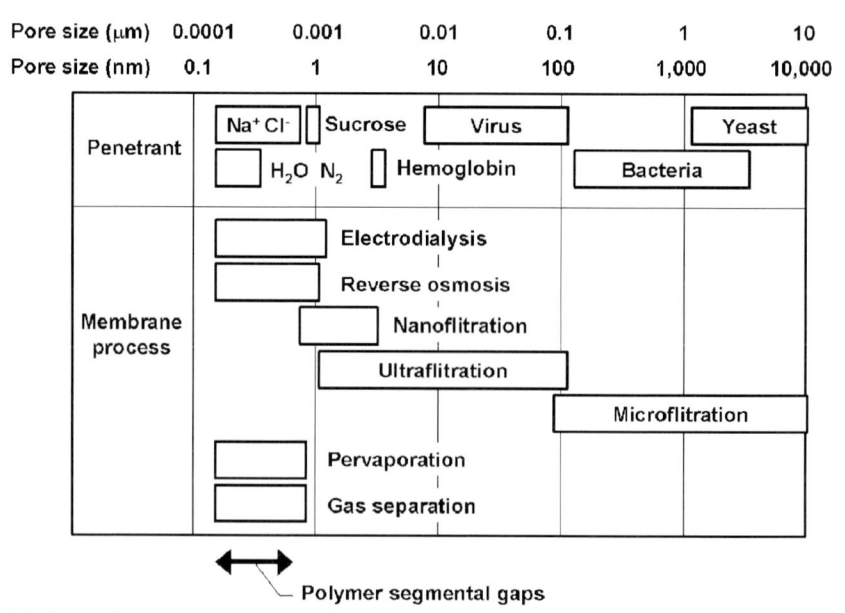

Fig. 12.7 Schematic representation of pore sizes of membranes and penetrant size.

separation membranes decreases in the order microfiltration > ultrafiltration > nanofiltration > electrodialysis > reverse osmosis = pervaporation.

The separation mechanism of microfiltration and ultrafiltration relies on molecular sieving through the pores. Microfiltration is a process to filter fine particles and bacteria, which have diameters between 0.1 and 10 µm, from water [1], whereas ultrafiltration membranes have pore diameters of 1–100 nm to separate water and microsolutes (e.g. proteins) from colloids. The pore sizes of nanofiltration membranes are intermediate between those of the membranes used for ultrafiltration and reverse osmosis. In contrast, reverse osmosis, electrodialysis and pervaporation membranes make use of both the polymer segmental gaps, about 0.3–0.5 nm in size and the micropores up to 1 nm. Reverse osmosis is used to bring about the separation of ions such as Na^+ and Cl^- and its separation mechanism is based on the solution–diffusion model. Electrodialysis makes use of ion-exchange membranes to effect separation and obeys the Donnan effect, i.e. the presence of impermeable ions on one side of the membrane affects the distribution of the permeable ions in a certain way. Pervaporation follows the solution–diffusion mechanism, using polymer segmental gaps for the separation of liquid mixtures. The pervaporation process provides the advantage of avoiding the azeotropic distillation of liquid mixtures and its fundamental theory is applicable to the evaluation of methanol crossover behavior, in direct methanol fuel cell applications [73].

According to the above-mentioned transport and separation mechanisms, pore size is a dominant factor for water purification by microfiltration, ultrafiltration and nanofiltration, regardless of the polymer structure. However, polymer characteristics are important to prevent the fouling of these membranes. Among water purification membranes, pervaporation is primarily dependent on the polymer characteristics; furthermore, a new strategy for the design of nanofiltration membranes involves an extensive dependence on the polymer properties.

12.3.2
Non-porous Nanofiltration Polymer Membranes

A non-porous membrane designed as a new type of nanofiltration membrane is illustrated in Fig. 12.8. Non-porous, microphase-separated block copolymers composed of fluorinated hydrophobic segments and hydrocarbon-based hydrophilic segments were synthesized for water purification applications. [74, 75] The fluorinated hydrophobic segments [e.g. poly(1,1-dihydroperfluorooctyl methacrylate) (PFOMA), poly(1,1,2,2-tetrahydroperfluorooctyl acrylate) (PTAN)] provide low surface energy and minimum adhesive sites to reduce membrane fouling, while the hydrophilic segments [e.g. poly(2-dimethylaminoethyl methacrylate) (PDMAEMA)] self-assemble to form hydrophilic channels for a high rate of water permeation through the membrane. A diblock copolymer, PDMAEMA-b-PFOMA, possesses irregular cylindrical morphologies, while triblock copolymers, PTAN-b-PDMAEMA-b-PTAN and PFOMA-b-PDMAEMA-b-PFOMA, have lamellar and cylindrical morphologies, respectively (Fig. 12.9) [74]. Water can ex-

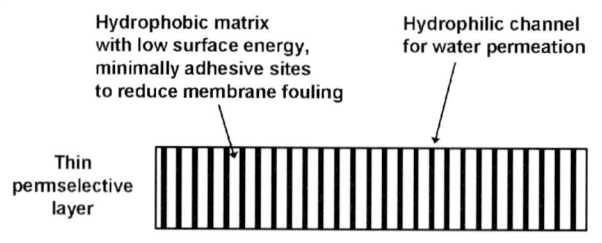

Fig. 12.8 Non-porous microphase-separated block copolymer nanofiltration membranes.

Labels: Hydrophobic matrix with low surface energy, minimally adhesive sites to reduce membrane fouling; Hydrophilic channel for water permeation; Thin permselective layer.

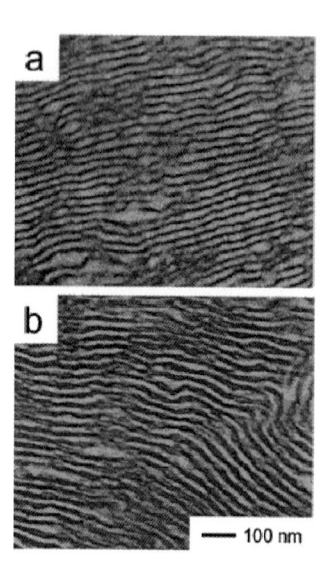

Fig. 12.9 Transmission electron micrographs of the PTAN-b-PDMAEMA-b-PTAN triblock copolymer with 69 mol% PDMAEMA (a) before and (b) after exposure to water [74]. An ordered lamellar morphology is observed in both images. Exposure to water appears to induce swelling of the PDMAEMA lamellae in (b). Reproduced with permission from the American Chemical Society © 2002.

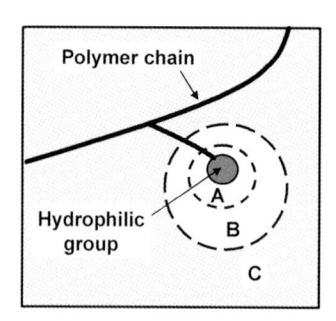

A : Non-freezing water
B : Bound water ⎫
C : Free water   ⎬ Freezing water

Fig. 12.10 States of water sorbed in a hydrophilic polymer membrane.

ist in multiple states after being sorbed in a polymer material. Based on the concept of two states of water as shown in Fig. 12.10, sorbed water is classified into *non-freezing water*, which interacts strongly with hydrophilic groups in a polymer and *freezing water*, which experiences rare interaction with these groups. Irrespective of the polymer morphology, salt partition and diffusion coefficients of these phase-separated block copolymers increase monotonically with the amount of freezing water in the hydrophilic domains, suggesting that the state of water in the phase-separated block copolymers is an important factor influencing their salt uptake and transport properties [75]. These copolymer membranes show less water flux decline due to fouling than the commercial membranes (e.g. Hydronautics NTR-7450) [76].

12.3.3
Pervaporation Polymer Membranes

As described previously, the mechanism of water purification using a membrane-based process involves a size-sieving filtration for the removal of small solids from water, while pervaporation is based on a solution–diffusion mechanism for the removal of volatile organic compounds (VOCs) from water. VOCs include BTX (benzene, toluene, xylene) and its analogues [77–80], halogen-containing compounds (e.g. trichloroethylene, chloroform) [79–86], alcohols (e.g. methanol, ethanol) [87–89], phenol [90–93], oxygen-containing compounds (e.g. acetic acid, butanone, 1,4-dioxane) [77, 94, 95], nitrogen-containing compounds (e.g. acrylonitrile) [77, 94] and endocrine disruptors [96, 97].

The permeability of a component A through a membrane can be expressed in terms of the diffusivity and solubility factors as follows:

$$P_A = K_A D_A \tag{9}$$

where D_A is the diffusion coefficient and K_A is the partition coefficient. The selectivity of component A over component B is defined as the ratio of the permeabilities of components A and B, P_A/P_B and can be expressed as

$$a_{(A/B)} = \frac{P_A}{P_B} = \left(\frac{K_A}{K_B}\right)\left(\frac{D_A}{D_B}\right) \tag{10}$$

where first term represents solubility selectivity and second one is concerned with diffusivity selectivity, hence an increase in solubility and/or diffusivity selectivity is a prerequisite for achieving an increase in overall selectivity.

In the case of polymer membranes, it is a general tendency for a family of polymers to exhibit either water selectivity [i.e. $(P_{VOC}/P_{H_2O}) < 1$] or VOC selectivity [i.e. $(P_{VOC}/P_{H_2O}) > 1$]. As the size of VOC molecules is larger than water, $(D_{VOC}/D_{H_2O}) < 1$ for the polymers in which separation is affected by the size sieving diffusion. On the other hand, the partition coefficient (K) is correlated with the solubility parameters of polymers and liquids (VOC/water). The affinity of a liquid (VOC/water) for a polymer membrane is determined by the extent of congruence between the solubility parameters of the two, hence hydrophilic polymers normally show $(K_{VOC}/K_{H_2O}) < 1$, whereas hydrophobic polymers tend to have $(K_{VOC}/K_{H_2O}) > 1$ (i.e. VOC-selectivity). In cases where diffusivity selectivity is the dominant contributing factor, $(K_{VOC}/K_{H_2O}) < (D_{VOC}/D_{H_2O})$, membrane show water-selective behavior even though $(K_{VOC}/K_{H_2O}) > 1$.

In the process of pervaporation, permeation through membrane is the rate-determining step; therefore, a minor component of the mixture must be removed by selective permeation through the membrane to minimize the amount of permeate liquid. Preferential water permeation (water selectivity) is the characteristic feature of most polymers and there have been only a few reports about VOC selective polymers. Substituted polyacetylenes furnish a remarkably interesting example of a class of polymers which exhibit both VOC and water selectivity [36].

Table 12.4 summarizes the total flux and ethanol/water selectivity of various VOC- and water-selective substituted polyacetylenes [88, 89]. Poly(TMSP) is a VOC-selective membrane with an ethanol/water selectivity of 4.5, whereas poly(1-phenyl-2-chloroacetylene) is water selective with a selectivity of 0.21. Poly(TMSP) shows $(K_{EtOH}/K_{H_2O}) > 1$ and $(D_{EtOH}/D_{H_2O}) < 1$ but, as the solubility selectivity is the factor which dominates the overall permeability selectivity in this case, $a_{EtOH/H_2O} = P_{EtOH}/P_{H_2O} > 1$ for poly(TMSP). Some of the polymers exhibit water selectivity despite their hydrophobic nature, as shown in Table 12.4. They preferentially sorb ethanol, hence the solubility selectivity would be $(K_{EtOH}/$

Table 12.4 Permeation rate (P) and separation factor [a(EtOH/H$_2$O)] of substituted polyacetylenes in ethanol–water pervaporation (EtOH feed concentration 10 wt.%) at 30 °C.

$-(CR=CR')_n-$		Downstream pressure (mmHg)	Thickness (μm)	[a(EtOH/H$_2$O)]	$P \times 10^3$ (g m m^{-2} h^{-1})	Ref.
R	R'					
(a) EtOH-selective						
Me	SiMe$_3$	1.0	~20	12	4.5	88
Ph	C$_6$H$_4$-p-SiMe$_3$	2.0	53	6.9	4.2	89
Ph	Ph	2.0	46	6.0	5.9	89
β-Naphthyl	C$_6$H$_4$-p-SiMe$_3$	2.0	32	5.3	6.9	89
Ph	β-Naphthyl	2.0	45	3.4	14	89
Cl	n-C$_6$H$_{13}$	1.0	~20	1.1	0.41	88
(b) Water-selective						
Me	n-C$_5$H$_{11}$	1.0	~20	0.72	0.57	88
H	t-Bu	1.0	~20	0.58	0.65	88
H	CH(n-C$_5$H$_{11}$)SiMe$_3$	1.0	~20	0.52	0.40	88
Me	Ph	1.0	~20	0.28	0.24	88
Cl	Ph	1.0	~20	0.21	0.23	88

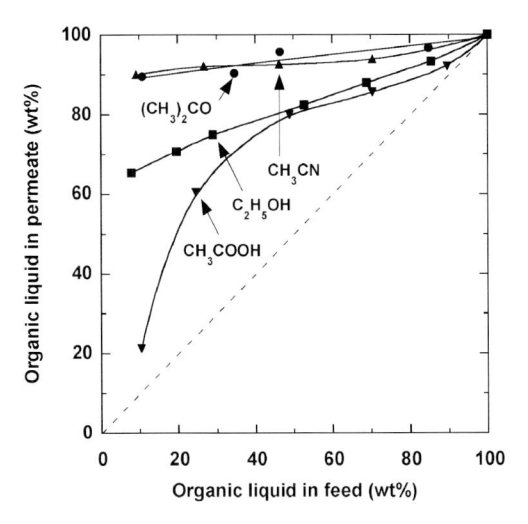

Fig. 12.11 Permeate composition curves for a poly(TMSP) membrane used for organic liquid–H$_2$O pervaporation at 30 °C [94]. Organic liquids: (CH$_3$)$_2$CO (●), CH$_3$CN (▲), C$_2$H$_5$OH (■) and CH$_3$COOH (▼). Reproduced with permission from Elsevier © 1990.

$K_{H_2O}) > 1$ but, as the overall permeation selectivity $(P_{EtOH}/P_{H_2O}) < 1$, the diffusivity effect is probably of greater significance in this case, $(K_{EtOH}/K_{H_2O}) < (D_{EtOH}/D_{H_2O})$. Poly(TMSP) displays faster permeation rates for acetone, acetonitrile and acetic acid, in comparison with water, as presented in Fig. 12.11 [94].

Grafting of a highly permeable rubbery polymer, poly(dimethylsiloxane), on to the α-methyl carbon of poly(TMSP) is the most effective way to enhance the ethanol/water separation efficiency of poly(TMSP) [98], e.g. in the separation of a 7 wt.% ethanol–water solution at 30 °C, poly(TMSP) has a flux of 1.2×10^{-3} g m m^{-2} h^{-1} and an ethanol/water selectivity of 11; however, in the presence of a 12 mol% poly(dimethylsiloxane) content in the graft copolymer, the flux increases to 2.5×10^{-3} g m m^{-2} h^{-1} and the ethanol/water selectivity is also enhanced to 28.

In the pervaporative separation of mixtures of water and various organic liquids (e.g. ethanol, toluene and trichloroethylene), poly(dimethylsiloxane) always exhibits organic liquid selectivity. As poly(dimethylsiloxane) has weak mechanical strength in comparison with the glassy polymers, it can be modified by the introduction of glassy polymer segments in the membrane, hence the permeate flux and separation properties are strongly influenced by the composition and the phase-separated structures. For example, phase-separated graft polymer membranes composed of poly(dimethylsiloxane) and poly(methyl methacrylate) segments have a transition point of their flux and benzene/water selectivity for a 0.05 wt.% benzene concentration in the mixture, at a poly(dimethylsiloxane) content of about 40 mol% in the membrane [79]. At this content, the continuous phase in the membrane changes from poly(methyl methacrylate) to poly(dimethylsiloxane) and, consequently, the flux and benzene/water selectivity undergo dramatic increases. Among the graft polymer membranes, one with a 68 mol% poly(dimethylsiloxane) content displays the highest benzene/water selectivity of 3730.

Furthermore, the pervaporation characteristics are significantly affected by the operating temperature. In the separation of an aqueous phenol solution through polyurethane membranes and of a 10 ppm 1,2-dibromo-3-chloropropane solution through cross-linked poly(dimethylsiloxane) membranes, the separation factor attains the maximum value at 60–70 °C [92, 97]. The water vapor pressure undergoes a drastic increase above 60–70 °C, leading to an increase in water diffusion and decreased (D_{VOC}/D_{H_2O}), thus resulting in an overall decline in (P_{VOC}/P_{H_2O}).

12.4
Membranes for Fuel Cells

12.4.1
Conventional Technologies

Up to now, sulfonated polymers have been used in significant technical applications such as in ion-exchange resins [99, 100], electrodialysis [101], bipolar membranes [102], sensors [103] and pervaporation [104]. Sulfonated polymer membranes have recently been studied extensively in the fuel cell field as potential polymer electrolytes. Unquestionably, the polymer electrolyte membrane fuel cell (PEMFC) and the direct methanol fuel cell (DMFC) are alternative power sources for portable applications, remote power generation (RPG) and transportation vehicles. This is mainly because of their power density, simplicity, flexible size and environmental benefits. Sulfonated polymer membranes and their derivatives are a key element of the fuel cell system. In order to realize quickly the practical applications of this promising technology, the sulfonated polymer membrane should be ameliorated with high priority.

To date, perfluorinated sulfonated ionomers (PFSA), such as Nafion, Aciplex and Flemion, have been perceived as the only prevalent proton-exchange membranes that meet the various technical requirements in fuel cell applications. The perfluorinated ionomers show high proton conductivity in a fully hydrated state and excellent chemical and mechanical stability even under harsh fuel cell conditions. However, the PFSA membranes are expensive (ca. US$ 500 m^{-2}), which contributes to the total cost of a fuel cell system. Furthermore, the PFSA membranes suffer from decreased proton conductivity at elevated temperatures (>80 °C), high fuel permeability, oxidative degradation under dry conditions and a high electro-osmotic drag.

Despite the fact that the first hydrocarbon-based-polymer membranes containing acid functional groups (i.e. sulfonated phenol–formaldehyde resin or sulfonated styrene) [99, 100, 105] did not surpass the high chemical and morphological stability of the PFSA membranes, the availability of sulfonated aromatic polymers with excellent oxidative resistance and low production cost brought non-fluorinated polymer membranes back into focus. The initial motivation was to mimic the properties of the PFSA membranes using cheap polymer materials. The distinct chemical and microstructural properties of this class of polyarylenes give rise to certain characteristic limitations and challenges in obtaining optimal membrane performance.

12.4.2
Membrane Performance Relevant for Fuel Cell Applications

Proton-exchange membranes (PEMs) should be an effective barrier to fuel molecules such as methanol and hydrogen. This implies chemical and morphological stability and a low gas or liquid permeability. Additionally, PEMs should also transport positive ions such as a proton or hydronium ion (~ 0.1 S cm^{-1} in the hydrated state). That is, the second requirement is high proton conductivity, which is closely related to the water content of the membranes. These functions of a successful PEM should be maintained under a variety of operating conditions and over the entire lifetime of the fuel cell. Moreover, the critical criteria for appropriate PEMs include excellent hydrolytic stability under acidic conditions, high mechanical strength (≥ 20 MPa in the dry state), high dimensional stability (<5%) and thermal stability at elevated temperatures (at least up to 200 °C). These requirements depend significantly on the operating conditions such as temperature, fuel type, humidity, fuel flow-rate, the interfacial properties of the membrane–electrode assembly (MEA), the transport in the gas diffusion layer (GDL) and the electrical current drained from the fuel cell [106]. Accordingly, proper polymer materials should be carefully selected considering their characteristic properties and the final sulfonated polymers should be fabricated using a modification of the chemical structure of the selected polymers to maintain an acceptable membrane performance under harsh conditions.

12.4.3
Aromatic Polymers with Sulfonic Acid Groups in the Their Main Chains

Among the promising membrane materials used for fuel cells, sulfonated aromatic polymers have attracted attention because of their high proton conductivity, high processability and excellent thermal, chemical and mechanical stability. These sulfonated aromatic polymers and their related derivatives include sulfonated polysulfone [107], sulfonated polyphenylsulfone, sulfonated polyarylene ether sulfone [108], sulfonated poly(ether ketone) [109], sulfonated poly(phenoxybenzoylphenylene) [110, 111], sulfonated polyimide [112–114], sulfonated polyphosphazene [115, 116] and sulfonated polybenzimidazole [117]. Even though the properties of these polymers, such as proton conductivity, fuel permeability and membrane stability, were measured under conditions different from those of a real fuel cell, the results can be taken as guidelines for the selection of a feasible polymer backbone.

12.4.3.1 Processability

Sulfonated aromatic polymers are fabricated by combining the hydrophobic backbone with highly hydrophilic groups (–SO$_3$H). While most sulfonated polyarylene polymers are soluble in common organic solvents, irrespective of the presence of salts in the polymers, polyimides and polybenzimidazoles with high T_g have a relatively low solubility even in the salt form. Their low solubility is related to poor processability, indicating a limitation on the mass production of such polymer membranes and their integration with a catalyst layer – MEA. According to the literature relating to sulfonated polyimides (SPIs) [112–114], the solubility of the sulfonated polymers may be changed through a modification of the chemical structure containing hydrophilic and hydrophobic moieties, as shown in Fig. 12.12. Bulky groups and flexible chains in SPIS can improve the solubility and random copolyimides show better solubility than block copolyimides. A homopolymer composed of only hydrophobic moieties is insoluble in

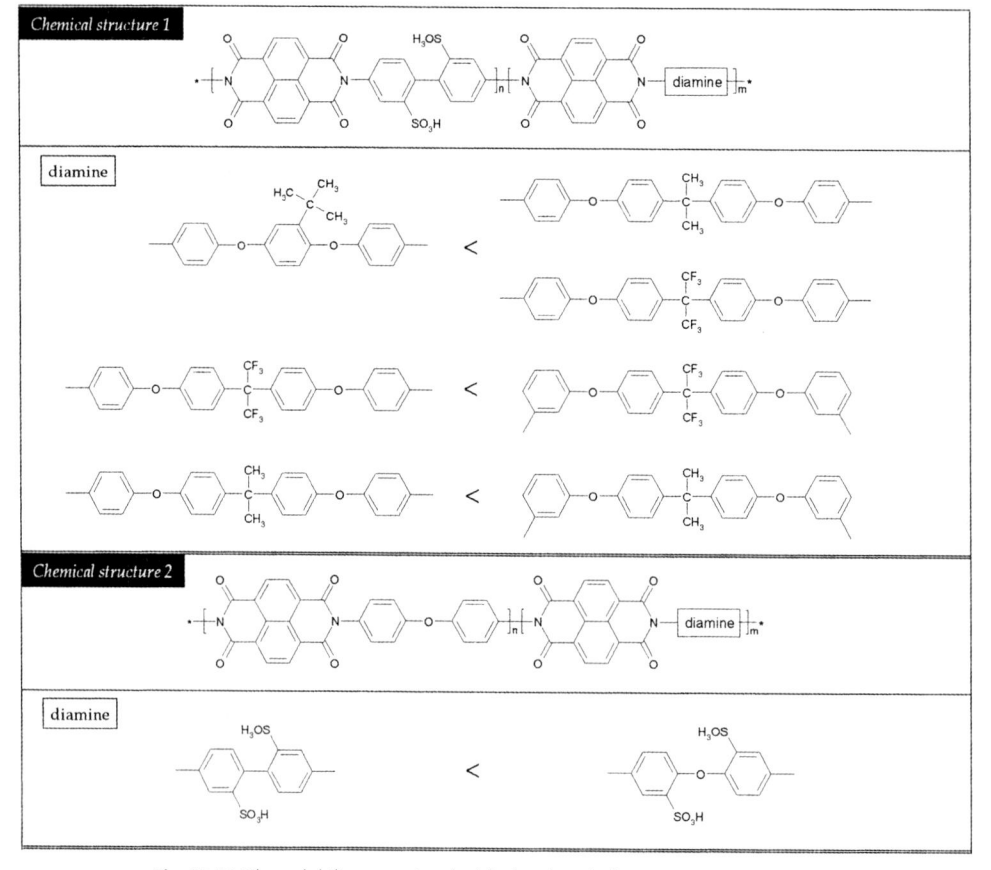

Fig. 12.12 The solubility associated with the chemical structures of SPIs in various organic solvents [112–114].

organic solvents, whereas block copolyimides which contain hydrophilic moieties show enhanced solubility to some extent. The results indicate that hydrophobic sequences are an important factor that affects the solubility of the block copolyimides. Furthermore, solubility studies revealed that SPIs derived from meta-substituted monomers exhibit better solubility than those derived from para-substituted monomers. Most of the sulfonated polymers described above have a salt form such as triethylammonium. Protonation of the sulfonic acid groups leads to an improvement in polymer solubility.

12.4.3.2 Water Sorption Behavior

The existence of water molecules in the sulfonated polymer membranes considerably affects the proton transport along with the formation of a hydrated structure around the negatively charged fixed ions, including sulfonic acid groups. Sufficient water uptake and its successful preservation in the membrane are a requirement for consecutive proton conduction because of the strong dependence of proton conductivity on the connectivity of the hydrophilic domains. Sulfonated polymers with a high degree of sulfonation or a high IEC show excellent proton conductivity above 10^{-2} S cm^{-1}. However, the highly sulfonated polymers simultaneously exhibit weak mechanical properties and poor hydrolytic stability, which makes MEA formation difficult. Hence it is necessary to maintain appropriate water uptake using hydrophilic domains responsible for the transport of protons and water molecules and hydrophobic domains that provide morphological stability and prevent dissolution in water.

The effect of humidity on water uptake in the sulfonated polymers should also be carefully considered because the proton conductivity across sulfonated polymers is strongly dependent upon the amount of water uptake. If the water uptake fluctuates in membranes due to even a slight change of humidity, it will considerably affect fuel cell performance. The hydration behavior of sulfonated aromatic polymers at low relative humidity (RH) is similar to that of Nafion. However, at 100% RH, Nafion has a high degree of water uptake owing to the high polarity of the sulfonic acid groups. This peculiar water uptake behavior of Nafion induces a higher proton conductivity than in other sulfonated polymers with the same IEC (0.91 mequiv. g^{-1}) at high RH, including liquid water. As systematically illustrated in Fig. 12.13 [118], the microstructure of the sulfonated poly(ether ether ketone) (SPEEK) is different from that of Nafion: SPEEK has narrow hydrophilic channels which are less separated and more branched with more dead-end water pockets than those of Nafion. These properties resulted in a large interface and a large micro-separation derived from the hydrophilic and hydrophobic domains [119]. Fixed-charged ions such as –SO$_3$H gave rise to a strong interaction with the water molecules in the narrow channels of the sulfonated aromatic polymers, which maintain a constant water retention level in their water pockets and compensate for the reduction of the proton conductivity owing to water vaporization at elevated temperatures.

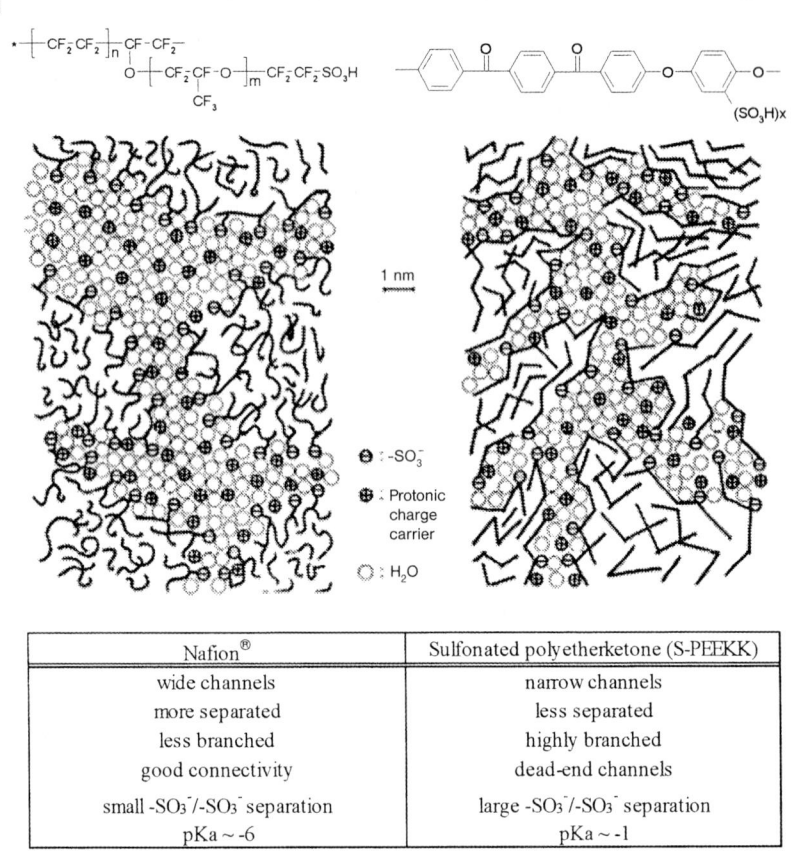

Nafion®	Sulfonated polyetherketone (S-PEEKK)
wide channels	narrow channels
more separated	less separated
less branched	highly branched
good connectivity	dead-end channels
small -SO_3^-/-SO_3^- separation	large -SO_3^-/-SO_3^- separation
pKa ~ -6	pKa ~ -1

Fig. 12.13 Schematic diagram of the microstructure of Nafion and a sulfonated poly(ether ether ketone) (SPEEK) derived from SAXS measurements [118].

The state of the water and also the water uptake should also be considered to improve proton conduction across the sulfonated polymer membranes. Generally, the state of the water in the sulfonated polymer can be categorized as either free water or bound water. Bound water can be classified into non-freezing bound water and freezing bound water [120]. Non-freezing bound water is derived from a strong interaction between the water molecules and the polymer matrix. The maximum volume of the non-freezing bound water is typically dependent on the polarity and the content of ionic groups in the sulfonated polymer. On the other hand, freezing bound water is defined as water that arises from a weak interaction with the hydrophilic polymer and free water as bulk water unaffected by the polymer matrix. The amount of freezing bound water and free water is obtained from an integration of the DSC endothermic peak area at around 0 °C. Then, the amount of bound water arising from hydrogen bonding with the hydrophilic groups in the sulfonated polymers is calculated using the difference of the total water uptake and the freezable water (freezing bound water and free water). Consequently, the state of water in the sulfonated polymers is significantly affected by total water uptake, which is strongly dependent on the polymer structure. The rigid-rod polymer structure in the sulfonated polymers causes a reduction in the free volume capable of containing water molecules [121]. According to the vehicle mechanism, the free water acts as a proton carrier and evaporates from the membranes at elevated temperatures. The existence of non-volatile bound water affects an increase of proton conductivity at temperatures above 80 °C. The ratio of bound water to the total sorbed water increases with a higher content of ionic groups in the sulfonated polymers. Furthermore, the incorporation of inorganic fillers including high-polarity silica can contribute to an enhancement of the amount of hydrogen-bonded water molecules – the non-freezing bound water – in sulfonated, polymer-based organic–inorganic composite membranes [122, 123]. The free water content is an important factor that determines the methanol permeation properties [127]. A low fraction of free water in the membranes generally leads to a low electro-osmotic drag and consequently to a low methanol permeability. In other words, methanol permeability can be lowered by a reduction of the free water content in the membranes.

12.4.3.3 Proton Conductivity

Membrane hydration is a critical factor in determining the proton conductivity. A high water content in the sulfonated polymers generally guarantees excellent proton conductivity. At temperatures exceeding 80 °C, the rapid rate of water vaporization reduces the proton conductivity. Therefore, sulfonated polymers with high water uptake are desirable as they maintain good proton conductivity. However, high water uptake can induce unexpected side-effects: (1) low mechanical strength, (2) poor hydrolytic stability, (3) low dimensional stability and (4) high methanol permeability in DMFC applications. Maintaining an appropriate level of hydration is necessary to ensure proton conductivities over 10^{-2} S cm^{-1} and balancing the unfavorable effects of excessive hydration at the same time.

Since the transport of water and protonic charge carriers (H_3O^+ and $H_5O_2^+$) takes place in the hydrated hydrophilic domains, the resultant proton transport behavior should be understood qualitatively on the basis of the microstructural data. In a heterogeneous system, the transport behavior of the protons is dependent on length at a molecular scale (<1 nm) and a single hydrophilic channel (>1 nm) [106]. On a scale smaller than the size of the channels (<1 nm), the diffusion of water and protonic charge carriers is reminiscent of the situation found in aqueous acid solutions. Namely, the structural diffusion occurs along with the characteristic proton mobility including the intermolecular proton transfer and the hydrogen bond breaking and forming processes [128–130]. On a single channel scale (>1 nm), the fixed charge anion ($-SO_3^-$) is distributed at the hydrophobic/hydrophilic interface of the polymer microstructure, which forms space charge layers – hydrophilic channels along with the interface composed of the immobile anions and the mobile protonic charge carriers. The electrical field around the fixed charged anions may localize even the dissociated protonic charge carriers including the bound water within the vicinity of their anionic counter ions. The enhanced localization of dissociated protonic charge carriers maintains the water retention level. It also prohibits any reduction of proton conductivity through fast vaporization of the water molecules in the sulfonated polymer at high temperature and/or low humidity. This effect additionally contributes the activation energy for the proton conductivity (E_a) at elevated temperatures. As a result, sulfonated polymers with a high content of localized water have a relatively low dependence of the proton conductivity on temperature. PFSA membranes exhibit a remarkable decrease in proton conductivity over 80 °C because the drying-out of the PFSA membrane leads to the collapse of the hydrophilic channels and impedes proton transport [124]. However, most of the sulfonated aromatic polymers exhibit a slight increase in proton conductivity even at elevated temperatures under low relative humidity (about 50% RH) [131]. This phenomenon is attributable to the relatively higher content of bound water in the aromatic sulfonated polymers than those in the PFSA membranes.

12.4.3.4 Fuel Permeability

The proton transport phenomena occurring in the sulfonated polymer membranes are very complicated, but two theories can be distinguished – the vehicle mechanism and the hopping mechanism. Both can explain proton conduction in the hydrated membrane system. According to the vehicle mechanism, protons can combine with molecules such as water and methanol and then form complex ions such as H_3O^+ and $CH_3OH_2^+$. After that, the protons are transported along with the complex ions. Moreover, protons can be transported between ionic sites such as sulfonic acids via the hopping mechanism. On the other hand, large molecules including methanol can permeate through the relatively broad hydrophilic channels [125]. Generally, water–methanol separation at a low water concentration can be effectively achieved using hydrophilic polymer membranes such as poly(vinyl alcohol) (PVA) [132], poly(acrylic acid) (PAAc) [133] and chitosan [134]. However, an increase in the hydrophilicity in the polymers results in excessive water swelling, which eventually leads to a low water–methanol separation factor. High methanol permeation is a serious obstacle in proton exchange membranes for DMFC. Methanol transport through hydrophilic polymer membranes can be controlled by factors such as hydrophilic channel size, water uptake, membrane compaction and other operational conditions [135]. The overall methanol permeation behavior is related to the water sorption and the proton conduction behavior. In other words, highly methanol-permeable membranes exhibit excellent proton conductivity owing to their strong hydrophilicity, whereas membranes with methanol barrier properties never guarantee high proton conductivity. To achieve desirable membrane performances for DMFC, the membrane morphology and microstructure should be controlled considering the trade-off relationship between proton and methanol transport.

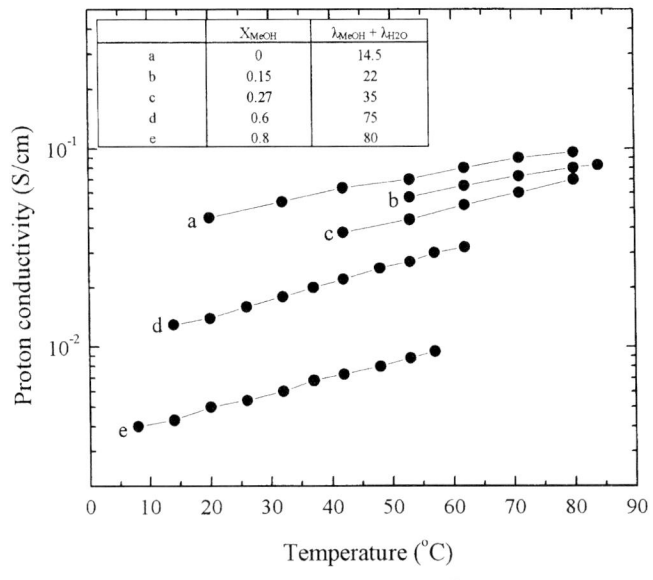

Fig. 12.14 Proton conductivity of Nafion 117 at different temperatures after soaking in water–methanol mixtures of various methanol contents at 60 °C [106]. Reproduced with permission from John Wiley & Sons © 2003.

	X_{MeOH}	$\lambda_{MeOH} + \lambda_{H2O}$
a	0	14.5
b	0.15	22
c	0.27	35
d	0.6	75
e	0.8	80

The swelling behavior as a function of the water–methanol mixture shows the tremendous effect of methanol on the membrane's morphology. For sulfonated polymers, including sulfonated poly(ether ketone) and Nafion, an increase in the methanol concentration in the water–methanol mixture leads to a high methanol uptake and subsequent swelling of the membrane. The methanol molecules, which contain the hydrophilic –OH group and the hydrophobic –CH_3 group, play an important role as a surfactant in improving the wettability of the sulfonated polymers by opening up the hydrophilic microstructure. Highly sulfonated polymers with a high IEC (> ∼ 1.4 mequiv. g^{-1}) may even be dissolved in the water–methanol mixture. This increased swelling significantly affects the electro-osmotic drag. While sulfonated polymers exhibit a higher swelling ratio than Nafion, their electro-osmotic drag increases with higher methanol concentrations [106]. Interestingly, as shown in Fig. 12.14, a high methanol content gives rise to the opposite effect on the proton conduction through the membrane. Methanol prohibits the dissociation of the acidic functional group (–$SO_3^-H^+$) in the sulfonated polymers, which reduces their proton conductivity. The decrease in the proton conductivity becomes marked as the mole fraction of methanol (X_{MeOH}) increases. This tendency becomes severe in most of the sulfonated aromatic polymers owing to the low acidity of the sulfonic groups in the aromatic rings.

12.4.3.5 Hydrolytic Stability and Oxidative Radical Stability

The hydrolytic stability of the PEMs can be one of the barometers used to estimate membrane durability in relation to the maintenance of their electrochemical and chemical performances during fuel cell operation. The membrane stability against attack by hydrolysis is examined by measuring the change in their intrinsic properties such as proton conductivity, IEC and methanol permeability under operational conditions similar to those in a PEMFC or DMFC [125]. As described above, PEMs should simultaneously be effective proton conductors and fuel barriers. A high degree of sulfonation is required to improve the proton conductivity of the PEMs owing to a large fixed ion and proton carrier concentration. Unfortunately, this makes the PEM excessively swollen or soluble in aqueous solutions. This dimensional instability of the PEMs brings about difficulties in MEA fabrication. In addition, highly sulfonated polymers also show poor hydrolytic stability and result in brittleness or cracking within a few days due to chain scission after immersion into acidic water. Even in the MEAs based on highly sulfonated aromatic polymers, the reliability of the electrochemical performances decreases with increasing contact resistance derived from the delamination problem in the interface between the catalytic layers and the membrane and the reduction of the molecular weight and acid functionality in the polymers. These phenomena can be easily observed in most of the highly sulfonated polyarylenes. It has been reported that sulfonated polyimides with five-membered phthalic imide structures have lower hydrolytic stability than other sulfonated aromatic polymers. The poor hydrolytic stability of the sulfonated polyimides limits their long-term operation in fuel cells. However, this problem can be solved to some extent via a chemical modification of the polymer backbones using hydrolytically durable monomers such as the six-membered naphthalenic dianhydride [112, 113, 136], sulfonated and non-sulfonated diamines containing a swivel group (–O– and –S–) and diamines which have sulfonic acid groups attached to their bridged phenyl rings [114, 137].

The sensitivity of polymer chains to free radical cleavage, which occurs under the aggressive conditions that exist at the electrode interfaces, can be another gauge which indicates the membrane stability with respect to oxidation. Oxidative stability is characterized by measuring the change in a membrane's properties, including proton conductivity, mechanical strength and color after immersion in Fenton's reagents (0.1 wt.% $FeSO_4$ in 30% H_2O_2). The earliest PEM, sulfonated polystyrene, had a relatively short lifetime mainly because of the low resistance of its aliphatic carbon-carbon backbone bonds to free radical attack [138]. The PFSA membranes are stable with respect to free radical skeletal cleavage reactions as compared with most of the sulfonated aromatic polymers. The inorganic components contribute significantly to an improvement in the oxidative radical stability of the sulfonated polymers. Sulfonated polyphosphazenes with an inorganic backbone are particularly stable to active peroxide radicals ($^\cdot OH$) [139]. The presence of nanosized fumed silica can compensate for the restricted oxidative stability of the pristine membranes required for the continuous electrochemical performances in a fuel cell [122, 123]. Additionally, sulfonated polymers containing flexible groups display fairly good stability to degradative peroxide radicals as compared with other rigid group-based sulfonated polymers, whereas the block copolymers show rather poor oxidative stability [114, 137].

12.4.4
Modified Aromatic Polymer Containing Sulfonic Acid Groups

Based on the difference in chemical structure and/or microstructure of the PFSAs and aromatic sulfonated polymers along with the membrane performances, chemical modifications such as blending, cross-linking and grafting have been envisaged to overcome the limitations of the intrinsic properties of sulfonated aromatic polymers. Cross-linking, in particular, can be a simple and powerful way to control the indispensable properties in PEM, including fuel

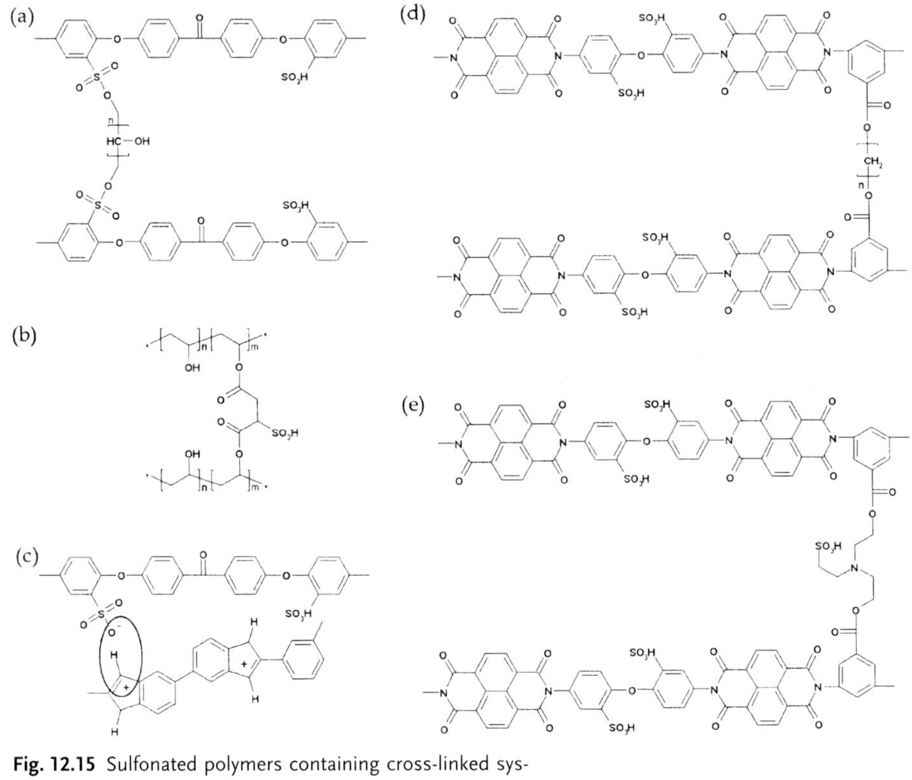

Fig. 12.15 Sulfonated polymers containing cross-linked systems derived from (a) SPEEK and polyatomic alcohols [141], (b) PVA and sulfosuccinic acid [144], (c) SPEEK and polybenzimidazole (PBI) [146], (d) sulfonated polyimide and aliphatic alcohols of different chain lengths [145] and (e) sulfonated polyimide and BES [125].

permeability, mechanical properties and dimensional stability. So far, there have been several investigations into the cross-linking of sulfonated polymers, including the thermally activated cross-linking of polyatomic alcohols and sulfonic acid groups in the sulfonated poly(ether ether ketone) (SPEEK) [140, 141], the UV-assisted photo-cross-linking of sulfonated polyphosphazene (PPh) [116, 142, 143], the chemical cross-linking of poly(vinyl alcohol) (PVA) and sulfosuccinic acid [121, 144], the thermochemical cross-linking of simple diols of different chain lengths or N,N-bis(2-hydroxylethyl)-2-aminoethanesulfonic acid (BES) with carboxylic acid groups in the sulfonated polyimide [122–126, 145] and the ionic cross-linking of acid–base blend ionomers [146, 147]. In most cases, as shown in Fig. 12.15, the cross-linking resulted in a significant improvement in the chemical and mechanical stability, but the proton conductivity and the methanol permeability decreased with a high degree of cross-linking owing to the increased cross-linking density and reduced water uptake.

Most of the membrane performances in cross-linked polymer systems are strongly related to the chain length of the cross-linker and cross-linker type [125, 145]. Originally, it was expected that longer cross-linked chains would create large vacancies between the rigid polymer chains and might lead to a higher water uptake if the degree of cross-linking remained constant. However, the water sorption in the cross-linked sulfonated polyimide (XSPI) showed a V-shape as the number (n) in the diol [HO–$(CH_2)_n$–OH, $n=2$–10] cross-linker increased [145]. Here, the water uptake decreased up to $n=4$ and then increased

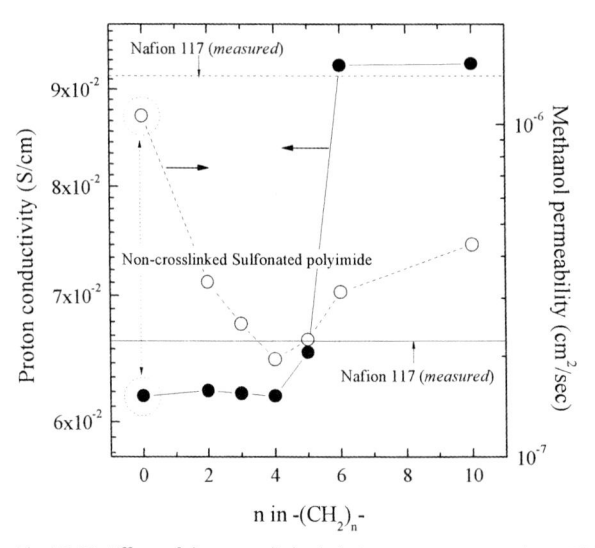

Fig. 12.16 Effect of the cross-linked chain length on the proton conductivity (solid line, measured at 30 °C and 90% RH using four-point probe a.c. impedance spectroscopy) and the methanol permeability (dashed line, measured at 25 °C using the two-chamber diffusion cell method) in sulfonated polyimide membranes containing a diol [HO–$(CH_2)_n$–OH] as a cross-linker [145].

from $n=4$ to 10. This trend in the proton conductivity and methanol permeability mirrored the characteristic water uptake behavior. Figure 12.16 shows the effect of the cross-linked chain length on the proton conductivity and methanol permeability in the XSPI membranes. The proton conductivity of the XSPI membranes remained constant or decreased very slightly up to $n=4$. However, an increase in the cross-linked chain length ($n>4$) considerably influenced the change in the proton conductivity of the XSPI. The proton conductivity increased drastically up to 9.3×10^{-2} S cm^{-1} at $n=6$. The use of longer cross-linked chains ($n>4$) improved their proton conductivities to greater than that of the Nafion 117 membrane. This indicates that the introduction of a flexible and long cross-linker between the rigid polymer chains can significantly affect the formation of hydrophilic channels in which protons can be effectively transported and thus improve the overall proton conductivity. The effect of the cross-linked chain length is distinctly observed in the methanol permeation properties. As in the other cases, the methanol permeability decreased slightly with respect to increasing n at low values but again increased with cross-linked chain length over $n=4$. Interestingly, the proton conductivity of the XSPI membranes with the longest cross-linked chain ($n=10$) was higher than that of the non-cross-linked SPI, whereas the methanol permeability was confined to a lower value than that of the non-cross-linked SPI. Furthermore, in spite of the ester bonding, the incorporation of diol cross-linkers also provided the resultant membrane with enhanced durability to hydrolysis and peroxide radical attack.

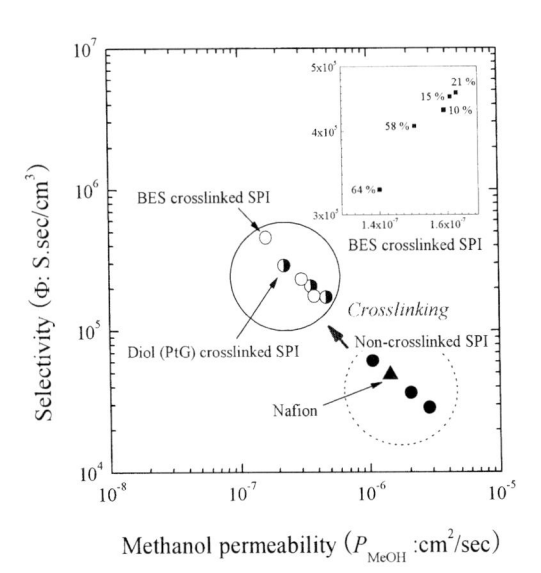

Fig. 12.17 The selectivity ($\Phi = \sigma/P_{MeOH}$) of non-cross-linked SPI, diol cross-linked SPI, BES cross-linked SPI and Nafion 117 membranes [125, 145]. Reproduced with permission from the American Chemical Society © 2005.

This is a good example which demonstrates the possibility of improving the proton conductivity and simultaneously reducing the methanol permeability with a morphological change due to cross-linking.

Another example of maintaining or increasing the proton conductivity and of reducing the methanol permeability is the use of a sulfonic acid-containing compound such as BES as a cross-linker and a proton carrier [125, 126]. The primary intent is to compensate for the probable loss of proton conductivity due to cross-linking by using a cross-linker that includes a fixed charge group ($-SO_3H$). Figure 12.17 shows the selectivity (Φ) or the ratio of the proton conductivity (σ) to the methanol permeability (P_{MeOH}) of Nafion 117 and sulfonated polyimides. Here, the selectivity is a characteristic factor used to evaluate the membrane performance considering both the proton conductivity and methanol permeability. Careful attention should be paid to use the selectivity plot because low methanol permeability and low proton conductivity can yield a high selectivity. Accordingly, the selectivity should be employed only in the comparison of membrane materials with proton conductivities that are of the same order of magnitude. In this case, the selectivity can be used as an indicator to develop the best proton conductive polymer membrane which also possesses reduced methanol permeability. A possible value in the current technical level will be a proton conductivity over 10^{-2} S cm^{-1} and a methanol permeability of below 10^{-6} cm^2 s^{-1} for DMFC application. The Nafion 117 and non-cross-linked SPI membranes had very similar proton conductivities and methanol permeabilities. On the other hand, the cross-linked SPI membranes displayed Φ values about 10 times higher than Nafion 117, irrespective of cross-linker type. Among the cross-linked SPI membranes, the BES-cross-linked SPI showed the highest selectivity due to the high proton conductivity and low methanol permeability of this membrane. As shown in the inset in Fig. 12.17, the selectivity varied with respect to the BES content and here the optimal BES content (21 mol%) existed in terms of the selectivity and the proton conductivity. In addition to the methanol permeability and proton conductivity, the brittleness of such membranes in the dry state should be considered in this cross-linking system.

In contrast to cross-linking via covalent bond formation, ionic cross-linking may reinforce the mechanical strength of the membrane [146, 147]. Ionic cross-linking occurs by blending a sulfonated polymer in the salted state such as the Na form with basic components. After the subsequent acidification of the Na-sulfonated polymer, the $-SO_3^-$ groups in the acidic polymer react with the basic groups including the H–N– group in the basic components. The interaction between the acid polymers and the basic components can be interchangeable with respect to the amount and the basicity of the basic components. Here, the basic components include weakly basic polymers, strongly basic polymers and basic materials such as imidazole and pyrazole with a low molecular weight (unpublished results from the authors' laboratory). Weakly basic polymers (e.g. post-sulfone-o-sulfonediamines) play a role only as proton acceptors and form physical cross-linking via hydrogen bonds with the acidic polymers (e.g. highly sulfonated Udel and Victrex). On the other hand, strongly basic polymers [e.g. PBI

(Celazole)] may be fully protonated by the acidic polymers and form ionic cross-linkages via ionic interactions between the blend components. When there is an excess of acidic sites over basic sites in the blended system, ionic or physical cross-linking coexists along with unsaturated sulfonic acid functional groups, which may act as proton carriers in the hydrated state. This type of blending converts the water-soluble acidic polymers into water-insoluble polymers. The high swelling in the water and the brittleness in the dry state are significantly reduced compared with the pristine polymers of similar IEC. In spite of even the best blend systems such as sulfonated PEEK and PBI with high morphological stability, membrane swelling is still higher than that in the covalent bond-based cross-linking system and even ionic cross-linking can be hydrolyzed in water at temperatures above about 80 °C [106].

12.4.5
Modified Aromatic PBIs

Aromatic PBIs are well known as high-temperature membrane materials owing to their remarkable thermal, mechanical, chemical and hydrolytic stability even at high temperatures approaching 200 °C. They also have excellent vapor barrier properties and acid–base complexation with strong protonic acids such as phosphoric acid (H_3PO_4), sulfuric acid (H_2SO_4), hydrochloric acid (HCl), nitric acid (HNO_3), hydrobromic acid (HBr) and perchloric acid ($HClO_4$) [148–151], organic acids such as CH_3SO_3H and $C_2H_5SO_3H$ [152], aromatic phosphoric acid [153] and polymeric acids [154]. The basic polymers establish hydrogen bonds with acidic molecules which undergo dissociation to some extent. For this acid–base complexation, PBI membranes were fabricated from a PBI solution with DMAc as the solvent and then immersed in a highly concentrated acid solution that included H_3PO_4 for a set period of time. Although the acid-doped PBI membranes exhibited proton conductivity up to $(1–4) \times 10^{-2}$ S cm^{-1} from 130 to 190 °C, compared with the non-modified PBI without any acid molecules [$(2–8) \times 10^{-4}$ S cm^{-1} at 0–100% RH], the proton conductivity of the acid-doped PBI was still too low for high-temperature fuel cell applications. Therefore, several methods were tried in order to improve the proton conductivity via the enhanced complexation of H_3PO_4 to PBI. These methods include the acid-solvent casting method using trifluoroacetic acid (TFA) [155] and the fabrication method of the porous PBI membrane for the entrapment and impregnation of strong acid molecules using phase inversion [156]. In spite of a high acid-doping level up to 70 wt.% and enhanced proton conductivity [$(4–8) \times 10^{-2}$ S cm^{-1}] under the same measurement conditions, these types of PBI membrane have a severe problem, so-called "leaching out of doped acids", which is the main reason for the reduction in the performances of the PEMFC at high temperature. Recently, a new sol–gel process was reported in relation to the PBI-based membranes with excellent proton conductivity and durability [157, 158]. In this process, termed the PPA process, polyphosphoric acid was used as the condensing agent for the polymerization and the membrane casting solvent. After casting, spontaneous

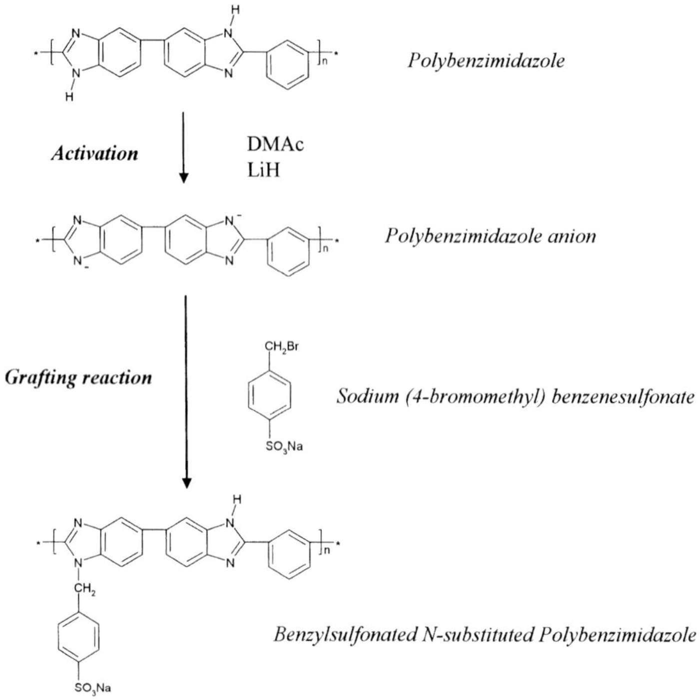

Fig. 12.18 Synthetic scheme of benzylsulfonated N-substituted polybenzimidazole [159, 160].

water absorption from the atmosphere led to the hydrolysis of the polyphosphoric acid to phosphoric acid. The conversion in the solvent induces a sol–gel transition and results in a specific three-dimensional network composed of PBI. The morphological change gives a high loading of H_3PO_4, excellent resistance to the problem of leaching and suitable physical properties in the PBI membranes.

In addition, aromatic PBIs have been considered as the main chain of sulfonated polymers with the proton conductivity in the hydrated state. The imidazole hydrogen can be replaced using sulfonated aryl or alkyl substituents [159, 160], as shown in Fig. 12.18. This method contributes to the improvement of the chemical stability of PBI because of the introduction of less reactive groups than the imidazole hydrogen into the imidazole ring. For instance, benzylsulfonated N-substituted PBI is obtained via the sequence of (1) the formation of the PBI anion using a soluble base (e.g. an alkali metal hydride) and (2) reaction with sodium (4-bromomethyl)benzenesulfonate. Interestingly, the high flexibility of the grafted chains reduces the brittleness of the membranes and leads to a high water uptake. The benzylsulfonated grafted PBI displays a hydration number (i.e. number of water molecules associated with each sulfonic acid group) of about 7 compared with 5 in Nafion or sulfonated poly(ether ether ketone) [130]. Unfortunately, the proton conductivity of the grafted PBI does not exceed 2×10^{-2} S cm^{-1}, which is lower than that (3×10^{-2} to 1×10^{-1} S cm^{-1}) of Nafion at 25 °C in aqueous phosphoric acid.

The development of high performance-membrane materials with long-term durability under harsh conditions including hydrolysis and free radical attacks using the described structure property relationship is under way.

12.5
Conclusion

Membrane separation technology has experienced substantial prominence, contributing to sustainable chemical processing in the past two decades, and is being exploited in a wide array of commercial applications in the medical, biochemical and chemical sectors. Some of the synthetic membrane materials display a better overall performance than their biological counterparts in industrially established operations. Since membrane functions are governed by both the primary and secondary structure of polymers, the rational design of highly efficient and economically viable membrane materials requires the elucidation of the relationship between transport properties and molecular structure of polymers.

Acknowledgment

We thank Ms. Fareha Zafar Khan for editing and typographic assistance.

References

1 Baker, R. W. *Membrane Technology and Applications*, McGraw-Hill, New York, **2000**.
2 Koros, W. J., Fleming, G. K. *J. Membr. Sci.* **1993**, *83*, 1.
3 Paul, D. R., Yampol'skii, Y. P. In *Polymeric Gas Separation Membranes*, Paul, D. R., Yampol'skii, Y. P. (Eds.), CRC Press, Boca Raton, FL, **1994**, p. 1.
4 Stern, S. A. *J. Membr. Sci.* **1994**, *94*, 1.
5 Henis, J. M. S. In *Polymeric Gas Separation Membranes*, Paul, D. R., Yampol'skii, Y. P. (Eds.), CRC Press, Boca Raton, FL, **1994**, p. 441.
6 Nakagawa, T. In *Polymeric Gas Separation Membranes*, Paul, D. R., Yampol'skii, Y. P. (Eds.), CRC Press, Boca Raton, FL, **1994**, p. 399.
7 Nakagawa, T. In *Membrane Science and Technology*, Osada, Y., Nakagawa, T. (Eds.), Marcel Dekker, New York, **1992**, p. 239.
8 Freeman, B. D., Pinnau, I. In *Polymer Membranes for Gas and Vapor Separation: Chemistry and Materials Science*, Freeman, B. D., Pinnau, I. (Eds.), American Chemical Society, Washington, DC, **1999**, p. 1.
9 Tsujita, Y. In *Membrane Science and Technology*, Osada, Y., Nakagawa, T. (Eds.), Marcel Dekker, New York, **1992**, p. 3.
10 Sato, S., Nagai, K. *Membrane* **2005**, *30*, 20.
11 Nagai, K. *Jpn. J. Polym. Sci. Technol.* **2004**, *61*, 420.
12 Poling, B. E., Prausnitz, J. M., O'Connell, J. P. *The Properties of Gases and Liquids*, 5th edn., McGraw-Hill, New York, **2000**.

13 Hirata, T., Sato, S., Nagai, K. *Sep. Sci. Technol.* **2005**, *40*, 2819.
14 Lin, H., Freeman, B. D. *J. Mol. Struct.* **2005**, *739*, 57.
15 Dixon-Garrett, S. V., Nagai, K., Freeman, B. D. *J. Polym. Sci., Part B: Polym. Phys.* **2000**, *38*, 1461.
16 Robeson, L. M. *J. Membr. Sci.* **1991**, *62*, 165.
17 Freeman, B. D. *Macromolecules* **1999**, *32*, 375.
18 Koros, W. J., Mahajan, R. *J. Membr. Sci.* **2000**, *175*, 181.
19 Kanehashi, S., Nagai, K. *J. Membr. Sci.* **2005**, *253*, 117.
20 van Krevelen, D. W. *Properties of Polymers, Their Correlation with Chemical Structure, Their Numerical Estimation and Prediction from Additive Group Contributions*, Elsevier, Amsterdam, **1990**.
21 Nagai, K. *Jpn. J. Polym. Sci. Technol.* **2003**, *60*, 468.
22 Lee, W. M. *Polym. Eng. Sci.* **1980**, *20*, 65.
23 Freeman, B. D., Hill, A. J. *ACS Symp. Ser.* **1998**, *710*, 1.
24 Nagai, K., Freeman, B. D., Hill, A. J. *J. Polym. Sci., Part B: Polym. Phys.* **2000**, *38*, 1222.
25 Yampolskii, Y. P., Korikov, A. P., Shantarovich, V. P., Nagai, K., Freeman, B. D., Masuda, T., Teraguchi, M., Kwak, G. *Macromolecules* **2001**, *34*, 1788.
26 Miyata, S., Sato, S., Nagai, K., Nakagawa, T., Kudo J. *Polym. Sci., Part B: Polym. Phys.*, submitted.
27 Nagai, K., Higuchi, A., Nakagawa, T. *J. Polym. Sci., Part B: Polym. Phys.* **1995**, *33*, 289.
28 Merkel, T. C., Bondar, V., Nagai, K., Freeman, B. D., Yampol'skii, Y. P. *Macromolecules* **1999**, *32*, 8427.
29 Merkel, T. C., Bondar, V. I., Nagai, K., Freeman, B. D., Pinnau, I. *J. Polym. Sci., Part B: Polym. Phys.* **2000**, *38*, 415.
30 Tanaka, K., Okano, M., Toshino, H., Kita, H., Okamoto, K. *J. Polym. Sci., Part B: Polym. Phys.* **1992**, *30*, 907.
31 Nagai, K., Nakagawa, T. In *Advanced Materials for Membrane Separations*, Pinnau, I., Freeman, B. D. (Eds.), American Chemical Society, Washington, DC, **2004**, p. 139.
32 Nagai, K., Freeman, B. D., Cannon, A., Allcock, H. R. *J. Membr. Sci.* **2000**, *172*, 167.
33 Allen, S. M., Fujii, M., Stannett, V., Hopfenberg, H. B., Williams, J. L. *J. Membr. Sci.* **1977**, *2*, 153.
34 Arnold, M. E., Nagai, K., Freeman, B. D., Spontak, R. J., Betts, D. E., DeSimone, J. M., Pinnau, I. *Macromolecules* **2001**, *34*, 5611.
35 Masuda, T., Sanda, F. In *Handbook of Metathesis*, Grubbs, R. H. (Ed.), Wiley-VCH, Weinheim, **2003**, p. 375.
36 Nagai, K., Masuda, T., Nakagawa, T., Freeman, B. D., Pinnau, I. *Prog. Polym. Sci.* **2001**, *26*, 721.
37 Masuda, T., Teraguchi, M., Nomura, R. *ACS Symp. Ser.* **1999**, *733*, 28.
38 Odani, H., Masuda, T. In *Polymers for Gas Separation*, Toshima, N. (Ed.), VCH, New York, **1992**.
39 Masuda, T., Higashimura, T. *Adv. Polym. Sci.* **1987**, *81*, 121.
40 Kesting, R. E., Fritzsche, A. K. *Polymeric Gas Separation Membranes*, Wiley, New York, **1993**.
41 Masuda, T., Isobe, E., Higashimura, T., Takada, K. *J. Am. Chem. Soc.* **1983**, *105*, 7473.
42 Masuda, T., Isobe, E., Hamano, T., Higashimura, T. *J. Polym. Sci., Part A: Polym. Chem.* **1987**, *25*, 1353.
43 Robeson, L. M., Burgoyne, W. F., Langsam, M., Savoca, A. C., Tien, C. F. *Polymer* **1994**, *35*, 4970.
44 Savoca, A. C., Surnamer, A. D., Tien, C. F. *Macromolecules* **1993**, *26*, 6211.
45 Takada, K., Matsuya, H., Masuda, T., Higashimura, T. *J. Appl. Polym. Sci.* **1985**, *30*, 1605.
46 Kwak, G. S., Masuda, T. *J. Polym. Sci., Part A: Polym. Chem.* **2000**, *38*, 2964.
47 Morisato, A., Pinnau, I. *J. Membrane. Sci.* **1996**, *121*, 243.
48 Tsuchihara, K., Masuda, T., Higashimura, T. *J. Am. Chem. Soc.* **1991**, *113*, 8548.
49 Tsuchihara, K., Masuda, T., Higashimura, T. *Macromolecules* **1992**, *25*, 5816.
50 Sakaguchi, T., Yumoto, K., Shiotsuki, M., Sanda, F., Yoshikawa, M., Masuda, T. *Macromolecules* **2005**, *38*, 2704.

51. Ito, H., Masuda, T., Higashimura, T. *J. Polym. Sci., Part A: Polym. Chem.* **1996**, *34*, 2925.
52. Kouzai, H., Masuda, T., Higashimura, T. *J. Polym. Sci., Part A: Polym. Chem.* **1994**, *32*, 2523.
53. Sakaguchi, T., Kwak, G., Masuda, T. *Polymer* **2002**, *43*, 3937.
54. Sakaguchi, T., Shiotsuki, M., Sanda, F., Freeman, B.D., Masuda, T. *Macromolecules* **2005**, *38*, 8327.
55. Hayakawa, Y., Nishida, M., Aoki, T., Muramatsu, H. *J. Polym. Sci., Part A: Polym. Chem.* **1992**, *30*, 873.
56. Aoki, T., Nakahara, H., Hayakawa, Y., Kokai, M., Oikawa, E. *J. Polym. Sci., Part A: Polym. Chem.* **1994**, *32*, 849.
57. Pauly, S. In *Polymer Handbook*, 4th edn., Brandrup, J., Immergut, E.H., Grulke, E.A. (Eds.), Wiley, New York, **1999**, p. VI543.
58. Teraguchi, M., Masuda, T. *Macromolecules* **2002**, *35*, 1149.
59. Toy, L.G., Nagai, K., Freeman, B.D., Pinnau, I., He, Z., Masuda, T., Teraguchi, M., Yampolskii, Y.P. *Macromolecules* **2000**, *33*, 2516.
60. Sakaguchi, T., Shiotsuki, M., Masuda, T. *Macromolecules* **2004**, *37*, 4104.
61. Shida, Y., Sakaguchi, T., Shiotsuki, M., Sanda, F., Freeman, B.D., Masuda, T. *Macromolecules* **2005**, *38*, 4096.
62. Kazama, S., Teramoto, T., Haraya, K. *J. Membr. Sci.* **2002**, *207*, 91.
63. Pinnau, I., He, Z., Morisato, A. *J. Membr. Sci.* **2004**, *241*, 363.
64. Pinnau, I., Toy, L.G. *J. Membr. Sci.* **1996**, *116*, 199.
65. Srinivasan, R., Auvil, S.R., Burban, P.M. *J. Membr. Sci.* **1994**, *86*, 67.
66. Anand, M., Langsam, M., Rao, M.B., Sircar, S. *J. Membr. Sci.* **1997**, *123*, 17.
67. Schultz, J., Peinemann, K.V. *J. Membr. Sci.* **1996**, *110*, 37.
68. Maxwell, C. *Treatise on Electricity and Magnetism*, Vol. 1, Oxford University Press, London, **1873**.
69. Merkel, T.C., Freeman, B.D., Spontak, R.J., He, Z., Pinnau, I., Meakin, P., Hill, A.J. *Science* **2002**, *296*, 519.
70. Nagai, K. *Jpn. J. Polym. Sci. Technol.* **2003**, *60*, 482.
71. Nagai, K. *Jpn. J. Polym. Sci. Technol.* **2005**, *62*, 564.
72. Toyomoto, K., Higuchi, A. In *Membrane Science and Technology*, Osada, Y., Nakagawa, T. (Eds.), Marcel Dekker, New York, **1992**, p. 289.
73. Akiba, C., Watanabe, K., Nagai, K., Hirata, T., Nguyen, Q.T. *J. Appl. Polym. Sci.* **2006**, *100*, 1113.
74. Arnold, M.E., Nagai, K., Freeman, B.D., Spontak, R.J., Leroux, D., Betts, D.E., DeSimone, J.M., DiGiano, F.A., Stebbins, C.K., Linton, R.W. *Macromolecules* **2002**, *35*, 3697.
75. Nagai, K., Tanaka, S., Hirata, Y., Nakagawa, T., Arnold, M.E., Freeman, B.D., LeRoux, D., Betts, D.E., DeSimone, J.M., DiGiano, F.A. *Polymer* **2001**, *42*, 9941.
76. DiGiano, F.A., Roudman, A., Arnold, M.E., Freeman, B.D., Preston, J., Nagai, K., DeSimone, J.M. *Laboratory Tests of New Membrane Materials*, AWWA Research Foundation, Denver, Colorado, **2001**.
77. Mishima, S., Nakagawa, T. *J. Membr. Sci.* **2004**, *228*, 1.
78. Ohshima, T., Miyata, T., Uragami, T., Berghmens, H. *J. Mol. Struct.* **2005**, *739*, 47.
79. Uragami, T., Yamada, H., Miyata, T. *J. Membr. Sci.* **2001**, *187*, 255.
80. Ohshima, T., Kogami, Y., Miyata, T., Uragami, T. *J. Membr. Sci.* **2005**, *260*, 156.
81. Hoshi, M., Saitoh, T., Yoshioka, C., Higuchi, A., Nakagawa, T. *J. Appl. Polym. Sci.* **1999**, *74*, 983.
82. Hoshi, M., Kobayashi, M., Saitoh, T., Higuchi, A., Nakagawa, T. *J. Appl. Polym. Sci.* **1998**, *69*, 1483.
83. Mishima, S., Kaneoka, H., Nakagawa, T. *J. Appl. Polym. Sci.* **1999**, *71*, 273.
84. Mishima, S., Nakagawa, T. *J. Appl. Polym. Sci.* **1999**, *73*, 1835.
85. Mishima, S., Nakagawa, T. *J. Appl. Polym. Sci.* **2000**, *75*, 773.
86. Nakagawa, T., Arai, T., Ookawara, Y., Nagai, K. *Sen'i Gakkaishi* **1997**, *53*, 423.
87. Ulutan, S., Nakagawa, T. *J. Membr. Sci.* **1998**, *143*, 275.
88. Masuda, T., Tang, B.Z., Higashimura, T. *Polym. J.* **1986**, *18*, 565.

89 Sakaguchi, T., Yumoto, K., Kwak, G., Yoshikawa, M., Masuda, T. *Polym. Bull.* **2002**, *48*, 271.
90 Hoshi, M., Ieshige, M., Saitoh, T., Nakagawa, T. *J. Appl. Polym. Sci.* **2000**, *76*, 654.
91 Hoshi, M., Ieshige, M., Saitoh, T., Nakagawa, T. *J. Appl. Polym. Sci.* **1999**, *71*, 439.
92 Hoshi, M., Kogure, M., Saitoh, T., Nakagawa, T. *J. Appl. Polym. Sci.* **1997**, *65*, 469.
93 Mishima, S., Nakagawa, T. *J. Appl. Polym. Sci.* **2004**, *94*, 461.
94 Masuda, T., Takatsuka, M., Tang, B. Z., Higashimura, T. *J. Membr. Sci.* **1990**, *49*, 69.
95 Mishima, S., Nakagawa, T. *J. Appl. Polym. Sci.* **2002**, *83*, 1054.
96 Yoon, B.-O., Koyanagi, S., Asano, T., Hara, M., Higuchi, A. *J. Membr. Sci.* **2003**, *213*, 137.
97 Higuchi, A., Yoon, B.-O., Asano, T., Nakaegawa, K., Miki, S., Hara, M., He, Z., Pinnau, I. *J. Membr. Sci.* **2002**, *198*, 311.
98 Nagase, Y., Takamura, Y., Matsui, K. *J. Appl. Polym. Sci.* **1991**, *42*, 185.
99 Grubb, W. T., *US Patent 2 913 511*, **1959**.
100 Grubb, W. T. *J. Elecrochem. Soc.* **1959**, *106*, 275.
101 Vallejo, E., Pourcelly, G., Gavach, C., Mercier, R., Pineri, M. *J. Membr. Sci.* **1999**, *160*, 127.
102 Eto, E., Tanioka, A. *J. Colloid Interface Sci.* **1998**, *200*, 59.
103 Sakai, Y., Matsuguchi, M., Yonesato, N. *Electrochim. Acta* **2001**, *46*, 1509.
104 Kirsh, Y. E., Fedotov, Y. A., Semenova, S. I., Vdovin, P. A., Valuev, V. V., Zemlianova, O. Y., Timashev, S. F. *J. Membr. Sci.* **1995**, *103*, 95.
105 Liebhafsky, H. A., Caims, E. J., *Fuel Cell and Fuel Batteries*, Wiley, New York, **1968**.
106 Kreuer, K. D. *Handbook of Fuel Cells – Fundamental Technology and Applications*, Wiley, New York, **2003**.
107 Lufrano, F., Gatto, I., Staiti, P., Antonucci, V., Passalacqua, E. *Solid State Ionics* **2001**, *145*, 47.
108 Manea, C., Mulder, M. *J. Membr. Sci.* **2002**, *206*, 443.
109 Alberti, G., Casciola, M., Massinelli, L., Bauer, B. *J. Membr. Sci.* **2001**, *185*, 73.
110 Kobayashi, T., Rikukawa, M., Sanui, K., Ogata, N. *Solid State Ionics* **1998**, *106*, 219.
111 Bae, J. M., Honma, I., Murata, M., Yamamoto, T., Rikukawa, M., Ogata, N. *Solid State Ionics* **2002**, *147*, 189.
112 Genies, C., Mercier, R., Sillion, B., Cornet, N., Gebel, G., Pineri, M. *Polymer* **2001**, *42*, 359.
113 Genies, C., Mercier, R., Sillion, B., Petiaud, R., Cornet, N., Gebel, G., Pineri, M. *Polymer* **2001**, *42*, 5097.
114 Fang, J., Guo, X., Harada, S., Wateri, T., Tanaka, K., Kita, H., Okamoto, K. *Macromolecules* **2002**, *35*, 9022.
115 Fedkin, M. V., Zhou, X., Hofmann, X. A., Chalkova, E., Weston, J. A., Allcock, H. R., Lvov, S. N. *Mater. Lett.* **2002**, *52*, 192.
116 Guo, Q., Pintauro, P. N., Tang, H., O'Connor, S. *J. Membr. Sci.* **1999**, *154*, 175.
117 Staiti, P., Lufrano, F., Acrico, A. S., Passalacqua, E., Antonucci, V. *J. Membr. Sci.* **2001**, *188*, 71.
118 Ise, M., *Polymer Elektrolyt Membranen*, PhD Thesis, University of Stuttgart, **2000**.
119 Kreuer, K. D. *J. Membr. Sci.* **2001**, *185*, 29.
120 Karlsson, L. E., Wesslen, B., Jannasch, P. *Electrochim. Acta* **2002**, *47*, 4589.
121 Kim, D. S., Park, H. B., Rhim, J. W., Lee, Y. M. *J. Membr. Sci.* **2004**, *240*, 37.
122 Lee, C. H., Park, C. H., Park, H. B., Lee, Y. M., Mulmi, S., Kim, J. Y. Presented at the International Congress on Membranes and Membrane Processes, Seoul, Korea, 21–26 August 2005.
123 Lee, C. H., Park, H. B., Lee, Y. M., Kim, J. Y. Presented at the International Congress on Membranes and Membrane Processes, Seoul, Korea, 21–26 August 2005.
124 Lee, C. H., Park, H. B., Lee, Y. M., Lee, R. D. *Ind. Eng. Chem. Res.* **2005**, *44*, 7617.
125 Lee, C. H., Park, H. B., Chung, Y. S., Lee, Y. M., Freeman, B. D. *Macromolecules* **2006**, *39*, 755–764.
126 Lee, Y. M., Park, H. B., Lee, C. H., US Patent 2004/0236038 A1
127 Kim, D. S., Park, H. B., Jang, J. Y., Lee, Y. M. *J. Polym. Sci., Part A: Polym. Chem.* **2005**, *43*, 5620.

128 Agman, N. *Chem. Phys. Lett.* **1995**, *244*, 456.
129 Tuckerman, M.E., Laasonen, K., Sprike, M., Parrinello, M. *J. Chem. Phys.* **1995**, *103*, 150.
130 Kreuer, K.D. *Chem. Mater.* **1998**, *8*, 610.
131 Rikukawa, M., Sanui, K. *Prog. Polym. Sci.* **2000**, *25*, 1463.
132 Uragami, T., Okazaki, K., Matsugi, H., Miyata, T. *Macromolecules* **2002**, *35*, 9156.
133 Uragami, T., Matsugi, H., Miyata, T. *Macromolecules* **2005**, *38*, 8840.
134 Lin, Y.L., Hsu, C.Y., Su, Y.H., Lai, J.Y. *Biomacromolecules* **2005**, *6*, 368.
135 Won, J.O., Park, H.H., Kim, Y.J., Choi, S.W., Ha, H.Y., Oh, I.H., Kim, H.S., Kang, Y.S., Ihn, K.J. *Macromolecules* **2003**, *36*, 3228.
136 Rusanov, A.L. *Adv. Polym. Sci.* **1994**, *111*, 115.
137 Guo, X., Fang, J., Wateri, T., Tanaka, K., Kita, H., Okamoto, K. *Macromolecules* **2002**, *35*, 6707.
138 Hubner, G., Roduner, E. *J. Mater. Chem.* **1999**, *9*, 409.
139 Allcock, H.R., Hofmann, M.A., Ambler, C.M., Lvov, S.N., Zhou, X.Y., Chalkova, E., Weston, J. *J. Membr. Sci.* **2002**, *201*, 47.
140 Xing, P., Roberson, G.P., Guiver, M.D., Mikhailenko, S.D., Wang, K., Kaliaguine, S. *J. Membr. Sci.* **2004**, *229*, 95.
141 Mikhailenko, S.D., Wang, K., Kaliaguine, S., Xing, P., Robertson, G.P., Guiver, M.aD. *J. Membr. Sci.* **2004**, *233*, 93.
142 Wycisk, R., Pintauro, P.N., Wang, W., O'Connor, S. *J. Appl. Polym. Sci.* **1996**, *59*, 1607.
143 Graves, R., Pintauro, P.N. *J. Appl. Polym. Sci.* **1998**, *68*, 827.
144 Rhim, J.W., Park, H.B., Lee, C.S., Jun, J.H., Kim, D.S., Lee, Y.M. *J. Membr. Sci.* **2004**, *238*, 143.
145 Park, H.B., Lee, C.H., Lee, Y.M., Freeman, B.D., Kim, H.J. *J. Membr. Sci.* **2006**, *285*, 432.
146 Kerres, J., Ullrich, A., Meier, F., Haring, T. *Solid State Ionics* **1999**, *125*, 243.
147 Hasiotis, C., Deimede, V., Kontoyannis, C. *Electrochim. Acta* **2001**, *46*, 2401.
148 Glipa, X., Bonnet, B., Mula, B., Jones, D.J., Roziere, J. *J. Mater. Chem.* **1999**, *9*, 3045.
149 Bouchet, R., Siebert, E. *Solid State Ionics* **1999**, *118*, 287.
150 Jones, D.J., Roziere, J. *J. Membr. Sci.* **2001**, *185*, 41.
151 Li, Q., He, R., Beg, R.W., Hjuler, H.A., Bjerrum, N.J. *Solid State Ionics* **2004**, *168*, 177.
152 Kawahara, M., Morita, J., Rikukawa, M., Sanui, K., Ogata, N. *Electrochim. Acta* **2000**, *45*, 1395.
153 Akita, H., Ichikawa, M., Nosaki, K., Oyanagi, H., Iguchi, M., *US Patent 6 120 060*, **2000**.
154 Kerres, J.A. *J. Membr. Sci.* **2001**, *185*, 3.
155 Savinell, R.F., Litt, M.H., *US Patent 5 525 436*, **1996**.
156 Sansone, M.J., Onorato, F.J., French, S.M., Marikar, F., *US Patent 6 187 231*, **2001**.
157 Xiao, L., Jana, T., Chen, R., Zhang, H., Opper, K., Scanlon, G., Yu, S., Choe, E.W., Benicewicz, B.C. Presented at Advances in Materials for Proton Exchange Membrane Fuel Cell Systems, Pacific Grove, CA, 2005.
158 Benicewicz, B.C. Presented at the Pacific Polymer Conference IX, Maui, HI, 11–14 December 2005
159 Glipa, X., Haddad, M.E., Jones, D.J., Roziere, J. *Solid State Ionics* **1997**, *97*, 323.
160 Gieselman, M., Reynolds, J.R. *Macromolecules* **1992**, *25*, 4832.

13
Utilization of Polymers in Sensor Devices

Basudam Adhikari and Alok Kumar Sen

13.1
Introduction

The function of a sensor is to provide information on our physical, chemical and biological environments. A chemical sensor furnishes information about its environment and consists of a physical transducer and a chemically selective layer. A biosensor contains a biological entity such as enzyme, antibody, bacteria, tissue, etc., as recognition agent but a chemical sensor does not contain such agents. Sensor devices have been made from classical semiconductors, solid electrolytes, insulators, metals and catalytic materials. Polymers, in general, have acquired a major position as materials in various sensor devices, among other materials. Intrinsically conducting polymers as coating or encapsulating materials on an electrode surface or non-conducting polymers for immobilization of specific receptor agents on sensor devices are being used extensively.

The electrical properties of solid-state sensor devices are influenced by the presence of gas phase or liquid phase species. Semiconducting metal oxide sensors, such as SnO_2 in the form of pressed powder or thin film, are active or are made active by adding catalysts. "Solid-state sensors" have been constructed from classical semiconductors, solid electrolytes, insulators, metals and catalytic materials, as well as from different organic membranes.

Solid electrolytes play an important role in commercial gas and ion sensors since their conductivity is dominated by the mobility of one type of ion only. Solid electrolytes used in commercial gas and ion sensors are yttria (Y_2O_3), stabilized zirconia (ZrO_2), an O^{2-} conductor at high temperature (>300 °C), for determination of oxygen in exhaust gases of automobiles, boilers or steel melts and LaF_3 for the determination of F^- even at room temperature. Nafion, a perfluorinated hydrophobic ionomer with ionic clusters, is a good example of a solid polymer electrolyte for a variety of room temperature electrochemical sensors [1].

Macromolecular Engineering. Precise Synthesis, Materials Properties, Applications.
Edited by K. Matyjaszewski, Y. Gnanou, and L. Leibler
Copyright © 2007 WILEY-VCH Verlag GmbH & Co. KGaA, Weinheim
ISBN: 978-3-527-31446-1

13.2
Different Analytes and Their Sensing Principles

13.2.1
Chemical Analytes

13.2.1.1 Gases and Volatile Chemicals

Gaseous pollutants, e.g., SO_2, NO_x and other toxic gases, in the environment are a global problem. Suitable sensor devices can detect and measure the concentration of such pollutants. Examples of some sensors and their applications are provided in Table 13.1. Table 13.2 contains some examples of different polymers used in gas sensors and their characteristics based on different working principles. Conducting polymers and their composites with other polymers such as PVC and PMMA having active functional groups and solid polymer electrolytes are being used extensively for sensing gases with acid–base or oxidizing characteristics.

Table 13.1 Sensors and their applications.

Sensor type	Polymer used	Fields of application	Special features	Ref.
Biosensor	Polyaniline (PANI)	Glucose, urea and triglyceride determination	Electrochemical deposition of polymer and enzyme immobilization	139
	Poly(o-aminophenol)	Glucose estimation	Immobilization on platinized glassy carbon electrode	137
	Polytyramine	Estimation of L-amino acids	Enzyme immobilization by electropolymerization	166
	Redox polymer	Detect ion of glucose, lactate, pyruvate	Glucose, lactate, pyruvate biosensor array based on enzyme–polymer nanocomposite film	146
	Polysiloxane	Blood glucose determination	Composite membrane was formed by condensation polymerization of dimethyldichlorosilane at the surface of a host porous alumina membrane	142
Chemical sensor	Poly(vinyl chloride)	Estimation of pethidine hydrochloride in injections and tablets	Pethidine phosphate–tungstate ion association as electroactive material	97
	Divinylstyrene and isoprene polymer	Environmental control of trace organic contaminants	Piezoelectric	33
	Nafion	Detection of dissolved O_2 in water	Gold–solid polymer–electrolyte sensor	6
	PVC	Determination of phentermine	PVC with tris(2-ethylhexyl) phosphate as solvent mediator and NaHFPB as ion exchanger	98

13.2 Different Analytes and Their Sensing Principles

Table 13.1 (continued)

Sensor type	Polymer used	Fields of application	Special features	Ref.
	PANI (emeraldine base)	Sensing humidity, NH_3, NO_2 and fabrication of other molecular devices	Nanocomposite ultra-thin films of polyaniline and isopoly-molybdic acid	25
	PANI and its derivatives	Sensing aliphatic alcohols	Extent of change governed by chain length of alcohol	91
	Epoxy resin	Li^+ detection	ΛE^+-MnO_2-based graphite–epoxy electrode	67
	PVC	Detection of phosphate ions	Plasticized PVC membrane containing uranyl salophene derivative	83
	PANI	pH sensing	Optical method	56
Odor sensor	Poly(4-vinylphenol) Poly(N-vinylpyrrolidone) Polysulfone Poly(methyl methacrylate) Polycaprolactone Poly(ethylene-co-vinyl acetate) Poly(ethylene oxide) Poly(vinylidene fluoride)	Odor detection	Array of conducting polymer composites	185
Gas sensor	Copolymers of poly-(EDMA-co-MAA)	Detection of terpene in atmosphere	Piezoelectric sensor coated with molecular imprinted polymer	32
	PANI PANI and acetic acid mixed film PANI–poly(styrenesulfonic acid) composite film	NO_2 was detected	Layers of polymer films formed by LB and self-assembly techniques	11
	PVC	Detection of gaseous NO_2 in air	A solid polymer electrode of 10% PVC is present in the sensor	12
	PPy nanocomposite	Sensing CO_2, N_2, CH_4 gases at various pressures	Nanocomposite of iron oxide PPy were prepared by simultaneous gelation and polymerization process	128
	Propylene–butyl copolymer	Detection of toluene, xylene gas	Polymer film-coated quartz resonator balance	50
Humidity sensor	PVA	Optical humidity sensing	Crystal Violet and Methylene blue are incorporated in PVA–H_3PO_4	127
Optical sensor	PVA	Optical sensing of nitro-aromatic compounds	Fluorescence quenching of benzo[k]fluoranthene in PVA film	111
Immuno-sensor	PMMA	Can detect RDX	Capillary-based immunosensors	190

Table 13.2 Polymers used in various gas sensors.

Gas	Device/techniques/principles	Polymer	Sensor characteristics	Ref.
NH_3	Change in optical transmittance using a 2-nm laser (He–Ne) source	PANI–PMMA	Sensitivity of PANI–PMMA coatings are ~10–4000 ppm, reversible response	26
	Electrical property measurement	PPy	Response time <20 s, recovery time ~60 s	28
	Electronic property of the film played the part in NH_3 sensing	PPY–PVA composite	Resistance increases with NH_3 concentration but becomes irreversible beyond 10% NH_3	29
	Electrical property measurement	PANI–isopolymolybdic acid nanocomposite	Resistance increases with NH_3 concentration and is reversible up to 100 ppm NH_3	25
	Electrical property measurement	Acrylic acid-doped polyaniline	Highly sensitive to even 1 ppm of NH_3 at room temperature and shows stable responses up to 120 days	27
NO_2	Electrical property measurement	PANI–isopolymolybdic acid nanocomposite	Resistance increases with NO_2 concentration	25
	An amperometric gas sensor based on Pt–Nafion electrode	Nafion	Electrode shows sensitivity of 0.16 $\mu A\ ppm^{-1}$ at room temperature, response time of 45 s and recovery time of 54 s, a long-term stability >27 days	10
	Amperometric gas sensor	Solid polymer electrolyte (10% PVC, 3% tetrabutyl-ammonium hexafluoro-phosphate, 87% 2-nitro-phenyl octyl ether)	Sensitivity 277 $nA\ ppm^{-1}$, recovery time 19 s	12
NO	Amperometric gas sensor	Polydimethylsiloxane (PDMS)	Shows sensitivity to 20 nM gas, high performance characteristics in terms of response time and selectivity	13
O_2	A mperometric gas transducer	PDMS	Analyte can be measured up to 1.2 mM	13
	Electrical property measurement	Nafion	Sensitivity 38.4 $\mu A\ ppm^{-1}$, lowest limit 3.8 ppm, stability excellent (30 h)	6
SO_2	QCM-type gas sensor	Amino-functional poly-(styrene-co-chloromethyl-styrene) derivatives	DPEDA functional copolymer with 5 wt% of siloxane oligomer shows response time 11 min and good reversibility even near room temperature (50 °C)	42
HCl	Optochemical sensor	5,10,15,20- Tetra(4'-alkoxy-phenyl)porphyrin [TP(OR)PH$_2$] embedded in poly(hexyl acrylate), poly(hexylmethacrylate), poly(butyl methacrylate)	Reversibly sensitive to sub-ppm levels of HCl	2
H_2S	Electrochemical detection	Nafion	High sensitivity (45 ppb, v/v), good reproducibility, short response time (0.5 s)	40

HCl in the environment Composites of alkoxy-substituted tetraphenylporphyrin–polymer composite films were developed by Nakagawa et al. [2] for the detection of sub-ppm levels of HCl in the environment. The basicity of the alkoxy group in the polymer increases the sensitivity to HCl. A high selectivity to sub-ppm levels of HCl gas using a 5,10,15,20-tetra(4′-butoxyphenyl)porphyrin–butyl methacrylate [TP(OC$_4$H$_9$)PH$_2$–BuMA] composite film was achieved.

Oxygen and ozone Luminescent sensors based on transition metal complex–polymer composites have attracted attention as oxygen sensors for both biomedical and barometric applications. Phosphorescent dyes were dispersed in a polymer matrix of high gas permeability. Using a sulfur–nitrogen–phosphorus (S-N-P) polymer, Pang et al. [3] controlled the sensitivity of the sensor over a wide range. Homogeneous polymer layers containing oxygen-quenchable luminescent dyes might lead to promising applications in oxygen sensing. Amao et al. [4] prepared an aluminum 2,9,16,23-tetraphenoxy-29H,31H-phthalocyanine hydroxide [AlPc(OH)]–PS film and measured its photophysical and photochemical properties. Based on the fluorescence quenching of the AlPc(OH)–PS film by oxygen, they developed an optical oxygen sensor. Later, Amao et al. [5] developed an optical oxygen sensor based on the luminescence intensity changes of tris(2-phenylpyridine anion)–iridium(III) complex [Ir(ppy)$_3$] immobilized in a fluoropolymer, poly(styrene-co-2,2,2-trifluoroethyl methacrylate) [poly(styrene-co-TFEM)] film. Its luminescence intensity decreased with increasing oxygen concentration.

Chou et al. [6] prepared a good and stable sensor for detecting dissolved oxygen (DO) in water with a lowest limit of 3.8 ppm using a gold–solid polymer electrolyte (Au–SPE) with Nafion as the SPE. An optochemical ozone sensor [7] was manufactured by immobilization of novel soluble indigo derivatives in permeable transparent polymeric films of polydimethylsiloxane–polycarbonate copolymer. 4,4′,7,7′-Tetraalkoxyindigo showed optimal sensitivity and specificity for ozone detection.

NO, NO$_2$, ammonia and alcohol A challenge is seen in the rapid and sensitive detection of nitric oxide (NO). There is increasing interest in the determination of NO because of its role in intra- and intercellular signal transduction in tissues. Ichimori et al. [8] introduced a commercially available amperometric NO-selective electrode. The Pt–Ir electrode was modified with an NO-selective nitrocellulose membrane and a silicone-rubber outer layer. The electrode was linearly responsive in the nanomolar concentration range with a time constant of ~1.5 s. There was an ~3-fold increase in sensitivity on raising the temperature from 26 to 37 °C. Such an electrode was used for *in vivo* applications for measuring NO in rat aortic rings under acetylcholine stimulation. Friedemann et al. [9] utilized a carbon-fiber electrode modified with an electrodeposited *o*-phenylenediamine coating for the detection of NO. In this device, a Nafion underlayer provided good sensitivity to NO and a three-layer overcoat of the Nafion optimized the selectivity against nitrite. Ho and Hung [10] developed an amperometric NO$_2$ gas sensor based on a Pt–Nafion electrode for the NO$_2$ concentration range from 0 to 485 ppm, where Nafion was used as a supporting membrane. Xie et al. [11] prepared polyaniline (PANI)-based gas sensors by ultra-thin film technology (Table 13.1). They prepared a pure PANI film, PANI and acetic acid mixed films and PANI and poly(styrenesulfonic acid) (PSSA) composite films by Langmuir–Blodgett (LB) and self-assembly (SA) techniques. PANI films prepared by the LB technique showed good sensitivity to NO_2, whereas SA films showed faster recovery. Contact of NO_2 with the π-electron network of PANI transfers an electron from the polymer to the gas making the polymer positively charged. This charge carrier increases the conductivity of the film. Reticulated vitreous carbon (RVC) [12] was tested as a material for the preparation of an indicator electrode in solid-state gas sensors for sensing NO_2 in air. The sensor contained an RVC indicator, a platinum auxiliary and a Pt–air reference electrode, with a solid polymer electrolyte of 10% PVC, 3% tetrabutylammonium hexafluorophosphate (TBAHFP) and 87% 2-nitrophenyl octyl ether (NPOE).

A hydrophobic polymer layer with a porous structure was used for the selective permeation of gases. Mizutani et al. [13] fabricated amperometric sensors with a permselective polydimethylsiloxane (PDMS) membrane for the determination of dissolved oxygen and nitric oxide. They were able to measure very low concentrations of nitric oxide (20 nM–50 µM) with these sensors at 0.85 V versus Ag/AgCl without serious interference from oxidizable species, such as L-ascorbic acid, uric acid and acetaminophen. They prepared the electrode by dip coating from a PDMS emulsion. The polymer coating, being permselective, is capable of discriminating between gases and hydrophobic species, which coexist in the analyte samples. Nitric oxide (NO) plays critical roles in vasodilation, neurotransmission, neurodegenerative diseases and metastases. Such vital biological roles have led to the development of a variety of strategies for imaging this small, seemingly simple, inorganic molecule. A number of promising conjugated polymer-based fluorescence turn-on sensors have been developed recently by Smith and Lippard [14] for bioimaging.

The conductivity of polyacetylene, the first organic conducting polymer (OCP), is altered by up to 11 orders of magnitude by exposure to iodine vapor [15, 16]. Charge transfer occurs from the polyacetylene chain (donor) to the iodine (acceptor), which leads to the formation of charge carriers. Above approximately 2% doping, the carriers are free to move along the polymer chains, resulting in metallic behavior. Later, heterocyclic polymers with π-electron conjugation were developed as attractive candidates for gas-sensing elements. Examples of heterocyclic OCPs are polyfuran (X=O), polythiophene (X=S) [17] and polypyrrole (X=NH). These are intrinsically conducting π-conjugated macromolecules that show electrical and optical property changes on doping/dedoping by some chemical agents. Change in electrical properties of these polymers can be observed at room temperature when exposed to lower concentrations of the gases, such as NO, NO_2, H_2S and NH_3, to be sensed.

Conducting polymer sensor arrays for gas and odor sensing based on substituted polymers of pyrrole, thiophene, aniline, indole and others have been reported [18]. It was observed that nucleophilic gases (ammonia, methanol and ethanol vapors) decrease conductivity, whereas electrophilic gases (NO_x, PCl_3 and SO_2) show the opposite effect [19]. In gas sensing applications, widely studied conducting polymers are polythiophene and its derivatives [20], polypyrroles [21] for NH_3, H_2 and CO and polyaniline (PANI) and their composites [22] for methanol. Electrically conducting polyacrylonitrile (PAN)–polypyrrole (PPY) [23], polythiophene–polystyrene, polythiophene–polycarbonate, polypyrrole–polystyrene and polypyrrole–polycarbonate [24] composites were prepared by electropolymerization of the conducting polymers into the matrix of the insulating polymers PAN, polystyrene and polycarbonates, respectively.

The conductivity of doped PANI has been exploited in gas sensor devices for the detection of NH_3 and other gases. Li et al. [25] fabricated electroactive nanocomposite ultra-thin films of PANI and isopolymolybdic acid (PMA) (dopant) for sensing humidity, NH_3 and NO_2 gases by alternate deposition following LB and self-assembly techniques. The NH_3 sensing occurs due to dedoping of PANI by basic ammonia, since the conductivity is strongly dependent on the doping level. In NO_2 sensing, NO_2 played the role of an oxidative dopant, causing an increase in the conductivity of the polymer film.

Nicho et al. [26] observed changes in the optical and electrical properties of PANI due to interaction of the emeraldine salt (ES) with NH_3 gas. In fact, NH_3 gas molecules decrease the polaron density in the bandgap of the polymer. They observed that PANI–PMMA composite coatings are sensitive to very low concentrations of NH_3 (<10 ppm). Chabukswar et al. [27] prepared acrylic acid (AA)-doped PANI for use as an NH_3 vapor sensor over a broad concentration range (1–600 ppm). They observed that d.c. electric resistance increased linearly up to 58 ppm of NH_3 concentration and saturated thereafter. The variation of the relative response of the NH_3 gas sensor with increase in NH_3 concentration is shown in Fig. 13.1. AA-doped PANI showed a sharp increase in relative response at ~10 ppm NH_3 and subsequently remained constant above 500 ppm, whereas PANI–PMA nanocomposite showed a decrease in relative response with increase in NH_3 concentration. A submicrometer polypyrrole film was found to be sensitive to NH_3 and its sensitivity was detected by the change in resistance of the polypyrrole film [28]. The change in resistance of the film was interpreted in terms of the formation of an electric barrier of NH_4^+ ions in the polypyrrole film. The electrons in NH_3 molecules act as donor to the p-type polypyrrole and reduce its number of holes and increase the resistivity of the thin film. A polypyrrole–poly(vinyl alcohol) (PVA) composite film prepared by electropolymerizing pyrrole in a cross-linked matrix of PVA also possesses significant NH_3 sensing ability [29].

Chemical vapors The quartz crystal microbalance (QCM) sensor, a piezoelectric quartz crystal with a selective coating on the surface, is a very stable device and can measure extremely small changes in mass. A change in mass by the adsorption of molecules on the coating disturbs the natural resonant frequency of the QCM. For example, a shift in resonance frequency of 1 Hz was measured for an AT-cut quartz plate with a resonance frequency of 5 MHz corresponding

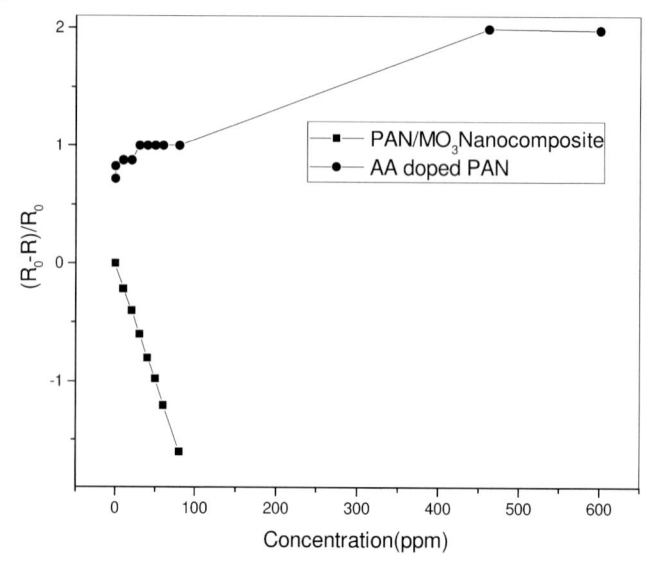

Fig. 13.1 Variation of $(R_0-R)/R_0$ of PAN–MO3 nanocomposite and AA-doped PAN with NH_3 concentrations. Adapted from [25, 27].

Table 13.3 Adsorption of strychnine (a pesticide) and β-ionene (an odor) at 45 °C by various films on a QCM surface [a].

Immobilized film	Δm (ng)	
	Strychnine	β-Ionene
Uncoated	2	2
$2C_{18}N_2^+C_1$–poly(styrene sulfonate) (PSS)	533	610
Dimyristoylphosphatidylethanolamine (DMPE)	560	540
Poly(vinyl alcohol)	4	4
Poly(methyl glutamate)	5	6
Polystyrene	7	7
Bovine plasma albumin cross-linked with glutaraldehyde	5	6
Keratin	7	6

a) Adapted from [30].

to a change in mass of just 17 ng cm^{-2} [30]. Table 13.3 shows how exposure to 19 ppm strychnine (a pesticide) or β-ionine (an odor) affects the absorption masses of QCM coated with various chemically sensitive films. It was observed that polymer-coated QCMs are sensitive towards volatiles possessing a complementary physicochemical character, e.g. acidic volatiles capable of hydrogen bond formation are best detected by hydrogen bond-forming basic polymers [31]. Alkanes and alkenes were differentiated by using hydrogen bond-forming acidic polymers that could interact with the weak hydrogen bond basicity of the alkenes, but the alkanes have no such hydrogen bonding capacity.

Percival et al. [32] showed that cross-linked poly(ethylene glycol dimethacrylate-co-methacrylic acid), a highly specific noncovalently imprinted polymer (MIP), coated on the surface of a gold-coated quartz crystal microbalance (QCM) sensor, could selectively detect the monoterpene menthol in the liquid phase with a sensitivity of 200 ppb and also could distinguish between the D- and L-enantiomers of menthol owing to the enantioselectivity of the imprinted sites. Pavlyukovich et al. [33] showed the utility of divinylstyrene polymer or polyisoprene in piezoelectric chemical sensors for environmental control of trace organic toxicants. Matsuno et al. [34] developed an odor sensor using an array of QCMs coated with a mixture of lipids and poly(vinyl chloride) for the identification of odorants such as amyl acetate, ethanol, acetic acid and water. An epoxy-coated quartz resonator [35] was used for sensing aromas from three wines (red, white and rosé). Hierlemann et al. [36] detected hydrocarbons, chlorinated compounds and alcohols using an array of QCMs coated with side-chain modified polysiloxanes.

Covington and coworkers [37] developed a novel chemFET sensor array using carbon black polymer composite for the detection of organic vapors such as ethanol and toluene. They used three vapor-sensitive carbon black polymer composites with poly(ethylene-co-vinyl acetate), poly(styrene-co-butadiene) and poly (9-vinylcarbazole) as the gate materials in FET devices.

A piezoelectric substance, incorporated in an oscillating electronic circuit, generates a surface acoustic wave (SAW) across the substance. Any change in velocity of the wave, caused by a change in mass of the coating on the piezoelectric substance due to absorbed species, will alter its resonant frequency [38]. Generally, the oscillations are applied to the sensor through a set of metallic electrodes formed on the piezoelectric surface over which a selective coating is formed. The acoustic wave is created by an a.c. voltage applied to a set of interdigited electrodes at one end of the device. The electric field distorts the lattice of the piezoelectric material beneath the electrode, causing an SAW to propagate towards the other end through the acoustic aperture. At the other end, a duplicate set of interdigitated electrodes generates an a.c. signal as the acoustic wave passes underneath them. Similarly to QCM sensors, the selectivity of the SAW device depends on the coating on the sensor, for example, fluoropolymers for sensing of organophosphorus pollutant gas [39].

H_2S and SO_2 A highly sensitive, fast-responding ion-exchange membrane (solid polymer electrolyte) in an electroanalytical sensor [40] was used for the determination of H_2S gas by amperometric monitoring, cathodic stripping measurement and flow injection analysis. A novel electrochemical sensor was developed by Shi et al. [41] for the detection of SO_2 in both gas and solution. They constructed the chemically modified electrode by polymerizing 4-vinylpyridine (4-VP) on to a platinum microelectrode in the presence of tetra-n-butylammonium perchlorate (TBAP) followed by electrodeposition of palladium and iridium oxide (PVP–Pd–

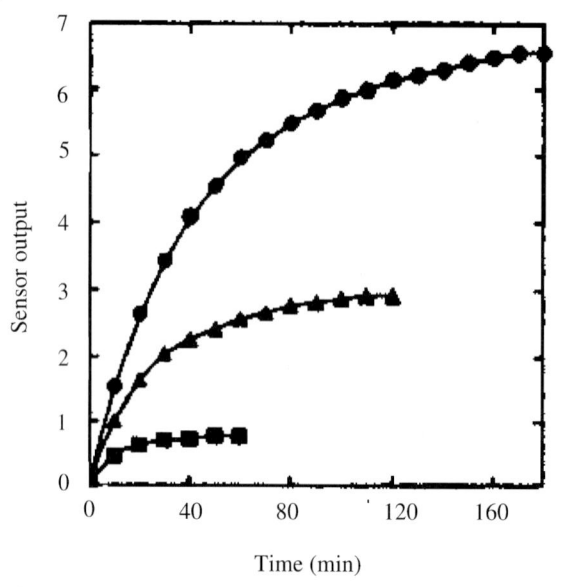

Fig. 13.2 Response characteristics of the sensor using amino-functional copolymers measured for 50 ppm of SO_2 at 30 °C: (●) DMEDA; (▲) DMPDA; (■) DPEDA [42]. Reproduced from Matsuguchi, M., Tamai, K. and Sakai, Y., *Sens. Actuators B*, 2001; 77: 363–367, by permission of Elsevier Science Ltd.

IrO_2). The sensor showed high current sensitivity, a short response time, good reproducibility in SO_2 detection and good potential for use in environmental monitoring and control. Matsuguchi et al. [42] fabricated QCM-type SO_2 gas sensors using an amino-functional poly(styrene-co-chloromethylstyrene) derivative (**I**) on the quartz surface. They used N,N-dimethylethylenediamine (DMEDA), N,N-dimethylpropanediamine (DMPDA) and N,N-dimethyl-p-phenylenediamine (DPEDA) to attach amine groups to the copolymer backbone. The basic amino group absorbs SO_2, being a strong Lewis acid gas. The sensor containing DPEDA functional copolymer showed the shortest response time and complete reversibility, even at 50 °C. Figure 13.2 shows the response characteristics of this SO_2 gas sensor using various amino-functional copolymers.

	R
DMEDA	$-(CH_2)_2-$
DMPDA	$-(CH_2)_3-$
DPEDA	$-CH_{64}-$

Amino-functional poly(styrene-co-chloromethylstyrene) derivate (**I**)

13.2 Different Analytes and Their Sensing Principles

An optical chemical sensor (OCS) was developed for controlling SO_2 in the air of a working area using films of polydimethylsiloxane copolymers with ionically bound Brilliant Green cations [43]. A detection limit of 3 mg m^{-3} indicated the possibility of using the proposed OCS to monitor workplace air.

CO, CO_2 and H_2 Otagawa et al. [44] fabricated a miniaturized electrochemical CO sensor containing three Pt electrodes (sensing, counter and reference) and solution-cast Nafion as a solid polymer electrolyte. They found a linear response with CO concentration in air, a sensitivity of about 8 pA ppm^{-1} and a 70-s response time. Wang et al. [45] reported a new type of solid polymer electrolyte (SPE)-based electrochemical sensor for the detection of CO_2.

Torsi et al. [46] doped electrochemically synthesized polypyrrole (PPy) and poly-3-methylthiophene with copper and palladium. Exposure of PPy and Cu-doped PPy sensors to H_2 and CO reducing gases showed the expected increase in the film resistance. However, the Pd–PPy sensor showed a drop in resistivity (Fig. 13.3a) when exposed to H_2 and CO and ammonia exposure produced a resistivity enhancement (Fig. 13.3b). Moreover, the responses of the Pd–PPy sensor to CO and H_2 were highly reversible and reproducible. Roy et al. [47] reported the hydrogen gas sensing characteristics of doped polyaniline and polypyrrole films. Nafion 117, a solid polymer electrolyte, was used in a compact hydrogen gas sensor based on the amperometric technique at room temperature [48]. Electrodes were prepared chemically by using the impregnation–reduction method, where the polymer was platinized by using a platinum chloroamino complex and sodium tetrahydroborate as impregnating salt and reducing agent, respectively. A maximum sensitivity of about 0.01 µA cm^{-2} ppm^{-1} and a response time of 10–50 s were obtained. A reproducible response with an approximately linear plot of current vs. partial pressure of H_2 was observed for the hydrogen concentration range 1–10%.

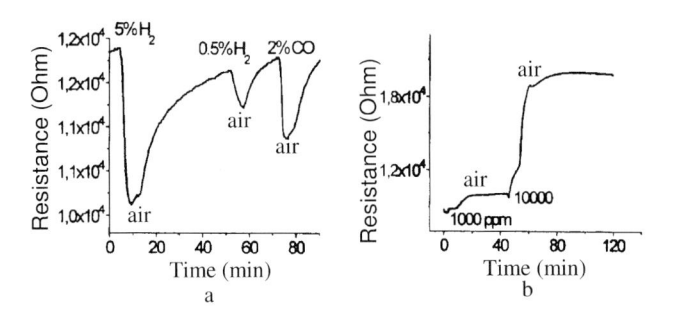

Fig. 13.3 Responses of Pd–Ppy-based gas sensor to different reducing gases: (a) H_2 and CO; (b) NH_3 [46]. Reproduced from Torsi, L., Pezzuto, M., Siciliano, P., Rella, R., Sabbatini, L., Valli, L. and Zambonin, P.G., Sens. Actuators B, 1998; 48: 362–367, by permission of Elsevier Science Ltd.

Hydrocarbons Methane, ethane, propane and butane gases were determined via pre-adsorption on a dispersed platinum electrode backed by Nafion in contact with 10 M sulfuric acid [49]. Nanto et al. [50] used copolymerized propylene–butyl film coated on a QCM for sensing harmful gases such as toluene, xylene, diethyl ether, chloroform and acetone. This gas sensor exhibited high sensitivity and excellent selectivity for these gases, especially toluene and xylene vapors.

A novel chemical gas sensor array composed of polymer–carbon black has recently been fabricated using poly(4-vinylphenol), poly(ethylene oxide) and polycaprolactone mixed with carbon black 2000 as sensing materials immobilized on an array of interdigitated microelectrode pairs [51] for the detection of acetone, toluene and tetrahydrofuran and their mixtures at low concentrations.

A trend has been observed for increased selectivity, response speed and sensitivity of various sensor devices in chemical and biological determinations of gases and liquids. Very recently, Potyrailo [52] nicely summarized the state-of-the-art, the development trends and the remaining knowledge gaps in the area of combinatorial polymeric sensor materials design.

13.2.1.2 Hydrogen Ion

Measurement and control of pH are very important in chemistry, biochemistry, clinical chemistry and environmental science. Polyaniline was found to be suitable for pH sensing in an aqueous medium [53]. Demarcos and Wolfbeis [54] developed an optical pH sensor using PPy prepared by oxidative polymerization of pyrrole. Others [55] have also developed polyaniline-based optical sensors for the measurement of pH in the range 2–12. Jin et al. [56] reported an optical pH sensor based on polyaniline (Table 13.1). The film showed a rapid reversible color change upon pH change. The effect of pH on the changes in the electronic spectrum of polyaniline was interpreted via the different degrees of protonation of the imine nitrogen atoms in the polymer chain [57]. The optical pH sensors do not show any change in sensor performance even when kept exposed in air for over 1 month.

Ferguson et al. [58] used a poly(hydroxyethyl methacrylate) hydrogel with acryloyl fluorescein as pH indicator. Other polymer-based pH sensor devices were also developed [55, 59]. Hydrogen ion-selective solid contact electrodes based on N,N'-dialkylbenzylethylenediamine were prepared. The response range and slopes of these electrodes were influenced by the alkyl chain length. Their 90% response time was <2 s and the electrical resistance varied in the range 2.37–2.76 MΩ. Solid contact electrodes with N,N'-didecylbenzylethylenediamine showed the best selectivity and reproducibility of e.m.f. [59]. Pandey and Singh [60] reported the pH-sensing function of polymer-modified electrodes obtained from electrochemical polymerization of aniline in dry acetonitrile containing 0.5 M tetraphenyl borate. They determined acetic acid in both aqueous and dry acetonitrile and ammonium hydroxide only in aqueous medium. Do et al. [61] investigated solid-contact pH sensors using polyaniline film as an internal solid-contact layer. The resulting sensors exhibited excellent selectivity of pH between 2.5 and 11.3 and a 30-s response time. The performances of five different potentiometric pH sensors based on amino group-containing polymer films obtained from electropolymerization of their respective monomers, 1,3-diaminopropane, diethylenetriamine, pyrrole, p-phenylenediamine and aniline, were compared [62]. Prissanaroon et al. [63] fabricated patterns of PPy on flexible PTFE films using a combination of micro-contact printing, electroless deposition of copper and electropolymerization of pyrrole for pH sensing.

13.2.1.3 Selective Ions

Ion-selective electrodes are suitable for the determination of some specific ions in a solution in the presence of other ions. The quantitative analysis of ions in solutions using ion-selective electrodes (ISEs) is a widely used analytical method. Commercial potentiometric devices of varying selectivity for both cations and anions are used in many laboratories. Ion-sensitive chemical transduction takes place due to ion selectivity conveyed by species such as ionophores – ion-exchange agents, charged carriers and neutral carriers – doped in polymer membranes. Organic salts and several macrocyclics such as antibiotics, crown ethers and calixarenes are used as neutral carriers [64, 65]. Structures of some ionophores are included in Table 13.4. The polymer-based ion-selective device is constructed using an internal electrode and reference solution, a selective membrane across which a potential difference develops and an external reference electrode. The composition of the membrane is mainly responsible for the selectivity and response of the device. A typical membrane for the usual cations and anions consists of ~33 wt% polymer, ~65 wt% plasticizer, ~1–5 wt% ion carrier and ~0–2 wt% ionic additives using PVC as the most common polymer matrix [66]. A list of ionophores and their structures that are commonly used in ion-selective sensor devices by entrapping in polymer films for the detections of different ions are shown in Table 13.5.

Selective cations For the detection of Na^+ in body fluids, a silicone-rubber-based membrane [65] containing a modified calix(4)arene was used. Teixeira et al. [67] used a λ-MnO_2-based graphite–epoxy electrode for the determination of lithium ions. The best potentiometric response was obtained with an electrode containing 35% λ-MnO_2, 15% graphite and 50% epoxy resin with a response time of < 30 s and a lifetime > 6 months.

A new Ca^{2+}-selective PANI-based membrane has been developed [68] for all-solid-state sensor applications. The membrane is made of PANI containing bis[4-(1,1,3,3-tetramethylbutyl)phenyl]phosphoric acid (DTMBP-PO_4H), dioctyl phenylphosphonate (DOPP) and cationic [tridodecylmethylammonium chloride (TDMACl)] or anionic [potassium tetrakis(4-chlorophenyl) borate (KTpClPB)] as lipophilic additives. Artigas et al. [69] described the fabrication of a Ca^{2+}-sensitive electrochemical sensor consisting of a photocurable polymer membrane based on aliphatic diacrylated polyurethane instead of PVC. These polymers are

Table 13.4 Structures of some ionophores used in ion-selective sensor devices.

Ionophore	Structure
Calix[n]arene	phenol with OH, -CH$_2$-, repeat unit n
Dioctyl phenylphosphonate R=(CH$_2$)$_7$CH$_3$ R$_1$ = phenyl	O=P(OR)(OR)-R$_1$
Tridodecylmethylammonium chloride	CH$_3$(CH$_2$)$_{11}$-N$^+$(CH$_3$)(CH$_2$(CH$_2$)$_{11}$)-CH$_3$(CH$_2$)$_{11}$ Cl$^-$
8-Hydroxyquinoline	quinoline with OH at 8-position
Sodium tetraphenylborate	B(C$_6$H$_5$)$_4^-$ Na$^+$
Ethyl-2-benzoyl-2-phenylcarbamoyl acetate	CH$_3$-COO-CH$_2$-CH(C$_6$H$_4$OCNH$_2$)(OC$_6$H$_4$)

compatible with photolithographic fabrication techniques and provide better adhesion to silanized semiconductor surfaces, such as the gate surfaces of ion-selective field-effect transistors (ISFETs).

A Be^{2+}-selective PVC-based membrane electrode was prepared [70] using 3,4-di[2-(2-tetrahydro-2H-pyranoxy)]ethoxystyrene–styrene copolymer (**II**) as a suitable ionophore for the determination of Be^{2+} in mineral samples. The membrane consists of oleic acid and sodium tetraphenylborate as anionic additives and dibutyl phthalate, dioctyl phthalate, acetophenone and nitrobenzene as plasticizing solvent mediators. The best performance was observed for the mem-

Table 13.5 Polymers used in different ion-selective sensors.

Ion	Polymer	Membrane components [a]	Sensor properties	Ref.
Ca^{2+}	Aliphatic diacrylated polyurethane, epoxy resin	1. Ionophore: bis[4-(1,1,3,3-tetramethylbutyl)phenyl] phosphate ionophore 2. Plasticizers: DOPP, TOP, o-NPOE	Quasi-Nernstian Sensitivity (26–27 mV decade^{-1}) in the range 5×10^{-6}–8×10^{-2} M, stable for more than 8 months	69
Zn^{2+}	PVC	1. Ionophore: dimethyl-8,13-divinyl-3,7,12,17-tetramethyl-21H,23H-porphine-2,18-dipropionate (proto-porphyrin IX dimethyl ester) 2. Anion excluder: NaTPB 3. Plasticizer: DOP	Working concentration range: 1.5×10^{-5}–1.0×10^{-1} M with a slope of 29.0 ± 1 mV decade^{-1} of activity, fast response time (10 s), stability >5 months	82
Ni^{2+}	PVC	Neutral carrier: DBzDA18C6	Nernstian response over a wide concentration range (5.5×10^{-3}–2.0×10^{-5} M), fast response time, stability of at least 6 weeks, good selectivity	81
Ca^{2+}, Mg^{2+}	Lipophilic acrylate resin	Calcium salt of bis[4-(1,1,3,3-tetramethylbutyl)phenyl] phosphate as ionophore, 1-decyl alcohol as plasticizer	Nernstian response with a slope of 29 mV decade^{-1} in the concentration range 10^{-5}–10^{-1} M, stability 1 year	79
$H_2PO_4^-$	PVC	o-NPOE (plasticizer), uranyl salophene III (ionophore), TDAB (lipophilic salt)	Linear response in the range 1–4 of $pH_2PO_4^-$ with a slope of 59 mV decade–1	83

DOPP, dioctyl phenylphosphonate; TOP, tris(2-ethylhexyl)phosphonate; o-NPOE, o-nitrophenyl octyl ether; DOP, dioctyl phthalate; NaTPB, sodium tetraphenylborate; DBzDA18C6, 1,10-dibenzyl-1,10-diaza-18-crown-6; TDAB, tetradecylammonium bromide.

brane having the composition 3% PVC, 55% nitrobenzene, 10% **II** and 5% oleic acid, which works well over the concentration range 1.0×10^{-6}–1.0×10^{-3} M. The detection limit of the electrode was 8.0×10^{-7} M (7.6 ng mL^{-1}). An excellent discrimination towards Be^{2+} ion with regard to alkali, alkaline earth, transition and heavy metal ions was obtained with this electrode. Liu et al. [71] developed a fast and simple analytical method for the selective determination of silver ions in electroplating wastewater using PVC membrane electrodes with 5% bis (diethyl dithiophosphates) ionophore and 65% 2-nitrophenyl octyl ether (o-NPOE) plasticizer. This electrode exhibited a linear response over the concentration range 10^{-1}–10^{-6} mol L^{-1} Ag^+. A PVC membrane sensor [72] containing 7-ethylthio-4-oxa-3-phenyl-2-thioxa-1,2-dihydropyrimido[4,5-d]pyrimidine (ETPTP) ionophore exhibited a good potentiometric response for Al^{3+} over a wide concentration range (10^{-5}–10^{-1} M). The sensor provided a stable response for at least 1 month, good selectivity for Al^{3+} in comparison with alkali, alkaline earth, transition and heavy metal ions and minimal interference from Hg^{2+} and Pb^{2+}.

3,4-[Di-(2-tetrahydro-2H-pyranoxy) ethoxy styrene-styrene copolymer (II)

Gupta and Mujawamariya [73] studied the potential response of a Cd^{2+} ISE based on cyanocopolymer matrices and 8-hydroxyquinoline as ionophore by varying the amount of ionophore, the plasticizer and the cyanocopolymer molecular weight. A significant dependence of sensitivity, working range, response time and metal ion interference on the concentration of ionophore, plasticizer and molecular weight of cyanocopolymers was observed. The cyano groups of the copolymers enhanced the selectivity of the electrode, e.g. selectivity for Cd^{2+} ions in the presence of alkali and alkaline earth metal ions in the pH range 2.5–6.5. The electrode had an average life of 6 months and did not show any leaching of membrane ingredients. New lipophilic tetraesters of calix(6)arene and calix(6)diquinone were investigated [74] as Cs^+ ion-selective ionophores in PVC membrane electrodes. The selectivity coefficients for Cs^+ over alkali, alkaline earth and ammonium ions were evaluated. The PVC membrane electrode based on calix(6)arene tetraester showed good detection limits, an excellent selectivity coefficient in pH 7.2 buffer solution and a linear response for Cs^+ ion concentrations of 1×10^{-6}–1×10^{-1} M.

An Ni^{2+}-selective electrode was developed by incorporating 5,7,12,14-tetramethyldibenzotetraazaannulene (Me_4BzO_2TAA) as an electroactive material in a PVC-based membrane [75]. Such electrodes are used for the quantitative determination of Ni^{2+} in chocolates and as an indicator electrode in the potentiometric titration of Ni^{2+} against EDTA. One such membrane prepared from Me_4BzO_2TAA, sodium tetraphenylborate (NaTPB) and PVC in the optimum ratio 2:1:97 (w/w) performed best in the working concentration range (7.9×10^{-6}–1.0×10^{-1} M) of the analyte at pH 2.7–7.6. The sensor exhibited a fast response time of 15 s and good selectivity for nickel(II) over a number of mono-, di- and trivalent cations. A heavy metal ion (Cu^{2+} and Ni^{2+}) sensor was demonstrated for drinking water analysis using a conducting polymer nanojunction array. Each nanojunction was formed by bridging a pair of nanoelectrodes separated with a small gap (<60 nm) with electrodeposited peptide-modified polyanilines [76].

Hassan et al. [77] developed a Hg(II) ion-selective PVC membrane sensor containing ethyl-2-benzoyl-2-phenylcarbamoyl acetate as a novel sensing material. The sensor showed good selectivity for Hg(II) ion in comparison with alkali, alkaline earth, transition and heavy metal ions and was used for the determination of Hg(II) in some amalgam alloys. Mahajan and Parkash [78] observed a high selectivity of a PVC membrane containing bis-pyridine tetraamide macrocycle for Ag^+ ions over that for Na^+, K^+, Ca^{2+}, Sr^{2+}, Pb^{2+} and Hg^{2+}. The electrode has a relatively fast response time and good stability for more than 5 months without any change in response. Numata et al. [79] developed a divalent cation-selective electrode, which utilizes a lipophilic acrylate resin as a matrix for the sensing membrane with long-term stability. The acrylate resin was impregnated with a solution of 1-decyl alcohol and calcium bis[4-(1,1,3,3-tetramethylbutyl)phenyl]phosphate at concentrations of 0.08 g mL^{-1} each. Having equal selectivity to Ca^{2+} and Mg^{2+} ions, such an electrode could be used as a water hardness sensor. The electrode maintained its initial performance for 1 year in a lifetime test conducted in tap water at a continuous flow-rate of 4 mL min^{-1}. The long-term stability of the electrode was attributed to the strong affinity of 1-decyl alcohol to the lipophilic acrylate resin.

Hassan et al. [80] described two novel PVC matrix membrane sensors responsive to uranyl ion. The first sensor contains tris(2-ethylhexyl) phosphate as both electroactive material and plasticizer and sodium tetraphenylborate as an ion discriminator. The sensor displays a rapid and linear response for UO_2^{2+} ions over the concentration range $1 \times 10^{-1} - 2 \times 10^{-5}$ mol L^{-1} at pH 2.8–3.6 and a life span of 4 weeks. The second sensor contains O-(1,2-dihydro-2-oxo-1-pyridyl)-N,N,N',N'-bis(tetramethylene)uranium hexafluorophosphate as a sensing material, sodium tetraphenylborate as an ion discriminator and dioctyl phenylphosphonate as a plasticizer. A linear and stable response for $1 \times 10^{-1} - 5 \times 10^{-5}$ mol L^{-1} UO_2^{2+} was obtained for the sensor at pH 2.5–3.5 with a life span of 6 weeks. Direct potentiometric determination of as little as 5 µg mL^{-1} of uranium in aqueous solution showed an average recovery of 97.2 ± 1.3%.

Using 1,10-dibenzyl-1,10-diaza-18-crown-6 as a neutral carrier, Mousavi et al. [81] constructed a PVC membrane nickel(II) ion-selective electrode, which exhibited a Nernstian response for Ni(II) ions over a wide concentration range ($5.5 \times 10^{-3} - 2.0 \times 10^{-5}$ M). The sensor exhibited good selectivity for Ni(II) over other metal ions and could be used in the pH range 4.0–8.0. It was used as an indicator electrode in the potentiometric titration of nickel ions (Fig. 13.4). Gupta et al. [82] constructed an ion-selective sensor with a PVC-based membrane containing dimethyl 8,13-divinyl-3,7,12,17-tetramethyl-21H,23H-porphine-2,18-dipropionate (protoporphyrin IX dimethyl ester) as the active material, sodium tetraphenylborate (NaTPB) as an anion excluder and dioctyl phthalate (DOP) as solvent mediator, in the ratio 15:100:2:200 (w/w) (I:DOP:NaTPB:PVC). It was used as an indicator electrode for end-point determination in the potentiometric titration of Zn^{2+} against EDTA. The sensor properties are presented in Table 13.5.

Selective anions Wróblewski et al. [83] reported the phosphate ion selectivity of plasticized PVC membranes containing uranyl salophene derivatives. The phosphate selectivity depends on the ionophore structure, dielectric constant and structure of the plasticizer and the amount of incorporated ammonium salt. The highest selectivity for $H_2PO_4^-$ over other anions tested was obtained for lipo-

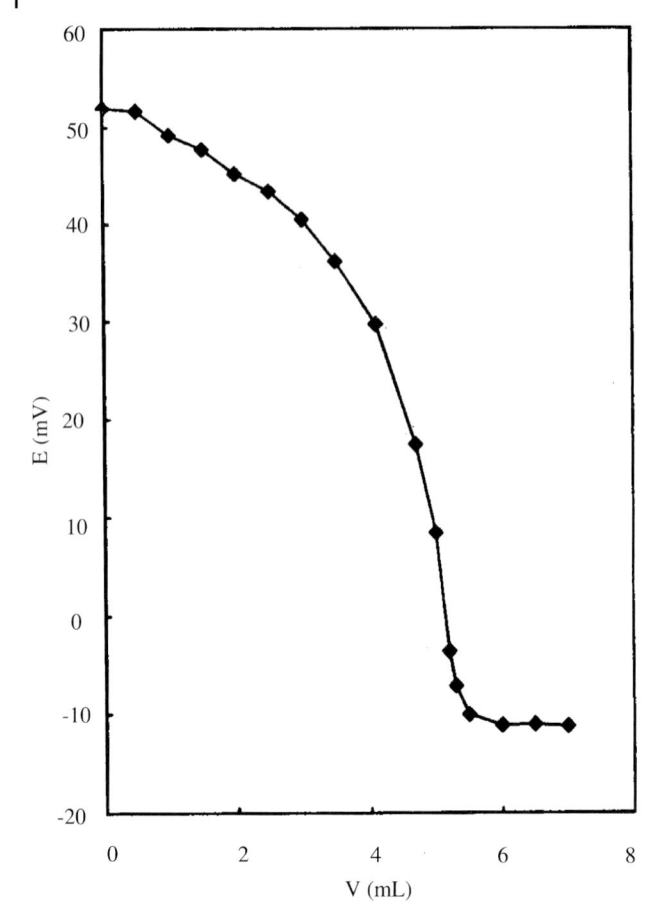

Fig. 13.4 Potentiometric titration curve of 20 mL of 0.01 M Ni(II) solution with 0.04 M EDTA in Tris buffer (pH 8), using the proposed sensor as an indicator electrode [81]. Reproduced from Mousavi, MF., Alizadeh, N., Shamsipur, M. and Zohari, N., Sens. Actuators B, 2000; 66: 98–100, by permission of Elsevier Science Ltd.

philic uranyl salophene III (without ortho substituents) in a PVC–o-nitrophenyl octyl ether membrane containing 20 mol% of tetradecylammonium bromide.

A PVC membrane electrode based on chloro[tetra(m-aminophenyl)porphinato]manganese [T(m-NH$_2$)PPMnCl] and 2-nitrophenyl octyl ether (o-NPOE) in the ratio 3:65:32 [T(m-NH$_2$)PPMnCl:o-NPOE:PVC] was prepared by Zhang et al. [84] for the determination of fluoroborate in electroplating solution. They obtained a Nernstian response to fluoroborate ion in the concentration range 5.1×10^{-7}–1.0×10^{-1} mol L^{-1}, with a pH range of 5.3–12.1 and a response time of 15 s. An improved selectivity towards BF$_4^-$ with respect to common coexisting ions was obtained. For anion-selective electrodes, Torres et al. [85] developed five different types of membranes by solubilizing poly(ethylene-co-vinyl acetate) copolymer (EVA) and tricaprylyltrimethylammonium chloride (Aliquat-336S) in chloroform followed by film casting. These ion-selective membrane electrodes are useful for iodide, periodate, perchlorate, salicylate and nitrate determinations in the concentration range 10^{-5}–10^{-1} mol L^{-1}. These systems were employed for the determination of salicylate and iodide in pharmaceutical samples.

Espadas-Torre and Meyerhoff [86] compared the thrombogenic properties of various polymer matrices used as implantable ion-selective membrane electrodes by *in vitro* platelet adhesion studies. A high molecular weight block copolymer of poly(ethylene oxide) and poly(propylene oxide) within ion-selective membranes reduces platelet adhesion. A thin photo-cross-linked poly(ethylene oxide) coating on a plasticized PVC membrane decreased platelet adhesion. Such surface-modified membranes retained selectivity and response time essentially equivalent to those for the untreated membranes.

13.2.1.4 Alcohols

In industrial and clinical analyses and also in biochemical applications, the determination of alcohols is essential. Due to the increasing popularity of conducting polymers as competent sensor materials for organic vapors, polyaniline has been used as a sensor material for alcohol vapors, such as methanol, ethanol and propanol [87, 88]. Good responses for alcohol vapors were also observed with polyaniline doped with camphorsulfonic acid [89, 90]. The sensing mechanism was better understood on the basis of the crystallinity of polyaniline. Athawale and Kulkarni [91] reported that polyaniline and its derivatives (**III**) such as poly(o-toluidine), poly(o-anisidine), poly(N-methylaniline), poly(N-ethylaniline), poly(2,3-dimethylaniline), poly(2,5-dimethylaniline) and polydiphenylamine are sensitive to methanol, ethanol, propanol, butanol and heptanol vapors (Table 13.1). These polymers respond to the saturated alcohol vapors by undergoing a change in electrical resistance of the polymer film. While short-chain alcohols,

Polymer	R_1	R_2	R_3	R_4
Polyaniline	H	H	H	H
Poly(o-toluidine)	H	CH$_3$	H	H
Poly(o-anisidine)	H	OCH$_3$	H	H
Poly(N-methyl aniline)	CH$_3$	H	H	H
Poly(N-ethyl aniline)	C$_2$H$_5$	H	H	H
Poly(2,3-dimethylaniline)	H	CH$_3$	CH$_3$	H
Poly(2,5-dimethyaniline)	H	CH$_3$	H	CH$_3$
Poly(diphenyl amine)	C$_6$H$_5$	H	H	H

Polyaniline derivatives (**III**)

viz. methanol, ethanol and propanol, decreased the resistance, an opposite trend in resistance change was observed with butanol and heptanol vapors. They attributed the change in resistance of the polymers exposed to different alcohol vapors to their chemical structure, chain length and dielectric nature. They explained the results based on the vapor-induced change in the crystallinity of the polymer. Polypyrrole was also studied as a sensing layer for alcohols. Polypyrrole [92] containing dodecylbenzenesulfonic acid (DBSA) and ammonium persulfate (APS) showed a linear change in resistance on exposure to methanol vapor (87–5000 ppm). Bartlett and Ling-Chung [93] also detected methanol vapor by the change in resistance of polypyrrole film.

The use of a poly(hydroxyethyl methacrylate) film as a sensor to measure alcohol-induced thickness changes was reported [94]. Another effort was made to sense aqueous ethanol by Blum et al. [95] using two lipophilic derivatives of Reichardt's phenolbetaine dissolved in thin layers of plasticized poly(ethylene–vinyl acetate) copolymer coated with microporous white PTFE in order to facilitate reflectance measurements. In fact, the sensor layers detected aqueous ethanol via the change in color from green to blue with increasing ethanol content in the solution. For the detection of ethanol and methanol at low concentrations, a novel gas sensor array of polymer–carbon black has recently been fabricated using poly(4-vinylphenol), poly(ethylene oxide) and polycaprolactone mixed with carbon black 2000 as the sensing materials immobilized on an array of interdigitated microelectrode pairs [51].

13.2.1.5 Drugs

El-Ragehy et al. [96] reported PVC membrane sensors for the determination of fluphenazine hydrochloride and nortriptyline hydrochloride based on the formation of ion-pair complexes between the two drug cations and sodium tetraphenylborate (NaTPB) or tetrakis(4-chlorophenyl) borate (KtpClPB). A novel PVC membrane electrode was developed by Liu et al. for the determination of pethidine hydrochloride drug in injections and tablets [97] (Table 13.1). For sensing stimulants such as phentermine, a PVC membrane electrode in combination with sodium tetrakis[3,5-bis(2-methoxyhexafluoro-2-propyl)phenyl]borate (NaHFPB), an ion exchanger, was developed [98]. The structure of phentermine is similar to that of amphetamine, a phenylalkylamine (**IV**). This electrode could discriminate between phentermine and analogous compounds.

Phentermine   Amphetamine   Phenylethylamine

Structures of a series of phenylalkylamines (**IV**)

13.2.1.6 Amines

For optical sensing of dissolved aliphatic amines, the absorbance-based chromoreactand 4-(N,N-dioctylamino)-4'-trifluoroacetylazobenzene (ETH^T 4001) was investigated [99] in different polymer matrices such as plasticized PVC, copolymers of acrylates, polybutadiene and silicone. Sensor layers with ETH^T 4001 and different polymer materials showed a decrease in absorbance at around 500 nm and an increase in absorbance at around 420 nm on exposure to dissolved aliphatic amines. Conversion of the trifluoroacetyl group of the reactant into a hemiaminal or a zwitterion caused a change in absorbance. The polarity of the polymer strongly influences the diol formation caused by conditioning in water and the absorbance maximum of the solvatochromic reactant. However, the selectivity of the sensor layers remains nearly unaffected by the polymer matrix for primary, secondary and tertiary amines. A conjugated polymer-based sensor array using regioregular poly(thiophene-3-propionic acid) was developed for the differential detection and quantification of biogenic diamines generated from the spoilage of foods [100].

13.2.1.7 Surfactant

There is a need for the determination of surfactant concentrations in industrial samples and food samples and in the environment. Comprehensive reviews are available on surfactant analysis [101]. An electrochemically prepared or solution-cast conducting polymer layer as a charge-transfer mediator between the ISE membrane and the solid substrate formed stable neutral carrier-type ISEs [102]. Bobacka et al. [103] also made a single-piece all-solid-state electrode by dissolving an appropriate conducting polymer in the PVC matrix of the ISE. Using teflonized graphite rods coated with an electrochemically deposited polypyrrole film as the electric connector support, an all-solid-state anionic surfactant electrode was developed [104]. Due to charge transfer at the graphite–polypyrrole–membrane interfaces, the surfactant electrode showed good stability.

13.2.1.8 Insecticides and Fungicides

Thin films of molecularly imprinted polymers on the surface of polypropylene membranes and on hydrophobized gold electrodes were prepared by graft polymerization for the detection of herbicides such as desmetryn [105]. On irradiation with UV light, an adsorbed layer of benzophenone initiated a radical polymerization near the surface. The electrodes coated with the molecularly imprinted polymers displayed specific binding of desmetryn. Only small capacitive effects were observed on addition of terbumeton or atrazine, whereas metribuzine displayed a capacitance decrease similar to that of desmetryn.

A new bulk acoustic wave (BAW) sensor modified with a cross-linked molecularly imprinted polymer (MIP) of methacrylic acid monomer using pyrimethamine as template was fabricated and applied for the determination of pyrimethamine (2,4-diamino-5-p-chlorophenyl-6-ethylpyrimidine), an effective antimicrobial agent, in serum and urine media with high selectivity and sensitive response [106]. The determination limit was 2.0×10^{-7} M. The sensor exhibited long-term stability in harsh chemical environments such as high temperature, organic solvents, bases, acids, etc. An imprinted polymer was prepared by reacting p-aminobenzoic acid and 1,2-dichloroethane in the presence of a typical diquat herbicide, 5,6-dihydropyrazine[1,2,3,4-lmn]-1,10-phenanthrolinium dichloride (DHPPCl$_2$) [107]. Such an imprinted polymer-modified hanging mercury drop electrode (HMDE) in a Model 303A system in conjunction with a PAR Model 264A polarographic analyzer/stripping voltammeter was used for the selective analysis of diquat herbicide in the differential-pulse cathodic stripping voltammetry mode. Drinking water and agricultural soil suspensions, spiked with a diquat herbicide, were directly analyzed by the adsorptive accumulation of the analyte over the working electrode. The limit of detection for diquat herbicide was found to be 0.34 nmol L^{-1} (0.1 ppb, RSD 2%, S/N=2). Jenkins et al. [108] used cross-linked polystyrene as molecularly imprinted polymer for specific sequestering of non-hydrolyzed organophosphate-containing pesticides and insecticides. These polymers were coated on optical fibers and used as optical sensors for the detection of these species in aqueous environments. The superior stability, sensitivity, selectivity and reversibility of such materials provided a real-time sensing application for molecularly imprinted polymers. Detection limits for these MIP sensors are reported to be less than 10 parts per trillion (ppt) with long linear dynamic ranges (ppt to ppm) and response times of less than 15 min. Li et al. [109] fabricated a novel organophosphate pesticide (OP) parathion sensor based on p-tert-butylcalix[6]arene-1,4-crown-4 as functional monomer by a sol–gel method using molecular imprinting technology. They characterized the electrochemical behavior of parathion at the imprinted p-tert-butylcalix[6]arene-1,4-crown-4 sol–gel film sensor by cyclic voltammetry, linear sweep voltammetry, chronoamperometry and a.c. impedance spectroscopy. A fast response of parathion was obtained after incubation in 0.1 M phosphonate buffer solution containing an appropriate amount of parathion for 20 min with a detection limit of 1.0×10^{-9} M (S/N=3). Thin films of a molecularly imprinted sol–gel polymer of tetraethyl orthosilicate (99%, TEOS), phenyltrimethoxysilane (99%, PTMOS) and aminopropyltriethoxysilane with specific binding sites for parathion were developed by Marx et al. [110]. The films cast on glass substrates and on glassy carbon electrodes were used to detect parathion in aqueous solutions. While the gas-phase binding measurements were performed on coated QCM resonators, the binding of parathion to the imprinted films in the liquid phase was investigated by GC/FPD and cyclic voltammetry. The imprinted films showed high selectivity towards parathion in comparison with similar organophosphates. The MIPs were based on hydrogen bonding as a means of enhancing specific recognition.

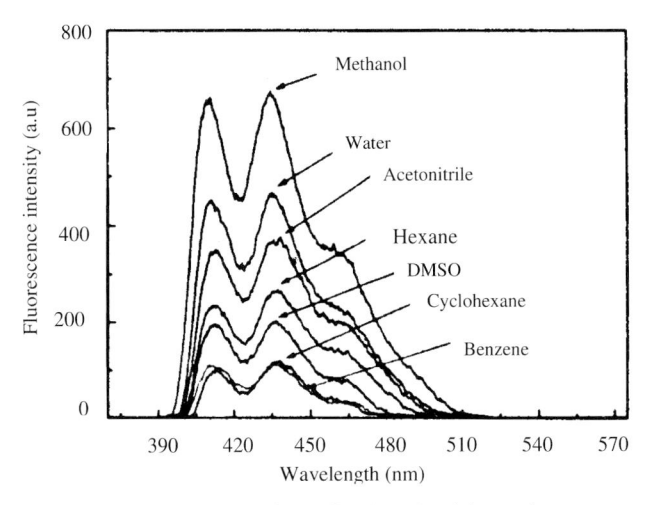

Fig. 13.5 Emission spectra of PVA film doped with benzo[k]-fluoranthene in different solvent media (λ_{ex}=308 nm) [111]. Reproduced from Patra, D. and Mishra, A.K., *Sens. Actuators B*, 2001; *80*: 278–282, by permission of Elsevier Science Ltd.

13.2.1.9 Aromatic Compounds

An optical sensor was developed for the detection of nitrobenzene, *m*-dinitrobenzene, *o*-nitrotoluene, *m*-nitrotoluene, *p*-nitrobromobenzene, *o*-nitroaniline, *p*-nitrophenol, etc., by fluorescence quenching of benzo[*k*]fluoranthene (BkF) in poly(vinyl alcohol) (PVA) film [111]. The fluorescence spectra of BkF-doped PVA films in various solvents shown in Fig. 13.5 demonstrate that the sensor film gives a good fluorescence quantum yield in methanol compared with other solvents, because of enhanced swelling of the film in methanol. A good selective response to aromatic hydrocarbons was observed with polypyrrole–nitrotoluene copolymer [112].

13.2.1.10 Hydrazine

For the detection of hydrazine and other compounds in flow injection analysis (FIA), Wang and Li [113] reported a novel composite electrode coating with a mixture of cobalt phthalocyanine and Nafion, which showed better properties than either of the two components alone. The analytical utility of the sensor was established by selective flow injection measurements of hydrazine with a detection limit of 5.7 ng or hydrogen peroxide in the presence of oxalic or ascorbic acids, respectively. Hou and Wang [114] developed a Prussian Blue-modified glassy carbon electrode coated with Nafion film for the detection of hydrazine in FIA with a detection limit of 0.6 ng. A Nafion–ruthenium oxide pyrochlore ($Pb_2Ru_{2-x}Pb_xO_{7-x}$)-modified glassy carbon electrode exhibited excellent electrocatalytic activity in the oxidation of hydrazine in neutral medium. Zen and Tang

[115] prepared the catalyst directly inside a spin-coated Nafion thin-film matrix on a glassy carbon electrode.

13.2.1.11 Humidity

The measurement and control of humidity are important in the paper, food and electronic industries, the domestic environment (air conditioners), medical areas (respiratory equipment), etc. Hence humidity sensors are useful for the detection of relative humidity (RH) in those environments. Polymers with hydrophilic properties are suitable for making humidity sensor devices. Table 13.6 describes some polymer-based humidity sensors and their properties. In order to measure the variation of electrical conductivity with water vapor, ion conducting polymers are used in humidity sensor devices. Apart from this type of polymer, electrolytes and mixtures or complexes of inorganic salts with polymer are the major materials for humidity sensors. Ionic copolymers from a variety of ionic and nonionic monomers, such as methyl methacrylate, styrene, methyl acrylate and 2-hydroxyethyl methacrylate as the nonionic monomers and sodium styrenesulfonate, sodium 2-acrylamide-2-methylpropanesulfonate (V), sodium methacrylate, ethacrylolyloxyethyldimethylammonium chloride, methacryloyloxyethyltrimethylammonium chloride and methacryloyloxyethyldimethyloctylammonium chloride as the ionic monomers were used for the preparation of humidity sensors [116]. The response time of this type of sensor varied in the order $-SO_3^- > -CO_2^- > -C_2H_4N^+H(CH_3)_2 > -C_2H_4N^+(CH_3)_3 > -C_2H_4N^+(C_8H_{17})(CH_3)_2$ with a rapid change of relative humidity from 40 to 60%.

$$\left[CH_2-CH \right]_n$$
$$\quad\quad\quad |$$
$$CONH_2C(CH_3)_2SO_3Na$$

Sodium salt of poly(2-acrylamide-2-methyl propane sulfonic acid) (V)

Table 13.6 Polymers used in humidity sensors

Polymer	Principles	Sensor properties	Ref.
2-Acrylamido-2-methyl propanesulfonate modified with tetraethyl orthosilicate	Electrical property measurement	Less hysteresis <2%, good linearity, 30–90% working range humidity, long-term stability of at least 31 days	126
Iron oxide–PPy nano-composite	Electrical property measurement	Sensitivity increases with increasing concentration of polypyrrole	128
Quaternized and cross-linked poly(chloromethyl styrene)	QCM sensor	Degree of hysteresis decreases with increase in quaternization	129
Crystal Violet and Methylene Blue incorporated in PVA–H_3PO_4 solid polymer electrolyte	Optical humidity sensor	Shows linearity of response	127

Humidity-sensitive copolymers were prepared by grafting hydrophilic monomers on hydrophobic polymers such as grafting of polystyrene on PTFE followed by sulfonation of the polystyrene branch and grafting of 4-vinylpyridine on PTFE film followed by quaternization with alkyl halide [117]. Microporous polyethylene (100 µm thick) film with 70% porosity was also used as the base polymer. 2-Acrylamide-2-methylpropanesulfonic acid (AMPS) or 2-hydroxy-3-methacryloyloxypropyltrimethylammonium chloride (HMPTAC) [118] was grafted on to the microporous polyethylene by benzoyl peroxide initiation or ultraviolet irradiation using benzophenone as a sensitizer.

Hydrophobic polymers were chemically modified by generating ionic groups to obtain a humidity-sensitive material. Sulfonation of polysulfone [119] showed good sensitivity to humidity. Plasma polymerization of organosilicones containing amino or amine oxide groups such as trimethylsilyldimethylamine, tetramethylsilane plus ammonia, bis(dimethylamino)methylsilane or bis(dimethylamino)methylvinylsilane, followed by treatment with methyl bromide, was used to fabricate humidity-sensing devices [120]. Various silane derivatives are shown in structures **VI**.

	R_1	R_2	R_3	R_4
Silane	H	H	H	H
Trimethylsilyldimethyl amine	CH_3	CH_3	CH_3	$(CH_3)_2NH$
Tetramethylsilane	CH_3	CH_3	CH_3	CH_3
Bis(dimethyl amino)methyl silane	CH_3	$(CH_3)_2NH$	$(CH_3)_2NH$	H
Bis(dimethyl amino)methyl-vinyl silane	CH_3	$(CH_3)_2NH$	$(CH_3)_2NH$	$CH_2=CH$

Silane derivatives (**VI**)

Surface-functionalized polymers such as sulfonated, lithiated and gold nanocluster-attached polyethylene and polypropylene have been found to be suitable as humidity sensors [121]. These polymers showed short response times and promising humidity-dependent resistance changes (10^9–10^6 Ω with change in RH from 30 to 95%).

Simultaneous quaternization and cross-linking of poly-4-vinylpyridine with an α,ω-dichloroalkane showed humidity-sensing properties [122]. Likewise, simultaneously cross-linked and quaternized poly(chloromethylstyrene) film by N,N,N',N'-tetramethyl-1,6-hexanediamine showed humidity-sensing behavior [123]. Rauen et al. [124] reported the humidity-sensing behavior of a spin-coated ionically conductive poly(dimethyldiallylammonium chloride) on a ceramic wafer. The hydroxy group of poly(2-hydroxy-3-methacryloxypropyltrimethylammonium chloride) (HMPTAC) (**VII**) was cross-linked by a diisocyanate to fabricate a humidity sensor [125]. An IPN film of cross-linked HMPTAC and ethylene

Poly(2-hydroxy-3-methacryloxypropyltrimethyl ammonium chloride) (**VII**)

glycol dimethacrylate (EGDMA) polymers was formed on a substrate with interdigitated electrodes [125]. The impedance of the sensor thus prepared decreased from 10^7 to 10^2 Ω from 0 to 90% RH.

Su et al. [126] fabricated a humidity sensor by thick-film deposition of poly(2-acrylamido-2-methylpropane sulfonate) modified with tetraethyl orthosilicate (TEOS) as the sensing material. They investigated the effect of adding triethylamine or diethylamine and, the dosage of TEOS on the capacity resistance to high humidity atmospheres and the thickness of the film. Certain dyes such as Crystal Violet and Methylene Blue can serve as excellent candidates for optical humidity sensing when incorporated in solid polymer electrolytes such as poly(vinyl alcohol) (PVA)–H_3PO_4, which is a good proton conductor [127]. Methylene Blue forms a charge-transfer complex with the PVA–H_3PO_4 and thereby changes its optical properties due to the change in humidity.

Nanocomposite pellets of iron oxide and polypyrrole were prepared for humidity and gas sensing (CO_2, N_2 and CH_4) by a simultaneous gelation and polymerization process [128]. Variation of the sensitivity of this sensor with change in relative humidity increased with increasing polypyrrole concentration (Fig. 13.6). Using a cross-linked and quaternized poly(chloromethylstyrene) (PCMS) (**VIII**) film, Sakai et al. [129] prepared a resistive-type humidity sensor. The sorption isotherm curves of water vapor in various cross-linked films were obtained using a QCM. The percentage hysteresis depended on the density of

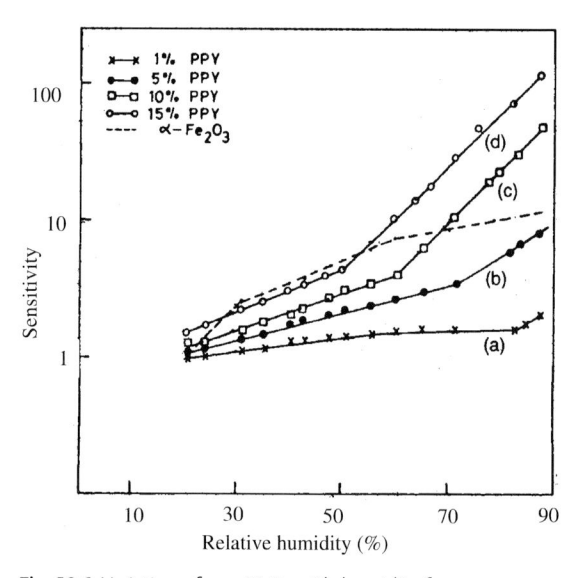

Fig. 13.6 Variation of sensitivity with humidity for sensors having (a) 1, (b) 5, (c) 10 and (d) 15% polypyrrole [128]. Reproduced from Suri, K., Annapurni, S., Sarkar, A.K. and Tandon, R.P., *Sens. Actuators B*, 2002; *81*: 277–282, by permission of Elsevier Science Ltd.

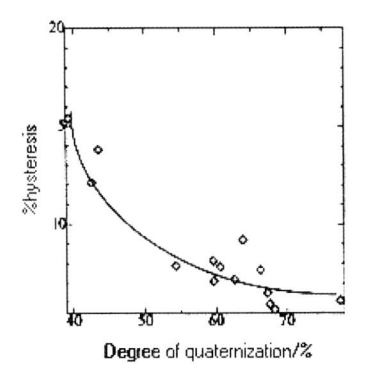

Fig. 13.7 Plot of % hysteresis at 40% RH as a function of degree of conversion [129]. Reproduced from Sakai, Y., Matsuguchi, M. and Hurukawa, T., *Sens. Actuators B*, 2000; 66: 135–138, by permission of Elsevier Science Ltd.

the quaternary ammonium groups, which affects the diffusion coefficient of the water molecules in the film (Fig. 13.7). Sun and Okada [130] investigated the interaction between methanol, water and Nafion (Ag) and determined the concentration of methanol and water (RH) using a QCM coated with Nafion film recast from Nafion (Ag) complex solution.

Cross-linked and quaternized poly(chloromethyl styrene) (VIII)

13.2.2
Biosensor Analytes

A biosensor, which transforms a biological event into an electrical signal, is a combination of a bioreceptor, the biological component and a transducer. Applications of different sensors are summarized in Table 13.1. The estimation of organic compounds is necessary for the control of food manufacturing processes and for the evaluation of food quality. Enzyme sensors are used for the direct measurement of such compounds, including organic pollutants for environmental control. Since hydrogen peroxide is used in the food, textile and dye industries for bleaching and sterilization purposes, can directly measure it:

$$H_2O_2 \xrightarrow{\text{Catalase}} H_2 + \tfrac{1}{2}O_2$$

Biosensors are based on the coimmobilization of an enzyme in a conducting polymer, namely polypyrrole [131, 132], and polyaniline [133]. Ichimura [134] photochemically immobilized enzymes in photo-cross-linkable poly(vinyl alcohol) (**IX**) bearing stilbazolium groups. Such a film of the photosensitive polymer containing the enzyme showed high enzyme activity.

Photocrosslinkable poly(vinyl alcohol) bearing stilbazolium group (**IX**)

Synthetic polymers having specific receptor structures are required in biology, medicine and biotechnology. Methacrylic polymers are selective for nucleotides, amino acids and herbicides [135]. Their selectivity depends on the extent and nature of the interactions between the substrate and the stationary polymer phase, the shape of the imprinting molecules and the polymer cavities. Peng et al. [136] fabricated a new bulk acoustic wave (BAW) sensor modified with a molecularly imprinted polymer (MIP), poly(methacrylic acid-*co*-ethylene glycol dimethacrylate), for the determination of pyrimethamine in serum and urine media and obtained high selectivity and a sensitive response to pyrimethamine.

The combination of a transducer and a thin enzymatic layer, which normally measures the concentration of a substrate, forms an enzyme sensor. The external surface of the enzymatic layer is kept immersed in a solution containing the substrate under study. The substrate, after migrating to the inner layer, reacts with the immobilized enzyme. In biosensors, polymers are the most suitable materials to immobilize the enzyme, the sensing component, and hence increase the sensor stability. In Table 13.7, polymers used in different enzyme sensors for the determination of glucose, urea and amino acids are shown.

Table 13.7 Polymers used in different enzyme biosensors for determination of glucose, amino acids, urea, etc.

Analyte	Polymer	Sensing elements	Sensor properties	Ref.
Glucose	Poly(o-aminophenol)	GOD	Response time < 4 s, lifetime > 10 months	137
	5-(1-Pyrrolyl)pentyl-2-(trimethylammonium)-ethyl phosphate, Nafion or poly(o-phenylenediamine) (PPD) as inner membrane	GOD	Influence of ascorbic acid eliminated, stability of 200 days	150
L-Amino acids	Polytyramine	L-Amino acid oxidase (L-AAOD)	Lower limit of detection 0.07 mM; stability > 1 month	166
Glucose, urea, triglycerides	PANI	GOD, urease, lipase	–	139

13.2.2.1 Glucose

Glucose determination is a routine analysis in clinical chemistry and in the microbiological and food industries. Diabetes has attracted continuous interest with respect to the development of an efficient glucose sensor (Table 13.1). An artificial pancreas dynamically responds to glucose level and controls insulin release based on the sensor's response. Zhang et al. [137] made a glucose biosensor by immobilizing glucose oxidase in an electropolymerized o-aminophenol film on a platinized glassy carbon electrode. Jung and Wilson [138] also developed a miniaturized (1.12 mm^2 sensing area) and potentially implantable amperometric glucose sensor based on an enzymatically catalyzed reaction. Polyaniline-based microsensors and microsensor arrays were fabricated for the estimation of glucose, urea and triglycerides [139] using microelectronics technology and electrochemical deposition of polymer and immobilization of enzyme (glucose oxidase, urease and lipase).

A glucose biosensor [140] was made with a composite membrane of sol–gel enzyme film and electrochemically generated poly(1,2-diaminobenzene) film with improved selectivity of the sol–gel enzyme sensors. Reliable results were obtained in assays of glucose in controlled human sera. Similar sensors were also developed for galactose and cholesterol, although their sensitivity was inferior to that of the glucose sensor.

Glucose-sensing enzyme electrodes with a low interference level were constructed using polydimethylsiloxane (PDMS) [141] by placing an enzyme [glucose oxidase (GOx) bilayer] on a PDMS-coated electrode. Another enzyme electrode with a PDMS layer containing lipid-modified GOx was prepared by placing aqueous dispersion of the polymer and the modified enzyme on the base electrode surface followed by drying. The modified GOx electrode is better for measuring glucose in high concentrations (around 5 mM) compared with the bilayer system. Myler et al. [142] used an ultra-thin polydimethyldichlorosiloxane film composite membrane as the outer covering barrier in an amperometric glucose oxidase enzyme electrode biosensor.

A novel amperometric biosensor was developed [143] for the cathodic detection of glucose using GOx and horseradish peroxidase (HRP). This was constructed by electrochemical deposition of polypyrrole (PPy) in presence of GOx on the surface of an HRP-modified sol–gel-derived mediated ceramic carbon electrode, using ferrocenecarboxylic acid as mediator to transfer electron between the enzyme and electrode. Miniaturized, disposable and interference-free amperometric biosensors [144] were developed from PPy film immobilizing GOx for glucose determination in serum. GOx was immobilized, either by entrapment in the electropolymerized film by an 'all-electrochemical' procedure (PPYox–GOx) or by gel entrapment over the PPYox-modified electrode by co-cross-linking with glutaraldehyde–bovine serum albumin (PPYox–GOx gel). A comparison of the glucose sensitivity showed that both sensors were linear up to 10 mM, although the Pt–PPYox–GOx gel sensor is the more sensitive of the two (168±15 versus 53±7 nA mM^{-1}). Copolymers of 3,4-ethylenedioxythiophene (EDOT) and modified EDOT-containing hydroxyl groups were electrochemically prepared by Kros et al. [145]. These copolymers bind proteins at the surface through the covalent coupling of GOx and were able to detect glucose amperometrically under aerobic and anaerobic conditions. Revzin et al. [146] prepared glucose, lactate and pyruvate biosensor arrays using redox polymer–oxidoreductase nanocomposite thin films deposited on photolithographically patterned gold microelectrodes.

Biodegradable synthetic polymers containing hydroxyl (**X**) or gluconamide (**XI**) groups are good immobilization matrixes for GOx [147]. The responses of these biosensors to glucose were measured by potentiostating the electrodes at 0.6 V vs. SCE in order to oxidize the hydrogen peroxide generated by the enzymatic oxidation of glucose in the presence of oxygen. The response of such electrodes was evaluated as a function of film thickness, pH and temperature. The polymers **X** and **XI** were synthesized by emulsion polymerization [148].

Poly(5-hydroxymethyl-bicyclo-2,2,1-hept-2-ene) (**X**)

Poly(N-methylnorbornylgluconamide) (**XI**)

An amperometric glucose sensor [149] was developed with electrodes coated with a four-layered membrane, 3-aminopropyltriethoxysilane, Nafion, GOx and perfluorocarbon polymer, for accurate and successive determination of glucose concentrations between 2.8 and 167 mM, over a 66-day period with no increase in response time.

New hemocompatible glucose sensors were developed using electropolymerized pyrrole derivatives with phosphatidylcholine, 5-(1-pyrrolyl)pentyl-2-(trimethylammonium)ethyl phosphate in the presence of GOx, and an inner membrane of Nafion or poly(o-phenylenediamine) [150]. These membranes not only eliminated the influence of ascorbic acid on the sensor response, but also increased the electrode stability. Selective amperometric glucose sensors were prepared by immobilization of GOx in a polyaniline layer. The polyaniline was prepared by electropolymerization of aniline using phosphate buffer on a Prussian Blue-modified platinum electrode. This sensor completely eliminated the influence of ascorbic acid and acetaminophen due to the impermeability of polyaniline to these substances [151]. A simple electropolymerization process [152] was followed for the fabrication of an ultra-thin PPy–GOx film (~55 nm) in a supporting electrolyte-free monomer solution for potentiometric biosensing of glucose. It was reported that this biosensor could be used repeatedly for over 2 months.

A new dipyrrolic derivative linked with a long hydrophilic spacer has been synthesized [153] by electrooxidation in water to form a polymer film on the electrode surface and GOx was immobilized in it. The amperometric response of the resulting biosensor to glucose was studied through the oxidation of the H_2O_2 generated. A new miniaturized glucose sensor with good selectivity and stable current response was fabricated by simultaneous electropolymerization of m-phenylenediamine and entrapment of Gox as an inner layer covered by a poly(m-phenylenediamine) film without GOx. The inner layer was covered with an outer bilayer of polytetrafluoroethylene and polyurethane films [154].

Teh and Lin [155] utilized conductive polymer-based nanocomposite as a MEMS sensing material via a one-step, selective, on-chip deposition process at room temperature. A doped PPy variant synthesized by incorporating multi-walled carbon nanotube (MWCNT) into electropolymerized PPy improved the sensing performance. The dodecylbenzenesulfonate (DBS)-doped PPy–MWCNT nanocomposites were responsive to H_2O_2, which was extended to glucose detection. The oxidant sensing effect was demonstrated by subjecting a glucose oxidase (GOx)-laden PPy–MWCNT nanocomposite film to various concentrations of glucose solution. Experimental GOx-laden, doped PPy–MWCNT was found to be sensitive to glucose concentrations up to 20 mM, covering the physiological range of 0–20 mM for diabetics. A highly sensitive and fast glucose biosensor was developed using glucose oxidase immobilized in mesocellular carbon foam of a Nafion membrane [156].

Bridge and Higson [157] used a polyacrylonitrile thin-film composite membrane as the outer covering barrier in an amperometric GOx enzyme electrode. Glucose determinations within patients' whole blood samples performed using this enzyme electrode sensor compared well with standard hospital analyses. A disposable glucose sensor based on a differential amperometric measurement was fabricated on activated carbon paste electrodes, using GOx immobilized in a nitrocellulose membrane [158]. Two identical three-electrode cells were screen-printed symmetrically on both faces of a single polyester substrate. Both electrodes of the two-sided sensor strip were covered with nitrocellulose membranes incorporating a mediator $\{K_3[Fe(CN)_6]\}$, one membrane with GOx and the other with bovine serum albumin (i.e. no GOx). The former served as a glucose-sensing cell and the other as a reference cell.

Kobayashi and Suzuki [159] utilized the swelling and shrinking of poly(N-isopropylacrylamide) aqueous gel in response to a variation in temperature to construct a micro glucose sensor based on the detection of hydrogen peroxide. After the gel was shrunk at 40 °C making contact with a buffer solution, the inlet of the sampling mechanism was placed in contact with a sample solution at 30 °C, which was introduced and moved into the interior of the system following the swelling of the gel.

13.2.2.2 Urea

Urea is the end product of protein degradation in the body and its content in blood serum depends on protein catabolism, nutritive protein intake and its renal excretion. Urea estimation in blood is important in clinical chemistry for monitoring kidney functioning. Mizutani et al. [160] prepared an amperometric urea sensor by immobilizing pyruvate oxidase (PyOx), pyruvate kinase and urea amidolyase on a polydimethylsiloxane-coated electrode. They monitored the oxygen consumption without interference from the PyOx reaction product, H_2O_2, and determined urea in a test solution (5 µM–0.35 mM) containing adenosine 5′-diphosphate and phosphoenolpyruvate with the trienzyme system. Pandey and Singh [161] used a polymer-modified electrode for the construction of a urea biosensor by immobilizing urease in a poly(vinyl alcohol) matrix and also in organically modified sol–gel glass on a polymer-modified electrode surface. This urea sensor showed a 160 mV response at 25 °C with a detection limit of 20 µM urea.

Komaba et al. [162] prepared a urea biosensor by immobilizing urease in an electropolymerized PPy on a Pt electrode, which showed a stable response potential to urea based on the pH response of the electroinactive PPy film electrode. A highly sensitive and rapid flow injection system for urea analysis was constructed with a composite film of electropolymerized inactive PPy and a polyion complex incorporating urease [163]. This system showed a sensitivity of 120 mV decade^{-1} and a lifetime of more than 80 assays.

13.2.2.3 Creatinine

For the diagnosis of renal and muscular function, measurement of the creatinine level in serum and determination of the renal clearance of creatinine are widely used. The first reversible chemosensor for creatinine was reported by Panasyuk-Delaney et al. [164] based on artificial chemoreceptors prepared by photo-polymerization of acrylamidomethylpropanesulfonic acid and methylenediacrylamide. They detected the creatinine binding by a decrease in the electrode capacitance. The detection limit for creatinine was 10 µM, which is optimum for medical applications.

Subrahmanyam et al. [165] reported a new 'bite-and-switch' approach for the design of an imprinted polymer recognition material useful for the development of a sensor for creatine and explored the ability of polymerizable hemithioacetal to react with primary amines to form an isoindole complex, which can give a fluorescence signal. They claimed that this sensor can be used for the detection of a broad range of amino compounds.

13.2.2.4 Amino acids

The concentration of amino acids is used as a measure of the nutritive value of a food. Single amino acids are measured to gain access to particular enzyme activities, e.g. transaminases and peptidases, and the determination of total serum amino acids also provides valuable clinical information. Although reports on the determination of amino acids are available, no mention has been made of the use of polymers in sensor devices. As a general strategy, either enzymes are physically immobilized in a polymer matrix or a series of chemical/electrochemical modifications are made to generate amine groups on a polymer for covalent attachment of the enzyme.

Cooper and Schubert [166] electrodeposited polytyramine [poly-4-(2-aminoethyl)phenol] (**XII**) for electrode modification through generating amine groups on which L-amino acid oxidase (L-AAOD) was covalently bound (Table 13.1). They used an L-AAOD–polytyramine electrode for the detection of L-amino acids, via current due to oxidation of enzymatically produced hydrogen peroxide.

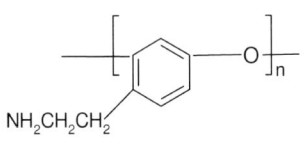

Polytyramine (**XII**)

13.2.2.5 Pyruvate

Situmorang and coworkers [167] constructed a biosensor by covalent attachment of pyruvate oxidase (PyOx) on the surface of electrodeposited polytyramine. This biosensor is sensitive to pyruvate (0.1–3.0 mM) with a 0.05 mM detection limit. An amperometric sensor [168] was prepared by modifying the tip of a 0.25-mm gold wire with a layer of electrically "wired" recombinant PyOx for the detection of pyruvate in biological fluids. The sensor did not require O_2 for its operation.

The electroactive area of the microwire tip was increased by electrodeposition of platinum black. The PyOx was adsorbed on the platinum black and then wired with the subsequently deposited cross-linked poly(4-vinylpyridine). When the electrode was set at 0.4 V vs. Ag/AgCl, the sensitivity at pH 6 was 0.26 A cm^{-2} M^{-1} and the current increased linearly with the pyruvate concentration through the 2–600 µM range. In calf serum, the detection limit was 30 µM, suggesting that the electrode might be used in the continuous monitoring of pyruvate in hypoxic organs. Mizutani et al. [169] prepared an amperometric pyruvate-sensing electrode by immobilizing PyOx on a polyion complex membrane deposited from aqueous solutions of poly-L-lysine and poly(4-styrenesulfonate) on a mercaptopropionic acid-modified gold surface. A photo-cross-linked poly(vinyl alcohol) layer containing PyOx was subsequently formed on the poly-L-lysine–poly(4-styrenesulfonate) complex layer. The polyion complex layer can eliminate electrochemical interferences of L-ascorbic acid, uric acid, L-cysteine and acetaminophen. A high sensitivity (detection limit 50 nM) and a low interference level were obtained.

13.2.2.6 Cholesterol

A cholesterol biosensor was constructed [170] by entrapping cholesterol oxidase (ChOx) within a composite poly(2-hydroxyethyl methacrylate)–PPy membrane. Pt electrode-supported polymer film was prepared by UV polymerization of the hydrogel component containing dissolved enzyme followed by electrochemical polymerization of entrapped pyrrole within the preformed hydrogel network. A linear response between 0.5 and 15 mM, detection limit 120 µM and response time 30 s were obtained for cholesterol. A cholesterol biosensor was also constructed by entrapment of cholesterol oxidase within a polypyrrole (PPy) film electropolymerized in a flow system [171].

A ChOx-based biosensor with enhanced sensitivity and stability was constructed by electropolymerization of a hydrophilic laponite nanoparticle–amphiphilic pyrrole derivative–enzyme mixture preadsorbed on the electrode surface [172]. By a simple process, Yon Hin and Lowe [173] constructed a bienzyme electrode for the detection of total cholesterol by incorporating cholesterol esterase and cholesterol oxidase in PPy films. The sensor showed a fast amperometric response to cholesterol and good storage stability. Kajiya et al. [132] also immobilized cholesterol oxidase and ferrocenecarboxylate ions in PPy films by electropolymerization of pyrrole in an aqueous solution containing these substances.

13.2.2.7 Peroxide

The determination of hydrogen peroxide and organic peroxides in clinical samples and in the environment is rapidly gaining practical importance. Measurement of lipid peroxides in food products and biological tissues is necessary to establish a relationship between diseases such as breast cancer and the level and type of fat in the diet [174].

García-Moreno et al. [175] prepared a biosensor by immobilizing horseradish peroxidase (HRP) enzyme during the electropolymerization of N-methylpyrrole for use in the determination of organic peroxides, e.g. 2-butanone peroxide and *tert*-butyl hydroperoxide, in a predominantly nonaqueous medium, such as reversed micelles. They used the poly-(N-methylpyrrole)–HRP amperometric biosensor for the determination of the organic peroxide content in body lotion samples, employing 2-butanone peroxide as a standard.

Wang and Dong [176] fabricated a hydrogen peroxide biosensor by coating a sol–gel peroxidase layer on a Nafion–Methylene Green (MG)-modified electrode, where the MG was immobilized by the electrostatic force between MG^+ and the negatively charged sulfonic acid groups in Nafion polymer. The sensor performance was evaluated in terms of response time, sensitivity and operational stability.

13.2.2.8 Lactate

Although lactate determination is not frequently used in clinical chemistry, its popularity in the diagnosis of shock and myocardial infarction and in neonatology and sports medicine is increasing. Strong efforts are being made to develop sensor-based lactate analyzers for quick use at the bedside.

Weigelt et al. [177] immobilized lactate monooxygenase in gelatin and fixed to a Clark-type oxygen probe. The resulting enzyme sensor could detect L-lactate between 0.01 and 60 mM, with a stability of 55 days. For lactate determination in dilute serum by flow injection analysis (FIA), lactate oxidase was electrochemically immobilized in a poly-*o*-phenylenediamine film for the one-step and all-chemical construction of a lactate amperometric biosensor, prepared *in situ* by simply injecting a plug of a solution containing the monomer and the enzyme [178]. Results obtained by FIA–amperometric detection compared well with those obtained by a standard enzymatic colorimetric assay. Chaube et al. [179] co-immobilized lactate oxidase and lactate dehydrogenase on conducting polyaniline films by physical adsorption for the estimation of L-lactate in cells and fermentation. Later Chaube et al. [180] immobilized lactate dehydrogenase (LDH) on electrochemically polymerized polypyrrole–poly(vinyl sulfonate) (PPy–PVS) composite films via a cross-linking technique using glutaraldehyde for application to lactate biosensors. These PPy–PVS–LDH electrodes exhibited detection limit 0.1 mM, response time 40 s and shelf-life about 2 weeks for 0.5–6 mM L-lactate estimation.

13.2.2.9 Fructose

Fructose in fruits or juices and lactulose (enzymatic hydrolyzed product of milk) are detected with a fructose biosensor. A ferrocene–Nafion-modified cellulose acetate (Fc-CA) membrane is a useful fructose biosensor [181]. Five cellulose acetate membranes were prepared, four containing 1.8, 5.3, 8.5 and 20.0% ferrocene and one containing 1.8% of ferrocene and 0.05% Nafion in the matrix. The biosensor containing 20.0% ferrocene in the matrix exhibited the lowest detection limit (3 µM), shortest response time (45 s) and highest sensitivity. The membrane containing 1.8% ferrocene and 0.05% Nafion showed better stability characteristics and retained almost 40% of the initial response after 8 h of continuous use. Garcia et al. [182] prepared a fructose biosensor by immobilizing D-fructose 5-dehydrogenase in PPy film. They used this biosensor for fructose determination in three different samples of dietetic products, with 200 analyses performed in 2 weeks of continuous use.

13.2.2.10 Foodstuffs and Drinks

Detection and maintenance of freshness of food products need prime attention. Compounds such as amines, carboxylates, aldehydes, ammonia, sulfur compounds and carbon dioxide are liberated during putrefaction, especially for fish, meat and green vegetables. Conventional methods of measurement involve extraction, centrifugation, steam distillation and titration, which lack reproducibility, whereas an enzyme sensor is easily capable of measuring the freshness of food products by estimating putrefaction products. A biosensor [183] was developed for the quantitative measurement of fish freshness by determining the ATP degradation products, hypoxanthine (HX), inosine (INO) and inosine monophosphate (IMP) metabolites in fish tissue. This biosensor consists of ferrocenecarboxylic acid mediator incorporated conducting polypyrrole enzyme electrodes with immobilized xanthine oxidase, nucleoside phosphorylase and nucleotidase enzymes for quantitative measurement of HX, INO and IMP, respectively, by an amperometric method.

Aromas of certain foods and beverages were discriminated by the use of sensor arrays coupled with pattern recognition [184]. Doleman and Lewis [185] compared an electronic nose and mammalian olfaction to reach the goal of designing an electronic analogue of mammalian olfactory sense. Guadarrama et al. [186] reported an electronic nose consisting of an array of electrodeposited conducting polymer for the organoleptic characterization of olive oil. The sensor was able not only to distinguish among olive oils of different qualities (extra virgin, virgin, ordinary and lampante) but also among Spanish olive oils prepared from different varieties of olives and even different geographic origins. Guadarrama et al. [187] also reported a set of 12 polymeric sensors as an artificial olfactory system for identification of wines from different sources. They prepared the sensors by electrochemical deposition of polypyrrole, poly-3-methylthiophene and polyaniline.

Some off-flavor compounds, such as 2-methylisoborneol (MIB) and geosmin, are the cause of odor in drinking water and fish. Ji et al. [188] developed piezoelectric odor sensors that exhibit selectivity for MIB with a molecularly imprinted polymer (MIP) as the recognition element. They first coated the piezoelectric quartz crystals with a layer of nylon by an LB technique, to provide increased sensitivity, followed by a layer of a cross-linked poly(methacrylic acid-*co*-ethylene glycol dimethacrylate) imprinted with MIB to endow the device with selectivity. A new, low-cost nitrite sensor [189] was developed by immobilizing a direct indicator dye in an optical sensing film for food and environmental monitoring. This sensor was fabricated by binding gallocyanine to a cellulose acetate film that had previously been subjected to an exhaustive base hydrolysis.

13.2.2.11 Environmental Pollutants

Environmental monitoring is conducted using immunosensors, which are based on the recognition of antigen–antibody interaction, in which immunoagents are immobilized in a polymer matrix such as PVC or polyacrylamide gel. Due to the antibody–antigen interaction, variations in electric charge, mass or optical properties are detected. Holt et al. [190] reported a capillary immunosensor in poly(methyl methacrylate) for environmental monitoring and remediation, which can provide on-site, real-time measurement of contaminant levels. The military sector needs such devices for monitoring environmental levels of 2,4,6-trinitrotoluene (TNT) and hexahydro-1,3,5-trinitro-1,3,5-triazine (RDX) introduced into soils and water supplies and absorbed by plants, causing toxicity to animal and human life. These sensors exhibited sensitivity to low RDX concentrations ($\mu g\ L^{-1}$) and peak-to-peak signal variations that were generally less than 10% for multiple injections at a single RDX concentration. The useful lifetime of the coupons in these experiments was more than 10 h (Table 13.1).

A quartz crystal thickness–shear mode (TSM) biomimetic sensor was described using a molecularly imprinted polymer (MIP) coating for the determination of nicotine (NIC) in human serum and urine [191]. The MIP was synthesized using NIC as the template molecule and methacrylic acid as the functional monomer. The sensor showed high selectivity and a sensitive response to NIC in aqueous systems. Konry et al. [192] described an optical microbiosensor for the diagnosis of hepatitis C virus (HCV) by using a novel photoimmobilization methodology based on a photoactivable electrogenerated polymer film deposited on surface-conductive fiber optics, which are then used to link a biological receptor to the fiber tip through light mediation. They claimed that the detection of anti-E2 antibodies with this microbiosensor might enhance significantly HCV serological standard testing, especially among patients during dialysis who were diagnosed as HCV negative by standard immunological tests but were known to carry the virus.

13.2.2.12 DNA-sensitive Infectious Agents and Pollutants

Considerable interest has been shown in DNA sensors for the analysis of unknown or mutant genes, diagnosis of infectious agents in various environments and detection of drugs, pollutants, etc., which interact with DNA. Dupont-Filliard et al. [193] described a versatile and reversible DNA sensor in which they immobilized single-stranded DNA to provide high sensitivity, versatility and ease of use. This system is based on biotin grafting-units, covalently linked to a PPy matrix, able to anchor large biomolecules due to biotin/avidin affinity. Wang and Jiang [194] described a biocomposite material based on the incorporation of nucleic acid dopants within an electronically conducting PPy network. The growth patterns and ion-exchange properties of this electropolymerized polypyrrole–oligonucleotide (PPy–ODN) film was characterized using an *in situ* electrochemical quartz crystal microbalance (EQCM). Such biomaterials open up new opportunities, including genoelectronic devices, composite materials, bioactive interfaces, genetic analysis or probing of DNA charge transfer.

A conjugated polymer in the form of a three-dimensional network of intrinsically conducting macromolecular wires is able to transport electric signals. Functionalization of such molecules with prosthetic groups shows recognition properties and such polymer architectures mimic the nervous system in living species. Using this idea, Garnier et al. [195] prepared new electrochemical sensors with electroactive PPy functionalized with oligonucleotide (ODN). The functionalization of PPy involved a PPy precursor bearing an easy leaving ester group on which an amino-labeled ODN could be directly substituted. The electrochemical response of this ODN-functionalized PPy electrode was analyzed in various aqueous media, containing either complementary or noncomplementary ODN targets. The functionalized PPy acts as macromolecular wires and transduces biological information into molecular signals.

13.3
Measurement of Sense of Taste

The senses of taste and smell are the chemical senses induced when chemical substances interact with the tongue and the nasal cavity, respectively [196]. It is generally assumed that the primary process of chemoreception in the mammalian olfactory system takes place at the cell membrane of sensory neurons. The mammalian sense of taste occurs as a result of complex chemical analyses that are completed in parallel at a series of chemically active sites called taste buds. These taste buds are located in depressions in the tongue, where the molecular and ionic analytes become restricted to allow time for their identification [197]. There are five primary tastes: sweet (carbohydrate based), sour (acidity), salty (ionic), umami or savory (amino acids) and bitter (quinine and other alkaloids). Sense of taste occurs as a result of interaction between the taste buds of the tongue and taste substances. Different lipid molecules in the taste buds play the key role in sensing tastes of food materials [198, 199]. Taste sensor fabrication by incorporating lipids in a PVC matrix has been reported. Hayashi et al. [200] utilized lipid membranes to mimic some of the functionality of mammalian taste bud cells. A multichannel taste sensor containing lipid membranes immobilized with PVC was developed [200–202]. They used *n*-decyl alcohol, oleic acid, dioctyl phosphate, trioctylmethylammonium chloride and oleylamine as lipids immobilized in PVC for the sensing of the tastes sourness, saltiness, bitterness, sweetness and umami.

In addition to the five primary taste substances, the multichannel taste sensor can also discriminate different amino acids, beers, mineral waters, various kinds of sakés, vegetables such as tomatoes and coffee [203]. The tastes of both raw and cooked samples of brown rice and milled rice, with different milling yields, were distinguished by a lipid membrane taste sensing system [204]. The consumption of Kimchi, a Korean traditional pickle fermented with lactic acid bacteria, is expanding worldwide [205]. The fermentation control during the preparation of Kimchi has routinely been done by measuring titratable acidity and pH. An eight-channel taste evaluation system was used in monitoring Kimchi fermentation. Eight polymer membranes which individually responded to cationic or anionic substances were prepared by mixing electroactive materials such as tri-*n*-octylmethylammonium chloride, bis(2-ethylhexyl) sebacate as the plasticizer and poly(vinyl chloride) in the ratio 1:66:33. Each membrane prepared was separately installed on the sensitive area of an ion-selective electrode to produce the respective taste sensor. A taste sensor fabricated with six ion-selective electrodes was used in distinguishing taste of fruits such as banana, oranges and apples [206]. Before study with this sensor, the fruits were crushed and dissolved in water, followed by sensing with the ion-selective electrodes. It appears that different polymers have been used as supporting materials for immobilizing lipids, the sensing element of the taste sensor.

Recently, Adhikari's group [207–211] exploited the tailorability of polymers to act as taste sensing materials without the help of lipid molecules. They modified different polymers to generate active functional groups which are responsible for taste sensing as observed with lipid molecules. They developed crosslinked and phosphorylated poly(vinyl alcohol) [207], xellophane phosphate [208], polyacrylic acid-grafted cellulose [209], polyacrylamide-grafted cellulose [210] and poly(vinyl alcohol)–cellulose composite [211] membranes for sensing the basic tastes, viz. sweetness of glucose, fructose and glycine, sourness of inorganic and organic acids, saltiness of NaCl, KBr and NaHCO3, bitterness of quinine hydrochloride, $MgSO_4$ and $MgCl_2$ and umami taste of monosodium glutamate, etc. Sense of taste was evaluated in terms of electric potential across the functionalized polymer membrane with the help of two Ag/AgCl electrodes. Responses in terms of electric potential of maleic acid, cross-linked and phosphorylated PVA to HCl, NaCl, sucrose, monosodium glutamate and quinine hydrochloride are shown in Fig. 13.8 [212]. Likewise, all other polymer membranes showed good taste-recognizing ability with all the substances mentioned. Threshold values of response of the membranes for different taste substances were below human threshold values. Riul et al. [213] constructed an electronic tongue using polyaniline oligomers (16-mer) and polypyrrole (PPy), which was able to distinguish salt, sweet, bitter and acidic solutions. The sensor was also capable of distinguishing different brands of mineral water, tea and coffee and could differentiate tastants below the human detection threshold, proving that conducting polymers are useful sensing materials and transducers for this sort of application. A three-channel sour taste sensor based on electrodes with polymer–lipid membranes containing elaidic acid, hexadecylamine and benzylcetyldimethylammonium chloride monohydrate has been prepared [214]. Although the sensor is very sensitive to citric acid concentration, it is less sensitive to acetic acid.

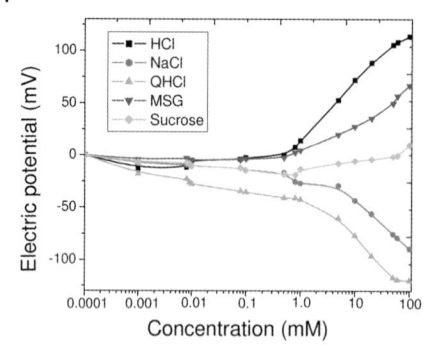

Fig. 13.8 Responses of maleic acid cross-linked and phosphorylated PVA to HCl, NaCl, sucrose, monosodium glutamate (MSG) and quinine-HCl (Q-HCl) [212].

13.4
Trends in Sensor Research

The state-of-the-art in sensor research indicates that various sensors have been developed for analytes, either as sensor arrays for multiple analytes or for specific analytes. An assessment of the successes and failures in these developments can provide a useful guide for further research. In that context, the following trends are notable for shaping the future of sensor research:
- improvement in the immobilization strategy of receptor components;
- innovations for multianalyte sensors, sensing arrays and chemometric approaches for non-selective or partially selective sensing;
- miniaturization and integration of components for combined separation and detection;
- use of conducting polymers and composites in modern sensor devices for the fastest response, highest sensitivity and lowest detection limit.

Significant progress in chemical modification of biosensor devices is apparent in four major areas: the molecular design of chemically selective and biospecific agents, improved immobilizations for molecular recognition, studying the actual interfaces between sample and energy transducer and designing the energy transducers. Very recently, Potyrailo [52] summarized the state-of-the-art, the development trends and the remaining knowledge gaps in the area of combinatorial polymeric sensor materials design. Increased selectivity, response speed and sensitivity in the chemical and biological determinations of gases and liquids are of great interest. Particular attention is paid to polymeric sensor materials, which are applicable to sensors exploiting various energy transduction principles, such as radiant, electrical, mechanical and thermal energy. Ideally, numerous functional parameters of sensor materials can be tailored to meet specific needs using rational design approaches. However, increasing the structural and functional complexity of polymeric sensor materials makes it more difficult to predict the desired properties. Combinatorial and high-throughput methods have had an impact on all areas of research on polymer-based sensor materials, including homo- and copolymers, formulated materials, polymeric structures with engineered morphology and molecular shape-recognition materials.

Novel smart textiles based on conducting polymer (PPy) coatings deposited on a foam substrate have been developed [215]. Being sensitive to pressures exerted from all three dimensions, it may be attractive for use in wearable sensors for sports and medical applications as a smart insole for patients with diabetes mellitus, who require constant monitoring of the pressure exerted underfoot during walking or standing in order to reduce the risk of damaging their feet due to excess pressure being applied. Future applications for this material may also lie in the area of wearable electronic components, whereby the material can be fabricated to produce resistors, capacitors, etc. An excellent approach [216] has been made for undergraduate education, which can provide experience in smart materials and electrochemistry with the help of a simple laboratory experiment using conducting polymers to convert electrical energy into mechanical energy at low voltage or current. Although a ClO_4-doped PPy-based actuator has been used for demonstration, other conjugated conducting polymers can also be used and evaluated as actuating and sensing materials.

Travas-Sejdic et al. [217] projected polymer electronics for the 21st century by addressing their research and development in the following areas: conducting polymer-based sensors for the identification of gene fragments; spectroscopic studies of doped polyanilines and their radical scavenging properties; *in situ* spectroscopic study of the polypyrrole doping–undoping cycle; investigation of the doping levels of poly(3,4-ethylenedioxythiophene); characterization of multilayer composite films containing conductive polymers (CP); and a.c. impedance studies of CPs and the improvement of the corrosion resistance by polyaniline coatings on Al surfaces.

13.5 Conclusion

Due to their excellent tailorability, polymers have found extensive use in sensor devices with definite roles, either in the sensing mechanism or through immobilizing the species responsible for sensing analyte components. It appears from the trend in the suitability of polymers in sensor devices that doped or undoped conducting polymers or polymers with conjugated double bonds are most promising with respect to their sensing roles. Polymer thin-film deposition technology and the design of more active and sensor-specific polymers will lead to successful miniature, multiple sensor arrays. The collaboration of polymer scientists and technologists with physicists and biotechnologists will accelerate the availability of efficient and economic artificial sensor devices for human adoption.

References

1 Madou MJ, Morrison SR. *Chemical Sensing with Solid State Devices*. Academic Press, London, **1989**, pp. 1–9.
2 Nakagawa K, Sadaoka Y, Supriyatno H, Kubo A, Tsutsumi C, Tabuchi K. *Sens Actuators B* **2001**; *76*: 42–46.
3 Pang Z, Gu X, Yekta A, Masoumi Z, Coll JB, Winnik MA, Manners I. *Adv Mater* **1996**; *8*: 768–771.
4 Amao Y, Asai K, Okura I. *Anal Chim Acta* **2000**; *407*: 41–44.
5 Amao Y, Ishikawa Y, Okura I. *Anal Chim Acta* **2001**; *445*: 177–182.
6 Chou TC, Ng KM, Wang SH. *Sens Actuators B* **2000**; *66*: 184–186.
7 Alexy M, Voss G, Heinze J. *Anal Bioanal Chem* **2005**; *382*: 1628–1641.
8 Ichimori K, Ishida H, Fukahori M, Nakazawa H, Murakami E. *Rev Sci Instrum* **1994**; *65*: 1–5.
9 Friedemann MN, Robinson SW, Gerhardt GA. *Anal Chem* **1996**; *68*: 2621–2628.
10 Ho K-C, Hung W-T. *Sens Actuators B* **2001**; *79*: 11–16.
11 Xie D, Jiang Y, Pan W, Li D, Wu Z, Li Y. *Sens Actuators B* **2002**; *81*: 158–164.
12 Hrnèirová P, Opekar F, Stulík K. *Sens Actuators B* **2000**; *69*: 199–204.
13 Mizutani F, Yabuki S, Sawaguchi T, Hirata Y, Sato Y, Iijima S. *Sens Actuators B* **2001**; *76*: 489–493.
14 Smith RC, Lippard SJ. In *Abstracts of Papers, 230th ACS National Meeting, Washington, DC, August 28–September 1, 2005*. American Chemical Society, Washington, **2005**, DC, INOR-462.
15 Chiang CK, Park YW, Heeger AJ, Shirakawa H, Louis EJ, MacDiarmid AG. *J Chem Phys* **1978**; *69*: 5098–5104.
16 Shirakawa H, Louis EJ, MacDiarmid AG, Chiang CK, Heeger AJ. *J Chem Soc, Chem Commun* **1977**; 578–580.
17 Roncali J. *Chem Rev* **1992**; *92*: 711–738.
18 Amrani MEH, Ibrahim MS, Persaud KC. *Mater Sci Eng* **1993**; *C1*: 17–22.
19 Slater JM, Watt EJ. *Analyst* **1991**; *116*: 1125–1130.
20 Marsella MJ, Carroll PJ, Swager TM. *J Am Chem Soc* **1995**; *117*: 9832–9841.
21 Torsi L, Pezzuto M, Siciliano P, Rella R, Sabbatini L, Valli L, Zambonin PG. *Sens Actuators B* **1998**; *48*: 362–367.
22 Unde S, Ganu J, Radhakrishnan S. *Adv Mater Opt Electron* **1996**; *6*: 151–157.
23 Park YH, Han MH. *J Appl Polym Sci* **1992**; *45*: 1973–1982.
24 Wang HL, Toppare L, Fernandez JE. *Macromolecules* **1990**; *23*: 1053–1059.
25 Li D, Jiang Y, Wu Z, Chen X, Li Y. *Sens Actuators B* **2000**; *66*: 125–127.
26 Nicho ME, Trejo M, García-Valenzuela A, Saniger JM, Palacios J, Hu H. *Sens Actuators B* **2001**; *76*: 18–24.
27 Chabukswar VV, Pethkar S, Athawale AA. *Sens Actuators B* **2001**; *77*: 657–663.
28 Yadong J, Tao W, Zhiming W, Dan L, Xiangdong C, Dan X. *Sens Actuators B* **2000**; *66*: 280–282.
29 Gangopadhyay R, De A. *Sens Actuators B* **2001**; *77*: 326–329.
30 Forster RJ. Miniaturized chemical sensors. In Diamond D (Ed.), *Principles of Chemical and Biological Sensors*. Wiley, New York, **1998**. p. 243.
31 Slater J M, Paynter J. *Analyst* **1994**; *119*: 191–195.
32 Percival CJ, Stanley S, Galle M, Braithwaite A, Newton MI, McHale G, Hayes W. *Anal Chem* **2001**; *73*: 4225–4228.
33 Pavlyukovich NG, Murashov DA, Dorozkina GN, Rozanov IA. *J Anal Chem* **2000**; *55*: 469–473.
34 Matsuno G, Yamazaki D, Ogita E, Mikuriya K, Ueda T. *IEEE Trans Instrum Meas* **1995**; *44*: 739–742.
35 Nanto H, Tsubakino S, Ikeda M, Endo F. *Sens Actuators B* **1995**; *25*: 794–796.
36 Hierlemann A, Weimar U, Kraus G, Gauglitz G, Goepel W. *Sens Mater* **1995**; *7*: 179–189.
37 Covington JA, Gardner JW, Briand D, de Rooij NF. *Sens Actuators B* **2001**; *77*: 155–162.
38 Buff W. *Sens Actuators A* **1992**; *30*: 117–121.
39 Déjous C, Rebière D, Pistré J, Tiret C, Planade R. *Sens Actuators B* **1995**; *24*: 58–61.

40 Schiavon G, Zotti G, Toniolo R, Bontempelli G. *Anal Chem* **1995**; *67*: 318–323.
41 Shi G, Luo M, Xue J, Xian Y, Jin L, Jin Ji-Ye. *Talanta* **2001**; *55*: 241–247.
42 Matsuguchi M, Tamai K, Sakai Y. *Sens Actuators B* **2001**; *77*: 363–367.
43 Soborover EI, Tverskoi VA, Tokarev SV. *J Anal Chem* **2005**; *60*: 447–453.
44 Otagawa T, Madou M, Wing S, Rich-Alexander J, Kusanagi S, Fujioka T, Yasuda A. *Sens Actuators B* **1990**; *1*: 319–325.
45 Wang Y, Yan H, Liu J. *Anal Lett* **2005**; *38*: 2057–2065.
46 Torsi L, Pezzuto M, Siciliano P, Rella R, Sabbatini L, Valli L, Zambonin PG. *Sens Actuators B* **1998**; *48*: 362–367.
47 Roy S, Sana S, Adhikari B, Basu S. *J Polym Mater* **2003**; *20*: 173–180.
48 Sakthivel M, Weppner W. *Sens Actuators B* **2006**; *113*: 998–1004.
49 Jacquinot P, Müller B, Wehrli B, Hauser PC. *Anal Chim Acta* **2001**; *432*: 1–10.
50 Nanto H, Dougami N, Mukai T, Habara M, Kusano E, Kinbara A, Ogawa T, Oyabu T. *Sens Actuators B* **2000**; *66*: 16–18.
51 Xie H, Yang Q, Sun X, Yang J, Huang Y. *Sens Actuators B* **2006**; *113*: 887–891.
52 Potyrailo RA. *Angew Chem Int Ed* **2006**; *45*: 702–723.
53 McQuade DT, Pullen AE, Swager TM. *Chem Rev* **2000**; *100*: 2537–2574.
54 Demarcos S, Wolfbeis OS. *Anal Chim Acta* **1996**; *334*: 149–153.
55 Grummant UW, Pron A, Zagorska M, Lefrant S. *Anal Chim Acta* **1997**; *357*: 253–259.
56 Jin Z, Su Y, Duan Y. *Sens Actuators B* **2000**; *71*: 118–122.
57 Chiang J-C, MacDiarmid AG. *Synth Met* **1986**; *13*: 193–205.
58 Ferguson JA, Healey BG, Bronk KS, Barnard SM, Walt DR. *Anal Chim Acta* **1997**; *340*: 123–131.
59 Han W-S, Park M-Y, Chung K-C, Cho D-H, Hong T-K. *Talanta* **2001**; *54*: 153–159.
60 Pandey PC, Singh G. *Talanta* **2001**; *55*: 773–782.
61 Do PQ, Vu TH, Nguyen MT, Tu VN, Pharm HV. *Tap Chi Phan Tich Hoa, Ly Va Sinh Hoc* **2005**; *10*: 72–79.
62 Lakard B, Herlem G, Lakard S, Guyetant R, Fahys B. *Polymer* **2005**; *46*: 12233–12239.
63 Prissanaroon W, Brack N, Pigram PJ, Hale P, Kappen P, Liesegang J. *Synth Met* **2005**; *154*: 105–108.
64 Suzuki K, Watanabe K, Matsumoto Y, Kobayashi M, Sato S, Siswanta D,Hisamoto H. *Anal Chem* **1995**; *67*: 324–334.
65 Tsujimura Y, Yokoyama M, Kimura K. *Anal Chem* **1995**; *67*: 2401–2404.
66 Cunningham AJ. *Introduction to Bioanalytical Sensors*. Wiley, New York, **1998**, p. 113.
67 Teixeira MF de S, Fatibello-Filho O, Ferracin LC, Rocha-Filho RC, Bocchi N. *Sens Actuators B* **2000**; *67*: 96–100.
68 Lindfors T, Ivaska A. *Anal Chim Acta* **2001**; *437*: 171–183.
69 Artigas J, Beltran A, Jiménez C, Bartrolí J, Alonso J. *Anal Chim Acta* **2001**; *426*: 3–10.
70 Shamsipur M, Ganjali MR, Rouhollahi A, Moghimi A. *Anal Chim Acta* **2001**; *434*: 23–27.
71 Liu D, Liu J, Tian D, Hong W, Zhou X, Yu JC. *Anal Chim Acta* **2000**; *416*: 139–144.
72 Saleh MB, Hassan SSM, Abdel Gaber AA, Abdel Kream NA. *Anal Chim Acta* **2001**; *434*: 247–253.
73 Gupta KC, Mujawamariya JD. *Talanta* **2000**; *52*: 1087–1103.
74 Oh H, Choi EM, Jeong H, Nam KC, Jeon S. *Talanta* **2000**; *53*: 535–542.
75 Gupta VK, Prasad R, Kumar P, Mangla R. *Anal Chim Acta* **2000**; *420*: 19–27.
76 Aguilar AD, Forzani ES, Li X, Tao N, Nagahara LA, Amlani I, Tsui R. *Appl Phys Lett* **2005**; *87*: 193108/1–193108/3.
77 Hassan SSM, Saleh MB, Abdel Gaber AA, Mekheimer RAH, Abdel Kream NA. *Talanta* **2000**; *53*: 285–293.
78 Mahajan RK, Parkash O. *Talanta* **2000**; *52*: 691–693.
79 Numata M, Baba K, Hemmi A, Hachiya H, Ito S, Masadome T, Asano Y, Ohkubo S, Gomi T, Imato T, Hobo T. *Talanta* **2001**; *55*: 449–457.
80 Hassan SSM, Ali MM, Attawiya AMY. *Talanta* **2001**; *54*: 1153–1161.
81 Mousavi MF, Alizadeh N, Shamsipur M, Zohari N. *Sens Actuators B* **2000**; *66*: 98–100.

82 Gupta VK, Kumar A, Mangla R. *Sens Actuators B* **2001**; *76*: 617–623.
83 Wróblewski W, Wojciechowski K, Dybko A, Brzózka Z, Egberink RJM, Snellink-Rueël BHM, Reinhoudt DN. *Sens Actuators B* **2000**; *68*: 313–318.
84 Zhang XB, Guo CC, Jian LX, Shen GL, Yu RQ. *Anal Chim Acta* **2000**; *419*: 227–233.
85 Torres KYC, Garcia CAB, Fernandes JCB, de Oliveira Neto G, Kubota LT. *Talanta* **2001**; *53*: 807–814.
86 Espadas-Torre C, Meyerhoff ME. *Anal Chem* **1995**; *67*: 3108–3114.
87 Sukeerthi S, Contractor AQ. *Indian J Chem, Sect A* **1994**; *33*: 565–571.
88 Hatfield V, Neaves P, Hicks PJ, Persaud K, Travers P. *Sens Actuators B* **1994**; *18/19*: 221–228.
89 Xia Y, Wiesinger JM, MacDiarmid AG, Epstein AJ. *Chem Mater* **1995**; *7*: 443–445.
90 MacDiarmid AG, Epstein AJ. *Synth Met* **1995**; *69*: 85–92.
91 Athawale AA, Kulkarni MV. *Sens Actuators B* **2000**; *67*: 173–177.
92 Jun H-K, Hoh Y-S, Lee B-S, Lee S-T, Lim J-O, Lee D-D, Huh J-S. *Sens Actuators B* **2003**; *96*: 576–581.
93 Bartlett PN, Ling-Chung SK. *Sens Actuators* **1989**; *19*: 141–150.
94 Mayes AG, Blyth J, Kyrolainen-Reay M, Millington RB, Lowe CR. *Anal Chem* **1999**; *71*: 3390–3396.
95 Blum P, Mohr GJ, Matern K, Reichert J, Spichiger-Keller UE. *Anal Chim Acta* **2001**; *432*: 269–275.
96 El-Ragehy NA, El-Kosasy AM, Abbas SS, El-Khateeb SZ. *Anal Chim Acta* **2000**; *418*: 93–100.
97 Liu ZH, Wen ML, Yao Y, Xiong J. *Sens Actuators B* **2001**; *77*: 219–223.
98 Katsu T, Ido K, Katoaka K. *Sens Actuators B* **2002**; *81*: 267–272.
99 Mohr GJ, Nezel T, Spichiger-Keller UE. *Anal Chim Acta* **2000**; *414*: 181–187.
100 Nelson TL, Maynor MS, Lavigne JJ. In *Abstracts, 57th Southeast/61st Southwest Joint Regional Meeting of the American Chemical Society, Memphis, TN, November 1–4, 2005*. American Chemical Society, Washington, DC, **2005**, NOV04-164.
101 Cullum DC. *Introduction to Surfactant Analysis*. Blackie, London, **1994**.
102 Bobacka J, Lindfors T, McCarrick M, Ivaska A, Lewenstam A. *Anal Chem* **1995**; *67*: 3819–3823.
103 Bobacka J, McCarrick M, Lewenstam A, Ivaska A. *Analyst* **1994**; *119*: 1985–1991.
104 Kovács B, Csóka B, Nagy G, Ivaska A. *Anal Chim Acta* **2001**; *437*: 67–76.
105 Panasyuk-Delaney T, Mirsky VM, Ulbricht M, Wolfbeis OS. *Anal Chim Acta* **2001**; *435*: 157–162.
106 Peng H, Liang C, He D, Nie L, Yao S. *Talanta* **2000**; *52*: 441–448.
107 Prasad BB, Arora B. *Electroanalysis* **2003**; *15*: 108–114.
108 Jenkins AL, Yin R, Jensen JL. *Analyst* **2001**; *126*: 798–802.
109 Li C, Wang C, Guan B, Zhang Y, Hu S. *Sens Actuators B* **2005**; *107*: 411–417.
110 Marx S, Zaltsman A, Turyan I, Mandler D. *Anal Chem* **2004**; *76*: 120–126.
111 Patra D, Mishra AK. *Sens Actuators B* **2001**; *80*: 278–282.
112 Josowicz M, Janata J, Ashley K, Pons S. *Anal Chem* **1987**; *59*: 253–258.
113 Wang J, Li R. *Talanta* **1989**; *36*: 279–284.
114 Hou W, Wang E. *Anal Chim Acta* **1992**; *257*: 275–280.
115 Zen J-M, Tang J-S. *Anal Chem* **1995**; *67*: 208–211.
116 Tsuchitani S, Sugawara T, Kinjo N, Ohara S, Tsunoda T. *Sens Actuators* **1988**; *15*: 375–386.
117 Sakai Y, Sadaoka Y, Fukumoto H. *Sens Actuators* **1988**; *13*: 243–250.
118 Sakai Y, Sadaoka Y, Matsuguchi M, Rao VL, Kamigaki M. *Polymer* **1989**; *30*: 1068–1071.
119 Xin Y, Wang S. *Sens Actuators A* **1994**; *40*: 147–149.
120 Inagaki N, Suzuki K. *Polym Bull* **1984**; *11*: 541–544.
121 Maddanimath T, Mulla IS, Sainkar SR, Vijayamohanan K, Shaikh KI, Patil AS, Vernekar SP. *Sens Actuators B* **2002**; *81*: 141–151.
122 Sakai Y, Sadaoka Y, Matsuguchi M. *J Electrochem Soc* **1989**; *136*: 171–174.
123 Sakai Y, Sadaoka Y, Matsuguchi M, Sakai H. *Sens Actuators B* **1995**; *25*: 689–691.

124 Rauen KL, Smith DA, Heineman WR, Johnson J, Seguin R, Stoughton P. *Sens Actuators B* **1993**; *17*: 61–68.
125 Sakai Y, Matsuguchi M, Sadaoka Y, Hirayama K. *J Electrochem Soc* **1993**; *140*: 432–436.
126 Su Pi-Guey, Chen IC, Wu Ren-Jang. *Anal Chim Acta* **2001**; *449*: 103–109.
127 Somani PR, Viswanath AK, Aiyer RC, Radhakrishnan S. *Sens Actuators B* **2001**; *80*: 141–148.
128 Suri K, Annapurni S, Sarkar AK, Tandon RP. *Sens Actuators B* **2002**; *81*: 277–282.
129 Sakai Y, Matsuguchi M, Hurukawa T. *Sens Actuators B* **2000**; *66*: 135–138.
130 Sun LX, Okada T. *Anal Chim Acta* **2000**; *421*: 83–92.
131 Schuhmann W, Lammert R, Uhe B, Schmidt H-L. *Sens Actuators B* **1990**; *1*: 537–541.
132 Kajiya Y, Tsuda R, Yoneyama H. *J Electroanal Chem* **1991**; *301*: 155–164.
133 Mu S, Xue H, Qian B. *J Electroanal Chem Interfacial Electrochem* **1991**; *304*: 7–16.
134 Ichimura K. *J Polym Sci, Part 1: Polym Chem* **1984**; *22*: 2817–2828.
135 Piletsky SA, Parhometz YP, Lavryk NV, Panasyuk TL, El'skaya AV. *Sens Actuators* **1994**; *19*: 629–631.
136 Peng H, Liang C, He D, Nie L, Yao S. *Talanta* **2000**; *52*: 441–448.
137 Zhang Z, Liu H, Deng J. *Anal Chem* **1996**; *68*: 1632–1638.
138 Jung S-K, Wilson GS. *Anal Chem* **1996**; *68*: 591–596.
139 Sangodkar H, Sukeerthi S, Srinivasa RS, Lal R, Contractor AQ. *Anal Chem* **1996**; *68*: 779–783.
140 Yao T, Takashima K. *Biosens Bioelectron* **1998**; *13*: 67–73.
141 Mizutani F, Yabuki S, Iijima S. *Electroanalysis* **2001**; *13*: 370–374.
142 Myler S, Collyer SD, Bridge KA, Higson SPJ. *Biosens Bioelectron* **2002**; *17*: 35–43.
143 Tian. F, Zhu G. *Anal Chim Acta* **2002**; *451*: 251–258.
144 Quinto M, Losito I, Palmisano F, Zambonin CG. *Anal Chim Acta* **2000**; *420*: 9–17.
145 Kros A, Nolte RJM, Sommerdijk NAJM. *J Polym Sci, Part A: Polym Chem* **2002**; *40*: 738–747.
146 Revzin AF, Sirkir K, Simonian A, Pishoko MV. *Sens Actuators* **2002**; *81*: 359–368.
147 Cosnier S, Szunerits S, Marks RS, Novoa A, Puech L, Perez E, Rico-Lattes I. *Talanta* **2001**; *55*: 889–897.
148 Puech L, Perez E, Rico-Lattes I, Bon N, Lattes A. *Colloids Surf A* **2000**; *167*: 123–130.
149 Matsumoto T, Ohashi A, Ito N, Fujiwara H, Matsumoto T. *Biosens Bioelectron* **2001**; *16*: 271–276.
150 Yasuzawa M, Matsuki T, Mitsui H, Kunugi A, Nakaya T. *Sens Actuators B* **2000**; *66*: 25–27.
151 Garjonyte R, Malinauskas A. *Biosens Bioelectron* **2000**; *15*: 445–451.
152 Adeloju SB, Moline AN. *Biosens Bioelectron* **2001**; *16*: 133–139.
153 Mousty C, Galland B, Cosnie S. *Electroanalysis* **2001**; *13*: 186–190.
154 Yang H, Chung TD, Kim YT, Choi CA, Jun CH, Kim HC. *Biosens Bioelectron* **2002**; *17*: 251–259.
155 Teh K-S, Lin L. *J Micromech Microeng* **2005**; *15*: 2019–2027.
156 Lee D, Lee J, Kim J, Kim J, Na HB, Kim B, Shin C-H, Kwak JH, Dohnalkova A, Grate JW, Hyeon T, Kim H-S. *Adv Mater* **2005**; *17*: 2828–2833.
157 Bridge KA, Higson SPJ. *Electroanalysis* **2001**; *13*: 191–198.
158 Cui G, Yoo JH, Yoo J, Lee SW, Nam H, Cha GS. *Electroanalysis* **2001**; *13*: 224–228.
159 Kobayashi K, Suzuki H. *Sens Actuators B* **2001**; *80*: 1–8.
160 Mizutani F, Sato Y, Hirata Y, Iijima S. *Anal Chim Acta* **2001**; *441*: 175–181.
161 Pandey PC, Singh G. *Talanta* **2001**; *55*: 773–782.
162 Komaba S, Seyama M, Momma T, Osaka T. *Electrochim Acta* **1997**; *42*: 383–388.
163 Osaka T, Komaba S, Fujino Y, Matsuda T, Satoh I. *J Electrochem Soc* **1999**; *146*: 615–619.
164 Panasyuk-Delaney T, Mirsky VM, Wolfbeis OS. *Electroanalysis* **2002**; *14*: 221–224.

165 Subrahmanyam S, Piletsky SA, Piletska EV, Chen B, Day R, Turner APF. *Adv Mater* **2000**; *12*: 722–724.
166 Cooper JC, Schubert F. *Electroanalysis* **1994**; *6*: 957–961.
167 Situmorang M, Gooding JJ, Hibbert DB, Barnett D. *Electroanalysis* **2002**; *14*: 17–21.
168 Gajovic N, Beinyamin G, Warsinke A, Scheller FW, Heller A. *Anal Chem* **2000**; *72*: 2963–2968.
169 Mizutani F, Yabuki S, Sato Y, Sawaguchi T, Iijima S. *Electrochim Acta* **2000**; *45*: 2945–2952.
170 Brahim S, Narinesingh D, Guiseppi-Elie A. *Anal Chim Acta* **2001**; *448*: 27–36.
171 Vidal JC, García E, Castillo JR. *Anal Chim Acta* **1999**; *385*: 213–222.
172 Besombes JL, Cosnier S, L'Abbé P. *Talanta* **1997**; *44*: 2209–2215.
173 Yon Hin BFY, Lowe CR. *Sens Actuators B* **1992**; *7*: 339–342.
174 Rose DP, Hatala MA, Connolly JM, Rayburn J. *Cancer Res* **1993**; *53*: 4686–4690.
175 García-Moreno E, Ruiz MA, Barbas C, Pingarrón JM. *Anal Chim Acta* **2001**; *448*: 9–17.
176 Wang B, Dong S. *Talanta* **2000**; *51*: 565–572.
177 Weigelt D, Schubert F, Scheller F. *Analyst* **1987**; *112*: 1155–1158.
178 Palmisano F, Centonze D, Zambonin PG. *Biosens Bioelectron* **1994**; *9*: 471–479.
179 Chaube A, Pande KK, Singh VS, Malhotra BD. *Anal Chim Acta* **2000**; *407*: 97–103.
180 Chaube A, Gerard M, Singhal R, Singh VS, Malhotra BD. *Electrochim Acta* **2001**; *46*: 723–729.
181 Tkáč J, Voštiar I, Šturdík E, Gemeiner P, Mastihuba V, Annus J. *Anal Chim Acta* **2001**; *439*: 39–46.
182 Garcia CAB, de Oliveira Neto G, Kubota LT. *Anal Chim Acta* **1998**; *374*: 201–208.
183 Ghosh (Hazra) S, Sarker D, Misra TN. *Sens Actuators B* **1998**; *53*: 58–62.
184 Agbor NE, Petty MC, Monkman AP. *Sens Actuators B* **1995**; *28*: 173–179.
185 Doleman BJ, Lewis NS. *Sens Actuators B* **2001**; *72*: 41–50.
186 Guadarrama A, Rodríguez-Méndez ML, Sanz C, Ríos JL, de Saja JA. *Anal Chim Acta* **2001**; *432*: 283–292.
187 Guadarrama A, Fernández JA, Íñiguez M, Souto J, de Saja JA. *Sens Actuators B* **2001**; *77*: 401–408.
188 Ji H-S, McNiven S, Lee K-H, Saito T, Ikebukuro K, Karube I. *Biosens Bioelectron* **2000**; *15*: 403–409.
189 Kazemzadeh A, Daghighi S. *Spectrochim Acta, Part A* **2005**; *61*: 1871–1875.
190 Holt DB, Gauger PR, Kusterbeck AW, Ligler FS. *Biosens Bioelectron* **2002**; *17*: 95–103.
191 Tan Y, Yin J, Liang C, Peng H, Nie L, Yao S. *Bioelectrochemistry* **2001**; *53*: 141–148.
192 Konry T, Novoa A, Shemer-Avni Y, Hanuka N, Cosnier S, Lepellec A, Marks RS. *Anal Chem* **2005**; *77*: 1777–1779.
193 Dupont-Filliard A, Roget A, Livache T, Billon M. *Anal Chim Acta* **2001**; *449*: 45–50.
194 Wang J, Jiang M. *Langmuir* **2000**; *16*: 2269–2274.
195 Garnier F, Korri-Youssoufi H, Srivastava P, Mandrand B, Delair T. *Synth Met* **1999**; *100*: 89–94.
196 Toko K. *Biomimetic Sensor Technology*. Cambridge University Press, Cambridge, **2000**.
197 Schmale H, Ahlers C, Blaker M, Kock K, Spielman AI. In Chadwick D, Marsh J, Goode J (Eds.), *The Molecular Basis of Smell and Taste Transduction*. Wiley, Chichester, **1993**, p. 167.
198 Kurihara K, Yoshii K, Kashiwayanagi M. *Comp Biochem Physiol* **1986**; *85A*: 1–22.
199 Miyake M, Kamo N, Kurihara K, Kobatake Y. *Biochim Biophys Acta* **1976**; *436*: 856–862.
200 Hayashi K, Yamanaka M, Toko K, Yamafugi K. *Sens Actuators B* **1990**; *2*: 205–213.
201 Kikkawa Y, Toko K, Matsuno T, Yamafuji K. *Jpn J Appl Phys* **1993**; *32*: 5731–5736.
202 Mikhelson KN. *Sens Actuators B* **1994**; *18*: 31–37.
203 Toko K. *Mater Sci Eng C* **1996**; *4*: 69–82.
204 Tran TU, Suzuki K, Okadome H, Homma S, Ohtsubo K. *Food Chem* **2004**; *88*: 557–66.

205 Kim N, Park K, Park I-S, Cho Y-J, Bae YM. *Biosens Bioelectron* **2005**; *20*: 2283–2291.
206 Tadayoshi Y, Shuichi M, Yoshihiro FA. *Denki Kagakkai Gijutsu, Kyoiku Kenkyu Robunshi* **1996**; *5*: 59–64.
207 Majumdar S, Adhikari B. *Sens Actuators B* **2006**; *114*: 747–755.
208 Majumdar S, Adhikari B. *Anal Chim Acta* **2005**; *554*: 105–112.
209 Majumdar S, Dey J, Adhikari B. *Talanta* **2006**; *69*: 131–139.
210 Majumdar S, Adhikari B. *J Sci Ind Res* **2006**; *65*: 237–243.
211 Majumdar S, Adhikari B. *Bull Mater Sci* **2005**; *28*: 703–712.
212 Majumdar S. *PhD Thesis*, IIT, Kharagpur, **2005**, Chapt. 3, p. 66.
213 Riul A Jr, Gallardo Soto AM, Mello SV, Bone S, Taylor DM, Mattoso LHC. *Synth Met* **2003**; *132*: 109–116.
214 Szwacki J, Szpakowska M. *Chemik* **2005**; *58*: 425–428.
215 Brady S, Lau KT, Megill W, Wallace GG, Diamond D. *Synth Met* **2005**; *154*: 25–28.
216 Cortes MT, Moreno JC. *J Chem Educ* **2005**; *82*: 1372–1373.
217 Travas-Sejdic J, Bowmaker GA, Cooney RP. *Chem N Z* **2004**; *68*: 10–12.

14
Polymeric Drugs

Tamara Minko, Jayant J. Khandare, and Sreeja Jayant

14.1
Introduction

A drug usually is defined as an agent intended for use in the diagnosis, mitigation, treatment, cure or prevention of disease in humans or animals [1]. A drug, as a pure chemical, is seldom used in its native form. Typically, a drug is added to a dosage form in order to formulate a usable product. The most advanced dosage form is a drug delivery system (DDS). A DDS is a formulation or device that safely brings a therapeutic agent to a specific body site at a certain rate in order to achieve an effective concentration at the site of drug action. A DDS is used to carry, distribute or transport drugs into and throughout the body. A DDS might be used to deliver drug(s) to a desired body location, organ, tissue or to specific types of cells, intracellular organelles or molecules (i.e., targeted DDSs) and/or provide a preprogrammed drug release profile (i.e., controlled release DDSs). A DDS should also protect the therapeutic agent(s) during transport to the site of action. In addition, advanced DDSs may contain several active components to enhance delivery or body drug retention or increase the efficacy of treatment by suppressing cellular or systemic mechanisms that may limit the therapeutic response of the drug. The development of DDSs is focused on the selection and/or fabrication of the most appropriate dosage form for a drug that allows patients to take their medication in a convenient and effective manner.

Drug delivery offers several benefits to patients. The two primary objectives are to increase patient compliance and to enhance drug efficacy and tolerance. Development of more "user friendly" dosage forms ultimately increases dosing convenience for the patient. This may involve reducing dosage frequency from three or more times daily to once daily or developing an oral formulation of a drug that traditionally is given by the parenteral route. Drug delivery also enables medicines to be administered in a more palatable format such as "fast melt" tablets that dissolve rapidly in the mouth, effervescent sachets that produce pleasant-tasting medicines or transdermal patches that avoid the need for taking medications orally. Currently, almost one billion prescriptions per year are taken incorrectly by pa-

Macromolecular Engineering. Precise Synthesis, Materials Properties, Applications.
Edited by K. Matyjaszewski, Y. Gnanou, and L. Leibler
Copyright © 2007 WILEY-VCH Verlag GmbH & Co. KGaA, Weinheim
ISBN: 978-3-527-31446-1

tients. A significant percentage of hospitalizations and nursing home admissions results from the incorrect usage of prescription medications. Providing medicine in a more acceptable manner increases patient compliance and the effectiveness of therapy. Contemporary DDSs can achieve these goals and also significantly increase the efficacy of a drug while limiting its side-effects. A DDS in its simplest form usually includes a drug and a carrier. More advanced DDSs often include several drugs and other active components with different mechanisms of action, or/and targeting moieties and other components. The carrier binds the various components together and provides for the required degree of solubility, molecular weight, size and other important characteristics of the whole DDS.

The quality, behavior in the body, toxicity, side-effects and overall efficiency of the whole DDS depend to a great extent on the type and properties of its carrier. Polymers are the most often used type of carriers for DDSs. Both naturally occurring and synthetic polymers are used as vehicles for drug delivery.

14.2
Polymeric Drug Delivery

Water-soluble polymers have been used in the last two decades to modify the pharmacokinetics and physicochemical properties of therapeutic agents. Polymers, whether synthetic or natural, have the property of encapsulating a diverse range of molecules of biological interest and bear distinct therapeutic advantages such as controlled release of drugs, protection against the premature degradation of drugs and reduction in drug toxicity [2].

14.2.1
Water-soluble Polymers

The use of water-soluble polymeric conjugates as drug carriers offers several possible advantages. These advantages include, but are not limited to: (1) improved drug pharmacokinetics, (2) decreased toxicity to healthy organs, (3) possible facilitation of accumulation and preferential uptake by targeted cells and (4) offering a programmed profile of drug release. Many soluble polymer conjugates have been designed as drug carriers. However, two of them are most widely used in several modifications for drug delivery: *N*-(2-hydroxypropyl)-methacrylamide (HPMA) copolymer and poly(ethylene glycol) (PEG).

HPMA copolymer is a water-soluble, biocompatible, nontoxic, nonimmunogenic polymer which is frequently used in different DDSs [3–21]. HPMA copolymers have been most intensively studied at the Center for Controlled Chemical Delivery (USA), at the Institute of Macromolecular Chemistry (Czech Republic), at the Center for Polymer Therapeutics at the University of Cardiff (UK) and at the Utrecht Institute for Pharmaceutical Sciences (The Netherlands). The main advantage of HPMA as a drug carrier is the existence of several sites for potential conjugation of drugs, targeting moieties and other components of a

DDS. The HPMA copolymer main chain is not biodegradable, so all conjugates developed clinically have been limited to a molecular weight of less than 40 000 Da to ensure eventual renal elimination. In most cases, an increase in molecular weight of HPMA copolymers results in prolonged circulation times of HPMA-bound drugs [22].

PEG is a water-soluble, nonionic polymer approved by the US Food and Drug Administration (FDA) for pharmaceutical applications [23–31]. Due to its non-toxic character, it is widely used in many biochemical, cosmetic, pharmaceutical and industrial applications. It is also important that PEG polymers show low antigen activity and, in most cases, decrease the antigenicity of active ingredients conjugated to them [32]. The major disadvantage of a linear PEG polymer is that it has only two binding sites. This allows binding of only two DDS components directly to the polymer. PEG is commercially available from major chemical suppliers.

It should be mentioned that, in addition to water-soluble polymers, a DDS may be constructed from many non-water-soluble polymers, polymer derivatives or other lipophilic or amphiphilic chemicals (surfactants, lipids, etc.) in the forms of liposomes, micelles, nanoparticles, etc. Many of them might include water-soluble polymers in their hydrophilic shell. Lipid-core micelles designed in Torchilin's laboratory are an example of a DDS which combines soluble copolymers with lipids (such as PEG–phosphatidylethanolamine conjugate, PEG–PE) to form self-assembling nanosized lipid-core micelles with a hydrophobic core and hydrophilic shell [33]. These micelles can effectively solubilize a broad variety of poorly soluble drugs (anticancer drugs in particular) and diagnostic agents. Drug-loaded lipid-core micelles can spontaneously target body areas with compromised vasculature (tumors, infarcts) via the enhanced permeability and retention (EPR) effect. Lipid-core mixed micelles containing certain specific components (such as positively charged lipids) are capable of escaping endosomes, thus delivering incorporated drugs directly into the cell cytoplasm. Various specific targeting ligand molecules (such as antibodies) can be attached to the surface of the lipid-core micelles and bring drug-loaded micelles to and into target cells. Lipid-core micelles carrying various reporter (contrast) groups may become the imaging agents of choice in different imaging modalities. The concept of targeted polymeric micelles was first introduced by Kataoka and co-workers [34–36]. Their experimental studies have successfully progressed towards actual applications of polymeric micelles. Such polymeric DDSs will not be reviewed in the present chapter, which is aimed mainly at describing water-soluble polymeric drugs.

14.2.2
Classification of Water-soluble Polymeric Drugs

Many different variants of polymeric drugs based on the use of water-soluble polymers as carriers have been developed and evaluated. Such delivery systems can be classified by different parameters, for instance, type of polymeric carrier,

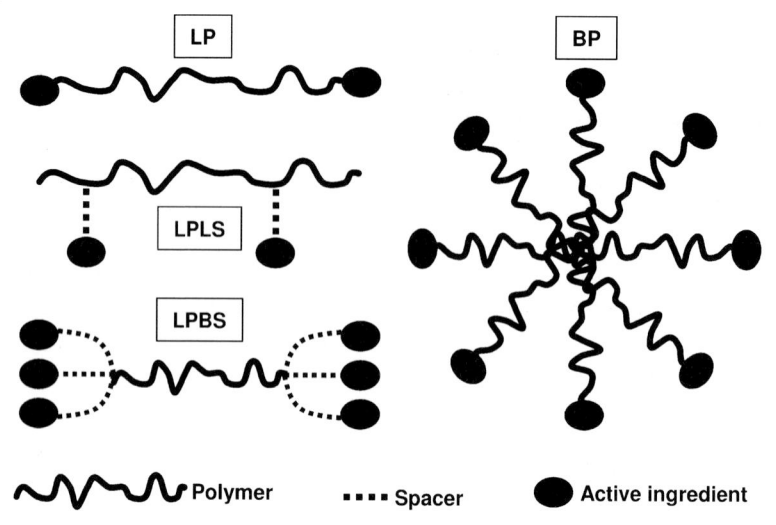

Fig. 14.1 Four main types of water-soluble polymeric DDSs:
LP, linear polymer; LPLS, linear polymer–linear spacer;
LPBS, linear polymer–branched spacer; BP, branched spacer,
represented by a star-like polymer.

general configuration of the system (linear, branched, star-like, etc.), types of linkage between components of the DDS (biodegradable, nonbiodegradable, etc.), the main therapeutic application (anticancer, anti-inflammatory, etc.) and so on. Here we select four main types of soluble polymeric drugs: LP (linear polymer systems); LPLS (linear polymer–linear spacer systems); LPBS (linear polymer–branched spacer systems); and BP (branched polymer systems) (Fig. 14.1). It should be stressed that more complex systems, for example BPLP (branched polymer–linear spacer) or even BPBP (branched polymer–branched spacer), theoretically can be designed. We will leave such complex systems out of the scope of the present chapter because the major types of DDSs that are currently in use can be classified, in most cases, under the four major types presented in Fig. 14.1. Below we briefly describe each of the above-mentioned types of DDSs.

14.2.2.1 Linear Polymer (LP) Systems

This type of DDS is the simplest and most often used type of delivery system when PEG polymer is used as a carrier. In this case, active ingredients are conjugated to the distal ends of a polymeric carrier. This requires certain modifications of polymer, drug or both. For example, to synthesize PEG prodrugs of cyclin-dependent kinase inhibitor, two methods were devised to conjugate PEG to alsterpaullone via the N atom of the indole ring portion of the molecule. In the first approach, activation of the indole was accomplished by reaction with *p*-nitrophenyl chloroformate to produce a reactive carbamate that was then condensed with a mono-blocked diamine to form a urea bond followed by deblocking and conjugation to PEG. The second route used the anion of the indole and produced a carbamate bond [27]. Native PEG has only two sites available for the conjugation and therefore only two drug molecules or one drug molecule and one other active ingredient (i.e. targeting moiety) may be conjugated to it. This limits the loading capacity of such linear polymeric carriers. Nevertheless, such a "simple" DDS is widely used and substantially enhances the properties of the drug. The most obvious improvement is the increase in solubility of the drug, which in turn increases cellular drug availability and enhances its specific activity. An example of such a soluble DDS constructed with linear PEG polymer and water-insoluble drug is camptothecin–PEG conjugate (CPT–PEG) [25–27, 29]. In most cases the conjugation of a drug to a PEG polymer also substantially improves its pharmacokinetics (PK). After the conjugation, the total plasma clearance rate usually is substantially decreased and the biological half-life is lengthened several-fold compared with the unconjugated compound [27]. However, improvement in PK is usually accompanied by a decrease in the specific activity of the drug. Such an effect is related to the increase in molecular weight of a polymeric drug, which in turn changes the route of drug penetration inside the cell. Low molecular weight drugs penetrate cellular membranes by diffusion. High molecular weight drugs are internalized by endocytosis, which is substantially slower than diffusion and therefore requires significantly higher concentrations of drug outside the cells. An increase in the drug load (the number of drug copies per one molecule of polymer) and/or the addition of a targeting moiety (penetration enhancer) are required. For instance, conjugation of CPT–PEG with biotin as a moiety to enhance nonspecific and/or targeted uptake via the sodium-dependent multivitamin transporter significantly increases the cellular availability of the camptothecin and therefore substantially enhances its toxicity to cancer cells [29]. However, under certain conditions, an increase in molecular weight of a drug might enhance its accumulation in targeted cells. Such a situation takes place in solid tumors, where polymeric drugs are accumulated preferentially in the tumor, increasing the local concentration of anticancer drugs and preventing drug accumulation in normal healthy organs. Such a phenomenon was first discovered by Matsumura and Maeda and was termed the "enhanced permeability and retention (EPR) effect" [37, 38]. This effect is explained by the increased blood circulation in the tumor and permeability of blood vessels, which enhances the penetration of high molecular weight drugs into the tumor, and by the poor lymphatic drainage of the tumor, which prevents the elimination of the drug.

14.2.2.2 Linear Polymer–Linear Spacer (LPLS) Systems

Certain polymers (similar to HPMA) allow conjugation of several active ingredients to one polymeric backbone directly or through so-called "spacers" (Fig. 14.1). Such spacers may play several roles in a DDS. First, they provide for binding structurally and chemically different compounds (polymer, drug, targeting moiety, etc.) in one complex relatively stable chemical structure. Second, they physically separate the polymer from the active ingredients, so preventing their negative influence on the specific activity of the drug(s) used in the DDS. Third, the specific construction of a spacer allows regulation of the degree of binding of drugs or other active ingredients to the polymer backbone and therefore predetermining the rate and specific place of drug release (if any). HPMA is generally a copolymer with a pendent reactive functional group and can be modified to obtain amides by reaction with carboxylic acids or amino acids. Small peptides might be considered as pendent spacer molecules with HPMA due to its hydrolysis and degradation ability. In general, the bond between such spacers and active ingredients can be non-biodegradable or biodegradable. Biodegradability, in this case, reflects the ease and rate of separation of the active ingredient (drug) in targeted organs, tissues or cells. An overall rule for spacer design is that it should provide stability of the whole DDS during its transfer to the site of action and then release the drug in the targeted site. Triggering of drug release can be initiated by changes in environment (e.g. pH), certain enzymes or microorganisms that cleave a spacer or even external stimuli [ultrasound, magnetic or electric field, heat (local temperature), injection of other chemicals, etc.]. Such types of drug release help to maintain a DDS in a non-effective state during transfer and attenuate the toxicity associated with drug activity at non-targeted tissues. Whole ranges of different DDS forms have been developed to provide different types of drug release from simple hydrolysis to such complicated and exotic approaches such as "ATTEMPTS" (Antibody Targeted Triggered Electrically Modified Prodrug Type Strategy) [39]. Different types of conjugation of active components of DDS to the polymer backbone which provide for the desired profile of drug release will be reviewed later in this chapter. In addition to the drug, such DDSs often contain a targeting moiety, providing accumulation and degradation of the delivery system in a specific organ, tissue or cell [3, 4, 17, 40]. Recently, targeting moieties have been developed to target a DDS even to particular molecules inside cells in order to suppress or activate some cellular processes, for instance to inhibit a cellular defense against the drug used [25, 41–47]. Despite clear advantages of an LPLS type of DDS, the drug load of such systems is still limited.

14.2.2.3 Linear Polymer–Branched Spacer (LPBS) Systems
Greater enhancement of drug loading might be achieved by branching the spacer (Fig. 14.1c). Such spacers are currently being developed in our laboratory [7]. The desired properties of such a DDS include the following. The spacer should be designed to have several active sites. One of these sites is used to conjugate to the polymeric carrier. This bond should be stable in order to prevent breakage during the delivery to the site of action. Other sites on the spacer are used to bind active ingredients of the DDS (drugs, targeting moieties, etc.). The biodegradability of such bonds should be based on the considerations discussed

above. LPBS DDSs have substantial advantages over linear systems, including, but not limited to, an increase in the drug load of the DDS and therefore increased overall specific activity of the whole DDS and prevention of strong steric hindrance at the macromolecular level. Moreover, each branched multi-armed spacer may in turn be bonded to another branched spacer. Therefore, the payload of the DDS might further be increased. It seems that the maximum number of active ingredients that can be included in such branched spacers is limited mainly by the solubility of the components. For example, preliminary data showed that only three non-water-soluble molecules of the anticancer drug camptothecin can be conjugated to such an LPBS system while maintaining a satisfactory solubility of the whole complex. Since branched spacers have a relatively low molecular weight (compared with the polymer), the addition of new active ingredients to the DDS is not associated with a substantial increase in the molecular weight of the whole DDS.

14.2.2.4 Branched Polymer (BP) Systems

In order to increase the load of the DDS and to provide the possibility of including active ingredients other than drugs, branched polymers can be used. An example of such BP polymers is shown in Fig. 14.1d and represents so-called "star-like" polymers. Several types of branched polymers have been designed and evaluated [19, 24, 48–52]. However, such types of DDSs include several copies of polymer and therefore might have an excessive molecular weight, which in turn can decrease the efficacy of the drug. Consequently, a compromise number of branches should be found. An alternative way of increasing the specific activity of drug in a BP DDS is by the incorporation of targeting moieties to enhance cellular uptake of the DDS. The field of biomedical dendrimers is still in its infancy, but the interest in dendrimers and dendronized polymers is currently exploding [53]. Dendrimers are attracting attention as active therapeutic agents, vectors for targeted delivery of drugs, peptides and oligonucleotides and permeability enhancers able to promote oral and transdermal drug delivery.

14.2.3
Advantages of Polymeric Drugs

Several different types of DDSs based on the use water-soluble polymer conjugates as carriers have been developed. The main advantage of such DDSs is the possibility to program a desired profile of drug pharmacokinetics, accumulation in the targeted site and drug release. In most cases, targeted polymeric delivery systems provide effective prevention of adverse side-effects to normal tissues and facilitate uptake in the targeted organ, tissues and cells. The main disadvantages of such systems include their relative complexity and high molecular weight. In addition to previously discussed types of polymeric DDSs, different approaches based on the use of other than the described classes of molecules (i.e. amino acids, peptides, proteins, etc.) are currently being developed. Such delivery systems are even more complicated than described in this chapter and present a number of challenges which still remain to be addressed. One can project that the cost of such polymeric DDSs will be substantially higher than that of free drugs. Comparing the main types of DDSs, the linear polymer-branched spacer (LPBS) type of polymeric system seems to be the most promising.

14.2.4
Future Directions

The following outstanding issues should be tackled for the future progress of polymeric drugs: the limited loading capacity of most currently developed polymeric systems, the possible interferences between active components of the DDS and the projected high cost of polymeric DDSs. In our opinion, the most promising direction in overcoming these obstacles is the use of branched types of polymeric drugs which utilize LPBS or BP approaches. If an increase in the number of branches of a spacer or polymer would lead to limited solubility of the whole resulting structure, micellar nanoparticles can be formed from LPBS or BP structured polymers.

14.3
Prodrug Approach

In most cases, polymeric drugs represent a special type of a more generic class – prodrugs. A prodrug is an inactive precursor of a drug which is converted into the drug in the blood, targeted organ, tissue or cell. A prodrug remains inactive during its transfer to the site of drug action. Advantages of the prodrug approach include, but are not limited to, the prevention of systemic side-effects, an increase in the bioavailability of modified drugs and the possibility of targeting specific organs, tissues or cell conditions to convert the inactive prodrug into its active form. The examples of different polymeric prodrugs and mechanisms of controlling their properties and functions as drug delivery vehicles will be presented in the Section 14.4. Here, we will only briefly mention our classification of prodrugs.

14.3.1
Prodrug Approach First Type Prodrugs

Two classes of prodrugs are generally used (Fig. 14.2a and b). Prodrugs of the first type (i.e., classical prodrugs) undergo reconversion to the active drug inside cells (Fig. 14.2a). Reconversion typically occurs by the cleavage of a chemical bond between the "pro" group and the drug moiety. To date, classical prodrugs

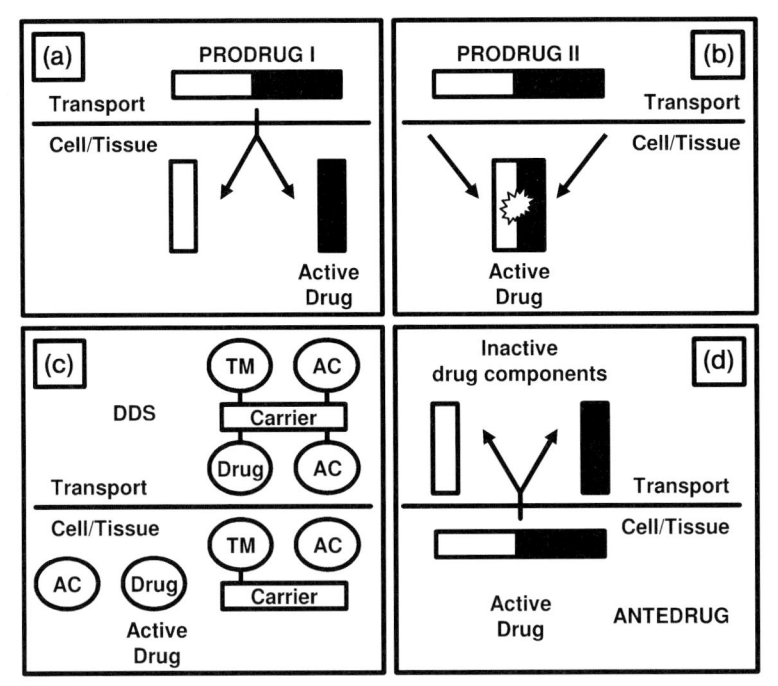

Fig. 14.2 Main types of prodrugs used for targeted drug delivery to the colon. Prodrugs of the first type (a) are broken inside cells to form active substance or substances. The second type of prodrug (b) is usually the combination of two or more substances. Under specific intracellular conditions, these substances react forming the active drug. A targeted DDS (c) usually includes several components: a targeting moiety (TM), a carrier, a drug, and might contain other active components (AC). Antedrug (d) is defined as an active derivative which is applied locally and undergoes biotransformation to the readily excretable inactive form upon entry into the systemic circulation.

have attracted the vast majority of research and development efforts and have had the most commercial success of any prodrug type. Examples of this type of prodrug will be discussed in Section 14.4.

14.3.2
Second Type Prodrugs

In contrast, a second type of prodrug usually involves the combination of two or more substances. Under specific intracellular conditions, these substances react forming the active drug (Fig. 14.2b). The combination of indole-3-acetic acid with horseradish peroxidase is a good example of the second type of prodrug approach [54]. Indole-3-acetic acid and some of its derivatives are oxidized by horseradish peroxidase, forming a radical-cation that rapidly fragments to form cytotoxic products. No toxicity is seen when either indole-3-acetic acid or horseradish peroxidase is incubated alone at concentrations that, in combination, form potent cytotoxins. The fluorinated prodrug–peroxidase combination also shows potent cytotoxic activity in human and rodent tumor cell lines. Cytotoxicity is thought to arise in part from the formation of 3-methylene-2-oxindole (or its analogues), which can conjugate with thiols and, probably, DNA and other biological nucleophiles. [54].

14.3.3
Tumor-activated Prodrugs (TAP)

A specific class of prodrugs, tumor-activated prodrugs, which remain inactive during their transfer in systemic and local circulation and are selectively activated in tumor tissue, has been identified [55]. Such prodrugs can be of either type I or II prodrugs. There are several mechanisms potentially exploitable for the selective activation of TAP. Some utilize unique aspects of tumor physiology, such as selective enzyme expression or hypoxia. Others are based on tumor-specific delivery techniques, including activation of prodrugs by exogenous enzymes delivered to tumor cells via monoclonal antibodies (ADEPT) or generated in tumor cells from DNA constructs containing the corresponding gene (GDEPT). Whichever activating mechanism is used, only a small proportion of the tumor cells are likely to be competent to activate the prodrug. Therefore, TAP need to exploit fully these "activator" cells by being capable of killing activation-incompetent cells as well as via a "bystander effect." A wide variety of chemistries have been explored for the selective activation of TAP [55]. Examples of the most important of them are: the reduction of quinones, *N*-oxides and nitroaromatics by endogenous enzymes or radiation, the cleavage of amides by endogenous peptidases and a hydrolytic metabolism by a variety of exogenous enzymes, including phosphatases, kinases, amidases and glycosidases.

14.3.4
Drug Delivery Systems

The more advanced type of prodrug, DDS, might include a targeting moiety and supplementary active ingredients in addition to the carrier and drug (Fig. 14.2c). The carrier binds all components of the DDS together and provides required characteristics of the whole DDS, i.e. solubility, molecular weight, size, etc. A targeting moiety enforces the specific delivery of a drug to the targeted organ, tissue or cell. It often also enhances the uptake of the complete DDS by targeted cells by acting as a penetration enhancer. Active ingredients are included in a complex DDS to augment the action of a main drug. They might be represented by drugs with mechanisms of action different to that of a main drug or might provide an independent influence on cellular functions to promote effects of the drug. Separate components of the DDS are bound together by means of spacers. Such spacers play an important role by providing release of DDS components under certain conditions or at specific time points. In most cases, the spacer which binds a targeting moiety to the carrier is made stable and is nondegradable in body fluids or inside targeted cells. In contrast, spacers, which connect a drug and other active components of the DDS, are biodegradable. The breakage of such spacers is normally caused by specific conditions inside the targeted organ, cell or cellular organelles, providing additional "targeting" to specific intracellular components, such as cellular cytoplasm, mitochondria or nuclei. The majority of DDSs are fabricated as prodrugs of the first type, where the degradation of spacers releases an active ingredient at the site of its action.

14.3.5
Antedrugs

The antedrug concept was introduced in designing potent, yet safer, locally active, anti-inflammatory steroids [56]. An antedrug is defined as an active synthetic derivative that is designed to undergo biotransformation to the readily excretable inactive form upon entry into the systemic circulation. Antedrugs are used locally (topical delivery) as effective drugs which are hydrolyzed or undergo other types of degradation to inactive components upon entering system circulation (Fig. 14.2d) and so minimizing systemic side-effects and increasing the therapeutic indices. Therefore, antedrugs can be considered as specific types of prodrugs, which are delivered locally as active drugs and undergo inactivation after entering the bloodstream. Steroid-21-oate esters have been developed as the first of this kind, which are quickly hydrolyzed to the corresponding inactive steroid acids, thereby exerting minimal systemic side-effects. Since then, other steroidal and nonsteroidal antedrugs have also been developed. The cost, complexity and length of development of a new therapeutic drug, in addition to the adverse drug reactions and the potential risk factors causing the withdrawal of the drugs from the market, have made the approach of antedrug design unique and attractive to those involved in drug design. Considering these factors, the concept by which a drug is strictly designed to eliminate deleterious adverse systemic effects is becoming an affordable approach in the drug development arena.

14.4
Controlling Properties and Function of Polymeric Drugs by Molecular Architecture

The primary step of identifying an appropriate polymer to form a prodrug remains critical and is reliant upon chemical structure, molecular weight and the chemico-physical properties of the polymer. Ever since the unique characteristics of biopolymers were identified, they have been used in prodrug form for many biological applications. The architecture of the polymer and its prodrug form can affect both pharmacodynamic and pharmacokinetic properties. Therefore, it is critical to design a prodrug which subsequently will release the drug from its conjugated form at the site of action and will not produce untoward effects before its elimination from the body. Only a few polymers, such as PEG and poly(lactide-*co*-glycolide) (PLGA) copolymers, have been approved by the FDA and are being successfully used in current clinical therapeutics. Clinical outcomes for many others, including HPMA and polyamines, are keenly awaited.

Varieties of polymers and novel delivery techniques have been developed to deliver drugs, genetic material, imaging compounds and other therapeutic agents. Based on the biological application, polymers can be used to deliver drugs by the following methods:

1. encapsulation of drugs in polymers, e.g. nano-particles, micro-particles, controlled drug delivery devices, liposomes;
2. complexation of drugs with polymers, e.g. dendritic, cyclodextrin complexes and poly-lysosomal DDSs;
3. conjugations, e.g. covalent conjugation of peptides, drugs and imaging agents.

Polymeric prodrugs have been extensively used to deliver drugs and other biomolecules; the process involves linking of components to form a stable conjugate. The prodrug bioconjugation process involves coupling of two or more biological components to form a unique molecule possessing the collective properties of its individual components. Jatzkewitz, in 1955, reported on the peptamin-polyvinylpyrrolidone (PVP) prodrug to improve the efficacy of the drug [57]. However, for the next 20 years, the biological aspects of prodrugs and their implications in therapeutics were not taken into account. Slowly the prodrug conjugation became more applied and a rational model for pharmacologically active polymers was proposed by Ringsdorf in 1975. Therefore, he may be considered the pioneer of prodrug research [58]. Table 14.1 shows examples of monomers; their polymeric architectures with chemical structures and their therapeutic applications. Homopolymers and copolymers are used extensively in drug delivery; their prodrug conjugates will be discussed in the next section. Delivery vehicles used in prodrug conjugates can be classified on the basis of their: (1) *origin and source* (either natural polymers or synthetic polymers) – polysaccharides, poly(α-amino acids), etc.; (2) *chemical nature* – vinylic or acrylic polymers; (3) *degradation* – biodegradable and nonbiodegradable; and (4) *degree of polymerization/molecular weight* – oligomers, macromers and polymers.

14.4.1
Design and Synthesis of Polymeric Prodrugs

The design, synthesis and characterization of conjugates are critical steps, since most of the conjugates have to undergo biological evaluations. In general, prodrug conjugation synthetic methodologies involve multiple chemical reactions including protections and deprotections of the functional groups during rigorous purification steps. In order to synthesize a prodrug conjugate, the polymer, and also the biocomponent, must possess reactive functional groups such as – COOH, –OH, –SH and –NH$_2$. Various methods have been reported for synthe-

Table 14.1 Monomers, their polymeric forms and therapeutic applications.

	Monomeric unit	Polymer architecture	Applications
1	$CH_2=CH-OH$ Vinyl alcohol	$-(CH_2-CH(OH))_n-$ Poly(vinyl alcohol) (PVA)	Copolymerization with vinyl acetate is used in microspheres and bioadhesive hydrogels
2	$CH_2=CH_2$ Ethylene	$-(CH_2-CH_2)_n-$ Polyethylene	Controlled delivery of drugs
3	$CH_2=C(R)-COOH$ R=H, acrylic acid; R=CH_3, methacrylic acid	$-(CH_2-C(R)(COOH))_n-$ Poly(acrylic acid) (PAA)	Stimuli-sensitive polymer – pH sensitive Reversible swelling hydrogels
4	$CH_2=CH-C(=O)-NH-CH(CH_3)_2$ N-Isopropylacrylamide	$-(CH_2-CH(C(=O)NH-CH(CH_3)_2))_n-$	Stimuli-sensitive polymer – temperature sensitive. Used in biomolecule recovery and hydrogels
5	$CH_2=C(CH_3)-C(=O)-NH-CH_2-CH(OH)-CH_3$ N-(2-Hydroxypropyl)-methacryl amide (HPMA)	$-(CH_2-C(CH_3)(C(=O)NH-CH_2-CH(OH)-CH_3))_n-$ Poly[N-(2-hydroxypropyl)-methacryla mide] p(HPMA)	Prodrug conjugates with doxorubicin, paclitaxel and other anticancer drugs
6	H_2C-CH_2 (epoxide, with O bridge) Ethylene oxide	$-(H_2C-CH_2-O)_n-$ Poly(ethylene glycol) (PEG)	Polymer–drug conjugates with increased aqueous solubility of drugs

Table 14.1 (continued)

	Monomeric unit	Polymer architecture	Applications
7	HOCH$_2$-C(=O)-OH Glycolic acid Glycolide Lactic acid Lactide	Polyglycolide Polylactide Poly(lactic-co-glycolic acid) (PLGA) copolymer	Nanoparticle and microparticle delivery of drugs and proteins and peptides
8	Amino acid, e.g. lysine	Poly(amino acid) (PAA), e.g. poly(lysine)	Formulation stabilizer and bioconjugation
9		R = NH$_2$/COOH, OH, e.g. poly(amidoamines) (PAMAM) dendrimers	Nanovehicle for delivery of high payload of drugs, imaging and targeting agents

sizing such conjugates; however, the selection of a particular method depends on the type of polymer and drug and their functional groups and may involve the use of reactants and reagents [59–61]. Below are listed the most common reagents that are used in synthesizing bioconjugates.

1. *Zero-order cross-linkers:* e.g. carbodiimides such as N,N'-dicyclohexylcarbodiimide (DCC), 1-ethyl-3-(3-dimethylaminopropyl)carbodiimide (EDC), EDC with N-hydroxysuccinimide (NHS), N-cyclohexyl-N-(4-methylmorpholinium)ethylcarbodiimide (CMC) and N,N-diisopropylcarbodiimide (DIC).
2. *Heterofunctional cross-linkers:*
   a. amine reactive, sulfhydryl reactive cross-linkers, m-maleimidobenzoyl-N-hydroxysuccinamide esters (MBS);
   b. amine reactive cross-linkers such as N-hydroxysuccinimidyl-4-azidosalicylic acid (NHS-ASA) esters, N-hydroxysuccinimidyl-6-(4'-azido-2'-nitrophenylamino)hexanoate cross-linkers;
   c. carboxylate reactive cross-linkers: 4-(p-azidosalicylamido)butylamine (ASBA).
3. *Homobifunctional cross-linkers:* NHS esters, glutaraldehyde, photoreactive cross-linkers, maleimides, sulfhydryl cross-linkers, ethylene glycol bis(succinimidyl)-succinate (EGS).
4. *Trifunctional/multifunctional cross-linkers:* 4-azido-2-nitrophenylbiocytin-4-nitrophenyl ester, sulfosuccinimidyl-2-[6-(biotinamido)-2-(p-azidobenzamidohexanoamido]ethyl 1,3-dithiopropionate (sulfo-SBED).
5. *Cleavable reagents:* hydroxylamine cleavable esters, dithionite cleavable diazo bonds.

14.4.2
Polymeric Bioconjugates in Drug Delivery

Below are given examples of polymers and their conjugates used as prodrugs.

14.4.2.1 Dextran and Its Prodrug Conjugates

Natural biopolymers, e.g. dextran and chitosan, have been used extensively in prodrug conjugate research (Table 14.2). Dextrans are natural macromolecules and consist of linear units of a-D-1,6-glucose-linked glucan with side-chains 1–3 linked to the backbone units of the dextran biopolymer (Fig. 14.3). Dextran is a complex branched polysaccharide made of many glucose molecules, generally produced by enzymes from certain strains of *Leuconostoc* or *Streptococcus*. The most common forms of dextrans used in drug delivery are native hydroxyl form (dextran-OH), aldehyde form (dextran-CHO) or amino form (dextran-NH_2). Dextrans possess multiple primary and secondary hydroxyl groups and therefore can be easily conjugated with drugs and proteins having reactive groups either by direct conjugation or by incorporation of a spacer arm. Dextran has the most compact structure due the extensive branching, i.e. one branch point on every 15th glucose unit. It is an ideal candidate for prodrug conjugate due to the following properties: (1) water solubility and low toxicity, (2) high stability due to

14 Polymeric Drugs

Table 14.2 Biopolymers with monomeric units: chemical structures and types.

Biopolymer	Monomer units	Structure	Type
Hydrocarbons/ lipids	CH_2 units in lipids or hydrocarbons	$-(CH_2-CH_2)_n-$ Polyethylene	Homopolymer
Proteins	Amino acids: lysine, serine	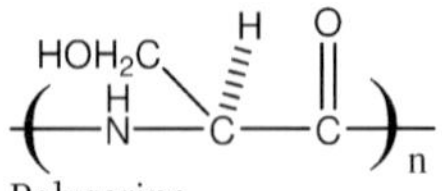 Polyserine	Heteropolymer
Polynucleotides	Nucleotides: adenine	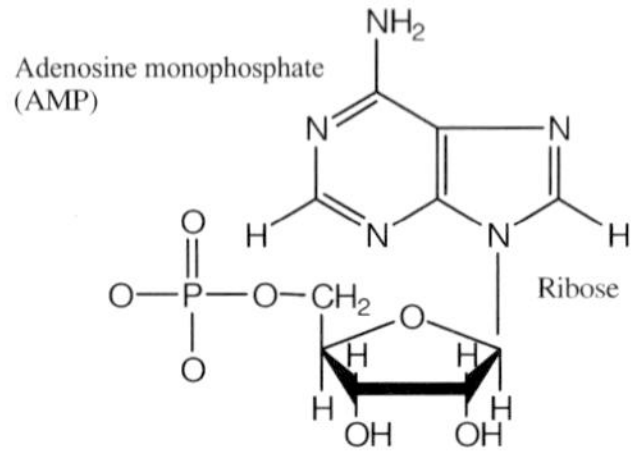 Adenosine monophosphate (AMP)	Heteropolymer
Polysaccharides	Sugars: glucose, fucose, etc.	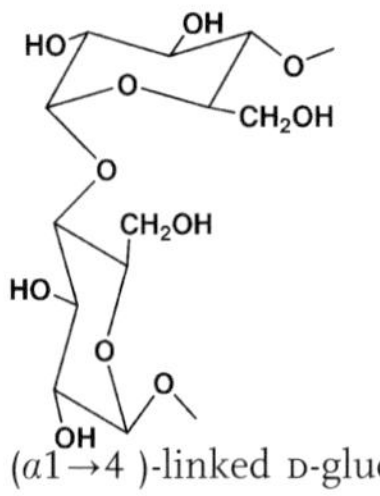 ($\alpha 1 \rightarrow 4$)-linked D-glucose units	Homo- and heteropolymers

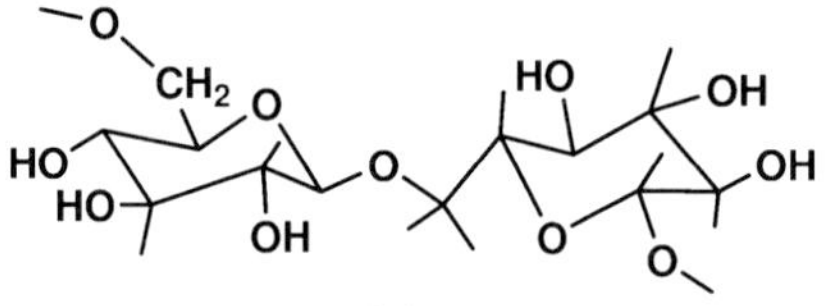

Fig. 14.3 Structure of dextran.

glycosidic bonds, (3) presence of multiple hydroxyl groups for bioconjugation and (4) its availability in different molecular weight ranges.

Most of the applications of dextrans as polymeric carriers are through injectable routes, whereas with systemic administration the pharmacokinetics of the dextran–therapeutic agent conjugates are significantly affected. Animal and human studies have shown that the distribution and elimination of dextrans are dependent on molecular weight and the net charge present on the polymer. Recently, a dextran–peptide–methotrexate conjugate was reported to achieve tumor-targeted delivery of chemotherapeutic agents [62]. Dextran is considered a prospective polymeric carrier because of its biocompatibility and biodegradability. It has been well evaluated as a polymeric carrier for delivery of anticancer drugs to the tumor by a passive accumulation of the dextran–anticancer conjugates in tumor tissues. In addition, the peptide linker was optimized to allow drug release in the presence of matrix-metalloproteinase-2 (MMP-2) and matrix-metalloproteinase-9 (MMP-9), two critical tumor-associated enzymes [62]. *In vivo* activity of the conjugate was compared with that of free methotrexate (MTX) having an MMP-insensitive linker. The MMP-sensitive conjugate showed tolerable *in vivo* side-effects and effective inhibition of tumor growth by 83% in each of the two separate models that overexpress MMP (HT-1080 and U-87). On the other hand, free MTX demonstrated no significant tumor reduction in the same models. Neither free MTX nor the conjugate possess any tumor inhibition in mice bearing RT-112, a slower growing model that does not overexpress MMP. MMP-insensitive conjugates, although able to inhibit tumor growth, caused toxicity in the small intestine and bone marrow.

Various chemotherapeutic drugs have been conjugated to dextran due to its non toxicity and aqueous solubility. Daunorubicin–dextran conjugate (Dau–Dex) was compared with free daunorubicin for acute and subacute toxicity and efficacy in tumor chemotherapy [63]. The LD_{50} of Dau–Dex was found to be threefold higher than that of the free drug. In addition, Dau–Dex conjugate showed no damage to heart tissue during the 2 months following four injections of the therapeutic dose and caused no change in the differential count of bone marrow cells. The subacute toxicity of Dau–Dex is much lower than that of free daunorubicin, as expressed by histological damage to all organs which were examined and compared with massive atrophy of spleen and bone marrow affected by the free drug.

The prodrugs of mitomycin C (MMC) and MMC–dextran conjugate (MMC–D) were evaluated against various murine tumors. MMC–D conjugate demonstrated enhanced antitumor activity against B16 melanoma, Ehrlich ascites carcinoma and P388 leukemia when compared with free MMC [64]. Administration of MMC–D at 24 h prior to tumor inoculation resulted in a significant increase in the life span of mice bearing L1210 leukemia. It was concluded that the conjugates of dextran with MMC show sustained pharmacological activity. The enhanced activity of MMC–D in various tumor systems was a result of the improved biopharmaceutical properties of MMC–D resulting from the modification of MMC into a polymeric drug.

Corticosteroids have been conjugated with dextrans to evaluate local delivery of steroids in the colon as anti-inflammatory agents [65]. A prodrug consisting of methylprednisolone (MP) was prepared by conjugating MP with dextran using succinic acid (SA) as a spacer [66]. MP–succinate (MPSA) conjugate was synthesized using 1,1′-carbonyldiimidazole as a coupling agent with SA as a spacer (Fig. 14.4). The hydrolysis of dextran–MP (DMP) conjugate in rat blood was achieved with a half life of ∼25 h and was proved to occur in the liver lysosomal fraction, but not in the control samples lacking lysosomes.

A novel macromolecular prodrug of Tacrolimus (FK506), FK506–dextran conjugate, was developed and its physico-chemical, biological and pharmacokinetic characteristics were studied [67]. The molar ratio of components (dextran to

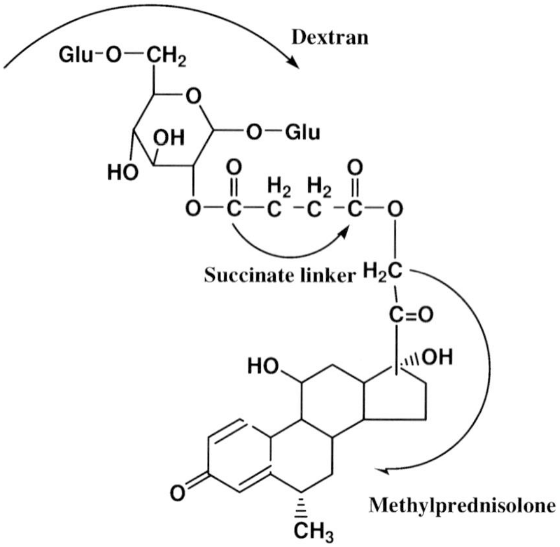

Fig. 14.4 Schematic representation of methylprednisolone–dextran conjugates with succinic acid as a linker. Reproduced with permission from [66].

FK506) was approximately 1:1. The dextran–FK506 conjugate was synthesized using a solution of carboxy-n-pentyldextran (C6D-ED) in phosphate buffer with activated ester of FK506 in dioxane. Conjugation of drugs with dextrans exhibited a prolonged effect and reduced toxicity and immunogenicity. Most of the studies have been carried out in animals, with only a few being extended to humans [68].

Among other polysaccharides, chitosan has been extensively investigated in drug delivery. Chitosan is a linear polysaccharide composed of randomly distributed β-(1–4)-linked D-glucosamine (deacetylated unit) and N-acetyl-D-glucosamine (acetylated unit). It is obtained commercially by deacetylation of chitin, which is the structural element in the exoskeleton of crustaceans (crabs, shrimp, etc.). Chitosan is the second most abundant biomaterial and is a natural cationic polymer with unique biological properties including biodegradability, biocompatibility and bioactivity. Therefore, it has recently emerged as one of the most promising biopolymers for a variety of potential applications in the biomedical and pharmaceutical fields. However, the solubility of non-cross-linked chitosan in weak acid solutions restricts its utility in DDSs.

14.4.2.2 PEG and Its Prodrug Conjugates

Most commonly used poly(ethylene glycol) (mPEG-OH) has either mono or bis terminal hydroxyl functional groups (Fig. 14.5a and b). PEG is a linear or branched polyether and is synthesized by anionic ring-opening polymerization

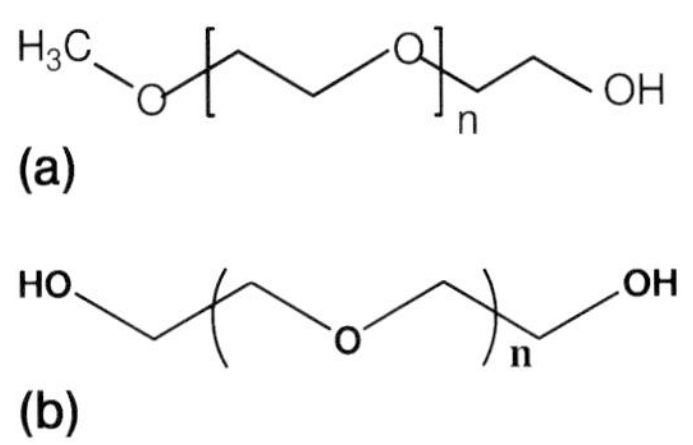

14.4 Controlling Properties and Function of Polymeric Drugs by Molecular Architecture

Fig. 14.5 Structures for monomethoxy-poly(ethylene glycol) (a) and dihydroxy functional PEG polymer (b).

of ethylene oxide initiated by nucleophilic attack of a hydroxide ion on the epoxide ring. The polymer is soluble in both aqueous and organic solvents and, therefore, is the most preferred candidate in prodrug conjugates.

Several PEGylated enzymes (adenosine deaminase, L-asparaginase) and cytokines (including interferon α and G-CSF) are routinely used in therapeutics. PEG-modified adenosine deaminase (Adagen) and PEG–L-asparaginase (Oncaspar) were the first PEG modified enzymes that came on the market in the early 1990s. Figure 14.6a presents multiple components of mPEG-OH polymer conjugates [69]. Derivatives **1** and **2** contain a reactive aryl chloride residue, which is displaced by a nucleophilic amino group by a reaction with peptides or proteins, as shown in Fig. 14.6b. Derivatives **1** and **2** are arylating reagents whereas derivatives **3–11** contain reactive acyl groups and are referred to as acylating agents. Protein modification with all of these agents results in acylated amine-containing linkages: amides, derived from active esters **3–6** and **11**, or carbamates, derived from **7–10**. Alkylating reagents (**12** and **13**) both react with proteins forming secondary amine conjugation with amino-containing residues. As represented in Fig. 14.6b, tresylate (**12**) alkylates directly, whereas acetaldehyde (**13**) is used in reductive alkylation reactions. The numbers **1–13** represent to the order in which these activated polymers were introduced [69].

Due to the presence of either one or a maximum of two functional hydroxyl groups, PEG has a limitation to its conjugation capacity. Recently, this limitation was overcome by coupling amino acids such as bicarboxylic amino acid and aspartic acid to PEG [70, 71]. The method doubled the number of active groups of the original molecule of PEG and it became possible to achieve a dendrimeric structure at each PEG extremity. However, a lower reactivity of the bicarboxylic acid groups towards drug binding was observed. It was concluded that the low reactivity was due to the presence of steric hindrance resulting from two molecules of 1–D-arabinofuranosilcytosyne (Ara-C)d-arabinofuranosilcytosyne (Ara-C)",4,1> when they are conjugated to the neighboring carboxylic moieties. The steric hindrance was substantially reduced by incorporating the dendrimer arms with an amino alcohol [$H_2N–(CH_2–CH_2–O)_2–H$].

PEG polymers with hydroxyl functional groups can be modified easily using small amino acids or other aliphatic chain molecules. For instance, linear or branched PEGs of varying molecular weight, PEG–Ara-C conjugates for con-

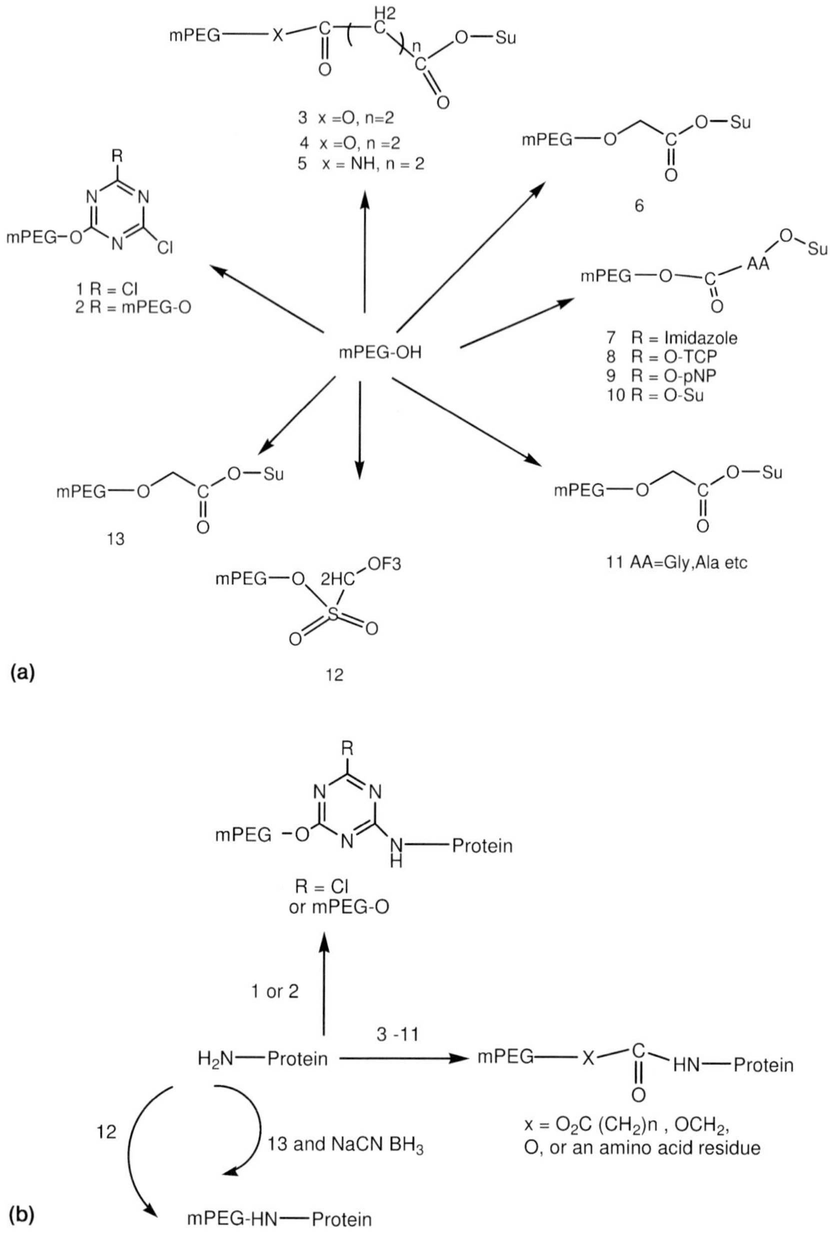

Fig. 14.6 mPEG and various protein-modifying methods (a). Protein modification with all of these agents results in acylated amine-containing linkages: amides, derived from active esters **3–6** and **11** or carbamates, derived from **7–10**. Alkylating reagents (**12** and **13**) both react with proteins forming secondary amine conjugation with amino-containing residues. Tresylate (**12**) alkylates directly, while acetaldehyde (**13**) is used in reductive alkylation reactions (b). **1–13** represent the order in which these activated polymers were introduced. Reproduced with permission from [69].

trolled release, were synthesized [72]. The antitumor agent Ara-C was covalently conjugated to varying molecular weight OH-terminal PEGs through an amino acid spacer to improve *in vivo* stability and blood residence time. PEG conjugates were synthesized using one or two hydroxyl groups at the polymer's termini. In order to increase the drug loading, the OH group of PEG was functionalized with a bicarboxylic amino acid to form a tetrafunctional derivative. Conjugates with four or eight Ara-C molecules for each PEG chain were prepared. The authors also investigated steric hindrance in PEG–Ara-C conjugates using molecular modeling to investigate the most suitable bicarboxylic amino acid with the least steric hindrance. Computer-aided analysis showed that aminoadipic acid was most suitable, because the carboxylic groups are sufficiently separated to accommodate Ara-C and so reduce the necessity to incorporate spacer arms. The theoretical observations were supported by experimental conjugation results. Ara-C was conjugated to PEG using an amino acid spacer (either norleucine or lysine). Hydroxyl groups in PEG were activated by *p*-nitrophenyl chloroformate, which forms a stable carbamate linkage between PEG and the amino acid. The degree of PEG hydroxyl group activation with *p*-nitrophenyl chloroformate was determined by UV analysis of the *p*-nitrophenol released from PEG–*p*-nitrophenyl carbonate after alkaline hydrolysis. Activated PEG was further coupled with amino acid and the intermediate PEG–amino acid was linked to Ara-C using the EDC–NHS activation method.

Many water-insoluble anticancer drugs are routinely conjugated with polymers to enhance aqueous solubility. Recently, camptothecin (CPT) was conjugated to PEG (M_w 3400 Da) [73]. The resulting drug–polymer conjugates were water soluble and could hydrolyze at a pH-dependent rate to release CPT. Further, the conjugates were evaluated *in vitro* for cytotoxicity using gliosarcoma cells and it was observed that PEGylation enhanced achievable drug concentrations with an increase in distribution of CPT. A wide range of PEGs of varying molecular weights and end functionality are routinely being used in prodrug conjugates. Higher molecular weight PEG (>20 000 Da) has a plasma circulating half-life of ~8–9 h in the mouse. Conjugation of small biomolecules with such high molecular weight PEG has proved to be a successful polymeric prodrug candidate [70, 71, 74].

14.4.2.3 Prodrug Conjugates of *N*-(2-Hydroxypropyl)methacrylamide (HPMA) Copolymer

HPMA drug conjugates are known to induce endocytosis for the delivery of drugs into the cells. Currently, antitumor agent HPMA–doxorubicin prodrug conjugate is under clinical trials. HPMA homopolymer was designed and synthesized in Czechoslovakia as a plasma expander [75]. Since 1973, HPMA has been the most investigated and advanced polymeric system; it is used widely in therapeutics due to its versatility as a vehicle in delivering active molecules. HPMA is hydrophilic and therefore can increase water solubility of the drugs and has proven to be nontoxic in rats. HPMA copolymers have proven to

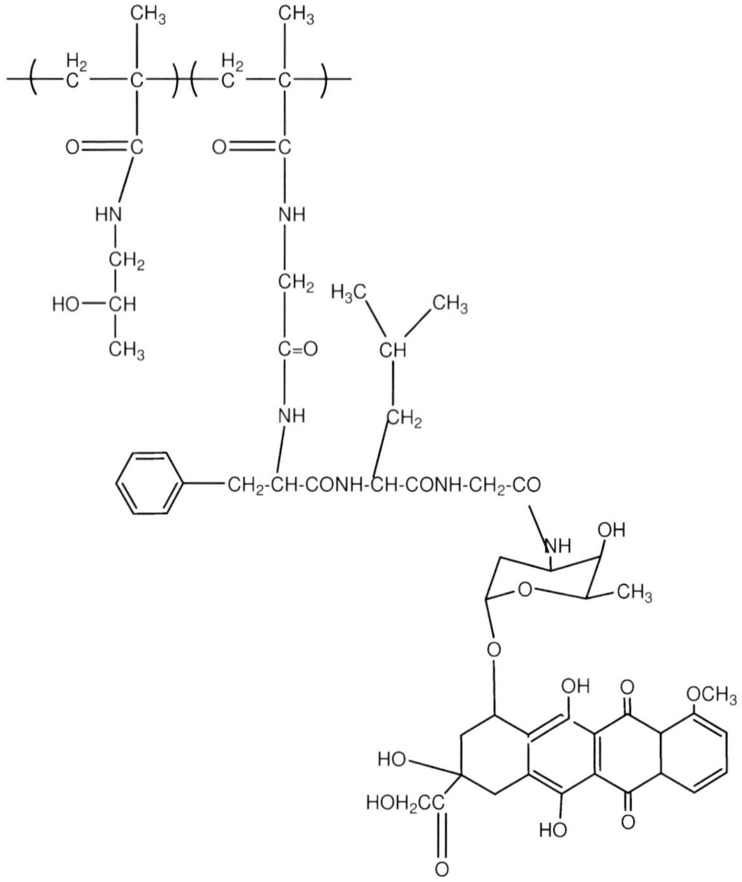

Fig. 14.7 Structure of adriamycin–HMPA copolymer conjugate (PK1). Reproduced with permission from [15].

be biocompatible, non-immunogenic and nontoxic [76]. In addition, their body distribution is well characterized and documented [77]. HPMA has shown great potential as a carrier for targeted delivery of antibodies, peptides and antisense oligonucleotides and controlled release of small molecules [78]. Moreover, HPMA copolymer–anticancer drug conjugates are highly water soluble and can deliver anticancer drugs to targeted carcinomas [79].

In most instances, an amide bond in the polymer is used to link to the drug via a peptide linkage or a spacer [10, 80]. HPMA copolymer is coupled to doxorubicin (Dox) by incorporation of the peptidyl linker Gly–Phe–Leu–Gly (PK1) (Fig. 14.7) [81]. Poly(HPMA) hydrazides were modified using N-succinimidyl 3-(2-pyridyldisulfanyl)propanoate (SPDP) to incorporate pyridyldisulfanyl groups for subsequent conjugation with a modified antibody (Fig. 14.8). The anticancer drug Dox was linked to the remaining hydrazide groups via an acid-labile hydra-

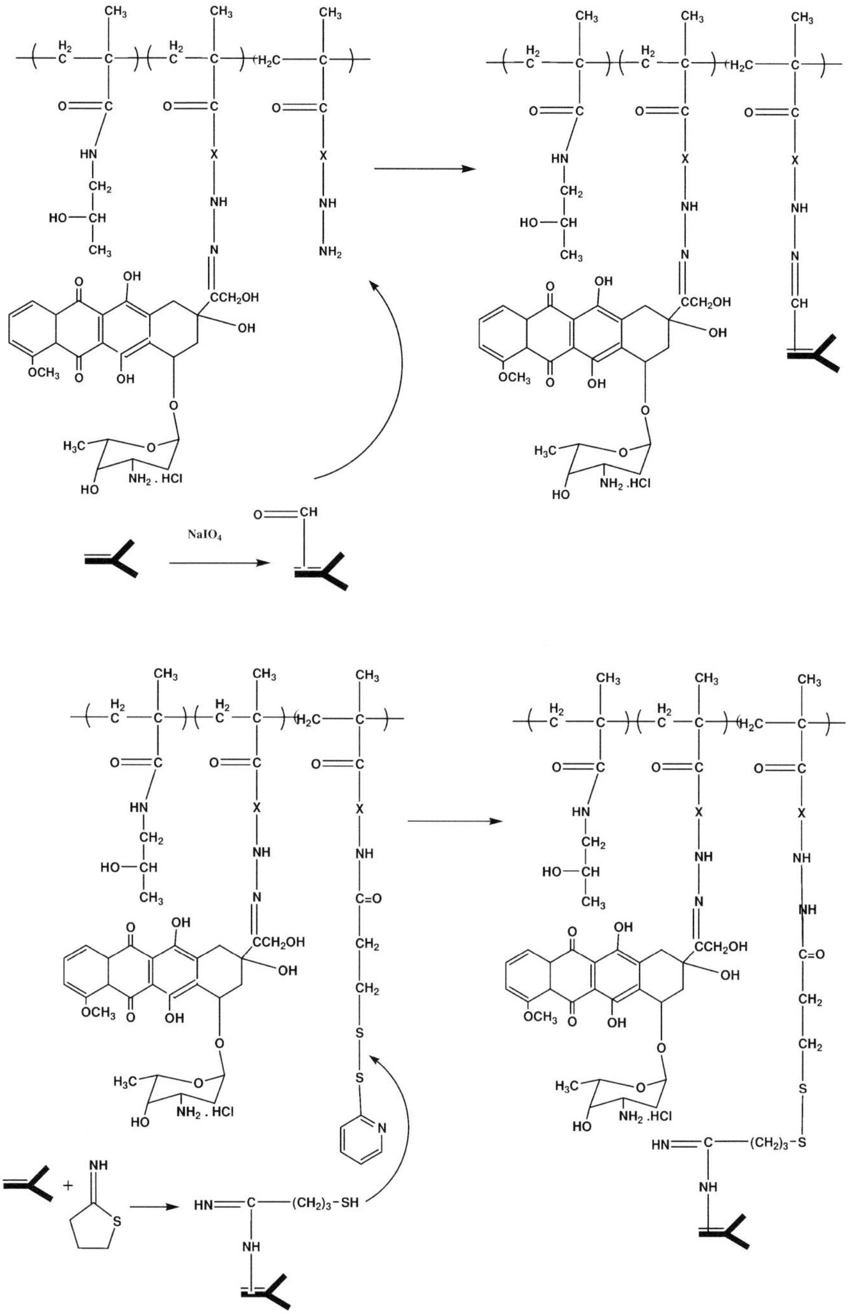

Fig. 14.8 Synthetic scheme for HPMA co-polymer–Dox-antibody bioconjugates. PHPMA hydrazides were modified with *N*-succinimidyl 3-(2-pyridyldisulfanyl)propanoate (SPDP) to introduce the pyridyldisulfanyl groups and for subsequent conjugation with modified antibody Reproduced with permission from [82].

zone bond [82, 83]. Human immunoglobulin G (IgG) was modified with 2-iminothiolane by conjugating it to the HPMA polymer through substituting the 2-pyridylsulfanyl groups of the polymer with –SH groups of the antibody. It was shown that HPMA–adriamycin conjugates were less toxic than the free drug and could accumulate inside solid tumors.

Various advanced studies using acid-sensitive Dox–HPMA copolymer conjugates have been reported [84, 85]. Polymer–Dox conjugates containing sidechains of hydrazone-bound Dox moieties were attached via single amino acid or longer oligopeptide spacers. Enzymatically degradable Gly–Phe–Leu–Gly or non-degradable Gly, Gly–Gly, β-Ala, 6-aminohexanoyl (AH) or 4-aminobenzoyl (AB) spacers were used. HPMA-based conjugates with Dox attached through Gly–Phe–Leu–Gly, Gly–Gly and AH spacers containing a *cis*-aconityl residue at the spacer end were also synthesized and studied. It was shown that the rate of Dox release from all the conjugates under study was pH dependent, with the highest release rates obtained at pH 5.

14.4.2.4 Dendrimers and Their Conjugates

Polymeric architecture, including linearity, polydispersity, molecular weight, solubility, size and functionality, has many implications in prodrug bioconjugates. Advances in polymer science have introduced the world to nanosized hyperbranched polymers (HBPs), e.g. dendrimers. Dendrimers (Greek *dendra* = tree) are hyperbranched macromolecules which can be chemically designed and synthesized to possess precise structural characteristics. These polymers are built up from monomeric units, with new branches being added in steps until a uniform tree-like structure has been formed. Dendritic polymers have emerged as unique drug delivery vehicles because they provide nanosized molecules with a significant amount of tailorability of large density functional groups [50, 86–91]. The following physical and chemical characteristics of dendrimers make them unique nano-vehicles in delivery of drugs: (1) monodispersion, (2) nanometer size ($\sim$ 20 nm), (3) multiple functional groups at the termini (generation 2–8), (4) possession of end-group tailorability such as –NH$_2$, –OH or –COOH, (5) aqueous and organic solvent solubility and (6) biocompatibility.

Theoretically, generation 3 poly(amidoamine) (PAMAM) dendrimers possess 32 end-groups on their surfaces, generation 4 consists of 64 and the generation 5 PAMAM dendrimer has 128 primary amine groups and 126 tertiary amine groups. The practicability of PAMAM dendrimers for cancer treatment is under critical investigation, since these vehicles currently serve in the delivery of targeted drug components, therapeutic agents and imaging agents [92, 93]. The characteristic nontoxicity of PAMAM dendrimers to biological systems makes their biocompatibility much greater than that of many other materials currently researched for use as controlled, chemotherapeutic DDSs [89]. Polyglycerol dendrimers with 4–5 generations were synthesized and used to investigate the effect of dendritic architecture and its generation on aqueous solubilization of paclitaxel [94].

Recently, polyether dendrimers were conjugated to folate residues on their surface for potential tumor cell specificity [86]. Baker et al. reported the conjugation of fluorescein isothiocyanate (FITC), folic acid (FA) and methotrexate (MTX) to a generation 5 PAMAM dendrimer [95]. The dendritic device synthesized was targeted to overexpressed membrane-associated folate receptors with FA and demonstrated cellular cytotoxicity. The drug MTX was conjugated to the PAMAM dendrimer to form an ester bond instead of an amide bond.

Although dendrimers do have the capability to conjugate high amounts of drug molecules, the conjugation ratio achieved has been substantially lower than the maximal theoretically possible ratio. On average, 4–5 molecules of a drug are conjugated per mole of dendrimer. The main reasons for poor conjugation ratios being obtained with dendrimers included (1) their small radius of gyration (R_h), (2) the high steric hindrance exhibited by the biomolecule as well as at the peripheral functional groups of the dendrimer, (3) the low reactivity of terminal functional groups for chemical conjugation with biocomponent and (4) the crowding effect of the end-groups. Steric hindrance is repulsion between the electron clouds on bulky groups of a molecule or between molecules, whereas the crowding effect is exerted due to the presence of multiple functional groups at the dendrimer peripherals in a smallest radius.

Various applications of dendrimers still remain unexplored in a broad range of therapeutic areas. The approach of simultaneous conjugation of drug and targeting moiety on dendrimers has been most widely used. In ester-terminated polyether dendrimers, hydrazide groups were introduced on to the surface of the dendrimers by reaction with the hydrazine moiety. In addition, folate residues were conjugated to the hydrazide chain ends of the dendrimers by direct condensation with folic acid in the presence of a condensing agent or by reaction with an active ester derivative of folic acid. Complete fictionalization of the terminal hydrazide groups was achieved for both the first- and second-generation dendrimers with four and eight hydrazide groups.

Most often only about 4–5 biomolecules are being conjugated per mole of dendrimer. In one attempt, a high payload, averaging ~50 ibuprofen molecules, was conjugated per mole of poly(amidoamine) PAMAM G4 hydroxyl-terminal dendrimers [51]. By decreasing the steric hindrance and increasing the reactivity of a drug, as many as 12 methylprednisolone (MP) molecules were conjugated to a PAMAM G4 hydroxyl-terminated dendrimer (Fig. 14.9). The spacer molecule glutaric acid (GA) was coupled to MP to enhance the reactivity of the drug. The resulting MP–GA–COOH moiety was further conjugated with a hydroxyl-terminal dendrimer using DCC as a coupling agent [96]. The conjugate demonstrated comparable therapeutic activity to the free drug, even over short intervals of time.

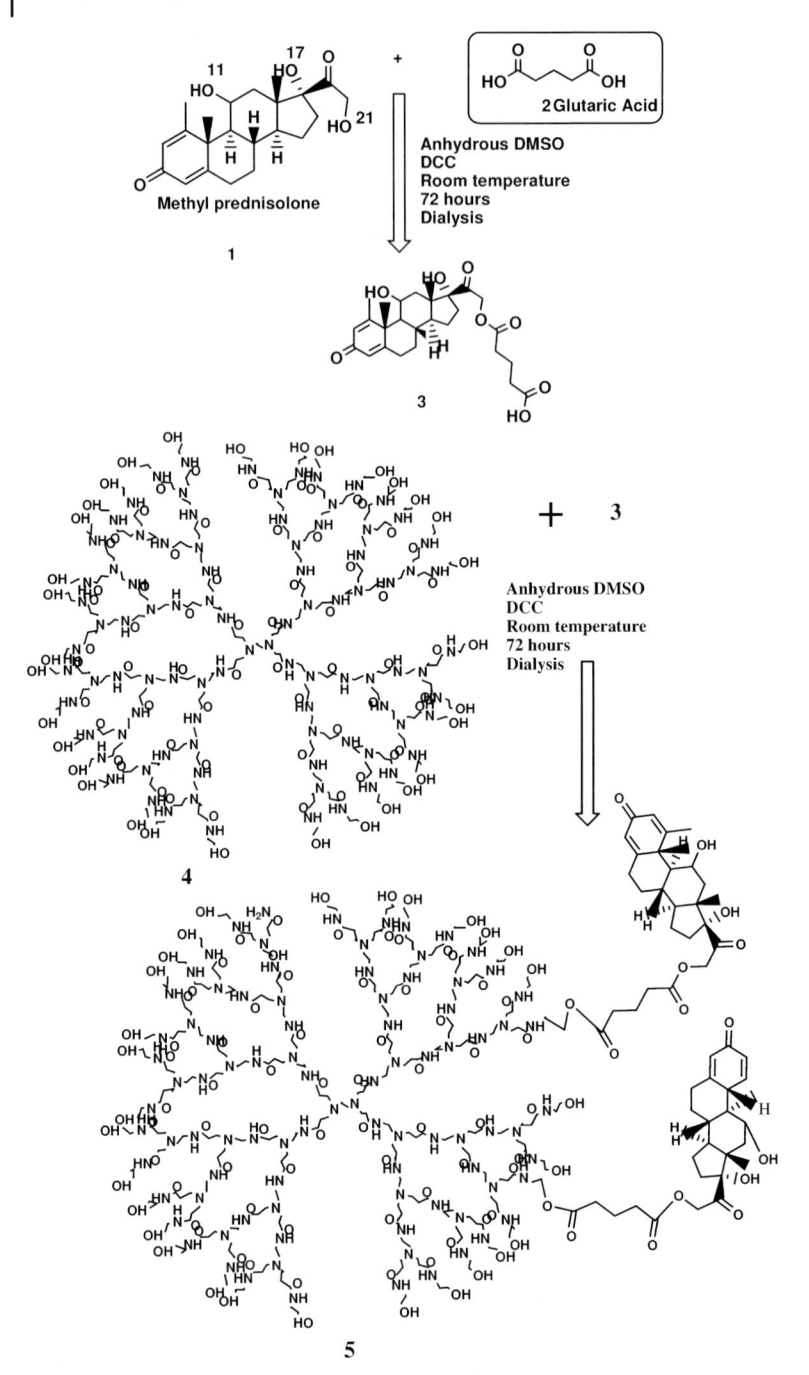

Fig. 14.9 Synthesis scheme for dendrimer–glutaric acid–methylprednisolone conjugate. An average of 12 methylprednisolones (MPs) were conjugated per molecule of generation 4 polyamidoamine OH-terminated dendrimer (PAMAM). Modified from [96].

14.4.2.5 Applications of Molecular Modeling in Prodrug Conjugates

Conformational structures and spacer chain lengths in prodrug conjugates are being elucidated using molecular modeling techniques [30, 72]. Due to high molecular weight constraints, the three-dimensional conformation of polymers along with conjugated biomolecules is not routinely studied. However, a few recent studies suggest applications of modeling in prodrug bioconjugates, such as involving incorporation of spacers and strategies to reduce the steric hindrance in prodrug synthesis [72]. The authors demonstrated steric hindrance associated in PEG–Ara–C conjugation by a molecular modeling method to determine the most suitable bicarboxylic amino acid that possesses the least steric hindrance. The model suggested that aminoadipic acid was most favorable, because the carboxylic groups are sufficiently separated to accommodate Ara-C without the necessity for incorporating spacer arms. The modeling results were confirmed and supported by the synthetic results. PEG conjugates with Ara-C were prepared through an amino acid spacer (e.g. norleucine or lysine) and hydroxyl groups of PEG were activated by *p*-nitrophenyl chloroformate, to form a stable carbamate linkage between PEG and the amino acid.

Molecular models were established for the quantitative description of the spacer length required for optimal binding with the receptor (Fig. 14.10). The structure of the bioconjugate without PEG, CPT–glycine (Gly)–folate and that of the bioconjugate with 3.4-kDa PEG, CPT–Gly–PEG–folate were built using SYBYL (Tripos) on a Silicon Graphics INSIGHT Workstation. The length of the spacer's Gly or Gly–PEG was determined using SYBYL. The bioconjugate with PEG was built to represent only one ethylene repeat ($-CH_2-O-CH_2-$). The number of repeats for a 3.4-kDa PEG was calculated to be 80. Based on the length of a single ethylene unit, the total length of the Gly–PEG spacer was calculated. The length of the Gly spacer was reported to be 3.7 Å, whereas the

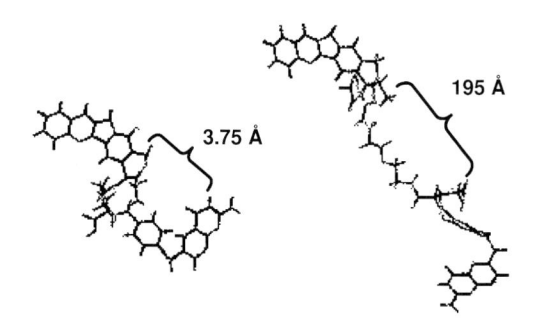

Fig. 14.10 Spacer length in PEG–prodrug conjugate: The structures were built in SYBYL and the linker lengths were calculated. The length of the glycine spacer ∼ 3.7 Å, whereas the length of the glycine–PEG spacer was observed to be 195 Å (one ethylene unit corresponds to 2.391 Å). Modified from [30].

Table 14.3 Molecular modeling studies for camptothecin (CPT) and bis-PEG conjugate.

	Prodrug conjugate	C-1–C-max	Bond length for C-1–C-max (Å)
1	CPT	C-1-C-20	10.54
2	CPT min. energy	C-1-C-20	11.14
3	CPT mol. dynamics	C-1-C-20	11.04
4	CPT bis-PEG conjugate	C-1-C-44	29.40
5	CPT bis-PEG min. energy	C-1-C-44	11.04
6	CPT bis-PEG mol. dynamics	C-1-C-44	5.40

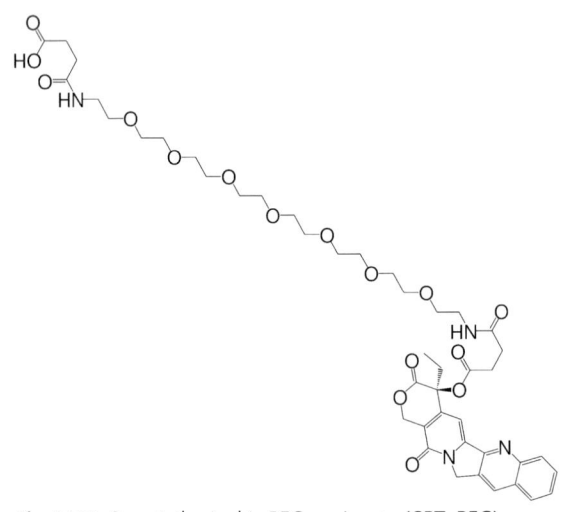

Fig. 14.11 Camptothecin–bis-PEG conjugate (CPT–PEG).

length of the Gly–PEG spacer was observed to be 195 Å (one ethylene unit corresponds to 2.391 Å).

We recently evaluated energy minima and molecular dynamics for synthesized anticancer-polymeric conjugates (Table 14.3). The minima represented the energies (kcal) indicating degree of conformational freedom for seven repeating units of bis-PEG polymeric chain conjugates containing one copy of anticancer drugs CPT (Fig. 14.11). Figure 14.11 show energy minima confirmation for CPT; the distance was observed to be 10.54 Å (C-1 to C-20). On the other hand, the molecular dynamics structure shows a distance of 11.04 Å. The distance between C-1 and C-44 for molecular dynamic CPT–bis-PEG prodrug conjugate was observed to be 5.40 Å. These results indicate that the longer drug–polymeric chains collapse, thereby decreasing the total length of the conjugate [97].

14.4.3
Crucial Aspects in Bioconjugation

Conjugation of small or large active biomolecules to polymers improves the delivery of biologically active components by increasing aqueous solubility and decreasing systemic toxicity. However, synthesizing bioconjugates using appropriate polymers and selecting an appropriate methodology are very critical. The following are the major crucial areas that must be considered when developing a prodrug bioconjugate.

14.4.3.1 Reactivity of the Polymer and Biocomponent

Every chemical modification or conjugation process involves the reaction of one functional group with another, resulting in the formation of a covalent bond. In general, polymers used for conjugation possess high molecular weights (from 1000 to 80 000 Da) and may undergo structural changes in the presence of differing pH ranges, solvents and other reactants. In addition, the reactivity of polymers and the biocomponent may not be sufficient for activation (e.g. in coupling reactions with use of homo- or heterobifunctional cross-linking agents). Therefore, most of the strategies involve incorporation of a spacer molecule between the polymers and active component, which would improve the reactivity for conjugation with a second component. It is well known that incorporating spacer arms enhances ligand–protein binding [98]. In addition, a spacer moiety or a linker in prodrug conjugate provides (1) stability at physiological pH and (2) release of the bioactive agent at an appropriate site of action. Amino acid spacers, e.g. glycine, alanine, serine and small peptides, are generally preferred due to their chemical versatility for covalent conjugation and biodegradability. In the past, various spacers have been incorporated along with the polymers and copolymers to decrease the crowding effect and steric hindrance [99]. Heterobifunctional coupling agents containing succinimidyl have also been used extensively as spacers. Linking of polymer to the drug is a key step in regulating the biological activity of the drug and for supporting a trigger for activation of the macromolecular prodrug. Three types of linkers are mainly used in polymeric prodrugs: (1) those susceptible to neutral pH, e.g. ester linkage; (2) those which react through acid-catalyzed cleavage, e.g. *cis*-aconityl groups; and (3) those prone to cleave by exposure to targeted enzyme, such as cathepsins and esterase.

14.4.3.2 Use of an Appropriate Method for Preparation of Bioconjugates

The method of preparation of a conjugate depends on two interrelated chemical reactions: the reactive functional groups on the different cross-linking or derivatizing reagents and functional groups present on the target biomolecule to be modified. Most bioconjugate reactions involve coupling of amine-containing molecules with *N*-hydroxysuccinimide (NHS) ester-activated polymers [26]. The primary coupling reactions for amine modifications are usually of two types: acylation or alkylation. Such reactions are efficient, rapid and give high yields with stable amide or secondary amine bonds. Carbodiimide coupling reactions or zero-length cross-linkers are widely used for coupling or condensation reactions. Coupling agents mediate the conjugation of the two molecules by forming a bond with no additional spacer atom. Therefore, one atom of the molecule is covalently linked to an atom of the second molecule with no additional linker or spacer. Carbodiimides are most commonly used as coupling reagents to obtain an amide linkage between a carboxylate and an amine or a phosphoramidate linkage between a phosphate and an amine. They are unique due to their efficiency and versatility since they can form a conjugate between two polymers, protein molecules, between a peptide and a drug molecule, between a peptide and a protein or any combination of these small molecules. Additionally, biomolecules containing phosphate groups (such as the 5′-phosphate of oligonucleotide) can be coupled to amine-containing molecules by using carbodiimide reactants (e.g. EDC). The carbodiimide activates the phosphate to an intermediate phosphate ester, similarly to its reaction with carboxylates. In the presence of an amine, polymers containing $-NH_2$ terminal groups can be conjugated to form a stable phosphoramidate bond. Other coupling procedures involve the use of heterobifunctional reagents to couple via modified lysine residues on one protein to sulfhydryl groups on the second protein. The modification of lysine residues involves the use of a heterobifunctional reagent comprising an NHS functional group together with a maleimide or protected sulfhydryl group. The linkage formed is either with a disulfide bridge or with a thioether bond, depending on whether the group introduced is a sulfhydryl or maleimide, respectively. The thiol group on the second protein may be an endogenous free sulfhydryl or chemically introduced by modification of lysine residues.

14.4.3.3 Steric Hindrance

Steric hindrance represents an interference of one molecular group with other groups in the structure or other molecules during chemical conjugation. This is due to the interactions of molecules governed by shape, size, structural conformation and/or spatial relationships. Molecules that have an affinity for one another may not be at an appropriate distance to attract each other or may have other atoms blocking them. Steric hindrance drives chemical conformations and may affect the chemical conjugation with bulkier and unstable molecules. Therefore, a conjugation reaction involving polymers, peptides and unstable molecules requires methodologies which would reduce this effect. The most preferred method to decrease steric hindrance has been to alter the synthetic approach, either by incorporating spacer arms or by increasing the reactivity of the polymer or biomolecules [96].

14.4.4
Future Directions

Polymeric prodrugs seem to have great applications in the delivery of therapeutic agents. There is a need to design and synthesize new polymer candidates based on individual biological applications. Thorough screening and evaluation of new polymer candidates along with their conjugate forms take many years. Therefore, polymer research is mainly dependent on well-established polymers. A variety of molecular architectures of polymers can be synthesized based on the properties and functions of drugs. However, the ideal polymer conjugate should selectively accumulate the drug in targeted organs or tissues. In addition, such conjugates must reduce accumulation in normal tissues, thereby decreasing drug-related adverse side-effects. Further, the polymer, and also its bioconjugate, should enhance the aqueous solubility and stability of the drug.

14.5
Controlled Drug Release

The realization that the therapeutic efficacy of certain drugs can be affected dramatically by the ways in which they are delivered has created immense interest in controlled DDSs. "Controlled" drug release from a DDS is defined as the release of a drug or other active agent in a predesigned manner. The release of the active agent may be constant over a long period, it may be cyclic over a long period or it may be triggered by the environment or other external events (Fig. 14.12). Polymers are very often used to design controlled release DDSs. The simple polymeric controlled release DDS might contain a polymer, whether natural or synthetic, combined with a drug or other active agent. More complex controlled release DDS might represent a complex device containing a microchip which controls the drug release. For instance, advances in microelectromechanical systems (MEMS) have allowed the microfabrication of polymeric substrates and the development of a novel class of controlled delivery devices [100]. Despite differences in the construction and complexity of controlled release DDSs, their common feature is that they all are intended to provide the release of an active agent(s) in a predesigned manner.

14.5.1
Advantages of Controlled Drug Release

Advantages of controlled drug release include more effective therapies, eliminating the potential for both under- and over-dosing, the maintenance of drug levels within a desired range, the need for fewer administrations, optimal use of the drug in question and increased patient compliance. At the same time, controlled release devices or DDSs might have certain disadvantages. Disadvantages of controlled drug release include the possible toxicity or nonbiocompatibility of

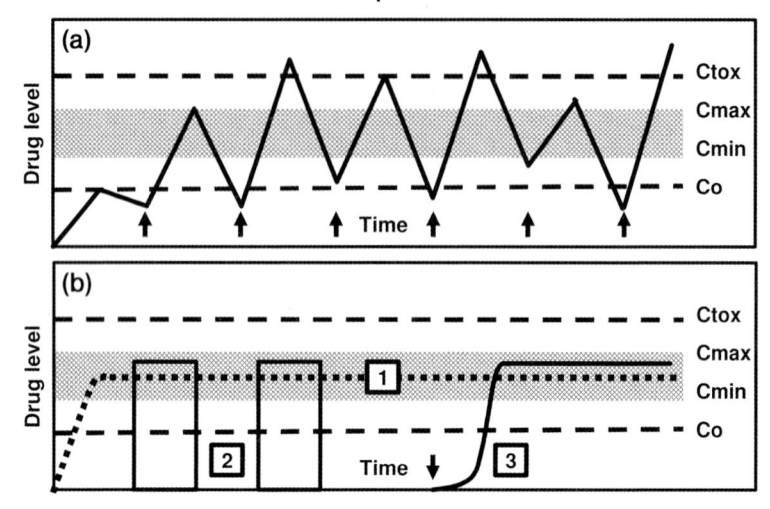

Fig. 14.12 Undesirable sawtooth kinetic profile under conditions of normal dosing (a) and optimum therapeutic profile obtainable with controlled release devices (b). The main types of controlled drug release kinetic profiles: (1) Constant over a long period; (2) cyclic over a long period; (3) triggered by the environment or other external events. Arrows in (a) show time points of repeated administration of the drug. A triggered environmental event (e.g. change in blood pH, glucose level) in (b) is indicated by an arrow.

the materials used in the controlled release device or DDS, undesirable by-products of degradation, any surgery required to implant or remove the system, the chance of patient discomfort from the delivery device and the higher cost of controlled release systems compared with traditional pharmaceutical formulations. Based on these advantages and disadvantages of controlled release, one might formulate requirements for the "ideal" controlled release DDS as follows: such systems should be inert, biocompatible, mechanically strong, comfortable for the patient, capable of achieving high drug loading, safe from accidental drug release, simple to administer and remove and easy to fabricate and sterilize. This system should certainly provide the desired profile of drug release.

14.5.2
Types of Controlled Drug Release

Three general types of controlled drug release systems are used, providing different profiles of plasma levels that are depicted in Fig. 14.12. Type (1) provides sustained release of the active ingredient at a constant rate over an extended period of time. This type of controlled release is used when the blood level of a drug must be maintained at a stable therapeutic level over a long period. Type (2) provides a cyclic pattern of release over a long period. Drug delivery systems that provide this type of drug release kinetics might be used for treatment that requires an episodic increase in drug concentration followed by a "rest" period (i.e. when the drug concentration rapidly decreases below the therapeutic level). Type (3) controlled drug release is triggered by the environment or other external events. Changes in pH, temperature or concentration of certain biologically active substances might serve as such stimuli that trigger drug release. For example, in diabetes treatment it is important to release insulin into the bloodstream only when the blood glucose concentration rises higher than a certain critical level. Therefore, a drug delivery device of such type should have a of blood glucose concentration sensor that would then trigger the release of insulin.

The most often used type of controlled drug release is sustained drug release. The rationale for the sustained controlled delivery of drugs is to promote therapeutic benefits while at the same time minimizing toxic effects. Normal drug dosing may follow a "sawtooth" kinetic profile (Fig. 14.12a) in which the dose first greatly exceeds the desired therapeutic level, then falls to a subclinical level and on subsequent dosing rises to dangerously high values, falling again to ineffective concentrations, in continuous cycles of excessive–ineffective levels. Controlled, sustained drug delivery can reduce the undesirable fluctuation of drug levels, enhancing therapeutic action and eliminating dangerous side-effects (Fig. 14.12b). Polymers are often used to provide a sustained release of the payload of DDSs into the desired body site. For instance, sustained drug delivery using implantable polymeric devices has been successfully used for local intraocular delivery of therapeutics. The eye is uniquely shielded from foreign substance penetration by its natural anatomical barriers. This makes effective drug delivery to the inside of the eye difficult. Two main barriers protect the eye: (1) the cornea, which protects the front of the eye, and (2) the blood–retina barrier, which protects the back of the eye. Hydrophobic or hydrophilic polymers shaped into a sheet, disc, rod, plug or a larger device can be implanted into the subretinal space, intrascleral space, vitreous space or peribulbar space or at the pars plana [101]. Drugs embedded in such polymeric devices can be released over a prolonged period, providing continuous treatment. Intraocular sustained drug release using implantable devices has been investigated to treat vitreoretinal diseases. Possible targeted diseases include those in which repeated intraocular injections are effective (cytomegalovirus retinitis, uveitis), diseases requiring surgery (proliferative vitreoretinopathy) and chronic diseases (AMD, macular edema, retinitis pigmentosa) [101]. Another example of a sustained release polymeric drug is a dry powder-inhaled PEG formulation for delivering peptides (e.g. insulin) efficiently across the lungs and to promote prolonged serum concentration of the peptide [102]. A sustained-release system of sparfloxacin was developed for use in the treatment of periodontal disease [103]. It was formulated, based on ethylcellulose 10 cP, PEG 4000 and diethyl phthalate (DEPh). This device is referred to as the sparfloxacin chip. The chip has dimensions of 10 mm length, 2 mm width and 0.5 mm thickness. The *in vitro* drug release pattern and clinical evaluation of the formulations were studied. Reports of the short-term clinical study show that the use of the chip may cause complete eradication of the pathogenic bacteria in the periodontal pockets of patients who have chronic generalized periodontitis. Significant improvements were observed in many parameters of the treatment group compared with the placebo group.

More "exotic" types of controlled drug release include the "release by demand" type of devices and other polymeric DDSs. Such DDSs can react to changes in homeostasis to release a drug into the bloodstream or specific tissues in order to correct the disturbances. For instance, one can imagine an implanted microdevice which senses the glucose concentration in the blood. If the concentration of glucose increases higher than a certain critical level, the device would release insulin until the glucose concentration returned to a normal value. Such "intelligent polymers", also referred to as "stimuli-responsive polymers" undergo strong property changes (in shape, surface characteristics, solubility, etc.) when only small changes in their environment occur (changes in temperature, pH, ionic strength, light, electrical and magnetic field, etc.). Conducting polymers (electroactive conjugated polymers) have emerged as one of the most promising transducers for chemical sensors. This is related to the unique electrical, electrochemical and optical properties of conjugated polymers that can be used to convert chemical information (concentration, activity, partial pressure) into electrical or optical signals in the solid state. Different approaches to the application of conducting polymers in chemical and biochemical sensors have been extensively studied since the mid-1980s and the achievements have been treated in several reviews [104, 105]. Stimuli-responsive polymers have been used in several novel applications, DDSs, tissue engineering scaffolds, bioseparation, biomimetic actuators and so forth [106]. The most popular of these type of polymers is poly(N-isopropylacrylamide) which exhibits a temperature-sensitive characteristic in which the polymer chains change from water-soluble coils to water-insoluble globules in aqueous solution as the temperature increases above the lower critical solution temperature of the polymer. Copolymerization of N-isopropylacrylamide with acrylic acid allows the synthesis of both pH- and temperature-responsive copolymers.

An interesting polymeric peptide delivery system with a controlled delay time and a burst-free pre-release phase has been described recently [107]. In general, the system consists of a blend of a tyrosine-derived polyarylate and a fast-degrading copolymer of lactic and glycolic acid (PLGA). Due to the peptide-like structure of the polyarylate backbone, peptide–polymer interactions prevent the release of peptide from neat polyarylate films. The addition of PLGA acts as a "delayed" excipient. As PLGA degrades, it generates acidic degradation products that cause a drop in the internal pH of the polyarylate matrix. This drop in pH weakens the peptide–polymer interactions and causes the release of peptide to commence. The initial molecular weight of PLGA can be used to control the length of time before degradation occurs. Consequently, this parameter can also be used to control the duration of the delay period prior to peptide release. As a specific model system, blends of poly(DTH adipate) with three different copolymers of lactic and glycolic acid were prepared and used for the delayed release of Integrilin, a synthetic water-soluble heptapeptide (used clinically in antithrombic injections) that acts as a highly potent glycoprotein IIb/IIIa antagonist. Blends composed of a 1:1 weight ratio of poly(DTH adipate) and PLGA and containing Integrilin (15%, w/w) were prepared. *In vitro* release studies were conducted in a phosphate-buffered solution at 37 °C and the release of Integrilin was followed by HPLC. As the initial molecular weight of PLGA varied from 12 000 to 62 000, the duration of the delay period prior to release increased from 5 to 28 days. PLGA copolymers have been also used for sustained release of other drugs, including 5-fluorouridine [108], carboplatin [109] and gentamicin [110].

14.5.3
Future Directions

The most exciting opportunities in controlled drug delivery are in the arena of responsive delivery systems. With this type of system, it will be possible to deliver drugs in response to a change in the environment precisely to a targeted site. Much of the development of novel materials in controlled drug delivery is focused on the preparation and use of these responsive systems with specifically designed macroscopic and microscopic structural and chemical features. While much previous work in drug delivery focused on achieving sustained drug release rates over time, a more recent trend is to make devices that allow the release rate to be varied over time. Advances in microfabrication technology have made an entirely new type of drug delivery device possible. Proof-of-principle experiments have shown that silicon microchips have the ability to store and release multiple chemicals on demand [111]. Future integration of active control electronics, such as microprocessors, remote control units or biosensors, could lead to the development of a "pharmacy on a chip", for instance, "smart" microchip implants or tablets that release drugs into the body automatically when needed.

14.6
Anticancer Polymeric Drugs

Anticancer drugs form a very important subset of polymeric drugs and DDSs. The main common feature of such drugs is their cell death-inducing activity. Current anticancer pharmacy possesses a wide range of highly toxic drugs which already are being used in cancer treatment or undergoing preclinical and clinical trials. The main goal in modern development of anticancer drugs is to prevent adverse side-effects of such toxic compounds to normal tissues, mainly by delivering anticancer drugs specifically to the tumor site, and to overcome or suppress intrinsic or acquired drug resistance of the cancer cells due to the treatment.

14.6.1
General Considerations

An "ideal" anticancer DDS should fulfill the following requirements: (1) effectively kill tumor cells; (2) be nontoxic for other healthy organs, tissues and cells; (3) not induce multidrug resistance and overcome existing drug resistance; (4) inhibit existing drug resistance; and (5) provide for the particular treatment scheme's required profile of drug release. It should be stressed that despite intensive research in this area, there are no clinically approved anticancer DDSs that comply with all of these requirements. However, extensive research efforts in the area of anticancer drug delivery have resulted in several novel approaches to cancer therapy that have substantially come close to achieving this goal.

14.6.2
Tumor-targeted Anticancer Polymeric Drugs

Targeting of anticancer drugs to the tumor provides several advantages over nontargeted drugs. The main advantages are the prevention of side-effects of drugs on healthy tissues and enhancement of the drug uptake by the targeted cells. Side-effects, which invariably impose dose reduction, treatment delay or even discontinuance of chemotherapy, should be avoided, especially with modern anticancer drugs with exceptionally high specific activities. Targeting a drug specifically to cancer cells and/or specific organelles inside the cells permits the internalization of substances with low cellular permeability by endocytosis and drug release in targeted organelles (e.g. lysosomes, mitochondria, nucleus). As a result, the targeted drug acts like a "magic bullet", selectively killing the villain and sparing the innocent. Targeting to cancer cells is especially important in the case of anticancer DDSs. Such systems usually contain one or more highly toxic components and have a relatively large size and high molecular weight compared with most low molecular weight anticancer drugs. Cellular uptake of macromolecular DDSs by diffusion is limited; specific mechanisms of cellular internalization are required. Two approaches are generally used to target the drug to the tumor and cancer cells: (1) passive and (2) active targeting [43].

14.6.2.1 Passive Targeting
Passive targeting approaches include (1) the EPR effect, (2) exploiting environmental conditions in tumors or tumor-bearing organs and (3) topical delivery directly to the tumor [43].

As a result of the increased permeability of the tumor vascular endothelium to circulating macromolecules combined with limited lymphatic drainage from the tumor interstitium, low molecular weight drugs coupled with high molecular weight carriers are inefficiently removed by lymphatic drainage and therefore accumulate in tumors. Theoretically, any high molecular water-soluble drug carrier should show such an effect, including water-soluble polymers, liposomes, etc. The existence of the EPR effect has been experimentally confirmed for many types of macromolecular anticancer DDSs [112].

In most cases when analyzing the EPR effect of an anticancer DDS, researchers look closely at the characteristics of the carrier (molecular weight, size, etc.), often disregarding the influence of the anticancer drug itself. However, Minko and coworkers [13, 14] showed that the presence of an anticancer drug in a DDS significantly influences the drug distribution in the tumor targeted by the EPR effect. The most pronounced effect was observed for cytotoxic drugs (such as HPMA–copolymer bound doxorubicin) that decrease the membrane permeability of tumor capillaries. In this case, a more uniform distribution of a drug within the tumor was found.

The second passive targeting approach utilizes specific conditions in an organ bearing a tumor or in the tumor environment itself to facilitate drug release from the DDS. These conditions include, but are not limited to, a particular pH, the existence of certain enzymes or microflora in a specific organ or tumor. For example, drug delivery to the colon might be targeted by preparing tablets with a specific coating that is destroyed in the colon by colon-specific pH and/or colon-specific bacteria [113–115]. Other colon-targeting approaches have been discussed in several reviews [116–118]. The limitation of this approach is the targeting of the whole organ, not the tumor itself. This potentially opens the door to severe organ cytotoxicity. Therefore, a passive targeting strategy based on specific tumor conditions is better in the sense that the specific tumor is being targeted rather than the whole organ.

One approach to passive targeting of cancer cells is based on the use of metalloproteases, enzymes overexpressed in certain tumors. It was found that the progression of malignant melanoma is characterized by overexpression of a number of matrix metalloproteinases (MMPs), especially MMP-2, which play a critical role in the degradation of basement membranes and the extracellular matrix. Consequently, a drug targeting strategy has been developed in which the protease activity of MMP-2 is exploited to release an anticancer agent from a macromolecular carrier [119]. For this purpose, a water-soluble maleimide derivative of doxorubicin incorporating an MMP-2 specific peptide sequence (Gly–Pro–Leu–Gly–Ile–Ala–Gly–Gln) was developed. This compound binds rapidly and selectively to the circulating albumin using the latter as a carrier. The albumin-bound form of this prodrug was efficiently and specifically cleaved by MMP-2, liberating a doxorubicin tetrapeptide (Ile–Ala–Gly–Gln–Dox) and subsequently doxorubicin.

Bae et al. [120] proposed a novel polymeric cancer-targeting DDS in which the enzymatic activity of type IV collagenases is used to cleave the inactive drug conjugate (prodrug), thereby activating drug fragments. The anticancer drug conjugate (mPEG–GPLGV–Dox) was synthesized by conjugating doxorubicin with Gly–Pro–Leu–Gly–Val (GPLGV) peptide and PEG methyl ether (mPEG). GPLGV pentapeptide was used as a substrate for MMP-2 and MMP-9, where the cleavage of the Gly–Val bond by MMP was expected. In addition, mPEG was grafted to the peptide–doxorubicin conjugate to increase the circulation time in the body and to reduce the cytotoxicity of the anticancer drug. The experiments showed that the proposed anticancer DDS increased the therapeutic efficiency and decreased the drug toxicity of the established anticancer drug by optimizing the degradation rate of the peptide link by MMP and circulation time in the body. Several chemical approaches have recently been applied to exploit the pH and redox potential triggers to release active pharmaceutical ingredients in the appropriate biological location [121].

The third passive tumor targeting approach is local (topical) delivery of anticancer agents directly into the tumor. This drug delivery technique has the obvious advantage of excluding drug delivery through the systemic circulation. Therefore, this approach potentially could limit adverse drug effects on healthy organs and tissues, as long as topically delivered drugs are retained in the tumor and do not enter the circulation. While topical delivery for some tumors might be achieved by injections or surgical procedures, other tumors, for instance lung cancers, are difficult to access for local drug delivery. To overcome this, several anticancer aerosols have been developed for lung cancer treatment [122, 123].

While all three main approaches to passive targeting of DDSs to tumors might be utilized to enhance cancer-specific delivery of pharmaceutical preparations, they are rarely used as a major targeting method in cancer therapy. The more frequently employed technique is active targeting.

14.6.2.2 Active Targeting

Active targeting of a DDS is usually achieved by adding to the DDS a targeting component that provides preferential accumulation of the whole system or drug in an organ bearing a tumor, in the tumor itself, cancer cells, intracellular organelles or certain molecules in specific cancer cells (Fig. 14.13). The active targeting approach is based on the interactions between a ligand and a receptor or between specific biological pairs (e.g. avidin–biotin, antibody–antigen, lectin–car-

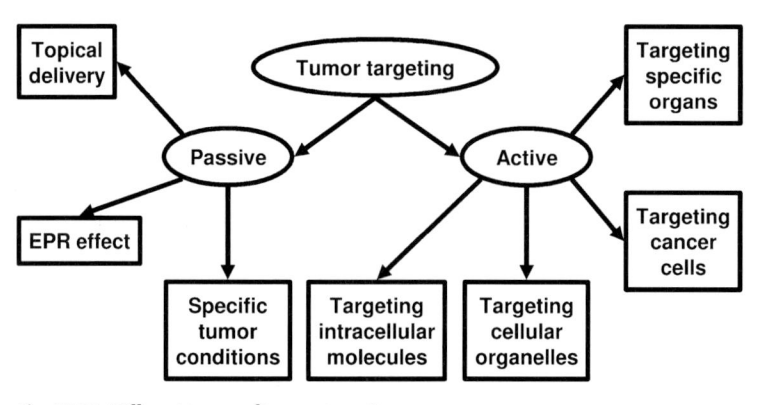

Fig. 14.13 Different types of tumor targeting.

bohydrate). In most cases, a targeting moiety in a DDS is focused on the specific receptor or antigen overexpressed in the plasma membrane or intracellular membrane in beleaguered cells.

Active targeting of specific organs is usually achieved by adding to the DDS a targeting moiety specific to the receptors or molecules overexpressed at the surface of the plasma membrane of cells in the targeted organ. Lectin–carbohydrate is one of the classic examples of a biological pair that can be used for targeted drug delivery. Lectins are proteins of nonimmunological origin, capable of recognizing and binding to glycoproteins expressed on cell surfaces. Lectin interactions with certain carbohydrates are very specific. This interaction is as specific as an enzyme with a substrate or an antigen with an antibody. Carbohydrate moieties can be used to target the DDS to lectins (direct lectin targeting) and *vice versa*; lectins can be used as targeting moieties to target cell surface carbohydrates (reverse lectin targeting) [117]. Several DDSs utilizing lectins or carbohydrates have been developed to target different organs [124]. This approach has been used most often for targeted drug delivery to normal and malignant colon cells and to deliver anticancer drugs. Some examples of such DDSs are N-(2-hydroxypropyl)methacrylamide copolymer-conjugated wheat germ agglutinin and peanut agglutinin [125], complex polymeric dendrimeric structures – glycodendrimers [126] – and liposomes targeted to cell surface galectins [127].

In contrast to other organs, serious difficulties arise when the brain is the target for drug delivery. The presence of the blood–brain barrier prevents the penetration of many molecules into the brain. Overcoming this problem might have a profound effect on the treatment of many brain disorders, including cancer. Transport vectors, such as endogenous peptides, modified proteins or peptidomimetic monoclonal antibodies, are promising tools to transfer large water-soluble DDSs to the brain [128].

The main drawback of targeting an anticancer DDS to a whole organ, not to cancer cells in the tumor, is the likelihood of adverse side-effects to healthy cells in the targeted organ. To overcome this, several strategies were developed to ensure the targeting of anticancer DDSs specifically to cancer cells.

The most obvious way to target a DDS specifically to cancer cells is to target receptors or antigens overexpressed on the surface of plasma membrane of cancer cells. The reaction of a ligand with a corresponding plasma membrane receptor (or antibodies with cell surface antigens) enhances the uptake of the entire DDS into the cell. Several "biological pairs" that have the ability to bind specifically might be used to target a DDS to cancer cells. Antibodies corresponding to antigens are one example of such pairs. Several targeted systems which utilize antibodies or their fragments have been developed. Polymer-based DDSs which target anticancer drugs to cancer cells using antibodies (or their recognition fragments) are examples of such targeted systems [9, 80, 129, 130]. Potential targets and targeting moieties for targeted DDSs will be discussed later in this chapter.

An interesting targeting approach of a DDS to cancer cells is based on exploiting their drug resistance. Antibodies to P-glycoprotein overexpressed in plasma membranes of cancer cells resistant to chemotherapy might be a potential targeting moiety for a targeted anticancer DDS [131]. Another approach to killing selectively resistant cancer cells has been proposed by Blagosklonny [132]. This approach is based on the temporary increase in the resistance of sensitive cells against certain anticancer drugs by specific protectors, such as pharmacological inhibitors of apoptosis. These protectors will be pumped out by multidrug-resistant cells while increasing the resistance in sensitive cells which do not have active drug efflux pumps. After applying a cytotoxic drug, sensitive cells will be protected and survive the exposure, while unprotected resistant cells will be killed. By abolishing several dose-limiting side-effects of chemotherapy, this strategy might provide a means to treat selectively the most deranged, aggressive and resistant cancers. A chemotherapy regimen based solely on targeting of anticancer drugs specifically to multidrug resistant cancer cells has at least two major drawbacks. First, a DDS targeted exclusively to resistant cancer cells will kill only such cells, leaving sensitive cancer cells viable. Therefore, it will not prevent tumor growth, especially in the initial stages of cancer treatment when most of the cancer cells have not yet developed multidrug resistance. Second, such a DDS will effectively kill normal cells, such as bone marrow hematopoietic cells, which have overexpressed P-glycoprotein active transporters [133]. Other cell types such as immune cells [134], cells of the blood–brain barrier [135], liver [136], lung alveolar epithelium [137], retinal epithelium [138], intestine, kidney, testis, adrenal gland and the pregnant uterus [139] may fall into this category.

Mitochondria and nuclei appear to be two intracellular organelles that might be considered the most important intracellular targets for anticancer drugs. They both contain genetic material and their damage is the main cause of cellular death by apoptosis [140]. Most gene delivery systems have been targeted to the nuclear genome, while the mitochondrial genome has been neglected for a long time [141, 142]. Although the targeting of anticancer DDSs to mitochondria could open new perspectives in cancer chemotherapy, this type of drug targeting requires further research. In addition to direct targeting to nuclei, anticancer DDSs designed to provide drug release in lysosomes after internalization of the DDSs by endocytosis might provide indirect nuclear targeting. This is possible because the DDS is designed so that after internalization and incorporation in the lysosome, the GFLG spacer degrades and the drug is released, diffusing through the lysosomal membrane into the cytoplasm and ultimately accumulating in the cell nuclei [143, 144]. Incorporation of acid-sensitive spacers between the drug and carrier allows the release of an active drug from the carrier in a tumor tissue, either in slightly acidic extracellular fluids or, after endocytosis, in endosomes or lysosomes of cancer cells [145].

In addition to targeting sites and binding ligands that permit the DDS to be directed to specific cancer cells or cellular organelles, an anticancer DDS might be targeted to several other molecular targets. Such targeting could provide a synergic effect that would ultimately increase the efficacy of the main anticancer agent. For example, this second target might include proteins responsible for the drug efflux from cancer cells, cellular detoxification mechanisms or mechanisms of cellular defense against apoptosis, etc. If, in addition to cell death induction by the anticancer agent, these mechanisms were to be suppressed, the efficacy of cancer treatment would be substantially increased, the concentration of the drug inside the cancer cells would be increased, the drug would not lose its anticancer activity and its ability to induce apoptosis would be significantly augmented. Such complex advanced anticancer DDSs and their possible molecular targets will be discussed later in the chapter.

14.6.3
Proapoptotic Anticancer Polymeric Drugs

The main task of anticancer treatment is the killing of tumor cells. The majority of anticancer drugs kills cells by inducing apoptosis – a programmable cell death, which is actively initiated by cancer cells themselves as a response to external stimuli or anticancer drugs, internal changes in cellular state or metabolism, DNA damage and so forth. In viable cells, including cancer cells, the apoptotic signal is suppressed by a special, multi-protein, complex, antiapoptotic system. This system plays an important role in the suppression of apoptosis and so protects cells from premature death. The same antiapoptotic system is activated after exposure to an anticancer drug, limiting and even totally preventing cell death as a response to anticancer drug exposure (Fig. 14.14). In addition to the antiapoptotic system, many cells, including cancer cells, have developed a mechanism of defense against xenobiotics, including anticancer drugs. The activation of such a system during chemotherapy leads to so-called "multidrug resistance", wherein initially drug responsive cells quickly become resistant to many anticancer drugs after the treatment with one agent. The main mechanism of multidrug resistance in cancer cells is the activation of membrane-bound active drug efflux pumps, which pump an anticancer drug out of the cell and therefore protect cancer cells from drug exposure. We termed this type of cancer cell resistance "pump" resistance [43–45]. In contrast, antiapoptotic defense, cellular detoxification mechanisms and other processes which increase

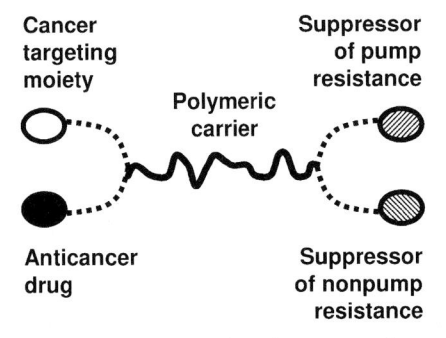

Fig. 14.14 Intracellular fate of anticancer drug.

cellular resistance to drugs and do not require an activation of drug efflux pumps can be termed as "nonpump" resistance. Therefore, to kill cancer cells effectively, an anticancer drug should not only induce cell death, but also simultaneously prevent activation or better suppress both pump and nonpump resistance. Unfortunately, it is not possible to fulfill this complex task of simultaneous cell death induction and suppression of cellular defenses by one simple anticancer drug or even by combined treatment with several drugs. The solution is a complex DDS which can protect the drug from both pump and nonpump resistance or even suppress them. This complex task can be achieved by including in a DDS, in addition to one or more copies of an anticancer drug, active ingredients which can suppress pump and nonpump resistance. Several different systems that can fulfill the task have recently been developed and tested [25, 43–45, 117, 146]. It was found that such DDSs are extremely toxic, because they almost completely suppress cellular resistance against exogenous drugs. Therefore, to prevent severe adverse side-effects to healthy tissues, the DDS should be targeted specifically to cancer cells minimizing the exposure of normal cells.

Several polymeric DDSs that fulfill some of the special requirements have been developed recently. Macromolecular water-soluble carriers have many of the characteristics of an "ideal" DDS. One example developed by Kopecek and coworkers is based on water-soluble N-(2-hydroxypropyl)methacrylamide (HPMA) copolymers [9, 10, 12, 14, 147]. HPMA copolymer-bound drugs enter cells by endocytosis in membrane-limited organelles that ultimately fuse with lysosomes to form secondary lysosomes. This allows the drug to escape drug efflux pumps located in the plasma membrane and protects it from the action of cytoplasmic detoxification enzymes. If a drug is bound to the polymer via a lysosomally degradable spacer, it is released from the secondary lysosomes into the cytoplasm in the perinuclear region. Cancer targeting is achieved by the EPR effect and/or by using targeted moieties including polymerizable antibody fragments, which also increase cytotoxicity by initiating receptor-mediated endocytosis [80]. This DDS provides targeted and protected drug delivery, overcomes pump resistance, at a certain level overcomes nonpump resistance and prevents the activation of antiapoptotic cellular defense. Therefore, the targeted HPMA copolymer–drug conjugate satisfies most of the demands of anticancer DDSs.

Sterically stabilized phospholipid mixed micelles have been manufactured and evaluated as novel carriers for water-insoluble drugs, (e.g. paclitaxel) [148]. Although this type of DDS cannot be considered as a strictly polymeric drug, it contains a polymer as an external shell of the vesicles. These micelles are composed of PEG 2000-grafted distearoylphosphatidylethanolamine and egg-phosphatidylcholine and can be used as an improved lipid-based carrier for water-insoluble anticancer drugs. Immunomicelles, new targeted carriers for poorly soluble anticancer pharmaceuticals, have been developed recently to attach antibodies chemically to reactive groups incorporated in the corona of polymeric micelles made of PEG–phosphatidylethanolamine conjugates [149]. Micelle-attached antibodies retained their ability to interact specifically with their antigens. Immunomicelles with attached antitumor monoclonal antibodies effectively recognized and bound various cancer cells *in vitro* and showed an increased accumulation in experimental tumors in mice when compared with nontargeted micelles.

Pluronic block copolymer P85 was shown to inhibit the P-glycoprotein (Pgp) drug efflux transporter and to increase the permeability of a broad spectrum of drugs in the blood–brain barrier. It was found that lipophilic Pluronics with an intermediate length of propylene oxide block (from 30 to 60 units) and Hydrophilic–Lipophilic Balance (HLB) <20 are the most effective at inhibiting Pgp efflux [150]. It has been discovered that Pluronic block copolymers interact with multidrug-resistant cancer tumors resulting in drastic sensitization of these tumors with respect to various anticancer agents, particularly anthracycline antibiotics. Furthermore, Pluronics affect several distinct drug resistance mechanisms including inhibition of drug efflux transporters and abolishing drug sequestration in acidic vesicles, as well as inhibiting the glutathione/glutathione *S*-transferase detoxification system. All of these mechanisms of drug resistance are energy dependent and therefore ATP depletion induced by Pluronic block copolymers in multidrug-resistant cells is considered to be one mechanism of chemosensitization of these cells. Further mechanistic studies and clinical evaluations of a formulation containing a mixture of doxorubicin and Pluronic (L61 and F127), SP1049C, are in progress [151].

14.6.4
Future Directions in Targeted Proapoptotic Anticancer Polymeric Drugs

DDSs provide "passive" and "active" targeting and, in most cases, a "passive" way to suppress pump and/or nonpump cellular resistance. A novel anticancer proapoptotic DDS that provides active targeting and actively suppresses both pump and nonpump cellular resistance has recently been developed by Minko and coworkers [25, 43–45, 146]. This system includes five main components: (1) an apoptosis-inducing agent (i.e. an anticancer drug), (2) a suppressor of pump resistance, (3) a targeting moiety/penetration enhancer, (4) an inhibitor of nonpump resistance and (5) a carrier (Fig. 14.15). The carrier binds the DDS components together and facilitates the solubility of the whole complex. PEG polymer or liposomes have been used as a carrier for the DDS.

Recently, CPT–PEG conjugate with a synthetic analogue of luteinizing hormone-releasing hormone (LHRH) as a targeting moiety and a synthetic BLC2 homology 3 (BH3) peptide as a suppressor of nonpump resistance was synthesized in our laboratory (Fig. 14.16) [25]. A hydroxyl group at position 20 in CPT was coupled with amino acid to form an ester bond using Boc-Cys (Trt) amino acid. DIPC was used as coupling agent and protected groups were removed using trifluoroacetic acid (TFA) in methylene chloride. The resulting CPT–cysteine ester conjugate had two potential orthogonal conjugation sites – the amino group and the thiol group (Fig. 14.16a). On the other hand, the CPT–Gly ester was first reacted with NHS–PEG–vinyl sulfone (VS). VS has an amino group and forms an amide bond with the active ester (*N*-hydroxysuccinimide ester) of

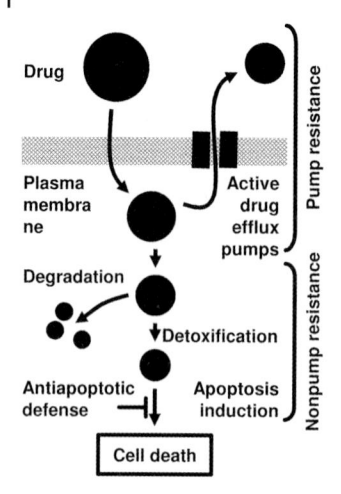

Fig. 14.15 Scheme of a novel advanced targeted proapoptotic anticancer DDS.

PEG. An analogue of BH3 (Ac–Met–Gly–Gln–Val–Gly–Arg–Gln–Leu–Ala–Ile–Ile–Gly–Asp–Asp–Ile–Asn–Arg–Arg–Tyr– Cys–NH$_2$), containing an extra residue of cysteine at the C-terminus, was synthesized by the solid-phase peptide method. The previous reaction mixture was coupled to the thiol group of BH3 and formed a thioether bond with the VS group on the PEG. The LHRH analogue LHRH–Lys6–des-Gly10–Pro9–ethylamide (Gln–His–Trp–Ser–Tyr–D-Lys–Leu–Arg–Pro–NH–Et), which possesses a reactive amino group on the side-chain of the lysine at position 6, was first coupled with one mole equivalent of NHS–PEG–VS using dimethylformamide (Fig. 14.16 b). CPT–Cys was then added to achieve thioether bond formation between the VS group and the thiol group.

The targeting moiety/penetration enhancer should substantially increase the internalization of active components specifically into cancer cells, enhancing the toxicity of the whole DDS to cancer cells and decreasing adverse side-effects on healthy tissues. We selected a synthetic analogue of LHRH as a targeting moiety to deliver the DDS specifically to cancer cells and facilitate its cellular uptake [25, 41, 152]. Experimental testing of synthetic LHRH decapeptide as a targeting moiety *in vitro* on human ovarian cancer cells and *in vivo* on mice bearing xenografts of human ovarian tumors showed that such targeting effectively prevented cell death induction in normal organs and enhanced apoptosis in tumors (Fig. 14.17), In addition, LHRH substantially enhanced internalization of the targeted DDS by cancer cells and its cytotoxicity (Fig. 14.18).

The apoptosis-inducing agent and the suppressor of antiapoptotic cellular defense can work concurrently to induce cell death and suppress cellular defense. Camptothecin and Dox have been tested as apoptosis-inducing agents. The BCL-2 protein family, characterized by specific regions of homology termed BCL-2 homology (BH1, BH2, BH3, BH4) domains, are the main players in the

Fig. 14.16 Synthesis of CPT–PEG–BH3 and LHRH–PEG–CPT conjugates. Modified from [25].

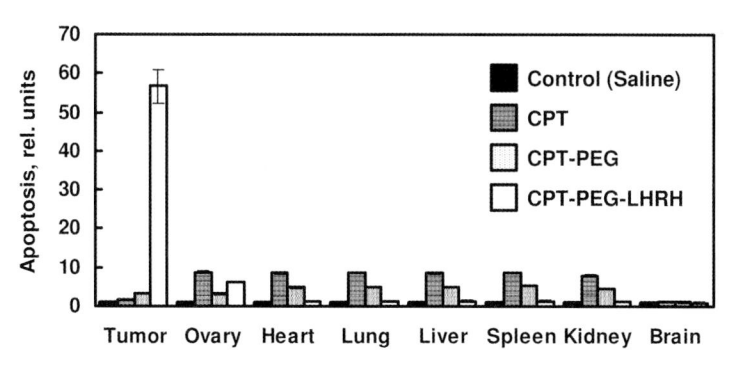

Fig. 14.17 Apoptosis induction after treatment with saline (control), CPT, CPT–PEG and CPT–PEG–LHRH in tumors and different tissues in mice bearing xenografts of A2780 human ovarian carcinoma. Means±SD from 4–5 independent measurements are shown.

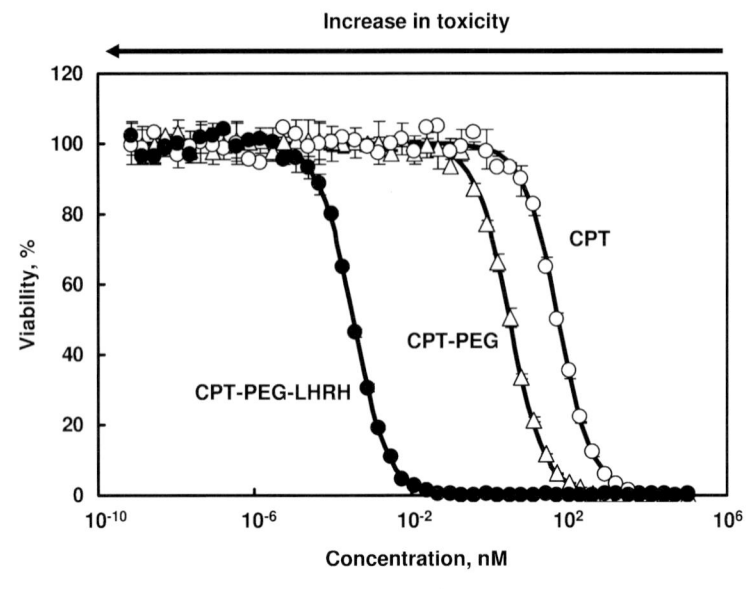

Fig. 14.18 Cytotoxicity of free CPT, CPT–PEG and CPT–PEG–LHRH conjugates. Viability of human A2780 ovarian carcinoma cells was measured by modified MTT assay as previously described [10] after 48 h of incubation with 48 different concentrations of drugs.

antiapoptotic cellular defense of cancer cells. These domains are critical to the functions of these proteins, including their impact on cell survival and their ability to interact with other family members and regulatory proteins [153, 154]. It was found that the BH3 domain of proapoptotic proteins from the BCL-2 family is responsible for the induction of apoptosis. Furthermore, expression of small-truncated derivatives of BAK protein containing the BH3 domain was sufficient for cell killing activity. Moreover, it was found that short synthetic peptides, corresponding to the minimal sequence of BH3 domain when bound to the antiapoptotic BCL-2 family proteins, suppress the cellular antiapoptotic defense [25, 41, 42, 153–156]. Therefore, the BH3 peptide has been tested in anticancer proapoptotic DDSs as an inhibitor of nonpump resistance. We found that the inclusion of synthetic BH3 peptide into the polymeric DDS containing camptothecin (CPT) as an anticancer drug prevented activation of antiapoptotic cellular defenses and therefore substantially enhanced cytotoxicity of the polymeric drug (Fig. 14.19).

In addition, antisense oligonucleotides against major antiapoptotic members of the BCL-2 protein family were also tested as inhibitors [44, 45]. Antisense oligonucleotides targeted to mRNA encoding the primary drug efflux pump's P-glycoprotein and multidrug resistance-associated proteins have been tested as suppressors of pump resistance [44, 45, 157–161].

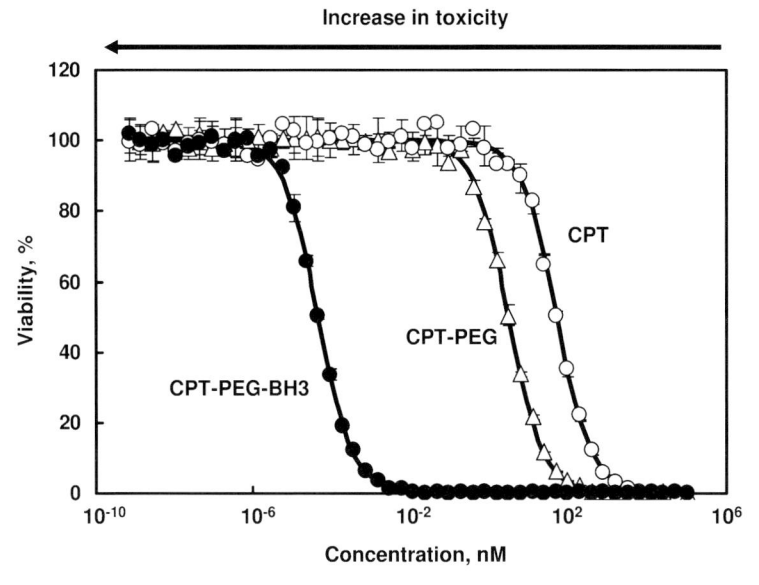

Fig. 14.19 Cytotoxicity of free CPT, CPT–PEG and CPT–PEG–BH3 conjugates. Viability of human A2780 ovarian carcinoma cells was measured by modified MTT assay as previously described [10] after 48 h of incubation with 48 different concentrations of drugs.

Experimental testing of proposed targeted proapoptotic DDSs and their components confirmed the feasibility of the approach and demonstrated that the use of such DDSs significantly enhanced the activity of traditional anticancer drugs to an extent that cannot be achieved by individual components applied separately.

Therefore, targeted anticancer proapoptotic DDSs utilize a novel three-tier approach, simultaneously attacking multiple targets: (1) the gene encoding drug efflux pumps to suppress cellular pump resistance and enhance drug retention by cancer cells; (2) mechanisms of apoptosis initiation in order to induce a cell death; and (3) controlling the mechanisms of apoptosis in order to suppress a cellular antiapoptotic defense. This is a three-pronged cellular–molecular attack on cancer. It targets, prevents multidrug resistance and promotes the cellular pathways leading to death in cancer cells.

References

1 L. V. J. Allen, N. G. Popovich, H. C. Ansel, *Pharmaceutical Dosage Forms and Drug Delivery Systems*, 8th edn., Lippincott Williams & Wilkins, Baltimore, **2005**.

2 R. Pandey, G. K. Khuller, Polymer based drug delivery systems for mycobacterial infections, *Curr Drug Deliv* **2004**, *1*, 195–201.

3 A. David, P. Kopeckova, J. Kopecek, A. Rubinstein, The role of galactose, lactose and galactose valency in the biorecognition of N-(2-hydroxypropyl)methacrylamide copolymers by human colon adenocarcinoma cells, *Pharm Res* **2002**, *19*, 1114–1122.

4 A. David, P. Kopeckova, T. Minko, A. Rubinstein, J. Kopecek, Design of a multivalent galactoside ligand for selective targeting of HPMA copolymer–doxorubicin conjugates to human colon cancer cells, *Eur J Cancer* **2004**, *40*, 148–157.

5 M. Demoy, T. Minko, P. Kopeckova, J. Kopecek, Time- and concentration-dependent apoptosis and necrosis induced by free and HPMA copolymer-bound doxorubicin in human ovarian carcinoma cells, *J Control Release* **2000**, *69*, 185–196.

6 Y. Kasuya, Z. R. Lu, P. Kopeckova, T. Minko, S. E. Tabibi, J. Kopecek, Synthesis and characterization of HPMA copolymer–aminopropylgeldanamycin conjugates, *J Control Release* **2001**, *74*, 203–211.

7 M. Kovar, L. Kovar, V. Subr, T. Etrych, K. Ulbrich, T. Mrkvan, J. Loucka, B. Rihova, HPMA copolymers containing doxorubicin bound by a proteolytically or hydrolytically cleavable bond: comparison of biological properties *in vitro*, *J Control Release* **2004**, *99*, 301–314.

8 K. Kunath, P. Kopeckova, T. Minko, J. Kopecek, HPMA copolymer–anticancer drug–OV-TL16 antibody conjugates. 3. The effect of free and polymer-bound adriamycin on the expression of some genes in the OVCAR-3 human ovarian carcinoma cell line, *Eur J Pharm Biopharm* **2000**, *49*, 11–15.

9 Z. R. Lu, J. G. Shiah, S. Sakuma, P. Kopeckova, J. Kopecek, Design of novel bioconjugates for targeted drug delivery, *J Control Release* **2002**, *78*, 165–173.

10 T. Minko, P. Kopeckova, J. Kopecek, Comparison of the anticancer effect of free and HPMA copolymer-bound adriamycin in human ovarian carcinoma cells, *Pharm Res* **1999**, *16*, 986–996.

11 T. Minko, P. Kopeckova, J. Kopecek, Chronic exposure to HPMA copolymer-bound adriamycin does not induce multidrug resistance in a human ovarian carcinoma cell line, *J Control Release* **1999**, *59*, 133–148.

12 T. Minko, P. Kopeckova, J. Kopecek, Efficacy of the chemotherapeutic action of HPMA copolymer-bound doxorubicin in a solid tumor model of ovarian carcinoma, *Int J Cancer* **2000**, *86*, 108–117.

13 T. Minko, P. Kopeckova, J. Kopecek, Preliminary evaluation of caspases-dependent apoptosis signaling pathways of free and HPMA copolymer-bound doxorubicin in human ovarian carcinoma cells, *J Control Release* **2001**, *71*, 227–237.

14 T. Minko, P. Kopeckova, V. Pozharov, K. D. Jensen, J. Kopecek, The influence of cytotoxicity of macromolecules and of VEGF gene modulated vascular permeability on the enhanced permeability and retention effect in resistant solid tumors, *Pharm Res* **2000**, *17*, 505–514.

15 T. Minko, P. Kopeckova, V. Pozharov, J. Kopecek, HPMA copolymer bound adriamycin overcomes MDR1 gene encoded resistance in a human ovarian carcinoma cell line, *J Control Release* **1998**, *54*, 223–233.

16 C. M. Peterson, J. G. Shiah, Y. Sun, P. Kopeckova, T. Minko, R. C. Straight, J. Kopecek, HPMA copolymer delivery of chemotherapy and photodynamic therapy in ovarian cancer, *Adv Exp Med Biol* **2003**, *519*, 101–123.

17 K. Ulbrich, T. Etrych, P. Chytil, M. Jelinkova, B. Rihova, Antibody-targeted polymer–doxorubicin conjugates with pH-controlled activation, *J Drug Target* **2004**, *12*, 477–489.

18 K. Ulbrich, T. Etrych, P. Chytil, M. Pechar, M. Jelinkova, B. Rihova, Polymeric anticancer drugs with pH-controlled activation, *Int J Pharm* **2004**, *277*, 63–72.

19 D. Wang, J. P. Kopeckova, T. Minko, V. Nanayakkara, J. Kopecek, Synthesis of starlike N-(2-hydroxypropyl)methacrylamide copolymers: potential drug carriers, *Biomacromolecules* **2000**, *1*, 313–319.

20 D. Wang, S. Miller, M. Sima, P. Kopeckova, J. Kopecek, Synthesis and evaluation of water-soluble polymeric bone-targeted drug delivery systems, *Bioconjug Chem* **2003**, *14*, 853–859.

21 A. M. Funhoff, C. F. van Nostrum, A. P. Janssen, M. H. Fens, D. J. Crommelin,

W. E. Hennink, Polymer side-chain degradation as a tool to control the destabilization of polyplexes, *Pharm Res* **2004**, *21*, 170–176.

22 T. Lammers, R. Kuhnlein, M. Kissel, V. Subr, T. Etrych, R. Pola, M. Pechar, K. Ulbrich, G. Storm, P. Huber, P. Peschke, Effect of physicochemical modification on the biodistribution and tumor accumulation of HPMA copolymers, *J Control Release* **2005**, 103–118.

23 L. C. Chang, H. F. Lee, M. J. Chung, V. C. Yang, PEG-Modified protamine with improved pharmacological/pharmaceutical properties as a potential protamine substitute: synthesis and *in vitro* evaluation, *Bioconjug Chem* **2005**, *16*, 147–155.

24 Y. H. Choe, R. B. Greenwald, C. D. Conover, H. Zhao, C. B. Longley, S. Guan, Q. Zhao, J. Xia, PEG prodrugs of 6-mercaptopurine for parenteral administration using benzyl elimination of thiols, *Oncol Res* **2004**, *14*, 455–468.

25 S. S. Dharap, B. Qiu, G. C. Williams, P. Sinko, S. Stein, T. Minko, Molecular targeting of drug delivery systems to ovarian cancer by BH3 and LHRH peptides, *J Control Release* **2003**, *91*, 61–73.

26 R. B. Greenwald, PEG drugs: an overview, *J Control Release* **2001**, *74*, 159–171.

27 R. B. Greenwald, H. Zhao, J. Xia, D. Wu, S. Nervi, S. F. Stinson, E. Majerova, C. Bramhall, D. W. Zaharevitz, Poly(ethylene glycol) prodrugs of the CDK inhibitor, alsterpaullone (NSC 705701): synthesis and pharmacokinetic studies, *Bioconjug Chem* **2004**, *15*, 1076–1083.

28 S. Gunaseelan, O. Debrah, L. Wan, M. J. Leibowitz, A. B. Rabson, S. Stein, P. J. Sinko, Synthesis of poly(ethylene glycol)-based saquinavir prodrug conjugates and assessment of release and anti-HIV-bioactivity using a novel protease inhibition assay, *Bioconjug Chem* **2004**, *15*, 1322–1333.

29 T. Minko, P. V. Paranjpe, B. Qiu, A. Lalloo, R. Won, S. Stein, P. J. Sinko, Enhancing the anticancer efficacy of camptothecin using biotinylated poly(ethylene glycol) conjugates in sensitive and multidrug-resistant human ovarian carcinoma cells, *Cancer Chemother Pharmacol* **2002**, *50*, 143–150.

30 P. V. Paranjpe, Y. Chen, V. Kholodovych, W. Welsh, S. Stein, P. J. Sinko, Tumor-targeted bioconjugate based delivery of camptothecin: design, synthesis and *in vitro* evaluation, *J Control Release* **2004**, *100*, 275–292.

31 H. Zhao, C. Lee, P. Sai, Y. H. Choe, M. Boro, A. Pendri, S. Guan, R. B. Greenwald, 20-*O*-acylcamptothecin derivatives: evidence for lactone stabilization, *J Org Chem* **2000**, *65*, 4601–4606.

32 P. Caliceti, O. Schiavon, F. M. Veronese, Immunological properties of uricase conjugated to neutral soluble polymers, *Bioconjug Chem* **2001**, *12*, 515–522.

33 V. P. Torchilin, Lipid-core micelles for targeted drug delivery, *Curr Drug Deliv* **2005**, *2*, 319–327.

34 M. Oishi, F. Nagatsugi, S. Sasaki, Y. Nagasaki, K. Kataoka, Smart polyion complex micelles for targeted intracellular delivery of PEGylated antisense oligonucleotides containing acid-labile linkages, *ChemBioChem* **2005**, *6*, 718–725.

35 D. Wakebayashi, N. Nishiyama, Y. Yamasaki, K. Itaka, N. Kanayama, A. Harada, Y. Nagasaki, K. Kataoka, Lactose-conjugated polyion complex micelles incorporating plasmid DNA as a targetable gene vector system: their preparation and gene transfecting efficiency against cultured HepG2 cells, *J Control Release* **2004**, *95*, 653–664.

36 N. Nishiyama, S. Okazaki, H. Cabral, M. Miyamoto, Y. Kato, Y. Sugiyama, K. Nishio, Y. Matsumura, K. Kataoka, Novel cisplatin-incorporated polymeric micelles can eradicate solid tumors in mice, *Cancer Res* **2003**, *63*, 8977–8983.

37 H. Maeda, J. Wu, T. Sawa, Y. Matsumura, K. Hori, Tumor vascular permeability and the EPR effect in macromolecular therapeutics: a review, *J Control Release* **2000**, *65*, 271–284.

38 Y. Matsumura, H. Maeda, A new concept for macromolecular therapeutics in cancer chemotherapy: mechanism of tumoritropic accumulation of proteins and the antitumor agent smancs, *Cancer Res* **1986**, *46*, 6387–6392.

39 S. S. Naik, J. F. Liang, Y. J. Park, W. K. Lee, V. C. Yang, Application of

"ATTEMPTS" for drug delivery, *J Control Release* **2005**, *101*, 35–45.

40 A. Mitra, J. Mulholland, A. Nan, E. McNeill, H. Ghandehari, B. R. Line, Targeting tumor angiogenic vasculature using polymer–RGD conjugates, *J Control Release* **2005**, *102*, 191–201.

41 S. S. Dharap, T. Minko, Targeted proapoptotic LHRH-BH3 peptide, *Pharm Res* **2003**, *20*, 889–896.

42 T. Minko, S. S. Dharap, A. T. Fabbricatore, Enhancing the efficacy of chemotherapeutic drugs by the suppression of antiapoptotic cellular defense, *Cancer Detect Prev* **2003**, *27*, 193–202.

43 T. Minko, S. S. Dharap, R. I. Pakunlu, Y. Wang, Molecular targeting of drug delivery systems to cancer, *Curr Drug Targets* **2004**, *5*, 389–406.

44 R. I. Pakunlu, T. J. Cook, T. Minko, Simultaneous modulation of multidrug resistance and antiapoptotic cellular defense by MDR1 and BCL-2 targeted antisense oligonucleotides enhances the anticancer efficacy of doxorubicin, *Pharm Res* **2003**, *20*, 351–359.

45 R. I. Pakunlu, Y. Wang, W. Tsao, V. Pozharov, T. J. Cook, T. Minko, Enhancement of the efficacy of chemotherapy for lung cancer by simultaneous suppression of multidrug resistance and antiapoptotic cellular defense: novel multicomponent delivery system, *Cancer Res* **2004**, *64*, 6214–6224.

46 Y. Wang, T. Minko, A novel cancer therapy: combined liposomal hypoxia inducible factor 1 alpha antisense oligonucleotides and an anticancer drug, *Biochem Pharmacol* **2004**, *68*, 2031–2042.

47 Y. Wang, R. I. Pakunlu, W. Tsao, V. Pozharov, T. Minko, Bimodal effect of hypoxia in cancer: the role of hypoxia inducible factor in apoptosis, *Mol Pharm* **2004**, *1*, 156–165.

48 H. H. Chung, G. Harms, C. M. Seong, B. H. Choi, C. Min, J. P. Taulane, M. Goodman, Dendritic oligoguanidines as intracellular translocators, *Biopolymers* **2004**, *76*, 83–96.

49 A. Guiotto, M. Canevari, P. Orsolini, O. Lavanchy, C. Deuschel, N. Kaneda, A. Kurita, T. Matsuzaki, T. Yaegashi, S. Sawada, F. M. Veronese, Synthesis, characterization and preliminary *in vivo* tests of new poly(ethylene glycol) conjugates of the antitumor agent 10-amino-7-ethylcamptothecin, *J Med Chem* **2004**, *47*, 1280–9.

50 S. Kannan, P. Kolhe, V. Raykova, M. Glibatec, R. M. Kannan, M. Lieh-Lai, D. Bassett, Dynamics of cellular entry and drug delivery by dendritic polymers into human lung epithelial carcinoma cells, *J Biomater Sci Polym Ed* **2004**, *15*, 311–330.

51 P. Kolhe, J. Khandare, O. Pillai, S. Kannan, M. Lieh-Lai, R. Kannan, Hyperbranched polymer-drug conjugates with high drug payload for enhanced cellular delivery, *Pharm Res* **2004**, *21*, 2185–2195.

52 H. Turk, R. Haag, S. Alban, Dendritic polyglycerol sulfates as new heparin analogues and potent inhibitors of the complement system, *Bioconjug Chem* **2004**, *15*, 162–167.

53 R. Duncan, L. Izzo, Dendrimer biocompatibility and toxicity, *Adv Drug Deliv Rev* **2005**, *57*, 2215–2237.

54 L. K. Folkes, O. Greco, G. U. Dachs, M. R. Stratford, P. Wardman, 5-Fluoroindole-3-acetic acid: a prodrug activated by a peroxidase with potential for use in targeted cancer therapy, *Biochem Pharmacol* **2002**, *63*, 265–272.

55 W. A. Denny, Tumor-activated prodrugs – a new approach to cancer therapy, *Cancer Invest* **2004**, *22*, 604–619.

56 M. O. Khan, K. K. Park, H. J. Lee, Antedrugs: an approach to safer drugs, *Curr Med Chem* **2005**, *12*, 2227–2239.

57 H. Jatzkewitz, Peptamin (glycyl-L-leucyl-mescaline) bound to blood plasma expander (polyvinylpyrrolidone) as a new depot form of a biologically active primary amine (mescaline). *Z Naturforsch* **1955**, *10*, 27–31.

58 H. Ringsdorf, Structure and properties of pharmacologically active polymers, *J Polym Sci Symp* **1975**, *51*, 135–153.

59 R. Duncan, M. J. Vicent, F. Greco, R. I. Nicholson, Polymer–drug conjugates: towards a novel approach for the treatment of endrocine-related cancer, *Endocr Relat Cancer* **2005**, *12* (Suppl 1), S189–199.

60 R. B. Greenwald, H. Zhao, J. Xia, Tripartate poly(ethylene glycol) prodrugs of the

open lactone form of camptothecin, *Bioorg Med Chem* **2003**, *11*, 2635–2639.
61 A. Kozlowski, S. A. Charles, J. M. Harris, Development of pegylated interferons for the treatment of chronic hepatitis C, *BioDrugs* **2001**, *15*, 419–429.
62 Y. Chau, R. F. Padera, N. M. Dang, R. Langer, Antitumor efficacy of a novel polymer–peptide–drug conjugate in human tumor xenograft models, *Int J Cancer* **2005**, *118*, 1519–1526.
63 F. Levi-Schaffer, A. Bernstein, A. Meshorer, R. Arnon, Reduced toxicity of daunorubicin by conjugation to dextran, *Cancer Treat Rep* **1982**, *66*, 107–114.
64 M. Hashida, A. Kato, T. Kojima, S. Muranishi, H. Sezaki, N. Tanigawa, K. Satomura, Y. Hikasa, Antitumor activity of mitomycin C-dextran conjugate against various murine tumors, *Gann* **1981**, *72*, 226–234.
65 A. D. McLeod, D. R. Friend, T. N. Tozer, Glucocorticoid–dextran conjugates as potential prodrugs for colon-specific delivery: hydrolysis in rat gastrointestinal tract contents, *J Pharm Sci* **1994**, *83*, 1284–1288.
66 R. Mehvar, R. O. Dann, D. A. Hoganson, Kinetics of hydrolysis of dextran-methylprednisolone succinate, a macromolecular prodrug of methylprednisolone, in rat blood and liver lysosomes, *J Control Release* **2000**, *68*, 53–61.
67 H. Yura, N. Yoshimura, T. Hamashima, K. Akamatsu, M. Nishikawa, Y. Takakura, M. Hashida, Synthesis and pharmacokinetics of a novel macromolecular prodrug of Tacrolimus (FK506), FK506–dextran conjugate, *J Control Release* **1999**, *57*, 87–99.
68 R. Mehvar, Dextrans for targeted and sustained delivery of therapeutic and imaging agents, *J Control Release* **2000**, *69*, 1–25.
69 S. Zalipsky, Chemistry of polyethylene glycol conjugates with biologically active molecules, *Adv Drug Deliv Rev* **1995**, *16*, 157–182.
70 Y. H. Choe, C. D. Conover, D. Wu, M. Royzen, Y. Gervacio, V. Borowski, M. Mehlig, R. B. Greenwald, Anticancer drug delivery systems: multi-loaded N4-acyl poly(ethylene glycol) prodrugs of ara-C. II. Efficacy in ascites and solid tumors, *J Control Release* **2002**, *79*, 55–70.
71 Y. H. Choe, C. D. Conover, D. Wu, M. Royzen, R. B. Greenwald, Anticancer drug delivery systems: N4-acyl poly(ethylene glycol) prodrugs of ara-C. I. Efficacy in solid tumors, *J Control Release* **2002**, *79*, 41–53.
72 O. Schiavon, G. Pasut, S. Moro, P. Orsolini, A. Guiotto, F. M. Veronese, PEG–Ara-C conjugates for controlled release, *Eur J Med Chem* **2004**, *39*, 123–133.
73 A. B. Fleming, K. Haverstick, W. M. Saltzman, In vitro cytotoxicity and *in vivo* distribution after direct delivery of PEG–camptothecin conjugates to the rat brain, *Bioconjug Chem* **2004**, *15*, 1364–1375.
74 R. B. Greenwald, Y. H. Choe, J. McGuire, C. D. Conover, Effective drug delivery by PEGylated drug conjugates, *Adv Drug Deliv Re.* **2003**, *55*, 217–250.
75 J. Kopecek, H. Bazilova, Poly(N-(hydroxypropyl)-methacrylamide) radical polymerization and copolymerization, *Euro Polym J* **1973**, *9*, 7–14.
76 B. Rihova, M. Bilej, V. Vetvicka, K. Ulbrich, J. Strohalm, J. Kopecek, R. Duncan, Biocompatibility of N-(2-hydroxypropyl) methacrylamide copolymers containing adriamycin. Immunogenicity and effect on haematopoietic stem cells in bone marrow *in vivo* and mouse splenocytes and human peripheral blood lymphocytes *in vitro*, *Biomaterials* **1989**, *10*, 335–342.
77 L. W. Seymour, K. Ulbrich, J. Strohalm, J. Kopecek, R. Duncan, The pharmacokinetics of polymer-bound adriamycin, *Biochem Pharmaco.* **1990**, *39*, 1125–1131.
78 L. Wang, J. Kristensen, D. E. Ruffner, Delivery of antisense oligonucleotides using HPMA polymer: synthesis of a thiol polymer and its conjugation to water-soluble molecules, *Bioconjug Chem* **1998**, *9*, 749–757.
79 A. Nan, H. Ghandehari, C. Hebert, H. Siavash, N. Nikitakis, M. Reynolds, J. J. Sauk, Water-soluble polymers for targeted drug delivery to human squamous carcinoma of head and neck, *J Drug Target* **2005**, *13*, 189–197.
80 J. Kopecek, P. Kopeckova, T. Minko, Z. Lu, HPMA copolymer–anticancer drug conjugates: design, activity and mecha-

nism of action, *Eur J Pharm Biopharm* **2000**, *50*, 61–81.
81 R. Duncan, H. C. Cable, J. B. Lloyd, P. Rejmanova, J. Kopecek, Degradation of side-chains of N-(2-hydroxypropyl)methacrylamide copolymers by lysosomal thiol-proteinases, *Biosci Rep* **1982**, *2*, 1041–1046.
82 K. Ulbrich, T. Etrych, P. Chytil, M. Pechar, M. Jelinkova, B. Rihova, Polymeric anticancer drugs with pH-controlled activation, *Int. J. Pharmaceutics* **2004**, *277*, 63–72.
83 O. Hovorka, M. St'astny, T. Etrych, V. Subr, J. Strohalm, K. Ulbrich, B. Rihova, Differences in the intracellular fate of free and polymer-bound doxorubicin, *J Control Release* **2002**, *80*, 101–117.
84 T. Etrych, M. Jelinkova, B. Rihova, K. Ulbrich, New HPMA copolymers containing doxorubicin bound via pH-sensitive linkage: synthesis and preliminary *in vitro* and *in vivo* biological properties, *J Control Release* **2001**, *73*, 89–102.
85 B. Rihova, T. Etrych, M. Pechar, M. Jelinkova, M. Stastny, O. Hovorka, M. Kovar, K. Ulbrich, Doxorubicin bound to a HPMA copolymer carrier through hydrazone bond is effective also in a cancer cell line with a limited content of lysosomes, *J Control Release* **2001**, *74*, 225–232.
86 H. R. Ihre, O. L. Padilla De Jesus, F. C. Szoka, Jr., J. M. Frechet, Polyester dendritic systems for drug delivery applications: design, synthesis and characterization, *Bioconjug Chem* **2002**, *13*, 443–452.
87 R. J. Christie, D. W. Grainger, Design strategies to improve soluble macromolecular delivery constructs, *Adv Drug Deliv Rev* **2003**, *55*, 421–437.
88 R. Jevprasesphant, J. Penny, D. Attwood, N. B. McKeown, A. D'Emanuele, Engineering of dendrimer surfaces to enhance transepithelial transport and reduce cytotoxicity, *Pharm Res* **2003**, *20*, 1543–1550.
89 N. Malik, R. Wiwattanapatapee, R. Klopsch, K. Lorenz, H. Frey, J. W. Weener, E. W. Meijer, W. Paulus, R. Duncan, Dendrimers: relationship between structure and biocompatibility *in vitro* and preliminary studies on the biodistribution of ^{125}I-labelled polyamidoamine dendrimers *in vivo*, *J Control Release* **2000**, *65*, 133–148.
90 A. Quintana, E. Raczka, L. Piehler, I. Lee, A. Myc, I. Majoros, A. K. Patri, T. Thomas, J. Mule, J. R. Baker, Jr., Design and function of a dendrimer-based therapeutic nanodevice targeted to tumor cells through the folate receptor, *Pharm Res* **2002**, *19*, 1310–1316.
91 F. Tajarobi, M. El-Sayed, B. D. Rege, J. E. Polli, H. Ghandehari, Transport of poly amidoamine dendrimers across Madin–Darby canine kidney cells, *Int J Pharm* **2001**, *215*, 263–267.
92 A. K. Patri, I. J. Majoros, J. R. Baker, Dendritic polymer macromolecular carriers for drug delivery, *Curr Opin Chem Biol* **2002**, *6*, 466–471.
93 J. C. Roberts, Y. E. Adams, D. Tomalia, J. A. Mercer-Smith, D. K. Lavallee, Using starburst dendrimers as linker molecules to radiolabel antibodies, *Bioconjug Chem* **1990**, *1*, 305–308.
94 T. Ooya, J. Lee, K. Park, Hydrotropic dendrimers of generations 4 and 5: synthesis, characterization and hydrotropic solubilization of paclitaxel, *Bioconjug Chem* **2004**, *15*, 1221–1229.
95 I. J. Majoros, T. P. Thomas, C. B. Mehta, J. R. Baker, Jr., Poly(amidoamine) dendrimer-based multifunctional engineered nanodevice for cancer therapy, *J Med Chem* **2005**, *48*, 5892–5899.
96 J. Khandare, P. Kolhe, O. Pillai, S. Kannan, M. Lieh-Lai, R. M. Kannan, Synthesis, cellular transport and activity of polyamidoamine dendrimer-methylprednisolone conjugates, *Bioconjug Chem* **2005**, *16*, 330–337.
97 J. J. Khandare, P. Chandna, Y. Wang, V. P. Pozharov, T. Minko, Novel polymeric prodrug with multivalent components for cancer therapy, *J Pharmacol Exp Ther* **2006**, *317*, 929–937.
98 J. Khandare, M. Kulkarni, Polymerizable monomers and process of preparation thereof, *US Patent 6 822 064*, **2004**.
99 A. A. Vaidya, B. S. Lele, M. G. Kulkarni, R. A. Mashelkar, Enhancing ligand–protein binding in affinity thermoprecipitation: elucidation of spacer effects, *Biotechnol Bioeng* **1999**, *64*, 418–425.

100 S. L. Tao, T. A. Desai, Micromachined devices: the impact of controlled geometry from cell-targeting to bioavailability, *J Control Release* **2005**, *109*, 127–138.

101 T. Yasukawa, Y. Ogura, E. Sakurai, Y. Tabata, H. Kimura, Intraocular sustained drug delivery using implantable polymeric devices, *Adv Drug Deliv Rev* **2005**, *57*, 2033–2046.

102 Insulin inhalation – Pfizer/Nektar Therapeutics: HMR 4006, inhaled PEG–insulin – Nektar, PEGylated insulin – Nektar, *Drugs R D* **2004**, *5*, 166–170.

103 V. Parthasarathy, R. Manavalan, R. Mythili, C. T. Siby, M. Jeya, Ethyl cellulose and polyethylene glycol-based sustained-release sparfloxacin chip: an alternative therapy for advanced periodontitis, *Drug Dev Ind Pharm* **2002**, *28*, 849–862.

104 J. Bobacka, A. Ivaska, A. Lewenstam, Potentiometric ion sensors based on conducting polymers, *Electroanalysis* **2003**, *15*, 366–374.

105 A. Giuseppi-Elie, G. G. Wallace, T. Matsue, in *Handbook of Conducting Polymers*, 2nd edn., T. A. Skotheim, R. L. Elsenbaumer, J. R. Reynolds (Eds.), Marcel Dekker, New York, **1998**, pp. 963–991.

106 E. Piskin, Molecularly designed water soluble, intelligent, nanosize polymeric carriers, *Int J Pharm* **2004**, *277*, 105–118.

107 D. M. Schachter, J. Kohn, A synthetic polymer matrix for the delayed or pulsatile release of water-soluble peptides, *J Control Release* **2002**, *78*, 143–153.

108 M. J. Dorta, A. Santovena, M. Llabres, J. B. Farina, Potential applications of PLGA film-implants in modulating *in vitro* drugs release, *Int J Pharm* **2002**, *248*, 149–156.

109 W. Chen, D. R. Lu, Carboplatin-loaded PLGA microspheres for intracerebral injection: formulation and characterization, *J Microencapsul* **1999**, *16*, 551–563.

110 W. Friess, M. Schlapp, Modifying the release of gentamicin from microparticles using a PLGA blend, *Pharm Dev Technol* **2002**, *7*, 235–248.

111 J. T. Santini, Jr., A. C. Richards, R. A. Scheidt, M. J. Cima, R. S. Langer, Microchip technology in drug delivery, *Ann Med* **2000**, *32*, 377–379.

112 M. Thanou, R. Duncan, Polymer–protein and polymer–drug conjugates in cancer therapy, *Curr Opin Investig Drugs* **2003**, *4*, 701–9.

113 M. Z. Khan, H. P. Stedul, N. Kurjakovic, A pH-dependent colon-targeted oral drug delivery system using methacrylic acid copolymers. II. Manipulation of drug release using Eudragit L100 and Eudragit S100 combinations, *Drug Dev Ind Pharm* **2000**, *26*, 549–554.

114 P. Nykanen, S. Lempaa, M. L. Aaltonen, H. Jurjenson, P. Veski, M. Marvola, Citric acid as excipient in multiple-unit enteric-coated tablets for targeting drugs on the colon, *Int J Pharm* **2001**, *229*, 155–162.

115 V. R. Sinha, R. Kumria, Binders for colon specific drug delivery: an *in vitro* evaluation, *Int J Pharm* **2002**, *249*, 23–31.

116 M. K. Chourasia, S. K. Jain, Pharmaceutical approaches to colon targeted drug delivery systems, *J Pharm Pharm Sci.* **2003**, *6*, 33–66.

117 T. Minko, Drug targeting to the colon with lectins and neoglycoconjugates, *Adv Drug Deliv Rev* **2004**, *56*, 491–509.

118 M. A. Shareef, R. K. Khar, A. Ahuja, F. J. Ahmad, S. Raghava, Colonic drug delivery: an updated review, *AAPS PharmSci* **2003**, *5*, E17.

119 A. M. Mansour, J. Drevs, N. Esser, F. M. Hamada, O. A. Badary, C. Unger, I. Fichtner, F. Kratz, A new approach for the treatment of malignant melanoma: enhanced antitumor efficacy of an albumin-binding doxorubicin prodrug that is cleaved by matrix metalloproteinase 2, *Cancer Res* **2003**, *63*, 4062–4066.

120 M. Bae, S. Cho, J. Song, G. Y. Lee, K. Kim, J. Yang, K. Cho, S. Y. Kim, Y. Byun, Metalloprotease-specific poly(ethylene glycol) methyl ether–peptide–doxorubicin conjugate for targeting anticancer drug delivery based on angiogenesis, *Drugs Exp Clin Res* **2003**, *29*, 15–23.

121 X. Guo, F. C. Szoka, Jr., Chemical approaches to triggerable lipid vesicles for drug and gene delivery, *Acc Chem Res.* **2003**, *36*, 335–341.

122 A. Gautam, N. Koshkina, Paclitaxel (taxol) and taxoid derivatives for lung cancer

123 C. Khanna, D. M. Vail, Targeting the lung: preclinical and comparative evaluation of anticancer aerosols in dogs with naturally occurring cancers, *Curr Cancer Drug Targets* **2003**, *3*, 265–273.

treatment: potential for aerosol delivery, *Curr Cancer Drug Targets* **2003**, *3*, 287–296.

124 N. Yamazaki, S. Kojima, N. V. Bovin, S. Andre, S. Gabius, H. J. Gabius, Endogenous lectins as targets for drug delivery, *Adv Drug Deliv Rev* **2000**, *43*, 225–244.

125 S. Wroblewski, B. Rihova, P. Rossmann, T. Hudcovicz, Z. Rehakova, P. Kopeckova, J. Kopecek, The influence of a colonic microbiota on HPMA copolymer lectin conjugates binding in rodent intestine, *J Drug Target* **2001**, *9*, 85–94.

126 W. B. Turnbull, J. F. Stoddart, Design and synthesis of glycodendrimers, *J Biotechnol* **2002**, *90*, 231–255.

127 M. Singh, N. Kisoon, M. Ariatti, Receptor-mediated gene delivery to HepG2 cells by ternary assemblies containing cationic liposomes and cationized asialoorosomucoid, *Drug Deliv* **2001**, *8*, 29–34.

128 W. M. Pardridge, Drug and gene targeting to the brain with molecular Trojan horses, *Nat Rev Drug Discov* **2002**, *1*, 131–139.

129 M. F. Fromm, The influence of MDR1 polymorphisms on P-glycoprotein expression and function in humans, *Adv Drug Deliv Rev* **2002**, *54*, 1295–1310.

130 X. Liu, Y. Zhen, Antitumor effects of monoclonal antibody Fab' fragment-containing immunoconjugates, *Chin Med Sci J* **2002**, *17*, 1–6.

131 T. Tsuruo, Molecular cancer therapeutics: recent progress and targets in drug resistance, *Intern Med* **2003**, *42*, 237–243.

132 M. V. Blagosklonny, Targeting cancer cells by exploiting their resistance, *Trends Mol Med* **2003**, *9*, 307–312.

133 Z. Chen, A. Takeshita, P. Zou, Z. Liu, M. Kozaka, Y. You, S. Song, K. Ohnishi, R. Ohno, Multidrug resistance P-glycoprotein function of bone marrow hematopoietic cells and the reversal agent effect, *J Tongji Med Univ* **1999**, *19*, 260–263.

134 R. Ruiz-Soto, Y. Richaud-Patin, X. Lopez-Karpovitch, L. Llorente, Multidrug resistance-1 (MDR-1) in autoimmune disorders III: increased P-glycoprotein activity in lymphocytes from immune thrombocytopenic purpura patients, *Exp Hematol* **2003**, *31*, 483–487.

135 A. W. Abu-Qare, E. Elmasry, M. B. Abou-Donia, A role for P-glycoprotein in environmental toxicology, *J Toxicol Environ Health B Crit Rev* **2003**, *6*, 279–288.

136 M. Daoudaki, I. Fouzas, V. Stapf, C. Ekmekcioglu, G. Imvrios, A. Andoniadis, A. Demetriadou, T. Thalhammer, Cyclosporine a augments P-glycoprotein expression in the regenerating rat liver, *Biol Pharm Bull* **2003**, *26*, 303–307.

137 L. Campbell, A. N. Abulrob, L. E. Kandalaft, S. Plummer, A. J. Hollins, A. Gibbs, M. Gumbleton, Constitutive expression of P-glycoprotein in normal lung alveolar epithelium and functionality in primary alveolar epithelial cultures, *J Pharmacol Exp Ther* **2003**, *304*, 441–452.

138 B. G. Kennedy, N. J. Mangini, P-glycoprotein expression in human retinal pigment epithelium, *Mol Vis* **2002**, *8*, 422–430.

139 Y. Tanigawara, Role of P-glycoprotein in drug disposition, *Ther Drug Monit* **2000**, *22*, 137–40.

140 V. Weissig, Mitochondrial-targeted drug and DNA delivery, *Crit Rev Ther Drug Carrier Syst* **2003**, *20*, 1–62.

141 V. P. Torchilin, PEG-based micelles as carriers of contrast agents for different imaging modalities, *Adv Drug Deliv Rev* **2002**, *54*, 235–252.

142 V. Weissig, V. P. Torchilin, Mitochondriotropic cationic vesicles: a strategy towards mitochondrial gene therapy, *Curr Pharm Biotechnol* **2000**, *1*, 325–346.

143 V. Omelyanenko, C. Gentry, P. Kopeckova, J. Kopecek, HPMA copolymer–anticancer drug–OV-TL16 antibody conjugates. II. Processing in epithelial ovarian carcinoma cells *in vitro*, *Int J Cancer* **1998**, *75*, 600–608.

144 V. Omelyanenko, P. Kopeckova, C. Gentry, J. Kopecek, Targetable HPMA copolymer–adriamycin conjugates. Recognition, internalization and subcellular fate, *J Control Release* **1998**, *53*, 25–37.

145 K. Ulbrich, V. Subr, Polymeric anticancer drugs with pH-controlled activation, *Adv Drug Deliv Rev* **2004**, *56*, 1023–1050.

146 S. S. Dharap, Y. Wang, P. Chandna, J. J. Khandare, B. Qiu, S. Gunaseelan, S. Stein, A. Farmanfarmaian, T. Minko, Tumor-specific targeting of an anticancer drug delivery system by LHRH peptide, *Proc Natl Acad Sci USA* **2005**, *102*, 12962–12967.

147 J. Kopecek, P. Kopeckova, T. Minko, Z. R. Lu, C. M. Peterson, Water soluble polymers in tumor targeted delivery, *J Control Release* **2001**, *74*, 147–158.

148 A. Krishnadas, I. Rubinstein, H. Onyuksel, Sterically stabilized phospholipid mixed micelles: in vitro evaluation as a novel carrier for water-insoluble drugs, *Pharm Res* **2003**, *20*, 297–302.

149 V. P. Torchilin, A. N. Lukyanov, Z. Gao, B. Papahadjopoulos-Sternberg, Immunomicelles: targeted pharmaceutical carriers for poorly soluble drugs, *Proc Natl Acad Sci USA* **2003**, *100*, 6039–6044.

150 E. V. Batrakova, S. Li, V. Y. Alakhov, D. W. Miller, A. V. Kabanov, Optimal structure requirements for pluronic block copolymers in modifying P-glycoprotein drug efflux transporter activity in bovine brain microvessel endothelial cells, *J Pharmacol Exp Ther* **2003**, *304*, 845–854.

151 A. V. Kabanov, E. V. Batrakova, V. Y. Alakhov, Pluronic block copolymers for overcoming drug resistance in cancer, *Adv Drug Deliv Rev* **2002**, *54*, 759–779.

152 J. J. Khandare, S. Jayant, A. Singh, P. Chandna, Y. Wang, N. Vorsa, T. Minko, Dendrimer versus linear conjugate: influence of polymeric architecture on the delivery and anticancer effect of paclitaxel, *Bioconjugate Chem.* **2006**, *17*, 1464–1472.

153 E. P. Holinger, T. Chittenden, R. J. Lutz, Bak BH3 peptides antagonize Bcl-xL function and induce apoptosis through cytochrome c-independent activation of caspases, *J Biol Chem* **1999**, *274*, 13298–13304.

154 H. L. Vieira, P. Boya, I. Cohen, C. El Hamel, D. Haouzi, S. Druillenec, A. S. Belzacq, C. Brenner, B. Roques, G. Kroemer, Cell permeable BH3-peptides overcome the cytoprotective effect of Bcl-2 and Bcl-X(L), *Oncogene* **2002**, *21*, 1963–1977.

155 S. C. Cosulich, V. Worrall, P. J. Hedge, S. Green, P. R. Clarke, Regulation of apoptosis by BH3 domains in a cell-free system, *Curr Biol* **1997**, *7*, 913–920.

156 R. J. Lutz, Role of the BH3 (Bcl-2 homology 3) domain in the regulation of apoptosis and Bcl-2-related proteins, *Biochem Soc Trans* **2000**, *28*, 51–56.

157 S. K. Alahari, R. DeLong, M. H. Fisher, N. M. Dean, P. Viliet, R. L. Juliano, Novel chemically modified oligonucleotides provide potent inhibition of P-glycoprotein expression, *J Pharmacol Exp Ther* **1998**, *286*, 419–428.

158 M. V. Corrias, G. P. Tonini, An oligomer complementary to the 5′ end region of MDR1 gene decreases resistance to doxorubicin of human adenocarcinoma-resistant cells, *Anticancer Res* **1992**, *12*, 1431–1438.

159 E. V. Kostenko, P. P. Laktionov, V. V. Vlassov, M. A. Zenkova, Downregulation of PGY1/MDR1 mRNA level in human KB cells by antisense oligonucleotide conjugates. RNA accessibility in vitro and intracellular antisense activity, *Biochim Biophys Acta* **2002**, *1576*, 143–147.

160 D. R. Mercatante, J. L. Mohler, R. Kole, Cellular response to an antisense-mediated shift of Bcl-x pre-mRNA splicing and antineoplastic agents, *J Biol Chem* **2002**, *277*, 49374–49382.

161 S. Motomura, T. Motoji, M. Takanashi, Y. H. Wang, H. Shiozaki, I. Sugawara, E. Aikawa, A. Tomida, T. Tsuruo, N. Kanda, H. Mizoguchi, Inhibition of P-glycoprotein and recovery of drug sensitivity of human acute leukemic blast cells by multidrug resistance gene (mdr1) antisense oligonucleotides, *Blood* **1998**, *91*, 3163–3171.

15
From Biomineralization Polymers to Double Hydrophilic Block and Graft Copolymers*

Helmut Cölfen

15.1
Biomineralization

Biomineralization is the process by which living organisms secrete inorganic minerals in the form of skeletons, shells, teeth, etc. It is already a rather old process in the development of life, which was adapted by living beings probably at the end of the Precambrian more than 500 million years ago [1]. Biominerals are highly optimized materials with remarkable properties, which have attracted a lot of recent attention not just among materials chemists because they are considered as natural archetypes for future materials. Nacre (mother of pearl), for example, is a layered material, where inorganic $CaCO_3$ platelets in the aragonite modification with a thickness between 300 and 800 nm are interspaced by a ca. 50-nm thick organic matrix composed of β-chitin layers and attached proteins (Fig. 15.1 a).

This material is 3000 times more fracture resistant than the brittle aragonite, although the mineral content in the biomineral exceeds 95%. The reason for this can be seen in the advanced material technology of a layered material between soft and hard components. Whereas the hard but brittle aragonite platelets add hardness to the composite material, the soft chitin/protein layers add elasticity; especially the silk fibroin, which is the main protein of spider and worm threads, and lustrin A, a multifunctional protein which behaves like a chain of springs elongating one after another when an increasing stretching force is applied [3]. The biopolymers enable the possibility of distributing energy upon nacre fracture as they are energy-absorbing fillers. Each mineral tablet is completely surrounded by a soft biopolymer layer (Fig. 15.1b). If nacre is fractured, a mineral tablet breaks but part of the energy is already dissipated by the next polymer layer until it breaks and so on, so that a crack cannot easily propagate through this organic–inorganic composite material. The increased fracture resistance corresponds to our every day experience. It is very difficult to break a

---

* A List of Abbreviations can be found at the end of this chapter.

Macromolecular Engineering. Precise Synthesis, Materials Properties, Applications.
Edited by K. Matyjaszewski, Y. Gnanou, and L. Leibler
Copyright © 2007 WILEY-VCH Verlag GmbH & Co. KGaA, Weinheim
ISBN: 978-3-527-31446-1

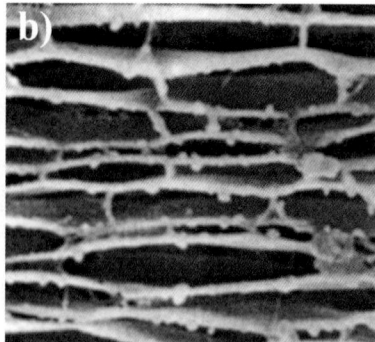

Fig. 15.1 (a) The structure of nacre (mother of pearl) from the bivalve *Mytilus californianus* as revealed by its fracture surface observed with scanning electron microscopy (1000-fold magnification) and (b) the same nacre cross-section after removal of the nacre mineral component. Copyright 1986 from Organization of extracellularly mineralized tissues: A comparative study of biological crystal growth by Stephen Weiner. Reproduced from [2] with permission of Taylor & Francis Group, LLC., http:/www.taylorandfrancis.com.

thin mussel shell, whereas this is easily possible with a much thicker $CaCO_3$ platelet or old mussel shells where the biopolymers were already degraded. This increased fracture resistance adds very favorably to the function of a mussel shell to protect the animal against predators.

In addition to their often improved mechanical or optical properties, many biominerals also exhibit a hierarchical structure – often over many hierarchy levels. The most prominent example of such material is bone, which has seven hierarchy levels from the nanometer to the macroscopic scale [4], as illustrated in Fig. 15.2. At each of these hierarchy levels, perfect control needs to be achieved over the organic and inorganic material deposition and assembly. The lowest structural level is characterized by the collagen molecule, the carbonated apatite crystals and water. The collagen molecules self-assemble into the collagen triple helices and form collagen fibrils (level 1). The next level represents the start of the crystallization in the fibrils (level 2), where the crystallization of the hydroxyapatite nano platelets starts in the gap zones between the collagen fibers in the fibril. These fibrils form bundles aligned along their long axes (level 3), which can form various patterns (level 4). Such self-assembly processes only work up to the micrometer range, as above this material transport can no longer be facilitated by a diffusion process. Therefore, cells (osteoblasts) take over the secretion of the bone-building materials at a higher structural level, which are arranged in concentric layers of collagen sheets (the osteon, level 5). Various osteon-based microstructures are formed with sizes up to the millimeter scale (level 6) until the bone macrostructure is formed with macroscopic size (level 7).

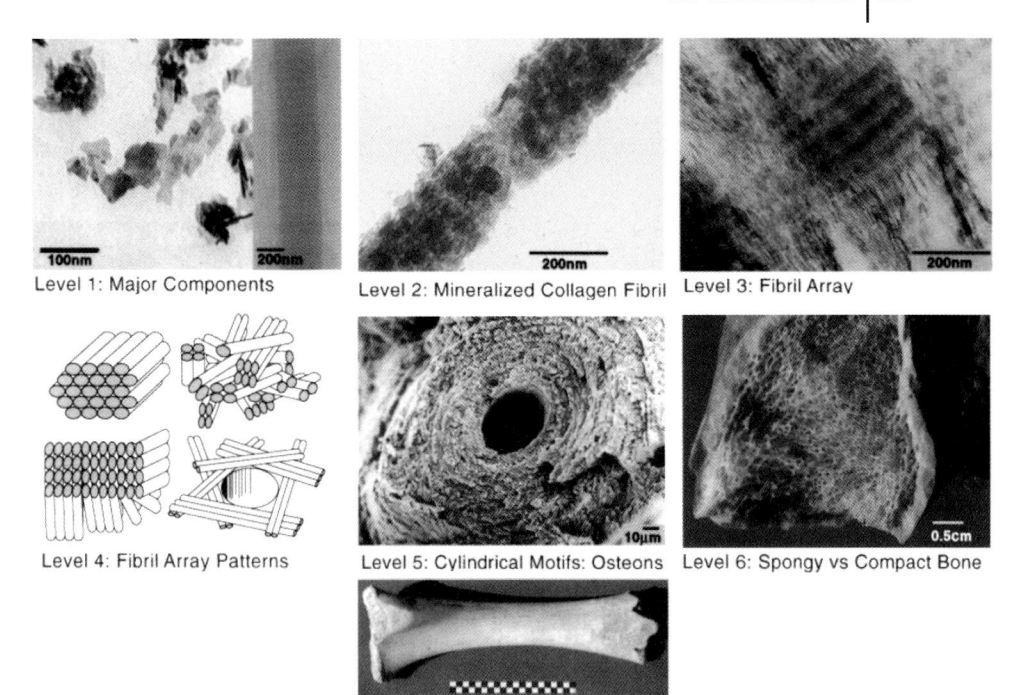

Fig. 15.2 The seven hierarchical levels of organization of the bone family of materials. Level 1: isolated crystals from human bone (left side) and part of an unmineralized and unstained collagen fibril from turkey tendon observed in vitreous ice in the TEM picture (right side). Level 2: TEM of a mineralized collagen fibril from turkey tendon. Level 3: TEM of a thin section of mineralized turkey tendon. Level 4: four fibril array patterns of organization found in the bone family of materials. Level 5: SEM of a single osteon from human bone. Level 6: light micrograph of a fractured section through a fossilized (about 5500 years old) human femur. Level 7: whole bovine bone (scale: 10 cm). Reproduced from [4] with permission from the Annual Review of Materials Science Volume 28 © 1998 Annual Reviews.

This example shows that perfect control is not only achieved over the mineral deposition site but also over the time of mineralization and demineralization as bone is a living material, which is continuously mineralized and demineralized. Usually, spatial control is achieved by biopolymers, which can build a scaffold for the mineralization site as collagen in the above example of bone. This is called the structural or insoluble matrix. However, the mineralization event itself is also highly controlled by biopolymers, which are usually soluble and therefore are called a soluble or functional matrix. It is clear that the functioning of the insoluble matrix and the various components of the soluble matrix is a highly synergetic process and only in some cases are the functions of the individual components of the soluble matrix already revealed. This makes the mimicking of biomineralization processes very difficult, as often only little is known about the formation of a biomineral archetype. Furthermore, the assembly of the structural matrix

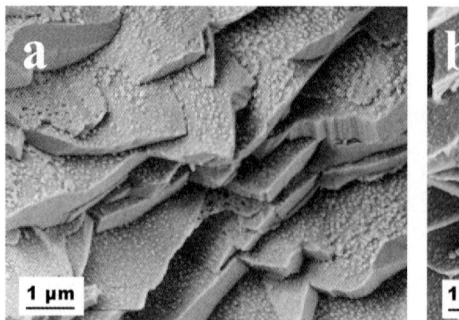

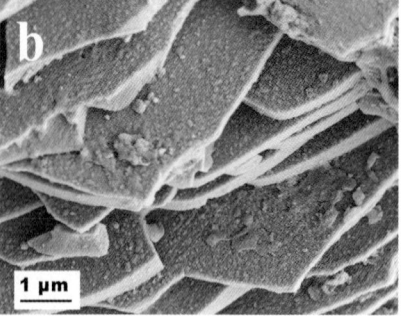

Fig. 15.3 Comparison of fracture surfaces of synthetically generated nacre (a) and natural nacre (b). Reproduced from [5] with permission of the American Chemical Society.

over several hierarchy levels is difficult to achieve as the synthetic chemist does not have cells at hand that do this job in Nature. However, biomimetic mineralization can be successful even if the formation of the natural archetype is not yet completely understood, as in the nacre example mentioned above. Our group was able to apply the demineralized structural matrix of nacre (Fig. 15.1 b) as scaffold for the deposition of $CaCO_3$ via amorphous precursor particles, which were generated in the presence of polyaspartic acid as a synthetic mimic of the natural soluble matrix of nacre [5]. These amorphous particles completely filled the compartments of the structural matrix and after crystallization resulted in a material which was indistinguishable from the natural archetype by both scanning and transmission electron microscopy (Fig. 15.3).

The above example shows that synthetic polymers are well able to take over some of the biopolymer functions in natural systems. Therefore, it makes sense to look at polymers that play an active role in biomineralization and to extract structural features, which can be transferred to synthetic polymers, in order to synthesize polymers with improved functions. This overview deals with this subject and ranges from the non-exhaustive description of some functional biomineralization polymers and their structures to their synthetic mimics, the so-called double hydrophilic block and graft copolymers, without the intention of being fully comprehensive.

15.2
Active Biomineralization Polymers

Especially the functional biopolymer matrix in biominerals is a multicomponent mixture with the possibility of all kinds of interactions. This makes the analysis of the individual polymer components difficult. Also, the biomineralization proteins are often polydisperse, have a non-globular shape and multiple charges and post-translational modifications, which further hampers protein separation.

Even more difficult is to reveal their function, as they are not only present in a polymer mixture with the associated possibility of polymer interactions, but are furthermore often only active for a certain time. For example, in calcified parts of crustacean cuticles, already 33 different proteins have been identified [6].

15.2.1
Proteins

Many of the functional proteins in biomineralization reveal a multifunctional structure. They can, for example, bind to one phase in the mineralization environment such as chitin in mollusks of crustaceans but also to a mineral and can furthermore even exhibit a crystallization-inhibiting moiety as found in the CAP-1 and CAP-2 polypeptides from the exoskeleton of the crayfish *Procambarus clarkii* [7,9]. These proteins can be linked to chitin and act as a nucleator

Table 15.1 Biomineralization proteins and genes.

Protein/gene	Species	Origin	Known characteristics	Refs.
CAP-1	Crayfish *Procambarus clarkii*	Cuticle	8.7 kDa, anti-calcification activity, chitin binding	7, 8
CAP-2	Crayfish *Procambarus clarkii*	Cuticle	7.4 kDa, anti calcification activity, chitin binding	9
Statherin	Human	Saliva	Asp–Ser–Ser–Glu–Glu binding unit, rich in proline, two-thirds of the molecule hydrophobic	12
Phosphoryn		Dentin/bone	Highly phosphorylated, $Asp[Ser(P)]_n$ repeat units, $n=1-3$	16
Casein		Milk	Phosphorylated acidic repeat units such as Ser–Ser–Ser–Glu–Glu and PSer–Gly–Ser–Ser–Glu–Glu, rich in polar and charged residues	16
Osteopontin		Bone/milk	Phosphorylated acidic repeat units such as Ser–Ser–Ser–Glu–Glu and PSer–Gly–Ser–Ser–Glu–Glu, rich in polar and charged residues	16
Bone sialoprotein		Bone/dentin	Phosphorylated acidic repeat units such as Ser–Ser–Ser–Glu–Glu and PSer–Gly–Ser–Ser–Glu–Glu, rich in polar and charged residues	16
Caspartin	Fan mussel *Pinna nobilis*	Calcite prisms	Long PAsp domain, Asx and Gly are 77% of AA residues	17
MSP-1	Scallop *Patinopecten yessoensis*	Foliated calcite shell	PI = 3.4; domains: 4 Gly–Ser, 4 Asp-rich, 1 basic, 3 Gly-rich	18, 19

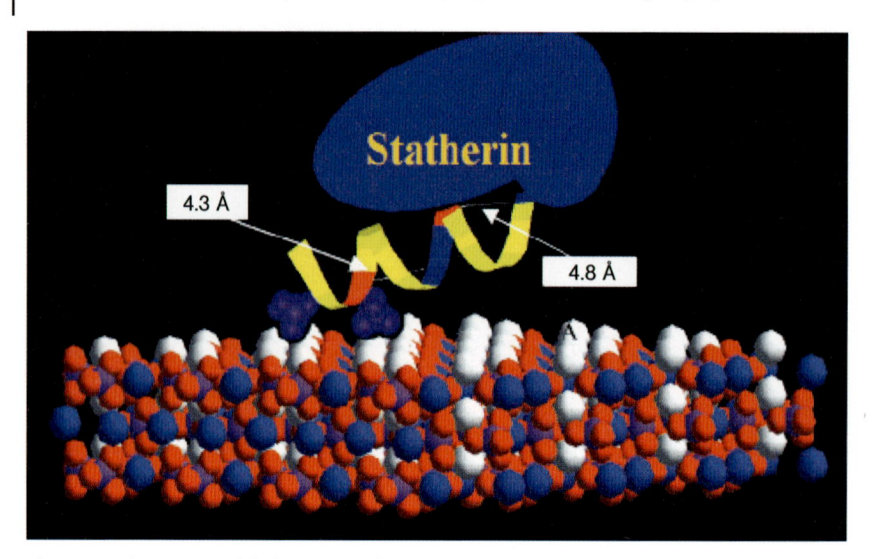

Fig. 15.4 Schematic model showing statherin interacting with the (001) face of HAP, with calcium ions in white. The interacting N-terminus is helical. The remaining portion of the protein (blue) is shown as a nondescript shape to indicate unknown surface-adsorbed structure. Reproduced from [11] with permission of the American Chemical Society.

for $CaCO_3$ deposition [9, 10] (see Table 15.1). The proteins DD4, DD5, DD9A and DD9B extracted from prawn cuticles partly have chitin-binding ability and are acidic, as expressed by their low isoelectric point, a feature which is common for the listed proteins (Table 15.1).

Another acidic protein with a multi-domain functionality is statherin [11], which has a hydroxyapatite (HAP) binding motif and another domain, which is water soluble (Fig. 15.4). Its natural function is HAP nucleation and growth inhibition in the supersaturated environment of saliva [12–14]. The helical N-terminus contains a highly acidic pentapeptide sequence (Asp–Ser–Ser–Glu–Glu) where both serine units are phosphorylated. This moiety binds to HAP whereas the other hydrophilic part has lubricating and hydrating properties for oral surfaces. The rest of the molecule (two-thirds) is hydrophobic.

Also, mollusk shell proteins have a modular primary structure [15]. Other acidic proteins exhibit highly phosphorylated serine units, such as phosphoryn, casein, osteopontin and bone sialoprotein [16], or long acidic PAsp domains as found in caspartin, which spontaneously forms multimers in solution [17] (Table 15.1).

Another interesting biomineralization protein is MSP-1, which can be extracted from a scallop calcite shell microstructure [18, 19]. This protein has a multi-domain structure with loops formed by four Gly–Ser domains and is suspected to serve for Ca binding (Asp-rich domains) and anchorage via the basic domain [15]. It has a short basic domain close to the N-terminus and two Gly–

Ser domains alternating with Asp-rich domains. As for the above-mentioned proteins, MSP-1 is enriched in Ser residues, which are likely phosphorylated to give a good $CaCO_3$ binding site.

15.2.2
Polysaccharides

Polysaccharides are far less studied than proteins as biomineralization polymers. They are of importance as a structural matrix such as chitin [20, 21] in mollusks and glycoaminoglycans (GAGs) [22], which are polysaccharides bound to a protein core found in vertebrate connective tissues. GAGs are alternating copolymers of a hexosamine and either a galactose or an alduronic acid and individual GAGs differ from each other by the type of hexosamine or alduronic acid (or hexose), the position and configuration of the glycosidic bonds and the degree and pattern of sulfation. The main acidic building blocks of GAGs are keratin sulfate, hyaluronic acid, chondroin-4-sulfate, chondroin-6-sulfate, dermatan sulfate, heparin sulfate and heparin [23].

Thus, polysaccharides are usually of importance when they bind covalently to proteins (glycoproteins, for example, in eggshell mineralization; see below). They only rarely occur unbound as a soluble matrix in biomineralization and often have an alternating structure, usually of the AB copolymer type. They are therefore closer to the structure of polyelectrolytes than to that of the above-discussed proteins/peptides with multiple domain structures and functionalities. Therefore, biomineralization polysaccharides have not yet served as archetypes for the development of synthetic functional polymers to a major extent and are thus not discussed further in this context.

15.2.3
Glycoproteins

Glycoproteins are also not too well studied as biomineralization polymers, although a number of biomineral-associated proteins are glycosylated [24–27] and evidence is reported that oligosaccharide chains interact with biominerals and/or lectin-like domains of other biomineralization proteins [28, 29]. For example, the calcite-binding protein lithostathine consists of a folded C-type lectin core domain and a random coil glycosylated mineral interaction domain consisting of 11 amino acids: pyro-(Glu–Glu–Asp–Gln–Thr–Glu–Leu–Pro–Gln–Ala–Arg) [24]. However, many biomineralization proteins offer potential glycosylation sites so that glycoproteins are certainly relevant for biomineralization. Glycoproteins are also of importance in antifreeze proteins, which are ice crystallization inhibition proteins found in a large variety of organisms where they can show an extended glycosylated helix as found in type IV antifreeze proteins [30, 32]. A very interesting glycoprotein is mucoperlin [33], an acidic polymer extracted from *Pinna nobilis* nacre with a short N-terminus, a long set of 13 almost identical tandem repeats enriched in serine and proline residues and a

C-terminus with short acidic motifs, three cysteine residues and two potentially sulfated tyrosine residues. The tandem repeat domain offers 27 sites for O-glycosylation at the serine units. Its function could be $CaCO_3$ inhibition and Ca binding. Glycoproteins are also found in the biomineralization process of avian eggshells. For example, on the outer sites of the eggshell membranes, discrete masses of organic material are deposited, the so-called mammillae, whose surface contain mammillan, a calcium-binding keratin sulfate-rich glycoprotein, which is involved in the nucleation of the first eggshell calcite crystals [34–36]. Another glycoprotein, which accompanies the growth of the crystalline columns of calcite, is ovoglycan, a dermatan sulfate-rich glycoprotein [34–37].

15.3
The Role of Biomineralization Polymers and Reduction to Their Functional Units

It is clear that polymers involved in biomineralization have to fulfill multiple tasks. It became obvious for the above-listed biomineralization protein and glycoprotein examples that they have a modular structure [15], are acidic and usually of low molar mass. For the example of mollusk shell proteins, the protein domains can be classified into four groups: (1) enzymatic domains, (2) structural domains, (3) domains interacting with $CaCO_3$ and (4) receptor or binding domains [15]. This feature enables the proteins to perform different functions with a single molecule. This is clearly a desirable feature for a synthetic polymer.

A molecule which has multiple functions such as simultaneously binding to calcium and chitin (e.g. CAP-1 and CAP-2 protein in Table 15.1) is amphiphilic in the sense of the word, which means "loving both". An amphiphilic molecule is therefore not only a surfactant with a hydrophilic and hydrophobic part, as is commonly associated with this term, but can be a molecule that can interact with two phases of whatever kind. Examples of such phases are minerals, metals, solvent, organic matrix, etc., and the amphiphile acts as a phase compartilizer.

If this inspiration from Nature's biomineralization polymer design is transferred to synthetic polymers, the strategy for the design of a new class of amphiphiles is to synthesize (multi)block or graft copolymers, with blocks that interact with the desired phases. The block-like polymer structure is sometimes also directly evident in biomineralization polymers such as statherin in Fig. 15.4. The advantage of this strategy is that tailor-made synthetic molecules become available, where each of the blocks can be highly optimized for its task so that different functions are separated within one molecule [38, 39]. For example, a drug carrier can be designed on the basis of a block copolymer, where one block binds the drug and the other one maintains water solubility so that self-assembly in an aqueous environment leads to micelle formation with the drug in the core [40, 41]. Similar considerations apply to the stabilization of mineral or metal nanoparticles in a solvent. However, one could also think of dispersing metal or mineral particles in a solid phase such as a polymer via a solution by applying a suitable block copolymer, and many other applications are possible.

15.4 Double-hydrophilic Block Copolymers (DHBCs) and Double-hydrophilic Graft Copolymers (DHGCs)

15.4.1 Concept of DHBCs and DHGCs

As discussed above, the important phases in biomineralization processes are an organic matrix, a mineral phase and the aqueous solvent. Therefore, the design of a block copolymer, which has a block interacting with a mineral (a polyelectrolyte or complexing block) and another block maintaining water solubility, mimics some biomineralization polymers as the simplest model case and should be a highly effective tool to generate mineral nanoparticles in aqueous solution (Fig. 15.5). Such block copolymers are called double-hydrophilic block copolymers (DHBCs), a new class of polymers which has attracted rapidly increasing interest in recent years [42].

In solution, DHBCs behave like normal polymers and polyelectrolytes and show no characteristics of an amphiphile such as micelle formation or lowered surface tensions of their solutions. The amphiphilicity is induced as soon as the molecules come into contact with a substrate, for example a mineral surface leading to an interaction and subsequent complex or superstructure formation. Alternatively, temperature, ionic strength or pH variation as well as a complexation reaction can completely reverse the hydrophilicity of one block into hydrophobicity [42]. In this case, DHBCs represent switchable amphiphiles with new applications. A special case of DHBCs are polyampholytes consisting of a polycation and polyanion block [43], which can act as their own substrate so that they show pH-dependent complex formation.

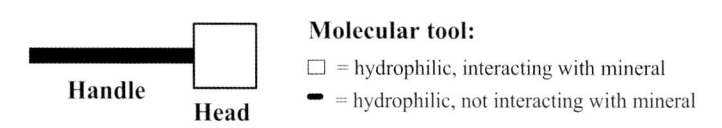

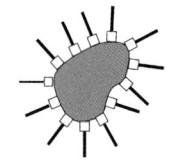

Fig. 15.5 DHBC design for the example of a polymer targeting mineral surfaces in water.

DHBC applications so far include drug carrier systems and gene therapy [44], switchable amphiphiles [45], mineralization templates [46] and crystal growth modifiers [47], induced nanoreactors for metal colloid synthesis [48], particle stabilizers [211] and desalination membranes [43], to list just a few, and it is likely that more applications will follow. This is because:
1. Ecology and society demand the use of environmentally friendly substances. Therefore, water as a solvent gains increasing importance for many kinds of technical processes and thus molecules are required which do the same job in water as other substances did previously in organic solvents (green chemistry).
2. The DHBC design is flexible with respect to the functional block and can be adapted to nearly every type of substrate. Often, biocompatible building blocks can be applied and such polymers can even be used for medical purposes.

From the viewpoint of particle stabilization, DHBCs have an optimized molecular design combining the advantages of electrostatic particle stabilization with those of steric particle stabilization by polymers. In addition, they are able to adsorb selectively on certain crystal faces and can therefore control particle shape upon further growth. This can also be achieved with ions or low molar mass additives (Fig. 15.6a), but then the particle is not sterically stabilized. On the other hand, polymers can be adsorbed and stabilize the particle (Fig. 15.6b), but if the polymers are long, they do not adsorb selectively but cover all crystal faces. A DHBC, however, with a short sticking block combines the advantages of face-selective adsorption with that of particle stabilization due to the longer stabilizing block (Fig. 15.6c).

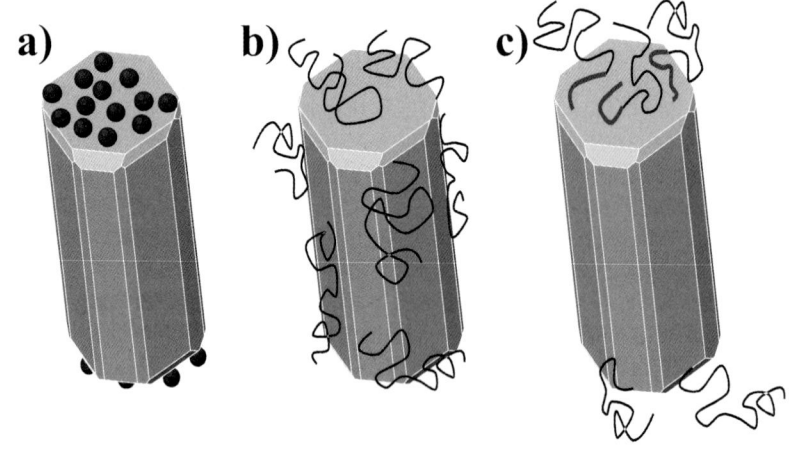

Fig. 15.6 Face-selective adsorption of ions or low molar mass additives (a), particle stabilization by polymers (b) and face-selective adsorption and particle stabilization by DHBC (c). Reproduced from [50] with permission of the Royal Society of Chemistry.

15.4 Double-hydrophilic Block Copolymers (DHBCs) and Double-hydrophilic Graft Copolymers

As their natural archetypes, DHBCs are typically rather small with block lengths between 10^3 and 10^4 g mol^{-1}. This also has practical reasons, as a too long interacting block might stick to more than one mineral nanoparticle and would thus lead to particle aggregation and bridging flocculation [49]. On the other hand, a too long stabilizing block has too high space demands, so that fewer DHBCs can adsorb on a particle surface and the stabilization effect is decreased. This is outlined in Fig. 15.7. In Fig. 15.7a, the sticking block is too long and the stabilizing block too short, so that the sticking block demands too much space on the particle surface, leading to a low polymer coverage and bad stabilization and, in extreme cases of polymer adsorption on several particle surfaces, even to bridging flocculation. If the length ratio between stabilizing and sticking block is balanced, good particle stabilization can be achieved (Fig. 15.7b), whereas a too long stabilizer block also leads to poor particle stabilization because its space requirements lead to a low polymer coverage of the particles (Fig. 15.7c). In an extreme case, the sticky block is far too short and the DHBC no longer adsorbs on the particle surface. In typical DHBCs for mineral stabilization, the polyelectrolyte sticking block is between 500 and 5000 g mol^{-1}, whereas the stabilizing block is between 3000 and 10000 g mol^{-1}. According to the above considerations, the length of the stabilizer block can be used to tune particle stabilization, so that mineral nanoparticles with only a temporary stabilization can also be generated, and which can be used as building units for complex superstructures via a self-assembly mechanism. The advantage of block copolymers as compared with homopolymers for the stabilization of particles is shown for the example of CaCO$_3$, which was precipitated in the presence of a PMAA homopolymer and a PMAA DHBC with a stabilizing PEO block

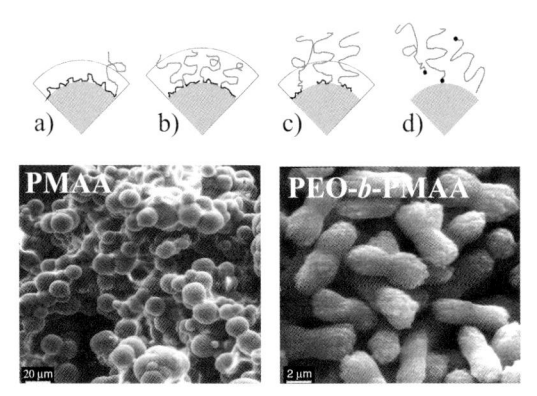

Fig. 15.7 Stabilization of a particle surface by a DHBC with a polyelectrolyte (PE) sticky block and a stabilizing (ST) block to illustrate the necessity of an appropriate block length ratio. (a) PE block too long and ST block too short; (b) PE and ST block in good ratio; (c) PE block too short, ST block too long; and (d) PE block far too short and ST block too long. The SEM pictures illustrate the difference of a PMAA polymer additive as homopolymer and DHBC with stabilizing PEO block for the example of CaCO$_3$ crystallization. Reprinted from [287] with permission of Wiley-VCH.

(Fig. 15.7). For the homopolymer, spherical particle aggregates are obtained, which are glued together by PMAA molecules adsorbing on several particle surfaces. This does not happen for the DHBC, as the stabilizing PEO shell prevents gluing of the dumbbell-shaped microparticles, which remain dispersed.

All above advantages of the DHBC design also apply to the corresponding graft copolymers (DHGCs), with the further advantage of an easier synthetic accessibility of graft copolymers. In the following, the synthesis and main applications of DHBCs and DHGCs are described. For a more extensive treatment of DHBC synthesis and applications, especially as additives for crystallization control, the reader is referred to the specialized literature [42, 50].

15.4.2
Synthesis of Double Hydrophilic Block Copolymers

The described synthesis strategies are mainly restricted to AB block copolymers, although some syntheses of ABA block copolymers are also described. Several synthetic strategies exist for DHBCs. The most defined block copolymers with adjustable block lengths can be obtained via living polymerization (anionic, cationic, group transfer and living radical polymerization). However, the range of suitable monomers is limited so that the possibilities of coupling two readily synthesized blocks or of modifying one block of the block copolymer by polymer analogous reactions are also advantageous synthetic strategies. In numerous cases, poly(ethylene oxide) (PEO) is used as one hydrophilic block of the DHBC so that many syntheses rely on the chemical nature of this polymer. Table 15.2 summarizes some common repeat units used to build up DHBCs as well as the polymerization reactions.

Table 15.2 gives a good overview of the synthesis of the different blocks in DHBCs. It is not possible to describe the syntheses in detail here, so the reader is either referred to an earlier review [42] or the cited primary literature. However, it becomes obvious that living polymerization methods are the most important methods to build up DHBCs. This is no natural limitation, as classical radical polymerization with appropriate macro initiators or coupling of two blocks possibly with a subsequent polymer analogous reaction are certainly also synthesis strategies for DHBCs, although they yield less defined polymers compared with the living methods. Nevertheless, the synthesis is much easier and can be scaled up.

15.4.2.1 Living Anionic Polymerization
Anionic polymerization of a polyanionic block is not directly possible so that esters have to be polymerized with subsequent hydrolysis after polymerization. Alternatively, blocks with reactive double bonds can be prepared, which can then be converted to a hydrophilic block via a polymer analogous reaction. Many synthesis routes apply hydrophobic precursor blocks resulting from monomers such as isoprene or butadiene [55, 58, 75], trimethylsilyl methacrylate [43, 70],

15.4 Double-hydrophilic Block Copolymers (DHBCs) and Double-hydrophilic Graft Copolymers

Table 15.2 Examples of typical blocks used in DHBCs and their functions and methods of synthesis [a].

Block	Structure	Remarks	Polymerization reaction	Refs. for synthesis
Solvating blocks				
Poly(ethylene oxide) (PEO)	$-[CH_2-CH_2-O]_n-$	Temperature-responsive block, hydrophilic at $T \sim 80\,°C$, strongly hydrated, mainly not interacting with substrates	Anionic	44, 47, 51–53
Dihydroxypropyl methacrylate (HMA)	(methacrylate with $-O-C(=O)-CH_2-CH(OH)-CH_2OH$ side chain)	Hydrophilic, mainly not interacting with substrates	Anionic via acetal precursor	54
Poly(vinyl alcohol) PVA)	$-[CH_2-CH(OH)]_n-$	Hydrophilic, mainly not interacting with substrate	Cationic via vinyl ether precursor	55–57
Poly(vinyl ether)s	$-[CH_2-CH(OR)]_n-$	Temperature-responsive block. Hydrophilic at: $R=(CH_2)_2OC_2H_5$: PEOVE, $T<20\,°C$; $R=(CH_2)_2OCH_3$: PMOVE, $T<70\,°C$; $R=CH_3$: PMVE, $T<34\,°C$; $R=C_2H_5$: PEVE, $T<\approx 15$–$20\,°C$; $R=[(CH_2)_2O]_3CH_3$: PMTEGVE, $T<83.5\,°C$	Cationic	55–61
Poly(oligoethylene glycol) methacrylate (POEGMA)	(methacrylate with $-O-C(=O)-O-((CH_2)_2-O)_6-CH_3$ side chain)	Hydrophilic, salt-responsive block	GTP	62
Poly(N-isopropyl-acrylamide) (PNIPAAm)	$-[CH_2-CH(C(=O)NH-CH(CH_3)_2)]_n-$	Temperature-responsive block, hydrophilic at $T<30.9\,°C$	Radical	63–65

Table 15.2 (continued)

Block	Structure	Remarks	Polymerization reaction	Refs. for synthesis
Poly(4-vinylbenzyl alcohol) (PVBA)		Hydrophilic, mainly not interacting with substrates	TEMPO free radical	66
Functional blocks				
Polycationic				
Poly(2-vinylpyridine) (P2VP in quaternized form)		Strong polyelectrolyte, very hydrophilic R=H, Me, Et, Bz X=Br, I	Anionic, TEMPO free radical	43–46, 52, 66, 68
Poly(4-vinylpyridine) (P4VP in quaternized form)		Strong polyelectrolyte, very hydrophilic R=H, Me, Et, Bz X=Br, I	Anionic	47, 48, 69–72
Poly(aminoethyl methacrylate)		Polyelectrolyte R=CH_3: PDMAEMA R=C_2H_5: PEMAEMA R=CH_3CHCH_3: PDPAEMA	Anionic, GTP, ATRP, RAFT	54, 73, 83, 133, 141
Poly(2-ethyloxazoline); poly(2-methyloxazoline)		Polyelectrolyte R=CH_3, C_2H_5	Cationic	84–88
Poly(4-dimethylamino-methylstyrene)		Polyelectrolyte	TEMPO free radical	66

15.4 Double-hydrophilic Block Copolymers (DHBCs) and Double-hydrophilic Graft Copolymers

Table 15.2 (continued)

Block	Structure	Remarks	Polymerization reaction	Refs. for synthesis
Polyanionic				
Poly(methacrylic acid) (PMAA); poly(acrylic acid) (PAA)	−[CH$_2$−CR(COOH)]$_n$−	Polyelectrolyte R=H, Me; also Na salts	Anionic via ester precursor, GTP via ester precursor, radical, ATRP of Na salt or ester	43, 45, 47, 48, 51, 53, 72, 74, 76–80, 89–97, 132
Poly(styrenesulfonic acid) (PSS)	−[CH$_2$−CH(C$_6$H$_4$SO$_3$H)]$_n$−	Strong polyelectrolyte, very hydrophilic, also Na, K salts	Anionic via polystyrene precursor, TEMPO free radical	66–68
Poly(vinyloxy-4-butyric acid) (PVOBA)	−[CH$_2$−CH(O−(CH$_3$)$_3$−COOH)]$_n$−	Polyelectrolyte, hydrophilic, also Na salts	Cationic via ester precursor	61
Poly(sodium 4-vinyl-benzoate) (PNaVBA)	−[CH$_2$−CH(C$_6$H$_4$COONa)]$_n$−	Polyelectrolyte, pH responsive, as Na salt	TEMPO free radical, ATRP	66, 98, 137
Other				
Peptides	−[C(=O)−CH(R)−NH]$_n$−	From hydrophilic to cationic and anionic depending on the composition of the residues R=CH$_2$COOH: Asp R=(CH$_2$)$_2$COOH: Glu R=(CH$_2$)$_4$NH$_2$: Lys	Solid-phase synthesis, ring opening via NCA	84, 99–117

a) The given references are restricted to the synthesis of a DHBC. This list of blocks is not exhaustive and further blocks can be found in the description of the different synthesis methods for DHBCs.

various alkyl methacrylates and acrylates (methyl, isopropyl and *tert*-butyl) [43, 71], methyl methacrylate (MMA) [51, 100, 118–120] and *tert*-butyl methacrylate (TBMA) [52, 54, 67, 68, 72–74, 121–123].

Blocks other than the polyanionic type can, however, be directly built up and often the DHBC can be directly synthesized. Examples include PEO-*b*-P2VP [59, 69], PEO-*b*-P4VP [72], PHMA-*b*-PDMAEMA [60], PDMAEMA-*b*-PMMA and PDMAEMA-*b*-PTBMA, which can be hydrolyzed to the PMAA block copolymers [56, 57], and PEO-*b*-PDMAEMA [61].

15.4.2.2 Living Cationic Polymerization
Cationic polymerization is also suitable for direct DHBC synthesis. Reported DHBCs include various poly(vinyl ether)s, PMOVE-*b*-PEOVE [76, 84], PVA blocks via deprotection of the vinyl ethers (benzyl, *tert*-butyl and trialkylsilyl) [76, 80], poly(vinyl ether) blocks (PMVE, PMTEGVE and PEVE) [77, 78], PBzVE, which can be deprotected to yield PVA [79, 80], and PVOBA [81]. Also, polyoxazoline blocks were prepared by cationic polymerization [82].

15.4.2.3 Group Transfer Polymerization (GTP)
GTP is an excellent method for the synthesis of well-defined methacrylate-based block copolymers of low polydispersity, but is more or less limited to acrylates and methacrylates in the choice of monomers. GTP was used for the preparation of PDMAEMA-*b*-PMAA via various ester precursors [62, 63, 89, 90, 124], PDMAEMA-*b*-PDEAEMA [64, 65], PDMAEMA-*b*-PDPAEMA) [65], PDEAEMA-*b*-PMEMA [65, 125] and POEGMA-*b*-PBzMA and POEGMA-*b*-PTHPMA [66]. Branched PDMA-*b*-PEGDMA DHBCs were also reported to be available via sequential monomer addition in GTP [126]. Interestingly, no acrylate-based DHBCs have been reported to be synthesized via GTP.

15.4.2.4 Radical Polymerization
Recent developments in controlled radical polymerization permit radical polymerizations with living character (ATRP, RAFT, TEMPO-mediated free radical polymerization). However, for the synthesis of DHBCs, radical polymerization was applied to build up either macro initiators or homopolymers, which are coupled to a block copolymer later.

15.4.2.5 Polymerization with Macro Initiators
Almost all macro initiator syntheses of DHBCs reported so far apply an appropriately modified monofunctional PEO. Therefore, many of the polymerization reactions described above can also be performed using PEO as macro initiator for radical, anionic, cationic and group transfer polymerization.

15.4 Double-hydrophilic Block Copolymers (DHBCs) and Double-hydrophilic Graft Copolymers

Radical polymerization This method was applied to synthesize PEO-b-PMMA [91, 92, 98], PEO-b-PAA [127] and PEO-b-PNIPAM [93, 94]. It also proved useful for synthesizing highly phosphonated block copolymers with a PEO macro initiator from the phosphonated acrylate methyl ester as well as via a PEO chain transfer agent [128]. Furthermore, a PSS macro initiator with terminal hydroxyl functionality could be used for the Ce^{4+}-initiated radical polymerization of methacrylates, which could be hydrolyzed to PMAA blocks [129]. However, the PSS-b-PMAA DHBC did not prove to be molecularly soluble in water due to an incompatibility between the two polyacid blocks. The same route could be applied for the polymerization of a second PDADMAC block [130]. An azo functionalized PEO could be used for the radical polymerization of PAMPS as second block as well as PDADMAC [130].

TEMPO-mediated free radical polymerization Recently, TEMPO-mediated living free radical polymerization of styrene derivatives was reported to be a suitable method to yield DHBCs directly without the protection–deprotection chemistry needed for living cationic, living anionic or GTP. By this method, PSS block copolymers with P2VP, PVBA, poly[4-(dimethylamino)methylstyrene] and poly(4-styrenecarboxylic acid) were obtained [95].

Atom transfer radical polymerization (ATRP) ATRP is another radical polymerization technique of steadily increasing importance, which can yield defined DHBCs via an appropriate macro initiator [131]. Examples include PEO-b-PMAA [132] or its sodium salt [96], PDMAEMA-b-PHEMA [133], PEO block copolymers with PHEMA, PDMAEMA, PDMEABr or PTMATf [134], PMeDMA block copolymers with PEO, BzDMA, GMA, MPC or SBMA [135], PAAM-b-PEO-b-PAAM [136] and PEO-b-PNaVBA [137], or with a second PBA [138] or P4VP [139] block. ATRP could also be applied with a three-arm PEO macro initiator to polymerize *tert*-butyl acrylate on to the three arms, which led to a star-like PEO_3-b-PAA_6 DHBC [140].

Reversible addition–fragmentation chain transfer (RAFT) Via DCC coupling, PEO monomethyl ether can be functionalized into a dithio ester RAFT macro initiator, which was used to polymerize a second PDMAEMA block [141]. A similar process to RAFT is macromolecular design via interchange of xanthates (MADIX), which uses xanthate chain transfer agents instead of the thiocarbonylthio compounds that are used for RAFT. MADIX was reported for the synthesis of acrylic acid/acrylamide DHBCs in aqueous solution, where the macro chain transfer agent was first synthesized with acrylic acid followed by subsequent polymerization of the second acrylamide block [142].

Anionic polymerization Anionic polymerization with appropriate macro initiators can also yield a variety of DHBCs partially via their precursors such as PEO-b-PMMA [53, 119, 120, 143, 144] and PEO block copolymers with PDMAEMA, PMEMA and PDPAEMA [101].

Ring-opening polymerization of *N*-carboxyanhydrides (NCAs) The ring-opening polymerization of NCAs of amino acids is especially promising for the synthesis of biocompatible DHBCs. The ring-opening polymerization of NCAs with amine initiators yielding block copolymers of two amino acids was initially applied to synthesize block copolymers between mainly nonpolar amino acids such as valine, isoleucine, alanine and phenylalanine or their block copolymers with polar or basic amino acids such as threonine or lysine. Chain breaking, transfer and termination reactions led to polymodal molar mass distributions. The application of organonickel initiators was reported to avoid these side-reactions and led to well-defined block copolymers ($M_w/M_n < 1.13$) such as PLys-*b*-PGlu [102]. Another way of obtaining DHBCs is the termination of living NCA polymerizations with an appropriate polymer such as living electrophilic polyoxazoline from 2-methyl-2-oxazoline [103].

However, most DHBCs with peptide blocks were so far polymerized with an amino monofunctionalized macro initiator which polymerizes the NCAs of the amino acids by ring opening yielding DHBCs such as PEO-*b*-PAsp [104–112], PGlu-*b*-PEO-*b*-PGlu [113–116], PEO-*b*-PGlu [117], PEO-*b*-PLys [85–87, 105, 106] and PMOX-*b*-PGlu [88].

Cationic polymerization Like anionic polymerization, cationic polymerization can also be carried out with an appropriately functionalized PEO macro initiator to yield polymers such as PEO DHBCs with 2-substituted-2-oxazoline [97, 145], PEO-*b*-PMOX [146] and PMVE-*b*-PEOX [147].

Group transfer polymerization A PEO macro initiator was reported to initiate the GTP of MMA yielding the PEO-*b*-PMMA DHBC precursor [148]. However, more complicated DHBC structures such as star polymers were also available via GTP, first polymerizing DMAEMA macro initiators, which were used for the subsequent polymerization of the second HEGMA block [149].

15.4.2.6 Coupling of Two Polymeric Blocks

This is the probably most variable route towards DHBCs. In principle, all kinds of organic chemistry reactions can be used. Practically, amide or urethane bonds, fairly stable to hydrolysis, have proven to be well suited as coupling units, in contrast to ester groups. However, polymers prepared by this synthesis route are usually polydisperse as technical polymers are mostly applied as precursors. Nevertheless, the advantage of this synthetic strategy is the use of common and well-established organic chemistry, which allows the scale-up of such reactions even for technical synthesis. In the following, the coupling of a PEO block as one of the most important hydrophilic blocks in DHBCs to other polymer blocks is described. A large variety of reactions exist, which allow the transformation of the hydroxyl group in PEO into reactive groups such as succinimidyl succinate, various carbonate and chloroformate derivatives or cyanuric chloride. The so activated PEO can subsequently be coupled to another polymeric block. In the past, this was mainly applied to proteins and liposomes and reviews are recommended which summarize the available chemistry for the modification of PEO and coupling reactions to other molecules [150, 151]. As most of the described coupling strategies for PEO should also work for all other blocks with similar functionalities, it is clear that a very large number of DHBCs are accessible via the coupling route of two polymeric blocks.

Coupling was reported via amide linkages to yield PEO-b-PEI [152], via an acid chloride with alcohols or amines as reported for PEI and a PAsp block [152, 153], via a succinimidyloxycarbonyl group yielding PHPMA or poly{N'-2-[(tert-butyloxycarbonyl)aminoethyl]-N-methacryloylglycinamide} DHBCs with poly[2-(trimethylammonio)ethyl] methacrylate or PHPMA-b-(PTMAEMCl), PHPMA-b-[PMA-Gly-NH(CH$_2$)$_2$NH$_2$] [154, 155] and PHPMA-b-PTMAEM [156–158] ester linkages via transesterification [159] or linkages via amide formation with a protected PMAA block [127] to yield PEO-b-PMMA and PMAA-b-PMMA, urethane bonds via 1,1′-carbonyldiimidazole reaction [160] to yield PEO–PSP and PEO–PEI [161] or via a diisocyanate to give PAsp-b-PEO-b-PAsp [162].

15.4.2.7 Polymer Analogous Reactions

The possibilities of modifications of existing AB diblock copolymers by polymer analogous reactions are numerous and restricted only by the necessity to find a suitable organic reaction for the desired modification. With this approach, one polymer can be modified into a whole family of molecules with different functionalities but identical polymer backbone. For example, PEO–PEI is a very useful polymer for polymer analogous reactions, offering the possibility of the application of most of the reactions described in the following, but many other polymer precursors are also available for post-functionalization. Hence it is impossible even to outline the possibilities offered by this approach for the synthesis of DHBCs and only a few examples of literature-reported reactions leading to new DHBCs can be given here such as acylation of PEI to a PEDTA block [152], betainization [65, 152], sulfonation [163], combined introduction of sulfamate and carboxyl groups [164], phosphonation [153], phosphorylation [165], amidation [47], hydroxylation [165], quaternization [69, 70, 121, 163] and hydrolysis [43, 166–168].

15.4.3
Synthesis of Double Hydrophilic Graft Copolymers (DHGCs)

Syntheses of DHGCs have only rarely been reported so far, although their synthesis does not require the demanding conditions required for the synthesis of DHBCs via living polymerization. Thus, DHGCs are also interesting from an industrial viewpoint as they are comparatively easy to synthesize by grafting-from or grafting-to polymer reactions and have the same separation of functional blocks like a DHBC. One possibility for preparing DHGCs was reported for a polyacetal backbone where the ester side-groups were partially replaced by a jeffamine forming an amide bond and thus linking PEO chains to the backbone. Subsequent hydrolysis of the remaining ester groups on the polymer backbone yielded carboxyl functionalities, which are interesting as additives for crystallization control [169]. Another reported synthetic pathway was the partial esterification of a PMAA with PEO [170] or the PEO grafting on to PVCL [171]. An interesting hybrid between a DHBC and DHGC was reported as a result of a RAFT copolymerization of AA and a PEO-derivatized methacrylate showing that the transition between DHBC and DHGC architecture can be continuous [172]. A similar hybrid is the PEO-grafted PAA-b-PMAA synthesized via PEO esterification of PAA and subsequent block copolymerization [173].

15.4.4
Applications of DHBCs and DHGCs

Multiple applications were reported for DHBCs and DHGCs. Those for DHGCs are in principle the same as for the DHBCs, as shown for the example of $CaCO_3$ [174]. Therefore, this review focuses on DHBC applications. Nevertheless, it should be mentioned that DHGCs offer more possibilities for structure tuning towards specific applications by changing the number and distribution of ionic groups on the polymer backbone as well as the design of the sidechains. Here we will focus on the application of DHBCs as crystallization additive and only briefly mention other applications. More extensive reviews can be found elsewhere [42, 50].

15.4.4.1 DHBC Superstructures

Induction of water insolubility and micelle/vesicle formation A considerable amount of literature concerning the application of DHBCs is related to the induction of water insolubility in one block or micelle formation so that water becomes a selective solvent. The DHBCs are applied as switchable amphiphiles, with one block made water insoluble by pH, temperature or ionic strength variation and also complexation of countercharged molecules. Especially the last is of practical importance as it allows the design of drug carrier systems as described by Kataoka in this volume.

Temperature variation Usually, DHBCs, which show a reversible micellization in water upon temperature variation, exhibit one block of a polymer with a lower critical solution temperature (LCST) above room temperature so that this polymer becomes insoluble upon heating, such as PMVE, PNIPAAm, PDEAEMA, PDPAEMA, PVCL or PMEMA. This was exploited for the reversible micelle formation of PEOVE-b-PMOVE [76, 84], PMVE-b-PVA [80], PMVE-b-PMTEGVE [78], PEO-b-PNIPAAm [93], PEO-b-PVCL [175], DHBCs prepared by selective betainization of the PDMAEMA block of PDMAEMA-b-PDEAEMA and PDMAEMA-b-PDPAEMA or PDMAEMA-b-PMEMA by propane sultone treatment [65]

15.4 Double-hydrophilic Block Copolymers (DHBCs) and Double-hydrophilic Graft Copolymers

and PDMAEMA-b-PMAA [62, 89]. The block copolymer cloud points depend strongly on the relative lengths of the two blocks and can thus be adjusted between the values for both homopolymers so that the micelle formation can be tailored in a limited experimental range.

pH variation If DHBCs are built up with a polyacid or polybase block, pH-induced reversible micellization can be observed if the polyacid becomes fully protonated or the polybase fully deprotonated so that these blocks become hydrophobic whereas the second block is still water soluble. Examples of this kind of hydrophobization were reported for combinations of a nonionic and an ionic block and also for combinations of two acidic or basic groups with different pK_as of both blocks. Reported examples include PEO-b-P2VP [59], PDMAEMA-b-PDEAEMA [64] also with a betainized PDMAEMA block [65], PSSNa-b-PSCOONa [95], PNaVBA-b-POEGMA [137], P2VP-b-PAA [71], P4VP-b-PMAA [72] and PDMAEMA-b-PMAA (Na salt) [176], PDEAEMA block copolymers with PEO, PDMAEMA and quaternized PDMAEMA [177] and PEO-b-PMAA [127]. However, other interactions such as hydrogen bonding can also account for aggregation, as evidenced for the interaction between PAA and PEO blocks in a copolymer at an appropriate pH [172].

Variation of the ionic strength The weakly basic PMEMA homopolymer is known to be completely soluble in acidic solution due to protonation but it can be relatively easily precipitated by the addition of electrolytes. Hence a DHBC with a PMEMA block can form micelles on addition of salts such as Na_2SO_4. This was demonstrated for PMEMA-b-PDMAEMA, which showed both micelle and reversible micelle formation upon pH adjustment or salt addition in aqueous media [125].

Complexation Polyelectrolyte blocks in DHBCs can also become hydrophobic due to charge neutralization upon complexation with an oppositely charged molecule. In such cases, so-called polyion complex (PIC) micelles can be formed due to electrostatic interaction. PIC micelles often show a narrow size distribution and small size. They combine the properties of block copolymers (micelle formation, enhanced solubility) with those of polyelectrolyte complexes. The general feature of PIC micelle cores is sensitivity to pH, salts and polyion exchange reactions). Thus, PIC micelles often have an impact on drug, DNA or oligonucleotide delivery as they are readily soluble, in contrast to their polyelectrolyte complex counterparts (for a review of these applications, see [178]). The micelle core is both pH (conversion degree of polyelectrolyte complex formation) and salt (charge shielding) sensitive. This is a general feature of PIC micelles. There is no clear transition to the polyelectrolyte complexes formed with DHBCs or to drug delivery systems, which are discussed later. The complexation can take place between DHBCs with oppositely charged polyelectrolyte blocks such as the pair PEO-b-PLys and PEO-b-PAsp with the same polymerization degree and 1:1 stoichiometry [107]. It was shown that chain length recognition exists between DHBCs with a polycationic or polyanionic block, indicating that the PIC micelle formation process is strictly driven through a phase separation process [110]. Even in mixtures with different block lengths of PEO-b-PLys or PEO-b-PAsp, micelles with extremely narrow size distribution formed with matched chain lengths of the charged segments [110]. The micelle stability can be enhanced by cross-linking [179].

Also, DHBCs can form PIC micelles with a polycation, as reported for PEO-b-PAsp with an equimolar ratio of PLys homopolymer to the Asp units [180], an oligopeptide model drug bearing basic arginine moieties [112] and lysozyme [181, 182]. The last PIC micelles were found to be an intelligent bioreactor through reversible release of lysozyme through micelle dissociation upon salt addition [183]. PIC micelles were also formed by binding a cationic drug (doxorubicin) to PEO–PMANa, which disintegrated in the presence of salts like all other PIC micelles [184]. PEO-b-PMAA formed PIC micelles with lysozyme [185]. Soluble polyelectrolyte complexes were reported for PEO-b-PAMPS with PDADMAC [130]. A DHGC can also be used for polyelectrolyte complex formation, as reported for a partially PEO-grafted PMAA with a PDADMAC polycation [170].

The complex formation of a DHBC with a polyanion is also possible. This type of complexation reaction is especially interesting for DNA or oligonucleotide delivery, which has become increasingly important in recent years. Major advantages of these complexes are their stability and decreased interaction with serum proteins. Recent reviews discuss this important application [186–188]. PEO-b-PLys and PHPMA-b-PTMAEM, PHPMA-b-[PMA-Gly-NH(CH$_2$)$_2$NH$_2$] and PEO block copolymers with polylysine/glycine and poly(1,3-butane-diol-1,4-diaminobutylphosphonate) as polycationic blocks spontaneously form micelles with DNA [87, 108, 154–158, 189–191]. PDMAEMA-b-PHMA forms micelles upon PAA addition [192]. Stable micelles with low polydispersity formed between PEO-b-PEI or PEO-b-PSP as DHBCs with a polycationic block and a negatively charged 24-mer oligonucleotide [161, 193]. PEO-b-PSP micelles with antisense oligonucleotides [160] were reported to form over a wide range of pH with the PEO–PSP-based micelles being less stable in the presence of electrolytes compared with PEO-b-PEI [161, 193]. Medical applications of PIC micelles were also reported. For example, phosphodiester oligonucleotides complexed via PIC micelles with PEO-b-PSP were used to transfect primary neurons [194] and phosphothioate oligonucleotides complexed in PIC micelles were successfully applied to modulate gene expression in the rat retina *in vivo* [195] or for the specific fibronectin expression reduction in retinal vascular cells [196]. Soluble polyelectrolyte complexes between a cationic PEO-b-PDADMAC DHBC and PSS as polyanion were also reported [130].

Another field of medical applications was found for the complexation of metal complexes with DHBCs with a polyanionic block. An example is the complexation of the anti-tumor active *cis*-dichlorodiammineplatinum(II) with PEO-b-PAsp by ligand exchange of the two chloride ligands by the carboxylate groups of PAsp [197, 198]. As the carboxylate ligands show low nucleophilicity, the complex formation is reversible, thus making this system a promising candidate for the release of the platinum complex as an anti-tumor drug.

DHBCs can also form complexes with surfactants. These complexes are water soluble and combine the virtues of block copolymers and polyelectrolyte-surfactant complexes. The first described studies concerned the complexes between the sodium salt of PEO-b-PMAA and the cationic single-tail surfactants dodecyltrimethylammonium bromide, tetradecyltrimethylammonium bromide, cetyltrimethylammonium bromide, dodecylpyridinium chloride and cetylpyridinium bromide [199, 202] and also the cationic double- and triple-tail surfactants DDDAB, DODAB and TMAB [202]. P1ME4VPCl-b-PMAA was also used for the preparation of polyelectrolyte surfactant complexes with the cationic surfactant CTA, the anionic surfactant SDS and the dye ANS in aqueous solution [203]. For the single-tail surfactants interacting with PEO-b-PMAA, spontaneous vesicle formation with low polydispersity was reported [199, 201] whereas complexation of all-*trans*-retinoic acid and PEO-b-PLys resulted in core–shell micelle formation with the core formed by a smectic PLys retinoate complex [204]. Core–shell micelles were reported with a core built up by densely packed dodecyltrimethylammonium bromide (DTAB) surfactant micelles complexed by the oppositely charged polyelectrolyte block with a shell formed by the neutral hydrophilic block of a PANa–PAM DHBC [205]. This was evidenced by a SANS study of poly(acrylamide)-b-poly(acrylic acid) sodium salt, which was prepared by controlled radical polymerization [206]. Core–shell corona micelles can be formed by complexation of a DHBC micelle with a charged corona by the DHBC with a counter-charged polyelectrolyte block, as evidenced for PSS-b-PAA complexed by PEO-b-P4VP in ethanol [139].

DHBCs for chemical reactions in confined volumes

Induced micelles as nanoreactors As DHBCs are able to complex metal ions, they offer the opportunity to form aggregates/micelles upon interaction with metal ions, which can be exploited for the formation of a nanoreactor in aqueous systems and for subsequent transformation of the metal ions. The first example of such an application was reported for the interaction of PEO-b-PEI with $AuCl_3$, $PdCl_2$ and $H_2PtCl_6 \cdot 6H_2O$ [207] and additionally $Na_2PtCl_6 \cdot 6H_2O$, K_2PtCl_4 and Na_2PdCl_4, the last only forming micelles/aggregates after additional protonation of the polymer [48, 207]. All systems based on PEO-b-PEI were very sensitive to the reaction conditions and the number of metal compounds which can induce micelle formation is limited. PEO-b-PEI was also reported to be a very effective stabilizer for high-quality crystalline CdS nanoparticles in water and methanol with tunable sizes via the polymer concentration in a one-step synthesis with branched PEI being superior to linear PEI or a corresponding dendritic structure [208].

PEO-b-P2VP was found to be better suited for nanoreactor application [209] because its micellization in water can already be introduced by pH changes [59]. Unlike the block copolymer micelles, which fall apart at pH<5, the metal-coordinated micelles remained stable even at a pH as low as 1.8 [209]. PAA-b-PAM and PAA-*b*-PHEA were used as a stabilizer for the formation of hairy needle-like colloidal lanthanum hydroxide through the complexation of lanthanum ions in water and subsequent micellization and reaction, where the polyacrylate blocks induced the formation of star-shaped micelles stabilized by the PAAM or PHEA blocks [210]. The size of the sterically stabilized colloids was controlled by simply adjusting the polymer-to-metal ratio. The concept of induced micellization of anionic DHBCs by cations was also applied in a systematic study of the direct synthesis of highly stable metal hydrous oxide colloids of Al^{3+}, La^{3+}, Ni^{2+}, Zn^{2+}, Ca^{2+} or Cu^{2+} via hydrolysis and inorganic polycondensation in the micelle core [211, 212]. The Al^{3+} colloids were characterized in detail by TEM [213], in addition to the intermediate species in the hydrolysis process by SANS, DLS and cryo-TEM [214].

15.4.4.2 DHBCs as Novel Surfactants

Although the examples for DHBC applications in various areas have increased significantly in the last years, the potential of DHBCs is still not fully exploited. For example, a large variety of mineral particles should be addressable by DHBCs in addition to biominerals, which are in the current focus of basic research on DHBCs [42, 50].

DHBCs at interfaces Due to the DHBC design (Fig. 15.5), the adsorption of DHBCs on inorganic particles alters the particle surface properties and can thus be used to optimize the particles for a particular use, as reported for the adsorption of PSS-*b*-P2VP on porous silica beads for chromatographic applications [163, 215]. The strong adsorption was demonstrated by an unaltered ion-exchange capacity of the coated silicas and showed the use of silica surfaces modified by polymer adsorption as an ion exchanger for liquid chromatography. Separations were reported for mixtures of histones [215] and nucleoproteins [216]. The easy polymer adsorption process was shown to be an alternative to the complex procedure of covalent attachment of ligands and the ion-exchange capacities of the modified silicas were found to be of the order of those reported for HPLC ion-exchange resins [163].

The adsorption behavior of charged block copolymers such as PDMAEMA-*b*-PHMA on silicon wafers coated with thin layers of TiO_2 and SiO_2 was studied with reflectometry as a function of pH, salt concentration and polymer composition [217, 218], suggesting an optimum composition between the charged and the neutral block of the DHBC as already discussed in Fig. 15.7, which was manifested in the experimentally found maximum in the amount of adsorbed block copolymer [219]. These systems are useful for studying DHBC–mineral interactions and their dependence on external parameters such as pH or ionic strength [217] as well as the polymer conformation under various conditions [220]. Similar studies were carried out on the adsorption of PDMAEMA-*b*-PMAA on oxidized silicon surfaces [221].

15.4 Double-hydrophilic Block Copolymers (DHBCs) and Double-hydrophilic Graft Copolymers

Particle stabilization DHBCs can advantageously be used for the stabilization of hydrophilic matter in aqueous environments, including the large area of inorganic particle stabilization. However, these applications are not yet commonly recognized. Therefore, so far, the literature is under-represented with respect to the general importance of the field. So far reported particle stabilization examples include PHMA-b-PDMAEMA and PEO-b-PMOx and PEO-b-PMVE for the stabilization/destabilization of colloidal silica [60, 146, 218, 222–224]. In these studies, a clear correlation between the amount of adsorbed polymer and the stabilization was found. a-Fe$_2$O$_3$ (hematite) particle stabilization was achieved using PMVE-b-PVOBA in water [81, 225, 226], but also various organic solvents ranging from polar alcohols such as 2-propanol via esters and ketones to nonpolar aromatic hydrocarbons such as toluene due to the good solubility of PMVE in these solvents [81]. Even the transfer of the stabilized pigment from water into methyl ethyl ketone was successfully performed without particle destabilization [81]. These results outline the possibilities of appropriately designed DHBCs as powerful stabilizers. CeO$_2$ is also among the examples of successfully DHBC-stabilized functional particles [227].

PDMAEMA-b-PMAA block and tapered polymers with a major PMAA stabilizing block in the form of the PMAA ammonium salt were found to be better stabilizers for iron oxide, pyrrolopyrrole and Cu-phthalocyanine pigments compared with the corresponding random copolymers [57]. The DHBCs tolerated even higher pigment contents than the so far applied stabilizers in organic solvents or with organic cosolvents showing the potential for water-based high-solid paint formulations. The color characteristics of the solvent-free paints were reported to be very favorable. Particles can also be stabilized and used for further synthesis steps as described for magnetite/maghemite nanoparticles initially synthesized and stabilized in water by PEO-b-PMAA, then repeptized into a HEMA–MAA mixture, which could be emulsified in decane with subsequent polymerization yielding superparamagnetic latexes [228].

PEO-b-PMAA was also reported to be a highly efficient dispersant for a-Al$_2$O$_3$ as PMAA acted as anchoring block whereas PEO provided steric stabilization, resulting in strongly repulsive interparticle potentials [73, 229]. High-solid dispersions (80 wt.%) of alumina-coated TiO$_2$ in water were successfully stabilized with various PEO, P2VP, P4VP and PMAA DHBCs [72]. DHBCs were found to be at least as efficient as or superior in stabilization to equivalent triblock copolymers with the outer blocks as stabilizing moiety, and also far better than the corresponding homopolymers or random copolymers. Also, the higher efficiency of electrosteric compared with steric stabilization was pointed out by comparison of different stabilizing blocks, which limits the use of PEO-based dispersants for high-solid dispersions [72]. On the other hand, a decrease in the anchoring ability of the diblocks drastically reduced the dispersion stability. BaTiO$_3$ can also be stabilized with a DHBC. Using AFM equipped with a BaTiO$_3$ sphere on the cantilever, the interactions between the sphere and a plane BaTiO$_3$ surface became accessible in the form of surface force versus distance curves [230, 231]. Hence a deeper insight into the interparticle forces, the polymer conformation on the particle surface and thus on the particle stabilization could be obtained. These investigations showed that PEO-b-PMAA could sterically stabilize a BaTiO$_3$ dispersion independently of the pH, whereas a PAA homopolymer showed a strongly pH- and ionic strength-dependent stabilization, in agreement with the DHBC concept [230–232]. Gold nanoparticles can also be stabilized by DHBCs such as PEO-b-PDMA, which can be subsequently cross-linked to give a pH-responsive hybrid gold nanoparticle with much enhanced colloidal stability compared with the uncross-linked shell [141].

Conjugation and complexation
Drug delivery systems Various applications of DHBCs have been reported for drug delivery purposes as reviewed in [233–235] and by Kataoka in this volume. Due to the requirement for *in vivo* application, the choice of DHBCs for drug delivery applications is limited and usually, the stabilizing block is PEO and the interacting block a charged amino acid. Examples are so far almost exclusively known for the delivery of anti-tumor drugs and include sulfido derivatives of cyclophosphamide fixed to PEO–PLys [236], adriamycin (ADR), which can be conjugated to PEO-b-PAsp [104–107, 111, 237–241] and can even be designed as a missile drug (a drug which selectively finds its target) by conjugation of the antibody immunoglobulin (IgG) to ensure binding of the drug conjugate to the target cell [104, 105]. First *in vivo* studies of the anticancer activity against mouse leukemia revealed a lower toxicity compared with free ADR [106] and high cancer activity [111, 237, 239, 242–247]. Physically entrapped and thus readily available ADR played a major role in the antitumor activity of the micelles, in contrast to the conjugated ADR, and correlations between the *in vivo* activity, ADR oligomerization, selective delivery and micelle composition were obtained [248, 249]. A biodistribution analysis revealed that the physically entrapped ADR accumulated at tumor sites in a highly selective manner, so that the role of the polymeric micelle can be seen as a selective drug carrier [249]. PEO-b-PHAA was also applied as a drug carrier system with MTX as anti-tumor agent [250, 251].

Surface recognition and controlled crystallization This DHBC application is in the focus of this overview section, as it is the application that is most closely related to the function of the natural biomineralization polymer archetypes for DHBCs. It can be shown that the DHBC design offers a versatile and powerful handle to control the size and shape of inorganic minerals through selective DHBC interaction with crystal faces, although the precise mechanism of additive interaction and control of the crystallization process of the inorganic material often remains an open question. In addition, DHBCs showed activity even in the deposition of amorphous minerals such as silica [252].

Control of crystal modification, stabilization of kinetic phases and chiral recognition DHBCs have been found to exert remarkable effects on the control of metastable crystal modifications and stabilization of kinetic phases. It is an established concept from biomineralization systems that the crystal modification of a developing crystal can be controlled by a polymeric matrix/template through epitaxy. One such example is the nacreous shell layer of *Nautilus repertus*, where $CaCO_3$ is crystallized in the thermodynamically metastable modification aragonite via epitaxy to a β-sheet of acidic proteins adsorbed on an insoluble chitin matrix [253, 254]. Here, a two-dimensional geometric match between the carboxylic acid groups and the Ca^{2+} ions of the aragonite a–b plane is postulated to account for the nucleation and stabilization of the aragonite phase, which is usually metastable under the applied ambient conditions. However, this epitaxial model was questioned, one reason being the molecular dynamics even in a macromolecular network (p. 45 in [255]).

Such argumentation is especially true for soluble macromolecules, which are also able to stabilize kinetic phases. An example is a flexible DHBC containing multiple strongly chelating EDTA functional groups [47]. The interesting fact here is not that a vaterite sphere consisting of nanocrystallites was formed, as this is common for the early stages of $CaCO_3$ crystallization, but that the vaterite modification was stabilized for at least 1 year under conditions where a transformation of the kinetically favored vaterite modification into the thermodynamically stable calcite would occur within 80 h [256]. The remarkable stabilization of a metastable crystal modification suggests that the functional pattern of the flexible PEDTA block adopted that of the initially formed vaterite surface and that the crystal was fixed at the surface by the adsorbed polymer. Hence a transformation into a thermodynamically stable crystal modification was no longer favored because too many electrostatic bonds would have to be opened for restructuring. Similarly prolonged stabilization of vaterite was observed with DHBCs with an acidic amino acid [257] or phosphoric acid functional groups [258]. The stabilization of usually metastable crystal modifications hints at the potential that DHBCs may have for crystal design purposes besides morphology control, as many crystals of technical importance are made at high temperatures to obtain the desired crystal modification. The same may be achieved by a polymer in water via a kinetic metastable phase if the functional block has the appropriate functional pattern, which requires a modular synthesis strategy towards DHBCs.

DHBCs can show strong interactions with water and can even change its structure in the liquid and solid state, as can be evidenced by water density, viscosity and additional enthalpic transitions close to the freezing point and changes of the ice unit cell [259]. Calorimetric investigations of the binding of these DHBCs to calcite surfaces revealed that the water structure plays an important but not yet fully recognized role in the control of the additive controlled mineralization process [260].

Another remarkable feature of DHBCs is their ability for chiral recognition of crystal enantiomers, provided that the interacting block is modified appropriately. Even a partial racemate separation was possible upon crystallization of racemic calcium tartrate in presence of chiral DHBCs with enantiomeric excess up to 40%, but only via a kinetic pathway [261].

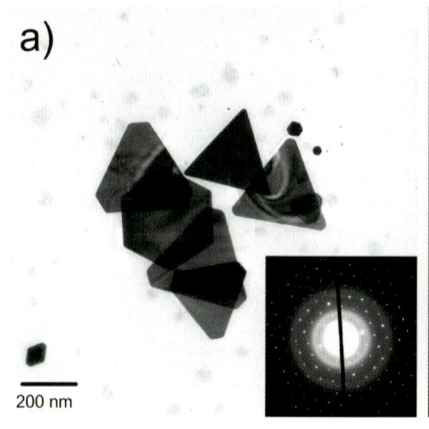

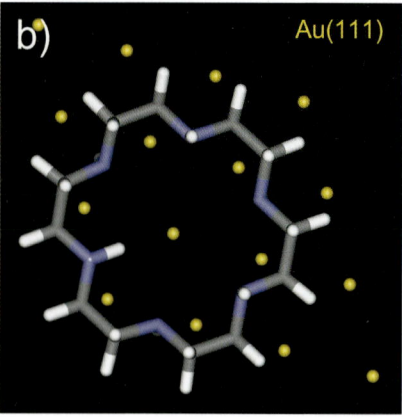

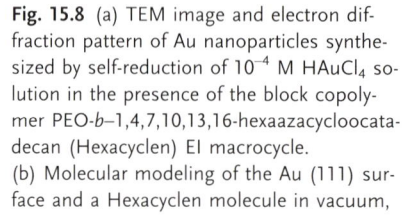

Fig. 15.8 (a) TEM image and electron diffraction pattern of Au nanoparticles synthesized by self-reduction of 10^{-4} M $HAuCl_4$ solution in the presence of the block copolymer PEO-b-1,4,7,10,13,16-hexaazacyclooctadecan (Hexacyclen) EI macrocycle. (b) Molecular modeling of the Au (111) surface and a Hexacyclen molecule in vacuum, which show the perfect match of this molecule to the hexagonal atom arrangement on Au (111). Yellow: Au; blue: N; gray: C; white: H. Figure drawn to scale. Reprinted with permission from S. H. Yu, H. Cölfen, Y. Mastai (2004) J. Nanosci. Nanotechnol. 4, 291. © American Scientific Publishers, http://www.aspbs.com.

Stabilization of specific crystal planes Recent reports show that DHBCs can be used for the stabilization of specific planes of some crystals for their oriented growth such as Au [262], ZnO [263, 265], $CaC_2O_4 \cdot xH_2O$ [266], $PbCO_3$ [267], $CdWO_4$ [268] and $BaSO_4$ [269]. If a DHBC is specifically adsorbed on a crystal face, the surface energy of this face will be lowered and the face becomes expressed upon further crystal growth, as this minimizes the total surface energy of the crystal. To achieve this, it can be expected that a geometric match of the polymer to a crystal surface needs to be established. This was demonstrated for Au nanoparticles with shapes such as triangles, truncated triangles or hexagons and a small fraction of smaller nearly spherical nanoparticles obtained in the presence of a PEO end-functionalized with a 1,4,7,10,13,16-hexaazacyclooctadecane (Hexacyclen) EI macrocycle, which selectively adsorbed on the Au (111) face due to a good geometric match, as revealed by computer modeling (Fig. 15.8) [262].

The same concept was applied for the control of ZnO morphology by DHBC adsorption. Addition of PEO-b-PMAA resulted in the strong decrease in the particle size distribution width and decreasing crystal lengths, suggesting a good stabilization and selective adsorption effects of the DHBCs for retardation of crystal growth perpendicular to (001) faces [263]. Taubert and coworkers showed that more rounded particles with a central grain boundary and a narrow size distribution were also produced in the presence of PEO-b-PMAA [264, 265, 270]. With the more strongly acidic PEO-b-PSS copolymers, a lamellar intermediate precipitate was found at the beginning, then this structure eventually dissolved and hexagonal prismatic crystals were produced. A striking effect was observed

15.4 Double-hydrophilic Block Copolymers (DHBCs) and Double-hydrophilic Graft Copolymers

when a subsequent growth process produced uniform "stack of pancakes"-shaped ZnO crystals. The formation of a metastable intermediate and the release of the polymer from the side faces of the core crystals, and the selective adsorption on the (001) ZnO faces display a collective effect in the determination of such "stack of pancakes" morphology [270]. A similar "stack of pancakes" morphology was also observed for $CaCO_3$ in an investigation to check if an epitaxial match between a PEO-b-Hexacyclen DHBC is necessary for selective adsorption on a specific crystal face [271]. However, the experimental data revealed that it is not only an epitaxial match which drives the polymer adsorption to a specific crystal face, but also charge/ion surface density, particle stabilization and the time for polymer rearrangement if it is initially adsorbed on an energetically less favorable face.

PEO-b-PMAA was found to favor the morphological transition of calcium oxalate dihydrate (COD) crystals from the default tetragonal bipyramids dominated by the (101) faces to elongated tetragonal prisms dominated by the (100) faces, which could be due to the preferential adsorption of the polymer on the (100) face of COD crystals. With increasing polymer concentration, the morphology of the COD crystals obtained gradually changed from tetragonal bipyramids dominated by the (101) faces to rod-like tetragonal prisms dominated by the (100) faces, which is a morphology adopted by some plant COD crystals but not previously obtained *in vitro* [266]. PEO-b-PMAA can also be used to stabilize a single crystal face of $PbCO_3$ crystals [267]. Disk-like nanoplatelets were formed due to selective polymer adsorption on the (001) face. Phosphonation of the precursor DHBC led to permanent stabilization of the interesting $PbCO_3$ platelet-like intermediates and a stiff PEO-b-{[2-(4-dihydroxyphosphoryl)-2-oxabutyl] acrylate ethyl ester} led to flat, uniaxially elongated quasi-hexagonal and single-crystalline nanoplates with a smooth surface due to the polymer adsorption on the (001) face [267]. Selective polymer adsorption can even lead to forbidden crystal symmetries, as demonstrated for flower-like $BaSO_4$ particles with a forbidden 10-fold symmetry, which can be produced by a multi-step growth process in the presence of a sulfonated PEO-b-PEI [269]. Flower-like $CaCO_3$ particles were also obtained in the presence of a DHGC with a PEO-grafted PAA block and a PMAA block [173].

In a complex and synergetic process, face-selective adsorption of PEO-b-PMAA with every second PMAA unit derivatized with a C_{12} chain to all hydroxyapatite (HAP) faces parallel to the HAP *c*-axis led to the development of very thin HAP whiskers, which, however, exhibited several branches as a probable result of inhomogeneous polymer adsorption (Fig. 15.9) [46].

The remarkable feature of this system is its synergetic nature. The polymer forms a loose aggregate in water due to the hydrophobic moieties. As the polymer controls HAP crystal growth along the *c*-axis by selective adsorption, but on the other hand the polymer aggregate structure becomes deformed by the growing crystal fibers, a feedback loop between polymer-controlled crystallization and crystallization-induced polymer aggregate deformation is established (Fig. 15.9) [46]. Interestingly, a similar nested structure was observed for calcium phos-

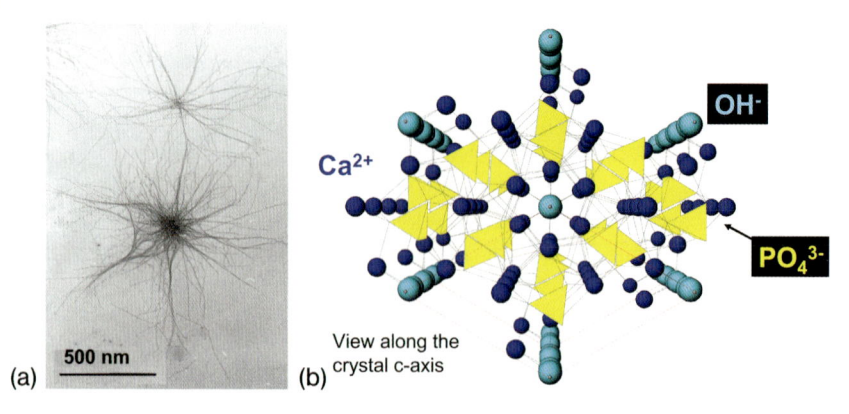

Fig. 15.9 (a) HAP whisker structures prepared within aggregates of C_{12} derivatized PEO-b-PMAA [46]. (b) View along the HAP crystal c-axis. The phosphate tetrahedra are shielded in this view so that adsorption of a negatively charged polymer to all faces parallel to the crystal c-axis is favored.

phate in the brushite modification precipitated in presence of the unmodified PEO-b-PMAA at low pH, which indicates that this complex morphology is not limited to HAP [132].

DHBC-controlled self assembly of mineral nanoparticles The controlled self-assembly of nanoparticles is certainly the most striking possibility for generating a complex crystal morphology. The flexible DHBC design enables one to design the stabilizing block short enough that no permanent stabilization of a crystalline nanoparticle is achieved, but only a temporary one. The simplest possibility to exploit this principle is the arrangement of crystalline nanoparticles around an external template, which can be as simple as a gas bubble. This was demonstrated for $CaCO_3$ with complex morphologies tunable by the surface tension of the crystallization solution [165, 258]. The template can also be a self-assembled structure of DHBC-coated metastable nanoparticles, which is sacrificed and dissolved to recrystallize into the thermodynamically stable polymorph. This was demonstrated for hollow spheres composed of calcite rhombohedra, which were growing on the surface of a sacrificial aggregate of vaterite nanoparticles formed in the presence of PEO-b-PMAA [267]. It is remarkable that hollow calcite spheres but with less expressed surface rhombohedra have also been reported when the same block copolymer was used together with sodium dodecyl sulfate (SDS) surfactant [272]. The inner part of the hollow structures obtained via sacrificial vaterite dissolution also consisted of nanoparticles.

The situation becomes even more complex if DHBCs are applied in interaction with other molecules such as surfactants, forming micelles or other superstructures, as shown for the PEO-b-PMAA–SDS system as control additive for $CaCO_3$ crystallization, which created further handles for the generation of new,

15.4 Double-hydrophilic Block Copolymers (DHBCs) and Double-hydrophilic Graft Copolymers

complex crystal morphologies such as hollow spheres, disks and pine cone-shaped $CaCO_3$ particles [273] or hollow silver spheres [274].

Another well-understood process exploits the face-selective DHBC adsorption with directed self-assembly in a process called "oriented attachment", as first described for the additive free crystallization of TiO_2 with subsequent fusion of the nanoparticles formed into chain-like single crystals [275]. In this process, nanoparticles crystallographically fuse together with their high energy faces, thus minimizing the surface energy of the crystal, leading to a highly directed particle fusion to an elongated single crystal. This was also found for the isomorphic $BaSO_4$ and $BaCrO_4$ systems. $BaSO_4$ and $BaCrO_4$ are often applied as mineral model systems as they only crystallize in the barite and hashemite modification, respectively, and the default platelet morphology in the case of $BaSO_4$ is not affected over the entire pH range so that the large influence of the DHBC functional group pattern on the final crystal morphology becomes evident [276–279].

If a phosphonated DHBC (PEO-b-PMAA–PO_3H_2) is adsorbed on all faces of primarily formed nanoparticles except (210) and its counter face, which do not exhibit attackable Ba^{2+} ions [50], primary nanoparticle building units are obtained, which exhibit only two high-energy faces [278, 280]. These primary nanoparticles undergo oriented attachment by crystallographic fusion of the (210) faces yielding defect-free single-crystalline fibers of about 30 nm in diameter and up to hundreds of micrometers in length without any branches as found for the HAP whiskers, which were grown via face-selective polymer adsorption (see above and [46]). The single-crystalline fibers form bundles of associated but individual fibers, which are likely interspaced by a PEO layer [278, 280]. It is not possible to obtain individual fibers as a result of the fiber attraction along their long axis, which can only be suppressed by adsorption of colloids on the sides of the fibers acting as a steric barrier for fiber bundle formation [281].

The fibrous structures always grow from a single starting point and from an aggregate of amorphous precursor particles, implying that they are nucleated on the glass wall or other substrates such as TEM grids, which obviously provide the necessary heterogeneous nucleation sites. In contrast, only spherical particles were obtained when the same reaction was carried out in polypropylene (PP) or plastic bottles [282]. The growth front of the fiber bundles is always very smooth, suggesting the homogeneous joint growth of all single nanofilaments with the ability to cure occurring defects, in line with the earlier findings for $BaSO_4$ [280]. Cones can also be formed and the opening angle of the cones is always rather similar, but depends on temperature, degree of phosphonation of the polymer and the polymer concentration [278]. Interestingly, secondary cones can nucleate from either the rim or defects on the cone. The cone-like superstructures tend to grow further into a self-similar, multi-cone "tree" or cone-in-a-cone "matrioshka" structures [278]. PEO-b-PMAA also appeared to induce the formation of Ag nanowires by oriented attachment, although the detailed formation mechanism could not be fully revealed [283].

It is interesting to note that a mixture of DHBCs, each of which results in defined $BaSO_4$ superstructures such as spheres (with a PEO-b-PMAA additive)

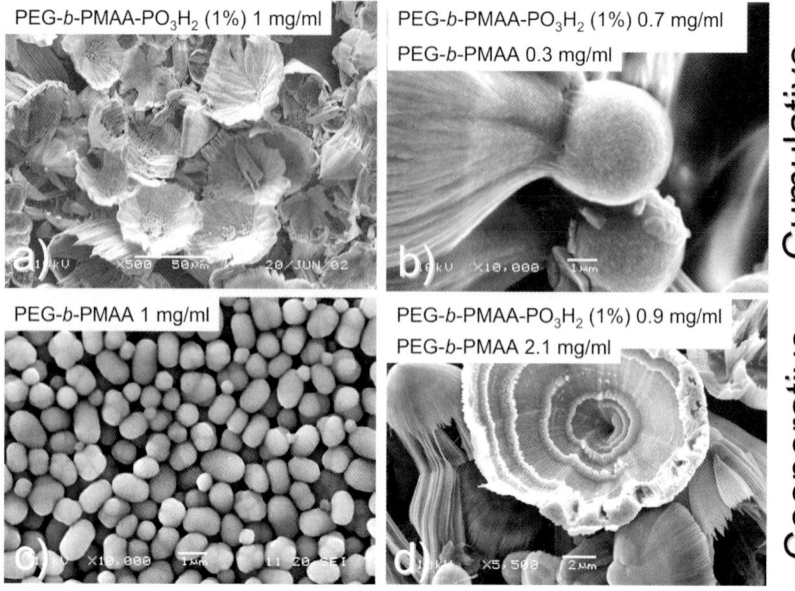

Fig. 15.10 BaSO$_4$ precipitated under the control of PEO-*b*-PMAA (c), its partially phosphonated derivative (a) and their mixtures at different total polymer concentration and mixing ratio. Reproduced from [284] with permission of the Royal Society of Chemistry.

and fibers (with its partially phosphonated derivative), can result in cumulative or cooperative morphogenesis mechanisms, depending on the DHBC mixing ratio and the total concentration (Fig. 15.10) [284].

For example, shuttlecock morphologies could be obtained by secondary fiber cone nucleation on initially precipitated spherical microparticles in a cumulative mechanism in a predictable manner, whereas a cooperative mechanism resulted in new complex fiber-related morphologies, which were not predictable from the functions of the individual DHBCs and depended strongly on the mixing ratio of the two functional DHBCs [284]. These results show that the range of morphologies obtainable can be significantly enhanced by mixing of functional DHBCs.

Whereas the arrangement of crystalline nanoparticles around templates or the DHBC-controlled oriented attachment are understandable, many self-organization processes of DHBC-coated nanoparticles are poorly or not yet understood. One of these morphogenesis sequences is certainly the rod–dumbbell–sphere transition, as first reported in detail for the growth of fluoroapatite in gelatin gels by Kniep and Busch [285, 286], which was observed for CaCO$_3$ [31, 32], but also for BaCO$_3$, MnCO$_3$ and CdCO$_3$ in the presence of DHBCs – often PEO-*b*-PMAA [267].

In the fluoroapatite case, rod-like particles are formed first, which can grow at their ends, resulting in dumbbell-like particles due to a dendritic growth [285,

286]. These dumbbells further grow into spherical particles. This morphology evolution process is qualitatively in agreement with the dumbbells and spheres and also the hollow spheres observed with $CaCO_3$, mineralized in presence of DHBCs. Here, however, the complex morphologies appear to result from a nanoparticle aggregation mechanism as WAXS data evaluated by the Scherrer equation reveal small primary nanoparticle building units with sizes around 20 nm [47]. A time-resolved electron microscopy study indeed revealed that dumbbells are the precursors to the much bigger spherical particles and that amorphous precursor nanoparticle aggregates are present in the initial experimental stages, which most likely act as a material depot [287]. This morphogenesis sequence has already been the subject of intense small-angle neutron scattering (SANS) investigations to characterize the structure formation on different size levels [289–291].

It appears that the polymer additive not only plays a structure-directing role, but also a control role for the generation and release of building material, consisting of amorphous particles. A variety of $CaCO_3$ intermediates of the rod–dumbbell–sphere mechanism with shapes such as rods, small spheres, peanuts, dumbbells and overgrown larger spherules, were also observed in the presence of PEO-b-PEIPA [292] and also in presence of mixed alcohol–water solvents in the presence of PEO-b-PMAA [293].

Despite the rod–dumbbell–sphere morphogenesis sequence, other morphologies were also obtained in the presence of DHBCs, but the exact role of the polymer and the self-organization mechanism of nanoparticles still remain unknown and sometimes it is not clear whether the observed crystals are aggregates of nanocrystals or not. Examples of further $CaCO_3$ morphologies include spheres [47, 257, 294, 295], hollow spheres [47, 153], twins and dumbbells [47], cocoon-like $CaCO_3$ crystals [294] and elongated overgrown calcite rhombohedra [296]. Some investigation of the very early stages in such DHBC-controlled $CaCO_3$ crystallization was possible by time-resolved small-angle X-ray scattering, showing that the DHBC architecture delays the aggregation of the primarily formed amorphous nanoparticles [297]. Interestingly, the random conformation of the functional block turned out to be more efficient in $CaCO_3$ crystallization control than its organized α-helix or β-sheet counterparts [294], which is in agreement with the findings in the literature about biomineralization proteins. These proteins tend to adopt the unfolded structures for their interacting motifs [29].

It is remarkable that a unifying overall scenario for the $CaCO_3$–PEO-b-PMAA system could be established, which allowed for an empirical morphology prediction if the pH and the polymer:mineral concentration ratio is known [287]. This result shows that already one particular polymer–mineral system can be tuned through a variety of particle morphologies, which are predominantly nanocrystal superstructures with complex morphologies, and that a uniform growth mechanism for several of these superstructures can be assumed. Therefore, it is possible to predict the crystal morphology for this system on an empirical basis by the choice of the appropriate experimental conditions.

A large variety of crystal morphologies have also been obtained for other crystal and DHBC systems such as $BaSO_4$ fibers, peanut-shaped and flower-like particles [276, 277] and also lozenge shapes [279], rod-like CaC_2O_4 particles [298], $BaHPO_4$ with irregular hexagonal or spherical shape [299], plate-like aggregates, ellipsoids and monodisperse $CaCO_3$ spheres [44] and, as is common for DHBC-controlled crystallization, other experimental parameters such as pH, polymer concentration and temperature play a morphology-determining role [277, 287]. This indicates the principal accessibility of many mineral systems for DHBC-controlled crystallization.

However, the most demanding task is to address selectively defined faces of nanocrystals to trigger their self-assembly into a defined superstructure. This also raises the question of whether crystallization only occurs via the addition of ions/molecules or can also be particle mediated with possible subsequent formation of single crystals [301, 302]. Striking examples were recently reported for a PEO-b-PEI, which was functionalized with an isobutyric acid group to control the crystallization of DL-alanine crystals [303], and also for a stiff phosphonated DHBC with PEO, which controlled the crystallization of $BaCO_3$ [304]. In the first case, selective polymer adsorption on the DL-alanine (001) face led to the development of platelets, which had a dipole moment along the crystal c-axis due to the crystal structure of DL-alanine. This finally resulted in a self-assembly of the primary nanoplatelets with interspacing polymer layers resulting in a highly aligned superstructure with external faces, as shown in Fig. 15.11 and evidenced by XRD analysis [303]. The nanoplatelets are almost perfectly aligned in three dimensions, as is demonstrated by a thin slice of the (010) face

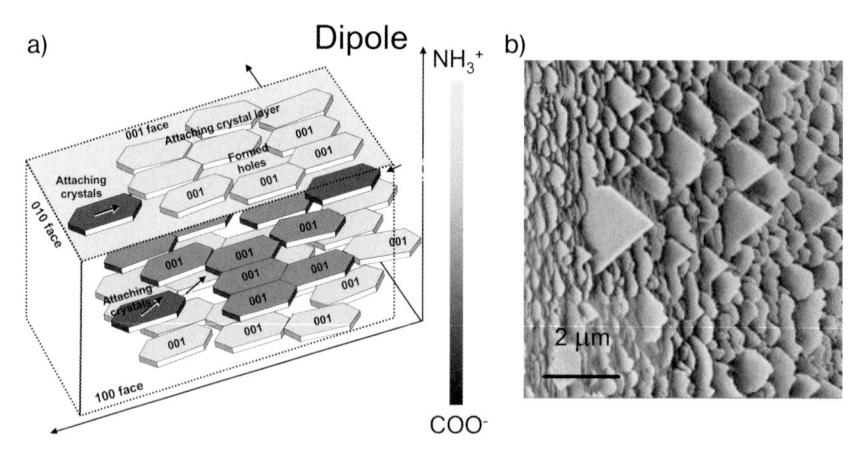

Fig. 15.11 (a) Schematic drawing of the self-structuring of primary nanocrystals to a mesocrystal for DL-alanine, with the externally exposed faces. Note that the rough face is (010), which exposes the nanocrystal platelet tips. Along the a and b directions, partial particle fusion occurs by oriented attachment. For simplicity, the polymer is not shown. (b) SEM image of DL-alanine crystals formed by crystallization of a supersaturated solution by the addition of the chiral anionic DHBC showing the semicut mesostructure as zoom. Reproduced from [303] with kind permission of Wiley-VCH.

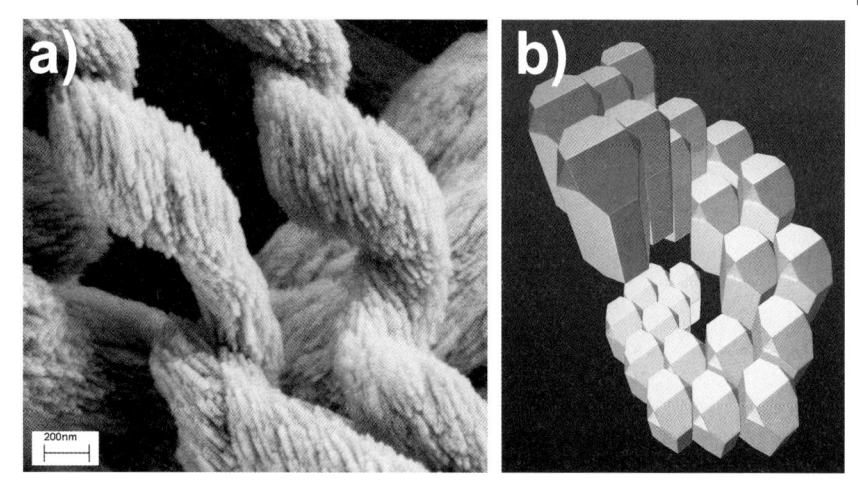

Fig. 15.12 (a) Tectonic arrangement of BaCO$_3$ nanocrystals into helices induced by a racemic DHBC and (b) schematic diagram of the nanoparticle assembly process into helices.

shown in Fig. 15.11b. This is supportive of particle-mediated crystallization processes as it is possible to fuse the nanoparticle building units together to form a single crystal.

However, nanoparticles can also be coded by face-selective adsorption of a sterically demanding DHBC so that the self-assembly of the primary crystals can be controlled by steric effects rather than directing dipole fields [304]. Strikingly, the crystallization of BaCO$_3$ in the presence of a stiff phosphonated DHBC yielded helical superstructures of elongated primary nanoparticles in approximate equal proportions between left- and right-handed helices [304] (Fig. 15.12a).

The two examples above show the high level of control over crystallization processes that can be achieved by using DHBC additives. Although, certainly, Nature's complexity is not yet reached with such bioinspired mineralization structures, already much can be learned about the mechanisms of how to arrange nanoparticles in a controlled manner by self-organization over several levels.

15.5
Conclusion

DHBCs are a class of amphiphilic diblock copolymers, which mimic biomineralization polymers in a very simplified way. The virtue of DHBCs lies in their wide applicability in aqueous systems, which makes them well suited for medical and environmentally friendly applications. A common feature of DHBCs is that their amphiphilicity in water can be induced by an external stimulus such as an inorganic surface, a polyelectrolyte, a temperature or pH change, etc., which leads to superstructure formation. The trigger limit for the induction of amphiphilicity can often be fine tuned via the molecular architecture so that DHBCs can be designed for applications such as selective stabilization or temperature-responsive drug carriers. One of the greatest advantages of DHBCs is their flexible design due to the separation of a functional and stabilizing unit so that they can even be tailor-made to interact with specific crystal faces. This offers a powerful handle for the controlled change of crystal morphology by face-selective DHBC adsorption followed by anisotropic growth, oriented attachment or controlled self-organization.

The toolbox of polymer chemistry is more or less available for DHBC synthesis and various strategies can be applied for the design of a tailor-made molecule. The synthesis toolbox is not limited only to living polymerization, as monodispersity is not a requirement for successful application, so that larger scale synthesis of DHBCs or DHGCs seems promising, for example via coupling of two homopolymer blocks or grafting a stabilizing block on to a polyelectrolyte backbone. DHBC applications as diverse as drug carriers and crystal growth modifiers show the versatility of DHBCs in aqueous systems and it is likely that more applications will follow, as DHBCs can be tailor-designed for a variety of substrates. This example shows that abstracting functional units from natural polymer archetypes can yield powerful synthetic model systems, which not only open up a whole variety of new applications, but also might be of help in understanding the complicated role of the natural archetypes in biomineralization processes.

List of Abbreviations

These abbreviations are predominantly adapted from the primary literature. Monomer abbreviations are listed with respect to the corresponding polymers (e.g. EO instead of PEO for the polymer). The standard three-letter amino acid abbreviations are used.

1ME4VPCl	1-methyl-4-vinylpyridinium chloride
2VP	2-vinylpyridine
4VP	4-vinylpyridine
AA	acrylic acid
AAM	acrylamide
ADR	adriamycin
AN	acrylonitrile
ANS	8-anilino-1-naphthalenesulfonic acid
Asp	aspartic acid
ATRP	atom transfer radical polymerization
BA	butyl acrylate
BzDMA	benzyl[2-(methacryloyloxy)ethyl]dimethylammonium chloride
BzMA	benzyl methacrylate

BzVE	benzyl vinyl ether
COD	calcium oxalate dihydrate
CTA	cetyltrimethylammonium chloride
DCC	dicyclocarbodiimide
DDDAB	didodecyldimethylammonium bromide
DEAEMA	diethylaminoethyl methacrylate
DHBC	double hydrophilic block copolymer
DHGC	double hydrophilic graft copolymer
DLS	dynamic light scattering
DMAEMA	dimethylaminoethyl methacrylate
DMEABr	2-(dimethylethylammonio)ethyl methacrylate bromide
DMF	dimethylformamide
DMT	4,4′-dimethoxytrityl
DODAB	dimethyldioctadecylammonium bromide
DPAEMA	2-(diisopropylamino)ethyl methacrylate
EDTA	ethylenediaminetetraacetic acid
EGDMA	ethylene glycol dimethacrylate
EI	ethylenimine
EO	ethylene oxide
EOVE	2-ethoxyethyl vinyl ether
EOX	ethyloxazoline
EVE	ethyl vinyl ether
GAG	glycoaminoglycan
Glu	glutamic acid
GMA	glycerol monomethacrylate
GTP	group transfer polymerization
HAA	2-hydroxyethylaspartamide
HAP	hydroxyapatite
HEA	hydroxyethylacrylate
HEGMA	hexa(ethylene glycol) methacrylate
HEMA	2-hydroxyethyl methacrylate
HMA	dihydroxypropyl methacrylate
HPLC	high-performance liquid chromatography
HPMA	N-(2-hydroxypropyl)methacrylamide
IBVE	isobutyl vinyl ether
IEP	isoelectric point
LCST	lower critical solution temperature
Lys	lysine
MAA	methacrylic acid
MANa	methacrylic acid sodium salt
MeDMA	2-(methacryloyloxy)ethyltrimethylammonium chloride
MEMA	2-(N-morpholinoethyl) methacrylate
MMA	methyl methacrylate
MOVE	2-methoxyethyl vinyl ether
MOX	methyloxazoline

MPC	2-(methacryloyloxy)ethylphosphorylcholine
MTEGVE	methyl tri(ethylene glycol) vinyl ether
MTS	1-methoxy-1-trimethylsilyloxy-2-methyl-1-propene
MTX	methotrexate
MVE	methyl vinyl ether
NaVBA	sodium-4-vinylbenzoate
NCA	N-carboxyanhydride
NiPAAm	N-isopropylacrylamide
OEGMA	oligo(ethylene glycol) methacrylate
PAMPS	poly(2-acrylamido-2-methyl-1-propanesulfonic acid)
PB	polybutadiene
PDADMAC	poly(diallyldimethylammonium chloride)
PEIPA	poly(ethylenimine)-poly(acetic acid)
PIC	polyion complex
PP	polypropylene
PVCL	poly(N-vinylcaprolactam)
SANS	small-angle neutron scattering
SBMA	[2-(methacryloyloxy)ethyl]dimethyl(3-sulfopropyl)ammonium hydroxide
SCOONa	sodium 4-styrenecarboxylate
SDS	sodium dodecyl sulfate
SP	spermine
SS	styrene sulfonate
TBABB	tetrabutylammonium bibenzoate
TBMA	*tert*-butyl methacrylate
TEM	transmission electron microscopy
TEMPO	2,2,6,6-tetramethyl-1-piperidinyloxy
THF	tetrahydrofuran
THPMA	tetrahydropyranyl methacrylate
TMAB	trioctylmethylammonium bromide
TMAEM	N-trimethylammonium ethyl methacrylate chloride
TMATf	2-(trimethylammonio)ethyl methacrylate trifluoromethanesulfonate
VA	vinyl alcohol
VBA	vinylbenzyl alcohol
VOBA	vinyloxy-4-butyric acid
WAXS	wide-angle X-ray scattering
XRD	X-ray diffraction

References

1. R. Wood, J. P. Grotzinger, J. A. D. Dickson, *Science* **2002**, *296*, 2383–2386.
2. S. Weiner, *CRC Crit. Rev. Biochem.* **1986**, *20*, 365–408.
3. B. L. Smith, T. E. Schäffer, M. Viani, J. B. Thompson, N. A. Frederick, J. Kindt, A. Belcher, G. D. Stucky, D. E. Morse, P. K. Hansma, *Nature* **1999**, *399*, 761–763.
4. S. Weiner, H. D. Wagner, *Annu. Rev. Mater. Sci.* **1998**, *28*, 271–298.
5. N. Gehrke, N. Nassif, N. Pinna, M. Antonietti, H. S. Gupta, H. Cölfen, *Chem. Mater.* **2005**, *17*, 6514–6516.
6. G. Luquet, F. Marin, *C.R. Palevol.* **2004**, *3*, 515–534.
7. H. Inoue, N. Ozaki, H. Nagasawa, *Biosci. Biotechnol. Biochem.* **2001**, *65*, 1840–1848.
8. H. Inoue, T. Ohira, N. Ozaki, H. Nagasawa, *Comp. Biochem. Physiol.* **2003**, *136B*, 755–765.
9. H. Inoue, T. Ohira, N. Ozaki, H. Nagasawa, *Biochem. Biophys. Res. Commun.* **2004**, *318*, 649–654.
10. G. Falini, S. Weiner, L. Addadi, *Calc. Tissue Int.* **2003**, *72*, 548–554.
11. J. R. Long, W. J. Shaw, P. S. Stayton, G. P. Drobny, *Biochemistry* **2001**, *40*, 15451–15455.
12. D. H. Schlesinger, D. I. Hay, *J. Biol. Chem.* **1977**, *252*, 1689–1695.
13. D. I. Hay, D. J. Smith, S. K. Schluckebier, E. C. Moreno, *J. Dent. Res.* **1984**, *63*, 857–863.
14. P. A. Raj, E. Marcus, D. K. Sukumaran, *Biopolymers* **1992**, *45*, 51–67.
15. F. Marin, G. Luquet, *C.R. Palevol.* **2004**, *3*, 469–492.
16. N. L. Huq, K. J. Cross, M. Ung, E. C. Reynolds, *Arch. Oral Biol.* **2005**, *50*, 599–609.
17. F. Marin, R. Amons, N. Guichard, M. Stigter, A. Hecker, G. Luquet, P. Layrolle, G. Alcaraz, C. Riondet, P. Westbroek, *J. Biol. Chem.* **2005**, *280*, 33895–33908.
18. I. Sarashina, K. Endo, *Am. Mineral.* **1998**, *83*, 1510–1515.
19. I. Sarashina, K. Endo, *Mar. Biotechnol.* **2001**, *3*, 362–369.
20. P. K. Dutta, J. Dutta, V. S. Tripathi, *J. Sci. Ind. Res.* **2004**, *63*, 20–31.
21. P. Jolles, R. A. A. Muzzarelli (Eds.), *Chitin and Chitinases*, Birkhäuser, Basel, 1999.
22. http://biol.lancs.ac.uk/gig/pages/toppage.htm.
23. J. L. Arias, A. Neira-Carrillo, J. I. Arias, C. Escobar, M. Bodero, M. David, M. S. Fernandez, *J. Mater. Chem.* **2004**, *14*, 2154–2160.
24. V. Gerbaud, D. Pignol, E. Loret, J. A. Bertrand, Y. Berlandi, J. C. Fontecilla-Camps, J. P. Canselier, N. Gabas, J. M. Verdier, *J. Biol. Chem.* **2000**, *275*, 1057–1064.
25. C. E. Kilian, F. H. Wilt, *J. Biol. Chem.* **1996**, *271*, 9150–9159.
26. F. H. Wilt, *J. Struct. Biol.* **1999**, *126*, 216–226.
27. M. MacDougall, D. Simmons, X. Luan, J. Nydegger, J. Feng, T. T. Gu, *J. Biol. Chem.* **1997**, *272*, 835–842.
28. K. Mann, I. M. Weiss, S. Andre, H. J. Gabius, M. Fritz, *Eur. J. Biochem.* **2000**, *267*, 5257–5264.
29. J. S. Evans, *Curr. Opin. Colloid Interface Sci.* **2003**, *8*, 48–54.
30. Z. Jia, P. L. Davies, *Trends Biochem. Sci.* **2002**, *27*, 101–106.
31. D. S. C. Yang, W. C. Hon, S. Bubanko, Y. Xue, J. Seetharaman, C. J. Hew, F. Sicheri, *Biophys. J.* **1998**, *74*, 2142–2151.
32. Z. Jiaz, C. I. DeLuca, H. Chao, P. L. Davies, *Nature* **1996**, *384*, 285–288.
33. F. Marin, P. Corstjens, B. De Gaulejac, E. De Vrind-De Jong, P. Westbroek, *J. Biol. Chem.* **2000**, *275*, 20667–20675.
34. J. L. Arias, D. A. Carrino, M. S. Fernandez, J. P. Rodriguez, J. E. Dennis, A. I. Caplan, *Arch. Biochem. Biophys.* **1992**, *298*, 293–302.
35. M. S. Fernandez, M. Araya, J. L. Arias, *Matrix Biol.* **1997**, *16*, 13–20.
36. M. S. Fernandez, A. Moya, L. Lopez, J. L. Arias, *Matrix Biol.* **2001**, *20*, 793–803.
37. J. E. Dennis, D. A. Carrino, K. Yamashita, A. I. Caplan, *Matrix Biol.* **2000**, *19*, 683–692.

38 G. Riess, P. Bahadur, Block copolymers, in *Encyclopedia of Polymer Science and Engineering*, M. F. Mark, N. M. Bikales, C. G. Overberger, G. Menges (Eds.), Wiley, New York, Vol. 2, pp. 324ff.
39 F. S. Bates, G. H. Fredrickson, *Annu. Rev. Phys. Chem.* **1990**, *41*, 525–557.
40 N. Nishiyama, K. Kataoka, *Adv. Polym. Sci.* **2006**, *193*, 67–101.
41 K. Osada, K. Kataoka, *Adv. Polym. Sci.*, **2006**, *202*, 113–153.
42 H. Cölfen, *Macromol. Rapid Commun.* **2001**, *22*, 219–252.
43 M. Kamachi, M. Kurihara, J. K. Stille, *Macromolecules* **1972**, *5*, 167.
44 A. V. Kabanov, V. A. Kabanov, *Adv. Drug Deliv. Rev.* **1998**, *30*, 49–60.
45 V. Bütün, N. C. Billingham, S. P. Armes, *J. Am. Chem. Soc.* **1998**, *120*, 11818–11819.
46 M. Antonietti, M. Breulmann, C. G. Göltner, H. Cölfen, K. K. W. Wong, D. Walsh, S. Mann, *Chem. Eur. J.* **1998**, *4*, 2493–2500.
47 H. Cölfen, M. Antonietti, *Langmuir* **1998**, *14*, 582–589.
48 L. M. Bronstein, S. N. Sidorov, A. Y. Gourkova, P. M. Valetsky, J. Hartmann, M. Breulmann, H. Cölfen, M. Antonietti, *Inorg. Chim. Acta* **1998**, *280*, 348–354.
49 D. H. Napper, *Polymeric Stabilization of Colloidal Dispersions*, Academic Press, London, **1989**.
50 S. H. Yu, H. Cölfen, *J. Mater. Chem.* **2004** *14*, 2124–2147.
51 E. Seiler, G. Fahrenbach, D. Stein (BASF AG), Ger. Pat. *2237954*, 1972; *Chem. Abstr.* **1974**, *81*, 13978x.
52 H. Reuter, I. V. Berlinova, S. Höring, J. Ulbricht, *Eur. Polym. J.* **1991**, *27*, 673–680.
53 T. Suzuki, Y. Murakami, Y. Takegami, *J. Polym. Sci., Polym. Lett. Ed.* **1979**, *17*, 241–244.
54 J. Wang, S. K. Varshney, R. Jerome, P. Teyssie, *J. Polym. Sci., Part A* **1992**, *30*, 2251–2261.
55 S. Förster, E. Krämer, *Macromolecules* **1999**, *32*, 2783–2785.
56 S. Creutz, P. Teyssie, R. Jerome, *Macromolecules* **1997**, *30*, 6–9.
57 S. Creutz, R. Jerome, G. M. P. Kaptijn, A. W. van der Werf, J. M. Akkerman, *J. Coat. Technol.* **1998**, *70*, 41–46.
58 J. Allgaier, A. Poppe, L. Willner, D. Richter, *Macromolecules* **1997**, *30*, 1582–1586.
59 T. J. Martin, K. Prochazka, P. Munk, S. E. Webber, *Macromolecules* **1996**, *29*, 6071–6073.
60 N. G. Hoogeveen, M. A. C. Stuart, G. J. Fleer, W. Frank, M. Arnold, *Macromol. Chem. Phys.* **1996**, *197*, 2553–2564.
61 K. Kataoka, A. Harada, D. Wakebayashi, Y. Nagasaki. *Macromolecules*, **1999**, *32*, 6892–6894.
62 A. B. Lowe, N. C. Billingham, S. P. Armes, *Macromolecules* **1998**, *31*, 5991–5998.
63 C. S. Patrickios, A. B. Lowe, S. P. Armes, N. C. Billingham, *J. Polym. Sci., Part A: Polym. Chem.* **1998**, *36*, 617–631.
64 V. Bütün, N. C. Billingham, S. P. Armes, *Chem. Commun.* **1997**, 671–672.
65 V. Bütün, C. E. Bennett, M. Vamvakaki, A. P. Lowe, N. C. Billingham, S. P. Armes, *J. Mater. Chem.* **1997**, *7*, 1693–1695.
66 V. Bütün, M. Vamvakaki, N. C. Billingham, S. P. Armes, *Polymer*, **2000**, *41*, 3173–3182.
67 R. C. Schulz, M. Schmidt, E. Schwarzenbach, J. Zöller, *Makromol. Chem., Makromol. Symp.* **1989**, *26*, 221–231.
68 S. Creutz, P. Teyssie, R. Jerome, *Macromolecules* **1997**, *30*, 6–9.
69 M. Ossenbach-Sauter, G. Riess, *C. R. Acad. Sci. Paris, Ser. C* **1976**, *283*, 269–272.
70 M. Kurihara, M. Kamachi, J. K. Stille, *J. Polym. Sci.* **1973**, *11*, 587–610.
71 N. P. Briggs, P. M. Budd, C. Price, *Eur. Polym. J.* **1992**, *28*, 739–745.
72 S. Creutz, R. Jerome, *Langmuir* **1999**, *15*, 7145–7156.
73 J. Orth, W. H. Meyer, C. Bellmann, G. Wegner, *Acta Polym.* **1997**, *48*, 490–501.
74 A. V. Kabanov, T. K. Bronich, V. A. Kabanov, K. Yu, A. Eisenberg, *Macromolecules* **1996**, *29*, 6797–6802.
75 M. A. Hillmyer, F. S. Bates, *Macromolecules* **1996**, *29*, 6994–7002.
76 S. Aoshima, E. Kobayashi, *Makromol. Chem., Makromol. Symp.* **1995**, *95*, 91–102

77 C. S. Patrickios, C. Forder, S. P. Armes, N. C. Billingham, *J. Polym. Sci., Polym. Chem.* **1997**, *35*, 1181–1195.
78 C. Forder, C. S. Patrickios, S. P. Armes, N. C. Billingham, *Macromolecules* **1996**, *29*, 8160–8169.
79 C. Forder, C. S. Patrickios, S. P. Armes, N. C. Billingham, *Macromolecules* **1997**, *30*, 5758–5762.
80 C. Forder, C. S. Patrickios, N. C. Billingham, S. P. Armes, *Chem. Commun.* **1996**, 883–884.
81 A. W. M. de Laat, H. F. M. Schoo, *Colloid Polym. Sci.* **1998**, *276*, 176–185.
82 W. H. Velander, R. D. Madurawe, A. Subramanian, G. Kumar, G. Sinai-Zingde, J. S. Riffle, C. L. Orthner, *Biotechnol. Bioeng.* **1992**, *39*, 1024–1030.
83 J. Furukawa, T. Saegusa, N. Mise, *Makromol. Chem.* **1960**, *38*, 244–247.
84 S. Aoshima, H. Oda, E. Kobayashi, *Kobunshi Ronbunshu* **1992**, *49*, 937–992.
85 M. K. Pratten, J. B. Lloyd, G. Hörpel, H. Ringsdorf, *Makromol. Chem.* **1985**, *186*, 725–733.
86 A. Harada, S. Cammas, K. Kataoka, *Macromolecules* **1996**, *29*, 6183–6188.
87 K. Kataoka, H. Togawa, A. Harada, K. Yasugi, T. Matsumoto, S. Katayose, *Macromolecules* **1996**, *29*, 8556–8557.
88 K. Tsutsumiuchi, K. Aoi, M. Okada, *Macromolecules* **1997**, *30*, 4013–4017.
89 A. B. Lowe, N. C. Billingham, S. P. Armes, *Chem. Commun.* **1997**, 1035–1036.
90 D. T. Wu, A. Yokoyama, R. L. Setterquist, *Polym. J.* **1991**, *23*, 709–714.
91 Y. Minoura, T. Kasuya, S. Kawamura, A. Nakano, *J. Polym. Sci., Part A* **1967**, *5*, 43–56.
92 K. K. Lee, J. C. Jung, M. S. Jhon, *Polym. Bull.* **1998**, *40*, 455–460.
93 M. D. C. Topp, P. J. Dijkstra, H. Talsma, J. Feijen, *Macromolecules* **1997**, *30*, 8518–8520.
94 M. D. C. Topp, I. H. Leunen, P. J. Dijkstra, K. Tauer, C. Schellenberg, J. Feijen, *Macromolecules* **2000**, *33*, 4986–4988.
95 L. I. Gabaston, S. A. Furlong, R. A. Jackson, S. P. Armes, *Polymer*, **1999**, *40*, 4505–4514.
96 E. J. Ashford, V. Naldi, R. O'Dell, N. C. Billingham, S. P. Armes, *Chem. Commun.* **1999**, 1285–1286.
97 C. I. Simionescu, I. Rabia, *Polym. Bull.* **1983**, *10*, 311–314.
98 J. C. Galin, M. Galin, P. Calme, *Makromol. Chem.* **1970**, *134*, 273–285.
99 J. C. Galin, *Makromol. Chem.* **1969**, *124*, 118–135.
100 D. Garg, S. Höring, J. Ulbricht, *Makromol. Chem. Rapid Commun.* **1984**, *5*, 615–618.
101 M. Vamvakaki, N. C. Billingham, S. P. Armes, *Macromolecules* **1999**, *32*, 2088–2090.
102 T. J. Deming, *Nature* **1997**, *390*, 386–389.
103 K. Tsutsumiuchi, K. Aoi, M. Okada, *Macromol. Rapid Commun.* **1995**, *16*, 749–755.
104 M. Yokoyama, S. Inoue, K. Kataoka, N. Yui, Y. Sakurai, *Makromol. Chem. Rapid Commun.* **1987**, *8*, 431–435.
105 M. Yokoyama, S. Inoue, K. Kataoka, N. Yui, T. Okano, Y. Sakurai, *Makromol. Chem.* **1989**, *190*, 2041–2054.
106 M. Yokoyama, M. Miyauchi, N. Yamada, T. Okano, Y. Sakurai, K. Kataoka, S. Inoue, *Cancer Res.* **1990**, *50*, 1693–1700.
107 M. Yokoyama, G. S. Kwon, T. Okano, Y. Sakurai, T. Sato, K. Kataoka, *Bioconjug. Chem.* **1992**, *3*, 295–301.
108 S. Katayose, K. Kataoka, PEG–poly(lysine) block copolymer as a novel type of synthetic gene vector with supramolecular structure, in *Advanced Biomaterials, in Biomedical Engineering and Drug Delivery Systems*, N. Ogata, S. W. Kim, J. Feijen, T. Okano (Eds.), Springer, Tokyo, **1996**, pp. 319*ff*.
109 A. Harada, K. Kataoka, *Macromolecules* **1995**, *28*, 5294–5299.
110 A. Harada, K. Kataoka, *Science* **1999**, *283*, 65–67.
111 M. Yokoyama, M. Miyauchi, N. Yamada, T. Okano, Y. Sakurai, K. Kataoka, S. Inoue, *J. Control. Release* **1990**, *11*, 269–278.
112 T. Aoyagi, K. I. Sugi, Y. Sakurai, T. Okano, K. Kataoka, *Colloids Surf. B* **1999**, *16*, 237–242.

113 G. Spach, L. Reibel, M. H. Loucheux, J. Parrod, *J. Polym. Sci., Part C* **1969**, *16*, 4705–4712.

114 T. Nishimura, Y. Sato, M. Yokoyama, M. Okuya, S. Inoue, K. Kataoka, T. Okano, Y. Sakurai, *Makromol. Chem.* **1984**, *185*, 2109–2116.

115 K. Kugo, A. Ohji, T. Uno, J. Nishino, *Polym. J.* **1987**, *19*, 375–381.

116 C. S. Cho, S. U. Kim, *J. Control. Release* **1988**, *7*, 283–286.

117 Z. Hruska, G. Riess, P. Goddard, *Polymer* **1993**, *34*, 1333–1335.

118 P. K. Seow, Y. Gallot, A. Skoulios, *Makromol. Chem.* **1975**, *176*, 3153–3166.

119 T. Suzuki, Y. Murakami, Y. Tsuji, Y. Takegami, *J. Polym. Sci., Polym. Lett. Ed.* **1976**, *14*, 675–678.

120 T. Suzuki, Y. Murakami, Y. Takegami, *Polym. J.* **1980**, *12*, 183–192.

121 E. A. Bekturov, V. A. Frolova, S. E. Kudaibergenov, R.C. Schulz, J. Zöller, *Makromol. Chem.* **1990**, *191*, 457–463.

122 S. Creutz, P. Teyssie, R. Jerome, *Macromolecules* **1997**, *30*, 5596–5601.

123 H. Reuter, S. Höring, J. Ulbricht, *Makromol. Chem. Suppl.* **1989**, *15*, 79–84.

124 C. S. Patrickios, W.R. Hertler, N. L. Abbott, T. A. Hatton, *Macromolecules* **1994** *27*, 930–937.

125 V. Bütün, N. C. Billingham, S. P. Armes, *J. Am. Chem. Soc.* **1998**, *120*, 11818–11819.

126 V. Bütün, I. Bannister, N. C. Billingham, D. C. Sherrington, S. P. Armes, *Macromolecules* **2005**, *38*, 4977–4982.

127 S. Holappa, M. Karesoja, J. Shan, H. Tenhu, *Macromolecules* **2002**, *35*, 4733–4738.

128 T. X. Wang, G. Rother, H. Cölfen, *Macromol. Chem. Phys.* **2005**, *206*, 1619–1629.

129 K. Tauer, V. Khrenov, N. Shirshova, N. Nassif, *Macromol. Symp.* **2005**, *226*, 187–201.

130 A. Zintchenko, H. Dautzenberg, K. Tauer, V. Khrenov, *Langmuir* **2002**, *18*, 1386–1393.

131 K. A. Davis, K. Matyjaszewski, *Adv. Polym. Sci.* **2002**, *159*, 1–13.

132 W. Tjandra, J. Yao, P. Ravi, K.C. Tam, A. Alamsjah, *Chem. Mater.* **2005**, *17*, 4865–4872.

133 X. P. Jin, Y. Q. Shen, S. P. Zhu, *Macromol. Mater. Eng.* **2003**, *288*, 925–935.

134 N. V. Tsarevsky, T. Pintauer, K. Matyjaszewski, *Macromolecules*, **2004**, *37*, 9768–9778.

135 Y. T. Li, S. P. Armes, X. P. Jin, S. P. Zhu, *Macromolecules* **2003**, *36*, 8268–8275.

136 M. Bednarek, T. Biedron, P. Kubisa, *Polimery* **2000**, *45*, 664–673.

137 X. S. Wang, R. A. Jackson, S. P. Armes, *Macromolecules* **2000**, *33*, 255–257.

138 M. Bednarek, T. Biedron, P. Kubisa, *Makromol. Rapid Commun.* **1999**, *20*, 59–65.

139 W. Q. Zhang, L. Q. Shi, L. C. Gao, Y. L. An, K. Wu, *Macromol. Rapid Commun.* **2005**, *26*, 1341–1345.

140 S. Hou, E. L. Chaikof, D. Taton, Y. Gnanou, *Macromolecules* **2003**, *36*, 3874–3881.

141 S. H. Luo, J. Xu, Y. F. Zhang, S. Y. Liu, C. Wu, *J. Phys. Chem.* **2005**, *109*, 22159–22166.

142 D. Taton, A. Z. Wilczewska, M. Destarac, *Macromol. Rapid Commun.* **2001**, *22*, 1497–1503.

143 M. Tomoi, Y. Shibayama, H. Kakiuchi, *Polym. J.* **1976**, *8*, 190–195.

144 D. Garg, S. Höring, J. Ulbricht, *Makromol. Chem., Rapid Commun.* **1984**, *5*, 615–618.

145 C. I. Simionescu, I Rabia, Z. Crisan, *Polym. Bull.* **1982**, *7*, 217–222.

146 H. D. Bijsterbosch, M. A. Cohen Stuart, G. J. Fleer, P. van Caeter, E. J. Goethals, *Macromolecules* **1998**, *31*, 7436–7444.

147 Q. Liu, M. Konas, R. M. Davis, J. S. Riffle, *J. Polym. Sci., Part A* **1993**, *31*, 1709–1717.

148 H. Budde, S. Höring, *Macromol. Chem. Phys.* **1998**, *199*, 2541–2546.

149 T. K. Georgiou, M. Vamvakaki, L. A. Phylactou, C. S. Patrickios, *Biomacromolecules* **2005**, *6*, 2990–2997.

150 S. Zalipsky, *Bioconjug. Chem.* **1995**, *6*, 150–165.

151 M. Sedlak, *Collect. Czech. Chem. Commun.* **2005**, *70*, 269–291.

152 M. Sedlak, M. Antonietti, H. Cölfen, *Macromol. Chem. Phys.* **1998**, *199*, 247–254.

153 M. Sedlak, H. Cölfen, *Macromol. Chem. Phys.* **2001**, *202*, 587–597.

154 C. Konak, L. Mrkvickova, O. Nazarova, K. Ulbrich, L. W. Seymour, *Supramol. Sci.* **1998**, *5*, 67–74.
155 D. Oupicky, C. Konak, K. Ulbrich, *Mater. Sci. Eng. C* **1999**, *7*, 59–65.
156 D. Oupicky, C. Konak, K. Ulbrich, *J. Biomater. Sci. Polym. Ed.* **1999**, *10*, 573–590.
157 D. Oupicky, C. Konak, P. R. Dash, L. W. Seymour, K. Ulbrich, *Bioconjug. Chem.* **1999**, *10*, 764–772.
158 M. A. Wolfert, E. H. Schacht, V. Toncheva, K. Ulbrich, O. Nazarova, L. W. Seymour, *Hum. Gene Ther.* **1996**, *7*, 2123–2133.
159 E. Esselborn, J. Fock, A. Knebelkamp, *Macromol. Chem., Macromol. Symp.* **1996**, *102*, 91–98.
160 A. V. Kabanov, S. V. Vinogradov, Y. G. Suzdaltseva, V. Y. Alakhov, *Bioconjug. Chem.* **1995**, *6*, 639–643.
161 S. V. Vinogradov, T. K. Bronich, A. V. Kabanov, *Bioconjug. Chem.* **1998**, *9*, 805–812.
162 M. Yokoyama, H. Anazawa, A. Takahashi, S. Inoue, *Makromol. Chem.* **1990**, *191*, 301–311.
163 E. Pefferkorn, Q. Tran, R. Varoqui, *J. Polym. Sci.* **1981**, *19*, 27–34.
164 S. Pispas, *J. Polym. Sci., Part A: Polym. Chem.* **2006**, *44*, 606–613.
165 J. Rudloff, M. Antonietti, H. Cölfen, J. Pretula, K. Kaluzynski, S. Penczek, *Macromol. Chem. Phys.*, **2002**, *203*, 627–635.
166 M. Tanahashi, T. Yao, T. Kokubo, M. Minoda, T. Miyamoto, T. Nakamura, T. Yamamuro, *J. Appl. Biomater.* **1994**, *5*, 339–347.
167 M. Tanahashi, T. Yao, T. Kokubo, M. Minoda, T. Miyamoto, T. Nakamura, T. Yamamuro, *J. Mater. Sci., Mater. Med.* **1995**, *6*, 319–326.
168 A. M. Nedeltscheva, S. Bencheva, E. Valcheva, V. Valchev, *Colloid Polym. Sci.* **1987**, *265*, 312–317.
169 M. Basko, P. Kubisa, *Macromolecules* **2002**, *35*, 8948–8953.
170 H. Dautzenberg, *Macromol. Chem. Phys.* **2000**, *201*, 1765–1773.
171 S. Verbrugghe, K. Bernaerts, F. E. Du Prez, *Macromol. Chem. Phys.* **2003**, *204*, 1217–1225.
172 E. Khousakoun, J. F. Gohy, R. Jerome, *Polymer* **2004**, *45*, 8303–8310.
173 F. Wang, G. Y. Xu, Z. Q. Zhang, L. Xiao, *Acta Chim. Sin.* **2003**, *61*, 1488–1491.
174 M. Basko, P. Kubisa, *J. Polym. Sci., Part A: Polym. Chem.* **2004**, *42*, 1189–1197.
175 S. Verbrugghe, A. Laukkanen, V. Aseyev, H. Tenhu, F. M. Winnik, F. E. Du Prez, *Polymer* **2003**, *44*, 6807–6814.
176 S. Creutz, J. van Stam, S. Antoun, F. C. de Schryver, R. Jerome, *Macromolecules* **1997**, *30*, 4078–4083.
177 A. S. Lee, V. Bütün, M. Vamvakaki, S. P. Armes, J. A. Pople, A. P. Gast, *Macromolecules* **2002**, *35*, 8540–8551.
178 A. V. Kabanov, V. A. Kabanov, *Adv. Drug Deliv. Rev.* **1998**, *30*, 49–60.
179 Y. Kakizawa, A. Harada, K. Kataoka, *J. Am. Chem. Soc.* **1999**, *121*, 11247–11248.
180 A. Harada, K. Kataoka, *J. Macromol. Sci., Pure Appl. Chem. A* **1997**, *34*, 2119–2133.
181 A. Harada, K. Kataoka, *Macromolecules* **1998**, *31*, 288–294.
182 A. Harada, K. Kataoka, *Langmuir* **1999**, *15*, 4208–4212.
183 A. Harada, K. Kataoka, *J. Am. Chem. Soc.* **1999**, *121*, 9241–9242.
184 T. K. Bronich, A. Nehls, A. Eisenberg, V. A. Kabanov, A. V. Kabanov, *Colloids Surf. B* **1999**, *16*, 243–251.
185 Y. Li, L. J. Ding, H. Nakamura, K. Nakashima, *J. Colloid Interface Sci.* **2004**, *264*, 561–564.
186 L. W. Seymour, K. Kataoka, A. V. Kabanov, Cationic block copolymers as self-assembling vectors for gene delivery, in *Self Assembling Complexes for Gene Delivery: from Laboratory to Clinical Trial*, A. V. Kabanov, P. L. Felgner, L. W. Seymour (Eds.), Wiley, Chichester, **1998**, pp. 219ff.
187 A. V. Kabanov, *Pharmaceutical Science and Technology Today* **1999**, *2*, 365–372.
188 P. Lemieux, S. V. Vinogradov, C. L. Gebhart, N. Guerin, G. Paradis, H. K. Nguyen, B. Ochietti, Y. G. Suzdaltseva, E. V. Bartakova, T. K. Bronich, Y. St-Pierre, V. Y. Alakhov, A. V. Kabanov, *J. Drug Target.* **2000**, *8*, 91–105.

189 P. R. Dash, V. Toncheva, E. Schacht, L. W. Seymour, *J. Control. Release* **1997**, *48*, 269–276.
190 S. Katayose, K. Kataoka, *J. Pharm. Sci.* **1998**, *87*, 160–163.
191 A. Kabanov, S. Vinogradov, Y. Suzdaltseva, V. Alakhov, *Pharm. Res.* **1996**, *13*, S-214.
192 M. A. Cohen Stuart, N. A. M. Besseling, R. G. Fokkink, *Langmuir* **1998**, *14*, 6846–6849.
193 T. Bronich, E. Batrakova, D. Miller, A. Kabanov, *Pharm. Res.* **1997**, *14*, S-641.
194 O. Mundigl, G. C. Ochoa, C. David, V. I. Slepnev, A. Kabanov, P. de Camilli *J. Neurosci.* **1998**, *18*, 93–103.
195 S. Roy, A. Kabanov, S. Vinogradov, K. Zhang, *Invest. Ophthal. Vis. Sci.* **1997**, *38*, 2086–2086.
196 S. Roy, K. Zhang, T. Roth, S. Vinogradov, R. S. Kao, A. Kabanov, *Nat. Biotechnol.* **1999**, *17*, 476–479.
197 M. Yokoyama, T. Okano, Y. Sakurai, S. Suwa, K. Kataoka, *J. Control. Release* **1996**, *39*, 351–356.
198 N. Nishiyama, M. Yokoyama, T. Aoyagi, T. Okano, Y. Sakurai, K. Kataoka, *Langmuir* **1999**, *15*, 377–383.
199 A. Kabanov, T. Bronich, V. Kabanov, K. Yu, A. Eisenberg, *Polym. Prepr.* **1997**, *38*, 648–649.
200 T. K. Bronich, A. V. Kabanov, V. A. Kabanov, K. Yu, A. Eisenberg, *Macromolecules* **1997**, *30*, 3519–3525.
201 A. V. Kabanov, T. K. Bronich, V. A. Kabanov, K. Yu, A. Eisenberg, *J. Am. Chem. Soc.* **1998**, *120*, 9941–9942.
202 T. K. Bronich, A. M. Popov, A. Eisenberg, V. A. Kabanov, A. V. Kabanov, *Langmuir* **2000**, *16*, 481–489.
203 E. A. Bekturov, S. E. Kudaibergenov, V. A. Frolova, R. E. Khamzamulina, R. C. Schulz, J. Zöller, *Makromol. Chem., Rapid Commun.* **1991**, *12*, 37–41.
204 A. F. Thünemann, J. Beyermann, H. Kukula, *Macromolecules* **2000**, *33*, 5906–5911.
205 J. F. Berret, J. Oberdisse, *Physica B* **2004**, *350*, 204–206.
206 P. Herve, M. Destarac, J. F. Berret, J. Lal, J. Oberdisse, I. Grillo, *Europhys. Lett.* **2002**, *58*, 912.
207 L. Bronstein, M. Sedlak, J. Hartmann, M. Breulmann, H. Cölfen, M. Antonietti, *Prepr. ACS, PMSE* **1997**, *76*, 54.
208 L. Qi, H. Cölfen, M. Antonietti, *Nano Lett.* **2002**, *1*, 61–65.
209 L. M. Bronstein, S. N. Sidorov, P. M. Valetsky, J. Hartmann, H. Cölfen, M. Antonietti, *Langmuir* **1999**, *15*, 6256–6262.
210 F. Bouyer, C. Gérardin, F. Fajula, J.-L. Putaux, T. Chopin, *Colloids Surf. A* **2003**, *217*, 179–184.
211 C. Gerardin, N. Sanson, F. Bouyer, F. Fajula, J. L. Putaux, M. Joanicot, T. Chopin, *Angew. Chem. Int. Ed.* **2003**, *42*, 3681–3685.
212 C. Gerardin, V. Buissette, F. Gaudemet, O. Anthony, N. Sanson, F. DiRenzo, F. Fajula, *Mater. Res. Soc. Symp. Proc.* **2002**, *726*, Q7.5.
213 N. Sanson, J. L. Putaux, M. Destarac, C. Gerardin, F. Fajula, *Macromol. Symp.* **2005**, *226*, 279–288.
214 N. Sanson, F. Bouyer, C. Gerardin, M. In, *Phys. Chem. Chem. Phys.* **2004**, *6*, 1463–1466.
215 R. Varoqui, E. Pefferkorn, Q. Tran, *Biol. Chem. Hoppe-Seyler* **1987**, *368*, 782.
216 E. Pefferkorn, Q. Tran, R. Varoqui, *J. Chim. Phys. Phys.-Chim. Biol.* **1981**, *78*, 551–553.
217 N. G. Hoogeveen, M. A. Cohen Stuart, G. J. Fleer, *Faraday Discuss. Chem. Soc.* **1994**, *98*, 161–172.
218 N. G. Hoogeveen, M. A. Cohen Stuart, G. J. Fleer, *Colloids Surf. A* **1996**, *117*, 77–88.
219 H. Walter, P. Müller-Buschbaum, J. S. Gutmann, C. Lorenz-Haas, C. Harrats, R. Jerome, M. Stamm, *Langmuir* **1999**, *15*, 6984–6990.
220 H. Walter, C. Harrats, P. Müller-Buschbaum, R. Jerome, M. Stamm, *Langmuir* **1999**, *15*, 1260–1267.
221 B. Mahltig, H. Walter, C. Harrats, P. Müller-Buschbaum, R. Jerome, M. Stamm, *Phys. Chem. Chem. Phys.* **1999**, *1*, 3853–3856.
222 H. D. Bijsterbosch, M. A. Cohen Stuart, G. J. Fleer, *J. Colloid Interface Sci.* **1999**, *210*, 37–42.

223 I. M. Solomentseva, M. A. Cohen Stuart, G. J. Fleer, *Colloid J.* **2000**, *62*, 218–223.
224 I. M. Solomentseva, M. A. Cohen Stuart, G. J. Fleer, *Colloid J.* **2000**, *62*, 218–223.
225 A. W. M. de Laat, H. F. M. Schoo, *J. Colloid Interface Sci.* **1997**, *191*, 416–423.
226 A. W. M. de Laat, H. F. M. Schoo, *J. Colloid Interface Sci.* **1998**, *200*, 228–234.
227 S. H. Yu, S. Cölfen, A. Fischer, *Colloids Surf. A* **2004**, *243*, 49–52.
228 K. Wormuth, *J. Colloid Interface Sci.* **2001**, *241*, 366–377.
229 L. M. Palmqvist, F. F. Lange, W. Sigmund, J. Sindel, *J. Am. Ceram. Soc.* **2000**, *83*, 1585–1591.
230 J. Sindel, W. Sigmund, B. Baretzky, F. Aldinger, Untersuchungen zu Wechselwirkungen zwischen Bindemittel und Pulverteilchen in wässrigen Bariumtitanatschlickern, in G. Ziegler, H. Cherdron, W. Hermel, J. Hirsch, H. Kolaska (Eds.), *Werkstoff- und Verfahrenstechnik, Proc. Werkstoffwoche*, Vol. 6, DGM Informationsgesellschaft Verlag, Oberursel **1996**, pp. 617–622.
231 W. M. Sigmund, J. Sindel, F. Aldinger, *Prog. Colloid Polym. Sci.* **1997**, *105*, 23–26.
232 J. Sindel, N. S. Bell, W. M. Sigmund, *J. Am. Ceram. Soc.* **1999**, *82*, 2953–2957.
233 K. Kataoka, G. S. Kwon, M. Yokoyama, T. Okano, Y. Sakurai, *J. Control. Release* **1993**, *24*, 119–132.
234 G. S. Kwon, K. Kataoka, *Adv. Drug Deliv. Rev.* **1995**, *16*, 295–309.
235 S. Cammas, K. Kataoka, Site specific drug carriers: polymeric micelles as high potential vehicles for biologically active molecules, in *Solvents and Self-organization of Polymers*, S. E. Webber et al. (Eds.), Kluwer, Dordrecht, **1996**, pp. 83ff.
236 H. Bader, H. Ringsdorf, B. Schmidt, *Angew. Makromol. Chem.* **1984**, *123/124*, 457–485.
237 M. Yokoyama, G. S. Kwon, T. Okano, Y. Sakurai, H. Ekimoto, K. Okamoto, H. Mashiba, T. Seto, K. Kataoka, *Drug Deliv.* **1993**, *1*, 11–19.
238 M. Yokoyama, T. Okano, Y. Sakurai, K. Kataoka, *J. Control. Release* **1994**, *32*, 269–277.
239 K. Kataoka, G. S. Kwon, M. Yokoyama, T. Okano, Y. Sakurai, Polymeric micelles as novel drug carriers and virus mimicking vehicles, in *Macromolecules*, J. Kahovec (Ed.), VSP, Utrecht, **1992**, pp. 267–276.
240 M. Yokoyama, T. Sugiyama, T. Okano, Y. Sakurai, M. Naito, K. Kataoka, *Pharm. Res.* **1993**, *10*, 895–899.
241 M. Yokoyama, G. S. Kwon, T. Okano, Y. Sakurai, M. Naito, K. Kataoka, *J. Control. Release* **1994**, *28*, 59–65.
242 M. Yokoyama, T. Okano, Y. Sakurai, H. Ekimoto, C. Shibazaki, K. Kataoka, *Cancer Res.* **1991**, *51*, 3229–3236.
243 G. Kwon, M. Yokoyama, T. Okano, Y. Sakurai, K. Kataoka, Pharmacokinetics and biodistribution of polymeric micelles based on AB block copolymers of polyethylene oxide and polyaspartic acid with bound adriamycin, in *Proceedings of the 1st International Conference on Intelligent Materials*, T. Takagi, K. Takahashi, M. Aizawa, S. Myata (Eds.), Tecnomic Pub. Co., Lancaster, PA, Naka-gun, Japan, **1992**, pp. 403–406.
244 G. S. Kwon, M. Yokoyama, T. Okano, Y. Sakurai, K. Kataoka, *Pharm. Res.* **1993**, *10*, 970–974.
245 G. S. Kwon, M. Yokoyama, T. Okano, Y. Sakurai, K. Kataoka, *J. Control. Release* **1994**, *28*, 334–335.
246 M. Yokoyama, G. S. Kwon, T. Okano, Y. Sakurai, H. Ekimoto, K. Kataoka, *J. Control. Release* **1994**, *28*, 336–337.
247 G. S. Kwon, S. Suwa, M. Yokoyama, T. Okano, Y. Sakurai, K. Kataoka, *J. Control. Release* **1994**, *29*, 17–23.
248 M. Yokoyama, S. Fukushima, R. Uehara, K. Okamoto, K. Kataoka, Y. Sakurai, T. Okano, *J. Control. Release* **1998**, *50*, 79–92.
249 M. Yokoyama, T. Okano, Y. Sakurai, S. Fukushima, K. Okamoto, K. Kataoka, *J. Drug Target.* **1999**, *7*, 171–186.
250 Y. Li, G. S. Kwon, *Colloids Surf. B* **1999**, *16*, 217–226.
251 Y. Li, G. S. Kwon, *Pharm. Res.* **2000**, *17*, 607–611.
252 L. M. Qi, *J. Mater. Sci. Lett.* **2001**, *20*, 2153–2156.
253 S. Weiner, W. Traub, *FEBS Lett.* **1980**, *111*, 311–316.
254 S. Weiner, W. Traub, *Philos. Trans. R. Soc. Lond. B* **1984**, *304*, 425–434.

255 S. Mann, J. Webb, R. J. P. Williams, *Biomineralization*, VCH, Weinheim, **1989**.
256 A. Richter, D. Petzold, H. Hofmann, B. Ulrich, *Chem. Tech.* **1996**, *48*, 271–275.
257 P. Kasparova, *PhD Thesis*, Potsdam University, **2002**.
258 J. Rudloff, H. Cölfen, *Langmuir*, **2004**, *20*, 991–996.
259 Y. Mastai, J. Rudloff, H. Cölfen, M. Antonietti, *Chem. Phys. Chem.* **2002**, *1*, 119–123.
260 R. Dimova, R. Lipowsky, Y. Mastai, M. Antonietti, *Langmuir* **2003**, *19*, 6097–6103.
261 Y. Mastai, M. Sedlak, H. Cölfen, M. Antonietti, *Chem. Eur. J.* **2002**, *8*, 2429–2437.
262 S. H. Yu, H. Cölfen, Y. Mastai, *J. Nanosci. Nanotechnol.* **2004**, *4*, 291–298.
263 M. Öner, J. Norwig, W. H. Meyer, G. Wegner, *Chem. Mater.* **1998**, *10*, 460–463.
264 A. Taubert, D. Palms, O. Weiss, M. T. Piccini, D. N. Batchelder, *Chem. Mater.*, **2002**, *14*, 2594–2601.
265 A. Taubert, D. Palms, G. Glasser, *Langmuir* **2002**, *18*, 4488–4494.
266 D. B. Zhang, L. M. Qi, J. M. Ma, H. M. Cheng, *Chem. Mater.*, **2002**, *14*, 2450–2457.
267 S. H. Yu, H. Cölfen, M. Antonietti, *J. Phys. Chem. B* **2003**, *107*, 7396–7405.
268 S. H. Yu, M. Antonietti, H. Cölfen, M. Giersig, *Angew. Chem. Int. Ed.* **2002**, *41*, 2356–2360.
269 H. Cölfen, L. M. Qi, Y. Mastai, L. Börger, *Cryst. Growth Des.*, **2002**, *2*, 191–196.
270 A. Taubert, C. Kübel, D. C. Martin, *J. Phys. Chem. B* **2003**, *107*, 2660–2666.
271 S. F. Chen, S. H. Yu, T. X. Wang, J. Jiang, H. Cölfen, B. Hu, B. Yu, *Adv. Mater.* **2005**, *17*, 1461–1465.
272 L. M. Qi, J. Li, J. M. Ma, *Adv. Mater.* **2000**, *14*, 300–303.
273 L. M. Qi, J. Li, J. M. Ma, *Adv. Mater.* **2002**, *14*, 300–303.
274 D. B. Zhang, L. M. Qi, J. M. Ma, H. M. Cheng, *Adv. Mater.* **2002**, *14*, 1499–1502.
275 R. L. Penn, J. F. Banfield *Geochim. Cosmochim. Acta* **1999**, *63*, 1549–1557.
276 L. M. Qi, H. Cölfen, M. Antonietti, *Angew. Chem. Int. Ed.* **2000**, *39*, 604–607.
277 L. M. Qi, H. Cölfen, M. Antonietti, *Chem. Mater.* **2000**, *12*, 2392–2403.
278 S. H. Yu, H. Cölfen, M. Antonietti, *Chem. Eur. J.* **2002**, *8*, 2937–2945.
279 K. L. Robinson, J. V. M. Weaver, S. P. Armes, E. D. Marti, F. C. Meldrum, *J. Mater. Chem.* **2002**, *12*, 890–896.
280 L. M. Qi, H. Cölfen, M. Antonietti, M. Li, J. D. Hopwood, A. J. Ashley, S. Mann, *Chem. Eur. J.* **2001**, *7*, 3526–3532.
281 S. H. Yu, H. Cölfen, M. Antonietti, *Adv. Mater.*, **2003**, *15*, 133–136.
282 S. H. Yu, M. Antonietti, H. Cölfen, J. Hartmann, *Nano Lett.* **2003**, *3*, 379–382.
283 D. B. Zhang, L. M. Qi, J. M. Ma, H. M. Cheng, *Chem. Mater.* **2001**, *13*, 2753–2755.
284 M. Li, H. Cölfen, S. Mann, *J. Mater. Chem.* **2004**, *14*, 2269–2276.
285 R. Kniep, S. Busch, *Angew. Chem. Int. Ed. Engl.* **1996**, *35*, 2624–2626.
286 S. Busch, H. Dolhaine, A. DuChesne, S. Heinz, O. Hochrein, F. Laeri, O. Podebrad, U. Vietze, T. Weiland, R. Kniep, *Eur. J. Inorg. Chem.* **1999**, *10*, 1643–1653.
287 H. Cölfen, L. M. Qi, *Chem. Eur. J.* **2001**, *7*, 106–116.
288 H. Cölfen, L. M. Qi, *Prog. Colloid Polym. Sci.*, **2001**, *117*, 200–203.
289 H. Endo, H. Cölfen, D. Schwahn, *J. Appl. Crystallogr.* **2003**, *36*, 568–572.
290 H. Endo, D. Schwahn, H. Cölfen, *J. Chem. Phys.* **2004**, *120*, 9410–9423.
291 H. Endo, D. Schwahn, H. Cölfen, *Physica B* **2004**, *350*, e943–e945.
292 S. H. Yu, H. Cölfen, J. Hartmann, M. Antonietti, *Adv. Funct. Mater.* **2002**, *12*, 541–545.
293 L. M. Qi, J. Li, J. M. Ma, *Chem. J. Chin. Univ.* **2002**, *23*, 1595–1597.
294 P. Kasparova, M. Antonietti, H. Cölfen, *Colloids Surf. A* **2003**, *250*, 153–162.
295 K. Kaluzynski, J. Pretula, G. Lapienis, M. Basko, Z. Bartczak, A. Dworak, S. Penczek, *J. Polym. Sci., Part A: Polym. Chem.* **2001**, *39*, 955–963.

296 J. M. Marentette, J. Norwig, E. Stockelmann, W. H. Meyer, G. Wegner, *Adv. Mater.* **1997**, *9*, 647–651.
297 J. Bolze, D. Pontoni, M. Ballauff, T. Narayanan, H. Cölfen, *J. Colloid Interface Sci.* **2004**, *277*, 84–94.
298 D. B. Zhang, L. M. Qi, J. M. Ma, H. M. Cheng, *Chem. Mater.* **2002**, *14*, 2450–2457.
299 F. Wang, G. Y. Xu, Z. Q. Zhang, *Mater. Lett.* **2005**, *59*, 808–812.
300 M. Lei, W. H. Tang, J. G. Yu, *Mater. Res. Bull.* **2005**, *40*, 656–664.
301 H. Cölfen, S. Mann, *Angew. Chem. Int. Ed.* **2003**, *42*, 2350–2365.
302 H. Cölfen, M. Antonietti, *Angew. Chem. Int. Ed.* **2005**, *44*, 5576–5591.
303 S. Wohlrab, N. Pinna, M. Antonietti, H. Cölfen, *Chem. Eur. J.* **2005**, *11*, 2903–2913.
304 S. H. Yu, H. Cölfen, K. Tauer, M. Antonietti, *Nat. Mater.* **2005**, *4*, 51–55.

16
Applications of Polymer Bioconjugates

Joost A. Opsteen and Jan C. M. van Hest

16.1
Introduction; Control at the Molecular Level Over Synthetic and Biological Macromolecules

In Nature, many biopolymers are employed, the (biological) activity of which is dependent on the level of structural control, that is unsurpassed by current synthetic materials. This level of control is a result of the well-defined order of arrangement of the biopolymer building blocks (e.g. nucleotides and amino acids) and the absolute control over molecular weight as observed in the case of DNA and proteins. However, this three-dimensional complexity also makes it often difficult to apply biopolymers in materials science, since they are susceptible to conformational changes, which can easily lead to loss of function.

Synthetic polymers have the advantage that they can be made in a wide range of different architectures and with a large variety of compositions. They can therefore be easily adapted to a specific application environment. However, since absolute control over composition and degree of polymerization cannot be achieved, the level of information and, hence, activity that can be incorporated in synthetic polymers remains limited in comparison with biopolymers. A logical approach which has recently experienced much interest in the materials community is to combine the natural structural control of biopolymers, leading to properties such as programmed assembly, recognition and bioactivity, with the versatility of synthetic polymers, in order to create a new class of hybrid macromolecules. Due to the synergistic effect of blending distinct properties in a single macromolecule, biohybrid polymers have many (potential) applications in medicine, nanotechnology and bioengineering.

The recent increase in activity in this field has much to do with the availability of new synthetic techniques that allow us to create polymeric building blocks of synthetic and biological origin and to construct well-defined hybrid architectures. This chapter will focus on well-defined covalently linked biohybrid polymers and, therefore, cross-linked systems and supramolecular conjugates will be omitted. First (Section 16.2), a short overview is given of the most important

Macromolecular Engineering. Precise Synthesis, Materials Properties, Applications.
Edited by K. Matyjaszewski, Y. Gnanou, and L. Leibler
Copyright © 2007 WILEY-VCH Verlag GmbH & Co. KGaA, Weinheim
ISBN: 978-3-527-31446-1

synthetic techniques that have surfaced in recent years. This is followed in the next three sections by a more detailed description of potential applications of polymer bioconjugates as they are described in the recent literature. In Section 16.3, pharmaceutical applications of hybrid polymers are described. Section 16.4 deals with the use of biohybrids for the preparation of bioactive surfaces. Finally, (Section 16.5), smart biohybrid materials, such as stimuli-responsive bioconjugates, are discussed.

16.2
Synthetic Methodologies

16.2.1
Controlled Polymerization Methods

The discovery of living anionic polymerization in 1956 by Szwarc et al. [1] enabled polymer chemists to have control over the polymerization process, which means that the degree of polymerization (DP) can be predetermined and the polydispersity index (PDI) is low. Due to exclusion of the termination process, every growing polymer chain remains active until all monomer is consumed [2, 3]. Until approximately 20 years ago, the only methods for performing controlled polymerizations were based on ionic mechanisms. Because of the sensitivity of this technique, stringent reaction conditions have to be applied and the number of monomers is limited. For this reason, polymer chemists have been striving for almost 40 years to introduce this high level of control in the free radical polymerization process. The advance of controlled or "living" radical polymerization (LRP) mechanisms in the past decade has led to methods that indeed combine the robustness of radical processes with the control first encountered only in anionic systems.

This level of control can be obtained in LRP by reducing the free radical concentration, as a result of which termination reactions are suppressed. The most widely used LRP techniques are reversible addition–fragmentation chain-transfer polymerization (RAFT) [4], atom-transfer radical polymerization (ATRP) [5, 6] and nitroxide-mediated polymerization (NMP) [7]. Considering the radical nature of these polymerization reactions, they are relatively easy to perform and, moreover, they are inert to many functional groups, permitting the polymerization of a myriad of monomers including biofunctional moieties such as peptides [8], nucleobases [9, 10] and oligosaccharides [11].

Due to the living character of anionic polymerizations and LRP, control over chain growth is achieved, which allows preparation of block copolymers by consecutive addition of different monomers. Inherently, this control over chain growth also implies control over polymer chain ends, which subsequently can be modified into numerous functional groups such as amines, azides, thiols and carboxylic acids [12, 13]. These functional moieties can be exploited specifically to conjugate polymers to biomolecules via their end-groups.

In addition to control over composition and functionalization, polymers of various topologies (comb, star, dendritic and so forth) can also be synthesized in a controlled fashion [14–17]. In Volume II, chain topologies are elaborately described. In conclusion, the advances in living radical polymerization have created many new opportunities for the construction of well-defined polymer bioconjugates.

The development of improved catalysts for metathesis reactions has paved the way for controlled ring-opening metathesis polymerization (ROMP) and acyclic diene metathesis polymerization (ADMET) [18, 19]. In ROMP, strained ring systems such as norbornenes can be polymerized by utilizing a transition metal catalyst. Initially, ill-defined polymers were obtained using $RuCl_3$; however, by using other ruthenium- [20] and molybdenum-based [21] catalysts developed by the groups of Grubbs and Schrock, respectively, well-controlled polymerizations can be conducted. Analogous to LRP, ROMP is highly tolerant to functional groups and, therefore, monomers containing biofunctionality can be polymerized in a controlled manner [22–26]. A more extensive overview of polymerization techniques and current state of the art on polymer synthesis can be found in Volume I.

16.2.2
Polypeptide Synthesis

The preparation of well-defined amino acid-based oligomers and polymers has always been a focal point of attention for synthetic chemists, since it allows the construction of bioactive moieties and molecules with a high level of control over three-dimensional structure. Also in this field much progress has been made in recent years.

Pioneering work by Merrifield [27], which was extended by Bayer and Mutter [28, 29], led to the synthesis of peptides via sequential coupling of amino acids on a solid-phase support. Solid-phase synthesis nowadays is still widely used to prepare a wide variety of peptides and oligonucleotides. This stepwise methodology, however, is limited to the preparation of peptides containing 40–50 amino acids. In order to obtain larger peptides with up to 200 amino acids, ligation methods can be used in which peptide sequences are "glued" together [30–32], as depicted in Scheme 16.1.

The most common way to prepare high molecular weight polypeptides is by polymerization of α-amino acid-N-carboxyanhydrides (NCAs). These NCA monomers, which are easily prepared in one step from commercially available amino acids, are capable of undergoing a ring-opening polymerization in the presence of nucleophiles or bases yielding high molecular weight polypeptides with preservation of chirality at the α-carbon center. By the development of nickel-based initiators, Deming succeeded in suppressing side-reactions, which resulted in the formation of polypeptides with controllable molecular weight and low PDI. Additionally, these living NCA polymerizations can be used to synthesize well-defined block copolypeptides [33, 34].

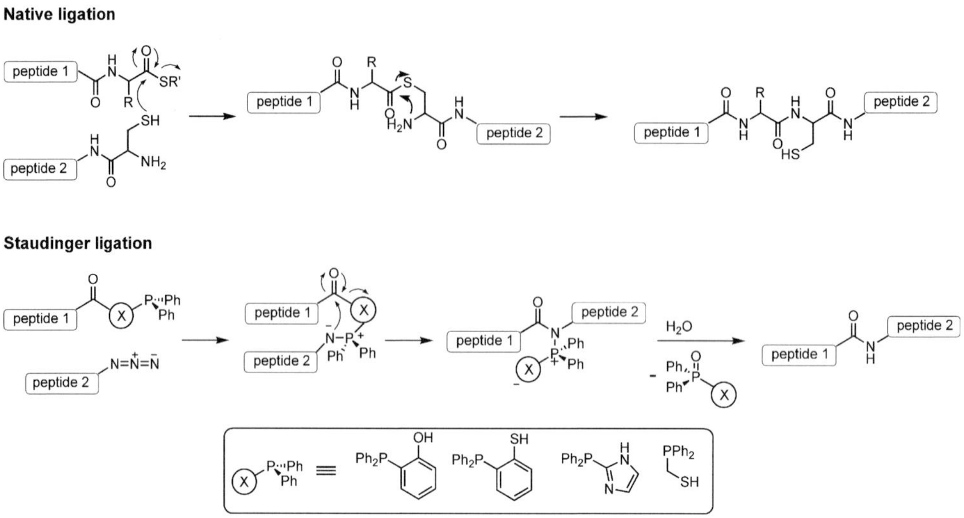

Scheme 16.1 Two examples of the chemical ligation process of peptides [30–32].

Albeit a very elegant technique to produce large quantities of block copolypeptides, no absolute control over the amino acid sequence in the peptides is possible. Control at the monomer level can be achieved by utilizing protein engineering. As in this case the protein production pathway of biological hosts, usually bacteria or yeast, is exploited for the synthesis, monodisperse polymers with predetermined chain lengths and primary amino acid sequences are prepared. These microorganisms can be programmed to produce the protein of interest by introducing the corresponding genetic information. Although first developed for molecular biology purposes, in the last few years protein engineering has become an important tool in materials design [35].

In comparison with other techniques, protein engineering seems to be limited to the ensemble of 20 amino acids. The introduction of functional groups in proteins that are absent in this series of proteinogenic amino acids, such as halides, alkynes and azides, would be useful in controlling properties or post-modification reactions. As a consequence, much effort has been applied to finding methods for incorporating unnatural amino acids. Nowadays two approaches are available that allow the extension of amino acid building blocks to non-proteinogenic species. The first approach utilizes rarely used genetic codons for the site-specific incorporation of unnatural amino acids [36]. The second multi-site replacement approach replaces one of the 20 proteinogenic amino acids with a structural analogue, which contains the functionality of interest [37–43]. Both approaches have been shown to be highly versatile and have led to a large extension of available amino acid building blocks.

The bio-related synthesis of polymers is extensively reviewed by Kiick and co-workers in Chapter 11 in Volume I.

16.2.3
Conjugation Methods

All aforementioned techniques exhibit control on the molecular level in the synthesis of both polymers and biomolecules. Moreover, functional groups can be introduced at predetermined locations in these macromolecules, allowing the construction of conjugates of synthetic polymers and biomolecules in a defined fashion. Nowadays, many methodologies are available to couple synthetic and biological macromolecules specifically via introduced functional groups [44, 45]. Due to the reduced reactivity in these large macromolecules and the presence of many functional groups, coupling chemistry has to be used, which is both efficient and also very specific. In addition to the previously shown ligation processes (Scheme 16.1), a myriad of conjugation reactions which meet these requirements is available, as depicted in Scheme 16.2 [46, 47].

Recently, Sharpless and coworkers improved Huisgen's 1–3-dipolar cycloaddition of azides and alkynes by means of copper(I) catalysis [48]. This so-called click chemistry process is very suitable for the synthesis of hybrid macromolecules due to the selectivity of the reactants and the efficiency of the reaction. Furthermore, azide and alkyne functionality can be readily introduced in synthetic polymers and in biomolecules [49–53]. Another strategy is to grow polymer chains from biopolymers using LRP techniques [54–58]. More on the synthesis of biohybrid polymers can be found in Chapter 12 in Volume I by Klok and Deming and Chapter 15 in Volume II by Antonietti et al.

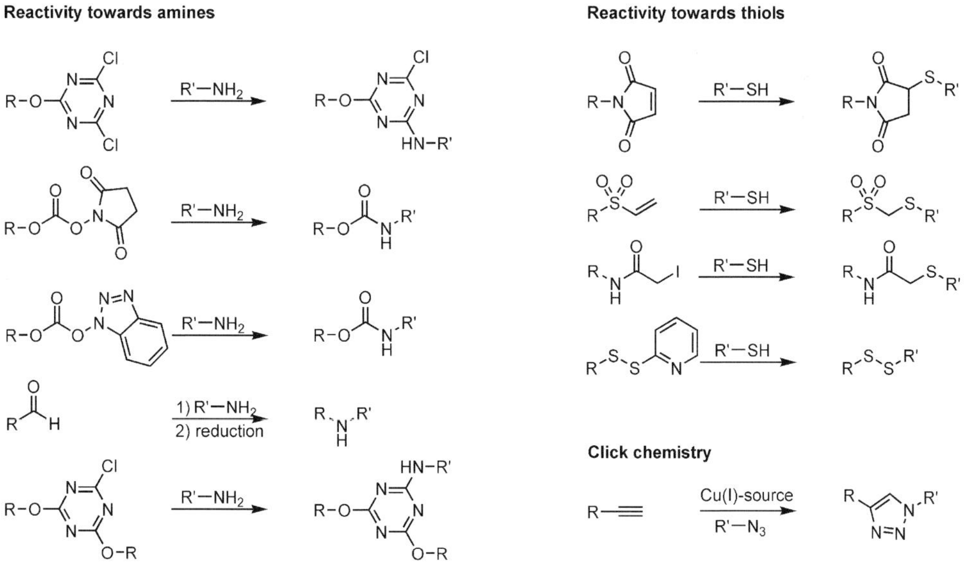

Scheme 16.2 Examples of conjugation methodologies [46, 47].

16.3
Pharmaceutical Applications

The usage of many potential pharmaceutical compounds is hampered by limitations such as harmful side-effects, nonspecific activity and short circulating half-life. Therefore, ideally, these compounds have to be modified in such a way that they can be transported to the site of action in an inactive form where they subsequently can be transformed into active species. In this respect, utilization of macromolecules as therapeutic agents has attracted much interest in the last decade. These "polymer therapeutics" comprise water-soluble macromolecules that are biologically active, polymeric micelles, polymer–drug and polymer–protein hybrid conjugates and supramolecular assemblies capable of forming multi-component polyplexes [59, 60] (Fig. 16.1).

As stated in the previous section, in this chapter the focus will be on polymer conjugates and therefore only polymer hybrid therapeutics will be discussed. Since polymeric drugs are extensively reviewed by Minko et al. elsewhere in this

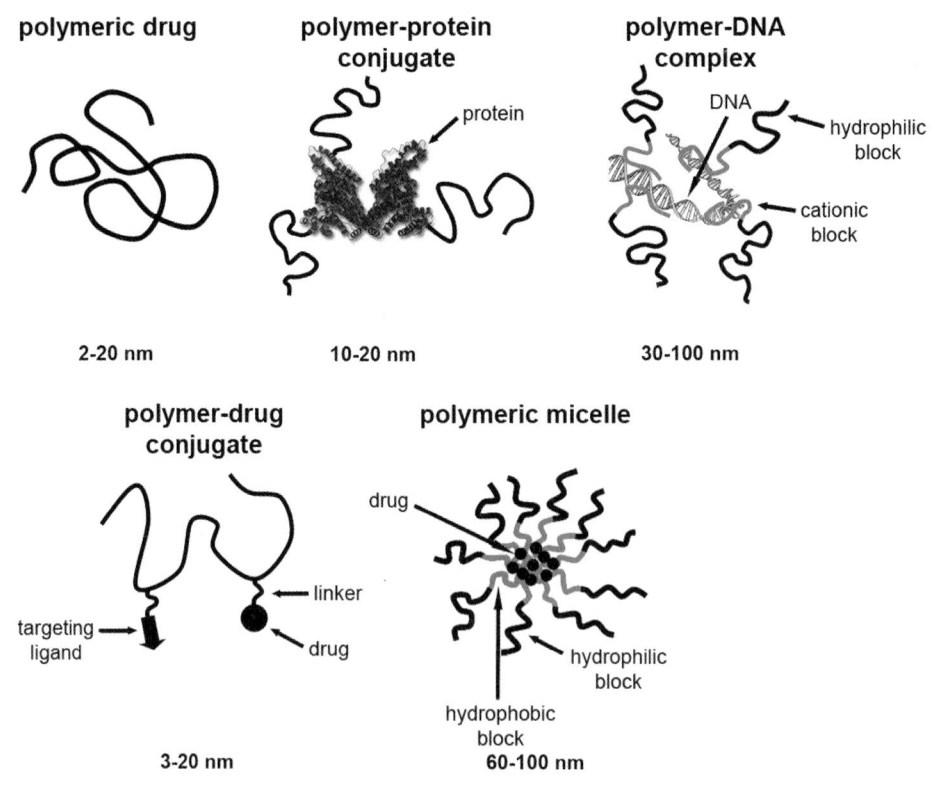

Fig. 16.1 Schematic representation of different classes of polymer therapeutics. Adapted from [60].

volume, this topic will be only briefly discussed and the added value of controlled architectures in designing polymer therapeutics will be emphasized.

16.3.1
Anti-cancer Therapeutics

As cancer cells in many respects are similar to normal host cells from which they are derived, chemotherapeutic treatment lacks selectivity, thereby causing adverse toxicity, which limits the dose of administrable drugs. Furthermore, many chemotherapeutic agents have further drawbacks such as low solubility and rapid degeneration and are sometimes rapidly excreted due to the presence of effective efflux pumps in tumor cells [61]. In order to overcome these problems, much research has been conducted on tumor cell targeting therapeutics. One possibility is the conjugation of drugs to polymer carriers. This concept of covalent linkage of polymers to drugs was first postulated by Ringsdorf and coworkers [62, 63].

An early example of such a macromolecular therapeutic is a conjugate of poly[styrene-co-(maleic anhydride)] (SMA) and neocarzinostatin (NCS), better known by the acronym SMANCS, which was developed by Maeda and coworkers [64–67]. Originally, they synthesized this conjugate in order to hydrophobize the antitumor protein NCS to allow dispersion in the phase contrast agent Lipiodol, which is used for patient imaging. While studying the pharmacokinetics of SMANCS using an *in vitro* tumor model, a liver tumor:blood ratio of >2500 was observed, which surpassed any existing targeting system at that time and, consequently, it was approved in Japan in 1993 as a treatment for hepatocellular carcinoma. Maeda and coworkers attributed this phenomenon to a combination of two factors, namely the hyperpermeability of tumor tissue, which allows the uptake of large polymers, and the ineffective tumor lymphatic drainage, which results in subsequent accumulation of these polymers. This passive form of drug targeting is nowadays referred to as the "enhanced permeation and retention (EPR) effect" [68, 69]. This EPR effect only functions properly when the molecular weight of the polymers exceeds the limit of 40 kDa in order to evade renal clearance.

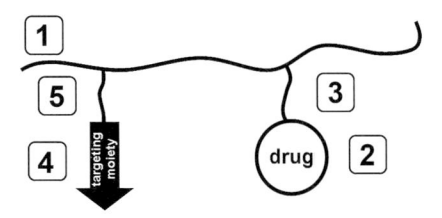

Fig. 16.2 Schematic construct of a polymer–drug conjugate. To the backbone of a water-soluble polymer (1), drugs (2) are conjugated via a biodegradable linker (3). Additionally, targeting moieties (4) can be attached via nondegradable linkers (5).

Many macromolecular drugs taking advantage of the EPR effect are currently in clinical use, in clinical trials or in development. As illustrated in Fig. 16.2, all these polymer–drug conjugates have to consist of at least three parts: (a) a (generally) water-soluble polymer; (b) a biodegradable linker between polymer and drug; and (c) a bioactive antitumor drug such as doxorubicin (Dox), paclitaxel and camptothecin.

The polymeric carrier, which is in general water soluble, can be either readily excreted from the body or biodegradable. These include both synthetic polymers such as N-(2-hydroxypropyl)methacrylamide (HPMA), poly(ethylene glycol) (PEG), poly(vinylpyrrolidone) (PVP) and poly(ethylenimine) (PEI) and natural polymers such as dextran (a-1,6-polyglucose), dextrin (a-1,4-polyglucose) and chitosans. Furthermore, synthetic poly(amino acids) such as poly(L-lysine) and poly(glutamic acid) (PGA) are used to aid drug delivery [70]. The most intensively used synthetic polymers for this purpose are (copolymers of) HPMA [71, 72] and PEG [73]. HPMA is a water-soluble nontoxic and nonimmunogenic

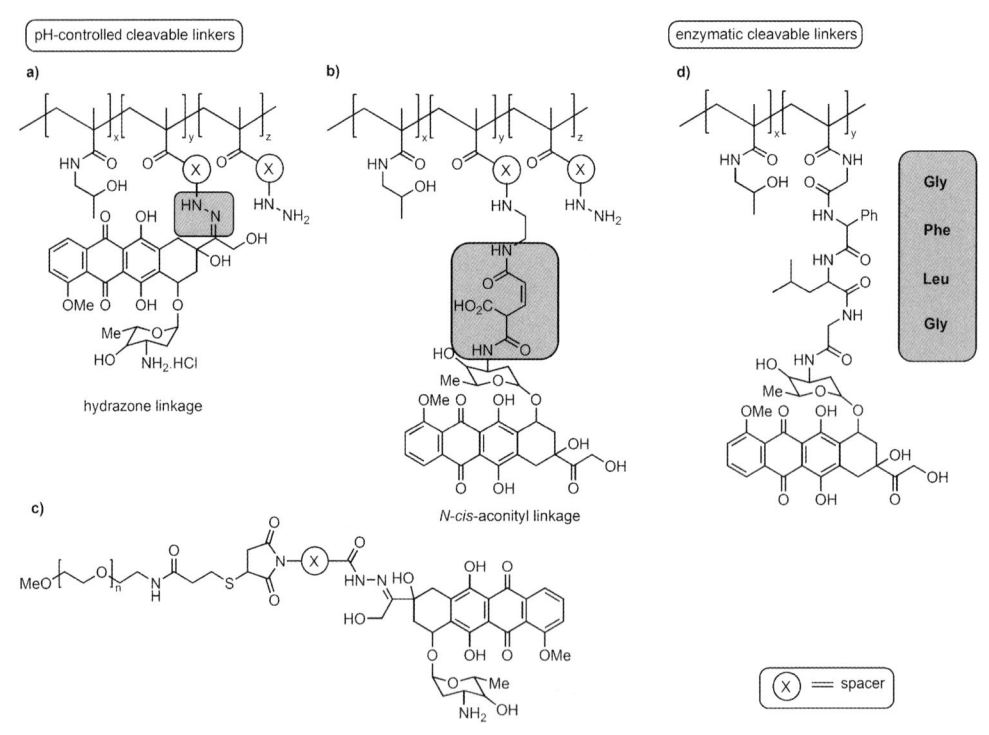

Fig. 16.3 Structures of some polymer–doxorubicin (Dox) conjugates containing biodegradable linkers. (a) Hydrazone-linked HPMA–Dox conjugate; (b) HPMA–Dox linked via an N-cis-aconityl spacer; (c) Dox attached to PEG via a maleimide–hydrazone spacer; (d) HMPA–Dox conjugate containing a Gly–Phe–Leu–Gly tetrapeptide linker.

polymer which has the advantage of the presence of many possible conjugation sites. On the other hand, in linear PEG only one or two binding sites at the chain ends are present. Nevertheless, PEG is a water soluble nonionic polymer which is approved by the FDA for application in pharmaceutics. Another important aspect is that these polymers exhibit low antigenicity and in most cases even decrease the antigenicity of conjugated compounds [74, 75].

Usually, as depicted in Fig. 16.2, the linkage between drugs and polymers is accomplished by a biodegradable linker. After administration, the drug attached to a polymer is transported in the body. The inactivated cytotoxic drug should be released from its polymeric carrier intracellularly in the lysosomes (lysosomotropic drug delivery) and/or intratumorally (tumoritropic drug delivery) [76]. In other words, cytotoxicity mainly occurs at the site of action. For these protected drugs, the term prodrugs is often used. In order to achieve controlled release from polymer–drug conjugates, linkers have to be chosen such that they can be cleaved hydrolytically or enzymatically at the correct location. Therefore, the development of spacers is mainly directed towards pH-sensitive spacers and oligopeptide spacers [77]. Some examples of polymeric drugs comprising biodegradable spacers are depicted in Fig. 16.3.

Polymeric drugs are taken up by cells through endocytosis and subsequently the conjugate is exposed to the acidic environment of the lysosome. Likewise, the pH in or in the vicinity of cancer cells is low compared with healthy tissue [78]. This relatively low pH (pH drop from physiological 7.4 to 5–6 in endosomes or 4–5 in lysosomes) can be exploited to apply spacers which are hydrolyzed at the location of tumor tissue and accordingly release the cytotoxic drug. In this respect, two types of spacers have been intensively studied and reviewed, namely the hydrazone and N-cis-aconityl linkages (Fig. 16.3 a and b, respectively) [79, 80]. Apart from the acidic environment, macromolecular drugs are also exposed to the degrading nature of lysosomal enzymes. Some of these lysosomal proteases, such as cathepsin B, play an important role in tumor growth and formation of metastasis [81]. By utilizing specific oligopeptide sequences recognized by lysosomal proteases, the drug can be released by these enzymes in or close to tumor tissue. Many peptide sequences have been tested and the tetrapeptide glycine–phenylalanine–leucine–glycine (Gly–Phe–Leu–Gly) appeared to be the most suitable degradable spacer and is, therefore, currently in the stage of clinical trials [82–87] (Fig. 16.3 d).

Hitherto, one aspect of the polymer–drug construct as depicted in Fig. 16.2 is not discussed. In addition to passive targeting via the EPR effect, additional targeting moieties can be attached to enhance the selective biorecognizability of the conjugate [88–90]. These targeting moieties range in complexity from carbohydrates and peptides to hormones and antibodies (Fig. 16.4) [91]. In targeting cancer there are two conceivable mechanisms, namely the targeting of tumor cells and the targeting of tumor vasculature. Tumor cell targeting is based on the recognition of receptors or antigens present in tumor cells. In Fig. 16.4a. an HPMA–Dox conjugate bearing pendant glucosamine units that function as targeting ligands is depicted. The aim of this asiaglycoprotein (ASGP) biomimetic

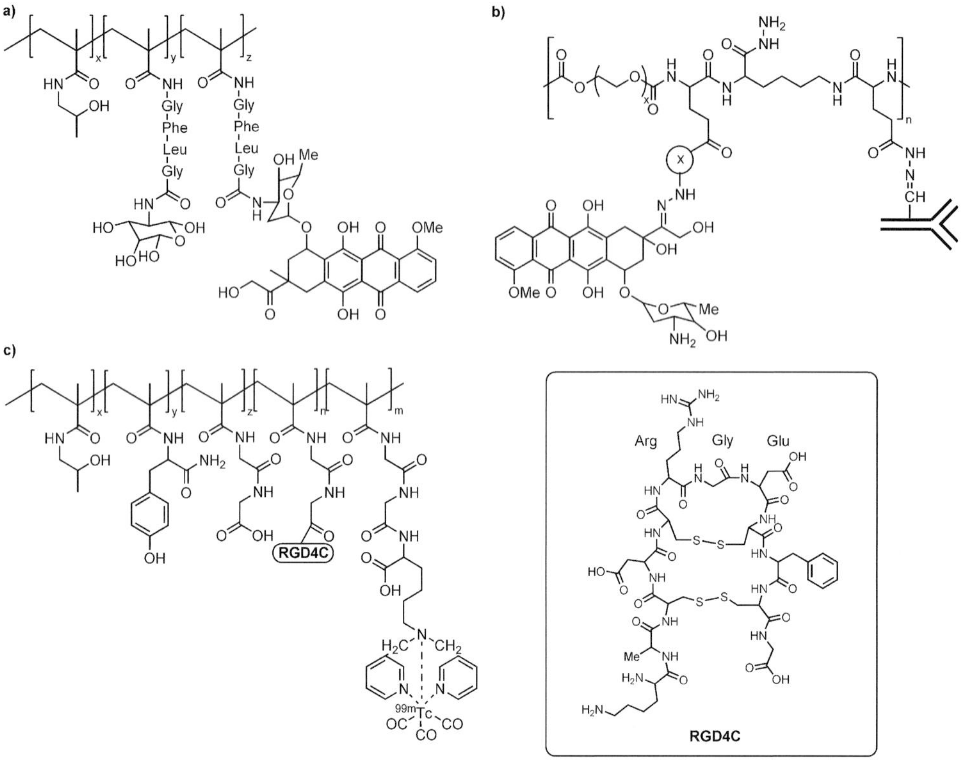

Fig. 16.4 Targeted polymeric drug delivery systems. (a) HPMA–Dox conjugate containing pendant N-acetylated galactosamine units (PK2) [92, 93]; (b) multiblock PEG–Dox conjugate embracing an antibody as targeting moiety [94–97]; (c) technetium–99m-labeled HPMA carrying a doubly cyclized Arg–Gly–Asp (RGD) binding motif [98].

is to target the hepatocyte and hepatocellular carcinoma ASGP receptor for the treatment of liver cancer. Up to now, this conjugate, better known as PK2, is the only conjugate containing a targeting moiety to enter the stage of clinical trials [92, 93]. Other markers for drug targeting are proteins present at the surface of tumor cells, some of which are specific for certain tumor types. Monoclonal antibodies can be used as targeting units which selectively bind to such tumor-specific antigens. An example of such a macromolecular drug with attached antibodies as targeting moieties is depicted in Fig. 16.4 b [94–97].

A second approach is targeting of tumor vasculature. An important control point in cancer is the formation of new capillary blood vessels from pre-existing vasculature, a process which is referred to as angiogenesis [99]. The process of angiogenesis involves many growth factors with accompanying receptors, cytokines, proteases and adhesion molecules [100, 101]. Therefore, many targeting possibilities for anti-angiogenic therapy for cancer exist [102, 103]. One such marker is the $a_V\beta_3$ integrin [104]. Attachment of the high affinity ligand Arg–Gly–Asp (RGD) to technetium-99m-labeled HPMA, depicted in Fig. 16.4c, displayed significant tumor localization of the conjugate *in vivo* [98].

16.3.2
Application of Polymer–Peptide/Protein/Antibody Conjugates as Therapeutics

A variety of novel peptides and proteins have emanated from the biotechnology revolution of the last two decades. Some of these peptides and proteins have become important new drugs for various diseases, including cancer, infectious diseases, autoimmune diseases and HIV/AIDS. Currently over 80 polypeptide drugs are marketed in the USA and, additionally, 350 are in the stage of clinical trials [105, 106]. Unfortunately, the application of peptides and proteins is hampered by some limitations, such as their susceptibility to denaturation by proteolytic enzymes, short circulating half-life, short shelf-life, low solubility, rapid kidney clearance and the tendency to generate negative immune responses.

Two distinct pathways can be followed in order to overcome the above-mentioned drawbacks, namely the encapsulation of peptides and proteins in stable carrier systems such as liposomes, polymer micelles and vesicles and, second, conjugation to synthetic polymers. Although the first method has proven to be very successful – noteworthy in this respect are, for instance, sterically stabilized or Stealth® liposomes, which have already been marketed for a decade [107] – it is beyond the scope of this chapter and therefore it will not be discussed further.

The stabilization of proteins by the attachment of synthetic polymers was already recognized by Davis and coworkers in the late 1970s [108–110]. They covalently linked PEG with molecular weights of 1.9 and 5 kDa to the protein bovine liver catalase. The conjugates obtained were injected into mice and exhibited a significantly enhanced circulating half-life, reduced immunogenicity and antigenicity while retaining their bioactivity to a large extent. The rationale behind this stabilization effect is the steric hindrance of the PEG shell, which prevents reaction of immune cells with the protein and protects it from degrading proteases. Interactions of cells and proteins with PEG–protein conjugates would involve the energetically unfavorable displacement of hydrated water molecules and the entropically unfavorable compression of the PEG chains [45].

The covalent attachment of PEG chains to peptides and proteins has since then been termed *PEGylation* and has been extensively reviewed [46, 47, 111–116]. The PEGylation of peptides and proteins nowadays is widely used in the pharmaceutical industry. One of the first PEGylated proteins that appeared on the market is a conjugate of PEG with bovine adenosine deaminase (Adagen®, Enzon Inc.) which received FDA approval in 1990. The enzyme bovine adenosine deaminase (ADA) is used to treat severe combined immunodeficiency disease [117], but is cleared rapidly from the plasma. However, the modification of this enzyme with multiple PEG chains with a molecular weight of 5 kDa extends the circulating half-life by a factor of 6.4 in rats [118].

Although the PEGylation of proteins has proven to be very valuable, many of the first-generation PEGylation products suffered from a severe loss in bioactivity, ranging from 20 to 95% [119]. This decrease in activity depends mainly on the chain length of the attached polymers and the site of the protein to which they are coupled [120]. Moreover, the attachment of multiple PEG chains leads to mixtures of isomers with different molecular weights, which makes it very difficult to reproduce drug properties from one batch to the next. For these reasons, it is of the utmost importance to have control over the conjugation process. The possibilities in order to gain control over the macromolecular architecture will be discussed in Section 16.3.3.

In addition to the development of peptide and protein drugs, there is growing interest in the utilization of antibodies and antibody fragments as therapeutic entities due to their high target binding specificity [121–124]. Furthermore, progress in genomic and proteomic technologies will lead to the identification of many more targets for antibody-based therapies. Additionally, antibodies and antibody fragments in particular can be engineered in order to fulfill specific needs. This implies that the antigen binding efficiency can be enhanced and new effector functions can be introduced via conjugation to toxins, enzymes or radionuclides. These antibodies and especially antibody fragments, however, suffer from stability problems *in vivo* in analogy with peptides and other proteins. These problems can also be circumvented by PEGylation [122]. Nevertheless, non-specific coupling of PEG chains can lead to a substantial loss in binding activity. After the attachment of three molecules of 5-kDa PEG to lysine residues of an antibody fragment, the binding activity was reduced to 62% compared with the unmodified fragment and it fell even further to 17% when six polymer chains were attached [125]. Presumably this is caused by PEGylation in the vicinity of the antibody binding domain preventing interaction with the antigen due to steric hindrance. Another plausible explanation of this decreased affinity of the antibody can be the interference with ionic interactions that are likely to initiate antigen binding [122]. Therefore, in accordance with other peptides and proteins, it is crucial to obtain control over the conjugation process at the molecular level. Moreover, in order to apply these conjugates in medicine, the synthesis has to be reproducible to preserve equal pharmacokinetic values.

16.3.3
Well-defined Macromolecular Therapeutic Architectures

Initial PEGylation processes were based on conjugation to the free amino group of lysine residues present in proteins, which usually results in statistical multiple site additions. Site-specific polymer attachment can be attained via conjugation to thiol groups present in cysteine residues. In Scheme 16.2, a number of coupling procedures to thiols are depicted. Free cysteines can be introduced at the surface of proteins either through reduction of disulfide bridges or by introduction of cysteine residues via protein engineering [126–129], as discussed in

Section 16.2.2. Another approach is to replace lysine residues by other amino acids, thereby limiting the number of conjugation sites [130].

Specific PEGylation to the amide group of glutamine and the hydroxyl functionality of serine and threonine can be accomplished under mild conditions using enzymes. Amine-functionalized PEG can be coupled to glutamine units using the enzyme transglutaminase [131]. For the site-specific PEGylation of serine and threonine a different methodology has been applied [132]. First, N-acetylgalactosamine (GalNAc) was specifically attached to serine and threonine residues using a recombinant O-GalNAc. Subsequently, to the glycosylated sites sialic acid-functionalized PEG was coupled enzymatically, which was accomplished by a sialyltransferase. Recently, a method was developed specifically to alkylate tyrosine residues using π-allylpalladium complexes [133].

As stated before, protein engineering can be exploited to introduce additional functionality in proteins via the incorporation of unnatural amino acids. This methodology makes it possible to apply coupling strategies which are orthogonal with respect to other functional groups present. By incorporation of unnatural amino acids bearing azido groups, the highly specific Staudinger ligation [134] and click reaction [51] can be employed for site-specific attachment of PEG in a very elegant fashion (Scheme 16.3).

Another elegant example of site-specific polymer attachment is the synthesis, using peptide chemistry and ligation methodologies, of "synthetic erythropoiesis protein" (SEP), which is a 51-kDa protein–polymer conjugate consisting of a 166 α-amino acid polypeptide chain and two coupled branched, monodisperse and negatively charged PEG-like polymer chains [135], as depicted in Fig. 16.5. Into the polypeptide chain, two noncoding levulinyl-functionalized lysine residues (Lys24 and -126) were introduced, which serve as conjugation sites for polymer attachment. This polymer bioconjugate displayed comparable activity *in vitro* but a prolonged half-life *in vivo* compared with human erythropoietin (Epo), which is a glycoprotein hormone controlling the proliferation, differentiation and maturation of erythroid cells. Kochendoerfer and coworkers extended this

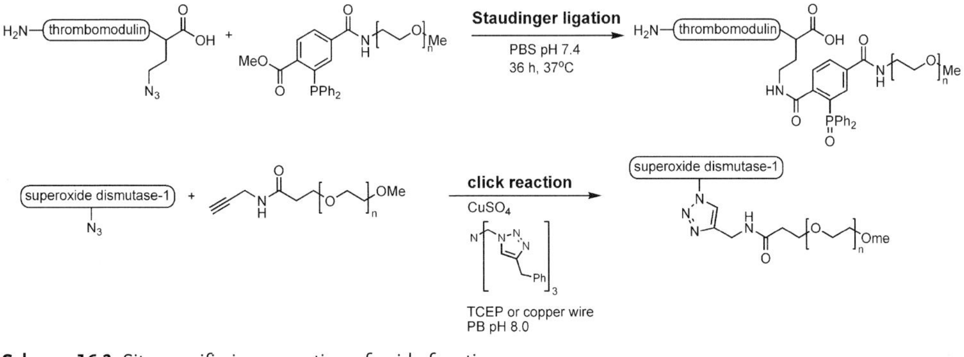

Scheme 16.3 Site-specific incorporation of azide-functionalized amino acids and subsequent ligation with PEG exploiting this azido group [51, 134].

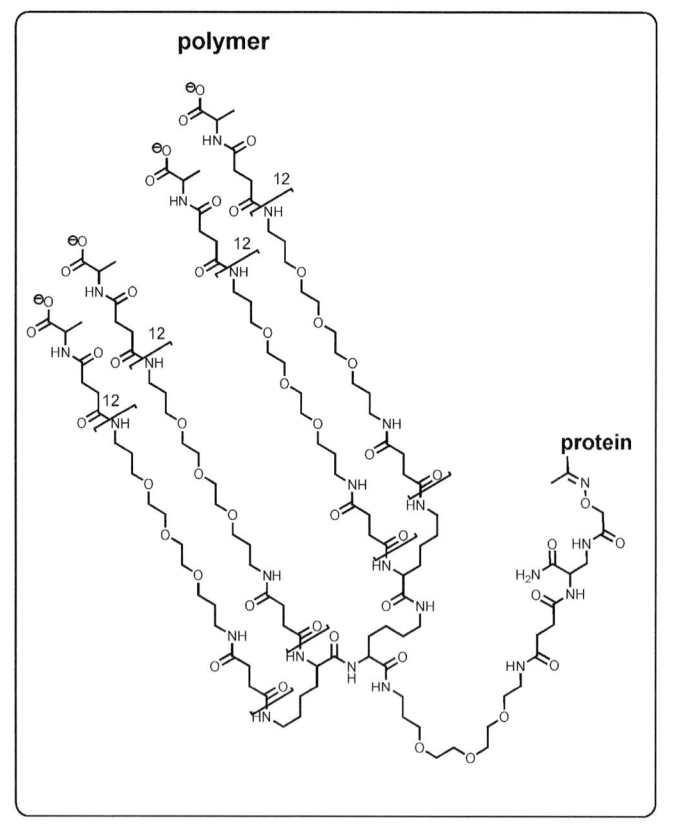

A¹
P
P
R-L-I-C-D-S-R-V-L-E-R-Y-L-L-E-A-K-E-A-E-K-I-T-T-G-C-A-E-H-C-S-L-N
                                                                E
    E-V-A-Q-Q-G-V-E-M-R-K-W-A-Y-F-N-V-K-T-D-P-V-T-I-K
    V
T-R-C-A-E   W
G   G   Q-G-L-A-L-L-S-E-A-V-L-R-G-Q-A-L-L-V-K-S-S-Q-P
D   T   W
R¹⁶⁸   Y   C-S
        L   L-A-R-L-L-T-T-L-S-R-L-G-S-V-A-K-D-H-V-L-Q-L-P
        K   G
        L   A
        K   Q
        G   K
        R   C-S
        L   A
        F   I-S-P-P-D-A-A-K-A-A-P-L
        N   R
        S-Y-V-R-F-L-K-R-F-T-D-A-T-I-T

Fig. 16.5 Schematic representation of the structure of SEP with two conjugated branched PEG-like polymer moieties. Adapted from [135].

methodology to a series of polymers with distinct architectures, showing that the duration of action *in vivo* is strongly dependent on the overall charge and size of the conjugate [136].

Solid-phase peptide synthesis is an elegant methodology to introduce PEG specifically at the N-terminus of peptides. Klok and coworkers used this technique to PEGylate a coiled coil-forming peptide [137]. The coiled coil is a protein folding motif consisting of two to five α-helical peptide strands which are intertwined into a superhelix [138]. In proteins these coiled coil structures are largely unfolded at neutral pH and fold into homo-oligomeric coiled coils at lower pH. The PEGylated peptides retained their folding ability, which makes these stimuli responsive conjugates interesting candidates for drug and gene delivery.

As mentioned in the first part of this chapter, progress in polymer science has resulted in the controlled synthesis of a myriad of polymers with numerous functional end-groups. Thereby the opportunities are increasing for protein conjugation. By application of ATRP, various polymers with N-succinimidyl ester [139], aldehyde [140], N-maleimido [141], pyridyl disulfide [142, 143] and azide [50] end-groups were synthesized and subsequently conjugated to amine, thiol and acetylene groups, respectively, present in proteins, as depicted in Scheme 16.4.

In order to simplify purification of the bioconjugate polymers obtained, solid-phase synthesis can be adopted, as shown by Kiessling and coworkers [144]. N-Succinimidyl ester- and maleimido-functionalized polymers prepared by ROMP were anchored to a Rink-type polystyrene resin via a Diels–Alder reaction, after which saccharides were tethered to the polymers. Subsequent cleavage of the desired product was accomplished using a retro-Diels–Alder reaction with furan.

The combination of controlled polymerization techniques with conjugation methodologies implies that we nowadays have access to tailor-made polymer biohybrids with all sorts of compositions and architectures. In addition to control over the molecular weight and composition of polymers, the chain topology can also be altered. Branched polymer structures have an increased surface shielding effect, which has a positive contribution to the protection of conjugated proteins against degradative proteases and antibodies [145]. Another advantage of branched polymers, graft polymers and star-shaped polymers is the presence of multiple sites which are available for drug attachment. This is especially valid for three-dimensional architectures such as dendrimers and dendronized polymers [146, 147]. Their surfaces are very suitable for a high loading of drugs, which make them good candidates as carriers for drug delivery [148–150]. Furthermore, the increased complexity of current macromolecular systems also allows multivalent interactions, which greatly enhances the biorecognition properties [151–155]. Maynard et al., for example, synthesized well-defined copolymers containing both multiple RGD integrin binding ligands and the pentapeptide Pro–His–Ser–Arg–Ans (PHSRN), which has a synergistic effect on RGD binding [156, 157] (Fig. 16.6a). Furthermore, Kiessling and coworkers studied in detail the potential use of saccharide-modified polymer architectures

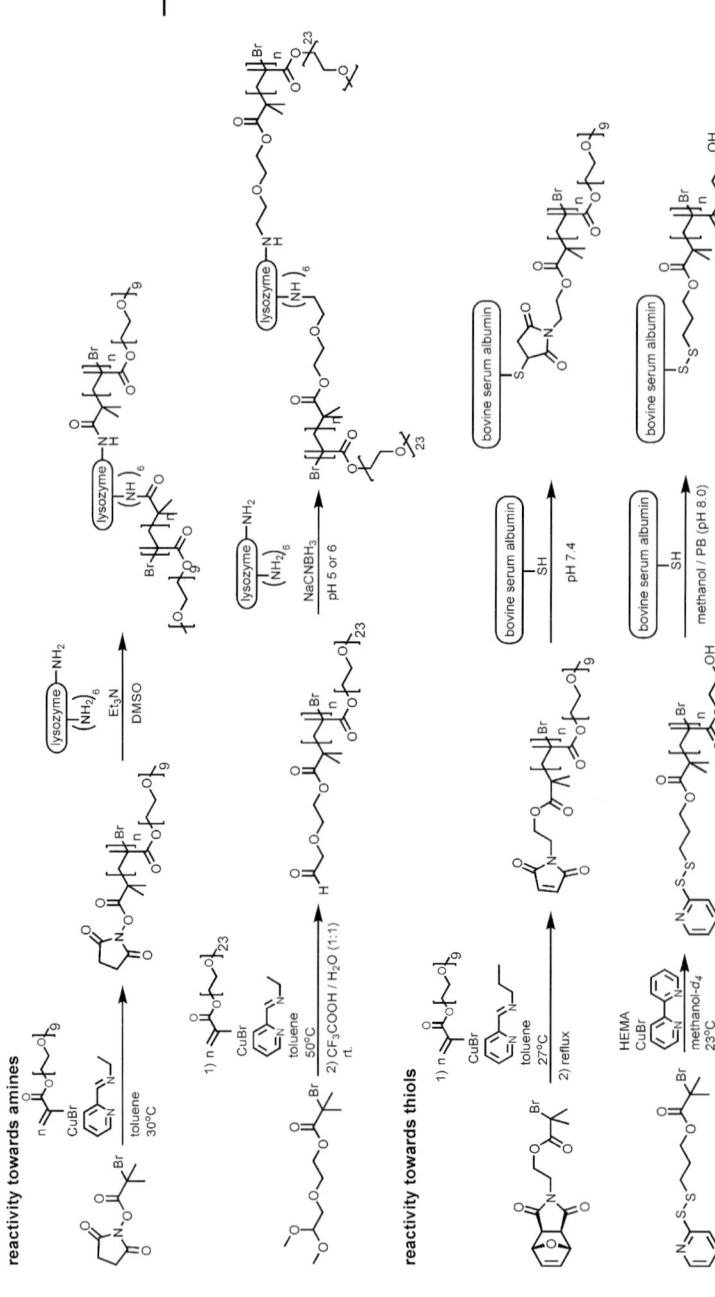

2660 *16 Applications of Polymer Bioconjugates*

Scheme 16.4 End-functionalized polymers prepared by ATRP which are ligated to proteins [50, 139–142].

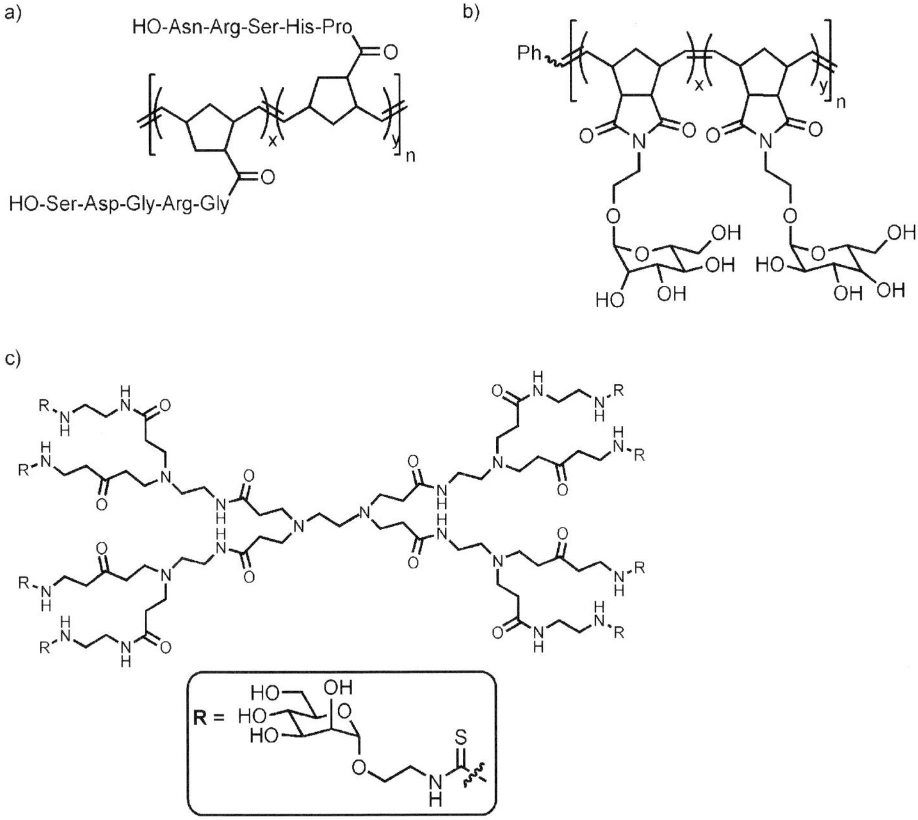

Fig. 16.6 Structural representation of well-defined multivalent ligand embracing polymer architectures. (a) Copolymer of the integrin binding motif RGD and pentapeptide PHSRN which synergistically enhances RGD binding [157]; (b) copolymer derivatized with galactose and mannose ligands [158, 159]; (c) first-generation PAMAM glycodendrimer [158, 159].

in multivalent ligand receptor binding [158, 159], both as powerful inhibitors [160–162] and as effectors eliciting a specific biological response [163, 164] (Fig. 16.6b and c). Although most polymer bioconjugates that are in clinical trials are still based on conventional synthetic approaches, the improvement of bioconjugate architectures via the application of new preparation methods will certainly lead to the construction of even more potent drugs and the future of tailor-made polymer therapeutics is therefore a bright one.

16.4
Bioactive Surfaces

The interaction of materials with the environment occurs via the interface. In recent years, much progress has been made in understanding and influencing these interfacial properties. Conjoined with the insights given by molecular biology in the molecular mechanisms and signaling processes which take place during the interaction of living cells with their surroundings, this has paved the way for the engineering of tailor-made polymer biohybrid materials which can communicate with a variety of biological entities [165, 166]. These advances have resulted in many applications in the fields of array development for diagnostics, cell culturing/tissue engineering and antibacterial- and non-thrombogenic coatings.

16.4.1
Bioconjugate Arrays

The discovery of many novel bioactive peptides, proteins and antibodies can be attributed to the progress in genomics and proteomics, which nowadays provide many insights into diseases on the DNA and protein expression level. As a consequence, there is a growing need for high-throughput screening of oligonucleotides and proteins and therefore for miniaturization of DNA and protein arrays in order to analyze hundreds or even thousands of markers simultaneously [167–170], which requires enhanced sensitivity of these arrays.

Initially, arrays were produced using robotic spotter techniques that physically apply the sample to the substrate using pins or needles, which limits the resolution attainable [171]. Photolithography [172, 173], ink-jet printing techniques [174, 175], dip-pen nanolithography [176, 177] and microcontact printing [178, 179] are examples of suitable alternatives for obtaining higher resolutions by spatial ordering of substrates at both the micrometer and nanometer length scales.

Biohybrid polymers constitute a class of materials highly suitable to be employed in diagnostic arrays. The utilization of polymers leads to the presence of multiple conjugation sites along the polymer backbone or at the polymer chain ends, which can result in improved sensitivity and resolution. The usefulness of biohybrid polymers in diagnostic tests has clearly been demonstrated with DNA microarrays. These arrays are based on specific hybridization of a known oligonucleotide sequence immobilized on a defined position and a complementary target sequence. Detection is based on signals arising from labeled targets either by fluorescence or via a specific enzyme which is able to convert substrate into a fluorescent product. Both oligonucleotide-functionalized polymers [180–182] and dendrimers [183, 184] have been used to enhance the sensitivity of DNA detection. Different strategies have been applied to synthesize these polymer–oligonucleotide conjugates, namely (a) deposition of polymers containing reactive groups and subsequent spotting and chemical binding of oligonucleotides [184],

(b) spotting of oligonucleotides bearing polymerizable handles which are anchored to the surface by copolymerization [180] and (c) covalent linkage of the oligonucleotide to the polymer prior to application to the surface [181]. An advantage of using controlled polymerization techniques is that block copolymers can also be synthesized. This latter method has been shown to enhance the sensitivity of the DNA array to a large extent [182].

In addition to improvements in sensitivity of the arrays due to a higher analyte capture capacity, the application of polymers has further advantages. Often spacers are required between the oligonucleotides or proteins and the support matrix to avoid steric hindrance, which limits the loading to the surface. A major problem in the development of protein arrays in particular is the denaturation of proteins on the surface. This problem is often caused by the hydrophobicity of the surface. Therefore, the surface chemistry for microarrays is still developing [185]. It has been shown that application of a thin layer of PEG between the solid support and the analyte introduces flexibility which reduces the steric hindrance, resulting in enhanced binding of proteins and oligonucleotides [186, 187]. PEG-modified surfaces furthermore not only lead to a higher capture capacity, but also prevent nonspecific binding of proteins to the surface. Therefore, these antifouling PEG coatings have a huge potential in microarray technology. Moeller and coworkers showed the successful deposition of 20mer oligonucleotides and green fluorescent protein on star PEG coatings containing isocyanate functionality [188, 189]. After covalent immobilization, no additional blocking of reactive groups was necessary, since isocyanates were completely hydrolyzed to amino groups by air humidity.

Hence the modification of solid supports with thin polymer films is a perfect tool to tailor surface properties. In this respect, controlled surface-initiated polymerization techniques such as NMP, RAFT and ATRP are very attractive because they allow the synthesis of polymer brushes of predetermined thickness, composition and density of reactive groups at the surface [14, 17, 190–193]. Klok and coworkers used surface-initiated ATRP to prepare well-defined brushes of biologically inert poly[oligo(ethylene glycol)methacrylate] (POEGMA), on to which fusion proteins containing O^6-alkylguanine-DNA-alkyltransferase (AGT) were immobilized selectively [194], as depicted in Scheme 16.5. AGT is a human DNA repair protein which irreversibly transfers the alkyl group of O^6-alkylguanine to one of its own cysteine residues [195]. However, the substrate specificity of AGT is low and therefore it also readily reacts with benzylguanine derivatives, which in this example was used to immobilize AGT fusion proteins.

The above-mentioned examples require accurate application of the analyte using techniques such as ink-jet printing and microcontact printing. Another possibility is to create selectively reactive groups at the polymer surface which serve as a handle for further functionalization. Maynard and coworkers deposited a film of poly(3,3'-diethoxypropyl methacrylate) (PDEPMA) on a silicon dioxide support. PDEPMA is a pH-responsive polymer containing acetal groups, which can be hydrolyzed under acidic conditions with formation of aldehydes. By employing a photoacid generator (PAG), a species that releases a proton

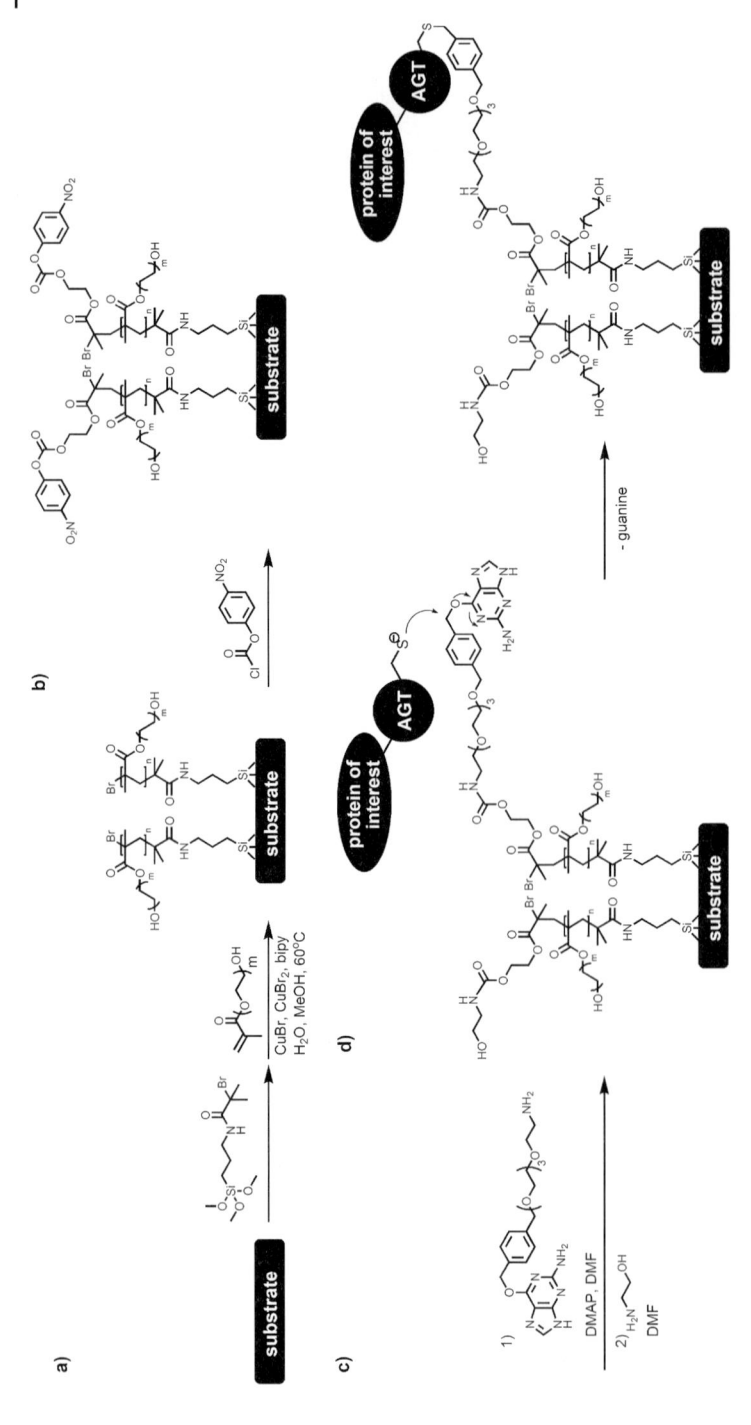

Scheme 16.5 Synthesis of protein-functionalized POEGMA brushes. (a) Grafting of initiator on to glass substrate and subsequent ATRP polymerization; (b) activation of hydroxyl groups by treatment with p-nitrophenyl chloroformate; (c) functionalization of polymer brush with benzylguanine (1) and blocking of residual reactive groups with 2-aminoethanol (2); (d) specific attachment of AGT fusion proteins to the surface. Adapted from [194].

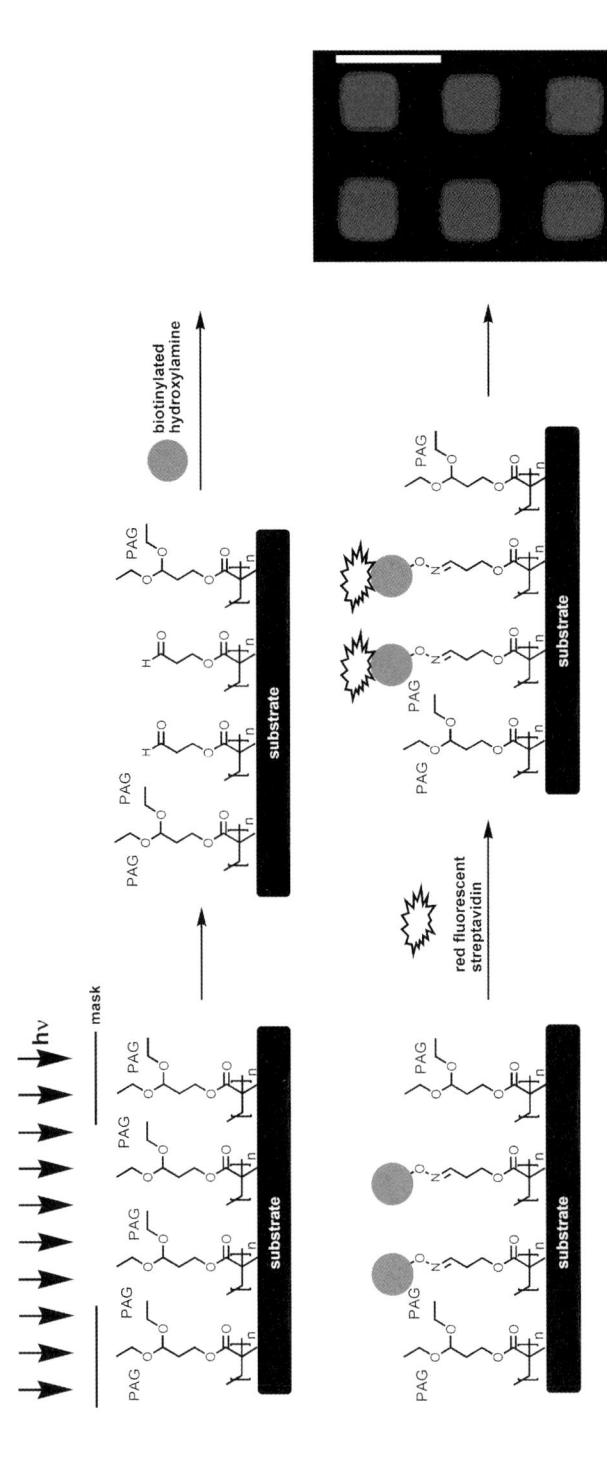

Scheme 16.6 Selective generation of aldehyde groups on PDEPMA film using PAG, UV irradiation and a mask. These reactive groups permitted selective biotinylation of the surface to which red fluorescent streptavidin has been coupled. The white scale bar represents 25 μm. Reprinted with permission from [196]. Copyright 2005 American Chemical Society.

upon UV exposure, aldehyde groups were formed on specific locations using UV irradiation and a mask. These so-formed conjugation sites were subsequently used to immobilize red fluorescent protein (Scheme 16.6) [196].

16.4.2
Cell Adhesive Coatings

The ability to decorate polymer films with proteins can also be applied to promote cell adhesion. Cells are normally surrounded by an extracellular matrix (ECM), which is a highly hydrated network of proteins consisting of fibrillar proteins, non-collagenous glycoproteins and hydrophilic proteoglycans. Furthermore, many soluble macromolecules such as growth factors, chemokines, cytokines and proteins at the surfaces of neighboring cells are present in the ECM. Cell functions such as differentiation, proliferation and migration are regulated by communication with the ECM effectors [197]. Mimicking the ECM permits directional cell growth, and functional substitutes for damaged tissue can be developed [198]. This field of research, called tissue engineering, will be discussed only briefly because it is extensively reviewed by Hubbell in this volume.

Many of the adhesive proteins present in the ECM contain the tripeptide Arg–Gly–Asp (RGD) serving as the interaction sequence for cell surface recep-

Fig. 16.7 Schematic representation of two categories of polymer–peptide hybrid materials which potentially can be used in tissue engineering, namely (1) soluble polymers bearing peptide moieties tethered to the polymer backbone (a), incorporated into the polymer backbone (b) or coupled to the chain ends (c); and (2) insoluble hybrid materials to which peptides either are immobilized on a polymer surface (a) or embedded in a cross-linked polymer matrix. Adapted from [119].

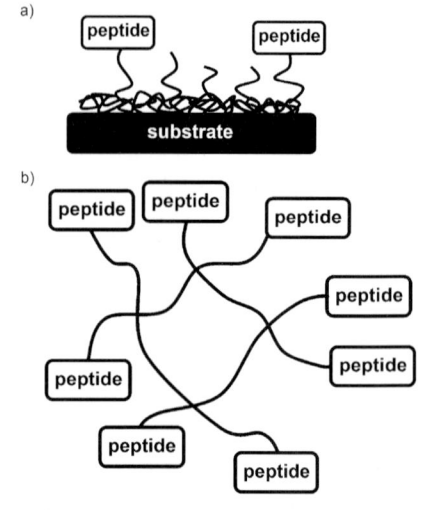

tors, called integrins [199, 200]. Polymers modified with this RGD motif can mimic the ECM, thereby promoting receptor-mediated cell adhesion. Additionally, binding of RGD sequences to integrins also inhibits cell functions such as angiogenesis, which is an important control point in cancer, and, therefore, these peptide sequences are also used for anti-cancer therapeutics, as discussed in Section 16.3.1. Owing to the cell binding properties of these polymer–peptide hybrids, they are excellent candidates to act as scaffolds or supports for tissue engineering. In this respect, many hybrid systems are being studied which can be roughly subdivided into two categories, as shown in Fig. 16.7, namely (1) soluble hybrid materials that direct cell growth either in solution or as thin deposited film and (2) peptide-modified polymer surfaces and cross-linked polymer matrices with embedded bioactive species [119].

The first category, soluble polymer–peptide hybrids, encompasses two distinct types of polymers. First, biodegradable copolymers such as polyglycolide [201, 202], polylactide [203], poly(lactide-*co*-glycolide) [204] and poly(lactide-*co*-lysine) [205–207] are used, which have the advantage of *in vivo* degradation after fulfillment of the desired task, thereby evading negative immune responses and retrieval of the polymer support. Additionally, graft and comb-type polymer topologies have been explored. Hubbell and coworkers, for example, functionalized poly(L-lysine)-*graft*-poly(ethylene glycol) (PLL-g-PEG) with an RGD sequence to promote cell adhesion (Fig. 16.8) [208]. The positively charged PLL block induced spontaneous adsorption from aqueous solution on to negatively charged surfaces, whereas the PEG block resisted nonspecific adsorption of extracellular and cell surface proteins.

Second, in addition to biodegradable polymers, stable polymers have also been utilized when long-term resistance in the body is required. For instance, polyurethanes with pendant RGD peptides have been prepared for this purpose [209, 210].

In addition to soluble hybrid materials, much research has been conducted in order to develop insoluble, bioinert polymer–peptide conjugates concerning tissue reconstitution. In this respect, inert surfaces composed of poly(vinyl chloride), polyurethane and polydimethylsiloxane with tethered RGD moieties have been exploited. These surfaces are either modified directly, if suitable functional groups are present at the surface, or subjected to a plasma treatment to induce conjugation sites [211]. Other insoluble polymeric matrices frequently used to mediate cell growth are cross-linked materials, particularly hydrogels [212]. Hydrogels have the ability to absorb and retain large amounts of water, thereby resembling soft tissue. Moreover, hydrogels are highly permeable to oxygen, nutrients and other water-soluble metabolites, which makes them ideal candidates to benefit cell growth for tissue engineering.

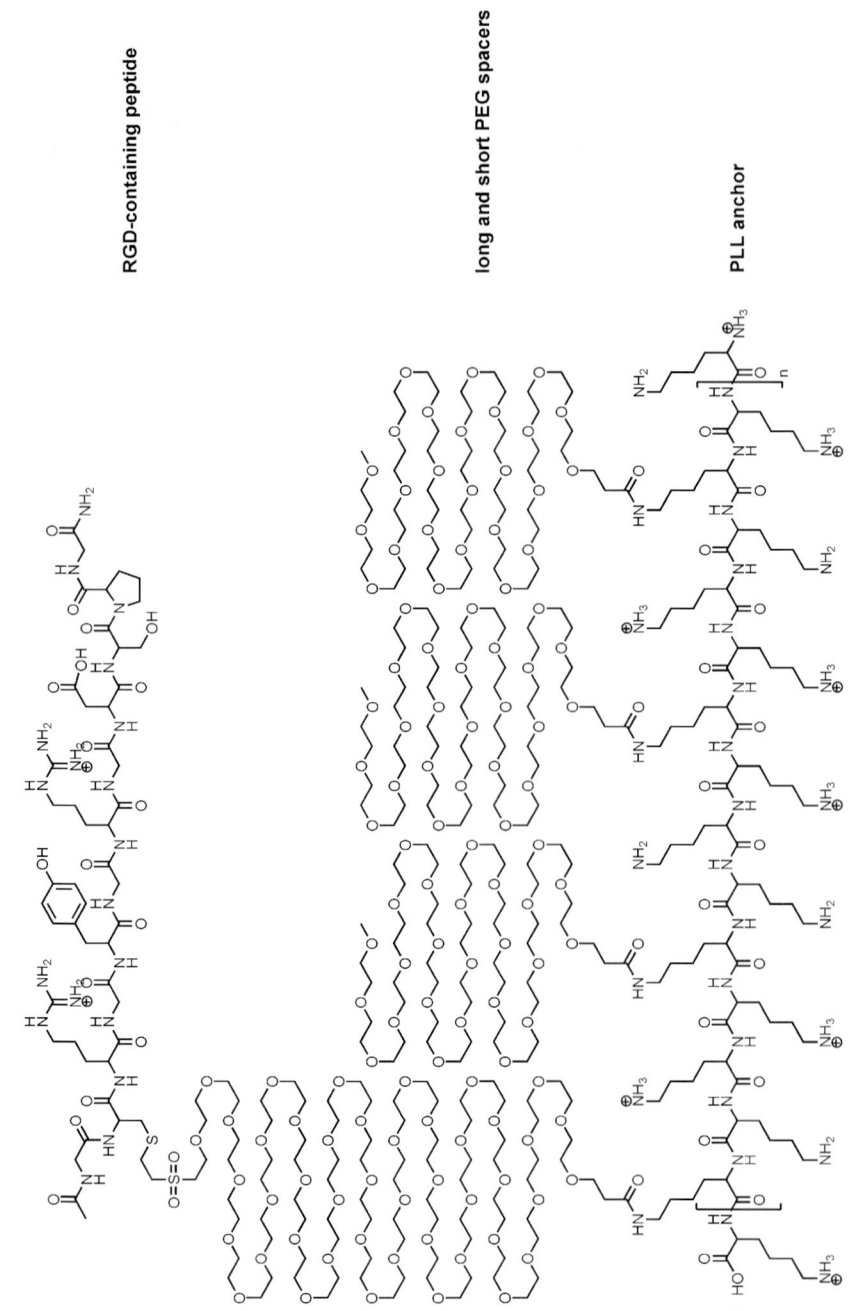

Fig. 16.8 Structure of a PLL-g-PEG copolymer laterally functionalized with the pentapeptide RGDSP. Adapted from [208].

16.4.3
Non-fouling Coatings

In the previous section, the deposition of cells on polymeric materials in order to perform cell culturing or to mediate tissue formation was described. However, in applications such as body implants, stents, soft contact lenses, biosensors and drug delivery vehicles, sedimentation of proteins and cells on the material surface must be circumvented. Unwanted protein adsorption causes blood

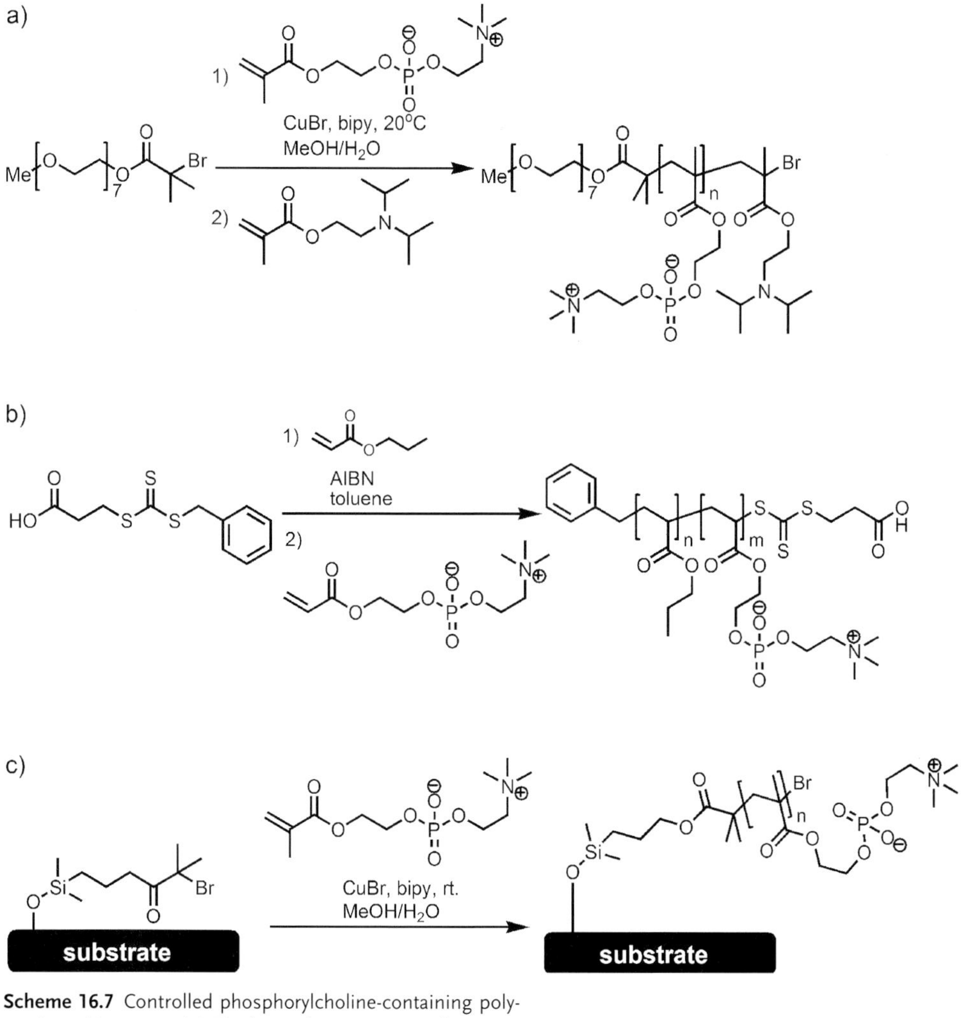

Scheme 16.7 Controlled phosphorylcholine-containing polymeric architectures. (a) Block copolymers prepared via ATRP [221]; (b) RAFT polymerization [222]; (c) biomimetic polymer brush [226].

platelet adhesion, thrombus formation, bacterial adhesion and other negative biological responses.

Biomembranes are regarded as the best surfaces for smooth interactions with blood components such as proteins and cells. Biomembranes are composed of a bilayered structure of amphiphilic phospholipids with proteins located on top of or within this bilayer. Mimicking biomembranes will avoid negative responses of the human body [213, 214]. By polymerization of phosphorylcholine-containing monomers [215–217], such a lipid like surface can be formed. These materials can find application, for instance, in the production of artificial vessels. Traditionally, these vessels are produced from poly(ethylene terephthalate) and polytetrafluoroethylene (PTFE). However, these materials can only be used to replace arteries with an internal diameter over 6 mm due to a required blood clotting pretreatment. Materials manufactured from segmented polyurethane (SPU) coated with phosphorylcholine polymers displayed significantly reduced protein adsorption and platelet adhesion *in vitro* [218]. In order to examine clot formation *in vivo*, an artificial heart made of SPU was coated with a phosphorylcholine polymer and implanted in a sheep. No clot formation on the surface was observed 1 month after implantation [214].

Phosphorylcholine functionalized polymers have already proven to be very successful as non-thrombogenic materials. To obtain more control over the surface properties of these materials, the application of controlled radical polymerization techniques has been explored [219, 220]. Consequently, this has led to the synthesis of well-defined block copolymers [221–224] and polymer brushes [225–227], thereby allowing tailoring of the properties (Scheme 16.7).

16.4.4
Antimicrobial Coatings

Bacterial adhesion to surfaces can lead to undesired situations, especially when food safety or medical devices, such as catheters, are concerned. The growing societal demand for healthy living conditions requires the design of materials that kill microorganisms on contact.

Cationic polymers comprising ammonium [228–233] or phosphonium [234] salts, polylysine [235], pyridinium salts [236] or polyguanidines and polybiguanidines [237] have proven to exhibit such biocidal properties. These polymers showed strong bactericidal activity against both Gram-positive and Gram-negative bacteria, owing to interactions of their positive charges with the negatively charged phospholipids and proteins present in the cytoplasm membranes [238].

Although these biocidal cationic polymers are very useful for many applications, no selectivity between bacterial and mammalian cells can be obtained. One way to overcome this lack of selectivity is to functionalize polymers with organism-specific antibiotics [236]. In Scheme 16.8a, an example of biodegradable polycaprolactone with the antibiotic ciprofloxacin included in the polymer backbone is depicted [239]. Furthermore, a large class of so-called host defensive peptides (HDPs) display broad spectrum antibiotic activity while retaining selec-

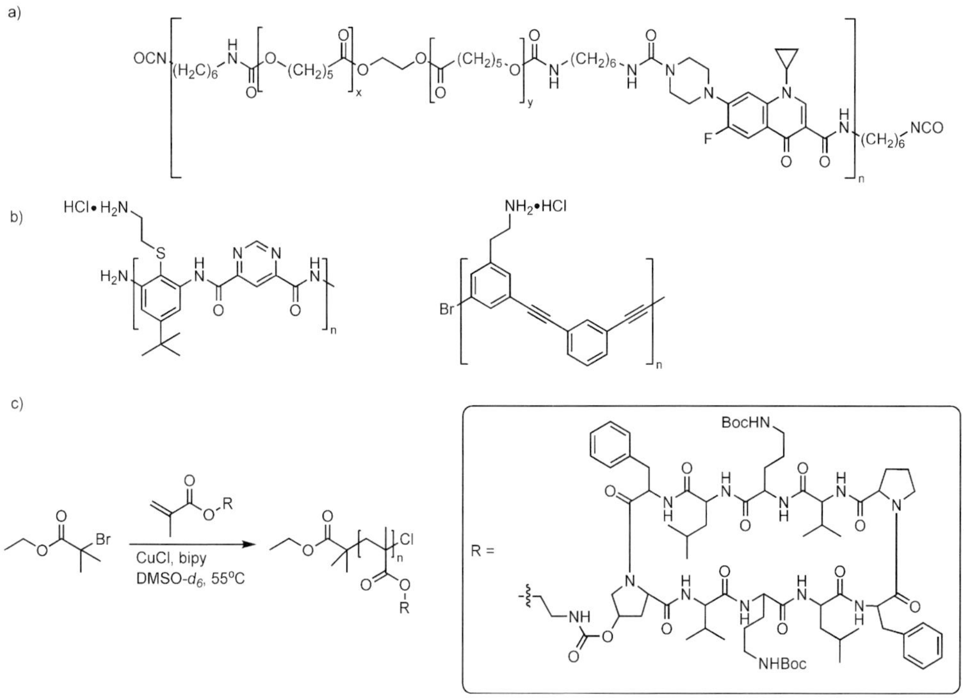

Scheme 16.8 Formation of polymers possessing antimicrobial properties. (a) Biodegradable polycaprolactone containing the antibiotic ciprofloxacin [236]; (b) antibiotic peptide mimicking facially amphiphilic polymer structures [246]; (c) preparation of a well-defined polymer with pendant gramicidin S moieties utilizing ATRP [248].

tivity between bacterial and mammalian cells [240]. The mode of action of these peptides stems from the ability to adopt a facially amphiphilic structure which disrupts the phospholipid membrane [241–243]. This structure can be mimicked by applying peptoids [244, 245] or synthetic polymers [246, 247], as depicted in Scheme 16.8 b.

Another possibility is to incorporate antimicrobial peptides into synthetic polymer materials. Van Hest and coworkers successfully polymerized the antibiotic cyclic peptide gramicidin S bearing a methacrylate handle using ATRP [248] (Scheme 16.8 c). The application of controlled polymerization techniques allows the combination of the versatility of synthetic polymers with the antibiotic properties of this type of peptide. Moreover, using controlled polymerization methodologies, a variety of topologies, such as block copolymers and surface-grafted polymers, can be prepared in order to fulfill specific requirements of distinct applications.

16.5
Smart Materials

In addition to uses in pharmaceutical applications and in bioactive coatings, biohybrid polymers can also be of interest for the construction of smart materials. The introduction of stimulus-responsive behavior in bulk materials such as hydrogels is particularly useful in the biomedical field for tissue engineering and controlled drug delivery. Stimulus-responsive behavior can also be introduced in non-cross-linked biohybrid polymers. Smart conjugates which can precipitate on demand have shown their potential, for example, in the affinity purification of biomolecules. The introduction of well-defined assembly behavior in proteins or DNA via conjugation with synthetic polymers can lead to the construction of (bio)active nanoscopic objects which could be used as building blocks for nanotechnology purposes. In this section, these different classes of smart conjugates are described, applications of which in some cases are somewhat more distant compared with the previous sections, but the potential of which is certainly appealing.

16.5.1
Smart Polymer Bioconjugates

Conjugates obtained by the attachment of stimulus-responsive synthetic polymers to biomolecules can be considered as "doubly smart" [249, 250]. The most extensively studied polymer in this respect is PNIPAAm. This polymer displays a lower critical solution temperature (LCST) of 32 °C. This implies that it is soluble in water up to 32 °C and with increasing temperature it precipitates abruptly [251]. This phase separation behavior is, furthermore, completely reversible.

Bioconjugates embracing PNIPAAm moieties also exhibit thermally induced phase separation. This behavior has been exploited to precipitate enzymes from their reaction solutions, hence facilitating both product recovery from the supernatant and recycling of the enzyme [252–254]. Moreover, poly(methacrylic acid) has been used to induce pH-dependent phase separation for enzyme recovery [255–258].

Likewise, this concept of reversible precipitation and solubilization of biohybrid polymers can be adopted in affinity precipitations to extract target molecules from complex mixtures [259, 260]. In the solution phase, specific proteins [261], antigens [262, 263] or oligonucleotide sequences [264, 265] are ligated to biohybrid polymers and isolated from solution by stimulus (temperature [261], pH [266] or light [267])-induced phase separation of the formed affinity complex. Furthermore, application of oligonucleotide sequences as anchors between proteins and smart polymers permits the selective incorporation of polymers with different stimuli responsiveness that are discriminated by the attached DNA sequences. Hence in this way, multiple affinity complexes can be selectively and sequentially separated from complex mixtures [268], thus opening up the possibility of performing immunoassays in the solution phase [250].

Analogous to polymer–protein drug conjugates discussed in Section 16.3.2, random conjugation of the above-mentioned smart polymers can impede the activity of proteins. Therefore, Hoffman and coworkers inserted specific reactive amino acids, such as cysteine, in proteins via protein engineering, to which they subsequently ligated PNIPAAm [269, 270]. Site specific introduction of PNIPAAm at a remote position from the active site of the protein hardly affected its activity [269].

Protein engineering techniques can also be used deliberately to attach responsive polymers in the vicinity of protein-active sites. In this case, environmental changes result in reversible blocking or unblocking of the protein's active site, which also can lead to a triggered release of bound ligands from the protein binding site [270–273]. This switching of enzyme activity is illustrated in Scheme 16.9. This methodology can also be exploited to achieve size-selective binding of proteins. Hoffman and coworkers therefore coupled poly(N,N-diethylacrylamide) (PDEAAm) to streptavidin approximately 20 Å from its active site [274]. Below the LCST, the polymer is in a random coil conformation and acts as a shield which prevents binding of large biotinylated proteins. Conversely, above the LCST, PDEAAm collapses and the binding site is exposed to large proteins. This distinction in binding size could be utilized in affinity separations, biosensors and diagnostics.

Another field where responsive biohybrid polymers could be very useful is antisense therapy. Antisense therapy prevents the expression of disease-related

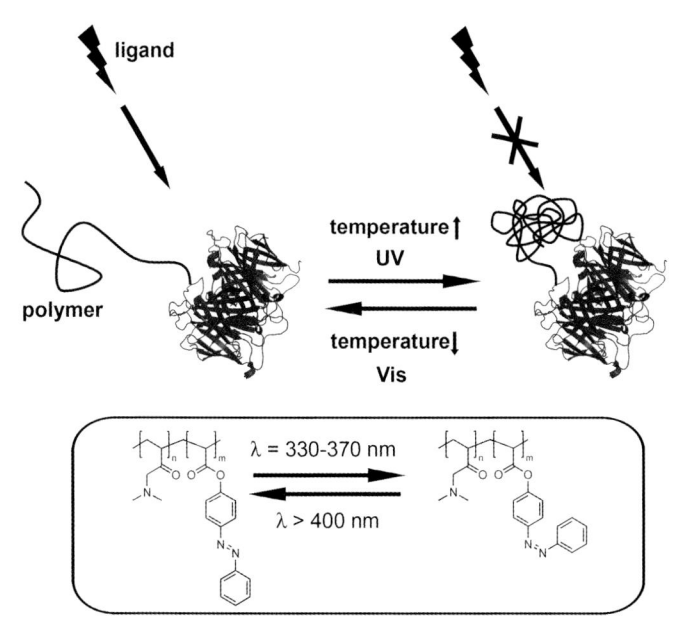

Scheme 16.9 Schematic representation of a temperature- [270, 271] or light-responsive [272, 273] enzyme switch.

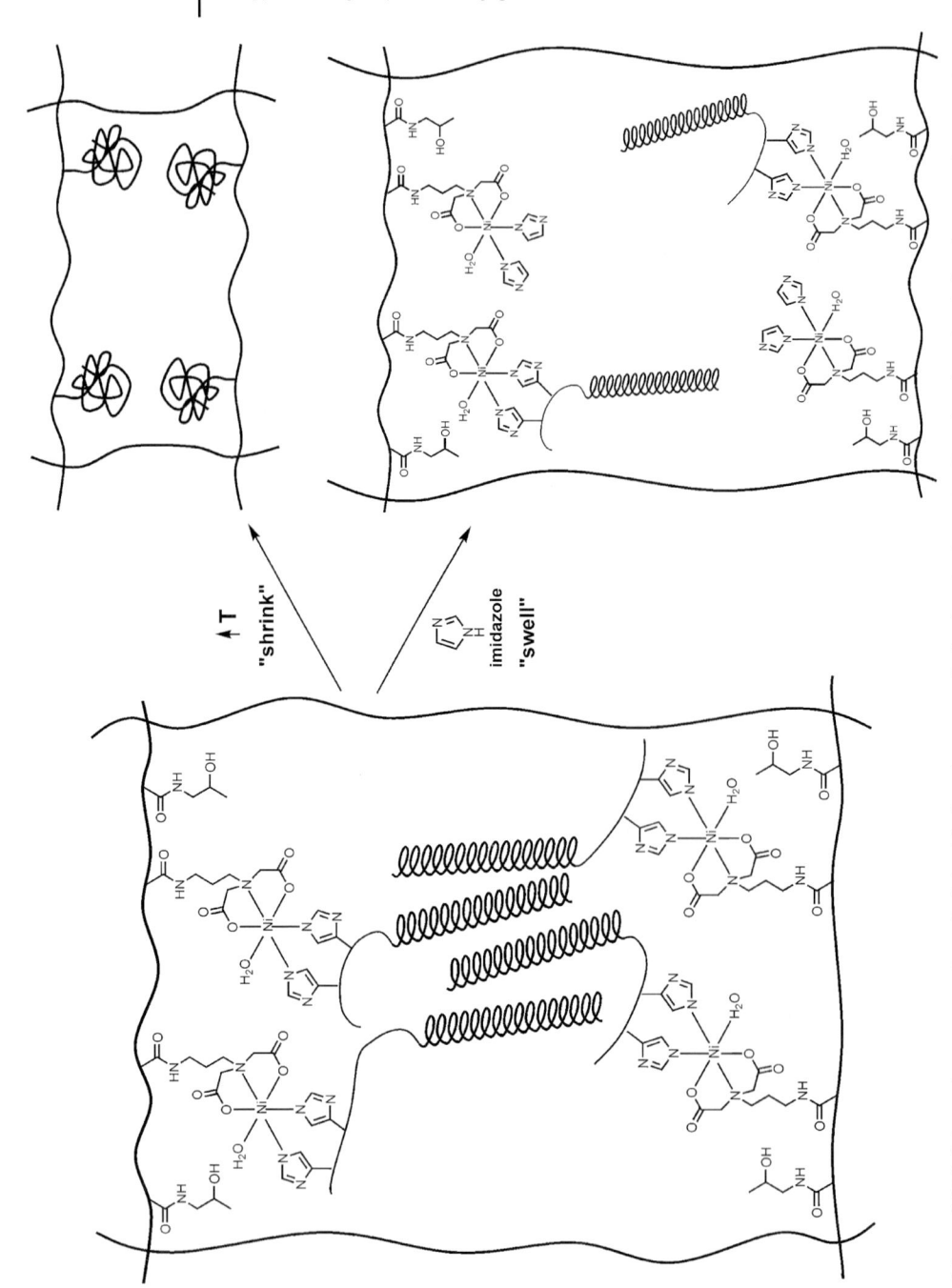

Scheme 16.10 Schematic representation of a hybrid hydrogel composed of poly(HPMA-co-DAMA) with tethered coiled coil protein domains. Temperature changes or imidazole addition induces a volume transition of this system. Adapted from [280].

proteins on the genetic level by blocking mutated or overactive genes. In this respect, Murata and coworkers synthesized PNIPAAm–oligodeoxynucleotide (ODN) conjugates [275–277]. Above the LCST, collapsed PNIPAAm chains serve as a shield preventing degradation of the ODN. With decrease in temperature, however, the antisense sequence is exposed, allowing binding to the target mRNA.

An example of a responsive biohybrid hydrogel worth mentioning was presented by Kopeček and coworkers, who prepared a hydrogel from a copolymer of HPMA and the metal-chelating monomer N-(N',N'-dicarboxymethylaminopropyl)methacrylamide (DAMA) [278–280]. Via the formation of a nickel complex of these pendant ligands and terminal histidine residues (His-tag) present in recombinant proteins, a hydrogel-containing chemical in addition to physical coiled coil cross-links was obtained (Scheme 16.10). In response to temperature, the proteins can undergo a conformational change from an extended coiled coil structure to a random coil, causing the hydrogel to shrink. Furthermore, addition of strong metal-chelating ligands such as imidazole leads to (partial) replacement of His-tags, inducing swelling of the hydrogel due to a decrease in cross-link density. It has been hypothesized that incorporation of distinct coiled coil motifs can result in assembled hydrogels which display several structural transitions upon temperature changes. In addition to proteins, thermoresponsive hydrogels based on physical cross-linking between complementary oligonucleotides have also been prepared [281].

16.5.2
Self-assembled Structures/Supramolecules; Towards Nanotechnology

The application of molecular self-assembly processes to construct functional nanometer-sized structures has become an attractive alternative for high-resolution lithography techniques. Particularly block copolymers have received much attention due to the ease of patterning via domain formation on the nanometer scale; the size and shape of these domains can be readily altered by changing the molecular weight and composition of the block copolymers [282–284].

Self-assembly processes occur frequently in Nature in order to build up all kinds of functional structures. One can think of the quaternary structure of proteins, the assembly of viruses or cell membranes. These well-defined three-dimensional structures are the result of a highly controlled hierarchical organization process, based on the information intrinsically stored in the biopolymer chain composition. By constructing biohybrid polymeric materials, a combination of the above-mentioned assembly procedures could be envisaged and therefore a class of materials with much potency in nanotechnology is to be expected [285].

In this respect, peptide–polymer hybrids are very attractive because of their folding behavior and synthetic accessibility. In general, research has been focused on two folding motifs, namely β-sheet- and α-helix-forming peptide sequences [45]. Meredith and coworkers were one of the first groups to report the

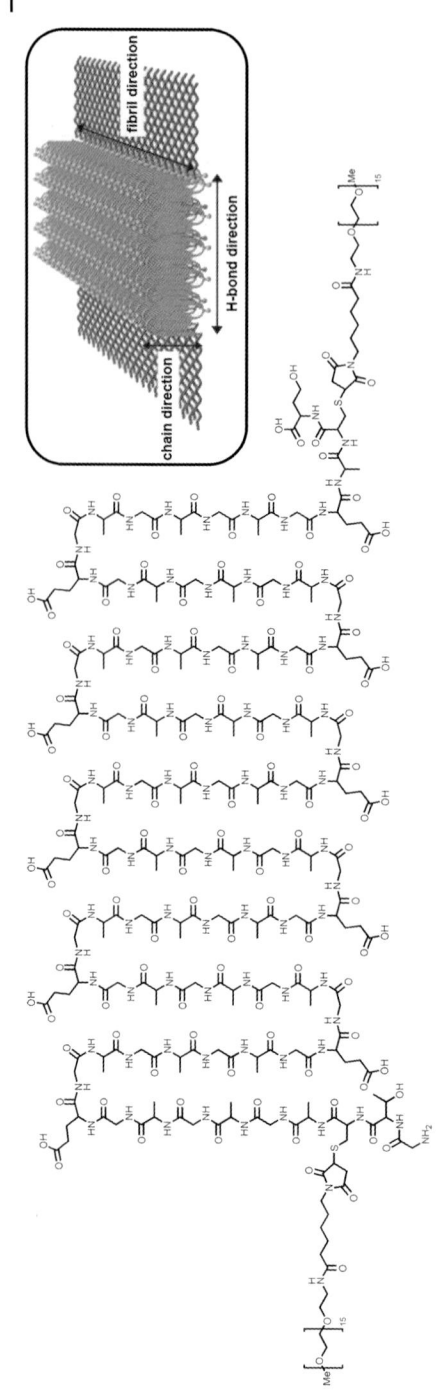

Fig. 16.9 ABA-type block copolymer containing a β-sheet segment flanked by two PEG chains, which is capable of forming well-defined fibrils (see inset). Reprinted with permission from [293]. Copyright 2005 Wiley-VCH.

conjugation of a synthetic polymer to a β-sheet-forming peptide [286]. PEG was attached to a β-amyloid peptide in order to affect fibril formation [287, 288], which was rendered reversible upon PEGylation. Inspired by the good mechanical properties of natural silk, Rathore and Sogah tethered β-sheet peptide crystalline segments to PEG [289, 290]. These β-sheet domains induce stiffness while the amorphous PEG phase prevents the material from becoming too brittle. In another example, Klok and coworkers showed that PEGylated peptides retain their β-sheet conformation upon aggregation, using infrared spectroscopy and X-ray scattering techniques [291, 292]. The attachment of PEG chains to β-sheet peptides can circumvent macroscopic crystallization and, moreover, forces peptides to assemble into micrometer-long fibers. This was shown by van Hest and coworkers by PEGylation of a β-sheet-forming peptide derived from *Bombyx mori* silk fibroin, which was prepared via protein engineering (Fig. 16.9) [293]. An attractive feature of these fibers is the presence of glutamic acid residues on top, allowing further modification.

As mentioned earlier, another interesting folding motif of peptides is the α-helical coil structure. Rod–coil block copolymers have attracted much attention due to their complex self-assembly and phase behavior characteristics [294, 295]. It has been shown that biohybrid block copolymers comprising α-helix-forming peptides also nicely form self-assembled structures [296–299]. Coiled coil structures composed of intertwined α-helices have also been used to direct aggregation of PEG-based biohybrid block copolymers [300, 301]. Using circular dichroism and analytical ultracentrifugation, Klok and coworkers showed the formation of uniform nanostructures composed of coiled coil cores with PEG shells closely wrapped around [302].

In addition to α-helical- and β-sheet-forming peptides, other peptide structures have also been exploited to direct aggregation of biohybrid polymers. Biesalski and coworkers showed the formation of hybrid nanotubes composed of a cyclic peptide core and a covalently coupled PNIPAAm shell [303].

By the attachment of hydrophobic polymers, such as polystyrene, to proteins, conjugates are formed which intrinsically are amphiphilic. Nolte and coworkers prepared such "giant amphiphiles" via the formation of biotin–streptavidin complexes [304] (Fig. 16.10a), via cofactor reconstitution in the enzyme horseradish peroxidase (HRP) [305] (Fig. 16.10b) and by coupling of polystyrene to a cysteine residue exposed on the outside of the lipase Cal B [50, 306] (Fig. 16.10c). These biohybrid amphiphiles displayed similar assembly behavior with respect to their low molecular weight counterparts, while retaining bioactivity, opening the door for functional enzyme assemblies.

Owing to advances in oligonucleotide synthesis, opportunities have arisen to utilize the unique recognition properties of DNA in materials science and nanotechnology [307–310]. The highly directional hydrogen-bonding interaction between complementary nucleobases is a perfect tool for reversible linear extension of polymer chains. These supramolecular polymers have interesting properties, such as temperature responsiveness, low melt viscosity and self-healing character [311]. Craig and coworkers functionalized monomers with oligonucleo-

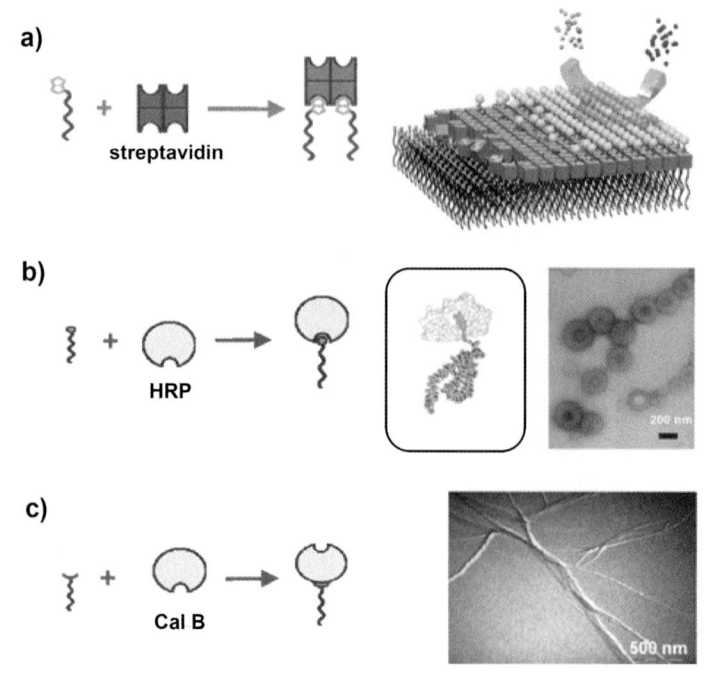

Fig. 16.10 Examples of the self-assembly process of "giant amphiphiles". (a) Via non-covalent linkage of biotinylated PS, a streptavidin functionalized surface is constituted to which the enzyme HRP was assembled, thereby obtaining a catalytic surface. Reprinted with permission from [304]. Copyright 2001 Wiley-VCH. (b) Vesicular aggregates formed by a polystyrene (PS)–HRP conjugate formed by cofactor reconstitution. Reprinted with permission from [305]. Copyright 2002 Wiley-VCH. (c) Covalent linkage of PS to the enzyme Cal B has led to the formation of micellar rods. Reprinted with permission from [306]. Copyright 2002 American Chemical Society. Adapted from [295].

tide strands which associated into linear chains [312, 313] and brushes [314, 315] via hydrogen bonding interactions. Rowan and coworkers showed that low molecular weight telechelic poly(tetrahydrofuran) terminated with single nucleobase moieties already self-assembled in the solid state with corresponding film- and fiber-forming characteristics [316, 317]. Due to weak interactions, these materials are extremely temperature sensitive, leading to the formation of very low-viscosity melts.

16.6
Conclusions and Outlook

Polymer bioconjugates have been recognized for many years as versatile materials for application in especially the field of drug delivery. Other applications, such as in diagnostics and the production of bioactive surfaces, are developing rapidly. The well-defined self-assembly properties of certain types of biohybrids are potentially useful for nanotechnology applications, albeit that this field of research is still on a more fundamental level than the other examples given.

Most of the currently applied polymer bioconjugates are made via traditional chemistry. Although impressive results have already been obtained with these macromolecules, it is to be expected that the recent advances in controlled production methods will lead to many opportunities for improvement of bioconjugate structure and, hence, functionality. Therefore, bioconjugate research will have as important theme within the coming years to implement these new synthetic schemes, leading to new and improved materials applications.

References

1 M. Szwarc, M. Levy, R. Milkovich, *J. Am. Chem. Soc.* **1956**, *78*, 2656–2657.
2 M. Szwarc, M. Van Beylen, *Ionic Polymerization and Living Polymers*, Chapman and Hall, New York, **1993**.
3 M. Szwarc, *Ionic Polymerization Fundamentals*, Hanser, New York, **1996**.
4 Y. K. Chong, T. P. T. Le, G. Moad, E. Rizzardo, S. H. Thang, *Macromolecules* **1999**, *32*, 2071–2074.
5 K. Matyjaszewski, J. H. Xia, *Chem. Rev.* **2001**, *101*, 2921–2990.
6 M. Kamigaito, T. Ando, M. Sawamoto, *Chem. Rev.* **2001**, *101*, 3689–3745.
7 C. J. Hawker, A. W. Bosman, E. Harth, *Chem. Rev.* **2001**, *101*, 3661–3688.
8 L. Ayres, K. Koch, P. Adams, J. C. M. van Hest, *Macromolecules* **2005**, *38*, 1699–1704.
9 A. Marsh, A. Khan, M. Garcia, D. M. Haddleton, *Chem. Commun.* **2000**, 2083–2084.
10 H. J. Spijker, A. J. Dirks, J. C. M. van Hest, *Polymer* **2005**, *46*, 8528–8535.
11 V. Ladmiral, E. Melia, D. M. Haddleton, *Eur. Polym. J.* **2004**, *40*, 431–449.
12 V. Coessens, T. Pintauer, K. Matyjaszewski, *Prog. Polym. Sci.* **2001**, *26*, 337–377.
13 G. Moad, Y. K. Chong, A. Postma, E. Rizzardo, S. H. Thang, *Polymer* **2005**, *46*, 8458–8468.
14 B. Zhao, W. J. Brittain, *Prog. Polym. Sci.* **2000**, *25*, 677–710.
15 N. Hadjichristidis, M. Pitsikalis, S. Pispas, H. Iatrou, *Chem. Rev.* **2001**, *101*, 3747–3792.
16 K. Inoue, *Prog. Polym. Sci.* **2000**, *25*, 453–571.
17 J. Pyun, T. Kowalewski, K. Matyjaszewski, *Macromol. Rapid Commun.* **2003**, *24*, 1043–1059.
18 J. E. Schwendeman, A. C. Church, K. B. Wagener, *Adv. Synth. Catal.* **2002**, *344*, 597–613.
19 T. W. Baughman, K. B. Wagener, *Adv. Polym. Sci.* **2005**, *176*, 1–42.
20 P. Schwab, R. H. Grubbs, J. W. Ziller, *J. Am. Chem. Soc.* **1996**, *118*, 100–110.
21 G. C. Bazan, R. R. Schrock, H. N. Cho, V. C. Gibson, *Macromolecules* **1991**, *24*, 4495–4502.
22 S. C. G. Biagini, V. C. Gibson, M. R. Giles, E. L. Marshall, M. North, *Chem. Commun.* **1997**, 1097–1098.
23 H. S. Bazzi, H. F. Sleiman, *Macromolecules* **2002**, *35*, 9617–9620.
24 R. G. Davies, V. C. Gibson, M. B. Hursthouse, M. E. Light, E. L. Marshall, M. North, D. A. Robson, I. Thompson, A. J. P. White, D. J. Williams, P. J. Williams, *J. Chem. Soc., Perkin Trans. 1* **2001**, 3365–3381.
25 S. C. G. Biagini, R. G. Davies, V. C. Gibson, M. R. Giles, E. L. Marshall, M. North, D. A. Robson, *Chem. Commun.* **1999**, 235–236.
26 C. Fraser, R. H. Grubbs, *Macromolecules* **1995**, *28*, 7248–7255.
27 R. B. Merrifield, *J. Am. Chem. Soc.* **1963**, *85*, 2149–2154.
28 E. Bayer, M. Mutter, *Nature* **1972**, *237*, 512–513.
29 M. Mutter, E. Bayer, *Angew. Chem. Int. Ed. Engl.* **1974**, *13*, 88–89.

30 T. Kimmerlin, D. Seebach, *J. Pept. Res.* **2005**, *65*, 229–260.
31 J. A. Prescher, C. R. Bertozzi, *Nat. Chem. Biol.* **2005**, *1*, 13–21.
32 M. Kohn, R. Breinbauer, *Angew. Chem. Int. Engl.* **2004**, *43*, 3106–3116.
33 T. J. Deming, *Nature* **1997**, *390*, 386–389.
34 T. J. Deming, *J. Polym. Sci., Part A: Polym. Chem.* **2000**, *38*, 3011–3018.
35 J. C. M. van Hest, D. A. Tirrell, *Chem. Commun.* **2001**, 1897–1904.
36 T. Hohsaka, D. Kajihara, Y. Ashizuka, H. Murakami, M. Sisido, *J. Am. Chem. Soc.* **1999**, *121*, 34–40.
37 E. Yoshikawa, M. J. Fournier, T. L. Mason, D. A. Tirrell, *Macromolecules* **1994**, *27*, 5471–5475.
38 S. Kothakota, T. L. Mason, D. A. Tirrell, M. J. Fournier, *J. Am. Chem. Soc.* **1995**, *117*, 536–537.
39 T. J. Deming, M. J. Fournier, T. L. Mason, D. A. Tirrell, *J. Macromol. Sci., Pure Appl. Chem.* **1997**, *A34*, 2143–2150.
40 J. C. M. van Hest, D. A. Tirrell, *FEBS Lett.* **1998**, *428*, 68–70.
41 J. C. M. van Hest, K. L. Kiick, D. A. Tirrell, *J. Am. Chem. Soc.* **2000**, *122*, 1282–1288.
42 K. L. Kiick, J. C. M. van Hest, D. A. Tirrell, *Angew. Chem. Int. Ed.* **2000**, *39*, 2148–2152.
43 A. J. Link, M. L. Mock, D. A. Tirrell, *Curr. Opin. Biotechnol.* **2003**, *14*, 603–609.
44 D. W. P. M. Löwik, L. Ayres, J. M. Smeenk, J. C. M. Van Hest, *Adv. Polym. Sci.* **2006**, *202*, 19–52.
45 H. A. Klok, *J. Polym. Sci., Part A: Polym. Chem.* **2005**, *43*, 1–17.
46 F. M. Veronese, *Biomaterials* **2001**, *22*, 405–417.
47 M. J. Roberts, M. D. Bentley, J. M. Harris, *Adv. Drug Deliv. Rev.* **2002**, *54*, 459–476.
48 V. V. Rostovtsev, L. G. Green, V. V. Fokin, K. B. Sharpless, *Angew. Chem. Int. Ed.* **2002**, *41*, 2596–2599.
49 J. A. Opsteen, J. C. M. van Hest, *Chem. Commun.* **2005**, 57–59.
50 A. J. Dirks, S. S. van Berkel, N. S. Hatzakis, J. A. Opsteen, F. L. van Delft, J. J. L. M. Cornelissen, A. E. Rowan, J. C. M. van Hest, F. P. J. T. Rutjes, R. J. M. Nolte, *Chem. Commun.* **2005**, 4172–4174.
51 A. Deiters, T. A. Cropp, D. Summerer, M. Mukherji, P. G. Schultz, *Bioorg. Med. Chem. Lett.* **2004**, *14*, 5743–5745.
52 B. S. Sumerlin, N. V. Tsarevsky, G. Louche, R. Y. Lee, K. Matyjaszewski, *Macromolecules* **2005**, *38*, 7540–7545.
53 B. Helms, J. L. Mynar, C. J. Hawker, J. M. J. Frechet, *J. Am. Chem. Soc.* **2004**, *126*, 15020–15021.
54 M. L. Becker, J. Q. Liu, K. L. Wooley, *Chem. Commun.* **2003**, 180–181.
55 M. L. Becker, J. Q. Liu, K. L. Wooley, *Biomacromolecules* **2005**, *6*, 220–228.
56 L. Ayres, P. Hans, J. Adams, D. Lowik, J. C. M. van Hest, *J. Polym. Sci., Part A: Polym. Chem.* **2005**, *43*, 6355–6366.
57 D. Bontempo, H. D. Maynard, *J. Am. Chem. Soc.* **2005**, *127*, 6508–6509.
58 Y. Mei, K. L. Beers, H. C. M. Byrd, D. L. Vanderhart, N. R. Washburn, *J. Am. Chem. Soc.* **2004**, *126*, 3472–3476.
59 R. Duncan, *Nat. Rev. Drug Discov.* **2003**, *2*, 347–360.
60 R. Satchi-Fainaro, R. Duncan, C. M. Barnes, *Adv. Polym. Sci.* **2006**, *193*, 1–65.
61 B. Twaites, C. D. Alarcon, C. Alexander, *J. Mater. Chem.* **2005**, *15*, 441–455.
62 H. Ringsdorf, *J. Polym. Sci. Polym. Symp.* **1975**, 135–153.
63 L. Gros, H. Ringsdorf, H. Schupp, *Angew. Chem. Int. Ed. Engl.* **1981**, *20*, 305–325.
64 K. Iwai, H. Maeda, T. Konno, *Cancer Res.* **1984**, *44*, 2115–2121.
65 H. Maeda, *Adv. Drug Deliv. Rev.* **1991**, *6*, 181–202.
66 H. Maeda, *Adv. Drug Deliv. Rev.* **2001**, *46*, 169–185.
67 H. Maeda, T. Sawa, T. Konno, *J. Control. Release* **2001**, *74*, 47–61.
68 H. Maeda, *Adv. Enzyme Regul.* **2001**, *41*, 189–207.
69 H. Maeda, K. Greish, J. Fang, *Adv. Polym. Sci.* **2006**, *193*, 103–121.
70 S. Brocchini, R. Duncan, *Encyclopedia of Controlled Drug Delivery*, Wiley, New York, **1999**, 786–816.
71 D. Putnam, J. Kopeček, *Adv. Polym. Sci.* **1995**, *122*, 55–123.
72 J. Kopeček, P. Kopečková, T. Minko, Z. R. Lu, *Eur. J. Pharm. Biopharm.* **2000**, *50*, 61–81.

73 R. B. Greenwald, C. D. Conover, Y. H. Choe, *Crit. Rev. Ther. Drug Carrier Syst.* **2000**, *17*, 101–161.

74 P. Caliceti, O. Schiavon, F. M. Veronese, *Bioconjug. Chem.* **2001**, *12*, 515–522.

75 T. Minko, *Drug Discov. Today* **2005**, *2*, 15–20.

76 K. Hoste, K. De Winne, E. Schacht, *Int. J. Pharm.* **2004**, *277*, 119–131.

77 A. J. M. D'Souza, E. M. Topp, *J. Pharm. Sci.* **2004**, *93*, 1962–1979.

78 A. J. Thistlethwaite, D. B. Leeper, D. J. Moylan, R. E. Nerlinger, *Int. J. Radiat. Oncol. Biol. Phys.* **1985**, *11*, 1647–1652.

79 K. Ulbrich, T. Etrych, P. Chytil, M. Pechar, M. Jelinkova, B. Rihova, *Int. J. Pharm.* **2004**, *277*, 63–72.

80 K. Ulbrich, V. Subr, *Adv. Drug Deliv. Rev.* **2004**, *56*, 1023–1050.

81 J. D. Vassalli, M. S. Pepper, *Nature* **1994**, *370*, 14–15.

82 R. Duncan, H. C. Cable, J. B. Lloyd, P. Rejmanova, J. Kopeček, *Biosci. Rep.* **1982**, *2*, 1041–1046.

83 R. Duncan, H. C. Cable, J. B. Lloyd, P. Rejmanova, J. Kopeček, *Makromol. Chem.* **1983**, *184*, 1997–2008.

84 P. Rejmanova, J. Kopeček, J. Pohl, M. Baudys, V. Kostka, *Makromol. Chem.* **1983**, *184*, 2009–2020.

85 J. Kopeček, *Biomaterials* **1984**, *5*, 19–25.

86 J. Kopeček, P. Rejmanova, R. Duncan, J. B. Lloyd, *Ann. N. Y. Acad. Sci.* **1985**, *446*, 93–104.

87 V. Subr, J. Strohalm, K. Ulbrich, R. Duncan, I. C. Hume, *J. Control. Release* **1992**, *18*, 123–132.

88 J. Kopeček, P. Kopečková, T. Minko, Z. R. Lu, C. M. Peterson, *J. Control. Release* **2001**, *74*, 147–158.

89 T. Minko, S. S. Dharap, R. I. Pakunlu, Y. Wang, *Curr. Drug Targets* **2004**, *5*, 389–406.

90 T. M. Allen, *Nat. Rev. Cancer* **2002**, *2*, 750–763.

91 B. Rihova, *Adv. Drug Deliv. Rev.* **1998**, *29*, 273–289.

92 R. Duncan, J. Kopeček, P. Rejmanova, J. B. Lloyd, *Biochim. Biophys. Acta* **1983**, *755*, 518–521.

93 R. C. Rathi, P. Kopečková, B. Rihova, J. Kopeček, *J. Polym. Sci., Part A: Polym. Chem.* **1991**, *29*, 1895–1902.

94 B. Rihova, J. Kopeček, *J. Control. Release* **1985**, *2*, 289–310.

95 B. Rihova, P. Kopečková, J. Strohalm, P. Rossmann, V. Vetvicka, J. Kopeček, *Clin. Immunol. Immunopathol.* **1988**, *46*, 100–114.

96 V. Omelyanenko, P. Kopečková, C. Gentry, J. G. Shiah, J. Kopeček, *J. Drug Target.* **1996**, *3*, 357–373.

97 Z. R. Lu, J. G. Shiah, S. Sakuma, P. Kopečková, J. Kopeček, *J. Control. Release* **2002**, *78*, 165–173.

98 A. Mitra, J. Mulholland, A. Nan, E. McNeill, H. Ghandehari, B. R. Line, *J. Control. Release* **2005**, *102*, 191–201.

99 D. Hanahan, J. Folkman, *Cell* **1996**, *86*, 353–364.

100 P. Carmeliet, R. K. Jain, *Nature* **2000**, *407*, 249–257.

101 P. Carmeliet, *Nature* **2005**, *438*, 932–936.

102 R. Satchi-Fainaro, C. M. Barnes, *Drug Deliv Companies Rep.* **2004** (Spring/Summer), 43–49.

103 R. Satchi-Fainaro, M. Puder, J. W. Davies, H. T. Tran, D. A. Sampson, A. K. Greene, G. Corfas, J. Folkman, *Nat. Med.* **2004**, *10*, 255–261.

104 P. C. Brooks, R. A. F. Clark, D. A. Cheresh, *Science* **1994**, *264*, 569–571.

105 J. M. Harris, R. B. Chess, *Nat. Rev. Drug Discov.* **2003**, *2*, 214–221.

106 S. Frokjaer, D. E. Otzen, *Nat. Rev. Drug Discov.* **2005**, *4*, 298–306.

107 T. M. Allen, C. B. Hansen, D. E. L. Demenezes, *Adv. Drug Deliv. Rev.* **1995**, *16*, 267–284.

108 A. Abuchowski, T. Vanes, N. C. Palczuk, F. F. Davis, *J. Biol. Chem.* **1977**, *252*, 3578–3581.

109 A. Abuchowski, J. R. McCoy, N. C. Palczuk, T. Vanes, F. F. Davis, *J. Biol. Chem.* **1977**, *252*, 3582–3586.

110 F. F. Davis, *Adv. Drug Deliv. Rev.* **2002**, *54*, 457–458.

111 C. Delgado, G. E. Francis, D. Fisher, *Crit. Rev. Ther. Drug Carrier Syst.* **1992**, *9*, 249–304.

112 S. Zalipsky, *Adv. Drug Deliv. Rev.* **1995**, *16*, 157–182.

113 Y. Kodera, A. Matsushima, M. Hiroto, H. Nishimura, A. Ishii, T. Ueno, Y. Inada, *Prog. Polym. Sci.* **1998**, *23*, 1233–1271.

114 P. Caliceti, F. M. Veronese, *Adv. Drug Deliv. Rev.* **2003**, *55*, 1261–1277.
115 F. M. Veronese, G. Pasut, *Drug Discov. Today* **2005**, *10*, 1451–1458.
116 F. M. Veronese, J. M. Harris, *Adv. Drug Deliv. Rev.* **2002**, *54*, 453–456.
117 M. S. Hershfield, R. H. Buckley, M. L. Greenberg, A. L. Melton, R. Schiff, C. Hatem, J. Kurtzberg, M L. Markert, R. H. Kobayashi, A. L. Kobayashi, A. Abuchowski, *N. Engl. J. Med.* **1987**, *316*, 589–596.
118 M. L. Nucci, R. Shorr, A. Abuchowski, *Adv. Drug Deliv. Rev.* **1991**, *6*, 133–151.
119 G. W. M. Vandermeulen, H. A. Klok, *Macromol. Biosci.* **2004**, *4*, 383–398.
120 G. E. Francis, D. Fisher, C. Delgado, F. Malik, A. Gardiner, D. Neale, *Int. J. Hematol.* **1998**, *68*, 1–18.
121 D. J. Hicklin, L. Witte, Z. P. Zhu, F. Liao, Y. Wu, Y. W. Li, P. Bohlen, *Drug Discov. Today* **2001**, *6*, 517–528.
122 A. P. Chapman, *Adv. Drug Deliv. Rev.* **2002**, *54*, 531–545.
123 J. S. Ross, K. Gray, G. S. Gray, P. J. Worland, M. Rolfe, *Am. J. Clin. Pathol.* **2003**, *119*, 472–485.
124 P. S. Chowdhury, H. Wu, *Methods* **2005**, *36*, 11–24.
125 A. P. Chapman, P. Antoniw, M. Spitali, S. West, S. Stephens, D. J. King, *Nat. Biotechnol.* **1999**, *17*, 780–783.
126 Y. Tsutsumi, M. Onda, S. Nagata, B. Lee, R. J. Kreitman, I. Pastan, *Proc. Natl. Acad. Sci. USA* **2000**, *97*, 8548–8553.
127 J. H. Wang, S. C. Tam, H. Huang, D. Y. Ouyang, Y. Y. Wang, Y. T. Zheng, *Biochem. Biophys. Res. Commun.* **2004**, *317*, 965–971.
128 M. S. Rosendahl, D. H. Doherty, D. J. Smith, S. J. Carlson, E. A. Chlipala, G. N. Cox, *Bioconjug. Chem.* **2005**, *16*, 200–207.
129 D. H. Doherty, M. S. Rosendahl, D. J. Smith, J. M. Hughes, E. A. Chlipala, G. N. Cox, *Bioconjug. Chem.* **2005**, *16*, 1291–1298.
130 Y. Yamamoto, Y. Tsutsumi, Y. Yoshioka, T. Nishibata, K. Kobayashi, T. Okamoto, Y. Mukai, T. Shimizu, S. Nakagawa, S. Nagata, T. Mayumi, *Nat. Biotechnol.* **2003**, *21*, 546–552.
131 H. Sato, *Adv. Drug Deliv. Rev.* **2002**, *54*, 487–504.
132 S. De Frees, D. Zopf, R. Bayer, C. Bowe, D. Hakes, X. Chen, *PCT int. appl.* **2004**, WO 2004099231 A2 20041118.
133 S. D. Tilley, M. B. Francis, *J. Am. Chem. Soc.* **2006**, *128*, 1080–1081.
134 C. S. Cazalis, C. A. Haller, L. Sease-Cargo, E. L. Chaikof, *Bioconjug. Chem.* **2004**, *15*, 1005–1009.
135 G. G. Kochendoerfer, S. Y. Chen, F. Mao, S. Cressman, S. Traviglia, H. Y. Shao, C. L. Hunter, D. W. Low, E. N. Cagle, M. Carnevali, V. Gueriguian, P. J. Keogh, H. Porter, S. M. Stratton, M. C. Wiedeke, J. Wilken, J. Tang, J. J. Levy, L. P. Miranda, M. M. Crnogorac, S. Kalbag, P. Botti, J. Schindler-Horvat, L. Savatski, J. W. Adamson, A. Kung, S. B. H. Kent, J. A. Bradburne, *Science* **2003**, *299*, 884–887.
136 S. Y. Chen, S. Cressman, F. Mao, H. Shao, D. W. Low, H. S. Beilan, E. N. Cagle, M. Carnevali, V. Gueriguian, P. J. Keogh, H. Porter, S. M. Stratton, M. C. Wiedeke, L. Savatski, J. W. Adamson, C. E. Bozzini, A. Kung, S. B. H. Kent, J. A. Bradburne, G. G. Kochendoerfer, *Chem. Biol.* **2005**, *12*, 371–383.
137 G. W. M. Vandermeulen, C. Tziatzios, R. Duncan, H. A. Klok, *Macromolecules* **2005**, *38*, 761–769.
138 Y. B. Yu, *Adv. Drug Deliv. Rev.* **2002**, *54*, 1113–1129.
139 F. Lecolley, L. Tao, G. Mantovani, I. Durkin, S. Lautru, D. M. Haddleton, *Chem. Commun.* **2004**, 2026–2027.
140 L. Tao, G. Mantovani, F. Lecolley, D. M. Haddleton, *J. Am. Chem. Soc.* **2004**, *126*, 13220–13221.
141 G. Mantovani, F. Lecolley, L. Tao, D. M. Haddleton, J. Clerx, J. Cornelissen, K. Velonia, *J. Am. Chem. Soc.* **2005**, *127*, 2966–2973.
142 D. Bontempo, K. L. Heredia, B. A. Fish, H. D. Maynard, *J. Am. Chem. Soc.* **2004**, *126*, 15372–15373.
143 K. L. Heredia, D. Bontempo, T. Ly, J. T. Byers, S. Halstenberg, H. D. Maynard, *J. Am. Chem. Soc.* **2005**, *127*, 16955–16960.
144 J. K. Pontrello, M. J. Allen, E. S. Underbakke, L. L. Kiessling, *J. Am. Chem. Soc.* **2005**, *127*, 14536–14537.

145 C. Monfardini, O. Schiavon, P. Caliceti, M. Morpurgo, J. M. Harris, F. M. Veronese, *Bioconjug. Chem.* **1995**, *6*, 62–69.

146 D. A. Tomalia, H. Baker, J. Dewald, M. Hall, G. Kallos, S. Martin, J. Roeck, J. Ryder, P. Smith, *Polym. J.* **1985**, *17*, 117–132.

147 J. M. J. Frechet, C. J. Hawker, I. Gitsov, J. W. Leon, *J. Macromol. Sci., Pure Appl. Chem.* **1996**, *A33*, 1399–1425.

148 U. Boas, P. M. H. Heegaard, *Chem. Soc. Rev.* **2004**, *33*, 43–63.

149 A. K. Patri, J. F. Kukowska-Latallo, J. R. Baker, *Adv. Drug Deliv. Rev.* **2005**, *57*, 2203–2214.

150 C. C. Lee, M. Yoshida, J. M. J. Frechet, E. E. Dy, F. C. Szoka, *Bioconjug. Chem.* **2005**, *16*, 535–541.

151 M. Mammen, S. K. Choi, G. M. Whitesides, *Angew. Chem. Int. Ed.* **1998**, *37*, 2755–2794.

152 A. David, P. Kopečková, A. Rubinstein, J. Kopeček, *Bioconjug. Chem.* **2001**, *12*, 890–899.

153 J. H. Griffin, M. S. Linsell, M. B. Nodwell, Q. Q. Chen, J. L. Pace, K. L. Quast, K. M. Krause, L. Farrington, T. X. Wu, D. L. Higgins, T. E. Jenkins, B. G. Christensen, J. K. Judice, *J. Am. Chem. Soc.* **2003**, *125*, 6517–6531.

154 A. David, P. Kopečková, T. Minko, A. Rubinstein, J. Kopeček, *Eur. J. Cancer* **2004**, *40*, 148–157.

155 P. Wu, M. Malkoch, J. N. Hunt, R. Vestberg, E. Kaltgrad, M. G. Finn, V. V. Fokin, K. B. Sharpless, C. J. Hawker, *Chem. Commun.* **2005**, 5775–5777.

156 H. D. Maynard, S. Y. Okada, R. H. Grubbs, *Macromolecules* **2000**, *33*, 6239–6248.

157 H. D. Maynard, S. Y. Okada, R. H. Grubbs, *J. Am. Chem. Soc.* **2001**, *123*, 1275–1279.

158 J. E. Gestwicki, C. W. Cairo, L. E. Strong, K. A. Oetjen, L. L. Kiessling, *J. Am. Chem. Soc.* **2002**, *124*, 14922–14933.

159 L. E. Strong, L. L. Kiessling, *J. Am. Chem. Soc.* **1999**, *121*, 6193–6196.

160 Y. C. Lee, R. T. Lee, *Acc. Chem. Res.* **1995**, *28*, 321–327.

161 L. L. Kiessling, N. L. Pohl, *Chem. Biol.* **1996**, *3*, 71–77.

162 J. J. Lundquist, E. J. Toone, *Chem. Rev.* **2002**, *102*, 555–578.

163 L. L. Kiessling, J. E. Gestwicki, L. E. Strong, *Curr. Opin. Chem. Biol.* **2000**, *4*, 696–703.

164 C. R. Bertozzi, L. L. Kiessling, *Science* **2001**, *291*, 2357–2364.

165 M. Tirrell, E. Kokkoli, M. Biesalski, *Surf. Sci.* **2002**, *500*, 61–83.

166 D. G. Anderson, J. A. Burdick, R. Langer, *Science* **2004**, *305*, 1923–1924.

167 G. Walter, K. Bussow, A. Lueking, J. Glokler, *Trends Mol. Med.* **2002**, *8*, 250–253.

168 D. S. Wilson, S. Nock, *Angew. Chem. Int. Ed.* **2003**, *42*, 494–500.

169 N. L. Rosi, C. A. Mirkin, *Chem. Rev.* **2005**, *105*, 1547–1562.

170 P. Mitchell, *Nat. Biotechnol.* **2002**, *20*, 225–229.

171 S. Choudhuri, *J. Biochem. Mol. Toxicol.* **2004**, *18*, 171–179.

172 H. Sorribas, C. Padeste, L. Tiefenauer, *Biomaterials* **2002**, *23*, 893–900.

173 P. M. Mendes, J. A. Preece, *Curr. Opin. Colloid Interface Sci.* **2004**, *9*, 236–248.

174 P. Calvert, *Chem. Mater.* **2001**, *13*, 3299–3305.

175 B. J. de Gans, E. Kazancioglu, W. Meyer, U. S. Schubert, *Macromol. Rapid Commun.* **2004**, *25*, 292–296.

176 D. S. Ginger, H. Zhang, C. A. Mirkin, *Angew. Chem. Int. Ed.* **2004**, *43*, 30–45.

177 J. M. Nam, S. W. Han, K. B. Lee, X. G. Liu, M. A. Ratner, C. A. Mirkin, *Angew. Chem. Int. Ed.* **2004**, *43*, 1246–1249.

178 M. Mrksich, G. M. Whitesides, *Trends Biotechnol.* **1995**, *13*, 228–235.

179 Y. N. Xia, G. M. Whitesides, *Annu. Rev. Mater. Sci.* **1998**, *28*, 153–184.

180 F. N. Rehman, M. Audeh, E. S. Abrams, P. W. Hammond, M. Kenney, T. C. Boles, *Nucleic Acids Res.* **1999**, *27*, 649–655.

181 V. A. Efimov, A. A. Buryakova, O. G. Chakhmakhcheva, *Nucleic Acids Res.* **1999**, *27*, 4416–4426.

182 B. de Lambert, C. Chaix, M. T. Charreyre, A. Laurent, A. Aigoui, A. Perrin-Rubens, C. Pichot, *Bioconjug. Chem.* **2005**, *16*, 265–274.

183 M. Beier, J. D. Hoheisel, *Nucleic Acids Res.* **1999**, *27*, 1970–1977.

184 V. Le Berre, E. Trevisiol, A. Dagkessamanskaia, S. Sokol, A. M. Caminade,

J. P. Majoral, B. Meunier, J. Francois, *Nucleic Acids Res.* **2003**, *31*, e88.
185 W. Kusnezow, J. D. Hoheisel, *J. Mol. Recognit.* **2003**, *16*, 165–176.
186 B. C. Weimer, M. K. Walsh, X. W. Wang, *J. Biochem. Biophys. Methods* **2000**, *45*, 211–219.
187 M. K. Walsh, X. W. Wang, B. C. Weimer, *J. Biochem. Biophys. Methods* **2001**, *47*, 221–231.
188 T. Ameringer, M. Hinz, C. Mourran, H. Seliger, J. Groll, M. Moeller, *Biomacromolecules* **2005**, *6*, 1819–1823.
189 J. Groll, W. Haubensak, T. Ameringer, M. Moeller, *Langmuir* **2005**, *21*, 3076–3083.
190 M. Husseman, E. E. Malmstrom, M. McNamara, M. Mate, D. Mecerreyes, D. G. Benoit, J. L. Hedrick, P. Mansky, E. Huang, T. P. Russell, C. J. Hawker, *Macromolecules* **1999**, *32*, 1424–1431.
191 Y. Tsujii, M. Ejaz, K. Sato, A. Goto, T. Fukuda, *Macromolecules* **2001**, *34*, 8872–8878.
192 M. Baum, W. J. Brittain, *Macromolecules* **2002**, *35*, 610–615.
193 S. Edmondson, V. L. Osborne, W. T. S. Huck, *Chem. Soc. Rev.* **2004**, *33*, 14–22.
194 S. Tugulu, A. Arnold, I. Sielaff, K. Johnsson, H. A. Klok, *Biomacromolecules* **2005**, *6*, 1602–1607.
195 A. Keppler, S. Gendreizig, T. Gronemeyer, H. Pick, H. Vogel, K. Johnsson, *Nat. Biotechnol.* **2003**, *21*, 86–89.
196 K. L. Christman, H. D. Maynard, *Langmuir* **2005**, *21*, 8389–8393.
197 M. P. Lutolf, J. A. Hubbell, *Nat. Biotechnol.* **2005**, *23*, 47–55.
198 R. Langer, J. P. Vacanti, *Science* **1993**, *260*, 920–926.
199 E. Ruoslahti, M. D. Pierschbacher, *Science* **1987**, *238*, 491–497.
200 E. Ruoslahti, *Annu. Rev. Cell Dev. Biol.* **1996**, *12*, 697–715.
201 L. E. Freed, J. C. Marquis, A. Nohria, J. Emmanual, A. G. Mikos, R. Langer, *J. Biomed. Mater. Res.* **1993**, *27*, 11–23.
202 L. E. Freed, A. P. Hollander, I. Martin, J. R. Barry, R. Langer, G. Vunjak-Novakovic, *Exp. Cell Res.* **1998**, *240*, 58–65.
203 C. R. Chu, J. S. Dounchis, M. Yoshioka, R. L. Sah, R. D. Coutts, D. Amiel, *Clin. Orthop.* **1997**, 220–229.
204 M. Sittinger, D. Reitzel, M. Dauner, H. Hierlemann, C. Hammer, E. Kastenbauer, H. Planck, G. R. Burmester, J. Bujia, *J. Biomed. Mater. Res.* **1996**, *33*, 57–63.
205 D. A. Barrera, E. Zylstra, P. T. Lansbury, R. Langer, *J. Am. Chem. Soc.* **1993**, *115*, 11010–11011.
206 D. A. Barrera, E. Zylstra, P. T. Lansbury, R. Langer, *Macromolecules* **1995**, *28*, 425–432.
207 A. D. Cook, J. S. Hrkach, N. N. Gao, I. M. Johnson, U. B. Pajvani, S. M. Cannizzaro, R. Langer, *J. Biomed. Mater. Res.* **1997**, *35*, 513–523.
208 S. VandeVondele, J. Voros, J. A. Hubbell, *Biotechnol. Bioeng.* **2003**, *82*, 784–790.
209 H. B. Lin, C. Garciaecheverria, S. Asakura, W. Sun, D. F. Mosher, S. L. Cooper, *Biomaterials* **1992**, *13*, 905–914.
210 H. B. Lin, Z. C. Zhao, C. Garciaecheverria, D. H. Rich, S. L. Cooper, *J. Biomater. Sci., Polym. Ed.* **1992**, *3*, 217–227.
211 D. Klee, H. Hocker, *Adv. Polym. Sci.* **1999**, *149*, 1–57.
212 K. Y. Lee, D. J. Mooney, *Chem. Rev.* **2001**, *101*, 1869–1879.
213 A. L. Lewis, *Colloids Surf. B* **2000**, *18*, 261–275.
214 Y. Iwasaki, K. Ishihara, *Anal. Bioanal. Chem.* **2005**, *381*, 534–546.
215 K. Ishihara, T. Ueda, N. Nakabayashi, *Polym. J.* **1990**, *22*, 355–360.
216 L. Ruiz, J. G. Hilborn, D. Leonard, H. J. Mathieu, *Biomaterials* **1998**, *19*, 987–998.
217 A. L. Lewis, Z. L. Cumming, H. H. Goreish, L. C. Kirkwood, L. A. Tolhurst, P. W. Stratford, *Biomaterials* **2001**, *22*, 99–111.
218 K. Ishihara, H. Hanyuda, N. Nakabayashi, *Biomaterials* **1995**, *16*, 873–879.
219 E. J. Lobb, I. Ma, N. C. Billingham, S. P. Armes, A. L. Lewis, *J. Am. Chem. Soc.* **2001**, *123*, 7913–7914.
220 I. Y. Ma, E. J. Lobb, N. C. Billingham, S. P. Armes, A. L. Lewis, A. W. Lloyd, J. Salvage, *Macromolecules* **2002**, *35*, 9306–9314.
221 Y. H. Ma, Y. Q. Tang, N. C. Billingham, S. P. Armes, A. L. Lewis, A. W. Lloyd, J. P. Salvage, *Macromolecules* **2003**, *36*, 3475–3484.
222 M. H. Stenzel, C. Barner-Kowollik, T. P. Davis, H. M. Dalton, *Macromol. Biosci.* **2004**, *4*, 445–453.

223 Y. Inoue, J. Watanabe, M. Takai, S. Yusa, K. Ishihara, *J. Polym. Sci., Part A: Polym. Chem.* **2005**, *43*, 6073–6083.

224 J. Z. Du, Y. P. Tang, A. L. Lewis, S. P. Armes, *J. Am. Chem. Soc.* **2005**, *127*, 17982–17983.

225 X. Y. Chen, S. R. Armes, *Adv. Mater.* **2003**, *15*, 1558–1562.

226 R. Iwata, P. Suk-In, V. P. Hoven, A. Takahara, K. Akiyoshi, Y. Iwasaki, *Biomacromolecules* **2004**, *5*, 2308–2314.

227 W. Feng, S. P. Zhu, K. Ishihara, J. L. Brash, *Langmuir* **2005**, *21*, 5980–5987.

228 J. C. Tiller, C. J. Liao, K. Lewis, A. M. Klibanov, *Proc. Natl. Acad. Sci. USA* **2001**, *98*, 5981–5985.

229 E. R. Kenawy, F. I. Abdel-Hay, A. El-Shanshoury, M. H. El-Newehy, *J. Polym. Sci., Part A: Polym. Chem.* **2002**, *40*, 2384–2393.

230 M. A. Gelman, B. Weisblum, D. M. Lynn, S. H. Gellman, *Org. Lett.* **2004**, *6*, 557–560.

231 S. B. Lee, R. R. Koepsel, S. W. Morley, K. Matyjaszewski, Y. J. Sun, A. J. Russell, *Biomacromolecules* **2004**, *5*, 877–882.

232 M. F. Ilker, K. Nusslein, G. N. Tew, E. B. Coughlin, *J. Am. Chem. Soc.* **2004**, *126*, 15870–15875.

233 C. J. Waschinski, J. C. Tiller, *Biomacromolecules* **2005**, *6*, 235–243.

234 A. Popa, C. M. Davidescu, R. Trif, G. Ilia, S. Iliescu, G. Dehelean, *React. Funct. Polym.* **2003**, *55*, 151–158.

235 R. E. W. Hancock, D. S. Chapple, *Antimicrob. Agents Chemother.* **1999**, *43*, 1317–1323.

236 T. Tashiro, *Macromol. Mater. Eng.* **2001**, *286*, 63–87.

237 M. Albert, P. Feiertag, G. Hayn, R. Saf, H. Honig, *Biomacromolecules* **2003**, *4*, 1811–1817.

238 B. L. Rivas, E. D. Pereira, M. A. Mondaca, R. J. Rivas, M. A. Saavedra, *J. Appl. Polym. Sci.* **2003**, *87*, 452–457.

239 G. L. Y. Woo, M. W. Mittelman, J. P. Santerre, *Biomaterials* **2000**, *21*, 1235–1246.

240 A. Tossi, L. Sandri, A. Giangaspero, *Biopolymers* **2000**, *55*, 4–30.

241 E. A. Porter, X. F. Wang, H. S. Lee, B. Weisblum, S. H. Gellman, *Nature* **2000**, *404*, 565–565.

242 E. A. Porter, B. Weisblum, S. H. Gellman, *J. Am. Chem. Soc.* **2002**, *124*, 7324–7330.

243 T. L. Raguse, E. A. Porter, B. Weisblum, S. H. Gellman, *J. Am. Chem. Soc.* **2002**, *124*, 12774–12785.

244 J. A. Patch, A. E. Barron, *J. Am. Chem. Soc.* **2003**, *125*, 12092–12093.

245 A. R. Statz, R. J. Meagher, A. E. Barron, P. B. Messersmith, *J. Am. Chem. Soc.* **2005**, *127*, 7972–7973.

246 J. Rennie, L. Arnt, H. Z. Tang, K. Nusslein, G. N. Tew, *J. Ind. Microbiol. Biotechnol.* **2005**, *32*, 296–300.

247 G. N. Tew, D. H. Liu, B. Chen, R. J. Doerksen, J. Kaplan, P. J. Carroll, M. L. Klein, W. F. DeGrado, *Proc. Natl. Acad. Sci. USA* **2002**, *99*, 5110–5114.

248 L. Ayres, G. M. Grotenbreg, G. A. van der Marel, H. S. Overkleeft, M. Overhand, J. C. M. van Hest, *Macromol. Rapid Commun.* **2005**, *26*, 1336–1340.

249 A. S. Hoffman, P. S. Stayton, V. Bulmus, G. H. Chen, J. P. Chen, C. Cheung, A. Chilkoti, Z. L. Ding, L. C. Dong, R. Fong, C. A. Lackey, C. J. Long, M. Miura, J. E. Morris, N. Murthy, Y. Nabeshima, T. G. Park, O. W. Press, T. Shimoboji, S. Shoemaker, H. J. Yang, N. Monji, R. C. Nowinski, C. A. Cole, J. H. Priest, J. M. Harris, K. Nakamae, T. Nishino, T. Miyata, *J. Biomed. Mater. Res.* **2000**, *52*, 577–586.

250 A. S. Hoffman, P. S. Stayton, *Macromol. Symp.* **2004**, *207*, 139–151.

251 H. G. Schild, *Prog. Polym. Sci.* **1992**, *17*, 163–249.

252 G. H. Chen, A. S. Hoffman, *Bioconjug. Chem.* **1993**, *4*, 509–514.

253 G. H. Chen, A. S. Hoffman, *J. Biomater. Sci., Polym. Ed.* **1994**, *5*, 371–382.

254 Z. L. Ding, G. H. Chen, A. S. Hoffman, *J. Biomed. Mater. Res.* **1998**, *39*, 498–505.

255 G. Q. Dong, R. Batra, R. Kaul, M. N. Gupta, B. Mattiasson, *Bioseparation* **1995**, *5*, 339–350.

256 R. Agarwal, M. N. Gupta, *Protein Expr. Purif.* **1996**, *7*, 294–298.

257 I. Roy, M. N. Gupta, *Curr. Sci.* **2000**, *78*, 587–591.

258 S. Sharma, P. Kaur, A. Jain, M. R. Rajeswari, M. N. Gupta, *Biomacromolecules* **2003**, *4*, 330–336.

259 M. N. Gupta, R. Kaul, D. Guoqiang, U. Dissing, B. Mattiasson, *J. Mol. Recognit.* **1996**, *9*, 356–359.
260 F. Hilbrig, R. Freitag, *J. Chromatogr. B* **2003**, *790*, 79–90.
261 J. E. Morris, A. S. Hoffman, R. R. Fisher, *Biotechnol. Bioeng.* **1993**, *41*, 991–997.
262 Y. G. Takei, M. Matsukata, T. Aoki, K. Sanui, N. Ogata, A. Kikuchi, Y. Sakurai, T. Okano, *Bioconjug. Chem.* **1994**, *5*, 577–582.
263 R. B. Fong, Z. L. Ding, A. S. Hoffman, P. S. Stayton, *Biotechnol. Bioeng.* **2002**, *79*, 271–276.
264 D. Umeno, T. Mori, M. Maeda, *Chem. Commun.* **1998**, 1433–1434.
265 T. Mori, D. Umeno, M. Maeda, *Biotechnol. Bioeng.* **2001**, *72*, 261–268.
266 M. A. Taipa, R. H. Kaul, B. Mattiasson, J. M. S. Cabral, *Bioseparation* **2000**, *9*, 291–298.
267 A. Desponds, R. Freitag, *Biotechnol. Bioeng.* **2005**, *91*, 583–591.
268 R. B. Fong, Z. L. Ding, C. J. Long, A. S. Hoffman, P. S. Stayton, *Bioconjug. Chem.* **1999**, *10*, 720–725.
269 A. Chilkoti, G. H. Chen, P. S. Stayton, A. S. Hoffman, *Bioconjug Chem.* **1994**, *5*, 504–507.
270 P. S. Stayton, T. Shimoboji, C. Long, A. Chilkoti, G. H. Chen, J. M. Harris, A. S. Hoffman, *Nature* **1995**, *378*, 472–474.
271 T. Shimoboji, E. Larenas, T. Fowler, A. S. Hoffman, P. S. Stayton, *Bioconjug Chem.* **2003**, *14*, 517–525.
272 T. Shimoboji, Z. L. Ding, P. S. Stayton, A. S. Hoffman, *Bioconjug Chem.* **2002**, *13*, 915–919.
273 T. Shimoboji, E. Larenas, T. Fowler, S. Kulkarni, A. S. Hoffman, P. S. Stayton, *Proc. Natl. Acad. Sci. USA* **2002**, *99*, 16592–16596.
274 Z. L. Ding, R. B. Fong, C. J. Long, P. S. Stayton, A. S. Hoffman, *Nature* **2001**, *411*, 59–62.
275 M. Murata, W. Kaku, T. Anada, N. Soh, Y. Katayama, M. Maeda, *Chem. Lett.* **2003**, *32*, 266–267.
276 M. Murata, W. Kaku, T. Anada, Y. Sato, M. Maeda, Y. Katayama, *Chem. Lett.* **2003**, *32*, 986–987.
277 M. Murata, W. Kaku, T. Anada, Y. Sato, T. Kano, M. Maeda, Y. Katayama, *Bioorg. Med. Chem. Lett.* **2003**, *13*, 3967–3970.
278 C. Wang, R. J. Stewart, J. Kopeček, *Nature* **1999**, *397*, 417–420.
279 C. Wang, J. Kopeček, R. J. Stewart, *Biomacromolecules* **2001**, *2*, 912–920.
280 A. Tang, C. Wang, R. J. Stewart, J. Kopeček, *J. Control. Release* **2001**, *72*, 57–70.
281 S. Nagahara, T. Matsuda, *Polym. Gels Netw.* **1996**, *4*, 111–127.
282 M. Shimomura, T. Sawadaishi, *Curr. Opin. Colloid Interface Sci.* **2001**, *6*, 11–16.
283 I. W. Hamley, *Angew. Chem. Int. Ed.* **2003**, *42*, 1692–1712.
284 C. Park, J. Yoon, E. L. Thomas, *Polymer* **2003**, *44*, 6725–6760.
285 S. G. Zhang, *Nat. Biotechnol.* **2003**, *21*, 1171–1178.
286 T. S. Burkoth, T. L. S. Benzinger, D. N. M. Jones, K. Hallenga, S. C. Meredith, D. G. Lynn, *J. Am. Chem. Soc.* **1998**, *120*, 7655–7656.
287 T. S. Burkoth, T. L. S. Benzinger, V. Urban, D. G. Lynn, S. C. Meredith, P. Thiyagarajan, *J. Am. Chem. Soc.* **1999**, *121*, 7429–7430.
288 P. Thiyagarajan, T. S. Burkoth, V. Urban, S. Seifert, T. L. S. Benzinger, D. M. Morgan, D. Gordon, S. C. Meredith, D. G. Lynn, *J. Appl. Crystallogr.* **2000**, *33*, 535–539.
289 O. Rathore, D. Y. Sogah, *Macromolecules* **2001**, *34*, 1477–1486.
290 O. Rathore, D. Y. Sogah, *J. Am. Chem. Soc.* **2001**, *123*, 5231–5239.
291 A. Rosler, H. A. Klok, I. W. Hamley, V. Castelletto, O. O. Mykhaylyk, *Biomacromolecules* **2003**, *4*, 859–863.
292 P. Parras, V. Castelletto, I. W. Hamley, H. A. Klok, *Soft Matter* **2005**, *1*, 284–291.
293 J. M. Smeenk, M. B. J. Otten, J. Thies, D. A. Tirrell, H. G. Stunnenberg, J. C. M. van Hest, *Angew. Chem. Int. Ed.* **2005**, *44*, 1968–1971.
294 H. A. Klok, S. Lecommandoux, *Adv. Mater.* **2001**, *13*, 1217–1229.
295 J. Elemans, A. E. Rowan, R. J. M. Nolte, *J. Mater. Chem.* **2003**, *13*, 2661–2670.
296 G. Floudas, P. Papadopoulos, H. A. Klok, G. W. M. Vandermeulen, J. Rodri-

guez-Hernandez, *Macromolecules* **2003**, *36*, 3673–3683.
297 S. Lecommandoux, M. F. Achard, J. F. Langenwalter, H. A. Klok, *Macromolecules* **2001**, *34*, 9100–9111.
298 H. Schlaad, B. Smarsly, M. Losik, *Macromolecules* **2004**, *37*, 2210–2214.
299 S. Ludwigs, G. Krausch, G. Reiter, M. Losik, M. Antonietti, H. Schlaad, *Macromolecules* **2005**, *38*, 7532–7535.
300 M. Pechar, P. Kopečková, L. Joss, J. Kopeček, *Macromol. Biosci.* **2002**, *2*, 199–206.
301 G. W. M. Vandermeulen, C. Tziatzios, H. A. Klok, *Macromolecules* **2003**, *36*, 4107–4114.
302 G. W. M. Vandermeulen, D. Hinderberger, H. Xu, S. S. Sheiko, G. Jeschke, H. A. Klok, *ChemPhysChem* **2004**, *5*, 488–494.
303 J. Couet, J. D. Jeyaprakash, S. Samuel, A. Kopyshev, S. Santer, M. Biesalski, *Angew. Chem. Int. Ed.* **2005**, *44*, 3297–3301.
304 J. M. Hannink, J. Cornelissen, J. A. Farrera, P. Foubert, F. C. De Schryver, N. Sommerdijk, R. J. M. Nolte, *Angew. Chem. Int. Ed.* **2001**, *40*, 4732–4734.
305 M. J. Boerakker, J. M. Hannink, P. H. H. Bomans, P. M. Frederik, R. J. M. Nolte, E. M. Meijer, N. Sommerdijk, *Angew. Chem. Int. Ed.* **2002**, *41*, 4239–4241.
306 K. Velonia, A. E. Rowan, R. J. M. Nolte, *J. Am. Chem. Soc.* **2002**, *124*, 4224–4225.
307 N. C. Seeman, *Angew. Chem. Int. Ed.* **1998**, *37*, 3220–3238.
308 J. J. Storhoff, C. A. Mirkin, *Chem. Rev.* **1999**, *99*, 1849–1862.
309 N. C. Seeman, *Trends Biochem. Sci.* **2005**, *30*, 119–125.
310 S. Sivakova, S. J. Rowan, *Chem. Soc. Rev.* **2005**, *34*, 9–21.
311 L. Brunsveld, B. J. B. Folmer, E. W. Meijer, R. P. Sijbesma, *Chem. Rev.* **2001**, *101*, 4071–4097.
312 E. A. Fogleman, W. C. Yount, J. Xu, S. L. Craig, *Angew. Chem. Int. Ed.* **2002**, *41*, 4026–4028.
313 J. Xu, E. A. Fogleman, S. L. Craig, *Macromolecules* **2004**, *37*, 1863–1870.
314 F. R. Kersey, G. Lee, P. Marszalek, S. L. Craig, *J. Am. Chem. Soc.* **2004**, *126*, 3038–3039.
315 J. Kim, Y. Liu, S. J. Ahn, S. Zauscher, J. M. Karty, Y. Yamanaka, S. L. Craig, *Adv. Mater.* **2005**, *17*, 1749–1753.
316 S. J. Rowan, P. Suwanmala, S. Sivakova, *J. Polym. Sci., Part A: Polym. Chem.* **2003**, *41*, 3589–3596.
317 S. Sivakova, D. A. Bohnsack, M. E. Mackay, P. Suwanmala, S. J. Rowan, *J. Am. Chem. Soc.* **2005**, *127*, 18202–18211.

17
Gel: a Potential Material as Artificial Soft Tissue

Yong Mei Chen, Jian Ping Gong, and Yoshihito Osada

17.1
Introduction

Hydrogels are polymer networks swollen with large amounts of water [1]. Normally, the water content is more than 50% of the total weight when one uses the term of "gel". Hydrogels are solids on the macroscopic scale: they have definite shapes and do not flow. At the same time, they behave like solutions on the molecular scale: water-soluble molecules can diffuse in hydrogels with various diffusion constants reflecting the sizes and shapes of the molecules.

Gel is a fascinating material for its unique properties, such as phase transition [2], chemomechanical behavior [3] and stimuli responsiveness [4], and for its possible wide application in many industrial fields. Recently, hydrogels have become especially attractive in the biological field due to their possible applications as soft artificial tissues. Hydrogels are expected to be used as scaffolds for repairing and regenerating a wide variety of tissues and organs or even as substitutes for tissues and organs since their three-dimensional network structure and viscoelasticity are similar to those of the macromolecular-based extracellular matrix (ECM) in biological tissue [5].

However, to be qualified for biological applications, significant problems with and shortages of hydrogels need to be solved, including the following. (1) Improving the mechanical toughness of the hydrogel: in the physiological condition, many tissues, such as blood vessels, articular cartilages, semi-lunar cartilages, tendons and ligaments, exist in a severe mechanical dynamic environment. For example, the articular cartilages sustain a daily compression of 10 MPa, while most of hydrogels are mechanically very weak. (2) Controlling the surface properties of hydrogels, such as surfaces with low sliding resistance against other tissues, surfaces with cellular viability and tissue compatibility, etc.: for example, the cartilages show a low resistance against sliding motion, and the endothelial cells that form a continuous monolayer in the inner surface of blood vascular protect procoagulant activity. (3) Controlling the transportation to ions, hormones and nutrition and improving the permeability to proteins.

Therefore, to replace natural tissues with hydrogels, the design and synthesis of hydrogels with some critical parameters, such as high mechanical strength, low surface friction and wear and high cellular viability, are important [6, 7]. Gel scientists have strived to design hydrogels with these critical parameters by using synthetic polymers in addition to natural polymers, because the chemical properties of synthetic gels are easily controllable and reproducible and the cost of synthetic gels is relatively low.

In this chapter, recent progress in the study and development of hydrogels with tough mechanical strength, low frictional coefficients and cellular viability is described and examples of applications of gels as substitutes for biological tissues are introduced.

17.2
Molecular Design of Robust Gels

Applications of hydrogels as mechanical devices are fairly limited due to the lack of mechanical strength. Reported values for the fracture energy of typical hydrogels fall within a range of 0.1–1 J m^{-2} [8–10], which are much lower than those in the usual rubbers [11]. Many researchers may have thought that the lack of mechanical strength of a gel is unavoidable because of its solution-like nature, i.e. low density of polymer chains and low friction between the chains. Furthermore, it is well known that in hydrogels synthesized from monomer solutions, inhomogeneity arises during the gelation [12]. This is also considered as a factor that decreases the mechanical strength. However, when we consider biological systems, we find some hydrogels with excellent mechanical performances. For example, cartilage exhibits a high mechanical strength [13]. It is a challenging problem in modern gel science to fill the gap between artificial gels and biological gels.

Recently, three new hydrogels with good mechanical performance have been developed [14–16]: "topological (TP) gel", "nanocomposite (NC) gel" and "double network (DN) gel". The TP gels have figure-of-eight cross-linkers that can slide along the polymer chains. Reflecting these flexible cross-linkers, the TP gels absorb large amounts of water and can be highly stretched without fracture. In the NC gels, polymer chains are cross-linked by inorganic clay slabs on the scale of a several tens of nanometers, instead of organic cross-linking agents. The NC gels are also highly stretchable and possess other favorable physical properties such as excellent optical transparency. The DN gels consist of two interpenetrated polymer networks: one is made of highly cross-linked rigid polymers and the other of loosely cross-linked flexible polymers. The DN gels, containing about 90 wt.% water, possess both hardness (elastic modulus 0.3 MPa) and toughness (fracture stress ~ 10 MPa). The development of these three kinds of novel gels not only has led to a major breakthrough in finding wider applications of gels in industrial and biomedical fields, but also approached fundamental problems in gel science. In this chapter we describe progress in the field, emphasizing the correlations between gel structures and mechanical properties.

17.2.1
Topological Gel

The TP gels have figure-of-eight cross-linkers; they can slide along polymer chains. A typical example is the polyrotaxane gel synthesized using a supramolecular chemistry technique by Okumura and Ito [14]; a polyrotaxane molecule consists of a poly(ethylene glycol) (PEG) chain, α-cyclodextrin (CD) circles threaded on the PEG chain and large end-groups trapping the α-CD rings; chemically cross-linking the α-CD in an aqueous solution of polyrotaxane results in the polyrotaxane gel.

The TP gels are highly water absorbent and stretchable. The gel swells to about 500 times its original (as-prepared) weight. A TP gel (in the as-prepared state) can be stretched to about 20 times its original length. Another unique nature of the TP gels appears in their macroscopic and equilibrium mechanical behavior when the cross-linking density is low: the stress–strain (S–S) curve of the TP gels under uniaxial elongation exhibits lower convexity over all the range of strain. This is different from the usual chemical gels, for which the S–S curve exhibits an upper convexity in a small strain regime.

The sliding motion of the figure-of-eight cross-linkers accounts for the special feature of the TP gels, which distinguishes them from the usual chemical and physical gels. In the chemical gel, the tension of each polymer chain is heterogeneous; scission of the polymer chains gradually occurs. On the other hand, in the TP gel, the relative sliding of the polymer chains by figure-of-eight cross-linkers occurs to equalize the tension of each polymer chain. Okumura and Ito refer to such an adjustment of the tensions by sliding as a "pulley effect". The sliding mode (pulley effect) is obviously related to the large water absorbance mentioned above.

17.2.2
Nanocomposite Ge

In the NC gels, the clay slabs work as multifunctional cross-linkers. A typical combination of chemicals is N-isopropylacrylamide (NIPA) for the polymer chains, a hectorite $[Mg_{5.34}Li_{0.66}Si_8O_{20}(OH)_4]Na_{0.66}$ as the clay slabs and potassium peroxodisulfate (KPS) as a radical initiator. According to Haraguchi and coworkers, the following two points are essential to obtain NC gels with excellent mechanical properties: (i) the gels must be prepared from mixed solutions of the clay slabs and NIPA *monomer* by the radical polymerization of NIPA (addition of the clay slabs to poly-NIPA solutions does not produce NC gels); (ii) a special radical initiator that can adsorb ionically on the clay surface must be used. Haraguchi and coworkers speculate that under the above polymerization conditions, a radical reaction occurs on the clay surfaces, which leads to strong adsorption of the NIPA chains on the clay surfaces [15].

The NC gels are highly stretchable: an NC gel (with an NIPA/water weight ratio of 0.1 and an elastic modulus of about 10^4 Pa) can be elastically stretched up to about 10 times its original length. Haraguchi and coworkers suggested that the following two points are responsible for the high stretchability: (i) the fluctuation in the cross-linking density of the NC gels is moderate compared with that of chemical gels; (ii) the polymer chains connected to the same clays can sustain the force cooperatively. Similar NC gels composed of another polymer [poly(N,N'-dimethylacrylamide)] are obtained [17, 18]. The shape of the S–S curves of NC gels is similar to that of the usual rubbers or chemical gels.

An important advantage of the NC gels is the simplicity of the synthesis technique. It is easy to induce new structures in the NC gels by modifying the technique: for example, (i) adding large clay particles ($\sim\mu$m) together with nanoscale clay slabs may improve the strength of the NC gels, because the large clay particles play a similar role to the filler particles in the particle-reinforced composites [19]; (ii) by synthesizing the DN gel on a substrate coated with clay, a strong joint may be formed between the NC gel and the substrate. On the other hand, from the viewpoint of fundamental research, even the gelation mechanism is not sufficiently understood at present. Physically, rotations of the clay slabs are characteristic modes of the NC gels.

17.2.3
Double Network Gel

The DN gel is synthesized via a two-step network formation: the first step is the formation of a rigid gel that is highly cross-linked and the second is the formation of a loosely cross-linked network in the first gel [16]. An optimal combination is poly(2-acrylamido-2-methylpropanesulfonic acid) (PAMPS) gel as the first network and polyacrylamide) (PAAm) gel as the second network.

Figure 17.1 shows a comparison of the behaviors of the PAMPS gel (a) and the PAMPS–PAAm DN gel prepared under the optimized conditions (b) under uniaxial compression. Figure 17.2 shows S–S curves for the PAMPS–PAAm, the PAMPS and the PAAm gels. The DN gel holds up to a stress of 17.2 MPa, where the vertical strain λ is about 92%. On the other hand, the PAMPS gel and PAA gel break at stresses of 0.4 MPa ($\lambda=41\%$) and 0.8 MPa ($\lambda=84\%$, respectively.

It should be emphasized that the DN gel exhibits high mechanical strength only when the second network is loosely cross-linked or even without cross-linking. When the cross-linking density of the first network is kept constant at 4 mol% and only that of the second network is changed systematically from 0 to 1.0 mol%, all the PAMPS–PAAm DN gels show a similar elastic modulus of 0.3 MPa, a water content of 90 wt% and a molar ratio of the second network to the first network of 20, regardless of the change in the cross-linking density of the second network. However, a dramatic change in the mechanical strength of DN gels is found, as shown in Fig. 17.3 [16, 20, 21]. The highest fracture strength and strain are obtained for uncross-linked samples [16, 20, 21].

When the second network, PAAm, is in the condition of an un-cross-linked state, the molecular weight of the second linear polymer (M_w) is a prior parame-

Fig. 17.1 Comparison of gel morphology of PAMPS gel (a) and PAMPS–PAAm DN gel (b) before and after uniaxial compression. Cross-linking density (ratio of cross-linker to monomer): 4 mol% for PAMPS, 0.1 mol% for PAAm. Fracture stress: PAMPS gel, 0.4 MPa; PAMPS–PAAm DN gel, 17.2 MPa. Reproduced with permission from [16].

ter related to the mechanical strength of PAMPS–PAAm DN gel [21, 22]. When the M_w is lower than 10^6, DN gels are as fragile as a PAMPS single network gel, the fracture strength and fracture energy being about 0.1 MPa and 1 J m^{-2}, respectively. However, when the M_w is higher than 10^6, the fracture strength and fracture energy increase dramatically to 20 MPa and 10^3 J m^{-2}, respectively, and saturates at $M_w = 3 \times 10^6$.

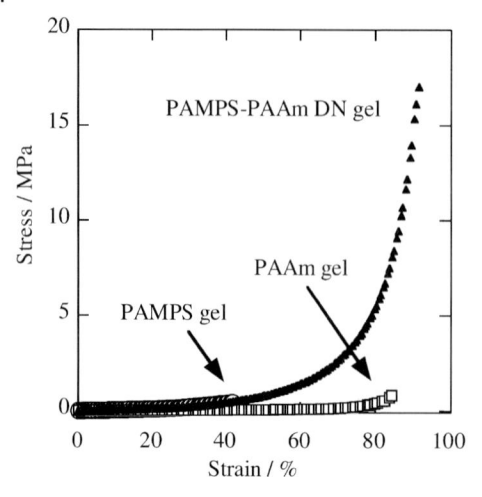

Fig. 17.2 Stress–strain curves for PAMPS–PAAm DN, PAMPS and PAAm gels. Cross-linking density: 4 mol% for PAMPS, 0.1 mol% for PAAm. Reproduced with permission from [16].

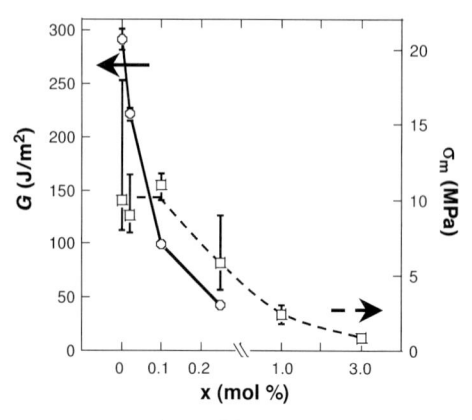

Fig. 17.3 Dependence of the fracture energy at $V = 0.5 \times 10^{-2}$ m s^{-1} (○) and the fracture stress (□) on the cross-linking density x of the second component, PAAm, of PAMPS–PAAm DN gels. Reproduced with permission from [21].

By dynamic light scattering (DLS) analysis, a heterogeneous structural model for PAMPS–PAAm DN gels has been proposed [20, 22]. As shown in Fig. 17.4, PAPMS networks (first component) are rigid and inhomogeneous and large "voids" exist due to the specific radical polymerization mechanism. When PAAm moieties are polymerized in the PAMPS network, some of them are interpenetrated in the PAMPS network and others are filled in the large "voids" of PAMPS gel, partially entangled with the PAMPS network. The linear or

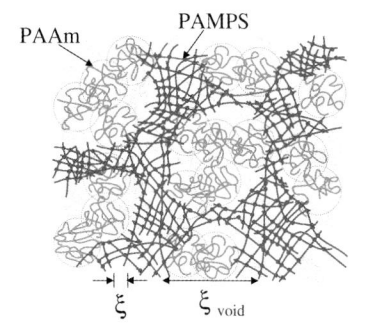

Fig. 17.4 Inhomogeneous structure model of the tough PAMPS–PAAm DN gels. The first PAMPS network represents as a rigid cross-linked polyelectrolyte and the second PAAm network represents an entangled flexible neutral polymer without cross-linker. For clarity, only some of the PAAm chains are highlighted by bolder lines. The second component PAAm chains are in a concentrated region and they are entangled both with each other and with the PAMPS network. The first network PAMPS has a mesh size ξ of several nanometers with large "voids" of size $\xi_{void} \gg \xi$. The voids that filled with concentrated PAAm act as a "crack-stop" to enhance the mechanical strength of the DN gels. Reproduced with permission from [22].

loosely cross-linked PAAm in the "voids" effectively absorbs the crack energy either by viscous dissipation or by large deformation of the PAAm chains, preventing crack growth to a macroscopic level. It was found that the critical molecular weight corresponds to the criteria of physical entanglement of PAAm chains with each other in the void of the PAMPS network with a size ξ_{void}, which is much larger than the radius of the PAAm chains. This indicates that the second PAAm chain is not only interpenetrated to the PAMPS network, but also forms a very soft, continuous network due to the physical entanglement.

The DLS study implies that the inhomogeneity correlates with a drastic improvement of the strength of the DN gels. It is well known in the field of biomechanics that inhomogeneity can improve mechanical strength; for example, the bone of animals is porous and the pores prevent crack propagation by suppressing the stress concentrations [23]. The DN gels suggest that a way to improve the strength of hydrogels is to optimize the inhomogeneities, rather than to remove them.

The DN gels possess both a high elastic modulus (hardness) and toughness. Adjusting the compositions of the first and second networks of the gel can independently control these two quantities. This is convenient for practical applications.

Randomness or inhomogeneity in the structure exists in all kinds of gels. The TP, NC and DN gels are concerned with this problem in different ways: The TP gels have adjustable cross-linking structures. The NC gels propose a cross-linking method effectively reducing the inhomogeneity. The DN gels reveal that the inhomogeneity may serve to improve the mechanical strength. The three new hydrogels described in this chapter introduce new concepts in gel science, and thus are stimulating both basic and applied research.

17.2.4
Robust Gels from Bacterial Cellulose

The discovery of a hydrogel with a good mechanical performance should enable it to find a wide application in industry, such as in fuel cell membranes, load-bearing water absorbents, separation membranes, the printing industry, optical devices and low friction gel machines, and in biomechanical fields, such as in artificial cartilages, tendons, blood vessels and other bio-tissues.

To realize biomedical applications, it is important to apply the concept of high-performance gels to biocompatible polymers. Hydrogels derived from natural polymers frequently demonstrate adequate biocompatibility and have been widely used in tissue engineering approaches [6]. On the basis of the design principle of tough DN gels, some progress has been made recently by combining bacterial cellulose (BC) and gelatin [24].

BC is a kind of extracellular cellulose, produced by bacteria of the genus *Acetobacter* and consisting of a hydrophobic ultra-fine fiber network stacked in a stratified structure [25]. Scanning electron microscopy (SEM) image shows that BC gel consists of alternating dense and sparse cellulose layers with a period of 10 µm, which contribute to the anisotropic mechanical properties. It has a high tensile modulus (2.9 MPa) along the fiber-layer direction but a low compressive modulus (0.007 MPa) perpendicular to the stratified direction. Due to its poor water-retaining ability, water is easily squeezed out from BC gel network, although an as-prepared BC contains 90% water and there is no further recovery in the swelling properties because of hydrogen-bond formation between cellulose fibers.

Gelatin is a polypeptide derived from an extracellular matrix, collagen. Gelatin gel can retain water and recovers from repeated compression. Due to its poor mechanical strength, a gelatin gel is easily broken into fragments under a modest compression of 0.12 MPa. By using the DN gel method, it is possible to create a strong gel consisting of BC and gelatin that has biocompatibility and high mechanical strength [24].

BC–gelatin DN gel is synthesized by immersing BC substrate in gelatin solution (30 wt%), then the gelatin is cross-linked with 1-ethyl-3-(3-dimethylaminopropyl)carbodiimide hydrochloride (EDC) (1 M). Wide-angle X-ray diffraction (WAXD) shows that the crystal pattern of BC is well maintained and the SEM image shows that the homogeneous gelatin textures are filled in the cellulose-poor layers in BC–gelatin gel. The structure of BC–gelatin indicates that anisotropic mechanical strength of BC is maintained after combination with gelatin. Therefore, mechanical strength against compression and elongation were performed perpendicular to and parallel to the stratified direction of BC–gelatin DN gels, respectively. Typical compressive and elongation S–S curves are shown in Fig. 17.5a and b, respectively. BC–gelatin DN gel shows a compressive elastic modulus of 1.7 MPa (Fig. 17.5a), which is more than 240 times higher than that of BC gel (0.007 MPa) and 11 times higher than that of gelatin gel (0.16 MPa). The fracture strength of BC–gelatin DN gel against compression in the direc-

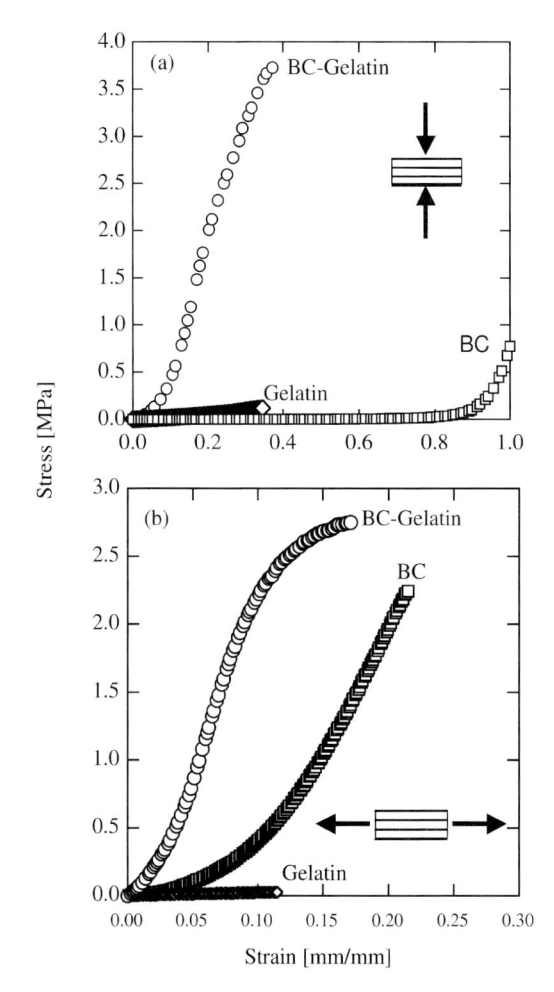

Fig. 17.5 Comparison of compressive (a) and elongational (b) stress–strain curves of BC–gelatin DN, gelatin and BC gels. The compression and elongation were performed perpendicular to and parallel with the stratified direction of BC and BC–gelatin DN gels, respectively. Concentration of gelatin in feed, 30 wt%; EDC concentration, 1 M. Reproduced with permission from [24].

tion vertical to the stratified structure reached 3.7 MPa. This strength is in the range of that of articular cartilage (1.9–14.4 MPa). This value is about 31 times higher than that of gelatin gel. Moreover, the BC–gelatin DN gel recovers well under a repeated compressive stress up to 30% strain. The elongation S–S curve (Fig. 17.5 b) shows that BC–gelatin DN gel can sustain nearly 3 MPa elongation stress with 23 MPa elastic modulus, which is 112 times larger than that of gelatin gel.

A similar improvement in the mechanical strength was also observed for the combination of BC with polysaccharides, such as sodium alginate, gellan gum and ι-carrageenan. For example, the compressive elastic modulus of BC–ι-carrageenan DN gel is 0.12 MPa, which is 17 and 13 times higher than that of individual BC and ι-carrageenan gels, respectively. In summary, the mechanical properties of synthetic and natural gels can be effectively improved by inducing a double network structure. The composition of the two kinds of networks can be adjusted according to different applications.

17.3
Surface Friction and Lubrication of Gels

Industrial and environmental problems caused by high frictional surfaces of materials always exist in our daily life. Looking for materials with a low surface friction has been one of the classical and everlasting research topics for materials scientists and engineers. Despite many efforts, it has been shown that surface modification or addition of a lubricant is not very effective in reducing the steady-state sliding friction between two solids, which show a frictional coefficient $\mu \approx 0.1$ even in the presence of lubricant [26].

In contrast to the artificial systems, one can observe fascinating surface behavior of bio-organs. For example, the cartilage of animal joints has a friction coefficient in the range 0.001–0.03, remarkably low even for hydrodynamically lubricated journal bearings [27]. It is not understood why the cartilage friction of the joints is so low even under conditions such that the pressure between the bone surfaces reaches as high as 3–18 MPa [28], and the sliding velocity is never greater than a few centimeters per second [27]. Under such conditions, the lubricating liquid layer cannot be sustained between two solid surfaces and the hydrodynamic lubrication does not work.

The cartilage, which consists of a three-dimensional collagen network filled with synovial fluid, provides the low friction of joints [27–30]. Cartilage cells synthesize a complex extracelluar matrix (ECM); the weight-bearing and lubrication properties of cartilage are associated primarily with this matrix and its high water content. The main macromolecular constituents of extracellular matrix are the proteoglycan, aggrecan and the cross-linked network of collagen fibrils [31, 32].

The tribological properties of polysaccharides, such as mucins, which comprise a family of high molecular weight, extensively glycosylated glycoproteins, play a crucial role in their biological activity, providing smooth lubrication of tissues and organs and protecting cell surfaces from damage. These facts suggest fascinating features of polysaccharides in tribology.

The role of the solvated polymer network that exists in the extracellular matrix in a gel state is critically important in the specific frictional behavior of cartilage. Believing that researchers should be able to find a solution to reduce the sliding friction of artificial systems by elucidating the low-friction secret of biological surfaces from viewpoint of hydrogels, the authors have been investigating the friction and lubrication of hydrogels for the past 10 years [33–39].

17.3.1
Complexity of Gel Friction

The friction between solids obeys Amonton's law (1699), $F = \mu W$, which says that the frictional force F is linearly proportional to load W and does not depend on the apparent contact area A of two solid surfaces and sliding velocity v. However, the gel friction does not simply obey this Amonton's law and shows rich and complicated features. Gel friction depends strongly on the chemical structure of gels, surface properties of opposing substrates and measurement conditions. In addition, normal load, sliding velocity, apparent contact area and the elasticity of the gel are important parameters in gel friction.

Figure 17.6 shows the frictional behavior of hydrogels with different chemical structures: poly(vinyl alcohol) (PVA), gellan, poly(2-acrylamido-2-methylpropane-sulfonic acid) (PAMPS) and its sodium salt (PNaAMPS) gels [34]. These hydrogels were slid on a glass substrate at a sliding velocity of 7 mm min^{-1}, using a tribolometer. Figure 17.6a shows the relationship between the normal load and the friction force. The friction coefficient μ, which is defined as the ratio of the frictional force to applied load, is shown in Fig. 17.6b. It is obvious that the frictional behavior of gels depends strongly on their chemical structure. The relationship between frictional force (F) and normal load can be classified into three types: (1) F almost does not depend on normal load, such as with gellan, a kind of polysaccharide; (2) F increases slightly with normal load, such as with PVA, a kind of nonionic gel; (3) F increases strongly with normal load, such as with PAMPS and PNaAMPS, two kinds of polyelectrolyte. The F of PNaAMPS gel shows a linear increase with the load with a value one order of magnitude lower than that of other gels.

The friction coefficient μ of these gels accordingly shows unique load dependences, which is completely different from those of solids. All the μ values of PVA, gellan and PAMPS gels decrease with an increase in load. On the other hand, the μ value of the PNaAMPS gel remains constant with change in load, similarly to rubber, but its value is as low as 0.002, which is two orders of magnitude lower than those of solids. PAMPS and PNaAMPS gels differ only in their counterions, but they show a striking difference in frictional behavior on a glass surface. In conclusion, in any one of these cases, the friction force of gels is 2–3 orders of magnitude lower than that of rubber. This behavior can be observed continuously for several hours.

The frictional behavior of gels is not only strongly dependent on the chemical structure of the gels, but also dependent on the surface properties of opposing substrates. When nonionic PVA gel is allowed to slide on a polytetrafluoroethylene (Teflon) plate, the behavior is the same as that on a glass surface. However, the behavior of strongly anionic PAMPS gel on Teflon changes substantially and becomes similar to that of PVA on glass. When a pair of polyelectrolyte gels carrying the same charges, e.g. PNaAMPS gel with PNaAMPS gel, are slid over with each other, the system shows a very low frictional force, which is comparable to or even lower than those between articular cartilages. On the other hand,

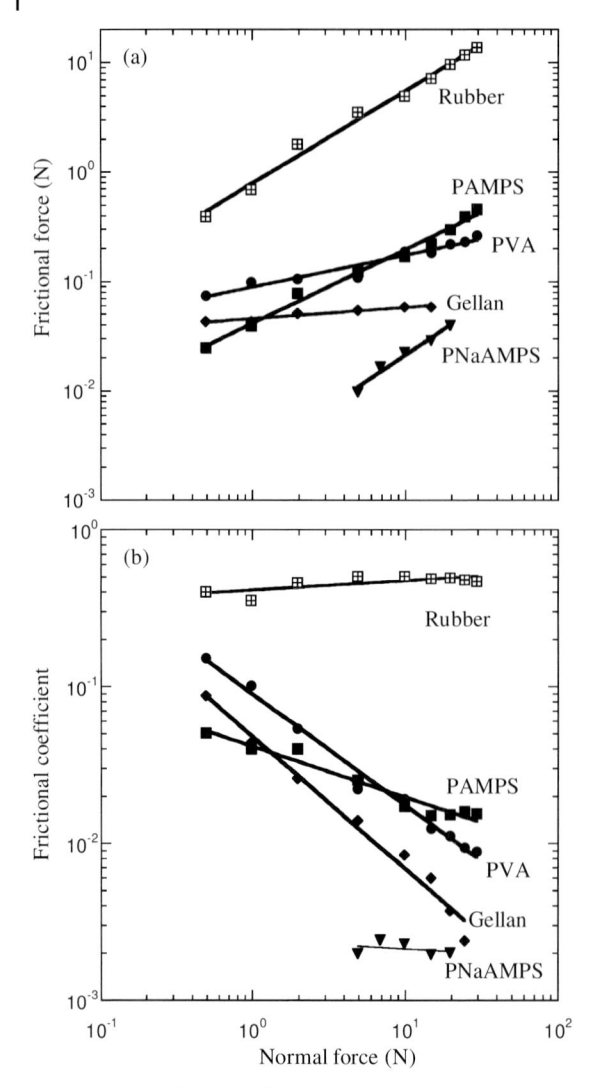

Fig. 17.6 Dependences of friction on load (a) and the coefficient of friction on load (b) for various kinds of hydrogels slid on a glass substrate. Sliding velocity: 7 mm min^{-1}. Sample sizes: for PVA, gellan and rubber, 3×3 cm; PAMPS and PNaAMPS, 2×2 cm. Compressive modulus E: PVA, 0.014 MPa; gellan, 0.06 MPa; PAMPS, 0.25 MPa; PNaAMPS, 0.35 MPa; rubber, 7.5 MPa. Degree of swelling q: PVA, 17; gellan, 33; PAMPS, 21; PNaAMPS, 15. The measurements were carried out in air. Reproduced with permission from [34].

when two polyelectrolyte gels carrying opposite charges are slid over each other, the adhesion between the two gels is so high that the gels are broken during the measurement [35]. This phenomenon indicates that the electrostatic interaction of gels with the sliding substrate is crucial in gel friction.

In order to describe the friction behavior of gels, a repulsion–adsorption model from the viewpoint of interfacial interaction between polymer and solid has been proposed [36]. When gels are slid on a solid substrate, the polymer network of the gel will be repelled from the surface if the polymer has a repulsive interaction with the substrate or will be adsorbed on the solid surface if the polymer has an attractive interaction with the solid surface. The former, for example, can be found for a gel carrying the same charge as that of a solid surface and the latter for a gel carrying the opposite charge to that of solid. In the former case, the viscous flow of solvent between the solid surface and the polymer network will make a dominant contribution to the frictional force. In the latter case, however, the adsorbing polymer chains will be stretched when the solid surface moves relative to the gel. The elastic force increases with the deformation and eventually detaches the adsorbing polymer chain from the substrate, which in turn appears as the frictional force.

The model predicts that (1) when the substrate is repulsive, the frictional force is lower than in the attractive case and it increases linearly with the normal pressure and velocity when the compressive strain (normal load) is not very high; (2) when the substrate is attractive, the frictional force increases with the attraction strength. For weak attraction, the pressure dependence of the frictional force is much weaker than in the repulsive case. It becomes stronger when the attraction strength increases. The frictional force exhibits a peak when the velocity is increased to a value such that the polymer blob relaxation is not quick enough for adsorption to occur on the substrate.

According to this model, the low frictional force and the strong load dependence for the PNaAMPS gel on a glass surface suggest a repulsive interaction between the gel and the solid. This repulsive interaction is attributed to the electrostatic repulsion between the anionic gel and the negatively charged glass surface in water. On the other hand, the much higher friction with less load dependence observed in PVA gel against a glass surface indicates an attractive interaction. The gellan gel might also have an attractive interaction but it should be even weaker than that of PVA, since it has a lower friction than that of PVA and the load dependence is also weaker than that of PVA. These suggestions from gel friction analysis were in agreement with the results obtained by colloidal probe atomic force microscopy (AFM) analysis [34].

As shown in Fig. 17.6, the friction of the PAMPS gel measured in air was much higher than that of the PNaAMPS gel. This should be attributable to the pH dependence of the surface property of glass. The surface SiO^- group density of glass decreases with decrease in pH and the glass surface becomes neutral at pH 3–4. Thus, the PAMPS gel has an electrostatic repulsion with SiO^- groups of the glass surface in the presence of large amounts of water. However, the proton of the PAMPS gel, which is a strong polyacid, might interact with the glass surface in air, which annihilates the negative charges of the glass surface. Hence the electrostatic repulsion is suppressed and the friction force increases. On the other hand, a repulsive force always exists between the PNaAMPS gel and the glass surface either in air or in water. These results again demonstrate the predominant importance of the interface interaction in the gel friction.

17.3.2
Template Effect on Gel Friction

It has been discovered that the surface friction of a hydrogel is strongly dependent on the substrate on which the gel is synthesized. Hydrogels that are synthesized between a pair of glass substrates have a mirror-like smooth surface. This is the same when a hydrogel is synthesized on other hydrophilic substrates, such as mica and sapphire. However, even for the same chemical structure, a hydrogel exhibits an eel-like slim surface when it is synthesized on hydrophobic substrates, such as Teflon and polystyrene (PS). The differences in the surface nature of the gels synthesized on different substrates are so obvious that they can be easily distinguished by touching with one's finger. If a hydrogel is synthesized between two plates, one of which is hydrophobic and the other hydrophilic, heterogeneous gelation occurs [37]. After equilibrated swelling in water, the gel exhibits significant curvature, as shown in Fig. 17.7, where the gel surface formed on the Teflon surface is always the outside of the curvature and that on the glass is the inside. This is because the gel surface close to the Teflon has a higher swelling whereas the next one has a lower value. The phenomenon indicates that the gel formed near the Teflon surface has lower cross-linking density. The gel formation at different positions of the reaction cell was analyzed by electronic speckle pattern interferometry. The result showed that the non-cross-linked polymer concentration increased with increasing distance near to the Teflon. In the cross-linking process, the gelation ability of the monomer on the side of the Teflon surface is lower than that on the side of the glass sur-

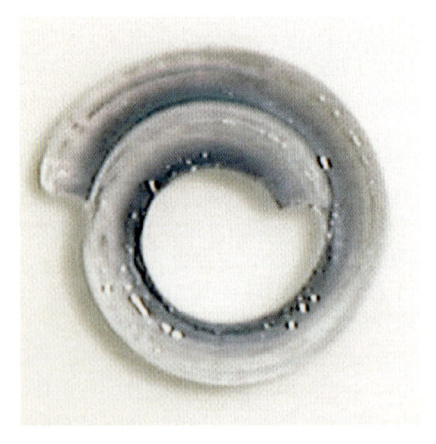

Fig. 17.7 Photograph of the PAMPS gel prepared between a Teflon plate and a glass plate after swelling in water. The gel was dyed after preparation for the clarity of image.

face, with the result that the monomer concentration near the Teflon side is high and monomers diffuse to the glass side where the monomers have already been consumed, leading to a gradient network density.

The lower cross-linking density of the gel close to the hydrophobic substrate is coincident with the fact that the surface layer synthesized on PS showed a lower compressive modulus of $\sim 10^3$ Pa whereas that in the bulk layer or the surface layer prepared on glass showed a value of $\sim 10^4$ Pa [38]. One can assume from this result that such a suppressed polymerization near the hydro-

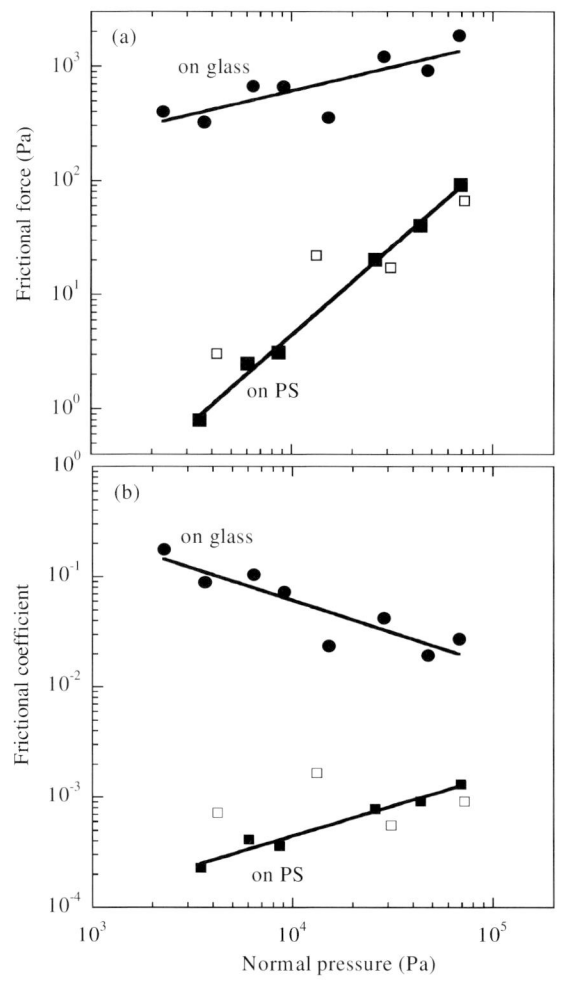

Fig. 17.8 Normal pressure dependence of the frictional force (a) and the frictional coefficient (b) of PAMPS gels slid against a glass plate in water at an angular velocity of 0.01 rad s^{-1}. (●) Prepared on glass, swelling degree, 21; (■) prepared on PS, swelling degree, 27; and (□) containing linear polymer chains, swelling degree, 15. Reproduced with permission from [39].

phobic substrate cannot make an efficient cross-linking reaction, but leads to the formation of branched dangling polymer chains.

The frictional force and frictional coefficient of PAMPS gel synthesized on a glass plate and on a PS plate are shown in Fig. 17.8 [39]. It shows that at low pressure, the frictional coefficient of the gel prepared on PS is as low as 10^{-4}, which at least two orders of magnitude lower than that of the gel synthesized on glass. Since the frictional force between two pieces of PAMPS gels in water increases only weakly with decrease in the degree of swelling of the gels [35], the large difference observed here should not be attributed to the small difference in the water content or the ionic strength of the two kinds of gels, but to the different surface natures of the gels. The decreased frictional coefficient of the gel prepared on a hydrophobic substrate is attributed to the presence of brush-like dangling chains on the gel surface, as revealed by the result for gels containing free linear polymer chains prepared on a glass plate, which showed the similar low friction coefficients (Fig. 17.8).

Such a substrate effect is observed in a wide variety of hydrogels prepared from water-soluble vinyl monomers, such as sodium styrenesulfonate, acrylic acid and acrylamide, on various hydrophobic substrates, such as Teflon, polyethylene, polypropylene, poly(vinyl chloride) and poly(methyl methylacrylate) (PMMA) [37].

17.3.3
Low Friction by Dangling Chains

The effect of polymer brushes on the reduction of sliding friction was also observed for solid friction by surface force apparatus (SFA) measurements [40, 41]. For example, Klein and coworkers reported massive lubrication between mica surfaces modified by repulsive polyelectrolyte brushes in water [42]. These results show that polymer dangling chains on solid or gel surfaces can dramatically reduce the surface friction if the polymer brush has a repulsive interaction with the sliding substrate.

The dramatic friction reduction effect caused by dangling chains has also been found in the negative load dependence of some physically cross-linked polysaccharide gels, such as gellan and κ-carrageenan. For example, the frictional force of gellan gel shows almost no dependence on normal load when the pressure is lower than a critical value [43]. However, the frictional force decreases with increase in normal load when the pressure is increased to a critical value (about 10^4 Pa). Polysaccharide gels are loosely cross-linked by salt. They undergo reversible "sol–gel" transition and transfer to the sol state under heating or pressure and gradually dissolve in water even at room temperature. Gellan and κ-carrageenan are loosened partially or dissolved slowly from the gel network under high pressure exceeding a critical value. The dissolved polymer chain forms a dangling-like architecture on the gel surface, leading to an increased hydrodynamic layer, which contributes to a decrease in frictional force beyond the critical load.

To investigate quantitatively the relationships between the friction coefficient and the polymer brush properties, such as polymer length, density and structure, hydrogels of poly(2-hydroxyethyl methacrylate) (PHEMA) with well-defined polyelectrolyte brushes of poly(sodium 4-styrenesulfonate) (PNaSS) of various molecular weights were synthesized, keeping the distance between the polymer brushes constant at ~20 nm [44]. The effect of polyelectrolyte brush length on the sliding friction against a glass plate, an electrorepulsive solid substrate, was investigated in water in the velocity range 7.5×10^{-5}–7.5×10^{-2} m s^{-1}. It was found that the presence of a polymer brush can dramatically reduce the friction when the polymer brushes are not very long. With an increase in the length of the polymer brush, this drag reduction effect works only at a low sliding velocity and a gel with long polymer brushes even shows a higher friction than that of a normal network gel at a high sliding velocity. The strong polymer length and sliding velocity dependence indicates a dynamic mechanism of the polymer brush effect [44].

It is still a question why a gel possessing branched dangling polymer chains on its surface exhibits such a low frictional coefficient. The reason might be associated with the enhanced hydrodynamic thickness of the solvent layer between branched dangling polymer chains and the sliding substrate [39]. As discussed in Section 17.3.1, when the interfacial interaction is repulsive between the gel and the substrate on which the gel is slid, a water layer is retained at the interface even under a large normal load to give a low friction. Under the same pressure, the static solvent layer thickness should be the same for both the chemically cross-linked gel and the gel having branched dangling polymer chains on its surface. However, under the shear flow which occurred in the process of friction measurement, the polymer brushes are deformed more easily than that of the cross-linked network and this would increase the effective thickness of the hydrodynamic layer, resulting in reduced shear resistance. This discovery should enable hydrogels to find wide application in many fields where low friction is required.

In conclusion, the frictional behavior of hydrogels is more complex than that of solids and dependent on the chemical structure of the gels, the surface properties of sliding substrates and the measurement conditions. In addition, the frictional coefficient of the gel reaches a value as low as 10^{-3}, which is much lower than that of solid materials. Two strategies that introduce branched dangling polymer chains and linear polyelectrolyte on the gel surface can effectively reduce the friction coefficient to as low as 10^{-4}–10^{-5}, which is comparable to that of animal joints. Studying friction behavior of hydrogels is helpful for understanding the low friction coefficient mechanism observed in biological systems and may be useful in finding novel approaches in the design of low-friction artificial organs. Soft and wet gel materials with both a high strength and an extremely low surface friction will promote the application of hydrogels in industrial and biomedical fields.

17.4
Polymer Gels as Scaffolds for Cell Cultivation

To mimic the macromolecular-based extracellular matrix in biological tissues, the cell adhesion and proliferation properties of hydrogels are another critical parameter. However, various hydrogels that originate from natural resources, such as alginate [45], chitosan [46, 47] and hyaluronic acid [48], and that are created synthetically, such as poly(N-isopropylacrylamide) (PNIPAAm) [49], poly(ethylene oxide) (PEO) [50], poly(vinyl alcohol) (PVA) [51] and poly(ethylene glycol) (PEG) [52], show poor cellular viability without modification with cell adhesive proteins or peptides, such as collagen, laminin, fibronectin and RGD sequence.

Recently, we found that bovine fetal aorta endothelial cells (BFAECs) can spread, proliferate and reach confluency on synthetic hydrogels with negative charges, such as poly(sodium 2-acrylamido-2-methylpropanesulfonate) (PNaAMPS), poly-(sodium p-styrenesulfonate) (PNaSS) and poly(acrylic acid) (PAA), without surface modification with any cell adhesive proteins or peptides [53]. The artificial cell scaffolds from synthetic sources have many advantages over natural sources such as collagen: the chemical properties of synthetic gels are easily controllable and reproducible, they are infection-free, withstand high-temperature sterilization and are relatively inexpensive. Therefore, the success in cultivating cells on negatively charged synthetic gel surfaces without any surface modification of adhesive proteins would substantially promote the application of gels for tissue engineering. Cell behaviors on various hydrogels and the effect of gel charge density on cell growth are described in this section.

17.4.1
Modulation of Cell Growth on Various Gels

The behavior of bovine fetal aorta endothelial cells (BFAECs) on several kinds of negatively charged hydrogels, such as poly(acrylic acid) (PAA), poly(methacrylic acid) (PMAA), poly(sodium p-styrenesulfonate) (PNaSS) and poly(sodium 2-acrylamido-2-methyl-1-propanesulfonate) (PNaAMPS) and on neutral hydrogels, such as polyacrylamide (PAAm) and poly (vinyl alcohol) (PVA) with different cross-linking density were investigated [53]. These polymers form strong hydrogels with a double network (DN) structure with proper combinations [16]. Understanding the cellular behavior of these gels helps in designing gels that have cell viability and also tough mechanical strength.

Phase-contrast micrographs of BFAECs on various kinds of gels cultured for 6 and 120 h are shown in Fig. 17.9. After 6 h of culture, BFAECs could be observed to adhere or spread on the gel surface, but not extending to the proliferation period. Therefore, cell morphology was analyzed at this time. The floating ratio represents the percentage of cells that remained in suspension solution and did not stick to the gel surface. The adhesive ratio represents the percentage of cells that stuck more or less tightly to the gel surface but remained in a

(I) (6h)   (II) (120h)

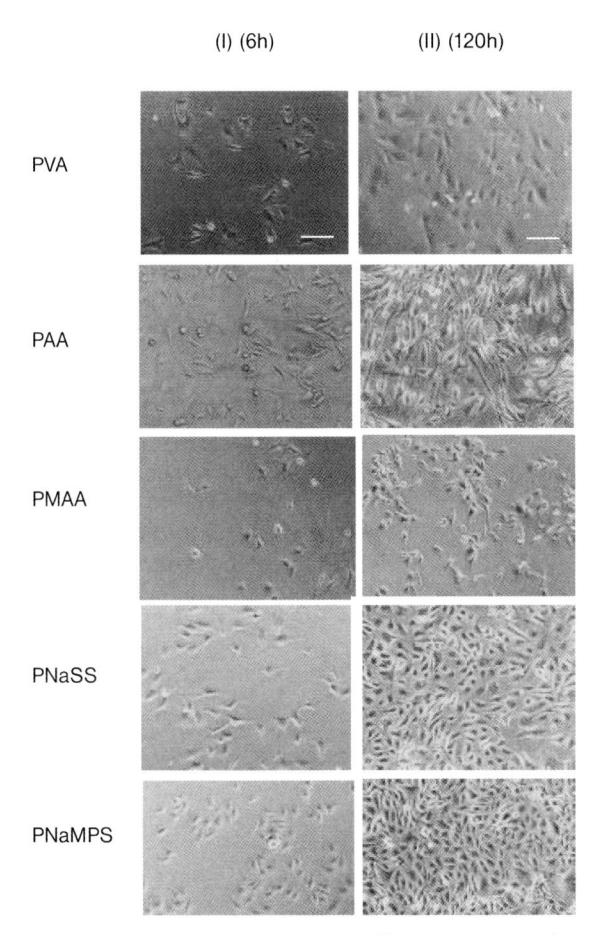

PVA

PAA

PMAA

PNaSS

PNaMPS

Fig. 17.9 Phase-contrast micrographs of bovine fetal aorta endothelial cell (BFAECs) on various gels at the initial stage (6 h, column I) and after a prolonged culture time (120 h, column II). Cross-linker concentration of gels: PVA, 6 mol%; PAA, 2 mol%; PMAA, 1 mol%; PNaSS, 10 mol%; PNaAMPS, 6 mol%. Original magnification: ×10. Scale bar: 100 µm. Reproduced with permission from [53].

spherical shape after lightly shaking the culture dish. The spreading ratio represents the percentage of cells that adhered on the gel surface with a change to fusiform or polygonal shape, concomitant with the extension of pseudopodia [54]. All kinds of hydrogels show substantially high cell adherence and the sum of adhesive ratio and spreading ratio is >50%. The results show that in the initial stage, the chemical structure and cross-linking concentration of gels do not have a remarkable influence on cell adherence. However, the spreading ratio obviously changes with chemical structure and cross-linking concentration of gels. Cells do not spread on non-ionic PAAm gels, except on a gel of 4 mol% cross-linking concentration with only about an 18% spreading ratio. Clearly, PAAm gels have a strong inhibiting effect on cell spreading, regardless of cross-linking concentration. It is considered that weak interaction between the proteins and the PAAm surface [55, 56] results in reduced cell activity.

However, about a 50% spreading ratio is observed on another nonionic gel, PVA, when the cross-linking concentration is 6 and 10 mol%, whereas it decreases to about 15% when the cross-linking concentration is decreased to 2 and 4 mol%. This result indicates that the cell viability on PVA gel is better than that on PAAm gel in the initial culture stage. When the cross-linking concentration of PVA gel is 2 and 4 mol%, cells do not proliferate with culture time, whereas when the cross-linking concentration is increased to 6 and 10 mol%, cells proliferate with culture time and reach subconfluency at about 168 h, with a cell density of 3.72×10^4 and 2.67×10^4 cells cm^{-2}, respectively.

The cell viability on PAA and PMAA gels, which are weak polyelectrolytes with carboxylic acid groups on the side-chains of the polymers, was also investigated. When the cross-linking concentration of PAA gel is 1 and 2 mol%, about 60% spreading cells is observed and cells proliferate with culture time, eventually reaching confluency at 144 h, with a cell density of 6.6×10^4 and 7.21×10^4 cells cm^{-2}, respectively. However, further increase in cross-linking concentration to 4 and 6 mol% induces a decline in the spreading cells to 46 and 23%, respectively, and the spreading cell number decreases to about half at 48 h and ultimately almost no spreading cells are observed.

The same phenomenon is observed on PMAA gels. The cell spreading ratio is about 50% when the cross-linking concentration is 1, 2 and 3 mol%. However, when the cross-linking concentration is increased to 4 mol%, the cell spreading ratio sharply decreases and almost no spreading cells are observed when the cross-linking concentration is 6 mol%. Cells cannot proliferate on PMAA gels except when the cross-linking concentration is 1 mol%, although PMAA has carboxylic acid groups as do PAA gels. On 1 mol% PMAA gel, cells reach subconfluency at 96 h with a cell density of 4.0×10^4 cells cm^{-2} and no further proliferation is observed after 96 h. The results show that in contrast to cell proliferation on PAA gel, although cells can proliferate on 1 mol% PMAA gel at 4 days, cells cannot proliferate after that. This indicates that cell proliferation is very sensitive to the chemical structure and hydrophobic/hydrophilic properties of gels. The above results indicate that cell spreading and proliferation are sensitive to the cross-linking concentration of weak polyelectrolyte gels. It has been reported that the carboxylic acid groups of the PAA-grafted surface have a negative effect on cell adhesion, spreading and growth [57]. Other research showed that there were only a few cells attached to the PAA-grafted surface and cell proliferation was very slow [58]. Our results indicate that if PAA is cross-linked under suitable conditions, PAA hydrogel is also able to be a cell scaffold.

On PNaSS and PNaAMPS gels, which are strong polyelectrolytes, both having sulfonate functional groups, distinct cell behaviors are observed compared with PAA and PMAA gels. When the cross-linking concentration of PNaSS gel is 4 mol%, the spreading ratio is 62% and it increases slowly with increasing cross-linking concentration, reaching 76% when the cross-linking concentration is 10 mol%. For PNaAMPS gel, when the cross-linking concentration is increased from 2 to 5 mol%, the spreading ratio is higher than 60%, which does not obviously change with cross-linking concentration. When the cross-linking concentration is 1 and 6 mol%, the spreading ratio decreases to 38 and 16.1%, respectively. The cell proliferation behaviors on PNaSS and PNaAMPS gels are very similar and independent of cross-linking concentration. Cells proliferate to confluency on PNaSS and PNaAMPS gels with a cell density higher than 1.1×10^5 cells cm^{-2} at 144 h. It is interesting that the cell spreading ratio on PNaAMPS gel at 6 h is only 16.2% when the cross-linking concentration is 6 mol%; however, cells proliferate normally and reach confluency with a prolonged culture time. The results indicate that cells can spread and proliferate on strong polyelectrolyte gels of negative charge over a wide range of cross-linker concentration. However, cells cannot proliferate on PNaAMPS gel with a 1 mol% cross-linking concentration although the experiment was repeated many times. Many chemical and physical properties of substrate affect cell compatibility, such as charge density [59, 60], interfacial energy, hydrophobicity/hydrophilicity balance, mobility of the polymer chain [38] and morphology of the material [61]. It is difficult to explain these phenomena at present.

Figure 17.10 shows the cell density on PAAm (1 mol%), PVA (6 mol%), PMAA (1 mol%), PNaSS (10 mol%) and PNaAMPS (6 mol%) gels as a function

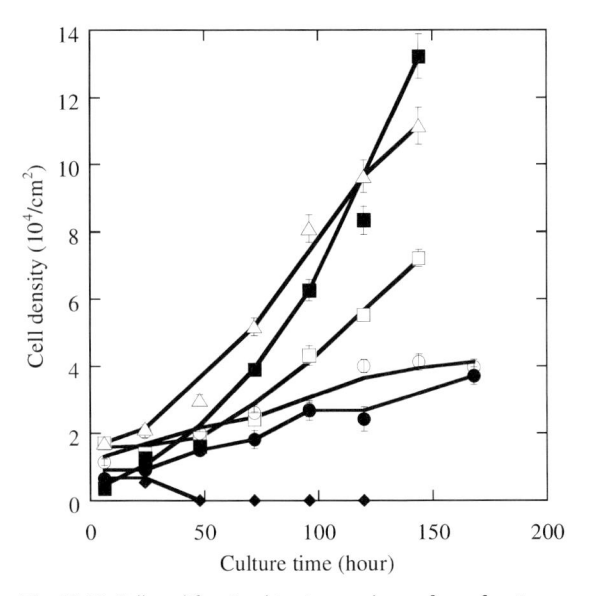

Fig. 17.10 Cell proliferation kinetics on the surface of various gels: (●) PVA (6 mol%); (◆) PAAm (4 mol%); (□) PAA (2 mol%); (○) PMAA (1 mol%); (△) PNaSS (10 mol%); (■) PNaMPS (6 mol%). Error ranges are standard deviations over n=4–8 samples. Reproduced with permission from [53].

of culture time. The cell proliferation rates on PNaSS gel with 4, 6 and 8 mol% cross-linking concentration are the same as that on PNaSS gel with 10 mol% cross-linking concentration and the cell proliferation rates on PNaAMPS gel with 2, 3, 4 and 5 mol% cross-linking concentration are the same as that on PNaAMPS gel with 6 mol% cross-linking concentration (data not shown). These results show that cell proliferation is fastest on PNaSS and PNaAMPS gels, which are strong polyelectrolytes with negative charges. The proliferation rate on these gels is very close to that on collagen gel (data not shown). The cell proliferation rate on PAA gel is lower than that on PNaSS and PNaAMPS gels, but higher than that on PMAA gel. This indicates that the cell proliferation rate is correlated with the charges of the gels. It has been reported that a PNaSS-grafted surface induces strong protein adsorption, which results in cell adhesion. It is argued that an aromatic ring close to the ionizable group is responsible for such a phenomenon [62]. The results suggested that the same protein adsorption from serum-containing medium on the PNaSS and PNaAMPS gel surface might occur to facilitate cell adhesion, spreading and proliferation. The study demonstrated that BFAECs can grow on PVA, PAA, PMAA, PNaSS and PNaAMPS gels without modification with any proteins or peptides, exhibiting higher levels of spreading and proliferation. The results also demonstrate that not only a positively charged environment promotes cell proliferation [63, 64], but also negatively charged surfaces.

17.4.2
Effect of Charge on Cell Growth

It has been reported that cell–substrate interactions are closely related to substrate surface charge [65]. However, so far, the correlation between the surface charge of a hydrogel and cell adherence and proliferation has scarcely been supported by experiment data. Our above study indicated that cell adherence, spreading and proliferation are correlated with the charge of the gels. Therefore, the zeta potentials of PAAm, PVA, PAA, PMAA, PNaSS and PNaAMPS gels with various cross-linking concentrations were measured, to analyze the effect of surface charge density on cell viability.

Figure 17.11 displays the zeta potential of hydrogels as a function of cross-linking concentration. It shows that the absolute value of the zeta potential is dependent on the chemical structure and the cross-linking density of the gels. Non-ionic gels, PAAm and PVA, show a negative zeta potential even with an electrically neutral structure, perhaps due to ionic adsorption in buffer solution. PAA and PMAA gels containing carboxylic acid groups dissociate at pH 7.4, inducing negative charges. When the cross-linking concentration of PAA hydrogel is 1 and 2 mol%, which show cell spreading and proliferation as described previously, the zeta potentials are −26.54 and −28.40 mV, respectively. On the other hand, for PAA gels with cross-linking concentrations of 4 and 6 mol%, on which cells show no proliferation, the zeta potentials are −16.05 and −13.19 mV, respectively. The zeta potential of PMAA is higher than that of PAA gel and de-

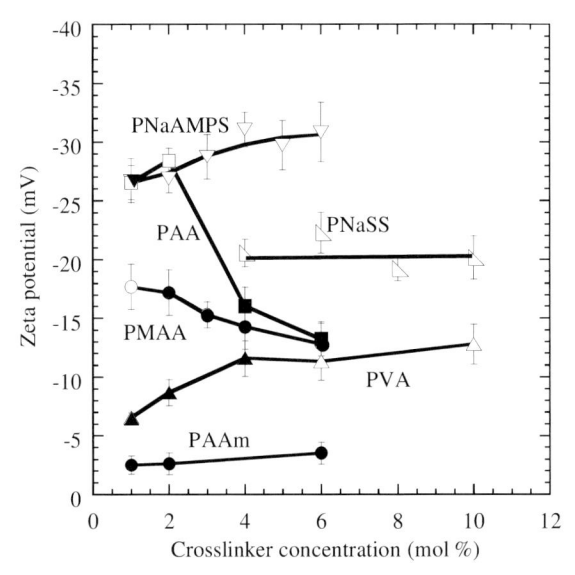

Fig. 17.11 Zeta potential of gels as a function of cross-linker concentration in HEPES buffer solution. Gels on which the cells showed a confluent or subconfluent are expressed by open symbols and those on which cells could not growth are by closed symbols. Reproduced with permission from [53].

creases slowly with increase in cross-linking concentration. This indicates that ionization of the carboxylic acid groups of weakly charged gel decreases with increasing cross-linking concentration. This is because the mesh size of the gel decreases with increasing cross-linking concentration, resulting in a decrease in the distance between polymer chains. To maintain a thermodynamically stable state in a small space, ionization of carboxylic acid groups is suppressed to decrease electrostatic repulsion in the gel network. This is in agreement with the result for cell proliferation on PAA and PMAA hydrogels, which is sensitive to the cross-linking concentration.

The zeta potential of strongly charged gels in water decreases dramatically with increase in cross-linking concentration, suggesting that the charge density of the gels increases with increase in cross-linking concentration (data is not shown). However, the zeta potential of PNaAMPS in HEPES buffer solution does not obviously change with cross-linking concentration and maintains a value of about −25 to −30 mV. The results indicate that in high ionic strength solution, the charge of gels is shielded by salt. Therefore, the zeta potential of strong polyelectrolyte gels is not sensitive to the cross-linking concentration. The same result is found in the case of PNaSS gel, which shows a constant zeta potential of −20 mV, regardless of the change in cross-linking concentration. This is in agreement with the result of cell proliferation on PNaAMPS and PNaSS hydrogels, which is not sensitive to the cross-linking concentration. The results suggested that the zeta potential of a gel in the range from about −30 to −20 mV is suitable for cell proliferation, regardless of their different chemical structures.

It is a question why cells can spread and proliferate on a protein-free gel surface with a negative zeta potential higher than 20 mV. Because the cell surface is negatively charged, gels that show good cell viability are also negative, therefore there is no direct electrostatic interaction between cells and gels. It is well known that adhesive protein plays an important role in the process of cell spreading and proliferation. A protein is an amphoteric polyelectrolyte with both acidic and basic peptides that behaves as a bridge between positive or negative groups on the cell or gel surface. Therefore, serum proteins in the culture medium might interact with the negatively charged gel by electrostatic interactions and can act as the bridges. Further work on protein absorption from serum medium on gel surfaces is necessary.

This study showed that bovine fetal aorta endothelial cells can adhere, spread and proliferate to confluency on PVA, PAAm, PAA, PMAA, PNaSS and PNaAMPS hydrogels without modification by any cell adhesive proteins or peptides, although these gels have different chemical structures. The results indicated that although gels have different properties, provided that they have a negative zeta potential higher than about 20 mV, they show cell viability in the environment of serum-containing medium.

17.5
Application of Gels as Substitutes for Biological Tissues

Most biological tissues are in a gel state, e.g. articular cartilage is a natural fiber-reinforced hydrogel composed of proteoglycans, type II collagen and ∼70% water. The normal cartilage tissue contributes substantially to the joint functions involving ultra-low friction, distribution of loads and absorption of impact energy. When the normal cartilage tissue is damaged, it is extremely difficult to regenerate these tissues with currently available therapeutic treatments. Therefore, it is important to develop substitutes for the normal cartilage tissue as a potential therapeutic option.

The most important characteristic of a materials to be used as a load-bearing tissue replacement is for its mechanical properties to be comparable to those of the native tissue. As described above, the DN hydrogels can withstand from a few to tens of MPa mechanical strength. The value is comparable to the severe loading conditions imposed on human articular cartilage (1.9–14.4 MPa) [66], satisfying the mechanical property requirement for an articular cartilage substitute.

17.5.1
Robust Gels with Low Friction – an Excellent Candidate as Artificial Cartilage

In Section 17.3.2, it was described that a hydrogel with polyelectrolyte branched dangling polymer chains on its surface can effectively reduce the surface sliding frictional coefficient to a value as low as 10^{-4}. Unfortunately, conventional hydrogels are mechanically too weak to be used practically in any stress- or strain-bearing applications, which hinders the extensive application of hydrogels as industrial and biomedical materials. The design and production of hydrogels with low surface friction and high mechanical strength are crucially important in the biomedical applications of hydrogels as contact lenses, catheters, artificial articular cartilage and artificial esophagus [67, 68]. As described in Section 17.2.3, we discovered a general method to obtain very strong hydrogels by inducing a double-network (DN) structure for various combinations of hydrophilic polymers. The DN hydrogel can exhibit fracture strengths as high as from a few to several tens of MPa [16].

Based on these studies, new soft and wet materials, with both lower friction and high strength, were synthesized by introducing a weakly cross-linked PAMPS network [to form a triple-network (TN) gel] or a non-cross-linked linear polymer chain (to form a DN-L gel) as a third component into the optimal tough PAMPS–PAAm DN gel [69]. The TN and DN-L gels were synthesized by UV irradiation after immersing the DN gels in a large amount of a third solution of 1 M AMPS and 0.1 mol% 2-oxoglutaric acid with (TN) and without (DN-L) the presence of 0.1 mol% MBAA. The mechanical properties of the gels are summarized in Table 17.1. After inducing cross-linked or linear PAMPS to DN gel, the fracture strength of TN and DN-L gels remains on the order of MPa and the elasticity of the gels is higher than that of DN gel (~ 2 MPa). In addition, the fracture strength of DN-L increases remarkably because PAMPS linear chains can effectively dissipate the fracture energy [16, 20].

Figure 17.12a and b show the frictional forces (F) and frictional coefficient (μ) of the three kinds of gels, respectively, as a function of normal pressure (P). The gels were slid on a glass plate in water. It clearly shows that the frictional coefficient decreases in the order DN > TN > DN-L, indicating that the introduction of PAMPS, especially linear PAMPS, as the third network component obviously reduces the frictional coefficient of the gels.

Table 17.1 Mechanical properties of DN, TN and DN-L gels.

Gel	Water content (wt.%)	Elasticity (MPa)	Fracture stress, σ_{max} (MPa)	Fracture strain, λ_{max} (%)
DN	84.8	0.84	4.6	65
TN	82.5	2.0	4.8	57
DN-L	84.8	2.1	9.2	70

Reproduced with permission from [69].

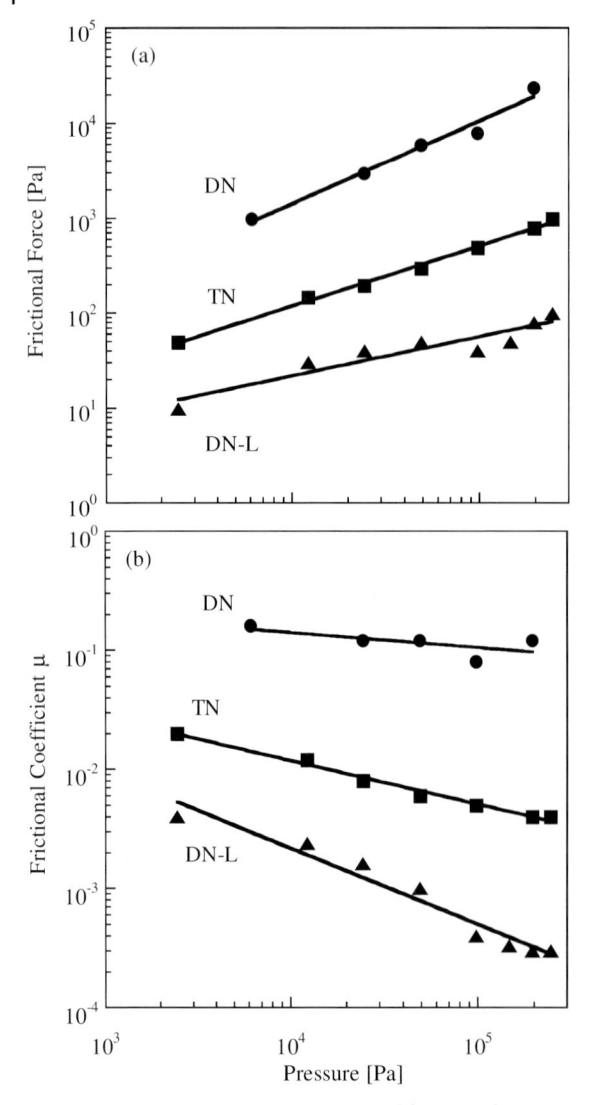

Fig. 17.12 Normal pressure dependence of frictional force (a) and frictional coefficient (b) of hydrogels against a glass plate in pure water. Sliding velocity: 1.7×10^{-3} m s^{-1}. Gels: (●) DN; (■) TN; and (▲) DN-L. Reproduced with permission from [69].

The DN gel has a relatively high frictional coefficient ($\sim 10^{-1}$) since the second network, nonionic PAAm, dominates the surface of the DN gel, which is adsorptive to the glass substrate.

However, when a PAMPS network is introduced to the DN gel as the third component, the frictional coefficient of the TN decreases to $\sim 10^{-2}$, which is two orders of magnitude lower than that of the DN gel, since the surface of TN gel is dominated by PAMPS, resulting in repulsive interaction with the glass substrate and a reduction in frictional force. Furthermore, when linear PAMPS chains are introduced to the surface of DN gel, the frictional coefficient is significantly reduced to $\sim 10^{-4}$, which is one to three orders of magnitude less than that of TN and DN gels, respectively. This demonstrates that the linear PAMPS chains on the gel surface reduce the frictional force due to further repulsive interaction with the glass substrate [39].

It should be emphasized that the lower friction coefficient of DN-L gel can be observed under a pressure range of 10^{-3}–10^5 Pa, which is close to the pressure exerted on articular cartilage in synovial joints. The results demonstrate that the linear polyelectrolyte chains are still effective in retaining lubrication even under an extremely high normal pressure.

17.5.2
Wearing Properties of Robust DN Gels

For application of DN gels as artificial articular cartilage, it is critical to evaluate the wear properties, because articular joints are subjected to high, rapid shear forces for millions of cycles over a lifetime. However, there are no established methods for evaluating the wear properties of gels. The pin-on-flat wear testing that has been used to evaluate the wear properties of ultra-high molecular weight polyethylene (UHMWPE), which is the only established rigid and hard biomaterial used in artificial joints, was used to evaluate the wear properties of DN gels. The wear properties of four kinds DN gels, composed of synthetic or natural polymers, PAMPS–PAAm, PAMPS–PDMAAm, BC–PDMAAm and BC–gelatin were evaluated [70].

Under 10^6 of friction cycles, equivalent to 50 km of friction (50 of mm×10^6), the maximum wear depths of the PAMPS–PAAm, PAMPS–PDMAAm, BC–PDMAAm and BC–gelatin gels were 9.5, 3.2, 7.8 and 1302.4 µm, respectively. It is amazing that the maximum wear depth of PAMPS–PDMAAm DN gel is similar to the value for UHMWPE (3.33 µm). In addition, although the maximum wear depth of PAMPS–PAAm DN and BC–PDMAAm DN gels is about 2–3 times higher than that of UHMWPE, these gels could withstand the 10^6 friction cycles. The results demonstrate that PAMPS–PAAm, PAMPS–PDMAAm and BC–PDMAAm DN gels are resistant to wear to a greater extent than conventional hydrogels and PAMPS–PDMAAm DN gel is potentially usable as a replacement material for artificial cartilage. On the other hand, BC–gelatin DN gel, which is composed of natural materials, shows extremely poor wear properties compared with the other DN gels. Some reasons for this may include the relatively low water content, higher friction coefficient and ease of roughening by abrasion.

For designing materials that potentially can be used as artificial cartilage, suitable viscoelasticity, high mechanical strength, durability to repetitive stress, low friction, high resistance to wear and resistance to biodegradation within the living body are required. It is difficult to develop a gel material that satisfies even two of these requirements at the same time. However, recent developments in synthesizing mechanically strong hydrogels have broken though the conventional gel concepts and opened up a new era of soft and wet materials as substitutes for articular cartilage and other tissues.

References

1 Y. Osada, K. Kajiwara, *Gels Handbook*, Academic Press, New York, **2001**.
2 Y. Tanaka, I. Nishio, S.T. Sun, S.V. Nishio, *Science* **1973**, *218*, 467–469.
3 Y. Osada, H. Okuzaki, H. Hori, *Nature* **1992**, *355*, 242–244.
4 Y. Osada, J.P. Gong, *Adv. Mater.* **1998**, *10*, 827–837.
5 Y. Osada, *Handbook of Biomimetics*, Academic Press, New York, **2000**.
6 K.Y. Lee, D.J. Mooney, *Chem. Rev.* **2001**, *10*, 1869–1879.
7 A.S. Hoffman, *Adv. Drug Del. Rev.* **2002**, *54*, 3–12.
8 J.J. Zarzycki, *Non-Cryst Solids* **1988**, *100*, 359–363.
9 Y. Tanaka, K. Fukao, Y. Miyamoto, *Eur. J. Phys. E* **2000**, *3*, 395–401.
10 D. Bonn, H. Kellay, M. Prochnow, K. Ben-Djemiaa, J. Meunier, *Science* **1998**, *280*, 265–267.
11 G.J. Lake, A.G. Thomas, *Proc. R. Soc. London* **1967**, *300*, 108–119.
12 H. Furukawa, K. Horie, R. Nozaki, M. Okada, *Phys. Rev. E* **2003**, *68*, 031406.1–014.
13 H. Abe, K. Hayashi, M. Sato, *Data Book on Mechanical Properties of Living Cells, Tissues and Organs*, Springer, Tokyo, **1996**.
14 Y. Okumura, K. Ito, *Adv. Mater.* **2001**, *13*, 485–487.
15 K. Haraguchi, T. Takehisa, *Adv. Mater.* **2002**, *14*, 1120–1124.
16 J.P. Gong, Y. Katsuyama, T. Kurokawa, Y. Osada, *Adv. Mater.* **2003**, *15*, 1155–1158.
17 K. Haraguchi, T. Takehisa, S. Fan, *Macromolecules* **2002**, *35*, 10162–10171.
18 K. Haraguchi, R. Farnworth, A. Ohbayashi, T. Takehisa, *Macromolecules* **2003**, *36*, 5732–5741.
19 D.C. Williams, Jr, *Materials Science and Engineering: an Introduction*, Wiley, New York, **2000**.
20 Y.-H. Na, T. Kurokawa, Y. Katsuyama, H. Tsukeshiba, J.P. Gong, Y. Osada, S. Okabe, M. Shibayama, *Macromolecules* **2004**, *37*, 5370–5374.
21 Y. Tanaka, R. Kuwabara, Y.-H. Na, T. Kurokawa, J.P. Gong, Y. Osada, *J. Phys. Chem. B*, **2005**, *109*, 11559–11562.
22 H. Tsukeshiba, M. Huang, Y.-H. Na, T. Kurokawa, R. Kuwabara, Y. Tanaka, H. Furukawa, Y. Osada, J.P. Gong, *J. Phys. Chem. B* **2005**, *109*, 16304–16309.
23 S.A. Wainwright, *Axis and Circumference: the Cylindrical Shape of Plants and Animals*, Harvard University Press, Cambridge, MA, **1988**.
24 A. Nakayama, A. Kakugo, J.P. Gong, Y. Osada, M. Takai, T. Erata, S. Kawano, *Adv. Funct. Mater.* **2004**, *14*, 1124–1128.
25 D. Klemm, D. Schumann, U. Udhardt, S. Marsch, *Prog. Polym. Sci.* **2001**, *26*, 1561–1603.
26 B.N.J. Presson, *Sliding Friction: Physical Principles and Applications*, 2nd edn., NanoScience and Technology Series, Springer, Berlin, **1998**.
27 C.W. McCutchen, *Wear* **1962**, *5*, 1–17.
28 W.A. Hodge, R.S. Fijian, K.L. Carlson, R.G. Burgess, W.H. Harris, R.W. Mann, *Proc. Natl. Acad. Sci. USA* **1986**, *83*, 2879–2883.
29 C.W. McCutchen, *Lubrication of Joints, the Joints and Synovial Fluid*, Academic Press, New York, **1978**.
30 G.A. Ateshian, H.Q. Wang, W.M. Lai, *J. Tribol.* **1998**, *120*, 241–251.
31 M.D. Buschmann, A.J. Grodzinsky, *J. Biomech. Eng.* **1995**, *117*, 179–192.

32 E. M. Wojtys, D. B. Chan, *Instrum. Course Lect.* **2005**, *54*, 323–330.

33 J. P. Gong, M. Higa, Y. Iwasaki, Y. Katsuyama, Y. Osada, *J. Phys. Chem. B* **1997**, *101*, 5487–5489.

34 J. P. Gong, Y. Iwasaki, Y. Osada, K. Kurihara, Y. Hamai, *J. Phys. Chem. B* **1999**, *103*, 6001–6006.

35 J. P. Gong, G. Kagata, Y. Osada, *J. Phys. Chem. B* **1999**, *103*, 6007–6014.

36 J. P. Gong, Y. Osada, *J. Chem. Phys.* **1998**, *109*, 8062–8068.

37 A. Kii, J. Xu, J. P. Gong, Y. Osada, X. M. Zhang, *J. Phys. Chem. B* **2001**, *105*, 4565–4571.

38 T. Narita, A. Hirai, J. Xu, J. P. Gong, Y. Osada, *Biomacromolecules* **2000**, *1*, 162–167.

39 J. P. Gong, T. Kurokawa, T. Narita, G. Kagata, Y. Osada, G. Nishimura, M. Kinjo, *J. Am. Chem. Soc.* **2001**, *123*, 5582–5583.

40 J. Klein, E. Kumacheva, D. Mahalu, D. Perahia, L. Fetters, *Nature* **1994**, *370*, 634–636.

41 G. S. Grest, *Adv. Polym. Sci.* **1999**, *138*, 149–183.

42 U. Raviv, S. Giasson, N. Kampf, J. F. Gohy, R. Jerome, J. Klein, *Nature* **2003**, *425*, 163–165.

43 J. P. Gong, Y. Iwasaki, Y. Osada, *J. Phys. Chem. B* **2000**, *104*, 3423–3428.

44 Y. Ohsedo, R. Takashina, J. P. Gong, Y. Osada, *Langmiur*, **2004**, *20*, 6549–6555.

45 J. A. Rowley, G. Madlambayan, D. J. Mooney, *Biomaterials* **1999**, *20*, 45–53.

46 Y. Y. Chang, S. J. Chen, H. C. Liang, H. W. Sung, C. C. Lin, R. N. Huang, *Biomaterials* **2004**, *25*, 3603–3611.

47 T.-W. Chung, Y.-F. Lu, S.-S. Wang, Y.-S. Lin, S.-H. Chu, *Biomaterials* **2002**, *23*, 803–4809.

48 Y. D. Park, N. Tirelli, J. A. Hubbell, *Biomaterials* **2003**, *24*, 893–900.

49 R. A. Stile, K. E. Healy, *Biomacromolecules* **2001**, *2*, 185–194.

50 J. L West, J. A. Hubbell, *Macromolecules* **1999**, *32*, 241–244.

51 R. H. Schmedlen, K. S. Masters, J. L. West, *Biomaterials* **2002**, *23*, 4325–4332.

52 M. H. Fittkau, P. Zilla, D. Bezuidenhout, M. P. Lutolf, P. Human, J. A. Hubbell, N. Davies, *Biomaterials* **2005**, *26*, 167–174.

53 Y. M. Chen, N. Shiraishi, H. Satokawa, A. Kakugo, T. Narita, J. P. Gong, Y. Osada, K. Yamamoto, J. Ando, *Biomaterials* **2005**, *26*, 4588–4596.

54 N. G. Maroudas, *J. Theor. Biol.* **1975**, *49*, 417–424.

55 J. K. Smetana, J. Vacík, D. Součková, Z. Krčová, J. J. Šulc, *Biomed. Mater. Res.* **1990**, *24*, 463–470.

56 Y. Nakayama, J. M. Anderson, T. Matsuda, *J. Biomed. Mater. Res.* **2000**, *53*, 584–591.

57 J. H. Lee, J. W. Lee, G. Khang, H. B. Lee, *Biomaterials* **1997**, *18*, 351–358.

58 S.-D. Lee, G.-H. Husiue, P. C.-T. Chang, C.-Y. Kao, *Biomaterials* **1996**, *17*, 1599–1608.

59 J. H. Lee, J. W. Lee, G. Khang, H. B. Lee, *Biomaterials* **1997**, *18*, 351–358.

60 G. B. Schneider, A. English, M. Abraham, R. Zaharias, C. Stanford, J. Keller, *Biomaterials* **2004**, *25*, 3023–3028.

61 A. Magnani, A. Priamo, D. Pasqui, R. Barbucci, *Mater. Sci. Eng. C* **2003**, *23*, 315–328.

62 A. Kishida, H. Iwata, Y. Tamada, Y. Ikada, *Biomaterials* **1991**, *12*, 786–792.

63 J. K. Smetana, J. Vacík, D. Součková, Z. Krčová, J. J. Šulc, *Biomed. Mater. Res.* **1990**, *24*, 463–470.

64 S. Hattori, J. D. Andrade, J. B. Hibbs, D. E. Gregonis, R. N. King, *J. Colloid Interface Sci.* **1985**, *104*, 72–78.

65 Y. Sugimoto, *Exp. Cell Res.* **1981**, *135*, 19–45.

66 G. Zheng-Qui, X. Jiu-Mei, Xiang-Hong Z., *J.Biomed. Mater. Engng.* **1998**, *8*, 75–81.

67 M. E. Freeman, M. J. Furey, B. J. Love, J. M. Hampton, *Wear* **2000**, *241*, 129–135.

68 U. Raviv, S. Giasson, N. Kampf, J. F. Gohy, R. Jerome, J. Klein, *Nature* **2003**, *425*, 163–165.

69 D. Kaneko, T. Tada, T. Kurokawa, J. P. Gong, Y. Osada, *Adv. Mater.* **2004**, *17*, 535–538.

70 K. Yasuda, J. P. Gong, Y. Katsuyama, A. Nakayama, Y. Tanabe, E. Kondo, M. Ueno, Y. Osada, *Biomaterials* **2005**, *26*, 4469–4475.

18
Polymers in Tissue Engineering

Jeffrey A. Hubbell

18.1
Introduction and Historical Perspective

Polymers play a very important role in the broad field of biomaterials science and engineering, with application in drug targeting, drug delivery, drug formulation, surgical reconstruction, cell transplantation and countless medical devices, to provide a non-exhaustive list. Some of these applications have been lumped into a relatively recently defined subfield of bioengineering and medicine, namely tissue engineering. This chapter introduces the field of tissue engineering, shows briefly what challenges within this field have been and are being met by polymer scientists and engineers and comments on the future of polymers in this field of biotechnology and medicine.

18.1.1
Tissue Engineering and Its Challenges for Macromolecular Engineering

Tissue engineering is a broadly defined field of bioengineering generally employing some combination of the following in a therapeutic, diagnostic or research product: cells, biomolecules and materials, each of which will be commented upon in the paragraphs below [1–4]. In the jargon of the field, the combination of at least two of these three components into a functional unit is referred to as a *tissue engineering construct*. These components are combined in a manner so as to produce constructs that are somehow active, either active due to the cells or active due to an incorporated biomolecule or active due to bioactive features incorporated in the material design (considered in detail in Section 18.4). A number of examples of tissue engineering constructs are also provided in the rest of this section, to illustrate the roles of the three components on the construct and to demonstrate concretely the role of the polymeric biomaterial in its overall function.

Immediately below, we refer back to the components of a construct and consider each one in individual detail. In particular, we also consider the requirements that each component places on the third, i.e. on the material that is employed:

18.1.1.1 Cells

Cells are frequently provided in tissue engineering as one component of a construct [5, 6]. The cells may arise by infiltration from a recipient, for example by invasion from the site of an implant, or they may be removed from the recipient's body, somehow processed and utilized within the construct; these cells, coming from a body and returning to it, are called *autologous* cells. Cells might be taken as a source from the body (e.g. a skin biopsy or a bone marrow aspirate), purified in culture (e.g. to purify stem cells from the skin or the bone marrow), multiplied in culture, differentiated in culture (e.g. to produce keratinocytes for mature skin or any number of cell types from bone marrow) and re-implanted within a biomaterial scaffold.

The use of autologous cells comes with great complexities, such as the typical need for the patient to visit the hospital at least twice, once to donate the cells and once to receive them in the final surgical procedure. Moreover, it is often not possible to treat acute emergency indications with autologous cells, since days and even weeks of preparation may be required and the patient may need treatment in hours. As an alternative, one which might allow for more off-the-shelf tissue engineering constructs, cells might also be taken from a donor and similarly processed for use in a patient of the same species; these cells are called *allogeneic* cells. Allogenic cells may be useful when immunotype matching is possible or when the cells being transplanted do not need to survive for a long time. An example in this case has already encountered clinical use, with cells isolated from the skin of a newborn (the foreskin, a tissue that is removed and discarded after many births), which are multiplied greatly in culture, seeded within biomaterial scaffolds and cultured to form skin-like substitute tissues used to treat patients with burns, as an alterative to skin grafting [7–9]. Since such allograft cells are not processed to order, but rather one batch of cells is grown to produce a large number of tissue engineering constructs, this cell source is much more practical for industrialization.

An even more attractive source is cells from non-human species, cells with might have some characteristics that are held very closely in common with human cells, even if they lack them all, including immunotype matching; these cells are referred to as *xenogeneic*. One such long-sought xenograft target is transplantation of islets of Langerhans, the glucose-sensing and insulin-producing cells of the pancreas, from animals such as pigs into humans [10]. Pigs regulate their glucose concentration very similarly to humans and porcine insulin is almost identical with human insulin, so it has long been hoped that one could protect such xenogeneic islets within a polymer membrane that is permselective, to protect the transplanted cells from the host's immune system. Although extensive research investment has been made towards this goal, it still remains elusive due to the difficulty of blocking immune rejection, i.e. due to the difficulty of distinguishing with a permselective polymer membrane between molecules that are recognized by the immune system (to reject at the membrane surface) and molecules that play important hormonal and nutritional roles (to pass through the membrane), while at the same time not restricting very much the permeation of glucose and oxygen.

The fact that cells are incorporated in tissue engineering constructs places a great deal of constraint on the materials used and their chemistries. Above all, the materials used must be non-cytotoxic and their processes of final formation or placement in the body must be non-cytotoxic. For example, if polymer matrices are injected as liquids in solvent, great care must be exercised in the selection of the solvent in view of its cytotoxicity and its route of elimination; e.g. degradable polyesters have been injected in ethyl acetate without negative toxicological consequences [11]. Other chemical schemes for *in situ* conversion from injectable precursors to solid implants have been devised, lacking in organic solvents, as is described in detail in Section 18.3.

18.1.1.2 Biomolecules

Biologically active, biological molecules constitute a second important class of components in tissue engineering constructs. These molecules are primarily peptides, proteins and DNA. Peptides and proteins are often included in constructs to serve signaling function, to induce cells in the construct to function or differentiate in certain ways. This function is played *in vivo* in large part by proteins of two classes: *adhesion proteins* and *growth factors*. Since formation of bioactive materials often includes incorporation of both of these kinds of molecules, it is important to develop a basic understanding of the molecules and their function.

Adhesion proteins play important roles in determining cell fate. Hundreds of proteins fall within the broad family of adhesion proteins, including fibronectin, vitronectin, laminin, collagen and elastin (although the last two are also structural proteins) [12, 13]. As indicated by their name, the adhesion proteins allow the cell to integrate mechanical stress from the cytoskeleton within the cell to the extracellular matrix and this is accomplished by the cell binding to the adhesion proteins via transmembrane receptors; the adhesion proteins are themselves bound within the extracellular matrix to structural proteins such as collagen and elastin, and also polysaccharide-based extracellular matrix molecules called proteoglycans. Together with intracellular proteins that link the receptor to the cytoskeleton, stress can be transmitted, allowing the cell to exert traction, and, when so signaled, to migrate [14, 15]. Hence adhesion proteins serve a primary role of providing traction to cells [16]. They also serve an important secondary role of signaling. The bound adhesion receptors cluster in the membrane of the cells into aggregates known as focal contacts and these focal contacts serve as sites for colocalization of many signaling proteins, which trigger a cascade of enzymatic reactions within the cell in response to its adherent state [17]. Thus the adhesion protein receptors with the signaling proteins function much like growth factor receptors, described below.

Table 18.1 The integrin family of adhesion receptors consist of a number of combinations of an α subunit and a β subunit and exhibit selective affinity for various adhesion proteins [13]. The affinity between integrins and their extracellular matrix adhesion protein ligands provides for traction between the cell and its environment. Peptide and protein ligands for the integrins can be engineered and incorporated into biomaterial surfaces used in tissue engineering constructs.

Integrin	Function
$\alpha 5\beta 1$, $\alpha 8\beta 1$, $\alpha v\beta 1$	RGD binding
$\alpha 3\beta 1$, $\alpha 6\beta 1$, $\alpha 7\beta 1$	Laminin binding
$\alpha 1\beta 1$, $\alpha 2\beta 1$, $\alpha 10\beta 1$, $\alpha 11\beta 1$	Fibrillar collagen binding
$\alpha v\beta 3$, $\alpha v\beta 5$, $\alpha v\beta 6$, $\alpha v\beta 8$	Many
$\alpha IIb\beta 3$	Platelet specific
$\alpha 6\beta 4$	Hemidesmosome
$\alpha L\beta 2$, $\alpha M\beta 2$, $\alpha X\beta 2$, $\alpha D\beta 2$	Immune specific
$\alpha 4\beta 7$, $\alpha E\beta 7$	Immune specific

The best studied of the adhesion protein receptors is the integrin family of receptors [13]. The integrins are heterodimers, consisting of one of several subunits noncovalently bound to one of several β subunits: a simplified list of several of the acceptable α–β combinations and their affinities is given in Table 18.1. Different cell types assemble different α–β combinations and this provides selective affinity for various adhesion proteins. Moreover, the same cell in different states of function or differentiation expresses different amounts of various α–β integrin combinations, allowing it also to express selective affinities. Still more interesting is the ability that cells have to respond to different adhesion molecules, sensed via their repertoire of integrin receptors, by adapting their function to the adhesion proteins displayed in their local extracellular matrix. This gives the biomaterials designer an ability to immobilize adhesion proteins, or their mimics, on a polymeric support and trigger different states of function or differentiation in the cells attached thereto. This concept is the basis for biofunctionalization with adhesion proteins, discussed in more detail in Section 18.4.

The second set of important signaling proteins are referred to as growth factors",4,1> [18]. Although one primary function of the growth factors is to signal cell growth, i.e. proliferation (the mitogen activity of the growth factor), this is by far not the only function of the growth factors. They signal in far more complicated ways and are responsible for telling the cell what type of function to display, such as whether to remain in one place or to migrate, whether to express its own extracellular matrix proteins or not or even whether to differentiate to become a bone cell or a cartilage cell. When stem cells are differentiated into specific cell types, one method that is commonly used is to expose them to certain combinations of growth factors over some predetermined time course.

Growth factors play this role *in vivo* and they can be manipulated in tissue engineering therapeutics to play it therapeutically also. For example, in wound healing, some members of the transforming growth factor beta (TGF-β) family signal for scar formation (TGF-β1 and -β2), whereas others signal against scar formation (TGF-β3) [19]. This plays an important role in adult versus fetal wound healing (TGF-β1 and -β2 dominate in the adult, whereas TGF-β3 dominates in the fetus) and by flooding the adult wound healing environment with the correct growth factor (here, TGF-β3) one can minimize scar formation in the adult [20]. Hence growth factors play important roles in tissue engineering, being very powerful regulators of biological function.

The fact that both adhesion proteins and growth factors are proteins places constraints on selections for materials and processes used in tissue engineering constructs. Most importantly, it means that the biomaterials designer must be careful with hydrophobicity and with organic solvents [21, 22]. Proteins are water-soluble polyamides consisting of hydrophilic and hydrophobic amino acids as co-monomers. The proteins fold into secondary and tertiary structures to minimize their free energy in their native solvent, i.e. in water. When this environment is altered, e.g. by the protein being placed near a hydrophobic surface or in an organic solvent, it can adopt a vastly different three-dimensional folding pattern, mostly likely destroying part all of its biological activity. This process is referred to as *denaturation* and denaturation of adhesion proteins or growth factors incorporated in biomaterials must be carefully avoided, lest their activity be lost. Even more so, the denatured proteins may be recognized by the immune system, leading to antibodies being formed against the denatured and potentially also the native protein; as such, denaturation not only renders the proteins inactive, but can also render them harmful.

Because of the denaturing activity of hydrophobic surfaces, highly hydrophobic materials in tissue engineering (but not all uses of biomaterials) are generally avoided, e.g. avoiding the use of perfluorinated polymers or at least surface modifying such materials to render them more hydrophilic. It is not the rule, but is the general trend, to use more hydrophilic polymers in tissue engineering. For surface modification, poly(ethylene glycol) (PEG) is frequently grafted to polymer surfaces to render them more hydrophilic, as described in Section 18.4.

Organic solvents are also generally avoided or, more specifically, mixtures of water and organic solvents. Since proteins are generally not soluble in organic solvents, they have no molecular mobility with which to denature – they are kinetically trapped in their native state or at least in something very close to their native state [21, 22]. Thus, if a protein powder is dispersed in an organic solvent, such as tetrahydrofuran (THF) or dichloromethane (DCM), processed and then dried before exposure to water, it will be perfectly active. However, if it is first dissolved in water and then if THF is added, the protein will be substantially denatured. As we will see in Section 18.4, addressing drug incorporation into polymers for tissue engineering, this will place considerable constraints on how protein drugs are incorporated into biomaterials.

18.1.1.3 Materials

Materials represent the third major component from which tissue engineering constructs are made – cells and materials, biomolecules and materials or cells, biomolecules and materials all together. Since this entire chapter is about the materials component of tissue engineering, an extensive introduction will not be provided here; rather, we will comment on the motivations to incorporate materials in the first place. Materials can organize and support cells, they can provide signaling cues via their mechanics [23] or via bound biomolecules [3] and they can release drugs over a prolonged period of time. Of course, these three classes of function are not the only functions that materials can play, but they are perhaps the major ones. Materials can also package a tissue engineering construct, they can protect cells to be transplanted from the immune system and so forth. In the paragraphs below on the main motivations, we do not attempt to be complete. Each of the main motivations is addressed briefly below before we move on in the chapter to the details.

First, materials can organize and support cells, whether we are considering cells that invade from tissues that neighbor an implant site or cells that are transplanted within a material from exogenous sources [24, 25]. Support and organization can occur on a number of scales. Most importantly, almost all mammalian cells must be adherent to an extracellular matrix or substrate in order to be normally functional; only a few cells, such as blood cells or sperm, are usually present within a liquid environment and most other cells eventually die when their supporting substrate is taken away. Thus, transplantation of cells in a liquid state, rather than attached to a polymer support, immediately challenges their function and viability. On a more subtle level, a material scaffold, or matrix can serve to organize and immobilize cells within a desired site of intervention. For example, if cells were merely placed into a wound area without a support, there would be little to retain them at that site. By contrast, a polymer support can organize them at their intended surgical site. Even at a more subtle scale, studies have demonstrated that cell shape [26] and substrate mechanics [23] play an important role in controlling cell function. For example, when cells are constrained in the area over which they can spread, by attachment to small spots, for example, they function differently then when they are able to spread entirely [26]. As such, it may be possible to organize cells in implants even on this very small length scale using nano- and micropatterning.

Second, materials can provide signaling cues. As mentioned above, biomolecules such as adhesion proteins and growth factor proteins are important signaling molecules. Some of these must be immobilized and others can be immobilized or free. For example, adhesion proteins must be immobilized in order to be functional – it is clear that traction between the cytoskeleton and the extracellular milieu cannot be achieved if the adhesion molecule is not somehow immobilized within the milieu. In fact, if it adhesion molecules are free, they will serve as competitive inhibitors and will serve the opposite function to that actually desired. Growth factors, by contrast, can function in either immobilized or free form. In nature, most growth factors that are of interest in tissue engineering bind to the extracellular matrix via electrostatic interactions by binding to polysaccharide-based molecules referred to as glycosaminoglycans [27]. In this sense, the extracellular matrix serves as a buffer to modulate the free concentrations of the growth factor. Synthetic materials can be thus designed to mimic this effect, as will be described in Section 18.4.

Third, materials can release drugs locally over prolonged periods of time. In tissue engineering, the drugs that are most of interest are protein drugs, mainly growth factors. As mentioned above, the fact that these drugs are hydrophilic macromolecules places extensive constraints upon the encapsulation and release mechanism. Ways in which these constraints are addressed are described in Section 18.4.

18.1.1.4 Examples of Tissue Engineering Constructs

Before launching into the details of polymeric materials and their use in tissue engineering, it is instructive to list characteristic examples in which cells, biomolecules and materials are combined in tissue engineering constructs:

Tissue engineered skin Dermal fibroblasts are isolated from the foreskin of a neonatal donor, cultured en mass in a processing center and seeded upon a polymer textile scaffold to product a skin substitute [7, 28]. The polymer provides a structure upon which the cells can grow and organize their own extracellular matrix and growth factors.

Skin regeneration matrix A dermal growth factor, such as platelet-derived growth factor, is bound to a polymer gel that is placed in chronic dermal wounds to enhance healing [29]. The polymer serves as a controlled release depot for the growth factor and as a scaffold upon which cells from the surrounding tissues can adhere as they infiltrate the wound area.

Bone regeneration matrix A bone growth factor, such as bone morphogenetic protein, is adsorbed on a biopolymer matrix such as type I collagen, which is placed in bone defects to enhance bone healing [30, 31]. The polymer serves as a controlled release depot for the growth factor.

Islet of Langerhans transplantation Islets of Langerhans are isolated from the pancreas of a cadaveric human donor and are encapsulated within a polymer hydrogel and implanted into the peritoneal cavity of a diabetic patient [10]. The polymer serves as a semipermeable membrane to help protect the transplanted cells against immune rejection by the recipient.

Chondrocyte transplantation Mesenchymal stem cells are isolated from bone marrow and multiplied in culture. A growth factor is exposed to the cells to differentiate them into chondrocytes. These cells are then placed in a polymer gel and implanted in cartilage defects to correct diseased cartilage [32, 33]. The polymer serves a structural role to retain the cells in this site of implantation.

The examples above illustrate only a few of the situations in which polymers are or might be used in tissue engineering, to show the roles that they play. Several other examples will be presented in the sections below.

18.1.2
History of Polymers in Tissue Engineering

Biopolymers and synthetic polymers were first used in tissue engineering as mechanical scaffolds upon which to culture cells prior to implantation or upon which cells could adhere as they infiltrate from surrounding tissues into a wound. These polymers were either biopolymeric (collagen or collagen and glycosaminoglycan blends) or synthetic degradable polyesters [poly(lactic acid) (PLA) and poly(lactic acid-co-glycolic acid) (PLGA)]. The polymers will be introduced in more detail below. The work was carried out by independently by three pioneers in the field, Robert Langer, Eugene Bell and Ioanis Yannas. Langer, working mostly with PLA and PLGA nonwoven textiles, together with surgeon collaborator Joseph Vacanti, demonstrated that cells would be isolated from a subject, seeded on polymer scaffolds and reimplanted to effect structural repair of a number of tissues and organs, such as cartilage and ureter; the technology was also demonstrated in a number of nonstructural tissues, such as liver [24, 25]. Bell demonstrated in early work that smooth muscle cells could be isolated from blood vessels and could be cultured in tubular collagen scaffolds to reconstruct tissues that closely resembled in form and function native blood vessels; similar work was also done with skin [34, 35]. Here, the polymer was native type I collagen isolated from tissues of animals. Yannas also employed the polymer collagen, but with much more sophisticated processing [36–39]. First, he sought polymers based on both collagen and glycosaminoglycans, to mimic more closely the biophysical characteristics of the natural extracellular matrix. He did this by isolating the two classes of extracellular matrix molecules from animal tissue, blending them and processing them to create a microporous material. He sought a nonuniform pore structure in which pores were aligned in a particular direction. This was obtained by directionally freezing the materials so as to grow ice crystals that were elongated in one direction. Lyophilization and cross-linking produced the desired matrices, which were useful in repair of skin and peripheral nerve.

These early experiments on the use of biopolymers and synthetic polymers as structural scaffolds in tissue engineering set the stage for countless experiments on the next steps – on more sophisticated approaches in materials bioactivity, on biomolecular grafting and release and on bioresponsiveness of more advanced materials. This chapter seeks to provide an introduction to those approaches in which materials have played a key role. Since the text is focused on macromolecule synthesis and engineering, the chapter focuses its attention towards synthetic polymers. For treatments of biological and biosynthetic polymers, the reader is directed elsewhere [2, 3].

18.2
Degradable Polymer Scaffolds

Degradable polymers have been used as scaffolds in tissue engineering. A scaffold is defined herein as a micro- or macroporous material formed as a nonwoven or woven textile or open cell foam. The material properties and processing conditions are selected so that the scaffold has mechanical characteristics that allow it so support the growth of cells that are cultured or transplanted within it or invasion of cells from the tissues that surround the implant site. The details of these mechanical characteristics depend on the application, of course [23]. In terms of supports for cell culture, to create a tissue *in vitro* for subsequent transplantation, an elastic modulus as low 250 Pa is sufficient. When these are later implanted in a tissue site, the mechanical requirements may be much higher, to prevent compaction or collapse due to forces from the surrounding tissues. Hence the scaffold serves principally a structural role, providing adhesion and traction to enable cells to organize into a tissue-like structure.

Scaffolds are micro- or macroporous, with minimum connected pore sizes as low as around 10 µm. A characteristic cell diameter is 10 µm, although cells may migrate through pores that are considerably smaller, say 3–5 µm. Hence if cells are placed in a material with connected porosity less than approximately 10 µm, the cells will be unable to migrate, infiltrate and assume normal biological function. This is true even with degradable materials, where the pore size may be changing with time – the time-scale of material degradation will be weeks to months, whereas that of cell migration will be hours to days. Hence the slowness of degradation of most materials used in tissue engineering makes it necessary that pores be added at the beginning, prior to implantation within the tissues.

By far the most commonly used polymers for tissue engineering scaffolds are the degradable polyesters poly(lactic acid) (PLA), poly(glycolic acid) (PGA), poly(ε-caprolactone) (PCL) and their co- and terpolymers (Fig. 18.1) [40–42]; the biologically derived poly(hydroxyalkanoates) are used to a lesser extent [43]. These polymers are formed by ring-opening polymerization of the cyclic dimers lactide (L,L to form a highly crystalline material or D,L to form a semicrystalline material), glycolide or ε-caprolactone (Fig. 18.2) [44]. The alternative, i.e. formation by condensation polymerization of the hydroxy acid, is notoriously difficult – the water that is produced during condensation hydrolyzes the ester links in the resulting polymer and high polymer molecular weights are not achieved. Rather, removal of water in the monomer stage, by condensation of two monomer molecules to yield a six-membered ring cyclic dimer in the case of lactide or glycolide or of one monomer to yield a seven-membered ring cyclic monomer in the case of ε-caprolactone, circumvents this problem. Water is simply distilled from the reaction mixture until the cyclic monomer or dimer is achieved in high purity and yield. The ring-opening polymerization is conducted in the presence of a basic metal catalyst such as tin octoate; the toxicity of this

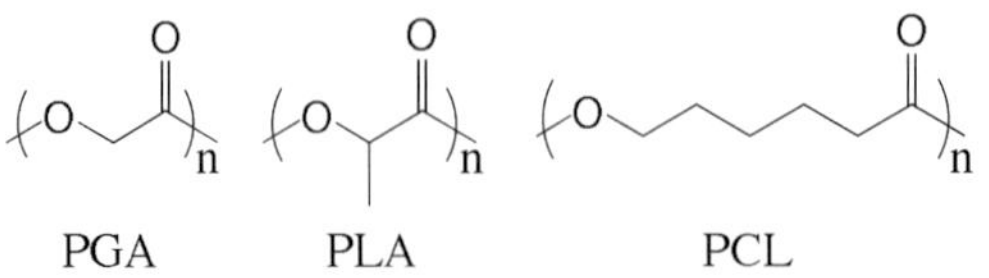

Fig. 18.1 The degradable polyesters poly(glycolic acid) (PGA), poly(lactic acid) (PLA) and poly(ε-caprolactone) (PCL), and their copolymers, are widely used in tissue engineering scaffolds. Scaffolds have been formed as textiles and as open-cell foams.

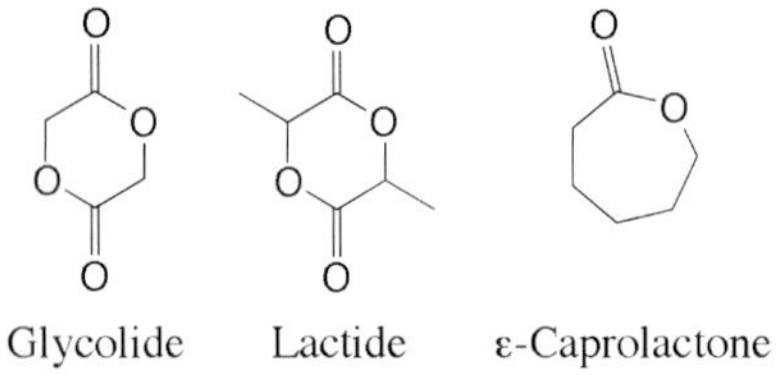

Fig. 18.2 PGA, PLA and PCL are formed by ring-opening polymerization of the cyclic dimers glycolide or lactide or the cyclic monomer ε-caprolactone.

Table 18.2 Physical characteristics of commonly utilized degradable polyesters [45].

Polymer[a]	T_m (°C)	T_g (°C)	Elastic modulus (GPa)	Degradation time (months)
PLLA	173–178	60–65	2.7	>24
PLA	Amorphous	55–60	1.9	12–16
85:15 PLA-co-PGA	Amorphous	50–55	2.0	5–6
75:25 PLA-co-PGA	Amorphous	50–55	2.0	4–5
65:35 PLA-co-PGA	Amorphous	45–50	2.0	3–4
50/50 PLA-co-PGA	Amorphous	45–50	2.0	1–2
PGA	Amorphous	35–40	7.0	6–12
PCL	58–63	−65 to −60	0.4	>24

a) PLLA, poly(L-lactic acid); PLA, poly(D,L-lactic acid); PLA-co-PGA, poly(lactic acid-co-glycolic acid); PGA, poly(glycolic acid); PCL, poly(ε-caprolactone).

catalyst is such that it need not be rigorously removed from the polymer before use in tissue engineering or other medical applications.

PLA, PGA and PCL degrade by hydrolysis in an acid-catalyzed manner. PCL degrades very slowly, because of its high hydrophobicity, which allows only very limited access of the polymer bulk to water. PLA is somewhat less hydrophobic. Poly(L,L-lactic acid) degrades very slowly, due to its highly crystalline nature, which also allows only very limited access of the bulk polymer to water.

Poly(D,L-lactic acid), which is semicrystalline, degrades more quickly and PGA more quickly still. The copolymers also degrade rather quickly, having even lower crystallinity and also relatively low glass transition temperatures, which allows higher penetration of water. Typical characteristics of PLA, PGA, PCL and their copolymers are given in Table 18.2 [45].

Polymer degradation is strongly influenced by acid catalysis; since lactic acid is produced during degradation, and since an acid end-group is produced during chain scission, the degradable polyesters degrade autocatalytically [40, 41, 46]. Compounding the polymer with inorganic bases may also lead to slower pH changes during degradation and this has been used in controlled release, as described in Section 18.4 [47–49].

In its native form, the degradable polyesters have no sites for functionalization, i.e. no sites for chemical grafting of biomolecules. Approaches have been presented for copolymerization with side-chain reactive monomers. For example, a cyclic dimer of lactic acid and lysine can be synthesized, of course with the lysyl amine protected, and this cyclic heterodimer can be copolymerized with the cyclic homodimer lactide to obtain a copolymer with a side-chain amino moiety for subsequent biofunctionalization [50, 51]. Schemes for copolymerization with side-chain protected hydroxyl-containing comonomers have also been presented [52, 53]. These approaches allow conjugation with biomolecules for biofunctionalization, as described in more detail in Section 18.4.

The formation of acidic degradation products can be a problem in the use of these polymers in tissue engineering. Inflammation may be triggered by the acidic degradation products, leading to a chronic foreign body encapsulation response as degradation proceeds *in vivo*. Clearly, if degradation is taking place within a tissue culture bioreactor to form a tissue engineering construct *in vitro* prior to implantation, the acidic degradation products can be swept away and neutralized by base in the culture medium. However, *in vivo*, mass transfer in the neighborhood of the implant may be much less and local acidification may ensue. Indeed, when degradable polyesters have been used in bone-forming applications, by local delivery of the growth factor bone morphogenetic protein-2 (BMP-2) from PLGA scaffolds, resorption of the newly formed bone was observed and was attributed to local acidification as the scaffold degraded [54].

More advanced degradable polymers have been proposed, for example polycarbonate structures designed to avoid local acidification during degradation [55, 56]. Key in many engineering polycarbonates as a monomer is the diphenolic bisphenol A. This structure has been mimicked to some extent in the toxicologically acceptable diphenolic compound desaminotyrosyltyrosine ethyl ester (Fig. 18.3), which can be esterified with ethyl, butyl, hexyl or octyl alcohols, for example. This monomer can be readily polymerized by exposure to phosgene to yield a polycarbonate of high molecular weight (Fig. 18.4). This material is a hydrophobic thermoplastic and can carry out many of the tasks that have been described above for the hydrophobic polyesters PLA, PGA and PLGA. By changing the ester group from ethyl to octyl, one can alter the glass transition temperature of these amorphous polymers from 81 °C (for ethyl) to 53 °C (for octyl) in a

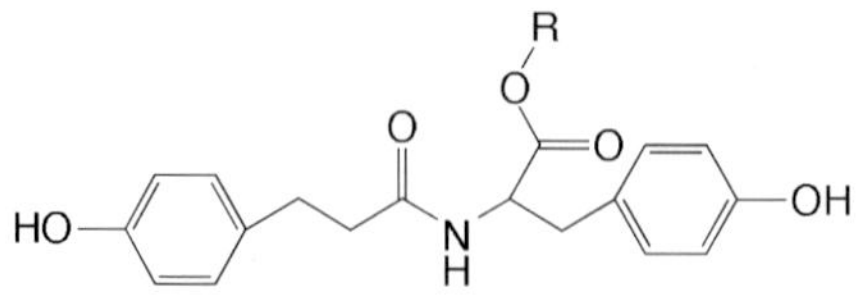

Fig. 18.3 Diphenol monomers have been developed based on tyrosines, here desaminotyrosyltyrosine, where the carboxyl of the tyrosine moiety has been esterified to an alkyl alcohol (R) to form desaminotyrosyltyrosine alkyl ester. The alkyl group can be an ethyl, butyl, hexyl or octyl group, for example.

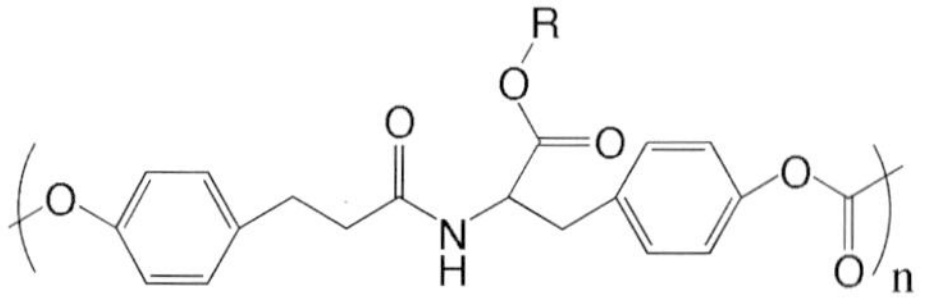

Fig. 18.4 Desaminotyrosyltyrosine alkyl ester can be polymerized into a high molecular weight polycarbonate by reaction with phosgene. The resulting hydrophobic, amorphous polymer degrades slowly in water by hydrolysis of the backbone carbonate. The side-chain ester is less sensitive to hydrolysis under physiological conditions.

gradated manner [56]. The materials degrade by hydrolysis of the carbonate linkage first and of the ester linkage second; the decreases in molecular weight and strength are due mainly to the cleavage of the backbone carbonate under physiological conditions [57]. Since these polymers do not generate acid as they degrade, one might expect their degradation to be more predictable, since it is not autocatalytic, and also that biocompatibility in environments where acid build-up would otherwise be detrimental, such as in bony implant sites. Indeed, both of these characteristics have been observed experimentally [58]. These polymers may assume an important role in the future as a replacement for the hydrophobic degradable polyesters.

In the above example of tyrosine-derived polycarbonates, the degradative characteristics of the polyesters were targeted for improvement. Other material designs have targeted the mechanical characteristics of the degradable polyesters, for example seeking degradable polymers that are elastomers. In one approach, glycerol was used as a multifunctional monomer and was condensation polymerized with sebacic acid at less than 1 mol equiv., resulting in a cross-linked network with residual alcohol moieties (Fig. 18.5). The mechanical characteristics of these thermosets were in a range such that the material could be useful for tissue engineering scaffolds.

To overcome the difficult processing that often accompanies the use of thermoset polymers, degradable elastomeric thermoplastics have also been sought. Nondegradable polyurethanes have been extensively utilized in medicine and re-

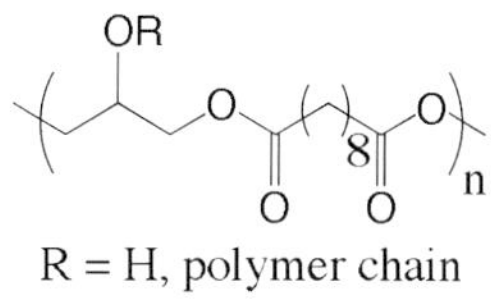

R = H, polymer chain

Fig. 18.5 Cross-linked polymer networks have been developed based on sebacic acid-cross-linked glycerol. Cross-link density was kept relatively low, in order to obtain networks that were not brittle, by adding the sebacic acid in less than equi-equivalent amounts. Moreover, residual hydroxyl groups are useful in biofunctionalization.

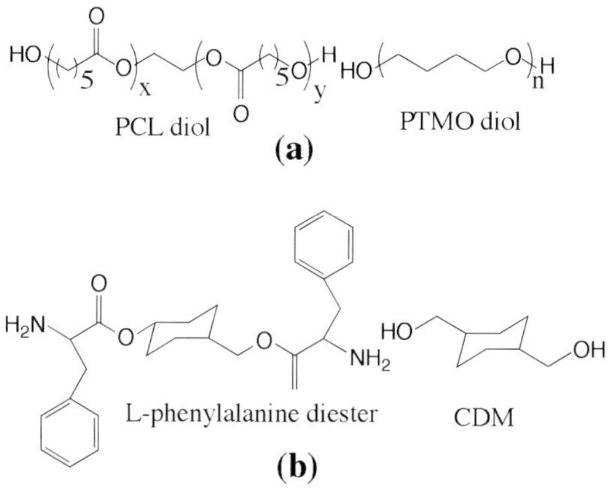

PCL diol   PTMO diol
(a)

L-phenylalanine diester   CDM
(b)

Fig. 18.6 Hydrolytically sensitive and enzymatically sensitive polyurethanes have been developed, (a) using in the former PCL domains as soft segments, shown next to a nonhydrolytic poly(tetramethylene oxide) (PTMO) soft segment and (b) in the latter amino acid components as chain extenders, shown next to an enzymatically insensitive chain extender of 1,4-cyclohexanedimethanol (CDM).

search groups, motivated by these uses, have sought to form polyurethane elastomers that are sensitive to hydrolytic degradation. As one approach to this, segmented polyurethanes have been developed comprising PCL as one soft segment component (Fig. 18.6). These materials degraded slowly after exposure to water. An alternative approach was also developed to obtain elastomers that degrade in response to enzymatic action, using aminio acid domains based on phenylalanine as the chain extender in the polyurethane polymerization reaction.

18.3
Degradable Polymer Matrices

Degradable polymers have also been used as matrices in tissue engineering. In contrast to our definition of a scaffold, here we define a matrix as a structural support that allows cell transplantation or cell infiltration but that possesses porosity that is less than the length scale of a typical cell process, i.e. less than a few micrometers (the definition of what is a matrix versus what is a scaffold is very loosely utilized in the literature; in this chapter, we adopt a more strict definition for clarity). Matrices are used for the same purposes as scaffolds reviewed above, with the exception that matrices could also be injectable, assuming that chemical schemes could be developed by which to convert liquid precursors into elastic gel matrices *in situ* [3, 59]. Such schemes will be discussed below.

In spite of the small porosity of hydrogel matrices, somehow cell migration and infiltration must be permitted; materials remodeling must be engineered to be on the same time-scale as cell migration. Either materials resorption could be particularly fast or the materials resorption could be tied to bioactive processes, such as proteolysis that is commonly associated with cell migration. Both concepts will be presented in this section.

Hydrolytically sensitive hydrogel formulations have been developed for conversion *in situ* from liquid precursors to elastic hydrogels [60]. These polymers have been used for engineering tissue healing [61–63] and for transplantation of slowly proliferating cells [64–66]. Water solubility, or at least extensive hydrophilicity, is the first requirement for a hydrogel matrix [60]. In this design, this was accomplished by building a polymer based on PEG. The hydroxy termini of linear PEG were used to ring-open lactide or other cyclic esters such as glycolide or -caprolactone. The degree of polymerization of the lactic acid block was kept low, so as to result in a polymer that was still highly soluble in water. The resulting hydroxy termini of the ABA block copolymer were then acrylated, to obtain the reactive macromonomer Ac-oligo(lactic acid)-block-PEG-block-oligo(lactic acid)-Ac (Fig. 18.7) [60].

Such macromonomers can be polymerized into cross-linked hydrogels *in situ*, i.e. in the presence of cells mixed into the precursors or in the presence of tis-

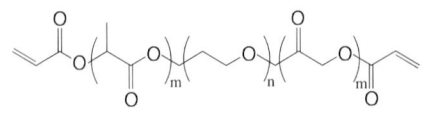

Fig. 18.7 Telechelically acrylated ABA block copolymers of PEG (B block) and oligo(lactic acid) have been synthesized to form reactive macromonomers that can be polymerized *in situ* to create biomaterial matrices for tissue engineering. These materials have been used to template tissue healing and to transplant relatively slowly growing cells such as chondrocytes.

sues directly. This has been accomplished, for example, by photopolymerization using UV-sensitive initiators [61, 63] or with eosin-yellowish in the presence of triethanolamine as a visible–sensitive initiation couple [67, 68]. Free radicals can be generated extracellularly in a manner that is sufficiently nontoxic to allow *in situ* photopolymerization. The resulting gel consists of formed polyacrylate chains, which are cross-linked by PEG chains, esterified to the polyacrylate side-chains via an intervening oligo(lactic acid) linker. The gel degrades as the linker is hydrolyzed to lactic acid, liberating also PEG and poly(acrylic acid). A limitation in the use of these gels is that they degrade relatively slowly compared with the rates of cell migration. Indeed, they have been used in tissue engineering applications in which cell migration is not desired or in which cell migration in slow. In one set of applications, these *in situ* polymerized materials were used as barriers to cell migration, to maintain healing in tissue planes (specifically, to prevent restenosis after balloon angioplasty [67] and to prevent scar tissue bridge formation between organs after surgery, [61, 63], both in animal models). In another set of applications using similar material, they were used as a matrix for culture on chondrocytes [64–66]. These cells proliferate and migrate only very slowly as they express their own extracellular matrix to create a cartilaginous tissue. This is to say, early during culture, the cells would be expected to express protein and proteoglycans, which would diffuse in to the PEG-based gel and be cross-linked by cell-produced enzymes to form an interpenetrating polymer network. As the PEG-based polymer degrades, the biopolymer network would remain.

As observed in the paragraph above, there exists a need somehow to tie the rate of degradation of a matrix to the rate of cell proliferation and migration within that material. One possible way to address this is to use slowly degrading materials for slowly forming or changing tissues. Another way would be somehow to connect the rate of material degradation with the rate of cell proliferation and cell migration. One approach that has been explored to accomplish this is to make analogous macromonomers as described above, but to replace the hydrolytically sensitive linkers with a proteolytic-sensitive linker, namely an oligopeptide that serves as a substrate for proteases that are expressed or activated on cell surfaces as they migrate. Nearly all cells activate plasminogen to plasmin (an enzyme that degrades the natural matrix material fibrin) and activate matrix metalloproteinases (MMPs) on their surfaces during spreading and migration [3]. Thus, if telechelic reactive ABA block copolymers are constructed with PEG (again as the B block) and peptide substrates for plasmin or MMPs (as the A block), then cells would be able to penetrate a material with sub-micrometer porosity (i.e. an amorphous, nonfibrillar gel) by proteolytically creating much larger pores, through which they could then migrate.

Other chemical schemes for *in situ* polymerization and for proteolytic penetration of hydrogel matrices have been devised, e.g. one using Michael-type addition reactions for cross-linking (Fig. 18.8) [69–74]. Here, as above, reactive PEG was copolymerized with reactive peptides, but in a cross-linking reaction. Specifically, branched PEGs were rendered reactive with terminal vinylsulfone moi-

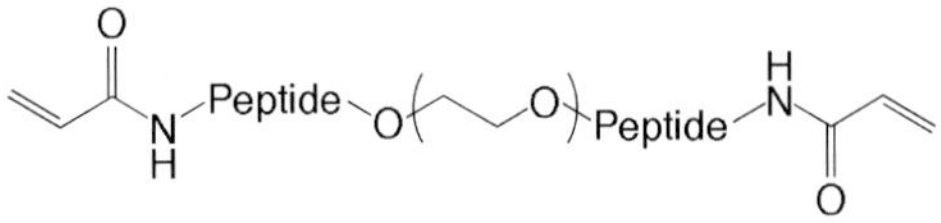

Fig. 18.8 ABA block copolymers have been developed using PEG as a central hydrophilic block to provide mechanical character, while using peptide blocks to provide enzymatic sensitivity. Peptides were synthesized with the C-termini attached to the PEG-OH groups and then the exposed peptide N-termini were acrylated. The precursor shown will polymerize into a cross-linked network that is sensitive to protease-mediated degradation. Given that all migrating cells activate proteases, this enables cells to migrate through materials with very low intrinsic porosity.

eties. In an extracellular environment, vinylsulfone end-groups are very non-cytotoxic, reacting mainly with the free cysteine residues available. However, the extracellular environment is oxidizing and almost all cysteine residues are present in disulfide-bonded form. Hence there are very few vinylsulfone-reactive functions *in vivo* extracellularly. The amine moieties on lysine or at the protein N-terminus is in principle reactive; however, at physiological pH very few of these functions are deprotonated and therefore sufficiently nucleophilic to be reactive. One should note that in the intracellular environment, there are numerous vinylsulfone-reactive functions, since the intracellular environment is much more reductive. However, with the large PEG chain attached to the vinylsulfone moieties, the rate of transport across the cell membrane is infinitesimal. In spite of the oxidative extracellular environment, reduced peptide co-reactant can be applied externally. This is, the vinylsulfone-derivatized PEG can be mixed with (for example) a peptide containing a reduced cysteine residue near each end. The result will be a cross-linked hydrogel, resulting from PEG and peptide cross-linking. If the peptide is designed as a substrate for plasmin or MMPs, cell penetration of the resulting hydrogel will be possible based on cell-mediated enzymatic degradation [70–75].

In the examples mentioned above, both hydrolytic and proteolytic degradation were highlighted. Indeed, hydrolytic degradation is a passive process, in which the concentration of water (essentially 55 M throughout) and the local pH (very close to 7.4 throughout) control gel degradation; gel degradation by hydrolysis is therefore close to homogeneous. By contrast, enzymatic degradation is an active process, in which processes at the surface of cells cleave bonds in the neighboring gel. This feedback between gel properties and cellular behavior, and vice versa, can be very heterogeneous and highly localized, resulting in the formation of micrometer-sized pores into which the smallest of cellular processes extent. In this sense, it is perhaps more meaningful to refer to degradation of the proteolytically-sensitive gels in an active voice: rather then degrading, they are degraded or even remodeled, by cells in contact with them.

Such materials have been used in the study of cell migration behavior, [75] and even in repairing complex tissues such as bone [70–74]. The bone morphogenetic protein-2 was entrapped within PEG- and peptide-based cross-linked gels in a manner such that passive release was very slow and active release (i.e. enzyme-dependent) dominated. When the polymeric matrices were placed in bony defects with the growth factor, cells that invaded from the defect site were shown to be capable of invading the matrix and releasing the growth factor. Those cells differentiated under the influence of the bone-forming growth factor and formed new bone within the defect, which was too large to heal on its own.

18.4
Biofunctionalization and Drug Release

Bioactivity is a key aspect of polymer design in forming tissue engineering constructs and this can be achieved by functionalizing biomaterial surfaces with biomolecules or by releasing biomolecules (drugs) from materials. This section addresses these two concepts.

As mentioned in the Introduction, most structural biomaterials are rather hydrophobic and most biomolecules of interest are very hydrophilic. There comes, then, a disconnect between worlds, that of the polymer and that of the biomolecule, which can lead to unfolding or *denaturation*, of the biomolecule, with resulting loss of function. To address this, many investigators have explored approaches by which to entrap, adsorb or graft water-soluble polymers such as PEG on biomaterial surfaces. To the hydrophilic polymer, then, biofunctional molecules can subsequently be grafted. Countless schemes have been presented in the literature by which to immobilize water-soluble polymers at polymer interfaces and only a few illustrative examples will be selected to be highlighted here.

Polymers such as PEG have been entrapped in polymer surfaces using dynamic physical processing, much like polymer membranes are sometimes formed. For example, thermoplastic polymers have been swollen in solvents that are also solvents for PEG [76]. As the polymer substrate slowly dissolves, PEG chains from the solution can slowly diffuse into the polymer interface. Rapidly exchanging the solvent to a solvent that is (a) miscible with the first solvent, (b) a nonsolvent for the substrate polymer and (c) a solvent for PEG leads to rapid collapse of the swollen polymer interface, entrapping the PEG as a blend within the substrate/polymer interface. Even if the blend is not thermodynamically stable, the rapid nature of the interface collapse can lead to the formation of a stable and phase-mixed polymer blend at the interface.

Polymer adsorption via electrostatics has also been extensively explored. For example, to treat anionic substrates, graft copolymers of polylysine-graft-PEG have been explored and optimized [77–79]. When these graft copolymers are exposed to anionic substrates, the polylysine central chain adsorbs to the substrate, thus immobilizing the PEG chains. If PEG is grafted to the polylysine only very sparsely, then a substantial cationic character is still present at the treated polymer substrate. If the grafting density is too high, then the PEG chains sterically block adsorption of the graft copolymer and no adlayer forms at all. At an optimal grafting ratio, a dense polymer brush of PEG is immobilized on the substrate surface, blocking the underlying negative charge and also the intervening positive charge used for immobilization.

Polymer adsorption via hydrophobic interactions has also been used. For example, hydrophobic polymer surfaces can be treated by adsorption with an ABA block copolymer of a water-soluble polymer, such as PEG, with a hydrophobic polymer, such as poly(propylene glycol) (PPG) [80–83]. Stable adlayers can be formed when the hydrophobic B block is sufficiently large, using processing that merely consists of dipping the substrate in a solution of the block copolymer, shaking off the excess fluid and allowing the substrate to dry.

Polymers to which PEG has been grafted can be very resistant to protein adsorption. Proteins adsorb primarily based on the hydrophobic effect and PEG's hydrophilicity provides very little hydrophobic driving force for protein adsorption. Proteins can also adsorb by electrostatic interactions and, of course, PEG's lack of charge provides no driving force for this mechanism of adsorption either. The biological result of the lack of protein adsorption is a corresponding lack of cellular adhesion – cells attach to biomaterial surfaces not directly, but rather indirectly via receptor–ligand interactions with proteins in the adlayer that forms on the biomaterial interface. If this adlayer fails to form, due to resistance imparted by PEG, then cells will not adhere.

The nonadhesive nature of PEG-grafted polymer surfaces can be an attractive interface for biofunctionalization. This is, if cells attach poorly to the substrate and then if a cell adhesion-promoting peptide or protein is immobilized to the PEG chain ends, for example, then cells will interact only with that intentionally grafted peptide. For example, hydrophobic surfaces treated with PEG-block-PPG-block-PEG are highly nonadhesive to cells, but those same surfaces treated with the same polymer to which an adhesion-promoting peptide is grafted can be highly adhesive to cells [82, 83]. Likewise, polylysine-graft-PEG adlayers are very resistant to cell adhesion; however, if adhesion peptides are grafted to the PEG, they can become very favorable substrates for cell adhesions [84]. To impart cell adhesion, certain integrin-binding peptides can be grafted, such as the well-studied Arg–Gly–Asp (RGD, in the single-letter nomenclature) sequence derived from fibronectin and a number of other adhesion proteins (Table 18.2) [3]. Peptides such as the RGD sequence bind to a wide variety of cell types; however, other peptide sequences can be more selective for specific types of cells. For example, when the fibronectin-derived sequence REDV was immobilized on otherwise nonadhesive surfaces, endothelial cell adhesion was strongly promoted [85]. Other chemical schemes have also been used for surface passivation and subsequent biomolecule functionalization, including formation of interpenetrating polymer networks at polymer surfaces, for the same ends [86, 87].

What kinds of molecules are most interesting for surface biofunctionalization? By and large, these are the adhesion proteins mentioned above, which provide traction to adhering cells and also serve a direct signaling function. These are proteins such as those mentioned in Table 18.2 and peptide fragments that have been derived therefrom, including the RGD sequence mentioned above.

Growth factor molecules, which are addressed in more detail further below, can also be interesting candidates for immobilization. These molecules bind to and activate receptors on the surfaces of cells. After activation, the receptors are usually internalized and either degraded or recycled. In early experiments with epidermal growth factor (EGF), it was demonstrated that surface-bound growth factor remained active, i.e. that the internalization steps were mainly off signals and that receptor bound to an immobilized ligand remained in an on state for a prolonged period of time [88, 89]. In this case, the EGF was chemically grafted to a reactive branched PEG via its N-terminal primary amine, thus creating a relatively passive interface (due to the PEG) attached to a homogeneously presented field of bound growth factor.

Many growth factors have evolved to function in an immobilized fashion, binding to the natural extracellular matrix by electrostatic interactions [27]. Many growth factors have a positively charged domain on their surface, which interacts with the polysaccharide-like molecules heparin and heparin sulfate, which are present in the extracellular milieu. This serves the primary role of creating stores of the growth factors in the extracellular matrix, ready for mobilization and release when the extracellular matrix is enzymatically remodeled. This affinity for heparin has been used in the binding of many growth factors, such as members of the TGF family [90] and the fibroblast growth factor family [91]. For example, in one approach in binding heparin-binding growth factors in biopolymeric gels, a heparin-binding peptide was covalently bonded to the hydrogel. Soluble heparin in the system was then bound to the immobilized heparin-binding peptide and the thus immobilized heparin serves as a site for binding of soluble fibroblast growth factor, nerve growth factor- and neurotrophin-3 in the system [91–95]. This provided for (reversibly) bound growth factor in the biomaterial system.

Covalent immobilization of biomolecules to surfaces provides a means of biofunctionalization, for example with adhesion proteins and growth factors as described above. At the other end of the spectrum, reversible immobilization provides both biofunctionalization (in the bound state) and controlled release (after dissociation), as mentioned in the paragraph immediately above. The remainder of this section addresses this end of that spectrum, that of drug release from biofunctional polymers – how protein drugs are entrapped within polymer scaffolds and matrices and how they are released.

Protein drugs may be entrapped within polymer scaffolds and matrices in a heterogeneous or homogeneous fashion, depending on the interactions between the drug and the polymer. Two cases will be considered here, hydrophobic polymers (such as PLA, PGA, PLGA and PCL) and hydrophilic polymers (such as PEG-based hydrogels). Only the case of the protein drug will be considered. As mentioned in the Introduction, protein molecules are hydrophilic macromolecules and this determines how the drug and the polymer will interact.

In hydrophobic polymers such as PLA, protein molecules will be present heterogeneously throughout the polymer matrix, either as solid particles or as highly concentrated pools of protein in aqueous solution. A number of methods have been developed to accomplish this, including formation of biomaterial scaffolds from the resulting polymer. Many of these methods involve processes with organic solvents, which are solvents for the polymer but nonsolvents for the protein. As mentioned in the Introduction, when proteins in water are mixed with water-miscible organic solvents, denaturation usually ensues. However, if proteins in the solid state are mixed in an organic solvent, since no molecular mobility is possible in the protein, no denaturation is possible either. This has been used for encapsulation. For example, micronized protein particles have been dispersed in degradable polyester solutions and polymer films formed by solution casting [96, 97]. Provided that the protein particles are very small, which can be obtained from lyophilization and subsequent ball-milling or by spray drying, for example, the resulting polymer films contain protein that is well dispersed.

Protein can be released from such solid dispersions by a number of means. If the protein concentration in the polymer is relatively high, protein particles will touch, allowing percolation of hydration in the polymer microphase throughout the protein–polymer network. Diffusion of protein throughout this tortuous pathway in the polymer results in slow, sustained release of protein from the scaffold material. If the protein concentration is too low to permit such percolation of hydration, then the protein will remain entrapped until the polymer is degraded to such an extent that the polymer layer in which the protein resides is exposed to the biomaterial surface.

Emulsion processes have also been used to encapsulate protein in degradable polyesters. In a typical process, referred to as the water-in-oil-in-water double emulsion method [98], the protein in one pot is dissolved in a buffer, the details of the buffer formulation being selected for optimal protein stability. In another pot, the polymer is dissolved in a water-immiscible organic solvent, usually DCM. The two pots are mixed to form a water-in-oil emulsion, the protein solution being dispersed as small droplets within the polymer solution. Polymer microparticles may be formed from this emulsion by further emulsifying the emulsion in a continuous water phase, with the aid of an emulsifying and stabilizing agent such as poly(vinyl alcohol), forming the water-in-oil-in-water double emulsion. Evaporation of the DCM under vacuum results in polymer microparticles containing nanodroplets of protein dissolved in the original buffer. This process has been used to develop a number of protein and peptide controlled-release formulations that are now in clinical use.

The degradation of the polyester into acid degradation products can create stability problems for the protein, for example acid-catalyzed deamidation of asparagine and glutamine residues. This difficulty has been partially addressed by co-encapsulation of poorly water-soluble inorganic bases, such as $Mg(OH)_2$, a common antacid [49].

Hydrogels have also been used for controlled release of proteins. For example, the *in situ* cross-linking chemistry of PEG gels by Michael addition described above has been used for protein entrapment and release. Protein to be encapsulated can be mixed in a branched PEG reactively functionalized with acrylate groups, which are good Michael acceptors for thiols under near-physiological conditions. If this mixture is exposed to a cross-linker, such as a PEG dithiol or dithiothreitol, then cross-linking to form a nanoporous gel rapidly ensues. If the PEG molecular weight is sufficiently low and the degree of branching of the PEG is sufficiently high, networks form that substantially hinder protein release [99]. These networks are not stable in water, due to designed hydrolysis of the ester link between the PEG and the acrylate moiety after reaction with the cross-linking thiol [100]. Hence, as the network degrades, protein is slowly released from the hydrogel in a manner that can be controlled via hydrogel architecture, i.e. via PEG acrylate architecture [99].

As illustrated above, polymers that are used to form biomaterial scaffolds (e.g. the hydrophobic polyesters) and matrices (e.g. the cross-linked PEG gels) can be used for protein encapsulation and subsequent release. Hence materials used in tissue engineering can be used for biofunctionalization across a complete spectrum of biomolecule immobilization through biomolecule entrapment and release.

18.5
Outlook

Polymeric biomaterials have been used extensively in tissue engineering, to serve both mechanical functions (as the platform in scaffolds and matrices) and biological functions (as a substrate for biofunctionalization and as a vehicle for controlled release of bioactive proteins and peptides). Early ventures using polymers in tissue engineering used those materials that were near at hand, having been developed for other purposes. It is clear that the future of polymers in tissue engineering lies in designing specifically for biological application, e.g. with chemical reactivity targeted towards functionalization with specific biological species or comprising directly biologically active moieties. This will require advances in both polymer chemistry and in biological chemistry and cell biology, on the one hand to develop schemes by which to incorporate new biological signals within materials and, on the other, to develop a deeper understanding of what those biofunctional agents should be to elicit designed responses from cells and tissues. From the pragmatic perspective of translation of basic laboratory advances in biomaterials in tissue engineering, the elegance of a design will also have to be balanced with practical constraints that may be placed on the material by demands of manufacturing, sterilization and cost. Even with these constraints considered, there exists substantial room for improvement of current technologies with more biomolecularly sophisticated polymer designs.

References

1 Hubbell, J. A. *Biotechnology (N. Y.)* **1995**, *13*, 565–576.
2 Langer, R., Tirrell, D. A. *Nature* **2004**, *428*, 487–492.
3 Lutolf, M. P., Hubbell, J. A. *Nature Biotechnol.* **2005**, *23*, 47–55.
4 Peppas, N. A., Langer, R. *Science* **1994**, *263*, 1715–1720.
5 Ahsan, T., Nerem, R. M. *Orthod. Craniofac. Res.* **2005**, *8*, 134–140.
6 Griffith, L. G., Naughton, G. *Science* **2002**, *295*, 1009–1014.
7 Ehrenreich, M., Ruszczak, Z. *Tissue Eng.* **2006**, in press.
8 Brouard, M., Barrandon, Y. *Curr. Opin. Biotechnol.* **2003**, *14*, 520–525.
9 Barrandon, Y., Green, H. *J. Invest. Dermatol.* **1988**, *91*, 315–318.
10 Gaglia, J. L., Shapiro, A. M., Weir, G. C. *Arch. Med. Res.* **2005**, *36*, 273–280.
11 Ravivarapu, H. B., Dunn, R. L. *J. Pharm. Sci.* **2000**, *89*, 732–741.
12 Kleinman, H. K., Philp, D., Hoffman, M. P. *Curr. Opin. Biotechnol.* **2003**, *14*, 526–532.
13 Bokel, C., Brown, N. H. *Dev. Cell* **2002**, *3*, 311–321.
14 Orsello, C. E., Lauffenburger, D. A., Hammer, D. A. *Trends Biotechnol.* **2001**, *19*, 310–316.
15 Lauffenburger, D. A., Griffith, L. G. *Proc. Natl. Acad. Sci. USA* **2001**, *98*, 4282–4284.
16 Maheshwari, G., Brown, G., Lauffenburger, D. A., Wells, A., Griffith, L. G. *J. Cell Sci.* **2000**, *113*, 1677–1686.
17 Cohen, L. A., Guan, J. L. *Curr. Cancer Drug Targets* **2005**, *5*, 629–643.
18 Ioannidou, E. *Curr. Pharm. Des.* **2006**, *12*, 2397–2408.
19 Ferguson, M. W., O'Kane, S. *Philos. Trans. R. Soc. Lond. Biol. Sci.* **2004**, *359*, 839–850.
20 Metcalfe, A. D., Ferguson, M. W. *Biochem. Soc. Trans.* **2005**, *33*, 413–417.
21 Carrea, G., Riva, S. *Angew. Chem. Int. Ed.* **2000**, *39*, 2227–2254.
22 Klibanov, A. M. *Nature* **2001**, *409*, 241–246.
23 Discher, D. E., Janmey, P., Wang, Y.-L. *Science* **2005**, *310*, 1139–1143.
24 Langer, R., Vacanti, J. P. *Science* **1993**, *260*, 920–926.
25 Langer, R., Vacanti, J. P. *Sci. Am.* **1995**, *273*, 130–133.
26 Sniadecki, N. J., Desai, R. A., Ruiz, S. A., Chen, C. S. *Ann. Biomed. Eng.* **2006**, *34*, 59–74.
27 Kreuger, J., Spillmann, D., Li, J. P., Lindahl, U. *J. Cell Biol.* **2006**, *174*, 323–327.
28 Mansbridge, J. *Expert Opin. Invest. Drugs* **1998**, *7*, 803–809.
29 Ehrbar, M., Djonov, V. G., Schnell, C., Tschanz, S. A., Martiny-Baron, G., Schenk, U., Wood, J., Burri, P. H., Hubbell, J. A., Zisch, A. H. *Circ. Res.* **2004**, *94*, 1124–1132.
30 Seeherman, H., Wozney, J. M. *Cytokine Growth Factor Rev.* **2005**, *16*, 329–345.
31 Seeherman, H., Wozney, J., Li, R. *Spine* **2002**, *27*, S16–S23.
32 Caplan, A. I. *Tissue Eng.* **2005**, *11*, 1198–1211.
33 Solchaga, L. A., Temenoff, J. S., Gao, J., Mikos, A. G., Caplan, A. I., Goldberg, V. M. *Osteoarthritis Cartilage* **2005**, *13*, 297–309.
34 Bell, E., Rosenberg, M., Kemp, P., Gay, R., Green, G. D., Muthukamaran, N., Nolte, C. *J. Biomech. Eng.* **1991**, *113*, 113–119.
35 Weinberg, C. B., Bell, E. *Science* **1986**, *231*, 397–400.
36 Deagalakis, N., Flink, J., Stasikelis, P., Burke, J. F., Yannas, I. V. *J. Biomed. Mater. Res.* **1980**, *14*, 511–528.
37 Yannas, I. V., Burke, J. F. *J. Biomed. Mater. Res.* **1980**, *14*, 65–81.
38 Yannas, I. V., Burke, J. F., Gordon, P. L., Huang, C., Rubenstein, R. H. *J. Biomed. Mater. Res.* **1980**, *14*, 107–132.
39 Yannas, I. V., Burke, J. F., Orgill, D. P., Skrabut, E. M. *Science* **1982**, *215*, 174–176.
40 Vert, M. *Biomacromolecules* **2005**, *6*, 538–546.
41 Vert, M., Schwach, G., Engel, R., Coudane, J. *J. Control. Release* **1998**, *53*, 85–92.
42 Albertsson, A. C., Varma, I. K. *Adv. Polym. Sci.* **2002**, *157*, 2–40.
43 Freier, T. *Adv. Polym. Sci.* **2006**, *203*, 1–61.

44 Stridsberg, K. M., Rzner, M., Albertsson, A. C. *Adv. Polym. Sci.* **2002**, *157*, 41–65.
45 Middelton, J. C., Tipton, A. J. *Biomaterials* **2000**, *21*, 2335–2346.
46 Vert, M., Mauduit, J., Li, S. *Biomaterials* **1994**, *15*, 1209–1213.
47 Ding, A. G., Shenderova, A., Schwendeman, S. P. *J. Am. Chem. Soc.* **2006**, *128*, 5384–5390.
48 Li, L., Schwendeman, S. P. *J. Control. Release* **2005**, *101*, 163–173.
49 Zhu, G. Z., Mallery, S. R., Schwendeman, S. P. *Nat. Biotechnol.* **2000**, *18*, 52–57.
50 Barrera, D. A., Zylstra, E., Lansbury, P. T., Langer, R. *J. Am. Chem. Soc.* **1993**, *115*, 11010–11011.
51 Barrera, D. A., Zylstra, E., Lansbury, P. T., Langer, R. *Macromolecules* **1995**, *28*, 425–432.
52 Saulnier, B., Coudane, J., Garreau, H., Vert, M. *Polymer* **2006**, *47*, 1921–1929.
53 Saulnier, B., Ponsart, S., Coudane, J., Garreau, H., Vert, M. *Macromol. Biosci.* **2004**, *4*, 232–237.
54 Fischgrund, J. S., James, S. B., Chabot, M. C., Hankin, R., Herkowitz, H. N., Wozner, J. M., Skhirkhoda, A. *J. Spinal Disord.* **1997**, *10*, 467–472.
55 Bourke, S. L., Kohn, J. *Adv. Drug Del. Rev.* **2003**, *55*, 447–466.
56 Ertel, S. I., Kohn, J. *J. Biomed. Mater. Res.* **1994**, *28*, 919–930.
57 Tangpasuthadol, V., Pendharkar, S. M., Kohn, J. *Biomaterials* **2000**, *21*, 2371–2378.
58 Tangpasuthadol, V., Pendharkar, S. M., Peterson, R. C., Kohn, J. *Biomaterials* **2000**, *21*, 2379–2387.
59 Hubbell, J. A. *Curr. Opin. Biotechnol.* **1999**, *10*, 123–129.
60 Sawhney, A. S., Pathak, C. P., Hubbell, J. A. *Macromolecules* **1993**, *26*, 581–587.
61 West, J. L., Chowdhury, S. M., Sawhney, A. S., Pathak, C. P., Dunn, R. C., Hubbell, J. A. *J. Reprod. Med.* **1996**, *41*, 149–154.
62 Sawhney, A. S., Pathak, C. P., Vanrensburg, J. J., Dunn, R. C., Hubbell, J. A. *J. Biomed. Mater. Res.* **1994**, *28*, 831–838.
63 Hill-West, J. L., Chowdhury, S. M., Sawhney, A. S., Pathak, C. P., Dunn, R. C., Hubbell, J. A. *Obstet. Gynecol.* **1994**, *83*, 59–64.
64 Rice, M. A., Anseth, K. S. *J. Biomed. Mater. Res.* **2004**, *70*, 560–568.
65 Bryant, S. J., Arthur, J. A., Anseth, K. S. *Acta Biomater.* **2005**, *1*, 243–252.
66 Elisseeff, J., Ferran, A., Hwang, S., Varghese, S., Zhang, Z. *Stem Cells Dev.* **2006**, *15*, 295–303.
67 West, J. L., Hubbell, J. A. *Proc. Natl. Acad. Sci. USA* **1996**, *93*, 13188–13193.
68 Cruise, G. M., Herge, O. D., Lamberti, F. V., Hager, S. R., Hill, R., Scharp, D. S., Hubbell, J. A. *Cell Transplant.* **1999**, *8*, 293–306.
69 Morpurgo, M., Veronese, F. M., Kachensky, D., Harris, J. M. *Bioconjug. Chem.* **1996**, *7*, 363–368.
70 Lutolf, M. P., Hubbell, J. A. *Biomacromolecules* **2003**, *4*, 713–722.
71 Lutolf, M. P., Lauer-Fields, J. L., Schmoekel, H. G., Metters, A. T., Weber, F. E., Fields, G. B., Hubbell, J. A. *Proc. Natl. Acad. Sci. USA* **2003**, *100*, 5413–5418.
72 Lutolf, M. P., Raeber, G. P., Zisch, A. H., Tirelli, N., Hubbell, J. A. *Adv. Mater.* **2003**, *15*, 888.
73 Lutolf, M. P., Tirelli, N., Cerritelli, S., Cavalli, L., Hubbell, J. A. *Bioconjug. Chem.* **2001**, *12*, 1051–1056.
74 Lutolf, M. P., Weber, F. E., Schmoekel, H. G., Schense, J. C., Kohler, T., Mèuller, R., Hubbell, J. A. *Nat. Biotechnol.* **2003**, *21*, 513–518.
75 Raeber, G. P., Lutolf, M. P., Hubbell, J. A. *Biophys. J.* **2005**, *89*, 1374–1388.
76 Desai, N. P., Hubbell, J. A. *Macromolecules* **1992**, *25*, 226–232.
77 Elbert, D. L., Hubbell, J. A. *Chem. Biol.* **1998**, *5*, 177–183.
78 Huang, N. P., Michel, R., Voros, J., Textor, M., Hofer, R., Rossi, A., Elbert, D. L., Hubbell, J. A., Spencer, N. D. *Langmuir* **2001**, *17*, 489–498.
79 Kenausis, G. L., Voros, J., Elbert, D. L., Huang, N. P., Hofer, R., Ruiz-Taylor, L., Textor, M., Hubbell, J. A., Spencer, N. D. *J. Phys. Chem. B.* **2000**, *104*, 3298–3309.
80 Amiji, M., Park, K. *J. Biomater. Sci. Polym. Edn.* **1993**, *4*, 217–234.
81 Amiji, M., Park, K. *Biomaterials* **1992**, *13*, 682–692.
82 Neff, J. A., Tresco, P. A., Caldwell, K. D. *Biomaterials* **1999**, *20*, 2377–2393.
83 Neff, J. A., Caldwell, K. D., Tresco, P. A. *J. Biomed. Mater. Res.* **1998**, *40*, 511–519.

84 VandeVondele, S., Voros, J., Hubbell, J. A. *Biotechnol. Bioeng.* **2003**, *82*, 784–790.
85 Hubbell, J. A., Massia, S. P., Desai, N. P., Drumheller, P. D. *Biotechnology (N.Y.)* **1991**, *9*, 568–572.
86 Bearinger, J. P., Castner, D. G., Healy, K. E. *J. Biomater. Sci. Polym. Edn.* **1998**, *9*, 629–652.
87 Bearinger, J. P., Castner, D. G., Golledge, S. L., Rezania, A., Hubchak, S., Healy, K. E. *Langmuir* **1997**, *13*, 5175–5183.
88 Kuhl, P. R., Griffith-Cima, L. G. *Nat. Med.* **1996**, *2*, 1022–1027.
89 Kuhl, P. R., Griffith-Cima, L. G. *Nat. Med.* **1997**, *3*, 93–93.
90 Bentz, H., Schroeder, J. A., Estridge, T. D. *J. Biomed. Mater. Res.* **1998**, *39*, 539–548.
91 Sakiyama-Elbert, S. E., Hubbell, J. A. *J. Control. Release* **2000**, *65*, 389–402.
92 Maxwell, D. J., Hicks, B. C., Parsons, S., Sakiyama-Elbert, S. E. *Acta Biomater.* **2005**, *1*, 101–113.
93 Sakiyama-Elbert, S. E., Hubbell, J. A. *J. Control. Release* **2000**, *69*, 149–158.
94 Sakiyama-Elbert, S. E., Panitch, A., Hubbell, J. A. *FASEB J.* **2001**, *15*, 1300–1302.
95 Taylor, S. J., McDonald, J. W., Sakiyama-Elbert, S. E. *J. Control. Release* **2004**, *98*, 281–294.
96 Langer, R., Folkman, J. *Nature* **1976**, *263*, 797–800.
97 Lu, H. H., Kofron, M. D., El-Amin, S. F., Attawia, M. A., Laurencin, C. T. *Biochem. Biophys. Res. Commun.* **2003**, *305*, 882–889.
98 Cohen, S., Yoshioka, T., Lucarelli, M., Hwang, L. H., Langer, R. *Pharm. Res.* **1991**, *8*, 713–720.
99 van de Wetering, P., Metters, A. T., Schoenmakers, R. G., Hubbell, J. A. *J. Control. Release* **2005**, *102*, 619–627.
100 Schoenmakers, R. G., van de Wetering, P., Elbert, D. L., Hubbell, J. A. *J. Control. Release* **2004**, *95*, 291–300.

IUPAC Polymer Terminology and Macromolecular Nomenclature

R. F. T. Stepto

Background

The important task of reviewing and maintaining an up-to-date, internationally agreed terminology for polymeric materials and nomenclature for macromolecules is the responsibility of the International Union of Pure and Applied Chemistry (IUPAC). The work is voluntary, carried out by established scientists in the various fields of polymer and macromolecular science and technology. It is organised by the Sub-Committee on Polymer Terminology of the Polymer Division of the IUPAC. The membership of the sub-committee is augmented when necessary to ensure that it has scientists with expertise appropriate to its current areas of work. In addition, in the field of macromolecular nomenclature, the sub-committee works in conjunction with the IUPAC Division of Chemical Nomenclature and Structure Representation.

The IUPAC Sub-Committee on Polymer Terminology presently (2006) has 33 members from 13 countries. Professor R. G. Jones (University of Kent) is Chairman and Dr. Michael Hess (University of Duisburg-Essen) is Secretary. As an indication of its size and the diversity of its work, it is currently working on 17 projects, 10 concerned with terminology, 5 with nomenclature and 2 of a general nature. Much of the sub-committee's work is carried out by e-mail correspondence. However, it also meets once a year to review projects and to discuss possibilities for new projects. Details of its projects and other activities can be found on the IUPAC Polymer Division web site, *www.iupac.org/divisions/IV.*

A history of the work of IUPAC on Polymer Terminology and Macromolecular Nomenclature is in the process of being published [1]. The history will not be repeated here as the emphases of the present chapter are the current work and recommendations of the IUPAC on Polymer Terminology and Macromolecular Nomenclature. Accordingly, this chapter concludes with the list of current members of the Sub-Committee on Polymer Terminology and the list of currently valid IUPAC publications emanating from the sub-committee and its predecessors. The recommendations listed will shortly be published as a compendium [1].

Macromolecular Engineering. Precise Synthesis, Materials Properties, Applications.
Edited by K. Matyjaszewski, Y. Gnanou, and L. Leibler
Copyright © 2007 WILEY-VCH Verlag GmbH & Co. KGaA, Weinheim
ISBN: 978-3-527-31446-1

Currently Valid IUPAC Recommendations on Polymer Terminology and Macromolecular Nomenclature

1. IUPAC Compendium of Macromolecular Terminology and Nomenclature (The Purple Book), 2nd Edition, eds. Edward S. Wilks, Michael Hess, W. Val Metanomski and Robert Stepto, Royal Society of Chemistry, Cambridge, *in press*.
2. Stereochemical definitions and notations relating to polymers (Recommendations 1980), *Pure Appl. Chem.* **53**, 733–752 (1981).
3. Note on the terminology of the molar masses in polymer science, *Macromol. Chem.* **185**, Appendix to No. 1 (1984); *J. Polym. Sci., Polym. Lett. Ed.* **22**, 57 (1984); *J. Colloid Interface Sci.* **101**, 277 (1984); *J. Macromol. Sci., Chem.* **A21**, 903–904 (1984); *Br. Polym. J.* **17**, 92 (1985).
4. Nomenclature for regular single-strand and quasi-single-strand inorganic and coordination polymers (Recommendations 1984), *Pure Appl. Chem.* **57**, 149–168 (1985). Reprinted as Chapter II-7 in *Nomenclature of Inorganic Chemistry II – Recommendations 2000*, Royal Society of Chemistry, Cambridge, 2001.
5. Source-based nomenclature for copolymers (Recommendations 1985), *Pure Appl. Chem.* **57**, 1427–1440 (1985).
6. Use of abbreviations for names of polymeric substances (Recommendations 1986), *Pure Appl. Chem.* **59**, 691–693 (1987).
7. Definitions of terms relating to individual macromolecules, their assemblies, and dilute polymer solutions (Recommendations 1988), *Pure Appl. Chem.* **61**, 211–241 (1989).
8. Definitions of terms relating to crystalline polymers (Recommendations 1988), *Pure Appl. Chem.* **61**, 769–785 (1989).
9. A classification of linear single-strand polymers (Recommendations 1988), *Pure Appl. Chem.* **61**, 243–254 (1989).
10. Compendium of Macromolecular Nomenclature (the "Purple Book"), prepared for publication by W. V. Metanomski, Blackwell Scientific Publications, Oxford (1991).
11. Nomenclature of regular double-strand (ladder and spiro) organic polymers (IUPAC Recommendations 1993), *Pure Appl. Chem.* **65**, 1561–1580 (1993).
12. Structure-based nomenclature for irregular single-strand organic polymers (IUPAC Recommendations 1994), *Pure Appl. Chem.* **66**, 873–889 (1994).
13. Graphic representations (chemical formulae) of macromolecules (IUPAC Recommendations 1994), *Pure Appl. Chem.* **66**, 2469–2482 (1994).
14. Basic classification and definitions of polymerization reactions (IUPAC Recommendations 1994), *Pure Appl. Chem.* **66**, 2483–2486 (1994).
15. Glossary of basic terms in polymer science (IUPAC Recommendations 1996), *Pure Appl. Chem.* **68**, 2287–2311 (1996).
16. Definition of terms relating to degradation, aging, and related chemical transformations of polymers (IUPAC Recommendations 1996), *Pure Appl. Chem.* **68**, 2313–2323 (1996).
17. Source-based nomenclature for non-linear macromolecules and macromolecular assemblies (IUPAC Recommendations 1997), *Pure Appl. Chem.* **69**, 2511–2521 (1997).
18. Definitions of terms relating to the non-ultimate mechanical properties of polymers (IUPAC Recommendations 1998), *Pure Appl. Chem.* **70**, 701–754 (1998).
19. Definitions of basic terms relating to low-molar-mass and polymer liquid crystals (IUPAC Recommendations 2001), *Pure Appl. Chem.* **73**, 845–895 (2001).
20. Generic source-based nomenclature for polymers (IUPAC Recommendations 2001), *Pure Appl. Chem.* **73**, 1511–1519 (2001). Errata, *Pure Appl. Chem.* **74**, 2019 (2002).
21. Definitions of basic terms relating to polymer liquid crystals (IUPAC Recommendations 2001), *Pure Appl. Chem.* **74**, 493–509 (2002).
22. Definitions relating to stereochemically asymmetric polymerizations (IUPAC

Recommendations 2001), *Pure Appl. Chem.* **74**, 915–922 (2002).
23 Nomenclature of regular single-strand organic polymers (IUPAC Recommendations 2002), *Pure Appl. Chem.* **74**, 1921–1956 (2002).
24 Definitions of terms relating to reactions of polymers and to functional polymeric materials (IUPAC Recommendations 2003), *Pure Appl. Chem.* **76**, 889–906 (2004).
25 Definitions of terms related to polymer blends, composites, and multiphase polymeric materials (IUPAC Recommendations 2004), *Pure Appl. Chem.* **76**, 1985–2007 (2004).

Subject Index

a

AB block copolymers 2165, 2174
- block efficiency 2174
- pigment dispersants 2165

AB diblock copolymers 1388
- theoretical phase diagram 1388

(AB)$_n$ linear multiblock copolymers 850
- condensation polymerization 850
- sequential addition of monomers 850

ABA block 550
- polysulfone 550

ABA block copolymers 550, 1118

ABA triblock copolymers 840, 842–844, 846–847, 849, 1991
- amphiphilic 843
- anionic polymerization 842
- biodegradable amphiphilic triblock copolymers 849
- cationic living polymerization 845
- controlled free radical polymerization 847
- 5-(N,N-diethylamino)isoprene (DAI) 844
- groups transfer polymerization 845
- ionomer-like 846
- nitroxide-mediated controlled radical polymerization 848
- PAA–P2VP–PAA triblock copolymer 844
- PDMA–PNIPAAm–PDMA copolymers 848
- physical gels 847
- PNIPAAm–PMPC–PNIPAAm block copolymer 848
- site-transformation techniques 849
- synthetic strategies 840

Abbe's equation 1650

ABC block copolymers 2169
- pigment dispersant 2169

ABC triblock copolymers 1412, 1949
- aggregate morphology 1412

- two-compartment micellar system 1949

ABC triblock polymers 1418
- ABC miktoarm copolymers 1420
- micellar morphologies 1418
- super-hydrophobicity 1420

ABC triblock terpolymers 850, 852–856
- anionic living polymerization 852
- double hydrophilic ABC terpolymers 852
- monohydrophilic 854
- controlled free radical polymerization 856
- PAA–PS–P4VP ABC triblock 853
- PEO–PDEAEMA–PHEMA 856
- PEOVE–PMOVE–PEOEOVE 855
- PS–P2VP–PBd triblock copolymer 852
- PS–P2VP–PEO 852
- PVBCl–PS–PVBFP 856
- structured nano-objects 854
- synthetic strategies 851

ABCA tetrablock terpolymers 858
- amphiphiles 858

ABCBA pentablock terpolymers 858
- PMMA–PS–PBd–PS–PMMA 858

ABCD quaterpolymers 859
- honeycomb morphology 859

absorption spectrum 645

Acacia catechu 417

Accelerator SLT100 synthesizer 1970

4-acetoxybenzoic acid 302, 309, 321, 333
- copolymerization 333
- fibrillar crystals 333
- microsphere 333
- self-polycondensation 321

acrylamide 420–421, 754
- HRP-mediated polymerization 421
- radical polymerization 754
- syndiotacticity 754

acrylates 558, 563, 657, 938, 945, 958

Macromolecular Engineering. Precise Synthesis, Materials Properties, Applications.
Edited by K. Matyjaszewski, Y. Gnanou, and L. Leibler
Copyright © 2007 WILEY-VCH Verlag GmbH & Co. KGaA, Weinheim
ISBN: 978-3-527-31446-1

- film photopolymerization   657
- polymerization   945, 958
- star polymers   938
acrylic copolymers   2011
acrylic polymers   589
activated monomer mechanism   423, 441, 583, 1308
activated monomer polymerization   583
- activated monomer mechanism   423, 441, 583, 1308
- transformation reactions   583
active chain-end mechanism   423
active site transformations   581
acyclic diene metathesis polymerization (ADMET)   2647
acyclic metathesis polymerization (ADMET)   697
Adagen®   2655
addition polymerization   526
addition polymerization reaction   953
addition–fragmentation transfer   796
- mechanism   796
additives   15, 168
- Lewis bases   61
- proton traps   61
- salts   61
adhesion protein receptors   2722
adhesion proteins   2721
adhesives   1
adriamycin (ADR)   2622
aerogels   2334
alcohol sensors
- polyaniline   2511
- polypyrrole   2512
aliphatic esters   1373
alkanethiol monolayers   1181
alkoxides   29, 118
alkoxyamine   569
alkyl (meth)acrylates   29
- polymerization   29
alkyl vinyl ethers (VE)   846
alkylazides   505
alkylboranes   312
"alligator clips"   2255
alpha-melanocyte stimulating hormone, α-MSH   1283
alpha-structure   1307
alternating copolymerization   829
- Lewis acid addition   829
alternating copolymers   813, 821, 828
- charge-transfer complexes   829
- copolymerization mechanism   828
- determining factor   821

alternating EP copolymers   237
- dual-site mechanism   237
- metallocene catalysts   237
amide–donor–acceptor interaction   360
amine mechanism   521, 1308
amine sensors   2513
amino acid N-carboxyanhydrides (NCA)   1308
α-amino acid N-carboxyanhydrides   519, 2647
- polymerization   519
amino acid analogues   503, 505, 507–508
- alkene-functionalized side-chains   507
- alkyne side-chains   507
- azide-functionalized side-chains   505
- halide-functionalized sde-chains   503
- ketone-functionalized side-chains   507
- photoreactive side-chains   508
- structural amino acid analogues   508
amino acids sensors   2525
amino-functional polyethers   2155
aminobenzoic esters   322
- chain-growth polycondensation   322
- meta-substituted   322
Amonton's law   2699
amorphous calcium carbonate (ACC)   1367
amorphous polymers   1788, 1793, 1802, 1804
- constitutive model   1793
- craze initiation   1802
- deformation   1804
- mechanical performance   1788
- neck formation   1804
amphiphilic initiators   1222
- peroxide surfactants   1222
amphiphilic surfactants   112
α-amylase   428
amylose   427, 462, 1320
- regioselective modification   462
analytical electron microscopy   1679, 1681
- energy-dispersive X-ray mapping   1679
- energy-filtered transmission electron microscopy   1679
anatase-TiO$_2$ 2120
- photocatalytic reactions   2120
anchoring groups
- acidic   2141
- aminic   2141
- electroneutral   2141
anchoring monomers   2147
Andrade–Eyring equation   367
angiogenesis   2654

angle light scattering apparatus (Rheo-SALS)   2105
anilines   418
- oxidative copolymerization   418
anion to cation transformation   578
anionic coordination polymerization   563
anionic copolymerization   38
- statistical copolymerization   38
anionic living polymerization (ALP)   842
anionic polymerization   7–10, 12–14, 16–18, 20, 22, 24, 26, 28, 30–32, 34–36, 38–42, 44–46, 48–50, 52, 54, 609, 632, 915, 1023, 1610, 2613, 9
- additives   15
- "associative" mechanism   37
- "ate" complexes   15
- bifunctional initiators   25
- butyllithium   19
- carbanionic active centers   11
- conjugated dienes   39
- "controlled"   7
- counterion   16
- degradation reactions   42
- dilithium initiators   28
- diluent   13
- dissociating agent   14
- electron transfer   18
- emulsion process   632
- experimental constraints   17
- general features   8
- graft copolymers   48
- group transfer   29, 37
- initiation   17
- ion pair stretching   15
- isoregulating mechanism   42
- kinetics of propagation   30–31
- "living"   7
- macromolecular engineering   8
- macromolecular synthesis   45
- (meth)acrylic monomers   36
- methyl methacrylate (MMA)   36
- monomers   12
- multifunctional initiators   25, 1023
- nucleophilic addition   19
- parameters   11
- persistence of active centers   42
- polymerization kinetics   14
- propagating species   14
- propagation step   30
- rate constants   35
- reactions   8
- reactivity   11
- reference monomers   10
- regioselectivity   39
- solvating agent   13
- solvents   13
- star-shaped polymers   48
- stereoregulating effects   24
- stereoselectivity   39
- stereospecific anionic polymerization   41
- suspension polymerization   609
- vinyl   8, 10, 12, 14, 16, 18, 20, 22, 24, 26, 28, 30, 32, 34, 36, 38, 40, 42, 44, 46, 48, 50, 52, 54
anionic polymerization to controlled radical polymerization transformation   556
anionic polymerization to radical transformation   555
anionic random copolymerization   863
anionic to ring-opening metathesis polymerization   585
anisotropic spin interactions   1939, 1941
- chemical shift   1939
- dipole–dipole interactions   1940
- manipulation   1941
- quadrupole interaction   1940
anisotropic structures, anisotropic mechanical properties   1814
- bridged bis(indenyl) catalysts   225
antidrug concept   2551
- steroid-21-oate esters   2551
anti-cancer therapeutics   2651
- "enhanced permeation and retention (EPR) effect"   2651
- neocarzinostatin (NCS)   2651
- tumor cell targeting   2651
anti-inflammatory drugs   1282
Antibody Targeted Triggered Electrically Modified Prodrug Type Strategy (ATTEMPTS)   2546
anticancer DDS   2576, 2578, 2580
- colon-targeting approaches   2577
- cytotoxic drugs   2577
- ectin–carbohydrate   2579
- EPR effect   2577
- mitochondria   2580
- molecular targets   2580
- requirements   2576
- targeted HPMA copolymer–drug conjugate   2582
- targeted systems   2579
- topical delivery   2578
anticancer polymeric drugs   2575
antimicrobial coatings   2670
antisense oligonucleotides   2586
antisense therapy   2673

apoptosis 2581
- antiapoptotic system 2581
apoptosis-inducing agents 2584
aptamers 2399
aqueous dispersed media 605
aragonite 2597
arborescent polymers 974
arbutin 408
ArF lithography 2320
- block copolymers 2320
"arm in–arm out" process 912
arm–arm coupling 939
- interstar 939
- intrastar 939
"arm-first" approach 911, 941, 946, 948–949, 958
"arm-first" method 943
"armored latexes" 1232
aromatic compound sensors 2515
aromatic monomers 405
- anilines 405
- phenolic compounds 405
aromatic polyamides 318, 336, 339, 953
- block copolymers 336, 339
- synthesis 953
- synthetic method 318
aromatic polyester 319
- polymerization 319
- substituent effects 319
arsonium ylides 313, 334
artificial cell scaffolds 2706
artificial urushi 414
aryl-substituted polymers 2239
arylates, block copolymers 558
arylazides 505
asphaltenes 2009
association constant 354, 357–358
association phenomena 1350
associative amphiphiles 846
Astramols® 144
ASW2000 synthesizer 1970, 1977
asymmetric block copolymers 1319
asymmetric stars 913–914
- functional group asymmetry 913
- molecular weight asymmetry 913
- topological asymmetry 914
atactic polypropylene 224
"ate" complexes 15, 33
atom transfer radical coupling (ATRC) 202, 791
atom transfer radical polymerization (ATRP) 177–178, 181–182, 186, 341, 529, 544, 607, 617, 618, 630, 682, 685, 784, 831, 936, 1111, 1142, 1146, 1185, 2053, 2171
- activators are generated by electron transfer (AGET) 181
- activators regenerated by electron transfer (ARGET) 181
- advantages 186
- block copolymers 682
- catalysts 177
- chain extensions 179
- conditions 180
- continuous activator regeneration (ICAR) 182
- controlled architectures 182
- copper-mediated surface-initiated ATRP 1143
- core–shell structures 1185
- Cu-mediated ATRP 180
- emulsion polymerizations 630
- functional groups 177
- gradient block copolymers 2171
- halogen exchange 178
- hydroxy-functional monomers 2171
- initiators 178, 2171
- kinetics 176
- limitations 186
- ligand functional homopolymers 682
- mechanism 176, 681
- metallomonomers 685
- miniemulsion polymerization 617
- monomers 177
- normal ATRP 181
- side-reactions 177
- simultaneous reverse and normal initiation (SR&NI) 618
- suspension polymerization 607
- transition metal complexes 179
- water-accelerated ATRP 1146
atomic force microscopy (AFM) 836, 1104, 1128, 1235, 1515–1522, 1525, 1532–1536, 1538–1539, 1541, 1544–1547, 1550–1551, 1554, 1556–1560, 1562, 1564, 2300, 2316
- adhesion measurement 1554
- applications 1515
- binding forces 1559
- block copolymer 1550
- block copolymer thin films 2316
- cantilevers 1519
- cantilever designs 1564
- capillary forces 1555
- chain-stretching experiments 1559
- chemical force microscopy (CFM) 1522, 1555

Subject Index

- chemical-binding potentials 1560
- "colloidal probe technique" 1554
- complementary techniques 1523
- compositional mapping 1526
- conformational transitions 1534
- contact AFM 1519
- contact area 1518
- contact mode 1521
- contrast mechanisms 1516
- crystalline superstructures 1538
- crystallization measurement 1541
- crystal morphology 1853
- dynamic properties measurement 1516
- electric force microscopy 1523
- energy dissipation imaging 1522
- environment-controlled AFM 1535
- flow mechanism 1546
- force modulation techniques 1557
- force–distance dependence 1518
- fractal dimension 1533
- frequency modulation AFM 1520
- friction behavior 1556
- height measurements 1539
- hydrophilic AFM 1555
- imaging conditions 1516
- imaging rate 1525
- instability principles 1547
- intermittent contact mode 1521
- jump-to-contact phenomenon 1519
- length distribution 1533
- measuring properties 1554
- microdomain observation 1551
- molecular resolution 1527
- molecular visualization 1526
- molecular weight measurement 1532
- monomer–monomer attraction 1559
- nanomechanical probing experiment 1557
- oscillations 1520
- persistence length 1533
- phase images 1521
- polymer blends examination 1545
- principles 1517
- probe calibration 1520
- pull-off forces 1555
- quantitative measurements 1554
- quasi-static contact AFM 1520
- rapid scanning AFM 1525, 1541
- resolution 1518
- scanning modes 1520
- self-assembled monolayers studies (SAMs) 1555
- semicrystalline polymers 1538
- shear modulation AFM 1558
- spectroscopic imaging 1524
- studies 1548
- surface morphology 1516
- surface-mediated chemical reactions 1536
- thermal properties measurement 1561
- thermoelectric actuation 1562
- thermomechanical surface patterning 1563
- thin films observation 1544
- topographic imaging 1526
- video-rate imaging 1526
- visualization molecular processes 1534
- wetting phenomena studies 1544

atom-transfer radical polymerization (ATRP) 2170
- pigment dispersants 2170

ATRP 550, 558, 573, 590, 785, 958–959
- initiators 785
- macroinitiator 550
- macromonomers 784
- mechanism 784

ATRP initiators 554, 561, 591
- hydroxy-functionalized 591

Au–polymer nanocomposites 2415

automated polymers synthesis
- 4,4′-azobis(4-cyanopentanoyl chloride) (ACPC) 563
- fully automated synthesis 1969
- semi-automated synthesis 1969
- standard laboratory techniques 1969

azo functionalized prepolymer 564
azo initiator-involved transformation 565
azo initiators 547
- bifunctional 547
azo-containing initiators 568
azo-containing monomer 565
azo-containing poly(cyclohexene oxide) 564
azobisisobutyronitrile (AIBN) 167

b

Bacillus sp. chitinase (ChiBs) 431
- reaction mechanisms 431
back-biting 140
bacterial cellulose (BC) 2696
- properties 2696
Bakelite 405
Baytron P 2401
BCl_3-based system, activity 62
BCP micelles 1406, 1409, 1420, 1424
- anomalous micellization 1410
- applications 1406

- artifactual micellar structures 1409
- branched worm-like micelles 1416
- corona free energy 1412
- crew-cut micelles 1412
- critical micelle temperature 1408
- crosslinking 1410
- drying process 1410
- homopolymer solvent 1420
- impurities 1410
- metallization 1425
- nonergodic dispersion 1416
- phase transiton 1411
- preparation methods 1409
- sample preparation 1409
- staining process 1410
- star-like micelle regimes 1412

BCPs 1388–1391, 1394–1395, 1412, 1415–1416
- BCP/inorganic phase systems 1394
- blending 1390
- bulk phase 1388
- concentrated systems 1413
- diluted systems 1416
- enthalpic interactions 1391
- intermaterial dividing surface 1388
- macromolecular nonionic surfactant 1390
- microphase separation 1389
- microstructural polymorphism 1415
- morphological transitions 1416
- morphology diagrams 1387
- nanoparticle-filled BCPs 1395
- phase behavior 1412
- self-assembly theories 1388
- self-consistent mean-field theory (SCFT) 1412
- solvent selectivity 1415
- spontaneous self-assembly 1387
- templates 1395
- weak segregation limit (WSL) 1389

benzyl bromides 554
γ-benzyl-L-glutamate 527
ε-benzyloxycarbonyl-L-lysine 527
Bessel function 1583
bicyclic polymers 894
- synthesis 894
bifunctional initiator 25, 339
- hydroxy-functionalized 591
- phenyl terephthalate 339
bifunctional monomers 303, 354
α,β-bifunctional polymers 74
bifunctional UPy monomers 366
bilirubin oxidase 419

binary homogeneous blends 1591
- scattering 1591
binary mixtures 1582, 1579
- Flory–Huggins interaction parameter 1583
- Form factors 1582
- Gaussian homopolymers 1583
- incompressibility constraint 1579
- intramolecular correlations 1580
- scattering length density 1580
- scattering volume 1578
bio-imaging 1071
bio-organs 2698
- surface behavior 2698
bioactive molecules 1282
bioactive proteins, multilayers 1279
bioactive surfaces 2662
- polymer biohybrid materials 2662
biocompatible polymers 1291, 2696
bioconjugate arrays 2662–2663
- DNA microarrays 2662
- pEG-modified surfaces 2663
bioconjugate polymers 2659
- macromolecular systems 2659
- purification 2659
- solid-phase peptide synthesis 2659
- stimuli responsive conjugates 2659
bioconjugates 2555
- coupling reactions 2570
- coupling reagents 2570
- reagents 2555
- steric hindrance 2570
bioconjugation 2569
- N-hydroxysuccinimide (NHS) ester-activated polymers 2569
- ligand–protein binding 2569
- reactivity 2569
- reparation 2569
- spacer molecule 2569
biodegradability 2118
- respirometric test 2118
biodegradable polyester system 450
- applications 450
- lipase-catalyzed ring-opening polymerization 450
biodegradable polymers 451, 2116
- aliphatic polyesters 2116
biofunctional coatings 1249
biofunctional films 1279, 1285
- anti-inflammatory response 1285
- cell adhesion 1279
- cell-protein communication 1289
- drug delivery systems 1287

- "fiber-like" protrusions 1286
- piroxicam–cyclodextrin complexes 1288
- piroxicam-β-cyclodextrin inclusion complexes 1287
- protein adsorption 1279
- pseudopods 1286
biofunctionalization 2735–2738
- acid-catalyzed deamidation 2738
- covalent immobilization 2737
- drug release 2735
- dynamic physical processing 2735
- emulsion processes 2738
- growth factor binding 2737
- polymer adsorption via electrostatics 2735
- polymer adsorption via hydrophobic interactions 2736
- protein adsorption 2736
- protein release 2738
- surface biofunctionalization 2736
biofunctionalized films, anti-inflammatory properties 1283
biofunctionalized multilayers, cell activity tests 1284
biogels 1689
biogenic biomacromolecules 1367
biohybrid polymers 2645, 2649, 2662, 2672–2673
- active site binding 2673
- affinity precipitations 2672
- antisense therapy 2673
- conjugation reactions 2649
- controlled polymerization 2646
- degree of polymerization (DP) 2646
- diagnostic arrays 2662
- polydispersity index (PDI) 2646
- polymer–oligonucleotide conjugates 2662
- polymer–protein drug conjugates 2673
- smart materials 2672
bioinspired synthetic materials 1365
biointerfaces 1167
- primary adsorption 1167
- secondary adsorption 1167
- tertiary adsorption 1167
biological tissues 2712
biologically active materials 1343
biomaterials 289, 1370
biomedical dendrimers 2547
biomembranes 2670
biomimetic materials chemistry 1365
biomimetic mineralization 2600
biomimetic polymers 1334
biomineralization 5, 2597, 2601

- functional proteins 2601
- glycoproteins 2603
- polysaccharides 2603
biomineralization polymers 2600, 2604
- amphiphilic molecule 2604
biomineralization protein 2602, 2604
- MSP-1 2602
- protein domains 2604
biominerals 2597–2598
- hierarchical structure 2598
biomolecules 2721
- adhesion proteins 2721
- growth factors 2721
biopolymers 2597, 2599, 2645
- mineralization 2599
biosensor analytes 2519
biosensors 2493, 2519–2521, 2526–2528
- amino acids sensors 2525
- aroma sensors 2528
- artificial olfactory system 2528
- ATP degradation products 2528
- bulk acoustic wave (BAW) sensor 2520
- cholesterol sensors 2526
- creatinine sensors 2524
- DNA sensors 2529
- environmental monitoring 2529
- food freshness 2528
- fructose biosensors 2527
- glucose sensors 2521
- immunosensors 2529
- lactate sensors 2527
- microbiosensor 2529
- molecularly imprinted polymer 2520
- peroxide sensors 2526
- pyruvate sensors 2525
- sense of taste 2530
- sol–gel enzyme sensors 2521
- urea sensors 2524
biosynthetic degradable materials 1373
- enzymatic synthesis 1373
- synthetic components 1373
biosynthetic hybrid materials 1365
biphenyl 18
bisphenol-A 410, 414
bisphenol-A bischloroformate 325
bisurea molecules 364
bitumen 2009
- modification 2009
block copolymer blends 1390
- AB/hA blends 1391, 1392
- ABC/hA phase transformations 1392
- BCP/nanoparticle blends 1394
- macrophase separation 1390

- order–disorder transition   1391
- order–order transition   1391
- phase behavior   1390
- phase diagram   1391
- temperature-induced structural transformations   1391

block copolymer brushes   1165
- surface morphologies   1165

block copolymer films   383
block copolymer host templates   1352
block copolymer lithography, solvents   2315
block copolymer micelles   1405, 1407
- critical micelle concentration   1407
- scaling relations   1407
- structural parameters   1407

block copolymer stabilizers   2209
- ω,ω'-hydroxy macromonomer   2210
- in situ formation   2209

block copolymer vesicles   1232
block copolymer–based adhesives   1733
- styrene–isoprene–styrene (SIS) triblock copolymer   1733
- tackifying resin   1733
- viscoelastic properties   1733

block copolymer-based adhesives   1736–1738, 1741–1742, 1744–1747, 1749
- acrylic-based block copolymers   1745
- adhesive properties   1736
- debonding mechanism   1746
- diblock content   1738
- dissipative properties   1749
- fibril extension stress   1745
- interfacial cracks   1747
- interfacial properties   1746
- large-strain behavior   1745
- nonlinear elastic properties   1742
- probe test method   1737
- small-strain modulus   1745
- statistical models   1742
- stress–strain curve   1737
- triblock:diblock ratio   1736
- uniaxial extension   1744
- uniaxial tensile tests   1741
- viscoelastic properties   1737

block copolymerization   195, 655
- dual initiators   195
- photopolymerization   655
- site transformation   195

block copolymers   4, 47, 80–83, 85, 87, 89–90, 92, 177, 194, 196–197, 204, 238, 258, 260–261, 334, 336, 451–452, 494, 525, 545–547, 549, 555, 558, 563, 570, 575, 580, 587–588, 631, 675, 678–679, 682, 684–685, 692, 694–695, 697, 701–703, 719, 836, 839–840, 885–886, 890, 897, 899, 901, 1199, 1307–1309, 1317, 1320–1321, 1333, 1343, 1345, 1347–1348, 1355, 1387, 1405, 1471–1472, 1475, 1480, 1484, 1492, 1495, 1502, 1549, 1551–1552, 1731, 1734, 1758, 1769, 1780, 1948, 1960, 2002–2003, 2005–2007, 2181, 2295, 2313, 2317–2318, 2320, 2322–2323, 2345, 2353, 2355, 2356, 2393, 2419, 2429, 2467, 2607, 2617
- (AB)n block copolymers   840
- (AB)$_n$ multiblock   542
- (AB)$_n$ multiblock copolymers   546
- AB diblock   542, 839
- AB-type   261, 570
- ABA block copolymers   547, 2005
- ABA triblock   542, 839
- ABA triblock amphiphiles   1355
- ABA-type   570, 575, 580
- ABABA pentablock   839
- ABC terpolymers   840
- ABC triblock copolymers   564
- ABC-type star-block   80
- ABCD quarterpolymers   840
- acrylates   558
- adhesive applications   1731
- advanced lithography   2318
- AFM   1549
- aggregates   1318
- alkenic-vinylic-type   587
- alkyne-functional block copolymers   679
- amorphous–crystalline   90
- amphiphilic radial block copolymer   2357
- applications   204, 897
- architectures   352, 1472
- atom transfer radical polymerization (ATRP)   260
- automotive application   2007
- biomimetic approaches   1307
- bio-organic segments   1308
- block copolymer   2318
- block copolymer microdomains   2313
- block copolypeptides   525
- boron-containing   2321
- t-BuMA block copolymers   92
- CdS–diblock copolymer nanocomposite   2429
- chromatographic separation   1917
- coil polymer   339
- coil–rod–coil triblock copolymers   1494
- coiled blocks   1495
- coiled coils   494
- colloidal building blocks   1333

- comb-type graft copolymers 80
- compatibilizers 1769
- conjugated blocks 1492, 1495
- conjugated rod–coil multiblock copolymers 1355
- coordinative block copolymers 1475
- coupling agents 85
- cyclic diblock copolymers 885
- cyclization 897
- dendritic-linear 2355
- diblock copolymer–small molecule additive system 2318
- a,ω-difunctional triblock copolymer 886
- direct polymerization 2317
- "double-ended block copolymers" 2003
- double-hydrophilic block copolymers (DHBCs) 2607
- electron transfer 1502
- emulsion polymerization 631
- ethylene oxide (EO) 555
- ethylene–propylene block copolymers 238
- fingerprint 836
- fluorene–bithiophene copolymer 2237
- footwear applications 2006
- functionalized stimuli-responsive materials 1347
- glassy block 1731
- graphoepitaxy 1551
- hierarchical ordering 1321
- hydrogenated block copolymers 2006
- IB 82
- indenofluorene–oligothiophene copolymer 2237
- interfacial interactions 1552
- intermediate capping reaction 83
- ionic multiblock copolymers 1475
- kinetic studies 86
- lamellar superstructures 1321
- lanthanide-containing block copolymers 684
- Lewis acidity tuning 83
- ligand exchange 83
- linear amylose segments 1309
- linear block copolymers 897
- linear coil–coil diblock copolymers 901
- linking reaction 87
- liquid crystalline block copolymers 586
- liquid crystalline blocks 563
- lithographic resists 2316
- lithographic techniques 2313
- lithography 2295
- living anionic polymerization 692
- living systems 239
- metal-containing block copolymers 675
- metalation 90
- metallo-linked block copolymers 675
- metalloblock copolymers 679
- micellar aggregates 1948
- micellar templates 2429
- micelle formation 2617
- micelles 897, 1405
- microdomains 1551
- monomer sequence control 1307
- monomers 1343
- morphology 901, 1405, 2315, 2353
- multisegmented block copolymers 196
- nanoparticles 2419
- "nanoreactors" 2419
- nanoscale morphologies 1960
- nanostructures 1343
- nanostructuring 1774
- nitroxide-mediated radical process 451
- NO-mediated polymerization 679
- nonlinear architecture 86
- oligosaccharide-terminated polymers 1309
- organic-inorganic hybrids 1199
- organometallic-*block*-inorganic diblock copolymers 719
- organotin block copolymers 702
- Pd-containing block copolymers 703
- periodic nanostructures 1471
- phase behavior 1318
- phase segregation 1355
- phenylene–thiophene copolymer 2237
- phosphino block copolymers 697
- photoresists 2315, 2323
- photovoltaic devices 2393
- poly(ethylene glycol)-*b*-poly(β-benzyl-L-aspartate) (PEG–PBLA) 528
- poly(ethylene)-*b*-polynorbornene copolymer 258
- poly(L-leucine)–poly(sodium L-glutamate)s 1318
- poly(St-*b*-IB-*b*-St) 92
- polybutadiene (PB) block copolymers 692
- polybutadiene–poly-(L-glutamate) diblock copolymers 1318
- polyethylene oxide (PEO) 555
- polyferrocenylsilane (PFS) block copolymers 695
- polymeric surfactant 2181

- polymerization mechanisms combination 89
- polynorbornene block copolymers 1333
- polynorbornene-*b*-poly(silyl vinyl ether) copolymer 258
- polypeptide-based block copolymers 1308
- polypropylene-based 588
- polystyrene-*b*-poly(ε-benzyloxycarbonyl-L-lysine) (PS–PZLL) 528
- polyvinylferrocene (PVFc) block copolymers 694
- polyvinylpyridine (PVP) 678
- post-assembly processing 1348
- post-polymerization modification 2317
- radical polymerization 92
- repulsive blocks 1472
- rhenium-containing block copolymer 685
- rod–coil diblock copolymers 1494
- ROMP 258, 697
- Ru-containing diblock copolymer 703
- rubbery block 1731
- secondary interaction motifs 1308
- secondary phase segregation 1345
- "selective" solvent 897, 899
- self-assembly 336, 899, 1343, 1780, 2315
- semifluorinated 452
- sequenced bio-organic segments 1309
- sequential block copolymerization 83
- sequestering agents 2322
- single-ion conducting block copolymer electrolytes 1484
- site-transformation method 89
- solution micelles 1549
- solution phase 1405
- spherical microdomains 1552
- star-block copolymers 87
- stereoblock copolymers 197
- stimuli-responsive diblock copolymers 82
- stimulus-responsive materials 1320
- structure formation 1317
- styrene 82, 555
- styrene–butadiene diblocks 2002
- styrene–butadiene–styrene triblock copolymer (SBS) 1196
- styrene–isoprene diblocks 2007
- styrenic block copolymers 2001
- sulfonated poly(aryl ether ketone) copolymers 1480
- surface micelles 1549
- surfactant-like additives 2323
- synthesis reactor 239
- synthetic methodologies 80
- synthetic-peptidic block copolymers 1343
- tackifying resin 1731
- thermoplastic elastomers (TPEs) 85
- thermotropic multiblock copolyesters 552
- thin films 2315
- thin-film characterization 2316
- time–temperature superposition principle 1734
- unimolecular end-to-end cyclization 890
- viscosity index improvers (VIIs) 2007
- water purification applications 2467
- zylindrical microdomains 1552

block copolymers, nonlinear 86
- (AB)$_n$-type star-block copolymers 86

block copolypeptides 495, 525–526, 536, 1309, 1321, 2648
- highly branched 536
- hydrophilic-hydrophobic 525
- hydrophilic-hydrophobic-hydrophilic 525
- poly(L-lysine)-*b*-poly(L-cysteine) block copolypeptides 526
- primary amine initiators 525
- "schizophrenic" vesicles 1321
- self-assembly 526
- solid-state structures 1321
- transition metal-mediated NCA polymerization 525

block macromonomers 1110
- diblock macromonomers 1110

block polymers 2383
- anionic living polymerization 2382
- controlled/"living" radical polymerization 2383
- convergent way 2382

blood–brain barrier 2579
blowing agents 2008
blue-shift solvatochromic effect 1354
Bombyx mori 481
bootstrap effect 830
boron neutron capture therapy 535
bottlebrushes 537
- polypeptide bottlebrushes 537
"bottom-up" construction 1344
"bottom-up" nanofabrications 1347
bovine fetal aorta endothelial cells (BFAECs) 2706
branched copolymers 1760

Subject Index

branched polymers 93, 973–974, 985, 994, 996–997, 1007, 1630, 1635, 1637, 1639, 1642, 1643, 2422, 2547
- arbitrary flow 1639
- arm retraction 1640
- back-biting 975
- branch-point withdrawal (BPW) 1635, 1640
- bulk properties 994
- Cayley tree 1631
- combs 973
- "controlled/living" radical polymerization 1007
- copolymerization 985
- curvilinear stretching 1640
- damping function 1637
- degree of branching, DB 974
- distribution of seniorities 1630
- drug delivery system (DDS) 2547
- dynamic dilution 1630
- elastic melt viscosity 997
- ethylenimine 975
- fingerprinting 1643
- glass transition temperature 996
- long-chain branching 1635
- mean-field gelation 1631
- melt dynamics 997
- metallocene ensembles 1633
- polymerization techniques 974
- pom-pom model 1642
- priority distribution 1636
- properties 973
- radical polymerization 191
- RAFT polymerization 985
- retraction dynamics 1630
- rheology 996
- scalar property 1643
- seniority distribution 1636
- stars 973
- stretch relaxation 1642
- synthesis 975
- templates/ligands 2422
- tensor property 1643
- time–strain factorability 1637
brush concentration profile 1163
brush copolymers 190, 1356–1360, 1362–1363
- amphiphilic cylindrical brush–linear block copolymers 1360
- ATRP 1358
- core–shell cylindrical brush copolymers 1359
- "grafting from" 1358
- "grafting on to" 1362
- "grafting through" 1360
- linear–brush–brush copolymer 1363
- molecular topologies 1357
- single-molecule templates 1356
brush–brush block copolymers 1118
brush–coil block copolymers 1116
brush–coil copolymers 1106
brush-*block*-brush copolymers 1106
brush-*block*-linear copolymers 1106
bulk heterojunction solar cells 1993
bulk materials 1654
- thin films 1654
bulk properties 355, 1126
bulk structures 1321
Buna synthetic rubbers 7
butadiene 621
- catalytic polymerization 621
butyl acrylate 823
- copolymerization 823
n-butyl acrylate (nBA) 1313
tert-butyl crotonate 767
β-butyrolactone (β-BL) 115, 122, 447
- lipase CA-catalyzed copolymerization 447
γ-butyrolactone, polymerization 107
butyllithium 16, 21
- order of reactivity 21

C

calcium carbonate 1366
calcium oxalate dihydrate (COD) 2625
- polymer adsorption 2625
calcium phosphate-based composite structures 1366
calixarenes 369
Camellia sinensis 417
camptothecin 2561, 2584
Candida antarctica lipase (lipase CA) 440
capillary viscosimetry 1983
capping agents 1141
- dithio ester 1141
- halogens 1141
- iodine 1141
- nitroxides 1141
ε-caprolactam 133
ε-caprolactone (ε-CL) 266, 439, 1191, 2043, 2343, 2345, 2348
- copolymerization 108
- derivative 2348
- *in situ* intercalative polymerization 2043

- lipase CA-catalyzed copolymerization 458
- lipase-catalyzed polymerization 452
- ring-opening polymerization 1191, 2345

capsules 1231–1233
- double emulsion technique 1233
- heterophase polymerization 1230
- interfacial polymerization 1232
- precipitation 1231
- submicron cross-linked hollow polymer capsules 1230
- surface adsorption 1232
- template method 1233

carbocationic polymerization 57
carbon nanorods 1123
carbon nanotubes 1196, 2059, 2249–2252, 2274
- "grafting from" 2059
- "grafting on to" 2059
- conjugated polymers 2251
- devices 2250
- electronic properties 2250
- graphene molecules 2252
- physical characteristics 2251
- pyrolysis 2251
- single-walled carbon nanotubes (SWNTs) 2274
- structures 2250
- synthesis 2251

carbon-doped oxides (CDOs) 2332
N-carboxyanhydrides (NCAs) 2614
- ring-opening polymerization 2614
N-carboxyanhydrides of α-amino acids (NCAs) 134–136
- amine mechanism 135
- high-vacuum technique 136
- ring-opening polymerization 134

cardanol 411
Carothers' equation 354
κ-carrageenan 2704
cartilage 2689, 2698
cashew nut shell liquid (CNSL) 411
- cardanol 411
cast films, PMMA films 1160
catalyst-transfer mechanism 322
catalytic chain transfer (CCT) 796
catalytic dendrimers 1080
- AIBN-mediated alkenylation 1083
- autocatalysis 1081
- catalytic turnover 1081
- chiral dendrimer catalysts 1084
- continuous-flow membrane reactor 1084
- cooperative effects 1080
- dendritic shell 1081
- enantioselective reactions 1084
- negative effects 1080
- organocobalt(III) species 1083
- pump effect 1082
- recyclable dendrimer catalysts 1084
- selectivity 1083
- types 1080

catalytic metal matrices 1348
catechin 417, 465, 466
- catechin-immobilizing polymer particles 466
- conjugation 465
- HRP-catalyzed polymerization 417
- polyallylamine–catechin conjugate 465

catechol 414
- oxidative polymerization 414
catechol derivatives 414–415
- in vitro enzymatic hardening reaction 415
- urushi 414

cation to radical transformation 566
cationic and anionic polymerization combination 89
cationic copolymers 2198
- amphiphilic copolymers 2198
- synthesis 2198

cationic initiator 1226
cationic living polymerization (CLP) 846
cationic photoinitiators 646, 647
- anionic photoinitiators 647
- diazonium salts 646
- metallocene derivatives 647
- onium salts 646

cationic polymerization 104, 335, 579, 609, 928, 1026, 2614
- cyclic amines 579
- multifunctional initiators 1026
- suspension polymerization 609

cationic polymerization to controlled radical transformation 568
cationic polymerization to conventional radical transformation 563
- azo-containing polymers 563

cationic ring-opening polymerization 136, 140, 965, 1983
- activated monomer (AM) mechanism 137, 140
- active chain end (ACE) mechanism 137
- chain transfer 138
- counterions 138
- cyclization 139

Subject Index | 2759

- mechanism 136
- propagation 137
cationic to ATRP transformation 571
- triblock copolymers 571
cell adhesive coatings 2666
cell cultivation 2706–2707
- adhesive ratio 2706
- floating ratio 2706
- spreading ratio 2707
cell migration behavior 2734
cells 2720–2721
- allogeneic cells 2720
- autologous cells 2720
- cytotoxicity 2721
- xenogeneic cells 2720
cellular detoxification mechanisms 2581
β-cellobiosyl fluoride (β-CF) 422
cellulose 422, 424, 462
- allomorphs 424
- *in vitro* synthesis 422
cellulose acetate 462
- lipase-catalyzed acylation 462
- transesterification 462
chain back-skip events 226
chain degradation mechanism 1036
chain extension 374
chain growth polymerizations 674
- mechanism 674
chain microstructure 1946
chain polymerization 103
chain topology 324
chain transfer agents (CTAs) 260, 753, 796, 1037
- mercaptans 1037
chain transfer agents (RAFT agents) 560
- macro RAFT agents 560
- PEO 560
chain transfer reactions 239
chain transfer species 2338
chain-end modification 80
chain-growth polycondensation 311, 316, 323, 953
- approaches 311
- reactive species transfer 311
chain-growth polymerization 311, 317, 980
- self-condensing vinyl polymerization (SCVP) 980
- self-condensing ring-opening polymerization (SCROP) 980
chain-growth self-condensing ring-opening polymerization 976
chain-growth self-condensing vinyl polymerization (SCVP) 976

chain-terminating agents 529
- isocyanates 529
"chain-walking" process 223
charge regulation 1164
charge-transfer complex 372
charge-transfer photopolymerization 660
- photoinitiator-free formulations 660
chemical analytes 2494
- gas sensors 2494
chemical composition distribution (CCD) 1882, 823
chemical force microscopy 1522
chemical modification 1756
chemical recycling 451
- industrial examples 451
- principle 451
chemical sensor 2493
chemical tagging 284
chemical transformations 1373
chemical vapor deposition (CVD) 2333
chemoenzymatic method 450
chemoselective oxidation polymerization 409
chemoselective polymerization 446
chemosensing 1353
chitin 430–431, 463
- application 431
- ChiBs-catalyzed polymerization 431
chitin spherulite 431
chitosan 463–464, 1291, 2558
- chitosan–catechin conjugate 464
- chitosan–gelatin gels 464
- enzymatic modification 464
- gelatin-grafted polymer 463
chlorosilane linking 918
chlorosilane reagents 1017
p-chlorostyrene (*p*-ClSt), living carbocationic polymerization 66
cholesterol biosensor 2526
chondroitin 436
- biological functions 436
- HAase-catalyzed ring-opening polyaddition 436
chondroitin sulfate 1370
chromatographic techniques 1983
chromatography 1881–1882
- retention volume 1882
chromatography at the critical point of adsorption 1912–1915
- copolymer separation 1915
- critical conditions 1911
- critical eluent composition 1912
- cyclic fractions 1915

- end-group separation   1913
- graft copolymer separation   1917
- liquid adsorption chromatography (LAC) mode   1912
- polymer blend separation   1915
- pore size   1913
- size-exclusion chromatography (SEC) mode   1911
- star separation   1914, 1911

chromophores   1499
chronic inflammatory response   1282
ε-CL   452, 458
Clausius–Mossotti equation   2457
clay layer   2091
- curved clay layer   2091
- flexibility   2090
- smectite clay   2090
- transmission electron microscope (TEM) images   2090

click chemistry   196, 791, 888, 1115, 1362, 2649
- transformations   505

click-coupling   1315
CMR   1983
- viscosimetry   1983

coarse-graining techniques   1455, 1458, 1460
- application   1460
- atomistic chain pair   1459
- atomistic model   1455
- coarse-grained level   1455
- conjugate gradient minimization   1459
- equilibrated starting configurations   1460
- force-matching scheme   1456
- gyration radii   1458
- inverse Boltzmann method   1457
- inverse mapping   1459
- mapping points   1458
- mapping scheme   1455
- melt equilibration   1460
- Metropolis criterion   1457
- Monte Carlo sampling   1457
- penalty function   1455
- slithering-like algorithms   1460
- spline interpolation   1457
- van der Waals equation-of-state analysis   1458

cobalt nanoparticles   2438–2439, 2441–2443
- block copolymer–cobalt nanocomposite   2439
- passivation   2442
- polymeric surfactants   2443
- small molecule surfactants   2441

- superparamagnetic colloids   2442
- surfactants   2439
- synthesis   2438

coil polymers   337, 339, 341
- block copolymers   337
- macroinitiator   341

coil–coil BCPs   1399
coil–coil block copolymers   1411
coil-to-globule transition   1120
coiled coil motif   494
coiled-coil folding motif   1330
- primary structure   1330

collagen   489, 1370, 2598
- tropocollagen   489

collagen-like polypeptides   489–490
- cell adhesion   490
- expression host   490
- gelatin-based protein polymer   490

collective properties   1441
colloidal crystals   1171–1173
- driving force   1173
- "fully stretched core-shell model"   1172
- interparticle potential   1171
- "semi-soft colloidal crystal"   1173
- shell model   1172

comb   1624, 1630
- polydispersity   1630
- relaxation modes   1629
- tube model   1630

comb synthesis   1611
Comb-burst® polymers   987
comb-like polymers   189, 581
- grafting from   189
- grafting on to   190
- grafting through   190

combinatorial material research (CMR)   1967, 1970
- automated synthesis robots   1970

combinatorial polymer research   1967, 1993
- fluorescence spectroscopy   1993

commodity polymers   2
comonomer feed compositions   826
comonomers   2344
- N,N-dimethylacrylamide   2344

compartmentalized multilayer films   1265–1267
- advantages   1271
- cell accessibility   1269
- cell degradation   1270
- cell fluorescence   1270
- confocal laser scanning microscopy   1266
- degradable barriers   1270

- degradation   1266
- efficiency   1266
- fluorescence recovery after photobleaching (FRAP)   1267
- PL^L diffusion   1266
- polylactide-co-glycolide layers   1270
compatibilization   1754
- strategies   1754
compatibilizer   1758
- architecture   1758, 1769
- branch/graft   1760
complementary binding motifs   378
complex functional degradable materials   1369
complex functional macromolecules   1341–1342
- macromolecular self-assembly   1342
composite particles   1209
composites   654
- photopolymerization   654
composition drift   823
compounding   2015
computer simulations   1431–1433, 1462–1464
- classical force fields   1450
- coarse-graining   1433, 1452, 1455
- concept of universality   1434
- conformational statistics   1432
- degrees of freedom   1433
- dynamics   1441
- examples   1462
- extendable nonlinear elastic interaction   1453
- generic models   1453
- Ising critical behavior   1453
- Lennard–Jones interactions   1450
- MacMillan–Mayer theory   1463
- molecular interactions   1446
- OPLS potential function   1450
- particle motion   1432
- phase diagrams   1464
- polyelectrolytes   1463
- polymer melt   1434
- potential energy functions   1447
- random walk   1434
- reactive force fields   1451
- reptation model   1443
- Rouse model   1441
- Ryckaert–Bellemans function   1449
- second-generation force fields   1451
- self-avoiding walk   1434
- simulation methods   1432
- small molecules   1462

- specificity   1433
- thermodynamic perturbation theory (PTP)   1464
- Zimm model   1445
computer-assisted tomography (CAT)   1671
concentrated brushes   1139, 1153, 1162
- dielectric properties   1162
- elastic properties   1162
- glass transition temperature   1160
- miscibility   1162
concentration-dependent transition   388
concurrent polymerizations   591
condensation polymer architecture   334
condensation polymerization   546, 1533
- Schulz–Flory theory   1533
condensation polymerization to anionic coordination-insertion polymerization transformation   552
condensation polymerization to controlled radical polymerization transformation   549
condensation polymerization to conventional radical polymerization transformation   547
condensation polymers   334, 337
- block copolymers   334, 337
conducting polymers   2369, 2493, 2574
- applications   2274, 2370
- biochemical sensors   2574
- electronic structure   2265
- electrostatic discharge protection   2274
conductivity   2265, 2369
conductors   1475
Cone calorimeter   2111
conformational transitions   1534
conjugated block copolymers   1351
conjugated block polymers   2381
- photovoltaic devices   2381
- rod–coil block copolymers   2381
conjugated polymer backbones   1350
- synthetic methodologies   1350
conjugated polymers   281, 1351, 1487–1489, 1491, 2225, 2228–2230, 2236, 2239, 2264, 2266–2267, 2270, 2272, 2370–2372, 2379, 2381, 2395, 2398, 2423, 2427
- bandgaps   2266
- block copolymers   2381
- chain packing   2236
- charge carrier mobility   1489, 2266
- charged derivatives   2264
- classes of molecules   2227
- conductivity   2227
- conformation   1487

- crystallinity 2267
- doping 2227, 2266
- effective conjugation length (ECL) 2229
- electron delocalization properties 2266
- exciplex formation 2379
- field-effect transistors 2395
- fine-tuning 2379
- "hairy rods" 1488
- head-to-head coupling 2230
- ketone defects 2239
- light-emitting diodes 2384
- optical properties 2230
- poly(aryleneethynylene)s 2372
- poly(arylenevinylene)s 2372
- polyacetylene 281, 2267
- polyalkylthiophenes 1489
- polyaniline 2272
- polyarenes 2228, 2270
- polyaryls 2372
- polyfluorenes 1491
- polymer lasers 2389
- processability 2228
- self-assembly 1487
- semiconductor nanoparticles 2427
- sensors 2398
- solid-state orientation 1351
- solution-state assemblies 1353
- structure 2264
- substituents 2229
- synthesis 2371
- transition metal-mediated cross-coupling reactions 2376

conjugation 1311
- methods 2649

constrained-geometry catalysts (CGCs) 222

continuous addition process 1218

continuous-stirred tank reactor (CSTR) 824, 1036

contour length 1125

controlled drug release 2574
- "release by demand" type 2574
- sustained drug release 2573
- types 2572

controlled free radical polymerization (CFRP) 634, 847, 2170
- microemulsion systems 634
- pigment dispersants 2170

controlled polymerizations 553, 675

controlled polymerized polymers 2026
- applications 2026

controlled radical polymerization (CRP) techniques 543, 784, 933, 1027, 1312–1313
- multifunctional precursors 1027
- peptide–polymer conjugates 1312

controlled radical polymerization processes 2055
- initiator anchoring 2055

controlled radical polymerization to anionic polymerization transformation 573

controlled/living free radical polymerization 607, 615, 629
- emulsion polymerization 629
- miniemulsion polymerization 615
- suspension polymerization 607

controlled/living radical polymerization 162, 171–172, 176, 185, 187–188, 190–193, 195, 199–201, 205–207, 831, 1014
- advantages 185
- applications 203, 813, 176
- brush topologies 190
- comb-like polymers 189
- combination of methods 195
- computational chemistry 206
- concept 171
- controlled topology 187
- copolymers 193
- cyclic polymers 192
- dendritic structures 192
- dormant chain end 202
- end-group Functionality 202
- functional monomers 201
- graft copolymers 197
- hyperbranched polymers 191
- kinetics 172
- limitations 185
- linear chains 188
- mechanisms 205
- microgels 192
- model reactions 205
- molecular architectures 206
- molecular hybrids 199
- persistent radical effect (PRE) 172
- polymer architectures 187
- polymer characterization 207
- polymer networks 192
- reversible transfer 171
- reversibly terminated 171
- side functional groups 202
- star-like 188
- structure–property correlation 207
- templated systems 200, 1110, 1139, 1141

- reversible activation process   1141, 203
conventional radical polymerization   555, 831, 833
conventional radical polymerization to anionic polymerization transformation   572
convergent approach   938
coordinating solvents   2425
coordination   218
- homogeneous catalysts   218
coordination polymerization   3, 217, 740
- active centers   111
- controlled/living radical polymerization   195
- enantiomer-selective site model   740
coordination polymers   355
coordination ring-opening polymerization   103
coordination ROP   116
- initiation   116
copolyesters   450
- AB_2 polyesters   450
copolymer, styrene–butyl acrylate copolymer   831
copolymer composition   824, 830
- chain-length dependence   830
copolymerization   165, 333, 1756
- degree of branching   975
- randomly functionalized polymers   1756
copolymerization models   813–814, 819
- comp-pen model   819
- complex dissociation model (CDM)   819
- complex participation model (CPM)   814, 819
- penultimate unit model   814
- terminal model   814
copolymers   193, 199, 831, 836, 2379
- alternating copolymers   2379
- controlled/living radical polymerization   199
- cross-coupling reactions (CCRs).   2380
- CRP techniques   199
- fluorene-based copolymers   2380
- gradient copolymers   199
- periodic copolymers   199
- properties   836
- random copolymers   2379
- segmented copolymers   193
- statistical copolymers   193
core-first stars   1023–1028, 1032
- atom transfer radical polymerization   1028
- cationic polymerization   1026

- condensative chain-growth polymerization   1032
- heterocyclics   1024
- initiators   1023
- organometallic routes   1032
- polyester-based dendrimers   1025
- reversible addition-fragmentation chain transfer   1030
- ring-opening methathesis polymerization   1031
- "ring-opening multi-branching polymerization"   1024
- stable free polymerization   1027
- terminations   1030core–shell brushes   1106
- vinylic monomers   1023
core–shell cylinders   1474
core–shell hybrid materials   2413
core–shell materials   2426
core–shell micelles   2619
core–shell molecular brushes   1122
core–shell nanoparticles   1180–1181, 1184–1185, 1187–1189, 1191–1192, 1194–1195, 2061, 2209
- alumina–polystyrene core–shell composites   1196
- ATRP (atom transfer radical polymerization)   1181, 1185
- CdS–SiO_2–poly(methyl methacrylate) core–shell–shell nanoparticles   1195
- chain–chain coupling   1186
- characterization methods   1184
- controlled radical polymerization methods   1181
- deactivating species   1187
- grafted polymer layers   1197
- "grafting from" method   1180
- "grafting from" technique   1180
- "grafting on to" technique   1180
- hollow spheres   1199
- initiator-modified Au nanoparticles   1182
- living anionic surface-initiated polymerization (LASIP)   1181
- magnetic cores   1194
- magnetic rings   1200
- metal nanoparticles   1191
- miscellaneous cores   1195
- nitroxyl radical-mediated polymerization   1188, 1181
- non-spherical cores   1196
- physical adsorption   1182
- polystyrene-coated Fe_3O_4 nanoparticles   1200

- reversible addition–fragmentation chain transfer   1189
- reversible addition–fragmentation chain transfer polymerization (RAFT)   1182
- secondary approach   1186
- semiconductor nanoparticles   1195
- surface functionality   1192
- synthesis   1180
- thermal behavior   1193
- transition temperature (T_g)   1193
- two-step functionalization process   1188

core–shell particles   1179, 1209–1210, 1212–1213, 1219–1225, 1230, 1233, 1239–1241
- AFM   1235
- amphiphilic core–shell particles   1223
- applications   1179, 1209
- 2,2′-azobis(2-amidinopropane) dihydrochloride (AIBA)   1226
- capsules   1230
- CdS–SiO$_2$ core–shell particles   1233
- conducting polymer   1225
- degree of phase separation   1238
- dispersion polymerization   1220
- electrically conductive core–shell particles   1225
- emulsifier-free polymerization   1222
- encapsulation efficiency   1226
- equilibrium morphology   1213
- filter experiments   1238
- fluorescence spectroscopy   1241
- fluorinated monomer   1224
- graft polymerization   1222
- heterophase polymerization   1230
- hollow polymer particles   1230
- inorganic core–polymer shell   1225
- interphase thickness   1239
- methods of characterization   1233
- microemulsion polymerization   1221
- MnFe$_2$O$_4$ 1226
- monomer hydrophilicity   1214
- NMR   1237
- optical properties   1234
- P(MMA-co-BA-co-MAA) particles   1226
- PA-PUR particles   1222
- physical properties   1209
- PBA–chitosan particles   1223
- [PBA–P(S-co-MMA)] core–shell particles   1213
- PBA–PS core–shell particles   1215
- PMMA–PMMA latex particles   1241
- polymer core–inorganic shell   1225
- polymer core–polymer shell   1219
- PPy–PMMA core–shell particles   1225
- PS–PANI core–shell latexes   1225
- PS–PANI particles   1221
- PS–PBA core–shell particles   1221
- PS–PIm core–shell particles   1221
- PS–PMA core–shell latexes   1241
- PS–PMMA core–shell particles   1241
- PS–PNIPAM core–shell latex particles   1240
- (PVAc–PBA) inverted core–shell particle   1213
- scattering methods   1240
- SEM   1234
- SiO$_2$–PMMA latex particles   1226
- spin-diffusion experiments   1239
- structural components   1179
- sub-micron size   1210
- synthesis   1219
- two-step emulsion polymerization process   1219
- WISE experiment   1238

core–shell particle morphology   1210–1211, 1217–1219
- addition mode   1218
- "confetti"-like structures   1215
- cross-linking agent   1217
- feed rate   1219
- "frozen" morphologies   1212
- "golf ball"-like morphology   1235
- initiator   1217
- inverse core–shell structure   1215
- kinetically controlled structures   1213
- "mushroom"-like structure   1217
- prediction   1210
- thermodynamic considerations   1211
- thermodynamically controlled structures   1213
- viscosity   1215
- weight ratio   1215

core–shell rubbers   1802
core–shell structure   628
core–shell systems   1044
"core-first" approach   911, 913, 915, 947, 949, 957
corticosteroids   2557
Cotton–Daoud model   2347
Coulombic interactions   360
coupling agents   85, 535, 841, 927, 930, 1194, 1611
- chlorosilanes   1611
- fullerene   927

- *m*-maleimidobenzoyl-*N*-hydroxysuccinimide ester (MBS)   535
- siloxane–ATRP coupling agent   1194
- silylenol ethers   930
- *N*-succinimidyl-3-(2-pyridyldithio)propionate (SPDP)   535

coupling reactions concurrent polymerizations   590
- oppositely charged macroions   590

covalent polymers   351
Cox–Merz relation   2102
creatinine sensors   2525
critical micelle concentration (CMC)   614
critical molecular weight   1606
cross-linking agents   1217
cross-linking films   1290
crotonates   768
- stereospecific polymerization   768

crown ethers   371
- dibenzylammonium ion-substituted   371
- methylviologen   371
- two-component system   371

CRP
- atom transfer radical polymerization (ATRP)   1313
- nitroxide-mediated radical polymerization (NMP)   1313
- reversible addition-fragmentation chain transfer radical polymerization (RAFT)   1313

cryogenic temperature (cryo-TEM)   1662, 1665–1669
- chemical-potential control   1666
- contrast reversal   1669
- direct imaging cryo-TEM   1662
- electron-beam damage   1668
- freeze-fracture-replication cryo-TEM   1662
- phase contrast   1667
- sample preparation   1662
- temperature-regulated effect   1669

Crystal Violet   1988
crystalline solids   355
crystallization polycondensation   299
- mechanisms   302
- morphology control   302

crystallization process   1835, 1862
- conformationally disordered (condis) phases   1835
- decoupling   1862
- intermediate phases   1835
- irreversible crystallization   1834
- irreversible melting   1834
- semicrystalline polymers   1835

CTAs   797
- design   796
- methacrylate-based CTA   797

cucurbit[*n*]urils (CB[*n*]s)   372
α-cyano-4-hydroxycinnamic acid (CAHA)   1975
cyclic alkene   880
- ring expansion polymerization   880

cyclic block copolymers   896, 898, 902
- morphology   898
- nano-organized films   902
- self-assembly   898

cyclic depsipeptides   453
- lipase-catalyzed ring-opening polymerization   453

cyclic ester polymerization   2348
cyclic esters   116, 119, 122, 2043
- catalyst   119
- coordination ROP initiators   116
- ring-opening polymerization   2043
- stereocontrolled polymerization   122

cyclic ethers   109, 140
- cationic ring-opening polymerization   140
- ROP   109

cyclic ketones   313
- peroxide oxidation   313

cyclic polyamides   325
cyclic polymers   192, 324–325, 875–876, 879, 881, 883, 887, 890, 895
- anionic polymerization   878
- back-biting/cyclization pathways   876
- bicyclic polymer   893
- cationic polymerization   878
- cyclic architectures   892
- cyclic homopolymers   887
- cyclic poly(dimethylsiloxane) (PDMS)   879
- cyclization efficiency   883
- display techniques   390
- end-to-end coupling   875
- end-to-end ring closure   875, 881
- glass transition temperature   875
- hemicyclic polymers   892
- α,ω-heterodifunctional polymers   887
- heterodifunctional polymer intermediate   886
- hydrodynamic volumes   875
- interfacial hydrolytic polycondensation   325
- physical properties   895
- polycondensation   324

- polycyclic polymers 893
- pre-cyclized systems 890
- propagation/depropagation equilibrium 876
- properties 875
- qualitative analysis 324
- ring expansion polymerization 879
- ring–linear chain equilibria 875
- scattering behavior 895
- synthesis 875–876
- tadpoles 892
- theta temperature 895
- unimolecular end-to-end cyclization 887

cyclic polymers, chain scrambling 877
cyclic sulfides, ring-opening polymerization 109
cyclobutene 254
- ROMP 254
β-cyclodextrin 1287
cyclohexene oxide (CHO) 546
ansa-cyclopentadienylamidotitanium catalysts 222
cyclooctadiene 276
- dispersion ROMP 276
cyclooctene 341, 1352
- polymerization (ROMP) 1352
- ring-opening metathesis polymerization 341
cyclopentene 585, 952
- block copolymerization 585
- ROMP 952

d

2D WISE experiment 1238
damping function measurements 1617
De Gennes' reptation concept 2
dead-end polymerization (DEP) 776
Debye function 895
Debye–Beuche equation 1589
decoupled segments 1862
defects 2235
degenerative transfer (DT) 182–183
- process 183
degradable polymer matrices 2732–2734
- cell-mediated enzymatic degradation 2734
- cross-linked hydrogels 2732
- hydrolytically sensitive hydrogel formulations 2732
- materials resorption 2732
- PEG-based polymer 2733
- proteolytic penetration 2733

- proteolytic-sensitive linkers 2733
- slowly degrading materials 2733
degradable polymer scaffolds 2727, 2729
- acidic degradation products 2729
- characteristics 2727
- degradable elastomeric thermoplastics 2730
- degradable polyesters 2729
- mechanical characteristics 2730
- polycarbonates 2729
- polymers 2727
- ring-opening polymerization 2727
degradation 1369
degradation–polymerization repetition 451
degree of branching 327
degree of crystallinity 1833
- two-phase model 1833
degree of polymerization 324, 352, 354
dehydration polymerization 455
- in water 455
- solvent-free 455
dendrigraft polymers 987
dendrimers 326, 973, 1058–1059, 1065, 1069, 1077, 1080, 2177, 2310, 2564–2565
- applications 2565
- catalytic dendrimers 1080
- characteristics 2564
- conjugates 2564
- conjugation ratio 2565
- convergent approach 1059
- divergent synthesis 1058
- features 1059
- light-harvesting dendrimers 1065
- morphologies 1059
- multiporphyrin arrays 1069
- payload 2565
- poly(amidoamine) (PAMAM) 2564
- polyether dendrimers 2565
- prodrug bioconjugates 2564
- redox-active dendrimers 1059
- self-assembled dendrimers 1084
- stimuli-responsive dendrimers 1077, 1079
- synthesis 1058
dendritic polymers 192, 1956
- supramolecular assembly 1956
density functional theory (DFT) 321, 2122
Derjaguin approximation 1152
desaminotyrosyltyrosine ethyl ester 2729
device applications, electronic properties 2226
dewetting 1548
- spinodal dewetting 1548

- studies 1548
dewetting phenomena 1544
- equilibrium thermodynamics 1544
Dexter-type energy transfer 1065
dextran 2557
- applications 2556
- daunorubicin–dextran conjugate 2557
- dextran–peptide–methotrexate conjugate 2556
- mitomycin C dextran conjugate 2557
- photoresists 2310
- prodrug conjugates 2555
- properties 2555
dialkyl esters 455
- lipase-catalyzed polycondensation 455
diamines 304, 306
- 2-(4-aminophenyl)ethylamine 306
diazonaphthoquinones (DNQs) 2305
diblock copolymers 336, 1344, 1403, 2615
- modifications 2615
- poly(acrylic acid)-b-polystyrene (PAA-b-PSt) synthetic amphiphilic diblock copolymers 1344
- polynorbornene diblock copolymers 2430
- self-assembly 1344
dibromomethane 297
dicarboxylic acids 454
- dehydration polycondensation 454
dicationic in-chain initiator 930
2,2-dichloro-1,3-benzodioxole 298, 309, 333, 921, 1023, 1373
- copolymerization 333
- fibrillar crystals 333
- microsphere 333
4,4′-dichlorodiphenylsulfone 325
dielectric structures 1497
- dielectric contrast 1497
- dielectric mirrors 1497
- photonic crystals 1497
- quarter-wave stacks 1497
dienes 32
- degree of aggregation 32
differential refractometer 1887
differential scanning calorimetry (DSC) 363, 1237, 1835, 1845, 1869
- heat flow-rate calibration 1837
- high-performance DSC 1869
- studies 363
- thin-film hot-stage 1869
differential thermal analysis (DTA) 1850
difunctional initiator (DI) 84–85, 842, 564, 840
- 1,3-bis(1-phenyl)benzene (PEB) 842

- Br–PMMA–Br macroinitiator 847
- sec-butyllithium (sBuLi) 842
difunctional monomers 93, 551, 911, 946–948
- DVB 927
- ethylene glycol dimethacrylate (EGDM) 947
- macrocyclic polymers 551
- methodology 928, 938, 943
- norbornadiene dimer 948
dihaloaromatic compounds 307
- polycondensation 307
dihalomethane 296
dihedral angle 1449
N,N-dihexadecyl,N'-3(triethoxysilyl)propylsuccinamide 1273
β-diketones 421
dimerization constant 357
dimethylsulfoxonium methylide 312, 334
dip pen nanolithography 416
- HRP-catalyzed polymerization 416
dip-pen nanolithography 1543, 1562–1563
- enzyme-assisted nanolithography 1563
- fountain pen technique 1563
- parallel DPN 1562
- tapping mode DPN 1562
direct imaging
- time sectioning 1665
- time-resolved cryo-TEM 1665
direct imaging cryo-TEM 1663
- blotting step 1664
- controlled-environment vitrification system (CEVS) 1664
- specimen preparation 1663
- thermal fixation 1663
- vitrification 1663
direct methanol fuel cell (DMFC) 2472
direct miniemulsion 1223
direct transformations 545
directed oligonucleotide hybridization 1331
directional complementary motif 354
discotic photoconductors 1957
- hexabenzocoronene (HBC) 1957
disordered isotropic systems 1589
- Ornstein-Zernike correlation function 1589
- scattering intensity profile 1589
- structure factor 1589
dispersion polymerization 276, 605, 1220–1221, 2204
- supercritical CO_2 1221
dispersion ring-opening polymerization 146–148

- mechanism 147
- particle nucleation 149

ditactic poly(α,β-disubstituted 3thylene)s 764
- disyndiotactic 767
- disyndiotactic ditacticity assignment 768
- erythrodiisotactic 767
- group transfer polymerization 767
- stereospecific polymerizations 764
- threodiisotactic 767

ditopic calixarene units 368
divinyl esters 457
- lipase-catalyzed polymerization 457
DNA expression 481
DNA recombinant technologie 519
DNA sensor 2529
DNA–polymer conjugates 1315, 1333, 1352
- DNA–poly(styrene) conjugate 1315
- hydrogels 1333
DNA-controlled structure formation 1332
dodecylbenzenesulfonic acid (DBSA) 1491
donor–acceptor block copolymers 2394
- C_{60}-modified PPV-PS block copolymer 2394
- collection efficiencies 2394
- photovoltaic performance 2394
- rod–coil polymer 2394
doping 2227
dormant species 1148
DOSY NMR spectroscopy 391
double hydrogen bond 356
double hydrophilic graft copolymers (DHGCs) 2615
- synthesis 2615
double network (DN) gels 2690, 2692, 2695, 2715
- BC–gelatin gel 2696
- inhomogeneity 2695
- PAMPS-PAAm gels 2692
- pin-on-flat wear testing 2715
- wear properties 2715
double quantum NMR 1942–1943
- sideband spectra 1943
double-head initiator 591
double-hydrophilic block copolymers (DHBCs)
- applications 2606, 2616
- block lengths 2607
- characteristics 2605
- complexation 2617
- crystal design 2623
- crystal enantiomers 2623
- crystal modification 2622
- crystal planes stabilization 2624
- design 2620
- DHBC-coated nanoparticles 2628, 2605–2608, 2614, 2616–2617, 2619–2620, 2623, 2628–2629
- drug delivery systems 2622
- EDTA functional groups 2623
- group transfer polymerization (GTP) 2612
- living anionic polymerization 2608
- living cationic polymerization 2612
- living NCA polymerizations 2614
- macro initiators 2612
- morphogenesis mechanisms 2628
- nanoreactors 2619
- "oriented attachment" 2627
- particle stabilization 2606, 2621
- pH-induced reversible micellization 2617
- PIC micelles 2617
- polymer adsorption 2620
- polymeric blocks coupling 2614
- radical polymerization 2612
- repeat units 2608
- rod–dumbbell–sphere morphogenesis 2629
- self-assembly of nanoparticles 2626
- solvent-free paints 2621
- surface recognition 2622
- surfactants 2619–2620
- synthesis 2608
double-hydrophilic block copolymers superstructure 2616
- micelle formation 2616
- water insolubility 2616
double-hydrophilic graft copolymers (DHGCs) 2605, 2616
- applications 2616
drug delivery 487–488, 1370
- applications 2622
- SELP hydrogels 488
drug delivery system (DDS) 1349, 2541–2543, 2547–2548, 2550, 2571, 2582
- biodegradability 2546
- camptothecin–PEG conjugate 2545
- carriers 2542
- controlled drug release 2571
- dendrimers 2547
- enhanced permeability and retention (EPR) effect 2545
- future progress 2548
- "ideal" 2582
- lipid-core micelles 2543

- loading capacity 2545
- polymeric controlled release 2571
- prodrug 2550
- solubility 2545
- spacer design 2546
- spacers 2545
- targeted polymeric delivery systems 2547
- water-soluble polymers 2543
drug delivery vehicles 2564
- dendrimers 2564
drug sensors 2512
- fluphenazine hydrochloride 2512
- nortriptyline hydrochloride 2512
- pethidine hydrochloride 2512
- phentermine 2512
dry brushes 1159
DSC
- heat flow-rate 1845
- latent heat 1845
- poly(ethylene-co-1-octene) 1845
dual core polymerization 660
dual functional initiators 553
dual living polymerizations 591
dumbbell polymers 973
Durham route 281
dynamic observables 1440
- mean-squared displacements 1440
dynamic properties 1441

e
Edwards' replica trick 2
elastic hydrogels 2732
elastin 484
- tropoelastin 484
elastin-like polypeptide copolymers 486
elastin-like protein polymers 485
- critical solution transition 485
- cross-linking 485
- thermal properties 485
elastin-like protein polymers (ELPs) 486, 488–489
- applications 488
- hydrogels 486
- purification processes 489
elastomers 1950
- properties 1950
elecron scattering 1652
- dark-field mode 1652
electric force microscopy 1523
- two-pass lift mode technique 1523
electroactive polymers 281
electrodialysis 2467
electroluminescence 2370

electron beam (EB) technology 665
- applications 665
electron beam lithography 2287
- molecular glass (MG) photoresists 2310
electron diffraction 1652
electron energy-loss spectroscopy (EELS) 1679–1680, 1682
- element-specific imaging 1682
- energy filter 1680
- structure-sensitive imaging 1681
- zero-loss imaging 1680
electron spin resonance (ESR) spectroscopy 821
electron transfer agent 18
electrorheological fluids (EFR)s 2109
- polyaniline/MMT nanocomposites 2110
electrostatic interactions 356
electrostriction 1162
elongation flow optorheometer 2105
emulsion polymerization 274, 626–633, 825, 1040, 1180
- ab initio batch emulsion polymerization 631
- ab initio emulsion polymerization 626–627
- applications 628
- atom-transfer radical polymerization 630
- cationic emulsion polymerization 633
- compartmentalization 628
- continuous monomer addition 629
- controlled/living free radical polymerization 629
- cross-linked polymers 633
- emulsion ROMP 274
- homogeneous nucleation mechanism 627
- homogeneous–coagulative nucleation 627
- ionic polymerization 632
- kinetics 628
- microprecipitation process 629
- molar mass distribution 628
- monomer droplet nucleation 629
- multifunctional monomers 1040
- nitroxide-mediated controlled free radical polymerization 629
- particle formation stage 627
- principle 626
- propylene sulfide 633
- reversible addition–fragmentation chain transfer 630
- ring-opening anionic emulsion polymerization 633
- ring-opening metathesis polymerization 632

- seeded emulsion polymerization   631, 627
- surfactant-free emulsion polymerization   631
- two-component initiating system   629

enantioselective membranes   532
- poly(L-glutamate) star polymers   532

enantioselective polymerization   446
encapsulating coatings   2410
end-capping agents   78, 110
end-functionalization   203, 1757
- ATRP polymers   203
- diphenyl chlorophosphate   110

end-functionalized polymers   580
- anionic polymerization   580

end-groups determination   110
end-to-end chain coupling
- bimolecular end-to-end cyclization   882
- unnimolecular end-to-end cyclization   886

energy-dispersive X-ray spectroscopy   1679
- composition analysis   1679

engineering plastics   139
engulfing   1211
- penetration   1211
- spreading   1211

enhancing agents   947
environmental monitoring   2529
enzymatic catalysis   401–402
- advantages   467
- characteristics   402
- disadvantages   467
- *in vitro* polymer synthesis   402

enzymatic polymerization   401
- *in vivo* enzymatic reactions   401

enzymatic synthesis   405, 1373
- advantages   405

enzyme model complexes   413
- tyrosinase model complex   413

enzyme-activated monomer (EM)   456
enzymes   404
- classification   404

epoxides, photopolymerization   658
epoxy resin   1655
equilibrium morphology   1211
2-ethyl-2-oxazoline   1978, 1983
- living cationic ring-opening polymerization   1978

ethylene   621, 632
- anionic oligomerization   12, 24
- catalytic polymerization   621, 632
- organic phase copolymerization   621

ethylene-based polymers   222
ethylene–propylene graft copolymers   241

ethylene–propylene rubber (EPR)   235
ethylene oxide (EO)   109
- polymerization   111

ethylene–propylene copolymers   235
- alternating copolymers   235
- EPR   235
- random copolymers   235

ethylene–propylene block copolymers   238
- ABA triblock copolymers   238
- heterogeneous catalyst   238
- properties   238

EUV lithography   2321
- block copolymer negative-tone resist   2321

exciplex emission   2379
excitons   2392
excluded volume interactions   358
experiments   1893, 1896
- low-angle light scattering (LALS)   1896
- particle scattering factor   1893
- scattering intensity   1893

exponentially growing films   1255, 1257–1260
- buildup   1255
- buildup mechanism   1255
- characteristic properties   1259
- cyclic voltammetry   1260
- diffusing polyelectrolyte   1255
- experimental facts   1257
- hyaluronan (HA)/poly(L-lysine) (PLL)   1255
- optical waveguide lightmode spectroscopy   1258
- swelling   1259
- viscoelastic properties   1260

extracellular matrix (ECM)   2666, 2698
extreme ultraviolet (EUV) lithography   2309
- line edge roughness (LER)   2309
- resists   2309

Eyring theory   1699

f

F–f curve   823
Fabry–Perot interferometer   2388
features   976
- step-growth polymerization   976

ferrocenophanes   693
- living photolytic anionic ROP   693

ferromagnetic nanoparticles   2437–2438, 2444
- cobalt nanoparticles   2437
- functionalization   2444
- iron nanoparticles   2438

fibrillogenesis 1325
fibrils 1324
– nanofibers 1324
– properties 1324
field-effect transistors (FETs) 1351
fine particles 1170–1171
– center-to-center distance D_p 1170
– colloidal crystals 1171
– gold nanoparticles (AuNP) 1171
– silica particle (SiP) 1168
– two-dimensional ordered arrays 1170
first-order phase transition 1833
fission–fusion process 1228
flavonoid polymers 414
flavonoids 417
– catechins 417
– oxidative polymerization 417
– pharmacological activity 417
Flory approximation 2
Flory distribution 303, 877
Flory–Erman theory 1712
Flory–Fox relation 996
Flory–Huggins interaction parameter 1388, 1344, 2315
Flory's scaling exponent 988
flow geometry 1615
– limiting cases 1615
fluorene 2238
– nematic liquid crystalline (LC) phases 2238
fluorescence detector 1356
fluorescence non-radiative energy transfer (NRET) 1241
fluoropolymers 2322
focused ion beam (FIB) sectioning 1654
foldamers 352
forced composition gradient copolymers 833
– control strategy 833
– *in situ* concentration measurements 834
Förster mechanism 2387
Förster-type energy transfer 1065
forward recoil spectrometry (FReS) 1545
Fourier transform infrared (FTIR) spectroscopy 1873, 1926
– chromatographic separation 1926
– commercial interface 1926
– temperature-resolved 1873
Fréchet-type dendritic azides 1362
free initiator 1144, 1169
– ethyl 2-bromoisobutylate (EBiB) 1169
free radical polymerization 3, 549, 568, 626

– emulsion polymerization 626
free radicals 162
freeze-fracture replication 1665–1666
– etching 1666
freezing water 2469
frequency modulation AFM 1528
friction force microscopy (FFM) 1556
fuel cells 1476, 2473, 2477, 2482, 2485
– aromatic PBIs 2485
– cross-linked polymer systems 2482
– membrane–electrode assembly (MEA) 2473
– membrane material 1476
– membranes 2472
– proton transport behavior 2477
– sulfonated aromatic polymers 2473
fullerene (C_{60}) 1019
fullerene-containing polymers 1360
functional biomimetic hybrid nanostructures 1364
functional brush copolymers 1356
functional capping agents 78
– 2-(dimethylamino)ethyl methacrylate 78
– 2-hydroxyethyl methacrylate (HEMA) 78
– 2-tributylstannylfuran 79
– allyltrimethylsilane (ATMS) 79
– organosilicon compounds 79
functional conjugated polymer assemblies 1349
functional dendrimers 1057
– architecture 1057
– generation number 1057
– molecular design 1057
– self-assembly 1057
functional initiator method 70, 76
– chlorosilyl functional initiators 71
– epoxide initiators 71
– ester functional initiators 71
– *p*-MeOSt 71
– *β*-pinene 71
– styrene 70
– vinyl ethers 70
functional initiators 202
functional macromolecules 1350, 1370
– assembly 1350
– degradability 1370
functional polyesters 459
– optically active polyester 459
– sugar-containing polyesters 459
– trehalose-containing polyester 459
α,ω-functional polymer precursor (FPP) 842
functional polymers 70, 139, 200, 1163

- application 70
- controlled/living radical polymerization 200
- end-functionalized polymers 70
- methods 70
functional self-assembled materials 1342
- applications 1347
functional terminator method 72
- functional PIBs 72
- hydrocarbon olefins 72
- *in situ* end functionalization 72
- *in situ* functionalization 73
- *p*-MeOSt 73
- β-pinene 77
- quantitative functionalization 72
- quenching 72
- vinyl ether polymerization 72
- vinyl ethers 77
functionality-type distribution (FTD) 1882
functionalization 46
functionalized polyesters 448
- epoxide-containing polyesters 461
- lipase-catalyzed polymerization 449
- reactive polyesters 461
- triblock polyesters 449
functionalized stars 913
- end-functionalized 913
- in-chain 913
fungicides sensors 2513
- desmetryn 2513
Fusarium oxysporum 1365

g
β-galactoside 426
gas chromatography (GC) 1983
gas sensors 2497–2499, 2501–2503
- alcohol 2497
- ammonia 2497
- amperometric 2498
- chemFET sensor array 2501
- chemical vapors 2499
- CO 2503
- CO_2 2503
- conducting polymer sensor arrays 2498
- conjugated polymer-based fluorescence turn-on sensors 2498
- H_2 2503
- H_2S 2501
- HCl 2497
- hydrocarbons 2504
- nitric oxide 2497
- NO_2 2497
- optical chemical sensor (OCS) 2503

- oxygen 2497
- ozone 2497
- polyaniline (PANI)-based gas sensors 2498
- polypyrrole film 2499
- QCM-type SO_2 gas sensors 2502
- quartz crystal microbalance (QCM) sensor 2499
- solid polymer electrolyte (SPE)-based electrochemical sensor 2503
- solid-state gas sensors 2498
- SO_2 2501
- surface acoustic wave (SAW) 2501
gas separation membranes 2452–2456, 2459–2461
- diffusion 2453
- diffusivity selectivity 2455
- dual-mode sorption model 2460
- gas permeability 2459
- gas permeation selectivity 2456
- gas transport model studies 2453
- separation mechanisms 2452
- solution–diffusion mechanism 2454
gas-barrier membranes 2463
gel friction 2699–2702
- colloidal probe atomic force microscopy analysis 2701
- electrostatic interaction 2700
- friction coefficient 2699
- opposing substrates 2699
- parameters 2699
- repulsion–adsorption model 2701
- template effect 2702
gel permeation chromatography (GPC) 1883–1884
gel permeation chromatography–multiple angle laser light scattering (GPC–MALLS) 327
gelatin 463, 465, 2696
- cross-linking 465
- gelatin–catechin conjugate 465
gellan gel 2704
Gibbs phase rule 1597, 1833
Gibbs–Duhem equation 1715
Gilch reaction 2372
Ginzburg region 1635
glass transition 1160
glass transition analysis 1862
glass transition temperatures 1209
glucose sensors 2521, 2523
- amperometric glucose oxidase enzyme electrode biosensor 2522
- aniline-based microsensors 2521

- conductive polymer-based nanocomposite 2523
- disposable glucose sensor 2524
- glucose-sensing cell 2524
- glucose-sensing enzyme electrodes 2521
- hemocompatible glucose sensors 2523
- interference-free amperometric biosensors 2522
- miniaturized glucose sensor 2523
- polyacrylonitrile thin-film composite membrane 2523
- redox polymer–oxidoreductase nanocomposite thin films 2522
glutenin 492
glycoproteins 2603
- lithostathine 2603
glycosaminoglycans (GAGs) 430, 435, 2603
- chondroitin (Ch) 430
- chondroitin sulfate (ChS) 430
- GAG families 435
- hyaluronan (hyaluronic acid; HA) 430
gold colloids 383
gold nanoparticles 2417, 2422, 2622
- block copolymers 2420
- dendritic surfactants 2422
- double-hydrophilic block copolymers (DHBCs) 2622
- functional ligands 2416
- micellar templates 2420
- polymer surfactants 2416
- stabilizers 2416
- surface-initiated polymerization 2417
- synthesis 2415–2416
- thiol-terminated polymers 2417
gradient brush/graft copolymers 1119
gradient chromatography 1918, 1920, 1931
- copolymers separation 1918
- critical conditions 1918
- ^{1}H NMR 1931
- polymer blend separation 1918
- standards 1920
- temperature gradient interaction chromatography (TGIC) 1920
gradient copolymer brushes 1106
gradient copolymers 199, 813, 830, 834, 836
- analysis 834
- controlled/living radical polymerization 830
- properties 836
A-g-B graft copolymers
- P2VP-g-PS copolymers 863
- pH-sensitive backbone 863

- PI-g-PS graft copolymers 868
- PMMA-g-PDMS graft copolymers 864
- PMMA-g-PS graft copolymers 863
- PnBA-g-PMMA graft copolymer 864
- PS-g-PMAA polymers 862
graft copolymerization 537
graft copolymers 47, 93, 189, 197–198, 204, 262–264, 266–267, 553, 572, 576, 581, 586–587, 589, 840, 862, 866, 1103, 1349, 1356, 1760, 1771–1772, 2604, 2608
- anionic polymerization 47
- anionic polymerization/ROMP 263
- applications 204, 1103
- ATRP/ROMP 265
- backbone 1103
- cationic polymerization/ROMP 265
- controlled/living radical polymerization 198
- double-hydrophilic graft copolymers (DHGCs) 2608
- grafting density 1772
- A-g-B graft copolymers 862
- "grafting from" method 262
- homopolymerization 198
- "macromonomer" method 262
- mechanical properties 198
- methacrylates 589
- molecular brushes 1103
- poly(chloroethyl vinyl ether) (PCEVE) 581
- poly(norbornene-g-methylmethacrylate) copolymers 265
- polyblend nanostructuring 1771
- polydispersity 1772
- polyethylene-based 866
- polypropylene (PP)-based 587
- PPP-type 553
- ROMP 262
- ROMP/ROMP 262
- ROMP/ROP/ATRP 267
- ROP/ROMP 266
- self-organization 1771
- side-chains 1103
- St graft propylene oxide (PO) copolymer 572
- supramolecular assemblies 1349
- synthesis 197
- unusual topologies 264
graft density 1137, 1139
- "concentrated" regime 1139
- "dilute" regime 1137
graft dormant 1142
graft polymerization 533, 1222

graft polymers 344, 1149
- compressibility 1152
- end-grafted polymers 1137
- interaction force 1152
- segmental density profile 1149
- surface-initiated polymerization 1137
graft polypeptides 533, 535
- block copolypeptide grafts 534
- degree of polymerization 534
- dendritic-graft polypeptides 535
- linear byproducts 534
- poly(L-lysine) backbone 535
graft polypropylene 233
- branched isotactic PP 234
graft terpolymer 2201
- synthesis 2201
graft-on-graft strategy 535
grafted polymer layer 1197
grafting approach 2372
"grafting from" 1107, 1363
- strategic combination 1363
"grafting from" methods 48, 1115, 1357
"grafting from" polymerization 1358
"grafting from" synthetic route 861
"grafting from" technique 94
"grafting on to" approach 1357
"grafting on to" methods 48, 860, 1362
grafting reactions 1188
- kinetics 1188
"grafting through" 1363
- strategic combination 1363
"grafting-through" approach 537
"grafting through" methods 1107, 1112, 1357, 1360
"grafting to" methods 1107, 1113
"grafting to" technique 94, 1137
"grafting-from" strategy 537
graphoepitaxy 1551
"green" chemistry process 611
green polymer chemistry 451, 467
Grignard metathesis (GRIM) method 1351
Grignard reagents 41, 324
Grignard thiophene 322, 337
- catalyst-transfer mechanism 323
- catalyst-transfer polycondensation 337
- polymerization study 322
GROMOS 96 force field 1448
group transfer polymerization (GTP) 29, 89, 589, 845, 945, 946, 2614
- advantages 845
- initiation process 2167
- mechanism 845, 2167
- silyl keten acetal activation 589

- tetrafunctional GTP initiator 946
growth factors 2723
- manipulation 2723
Grubbs catalyst 255, 278
Grubbs catalytic systems 948
Grubbs' third-generation catalyst 1361
guest molecules 1088
- Bengal Rose 1088
- p-nitrobenzoic acid 1088
Guinier equation 1589
gyroid structure 1475, 1477

h

^{1}H NMR spectroscopy 734, 736
- overlapped signals 736
- stereochemical configuration 734
2-hydroxyethyl methacrylate (HEMA) 1118, 1313
H-polymers 1624, 1626–1628
- arm fluctuation 1627
- dynamic processes 1626
- polydispersity 1628
- spectrum 1628
- tube model 1626
- volume fraction 1628
Hagahira–Sonogashira synthesis 2378
"hairy rod" polymers 1951
halogen functional polymer 789
halomethyl benzene reagents 1018
Hammett relation 9
head-to-head polyamide 305
- polycondensation 305
head-to-head polymer 304
head-to-tail poly(alkylthiophene) 308
head-to-tail polyamide 307
head-to-tail polymer 304, 307
head-to-tail polyurea 307
heat capacity 1841, 1848
- first-order phase transitions 1848
- frequency correction 1841
- large-amplitude motion 1848
- reversing heat capacity 1839, 1841, 1848
- thermodynamic contributions 1848
- total heat capacity 1839
heat flux calorimetry 1836
- calorimeters 1836
Heck reaction 2374
hematin 419
hematite 2621
- particle stabilization 2621
hemiisotactic polypropylene 226
- asymmetric catalyst 226
hemiisotactic PP homopolymerization 237

herbicide sensors
- imprinted polymer-modified hanging mercury drop electrode (HMDE) 2514
- MIP sensors 2514
- organophosphate-containing pesticides 2514

heteroarm star-like micelles 853
heteroarm stars 914
heterocyclic monomers 111, 878
- anionic polymerization 878
- cationic polymerization 878
- ring–linear chain equilibration reactions 878

a,ω-heterodifunctional polymer 887
heterogeneity 1881
- molar mass 1881
heterogeneous catalysts
- bis(arene) group IV catalysts 227
- group IV tetraalkyl complexes 227
- titanium tetrachloride catalyst 227
heterogeneous copolymers 822
heterogeneous polymerization 4, 605
heterogeneous solution process 462
heterogeneous Ziegler–Natta catalysts 218, 238
heterograft brush copolymers 1105, 1112
heterograft copolymers 1118
heterografted block copolymers 866
heterotactic polymers 758–760
- alternating copolymerization 759
- coheterotacticity 759
- cyclopolymerization 758
- definition 758
- heterotactic-specific radical polymerization 760
- preparation 758
- ring-opening polymerization 760
heterotactic-specific polymerization 758, 760–764
- ^{13}C NMR analysis 763
- concepts 761
- heterotactic living polymerization 760
- polymerization temperature 762
- propagating anions 762
- propagation process 761
- stereochemical defects 764
- stereoregulation 762
- supercritical fluid chromatographic (SFC) analysis 762
hexa-peri-hexabenzocoronene (HBC) 2242, 2244–2245, 2248–2249
- aligned 2244
- carbon nanotubes 2249

- dendronized HBC 2248
- HBC–peryleneimide hybrids 2248
- intracolumnar order, intracolumnar order 2245
- Langmuir–Blodgett films 2247
- zone-casting 2244
hexabenzocoronenes 1957
- columnar mesophases 1957
hexachlorocyclophosphazene 1019
hexafunctional initiators 937, 959
- Ru-centered complexes 959
high molecular weight polymer 161
high-density polyethylene (HDPE) 2342
high-performance dispersants 2160
high-performance liquid chromatography (HPLC) 1158, 1882
high-speed calorimetry 1868, 1870
- foil calorimeter 1869
- heating rates 1868
- super-fast chip calorimeter (SFCC) 1870
high-throughput screening 5, 1967, 1971–1972, 1983, 1985–1986, 1989–1990
- attenuated total reflection FTIR (ATR-FTIR) 1986
- chromatographic techniques 1983
- fiber-optic mid-IR 1985
- Fourier transform IR 1985
- MALDI-TOF-MS 1974
- matrices 1975
- mechanical properties 1990
- nanoindentation 1989
- optical methods 1983
- size-exclusion chromatography (SEC) 1972
high-vacuum techniques 524
highly oriented pyrolytic graphite (HOPG) 1129
Hofmeister series 1408
hollow spheres 1199
homogeneous binary homopolymer blends 1591
- binary phase diagrams 1593
- Flory–Huggins interaction parameters 1593
- scattering 1591
homogeneous catalysts 218, 227
- active species 218
- chain transfer mechanisms 220
- indenylfluorenyl metallocenes 228
- MAO 218, 227
- mechanism of initiation 218
- mechanism of propagation 219
- mechanisms of termination 220

homogeneous copolymers 822, 824
homogeneous multicomponent
   blends 1594
- composition 1594
- scattering profiles 1594
homogeneous Ziegler–Natta polymerization
   catalysts 217
homopolymers 89, 2371
- polyacetylene (PA) 2371
Horner–Wittig polycondensations 2373
horseradish peroxidase (HRP) 404
host defensive peptides (HDPs) 2670
host–guest chemistry 360, 1992
- axle component 360
- screening 1992
- wheel component 360
host–guest complexes 361
- crown ethers 361
- cucurbit[n]uril family 361
- cyclodextrin family 361
host–guest interactions 370
- main-chain polyrotaxanes 370
hot melt adhesives 2011
HPMA copolymer 2542
- drug carrier 2542
HRP-catalyzed polymerization 418
humidity sensors 2516–2518
- humidity-sensitive copolymers 2517
- ion conducting polymers 2516
- Methylene Blue 2518
- nanocomposite pellets 2518
- resistive-type humidity sensor 2518
- silane derivatives 2517
- surface-functionalized polymers 2517
hyaluronan (HA) 435, 1255, 1291
- exponentially growing films 1255
- HAase-catalyzed copolymerization 436
hyaluronidase 435
HYBRANES 2338
hybrid block copolymers 136, 523, 526
- peptide–synthetic 526
hybrid complex structures 1345
hybrid cure polymerization 660
hybrid graft copolymers 537
- poly(γ-benzyl-L-glutamate)-based 537
hybrid macromolecules 2645
hybrid materials 290
hybrid nano-objects 1106
- nanocylinders 1106
- nanowires 1106
hybrid nanocomposites 2411
hybrid organic inorganic objects 1179
hybrid organic–inorganic films 2360

hybrid polysaccharides 437
- cellulose–chitin hybrid 437
- ChiBs-catalyzed ring-opening polyaddition 437
- chitin–chitosan hybrid 437
- unnatural hybrid polysaccharides 437
hybrid sol–gel (HSG) materials 655
hybrid-type polysaccharides 428
- cellulose–chitin hybrid 429
- cellulose–xylan hybrid 429
hydrazine sensors 2515
hydroboration 312–313, 334
hydrogels 655, 1366, 2667, 2689–2690,
   2695–2696, 2698–2699, 2702, 2704–2713
- BC–gelatin double network (DN)
   gel 2696
- biomedical applications 2696, 2713
- κ-carrageenan 2704
- cell adherence 2707
- cell cultivation 2706
- cell density 2709
- cell growth modulation 2706
- cell spreading 2708
- cell–substrate interactions 2710
- cross-linking concentration 2708
- cross-linking process 2702
- "double network (DN) gel" 2690
- drag reduction effect 2705
- fracture energy 2690
- frictional behavior 2699, 2705
- gellan 2704
- hydrodynamic layer 2705
- inhomogeneity 2695
- load-bearing tissue replacement
   2712
- low-friction gels 2713
- lubrication 2698
- "nanocomposite (NC) gel" 2690
- mechanical toughness 2689
- photopolymerization 655
- PLA–PEO–PLA hydrogels 1375
- polymer brush properties 2705
- polymer dangling chains 2704
- properties 2689
- protein absorption 2712
- strong polyelectrolyte gels 2709, 2711
- substrate effect 2704
- substrates 2704
- surface charge density 2710
- surface friction 2698
- surface nature 2702
- surface properties 2689

- swelling 2702
- "topological (TP) gel" 2690
- weak polyelectrolyte gels 2708
- zeta potential 2710
hydrogen bond acceptor 356
hydrogen bond donor 356
hydrogen bonding 2240
hydrogen bonding groups 383
- diamidopyridine 383
- diaminotriazine 383
- thymine 383
hydrogen bonds 356
- energy 356
- forces 356
- thermodynamic stability 356
hydrogen-bonded complex 357
- free energies of complexation 357
- ureido-s-triazine (UTr) dimer 358
- ureidoguanosine 360
- ureidopyrimidinone (UPy) 358
hydrolases 422, 428–429
- genetically engineered 429
hydrolytic degradation 2734
hydrophobically modified polyelectrolytes 862
hydrosilylation 339
N-(2-hydroxypropyl)methacrylamide (HPMA) copolymers 2561, 2564
- N-(2-hydroxypropyl)methacrylamide (HPMA) copolymer–anticancer drug conjugates 2562, 2564
- Dox–HPMA copolymer conjugates 2564
- prodrug conjugates 2561
hydroxy telechelic PEB, functionalization 375
hydroxyapatite (HAP) 2625
face-selective adsorption 2625
hydroxyethylcellulose (HEC) 462
- lipase-catalyzed modification 462
12-hydroxystearic acid 2158
hyperbranched copolymers 1039
hyperbranched poly(ether ketone)s 330
hyperbranched polyamides 328
- amino-terminated 328
- carboxyl-terminated 328
- melt polycondensation 328
hyperbranched polycarbonates 327
hyperbranched polyesters 327
- AB_2 monomer 327
- aromatic-aliphatic polyester 327
hyperbranched polyethers 329
- aromatic–aliphatic 329
- poly(aryl ether)s 329

hyperbranched polymers 93–94, 191, 326, 581, 973, 976–979, 982, 987–990, 995–996, 998–999, 2349
- AB_2 condensation monomers 326
- addition reactions 978
- back-biting reaction 978
- chain-growth polymerization 976
- co-condensation 2349
- contraction factor 990
- core molecule 978
- degree of branching 990, 995
- dendrigraft polymers 987
- double-monomer methodologies 979
- draining parameter 988
- dynamics 1605
- elongational flows 998
- Flory–Fox parameter 992
- flow behavior 999
- glass transition 995
- grafting on to approach 976
- hyperbranched polyglycerol 986
- light scattering 992
- Mark–Houwink exponents 991
- Mark–Houwink plots 989
- molecular weight distribution 976, 999
- multiple grafting processes 987
- polyaddition 976
- polycondensation 976
- polydispersity 989
- power-law dependence 992
- properties 326
- radical polymerization 191
- relaxation processes 999
- rheology 1605
- SCROP 986
- self-condensing vinyl polymerization 94
- self-polycondensation 979
- side-reactions 978
- slow monomer addition 979
- solution behavior 987
- two monomer method 976
- types 977
hyperbranched polypeptides 533
hyperfunctional polymers 1335

i
IB 89
- block copolymers 82, 89
- sequential block copolymerization 83
imidazole 1955
- proton conductors 1955
immersion lithography 2289
immunoglobulin G (IgG) 2564

immunomicelles 2582
immunosensors, quartz crystal thickness–shear mode (TSM) biomimetic sensor 2529
imprint lithography 2290
in situ encapsulation 1180
in situ nanoparticle formation 2413
in vivo enzymatic reactions 402
- "key and lock" theory 402
- transition state 402
"in–out" procedure 926
indene 67
indenofluorene 2238–2239
- alkylation 2239
- nematic liquid crystalline phases 2238
indirect transformation 546
inductive effect 321
inelastic scattering 1653
inifer technique 57
iniferter radical polymerization 567
iniferter technique 562
inimers 982, 1188
- SCVP 1188
initiating dormant species 1148
initiating systems 61
initiator method 449
initiator monolayers 1185
initiator/co-initiator binary system 61
initiator/co-initiator system 60
initiators 20–21, 24, 27, 117, 119–121, 178, 1308, 2343
- Al(O^{i}Pr)$_3$ 119
- Al-based 117
- alkoxides 110
- *tert*-butyl hydroperoxide (*t*-BHP) 1217
- butyllithium 21
- carboxylates 110
- earth metal alkoxides 120
- ionic initiator 1217
- K$_2$S$_2$O$_8$ 1217
- lithium-based initiators 20, 27
- miktofunctional initiators 2344
- non-ionic initiator 1217
- primary amine hydrochlorides 523, 1308
- redox initiating system 1217
- R$_2$Mg initiators 24
- "single-site" initiators 117, 121
- Sn(Oct)$_2$ 119
- transition metal 1308
- transition metal complexes 522
initiators atom-transfer radical polymerization (ATRP) initiators 178
ink-jet printing 385

inorganic cores 1179, 1181
- gold (Au) nanoparticles 1181
- silica (SiO$_2$) nanoparticles 1181
inorganic hybrid micelles 1424
inorganic polymers 673
- metal-containing polymers 673
insecticide sensors 2513
- bulk acoustic wave (BAW) sensor 2513
instantaneous copolymer composition 822
insulating materials 2332–2333
- inorganic 2332
- organic high-temperature polymers 2332
- porous insulators 2333
integrins 2667
intelligent design 1341
inter-chain interactions 385
interaction chromatography 1906
interfacial initiator, PEGA200 1223
interpenetrating network (IPN) 2017
- morphology 2017
interpenetrating networks 1501
interpenetrating polymer networks 1724–1726
- phase separation 1725–1726
- properties 1726
- sequential method 1725
intrinsic behavior 1789
inverse miniemulsion 1223
iodine transfer polymerization (ITP) 803
ion-selective sensor devices 2505–2506
- Be^{2+}-selective PVC-based membrane electrode 2506
- Cs$^+$ ion-selective ionophores 2508
- cyanocopolymer matrices 2508
- divalent cation-selective electrode 2509
- heavy metal ion sensor 2508
- Hg(II) ion-selective PVC membrane sensor 2508
- ion-selective electrodes 2505
- ion-selective field-effect transistors (ISFETs) 2506
- ion-sensitive chemical transduction 2505
- ionophores 2506
- λ-MnO$_2$-based graphite–epoxy electrode 2505
- Ni^{2+}-selective electrode 2508
- PANI-based membrane 2505
- PVC matrix membrane sensors 2509
- PVC membrane nickel(II) ion-selective electrode 2509
- selective anions 2509
- selective cations 2505
ionic liquids 456

ionic polymerization 608, 622
- miniemulsion 622
- monomers 622
ionic ring-opening polymerization 103
ionic solvents 452
- 1-butyl-3-methylimidazolium salts 452
ionic surfactant 274
- dodecyltrimethylammonium bromide (DTAB) 274
- sodium dodecyl sulfate (SDS) 274
iron nanoparticles 2438
irreversible crystallization 1850
irreversible melting 1860
isobutene 62, 571, 845
- co-initiator 64
- kinetic studies 63
- polyisobutene (PIB) 63
- polymerization initiators 63
- protic initiation 63
isobutyl vinyl ether (IBVE), polymerization 566
isometric nanoparticles 2061
- covalent anchoring 2061
- homogeneous coating 2061
isometric or 0D nanoparticles 2034
isophthalic acid 305
isoprene 2002
N-isopropylacrylamide 1190, 2691
isotactic living polymerization 731
isotactic polypropylene 224
isotactic–stereospecific polymerizations 743, 746
- methacrylates 743
- propylene 746
isothermal microcalorimetry 360
isotropic solution 363
- supramolecular polymers 363
iterative tandem catalysis 447

j
Jacobson–Stockmayer dependence 1707
Jacobson–Stockmayer theory 139
Janus micelles 855, 1345
Janus-type structures 1106, 1119
joint confidence interval (JCI) 826
Jordharno's equation 1766

k
knitting pattern morphology 1661
Knoevenagel polycondensation 2373
Kraton copolymers 1661
Kuhn–Mark–Houwink–Sakurada scaling laws 987

l
laccases 419
lactam monomer 2040
- ring-opening polymerization (ROP) 2040
lactams 132–133
- activated monomer mechanism 133
- ring-opening polymerization 132
lactate biosensors 2527
lactides 115, 120, 122–123
- copolymerization 108
- polymerization 115, 120
- rac-LA 123
- stereocontrolled polymerization 123
- transesterification 123
lactones 115, 439–440, 442–443, 445–446, 448, 450, 457–458, 2145
- ε-caprolactone 439
- δ-valerolactone 439
- atom transfer radical polymerization (ATRP) 450
- enzymatic ring-opening polymerization (eROP) 450
- lipase-catalyzed copolymerization 457
- lipase CA-catalyzed copolymerization 443
- lipase CA-catalyzed enantioselective polymerization 446
- lipase CA-catalyzed polymerization 448
- lipase PF-catalyzed copolymerizations 445
- lipase–substrate interaction 458
- non-substituted 440
- polymerization 115
- reactivities 442
- ring-opening mechanism 2145
- ring-opening polymerization 439
β-lactosyl fluoride 426
Langmuir–Blodgett film techniques 2419
Langmuir–Blodgett monolayers 1128
Langmuir–Blodgett technique 655, 1143, 1491
large building blocks 353, 373
Lauritzen–Hoffman theory 1541
layer-by-layer self-assembly 419
layer-by-layer assembly 1249
layered nanoparticles 2057
- H-magadiite 2057
layered silicates 2038–2039
- interlayer spacing 2039
- phyllosilicates 2038
Leidenfrost effect 1664
Leuch's anhydride 1308
Lewis acids 15

ligand exchange methods, surface modification 2413
light absorption 645
light emitting diodes (LEDs), device parameters 2226
light scattering 1893, 1896
– off-line 1895
– on-line 1893
– Zimm plot 1893
light scattering experiment 1892, 1895, 1900
– copolymers 1900
– multi-angle light scattering (MALS) 1895
light source 644
– laser beams 644
light emitting diodes 2384
– development 2384
light-harvesting antenna 1065
light-harvesting dendrimers 1065
– conjugated dendrimers 1068
– cooperative energy migration 1071
– dendritic zinc porphyrin 1072
– electron transfer 1073
– energy gradient 1066
– energy transfer 1073
– fluorescence quantum yield 1068
– fluorescence-resonance energy transfer 1071
– intramolecular energy transfer 1065
– intramolecular hopping 1075
– low-energy photons 1071
– multichromophoric dendrimers 1065
– multiporphyrin dendrimers 1069
– non-conjugated dendrimers 1066
– photochemical hydrogen evolution 1076
– photodynamic therapy 1071
– photoinduced electron transfer 1072
– site-isolation effect 1068
– time-resolved fluorescence-anisotropy measurements 1069
– vibronic energy dissipation 1072
– zinc porphyrin-to-fullerene 1075
linear chains 188
linear diblock copolymers 81
– alkyl vinyl ethers 81
– crossover rate 81
linear high-density polyethylene 222
linear low-density polyethylene 222
linear main-chain polymers 353
– structural motif 353
linear multisegmental block copolymers 840, 859

– $(AB)_n$ multiblock copolymers 850
– ABA triblock 840
– ABC terpolymers 850
– ABCBA terpolymers 858
– ABCD quaterpolymers 859
linear polymers 973
linear rheology 1613
linear supramolecular polymers 362
linear triblock copolymers 84
– ABA-type 85
– difunctional initiation 84
– intermediate capping reaction 85
– poly(St-b-IB-b-St) triblock copolymer 85–86
– tensile strength 85
linear–ring chain equilibrates 876
linearly distributed molecular brushes 1106
linearly growing films 1257
– charge excess 1257
linking agents 859
– chloromethylphenylethenyldimethylchlorosilane 859
linking agent methodology 919
linking agents 917, 921–922, 926, 945
– 1,3-bis(1-phenylethenyl)benzene (MDDPE) 926
– 1,3,5-tris(bromomethyl)benzene 945
– bromomethylbenzene derivatives 921
– chloromethylbenzene derivative 921
– diphenylethylene derivatives 922, 926
– functionalized 1,1-diphenylethylene derivatives 925
lipase 439
– catalysis 401
– reaction mechanism 441
lipase PC 446
lipase-catalyzed ester hydrolysis 441
lipase-catalyzed polycondensation/transesterification 1373
lipase-catalyzed polymerization 441
lipase-catalyzed ring-opening polymerization 445
– mechanism 445, 442
– kinetics 442
lipid core micelles 2543
– applications 2543
liquid adsorption chromatography 1884, 1907, 1909–1910
– isocratic elution 1909
– liquid chromatography under critical conditions of adsorption 1884
– mass dependence 1884

- multiple attachment mechanism 1907
- non-uniformity 1908
- peak broadening 1910
- polydispersity 1908

liquid chromatography 1926, 1928, 1931
- FTIR spectroscopy 1926
- MALDI-TOF-MS 1931
- NMR spectroscopy 1928
- spectroscopic methods 1926

liquid core–polymer shell 1230
liquid crystalline block copolymers 564
liquid crystalline materials 380, 1092
- LC dendrimers 1092
- non-covalent side-chain functionalization 380

liquid crystalline (LC) phase transitions 1352
liquid crystalline polymers 280, 355
- nematic–isotropic transition 280
- side-chain liquid crystalline polymers (SC-LCPs) 280

liquid crystalline systems 1590
- scattering 1590

liquid-phase polymerization 1219
lithium hexamethyldisilazide (LHMDS) 318
lithographic techniques 1165
lithography 2295
- atomic force microscopy 1516
- "bottom-up" approach 2295
- electron beam technology 2287
- functional resists 2289
- immersion lithography 2289
- microlithographic applications 2282
- "top-down" approach 2295

living anionic polymerization 377, 557, 690–692, 2608
- block copolymers 691
- dianionic chain propagation 690
- ligand functional homopolymers 691
- mechanism 690
- metallomonomers 692

living cationic and radical polymerization combination 92
living cationic homopolymerization 92
living cationic polymerization 58, 61, 95, 338, 688
- additives 61
- commercialization 95
- equilibrium constants 59
- high-vacuum technique 60
- mechanism 688
- monomers 61, 689

- prior kinetics 59
- rate 59
- rigorous purification 61

living ionic polymerization 542, 777
- reaction scheme 777

living polyanions 1757
living polymer precursor 911
living polymerization 57, 312, 675, 1610, 1984
- diagnostic proof 58
- experimental criteria 58
- first-order kinetics 1984
- model polymers 58

living polymers 355
living radical polymerization 788, 2646
- atom-transfer radical polymerization (ATRP) 2646
- nitroxide-mediated polymerization (NMP) 2646
- quenchers 788
- reversible addition–fragmentation chain-transfer polymerization (RAFT) 2646

living sequential anionic polymerization 542
living/controlled polymerization 839
long-chain branches (LCBs) 223, 1946
lotus effect 1364
low-angle laser light scattering (LALLS) 329
low-conversion controlled/living radical polymerization 833
low-density polyethylene (LDPE) 222, 2342
lubricated compression 1784
luminescent rewritable paper 1090
luminescent sensors 2497

m

MA-type polymers 589
macro azo initiator 565
macro initiator syntheses 2612
macrocycles 273
- ROMP 273

macrocyclic molecular brushes 1114
macrocyclic polymerization to condensation polymerization transformation 551
macrocyclic polymers 49
- anionic polymerization 49

macroinitiator approach 849, 851
- polymerization mechanisms 851

macroinitiator method 341
macroinitiators 338, 528, 532, 558, 570–572, 579–580, 589-590, 841, 849, 856, 966, 1115–1116, 1120, 1185, 1313, 1360

- amine-terminated PMA macroinitiator 529
- ATRP-active halogen atoms 1116
- BPEM-based macroinitiators 1359
- brominated PIB-co-*p*-MeSt 571
- brush–linear macroinitiators 1363
- halo ester-based polyfunctional macroinitiators 1358
- initiator-modified particles 1185
- metallacycle substituted 528
- PBIPM 1359
- PDMS 579
- PEO–Br macroinitiator 856
- peptide macroinitiator 1313
- α,ω-polyethylene oxide 849
- polyferrocenylsilane-based ATRP macroinitiator 705
- polyfunctional anionic macroinitiator 1360
- poly(propylenimine) dendrimers 532
- poly(propylene oxide)-*p*-nitrobenzene sulfonate 580
- polystyrene-*b*-polyisoprene (PSt-*b*-PIP) 558
- polystyrene-based 528
- primary amine-functionalized poly(amidoamine) (PAMAM) dendrimer 532
- PS 966
- PSt 570, 558, 590
- silyl keten acetal 589
- star macroinitiators 1120

macrolides 441–442, 451
- 12-dodecanolide (DDL) 441
- 15-pentadecanolide (PDL) 441

macromolecular architecture 4, 2342
- porous structures 2342

macromolecular aromatic networks 2280
- SiLK 2280
- synthetic approaches 2280

macromolecular conjugates 1312

macromolecular engineering 1, 3, 46, 249, 839
- anionic polymerization 46
- applications 1
- cationic macromolecular engineering 57
- chemoselectivity 3
- degree of polymerization 46
- "functionalizing" reactions 46
- regioselectivity 3
- stereoselectivity 3

macromolecular therapeutic architectures 2656
- protein engineering 2656

macromolecules 126
- stereocomplexes 126

macromonomer technique 593

macromonomers 76–80, 241, 264, 266, 276, 410, 448, 451, 461, 537, 547, 775–777, 779–783, 785–787, 792–793, 796–797, 803–805, 851, 853, 861, 1109–1111, 1113, 2177
- addition–fragmentation processes 796
- alkenyl-type macromonomer 449
- allylic macromonomers 785
- anionic homopolymerization 1108
- ATRP 785
- cationic polymerizations 779
- chain-end halogen atom 786
- dendritic macromonomers 1111
- 1,1-diphenylethylene (DPE) 80
- direct chemical change 804
- functional initiators 779
- furan telechelic PIB 79
- G functionality 775
- heterografted brush macromolecules 782
- ω-hydroxy oligo polycaprolactone macromonomers 783
- induced deactivation 780
- initiators 785
- iodine transfer polymerization 803
- living polymerizations 777
- long-chain macromonomers 241
- methacryl-type polyester macromonomer 449
- methacrylate functional PIB macromonomers 77
- methacrylate functional poly(methylstyrene) 77
- methacrylate-based macromonomers 797
- methacryloyl-ended PEO macromonomers 1363
- ω-methacryloyloxy poly(styrene-*block*-vinyl-2-pyridine) macromonomers 781
- monofunctional chain polymers 547
- nitroxide-mediated radical polymerization (NMRP) 792
- norbornenyl macromonomers 1109
- norbornenyl-PEO macromonomer 276
- nucleophilic substitutions 788
- PEG macromonomers 1113
- PEO macromonomers 1109
- phenolic polymers 410
- β-pinene 77
- poly(ε-caprolactone) 266
- poly(ε-CL) 451

- poly(2-vinylpyridine) macromonomers 781
- poly(dimethylsiloxane) (PDMS) macromonomer 1113
- poly(ethylene oxide) (PEO) macromonomers 776
- poly(vinyl ether)s 78
- poly-THF macromonomers 783
- polydiene macromonomers 776
- polyester macromonomers 783
- polystyrene 77
- α position modification 792
- ω position modification 793
- PtBA macromonomer 853
- quencher agents 787
- R group 785
- radical addition 805
- radical addition reactions 787
- radical coupling 805
- radical homopolymerization 1110
- reversible addition–fragmentation transfer 798
- ring-opening metathesis polymerization 264, 1109
- styrene 79
- styrene end-functionalized PBLG 537
- synthesis 76, 775
- two-step procedures 783
macrophages 1264
macrophase morphology 378
macroinimers 985
magic angle spinning (MAS) 1941
- high-resolution MAS (HRMAS) 1941
magnetic materials 2434–2436
- antiferromagnetism 2435
- diamagnetism 2434
- ferrimagnetism 2435
- ferromagnetism 2434
- hysteresis 2435
- magnetization 2435
- paramagnetism 2434
- superparamagnetic materials 2436
magnetic nanoparticles 1194
- applications 1194
- iron oxide–polymer core–shell structures 1194
magnetic resonance imaging (MRI) 1671
magnetic rings 1200
main group element polymers 717
main-chain polymers 353
- bidirectional units 353
- linear main-chain polymers 353
- networks 353

main-chain supramolecular polymers 374
maleic anhydride (MAnh) 819
- copolymerizations 819
- cross-propagation 828
maltenes 2009
maltose 427
Mark–Houwink curves 989
Mark–Houwink–Sakurada equation 466
matrix-assisted laser desorption/ionization time-of-flight mass spectrometry (MALDI-TOF-MS) 1931, 1974–1975, 1979–1981
- chromatographic experiments 1931
- dried droplet (DD) 1974
- inkjet printing technology 1980
- matrix–solvent combinations 1981
- multiple-layer spotting technique 1974
- polymerization monitoring 1979
- seed-layer technique 1974, 324, 834
Maya blue 2072
Mayo–Lewis method 38
Mayo–Lewis model 814
- terminal model 814
mciroelectronics 2263
mean-field gelation 1612
mechanical network 1034
mechanical performance prediction 1783–1784, 1803–1804, 1806, 1812
- age-independent cavitation stres 1801
- ageing during processing 1796
- ageing kinetics 1786, 1794
- anisotropic crystal plasticity 1811
- anisotropic mechanical properties 1814
- craze-initiation 1799
- craze-initiation stress 1787
- critical cavitation stress 1799
- critical triaxial stress state 1787
- draw down ratios 1808
- elastic contribution 1785
- entanglement density 1786
- entropic spring 1790
- finite element analyses 1784
- geometrical softening 1803
- Hill's anisotropic definition 1815
- influence of flow 1808
- injection-molded product 1798
- intrinsic properties 1784
- intrinsic response 1789
- large-strain solid-state rheology 1803
- Leonov model 1812
- lubricated uniaxial compression tests 1806
- Maxwell element 1790
- mechanical rejuvenation 1792

- mold temperatures 1797
- molecular rheology 1812
- multi-level finite element method 1802
- multi-scale, micromechanical modelling approaches 1814
- multilevel finite element method 1799
- parallel contribution 1785
- PomPom model 1812
- rejuvenated state 1784
- slip-field theory 1799
- state parameter 1792
- strain-hardening modulus 1804
- strain-rate dependent response 1789
- stress-dependent viscosity 1789
- S_{J2} model 1812
- viscoelastic constitutive model 1812
- Weissenberg numbers 1812

melt blending 1753
melt polycondensation 327
melting temperature 1858
- Broadhurst equation 1858
- Gibbs–Thomson equation 1858
- macroconformation 1858

membrane separation technology 2451
p-MeOSt, end-functionalized polymers 71
a-MeS, star-block copolymers 931
aMeSt 85
mesomorphic molecular assemblies 380
meta-substituted monomers 321
metal template-assisted star polymer synthesis 956
- multifunctional metalloinitiators 956

metal-centered star copolymers 709
metal-containing block copolymers 673, 680, 702, 704, 712, 714
- applications 712
- nanolithography 714
- nanoporous thin film 712
- organic–organocobalt block copolymers 680
- phase-separated diblock copolymers 714
- Pt-containing block copolymers 702
- sequential polymerization 704

metal-containing diblock copolymer 705, 716
- SERS-active substrates 716
- silver-nanotextured surfaces 716

metallic nanoparticles 2414–2415
- gold nanoparticles 2415
- metal-to-insulator transition 2415

metallo block copolymers 679
ansa-metallocene systems 225

metallo-linked block copolymers 707
metallo-polymers 656
metallo-star polymer 709
metallocene catalysts 225
metallocene polymerization 543
metallocene-mediated olefin polymerization to anionic polymerization transformation 588

metallocenes 744, 747
- C_1-symmetric metallocenes 747
- C_2-symmetric ansa-metallocenes 747

metallocenophanes, anionic ROP 692
metalloinitiator method 958
metalloinitiators 709
- Zn-based metalloinitiators 709

metallomonomers 680, 684, 694, 702
- atom-transfer radical polymerization 684
- living anionic polymerization 694
- nitroxide-mediated radical polymerization 680

metallopolymers 673, 675
- "controlled" polymerization 675
- anionic polymerization 690
- applications 712
- atom-transfer radical polymerization 681
- chain growth polymerizations 674
- living cationic polymerizations 688
- living polymerizations 675
- metalloinitiators 689
- transition metal-containing polymer 689

metalloypolyerms, nitroxide-mediated radical polymerization 676
metathesis polymerization 696–697
- block copolymers 697
- metallomonomers 697

metathesis polymerization transformation 585
(meth)acrylic monomers 36, 41
- anionic polymerization 36
- stereospecific anionic polymerization 41

(meth)acrylic polymers 44
- termination reaction 44

(meth)acrylic-based copolymers 2208
- anionic polymerization 2208
- PMMA-b-PtBA block copolymers 2208

methacrylates 743, 745, 749–752, 754
- group transfer polymerization 751
- isotactic-specific living polymerization 743
- ligated anionic initiator system 745
- organotransition metal initiator 752
- radical polymerization mechanism 754
- stereoregular polymethacrylates 743

- syndiotactic-specific polymerization   749
- template polymerization   754
- trimethylsilyl methacrylate (TMSMA)   750

p-methoxystyrene (p-MeSt)   85, 546
p-methoxystyrene (p-MeOSt), living carbocationic polymerization   66
p-methoxystyrene (MeSt), block copolymers   545
3-methyl-4-oxa-6-hexanolide (MOHEL)   446
- enantioselective polymerization   446

a-methylenemacrolides   448
- chemoselective polymerization   448
- lipase-catalyzed polymerization   448

methyl methacrylate (MMA)   41, 89, 744, 1119, 1146, 2338
- isoregulation   41
- isotactic specific polymerization   744
- single-site polymerization catalysts   744
- surface-initiated ATRP   1146

methylaluminoxane (MAO)   218
methylsilsequioxane (MSSQ)   2335–2336, 2351
- phase diagram   2335
- prepolymers   2336

a-methylstyrene (aMeSt)   67
p-methylstyrene (p-MeSt), living carbocationic polymerization   65

micelles   1550
- AFM   1550
- block copolymer micelles   1550
- critical aggregate concentration (cac)   1949
- critical micellar concentration (cmc)   1949
- cryogenic TEM   1550
- morphology   1550
- properties   1550

micro-phase segregation   1351
micro-thermal analysis   1561
microarrays   2663
microcapsules   612–613, 654
- application   612
- inter-wall cross-linking   612
- kinetics   613
- photopolymerization   654
- porosity   613
- shell formation   612
- synthesis   612

microcontact printing (μCP)   2296–2297, 2299
- nanocomposites   2299
- nanocontact printing (nCP)   2300

- stamp material   2296
- substrate   2299
- surface-initiated polymerization (SIP)   2297
- "top-down" manner   2299

microelectromechanical systems   2571
microelectronic materials   2331
microemulsion polymerization   634, 635, 1221
- controlled free radical polymerization   634
- features   634
- ionic polymerization   635
- polycondensation   635
- vesicles   635

microemulsions   1041, 1589, 1597
- scattering   1597
- scattering profiles   1597
- Teubner–Strey intensity profile   1589

microfiltration   2467
microfluidic devices   2324
microgel nodules   911
microgels   192, 909, 1007–1008, 1022, 1032–1034, 1036–1038, 1040–1046, 1008
- anionic crosslinking copolymerization   1042
- applications   1046
- bulk process   1036
- characterization   1045
- core–shell structure   1040, 1044
- crosslinking density   1034
- doubly stimuli-responsive microgels   1044
- emulsion RCC   1040
- hydrophilic microgels   1044, 1046
- "initiator-fragment incorporation radical polymerization" (IFIRP)   1037
- in situ real-time monitoring   1045
- intramolecular crosslinking   1043
- inverse emulsion   1041
- ionizing radiation   1043
- "local molecular clusters"   1034
- microemulsion   1041
- monomer-free approach   1042
- "morphological map"   1037
- one-pot processes   1008
- radical crosslinking copolymerization (RCC)   1034
- rheology   1045
- shell-crossliked micelles   1033
- solution RCC   1037
- solvent-free processes   1036
- solvents   1037

- step-growth crosslinking copolymerization 1042
- stimuli-responsive microgels 1044
- surfactant-free precipitation 1041
- swollen microgels 1033
- synthesis 1033
- viscosimetry 1045

microlens arrays 1368
microlithography 2331
microphase separated multicomponent phases 1596
- compositions 1596
- Fourier analysis 1602
- neutron scattering 1596
- scattering intensity 1601
- scattering length density profiles 1599
- scattering profiles 1596
- SCFT calculations 1599
- self-consistent field theory (SCFT) 1599

microphase separation 378
microscopy 1234
- atomic force microscopy (AFM) 1234
- scanning electron microscopy (SEM) 1234
- transmission electron microscopy (TEM) 1234

microspheres, organic–inorganic hybrid microspheres 526
microsuspension polymerization 615
microwave polymerization 666
miktoarm polymers 2344
miktoarm stars 914
Millipede Project 1563
miniemulsion polymerizations 278, 613–615, 617–619, 620–625
- atom-transfer radical polymerization 617
- n-butyl cyanoacrylate (BCA) 624
- catalytic polymerization 621
- chain transfer 623
- cobalt-mediated miniemulsion polymerization 620
- colloidal state 623
- complex architectures 618
- controlled/living free radical polymerization 615
- copper–complex catalyst 618
- "critical DP limit" 623
- cross-linked nanocapsules 625
- cyclosiloxanes 624
- direct atom-transfer radical polymerization (ATRP) 618
- droplet nucleation 614
- free radical polymerization 613
- hydrolysis–condensation 625
- iodine-transfer polymerization (ITP) 619
- ionic polymerization 622
- kinetics 615
- metastable miniemulsion 614
- nitroxide-mediated controlled free radical polymerization 615
- olefin polymerization 621
- polycondensation 624–625
- polyaddition 625
- principle 613
- RAFT 619
- rate retardation 620
- reverse atom-transfer radical polymerization# (ATRP) 618
- reversible chain transfer 619
- ring-opening metathesis polymerization 620
- surfactant 623
- ultrahydrophobe 614

miniemulsion process 1226
- encapsulation 1226

mixed homopolymer brushes 1166
- morphologies 1166

mobile amorphous fraction (MAF) 1862
model discrimination 819, 821
- comonomer feed ratio 821

model graft-like block copolymers 868
- barbed-wire (hexafunctional) 868
- centipede (tetrafunctional) 868
- H-shaped $(PS)_2$–PEO–$(PS)_2$ 869

modified polysaccharides
- amylose 462
- carboxymethylcellulose 462
- cellulose acetate 462
- chitosan 463

modulated-temperature differential scanning calorimetry (M-TDSC) 1241
moisture vapor transition rate (MVTR) 1988
molar mass–chemical composition distribution (MMCCD) 830
molar weight distribution (MWD) 1882
Moldflow simulation 1796
molecular brushes 204, 860, 1103–1110, 1112–1114, 1116, 1118, 1120–1126, 1128–1130, 1132, 1134
- anionic homopolymerization 1108
- applications 204
- block copolymers 1116
- brush–substrate interactions 1129
- bulk properties 1126
- classifications 1104

- copolymerization 1112
- double-graft brushes 1126
- grafting from method 1115
- grafting to reactions 1113
- homopolymer side-chains 1121
- initiation efficiency 1124
- interfacial properties 1127
- linearly distributed copolymers 1105
- long-range/epitaxial ordering 1129
- macroinitiators 1116
- macroscopic gelation 1123
- precautions 1123
- radially distributed brushes 1105
- radical homopolymerization 1110
- rod-to-globule transitions 1125
- side-chain attachment 1113
- solution properties 1124
- surface conformation 1127
- synthesis 1107
- thin-film spreading 1129
- topology 1104
molecular chimeras 136, 1309
molecular crystals 355
molecular dynamics (MD) method 1435
- ensembles 1436
- straightforward approach 1435
- Verlet algorithm 1435
molecular dynamics (MD) simulation 1432
molecular electronics 2254–2255
- chemical sensors 2255
- prospects 2254
- transistor properties 2255
molecular force spectroscopy 1558
molecular glass (MG) photoresists 2310–2311
- synthesis 2311
molecular hybrids 199, 205
- applications 205
- atom-transfer radical polymerization (ATRP) 200
- functional groups 200
- organic–inorganic hybrids 199
molecular modeling 2237
molecular visualization 1526–1527, 1530, 1532, 1536
- dissociation reactions 1536
- molecular characterization 1532
- single molecules visualization 1529
- spreading behavior 1546
- structure verification 1530
- synthetic polymers 1529
- true molecular resolution 1527
molecular weight control 310

molecular wires 2255
Money–Rivlin dependence 1713
monoclonal antibodies 2654
monocytes 1286
monodispersed biopolymers 311
monofunctional chain stoppers 354
monofunctional initiators 21, 344, 841
- cumyl carbanions 21
monohydroxy polyacrylates 2146
monolithic supports 283
monomer feed composition 824
monomer purification 354
monomer reactivity control (stoichiometric-imbalanced polycondensation) 295
monomers 7, 12, 20, 61, 166, 174, 177, 184, 677
- anionic polymerization 7
- carbon–carbon double bond 7
- cationic 650
- charge density 12
- methacrylic monomers 20
- nitrogen-containing 177
- radical 650
- reversible addition-fragmentation transfer (RAFT) polymerization 184
- spin coating 1251
- spraying 1251
- styrenyl 677
Monte Carlo (MC) method 1435–1438
- classical Monte Carlo approach 1436
- local particle move 1438
- Markov process 1438
- Metropolis criterion 1437
Monte Carlo (MC) simulation 1432
montmorillonite 2038, 2041, 2074
- polymerization initiators 2041
Morton's equation 627
motif 1326
- β-sheet fibrils 1326
multi-angle laser light scattering (MALLS) 1533
multi-arm star (co)polymers 93
multi-block copolymers 378
- non-covalent interactions 378
multiarm star polymers 1009
- arm-first approach 1009
- macromonomer method 1009
- synthesis 1009
multiblock copolymers 1343
multiblock protein-based 481
multiblock stereocopolymers 124
- chain-end control mechanism (CEM) 124

- site-control mechanism (SCM) 124
multichain polyamino acids 533
multichain polypeptides 534
- random copolymers 534
multicomponent copolymers 565
- polyisobutene (PIB) 565
multicomponent materials 1475
- conducting channels 1475
multicomponent polymer mixtures 1584
- coherent scattering profile 1586
- Gaussian homopolymer chain 1586
- homogeneous multicomponent mixtures 1584
- homopolymer–diblock copolymer blend 1587
- incompressibility constraint 1585
- interaction matrix 1585
- structure factor 1585
- triblock copolymer 1585
multidrug resistance 2581
multidrug-resistant cells 2580
multifunctional antagonist reagents 1017
multifunctional ATRP initiators 938
multifunctional initiating systems 859
multifunctional initiator approach 915, 952
multifunctional initiator method 945
multifunctional initiators 25, 74, 86, 93, 533, 591, 911, 936, 960, 1027, 1116
- anionic multifunctional initiators 93
- arb-PIB macroinitiators 86
- calix[8]arene-based initiator 86
- epoxy-type initiators 1027
- hexaepoxysqualene (HES) 86
- multifunctional initiator method 93
- pentaerythritol 960
- poly(aMeSt-b-IB) macroinitiator 92
- poly(DL-ornithine) 533
- poly(L-lysine) 533
- polymer linking method 93
- 1,3,5-tris(a-methoxybenzyl)benzene 25
multifunctional linking agents 910
multifunctional macroinitiators 1024
- hyperbranched polyglycerols 1024
multifunctional monomers 354, 406
- oxidative polymerization 406
multifunctional polymers 74
multifunctional terminator 93
- multifunctional terminator method 93
multigraft copolymers 1769
- reactive blending 1769
multilayer devices 2227

multilayer films 1253–1254, 1282, 1287, 1290, 1296
- active agents 1283
- active drugs embedding 1287
- active peptide embedding 1282
- bioactive films 1265
- cell adhesive properties 1291
- cellular activity 1264
- chemical cross-linking 1292
- compartmentation 1265
- composition 1262
- cross-linking extent 1296
- cross-linked polysaccharide multilayers 1290
- D-molecules 1264
- exponentially growing films 1250
- film stiffness 1292
- film surface chemistry 1291
- functionalized coatings 1282
- gold nanoparticles 1253
- hyaluronan-containing films 1291
- layer-by-layer deposition 1253
- linearly growing films 1250
- multi-material films 1283
- nanoparticle templates 1253
- polyelectrolyte degradation 1280
- β-sheet content 1263
- solution coupling 1283
- solution dipping 1251
- surface functionalization technique 1283
multilayer formation 1249
- zeta potential 1249
multiscale micromechanical modeling 2088
multisegmental block/graft copolymers 839
multisegmental graft copolymers 860
- synthetic strategies 860
multisite mesogens 362
- main-chain supramolecular polymers 362
multiwall carbon nanotubes (MWNTs) 2059
"mushroom" conformation 1138
mushroom-to-brush crossover 1156
mussel adhesive plaque protein (MAP) 491
mutated enzymes 429, 432

n

nacre 2597, 2600
- synthetic 2600
Nafion 1478, 1988, 2475, 2493
nano building blocks 1179

Subject Index

nano-clay 2090
nano-dispersion 2334
nano-filler surfaces 2091
nano-lithography 1394
nano-objects 1343
nano-organization 901
nano-patterning 1394
nanobarriers 2066
nanocapsules 635, 1231–1232
- chitosan-containing nanocapsules 1231
- nanoprecipitation 1231
- polyalkylcyanoacrylate (PACA) nanocapsules 1232
nanocomposite gels 2690–2692
- synthesis 2692
nanocomposite foam 2124, 2126–2127
- cell wall thickness 2127
- interfacial tension 2125
- structure 2124
- bubble nucleation 2126
nanocomposites 5, 654, 1123, 1372, 2033–2034, 2036–2037, 2039, 2041–2044, 2046–2050, 2054, 2057, 2066–2067, 2071, 2093, 2333, 2338, 2351, 2423, 2431–2432
- applications 2065
- bicationic initiator-based clay platelets 2042
- CdSe-conjugated polymer nanocomposites 2432
- clay–polyamide 6 nanocomposite 2065
- composite morphologies 2043
- controlled grafting 2047
- controlled polymerization processes 2050
- definitions 2034
- exfoliated PCL organoclay nanocomposites 2046
- exfoliated structure 2039
- flame retardancy 2067
- free radical polymerization 2041
- gas permeation 2048
- "grafting through" technique 2041, 2054
- hydrothermal synthesis 2036
- in situ "coordination–insertion" polymerization 2056
- in situ ROP reactions 2041
- intercalated structure 2039
- ionic-like processes 2055
- masterbatch 2037, 2049
- melt blending 2036
- methylsilsequioxane (MSSQ) 2351
- monocationic initiator-based clay platelets 2042
- monomer-to-clay ratio 2044
- monomer-to-initiator ratio 2055
- montmorillonite (MMT) 2071
- Nylon-based nanocomposites 2093
- organomodifier 2053
- particulate fillers 2033
- phase-separated composite 2039
- photopolymerization 654
- PLA-coated cellulose whisker composites 1372
- polyamide–clay nanocomposites 2066
- polydithiafulvene–Au nanoparticle composites 2423
- polyester-grafted nanoclay 2049
- polymer–carbon nanotube nanocomposites 2059
- polymer/layered silicate nanocomposites 2071
- polypropylene–nanoclay nanocomposite 2065
- polysaccharides 1372
- preparation 2423
- production pathways 2036
- properties 2037
- PS–magadiite nanocomposites 2057
- quantum dot–P3HT nanocomposites 2431
- radical processes 2050
- silica-based PS nanocomposites 2061
- structures 2034
- tacticity control 2056
- thermoplastic polyolefin (TPO)-clay nanocomposites 2065
- types 2034
- water permeability 2048
nanocontainers 1349
nanofiltration polymer membranes 2467
- non-porous membrane 2467
nanogels 1039
- direct synthesis 1039
nanohybrids 1249, 2034, 2036–2037, 2044, 2046, 2049, 2336, 2355
- definitions 2034
- "grafting from" process 2037
- "grafting on to" process 2036
- masterbatches 2049
- organic–inorganic nanohybrid materials 2046
- PCL–montmorillonite nanohybrids 2044
- synthesis 2034
- thermoplastic matrices 2049
nanolithography 1348
nanoparticle systems 1395
- ex situ synthesis 1395

nanoparticle–polymer composites 2409
nanoparticle-filled BCP systems 1395, 1397, 1399
- Monte Carlo simulation 1399
- phase behavior 1397
nanoparticle-polymer composites 2418
nanoparticles 656, 1193, 1197, 1226, 1228–1230, 1253, 1354, 1424, 2033–2034, 2037–2038, 2352, 2409, 2413, 2420–2421, 2423, 2607, 2626, 2628, 2631
- adhesive properties 1199
- block copolymer-encapsulated nanoparticles 2420
- block copolymers 1199
- "bricks and mortar" approach 2421
- CdS–ZnS-coated CdSe quantum dots 1228
- CdSe nanoparticles 1193
- characteristics 2037
- conjugated polymers 2423
- core/shell nanoparticle suspensions 1253
- encapsulation 2411
- face-selective adsorption 2631
- Fe_3O_4 nanoparticle 1194
- functionalization 2412
- gold nanoparticles 1230, 2061
- hydrodynamic radii 2352
- hydrophilic $CaCO_3$ 1227
- hydroxyapatite particles 1197
- inorganic nanoparticles 2412
- *in situ* particle reactions 2412
- Laponite nanoparticles 1230
- layered silicates 2038
- ligand exchange 2412
- living anionic surface-initiated polymerization (LASIP) 1191
- magnetic nanoparticles 1226
- magnetite particles 1229
- "oriented attachment" 2627
- particle stabilization 2607
- PNB nanoparticles 279
- polyamidoamine (PAMAM) 2411
- polyethylene glycol (PEG)-coated silica particles 1197
- polymer surfactants 2411
- ring-opening metathesis polymerizations (ROMP) 1191
- self-assembly of nanoparticles 2626
- semiconductor nanoparticles 2413
- silica-coated gold nanoparticles 2061
- silver particles 1226
- stabilization 2626
- surface functionalization 1253
- surface-initiated 2412
- templates 1253
- TiO_2 particles 1227
nanoporous silica 2338
nanoreactor
nanoreactors 1958, 2411, 2619
- metal-coordinated micelles 2619
nanorods 364
- cyanuric acid–melamine motifs 364
nanostripes 2237
nanostructured polyblends 1771
- AB symmetric copolymer/A homopolymer type 1771
- co-continuous nanostructures 1773
- reactive blending 1771
nanostructures 897, 1343, 1345, 1347
- bamboo-like structures 1345
- bowl-shaped micelles 1345
- "smart" nanostructures 1347
- "tagged" nano-objects 1347
nanotribology 1556
nanotubes 364, 1355
- bidirectional cyclic peptides 364
- chiral polyaniline (PANI) nanotubes 1355
- hollow polypyrrole nanotubes 1355
nanovessels 1349
naphthalene 18
naphthalene dicarboxymonoimide (NDI) 1066
2-naphthol 412
- oxidative polymerization 412
native chemical ligation 496, 1311
natural rubber 226, 420, 2003
- synthesis 420
natural self-interaction 1307
- reversible, secondary interaction 1307
NCA polymerizations 520–522, 524, 529, 533, 538
- activated-monomer mechanism 521
- amine mechanism 521
- basic initiators 521
- controlled NCA polymerization technique 533
- history 520
- "living" polymerization 522
- primary amine-initiated 524
- techniques 538
- transition metal-mediated 522
NCA ring-opening polymerization 520, 522–523, 534, 536
- competing pathways 522

- end-functionalization reactions 536
- initiators 520
- multifunctional initiators 534
negative photoresists 2285
negative staining 1235
neoglycopolymers 286
Nephila clavipes 481
- spidroin 481
network formation theories 1704
- application 1709
- cascade theory 1706
- computer simulation 1705
- kinetic theories 1704
- perturbation methods 1706
- recursive theory 1706
- reference system 1706
- statistical theories 1704
nickel catalysts 223
nitrogenous porogens 2338
- dendritic amidoamines 2338
- hyperbranched polyester–amides 2338
- nanoporous silica 2338
nitroxide-based regulators, structure–activity relationships 2173
nitroxide-mediated controlled free radical polymerizations 615, 629, 867
- emulsion polymerization 629
- miniemulsion polymerization 615, 607, 616
- kinetics 616
- partition coefficient 616
- radical initiators 616
- suspension polymerization 607
nitroxide-mediated polymerizations (NMP) 173–175, 556, 792, 829, 831, 933–934, 1150, 2053
- clay nanoplatelets 2053
- conditions 175
- controlled architectures 175
- DEPN 174
- graft density 1150
- β-hydrogen-containing nitroxides 934
- initiators 174
- monomers 174
- nitroxides 173
- TEMPO 173
- TIPNO 174
nitroxide-mediated radical polymerizations 676–678, 680
- ligand functional block copolymers 678
- ligand functional homopolymers 677
- metallomonomers 680
- styrene 677, 676

nitroxyl radical-mediated polymerization (NMP) 1188
nitroxyl radicals 556, 558, 568
- side groups 558
- 2,2,6,6-tetramethylpiperidine-1-oxyl (TEMPO) 556
nitroxyl-mediated radical polymerization 2172–2173
- mechanism 2173
- polymerization regulators 2172
NMR spectroscopy 1937–1938, 1940–1941, 1944–1945, 1947, 1950, 1952, 1954–1955, 1962
- anisotropic spin interactions 1938
- applications 1945
- background 1938
- chain dynamics 1947
- chemical shift 1938, 1940
- J-coupling 1938
- dielectric spectroscopy 1952
- dipole–dipole interactions 1940
- end-groups 734
- features 1938
- heteronuclear dipolar decoupling 1941
- heteronuclear interaction 1940
- high-resolution NMR 1937
- homonuclear interaction 1940
- magic angle spinning (MAS) 1941
- Pake spectrum 1940
- α-process 1952
- β-process 1952
- proton conductors 1955
- quadrupole interaction 1940
- rotational-echo double-resonance (REDOR) technique 1941
- solid-state NMR 1937
- supramolecular systems 1954
- two-dimensional NMR spectroscopy 1944
- viscoelastic materials 1950
NMR spin-diffusion techniques 1239
noble metal nanocomposites 2414
non(homo)polymerizable bis-DPE compounds 75
non-covalent interactions 351, 355–356
- advantages 356
- Coulombic interactions 355
- hydrogen bonding 351
- hydrophilic interactions 351
- hydrophobic interactions 351
- ionic interactions 351
- limitations 356
- metal–ligand coordination 351

- solvophobic interactions 355
non-fouling coatings 2669
non-freezing bound water 2476
non-freezing water 2469
non-isothermal crystallization 1849
non-isothermal recrystallization 1853
non-linear polymeric architecture 932
non-linear rheology 1615
non-natural amino acid in corporation 496
- functional groups 496
- synthetic methodologies 496
non-natural amino acid incorporation 498–499, 500–501, 503
- $in\ vivo$ strategies 498
- aaRS mutation 501
- auxotrophic expression 500
- bacterial expression hosts 500
- codon reassignment 503
- methionine analogues 501
- multisite incorporation 499
- purified translation system 498
- requirements 499
- site-specific incorporation 498
- suppression-based methods 498
non-overlapping hard objects 1588
- non-interacting hard spheres 1588
- organization 1588
- scattering 1588
non-porous membranes 2453
nonaqueous heterogeneous polymerizations 2204
nonlinear optical materials 1499
- chromophores 1499
norbornene 253, 696, 1193
- nanoparticle surfaces 1193
- ROMP 253, 696–698
norbornene derivatives, fullerene C_{60}-based norbornene derivative 1360
norbornenylpolyphosphazenes 265
Novozyme 435 1373
nuclear magnetic resonance (NMR) 1237
- high-resolution NMR 1237
- solid-state NMR 1237
nuclear magnetic resonance (NMR) spectroscopy 1928
- liquid chromatography 1928
nucleation–elongation polycondensation 302
nucleophilic initiators, nucleophilicity/basicity ratio 521
nucleophilic substitution 789
nucleophilicity parameter 81

Nylon 6-based nanocomposite (N6CN3.7) 2092–2093
- growth process 2092
- heat distortion temperature 2093
- γ-phase 2092
- structure 2092
- transcrystallization 2093

o

observables computer simulations 1438
octadecyl methacrylate (ODMA) 1118
octafunctional initiators 937
- 2-bromopropionate-modified calixarene 937
Ohm's law 2265
oil gels 2024
olefin metathesis 249–250
- mechanism 250
olefin polymerization 219
- metal halide complexes 219
- neutral metal alkyl complexes 219
olefins 218
- coordination polymerization 218
α-olefins 312
oligo(ethylene glycol) methyl methacrylate (OEGMA) 1149
- surface-initiated ATRP 1149
oligonucleotides 369
oligopeptide macroinitiators 1313
on-chip wiring 2331
- back-end-of-the line (BEOL) 2331
- front-end-of-the line (FEOL) 2331
- insulator 2331
optical ellipsometry 1546
optical microscopy 1234
optical screening methods 1983
opto-electronic devices 1351, 1498
optoelectronics 2384
ordered arrays 1590
- order–disorder transition 1590
- scattering 1590
organic electronics 2225
- classes of molecules 2227
organic light-emitting diodes (OLEDs) 2263, 2270
"organic metal" 2272
organic nanoparticles 2352
- formation 2352
organic nanotubes 854
- (PI–PCEMA–PtBA) terpolymers 854
organic oxidants 584
- onium salts 584
organic pigments 2136

– classes 2136
organic radicals 162
organic semiconductors 2225
organic–inorganic hybrid materials 1959–1960, 2409
– aluminosilicate–PI-b-PEO hybrids 1960
– double quantum 1959
– NMR spin-diffusion experiments 1960
– ^{29}Si NMR correlation experiments 1959
organic–inorganic hybrids 1348
organic–inorganic thin films 2343
organoclay 2039, 2072
organosilicate thin films 2360
– Young's modulus 2360
organosilicates 2332, 2335
– nanoporous organosilicates 2335
– nucleation and growth process 2335
organotellurium-mediated radical polymerization (TERP) 1142
orientational memory 1953
Ornstein–Zernike scattering equation 1589
OsO$_4$ 1659
Ostwald ripening 278
oxanorbornene 700
– ROMP 700
oxazoline monomers 436
oxazolines 145, 1984
– cationic polymerization 145
– cationic ring-opening polymerization 1984
oxidoreductases 404
– laccases 404
– peroxidases 404
– tyrosinase 404
oxiranes 111, 113, 139, 1110
– anionic polymerization 1110
– branched polyoxiranes 129
– cationic ROP 139
– chiral oxiranes 113
– ring-opening polymerization 111, 129
– star-shaped polyoxiranes 129
– stereocontrolled polymerization 113
oxolane 779
oxyacid derivatives 458
– lipase-catalyzed polymerization 458
– methyl 6-hydroxyhexanoate 458
– ricinoleic acid 458

p

π-stacking 2242
P(CL-co-PLLA)-perfluoropolyether composite materials 1375
PAA-based copolymers 2193
– *ab initio* emulsion polymerization 2195
– chain transfer agent 2194
– P*n*BA-*b*-PAA copolymer 2195
– PS-*b*-PAA block copolymers 2196
– star-block copolymers 2196
palladium catalysts 223
PAMAM dendrimers 2564
para-substituted monomers 316
parallel polymer synthesis 1969
particle morphology 1212
– core–shell 1212
– hemispherical 1212
– sandwich 1212
particle replication in non-wetting templates (PRINT) 2302
particle–particle coupling 1187
PCBs, salt-induced phase transitions 1418
PCL 556
Pd-catalyzed polycondensation 298
PEGylation 2655, 2656
pentablock copolymers 530
– PEG-capped CABAC-type hybrid pentablock copolymers 529
penultimate unit model (PUM) 816–818
– conditional probabilities 817
– implicit penultimate unit effect (EPUE) 818
– propagation reactions 816
PEO macromonomers 782
– deactivation 782
– ω-hydroxypoly(ethylene oxide) macromonomers 782
– ionic initiation 778
PEO methacrylates 1196
PEO-based block copolymers 2192
PEO-based copolymers 2183
– anionic polymerization 2184
– free-radical polymerization 2184
– PEO-*b*-PPO block copolymers 2183
– PEO-*b*-PPO-*b*-PEO block copolymers 2183
– synthesis 2184
peptide coupling 535
peptide hairpins 1323
peptide ligation 1315
peptide synthesis 461
– protease-catalyzed polymerization 461
peptide–nucleic acids (PNA) 1316
– pharmacological potential 1316
– synthesis 1316
peptide–peptide interactions 1322
peptide–polymer conjugates 1313–1314
– (HPMA)–drug conjugate 1314

- CRP methods 1313
- kinetic investigations 1314
- multiple peptide segments 1314
- polymethacrylate-*graft*-oligopeptide 1314

peptide–synthetic hybrid block copolymers 526–528, 530
- poly(γ-benzyl-L-glutamate)-*b*-polyisocyanide rod-rod type 530
- polyurethane prepolymer 527
- synthetic polymer blocks 526
- transition metal complexes 528

peptide–synthetic hybrid copolymers 528
- (AB)$_n$-type 528

peptide-based hybrid polymers 1309
peptide-polymer conjugates 1329
- coiled-coil superstructure 1331
- DNA–DNA interactions 1331
- elastin mimics 1329
- peptide–PEO conjugates 1331

peptides 1309
perbranched polymers 993
- inter-particle interaction 993
- reduced osmotic modulus 993
- semi-dilute solutions 993
- specific viscosity 993

percolation statistics 1612
Percus–Yevick equation 1588
perfluorinated polyethers (PFPEs) 2302
perfluorinated sulfonated ionomers (PFSA) 2472, 2481
- chemical modifications 2481

peroxidase catalyst 414
- laccase 414

peroxidases 404
- horseradish peroxidase (HRP) 404
- soybean peroxidase (SBP) 404

peroxide sensors 2526
peroxides 166
persistent radical effect 1144
pervaporation 2467
pervaporation membranes 2470
- separation efficiency 2471
- water-selective behavior 2470

pervaporation polymer membranes 2469
- volatile organic compounds (VOCs) 2469

perylene dicarboxymonoimide (PDI) 1066
PEVE macromonomers 76
pH sensors
- hydrogen ion 2504
- hydrogen ion-selective solid contact electrodes 2504
- optical pH sensor 2504

phase diagrams 1464

phase transfer 2416
phase transitions 1827
phase-modulated microscopy 1546
phase transfer catalysis (PTC) 610
- PTC agent 610

phase transfer catalyst 962
- tetrabutylammonium bromide (TBAB) 962

phenol 406
- HRP-catalyzed polymerization 406
- polymer synthesis 407

phenol compounds 405
- bisphenol 296
- glucose–D-hydroquinone 408
- polyphenols 405
- substituted phenols 408
- tyrosine ester 409
- unsubstituted phenol 406
- vinyl monomers 410

phenol derivatives 418
- oxidative copolymerization 418

phenolic polymers 405
- phenol–formaldehyde resin 405
- poly(2,6-dimethyl-1,4-oxyphenylene) 405

phenols 413
- polymerization 413

phenyl 4-(octylamino)benzoate 317
- substituent effects 317

phenylalanyl-tRNA synthetase (A294G) 1375

o-phenylenediamine 418
- oxidative polymerization 418

PHIC 1404
- backbone structure 1403
- morphology 1404
- morphology diagram 1404

phospholipid micelles 2582
phospholipid vesicles 1272
- cyclic voltammetry 1276
- deposition kinetics 1274
- dissipation 1273
- embedding 1274
- enzyme embedding 1278
- ferrocyanide anions 1276
- modified large unilamellar vesicles (mLUVs) 1273
- molecular integrity 1276
- preparation 1273
- spontaneous vesicle rupture 1273
- unmodified large unilamellar vesicles (uLUVs) 1273

phosphoranimines 313, 338
- polymerization mechanism 315

photoacid generators (PAGs) 2305–2306, 2663
photocross-linking agents 508
photocross-linking reaction 644
photocuring 644
photoinduced charge-transfer polymerization 555
photoinduced electron transfer 1072
photoinduced polymerization reaction 644
photoinitiated polymerization reaction 649
– formulation 649
– monomers 649
– oligomers 649
photoinitiators 645, 648–549, 646–647, 649
– absorption spectra 647
– cationic photoinitiators 646
– co-initiator 646
– excited state reactivity 648
– photolatent compounds 647
– properties 649
– radical photoinitiators 646
– titanocene derivatives 646
photolithographic methods 2282
– photoresists 2282
photolithography 2283–2284, 2303–2304
– macromolecules 2303
– photoresist films 2304
– photoresists 2283, 2304
– resists 2303
– substrate 2284
photonic crystals 1497–1498
– holographic lithography 1498
– roll casting 1497
photopolymers 643
photopolymerization 643–644, 653, 655, 657–664
– acrylates 657
– bulk reactivity 664
– cationic monomers 650
– charge-transfer polymerization 660
– developments 650
– dual cure technique 660
– epoxides 658
– excited state processes 663
– frontal photopolymerization 662
– functional properties 655
– hybrid cure technique 660
– interaction rate constants 663
– kinetics 662
– light sources 644
– molecular modeling calculations 663
– monitoring 661
– photochemical/chemical reactivity 653

– photoinitiation quantum yield 663
– polymerization propagating radicals 664
– powder formulations 659
– processes 661
– pulsed laser polymerization 662
– radical monomers 650
– rate constants 663
– reactions 653
– reactivity 663
– solution reactivity 664
– stimuli 664
– structural evolution 662
– thiol–ene photopolymerization 658
– vinyl ethers 658
– water-borne light-curable systems 659
photopolymerization monomers
– new monomers 651
– properties 651
– reactive monomers 651
photoresists 2285–2286, 2304–2310, 2313
– 157-nm resists 2308
– 193-nm resists 2307
– acrylic photoresists 2307
– APEX-type resist 2306
– chemically amplified (CA) 2306
– CO_2-soluble photoresists 2313
– deep-UV (248-nm) resists 2305
– dendrimers 2310
– ESCAP-based resists 2306
– extreme UV resists 2309
– hyperbranched polymers 2310
– immersion lithography 2307
– KRS-type resists 2306
– molecular glass photoresists 2310
– negative photoresist 2285
– norbornene-based resists 2309
– Novolac-based photoresists 2305
– Ohnishi parameter 2304
– photoacid generators 2306
– polysilane-based resists 2309
– positive photoresists 2286
– ring parameter 2304
– solubility contrast 2306
photosensitizers 645, 648–649
– cationic photoinitiator/photosensitizer systems 649
– examples 648
– processes 648
– properties 649
– radical photoinitiator/photosensitizer systems 648
photovoltaic cells (PVs) 2392
photovoltaics 1500

phyllosilicates 2073
physical ageing 1785, 1792, 1795, 1797
- processing conditions 1798
- state parameter 1792
- stress-enhanced physical 1795
- temperature-enhanced ageing 1797
physical gels 840
PIB 71–72, 75, 565, 571, 582
- amphiphilic pentablock copolymers 571
- multicomponent block copolymers 565
- segmented graft copolymers 571
- star copolymers 582
PIC micelles 2618
- complex formation 2618
- intelligent bioreactor 2618
- medical applications 2618
piezoelectricity 1500
pigment dispersants 2136, 2139, 2142, 2148, 2156, 2165, 2168
- AB block copolymer dispersants 2165
- AB block dispersant synthesis 2175
- ABC block copolymer 2169
- advantages 2136
- application criteria 2136
- application performance 2175
- atom transfer radical polymerization 2170
- block copolymers 2165
- characteristic data 2139
- controlled free radical polymerization 2165
- controlled free radical polymerization 2170
- dendrimers 2177
- flocculation 2166
- gel permeation chromatography 2139
- gradient polymers 2172
- heterocyclic aromatic rings 2168
- hyperbranched polymers 2177
- "living" polymerization 2165
- manufacture 2137
- millbase rheology 2136
- MMA–GMA block copolymer 2168
- nitroxyl-mediated radical polymerization 2172
- pigment dispersion process 2137
- polyacrylate-based dispersants 2148
- polyester-based dispersants 2156
- polyether-based dispersants 2162
- polyurethane-based dispersants 2142
- radical addition and fragmentation transfer 2176
pigments 1227–1228

- carbon black 1228
- copper phthalocyanine 1227
- phthalocyanine blue 1227
β-pinene 71, 77
- copolymerization 64
- macromonomers 77
piroxicam 1287, 1290
- prototropic 1290
plasma polymerization 665
- mechanism 665
plasticizers 2014
plug and play polymers 382
PMAA-based copolymers 2193
- emulsion polymerization 2194
- micro-emulsion polymerization 2194
- PS-b-PMAA block copolymers 2196
- statistical copolymerization 2194
PMB 333
- microspheres 333
PMMA 582
- star copolymers 582
PMMA–SiP monolayers 1170–1171
- surface morphologies 1171
PMMA-b-PEO block copolymers 2191
PNA–polymer hbrids 1316
Poissonian molecular-weight distribution 1309
polyacetylenes 281, 1350, 2227, 2264, 2267, 2269, 2371–2372, 2460–2461, 2464, 2470
- catalysts 2269
- conductivity 2268
- crystallinity 2268
- functionalization 374
- grafting approach 2372
- oxygen permeability 2460
- cis-polyacetylene 2268
- $trans$-polyacetylene 2268
- post-polymerization reactions 2269
- precursor route 2372
- ROMP 2269
- size-sieving ability 2464
- substituted polyacetylenes 2461
- synthesis 2371
- water purification 2452, 2454, 2456, 2458, 2460, 2462, 2464, 2466, 2468, 2470, 2472, 2474, 2476, 2478, 2480, 2482, 2484, 2486, 2488, 2490
polyacrylamide (PAAM) 2692
poly(2-acrylamido-2-methylpropanesulfonic acid) (PAMPS) 2692
polyacrylate dispersants 2148
- architecture 2150

- free radical copolymerization 2150
- product classes 2150
polyacrylic dispersants 2151
- anchoring groups introduction 2153
- macromonomers 2152
- monomers 2152
- polyacrylate backbone 2151
- SMA resins 2154
- transesterification 2153
polyaddition 308, 625
polyalkenes 1946
- chain microstructure 1946
poly(n-alkyl methacrylates) 1952–1953
- backbone dynamics 1953
poly(alkylthiophene)s 322
poly(3-alkylthiophene)s (P3ATs) 2374–2376
- head-to-head configuration 2376
- head-to-tail configuration 2376
- polymerization process 2375polyallylamine 2157
polyamides 132, 309, 316, 328, 343
- diblock copolymer 341
- hyperbranched aromatic polyamides 328
- ring-opening polymerization 132
- sequence-controlled 309
- star polyamides 343
poly(amino acid) 466
- intermolecular oxidative coupling 466
polyanilines (PAs) 418–420, 2228, 2272–2273, 2288, 2400
- applications 2273
- conducting property 418
- electroactive polyaniline films 419
- electron beam applications 2288
- HRP-catalyzed polymerization 420
- oxidation states 2272
- oxidative coupling 419
- "PanAquas" 2273
- Pd-catalyzed aryl amination 2272
- post-polymerization 418
- reaction conditions 418
- water-soluble polyanilines 419
polyarenes 2270
- applications 2270
- bandgaps 2270
poly(aromatic)s, synthesis 404
poly(aryleneethynalene)s (PAEs) 2378
- alkyne metathesis 2378
- Hagihara–Sonogashira coupling 2378
polyarylenes 2227
poly(arylenevinylene) 2374
- cross-coupling 2374
poly(benzamide) 336

- block copolymer 336
poly(m-benzamide)s 322
- solubility 322
polybenzimidazole (PBI), proton conductivity 2485
polybenzocyclobutanes 2275, 2278–2279
- cyclotene 2278
- dielectric constants 2279
- polymerization mechanism 2279
polybenzoxazoles 2275, 2277
- processing 2277
poly(benzyl-L-glutamate) homopolymers 524
poly(γ-benzyl-L-glutamate) (PBLG) 1403
poly(butyl acrylate) 342
poly(butyl methacrylate) (PBMA) 385
poly(butylene succinate) (PBS) 2096
- di-isocyanate [OCN-(C_6H_{12})-NCO] type end-groups 2096
polycaps 368
poly(ε-caprolactone) (PCL) 119, 865
polycarbazole (PC) 2377
- synthesis 2377
poly(carbazolevinylene) (PCV) 2373
polycarbonate 453, 1785
- lipase-catalyzed ring-opening polymerization 453
- mechanical rejuvenation 1785
polycarbonates 127–128, 1376, 1461
- absorbable sutures 127
- coarse-graining 1461
- coordination initiators 128
- functionalization 375
- lipase-catalyzed degradation 1376
- ring-opening polymerization 127
- synthesis 128
polycatechin 418
polycatenanes 355
polycations 1291
poly(ε-CL) 451
- lipase CA-catalyzed degradation 451
polycondensates 1756
polycondensations 4, 295–298, 304, 308, 310–311, 316, 321–322, 334, 424, 454–455, 610, 625, 635
- catalyst-transfer mechanism 322
- cellulose 424
- chain-growth polycondensation 311
- condensation polymer architecture 334
- controlled/living radical polymerization 196
- degree of polymerization 295
- diad orientational arrangement 308

- hydrolysis–condensations 625
- inductive effect 321
- meta-substituted monomers 321
- molecular weight control 310
- palladium-catalyzed 298
- para-substituted monomers 316
- polydimethylsiloxane (PDMS) 625
- polydispersity control 310
- reactions 953
- α,α-sihalogenated monomers 296
- solvent-free system 455
- symmetrical monomer 304
- unsymmetrical monomers 304

polycrotonates 765

polycyclic aromatic hydrocarbons
 (PAHs) 2229, 2242–2243, 2245, 2247
- all-benzenoid PAHs 2242
- films 2245
- heterotropic alignment 2243
- hexa-peri-hexabenzocoronenes
  (HBCs) 2245, 2247
- homeotropic alignment 2243
- intercolumnar order 2245
- intracolumnar order 2245
- Langmuir–Blodgett films 2247
- liquid crystalline phases 2243
- ribbons 2246
- supramolecular assembly 2242
- supramolecular order 2243
- synthesis 2242

polycyclic polymers, synthesis 894

polycyclosiloxanes 624
- emulsion polymerization 624

poly(dichlorophosphazene) 338

polydicyclopentadienes 249

polydienes 39, 884
- cyclization 884
- degradation 44
- enchainment 39
- polymerization conditions 39

poly(3,4-diethyloxy)-thiophene
 (PEDOT) 2376

polydimethylsiloxane (PDMS) 374, 2296
- lithography 2296

poly(dimethylsiloxane) 339

poly(9,9-dimethyl-2-vinylfluorene) 885
- cyclization 885

poly(diphenylacetylene) 2461

polydipropylsiloxane (PDPS) 1543
- mesophase transition 1543

polydisperse precursors 1710

polydispersity control 310

polydispersity index 1144, 1610

polyelectrolyte brushes 1163
- poly(methacrylic acid) brush 1164
- strong polyelectrolyte brush 1164
- swelling behavior 1163
- weak polyelectrolyte brush 1164

polyelectrolyte multilayer (PEM)
 films 11249, 261, 1272
- active reservoir 1272
- deposited active compounds 1272
- phospholipid vesicles 1272
- polyelectrolyte mixture 1261
- protein embedding 1279
- thickness 1261

polyelectrolytes 4, 495, 1197, 1233, 1272,
 1463
- "carrier" polyelectrolyte 1272
- natural polyelectrolytes 1290
- osmotic coefficient 1463
- polyelectrolyte capsules 1272

poly(ester–ether) 453
- ring-opening polymerization 453

polyesters 114, 126, 131, 143, 147, 302,
 316, 326, 439, 454, 456–457, 2343, 2348,
 2350
- active centers 114
- aromatic polyester synthesis 455
- aromatic polyesters 457
- biocatalytic production 457
- cationic AM polymerization 143
- condensation polymerization 454
- cyclic polymers 326
- dispersion polymerization 147
- functionality 2350
- functionalization 375
- hyperbranched polyesters 326, 2343
- initiating system 131
- lipase-catalyzed polymerization 439
- lipase PF-catalyzed polymerization 456
- ring-opening polymerization 114, 439
- star-shaped polyesters 131, 2343
- stereocomplexes 126

polyester-based dispersants 2159, 2161
- carboxyl-functional polyesters 2158
- carboxylic acid-modified dispersants 2161
- comonomers 2160
- hydroxy-functional polyesters 2160
- polyallylamines 2160
- polyester side chains 2157
- polymeric amine anchoring blocks 2156
- synthesis 2157, 2159

polyether-based dispersants 2161, 2163–2164

- alkyl ethoxylates 2162
- bifunctional dispersants 2163
- HLB value 2162
- monoalkyl polyether dispersants 2164
- nonylphenol ethoxylates 2162
- polyester–polyether-based phosphate dispersants 2163

poly(ether imide) 331
- approaches 331
- hyperbranched aromatic poly(ether imide) 331

poly(ether ketone) 330
- hyperbranched poly(ether ketone)s 330

poly(ether ketones) (PEK) 378
- cyclic polymers 326

poly(ether sulfone)s 320, 325, 330
- chain-growth polycondensation 320
- polycondensation 325
- step-growth polycondensation 320

poly(ether)s 331
polyethers 316, 319–320, 329
- chain-growth polycondensation 319
- crystallinity 320
- functionalization 375
- hyperbranched aromatic polyethers 329
- step-growth polycondensation 320
- poly(ethylene-co-1-octene) 1873

poly(3,4-ethylenedioxythiophene) 2400
poly(ethylene glycol) (PEG) 337, 379, 2543, 2558–2559
- conjugation capacity 2559
- poly(ethylene glycol)–p-nitrophenyl carbonate 2561
- poly(ethylene glycol)–Ara-C conjugates 2559
- poly(ethylene glycol)lated enzymes 2559
- prodrug conjugates 2558

poly(ethylene glycol) monomethacrylate (PEGMA) 1110
polyethylenimine (PEI) 846, 2156
poly(ethylene naphthalate) (PEN) 2299
poly(ethylene oxide) (PEO) 112, 555, 1109, 1909, 2336, 2416, 2354
- diblock copolymer 2354

poly(ethylene oxide) macromonomers 778
- end-modification 778

polyethylenes (PP) 217, 1538, 222, 1846, 2342
- coordination polymerization method 217
- crystals 1538
- elastomeric PP 227
- hemiisotactic PP 226

- HDPE 222
- LDPE 222
- LLDPE 222
- low-density ethylene copolymers 1846
- particles 2208
- polymerization catalysts 222
- radical polymerization 169
- skeletal structure 2342
- ULDPE 222

poly(ethylene terephthalate) (PET) 2299
polyferrocenylenedivinylenes 701
polyfluorenes (PF) 1352, 2377
- blue light-emitting active layers 1352
- cross-coupling 2377

poly(L-glutamic acid) (PLGA) 493
polyglycerol 129
- anionic polymerization 129

polyhedral oligomeric silsesquioxane (POSS) 2312
polyheteroarylenes 2227
poly(hexyl isocynate) (PHIC) 1403
poly(3-hexylthienylvinyleneethynylene) (P3HTE) 2378
- regioregularity effect 2378

poly(hexylthiophene) 341
- block copolymers 341

polyhomologation 312
poly(β-hydroxyalkanoate)s (PHAs) 121
poly(β-hydroxybutyrate)s 122
- tacticity 122

polyimides 2275–2277
- anisotropy 2276
- cyclic polymers 326
- dielectric constants 2277
- fluorinated polyimides 2277
- synthesis 2276

polyion complex (PIC) 2617
poly(isobutylenes) (PIB) 378
polyisoprene(PI) 842, 885, 1614
- cyclization 885
- linear rheology 1614

poly(ε-lysine) 465
- conjugation 465

poly(L-lysine) (PLL) 1255
- linearly growing films 1255

poly(lactic acid) (PLA) 451
- lipase-catalyzed degradation 451

poly(lactic-co-glycolic) acid (PLGA) 1270
poly(L-lysine)–calcium phosphate hybrid coatings 1366
polymacromonomers 1108
polymer ageing 1783
polymer analogous reactions 2615

polymer bioconjugates   2645–2646, 2648,
    2650–2652, 2654–2658, 2660, 2662, 2664,
    2666, 2668, 2670, 2672, 2674, 2676–2678,
    2680, 2682, 2684, 2686
- amine-functionalized PEG   2657
- antibody-based therapies   2656
- bactericidal activity   2670
- biohybrid amphiphiles   2677
- biomembranes   2670
- cross-linked polymer matrices   2667
- direct aggregation   2677
- hydrogels   2667
- non-thrombogenic materials   2670
- PEG-based biohybrid block copolymers   2677
- PEG–protein conjugates   2655
- PEGylation   2655
- peptide–polymer hybrids   2675
- peptide-modified polymer surfaces   2667
- pharmaceutical applications   2650
- polymer hybrid therapeutics   2650
- protein engineering   2657
- protein–polymer conjugate   2657
- responsive biohybrid hydrogel   2675
- self-assembly processes   2675
- site-specific PEGylation   2657
- SMANCS   2651
- soluble polymer–peptide hybrids   2667
polymer blending   1753, 1755
- reactive compatibilization   1755
polymer blends   1545, 1771
- homogeneous coupling   1764
- interfacial adhesion   1766
- interfacial reaction   1761, 1763
- interfacial reaction conditions   1767
- phase dispersion   1761
- phase morphology   1760
- phase separation   1545
- reactive chain length   1765–1766
- relative melt viscosity   1766
- thin films   1545
- volume average diameter   1762
- volume fraction   1766
polymer brushes   94, 1103, 1138–1140,
    1152, 1159, 1163, 1167–1168, 2297, 2704
- applications   1163
- atomic force microscopy   1533
- biointerfaces   1167
- block-copolymer semi-dilute
    brushes   1152
- "bottom-up" patterning   2297
- "concentrated brushes"   1139
- diblock copolymer brushes   1140

- dry brushes   1159
- fine particles   1168
- high-density PMMA brush   1152
- morphological control   1165
- polyelectrolyte brush systems   1163
- semi-dilute brushes   1138
- structure   1152
- surface friction   2704
- surface-initiated polymerization   1140, 2297
- swollen brushes   1152
- "top-down" patterning   2297
poly(p-mercaptobenzoyl) (PMB), whisker   332
polymer defects, emission color   2239
polymer degradation   2729
a,ω-polymer dianions   882
polymer dispersions   605
polymer dynamics   1610
- branch point withdrawal   1635
- Cayley tree   1630
- complex topologies   1630
- contour-length fluctuations   1634
- cross-bar orientation   1641
- damping function   1617
- disentanglement transitions   1634
- dynamic dilution theory   1623
- dynamical degrees of freedom   1609
- entangled branched chains   1618
- entangled combs   1625
- extensional rheology   1616
- H-shaped polymer   1625
- linear response   1613
- long-chain branching   1635
- long-chain-branched metallocene catalysed ensemble   1632
- Maxwell-like responses   1615
- mean-field gelation   1632
- molecular-based constitutive
    equations   1617
- nonlinear response   1615
- nonlinear shear rheology   1616
- partial retraction   1638
- polydispersity   1610
- pom-pom architectures   1628
- priority distribution   1635
- reaction schemes   1612
- relaxation fronts   1634
- relaxation modulus   1613
- reptation time   1641
- seniority-2 segment   1636
- star polymers   1619
- step-strain response   1639

Subject Index | **2801**

- strain dependence 1617
- strain response 1636
- stress relaxation function 1613
- stress-growth coefficient 1616
- time–strain factorability 1638
- time–temperature superposition 1614
- topological renormalization 1643
- waiting time 1633

polymer electrolyte membrane fuel cell (PEMFC) 2472
- PFSA membranes 2472
- sulfonated polymer membranes 2472

polymer encapsulation 2409

polymer field-effect transistors 2395–2398
- bottom-contact transistors 2398
- charge carrier transport 2397
- chemical purity 2397
- current ratio 2396
- device performance 2397
- field-effect mobility 2396
- principle 2396
- semiconductor–metal contacts 2397
- n-type PFETs 2397

polymer films 1165, 1547, 2389
- amplified spontaneous emission 2389
- morphology 1165
- patterned graft layer 1165
- spreading behavior 1546
- spreading instability 1547

polymer gels 465

polymer immiscibility 1753

polymer laser 2389, 2391
- assembly 2389
- electrically pumped polymer laser 2391
- polymer blends 2391
- resonator structure 2389

polymer laser materials 2388
- principle 2388
- requirements 2389

polymer libraries 1971, 1985
- molecular weight screening 1973
- polymer composition screening 1985

polymer light-emitting diodes 2385–2387
- double-layer PLEDs 2386
- electroluminescence mechanism 2385
- electron-transporting layer (ETL) 2386
- hole-transporting layer (HTL) 2386
- indium–tin oxide (ITO) 2386
- polymer blend PLEDs 2387
- polymer matrix 2387
- single-layer PLEDs 2385

polymer loading
- critical triaxial stress state 1787

- triaxial loading 1787
- unidirectional loading 1787

polymer matrix 1162

polymer melts 1462, 1607, 1947
- coarse-grained model 1462
- entangled melt dynamics 1608
- extension hardening 1607
- heterogeneous melt 1947
- long-chain branched (LCB) melts 1607

polymer membranes 2452, 2454, 2461, 2463
- gas separation mechanisms 2452
- poly(diphenylacetylene) membrane 2461
- reverse-selective polymer membranes 2463

polymer modifications 402
- advantages 462
- enzymatic catalysis 461
- properties 462
- reactions 462

polymer morphology control 324, 332

polymer nanofibers 1325

polymer network formation 1693–1694
- carboxy–epoxide reaction 1693
- cyclotrimerization 1694
- epoxide group 1691
- epoxide-amine addition 1692
- etherification reaction 1692
- glycidyl group 1691
- isocyanate group 1693
- isophorone diisocyanate 1694
- OH–epoxide reaction 1692
- primary amine addition 1691
- reaction with water 1695
- transesterification 1693
- transurethanization 1695
- urethane bonds 1694

polymer networks 192, 1687–1690, 1696, 1699, 1702, 1711, 1713, 1718, 1722
- affine network 1712
- application 1688
- branch points 1687
- bulk phase separation mechanism 1720
- chain crosslinking (co)polymerization 1722
- chain polymerization 1690
- chemical reactions 1690
- connectivity 1687
- crosslinking kinetics 1699
- crosslinking reactions 1690
- cyclization 1707
- cyclization reactions 1690
- degree of swelling 1716

- diffusion control 1700
- dynamic mechanical properties 1713
- elastically active network chains (EANC) 1696, 1702
- equilibrium rubber elasticity 1711
- equilibrium swelling 1714
- features 1689
- Flory–Huggins mean-field model 1714
- functionality 1710
- gel point time 1714
- glass transition 1714
- glass transition temperature 1701
- graph 1688
- inhomogeneity 1722
- internal branching 1710
- interpenetrating polymer networks (IPN) 1724
- junction-fluctuation theory 1712
- liquid-crystalline order 1727
- lower critical solution temperature (LCST) 1716
- macrosyneresis 1718
- materials 1688
- microsyneresis 1718
- network buildup 1702
- network formation 1689
- network formation studies 1696
- network formation theories 1704
- off-stoichiometry 1710
- phantom network 1712
- phase separation 1711, 1718
- polydispersity 1710
- polyelectrolyte networks 1717
- polymer solvent interaction 1716
- precursor architectures 1696
- precursor molecule 1689
- precursors 1696, 1710
- reactivity 1710
- simultaneous method 1725
- static fluctuations 1711
- step growth reactions 1690
- stress–strain relations 1712
- structural features 1711
- substructures 1702
- swelling transition 1721
- topological cluster 1722
- topological limit 1700
- Trommsdorff effect 1700
- upper critical solution temperature (UCST) 1716
- vertex 1688
- viscosity 1713
- volume phase transition 1717, 1720

polymer photovoltaic materials 2392
- donor–acceptor concept 2392
- donor–acceptor heterojunctions 2393
- performance 2393
- photoconductivity 2392
- requirement 2392
polymer processing 1796
polymer property screening 1987
- AFM 1987
polymer self-assembly 1317, 1322
- non-specific interactions 1317
- specific interactions 1322
polymer surfactants 2416
- ferrocene-functionalized norbornenes 2418
- thiol-terminated polymers 2417
- water-soluble homopolymers 2416
polymer synthesis, primary structures 1308
polymer therapeutics 2650
polymer thin-film libraries 1987–1988, 1991, 1994
- contact angle measurements 1994
- polydimethylsiloxane microlens library 1988
- property screening 1987
- UV/visible spectroscopy 1991
polymer-based sensors 2398
- analyte–polymer interactions 2399
- colorimetric sensors 2398
- conductometric sensors 2398
- potentiometric sensors 2398
polymer–biopolymer conjugate 1334
polymer-bound sulfonium salts 576
polymer–DNA hybrids 1332
polymer–drug conjugates 2652–2653
- degradable spacer 2653
- linkers 2653
- polymeric carrier 2652
- selective biorecognizability 2653
- targeting moieties 2653
polymer-layered silicate nanocomposites 2039–2040, 2043
- clay delamination 2043
- controlled polymerization process 2043
- exfoliation–adsorption 2039
- in situ intercalative polymerization 2040
- melt intercalation 2040
- non-controlled polymerization process 2040
- ring-opening polymerization 2043
- surface-grafted diblocks 2046

polymer/layered filler nanocomposites
  (PLFNCs)   2071, 2073, 2076, 2079, 2084,
  2095, 2102, 2105, 2110–2111, 2114–2115,
  2119, 2121, 2124–2125, 2128–2129
- all-trans conformation   2077
- anti-thixotropy   2106
- atomic scale structure   2083
- biodegradability   2116
- burning   2128
- characterization   2079
- clays   2074
- compatibilizer concentration   2107
- compounding process   2083
- computer simulation   2121
- Cox–Merz relation   2102
- crystallization   2086
- 2D small angle X-ray scattering
  (SAXS)   2107
- deformation   2129
- degradation process   2116
- diffuse in mechanism   2087
- diffuse out mechanism   2087
- dynamic storage modulus   2096
- elastomer/MMT nanocomposites   2089
- electrorheology   2109
- electrospinning processing   2127
- elongational flow   2105
- exfoliated nanocomposites   2080
- fire retardant properties   2111
- flocculation   2079, 2096
- flocculation control   2095
- flow behavior   2098
- foam processing   2124
- frequency shift factor   2100
- gas barrier properties   2113
- gel permeation chromatography   2118
- heat barrier effect   2111
- higher-order structure development   2091
- *in situ* intercalative polymerization   2076
- *in situ* polymerization   2083, 2098
- intercalants   2077
- intercalated nanocomposites   2080
- intercalation   2075, 2086
- interfacial tension   2125
- interlayer structure   2077
- ion-exchange reactions   2074
- ionic conductivity   2115
- linear viscoelastic properties   2097
- lipophilic–hydrophobic balance   2075
- lipophilization   2072
- liquid crystalline ordering   2121
- macroscopic modulus   2089
- magnetic fluctuations   2082

- mean field theory   2076
- mechanical properties   2088
- melt intercalation   2076
- melt rheology   2097
- melt state linear dynamic oscillatory shear
  properties   2098
- micromechanical models   2089
- modified clays   2075
- modulus enhancement   2095
- molecular dynamics   2083
- multiscale modeling strategy   2088
- Nielsen model   2114
- Nylon 6/layered silicate composites   2072
- Nylon 6/MMT nanocomposite   2091
- O_2 gas permeability   2113
- OMLF structure   2077
- optical transparency   2115
- photodegradation   2119
- polylactide (PLA)-based nanocomposites   2095
- polymer penetration   2079
- polyvinylidene (PVDF)-based nanocomposites   2093
- porous ceramic material   2128
- PPCN4   2086
- PPCN7.5   2086
- pressure–volume–temperature (PVT)
  behavior   2121
- processing operations   2123
- relaxation process   2100
- rheo-optical device   2105
- rheological theory of suspension   2097
- rheopexy   2106
- $scCO_2$-mediated interaction process   2083
- semi-crystalline PLFNCs   2089
- solid-state nuclear magnetic resonance   2081
- solid-state shear processing   2085
- solvent-free electrolytes   2115
- steady shear flow   2102
- sticker/surface attraction   2123
- strain-induced hardening   2105
- structure   2073, 2079
- talc fillers   2085
- thermal stability   2110
- thermogravimetric analysis (TGA)   2110
- thermoplastic-based nanocomposites   2084
- thermorheological properties   2098
- "tortuous path"   2113
- transcrystallization behavior   2089
- ultrasound preparation   2084
- UV/vis transmission spectra   2120

- viscoelastic behavior 2102
- WAXD patterns 2080
- wide-angle X-ray diffraction 2079

polymerase chain reaction (PCR)-based methods 1310

polymeric bioconjugates
- cross-linkers 2555
- examples 2555

polymeric dielectrics 2275, 2281
- coefficient of thermal expansion (CTE) 2275
- dielectric constants 2275
- extended aromatic networks 2280
- fluorinated polymers 2281
- interlayer dielectrics 2276
- polybenzoxazoles 2277
- polyimides 2276
- polysilsesquioxanes 2280

polymeric dispersants 2135, 2140, 2167
- anchoring groups 2141
- architecture 2140
- electrostatic stabilization 2138
- group transfer polymerization 2167
- molecular weight 2138
- steric stabilization 2138

polymeric drug delivery 2542
- N-(2-hydroxypropyl)methacrylamide (HPMA) copolymer 2542
- poly(ethylene glycol) (PEG) 2542
- water-soluble polymers 2542

polymeric drugs 2541, 2548, 2551
- delivery techniques 2552
- delivery vehicles 2552
- molecular architecture 2551
- anticancer drugs 2575
- prodrug bioconjugation 2552
- prodrugs 2548

polymeric glycols 374

polymeric light-emitting diodes (PLEDs) 1195, 1352

polymeric membranes 2451–2452, 2454, 2456, 2458, 2460, 2462, 2464, 2466, 2468, 2470, 2472, 2474, 2476, 2478, 2480, 2482, 2484, 2486, 2488, 2490
- applications 2451
- free-standing membrane 2462
- impermeable filler-containing polymer membranes 2464
- poly(dimethylsiloxane) membranes 2465
- Si-containing polymer membranes 2462

polymeric nanogels 1032, 1034
- radical crosslinking copolymerization 1034

polymeric prodrugs 2552
- design 2552
- synthesis 2552

polymeric spacers 374

polymeric surfactants 2181–2183, 2200
- block copolymers 2181
- cationic copolymers 2198
- double hydrophilic block copolymers 2203
- electrostatic interaction 2182
- fluorinated copolymers 2213
- gemini-type surfactants 2190
- glycopolymers 2202
- heterogeneous polymerizations in scCO$_2$ 2211
- heterogeneous radical polymerization 2183
- in situ 2209
- nonaqueous dispersions 2205
- nonaqueous heterogeneous polymerizations 2204
- PAA block copolymers 2193
- PEO-based block copolymers 2205
- PEO-based copolymers 2183
- PMAA copolymers 2194
- PMMA-b-PEO copolymers 2191
- PMMA-based copolymers 2208
- poly(vinyl alcohol) block copolymers 2201
- polycarbonates 2215
- polyolefin-based copolymers 2206
- polysaccharides 2202
- PPO-based copolymers 2205
- PS-b-PEO block copolymers 2187
- silicon-based copolymers 2207
- silicon-based stabilizers 2211
- star-shaped stabilizers 2202
- steric stabilization 2182
- steric stabilizers 2205
- sulfonate block copolymers 2196
- surface activity 2181
- types 2410

polymerization 643
- stimuli 643
- UV curing 643

polymerization catalysts 401
polymerization kinetics 1983
polymerization medium 824
polymerization methods 960
- combination 960
polymerization processes 1610
polymerization-induced phase separation (PIPS) 385

polymers 1431
- physical properties 1431
polymethacrylates 2427
- semiconductor nanoparticles 2427
poly(meth)acrylates 170
- radical polymerization 170
poly(methyl acrylate) (PMA) 341
- block copolymers 341
poly(methyl methacrylate) (PMMA) 170, 706, 2286
- anionic polymerization 706
- graft polymers 344
- positive resist 2286
- radical polymerization 170
polymethylene 312
- hydroboration 312
poly(α-methylstyrene) 884
- cyclization 884
poly(p-methylstyrene) 1898
- size-exclusion chromatography/light scattering 1898
polymer–monomer equilibrium 108
polymer–peptide conjugates 1312
- coupling strategies 1312
- polymerization strategies 1312
polynorbornenes 249
- graft polymers 344
polyoctenamers 249
polyolefin catalysts 1970
- high-throughput screening methodology 1970
polyolefin-based copolymers 2206
- PS-b-PI block copolymer 2206
- PS-b-poly(ethylene-co-propylene) copolymers 2206
polyolefins 218, 384, 1756–1757, 2016
- model chains 1757
polyorthocarbonate 298
poly(p-oxybenzoyl) (POB) 332
- whisker 332
poly(1,4-oxyphenylene) (PPO) 413
polypeptide drugs 2655
polypeptide synthesis 2647–2648
- ligation methods 2647
- NCA monomers 2647
- protein engineering 2648
- solid-phase synthesis 2647
polypeptides 481, 519–520, 1309–1311, 1403
- applications 519
- chain length uniformity 519
- defined monomer sequences 519
- high molecular weight polypeptides 520

- homopolymeric bio-organic segments 1308
- ligation methods 1311
- macromolecular engineering 519
- polypeptide-based block copolymers 520
- recombinant DNA methods 1309
- solid-phase supported synthesis 1310
- solid-phase techniques 481
- structural transitions 1403
polyphenolic compounds 417
polyphenols 405–406, 414
- enzymatic synthesis 414
- HRP-catalyzed polymerization 406
polyphenylene 326
- hyperbranched polyphenylene 326
polyphenylene dendrimers 1958
poly(p-phenylene) (PPP) 2232, 2270, 2272, 2374
- condensation-type polymerizations 2272
- conjugation length 2233
- ladder-type PPPs (LPPPs) 2232
- optical properties 2232
- synthesis 553, 2233
poly(p-phenyleneterephthalamide), rigid-rodanalogues 325
poly(p-phenylenevinylene)s (PPVs) 2227–2228, 2234–2235, 2270–2271, 2370, 2372
- defects 2234
- excited states 2271
- LEDs 2235
- poly(2,5,2',5'-tetrahexyloxy-7,8'-dicyanodi-p-phenylenevinylene) (CN-PPV) 2373
- polycondensation reactions 2373
- poly[2-methoxy-5-(2'-ethylhexyloxy)-1,4-phenylenevinylene] (MEH-PPV) 2373
- synthesis 2234, 2372
- topology 2228
polyphosphazenes 313, 724–725, 335, 337–338, 343
- applications 724
- block copolymers 335, 337
- diblock copolymers 338
- graft polymers 344
- kinetics 314
- polymerization 724
- polymerization mechanism 314
- polyorganophosphazenes 724
- properties 724
- thermal condensation 725
- three-armed star polyphosphazene 343
- triblock copolymers 338
polypropylene (PP) 224, 1757, 2627
- atactic PP 224

- isotactic PP   224
- radical grafting   1757
- syndiotactic PP   225

poly(propylene oxide) (PPO)   2336

polypyrrole (PP)   2270, 2272
- electrochemical polymerization   2272

polyrotaxanes   355, 380
- side-chain type   380

polysaccharide gels   2704
- "sol–gel" transition   2704

polysaccharide PEM films   1293, 1295
- biodegradability   1293
- cell adhesion   1295
- in vivo experiments   1293

polysaccharide-based copolymers   1320
- polystyrene–amylose block copolymers   1320

polysaccharides   422, 1309, 1370, 2603, 2698
- biosynthesis   422
- collagenase-resistant scaffold material   1370
- in vitro production   422
- tribological properties   2698

polysilanes   720, 722–723
- anionic ROP   722
- dendritic polysilanes   723
- photodegradable polysilane cores   723
- properties   720

polysilastyrene   721

polysiloxanes   718–719
- living anionic ROP   719
- organic–inorganic block copolymer   719
- properties   718
- synthesis   719

polysilsesquioxanes   2276, 2280

polystyrene (PS)   77, 341–342, 421, 857, 882, 1108, 1462, 1479, 1799–1800, 1886, 2288, 2354, 2417
- nanoporous   857
- conducting sulfonated polystyrene   1479
- copolymerization   169
- craze initiation   1799
- cyclic polystyrene   882
- diblock copolymers   338, 341, 2354
- electron beam lithography   2288
- functionalization   374
- gold nanoparticles   2417
- graft polymers   344
- macroinitiators   523
- micro-indentation   1800
- polystyrene standards   1886
- polystyreneblock copolymers   341

- positronium lifetimes   1462
- radical polymerization   169
- ring closure reaction   882
- star   933
- triblock copolymers   338
- vitamin C functionalized polystyrene   421

poly(styrene–ethyl acrylate) copolymers   1930
- chemical composition distribution   1930

polystyrene (PS) macromonomers   777

polystyrene latexes   2198
- pH-dependent surface activity   2198

polystyrene-b-poly(ethylene oxide) copolymers   2186
- anionic polymerization   2186
- applications   2188
- controlled radical polymerization   2187
- dendrimer-like copolymers   2190
- in situ formation   2189
- PEO macro-initiator   2187
- shear rheological measurements   2189
- star block polymers   2190
- styrenyl end-capped PS-b-PEO   2189
- synthesis   2186

polystyrene-b-polybutadiene-b-polystyrene (PS–PBd–PS)   842

polystyrenes   343, 887, 891
- cyclization   887
- pre-cyclized polymer   891
- telechelic precursors   891

poly(styrenesulfonic acid) (PSS)   2401

polystyryllithium   32

polystyrylsodium   34

polysulfide   297

polysulfone   342
- telechelic OH polysulfone   342
- macroinitiator   342

polytetrafluoroethylene (Teflon, PTFE)   1872, 2281
- heating scan   1872

poly(tetrahydrofuran) (PTHF)   846

poly(thienylenevinylene) (PTV)   2373

polythiophenes (PT)   324, 337, 341, 1350–1351, 1491, 2228, 2231, 2236, 2270
- block copolymer   337
- catalyst-transfer polycondensation   341
- chain packing   2236
- conjugation   1351
- conjugation length   2231
- difunctionalized   324
- lamellar sheets   1351
- macroinitiator   341

- monofunctionalized   324
- regioregularity   2231
poly(TMSP)   2464
poly(1-trimethylsilyl-1-propyne)
  [poly(TMSP)]   2459–2460
- gas permeation mechanism   2460
polytyrosine   409
- peptide-type   409
polyurethanes   332
- cyclic polymers   326
- hyperbranched   332
- microspheres   2209
polyurethane-based dispersants   2142
- anchoring groups   2147
- anchoring groups incorporation   2148
- cross-linkers   2144
- monofunctional side-chains   2145
- polyisocyanate resins   2142
- polyurethane dispersant   2147
poly(vinyl alcohol) (PVA)   170, 1122
polyvinylamine   2157
poly(vinylcarbazole) (PVK)   2387
poly(vinyl chloride) (PVC)   170
- radical polymerization   170
poly(2-vinylnaphthalene)   885
- cyclization   885
poly(2-vinylpyridine)   885
- cyclization   885
polyvinylpyrrolidone (PVPy)   2416
poly(VPGVG)   1330
pom-pom model   1643
- application   1643
porogens   2334, 2337, 2339–2340, 2351, 2362
- IR studies   2339
- nitrogenous porogens   2337, 2339, 2341
- polycaprolactone porogens   2350
- polydispersity index   2340
- pore size   2341
- porogen branching   2340
- porogen decomposition   2362
- scattering length density   2341
- small-angle X-ray scattering studies   2340
porosity   2334
- polymer decompositionporosity   2334
- sol–gel processes   2334
porous ceramic materials   2128
porous membranes   2453
porphyrins   175
positive photoresists   2286
- Novolac resins   2286
- PMMA   2286
positive staining   1235

positron annihilation lifetime spectroscopy (PALS)   2456
positron emission tomography (PET)   1671
post-modification   2153
potassium methoxyethanolate   782
potato phosphorylase   428
potential energy function (PEF)   1447
- bonded interactions   1448
- external terms   1447
- internal terms   1447
- nonbonded interactions   1448
powder coatings   659
polyurea   332
pre-formed polymers, intramolecular cross-linking   1042
precision polymerization   401
precursor route   2372
- Wessling method   2372
pressure-induced polymerization   666
pressure-sensitive adhesives (PSAs)   1731–1732, 1734, 1736–1737, 1742, 1746–1747, 2011
- base polymer blends   1736
- classes   2011
- crack propagation   1747
- elastic modulus   1734
- interfacial properties   1746
- molecular characteristics   1736
- neo-Hookean model   1742
- neo-Mooney–Rivlin modell   1742
- order–disorder transition (ODT)   1731
- probe test method   1737
- relaxation process   1734
- requirements   2011
primary amine initiators   532
primary amine-initiated NCA polymerization   526, 534
primary crystallization   1849
primary structure   519
primer   1309
proapoptotic anticancer polymeric drugs   2581–2583
- CPT–PEG conjugate   2583
- immunomicelles   2582
- "nonpump" resistance   2582
- pluronic block copolymer P85   2583
- "pump" resistance   2581
proapoptotic proteins   2586
prodrug   2548–2549
- activating mechanism   2550
- antedrugs   2551
- "bystander effect"   2550
- design   2551

- drug delivery systems 2550
- first type prodrugs 2548
- monoclonal antibodies 2550
- second type prodrugs 2549
- tumor-activated prodrugs (TAP) 2550

prodrug conjugates 2567
- molecular dynamics 2568
- molecular modeling techniques 2567
- poly(ethylene glycol)–Ara–C conjugation 2567
- spacer length 2567

prodrug linkers 2569
proline analogue 509
- 4-fluoroproline 509
- 4-hydroxyproline 509

propylene 226, 587, 746–747
- block copolymers 226
- isotactic-specific polymerization 747
- living coordination polymerization 587

N-(n-propyl)-2-pyridylmethanimine 2340
propylene-based polymers 224
1,2-propylene oxide (PO) 109, 112
- polymerization 112
prosthetics 1366
protease 455
protecting groups 2350
protecting-group strategies 1314
protein 2601
Protein A 1279
protein polymer 504
- coiled coil 504
- fluorinated amino acids 504

protein polymers 492–496, 505
- azido-functionalized amino acids 505
- coiled coils 494
- *de novo* design 492
- helical protein polymers 495
- hydrogels 494
- liquid crystals 493
- non-natural amino acids 496
- non-structured protein polymers 495
- β-sheet-forming protein polymers 492

protein semisynthesis 1311
protein-based polymers 479, 481, 510–511
- applications 510
- biosynthesis 479
- commercial production 511
- genetic methods 481
- ligation methods 481
- prospects 510
- solid-phase techniques 481

proteins 519, 2723

- denaturation 2723
proteolytic degradation 2734
proton conductivity 1476–1477
- hydrogen-bonded compounds 1477
- proton hopping 1477
proton conductors 1955
proton-exchange membrane fuel cells (PEMFCs) 2452
proton-exchange membranes (PEMs) 2473
prototropy 358
PS 379, 385
PS macromonomers 778, 780–781
- ionic deactivation 780
- ionic initiation 777
- ω-methacryloyloxy PS macromonomers 780
- norbornene end-functionalized PS macromonomers 781
- α-norbornenyl PS macromonomers 778
- ω-norbornenyl PS macromonomers 778
- ω-undecenyl poly(styrene-*block*-isoprene) macromonomers 780

PSAs
- block copolymer systems 1732
- polystyrene–polyisoprene adhesive blends 1732
- properties 1732

pseudo-peptides 1311
pseudo-rotaxane inclusion complexes 371
Pseudomonas fluorescens lipase 442
- lipase PF 442
Pseudomonas oleovorans 1365
pseudopods 1286
PTHF 566
pulse-field-gradient NMR spectroscopy 390
pulsed laser photolysis (PLP) 164
PUM
- explicit penultimate unit effect (IPUE) 818
- reactivity ratios 817
pyruvate sensors 2525

q
Q–e scheme 827
quadruple hydrogen-bonded motifs 356
quantitative structure–property relationships (QSPR) 1967
quantum dots 2425
quartz crystal microbalance (QCM) sensor 2501
quinones 463
quinoxaline 329

r

rac-LA 124–125
- enantioselective polymerization 124
- stereoselective polymerization 125

radially distributed molecular brushes 1106

radical addition and fragmentation transfer (RAFT) 2176

radical back-biting reaction 222

radical copolymerization 813, 829
- solvent dependence 829

radical crosslinking copolymerization (RCC) 1034

radical initiators 166, 168, 174, 549
- azobisisobutyronitrile (AIBN) 549
- diazenes 166
- peroxides 166
- redox systems 168
- reversible addition-fragmentation transfer polymerization 184

radical polymerization 161–163, 169, 172, 2612–2613
- atom transfer radical polymerization 2613
- conditions 168
- copolymerization 165
- degenerative transfer processes 182
- double-hydrophilic block copolymers 2613
- features 162
- inhibitors 168
- initiation 163
- initiators 166
- kinetics 163–164
- limitations 161
- monomers 166
- NMP 173
- organic radicals 162
- poly(meth)acrylates 170
- polyethylene 169
- polystyrene 169
- propagation 163
- PVC 170
- RAFT 182
- reversible addition–fragmentation chain transfer 2613
- SFRP 173
- structures 163
- TEMPO-mediated free radical polymerization 2613
- termination 163
- transfer 164

radical polymerization to cationic polymerization transformation 574

radical to cationic transformation 584

radical to cationic transformation polymerizations 574

radiofrequency identification tags (RFIDs) 2403

RAFT 799–800, 1190
- S,S'-bis(2-hydroxyethyl-2'-butyrate)trithiocarbonate (BHEBT) 800
- chain-end functionality 799
- kinetics 1190
- mechanism 798
- pathways 799
- surface-initiated RAFT 1190
- trithioester transfer agent 800

RAFT polymerization 965

RAFT-mediated polymerization 829

random chain polymer 304

random copolymers 2379
- determining factor 821
- polycondensation 2379

ransferases 428

rapid scanning AFM 1525

RCC 1034–1036, 1038, 1040–1041
- bulk RCC 1035
- controlled RCCs 1038
- conventional RCC 1038
- emulsion RCC 1040
- precipitation RCC 1036, 1041
- surfactant-free RCC 1041

reaction solvents 451

reactive blending 1753

reactive compatibilization 1755, 1758, 1769
- compatibilization efficiency 1769
- compatibilization reactions 1758
- functional groups 1755
- imidization reaction 1758
- kinetics 1763
- reactive group content 1769
- reactive polymers 1755
- routes 1754

reactive polyesters 460
- cross-linkable polyester 460

reactive polymers 1755
- preparation 1755

reactive species transfer, categories 311

reactivity ratios 821, 825–828
- experimental determination 826
- Kelen–Tüdös method 826
- patterns of reactivity 828
- theoretical predictions 827

recognition-induced polymersomes (RIPs) 383

recombinant collagen 490

recombinant DNA methods   1309
recombinant synthesis   1365
redox telomerization   784
redox-active dendrimers   1060–1061, 1063
- biomimetic applications   1062
- dendrimer porphyrin complexes   1062
- dinuclear complexes   1063
- electron reservoirs   1063
- electronic properties   1060
- ferrocene-decorated dendrimers   1064
- heterogeneous electron transfer   1060
- iron–sulfur cluster focal core   1061
- oxygenation–deoxygenation cycle   1062
- redox-active core   1060
- spatial isolation   1060
- zinc porphyrin core   1060
refractive indices   1581
rejuvenated polymers   1785
- mechanical rejuvenation   1785
relative diffusion constants   390
reprecipitation   1354
reptation model   1443–1444
- melt properties   1443
resilin   491
resonance effect   316
responsive materials   351
reverse LRP   1142
reverse osmosis   2467
reversible activation reactions   1142
- atom transfer (AT)   1142
- degenerative chain transfer (DT)   1142
- dissociation–combination (DC   1142
reversible addition fragmentation chain transfer polymerization (RAFT)   940
- dendritic multifunctional RAFT agent   941
- thiocarbonyl agent   940
reversible addition fragmentation termination (RAFT) polymerization   686
- mechanism   686-687
- block copolymers   687
reversible addition-fragmentation chain transfer (RAFT)   1189
reversible addition-fragmentation chain transfer   630
- emulsion polymerization   630, 608, 796, 798
- suspension polymerization   608, 831, 984, 965
reversible addition-fragmentation chain transfer (RAFT) polymerization   1150–1151
- graft polymerization   1151

reversible addition-fragmentation transfer (RAFT) polymerization   2054
- clay platelets   2054
- styrene polymerization   2054
reversible addition-fragmentation chain transfer (RAFT) polymerization   1142
reversible addition-fragmentation transfer (RAFT) polymerization   183–186
- advantages   186
- controlled architectures   185
- initiators   184
- limitations   186
- mechanism   183
- monomer   184
- termination   184
- transfer agents   185, 559
- mechanism   559
reversible chain transfer reaction   619
- chain transfer agents   619
reversible coordination polymers   390
reversible crystallization   1849, 1851, 1853
- crystal morphology   1853
- local equilibria   1851
- reversible insertion–crystallization   1851
reversible melting   1849–1852, 1855–1856, 1858
- extended-chain crystal   1851
- folded-chain crystal   1852
- isotactic polypropylene   1856
- quantification   1855
- reversible fold-surface melting   1851
- semicrystalline polymers   1849
- specific reversibility   1856
- total reversibility of melting   1856
rheology   1606
rheometers   1615
- cone-and-plate rheometer   1616
- extensional rheometers   1616
Rieke zinc method   1351
rigid amorphous fraction   1861
rigid amorphous phase (RAF)   1862–1866, 1868
- crystal formation   1866
- glass transition   1866
- heat capacity   1866
- molecular architecture   1864
- poly(butylene terephthalate)   1866
- semicrystalline polymers   1863
- specific RAF   1865
- thermal analysis   1863
- transition temperature   1868
- vitrification   1866
rigid rod polymers   362

rigid rods   1471
ring expansion   313
ring opening polymerization   959
ring–chain equilibria   387
– supramolecular polymers   387
ring–chain equilibrium   388
– theoretical treatment   388
ring-opening addition   308
ring-opening metathesis polymerization (ROMP)   3, 249, 585, 608, 620, 632, 696, 879, 948, 1141, 1315, 2647
– cycloolefins   585
– emulsion polymerization   632
– metal–alkylidyne catalysts   696
– metal–carbene initiators   696
– miniemulsion polymerization   620
– miniemulsion ROMP   278
– polynorbornene-*graft*-DNA conjugates   1316
– suspension polymerization   608, 696
ring-opening polyaddition   430–431
ring-opening polyaddition–condensation   452
ring-opening polymerization (ROP)   90–91, 103, 106, 109–110, 114, 121, 126–127, 129, 133–134, 136, 141, 145–146, 439, 447, 519, 544, 551–554, 573, 591, 689, 943, 2145, 2614
– "activated monomer" mechanism   133
– active centers   110
– active chain end mechanism   141
– anionic   91
– anionic ring-opening polymerization   109
– basic mechanistic features   109
– ε-caprolactone   591
– cationic ROP   136
– cationic polymerization mechanism   689
– ceiling/floor temperatures   106
– controlled/living radical polymerization   195
– coordination ring-opening polymerization   109
– cyclic carbonates   127
– cyclic esters   114
– dispersion polymerization   146
– equilibrium monomer concentration   106
– initiators   110
– metal-free catalysts   118
– monohydroxy initiators   2145
– "multiple-site" metal alkoxides   118
– "multiple-site" metal carboxylates   118
– NCA polymerization   134
– oxazolines   145
– poly($β$-hydroxybutyrate)s   121
– polyamides   132
– propagation–depropagation   106
– $α$-ring opening   141
– $β$-ring openings   141
– single-site metal alkoxides   121
– stereocomplexes   126
– thermodynamics   106
– "trimolecular mechanism"   118
road marking   2010
robotic spotter techniques   2662
rod–coil BCPs   1399, 1400, 1402, 1404–1405, 1421
– anisotropic liquid crystal phases   1399
– crystalline rod phases   1399
– dendron rod–coil   1423
– domain periodicities   1404
– kinetics   1423
– liquid-crystalline inter-rod interactions   1402
– liquid-crystalline orders   1421
– microphase-separated morphologies   1399
– morphology   1400
– packing constraints   1405
– packing schemes   1400
– rod-like structures   1402
– self-assembly   1399, 1400
– types   1399
– solution phase behavior   1421
– thermal transitions   1400
rod–coil block copolymer systems   378
rod–coil block copolymers   1422, 2239–2240, 2677
– fluorene–ethylene oxide block copolymers   2239
– morphology   2240
– rod–coil BCP/solvent systems   1422
rod–coil copolymers   2381, 2384
– convergent way   2381
– divergent way   2381
– donor–acceptor materials   2384
rod-to-globule transitions   1128
ROMP   252, 258, 262, 268, 273–274, 277, 280–281, 284, 286
– advanced materials   280
– biological materials   286
– block copolymers   258
– dispersed medium   273
– dispersion ROMP   276
– "Durham route"   281
– emulsion ROMP   274
– initiators   252

- multivalent chiral catalysts 284
- raft copolymers 262
- star polymers 268
- suspension ROMP 277
- titanacyclobutane catalyst 258

ROMP initiators 252–253
- antalacyclobutane catalysts 254
- first generation 253
- first-generation Grubbs catalyst 255
- midotungsten complexes 254
- molybdenum complexes 254
- ruthenium alkylidenes 255
- titanacyclobutanes 253
- transition metals 252

ROMP–ATRP tandem reaction 260
roofing 2010
rotated Fischer projection 732
rotaxanes 360
Rouse model 997, 1441–1442, 1444
- Langevin dynamics 1442
- melt viscosity 1442
- power-law sequence 1444
- Rouse relaxation time 1444
- Rouse time 1442

artificial urushi 416
rubber elasticity 1711
- statistical-mechanical theory 1711
rubber elasticity theory 1606
RuO_4 1661
rushi 414
- urushi trees 414

S

sacrificial filler materials 1370
Samia cynthia ricini 481
Sauerbrey relation 1284
SBC compounds 2015–2016, 2021
- automotive applications 2020
- contour diagrams 2016
- flame retardants 2021
- formulating ingredients 2016
- hydrogenated block copolymers 2015
- medical applications 2021
- SBS-based formulations 2015
- SEBS compounds 2021
- SEBS-based formulations 2015
- soft-touch overmolding 2022
- strand cutting 2020
- twin screw extruder 2020
- two-component molding 2022
- ultra-soft compounding 2022
- under-water pelletization 2020
- unsaturated block copolymers 2015

scanning calorimetry 1827–1828, 1830, 1832, 1834, 1836, 1838, 1840, 1842, 1844, 1846, 1848, 1850, 1852, 1854, 1856, 1858, 1860, 1862, 1864, 1866, 1868, 1870, 1872, 1874, 1876, 1878, 1880
- high-speed calorimetry 1828, 1830, 1832, 1834, 1836, 1838, 1840, 1842, 1844, 1846, 1848, 1850, 1852, 1854, 1856, 1858, 1860, 1862, 1864, 1866, 1868, 1870, 1872, 1874, 1876, 1878, 1880
- temperature-modulated calorimetry 1827

scanning electron microscopy (SEM) 336, 1234, 1653, 2316
- block copolymer thin films 2316

scanning local acceleration microscopy 1561
scanning near-field optical microscopy (SNOM) 1524
scanning probe contact lithography (SPCL) 2302
scanning probe lithography (SPL) 1562, 2300–2301
- conducting polymers 2301
- dip pen lithography, DPN 2300
- "dip-pen" nanolithography (DPN) 1562
- direct patterning 2301
- instrumental techniques 2300
- "millipede" data-storage technology 1562
- nanoscale patterning 2301
- nanoshaving 2300
- resists 2301

scanning transmission electron microscopy (STEM) 1653
scanning tunneling microscopy (STM) 1517, 2300
- tip–sample interactions 1517

scattering 1575–1577
- binary mixtures 1577
- elastic scattering 1576
- incident radiation 1575
- refractive index 1581
- scattered intensity 1577
- scattering length density 1581
- scattering vector 1576

scattering equations 1591
- applications 1591

scattering experiments 1575
- wide-angle scattering experiment 1575

Schrock catalysts 254
Schrock catalytic systems 948
secondary crystallization 1849
secondary electrostatic interactions 357

secondary structure-induced assembly   1318
seeded swelling method   1218
segmented copolymers   541–544, 546, 548, 550–552, 554, 556, 558–560, 562, 564, 566, 568
- concurrent polymerizations   544
- coupling reactions   544
- direct transformation reactions   543
- direct transformations   545
- graft terpolymers, comb-like block copolymers   559
- indirect transformation   543, 546
- multiblock   551
- multiblock thermoplastic elastomers   541
- PEO-b-PMMA-b-PSt triblock segmented copolymers   556
- PMMA-b-PSt-b-PMMA triblock segmented copolymers   556
- poly(ether ester)s   541
- polycondensate-type   541
- polymer-ceramic hybrid materials   559
- polyurethanes   541
- synthesis   541
selective fragment coupling reactions   1311
self-assembled dendrimers   1084
- capsules   1090
- catenanes   1088
- chiroptical activity   1094
- dendrimer-encapsulated metal nanoparticles   1088
- dendritic box   1087
- dendritic metallocycles   1090
- dendron rod–coil molecule   1090
- electrostatic (acid–base) interactions   1087
- hierarchical self-assembly   1088
- high-resolution lithography   1094
- hydrogen-bonding arrays   1085
- light-emitting organogels   1090
- liquid crystalline dendrimers   1092
- metal–ligand interaction   1088
- metallophilic interactions   1090
- monolayers   1092
- multilayer thin films   1092
- physical gels   1088
- polyphenylazomethine dendrimers   1088
- rotaxanes   1088
- secondary structures   1086
- supramolecular chemistry   1085
- supramolecular electronic materials   1092
- supramolecular functionalization   1086
- urea adamantyl poly(propylenimine) dendrimers   1086
- vesicles   1090
self-assembled ionically conducting polymers   1483, 1486
- "hairy rods"   1486
- lithium bis(trifluoromethylsulfonyl)imide salt (LiTFSI)   1483
- matrix polymer   1486
- salt-in-polymer system   1483
self-assembled monolayers   1140, 1556, 2255
- AFM   1556
self-assembled polymer architectures   352
self-assembled protonically conducting membranes   1480, 1476
- block copolymers   1477
- conducting channel orientation   1479
- model material   1479
- perfluorosulfonic acids   1476
- polystyrene-b-polybutadiene-b-polystyrene (SBS)   1477
- sulfonated SBS   1478
- thin films   1480
self-assembled structures   900
- additives   900
self-assembled systems   1471
- electro-optical properties   1471
- transport   1471
self-assembling layer-by-layer technique   1230
self-assembly   1471–1472, 1475–1477, 1481, 1488–1489, 1491–1492, 1494–1497, 1499–1500, 2675
- block copolymers   1472
- charge carrier mobility   1489
- charge-transfer salts   1492
- colloidal self-assembly   1481
- conducting channels   1475
- conducting domains   1477
- conjugated polymers   1475, 1488
- hierarchically self-assembled systems   1483
- insulator–semiconductor layered structure   1494
- lamellar-within-lamellar self-assembly   1481
- nanowire elements   1496
- noncentrosymmetric assembly   1499
- noncentrosymmetric self-assembly   1500
- photonic crystals   1497
- polyaniline   1491
- polyelectrolyte–surfactant complex   1491

- properties 1476
- ribbon-like entities 1494
- sulfonic acid-doped polyaniline 1492
- template 1495
self-assembly polymer systems 1496
- optical properties 1496
self-assembly principle 932
self-assembly structures 1484
- phase transitions 1484
self-assembly techniques 1180
self-complementary arrays 357
self-complementary binding units 368
- polycaps 368
self-complementary motif 354
self-condensing ring-opening polymerization (SCROP) 986
self-condensing vinyl copolymerization (SCVCP) 984
- initial steps 984
- kinetics 984
- rate constants 984
self-condensing vinyl polymerization (SCVP) 980–985, 1188, 2349
- iniferter group 985
- inimers 981, 983
- kinetics 982
- living polymerization systems 984
- monomer-terminator 985
self-consistent field model (SCF) 2122
self-polycondensation 323, 341
semi-crystalline polymers 1783, 1803–1805, 1808–1810, 1814, 1816
- anisotropic properties 1810
- degree of crystallinity 1804
- geometrical softening 1803
- injection molding 1809
- intrinsic behavior 1803
- multi-scale modelling 1814
- neck formation 1804
- slow-induced nucleation 1808
- texture-induced hardening 1816
- yield stress 1805
semi-dilute brushes 1138
semiconducting polymers 2110, 2369–2370
- charge carriers 2370
- optoelectronic applications 2369
- properties 2370
- n-type semiconductor 2370
- p-type semiconductor 2370
semiconductor nanoparticles 1195, 2424–2425, 2427–2433
- applications 1195
- block copolymers 2429

- CdSe semiconductor nanoparticles 2425
- conjugated polymers 2430
- dendritic encapsulation 2433
- nanocomposite films 2430
- nanoparticle dispersion 2432
- passivation 2426
- polymer ligands 2427
- polymer-coated CdS nanoparticles 2427
- PPV-coated CdSe quantum dots 2433
- properties 2424
- quantum dots 2425
- size-selective precipitation 2426
- size-tunable photoluminescence emission 2425
- structure 2424
- surface-initiated controlled radical polymerization 2428
- surfactants 2427
- synthesis 2425
- types 2425
- n-type semiconductors 2431
semicrystalline polymers 1538–1539, 1541, 1863, 1867, 1947
- kinetic trapping 1538
- ATHAS Database 1863
- behavior 1538
- crystallization 1541
- crystallization kinetics 1541
- fold-controlling units 1541
- mobile-amorphous fractions 1863
- model molecules 1539
- morphology
- real-time AFM studies 1538
- thermal properties 1867
- thickness distribution 1538
seniority distribution 1630
sense of taste 2530
sensing microarrays 1356
sensitivity analysis 820
sensor array modular measurement system (SAMMS) 1988
sensor devices 2493
- solid-state sensor devices 2493
sequence control 303
sequence statistics 737, 739–741
- Bernoullian statistics 737
- enantiomer-selective site model 740
- first-order Markovian statistics 737
- higher order statistics 739
- isotactic polymers 741
- second-order Markovian statistics 737
sequential conventional free radical 590
sequential monomer addition (SMA) 841

sequential polymerization 704
- metal-containing block copolymers 704
sequential polymers 304, 308–309
- (–AA–B'B–AA–BB'–)$_n$ 304
- (–AA–BB'–)$_n$ 307
- (–AA'–BB'–)$_n$ 308
- (–B''B'–A'A–BB–AA'–B'B''–A''A''–)$_n$ 310
- (–B'B–AA–BB'–A'A'–)$_n$ 309
- sequential poly(acylhydrazide–imide) 308
- sequential poly(amide–thioether) 308
- symmetrical monomers 304, 309
- unsymmetrical monomers 304, 308–310
Serratia marcescens chitinase 438
shape-persistent nanoparticles 1958
β-sheet 1326
β-sheet motif 1322, 1324, 1327
- double β-sheets denoted ribbons 1327
- lamellar crystals 1327
- unisometric tape structures 1324
shell-cross-linked micelles 854
shell-cross-linked nanoparticles (SCKs) 2421
short-chain branches (SCBs) 1946
side-chain attachment 1115
- click chemistry 1115
- nucleophilic substitution 1113
side-chain polymers 353
side-chain supramolecular polymers 380
- benzoic acid–pyridine mesogenic complexes 380
- poly(4-hydroxystyrene) 381
- poly(4-vinylpyridine) (P4VP) 381
- side-chain homopolymers 381
[1,3]-sigmatropic rearrangement 313
silanolates 29
silica particle (SiP) 1169
- dispersibility 1169
- initiator-coated SiP 1169
silicate layers 2107
silicon-based copolymers 2207
silicon-based stabilizer
- PDMS-*b*-PMA stabilizer 2212
- polydimethylsiloxane (PDMS) macromonomer 2211
silicon-mediated polycondensation 552
silk 481–482, 484
- capture silk 482
- dragline silk 482
- mechanical properties 482
- protein sequences 481
- triggering process 484

silk analogues 483–484
- biological function 484
silk fibroin 2677
silk mimic 1323
- oligoalanine–PEG multiblock copolymer 1323
silk proteins 482–483
- expression 482
- expression hosts 483
- recombinant silk proteins 483
silk–elastin-like polypeptides (SELPs) 487
- ProLastin 47K 487
silkworm silk 483–484
- chimeric 483
- fibroin 484
- sericin 484
siloxanes 1182
simulation software
- DL_POLY 1452
- ESPResSo 1452
- GROMACS 1451
- NAMD 1451
single initiator 591
single-molecule spectroscopy (SMS) 2254
single-site catalysts 121, 746
- dialkylaluminum alkoxides 121
single-wall nanotube (SWNT) 1525
single-walled carbon nanotube (SWCNTs) 715
site epimerization 228
site-transformation technique 90
size-exclusion chromatography (SEC) 325, 1883–1886, 1888, 1892, 1897, 1901, 1972–1973
- branching ratio 1897
- calibration 1886
- calibration curve 1885
- distribution coefficien 1885
- exclusion limit 1886
- flow-rate 1973
- light scattering 1892
- linear columns 1886
- Mark–Houwink equation 1888
- molar mass distribution 1887
- molar mass-sensitive detection 1892
- separation limit 1886
- separation principle 1885
- viscometric detection 1901
size-exclusion chromatography (SEC)/light scattering 1894, 1896–1897, 1899–1900
- copolymers 1900
- detectors 1896
- MALS system 1897

- number-average molar mass 1900
- setup 1894
- separation mechanism 1899
- solvents 1900

size-exclusion chromatography (SEC)/viscometric detection 1901–1902
- differential pressure transducer 1901
- differential viscosimeters 1901
- intrinsic viscosity 1902
- Mark–Houwink plot 1902
- Mark–Houwink relation 1902
- universal calibration curve 1902

size-exclusion chromatography calibration 1888–1891
- broad molar mass distribution 1889
- calibration curve 1890
- integral calibration 1891
- universal calibration 1888

size-exclusion chromatography/viscometric detection 1903–1904
- copolymers 1904
- mechanism 1903

small molecule libraries 1971
small molecule surfactants, synthesis 2415
small-angle neutron scattering (SANS) 364, 425, 1240
small-angle X-ray scattering (SAXS) 1240
small-angle X-ray scattering under grazing incidence (GISAXS) 2317
small-molecule dyes 1356
smart coatings 1498
smart materials 288, 1077, 1343, 2672
smart polymer bioconjugates 2672
- multiple affinity complexes 2672
- PNIPAAm 2672

sodium montmorillonite lamellae 1196
soft tissue engineering 1370
software 1451
solar cells, device parameters 2226
solid electrolytes 2493
solid surfaces 1137, 1146, 1168
- Au surfaces 1182
- Au-coated substrate 1146
- azo initiator-bound silicate surfaces 1150
- end-grafted polymers 1137
- grafting-from method 1137
- grafting-to method 1137
- initiator-modified surface 1182
- montmorillonite clays 1182

- poly(lactic acid) (PLA)-grafted surface 1141
- protein adsorption behavior 1168
- silica particle surfaces 1182

solid-phase peptide synthesis (SPPS) 496, 519
solid-phase supported peptide synthesis (SPPS) 1310
solid-phase supported polymerization 1313
solid-phase techniques 480
solid-state assemblies 1351
solid-state NMR 1938
solid-state polycondensation 302
solution-state assemblies 1353
solvent effects 829
solvophobic interactions 360
soybean peroxidase (SBP) 404
sparfloxacin 2573
specialized nanofabrication 2295–2296
- microcontact printing (μCP) 2296
- scanning probe lithography (SPL) 2296
- soft lithography 2296

specific heat capacity 1827–1832, 1835, 1853, 1857
- ATHAS Database 1830
- bulk specific heat 1833
- excess heat capacity 1857
- extended-chain crystals 1853
- folded-chain crystals 1853
- heat-capacity baseline 1832
- latent heat 1832
- long-term reorganization 1828
- measurement 1835
- polyethylene 1829–1830
- reversing apparent specific heat capacity 1852
- skeletal vibrations 1831
- temperature-modulated DSC 1835
- vibrational heat capacity 1828

specific viscosity 388
spectroscopic imaging 1524
spider silk 483, 1323, 1365
spin-on methods 2333
spray deposition 1251
stable free radical polymerization (SFRP) 173–176, 186
- advantages 186
- catalytic chain transfer process 175
- limitations 186
- moderators 176
- SFRP systems 175

staining agents 1235, 1660, 1662
- electrophilic 1662

star architectures   912
star block copolymers   86
star copolymers   2422
- gold nanoparticles   2422
star oligomers   1021
star poly(ε-caprolactone) (PCL)   1372
star polymers   48, 129, 268, 272, 343, 531, 593, 709, 909–910, 915, 917, 919, 921, 925, 927, 933, 936, 938–941, 943, 949, 952–953, 958, 961–962, 1007, 1009–1022, 1027, 1029, 1615, 1619, 1624, 2024, 2346–2347
- A_2B miktoarm stars   925
- ABC-type miktoarm   593
- arm-first   531
- "arm-first" approach   1017
- "arm-first" technique   1013
- 12-arm PS star homopolymers   934
- asymmetric $PS(PS')_2$ 939
- atom transfer radical polymerization   1015
- barrier-hopping   1620
- catalytic scaffolds   1016
- categories   1009
- characterization   911
- chloromethylbenzene functions   1018
- chlorosilane derivatives   1017
- controlled/living radical polymerization   1014
- conventional radical polymerization   1014
- convergent approach   910
- "convergent" method   48
- "core-first" approach   1022
- core–shell architecture   1011
- core–shell structure   1039
- core-first approach   1009
- core-first methodology   531
- crosslinking reactions   1012
- dielectric relaxation   1624
- diffusive properties   1623
- diphenylethylene derivatives   1019
- divergent approach   911
- divinylic comonomer approach   1022
- dynamic dilution   1621
- dynamic tube dilation approximation   1624
- electrophilic reagents   1019
- experimental variables   1011
- Fe polyoxazoline star polymers   958
- functionalized stars   912
- group transfer polymerization (GTP)   1013
- halogen–lithium exchange   1018
- hydroxyl-terminated poly(ε-caprolactone) stars   943
- in-chain functionalized stars   949
- inimer   1030
- "in–out" approach   1016
- linear rheology   1615
- linking agents   917
- linking reagents   1019
- "living" anionic polymerization   1012
- "living" cationic polymerization   1013
- macrogelation   1012
- metal-centered   709
- metallo-supramolecular star polymers   1022
- microgel-type core   1010
- miktoarm $PS(PnBuA)_2$ stars   939
- miktoarm star   909, 960, 1009
- miktoarm star $PS(PI)_2$ 920
- miktoarm star quarterpolymers   921
- miktoarm star terpolymer   920
- Mitsunobu condensation   2346
- motional narrowing   1622
- multiarm PBd stars   919
- multiarm polyelectrolyte   1015
- multiarm star polymers   1009
- multifunctional RAFT agents   1020
- nitroxide-mediated polymerization   1015
- nodulus approach   1010
- nucleophilic reagents   1020
- oligomers   1021
- oligothiophene stars   1021
- PnBuA stars   943
- PEO stars   915, 938, 962
- PMMA star homopolymers   925
- PMMA stars   927
- $PMMA_nPnBuA_m$ stars   938
- polybutadiene (PBd) star   915
- polyethylene star polymers   271
- polyimide stars   953
- polyisobutene (PIB) star homopolymers   928
- polymerization techniques   909
- poly(methyl acrylate) stars   941
- polynorbornene stars   272, 952
- polyoxazolines stars   1027
- poly(1-pentenylene) stars   271
- polystyrene stars   958
- $(PS)(PtBA)_2(PMMA)_2$ miktoarms stars   966
- $(PS)_2(PEO)_2$ miktoarm stars   961
- PS star homopolymers   943
- PS stars   917, 941, 1012

- R-group approach   1020
- radial density distribution   2347
- RAFT polymerization   940
- reversible addition–fragmentation chain transfer   1016
- rheological relaxation times   1623
- ring-opening metathesis polymerization (ROMP)   268, 1013
- SANS studies   2348
- self-diffusion   1623
- stable free radical polymerization   1015
- star arm retraction   1622
- star homopolymers   927
- star-arm dynamics   1619
- star-block copolymers   909, 1009
- star-like PEOs   1013
- star polymethylene methanols   343
- star–star couplings   1012, 1015
- steric crowding   1017
- stress relaxation   1619
- supramolecular approach   1022
- synthesis   909–910, 915
- three-arm P2VP stars   921
- three-arm PMMA star homopolymers   936
- three-arm PS   925
- three-arm PS-d7   933
- three-miktoarm star terpolymers   963
- tube model   1619
- Z-group approach   1020

star polymethylene   343
- triol   343

star polypeptides   531, 533
- β-benzyl-L-aspartate stars   531
- controlled NCA polymerization technique   533
- core-first strategy   531
- many-arm star polypeptides   533
- poly(β-benzyl-L-glutamate) star   531
- poly(γ-stearyl-L-glutamate) star polypeptides   531

star polypropylene   233
star-block copolymers   912, 919, 931–932, 943, 950
- amphiphilic star-block copolymers   950
- 12-arm PS-P4VP star-block copolymers   934
- AA'B asymmetric star-block copolymers   87
- ABB' asymmetric star-block copolymers   87
- ABC asymmetric star-block copolymers   87
- A_nB_n hetero-arm star-block copolymers   87
- diblock PS-b-poly(N-isopropylacrylamide) copolymer   943
- four-arm amphiphilic block copolymer   932
- α-methylstyrene (α-MeS)   931
- miktoarm star copolymer   919, 930, 960
- PCL-b-PEO star-block copolymer   943
- $(PIB)_2(PMVE)_2$ miktoarm copolymer   931
- $(PIB-b-PS)_3$ star-block copolymer   930
- $(PnBuA-b-PS)_3$ star-block copolymers   936
- PS-b-poly(N-isopropylacrylamide) star-block copolymers   943

star-like polymers   188–189
- arm-first process   189
- atom-transfer radical polymerization (ATRP)   188
- reversible addition-fragmentation transfer (RAFT) polymerization   188

star-shaped copolymers   2356
- stimuli-responsive   2356

star-shaped polymers   144, 1607, 2344
- cationic ring-opening polymerization   144
- "core-in"/"core-out" approach   2344
- functional dendrons   2344
- miktoarm dendritic-linear star polymers   2344
- modification   2344
- star-shaped polyester   144
- star-shaped PTHF   144
- tracer diffusion constant   1607
- viscosity   1607

static observables   1438
- gyration tensor   1438
- hydrodynamic radius   1438
- mean-squared end-to-end distance   1438
- structure factor   1439

statistical copolymers   813, 821
- determining factor   821

Stealth® liposomes   2655
step growth polymerization   550
step polymerization   546
step-growth polycondensation   316, 976
step-growth polymerization   161, 295, 953
step-ladder polymers   2232
stereoblock polymers   756
- acrylamides   757
- aluminum bisphenoxides   756
- stereoregularity   756
- synthesis   756

- tertiary alkyl methacrylates   757
stereoblock polypropylene   226, 228, 230–232
- binary catalyst mixtures   231
- isotactic–atactic   228
- isotactic–hemiisotactic   229
- isotactic–syndiotactic   231
- non-metallocene Group IV catalysts   230
- syndiotactic-atactic   232
stereocomplexes   126
stereoisomerism   731
stereospecific living polymerization   731
steric stabilizers   276
stilbene   828
Stille reaction   2374
stimuli-responsive dendrimers   1077, 1079
- chiroptically sensing   1080
- dendroclefts   1079
- molecular recognition   1079
- pH-responsive shape changes   1078
- phosphorus dendrimer   1078
- photoinduced trans-to-cis isomerization   1078
- thermally induced spin transition   1078
stimuli-responsive polymers   2574
- poly(N-isopropylacrylamide)   2574
stimuli-responsive vesicles   1343
stoichiometric imbalanced polymerization   303
- folding-driven nucleation–elongation mechanism   303
stoichiometric imbalanced polycondensation   296
structure–property correlation   1124
- molecular brushes   1124
styrene   18, 33, 35, 43, 70, 79, 93, 421, 450, 571, 579, 677, 819, 823, 828, 915, 934, 937–938, 1140, 2188–2189, 2200
- ATRP polymerization   937
- block copolymerization   585
- block copolymers   82, 571, 579, 587
- copolymerization   819, 823
- cross-propagation   828
- Cu-catalyzed radical polymerization   450
- degradation reaction   43
- dispersion polymerization   2189
- electroactive polymer   587
- electron transfer   18
- emulsion polymerizations   2188
- half-lifetime   43
- intramolecular alkylation   67
- kinetic studies   33
- living carbocationic polymerization   65

- macromonomers   79
- mini-emulsion polymerization   2200
- NO-mediated polymerization   677
- polymerization   557
- rate constants   35
- star polymers   938
- star polystyrene   93
styrene-isobutene-styrene   2028
- applications   2028
- S–IB–S copolymers   2028
styrene–maleic anhydride copolymers   2154
- polymeric dispersants   2154
p-styrenesulfonic acid   421
styrenic block copolymers   2001, 2008
styrenic block copolymers (SBCs)   2001, 2008–2012, 2018–2019, 2023–2025
- additives   2018
- adhesive creep compliance   2012
- adhesives   2011
- bitumen modification   2009
- butadiene-based SBCs   2012
- compound morphology   2019
- compounding applications   2015
- cross-linkable adhesives   2014
- elastic modulus   2012
- endblock plasticizers   2017
- endblock resins   2014, 2017
- energy-absorbing mechanisms   2023
- fillers   2017
- history   2001
- hydrogenated SBCs   2019
- oil gels   2024
- order–disorder transition temperature (ODT)   2019
- paving materials   2009
- polar engineering thermoplastics   2023
- polymer modification   2023
- processing techniques   2018
- PVC replacement   2025
- recycling compatibilization   2025
- rheology modifiers   2024
- road marking paints   2010
- roofing membranes   2010
- SBS-based footwear   2008
- SBS block copolymers   2009
- sealants   2011
- silicone replacement   2025
- SIS polymers   2012
- tackifying resin   2012
- twin screw extruder   2020
- viscosity index improvers   2024
- waterproofing coatings   2010
submicronic reactors   1278

substituent effects   316
substituted phenols   408
– meta-substituted phenols   410
– oxidative polymerization   408
– para-substituted phenols   408
substituted polyacetylenes   2464
substrate-assisted mechanism   431
sulfonate block copolymers   2196–2197
– electrosteric stabilization   2197
– reactive stabilizers   2197
– synthesis   2197
sulfonated polymer membranes   2474–2481, 2484, 2486
– chemical modifications   2481
– cross-linked sulfonated polyimide (XSPI) membranes   2482
– cross-linking   2481
– electro-osmotic drag   2479
– fuel permeability   2478
– hopping mechanism   2478
– hydrolytic stability   2480
– ionic cross-linking   2484
– methanol permeability   2484
– organic–inorganic composite membranes   2477
– oxidative eadical atability   2480
– PBI membranes   2486
– processability   2474
– proton conductivity   2477, 2484
– sulfonated poly(ether ether ketone) (SPEEK)   2475
– sulfonated polyarylenes   2480
– sulfonated polyimides   2480
– sulfonated polyphosphazenes   2480
– swelling behavior   2479
– vehicle mechanism   2478
– water sorption behavior   2475
– water–methanol separation   2478
sulfonated polystyrene (SPS)   420
sulfone, hyperbranched poly(ether sulfone)s   331
super-fast chip calorimeter (SFCC)   1871
– nanophase structure   1871
super-fast chip calorimetry (SFCC)   1850
super-soft elastomers   1113, 1120, 1126
supercooling   1850
supercritical carbon dioxide (scCO$_2$)   451, 457, 1221, 2083, 2211, 2312
– lipase CA-catalyzed polymerization   452
supercritical CO$_2$ (scCO$_2$) technology   2322
– block copolymers   2322
supercritical CO$_2$ processing   2312
superparamagnetic latexes   2621

supported catalysts   284
supported films   1159, 1161
– cast film   1160
– glass transition temperature   1160
– graft film   1160
– PMMA films   1160
– surface molecular motion   1161
supramolecular assemblies   2225
supramolecular block copolymers   378
– AB diblock copolymers   379
– coil–rod–coil triblock type copolymer   380
– main-chain rod–coil block copolymer   378
supramolecular block polymers, comb–coil supramolecular diblock copolymer   381
supramolecular chemistry   351, 1471, 1954
supramolecular helical dendrimers   1957
supramolecular liquid crystals   355, 362
supramolecular main-chain polymers   364
– nanotubes   364
supramolecular polymer engineering   351
supramolecular polymers   351–355, 361–362, 369, 371, 392, 1954, 2229, 2240
– 2,6-bis(diamidopyridine)phenol   369
– classes   353
– cyclodextrin-based   371
– definition   353
– discotic molecules   355
– donor–acceptor–donor triads   2240
– general aspects   352
– hydrogen bonding   2240
– hydrogen bonds   1954
– isotropic solution   369
– large building blocks   353
– liquid crystalline monomers   362
– metal coordination bonding   354
– multiple hydrogen-bonded   352
– research objectives   392
– self-assembly   2240
– small building blocks   361
– stress relaxation mechanisms   367
– thermodynamic equilibrium   392
– ureidopyrimidone-based   355
supramolecular UPy materials   385
– applications   385
surface microscopy techniques   1127
surface-initiated ATRP   1146, 1148
– graft density   1146
– initiation efficiency   1146
– kinetics   1149
– volume effect   1148

surface-initiated graft polymerization
- nitroxide-mediated polymerization (NMP) 1150
- reversible addition–fragmentation chain transfer (RAFT) polymerization 1150

surface-initiated living polymerizations 1139
- LRP 1139

surface-initiated LRP 1165

surface-initiated polymerization 450, 1137, 1140–1142, 1180, 2428, 2663
- anionic polymerization 1140
- atom transfer radical polymerization (ATRP) 1142
- cationic polymerization 1140
- controlled/living radical polymerization (LRP) 1141
- initiation efficiency 1142
- polymer-encapsulated materials 2428
- ring-opening polymerization 1140

surface-initiated polymerization (SIP), "backfilling" technique 2297

surface-initiated polymerization methods 2413
- step-growth polymerization 2413

surface-initiated polymerization 94
- polymer brush 94

surfactant sensors 2513

surfactant systems 1227
- hydrophilic–lipophilic balance (HLB) value 1227

surfactant templates 633
- microemulsion polymerization 633
- surfactant–cosurfactant pair 633

surfactants 1346, 1376
- sugar-based gemini surfactants 1376

surfmers 2185

suspension polymerization 277, 606–612
- activated monomer mechanism 609
- aromatic polymer synthesis 611
- atom-transfer radical polymerization (ATRP) 607
- catalyst system 609
- cationic surfactants 611
- co-stabilizer 277
- controlled/living free radical polymerization 607
- free radical polymerization 606
- ionic polymerization 608
- kinetic 606
- microcapsules 612
- metallic catalyst 609
- morphology 606

- nitroxide-mediated controlled free radical polymerization (NMP) 607
- oil-in-water processes 608
- "pearl polymerization" process 609
- phase-transfer catalysis 610
- polycondensates 611
- polycondensation/polyaddition 610
- principle 606
- reversible addition–fragmentation chain transfer (RAFT) 608
- ring-opening metathesis polymerization 608
- water-in-oil processes 610

sustained drug delivery 2573
- intraocular sustained drug release 2573
- powder-inhaled PEG 2573

Suzuki coupling 2376
Suzuki coupling reaction 326
Suzuki polycondensations 553
Suzuki reaction 2380

swollen brushes 1156–1157
- brush height 1155
- compressibility 1153
- equilibrium thickness 1153
- force profile 1154
- friction 1156
- frictional property 1157
- graft density 1155
- hydration transition 1156
- hydrodynamic thickness 1156
- size-exclusion properties 1157
- "super lubrication" 1157
- terminal block 1154
- tribological Properties 1156
- wear resistance 1157

syndiotactic living polymerization 731
syndiotactic polypropylene 224
syndiotactic-specific polymerization 747, 749–752, 754–755
- aluminum phenoxides 750
- anionic polymerization 749
- C_s-symmetric catalysts 747
- group transfer polymerization 751
- methacrylates 749
- monomer-activation mechanism 749
- NIPAAm polymerization 755
- α-olefins 747
- organotransition metal initiator 752
- radical polymerization mechanism 754

synergistic UPy peptides 387
synthetic degradable materials 1375
synthetic erythropoiesis protein (SEP) 2657

synthetic metals 2369
synthetic polypeptide antigens 534
synthetic rubber 2002
- government rubber–styrene 2002
syringic acid 411

t

Tacrolimus 2557
- FK506-dextran conjugate 2557
tacticity 731–733, 737, 739, 741–742, 769
- definition 732
- diad 732
- end-group analysis 742
- isotactic structure 732
- NMR spectroscopy 733
- propagation mechanism 737, 739, 741
- sequence statistics 737
- stereochemical assignments 733
- stereochemical defects 741
- stereoregularity 739, 769
- syndiotactic structure 732
- triads 733
- uniformity 769
tadpoles 892–893
- poly(THF) 893
- polybutadiene 892
- polyisoprenes 893
- polystyrene 893
tail-to-tail polymer 304
tandem siloxane–initiator compounds 1182
tandem thiol–polymerization initiators 1182
targeted libraries 1995
taste sensors 2530–2531
- electronic tongue 2531
- multichannel taste sensor 2530
- sensing materials 2531
- three-channel sour taste sensor 2531
telechelic ionomers 846
telechelic oligomers 775–776
- telomerization process 776
telechelic polyelectrolytes 843, 845
telechelic polymers 74–75, 143, 188, 547, 891
- AM polymerization 143
- α,ω-asymmetric polymers 75
- bifunctional initiator 74
- coupling agents 74
- four-arm star polymers 75
- haloboration–initiation 75
- PIB 75
- pre-cyclized structures 891

- symmetric telechelic polymers 74
telechelics 27, 448–449, 461, 787, 790, 794, 796, 800, 803–805
- addition–fragmentation processes 796
- aminoxyl group 794
- chemical modifications 790
- "click" chemistry 791
- direct chemical change 804
- iodine transfer polymerization 803
- nucleophilic substitutions 788
- quencher agents 787
- radical addition 805
- radical addition reactions 787
- radical coupling 791
- synthesis 449
- thioester modification 801
- trithioester compounds 800
TEM 1234
temperature-modulated differential scanning calorimetry (TMDSC) 1835, 1837–1838, 1840, 1842, 1844, 1846, 1858
- annealing experiment 1846
- data evaluation 1838
- linear response 1840
- non-isothermal TMDSC 1842
- quasi-isothermal TMDSC 1838, 1842
- reversible melting 1858
- sawtooth modulation 1840
TEMPO 173, 616
terafunctional initiators 930
terminal model, steady-state assumption 815
terminal model (TM) 814
- conditional probabilities 814
- propagation reactions 814
- reactivity ratios 814
terminator method 449
terpyridine 677
terpyridine polymers 682
terylene tetracarboxydiimide (TDI) 1066
tetrabutylammonium fluoride (TBAF) 315
tetrafunctional initiators 937
tetramethylethylenediamine (TMEDA) 24
2,2,6,6-tetramethylpiperidinoxy (TEMPO)-mediated polymerization 847, 1119
tetramethyltetraazacyclotetradecane 32
Teubner–Strey equation 1597
theory of cyclization 388
- critical concentration 388
thermal analysis 1828, 1849, 1868, 1872
- additional experimentation 1872
- advance 1868
- dilatometry 1874

- dynamic mechanical analysis (DMA) 1874
- thermogravimetry (TGA) 1874
- thermomechanical analysis (TMA) 1874
thermal conductivity 1830
thermal iniferters 562
- substituted tetraphenylethanes 562
thermal properties 1828–1829
- glass transition temperature 1829
- heat capacity 1829
- melting temperature 1829
- volume expansivity 1828
thermodynamic functions of state 1827, 1832–1833
- enthalpy 1833
- entropy 1833
- polyethylene 1832
thermoplastic elastomers 27, 226, 571, 839–840, 1779, 2001, 2022–2023
- impact modifiers 2023
- soft-touch overmold material 2022
- styrenic thermoplastic elastomers (TPEs) 2001
thermoplastics 217, 1775, 1829
- nanostructures 1775
- reactive blending 1775
thermoreversible networks 363
thermosetting materials 2332
thin polymer films 2663
- surface-initiated ATRP 2663
thin-film conducting transparent polymer materials 2401
- applications 2401
- core–shell latex spheres 2401
- polymer blend method 2402
- polypyrrole-based materials 2401
- PPy-coated poly(butyl methacrylate) latex 2401
- spin casting 2402
thin-film transistors 1352
thiol–ene polymerization 658
- propagation mechanism 658
thiols 2656
- coupling procedures 2656
thymine derivatives 383
tie molecules 1862
tissue engineering 387, 487, 2666, 2719, 2721, 2723–2725, 2726–2727, 2732, 2735
- biofunctionalization 2735
- biomolecules 2721
- bone regeneration matrix 2725
- construct 2719
- chondrocyte transplantation 2725
- degradable materials 2727
- denaturing activity 2723
- islet of Langerhans transplantation 2725
- matrices 2732
- materials 2724
- polymers 2726
- scaffold 2727
- skin regeneration matrix 2725
- tissue engineered skin 2725
tissue engineering construct 2719–2720
- cells 2720
tissue engineering materials 2724
- polymer support 2724
- scaffold 2724
tissue engineering scaffold 1370, 2724
tissue regrowth 1366
titania (TiO_2) 1195
- anatase powders 1195
TMDSC, heat flow-rate 1844
tobacco mosaic virus (TMV) 352
tomography 1650
topological (TP) gels 2690, 2691
- polyrotaxane gel 2691
- pulley effect 2691
topological clusters 1723
topology 839
topology distribution 1882
transcription 479
transesterification 319, 455, 456, 457
transfer agents (TA) 164, 182, 185, 1190
- dithioester 1190
transformation
- amino telechelics 572
- hydroxy telechelics 572
transformation agents 547, 549
- azo initiators 547
- photoinitiators 549
transformation reactions 593
transition metal complex 179
- atom-transfer radical polymerization (ATRP) catalysts 179
transition metal complexes 522
- bipyNi(COD) 522
- $(PMe_3)_4Co$ 522
transition-state analogue substrate (TSAS) 403, 423
translation 479
transmission electron microscopy (TEM) 1234, 1649–1654, 1671
- background 1650
- bulk materials 1654
- co-continuous morphologies 1656
- distortions 1656

- elastic scattering 1652
- electron energy-loss spectroscopy (EELS) 1652
- energy-filtered transmission electron microscopy (EFTEM) 1652, 1680
- field emission guns (FEGs) 1652
- imaging mode 1652
- inelastic scattering 1653
- phase contrast 1657
- resolution 1651
- secondary polymer medium 1655
- sectioning 1655
- staining 1657
- staining agent 1659
- staining reactions 1659
- stigmator coils 1652
- TEMsim 1671
- units 1651

transmission electron microtomography (TEMT) 1670–1672, 1674–1678
- background 1670
- 3D energy-filtered (EF-TEMT) images 1678
- 3D TEMT reconstruction 1671
- image cross-correlation 1672
- global structural analysis 1677
- gyroid morphology 1674
- image motion strategy 1672
- isosurface 1674
- methodology 1671
- quantitative analysis 1676
- reconstructed volume elements 1675
- reconstruction fidelity 1673
- topological analysis 1676–1677
- r-weighted back-projection algorithm 1673

triad fractions 827
- measurement 827

triblock
- cationic living polymerization 855
- group transfer polymerization 855

triblock copolymers
- PEG–aromatic polyamide–PEG triblock copolymer 339
- polyamide–poly(THF)–polyamide triblock copolymer 340
- polyphosphazene–polysiloxane–polyphosphazene 339
- polysiloxane–polyphosphazene–polysiloxane 339

Trichoderma viride cellulose (CelTV) 423
- reaction mechanisms 423

trifunctional initiators 344, 930, 934, 937

- phenyl ester initiator 954
- trifunctional dichloroacetate initiator 937

2,4,6-trimethylstyrene (TMeSt) 66
triphenylmethyl crotonate 767
triple detection 1905–1906
- advantage 1906
- principle 1905

triple hydrogen-bonded dimers 356
tritrypticin 1343
Trommsdorff effect 169, 1700
tropoelastin 1329
true molecular resolution 1528
- polydiacetylene crystal 1528
- reproducible observations 1528

Tsuji-Trost reaction 298, 300
tube model 1607, 1609
tumor-targeted anticancer polymeric drugs 2576
- active targeting 2578
- cellular internalization 2576
- passive targeting 2576

two-dimensional (2D) spectroscopy 1944
- correlation techniques 1944
- exchange spectroscopy 1945
- separation spectroscopy 1944
- spin diffusion 1945

two-dimensional chromatography 1922–1923, 1925
- combined functionality type (FTD) 1922
- fully automated two-dimensional chromatography 1922
- molar weight distribution (MWD) 1922
- topology-based separation 1925
- types of separations 1923

two-dimensional polyphenylenes 2242
two-photon lithography 2323–2324
- patterning process 2324

two-stage seeded emulsion polymerization 1219
- core–shell latex particles 1219

tyrosinase model complex catalyst 414
tyrosinases 463

u

Ullman coupling 805
ultra-high-throughput screening (UHTS) 1967
ultra-low-density polyethylene (ULDPE) 222
ultrafiltration 2467
ultramicrotome 1655
uniform polymers 769

universal polymer backbone   384
UPy derivatives   367
– bulk viscoelastic properties   367
UPy poly(ethylene oxide–propylene oxide) copolymer   376
UPy synthon   375
UPy telechelic oligocaprolactones   386
UPy telechelic oligodimethylsilane   375
UPy telechelic polymers   377
– material properties   377
urea biosensor   2524
ureido-s-triazine (UTr) derivatives   367
ureidopyrimidinone (UPy) functionality   364
urushi   411, 415
– "artificial urushi"   411
– fast drying hybrid urushi   415
urushiols   415, 416
– analogues   416
– laccase catalysis   415
utilizing protein engineering   2648
UV curing   652, 657
– acrylates   657
– applications   652
– reactions   657
– "remote curing process"   661
– sterically hindered amine light stabilizers (HALSs)   657
– UV absorbers (UVAs)   657

v

vapor-diffusion method   1154
vapor-phase osmometry (VPO)   372
vibrational spectroscopy   1985
vinyl   7, 524
– anionic polymerization   7, 524
N-vinylcarbazole   68
vinyl (macro)monomers   1103
vinyl esters   456
– lipase-catalyzed acylation   456
vinyl ethers   68, 70, 77, 85
– copolymerization   69
– fast living polymerization   69
– homopolymerization   69
– living cationic polymerization   68
– macromonomers   77
– photopolymerization   658
– stability   68
vinyl monomers   420
– enzymatic polymerization   420
vinyl polymers   420
vinylferrocene   692
vinylic monomers   9–10
– "living" polymerizations   9
– polymerizability   10
– reactivity   10
– thermodynamic requirement   9
virus-based scaffolds   1365
viscoelastic windows concept   2013
viscoelasticity   1606
viscosity modification   2024
volume expansivity   1829
– thermal expansion   1829

w

water purification membranes   2451, 2466, 2471
– mechanisms   2466
– nanofiltration membranes   2467
– phase-separated graft polymer membranes   2471
– pore size   2466
– ultrafiltration membranes   2467
water-soluble polymeric drugs   2543
– branched polymer (BP) systems   2547
– classification   2543
– linear polymer (LP) systems   2544
– linear polymer–branched spacer (LPBS) systems   2546
– linear polymer–linear spacer (LPLS) systems   2545
wave–particle duality   1650
weight ratio   1215
wetting phenomena   1544
– equilibrium thermodynamics   1544
whisker   332
wide-angle X-ray scattering (WAXS)   1873
– temperature-resolved   1873
Wigner–Seitz cell   1392
Wilchinsky triangle   2108

x

X-ray scattering   1961
xerogels   2334
xylan   427
xylanase   429

y

Yamamoto coupling   2379
Yamamoto polycondensations   553
Yamamoto route   2375

z

zeolites   1196
zeta potential   2148
Ziegler–Natta polymerization   585–586

Zimm equation 1583
Zimm model 997, 1445
– hydrodynamic interactions 1445
– mobility tensor 1445
– Oseen tensor 1445
ZnO crystals 2625

Further of Interest

Elias, H.-G.

Macromolecules
Volume 4: Applications of Polymers

2008
ISBN 978-3-527-31175-0

Elias, H.-G.

Macromolecules
Volume 3: Physical Structures and Properties

2007
ISBN 978-3-527-31174-3

Elias, H.-G.

Macromolecules
Volume 2: Industrial Polymers and Syntheses

2007
ISBN 978-3-527-31173-6

Elias, H.-G.

Macromolecules
Volume 1: Chemical Structures and Syntheses

2005
ISBN 978-3-527-31172-9

Kemmere, M. F., Meyer, T. (Eds.)

Supercritical Carbon Dioxide
in Polymer Reaction Engineering

2005
ISBN 978-3-527-31092-0

Meyer, T., Keurentjes, J. (Eds.)

Handbook of Polymer Reaction Engineering
2 Volumes

2005
ISBN 978-3-527-31014-2